MOLECULAR SPECT

and

MOLECULAR STRUCTURE

IV. CONSTANTS OF DIATOMIC MOLECULES

By

K. P. Huber and G. Herzberg

National Research Council of Canada

VAN NOSTRAND REINHOLD COMPANY

NEW YORK CINCINNATI ATLANTA DALLAS SAN FRANCISCO
LONDON TORONTO MELBOURNE

Van Nostrand Reinhold Company Regional Offices:
New York Cincinnati Atlanta Dallas San Francisco

Van Nostrand Reinhold Company International Offices:
London Toronto Melbourne

Library of Congress Catalog Card Number: 50-8347
ISBN: 0-442-23394-9

Manufactured in the United States of America

Published by Van Nostrand Reinhold Company
135 West 50th Street, New York, N.Y. 10020

Published simultaneously in Canada by Van Nostrand Reinhold Ltd.

15 14 13 12 11 10 9 8 7 6 5 4 3 2 1

Library of Congress Cataloging in Publication Data
Herzberg, Gerhard, 1904-
Molecular spectra and molecular structure.

Vol. 4 by K. P. Huber and G. Herzberg.
Bibliography: p.
Includes indexes.
CONTENTS: 1. Spectra of diatomic molecules.
–4. Constants
of diatomic molecules.
1. Molecular spectra–Tables. 2. Molecular structure–Tables. I. Huber, Klaus-Peter, 1934–
II. Title.
QC451.H64 543'.085 50-8347
ISBN 0-442-23394-9 (v. 4)

PREFACE

Since the publication in 1950 of Vol. I, *Spectra of Diatomic Molecules* of *Molecular Spectra and Molecular Structure*, much progress has been made in the field. While there have been some important refinements in the theory of diatomic molecular spectra, most of the advances have been in the further exploration of individual spectra. Not only has the number of molecules about which some spectroscopic data are available been increased by a factor of 2 to 3, but also the spectroscopic information about the molecules known in 1950 has been vastly extended. This is due to the observation of new electronic states (about three times as many as known before), the enormous improvements in the accuracy of the constants of the states known in 1950, and the determination of higher order constants.

In view of the increasing use of spectroscopic information on diatomic molecules in other fields of physics, in chemistry, and in astrophysics, it appeared desirable to prepare an up-to-date version of the table of molecular constants in the appendix of Vol. I. This updating proved to be far more time-consuming than originally anticipated, and it is only now, 10 years after its initiation, that we are able to present such a table, which, instead of the original 80 pages (plus 30 pages of bibliography), now fills a volume of 700 pages.

In the interest of economy, and unlike the original version, the new table has been produced by photo-offset from the final manuscript. Thus, typographical errors, which would have been likely to occur in such a difficult typesetting job, were entirely eliminated. We have spared no effort to make the table as up-to-date as possible. The date of final revision of each part is indicated. While the principal constants of each state of each molecule are listed uniformly in the body of the table, additional constants that could not be fitted in are given in the footnotes. Also, many qualifications and other explanations are added in this way.

We hope that users of these tables will find them helpful in supplying up-to-date constants as well as references to the most recent literature.

We are greatly indebted to many colleagues who have supplied advance information on recent unpublished work. The tables edited by the late B. Rosen were of great help to us in finding some of the earlier literature on many of the molecules dealt with therein. We have also greatly profited by the special tables prepared by P. H. Krupenie on O_2 and CO and by Krupenie and A. Lofthus on N_2. But wherever possible we have checked with the original publications. Particularly valuable to us throughout the entire course of this work were the *Berkeley Newsletters* prepared by J. G. Phillips and S. P. Davis; without them, a great many of the publications listed in the tables and in the appendix might have escaped our attention.

Although the final manuscript was prepared by one of us (K.P.H.), many drafts of the tables were typed by M. P. Thompson, and much checking of numbers and references was done by her and I. Dabrowski. We are most grateful for their efforts.

Finally, we must acknowledge that the National Research Council of Canada supported this work throughout the protracted period of its preparation by permitting one of us to spend all of his time for more than 10 years—and the other, part of his time—on this project and by providing other necessary facilities.

K. P. Huber
G. Herzberg

Ottawa (Canada), May 1978

INTRODUCTION

In this volume, we present a compilation of available data for all diatomic molecules and ions. The format of most of the tables follows closely that used in Table 39 of Volume I[1] of this series; that is, the various known electronic states and their symmetry symbols appear in the first column, and in the adjacent columns the electronic energy (T_e), the vibrational constants (ω_e and $\omega_e x_e$), the rotational constants (B_e, α_e, D_e), and the internuclear distance (r_e) are given. In the last three columns, the observed transitions involving the particular electronic state, their ν_{00} values, and the references on which the information is based are listed. The columns "$\omega_e y_e$" and "Remarks" of the previous table have been omitted, but this information is given in the footnotes. In addition, in the footnotes, references about Franck-Condon factors, potential functions, and other constants are given such as the following:

- A Spin-orbit interaction parameter
- λ Spin-spin interaction parameter
- γ Spin-rotation interaction parameter (not to be confused with the rotation-vibration interaction constant, γ_e)
- τ Radiative lifetime
- f Oscillator strength (f value)
- μ_{el} Electric dipole moment in Debye units (1 D = 10^{-18} esu cm).
- g_J Rotational g factor in units of nuclear magnetons (μ_N)
- Λ-type doubling constants p, q, . . .
- Hyperfine structure (hfs) constants such as the magnetic coupling constants, a, b, c, d, and the electric quadruple coupling constant, eqQ.

At the top of each table, the reduced mass μ of the molecule is given, as well as the dissociation energy D_0^0 in the ground state and the ionization potential (I.P.). Dissociation energies for states other than the ground state are easily derived if the dissociation products are known. Normally they are not listed explicitly.

[1]G. Herzberg, *Molecular Spectra and Molecular Structure. I. Spectra of Diatomic Molecules*, 2nd ed., Van Nostrand (1950).

The data presented in the tables come from a wide variety of experimental and theoretical studies. Of these, the most important are electronic spectra in emission or absorption (extending from the infrared to the vacuum ultraviolet and, in a few cases, the X-ray region), rotation-vibration spectra in the near infrared, rotation spectra in the far infrared and especially in the microwave (centimeter, millimeter, and submillimeter wave) regions. Additional highly precise information comes for some molecules from molecular beam electric and magnetic resonance studies and electron spin resonance spectra. Still other information has been taken from photofragment spectra, photoelectron and photoion spectra, and Auger electron spectra. In addition, data from electron scattering (elastic and inelastic), atomic scattering, mass spectrometry, flame photometry, and thermochemical studies have been used. References to lifetime measurements by various methods are also included, as well as references to theoretical calculations, which may be compared with experimental data or may fill gaps in existing experimental information.

The molecules are listed in strict alphabetical order (e.g., BaBr, BaCl precede BBr and BCl, even though this separates the latter from B_2). Positive and negative ions, in this order, follow immediately after the corresponding neutral molecule. Constants for hydrides, deuterides, and tritiides are given separately; for all other molecules, they are given for only one, usually the most abundant, isotope or for the natural isotopic mixture. In the latter case, the mass number for the most abundant species of one or both of the constituent atoms appears in parentheses and the reduced mass μ, also in parentheses, has been calculated accordingly.[2]

As in Volume I, the footnotes are referred to by lower-case letters$^{a,b,c,\ldots}$ continuing where necessary with $^{a',b',c',\ldots}$. In tables that extend over several pages, the sequence of footnotes starts with ^{a} on each page.

The references to the original literature are numbered in chronological order for each molecule and follow immediately at the end of the corresponding table. In order to save space, they are given in abbreviated form, omitting the initials of the authors and using code names for the journals as well as for the monographs. An alphabetical list of these abbreviated publication titles may be found on pages 1 through 7.

Each table carries the date (month and year) of its last revision. Considering the inevitable delay between publication of a paper and its eventual digestion for the purpose of the table, we estimate that the information in the table can be regarded as complete up to an effective cutoff time of three or four months prior to the indicated date. When the date is followed by the letter A, it indicates that in the appendix on pages 690 through 716 a list and short description may be found of additional publications that came to our notice after completion of the particular table.

[2]For ions with unequal nuclear charges, the reduced mass is not given explicitly. An ambiguity arises here with respect to the calculation of μ, which we did not undertake to resolve; instead, we used the reduced mass of the corresponding neutral molecule for the evaluation of internuclear distances. This approximation will normally not significantly increase the uncertainty of the result.

The notation for spectra of diatomic molecules has been standardized for many years. We have followed this notation throughout this volume. There is one important change in the notation, which was internationally agreed upon after the publication of Volume I: the change from K to N for the angular momentum excluding electron spin and the corresponding quantum number. There are still authors who are not aware of this change, which was agreed upon because, in polyatomic molecules, K represents the component of J in the direction of the top axis, and both K and N are needed there when $S \neq 0$.

All numbers in the tables are in cm^{-1} units except where otherwise indicated. The conversion factors and fundamental constants which we have used are based on "The 1973 Least-Squares Adjustment of the Fundamental Constants" by Cohen and Taylor (*J. Phys. Chem. Ref. Data* **2** [4], 663–734 [1973]). The most important of these are the following:

$$1\ \text{eV} = 8065.47_9\ \text{cm}^{-1}$$
$$1\ \text{kcal/mol} = 349.75_5\ \text{cm}^{-1}$$
$$1\ \text{kJ/mol} = 83.5935\ \text{cm}^{-1}$$
$$c = 2.99792458 \times 10^{+10}\ \text{cm s}^{-1}$$
$$\frac{m(^{12}\text{C})}{12} = 1.6605655 \times 10^{-24}\ \text{g}$$
$$\frac{h}{8\pi^2 c} = 27.9932_0 \times 10^{-40}\ \text{g cm}^2\ \text{cm}^{-1}$$
$$= 16.8576_3\ \text{a.m.u. Å}^2\ \text{cm}^{-1}$$
$$\frac{k}{hc} = 0.695030\ \text{K}^{-1}\ \text{cm}^{-1}$$
$$1\ \text{K} = 0.000086173\ \text{eV}$$

The reduced masses are referred to $m(^{12}C) = 12.00000000$ and are calculated from the atomic masses given in the table by Wapstra and Gove (*Nucl. Data Tables* **9**, 265–301 [1971]).

All the data in the table have been reviewed critically. They are, however, presented without error limits. The addition of meaningful and uniformly evaluated error limits would have meant an enormous increase in the time required to complete these tables. Instead, we hope that the number of digits quoted may serve as a *very* rough indication of the estimated order of magnitude of the error, generally ± 9 units of the last decimal place. Where the last digit is given as a subscript, we expect that the uncertainty may considerably exceed ± 10 units of that last decimal place.

In almost all cases, and unless stated otherwise, the constants ω_e, $\omega_e x_e$, B_e, α_e, and D_e that are listed are effective constants; that is, apart from sign they correspond to the coefficients Y_{lm} in the Dunham series expansion for the term values

$$T_{vJ} = \sum_{lm} Y_{lm}(v + \tfrac{1}{2})^l J^m (J + 1)^m$$

The signs are defined as follows:

$$G(v) = \omega_e(v + \tfrac{1}{2}) - \omega_e x_e(v + \tfrac{1}{2})^2 + \omega_e y_e(v + \tfrac{1}{2})^3 + \omega_e z_e(v + \tfrac{1}{2})^4 + \ldots$$
$$F_v(J) = B_v J(J + 1) - D_v J^2(J + 1)^2 + H_v J^3(J + 1)^3 \ldots$$
$$B_v = B_e - \alpha_e(v + \tfrac{1}{2}) + \gamma_e(v + \tfrac{1}{2})^2 + \ldots$$
$$D_v = D_e + \beta_e(v + \tfrac{1}{2}) + \ldots$$

The higher order constants are given in the footnotes.

T_e is usually calculated from the observed transitions without taking into account the quantity Y_{00} in the upper or lower state. Exceptions are mentioned in the footnotes.

The dissociation energy D_0^0 is always defined as *the energy of the ground state atomic products relative to the lowest existing level of the molecule.* In almost all cases, there is a footnote after D_0^0, which explains the method used in its derivation. The description "thermochemical value" is used to indicate any determination (mass spectrometric, flame photometric, or other) that involves the evaluation of a thermochemical equilibrium. Where necessary, thermochemical values that have been derived from exchange reactions have been adjusted to take account of recent changes in the dissociation energies of the reference molecules.

In agreement with common practice, the first ionization potential (in eV) is taken as *the energy difference of the lowest existing level of the ion and the lowest existing level of the neutral system.* Analogous definitions apply to positive as well as negative ions although, for the latter, the I.P. is more commonly referred to as electron affinity of the neutral. Higher ionization potentials have in some cases been added in the table or in the footnotes. Electron impact appearance potentials have only rarely been included since their accuracy is usually low (typically ± 0.5 eV).

It appears nearly impossible to give a generally applicable definition of the band origins. Here, origins in singlet systems normally refer to the zero lines; that is, they include the J independent term $-B\Lambda^2$, which some authors prefer to include in the rotational energy expression. Similarly, for case "a" multiplet states, a corresponding definition applies to individual sub-bands; exceptions to these rules are usually indicated in the footnotes. For intermediate coupling or coupling close to case "b", we frequently refer to the zero-point of the Hill-Van Vleck (or equivalent) expression; an explanation is usually given in the footnotes. Multiplets very close to case "b" are often treated as singlets disregarding electron spin. In all cases where our definition of the origin deviates strongly from that used by the original author, we have indicated this in the accompanying footnote.

The magnitude and sign of the Λ-type doubling is indicated in many instances in footnotes by quoting either the difference $B(R,P) - B(Q)$, which is equivalent to $B(\Pi^+) - B(\Pi^-)$ for transitions involving a Σ^+ state, or the leading terms in the expression giving the observed splitting as a function of J. For the labeling of the parity doublet levels, we have adopted the recommendations of Brown et al. (*J. Mol. Spectrosc.* **55,** 500 [1975]):

integral J: e levels have parity $+(-1)^J$
f levels have parity $-(-1)^J$
half-integral J: e levels have parity $+(-1)^{J-\frac{1}{2}}$
f levels have parity $-(-1)^{J-\frac{1}{2}}$

The sign of the splitting is defined by

$$\Delta\nu_{ef}(J) = F_e(J) - F_f(J) = -\Delta\nu_{fe}(J)$$

For some case "b" $^2\Pi$ states we give $\Delta\nu_{ef}(N)$ and refer to the F_1 component.

The following symbols are used throughout the table:

H Data obtained from band head measurements
Z Data obtained from, or referring to, band origins
R Shaded towards longer wavelengths
V Shaded towards shorter wavelengths
( ) Uncertain data
[ ] Data refer to $v = 0$ or lowest observed level. T_e values in square brackets give the energy of this level relative to the minimum of the ground-state potential energy curve. Vibrational frequencies in square brackets correspond to $\Delta G(\frac{1}{2})$ or the lowest observed integral.
{ } Hypothetical levels
→ Emission
← Absorption
↔ Emission and absorption
(→, ←, ↔: According to international agreement the *upper* state always comes first.)
* Indicates a reference number where a good record or reproduction of the spectrum may be found

CONTENTS

MOLECULAR SPECTRA

and

MOLECULAR STRUCTURE

IV. CONSTANTS OF DIATOMIC MOLECULES

[illegible]ED PUBLICATION TITLES.

AA	Astronomy and Astrophysics.
AA(Suppl.)	Astronomy and Astrophysics, Supplement Series.
AANL	Atti della (Reale) Accademia Nazionale dei Lincei. Rendiconti, Classe di Scienze Fisiche, Matematiche e Naturali.
AAp	Annales d'Astrophysique.
AAQA	Anales de la Asociacion Quimica Argentina.
AC(Int. Ed. Engl.)	Angewandte Chemie. International Edition in English.
AD	Atomic Data.
AdC	Advances in Chemistry Series.
AdGp	Advances in Geophysics.
AdHTC	Advances in High Temperature Chemistry.
ADNDT	Atomic Data and Nuclear Data Tables.
AdRS	Advances in Raman Spectroscopy.
AF	Arkiv foer Fysik.
AGEP	Annales de Géophysique.
AMAF	Arkiv foer Matematik, Astronomi och Fysik.
AO	Applied Optics.
AP(Leipzig)	Annalen der Physik (Leipzig).
AP(Paris)	Annales de Physique (Paris).
APH	Acta Physica Academiae Scientiarum Hungaricae.
ApJ	Astrophysical Journal.
ApJ(Suppl.)	Astrophysical Journal, Supplement Series.
APL	Applied Physics Letters.
ApL	Astrophysical Letters.
APP	Acta Physica Polonica.
ARAA	Annual Review of Astronomy and Astrophysics.
ARSEFQ	Anales de la Real Sociedad Espanola de Fisica y Quimica.
AS	Applied Spectroscopy.
AZ	Astronomicheskii Zhurnal. - For English translation see SAAJ.

BAMS	Berkeley Analyses of Molecular Spectra. University of California Press.
	1 S. P. Davis, J. G. Phillips, "The Red System ($A\ ^2\Pi - X\ ^2\Sigma$) of the CN Molecule" (1963).
	2 J. G. Phillips, S. P. Davis, "The Swan System of the C_2 Molecule"; "The Spectrum of the HgH Molecule" (1968).
BAPS	Bulletin of the American Physical Society.
BAPS(MAP)	Bulletin de l'Académie Polonaise des Sciences. Série des Sciences Mathématiques, Astronomiques et Physiques.
BASPS	Bulletin of the Academy of Sciences of the USSR, Physical Series. - English translation of IANSF.
BBPC	Berichte der Bunsen-Gesellschaft für Physikalische Chemie.
BCSARB	Bulletin de la Classe des Sciences, Académie Royale de Belgique.
BSCB	Bulletin des Sociétés Chimiques Belges.
BSCF	Bulletin de la Société Chimique de France.
BSRSL	Bulletin de la Société Royale des Sciences de Liège.
CaP	Cahiers de Physique.
CC	Chemical Communications.
CCA	Croatica Chemica Acta.
CF	Combustion and Flame.
CJC	Canadian Journal of Chemistry.
CJP	Canadian Journal of Physics.
CJPS	Chinese Journal of Physics. (Chinese Physical Society) Shanghai.
CJR	Canadian Journal of Research.
CJS	Canadian Journal of Spectroscopy.
CP	Chemical Physics.
CPAS	Commentarii, Pontificia Academia Scientiarum.
CPL	Chemical Physics Letters.
CR	Comptes Rendus Hebdomadaires des Seances de l'Académie des Sciences. Paris.
CRev	Chemical Reviews.
CS	Current Science.
CSp	Canadian Spectroscopy.
DANS	Doklady Akademii Nauk SSSR. - For English translation see DC, DPC.
	Doklady Chemistry. - English translation of DANS (Chemistry Section).
	Discussions of the Faraday Society.
	A. G. Gaydon, "Dissociation Energies and Spectra of Diatomic Molecules", 3rd ed., Chapman and Hall (1968).
	Tables Internationales de Constantes Sélectionnées. 17. Données Spectroscopiques relatives aux Molécules Diatomiques (établi sous la direction de B. Rosen). Pergamon Press (1970).

DPC	Doklady Physical Chemistry. - English translation of DANS (Physical Chemistry Section).
FDCS	Faraday Discussions of the Chemical Society.
Fizika	Fizika (Zagreb).
FSCS	Faraday Symposia of the Chemical Society.
GCI	Gazzetta Chimica Italiana.
GHDB	Glasnik Hemijskog Drustva, Beograd. (Bulletin of the Chemical Society, Belgrade.)
HCA	Helvetica Chimica Acta.
HPA	Helvetica Physica Acta.
HT(USSR)	High Temperature. - English translation of TVT.
HTS	High Temperature Science.
IAMS	Identification Atlas of Molecular Spectra. U. of Western Ontario (1964/5), York University (1967/72).
	1 D.C. Tyte, R.W. Nicholls, "The AℓO $A\,^2\Sigma - X\,^2\Sigma$ Blue-Green System" (1964).
	2 D.C. Tyte, R.W. Nicholls, "The N_2 $C\,^3\Pi_u - B\,^3\Pi_g$ Second Positive System" (1964).
	3 D.C. Tyte, R.W. Nicholls, "The N_2^+ $B\,^2\Sigma_u^+ - X\,^2\Sigma_g^+$ First Negative System of Nitrogen" (1965).
	4 G.R. Hébert, S.H. Innanen, R.W. Nicholls, "The O_2 $B\,^3\Sigma_u^- - X\,^3\Sigma_g^-$ Schumann-Runge System" (1967).
	5 D.C. Tyte, S.H. Innanen, R.W. Nicholls, "The C_2 $A\,^3\Pi_g - X'\,^3\Pi_u$ Swan System" (1967).
	6 V. Degen, S.H. Innanen, G.R. Hébert, R.W. Nicholls, "The O_2 $A\,^3\Sigma_u^+ - X\,^3\Sigma_g^-$ Herzberg I System" (1968).
	7 J.A. Harrington, R.M. Seel, G.R. Hébert, R.W. Nicholls, "The VO $C\,^4\Sigma^- - X\,^4\Sigma^-$ Yellow-Green and $B\,^4\Pi - X\,^4\Sigma^-$ Red Systems" (1970).
	8 B. Brocklehurst, G.R. Hébert, S.H. Innanen, R.M. Seel, R.W. Nicholls, "The CN $A\,^2\Pi - X\,^2\Sigma^+$ Red System" (1971).
	9 B. Brocklehurst, G.R. Hébert, S.H. Innanen, R.M. Seel, R.W. Nicholls, "The CN $B\,^2\Sigma^+ - X\,^2\Sigma^+$ Violet System" (1972).
IANNM	Izvestiya Akademii Nauk SSSR, Neorganicheskie Materialy. - For English translation see IM.
IANSF	Izvestiya Akademii Nauk SSSR, Seriya Fizicheskaya. - For English translation see BASPS.
Icarus	Icarus.
ICB	Industrie Chimique Belge.
IDSPEC	R.W.B. Pearse, A.G. Gaydon, "The Identification of Molecular Spectra". Chapman and Hall, 3rd ed. 1963, 4th ed. 1976.
IJMSIP	International Journal of Mass Spectrometry and Ion Physics.
IJP	Indian Journal of Physics.
IJPAP	Indian Journal of Pure and Applied Physics.
IJQC	International Journal of Quantum Chemistry.
IM	Inorganic Materials. - English translation of IANNM.

INCL	Inorganic and Nuclear Chemistry Letters.
IPCR	See Sci. Pap. IPCR (Tokyo).
ISOANK	Izvestiya Sibirskogo Otdeleniya Akademii Nauk SSSR, Seriya Khimicheskikh Nauk. - For English translation see SCJ.
IVUZF	Izvestiya Vysshikh Uchebnykh Zavedenii, Fizika. - For English translation see SPJ.
IVUZK	Izvestiya Vysshikh Uchebnykh Zavedenii, Khimiya i Khimicheskaya Tekhnologiya.
JACS	Journal of the American Chemical Society.
JANAF	JANAF Thermochemical Tables. 2nd edition. NSRDS-NBS 37 (1971).
JAP	Journal of Applied Physics.
JAS	Journal of Applied Spectroscopy. - English translation of ZPS.
JATP	Journal of Atmospheric and Terrestrial Physics.
JCP	Journal of Chemical Physics.
JCPPB	Journal de Chimie Physique et de Physico-Chimie Biologique. Paris.
JCS	Journal of the Chemical Society.
JCS FT	Journal of the Chemical Society. Faraday Transactions.
JESRP	Journal of Electron Spectroscopy and Related Phenomena.
JGR	Journal of Geophysical Research.
JIC(USSR)	Journal of Inorganic Chemistry (USSR). - English translation of ZNK.
JINC	Journal of Inorganic and Nuclear Chemistry.
JJP	Japanese Journal of Physics.
JLTP	Journal of Low Temperature Physics.
JMS	Journal of Molecular Spectroscopy.
JMSt	Journal of Molecular Structure.
JOSA	Journal of the Optical Society of America.
JP	Journal of Physics.
JP(Paris)	Journal de Physique (Paris), and Journal de Physique (Paris), Lettres.
JPC	Journal of Physical Chemistry.
JPCRD	Journal of Physical and Chemical Reference Data.
JPhoC	Journal of Photochemistry.
JPR	Journal de Physique et le Radium.
JPSJ	Journal of the Physical Society of Japan.
JPUSSR	Journal of Physics (Moscow).
JQE	IEEE Journal of Quantum Electronics.
JQSRT	Journal of Quantitative Spectroscopy and Radiative Transfer.

	Journal des Recherches du Centre National de la Recherche Scientifique.
	Journal of Research of the National Bureau of Standards.
JRS	Journal of Raman Spectroscopy.
JSHU	Journal of Science of the Hiroshima University.
JSIR	Journal of Scientific and Industrial Research.
JSRBHU	Journal of Scientific Research of the Banaras Hindu University.
KEM	Kvantovaya Elektronika (Moscow). - For English translation see SJQE.
LNC	Lettere al Nuovo Cimento (della Società Italiana di Fisica).
LTS	Low Temperature Science, Series A: Physical Sciences. (Teion Kagaku, Butsuri-Hen.)
METOX	A. Gatterer, J. Junkes, E. W. Salpeter, B. Rosen, "Molecular Spectra of Metallic Oxides", Specola Vaticana (1957).
MOLSPEC	G. Herzberg, "Molecular Spectra and Molecular Structure". Van Nostrand Reinhold.
	1 Spectra of Diatomic Molecules. 2nd ed. (1950).
	2 Infrared and Raman Spectra of Polyatomic Molecules (1945).
	3 Electronic Spectra and Electronic Structure of Polyatomic Molecules (1966).
MP	Molecular Physics.
MSRSL	Mémoires de la Société Royale des Sciences de Liège (Collection en 8°).
MSRSL*	Mémoires de la Société Royale des Sciences de Liège, Volume hors série.
MUCB	Moscow University Chemistry Bulletin. - English translation of VMUK.
NARSSU	Nova Acta Regiae Societatis Scientiarum Upsaliensis.
Nature	Nature (London).
Nature PS	Nature (London), Physical Sciences.
Naturw.	Naturwissenschaften.
NBSM	National Bureau of Standards (U.S.), Monograph.
NC	Nuovo Cimento (della Società Italiana di Fisica).
NC(Suppl.)	Nuovo Cimento, Supplemento.
NDVSK	Nauchnye Doklady Vysshei Shkoly, Khimiya i Khimicheskaya Tekhnologiya.
NIM	Nuclear Instruments and Methods.
NSRDS-NBS	National Standard Reference Data Series, National Bureau of Standards (U.S.).
OC	Optics Communications.
OPA	Optica Pura y Aplicada.
OS	Optika i Spektroskopiya. - For English translation see OS(Engl. Transl.).
OS(Engl. Transl.)	Optics and Spectroscopy (USSR). - English translation of OS.
PASP	Publications of the Astronomical Society of the Pacific.

PBCS	Proceedings of the British Ceramic Society.
PC	Physics in Canada.
PCS	Proceedings of the Chemical Society, London.
PDAO	Publications of the Dominion Astrophysical Observatory, Victoria, British Columbia.
Physica	Physica.
PIAS	Proceedings of the Indian Academy of Sciences.
PKNAW	Proceedings of the Koninklijke Nederlandse Akademie van Wetenschappen, Amsterdam.
PL	Physics Letters.
PM	Philosophical Magazine.
PNASI	Proceedings of the National Academy of Sciences, India.
PNASU	Proceedings of the National Academy of Sciences of the United States of America (Washington).
PNISI	Proceedings of the National Institute of Sciences of India.
PP	Photochemistry and Photobiology.
PPMSJ	Proceedings of the Physico-Mathematical Society of Japan.
PPS	Proceedings of the Physical Society, London.
PR	Physical Review.
Pramāṇa	Pramāṇa (India).
PRIA	Proceedings of the Royal Irish Academy.
PRL	Physical Review Letters.
PRR(Suppl.)	Philips Research Reports, Supplements.
PRS	Proceedings of the Royal Society of London.
PS	Physica Scripta.
PSS	Planetary and Space Science.
PTRSL	Philosophical Transactions of the Royal Society of London.
PZ	Physikalische Zeitschrift.
PZS	Physikalische Zeitschrift der Sowjetunion.
QR	Quarterly Reviews, Chemical Society.
RIHTR	Revue Internationale des Hautes Températures et des Réfractaires. Paris.
RiSc	La Ricerca Scientifica.
RJIC	Russian Journal of Inorganic Chemistry. - English translation of ZNK.
RJPC	Russian Journal of Physical Chemistry. - English translation of ZFK.
[illegible]	Reviews of Modern Physics.
[illegible]	Revue d'Optique, Théorique et Instrumentale.

RPA	Revue de Physique Appliquée.
RR	Radiation Research.
RRP	Revue Roumaine de Physique.
RS	Ricerche Spettroscopiche, Laboratorio Astrofisico della Specola Vaticana.
SA	Spectrochimica Acta.
SA(Suppl.)	Spectrochimica Acta (Vol. 11), Supplement (1957). Proceedings of the 6th Colloquium Spectroscopicum Internationale, Amsterdam, May 14-19, 1956.
SAAJ	Soviet Astronomy - AJ. - English translation of AZ.
SAJS	South African Journal of Science.
Science	Science.
Sci. Pap. IPCR (Tokyo)	Scientific Papers of the Institute of Physical and Chemical Research, Tokyo.
SCJ	Siberian Chemistry Journal. - English translation of ISOANK.
SJQE	Soviet Journal of Quantum Electronics. - English translation of KEM.
SL	Science of Light (Tokyo).
SPJ	Soviet Physics Journal. - English translation of IVUZF.
SpL	Spectroscopy Letters.
SPU	Soviet Physics - Uspekhi. - English translation of UFN.
TA	Thermochimica Acta.
TFS	Transactions of the Faraday Society.
TVT	Teplofizika Vysokikh Temperatur. - For English translation see HT(USSR).
UFN	Uspekhi Fizicheskikh Nauk. - For English translation see SPU.
USIP	University of Stockholm, Institute of Physics. Reports.
VMUK	Vestnik Moskovskogo Universiteta (Seriya II), Khimiya. - For English translation see MUCB.
ZA	Zeitschrift für Astrophysik.
ZAMP	Zeitschrift für Angewandte Mathematik und Physik. [Journal for Applied Mathematics and Physics (ZAMP)].
ZE	Zeitschrift für Elektrochemie. Berichte der Bunsengesellschaft für Physikalische Chemie.
ZFK	Zhurnal Fizicheskoi Khimii. - For English translation see RJPC.
ZN	Zeitschrift für Naturforschung.
ZNK	Zhurnal Neorganicheskoi Khimii. - For English translation see JIC(USSR) [1956 - 1958] and RJIC [1959 +].
ZP	Zeitschrift für Physik.
ZPC	Zeitschrift für Physikalische Chemie.
ZPS	Zhurnal Prikladnoi Spektroskopii. - For English translation see JAS.
ZWP	Zeitschrift fur wissenschaftliche Photographie, Photophysik und Photochemie.

State	T_e	ω_e	$\omega_e x_e$	B_e	α_e	D_e $(10^{-7}cm^{-1})$	r_e (Å)	Observed Transitions Design.	Observed Transitions ν_{00}	References
$^{(107,109)}Ag_2$		(μ = 53.947829_3)		D_0^0 = 1.66 eV[a]						OCT 1974 A
E	40159.1	146.0_8 H	1.54					E← X, R	40135.7 H	(5)(7)(8)
D $^1\Pi_u$ (1_u)	39023.7	166.7 H^R	1.13_4					D← X, R	39010.7^b H^R	(5)(8)
C $^1\Pi_u$ (1_u)	37626.9	$172._9$ H^Q	1.0_7					C← X, R	$37617._0{}^c$ H^Q	(5)(7)(8)
B	35827.3	151.3 H	0.70					B↔ X, R	35806.7 H	(5)(8)
A	22996.4	154.6 H	0.587^d					A↔ X, R	22977.5_0 H	(1)(2)* (5)
X $^1\Sigma_g^+$	0	192.4 H	0.643^d							
$^{107}Ag^{27}Al$		μ = 21.5440781		D_0^0 = 1.9_5 eV[a]						OCT 1974
C $^1\Pi$	$31744._8$	$[221.06]^b$		$[0.1225]^c$		[1.6]	[2.525]	C← X, R	31727.33^c Z	(2)
B $^1\Sigma^+$	27459.1_7	[199.85] Z	(1.6_6)	0.11706^d	0.00112	1.6	2.5854	B← X, R	27432.32 Z	(2)
X $^1\Sigma^+$	0	[254.34] Z	(1.1_3)	0.12796	0.00076	1.27	2.4728			
$^{(107)}Ag^{197}Au$		(μ = 69.294807_5)		D_0^0 = 2.0_6 eV[a]						OCT 1974
		Four bands in thermal emission, possibly forming a progression with $\omega \approx 200$ cm^{-1}. No wavelengths given.								(1)
$^{109}Ag^{209}Bi$		μ = 71.594910_6								FEB 1975 A
B (0^+)	20753.0	145.29 H	0.335^a					B↔ X, R	20749.6 H	(1)(2)(3)*
A (0^+)	16364	144.0	0.53	$(0.0200)^b$	(0.0000551)	(0.0153)	(3.43)	A→ X, R	16360^c	(2)(3)*
								A→ X', R_V	12176.8 H^Q	(3)*
X'(1)	4185	148.5	0.50	$(0.0199)^b$	(0.0000483)	(0.0143)	(3.44)			
X (0^+)	0	152.14 H	0.405^d	$(0.0198)^b$	(0.0000435)	(0.0135)	(3.45)			

Ag_2: [a]Thermochemical value (mass-spectrom.)(3)(4)(6), recalculated (9).
[b]Q head at 38995.6 cm^{-1}.
[c]R head at 37628.2 cm^{-1}.
[d]Constants for $^{107}Ag^{109}Ag$; $\omega_e y_e(A) = +0.0023$, $\omega_e y_e(X) = +0.0003$.

(1) Ruamps, CR 238, 1489 (1954).
(2) Kleman, Lindkvist, AF 9, 385 (1955).
(3) Drowart, Honig, JCP 25, 581 (1956); JPC 61, 980 (1957).
(4) Schissel, JCP 26, 1276 (1957).
(5) Ruamps, AP(Paris) 4, 1111 (1959).
(6) Ackerman, Stafford, Drowart, JCP 33, 1784 (1960).
(7) Maheshwari, IJP 37, 368 (1963).
(8) Choong, Wang, Lim. Nature 209, 1300 (1966).
(9) Smoes, Mandy, Vander Auwera-Mahieu, Drowart, BSCB 81, 45 (1972).

AgAℓ: [a]Thermochemical value (mass-spectrom.)(1).
[b]From (2), no details.
[c]Constants refer to the normal $^{107}Ag/^{109}Ag$ isotopic mixture.
[d]Perturbations in v=0 and 1.

(1) Blue, Gingerich, 16th Annual Conference on Mass Spectrometry and Allied Topics, Pittsburgh (May 1968); paper 129.
(2) Clements, Barrow, TFS 64, 2893 (1968).

AgAu: [a]Thermochemical value (mass-spectrom.)(2).

(1) Ruamps, SA(Suppl.) 11, 329 (1957).
(2) Ackerman, Stafford, Drowart, JCP 33, 1784 (1960).

AgBi: [a]$\omega_e y_e = +0.00007$.
[b]Rotational structure not resolved. The rotational constants have been estimated (3) from the appearance of extra heads in the red and infrared systems. They appear doubtful considering the unexpectedly large values of r_e resulting from them.
[c]Head of the 0-0 band at 16394.5 cm^{-1}.
[d]$\omega_e y_e = +0.00044$.

(1) Houdart, Lochet, CR B 271, 38 (1970).
(2) Lochet, CR B 272, 44, 797 (1971); 274, 174 (1972).
(3) Lochet, JP B 7, 505, L543 (1974).

State	T_e	ω_e	$\omega_e x_e$	B_e	α_e	D_e ($10^{-8}cm^{-1}$)	r_e (Å)	Observed Transitions Design.	Observed Transitions ν_{00}	References
$^{109}Ag^{81}Br$		$\mu = 46.423560_9$		$D_0^0 = 3.1$ eV^a						OCT 1974
		Fragments of a band system at 24000 cm^{-1}, in absorption.								(3)
C	43537.4	205.0 H	0.74					C← X, R	43516.0 H	(4)
		Continuous absorption 29800 - 32300 cm^{-1}, maximum at 31400 cm^{-1}.								(1)(8)
B (0^+)	31280.4_3	180.8 H	4.45^b					B↔X, R	31246.0_2 H	(1)* (2)
X $^1\Sigma^+$	0	247.7_2 H	0.679_5	0.063407_3	$0.0002282_2{}^c$	1.7	2.39309	Microwave sp.^{d}		(7)(9)
$^{107}Ag^{35}Cl$		$\mu = 26.3497888$		$D_0^0 = 3.22$ eV^a						OCT 1974
D	(48800)	(290) Fragments of two progressions.						D← X, R		(4)
		Continuous absorption above 47500 cm^{-1}.								(4)
C	43525.7	294.1 H	1.70					C← X, R	43500.9^b H	(4)
B 0^+	31602.65	278.36 Z	4.047^c	0.11912_7	0.001492^d	9.40^e	2.3174_2	B↔X, R	31569.32 Z	(1)* (2)(3) (5)(8)
		Continuous absorption with maximum at 30300 cm^{-1}.								(9)
X $^1\Sigma^+$	0	343.49 Z	1.17	0.12298388	$0.00059540_5{}^f$	$6.305_4{}^g$	2.280792	Microwave sp.^{h}		(6)(7)(11)
$^{(107)}Ag^{(63)}Cu$		$(\mu = 39.612012_7)$		$D_0^0 = 1.7_6$ eV^a						OCT 1974
B	25851.6	178.5 H	0.50					B→X, R	25825.0 H	(3)
A	20836	171.5 H	<0.5					A→X, R	20806 H	(1)
X	0	231.8 H	0.80							
$^{107}Ag^{19}F$		$\mu = 16.1316107$		$D_0^0 = 3.6_4$ eV^a						OCT 1974
C		Unclassified bands, $\nu \geq 38400$ cm^{-1}.						C← X, R		(3)(5)
		Continuous absorption at 33000 cm^{-1}.								(3)(5)
B 0^+	(31663)	[373.6] H	b	$[0.25548_6]$	c	[47.2]	$[2.0224_4]$	B← X, R	31594.13 Z	(1)(3)(5)*
A 0^+	(29220)	[455.39] Z	d	[0.27268]	0.00593^d	$[37.1]^e$	[1.9576]	A← X, V	29250.87 Z	(3)(5)*
X $^1\Sigma^+$	0	513.447 Z	2.593	0.2657020	0.0019206^f	28.4^f	1.983179	Microwave sp.^{g}		(4)

AgBr: [a]Thermochemical value (5)(6).

[b]$\omega_e y_e = -0.060$.

[c]$\gamma_e = +1.6_0 \times 10^{-7}$. (9) give constants for $^{107}Ag^{79}Br$.

[d]For values of eqQ(^{79}Br, ^{81}Br) see (7)(9).

(1) Brice, PR 38, 658 (1931).
(2) Mulliken, PR 51, 310 (1937).
(3) Metropolis, Beutler, PR 55, 1113 (1939).
(4) Barrow, Mulcahy, Nature 162, 336 (1948).
(5) Brewer, Lofgren, JACS 72, 3038 (1950).
(6) Barrow, JCP 22, 573 (1954).
(7) Krisher, Norris, JCP 44, 974 (1966).
(8) Davidovits, Bellisio, JCP 50, 3560 (1969).
(9) Hoeft, Lovas, Tiemann, Törring, ZN 26 a, 240 (1971).

AgCℓ: [a]Thermochemical value (10).

[b]0-0 band not observed.

[c]$\omega_e y_e = -0.38_6$. The vibrational levels converge rapidly above v=4.

[d]$+[6.2(v+\frac{1}{2})^2 - 2.9(v+\frac{1}{2})^3] \times 10^{-5}$ (for $v \leq 4$).

[e]$-[0.0257(v+\frac{1}{2}) - 0.467(v+\frac{1}{2})^2] \times 10^{-8}$ (for $v \leq 4$).

[f]$\gamma_e = +6.28_1 \times 10^{-7}$.

[g]$\beta_e = -0.0018 \times 10^{-8}$; $H_e = -0.9_7 \times 10^{-14}$.

[h]$\mu_{e\ell} = 5.7_0$ D [(6), criticised by (11) for neglect of quadrupole hfs]. For value of eqQ($^{35}C\ell$) see (11).

(1) Brice, PR 35, 960 (1930).
(2) See ref. (1) of AgBr.
(3) See ref. (2) of AgBr.
(4) Jenkins, Rochester, PR 52, 1141 (1937).
(5) Barrow, Morgan, Wright, PCS (1959), p. 303.
(6) Krisher, Norris, JCP 44, 391 (1966).
(7) Pearson, Gordy, PR 152, 42 (1966).

AgCℓ (continued):

(8) Clements, Barrow, TFS 63, 2876 (1967).
(9) Davidovits, Bellisio, JCP 50, 2787 (1969).
(10) Hildenbrand, JCP 52, 5751 (1970).
(11) Hoeft, Lovas, Tiemann, Törring, ZN 26 a, 240 (1971).

AgCu: [a]Thermochemical value (mass-spectrom.)(2).

(1) Ruamps, SA(Suppl.) 11, 329 (1957).
(2) Ackerman, Stafford, Drowart, JCP 33, 1784 (1960).
(3) Joshi, Majumdar, PPS 78, 197 (1961).

AgF: [a]Thermochemical value (2). In agreement with an upper limit ($D_0^0 < 3.70$ eV) from predissociation in A 0^+ (3)(5).

[b]From band heads: $\Delta G(3/2, 5/2, 7/2) = 380.3, 373.2, 356.9$.

[c]Levels with $v \geq 1$ are predissociated.

[d]Predissociation in v=0 for $J > 85$ and in v=1 for $J > 40$. Levels with $v > 1$ are not observed.

[e]$D_1 = 67.3 \times 10^{-8}$; $H_0 = -3.2 \times 10^{-12}$.

[f]$\gamma_e = +3.8_0 \times 10^{-6}$; $\beta_e = +0.1 \times 10^{-8}$.

[g]$\mu_{e\ell}(v=0) = 6.2_2$ D.

(1) Joshi, Sharma, IJPAP 1, 86 (1963).
(2) Zmbov, Margrave, JPC 71, 446 (1967).
(3) Clements, Barrow, CC (1968), p. 27.
(4) Hoeft, Lovas, Tiemann, Törring, ZN 25 a, 35 (1970).
(5) Barrow, Clements, PRS A 322, 243 (1971).

State	T_e	ω_e	$\omega_e x_e$	B_e	α_e	D_e ($10^{-4}cm^{-1}$)	r_e (Å)	Observed Transitions		References
								Design.	ν_{00}	
$^{(107)}Ag^{(69)}Ga$		(μ = 41.906769_3)		$D_0^0 = 1.8_7$ eV[a]						OCT 1974
A	33061.1	153.2 H	0.8_2					A← X, R	33045.3 H	(1)*
X ($^1\Sigma$)	0	184.7 H	0.6_5							
$^{(107)}Ag^1H$		(μ = 0.99841289)		D_0^0 = 2.28 eV[a]						OCT 1974
$c_0(^3\Pi_0)$				[5.10][b]		[6.0]	$[1.82_0]$	c_0← X, R	46981.1 Z	(10)*
d $^3\Sigma^+$				[(3.83)][c]			$[(2.10_0)]$		(46875)[d]	(10)*
c_1 $^3\Pi_1$				[(4.95)][e]		[(6.0)]	$[(1.84_7)]$	c_1← X, R	(46600)[e]	(10)*
D $^1\Pi$	(46720)	[844.7] Z	(120)	5.23[f]	0.58	6.0	1.79_7	D← X, R	46360.9 Z	(10)*
b $(^3\Delta_1)$ 1				[4.805][g]		[3.5]	$[1.874_5]$	b← X,[h] R	44529.2 Z	(10)*
B $^1\Sigma^+$	(44512)	[(1089)][i]	(65)	(4.87)[i]	(0.31)	(3.8)	(1.86_2)	B← X, R	(44234)[i]	(10)
a $^3\Pi_r$	(41700)[j]	(1450)	(50)	(>6.3)[j]			(<1.64)			(10)
C $^1\Pi$	(41261)	(1589)[k]	(42)[k]	[6.54]	(0.31)[k]	[4.9]	$[1.60_7]$	C← X,	41173.6 Z	(10)*
A $^1\Sigma^+$	29959	[1489.6] Z	87.0[ℓ]	$[6.090_5]$	0.348_5[ℓ]	$[3.8_9]$	[1.6650]	A↔ X, R	29897.9_4 Z	(2)(3)* (5) (6)(8)
X $^1\Sigma^+$	0	1759.9 Z	34.06[m]	6.449[o]	0.201[m]	[3.44][n]	1.618_1			
$^{(107)}Ag^2H$		(μ = 1.97685802)		D_0^0 = 2.31 eV[p]						OCT 1974
d $^3\Sigma^+$				[(2.09)][c]			[(2.02)]		(47019)[d]	(10)
$c_2(^3\Pi_2)$	(47025)	[644.7] Z	(29)	2.36[q]	0.12	[1.6]	1.90	c_2← X,[r] R	46748.1 Z	(10)*
$c_1(^3\Pi_1)$	(46939)	[(716.0)][s]	(32)	(2.58)[s]	(0.13)	(1.6)[s]	(1.82)	c_1← X, R	(46700)[s]	(10)*
D $^1\Pi$		Constants for v=1:		2.35[f]		1.6	1.90	D← X, R	47160.3 Z	(10)
b $(^3\Delta_1)$ 1				[2.468][g]		[0.95]	[1.859]	b← X,[h] R	44600.5 Z	(10)
B $^1\Sigma^+$	(44476)	[(811)][i]	(36)	(2.52)[i]	(0.12)	(1.10)[i]	(1.84)	B← X, R	$(44287._0)$[i]	(10)*
a $^3\Pi_r$	(41700)[j]	(1040)	(25)	(>3.2)[j]			$(<1.6_3)$			(10)
C $^1\Pi$	(41269)	(1108)[k]	(25)	[3.335]	(0.118)[k]	[1.25]	$[1.599_1]$	C← X, V_R	41195.5 Z	(10)*
A $^1\Sigma^+$	29960.0_4	1160.82	31.73[ℓ]	3.154	0.100[ℓ]	0.97	1.644_3	A→ X, R	29911.22 Z	(4)(6)(7)*
X $^1\Sigma^+$	0	1250.70 Z	17.17	3.2572	0.0722	[0.859][n]	1.6180_4			

AgGa: [a]Thermochemical value (mass-spectrom.)(2).

(1) Biron, CR B 264, 1097 (1967).
(2) Gingerich, Blue, 18th Annual Conference on Mass Spectrometry and Allied Topics, San Francisco (June 1970); paper F2.

Ag^1H, Ag^2H:

[a]Graphical Birge-Sponer extrapolation of X $^1\Sigma^+$. Thermochemical value 2.4_9 eV (1).

[b]Strong perturbations. Constants valid near J=0 only.

[c]From perturbations in $c_1(v=0)$.

[d]Estimated energy of v'=0,N'=0 relative to X $^1\Sigma(v=0,N=0)$.

[e]The only observed level is strongly perturbed by more than one state. Constants refer to unperturbed region around J=19. Origin of 0-0 band (not observed) at 46692 cm^{-1}.

[f]Perturbations in v=0,1; bands going to v'=0 of Ag^2H have not been analyzed. Constants refer to unperturbed regions near J=0. Large Λ-type doubling;
Ag^1H: $\Delta v_{ef}(v=0) = +0.244\,J(J+1) - \ldots$
Ag^2H: $\Delta v_{ef}(v=1) = +0.077_2\,J(J+1)$.

[g]Constants refer to unperturbed region near J=0. Large Ω-type doubling;
Ag^1H: $\Delta v_{ef} = +0.156\,J(J+1) - \ldots$
Ag^2H: $\Delta v_{ef} = +0.033_7\,J(J+1) + \ldots$

[h]P branch very weak or absent.

[i]Approximate constants for the deperturbed state. Origins of the 0-0 bands at 44225.0 cm^{-1} (Ag^1H) and 44277.4 cm^{-1} (Ag^2H). v=1 is free of perturbations:
Ag^1H: $B_1 = 4.412$, $D_1 = 3.5 \times 10^{-4}$, $v_0(1-0) = 45322.8$ cm^{-1};
Ag^2H: $B_1 = 2.343$, $D_1 = 0.95 \times 10^{-4}$, $v_0(1-0) = 45097.8$ cm^{-1}.

[j]Constants estimated from perturbations in C $^1\Pi$, B $^1\Sigma^+$,

Ag^1H, Ag^2H (continued):

D $^1\Pi$ of Ag^1H, and b 1 of Ag^2H; vibrational numbering uncertain. $+150 \leq A \leq +200$.

[k]Approximate constants for the deperturbed state. Perturbations in v=0,1,2 are caused by $a\,^3\Pi$. $\Delta G(1/2, 3/2)$ = 1519, 1453 (Ag^1H); 1056, 987 (Ag^2H).

[l]Anomalous potential curve; see (6). Constants for higher vibrational levels of Ag^1H in (2)(3). The constants for Ag^2H ($\omega_e y_e = -1.94_2$, $\gamma_e = -0.009$) are valid for $v \leq 4$.

[m]Higher order terms are needed to represent levels with $v \geq 4$; see (2)(3)(5).

[n]$H_0 = +88 \times 10^{-10}$ (Ag^1H); $+6 \times 10^{-10}$ (Ag^2H).

[o]RKRV potential curve (9).

[p]From the value for Ag^1H.

[q]Constants refer to unperturbed regions near J=0. Small Λ-type doubling.

[r]R, P much weaker than Q branches.

[s]Approximate constants for the deperturbed state. v=0 strongly perturbed by more than one state. Origin of 0-0 band (not observed) at 46794 cm^{-1}.

(1) Farkas, ZPC B 5, 467 (1929).
(2) Bengtsson-Knave, Olsson, ZP 72, 163 (1931).
(3) Bengtsson-Knave, NARSSU (IV) 8(4) (1932).
(4) Koontz, PR 48, 138 (1935).
(5) Gerö, Schmid, ZP 121, 459 (1943).
(6) Learner, PRS A 269, 327 (1962).
(7) Ringström, AF 21, 145 (1962).
(8) Loginov, OS(Engl. Transl.) 16, 220 (1964).
(9) Singh, Rai, CJP 43, 1685 (1965).
(10) Ringström, Åslund, AF 32, 19 (1966).

State	T_e	ω_e	$\omega_e x_e$	B_e	α_e	D_e ($10^{-8}cm^{-1}$)	r_e (Å)	Observed Transitions Design.	Observed Transitions ν_{00}	References
$^{(107)}Ag^{165}Ho$		(μ = 64.862383)		$D_0^0 = 1.2_6$ eV[a]						NOV 1974
$^{107}Ag^{127}I$		$\mu = 58.024719_5$		$D_0^0 = 2.6$ eV[a]						NOV 1974
D	(45487)	(176)[b] H	(2.5)					D← X, R		(3)(6)
C	44721	$155._5$[c] H	1.7					C← X, R	44695 H	(3)(6)
		Continuous absorption above 42400 cm^{-1}.								(3)
		Continuous absorption 29800 - 33300 cm^{-1}, maximum at 31500 cm^{-1}.								(1)(2)(12)
B 0^+	31194.06	127.14 Z	4.255[d]	0.040716	0.000572[d]	[2.19][e]	2.6712	B⟷X, R	31153.42 Z	(1)* (2)(5) (7)(10)(11)
		Continuous absorption above 24000 cm^{-1}.								(4)
A	23906	151.2 H						A← X, R	23879 H	(4)
X $^1\Sigma^+$	0	206.52 Z	0.445	0.04486806	0.0001413_9[f]	0.847[g]	2.544621	Microwave sp.[h]		(13)
$^{(107)}Ag^{115}In$		$(\mu = 55.380129_3)$		$D_0^0 = 1.6_9$ eV[a]						NOV 1974
B	33535.52	140.66[b] H	0.462					B← X, R	33527.91[b] H	(2)
A	32471.41	133.20[c] H	0.377					A← X, R	32460.41[c] H	(1)
X ($^1\Sigma$)	0	155.54[d] H	0.423							
$^{(107)}Ag^{(7)}Li$		(μ = 6.5839134)		$D_0^0 = 1.81$ eV[a]						APR 1975
$^{(107)}Ag^{23}Na$		$(\mu = 18.920867_7)$		$D_0^0 = 1.40$ eV[a]						NOV 1974
$^{107}Ag^{16}O$		$\mu = 13.9132446_1$		$D_0^0 = 2.2_9$ eV[a]						NOV 1974 A
		Unclassified band system in the red; no details.								(1)
B $^2\Pi$ 3/2	x_2+28113.8	539.1[b] H	6.1_5	[0.3195]		[50]	$[1.947_4]$	B→X, V	28137.3[b] H	(2)*
B $^2\Pi$ 1/2	x_1+28072.3	535.7[b] H	6.3_4	[0.3178][bc]		[51][b]	$[1.952_6]$		28094.2[b] H	
A $^2\Pi$ 3/2	x_2+24540	[241.1][b] H	d	[0.2816][d]		[171]	[2.074]	A→X, R	24416.0[b] H	(2)*
A $^2\Pi$ 1/2	x_1+24370	[237.3][b] H	d	[0.2812][cd]		[168]	[2.076]		24244.5[b] H	

AgO (continued):

[e]Not quite certain that this is the ground state. It appears to be regular with $A \approx +135\ cm^{-1}$, contrary to expectations; see (3).

(1) Loomis, Watson, PR 48, 280 (1935).
(2) Uhler, AF 7, 125 (1954).
(3) Cheetham, Barrow, AdHTC 1, 7 (1967).
(4) Smoes, Mandy, Vander Auwera-Mahieu, Drowart, BSCB 81, 45 (1972).

AgS, AgSe, AgTe:

[a]Thermochemical value (mass-spectrom.)(2).
[b]Analysis uncertain.

(1) Maheshwari, PPS 81, 514 (1963).
(2) See ref. (4) of AgO.

AgSb: [a]According to (1) the bandhead measurements refer to $^{109}Ag^{121}Sb$.

(1) Lefebvre, Lochet, CR B 275, 85 (1972).

AgSn: [a]Thermochemical value (mass-spectrom.)(1).

(1) Ackerman, Drowart, Stafford, Verhaegen, JCP 36, 1557 (1962).

State	T_e	ω_e	$\omega_e x_e$	B_e	α_e	D_e $(10^{-7}cm^{-1})$	r_e (Å)	Observed Transitions Design.	ν_{00}	References
$^{27}Al_2$		μ = 13.4907703		$D_0^0 = 1.5_5$ eV[a]						NOV 1975
A $^3\Sigma_u^-$	17269.3_6	278.80[b] H	0.831[c]	0.1907	0.0013	3.9	2.560	A→X, R	17234.0_5 H	(1)* (2)*
X $^3\Sigma_g^-$	0	350.01[b] H	2.022[d]	0.2054	0.0012	3.0_7	2.466			
$^{27}Al^{79}Br$		μ = 20.1070892		$D_0^0 = 4.4_3$ eV[a]						NOV 1975 A
A $^1\Pi$	35879.5	297.2 H^Q	6.40[b]	0.1555[c]	0.00216[d]	1.8[e]	2.322	A↔X, R	35837.8 H^Q	(1)(2)* (3) (4)* (5)
a $^3\Pi_1$	23779.3	410.32 H^Q	1.75	0.164[f]	0.001		2.26	a→X, V	23795.3 H^Q	(4)(6)*
a $^3\Pi_0$	23647	411.2 (Z)	1.75						23663[g] (Z)	
X $^1\Sigma^+$	0	378.0 H^Q	1.28	0.15919713	0.00086044$_9$[h]	1.1285[i]	2.294807[j]	Microwave sp.[k]		(9)
AlC										NOV 1975
		The spectrum originally attributed to AℓC (1) was later shown (2) to be due to $A\ell_2$.								
$^{27}Al^{35}Cl$		μ = 15.2301459		$D_0^0 = 5.12$ eV[a]						NOV 1975 A
		Fragments of additional systems in the region 48200 - 54900 cm^{-1}.								(7)
b ($^3\Sigma$)	(43591)	(350)[b]		[0.226][c]			[2.21]	b→a, R	18847.40 H^Q 18911 (Z) 18975.00 H^Q	(5)*
A $^1\Pi$	38254.0	449.96 H^Q	4.37[d]	0.259[e]	0.006		2.06_7	A↔X, R_V	38237.7 H^Q	(1)* (2)(3) (4)
a $^3\Pi_r$	24658 24593.84 24528	524.35 H^Q	2.175	0.250[f]	0.002		2.10	a→X, V	24680 (Z) 24615.31 H^Q 24541.65 H^P	(5)*
X $^1\Sigma^+$	0	481.30 H^Q	1.95	0.24393012	0.00161113[g]	2.502_7[h]	2.130113[i]	Microwave sp.[j]		(9)(11)

Al_2: [a]Thermochemical value (mass-spectrom.)(5). Values in (3)(4) are somewhat higher.
[b]From (1) who wrongly attributed the spectrum to AlC.
[c]$\omega_e y_e = -0.010_4$.
[d]$\omega_e y_e = -0.010_5$.

(1) Zeeman, CJP 32, 9 (1954).
(2) Ginter, Ginter, Innes, ApJ 139, 365 (1964).
(3) Blue, Gingerich, 16th Annual Conference on Mass-Spectrometry and Allied Topics, Pittsburgh (1968); paper 129.
(4) Uy, Drowart, TFS 67, 1293 (1971).
(5) Stearns, Kohl, HTS 5, 113 (1973).

AlBr: [a]Thermochemical value (8); 4.58 eV from prediss. in $A\,^1\Pi$.
[b]$\omega_e y_e = -0.527$. This state may have a potential hump of $\sim$0.20 eV (7)(8).
[c]Predissociation for $v > 3$. According to (11) emission from v=2, 3 breaks off at J=93, 67, resp..
[d]$\gamma_e = -0.000175$.
[e]Rapidly increasing with v.
[f]Estimated from P and Q head separations.
[g]P head at 23657.9 cm^{-1}.
[h]$\gamma_e = +2.030 \times 10^{-6}$.
[i]$\beta_e = -0.0020_7 \times 10^{-7}$; $H_e = -1.119 \times 10^{-14}$.
[j]From the corrected $B_e = 0.15920431$.
[k]For values of eqQ see (10).

(1) Crawford, Ffolliott, PR 44, 953 (1933).
(2) Howell, PRS A 148, 696 (1935).
(3) Mahanti, IJP 9, 369 (1935).
(4) Miescher, HPA 8, 279 (1935); 9, 693 (1936).
(5) Jennergren, Nature 161, 315 (1948); AMAF A 35, No. 22 (1948).
(6) Sharma, ApJ 113, 219 (1951).
(7) Barrow, TFS 56, 952 (1960).
(8) Barrow, Nature 189, 480 (1961).

AlBr (continued):

(9) Wyse, Gordy, JCP 56, 2130 (1972).
(10) Hoeft, Törring, Tiemann, ZN 28 a, 1066 (1973).
(11) Ram, Upadhya, Rai, Singh, OPA 6, 38 (1973).

AlC: (1)(2) See ref. (1)(2), respectively, of Al_2.

AlCl: [a]Thermochemical value [see Appendix of ref.(10)]. See also (8).
[b]Estimated from observed isotope shifts.
[c]$v > 0$ probably predissociated.
[d]$\omega_e y_e = -0.216$. This state may have a potential hump of $\sim$0.26 eV (6)(8).
[e]Predissociation in v=10 (3).
[f]Estimated from P and Q head separations.
[g]$+ 4.69_7 \times 10^{-6}(v+\frac{1}{2})^2 - 5._7 \times 10^{-9}(v+\frac{1}{2})^3$.
[h]$\beta_e = -0.0053 \times 10^{-7}$; $H_e = -4.45_6 \times 10^{-14}$.
[i]From the corrected $B_e = 0.2439422_9$.
[j]$\mu_{el} = 1$ to 2 D (9). Values of eqQ in (9)(12).

(1) Bhaduri, Fowler, PRS A 145, 321 (1934).
(2) Mahanti, ZP 88, 550 (1934).
(3) Holst, ZP 93, 55 (1934); Dissertation (Stockholm, 1935).
(4) See ref. (4) of AlBr.
(5) Sharma, ApJ 113, 210 (1951).
(6) Barrow, JCP 22, 573 (1954).
(7) Reddy, Rao, CJP 35, 912 (1957).
(8) See ref. (7) of AlBr.
(9) Lide, JCP 42, 1013 (1965); 46, 1224 (1967) (erratum).
(10) Hildenbrand, Theard, JCP 50, 5350 (1969).
(11) See ref. (9) of AlBr.
(12) See ref. (10) of AlBr.

State	T_e	ω_e	$\omega_e x_e$	B_e	α_e	D_e ($10^{-7}cm^{-1}$)	r_e (Å)	Observed Transitions Design.	ν_{00}	References
$^{27}Al^{19}F$		μ = 11.1484740		$D_0^0 = 6.8_9$ eV[a]		I.P. = 9.8 eV[b]				MAR 1976 A
		Unclassified V shaded bands in the region 70900 - 74600 cm^{-1}; in absorption.								(6)*
H $^1\Sigma^+$	67320	958 H	$7._0$	[0.59214]		[8.3]	[1.5980]	H→B, V	13114.57 Z	(17)(26)
								H→A, V	23447.32[c] Z	(17)(26)
								H←X, V	67397.03 Z	(5)(6)*
g $^3\Sigma^+$	(66910)			[0.59544]		[9.51]	[1.5936]	g→b, V	22177.12 Z	(23)(26)
G $^1\Sigma^+$	66334.0	[931.46] Z	$8._0$ H	0.60490	0.00767[d]	[10.26]	1.5811	G→B, V	12123.34 Z	(26)
								G→A, V	22456.09[c] Z	(9)* (26)
								G←X, V	66405.81 Z	(5)(6)*
f $^3\Pi$	65803	[938.90] Z	(5.9)	0.59355[ef]	0.00480	[9.29]	1.5961	f→c, V	10853.84[c] Z	(26)
								f→b, V	21072.71[c] Z	(17)(23)(26)
								f→a, V	38576.1 38623.6 38670.9	(7)(10)
F $^1\Pi$	65795.6	955.33 Z	5.38	0.59281[gf]	0.00459	[8.78]	1.5971	F→B, V	11589.46[c] Z	(9)* (26)
								F→A, V	21922.22[c] Z	(9)* (26)
								F←X, V	65871.95[c] Z	(5)(6)*
e $^3\Sigma^+$	(65010)			[0.59464]		[8.4]	[1.5946]	e→c, V	10064.76 Z	(23)(26)*
								e→b, V	20283.63 Z	(17)(23)(26)
E $^1\Pi$	63689.4	923.02 Z	5.28	0.58709[h]	0.00464	[9.46]	1.6049	E→A, V	19799.95[c] Z	(17)(26)
								E←X, V	63749.68[c] Z	(5)(6)
d ($^3\Delta$) [i]	(63203)	$[930._2]$ (Z)						d→a, V	36017.6 (Z)	(7)(10)
D $^1\Delta$ [i]	61229.5	901.05 Z	6.11	0.58297	0.00502	[9.87]	1.6105	D→A, V	17328.85[c] Z	(4)(9)* (26)

AℓF: [a]Thermochemical value, see Appendix of ref. (20), also (21).

[b]Interpretation of Rydberg states (26). Electron impact appearance potentials vary from 9.5 to 10.1 eV (16)(19)(21).

[c]Band origins as defined in (26); add $B''\Lambda''^2 - B'\Lambda'^2$ to obtain zero lines.

[d]Constants from (26). Small discrepancy with the B_1 value in the same paper.

[e]Λ-type doubling $\Delta v_{fe} = + 0.00338 \times N(N+1)$.

[f]Interactions between levels of F $^1\Pi$ and f $^3\Pi$.

[g]Λ-type doubling $\Delta v_{fe}(v=0) = + 0.00559 \times J(J+1) - \ldots$, decreasing with increasing v.

[h]Λ-type doubling $\Delta v_{fe} = + 0.00025 \times J(J+1)$.

[i]Compare with ab initio calculations by (28).

(continued p. 22)

State	T_e	ω_e	$\omega_e x_e$	B_e	α_e	D_e $(10^{-7} cm^{-1})$	r_e (Å)	Observed Transitions		References
								Design.	v_{00}	
$^{27}Al^{19}F$ (continued)										
C $^1\Sigma^+$	57688.0	938.22 Z	5.09[j]	0.58992	0.00458[k]	9.23[ℓ]	1.6010	C→A, V	13806.16[c] Z	(2)(3)* (4) (26)
								C↔X, V	57755.89 Z	(5)(6)(11) (13)
c $^3\Sigma^+$	54957.7	933.66 Z	4.81	0.58861[m]	0.00457	9.8	1.6028	c→b, V	10218.89 Z	(23)(26)
								c→a, V	27722.2 27769.8 27817.1	(4)* (10) (26)
B $^1\Sigma^+$	54251.0	866.6_0 Z	7.4_5[n]	0.57968	0.00560	[10.49]	1.6151	B↔X, V	54282.46 Z	(5)(6)* (11) (13)(26)
b $^3\Sigma^+$ i	44813.2[o]	786.37 Z	7.64[p]	0.56280[mq]	0.00651[r]	11.5	1.6391	b→a, V_R	17503.4 17550.9 17598.2	(4)(23)(26)
A $^1\Pi$ i	43949.2	803.9_4 Z	5.99[s]	0.55640[tq]	0.00534[u]	10.56[v]	1.6485	A↔X, V_R	43949.73[c] Z	(1)* (2)(3) (4)* (5) (12)(25)
a $^3\Pi_r$ i	27241[w]	827.8 Z	3.9	0.5570_3	0.00453	[9.82]	1.6476	a→X,	27254	(29)
X $^1\Sigma^+$ i	0	802.2_6 Z	4.77	0.5524798	0.0049841[x]	10.464[y]	1.654369	Microwave sp.[z]		(18)(22)(24)

AℓF (continued):

[j] $\omega_e y_e = -0.017$.
[k] $\gamma_e = -0.000011$.
[ℓ] $H_e \approx -1.1 \times 10^{-12}$.
[m] Magnetic hyperfine structure; very small spin splitting.
[n] $\omega_e y_e = -0.04_5$.
[o] Relative energies of singlet and triplet states derived from the analysis of spin-forbidden perturbations, see [q].
[p] $\omega_e y_e = -0.009$.
[q] Interactions between levels of A $^1\Pi$ and b $^3\Sigma^+$ (26).
[r] $\gamma_e = -0.00002$.
[s] $\omega_e y_e = -0.050$. This state may have a potential maximum of ~0.35 eV; see e.g. (15).
[t] Λ-type doubling $\Delta v_{fe} = +0.00010\, J(J+1)$.
[u] $\gamma_e = -0.000043$.

AℓF (continued):

$^{v}H_e \approx -2.0 \times 10^{-12}$.

$^{w}A = +47\ cm^{-1}$.

$^{x} + 1.71_8 \times 10^{-5}(v+\frac{1}{2})^2 + 4._7 \times 10^{-8}(v+\frac{1}{2})^3$. From (24); slightly different constants in (22).

$^{y}\beta_e = -0.015 \times 10^{-7}$.

$^{z}\mu_{e\ell} = 1.5_3$ D (18). For eqQ(Aℓ) see (22)(24)(27). Additional constants derived from Zeeman effect measurements (g_J, mol. quadrupole moment, etc.) in (27).

(1) Rochester, PR 56, 305 (1939).
(2) Naudé, Hugo, PR 90, 318 (1953).
(3) Naudé, Hugo, CJP 31, 1106 (1953).
(4) Rowlinson, Barrow, PPS A 66, 437 (1953).
(5) Rowlinson, Barrow, PPS A 66, 771 (1953).
(6) Barrow, Rowlinson, PRS A 224, 134 (1954).
(7) Dodsworth, Barrow, PPS A 67, 94 (1954).
(8) Gross, Hayman, Levi, TFS 50, 477 (1954).
(9) Naudé, Hugo, CJP 32, 246 (1954).
(10) Dodsworth, Barrow, PPS A 68, 824 (1955).
(11) Naudé, Hugo, CJP 33, 573 (1955).
(12) Barrow, Johns, Smith, TFS 52, 913 (1956).
(13) Naudé, Hugo, CJP 35, 64 (1957).
(14) Witt, Barrow, TFS 55, 730 (1959).
(15) Barrow, TFS 56, 952 (1960).
(16) Porter, JCP 33, 951 (1960).
(17) Barrow, Kopp, Scullman, PPS 82, 635 (1963).
(18) Lide, JCP 38, 2027 (1963); 42, 1013 (1965).
(19) Ehlert, Blue, Green, Margrave, JCP 41, 2250 (1964).
(20) Hildenbrand, Murad, JCP 44, 1524 (1966).
(21) Murad, Hildenbrand, Main, JCP 45, 263 (1966).
(22) Hoeft, Lovas, Tiemann, Törring, ZN 25 a, 1029 (1970).
(23) Kopp, Barrow, JP B 3, L118 (1970).
(24) Wyse, Gordy, Pearson, JCP 52, 3887 (1970).
(25) Liszt, Smith, JQSRT 12, 947 (1972).
(26) Barrow, Kopp, Malmberg, PS 10, 86 (1974); USIP Report 74-15 (March 1974).
(27) Honerjäger, Tischer, ZN 29 a, 342 (1974).
(28) So, Richards, JP B 7, 1973 (1974).
(29) Rosenwaks, Steele, Broida, CPL 38, 121 (1976).

State	T_e	ω_e	$\omega_e x_e$	B_e	α_e	D_e ($10^{-4}cm^{-1}$)	r_e (Å)	Observed Transitions Design.	Observed Transitions ν_{00}	References
$^{27}Al^{1}H$		μ = 0.97153602		$D_0^0 < 3.06$ eV[a]						NOV 1975 A
c $^3\Pi_r$				b				c → a,	(R)(36950)	(15)*
b $^3\Sigma$				[6.759][c]		[4.36]	[1.602_2]	b → a,	26217 Z	(5)(9)
a $^3\Pi_r$	a[d]			[6.704]		[4]	[1.608_8]			
E $^1\Pi$				[5.620][e]		[10.0][e]	[1.757_1]	E → A, R	29512.2 Z	(7)(26)
								E ↔ X, R	52982.9 Z	(13)(26)*
D $^1\Sigma^+$				[6.56][f]		[6.1]	[1.62_6]	D ↔ X, V_R	49288 Z	(3)(11)(13)(17)*
C $^1\Sigma^+$	44676	1575.3[g]	125.5	6.664[h]	0.544	[5.5][i]	1.613_6	C → A, V	21127.0 Z	(1)(13)*
								C ↔ X, R_V	44597.9 Z	(11)(13)(17)*
A $^1\Pi$		[1082.76] Z	k	6.3869[jkℓ]	0.7323	[6.200][m]	1.648	A ↔ X, R_V	23470.93 Z	(2)(4)(6)*(8)(10)(13)*(14)(16)*(20)(25)
X $^1\Sigma^+$ n	0	1682.56 Z	29.09[o]	6.3907	0.1858[p]	3.565[q]	1.6478			
$^{27}Al^{2}H$		μ = 1.87419817		$D_0^0 < 3.09$ eV[a]						NOV 1975
c $^3\Pi_r$	r	(800)		s		s		c → a, R	36959.2 H	(15)*
a $^3\Pi_r$		[1237.4] Z		3.545	0.096	1.1	1.592_9			
E $^1\Pi$				[2.995][t]		[2.95][t]	[1.733_0]	E → A, R	29546.0 Z	(26)
								E → X, R	53082.8 Z	(26)*
G $^1\Sigma^+$				[3.40][u]			[1.62_6]	G → A, R	29517.9 Z	(26)
								G → X, R	53054.7 Z	(26)*
D $^1\Sigma^+$				[3.45]		[1.14]	[1.61_5]	D → X, V	49288 Z	(11)
C $^1\Sigma^+$	44686	1134.2[g]	65.1	3.438	0.176	1.8	1.617_5	C → X, R_V	44634.4 Z	(11)(19)
A $^1\Pi$	23653	1014.6 Z	86.0[k]	3.235[vk]	0.122[w]	[1.67_5][x]	1.667	A → X, R	23536.8 Z	(6)* (13)*
X $^1\Sigma^+$	0	1211.95 Z	15.14[y]	3.3186	0.0697[z]	[0.97]	1.6463			

$A\ell^1H$, $A\ell^2H$:

[a]From the predissociation in A $^1\Pi$ (10)(12). The most recent theoretical calculations for $A\ell^1H$ (28) recommend 3.05 eV.

[b]v=0 perturbed.

[c]Perturbation at N=19, predissociation at N > 25.

[d](22) estimate $a \approx 12000$ cm^{-1}. A = +40.2.

[e]Constants for the f component. Large Λ-type doubling, Δv_{fe} positive; J. W. C. Johns [see (26)] reports a new $^1\Sigma^+$ state just above E $^1\Pi$. Predissociation for J > 12 (7)(26).

[f]Perturbed at J=5, predissociated for J > 10.

[g]From $\Delta G(\frac{1}{2})$ of the hydride and deuteride.

[h]Perturbations.

[i]$D_1 = 10 \times 10^{-4}$.

[j]Λ-type doubling, $\Delta v_{ef}(v=0) = +0.0097\,J(J+1) - \ldots$.

[k]Predissociation by rotation in v=0,1 ($A\ell^1H$) and v=0,1,2 ($A\ell^2H$); this state has a potential maximum (12) of ~0.15 eV (18).

[ℓ]Zeeman effect (21).

[m]$D_1 = 11.20 \times 10^{-4}$; also higher order constants (16).

[n]For theoretical calculations concerning the ground state of AℓH see (23)(24)(27)(28).

[o]$\omega_e y_e = +0.239$.

[p]$\gamma_e = +0.0016_1$.

[q]$\beta_e = -0.047 \times 10^{-4}$.

[r]0 < A < 12.

[s]v=0 perturbed. v=1 ($B_1 = 2.938$, $D_1 = 3 \times 10^{-5}$) predissociated at N≈16 and perturbed in $^3\Pi_0$.

[t]Constants for the f component. The e component interacts with G $^1\Sigma^+$.

[u]Approximate deperturbed value; interaction with the e component of E $^1\Pi$.

[v]Λ-type doubling, $\Delta v_{ef} = +0.0020\,J(J+1)$.

[w]$\gamma_e = -0.05_2$.

[x]$D_1 = 2.2_1 \times 10^{-4}$, $D_2 = 6.9_6 \times 10^{-4}$.

[y]$\omega_e y_e = +0.098$.

[z]$\gamma_e = +0.00055$.

(1) Bengtsson, Hulthén, ZP 52, 275 (1929).
(2) Farkas, ZP 70, 733 (1931).
(3) Bengtsson-Knave, NARSSU 8, No. 4 (1932).
(4) Farkas, Levy, ZP 84, 195 (1933).
(5) Holst, ZP 86, 338 (1933).
(6) Holst, Hulthén, ZP 90, 712 (1934).
(7) Holst, ZP 90, 728 (1934).
(8) Holst, ZP 90, 735 (1934).
(9) Challacombe, Almy, PR 51, 930 (1937).
(10) Olsson, Dissertation (Stockholm, 1938).
(11) Grabe, Hulthén, ZP 114, 470 (1939).
(12) Herzberg, Mundie, JCP 8, 263 (1940).
(13) Nilsson, Dissertation (Stockholm, 1948); AMAF A 35, No. 19 (1948).
(14) Kleman, Lagercrantz, Uhler, AF 2, 359 (1950).
(15) Kleman, AF 6, 407 (1953); Dissertation (Stockholm, 1953).
(16) Zeeman, Ritter, CJP 32, 555 (1954)
(17) Khan, PPS 71, 65 (1958).
(18) Hurley, PRS A 261, 237 (1961).
(19) Khan, PPS 79, 745 (1962).
(20) Loginov, OS (Engl. Transl.) 16, 220 (1964).
(21) Klynning, Neuhaus, AF 28, 249 (1965).
(22) Grimaldi, Lecourt, Lefebvre-Brion, Moser, JMS 20, 341 (1966).
(23) Cade, Huo, JCP 47, 649 (1967).

(continued p. 27)

State	T_e	ω_e	$\omega_e x_e$	B_e	α_e	D_e ($10^{-5}cm^{-1}$)	r_e (Å)	Observed Transitions: Design.	Observed Transitions: ν_{00}	References
$^{27}Al^{1}H^{+}$										DEC 1975
A $^2\Pi_r$	$(27686)^a$ (27593)	(1770)		6.851^b	0.248	41	1.591_4	A→X, V	27760.2 Z 27667.8 Z	(1)* (2)
X $^2\Sigma^+$	0	(1620)		6.763^c	0.398	47	1.601_8			
$^{27}Al^{127}I$		$\mu = 22.2507434$		$D_0^0 = 3.7_7$ eV[a]						DEC 1975 A
A $^1\Pi$		Unstable; diffuse fluctuation bands with various v".						A←X,	31487^b	(1)*
a $^3\Pi_1$	22089.5	333.4 H	2.0					a↔X, V	22097.9 H	(1)*
a $^3\Pi_0$	21889.3	337.2 H	2.0					V_R	21899.6 H	
X $^1\Sigma^+$	0	316.1 H	1.0	0.11769985	0.000558586^c	d	2.53710_2	Microwave sp.[e]		(3)
$^{27}Al^{14}N$		$\mu = 9.2186914_7$								DEC 1975
A $^3\Pi_i$	a			$[0.5811]^{bc}$		[0.139]	[1.7739]	A→X, V	19727.37 Z	(1)*
(X) $^3\Pi_i$ [d]	e	[746.93] Z		0.5730^b	0.0056	[0.136]	1.7864			

$A\ell^1H$, $A\ell^2H$ (continued):

(24) Cade, Bader, Henneker, Keaveny, JCP 50, 5313 (1969).
(25) Huron, Physica 41, 58 (1969).
(26) Lagerqvist, Lundh, Neuhaus, PS 1, 261 (1970).
(27) Laws, Stevens, Lipscomb, JCP 54, 4269 (1971).
(28) Meyer, Rosmus, JCP 63, 2356 (1975).

$A\ell^1H^+$: [a]$A = +108$.
[b](2) finds breaking-off in P and R at N=28 (v=0).
[c]Spin splitting constant $\gamma \approx +0.06$.

(1) Almy, Watson, PR 45, 871 (1934).
(2) Holst, ZP 89, 40 (1934).

$A\ell I$: [a]From the analysis of fluctuation bands (2).
[b]Vertical transition from v"=0.
[c]$+ 1.047_7 \times 10^{-6}(v+\frac{1}{2})^2 - 2.46 \times 10^{-9}(v+\frac{1}{2})^3$.
[d]$D_v = [6.521_2 - 0.0095_7(v+\frac{1}{2})] \times 10^{-8}$; $H_e = -4.590 \times 10^{-15}$.
[e]Values of eqQ_v (v=0...3) in (4).

(1) Miescher, HPA 8, 279 (1935); 9, 693 (1936).
(2) Barrow, TFS 56, 952 (1960).
(3) Wyse, Gordy, JCP 56, 2130 (1972).
(4) Törring, Tiemann, Hoeft, ZN 28 a, 1062 (1973).

$A\ell N$: [a]$A_0 = -23._0$.
[b]All lines in the $^3\Pi_2 - {}^3\Pi_2$ and high J lines in the $^3\Pi_0 - {}^3\Pi_0$ subbands are broad (~ 0.15 cm^{-1}), probably on account of unresolved nuclear hyperfine structure.
[c]Predissociation near J=48.
[d]Not certain that this is the ground state.
[e]$A_0 = -33._0$, $A_1 = -34._5$.

(1) Simmons, McDonald, JMS 41, 584 (1972).

State	T_e	ω_e		$\omega_e x_e$	B_e	α_e	D_e ($10^{-6}cm^{-1}$)	r_e (Å)	Observed Transitions: Design.	Observed Transitions: ν_{00}	References
$^{27}Al^{16}O$		μ = 10.0419507$_1$			D_0^0 = 5.27 ± 0.04 eV[a] I.P. = 9.5$_3$ eV[b]						DEC 1975 A
		Additional states and transitions predicted by (43).									
F $^2\Sigma^+$	[47677.3]				[0.5088]		[2.15]	[1.816$_4$]	F→A, R	41843.5$_2$ Z 41972.3$_6$ Z	(44)*
E $^2\Delta_i$ c	45562[d] 45431	(503)			[0.4951] [0.4919]		[1.9]	[1.844$_4$]	E↔A, R	39979.81 Z 39977.17 Z	(4)(25)(28)* (31)*
D $^2\Sigma^+$ c	40266.7	819.6	Z	5.8	0.5652[e]	0.0046[f]	1.10	1.723$_4$	(D→B),[g] R	(19552)	(30)
									D↔A, V	34841.23 Z 34970.09 Z	(4)(25)(28)*
									D↔X, R	40187.2 Z	(23)* (25) (31)(32)* (39)(40)(45)
C $^2\Pi_r$ c	33153 33079	856	H	6	h				(C→B),[i] R_V	(12457) (12383)	(10)(18)*
									C↔X, R	33092 33018	(4)(10)* (12)(25) (26)* (29)
B $^2\Sigma^+$ j	20688.95	870.05	Z	3.52	0.6040$_8$[k]	0.0044$_7$	1.16	1.6670	B[ℓ]↔X, R	20635.22 Z	(1)(2)* (3)* (6)* (7)* (9) (11)(14)(15) (16)(17)* (21)* (24) (25)(30)(31)* (33)(36)(37)
A $^2\Pi_i$ j	5470.6[m] 5341.7	728.5	H	4.15	[0.5374][no] [0.5333][o]		[1.1]	[1.7708]	A↔X,[p] R	5346 5217	(33)(51)
X $^2\Sigma^+$ j	0	979.23	Z	6.97	0.6413$_6$[qr]	0.0058$_0$	1.08[s]	1.6179	IR[t] and ESR[u] sp.		

AlO: [a]Lower limit from the laser fluorescence study of AlO formed in the reaction Al + O_2 (49), upper limit from the re-interpretation (28)(41) of the long-wavelength limit of an absorption continuum (25). Good agreement with the most recent mass-spectrometric results (34)(42); slightly lower value by flame photometry (48). Further references reviewed in (49).

[b]Electron impact appearance potential (13)(22)(42).

[c]Theoretical calculations concerning these states in (43).

[d]A_0 = -64.8.

[e]Spin splitting constant γ = +0.0060 (50).

[f]γ_e = -0.00005.

$A\ell O$ (continued):

[g]Uncertain observation. See also (43).

[h]Rotational constants in (10) are unreliable; see (29). Perturbations.

[i]Uncertain identification (18). See also (43).

[j]Theoretical calculations concerning these states (46)(47).

[k]Spin splitting constant $\gamma = -0.0074$ (50), disagreeing with +0.020 in (6)(10).

[ℓ]Radiative lifetime $\tau = 102$ ns (49), 127 ns (35). The corresponding B–X oscillator strengths, $f_{00} = 0.027$ and 0.021, resp., are compared in (49) with additional experimental and theoretical (38)(46) results. Relative b. strengths (27).

[m]$A_0 = -127.8$.

[n]Λ-type doubling $\Delta v_{fe} = -0.0128(J+\frac{1}{2})$.

[o]Slightly different constants in (44).

[p]Theoretical oscillator strengths in (46).

[q]Spin splitting constant $\gamma_0 = +0.0050$ (50), smaller than earlier values (6)(10) and in better agreement with ESR (33) and theoretical (50) results.

[r]Perturbations (5)(9) by A $^2\Pi_i$ (28).

[s]$\beta_e = +0.02 \times 10^{-6}$.

[t]Absorption f-number for the IR fundamental band 0.000033 (52). Theoretical values in (46)(47)(52).

[u]In rare gas matrices at 4 K (33).

(1) Pomeroy, PR 29, 59 (1927).
(2) Sen, IJP 11, 251 (1937).
(3) Roy, IJP 13, 231 (1939).
(4) Coheur, Rosen, BSRSL 10, 405 (1941).
(5) Rosen, PR 68, 124 (1945).
(6) Lagerqvist, Nilsson, Barrow, PPS A 69, 356 (1956); AF 12, 543 (1957).
(7) Gatterer, Junkes, Salpeter, Rosen, METOX (1957).
(8) Gurvich, Veits, IANSF 22, 673 (1958). For engl. transl. see BASPS 22, 670 (1958).
(9) Shimauchi, SL 7, 101 (1958).
(10) Goodlett, Innes, Nature 183, 243 (1959).
(11) Loginov, OS(Engl. Transl.) 6, 67 (1959); 16, 220 (1964).
(12) Bécart, Declerck, CR 251, 2153 (1960).
(13) Drowart, De Maria, Burns, Inghram, JCP 32, 1366 (1960).
(14) Nicholls, JRNBS A 66, 227 (1962).
(15) Tawde, Korwar, PPS 80, 794 (1962).
(16) Bécart, Mahieu, CR 256, 5533 (1963); JP(Paris) 25, 873 (1964).
(17) Edse, Rao, Strauss, Mickelson, JOSA 53, 436 (1963).
(18) Tyte, Nature 202, 383 (1964).
(19) Hébert, Tyte, PPS 83, 629 (1964).
(20) Tyte, Hébert, PPS 84, 830 (1964).
(21) Tyte, Nicholls, IAMS 1 (1964).
(22) Burns, JCP 44, 3307 (1966).
(23) Krishnamachari, Narasimham, Singh, CJP 44, 2513 (1966).
(24) Sharma, JQSRT 7, 283, 289 (1967).
(25) Tyte, PPS 92, 1134 (1967).
(26) Mahieu, Bécart, CSp 13, 95 (1968).
(27) Linton, Nicholls, JQSRT 9, 1 (1969).
(28) McDonald, Innes, JMS 32, 501 (1969).
(29) McDonald, Innes, Goodlett, Tolbert, JMS 32, 511 (1969).
(30) Prasad, Narayan, IJPAP 7, 413 (1969).
(31) Singh, Narasimham, JP B 2, 119 (1969).
(32) Singh, PIAS A 71, 82 (1970).
(33) Knight, Weltner, JCP 55, 5066 (1971).
(34) Farber, Srivastava, Uy, JCS FT I 68, 249 (1972).
(35) Johnson, Capelle, Broida, JCP 56, 663 (1972).

(continued p. 31)

State	T_e	ω_e	$\omega_e x_e$	B_e	α_e	D_e $(10^{-7}cm^{-1})$	r_e (Å)	Observed Transitions Design.	ν_{00}	References
$^{27}Al^{31}P$		μ = 14.4200754		$D_0^0 = 2.2_0$ eV[a]						DEC 1975
$^{27}Al^{(32)}S$		(μ = 14.6327889)		$D_0^0 = 3.84$ eV[a]						DEC 1975
C $^2\Sigma^+$	(35797)	[430.0] H	(14)	0.2402[b]	0.0036	3.1	2.190	C← X, R	35714.9	(2)
B ($^2\Pi$)	(30104) (29986)	[530] Strong perturbations.						B← X, R	30061 H 29943 H	(2)(4)
A $^2\Sigma^+$	23433.80	510.9_1 Z	1.45	0.2461[cd]	0.0012	2.1	2.164	A↔X, R	23381.16 Z	(1)* (2)(4) (6)*
X $^2\Sigma^+$	0	617.1_2 Z	3.33	0.2799[e]	0.0018	2.2	2.029			
$^{27}Al^{(80)}Se$		(μ = 20.1712814)		$D_0^0 = 3.4_6$ eV[a]						DEC 1975
A ($^2\Sigma$)	23183.5	389.8 H	1.23					A↔X, R	23144.8 H	(2)
X ($^2\Sigma$)	0	467.6 H	2.08							
$^{27}Al^{(28)}Si$		(μ = 13.7351103)		$D_0^0 = 2.3_4$ eV[a]						DEC 1975
$^{27}Al^{(130)}Te$		(μ = 22.3412584)		$D_0^0 = 2.7_3$ eV[a]						DEC 1975

AℓO (continued):

(36) Gole, Zare, JCP 57, 5331 (1972).
(37) Liszt, Smith, JQSRT 12, 947 (1972).
(38) Michels, JCP 56, 665 (1972).
(39) Singh, JQSRT 12, 1343 (1972).
(40) Tawde, Tulasigeri, JP B 5, 1681 (1972).
(41) Drowart, private communication (1973).
(42) Hildenbrand, CPL 20, 127 (1973).
(43) Schamps, CP 2, 352 (1973).
(44) Singh, JP B 6, 521 (1973).
(45) Singh, Saksena, PIAS A 77, 139 (1973).
(46) Yoshimine, McLean, Liu, JCP 58, 4412 (1973).
(47) Das, Janis, Wahl, JCP 61, 1274 (1974).
(48) Frank, Krauss, ZN 29 a, 742 (1974).
(49) Dagdigian, Cruse, Zare, JCP 62, 1824 (1975).
(50) Mahieu, Jacquinot, Schamps, Hall, JP B 8, 308 (1975).
(51) Rosenwaks, Steele, Broida, JCP 63, 1963 (1975).
(52) Sulzmann, JQSRT 15, 313 (1975).

AℓP: [a]Thermochemical value (mass-spectrom.)(1).

(1) De Maria, Gingerich, Malaspina, Piacente, JCP 44, 2531 (1966).

AℓS: [a]Thermochemical value (mass-spectrom.)(3)(5).
[b]Predissociation for $v \geq 2$.
[c]Spin splitting constant $\gamma_0 \approx -0.0055$ (6).

AℓS (continued):
[d]Small perturbations (4)(6).
[e]Spin splitting constant $\gamma_0 \approx +0.0045$ (6).

(1) McKinney, Innes, JMS 3, 235 (1959).
(2) Mal'tsev, Shevelkov, Krupnikov, OS(Engl. Transl.) Suppl. No. 2, 4 (1966).
(3) Ficalora, Hastie, Margrave, JPC 72, 1660 (1968).
(4) Kronekvist, Lagerqvist, AF 39, 133 (1969).
(5) Uy, Drowart, TFS 67, 1293 (1971).
(6) Lavendy, Mahieu, Bécart, CJS 18, 13 (1973); Lavendy, Jacquinot, CJS 20, 141 (1975).

AℓSe: [a]Thermochemical value (mass-spectrom.)(1)(3).

(1) See ref. (3) of AℓS.
(2) Singh, Tewari, Mohan, JP B 2, 627 (1969); IJPAP 10, 386 (1972).
(3) See ref. (5) of AℓS.

AℓSi: [a]Thermochemical value (mass-spectrom.)(1).

(1) Stearns, Kohl, HTS 5, 113 (1973).

AℓTe: [a]Thermochemical value (mass-spectrom.)(1)(2).

(1) See ref. (3) of AℓS.
(2) See ref. (5) of AℓS.

State	T_e	ω_e	$\omega_e x_e$	B_e	α_e	D_e ($10^{-7}cm^{-1}$)	r_e (Å)	Observed Transitions Design.	ν_{00}	References
$^{40}Ar_2$		μ = 19.9811921 D_0^0 = 0.01051 eV[a]								DEC 1975 A
		Unclassified, mostly diffuse absorption bands in the regions 88100 - 90100, 106600 - 108200, 113000 - 113900, 114900 - 127200 cm^{-1}.								(8)(13)
		Fragments of a system of very diffuse absorption bands.							(117666)[b]	(13)
H		[128.4][bc] H		Long upper state progressions.				H← X, V	112033.9[bc] H	(13)
G		[170.5][bc] H						G← X, V	110930.9[bc] H	(13)*
F (0_u^+)		[134.2][b] H						F← X, V	108492.2[b] H	(13)
E		[170][b] H		All bands appear diffuse.				E← X, V	107330[b] H	(13)
D		[183.2][b] H		All bands appear diffuse.				D← X, V	106029.5[b] H	(13)
C (0_u^+)	95033.0	67.0 H	4.03[d]					C← X, V	95050.7 H	(13)*
B $(^1\Sigma_u^+)$ 0_u^+		[136.49][e] Z		[0.1057][e]		[7][e]	[2.82_5][e]	B← X, V	93241.26[e] Z	(13)* (24)
A $(^3\Sigma_u^+)$ 1_u		[133.7][b] H	f					A↔ X, V	92393.3[b] H	(13)*
		Emission continuum longward of 1067 Å (93750 cm^{-1}); maxima near 93000 and 79000 cm^{-1}, the former exhibiting oscillatory structure (2)(10)(23) extending to 67000 cm^{-1} (21). The emission originates from the lowest excited states [see (12)] and at 79000 cm^{-1} is characterized by radiative lifetimes of 3.2 μs (1_u?) and 4.2_0 ns (0_u^+?) (20); see also (22).						A,B → X		(1)* (3)* (5)(6)
		Additional continua at 65000, 53000, 47000 cm^{-1}(1)(4)(7)(9), and diffuse emission bands in the region 87000 - 89000 cm^{-1} (1)(2).								
X $^1\Sigma_g^+$	0	[25.74] Z	g	0.05975	0.00375[h]	[11.3][i]	3.758	j		(18)(24)
$^{40}Ar_2^+$		μ = 19.9810550 $D_0^0 \geq$ 1.049 eV[a]								DEC 1975 A
		Theoretical calculations, applied to scattering data (4)(4b)(4c)(5).								
X $^2\Sigma_u^+$	0						2.43 (theor.)			

Ar_2: [a]Calculated by (24) from spectroscopic data for the five lowest vibrational levels in X $^1\Sigma_g^+$ and using information from long-range forces and from the second virial coefficient. $D_e = 99.55$ cm^{-1}; compare with earlier results by (11)(14)(15)(16)(17)(19).

[b]Lowest observed level and interval; vibrational numbering unknown.

[c]The vibrational levels assigned to G and H may belong to a single state with $\omega_e \geq 176$, $\omega_e x_e \approx 2.85$.

[d]This state correlates with $4s'[\frac{1}{2}]_1 + {}^1S$.

[e]Constants for the lowest of four rotationally analyzed levels (24); vibrational numbering unknown, (21) estimate v=22. This state correlates with $4s[\frac{3}{2}]_1 + {}^1S$ and may have a potential hump of ~70 cm^{-1} (13).

[f]Rapid convergence to the limit $4s[\frac{3}{2}]_2 + {}^1S$.

[g]$\Delta G(\frac{3}{2},\ldots,\frac{9}{2})$ = 20.41, 15.60, 10.91, 6.78. G(0) = 14.80.

[h]$\gamma_e = -0.000313$.

[i]$D_1,\ldots,D_5(10^{-7}$cm$^{-1})$ = 16.6, 20, 33, 52, 200.

[j]Raman spectrum (15a)(21a)(22a).

(1) Tanaka, JOSA 45, 710 (1955).
(2) Tanaka, Jursa, LeBlanc, JOSA 48, 304 (1958).
(3) Tanaka, Huffman, Larrabee, JQSRT 2, 451 (1962).
(4) Strickler, Arakawa, JCP 41, 1783 (1964).
(5) Huffman, Larrabee, Tanaka, AO 4, 1581 (1965).
(6) Wilkinson, CJP 45, 1715 (1967).
(7) Verkhovtseva, Fogel, Osyka, OS(Engl. Transl.) 25, 238 (1968).
(8) Wilkinson, CJP 46, 315 (1968).
(9) Hurst, Bortner, Strickler, PR 178, 4 (1969).
(10) Michaelson, Smith, CPL 6, 1 (1970).
(11) Cavallini, Gallinaro, Meneghetti, Scoles, Valbusa, CPL 7, 303 (1970).
(12) Mulliken, JCP 52, 5170 (1970).

Ar_2 (continued):

(13) Tanaka, Yoshino, JCP 53, 2012 (1970).
(14) Barker, Fisher, Watts, MP 21, 657 (1971).
(15) Maitland, Smith, MP 22, 861 (1971).
(15a) Morgan, Frommhold, PRL 29, 1053 (1972).
(16) Parson, Siska, Lee, JCP 56, 1511 (1972).
(17) Le Roy, JCP 57, 573 (1972).
(18) Docken, Schafer, JMS 46, 454 (1973).
(19) Present, JCP 58, 2659 (1973).
(20) Keto, Gleason, Walters, PRL 33, 1365 (1974).
(21) Michaelson, Smith, JCP 61, 2566 (1974).
(21a) Frommhold, JCP 61, 2996 (1974); 63, 1687 (1975).
(22) Oka, Rao, Redpath, Firestone, JCP 61, 4740 (1974).
(22a) Frommhold, Bain, JCP 63, 1700 (1975).
(23) Verkhovtseva, Yaremenko, Fogel, Ovechkin, Katrunova, OS(Engl. Transl.) 38, 98 (1975).
(24) Colbourn, Douglas, JCP 65, 1741 (1976).

Ar_2^+: [a]Difference between the ionization potential of Ar and the appearance potential of Ar_2^+, the latter obtained as the longest-wavelength absorption line of Ar which produces photoionization (2). Similar results by electron impact; see ref. in (2)(3). Ar^+-on-Ar scattering data indicate D_e = 1.30 eV (4a)(4b), consistent with theoretical (4) and semiempirical (3) calculations.

(1) Aberth, Lorents, PR 144, 109 (1966).
(2) Huffman, Katayama, JCP 45, 138 (1966).
(3) See ref. (12) of Ar_2.
(4) Gilbert, Wahl, JCP 55, 5247 (1971).
(4a) Lorents, Olson, Conklin, CPL 20, 589 (1973).
(4b) Mittmann, Weise, ZN 29 a, 400 (1974).
(4c) Jones, Conklin, Lorents, Olson, PR A 10, 102 (1974).
(5) Sidis, Barat, Dhuicq, JP B 8, 474 (1975).

State	T_e	ω_e	$\omega_e x_e$	B_e	α_e	D_e (10^{-4} cm^{-1})	r_e (Å)	Observed Transitions: Design.	Observed Transitions: ν_{00}	References
$^{40}Ar^{(35)}Cl$	(μ = 18.6496153$_4$)									JAN 1976
	Continuous emission with strongest peak near 57250 cm^{-1} and additional maxima of decreasing intensity at higher energies; from chemiluminescent reactions (1)(2). Estimated constants for the strongly bound upper state: T ≈ 56800, ω ≈ 230 cm^{-1}, D[Ar($^3P_2$) — Cℓ] ≈ 4.51 eV.									(2)
	Additional weaker emission at longer wavelengths (2).									
$^{40}Ar^{19}F$	μ = 12.8767196									JAN 1976 A
	Continuous emission with strongest peak near 52550 cm^{-1} and additional peaks of decreasing intensity at higher energies; from chemiluminescent reactions (2). Estimated constants for the strongly bound upper state: T ≈ 54300, ω ≈ 280 cm^{-1}, D[Ar($^3P_2$) — F] ≈ 4.81 eV.									(2)
	Additional weaker emission in the region 35700 - 50000 cm^{-1} (2).									
$^{40}Ar^{19}F^+$	$D_0^0 \geq 1.67$ eV[a]									JAN 1976
$^{40}Ar^{1}H$	μ = 0.98303375									JAN 1976 A
B $^2\Pi$	a			[10.129][b]		[5.03]	[1.3012]	B→A,[c]	R 13024.5 Z	(1)*
A $^2\Sigma(+)$				[10.200][d]		[5.31]	[1.2966]			
	For a theoretical calculation of the ground state potential (D_e = 0.0042 eV, r_e = 3.57 Å) and a comparison with experimental scattering data see (2).									
$^{40}Ar^{2}H$	μ = 1.91746222									JAN 1976
B $^2\Pi$	a	(1990)		5.3262[b]	0.1310	1.500[e]	1.2848	B→A,[f]	R_V 13040.38 Z	(1)*
A $^2\Sigma(+)$		(2057)		5.3402[gd]	0.1399	[1.440][h]	1.2831			

State	T_e	ω_e	$\omega_e x_e$	B_e	α_e	D_e (10^{-} cm^{-1})	r_e (Å)	Observed Transitions Design.	ν_{00}	References
$^{40}Ar^{1}H^{+}$		$D_0^0 \geq 2.65$ eV[a]								JAN 1976
X $^1\Sigma^+$	0						1.28[b]			

ArCℓ: (1) Golde, Thrush, CPL 29, 486 (1974).
(2) Golde, JMS 58, 261 (1975).

ArF, ArF^+:

[a]From the observed exothermicity of the reaction $F_2^+ + Ar = ArF^+ + F$ (1).

(1) Berkowitz, Chupka, CPL 7, 447 (1970).
(2) See ref. (2) of ArCℓ.

Ar^1H, Ar^2H:

[a]$A_0 = (-)2.0$; see (1).
[b]Λ-type doubling; see (1). Small perturbations are evident in the hydride spectrum.
[c]0-0 band only; diffuse lines.
[d]v=0 of Ar^1H, and v > 0 of Ar^2H, are predissociated, probably by interaction with the unstable X $^2\Sigma^+$ ground state; see (1).
[e]From Q branches; $H_e = +1.34 \times 10^{-9}$.
[f]0-0 sequence only. Lines are sharp in the 0-0 band, but become more diffuse as v increases.
[g]Spin splitting $\Delta\nu_{12} = (+)[0.0369(N+\frac{1}{2}) - \ldots]$.
[h]$D_1 = 1.426 \times 10^{-4}$; $H_0 = +0.96 \times 10^{-9}$.

Ar^1H, Ar^2H (continued):

(1) Johns, JMS 36, 488 (1970).
(2) Wagner, Das, Wahl, JCP 60, 1885 (1974).

Ar^1H^+: [a]From the observed exothermicity of the reaction $H_2^+ + Ar = ArH^+ + H$ (1). A higher limit, $D_0^0 \geq 3.40$ eV, is suggested by a tentative interpretation (1) of chemi-ionization processes. H^+-on-Ar scattering data [(3)(5), and ref. therein] give $D_e = 4.17$ eV, in reasonable agreement with theoretical calculations (2)(4).
[b]From proton-argon scattering data [(3)(5), and ref. therein]. Slightly larger values from theoretical calculations (2)(4).

(1) Chupka, Russell, JCP 49, 5426 (1968).
(2) Roach, Kuntz, CC (1970), p. 1336.
(3) Klingbeil, JCP 57, 1066 (1972).
(4) Sidis, JP B 5, 1517 (1972).
(5) Weise, BBPC 77, 578 (1973).

State	T_e	ω_e	$\omega_e x_e$	B_e	α_e	D_e $(10^{-}\ cm^{-1})$	r_e (Å)	Observed Transitions Design.	Observed Transitions ν_{00}	References
$^{40}Ar^{(84)}Kr$		(μ = 27.0703037) D_0^0 [a]								JAN 1976 A
		Four V shaded absorption bands in the region 94138 - 94202 cm^{-1}.								(1)
		Two groups of diffuse absorption bands, 91108 - 91209 and 92296 - 92413 cm^{-1}.								(1)
		Two groups of very diffuse bands extending from the Kr resonance lines at 80918 and 85847 to 81200 and 86100 cm^{-1}, resp.; in emission and absorption.								(1)
		Continuous emission at 74000 cm^{-1}, tentatively attributed to ArKr.								(2)
X $^1\Sigma^+$	0	[a]								
$^{40}Ar^{(84)}Kr^+$										JAN 1976
		Large number of R shaded emission bands in the region 15500 - 16500 cm^{-1}; no analysis. The upper and the lower state of the system are believed to arise from $Ar^+(^2P_{\frac{1}{2}}) + Kr(^1S)$ and $Ar(^1S) + Kr^+(^2P_{\frac{3}{2}})$, respectively. Additional systems expected in the near IR.								(1)
$^{40}Ar^{16}O$		μ = 11.4229054$_7$								JAN 1976 A
		Unclassified emission bands associated with the $^1S_0 - {}^1D_2$ transition of oxygen at 17924 cm^{-1}.								(2)
		Continuous emission with long-wavelength peak at 67000 and extending to above 71000 cm^{-1}; in chemiluminescent reactions. The strongly bound upper state from $Ar(^3P_2) + O(^3P)$ is estimated to have $D_0 \lesssim 3.2$ eV. For the ground state preliminary scattering data (3) indicate $r_e \approx$ 3.31 Å and $D_e \approx 0.0101$ eV.								(4)
$^{40}Ar^{(132)}Xe$		(μ = 30.6703362)								JAN 1976 A
X $^1\Sigma^+$	0	Estimated constants by non-spectroscopic methods: $D_e \approx 0.0155$ eV, $r_e \approx 4.1_5$ Å. For references see Table III of (1). Translational spectrum (2).								

State	T_e	ω_e	$\omega_e x_e$	B_e	α_e	D_e (10^{-} cm^{-1})	r_e (Å)	Observed Transitions		References
								Design.	ν_{00}	
$^{40}Ar^{(132)}Xe^+$										JAN 1976
	Large number of emission bands in the regions (A) 30230 - 31180 cm^{-1},							R_V		(1)
	(B) 28840 - 29930 cm^{-1},							R_V		(1)
	(C) 28570 - 28840 cm^{-1},							V		(1)
	(D) 19730 - 19760 cm^{-1},							V		(1)
	(E) 18300 - 18500 cm^{-1},							V_R		(1)
	No analyses. The five systems correspond to transitions between upper and lower states arising from $Ar^+(^2P_{\frac{3}{2},\frac{1}{2}}) + Xe(^1S)$ and $Ar(^1S) + Xe^+(^2P_{\frac{3}{2},\frac{1}{2}})$, respectively.									

ArKr: [a]A large number of non-spectroscopic determinations give D_e^0 values in the range 0.0122 - 0.0156 eV and r_e values centred at 3.9 Å. See (3), and Table XIII of (1).

(1) Tanaka, Yoshino, Freeman, JCP 59, 5160 (1973).
(2) Verkhovtseva, Ovechkin, Fogel, CPL 30, 120 (1975).
(3) Gough, Matthews, Smith, Maitland, MP 29, 1759 (1975).

ArKr$^+$: (1) Tanaka, Yoshino, Freeman, JCP 62, 4484 (1975).

ArO: (1) Herman, Weniger, Herman, PR 82, 751 (1951).
(2) Cooper, Lichtenstein, PR 109, 2026 (1958).
(3) Aquilanti, Liuti, Vecchio-Cattivi, Volpi, FDCS 55, 187 (1973).
(4) Golde, Thrush, CPL 29, 486 (1974).

ArXe: (1) Kim, Gordon, JCP 61, 1 (1974).
(2) Marteau, Granier, Vu, Vodar, CR B 265, 685 (1967).

ArXe$^+$: (1) See ref. (1) of ArKr$^+$.

State	T_e	ω_e	$\omega_e x_e$	B_e	α_e	D_e ($10^{-8}cm^{-1}$)	r_e (Å)	Observed Transitions Design.	Observed Transitions ν_{00}	References
$^{75}As_2$		$\mu = 37.460800_2$		$D_0^0 = 3.96$ eV[a]						JAN 1976
N	(72137)	[319]						N← X,	72082	(8)
M	(69607)	[365]						M← X,	69575	(8)
J								J← X,	66133 66015	(8)
I								I← X,	65432 65308	(8)
H	(61726)	[364] H						H← X,[b]	R 61694 H	(8)
		Unclassified absorption bands in the region 47000 - 55000 cm^{-1}.								(9)
G	(54586)	[377][c] H						G← X,[d]	R 54560 H	(8)(9)
F	(52221)	[386] H						F← X,[e]	R 52200 H	(8)(9)
		Fragments of other electronic states in the region 42400 - 44500 cm^{-1}.								(11)
b ($^3\Pi_u$)		Only v=0 observed.						b→ X,	R 42006	(2)(3)(4)
B $^1\Sigma_u^+$	(40925)	[243.6][f] Z		[0.0712][f]			[2.5]	B↔ X,	R 40832.2 Z	(1)(2)(4)* (6)(7)(10)
A $^1\Sigma_u^+$	(40349)	[260.3][gh] Z		[0.07202][gh]			[2.50]	A↔ X,	R 40265.0 Z	(1)(2)(4)* (6)(7)(9) (10)(11)(12)
d ($^3\Pi_g$) 1_g	30818.8	336.7 H	1.36	0.09222	0.00033		2.209_0	d→c,[i] d→ X,	V 16185.6 H 16348.4 H R 30772.4 H	(4)* (5)* (6) (5)* (6)
a ($^3\Sigma_u^-$) 0_u^+	24641.2	337.0 H	0.83	0.08664	0.00030		2.279_0	(a→c),[j] a→X,[k]	V (10100) R 24595.0 H	(5)* (4)* (6)
e	19914.7	330.0 H	0.90					e→X,	R 19865.0[ℓ] H	(6)
c $^3\Sigma_u^+$ $\{0_u^-$	14644.4	314.3 H	1.17	0.08491	0.00035		2.302_1			
c $^3\Sigma_u^+$ 1_u	14481.6			0.08471[m]	0.00035		2.303_6	$c(1_u)$→X,	R 14424.0	(6)
X $^1\Sigma_g^+$	0	429.55 H	1.117[n]	0.10179	0.000333[o]		2.1026			

As_2: [a]From the predissociation limit in A $^1\Sigma_u^+$ assuming dissociation into $^4S + {}^2D$ without kinetic energy (4). Extrapolation of the ground state vibrational levels (observed to v=70, i. e. 80% of D_0^0) gives 3.93 eV (3)(4).

[b]System G← X of (8).

[c]Bands with v' ≥ 3 are diffuse.

[d]System E← X of (8).

[e]System f← X of (8).

[f]Additional $\Delta G(v+\frac{1}{2})$ and B_v values ($v \leq 7$) in (10). Strong perturbations.

[g]Additional $\Delta G(v+\frac{1}{2})$ and B_v values ($v \leq 17$) in (10)(11)(12). Strong perturbations produced by interaction with the B and, possibly, other states.

[h]In emission no bands with v' ≥ 10 are observed except those with v' = 14 (2). In absorption bands with v' ≥ 10 are observed (9)(11) but the lines in them are broad except those with v' = 14 which are sharp. The line width varies greatly as a function of J in bands with v' ≥ 10. There is clearly a predissociation limit at 42700 ± 100 cm^{-1}.

[i]Originally (4) attributed to As_2^+. Recognized as being due to neutral As_2 by (5). The lower states are the two case "c" components of c $^3\Sigma_u^+$ (6).

[j]Strong system in the infrared (5), not analysed in detail.

[k]System D→X of (4).

[ℓ]19855 in (6) appears to be erroneous.

[m]Ω-type doubling, $\Delta v_{fe} = (-)0.00018 \times J(J+1)$.

[n]$\omega_e y_e = +\ 0.000139$, $\omega_e z_e = -\ 0.00001958$.

[o]$\gamma_e = -\ 2.8 \times 10^{-7}$.

(1) Gibson, MacFarlane, PR 46, 1059 (1934).
(2) Almy, Kinzer, PR 47, 721 (1935).
(3) Almy, JPC 41, 47 (1937).
(4) Kinzer, Almy, PR 52, 814 (1937).
(5) Mrozowski, Santaram, JOSA 57, 522 (1967).
(6) Perdigon, D'Incan, CJP 48, 1140 (1970).
(7) Perdigon, Martin, D'Incan, JMS 36, 341 (1970).
(8) Donovan, Strachan, TFS 67, 3407 (1971).
(9) Topouzkhanian, Sibaï, SA A 28, 2197 (1972).
(10) Martin, Perdigon, D'Incan, JMS 50, 45 (1974).
(11) Sibaï, Perdigon, Topouzkhanian, ZN 29 a, 429 (1974).
(12) Martin, Figuet, Perdigon, JMS 57, 319 (1975).

State	T_e	ω_e	$\omega_e x_e$	B_e	α_e	D_e	r_e (Å)	Observed Transitions: Design.	Observed Transitions: ν_{00}	References
$^{75}As_2^+$?										JAN 1976

The spectrum tentatively assigned to As_2^+ in (1) has been shown to be due to the d→c transition of the neutral molecule; see [i] of As_2.

The authors of (2) have described two additional systems,

$$\nu = \begin{matrix}13701.6\\13403.4\end{matrix} + 365(v'+\tfrac{1}{2}) - 3.95(v'+\tfrac{1}{2})^2 - 317(v''+\tfrac{1}{2}) + 1.68(v''+\tfrac{1}{2})^2,$$

$$\nu = \begin{matrix}12697.6\\12365.8\end{matrix} + 365(v'+\tfrac{1}{2}) - 3.95(v'+\tfrac{1}{2})^2 - 354(v''+\tfrac{1}{2}) + 3.00(v''+\tfrac{1}{2})^2,$$

obtained under the same conditions at which the d→c bands of As_2 appear. Since they thought the latter were due to As_2^+ and since one of the new systems apparently has the same lower state (c) they concluded that both new systems belong to As_2^+. This conclusion is now very doubtful.

State	T_e	ω_e	$\omega_e x_e$	B_e	α_e	D_e	r_e (Å)	Observed Transitions: Design.	Observed Transitions: ν_{00}	References
$^{75}As_2^-$		I.P. = $0.1_0 \pm 0.1_8$ eV[a]								JAN 1976
$^{75}As^{(79)}Br$		($\mu = 38.434024_5$)								JAN 1976
b (0^+)		0-0 sequence only; $\omega' - \omega'' = 15.0$ cm^{-1}.						b→X,	V 12316.9 H	(1)*
X	0									
$^{75}As^{(35)}Cl$		($\mu = 23.8412199$)								JAN 1976
B	40826	[520] H						B←X,	V 40865 H	(1)
X	0	443 H	2							

As_2^+: (1) Herzberg, MOLSPEC Vol. 1 (1950).
(2) Rao, Lakshman, IJPAP 4, 259 (1966).

As_2^-: [a]Electron impact mass-spectrometry (1).
(1) Bennett, Margrave, Franklin, Hudson, JCP 59, 5814 (1973).

AsBr: (1) de Bie - Prévot, Thèse (Université Libre de Bruxelles, 1974)

AsCℓ: (1) Basco, Yee, CC (1967), p. 1255.

State	T_e	ω_e	$\omega_e x_e$	B_e	α_e	D_e ($10^{-7}cm^{-1}$)	r_e (Å)	Observed Transitions: Design.	Observed Transitions: ν_{00}	References
$^{75}As^{19}F$		μ = 15.1553535		D_0^0 = 4.2 eV[a]		I.P. = 9.4 eV[a]				JAN 1976 A
d $^1\Pi$				[0.3989][b]		[4.3]	[1.669_8]	d→b, V	37032.07 Z	(3)(5)
								d↔a, V	43628.60 Z	(3)(5)*
								d←X_1, V	50686.43 Z	(6)
C ($^3\Pi$)								C→X, V	49758 H	(6)
c $^1\Pi$	48672.5	817.30 Z	4.39	0.4004[cd]	0.0027	4.0	1.666_7	c→b, V	35083.56 Z	(3)(5)* (6)
								c↔a, V	41680.09 Z	(3)* (5)*
								c↔X_2, V	48599[e]	(6)
								X_1,	48737.92 Z	(6)*
B ($^3\Pi$)	(48138)	815.5[f] H	3.9_6	Very complicated rotational structure.				B↔X, V	48202.2[f] H	(6)*
c' $^1\Pi$	32479.5	399.38 Z	1.34[g]	0.2932[h]	0.0018	6.2	1.948	c'→b, R	18682.4	(7)*
								c'→a, R	25278.88 Z	(7)*
						Fragments of a system		c'→X_2, R	32198	(7)
A_4 $^3\Pi_r$ 2	27152	412.28 Z	1.43[i]	0.2918	0.0020			A_4→X_2, R	26877.0 Z	(10)*
A_3 $^3\Pi_r$ 1	26348	412.13 Z	1.44[j]	0.2920[k]	0.0020	6	1.954	A_3→X_1, R	26211.7 Z	(10)*
A_2 $^3\Pi_r$ 0^+	25751	(419.8) H		(0.2914)	(0.0018)			A_2→X_2, R	25480	(10)
A_1 $^3\Pi_r$ 0^-	25719	412.21 Z	1.32[j]	0.2904	0.0018			A_1→X_2, R	25444.5 Z	(10)*
b $^1\Sigma^+$	13648.6	697.34 Z	3.08	0.3719	0.0028	4.7	1.729_4	b→X_2, ℓ	13515.78 Z	(6)*
								X_1,	13654.4 Z	(6)
a $^1\Delta$	7053.5	694.44 Z	3.06	0.3707	0.0026	4.4	1.732_2			
X_2 $^3\Sigma^-$ 1	138.7	685.50 H[Q]	2.95	0.3691[mn]	0.0028	4.5	1.736_0[m]			
X_1 $^3\Sigma^-$ 0^+	0	685.78 Z	3.12	0.3648[on]	0.0024					

AsF: [a]Theoretical calculations, for D_0^0 supported by limited experimental data; see (8). The same paper gives theoretical values for the electron affinity (1.1 eV) and dipole moment (1.75 D) of AsF.

[b]Predissociated near J=30.

[c]Λ-type doubling, $\Delta\nu_{ef} \approx + 0.00005 \times J(J+1)$.

[d]Predissociation in v=1 at J≈60; v=2 observed in absorption only.

[e]Very weak band.

[f]The vibrational analysis of (6) assigns v=0 to the lowest level observed in emission. However, the 1-0 and 0-0 bands may have been observed in absorption (4) at 49001 and 48192 cm^{-1} [47192 in (4) seems to be a misprint] as members of a strong upper state progression which begins at 47381 cm^{-1} and may even include a diffuse band at 46570 cm^{-1}. No details.

[g]$\omega_e y_e = - 0.015$.

[h]Λ-type doubling, $\Delta\nu_{ef} \approx + 0.000015 \times J(J+1)$.

[i]$\omega_e y_e = - 0.031$.

[j]$\omega_e y_e = - 0.026$.

[k]Small Λ-type doubling.

[ℓ]Long 0-0 sequence of V shaded Q heads; R, P branches undegraded.

[m]B_e and r_e from the f component; $\Delta\nu_{ef} = + 0.0035 \times J(J+1)$.

[n]For a more detailed discussion of the X $^3\Sigma^-(0^+, 1)$ fine structure see (9).

[o]Effective value.

(1) Pannetier, Deschamps, Guillaume, CR 261, 3396 (1965).

(2) Pannetier, Deschamps, Guillaume, CR C 265, 710, (1967).

(3) Yee, Liu, Jones, JMS 35, 153 (1970).

(4) Chatalic, Danon, Pannetier, CR C 273, 874 (1971).

(5) Liu, Yee, Jones, JMS 38, 512 (1971).

(6) Liu, Jones, CJP 50, 1230 (1972).

(7) Chatalic, Danon, Iacocca, Pannetier, JCPPB 70, 1600 (1973).

(8) O'Hare, Batana, Wahl, JCP 59, 6495 (1973).

(9) Veseth, JP B 6, 1473, 1484 (1973).

(10) Chatalic, Danon, Pannetier, JCPPB 71, 243 (1974).

State	T_e	ω_e	$\omega_e x_e$	B_e	α_e	D_e $(10^{-5}cm^{-1})$	r_e (Å)	Observed Transitions Design.	Observed Transitions ν_{00}	References
$^{75}As^{1}H$		μ = 0.99444817		D_0^0 < 3.6 eV[a]						FEB 1976
A $^3\Pi_i$ 0	b	[1207.5][c] Z		[6.5623][df]	0.886[c]	[70.3]		A← X, R	30518.34 Z	(1)* (2)
A $^3\Pi_i$ 1				[6.4726][ef]		[62.4]	[1.6203][g]		29821.97 Z	
A $^3\Pi_i$ 2				[6.3387][f]		[55.4]			29282.16 Z	
X $^3\Sigma^-$	0[h]	(2130)		[7.199_8][h]		[$32._9$]	[1.5344]			
$^{75}As^{2}H$		μ = 1.96137497		D_0^0 = 2.7_6 eV[i]						FEB 1976
A $^3\Pi_i$ 0	j	[933.6] Z		[3.3467][kf]	0.264	[16.7]		A← X, R	30628.87 Z	(1)* (2)
A $^3\Pi_i$ 1		[934.8] Z		[3.3220][ℓf]	0.240	[16.1]	[1.6095][g]		29932.48 Z	
A $^3\Pi_i$ 2		[954.2] Z		[3.2881][f]	0.227	[14.5]			29389.55 Z	
X $^3\Sigma^-$	0[m]	(1484)		[3.6688][m]		[8.9_7][n]	[1.5306]			(3)
$^{75}As^{14}N$		μ = 11.7979936_0								FEB 1976 A
A $^1\Pi$	35999.7	[853.3][a] Z	8.24 H	0.501_8[bc]	0.009	0.066	1.687	A→X, R	35899.6 Z	(1)* (5)* (6)
($^1\Sigma^+$)		Single band.		[0.501_1]		[0.08_8]	[1.689]	($^1\Sigma$)→X, R	29124.9 Z	(3)*
X $^1\Sigma^+$	0	1068.54 Z	5.41[d]	0.54551[c]	0.003366	0.053	1.6184_3			

As^1H, As^2H:

[a]From the predissociation in A $^3\Pi$.

[b]$A_0 = -615.4$. Small J-dependence (2).

[c]$\Delta G(\frac{1}{2})$ and α for the $^3\Pi_0+$ component only.

[d]Λ-type doubling, $\Delta v_{ef}(v=0) = +44.72 + 0.0056 \times J(J+1)$.

[e]Λ-type doubling, $\Delta v_{ef}(v=0) = +0.0214 \times J(J+1)$.

[f]Lines are sharp for the $^3\Pi_0+$ state only. Line width increases in the order $^3\Pi_0+ < {}^3\Pi_0- < {}^3\Pi_1 < {}^3\Pi_2$, and also with increasing vibrational energy. The predissociation is due to interaction with the unstable $^5\Sigma^-$ state arising from $As(^4S) + H(^2S)$ (1).

[g]From the "true" B_0 values in (2).

[h]Spin splitting constants $\lambda_0 = +58.87$, $\gamma_0 = -0.281$.

[i]From a short extrapolation (1) of the $^3\Pi$ state to the limit $As(^2D) + H(^2S)$.

[j]$A_0 = -616.9$, small J-dependence (2); $A_1 = -599.8$.

[k]Λ-type doubling, $\Delta v_{ef}(v=0) = +43.39 + 0.00472 \times J(J+1)$.

[l]Λ-type doubling, $\Delta v_{ef}(v=0) = +0.0046 \times J(J+1)$.

[m]Spin splitting constants $\lambda_0 = +58.811$, $\gamma_0 = -0.147_5$.

[n]$H_0 = +5 \times 10^{-9}$.

(1) Dixon, Lamberton, JMS 25, 12 (1968).
(2) Veseth, JP B 5, 229 (1972).
(3) Lindgren, PS 12, 164 (1975).

AsN: [a]$\omega_e = 871.3$ from band heads (1).

[b]Strong perturbations.

[c]Independent rotational analyses by (2) and (4) gave different B values. The values of (6) are based on plates with higher resolution and are adopted here.

[d]$\omega_e y_e = +0.04$.

(1) Spinks, ZP 88, 511 (1934).
(2) D'Incan, Fémelat, CR B 264, 1261 (1967).
(3) D'Incan, Fémelat, CR B 267, 796 (1968).
(4) Dixit, Krishnamurty, Narasimham, PIAS A 71, 23 (1970).
(5) Jones, JMS 34, 320 (1970).
(6) Fémelat, Jones, JMS 49, 388 (1974).

State	T_e	ω_e	$\omega_e x_e$	B_e	α_e	D_e $(10^{-7}cm^{-1})$	r_e (Å)	Observed Transitions: Design.	Observed Transitions: ν_{00}	References
$^{75}As^{16}O$		μ = 13.1809344_6		$D_0^0 \leq$ 4.980 eV[a]						FEB 1976
B $^2\Sigma^+$	39866.0	1098.3 H[Q]	6.1	[0.51284][bc]	0.0036[d]	[4.57]	1.5764	B↔X, V	38905.88 Z 39931.36 Z	(1)(2)* (6)* (8)(14)
C $^2\Delta$ 5/2 3/2	38686 (38638)	655.7[e] Only v=0 observed.	4.53	0.4164[e] [0.4028]	0.0040	[14.1]	1.765	C→X, R	37506[e] 38483	(3)(15)(17)
C' $^2\Phi$ or $^2\Delta$	(38800)	(600)[f]		(0.3798)[f]	(0.0058)					(15)(17)
D $^2\Sigma^-$	37555.4	629.9[g] Z	3.79	0.3973[gh]	0.0034	6.5	1.794_2	D→X, R	36361.7 Z 37387.1 Z	(3)(6)(7)(8) (11)(13)(15) (17)
D' $^2\Sigma^-$	[37857][i]			[0.360][i]						(17)
H $^2\Pi$ 1/2 3/2	j 37053.7	606.92 Z	4.913	0.36539[kℓ]	0.00273	5.4	1.8709	H→X, R	j 35848.2 Z	(12)(17)
A $^2\Sigma^+$	31652.45	686.68 Z	10.78	0.46240[mn]	0.00710	8.56[o]	1.6631	A→X, R	30485.28 Z 31510.87 Z	(1)* (2)* (4)(6)* (8) (14)
		Fragments of additional systems of R shaded emission bands in the region 25000 - 30000 cm^{-1}.								(5)(17)
A' $^2\Pi_i$ 1/2[p] 3/2	26485.2[q] 26168.4	630.30 Z 633.23 Z	3.006[r] 2.895	0.37183[ksp] 0.37124[k]	0.002701 0.002622	5.0	1.8553	A'→X, R	26317.30 Z 24976.59 Z	(6)(10)* (11)(16)(17)
		Fragments of a weak system of R shaded emission bands in the region 15300 - 17300 cm^{-1}.								(16)
X $^2\Pi_r$ 3/2 1/2	1025.97[t] 0	965.90 Z 967.08 Z	4.909[u] 4.850[v]	0.48552[k] 0.48482[kw]	0.003320[u] 0.003299[v]	4.9	1.6236			
$^{75}As^{16}O^+$										FEB 1976
A $^1\Pi$	42594.2	780.8 Z	7.4	0.4491[a]	0.0047	5.0	1.688	A→X, R	42433.4 Z	(1)(2)(4)* (5)
X $^1\Sigma^+$	0	[1091.32] Z	5.0 H	0.5199[a]	0.0031	3.9	1.568			

AsO: [a]From the predissociation in B $^2\Sigma^+$.

[b]Spin splitting constant $\gamma_0 = +0.0043$.

[c]Weak predissociation above v=0, N=20 (23 for $As^{18}O$) (6) (8), attributed (6) to interaction with the unstable $^4\Sigma^+$ state arising from ground state atomic products.

[d]Estimated from band head separations.

[e]Deperturbed (15) constants. Strong perturbations by successive levels in both doublet components of a $^2\Phi$ or $^2\Delta$ state, and possibly by other states.

[f]From perturbations in C $^2\Delta_{5/2}$. Vibrational numbering unknown.

[g]v=0 and 1, formerly (12) attributed to a $^4\Sigma$ state, are perturbed by levels of a $^2\Sigma^-$ state and of H $^2\Pi_{\frac{1}{2},\frac{3}{2}}$ (17).

[h]The spin splitting constant γ_v increases from $\gamma_0 = +0.021$ to $\gamma_7 = +0.039$.

[i]Lowest observed level, from a perturbation in D(v=0). Vibrational numbering unknown.

[j]Bands of the $\frac{1}{2}-\frac{1}{2}$ transition have not been identified. One level of $^2\Pi_{\frac{1}{2}}$ (vibrational numbering unknown) may be responsible for a perturbation in D $^2\Sigma^-$(v=0).

[k]Effective constants.

[ℓ]Perturbations in v=1,2 by v=0,1, resp., of D $^2\Sigma^-$.

[m]Spin splitting constant $\gamma = -0.035$.

[n]v=0,1,2 perturbed by v=9...12 of A' $^2\Pi_{\frac{1}{2}}$. Additional perturbations by unidentified levels of Σ and Π symmetry (6)(17).

[o]$\beta_e = +0.18 \times 10^{-7}$; $H_e = -2.6 \times 10^{-12}$.

[p]The levels v=9...12 of $^2\Pi_{\frac{1}{2}}$ have been observed in perturbations of A $^2\Sigma^+$(v=0,1,2) and were previously (6) assigned to a state G $^2\Pi$.

[q]$A_e = -316.08$.

[r]$\omega_e y_e = +0.010$.

AsO (continued):

[s]Λ-type doubling $\Delta v_{fe} = -0.002(J+\frac{1}{2})$.

[t]$A_e = +1026.94$.

[u]$\omega_e y_e = -0.0034$; $\gamma_e = -4.6 \times 10^{-6}$.

[v]$\omega_e y_e = -0.0015$; $\gamma_e = -2.5 \times 10^{-6}$.

[w]Λ-type doubling $\Delta v_{fe} = +0.0179(J+\frac{1}{2})$.

(1) Connelly, PPS 46, 790 (1934).
(2) Jenkins, Strait, PR 47, 136 (1935).
(3) Lakshman, Rao, IJP 34, 278 (1960).
(4) Klynning, Naturw. 49, 252 (1962).
(5) Venkataramanaiah, Lakshman, IJP 38, 209 (1964).
(6) Callomon, Morgan, PPS 86, 1091 (1965).
(7) d'Incan, Goure, CR 261, 3086 (1965).
(8) Meyer, JMS 18, 443 (1965).
(9) d'Incan, Goure, Zgainsky, CR B 263, 1319 (1966).
(10) Mrozowski, Santaram, JOSA 56, 1174 (1966).
(11) Goure, d'Incan, CR B 268, 1311 (1969).
(12) d'Incan, Goure, CR B 268, 1647 (1969).
(13) Topouzkhanian, Goure, Figuet, d'Incan, CR B 270, 1676 (1970).
(14) Lakshman, Rao, JP B 4, 269 (1971).
(15) Goure, Figuet, Massot, d'Incan, CJP 50, 1926 (1972).
(16) Kushawaha, Asthana, Pathak, JMS 41, 577 (1972).
(17) Anderson, Callomon, JP B 6, 1664 (1973).

AsO^+: [a]Different constants in (3).

(1) See ref. (3) of AsO.
(2) Lakshman, PPS 89, 774 (1966).
(3) Shanker, Singh, Singh, CJP 47, 1601 (1969).
(4) Rao, Rao, JP B 3, 430 (1970).
(5) See ref. (14) of AsO.

State	T_e	ω_e	$\omega_e x_e$	B_e	α_e	D_e $(10^{-8} cm^{-1})$	r_e (Å)	Observed Transitions Design.	Observed Transitions ν_{00}	References
$^{75}As^{31}P$		$\mu = 21.9141220$								FEB 1976
A $^1\Pi$	32417.05	475.52 Z	2.12	0.1744[a]	0.0009	11	2.100	A→X, R	32352.76 Z	(1)(2)*
X $^1\Sigma^+$	0	604.02 Z	1.98	0.1925	0.0008	7.8	1.999			
$^{75}As^{32}S$		$\mu = 22.4091734$								FEB 1976 A
		Unclassified bands, most of them red-degraded, in the region 26000 - 31000 cm^{-1}.								(1)
A_2' $^2\Pi_{1/2}$	20474.9	[399.84] Z		0.14871[a]	0.00073	$8._1$[b]	2.2496	$A_2' \to X_1$, R	20392.13 Z	(4)*
A_1' $^2\Pi_{3/2}$	x + 19263.28	405.5_8 Z	1.11	0.14859	0.00069[c]	8.0[b]		$A_1' \to X_2$,	19183.22 Z	(2)* (3)* (5)*
X_2 $^2\Pi_{3/2}$	x	566.1_3 Z	1.96[d]	0.18492	0.00083	7.9[b]	2.0174			
X_1 $^2\Pi_{1/2}$	0	567.9_4 Z	1.97	0.18476[a]	0.00082	7.8[b]				
$^{75}As^{32}S^+$										FEB 1976
A $^1\Pi$	37359.7	441.18 Z	3.45	0.1732[a]	0.00123	10.0	2.084	A→X, R	37257.82 Z	(1)(2)*
X $^1\Sigma^+$	0	644.25 Z	2.09	0.1989	0.00089	5.7	1.9447			
$^{75}As^{(80)}Se$		$(\mu = 38.669248_6)$								FEB 1976
A $(^2\Pi)$								A→X, R	18717 H 18044 H	(1)
X $(^2\Pi)$	0	281 H	0.5							

AsP: [a]Small perturbations in both Λ-components [see (2)].

(1) Yee, Jones, CC (1969), p. 586.
(2) Harding, Jones, Yee, CJP 48, 2842 (1970).

AsS: [a]Λ-type splitting in the $^2\Pi_{\frac{1}{2}} - {}^2\Pi_{\frac{1}{2}}$ subbands $\Delta\nu \approx 0.037(J+\frac{1}{2})$.
[b]Values of $4B_e^3/\omega_e^2$, in satisfactory agreement with experimental values.
[c]$\gamma_e = +5 \times 10^{-6}$.
[d]$\omega_e y_e = -0.003_1$.

(1) Shimauchi, SL 18, 90 (1969).
(2) Shimauchi, CJP 49, 1249 (1971).
(3) Shimauchi, Sakaba, Kikuchi, SL 21, 1 (1972).
(4) Shimauchi, Iwata, Matsuno, Sakaba, Lee, Karasawa, SL 21, 145 (1972).
(5) Shimauchi, Karasawa, SL 22, 127 (1973).

AsS^+: [a]Small perturbations in both Λ-components [see (2)].

(1) See ref. (1) of AsS.
(2) Shimauchi, Karasawa, Ninomiya, SL 23, 72 (1974); Shimauchi, Karasawa, CJP 53, 831 (1975).

AsSe: (1) Vasudev, Jones, JMS 54, 144 (1975).

State	T_e	ω_e	$\omega_e x_e$	B_e	α_e	D_e (10^{-8}cm^{-1})	r_e (Å)	Observed Transitions Design.	ν_{00}	References
$^{197}Au_2$		μ = 98.483274		D_0^0 = 2.30 eV[a]						FEB 1975
B 0_u^+	25685.5	179.85 H	0.680[b]	[0.026961]	(0.0000963)	[0.260]	[2.5197]	B↔X, R	25679.87 Z	(2)(5)* (8)*
A 0_u^+	19668.1	142.3 H	0.445[c]	0.02595_8	0.0000903	0.35	2.567_9	A↔X, R	19643.8 H	(1)* (2)(5) (8)
X $^1\Sigma_g^+$	0	190.9[d] H	0.420[e]	0.028013[f]	0.0000723	[0.250]	2.4719			
$^{197}Au^{27}Al$		μ = 23.7307715		D_0^0 = 3.34 eV[a]						NOV 1974
C 1	24623	250 H	2	$[0.12332_2]$[b]		$[11.5_9]$	[2.4001]	C↔X, R	24581.28 Z	(1)(2)*
B 1	22490.3	291.83 H^Q	3.03	0.13131[c]	0.00130	$(10._5)$	2.3259	B↔X, R[d]	22469.22 H^Q	(1)(2)*
A 0^+	16265.06	347.96 Z	1.85_4	0.13645_5	0.00084_8	7.8_0[e]	2.2816_4	A↔X, V	16272.37 Z	(1)(2)* (3)
X $^1\Sigma^+$ (0^+)	0	333.00 Z	1.16_3	0.12991_3	0.00066_8	7.1_8[f]	2.3383_9			
$^{197}Au^{75}As$		$\mu = 54.276176_0$								NOV 1974
System D:		$\nu_H = 18471.2 + 228.1(v'+\frac{1}{2}) - 0.8(v'+\frac{1}{2})^2 - 254.8(v''+\frac{1}{2}) + 0.6(v''+\frac{1}{2})^2$						a	R $18457._8$ H	(1)
System C:		$\nu_H = 17442.1 + 242.2(v'+\frac{1}{2}) - 1.1(v'+\frac{1}{2})^2 - 257.7(v''+\frac{1}{2}) + 0.8(v''+\frac{1}{2})^2$						a	R 17434.3 H	(1)
System B:		$\nu_H = 16582.1 + 241.7(v'+\frac{1}{2}) - 0.9(v'+\frac{1}{2})^2 - 254.8(v''+\frac{1}{2}) + 0.6(v''+\frac{1}{2})^2$						a	R 16575.5 H	(1)
System A:		$\nu_H = 16473.5 + 226.9(v'+\frac{1}{2}) - 1.1(v'+\frac{1}{2})^2 - 239.3(v''+\frac{1}{2}) + 0.7(v''+\frac{1}{2})^2$						a	R 16467.2 H	(1)
$^{197}Au^{(11)}B$		$(\mu = 10.4265222_7)$		$D_0^0 = 3.6_5$ eV[a]						NOV 1974
$^{197}Au^{(138)}Ba$		(μ = 81.113786)								NOV 1974
		Unclassified R shaded bands in the region 13500 - 14100 cm^{-1}; assignment to AuBa uncertain.[a]								(2)
B	21804	137.5 H	0.5[b]					B→(X),[a] V_R	21808 H	(1)(2)
A'	13522	154.1 H	0.35					A'→(X),[ac] V	13535 H	(2)
A	12636.4	158.1 H	0.30					A→X,[a] V	12651.0 H	(1)(2)
X	0	128.8 H	0.18							

Au_2: [a]Thermochemical value (mass-spectrom.)(3)(4)(6), recalculated (9)(10), and (11).
[b]$\omega_e y_e = +0.003$.
[c]$\omega_e y_e = -0.0015$.
[d]From band origins $\Delta G(\frac{1}{2}) = 190.17_6$ (8).
[e]$\omega_e y_e = -0.0001$.
[f]Ground state potential curve (7).

(1) Kleman, Lindkvist, Selin, AF 8, 505 (1954).
(2) Ruamps, CR 238, 1489 (1954).
(3) Drowart, Honig, JCP 25, 581 (1956); JPC 61, 980 (1957).
(4) Schissel, JCP 26, 1276 (1957).
(5) Ruamps, AP(Paris) 4, 1111 (1959).
(6) Ackerman, Stafford, Drowart, JCP 33, 1784 (1960).
(7) Dolgushin, OS(Engl. Transl.) 19, 289 (1965).
(8) Ames, Barrow, TFS 63, 39 (1967).
(9) Smoes, Drowart, CC (1968), p. 534.
(10) Smoes, Mandy, Vander Auwera-Mahieu, Drowart, BSCB 81, 45 (1972).
(11) Kordis, Gingerich, Seyse, JCP 61, 5114 (1974).

AuAl: [a]Thermochemical value (mass-spectrom.)(4).
[b]Average of the two Ω-type doubling components; $\Delta\nu_{fe} = +0.00011\, J(J+1)$.
[c]From bandhead measurements, using ground state constants from A − X.
[d]Reversal of shading in the 0-0 band from V at low J to R at high J.

AuAl (continued):
[e]$\beta_e = +0.0_6 \times 10^{-8}$.
[f]$\beta_e = -0.0_1 \times 10^{-8}$.

(1) See ref. (5) of Au_2.
(2) Barrow, Travis, PRS A 273, 133 (1963).
(3) Lochet, CR B 272, 797 (1971).
(4) Gingerich, Blue, JCP 59, 185 (1973).

AuAs: [a]Thermal emission and absorption.

(1) Houdart, Bocquet, CR B 264, 1717 (1967).

AuB: [a]Thermochemical value (mass-spectrom.)(1)(2).

(1) Gingerich, ZN 24 a, 293 (1969); JCP 54, 2646 (1971).
(2) Vander Auwera-Mahieu, Peeters, McIntyre, Drowart, TFS 66, 809 (1970).

AuBa: [a]All systems in thermal emission.
[b]Analysis uncertain.
[c]It is assumed that the lower state of this system ($\omega_e'' = 128.8$, $\omega_e''x_e'' = 0.44$) is identical with the lower state of A − X.

(1) Schiltz, CR 253, 1777 (1961).
(2) Schiltz, AP(Paris) 8, 67 (1963).

State	T_e	ω_e	$\omega_e x_e$	B_e	α_e	D_e $(10^{-6} cm^{-1})$	r_e (Å)	Observed Transitions Design.	Observed Transitions ν_{00}	References
$^{197}Au^{9}Be$		$\mu = 8.6178730$								NOV 1974
B $\frac{1}{2}$ $(^2\Pi_{\frac{1}{2}})$	18945.98	628.9_5 Z^a	$3.22_5{}^b$	0.47944^c	0.00434	$[1.09]^d$	2.0199_1	B ↔ X,	V 18956.68 Z	(1)* (2)
A $\frac{1}{2}$ $(^2\Sigma^+)$	17171.04	655.4_4 Z^a	$3.59_5{}^e$	0.49264^f	0.00463	$[1.09]^g$	1.9926_6	A ↔ X,	V 17194.88 Z	(1)* (2)
X $^2\Sigma^+$	0	607.68 Z^a	3.53	0.46074	0.00400	$[1.04]^h$	2.0604_9			(1)(2)
$^{197}Au^{209}Bi$		$\mu = 101.397851$								NOV 1974
A	$17787._4$	$149._6$ H^a	0.34					A ← X,	R 17783.3 H	(1)*
X	0	$157._7$ H^a	0.25							
$^{197}Au^{(40)}Ca$		$(\mu = 33.2221432)$								NOV 1974 A
		Very complex system of R and V shaded bands in the region 16000 - 19000 cm^{-1}.[a]								(1)*
B	15024.8	220.2 H^Q	0.6					B → $(X)^b$,[a]	V 15024.1 H^Q	(1)*
A	14512.3	212.7 H^Q	0.10					A → X,[a]	V 14508.8 H^Q	(1)
X	0	220.0 H^Q	0.62^c							
$^{197}Au^{(140)}Ce$		$(\mu = 81.801686)$		$D_0^0 = 3.3_4$ eV^a						NOV 1974
$^{197}Au^{35}Cl$		$\mu = 29.6966066$		$D_0^0 = (3.5)$ eV						NOV 1974
B	19238.3	316.3 H	1.45					B → X,	R 19205.0 H	(1)*
A	19113.8	312.0 H	0.70					A → X,	R 19078.6 H	(1)*
X $(^1\Sigma^+)$	0	382.8 H	1.30							
$^{197}Au^{59}Co$		$\mu = 45.360994_8$		$D_0^0 = 2.2_2$ eV^a						NOV 1974

State	T_e	ω_e	$\omega_e x_e$	B_e	α_e	D_e (10^{-} cm^{-1})	r_e (Å)	Observed Transitions Design.	ν_{00}	References
$^{197}Au^{(52)}Cr$		(μ = 41.101859$_8$)		D_0^0 = 2.2$_5$ eV[a]						NOV 1974

AuBe: [a]Using calculated head-origin separations.
[b]$\omega_e y_e$ = -0.065$_8$.
[c]Average of the two Ω-type doubling components;
$\Delta\nu_{fe}(v=0) = -0.0189_5(J+\frac{1}{2}) + \ldots$
[d]$H_0 = -8.1 \times 10^{-13}$.
[e]$\omega_e y_e$ = -0.19$_3$.
[f]Average of the two Ω-type doubling components;
$\Delta\nu_{fe}(v=0) = +0.1057_9(J+\frac{1}{2}) + \ldots$
[g]$H_0 = -11.1 \times 10^{-13}$.
[h]$H_0 = -5.4 \times 10^{-13}$.

(1) Barrow, Gissane, Travis, Nature 201, 603 (1964); PRS A 287, 240 (1965).
(2) Singh, Pathak, JQSRT 10, 819 (1970).

AuBi: [a]From the variation of $\Delta G(v+\frac{1}{2})$ along the rows and columns of the Deslandres table it appears that the head-origin separations are very large.

(1) Houdart, Bocquet, CR B 263, 151 (1966).

AuCa: [a]Thermal emission.
[b]The lower state of this transition (ω_e'' = 221.85, $\omega_e''x_e''$ = 1.08, $\omega_e''y_e''$ = +0.042) is probably identical with the lower state of A→X.
[c]$\omega_e y_e$ = +0.01.

(1) Schiltz, CR 252, 1750 (1961); AP(Paris) 8, 67 (1963).

AuCe: [a]Thermochemical value (mass-spectrom.)(1).

(1) Gingerich, Finkbeiner, JCP 52, 2956 (1970); 54, 2621 (1971).

AuCl: (1) Ferguson, PR 31, 969 (1928).

AuCo: [a]Thermochemical value (mass-spectrom.)(1)(2).

(1) Kant, JCP 49, 5144 (1968).
(2) Smoes, Mandy, Vander Auwera-Mahieu, Drowart, BSCB 81, 45 (1972).

AuCr: [a]Thermochemical value (mass-spectrom.)(1).

(1) Ackerman, Stafford, Verhaegen, JCP 36, 1560 (1962).

State	T_e	ω_e	$\omega_e x_e$	B_e	α_e	D_e (10^{-4} cm^{-1})	r_e (Å)	Observed Transitions Design.	Observed Transitions ν_{00}	References
$^{197}Au^{(63)}Cu$		(μ = 47.692221_1)		$D_0^0 = 2.3_6$ eV[a]						NOV 1974
		Unclassified R shaded bands in the region 16900 - 19600 cm^{-1}.								(1)
D	(23699)	(182) H						D→X,	R (23665) H	(1)
C	(22176)	(231) H						C→X,	R (22167) H	(1)
B	(20655)	(250)[b] H						B→X,	R 20655.4 H	(1)
A	(20241)	[195.7][c] H						A→X,	R 20214 H	(1)
X	0	250 H	0.7							
$^{197}Au^{19}F$?		μ = 17.3271178								JAN 1975
		Stimulated IR emission in the range 420 - 950 cm^{-1}.								(1)
$^{197}Au^{(56)}Fe$		(μ = 43.563646_8)		$D_0^0 = 1.9_0$ eV[a]						NOV 1974
$^{197}Au^{69}Ga$		μ = 51.058425_8								NOV 1974
A (0^+)	18061.6	219.1 H	1.22[a]	b				A↔X,	V_R 18058.0 H	(1)(2)(3)
X (0^+)	0	226.0 H	0.61	b						
$^{197}Au^{(74)}Ge$		(μ = 53.749202_9)								NOV 1974
A $(^2\Sigma)$	13743.3	242.6[a] H	0.59					A↔X_2, A↔X_1,	R[b](12188) H R[b] 13739.7 H	(1)(2)* (3)
$X_2(^2\Pi_{3/2})$	(1552)									
$X_1(^2\Pi_{1/2})$	0	249.7[a] H	0.33							
$^{197}Au^{1}H$		μ = 1.00269470		$D_0^0 = 3.2_2$ eV[a]						NOV 1974
C $^1\Sigma^+$	(43350)	(1550)[b]		[(6.66)][b]			[(1.58_9)][b]	C←X,	R (42986)[b]	(5)*
c 0^-	42922	[1229] Z	(42)	5.96	0.27	[5.3][c]	1.68_0	c←X,[d]	R 42426 Z	(5)
b 1	42883	[1075.9] Z	(58)	5.627[e]	0.320	[3.5]	1.728_5	b←X,[e]	R 42323.0 Z	(5)
a 2	(42720)	(1020)	(45)	[5.523][f]	f	[3.9]	[1.744_7]	a←X,	R 42077.1 Z	(5)

AuCu: [a]Thermochemical value (mass-spectrom.)(2).
[b]Perturbations.
[c]Perturbations; irregular vibrational intervals.

(1) Ruamps, CR 239, 1200 (1954); SA(Suppl.) 11, 329 (1957).
(2) Ackerman, Stafford, Drowart, JCP 33, 1784 (1960).

AuF: (1) Rice, Beattie, CPL 19, 82 (1973).

AuFe: [a]Thermochemical value (mass-spectrom.)(1).

(1) Kant, JCP 49, 5144 (1968).

AuGa: [a]$\omega_e y_e = -0.004_9$.
[b]For rotational constants derived from incompletely resolved rotational structure in the 0-0 band see (4). Uncertain.

(1) Barrow, Gissane, Travis, Nature 201, 603 (1964).
(2) Bocquet, Houdart, CR B 265, 979 (1967).
(3) Lochet, CR B 272, 797 (1971).
(4) Bocquet, Lefebvre, Houdart, JP(Paris) 34, 317 (1973).

AuGe: [a]From (3). (1) give $\Delta G'(\frac{1}{2}) = 244$, $\Delta G''(\frac{1}{2}) = 251.7$ (from bandheads).
[b]Multiple heads. No analysis of $A \leftrightarrow X_2$.

(1) See ref. (1) of AuGa.
(2) Houdart, CR 261, 2609 (1965).
(3) Houdart, Schamps, JP B 6, 2478 (1973).

Au^1H, Au^2H:

[a]Based on the value for the common dissociation limit $Au(^2D_{3/2}) + H(^2S)$ of states B, a, b, and c (5). Agrees within error limits with an earlier thermochemical value by (1).
[b]Approximate constants for the deperturbed state. Perturbed values are $B_0 = 5.79$, $\nu_0(0-0) = 43105.5$. Rotational levels in v=0 are broadened; broadening increases with J. See [j].
[c]$D_1 = 7.6 \times 10^{-4}$.
[d]Q branches only having $J \geq 18$.
[e]Ω-type doubling $\Delta\nu_{ef}(v=0) \approx +0.006\, J(J+1) + 0.00007\, J^2(J+1)^2 - \ldots$ for $J < 17$. b 1(v=0) perturbed by B 0^+(v=3) at $J \approx 18$. Bands with $v' \geq 1$ consist of Q branches only.
[f]Ω-type doubling $\Delta\nu_{ef}(v=0) \approx -0.005\, J(J+1) + 0.00002\, J^2(J+1)^2$. Bands ascribed by (5) to v=4 and 5 ($B_4 = 4.60$, $B_5 = 3.63$) consist of Q branches only; line width increasing with J, see [j].

(continued p. 57)

State	T_e	ω_e	$\omega_e x_e$	B_e	α_e	D_e $(10^{-4}cm^{-1})$	r_e (Å)	Observed Transitions: Design.	Observed Transitions: ν_{00}	References
$^{197}Au^{1}H$ (continued)										
B 0^{+} g	(38545)	[1544.5] Z	$(74)^{h}$	5.849^{ij}	0.187^{h}	$[3.0]^{k}$	1.695_4	B↔X, R	38231.8 Z	(2)(5)* (6)
A 0^{+} g	27665.7	1669.55 Z	55.06^{ℓ}	6.0069	0.249^{ℓ}	3.24^{ℓ}	1.6729_7	A↔X, R	27344.5 Z	(1)(2)(3) (4)* (6)
X $^{1}\Sigma^{+}$	0	2305.01 Z	43.12^{m}	7.2401	0.2136	2.79^{m}	1.5238_5			
$^{197}Au^{2}H$		μ = 1.99371527		$D_0^0 = 3.2_6$ eV^{n}						NOV 1974
C $^{1}\Sigma^{+}$	(43350)	$(1100)^{o}$		$[(3.53)]^{o}$			$[(1.54_8)]^{o}$	C←X, R	$(43086)^{o}$	(5)*
b 1	(42838)	$(845)^{n}$	$(29)^{n}$	$[2.762]^{p}$		[0.93]	$[1.749_7]$	b←X, R	42440.7 Z	(5)
a 2		Identification uncertain.						a←X, R		(5)
B 0^{+}	38531.9	1187.4 Z	33.9^{q}	2.951^{r}	0.067^{q}	0.73^{q}	1.692_7	B↔X, R	38305.1 Z	(2)(5)*
A 0^{+}	27644.1	1195.24 Z	34.813	3.0352	0.0963^{s}	0.79^{s}	1.6690_6	A→X, R	27420.9 Z	(2)(3)*
X $^{1}\Sigma^{+}$	0	1634.98 Z	21.655^{t}	3.6415	0.07614	0.709^{t}	1.5237_9			
$^{197}Au^{165}Ho$		μ = 89.765242		$D_0^0 = 2.7$ eV^{a}						FEB 1975
$^{197}Au^{(115)}In$		(μ = 72.569304)								NOV 1974
		Red degraded bands in the region 15600 - 17900 cm^{-1}.								(1)
$^{197}Au^{139}La$		μ = 81.459119		$D_0^0 = 3.4_5$ eV^{a}						NOV 1974
$^{197}Au^{(7)}Li$		(μ = 6.7746885)		$D_0^0 = 2.92$ eV^{a}						APR 1975 A
$^{197}Au^{(175)}Lu$		(μ = 92.650724)		$D_0^0 = 3.4_1$ eV^{a}						NOV 1974
$^{197}Au^{(24)}Mg$		(μ = 21.3813864)								NOV 1974
		Additional unidentified R shaded absorption bands in the region 38230 - 37580 cm^{-1}.								(2)

Au^1H, Au^2H (continued):

[g] A and B correspond to $^1\Sigma*$ and $^1\Sigma**$, respectively, of (2).

[h] Vibrational levels observed up to v=4. Levels with $v \geq 2$ are strongly perturbed by C $^1\Sigma^+$. For details see (5).

[i] Broadening, increasing with J, of rotational levels in v=3 and 4. See [j].

[j] Predissociations in B 0^+, a 2, and C $^1\Sigma^+$ are due to interaction with an unstable 1 state arising from $Au(^2D_{5/2}) + H(^2S)$; see (5).

[k] $D_1 = 3.4 \times 10^{-4}$.

[ℓ] $\omega_e y_e = -3.93$; $\gamma_e = -0.027$; $\beta_e = +0.28 \times 10^{-4}$.

[m] $\omega_e y_e = -0.044$; $\beta_e = -0.0084 \times 10^{-4}$.

[n] From the value for Au^1H.

[o] Approximate constants for the deperturbed state. Perturbed values are $B_0 = 2.954$, $v_0(0-0) = 43011.0$. Constants for an additional level at 45122.8 cm^{-1} above X $^1\Sigma^+(v=0)$, assumed to be unperturbed and tentatively assigned as v=2, are $B_2 = 2.75$, $D_2 = 8.3 \times 10^{-4}$.

[p] Average of the two Ω-type doubling components; $\Delta v_{ef} \approx +0.0055\,J(J+1)$.

[q] Vibrational levels observed for $v \leq 7$, except v=6. Levels with $v > 2$ are strongly perturbed by C $^1\Sigma^+$ (5). $\beta_e \approx +0.06 \times 10^{-4}$.

[r] Line width in 5← 0 band increases with J.

[s] $\gamma_e = -0.0038$; $\beta_e = +0.071 \times 10^{-4}$.

[t] $\omega_e y_e = -0.0288$; $\beta_e = -0.0034 \times 10^{-4}$.

Au^1H, Au^2H (continued):

(1) Farkas, ZPC B 5, 467 (1929).
(2) Heimer, Naturw. 24, 78 (1936); ZP 101, 121 (1936); 104, 303 (1937); Dissertation (Stockholm, 1937).
(3) Imanishi, Sci. Pap. IPCR (Tokyo) 31, 247 (1937).
(4) Nilsson, Dissertation (Stockholm, 1948).
(5) Ringström, Nature 198, 981 (1963); AF 27, 227 (1964).
(6) Loginov, OS (Engl. Transl.) 16, 220 (1964).

AuHo: [a] Thermochemical value (mass-spectrom.)(1)(2)(3).

(1) Cocke, Gingerich, JPC 75, 3264 (1971).
(2) Gingerich, CPL 13, 262 (1972).
(3) Kordis, Gingerich, Seyse, JCP 61, 5114 (1974).

AuIn: (1) Barrow, Gissane, Travis, Nature 201, 603 (1964).

AuLa: [a] Thermochemical value (mass-spectrom.)(1).

(1) Gingerich, Finkbeiner, JCP 52, 2956 (1970); 54, 2621 (1971).

AuLi: [a] Thermochemical value (mass-spectrom.)(1).

(1) Neubert, Zmbov, JCS FT I 70, 2219 (1974).

AuLu: [a] Thermochemical value (mass-spectrom.)(1).

(1) See ref. (2) of AuHo.

AuMg: See p. 59.

State	T_e	ω_e		$\omega_e x_e$	B_e	α_e	D_e ($10^{-7}cm^{-1}$)	r_e (Å)	Observed Transitions		References
									Design.	ν_{00}	
$^{197}Au^{(24)}Mg$ (continued)											
D		Six R shaded bands in the region 34230 - 32720 cm^{-1}. No analysis.							D ← X, R		(2)
C	31058	242	H	2^a					C ← X, R	31025 H	(2)
B $\frac{1}{2}$ ($^2\Pi_{\frac{1}{2}}$)	19492.3	338.5	H^Q	1.46^b	$[0.14043]^c$		$[1.02]^d$	[2.3695]	B ↔ X, V	19507.52 Z	(1)(3)
A $\frac{1}{2}$ ($^2\Sigma^+$)	18392.7	341.7	H	3.3	$[0.14201]^e$		$[1.06]^f$	[2.3562]	A ↔ X, V	18409.05 Z	(1)* (3)
X $^2\Sigma^+$	0	307.9^g	H	1.1	0.13214	0.00073	$[1.02]^h$	2.4427			
$^{197}Au^{55}Mn$		$\mu = 42.956570_0$ $\quad D_0^0 = 1.8_8$ eV^a									NOV 1974
$^{197}Au^{(142)}Nd$		(μ = 82.482152) $\quad D_0^0 = 3.0_6$ eV^a									NOV 1974
$^{197}Au^{(58)}Ni$		($\mu = 44.767511_8$) $\quad D_0^0 = 2.5_3$ eV^a									NOV 1974
$^{197}Au^{16}O$		$\mu = 14.7935835_5$ $\quad D_0^0 = 2.3_3$ eV^a									NOV 1974 A
$^{197}Au^{(208)}Pb$		(μ = 101.160962)									NOV 1974
		A second system under low resolution near 8800 cm^{-1}, possibly due to $A \to X_2$.[a]									(2)
A	16357.6	152.7	H	0.9					$A \to X_1$,[a] R	16354.6 H	(1)*
X_1	0	158.6	H	0.6							
$^{197}Au^{(106)}Pd$		(μ = 68.872592) $\quad D_0^0 = 1.4_4$ eV^a									NOV 1974
$^{197}Au^{141}Pr$		μ = 82.143292 $\quad D_0^0 = 3.1_2$ eV^a									NOV 1974
$^{197}Au^{(32)}S$		(μ = 27.5070619) $\quad D_0^0 = 2.5_9$ eV^a									NOV 1974

AuMg: [a]$\omega_e y_e$ = - 0.1.
[b]$\omega_e y_e$ = + 0.009.
[c]Average of the two Ω-type doubling components;
$\Delta v_{fe}(v=0) = - 0.0039_3(J+\frac{1}{2}) + \ldots$
[d]$H_0 = + 1.54 \times 10^{-13}$.
[e]Average of the two Ω-type doubling components;
$\Delta v_{fe}(v=0) = + 0.0264_0(J+\frac{1}{2}) + \ldots$
[f]$H_0 = + 0.96 \times 10^{-13}$.
[g]From band origins $\Delta G(\frac{1}{2}) = 306.10$.
[h]$H_0 = + 1.56 \times 10^{-13}$.

(1) Ruamps, AP(Paris) 4, 1111 (1959).
(2) Schiltz, CR 251, 682 (1960).
(3) Barrow, Gissane, Travis, Nature 201, 603 (1964); PRS A 287, 240 (1965).

AuMn: [a]Thermochemical value (mass-spectrom.)(1).

(1) Smoes, Drowart, CC (1968), 534.

AuNd: [a]Thermochemical value (mass-spectrom.)(1).

(1) Gingerich, Finkbeiner, JCP 52, 2956 (1970); 54, 2621 (1971).

AuNi: [a]Thermochemical value (mass-spectrom.)(1)(2).

(1) Kant, JCP 49, 5144 (1968).
(2) Smoes, Mandy, Vander Auwera-Mahieu, Drowart, BSCB 81, 45 (1972).

AuO: [a]Thermochemical value (mass-spectrom.)(1).

(1) See ref. (2) of AuNi.

AuPb: [a]Thermal emission.

(1) Houdart, Carette, CR 260, 5746 (1965).
(2) Houdart, Schamps, JP B 6, 2478 (1973).

AuPd: [a]Thermochemical value (mass-spectrom.)(1).

(1) Ackerman, Stafford, Verhaegen, JCP 36, 1560 (1962).

AuPr: [a]Thermochemical value (mass-spectrom.)(1).

(1) See ref. (1) of AuNd.

AuS: [a]Thermochemical value (mass-spectrom.)(2). See, however, (1) who gives $D_0^0 = 4.2_4$ eV.

(1) Gingerich, CC (1970), 580.
(2) See ref. (2) of AuNi.

State	T_e	ω_e	$\omega_e x_e$	B_e	α_e	D_e	r_e (Å)	Observed Transitions Design.	ν_{00}	References
$^{197}Au^{(121)}Sb$		(μ = 74.917358$_6$)								NOV 1974
		Four systems of R shaded bands, presumably in thermal emission:								
System D:		$\nu_H = 16762.3 + 186.2(v'+\frac{1}{2}) - 0.6(v'+\frac{1}{2})^2 - 203\ (v''+\frac{1}{2}) + 0.4(v''+\frac{1}{2})^2$.								(1)
System C:		$\nu_H = 16108.6 + 192.8(v'+\frac{1}{2}) - 0.3(v'+\frac{1}{2})^2 - 204.7(v''+\frac{1}{2}) + 0.3(v''+\frac{1}{2})^2$.								(1)
System B:		$\nu_H = 15189.7 + 192.8(v'+\frac{1}{2}) - 0.3(v'+\frac{1}{2})^2 - 203.9(v''+\frac{1}{2}) + 0.3(v''+\frac{1}{2})^2$.								(1)
System A:		$\nu_H = 15047.9 + 186.2(v'+\frac{1}{2}) - 0.6(v'+\frac{1}{2})^2 - 204.7(v''+\frac{1}{2}) + 0.3(v''+\frac{1}{2})^2$.								(1)
$^{197}Au^{45}Sc$		μ = 36.601858$_6$		$D_0^0 = 2.8_7$ eV[a]						NOV 1974
$^{197}Au^{(80)}Se$		(μ = 56.850286$_8$)		$D_0^0 = 2.4_8$ eV[a]						NOV 1974
(A)	(15834)	(210)[b] H						(A)→(X),	(15804)[b] H	(1)
(X)	0	(270)[b] H								
$^{197}Au^{28}Si$		μ = 24.4973499		$D_0^0 = 3.2_4$ eV[a]						NOV 1974 A
A ($^2\Sigma$)	13632.7	389.5[b] H	2.22					A↔X_2,[c]	(12561) H	(1)(2)* (5)
								A↔X_1, R	13631.8 H	(1)(2)* (5)
X_2 ($^2\Pi_{3/2}$)	(1071)									
X_1 ($^2\Pi_{1/2}$)	0	390.9[b] H	1.32							
$^{197}Au^{(120)}Sn$		(μ = 74.531563$_9$)		$D_0^0 = 2.4_9$ eV[a]						NOV 1974
		Fragments of an unidentified system, possibly due to AuSn, in thermal emission from 25000 to 26300 cm^{-1}.								(2)
A ($^2\Sigma$)	13899.0	179.0 H	1.44					A↔X_2,[b]	(11348) H	(2)(3)
								A↔X_1, R	13893.3 H	(2)(3)
X_2 ($^2\Pi_{3/2}$)	(2545)									
X_1 ($^2\Pi_{1/2}$)	0	190.4 H	1.26							

AuSb: (1) Houdart, Bocquet, CR B <u>264</u>, 860 (1967).

AuSc: [a]Thermochemical value (mass-spectrom.)(1).

(1) Gingerich, Finkbeiner, Proc. 9th Rare Earth Res. Conf., Blacksburg Va. (October 1971). Edited by P. E. Field. Vol. II, 795.

AuSe: [a]Thermochemical value (mass-spectrom.)(2).
[b]Preliminary data only, no details.

(1) Joshi, JMS 8, 79 (1962).
(2) Smoes, Mandy, Vander Auwera-Mahieu, Drowart, BSCB 81, 45 (1972).

AuSi: [a]Thermochemical value (mass-spectrom.)(3)(4).
[b]$\Delta G'(\frac{1}{2}) = 386.0$, and $\Delta G''(\frac{1}{2}) = 391.2$, from band heads acc. to (1).
[c]Complex system of V and R shaded bands. No analysis.

(1) Barrow, Gissane, Travis, Nature 201, 603 (1964).
(2) Houdart, CR B 262, 550 (1966).
(3) Gingerich, JCP 50, 5426 (1969).
(4) Vander Auwera-Mahieu, Peeters, McIntyre, Drowart, TFS 66, 809 (1970).
(5) Houdart, Schamps, JP B 6, 2478 (1973).

AuSn: [a]Thermochemical value (mass-spectrom.)(1).
[b]Three sequences of bandheads.Low dispersion only.

(1) Ackerman, Drowart, Stafford, Verhaegen, JCP 36, 1557 (1962).
(2) Collette, Schiltz, CR 257, 2092 (1963).
(3) See ref. (5) of AuSi.

State	T_e	ω_e	$\omega_e x_e$	B_e	α_e	D_e $(10^{-}\ cm^{-1})$	r_e (Å)	Observed Transitions Design.	ν_{00}	References
$^{197}Au^{(88)}Sr$		($\mu = 60.779779_9$)								NOV 1974
		Unclassified bands from 14500 to 15200 cm^{-1}.[a]								(1)
C	14162.2	155.6_3 H	0.28					C→X,[a]	14163.3 H	(1)
B	13832.7	147.04 H^Q	0.92[b]					B→X,[a] V	13829.4 H^Q	(1)*
(A)		(140)[c] H						(A→X)[ac]		(1)
X ($^2\Sigma$)	0	153.33 H^Q	0.19[d]							
$^{197}Au^{159}Tb$		$\mu = 87.956432$		$D_0^0 = 3.0$ eV[a]						FEB 1975
$^{197}Au^{(130)}Te$		($\mu = 78.278717_7$)		$D_0^0 = 2.4_2$ eV[a]						NOV 1974
A	15481	$156._1$ H	0.1_0					A→X, R	15453 H	(1)*
(X)	0	$212._5$ H	0.5_0							
$^{197}Au^{238}U$		$\mu = 107.784312$		$D_0^0 = 3.2_5$ eV[a]						NOV 1974
$^{197}Au^{89}Y$		$\mu = 61.256283_4$		$D_0^0 = 3.0_8$ eV[a]						NOV 1974

AuSr: [a]Thermal emission.

[b]$\omega_e y_e = +0.06$.

[c]Uncertain analysis of bands in the region 11600 - 13200 cm^{-1}, possibly forming part of $B \rightarrow X$ (1).

[d]$\omega_e y_e = -0.007$.

(1) Schiltz, AP(Paris) 8, 67 (1963).

AuTb: [a]Thermochemical value (mass-spectrom.)(1)(2).

(1) Gingerich, CPL 13, 262 (1972).

(2) Kordis, Gingerich, Seyse, JCP 61, 5114 (1974).

AuTe: [a]Thermochemical value (mass-spectrom.)(2).

(1) Maheshwari, Sharma, PPS 81, 898 (1963).

(2) Smoes, Mandy, Vander Auwera-Mahieu, Drowart, BSCB 81, 45 (1972).

AuU: [a]Thermochemical value (mass-spectrom.)(1).

(1) Gingerich, Blue, JCP 47, 5447 (1967).

AuY: [a]Thermochemical value (mass-spectrom.)(1).

(1) Gingerich, Finkbeiner, Proc. 9th Rare Earth Res. Conf., Blacksburg Va. (October 1971); ed. Field. Vol. II, p. 795.

State	T_e	ω_e	$\omega_e x_e$	B_e	α_e	D_e $(10^{-}\ cm^{-1})$	r_e (Å)	Observed Transitions: Design.	Observed Transitions: ν_{00}	References
$^{11}B_2$		$\mu = 5.5046526_7$		$D_0^0 = 3.0_2$ eV[a]						SEP 1976
A $^3\Sigma_u^-$	30573.4	937.4 Z	2.6	1.160	0.011		1.625	A↔X, R	30518.10 Z	(1)* (2)(6)
X $^3\Sigma_g^-$ [b]	0	1051.3 Z	9.35	1.212	0.014		1.590			
$^{(138)}Ba^{79}Br$		$(\mu = 50.194041_5)$		$D_0^0 = 3.7_9$ eV[a]						FEB 1976 A
E $^2\Sigma^+$	26865.9	219.9 H	0.35					E↔X, V	26878.9 H	(2)(7)(8)
D $^2\Sigma^+$	25670.9	209.1 H	0.53					D↔X, V	25678.5 H	(2)(7)(8)
C $^2\Pi$	19192.5 18650.9	197.4 H	0.41					C^b↔X, R	19194.3 H 18652.7 H	(1)(2)(7)(8)
B $(^2\Sigma^+)$	11325[c]							B→X,		(7)
A $(^2\Pi)$	10604[c] 9980[c]							A→X,		(7)
X $^2\Sigma^+$	0	193.8 H	0.42							
$^{138}Ba^{35}Cl$		$\mu = 27.8953776$		$D_0^0 = 4.5_5$ eV[a]						FEB 1976 A
G $(^2\Sigma)$	32511.4	331.3	1.29					G←X, V	32537.3	(2)
F $(^2\Sigma)$	29493.6	331.8 H	1.30					F←X, V	29519.7 H	(2)
E $^2\Sigma^+$	27064.8	311.5 H	0.93					E↔X, V	27080.9 H	(1)* (2)
D $^2\Sigma^+$	25471.6	304.6 H	1.04					D↔X, V	25484.2 H	(1)* (2)
C $^2\Pi$	19450.1 19062.9	285.0 H^Q 280.2 H^Q	0.79 0.79					C^b↔X, R_V	19453.0 H^Q 19063.4 H^Q	(1)* (2)
B $(^2\Sigma^+)$	11880.0	255.25 H	0.83					B↔X, R	11868.0 H	(3)(14)
A $(^2\Pi)$	10995.3 10363	256.35 H	0.73					A↔X, R	10983.9 H 10351.1 H	(3)(14) (8)(14)
X $^2\Sigma^+$	0	279.3 H	0.89							

B_2: [a]Thermochemical value (mass-spectrom.); from (3), but using the higher value from (5) for the heat of sublimation of boron.

[b]Experimental evidence (6) supports a $^3\Sigma$ ground state; theoretical calculations (4) predict a $^5\Sigma_u^-$ state at approximately 1100 cm^{-1} below the lowest state of $^3\Sigma_g^-$ symmetry.

(1) Douglas, Herzberg, CJR A 18, 165 (1940).
(2) Nicholls, Fraser, Jarmain, CF 3, 13 (1959).
(3) Verhaegen, Drowart, JCP 37, 1367 (1962).
(4) Bender, Davidson, JCP 46, 3313 (1967).
(5) JANAF (1971).
(6) Graham, Weltner, private communication (1976); JCP 65, 1516 (1976).

BaBr: [a]Thermochemical value (flame photometry)(5). In agreement with a theoretical lower limit of 3.76 eV (3).

[b]Radiative lifetimes (6) for the upper (v=8) and for the lower (v=4) component of C $^2\Pi$ are $16._5$ and $17._9$ ns, respectively. A shorter lifetime of 8 ns was reported by (4).

[c]Preliminary data, no details.

(1) Hedfeld, ZP 68, 610 (1931).
(2) Harrington, Dissertation (U. of California, 1942).
(3) Krasnov, Karaseva, OS(Engl. Transl.) 19, 14 (1965).
(4) Capelle, Bradford, Broida, CPL 21, 418 (1973).
(5) Gurvich, Ryabova, Khitrov, FSCS No. 8, 83 (1973); Khitrov, Ryabova, Gurvich, HT(USSR) 11, 1005 (1973).
(6) Dagdigian, Cruse, Zare, JCP 60, 2330 (1974).
(7) Bradford, Jones, Southall, Broida, JCP 62, 2060 (1975).
(8) Joshi, Gopal, Pramāṇa 4, 276 (1975).

BaCℓ: [a]Thermochemical value (mass-spectrom.)(6)(7). In agreement with a theoretical lower limit (5) of 4.35 eV. Flame photometry (11)(12) gives 4.6_3 eV. A higher value, $D_0^0 \geq 4.7_7$ eV, was estimated (9) from the short-wavelength limit of the BaCℓ chemiluminescence spectrum.

[b]Radiative lifetimes (13) for the upper and the lower component of C $^2\Pi(v=0)$ are $16._6$ and $17._5$ ns, respectively. (10) give $\tau = 22$ ns.

(1) Parker, PR 46, 301 (1934).
(2) See ref. (2) of BaBr.
(3) Barrow, Crawford, Nature 157, 339 (1946).
(4) Gurvich, Ryabova, HT(USSR) 2, 190 (1964); 3, 604 (1965).
(5) See ref. (3) of BaBr.
(6) Zmbov, CPL 4, 191 (1969).
(7) Hildenbrand, JCP 52, 5751 (1970).
(8) Lagerqvist, quoted in DONNSPEC (1970).
(9) Jonah, Zare, CPL 9, 65 (1971).
(10) See ref. (4) of BaBr.
(11) See ref. (5) of BaBr.
(12) Ryabova, Khitrov, Gurvich, HT(USSR) 10, 669 (1973).
(13) See ref. (6) of BaBr.
(14) See ref. (7) of BaBr.

State	T_e	ω_e	$\omega_e x_e$	B_e	α_e	D_e (10^{-6}cm^{-1})	r_e (Å)	Observed Transitions Design.	Observed Transitions ν_{00}	References
$^{(138)}Ba^{19}F$		(μ = 16.6980158)		D_0^0 = 6.05 eV[a]		I.P. = 4.8_5 eV[b]				FEB 1976
I		(514)						I→X,	(33717)	(11)
H ($^2\Sigma$)	31582.3	508.8 H	2.00					H↔X, (V)	31602.2 H	(3)(11)
G ($^2\Sigma$)	31451.9	510.4 H	0.83					G↔X, (V)	31472.9 H	(3)(11)
F ($^2\Sigma$)	29411.3	529.9 H	2.00					F↔X,[c] (V)	29441.8 H	(3)(11)
E $^2\Sigma^+$	28139.7	538.4 H	1.90	[0.2290][d]	(0.0011)	[0.162]	[2.100]	E↔X, V	28174.45 Z	(3)(7)(11)
D' $^2\Sigma^+$	26227.0	504.9 H	1.54	0.2269[e]	(0.00099)	0.176[e]	2.109[e]	D'↔X, V	26245.0[f] Z	(2)* (3)(7)(9)(11)
D $^2\Sigma^+$	24156.8	508.4 H	1.88	[0.2273][g]	(0.0011)	[0.173]	[2.107]	D↔X,[h] V	24176.54 Z	(2)* (3)(7)(11)
C $^2\Pi$	20197 19998.2	$456._0$ H	1.6_7	(0.2148)[i] (0.2138)[i]	(0.0012)		(2.170)	C[j]↔X, R	20191 H 19991.8 H	(2)* (9)(12)(13)(15)
B $^2\Sigma^+$	14062.5	424.4 H	1.88	[0.2071][k]	(0.0012)	[0.190]	[2.208]	B↔X,[c] R	14040.21 Z	(1)(2)* (7)(9)
A $^2\Pi_r$	12278.2 11646.9	436.7 H^Q 435.5 H^Q	1.82 1.68	[0.2119] [0.2118][ℓ]	(0.0012) (0.0011)	[0.208]	[2.183]	A↔X, R	12262.09 Z 11630.20 Z	(1)(2)(7)(9)
X $^2\Sigma^+$	0	468.9 H^Q	1.79	[0.2158_5][m]	(0.0012)	[0.175]	[2.162_7]	ESR sp.[n]		(10)
$^{(138)}Ba^{1}H$		(μ = 1.00051337)		$D_0^0 \leq$ 1.95 eV[a]						FEB 1976
G $^2\Sigma$		Single band.		[3.65]			[2.15]	G←X, V	31645	(12)
F $^2\Sigma^+$				[3.626][b]		[114]	[2.156]	F←X, V	30747.91 Z	(8)* (12)
C $^2\Sigma^+$	23675	1282 Z	15	3.59[cde]	0.064	(100)	2.17	C↔X, V	23732[d] (Z)	(6)(7)(9)*
D $^2\Sigma^+$	21885[f]	428[f]	4.5	1.62[fgh]	0.017	(100)	3.22	D↔X,[i] R	21517[f]	(9)
E { $^2\Pi_{3/2}$	$15055._4$	$1228._6$ Z	16.9	3.560[j]	0.075	[122][k]	2.187	E→X, V	15084.9[m] Z	(1)* (2)
$^2\Pi_{1/2}$	14605	[$1186._7$] Z		3.486[j]	0.072	[110][ℓ]			14630.1[m] Z	
B $^2\Sigma^+$	11092.4_4	1088.9 Z	15.4	3.266[n]	0.070	111	2.308[o]	B→X, R	11052.51 Z	(3)(5)

BaF: [a]Thermochemical value (mass-spectrom.)(4)(5)(8). 6.3_7 eV by flame photometry (6).

[b]Electron impact appearance potential (5)(8).

[c]Double heads.

[d]Spin-doubling, $|\gamma' - \gamma''| = 0.184$.

[e]Constants for v=1. Spin-doubling, $|\gamma' - \gamma''| = 0.0091$.

[f]Calc. from the origin of the 1-0 band at 26746.86 cm^{-1}.

[g]Spin-doubling, $|\gamma' - \gamma''| = 0.0047$.

[h]The bands have normal structure, but the four branches have unusual intensities, the two strong branches being either R_1 and P_2 or R_2 and P_1.

[i]Estimates based on band head separations (12)(13).

[j]Radiative lifetime $\tau(v=0) = 23.6$ ns (14).

[k]Spin-splitting constant $\gamma = (-)0.263$.

[ℓ]Λ-type doubling, $\Delta\nu_{fe} = -0.258(J+\frac{1}{2})$.

[m]Spin-splitting constant $\gamma = +0.00278$.

[n]In Ne and Ar matrices at 4 K (10).

(1) Nevin, PPS 43, 554 (1931).
(2) Jenkins, Harvey, PR 39, 922 (1932).
(3) Fowler, PR 59, 645 (1941).
(4) Blue, Green, Ehlert, Margrave, Nature 199, 804 (1963).
(5) Ehlert, Blue, Green, Margrave, JCP 41, 2250 (1964).
(6) Gurvich, Ryabova, HT(USSR) 2, 366 (1964).
(7) Barrow, Bastin, Longborough, PPS 92, 518 (1967).
(8) Hildenbrand, JCP 48, 3657 (1968).
(9) Mohanty, Mohanty, Mohanty, IJPAP 8, 423 (1970).
(10) Knight, Easley, Weltner, Wilson, JCP 54, 322 (1971).
(11) Singh, Mohan, JP B 4, 1395 (1971); IJPAP 11, 918 (1973).
(12) Kushawaha, Asthana, Shanker, Pathak, SpL 5, 407 (1972); Kushawaha, SpL 6, 633 (1973).
(13) Cruse, Dagdidgian, Zare, FDCS No. 55, 277 (1973).
(14) See ref. (6) of BaBr.
(15) See ref. (7) of BaBr.

Ba^1H: [a]From the predissociation in C $^2\Sigma^+(v=1)$, assuming dissociation into $^3D_3 + {}^2S$; see (9).

[b]Diffuse lines.

[c]Spin-splitting constant $\gamma = +0.14$.

[d]v=0 interacts strongly with successive vibrational levels of D $^2\Sigma^+$. A third (unidentified) state affects the low J levels in v=0; see (9).

[e]Emission from v=1 breaks off above N=8; in absorption, the lines become broad for $N' \geq 10$. Similar predissociation effects in absorption to v'=2 and 3; see (9).

[f]Vibrational numbering uncertain.

[g]Spin-splitting constant $\gamma = +0.12$.

[h]Strong interactions with C $^2\Sigma^+(v=0)$. Weaker perturbations may be caused by a $^4\Sigma$ state; for details see Figs. 6 and 10 of (9).

[i]The P_2 and R_1 branches are about twice as strong as the P_1 and R_2 branches (9)(11).

[j]Small perturbations. Large Λ-type doubling in $^2\Pi_{\frac{1}{2}}$. For a refined treatment of the rotational structure see (15).

[k]$H_0 = +3.1 \times 10^{-9}$.

[ℓ]$H_0 = +2.0 \times 10^{-9}$.

[m]$\{J'=0\}$ relative to N"=0.

[n]Effective constants. Very large spin doubling ($\gamma_0 \approx -4.9$) due to interaction with A $^2\Pi$. "True" constants ($B_e = 3.164$, $\alpha_e = 0.061$) have been determined by (16) whose revised spin splitting parameters are in good agreement with "pure precession" values for $\ell=2$.

[o]From the "true" rotational constants; see [n] and [p].

(continued p. 68)

State	T_e	ω_e	$\omega_e x_e$	B_e	α_e	D_e ($10^{-6}cm^{-1}$)	r_e (Å)	Observed Transitions: Design.	Observed Transitions: ν_{00}	References
$^{(138)}Ba^1H$ (continued)										
A $^2\Pi_{3/2}$	9939.82	1109.9_8 Z	13.59	3.322^{pq}	0.082	$[132.2]^r$	2.249^o	A⟷X, R	9910.88^m Z	(4)(5)(10)
A $^2\Pi_{1/2}$	9457.45	1110.55 Z	15.29	3.27887^p	0.07283	$[121]^s$			9428.37^m Z	
H ($^2\Delta_{3/2}$)	$[10609]^t$	$[1023]^t$		$[2.97]^t$						(10)(16)
X $^2\Sigma^+$	0	1168.31 Z	14.50	3.38285^u	0.06599	$[112.67]^v$	2.23175	ESR sp.[w]		(14)
$^{(138)}Ba^2H$		(μ = 1.98510975)		$D_0^0 \leq 1.97$ eV[a]						FEB 1976
F $^2\Sigma^+$	(30708)	[860.6] Z		$[1.8376]^b$	c	[29.2]	[2.1497]	F←X, V	30725.25 Z	(1)(6)(7)
C $^2\Sigma^+$	23675	910	7.6	1.80^d	0.020		2.17	C←X, V	23715	(2)* (7)
D $^2\Sigma^+$	21884^e	304^e	(2.3)	0.827^{ef}	(0.008_2)		3.20	D←X, R	21622^e	(2)(7)
E $^2\Pi_{3/2}$	15059.32	872.19 Z	8.47	1.7874^{gh}	0.0259	$[30.4]^j$	2.186	E←X, V	15080.24^k Z	(7)*
E $^2\Pi_{1/2}$	14604.12	867.82 Z	8.78	1.7653^{gi}	0.0255	[28.4]			14622.78^k Z	
B $^2\Sigma^+$	11089.62	772.99 Z	7.72	1.6355^{ℓ}	0.0252	$[28.9]^m$	2.298^n	B←X, R	11061.13 Z	(3)* (7)
A $^2\Pi_{3/2}$	9938.34	791.23 Z	7.80	1.65983^o	0.02710	$[31.0]^p$	2.259^n	A←X, R	9918.95^k Z	(4)* (7)
A $^2\Pi_{1/2}$	9456.20	788.85 Z	7.67	1.6475^{oq}	0.0259	[29.4]			9435.65^k Z	
H ($^2\Delta_{3/2}$)	$[9840]^r$			$[1.591]^r$						(4)(7)(10)
X $^2\Sigma^+$	0	829.77 Z	7.32	1.7071^s	0.02363	$[28.77]^t$	2.2304			

Ba^1H (continued):

[p]Effective constants. Large Λ-type doubling due to interaction with B $^2\Sigma^+$:
$^2\Pi_{3/2}$, $\Delta\nu_{fe}(v=0) = -3.38 \times 10^{-4}(J-\frac{1}{2})(J+\frac{1}{2})(J+\frac{3}{2}) + \ldots$;
$^2\Pi_{1/2}$, $\Delta\nu_{fe}(v=0) = -5.307(J+\frac{1}{2}) + \ldots$.
(16) has evaluated "true" constants for v=0,1,2 (A_0 = +438.1, B_0 = 3.306, ...) and has shown that the Λ-type doubling agrees with "pure precession" values for ℓ=2.

[q]Perturbations by H $^2\Delta_{3/2}(v \geq 2)$ (10)(16).

[r]$H_0 = +6.6 \times 10^{-9}$.

[s]$H_0 = +4 \times 10^{-9}$.

[t]Constants (10) for the lowest observed level, probably v=2 (16). From perturbations in A $^2\Pi_{3/2}$.

Ba^1H (continued):

[u]Spin splitting constant $\gamma_0 = +0.1927$; also higher order terms (10)(13).

[v]$H_0 = +2.89 \times 10^{-9}$.

[w]In Ar matrix at 4 K (14).

(1) Fredrickson, Watson, PR <u>39</u>, 753 (1932).
(2) Funke, ZP <u>84</u>, 610 (1933).
(3) Watson, PR <u>43</u>, 9 (1933).
(4) Watson, PR <u>47</u>, 213 (1935).
(5) Koontz, Watson, PR <u>48</u>, 937 (1935).
(6) Grundström, ZP <u>99</u>, 595 (1936).
(7) Funke, Grundström, ZP <u>100</u>, 293 (1936).
(8) Edvinsson, Kopp, Lindgren, Åslund, AF <u>25</u>, 95 (1963).
(9) Kopp, Åslund, Edvinsson, Lindgren, AF <u>30</u>, 321 (1965).
(10) Kopp, Kronekvist, Guntsch, AF <u>32</u>, 371 (1966).
(11) Kopp, Hougen, CJP <u>45</u>, 2581 (1967).
(12) Khan, JP B <u>1</u>, 985 (1968).
(13) Veseth, JP B <u>3</u>, 1677 (1970).
(14) Knight, Weltner, JCP <u>54</u>, 3875 (1971).
(15) Veseth, JMS <u>38</u>, 228 (1971).
(16) Veseth, MP <u>20</u>, 1057; <u>21</u>, 287 (1971); <u>25</u>, 333 (1973).

Ba^2H: [a]See [a] of Ba^1H.

[b]See [b] of Ba^1H.

[c]v=1 perturbed, $B_1 \approx 1.855$, $D_1 \approx 41 \times 10^{-6}$.

[d]Strong interactions with $D\,^2\Sigma^+$. Only v=1 has been analyzed in detail. Lines are broad above v'=1,N'=22.

[e]See [f] of Ba^1H.

[f]Strong interactions with C $^2\Sigma^+$. v=8...11 have been analyzed, but only v=11 has been deperturbed.

[g]Effective constants. For a refined treatment of the ro-

Ba^2H (continued):

tational structure see (9).

[h]Λ-type doubling,
$\Delta\nu_{fe}(v=0) = +0.060 \times 10^{-4}(J-\frac{1}{2})(J+\frac{1}{2})(J+\frac{3}{2}) - \ldots$.

[i]Λ-type doubling, $\Delta\nu_{fe}(v=0) = +0.433(J+\frac{1}{2}) - \ldots$. v=0 is perturbed by B $^2\Sigma^+(v=5)$.

[j]$H_0 = +4.1 \times 10^{-10}$.

[k]See [m] of Ba^1H.

[l]Effective constants. Very large spin splitting, $\gamma_0 = -2.433$. "True" constants ($B_e = 1.608$, $\alpha_e = 0.024$) have been evaluated (10); see [n] of Ba^1H.

[m]$H_0 = +3.7 \times 10^{-10}$.

[n]From the "true" rotational constants; see [l] and [o].

[o]Effective constants. Large Λ-type doubling:
$^2\Pi_{3/2}$, $\Delta\nu_{fe}(v=0) = -0.524 \times 10^{-4}(J-\frac{1}{2})(J+\frac{1}{2})(J+\frac{3}{2}) + \ldots$;
$^2\Pi_{1/2}$, $\Delta\nu_{fe}(v=0) = -2.636(J+\frac{1}{2}) + \ldots$; see also (5).
"True" constants for v=0,1,2 ($A_0 = +433.6$, $B_0 = 1.652$, ...) have been evaluated (10); see [p] of Ba^1H.

[p]$H_0 = +5.3 \times 10^{-10}$.

[q]v=0 perturbed by H $^2\Delta_{3/2}(v=2)$.

[r]Single level, probably v=2 (10), observed in a perturbation of A $^2\Pi_{\frac{1}{2}}(v=0)$.

[s]Spin splitting constant $\gamma_0 = +0.0972$; also higher order terms (7)(8).

[t]$H_0 = +4.0 \times 10^{-10}$.

(1) See ref. (8) of Ba^1H.
(2) See ref. (9) of Ba^1H.
(3) Kopp, Wirhed, AF <u>32</u>, 307 (1966).
(4) See ref. (10) of Ba^1H.

(continued p. 71)

State	T_e	ω_e	$\omega_e x_e$	B_e	α_e	D_e $(10^{-}$ $cm^{-1})$	r_e (Å)	Observed Transitions Design.	Observed Transitions ν_{00}	References
$^{(138)}Ba^{127}I$		$(\mu = 66.088178_9)$ $D_0^0 = 4.42$ eV[a]								SEP 1976
		Additional emission bands in the region 22300 - 23300 cm^{-1}: $\nu_e = \frac{23224.8}{22664.7}$ $\omega_e' = \frac{173.0}{171.0}$ $\omega_e'x_e' = \frac{0.50}{0.50}$ $\omega_e'' = \frac{177.0}{177.0}$ $\omega_e''x_e'' = \frac{0.60}{0.65}$								(8)
E $(^2\Sigma^+)$	26753	176.0[b] H	0.30					E↔X, V	26765 H	(4)(5)(6)*
D $(^2\Sigma^+)$	25764	161.1[b] H	0.26					D↔X, V	25769[b] H	(4)(5)(6)*
C $(^2\Pi)$	18569 17814	150.05[c] H	0.275					C[d]↔X, V_R	18568 H 17813 H	(1)(2)(4)(5) (12)
B $(^2\Sigma)$	10417[e]							B→X,		(10)
A $(^2\Pi)$	9921[e] 9268[e]							A→X,		(10)
X $(^2\Sigma^+)$	0	152.30[c] H	0.270							

Ba^2H (continued):

(5) See ref. (11) of Ba^1H.
(6) See ref. (12) of Ba^1H.
(7) Kopp, Wirhed, AF 38, 277 (1968).
(8) See ref. (13) of Ba^1H.
(9) See ref. (15) of Ba^1H.
(10) See ref. (16) of Ba^1H.

BaI: [a]From the Ba + I_2 chemiluminescence spectrum, see (11). Lower bounds were earlier predicted by (3) and (7).

[b]Vibrational constants from (4). Similar results in (5). A different analysis of the D – X system has been suggested (6), placing the 0-0 band at 26101 cm^{-1}, and resulting in increased upper and lower state vibrational constants.

[c]Vibrational constants recalculated by (12) from unpublished data of M. M. Patel; for the upper state there is only moderate agreement with the constants in (4).

[d]Radiative lifetime $\tau = 16._5$ ns (9).

[e]Preliminary data, no details.

(1) Walters, Barratt, PRS A 118, 120 (1928).
(2) Mesnage, AP(Paris) 12, 5 (1939).
(3) See ref. (3) of BaBr.
(4) E. Morgan, unpublished, quoted in DONNSPEC (1970).
(5) Patel, Shah, IJPAP 8, 681 (1970).
(6) Reddy, Rao, JP B 3, 1008 (1970).
(7) Mims, Lin, Herm, JCP 57, 3099 (1972).
(8) Shah, Patel, Darji, JP B 5, L191 (1972).
(9) See ref. (6) of BaBr.
(10) See ref. (7) of BaBr.
(11) Dickson, Kinney, Zare, CP 15, 243 (1976).
(12) Dagdigian, Cruse, Zare, CP 15, 249 (1976).

State	T_e	ω_e	$\omega_e x_e$	B_e	α_e	D_e ($10^{-7}cm^{-1}$)	r_e (Å)	Observed Transitions Design.	ν_{00}	References
$^{138}Ba^{16}O$		μ = 14.3325559$_8$		D_0^0 = 5.79 eV[a]		I.P. = 6.9$_1$ eV[b]				SEP 1976 A
	[36490 - 38620]	Nineteen $^1\Sigma^+$ vibronic levels belonging to two or more perturbed electronic states.[c]								(35)
B ($^1\Pi$)	32866.$_4$	488 H	3.6					B↔X, R	32775 H	(5)*
A' $^1\Pi$ [d]	17691	443 H	1.66	0.2252[e]	0.0013$_0$	(2.3)	2.285	A'[f]↔X, R	17578 H	(32)(34)(39)* (40)
a $^3\Pi_{0^+}$ [d]	17586.$_5$	449.$_0$[g]	2.5	0.2254$_4$[g]	0.0013$_8$					
a $^3\Pi_1$ [d]	17442	[445][g]	(4.5)	[0.2254][g]			2.289			
a $^3\Pi_2$ [d]	17393	[g]		[g]						
A $^1\Sigma^+$	16807.$_2$	499.$_7$[h] Z	1.6$_4$	0.25832[h]	0.001070	2.8	2.1338	A[i]↔X,[j] R	16722.25 Z	(1)* (2)* (4)* (18) (31)
								Microwave optical double resonance		(22)(23)(26)
X $^1\Sigma^+$	0	669.76[k] Z	2.02$_8$[ℓ]	0.3126140[k]	0.0013921[m]	2.724	1.939692[n]	Microwave sp.[o]		(7)(15)(33)
								Mol. beam electric[p] and magn. reson.[q]		(6) (10)

BaO: [a]By extrapolation of the highest observed ground state levels (v=0,J) populated in the reaction Ba + CO_2 under single-collision conditions (31). Compatible with lower bounds obtained from the short-wavelength limits of the Ba + NO_2 (17)(21) and Ba + $C\ell O_2$ (40) chemiluminescence spectra. Flame photometric values (16)(27)(36), if corrected to a $^1\Sigma$ ground state, are near 5.6$_9$ eV. Earlier measurements reviewed in (12)(13).

[b]Electron impact appearance potential (28)(38).

[c]Optical-optical double resonance laser spectroscopy.

[d] $^3\Pi_2$, $^3\Pi_1$, $^3\Pi_0$, and $^1\Pi$ [see (29)] correspond to Q, Z, Y, and X, respectively, of (2)(3).

[e]Only v=12, 17, 18 have been rotationally analyzed (34). In addition, constants for v=2, 3, 4 have been derived (2)(3) from perturbations (29) in v=3, 4, 5, resp., of A $^1\Sigma^+$.

[f]Radiative lifetime τ(v=13...17) = 9 μs (34). Similar lifetimes observed by (18)(20) can be attributed to either A' $^1\Pi$ or a $^3\Pi_i$; see (30).

[g]From perturbations in A $^1\Sigma^+$ (2)(3)(29). (29) has adopted the following constants for a $^3\Pi_i$: T_e = 17483, A = -105, ω_e = 448.3, $\omega_e x_e$ = 2.39, B_e = 0.2244, α_e = 0.0014.

[h]Partially deperturbed constants. Numerous interactions with levels of a $^3\Pi_i$ and A' $^1\Pi$, except in v=0 which is unperturbed (2)(3)(29). Potential curve (11)(37).

BaO (continued):

[i] Radiative lifetime $\tau(v=0)$ = 0.356 μs; $f_{00}(A \leftarrow X)$ = 0.00026 (20). Slight variation of τ with v.

[j] Relative intensities (9)(14)(19)(25); Franck-Condon factors (8)(24)(37).

[k] The ground state constants have been derived (35) from a combination of microwave rotational constants (15)(33), electronic band origins (2), and data on higher vibrational levels ($v \leq 34$) obtained by optical-optical double resonance photoluminescence spectroscopy. They were used for the construction of the RKR potential curve up to v=40 (35).

[l] $\omega_e y_e = -0.003_5$, $\omega_e z_e = -6._3 \times 10^{-5}$.

[m] $\gamma_e = -4.3_3 \times 10^{-6}$.

[n] From the Dunham corrected microwave B_e value in (33).

[o] Values of eqQ(^{135,137}Ba) in (15)(33).

[p] $\mu_{el}(v) = [7.934 + 0.042(v+\frac{1}{2})]$ D.

[q] g_J = (-)0.103.

(1) Mahanti, PPS 46, 51 (1934).
(2) Barrow, Lagerqvist, Lind, Nature 164, 923 (1949); Lagerqvist, Lind, Barrow, PPS A 63, 1132 (1950).
(3) Kovács, Lagerqvist, JCP 18, 1683 (1950); AF 2, 411 (1950).
(4) Gatterer, Junkes, Salpeter, Rosen, METOX (1957).
(5) Parkinson, PPS 78, 705 (1961).
(6) Wharton, Kaufman, Klemperer, JCP 37, 621 (1962); 39, 240 (1963).
(7) Wharton, Klemperer, JCP 38, 2705 (1963).
(8) Ortenberg, Glasko, Dimitriev, SAAJ 8, 258 (1964).
(9) Vaidya, Desai, Bidaye, JQSRT 4, 353 (1964).
(10) Brooks, Kaufman, JCP 43, 3406 (1965).
(11) Singh, Rai, IJPAP 4, 102 (1966).
(12) Schofield, CRev 67, 707 (1967).
(13) Gaydon, DISSEN (1968), p. 241.
(14) Walvekar, Korwar, JP B 2, 115 (1969).
(15) Hoeft, Lovas, Tiemann, Törring, ZN 25 a, 1750 (1970).
(16) Kalff, Alkemade, JCP 52, 1006 (1970).
(17) Ottinger, Zare, CPL 5, 243 (1970).
(18) Sakurai, Johnson, Broida, JCP 52, 1625 (1970).
(19) Degen, Brown, Romick, PSS 19, 1625 (1971).
(20) Johnson, JCP 56, 149 (1972).
(21) Jonah, Zare, Ottinger, JCP 56, 263 (1972).
(22) Field, Bradford, Harris, Broida, JCP 56, 4712 (1972).
(23) Field, Bradford, Broida, Harris, JCP 57, 2209 (1972).
(24) Wentink, Spindler, JQSRT 12, 129 (1972).
(25) Best, Hoffman, JQSRT 13, 69 (1973).
(26) Field, English, Tanaka, Harris, Jennings, JCP 59, 2191 (1973).
(27) Kalff, Alkemade, JCP 59, 2572 (1973); 60, 1698 (1974).
(28) Panchenkov, Gusarov, Gorokhov, RJPC 47, 55 (1973).
(29) Field, JCP 60, 2400 (1974).
(30) Field, Jones, Broida, JCP 60, 4377 (1974).
(31) Dagdigian, Cruse, Schultz, Zare, JCP 61, 4450 (1974).
(32) Hsu, Krugh, Palmer, Obenauf, Aten, JMS 53, 273 (1974).
(33) Tiemann, Bojaschewsky, Sauter-Servaes, Törring, ZN 29 a, 1692 (1974).
(34) Pruett, Zare, JCP 62, 2050 (1975).
(35) Field, Capelle, Revelli, JCP 63, 3228 (1975).
(36) Van der Hurk, Hollander, Alkemade, JQSRT 15, 113 (1975).
(37) Tawde, Tulasigeri, APH 38, 299 (1975).
(38) Rauh, Ackermann, JCP 64, 1862 (1976).
(39) Wyss, Broida, JMS 59, 235 (1976).
(40) Engelke, Sander, Zare, JCP 65, 1146 (1976).

State	T_e	ω_e	$\omega_e x_e$	B_e	α_e	D_e $(10^{-7}cm^{-1})$	r_e (Å)	Observed Transitions Design.	Observed Transitions ν_{00}	References
$^{138}Ba^{32}S$		$\mu = 25.9547097$		$D_0^0 = 4.3_6$ eV[a]						MAR 1976 A
B $^1\Sigma^+$	27060.29	254.10 Z	0.438	0.08604	0.00044		2.747_5	B←X, R	26997.74 Z	(5)
A $^1\Sigma^+$	14493	$294._3$[b]	3.0_8	0.0935_3[b]	0.0007_2		2.635	A←X, R	14450	(3)(5)
a ($^3\Pi_{0^+}$)	(14570)	(235)[c]		(0.078_5)[c]			(2.88)			
X $^1\Sigma^+$	0	379.42 Z	0.8842	0.10331	0.0003188	0.306	2.5074	Mol. beam el. reson.[d]		(4)
$^{11}B^{79}Br$		$\mu = 9.6615016_1$		$D_0^0 = 4.49$ eV[a]						MAR 1976
A $^1\Pi$	33935.3	637.63 H^Q	17.58[bc]	[0.496][d]	(0.0090)	[12.8]	1.87	A^e↔X, V_R	33908.6 H^Q	(1)* (2)* (6)
a $^3\Pi_2$										
a $^3\Pi_1$	18851.4_8	757.10 H^Q	4.81[f]	0.5083[d]	0.0036	$[9._3]$	1.853	a→X, V	18887.5_5 Z	(5)* (7)*
a $^3\Pi_{0^+}$	18673.8_3	759.8_0	4.8_0	0.5062[d]	0.0036				18711.2_5 Z	
X $^1\Sigma^+$	0	684.31 H^Q	3.52	0.4894[d]	0.0035	10.0	1.888_2			
$^{(11)}B^{12}C$		$(\mu = 5.74166243)$		$D_0^0 = 4.6_0$ eV[a]						MAR 1976
		Theoretical calculations (2) predict a $^4\Sigma^-$ ground state.								
$^{11}B^{35}Cl$		$\mu = 8.3731666_4$		$D_0^0 = 5.5$ eV[a]						MAR 1976
A $^1\Pi$	36750.92	849.04 H^Q	11.37[b]	0.7054[c]	0.00820[d]	16.0[e]	1.689_4	A^f↔X, V_R	36754.30 H^Q	(1)(2)(3)* (5)(6)* (7) (10)
a $^3\Pi_1$	20200	911	5.7	0.698_6	0.004_7		1.698	a→X, V	20235.7_0 Z	(11)
X $^1\Sigma^+$	0	839.12 H^Q	5.11	0.6838[c]	0.00646	17.2[g]	1.715_9			

BaS: [a]Thermochemical value (mass-spectrom.)(2), corrected to a $^1\Sigma$ ground state.

[b]Extensive perturbations arising from successive vibrational levels of three perturbing states or sub-states of smaller B values.

[c]From perturbations in A $^1\Sigma^+$; vibrational numbering unknown.

[d]$\mu_{el}(v) = [10.853 + 0.021(v+\frac{1}{2})]$ D.

(1) Barrow, Gissane, Rose, PPS 84, 1035 (1964).
(2) Colin, Goldfinger, Jeunehomme, TFS 60, 306 (1964).
(3) Clements, Barrow, CC (1968), 1408.
(4) Melendres, Hebert, Street, JCP 51, 855 (1969).
(5) Barrow, Burton, Jones, TFS 67, 902 (1971).

BBr: [a]From the predissociation in A $^1\Pi$; see (3).

[b]$\omega_e y_e = + 1.10$, $\omega_e z_e = - 0.250$. This state may have a potential hump of $\sim$ 0.13 eV; see (3).

[c]Predissociation above v = 4 (1).

[d]The rotational constants refer to the normal $^{79,81}Br$ isotopic mixture.

[e]Radiative lifetime $\tau(v=0, 1) = 26$ ns (4), corresponding to an absorption f_{00} value of 0.10.

[f]$\omega_e y_e \approx - 0.004$.

(1) Miescher, HPA 8, 279 (1935); 9, 693 (1936).
(2) Rosenthaler, HPA 13, 355 (1940).
(3) Barrow, TFS 56, 952 (1960).
(4) Lutz, Hesser, JCP 48, 3042 (1968).
(5) Lebreton, Marsigny, Bosser, CR C 271, 1113 (1970).
(6) Wentink, Spindler, JQSRT 10, 609 (1970).
(7) Lebreton, JCPPB 70, 738 (1973).

BC: [a]Thermochemical value (mass-spectrom.)(1).

(1) Verhaegen, Stafford, Drowart, JCP 40, 1622 (1964).
(2) Kouba, Öhrn, JCP 53, 3923 (1970).

BCl: [a]Extrapolation of A $^1\Pi$; see (4).

[b]$\omega_e y_e = - 0.100$, $\omega_e z_e = - 0.0271$.

[c]RKR potential curve (8).

[d]$\gamma_e = - 0.00050$.

[e]$\beta_e = + 2.7 \times 10^{-7}$.

[f]Radiative lifetime $\tau(v=0, 1, 2) = 19._1$ ns (9); $f_{00}(A \leftarrow X) = 0.11$.

[g]$\beta_e = + 0.7 \times 10^{-7}$.

(1) Jevons, PRS A 106, 174 (1924).
(2) See ref. (1) of BBr.
(3) Herzberg, Hushley, CJR A 19, 127 (1941).
(4) See ref. (3) of BBr.
(5) Thrush, Nature 186, 1044 (1960).
(6) Verma, JMS 7, 145 (1961); CJP 40, 1852 (1962).
(7) Pannetier, Goudmand, Dessaux, Arditi, CR 258, 1201 (1964).
(8) Gélébart, Johannin-Gilles, CR B 267, 408 (1968).
(9) Hesser, JCP 48, 2518 (1968).
(10) See ref. (6) of BBr.
(11) Lebreton, Marsigny, Ferran, CR C 272, 1094 (1971).

State	T_e	ω_e	$\omega_e x_e$	B_e	α_e	D_e (10^{-6} cm^{-1})	r_e (Å)	Observed Transitions Design.	ν_{00}	References
$^9Be_2$		μ = 4.5060914								APR 1976
A ($^1\Sigma_u^+$)		502^a 483^b	6.8 4.5					A← X,	28551^a 27670^b	(2)
X $^1\Sigma_g^+$		Theoretical calculations give no evidence for a bound ground state (1).								
$^9Be^{40}Ar^+$		μ = 7.3534172^a								APR 1976 A
A $^2\Pi_r$	(24605) (24563)	[570] H		0.6127^b			1.934	A→X, V	24708 H 24666 H	(1)(2)*
X $^2\Sigma^+$	0	369 H	11	0.5271^b			2.085			
$^9Be^{(79)}Br$		(μ = 8.0885053)								APR 1976
A $^2\Pi_r$	26554 26353 a	695 H 702 H	5.2 4.4	$[0.5332]^b$		[1.31]	[1.976]	A→X, R	26543.3 H 26346.1 H	(1)(2)* (3)*
X $^2\Sigma^+$	0	715 H	3.8	$[0.5459]^c$		[1.31]	[1.953]			
$^9Be^{35}Cl$		μ = 7.1654906 D_0^0 a								SEP 1976
		R shaded emission bands in the region 37900 - 38300 cm^{-1} (3) are due to AlCl. Unidentified system (ν_e = 18686, ω_e' = 540, ω_e'' = 552) in emission (1); assignment to BeCl not confirmed.								
B $^2\Sigma^+$	(48773)	[952.5] Z	b	0.7751	0.0043	$[3.0]^c$	1.7422	B→X, V	48827.6 Z	(6)* (8)*
A $^2\Pi_r$	27992.0^d	822.11 Z	5.24	0.7094^e	0.0068	2.3	1.8211	A↔X, R	27979.63	(2)* (3)(5)*
X $^2\Sigma^+$	0	846.7 Z	4.8_5	0.7285	0.0069	2.5	1.7971			
$^9Be^{19}F$		μ = 6.1125850 D_0^0 = 5.85 or 6.26 eV^a								APR 1976 A
		R shaded double-headed emission bands in the regions 62900 - 64100 and 65800 - 66700 cm^{-1}.								(9)
C $^2\Sigma^+$	50364.0	1419.7 H	9.9	1.570	0.01_4		1.325	C→A, V	17253.7	(6)
								C→X, V	50440.86 Z	(9)*
B $^2\Sigma^+$	49563.9	1350.8 H	12.6	$[1.54_7]$			$[1.33_5]$	B→X, V	49605.6 Z	(9)*
A $^2\Pi_r$	33233.7^b	[1154.67] Z	8.78^c H^Q	$1.4202_4{}^d$	0.0175	$[8.40]^e$	1.3935	A↔X, R	33187.13^f Z	(1)* (2)* (3) (4)(5)(13)
X $^2\Sigma^+$	0	[1247.36] Z	9.12^c H^Q	1.4889_3	0.0176	8.28	1.3610			

Be_2: [a]In Ne matrix at 4 K.
[b]In Ar matrix at 4 K.

(1) Bender, Davidson, JCP 47, 4972 (1967).
(2) Brom, Hewett, Weltner, JCP 62, 3122 (1975).

$BeAr^+$: [a]Reduced mass of $Be^+ + Ar$.
[b]Preliminary results.

(1) Subbaram, Vasudev, Jones, JOSA 65, 318 (1975).
(2) Subbaram, Coxon, Jones, CJP 53, 2016 (1975).

BeBr: [a]$A_0 = +198.0$.
[b]Λ-type doubling in $^2\Pi_{\frac{1}{2}}$, $\Delta v_{ef}(v=0) = (+)0.030(J+\frac{1}{2})$; details in (3).
[c]Spin splitting constant $\gamma_0 = +0.0242$.

(1) Reddy, Rao, JP B 1, 482 (1968).
(2) Reddy, Reddy, Rao, JP B 3, L1 (1970).
(3) Carleer, Herman, Colin, CJP 53, 1321 (1975).

BeCℓ: [a]Thermochemical values for D_0^0 obtained by various methods have been summarized in (7), all being close to 4.51 eV and substantially higher than a mass-spectrometric value of 3.99 eV (4). The interpretation (8) of an inverse predissociation assumed to be responsible for the strikingly high intensity of $B \rightarrow X$ bands with v'=1 suggests an even lower value of 3.45 eV.
[b]A very weak head at 48502.7 cm^{-1} has tentatively been identified as 2-3 band; $\Delta G'(3/2) = 1212.7$ cm^{-1}.
[c]$D_1 = 1.9 \times 10^{-6}$.
[d]$A_v = +52.8 - 1.4(v+\frac{1}{2})$.
[e]The Λ-type doubling in the $\frac{1}{2}$ [$\Delta v_{ef}(v=0) \approx -0.011(J+\frac{1}{2})$] and $\frac{3}{2}$ components does not conform with the "pure precession" pattern frequently observed in regular $^2\Pi$ states; see (5).

BeCℓ (continued):

(1) Parker, PR 45, 752 (1934).
(2) Fredrickson, Hogan, PR 46, 454 (1934).
(3) Novikov, Tunitskii, OS(Engl. Transl.) 8, 396 (1960).
(4) Hildenbrand, Theard, JCP 50, 5350 (1969).
(5) Colin, Carleer, Prévot, CJP 50, 171 (1972).
(6) Burtin, Thèse (U. Libre de Bruxelles, 1974).
(7) Farber, Srivastava, JCS FT I 70, 1581 (1974).
(8) Carleer, Burtin, Colin, CJP 55, 582 (1977).

BeF: [a]Mass-spectrometric values (7)(16).
[b]$A_0 = +21.82$, $A_1 = +21.93$; slight J dependence. A negative value of A has been ruled out by the calculations of (14).
[c]Slightly different constants in (5).
[d]The Λ-type doubling in the $\frac{1}{2}$ and $\frac{3}{2}$ components (13) does not follow the "pure precession" pattern for a regular $^2\Pi$ state. An earlier explanation (10) of this "anomaly" has since been revised (14).
[e]$D_1 = 8.26 \times 10^{-6}$.
[f]$J'=\frac{1}{2}$ (average of F_1 and $\{F_2\}$) relative to N"=0.

(1) Jevons, PRS A 122, 211 (1929).
(2) Jenkins, PR 35, 315 (1930).
(3) Mulliken, PR 38, 836 (1931).
(4) Fowler, PR 59, 645 (1941).
(5) Tatevskii, Tunitskii, Novikov, OS 5, 520 (1958).
(6) Rao, Rao, IJPAP 3, 177 (1965).
(7) Hildenbrand, Murad, JCP 44, 1524 (1966).
(8) Singh, Rai, IJPAP 4, 102 (1966).
(9) Novikov, Gurvich, OS(Engl. Transl.) 23, 173 (1967).
(10) Walker, Richards, PPS 92, 285 (1967).
(11) Katti, Sharma, IJPAP 6, 458 (1968).

(continued p. 79)

State	T_e	ω_e	$\omega_e x_e$	B_e	α_e	D_e ($10^{-4}cm^{-1}$)	r_e (Å)	Observed Transitions Design.	ν_{00}	References
$^{9}Be^{1}H$		μ = 0.90645687		$D_0^0 = 2.03_4$ eV^{a}		I.P. = 8.21 eV^{b}				SEP 1976
G $^2\Pi$	58711	405.3 Z	22.7	5.02^{cd}	-0.556^{e}	$[18]^{f}$	1.92_5	G← X, R	57886.2 Z	(13)*
F $^2\Sigma^+(4p\sigma)$	(56606)	$(2153)^{g}$		$[10.576]^{h}$		$[8.0]^{f}$	[1.326]	F← X,	56661.24 Z	(13)*
E $^2\Sigma^+(4s\sigma)$	(54134)	$(1970)^{g}$		$[10.578]^{i}$		$[22.4]^{f}$	[1.326]	E← X,	54097.6 Z	(13)*
D $\{^2\Sigma^+, ^2\Pi, ^2\Delta\}$ (3d)	(54000)	Strong absorption, complex structure.						D← X,	54050	(13)*
B $^2\Pi$ (3pπ)	50882^{j}	2265.94 Z	71.52	$10.8495^{k\ell d}$	0.1016^{m}	$[10.35]^{f}$	1.3092	B↔X, V_R	50976.17 Z	(3)(13)*
A $^2\Pi_r(2p\pi)$	20033.19^{n}	2088.58^{o} Z	40.14^{p}	10.4567^{q}	0.3222^{r}	10.41^{s}	1.3336	A↔X,^{t} V^{u}	20045.81^{o} Z	(1)* (2)(9)*
X $^2\Sigma^+(2p\sigma)$	0	2060.78^{o} Z	36.31^{v}	10.3164^{w}	0.3030^{x}	10.221^{y}	1.3426	ESR sp.^{z}		(10)
		For theoretical calculations see references in (9)(11)(14).								

Be^1H: ^{a}From the predissociation by rotation (13) in the B' $^2\Pi$, v'+1 level (see $^{\ell}$), assuming dissociation into Be(^{1}P) + H(^{2}S). The experimental X $^2\Sigma^+$ well depth of 2.16_1 eV is in good agreement with the calculated values $D_e = 2.11_5$ eV (11) and D_e = 2.15 eV (14).

^{b}From the observation of Rydberg states in the absorption spectrum, and from *ab initio* calculations for BeH and BeH^+; see (12).

$^{c}\Lambda$-type doubling; for details see (13). R and P lines involving v'=0 and 1 are twice as broad as the corresponding Q lines. v=2 is strongly perturbed.

^{d}(13) suggest that the B and G states result from an avoided $^2\Pi$ - $^2\Pi$ crossing, the former (B $^2\Pi$) having a double minimum potential curve with a potential maximum corresponding to T_e = 56950 cm^{-1} at r = 1.94 Å.

$^{e}\gamma_e$ = -0.040.

^{f}For additional D_v and higher order constants see (13).

^{g}Using isotope relations.

^{h}Perturbation at low N. Higher vibrational levels are probably predissociated.

^{i}Line width increases with decreasing N; the first lines are not observed.

^{j}(13) give 50888.57, without explanation. A ≈ 0.

$^{k}\Lambda$-type doubling, $\Delta v_{ef}(v=0) = +0.217\,N(N+1)$.

$^{\ell}$The B↔X bands (v' ≤ 2, and fragments of the 3-3 band of Be^2H) consist of Q branches only except the 0-0 band which, in absorption, has P and R branches showing a marked broadening increasing with N. The predissociation mechanism involves both the unobserved 3pσ state and the first excited $^2\Sigma^+$ state which is unstable except at very large r

Be^1H (continued):

values; for details see (15). The Q branch lines of Be^1H (Be^2H) break off at N'=31, 24(38), 14(27), (14) in v'=0,1,2,3, respectively, owing to the presence of a maximum in the B $^2\Pi$ potential curve; see [d]. Two additional levels of Be^1H, very likely belonging to B $^2\Pi$ but called B' $^2\Pi$(v', v'+1) in (13), are situated above the potential energy maximum in B $^2\Pi$:

	T_e	B	D	
B' $^2\Pi$, v'+1	[58435.8]	6.01	187×10^{-4}	and higher or-
B' $^2\Pi$, v'	[58284.8]	8.34	129×10^{-4}	der constants.

Transitions to these levels from X $^2\Sigma^+$ consist of Q branches only, breaking off at N'=14 in v'+1.

[m] $\gamma_e = -0.1324$.

[n] Taking into account the usually neglected contributions Y'_{00} and Y''_{00} to the zero point energies of A and X. A (spin-orbit) = +2.1; for an ab initio calc. see (4).

[o] Derived (9) from pure vibronic energy separations which differ from the origins normally referred to in these tables.

[p] $\omega_e y_e = -0.47$.

[q] Λ-type doubling, $\Delta\nu_{ef}(v=0) = +0.0141\,N(N+1) - \ldots$; higher order constants in (9). Theoretical calculations (7)(8).

[r] $\gamma_e = -0.0042$.

[s] Higher order constants in (9).

[t] Theoretical absorption oscillator strengths (5)(6).

[u] Reversal of shading in some of the bands.

[v] $\omega_e y_e = -0.38$.

[w] Spin splitting constant $\gamma_0 = +0.005$.

[x] $\gamma_e = -0.0027$.

[y] $\beta_e = -0.039 \times 10^{-4}$; higher order constants in (9).

[z] In Ar matrix at 4 K (10).

(1) Watson, Parker, PR 37, 167 (1931).
(2) Olsson, ZP 73, 732 (1932).
(3) Watson, Humphreys, PR 52, 318 (1937).
(4) Walker, Richards, PR 177, 100 (1969).
(5) Henneker, Popkie, JCP 54, 1763 (1971).
(6) Popkie, JCP 54, 4597 (1971).
(7) Hinkley, Hall, Walker, Richards, JP B 5, 204 (1972).
(8) Hinkley, Walker, Richards, JP B 5, 2016 (1972).
(9) Horne, Colin, BSCB 81, 93 (1972).
(10) Knight, Brom, Weltner, JCP 56, 1152 (1972).
(11) Bagus, Moser, Goethals, Verhaegen, JCP 58, 1886 (1973).
(12) Colin, De Greef, Goethals, Verhaegen, CPL 25, 70 (1974).
(13) Colin, De Greef, CJP 53, 2142 (1975).
(14) Meyer, Rosmus, JCP 63, 2356 (1975).
(15) Lefebvre-Brion, Colin, JMS 65, 33 (1977).

BeF (continued):

(12) Mishra, Khanna, IJPAP 7, 63 (1969).
(13) Walker, Barrow, JP B 2, 102 (1969)
(14) Walker, Richards, JP B 3, 271 (1970).
(15) Rai, Singh, SpL 4, 129 (1971).
(16) Farber, Srivastava, JCS FT I 70, 1581 (1974).

State	T_e	ω_e	$\omega_e x_e$	B_e	α_e	D_e ($10^{-4}cm^{-1}$)	r_e (Å)	Observed Transitions Design.	ν_{00}	References
$^{9}Be^{2}H$		μ = 1.64619882		$D_0^0 = 2.06_6$ eV[a]						SEP 1976
G $^2\Pi$	58750	302.5 Z	23.6	[2.747][bc]		[10][d]	[1.931]	G← X, R	58134.9 Z	(3)
F $^2\Sigma^+$	(56606)	(1598)[e]		[5.879]		[3.01][d]	[1.319_8]	F← X,	56644.87 Z	(3)
E'($^2\Sigma^+$)	[54230.2][f]			[5.260][f]			[1.395][f]	E'← X,	50533.8[f] Z	(3)
E $^2\Sigma^+$	(54135)	(1460)[e]		[5.745][g]		[3.8]	[1.335]	E← X,	54104.9 Z	(3)
D {$^2\Sigma^+$, $^2\Pi$, $^2\Delta$}	(54000)	Strong absorption, complex structure.						D← X,	54050	(3)
B $^2\Pi$	50906	1646.22 Z	25.15[h]	5.9963[ij]	0.1217[k]	[3.114][d]	1.3068	B← X, V	50963.02 Z	(3)
A $^2\Pi_r$	20036.11[ℓ]	1551.13 Z	23.00	5.7614[m]	0.1344[n]	[3.181][o]	1.3332	A↔ X, V	20045.91[p] Z	(1)* (2)
X $^2\Sigma^+$	0	1530.32 Z	20.71	5.6872	0.1225[q]	3.138[r]	1.3419			
$^{9}Be^{3}H$		$\mu = 2.2597826_7$		D_0^0 = 2.08 eV[a]						APR 1976
A $^2\Pi_r$	20037.91[b]	1322[c]	16[c]	4.192	0.068[d]	[1.66][e]	1.334_0	A→X, V	20045.83[f] Z	(1)
X $^2\Sigma^+$	0	1305[c]	15[c]	4.142	0.064[g]	[1.67][e]	1.342_0			
$^{9}Be^{1}H^{+}$				D_0^0 = 3.14 eV[a]						APR 1976
A $^1\Sigma^+$	39417.0	1476.1 Z	14.8[b]	7.184	0.125[c]	6.1	1.608_9	A→X,[d] R	39050.4 Z	(1)(3)
X $^1\Sigma^+$	0	2221.7 Z	39.79[e]	10.800	0.294[f]	9.9	1.3122	Theoret. calc., pot. curves		(4)(5)(6)
$^{9}Be^{2}H^{+}$				D_0^0 = 3.18 eV[g]						APR 1976
A $^1\Sigma^+$	$39416._2$	1096.4 Z	8.5	3.971	0.057_8	1.9	1.605_9	A→X, R	39143.9 Z	(2)
X $^1\Sigma^+$	0	1647.6 Z	21.9	5.955	0.123_3	2.9	1.3113			

Be^2H: [a]From the value for Be^1H.

[b]Average from P,R and Q branches. Transitions to v'=1 and 2 consist of Q branches only, strongly perturbed for v'=2. The Q branch constants are given by $B'_v = 2.197 + 1.205(v+\frac{1}{2}) - 0.167(v+\frac{1}{2})^2$.

[c]See [d] of Be^1H.

[d]Additional D_v and higher order constants in (3).

[e]Using isotope relations.

[f]Unidentified single level, vibrational numbering unknown. In absorption from v"=2.

[g]See [i] of Be^1H.

[h]Inclusion of v=3 changes these constants to $\omega_e = 1607.29$, $\omega_e x_e = -5.32$, $\omega_e y_e = -6.77$.

[i]Λ-type doubling, $\Delta v_{ef}(v=0) = +0.074 \times N(N+1)$.

[j]See [d] and [ℓ] of Be^1H.

[k]$\gamma_e = -0.0089$ ($v \leq 2$); recalculated from (3).

[ℓ]See [n] of Be^1H. $A \approx +1.9$.

[m]Λ-type doubling, $\Delta v_{ef}(v=0) = +0.00419 \times N(N+1) - \ldots$; additional constants in (2).

[n]$\gamma_e = -0.0003$.

[o]For additional D_v and higher order constants see (2).

[p]Pure vibronic energy difference, not to be confused with the band origin as usually defined in these tables. See (2).

[q]$\gamma_e = -0.001$.

[r]$\beta_e = -0.022 \times 10^{-4}$; higher order constants in (2).

(1) Koontz, PR 48, 707 (1935).

(2) See ref. (9) of Be^1H.

(3) See ref. (13) of Be^1H.

Be^3H: [a]From the value for Be^1H.

[b]See [n] of Be^1H.

[c]From isotope relations. $\omega'_e - \omega''_e = 16.36$, $\omega'_e x'_e - \omega''_e x''_e = 0.84$, $\omega'_e y'_e - \omega''_e y''_e = -0.15$.

[d]$\gamma_e = -0.007$.

[e]Additional D_v and higher order constants in (1).

[f]See [p] of Be^2H.

[g]$\gamma_e = -0.006$.

(1) De Greef, Colin, JMS 53, 455 (1974).

Be^1H^+, Be^2H^+:

[a]$D_0^0(Be^1H) + I.P.(Be) - I.P.(Be^1H)$.

[b]$\omega_e y_e = -0.38$; -0.038 in (3) is obviously wrong.

[c]$\gamma_e = -0.0054$.

[d]Theoretical oscillator strength (7).

[e]$\omega_e y_e = -0.21$; -0.021 in (3) is obviously wrong.

[f]$\gamma_e = -0.0049$.

[g]From the value for Be^1H^+.

(1) Bengtsson-Knave, NARSSU (IV) 8, No. 4, 65 (1932).

(2) See ref. (1) of Be^2H.

(3) See ref. (3) of Be^1H.

(4) Reed, Vanderslice, Jenč, JCP 37, 205 (1962).

(5) Brown, JCP 51, 2879 (1969).

(6) Banyard, Taylor, JP B 8, L137 (1975).

(7) Stewart, Watson, Dalgarno, JCP 63, 3222 (1975).

State	T_e	ω_e	$\omega_e x_e$	B_e	α_e	D_e (10^{-6}cm^{-1})	r_e (Å)	Observed Transitions Design.	Observed Transitions ν_{00}	References
$^9Be^{127}I$		μ = 8.4146148								SEP 1976
A $^2\Pi_{3/2}$				[0.4182]		[0.80]	[2.188_7]	A→X, R	23900.83 Z	(1)* (2)
A $^2\Pi_{1/2}$	23544.7	603.8 H	2.1	[0.4216][a]		[0.73]	[2.179_9]		23540.59 Z	
X $^2\Sigma^+$	0	611.7 H	1.6	[0.4219][b]		[0.82]	[2.179_1]			
$^9Be^{(84)}Kr^+$		(μ = 8.1376909)[a]								APR 1976 A
A $^2\Pi$	23857 23692	557 H	3.5					A→X, V	23954 H 23789 H	(1)*
X $^2\Sigma^+$	0	364 H	5.4							
$^9Be^{16}O$		μ = 5.7643273_5 D_0^0 = 4.6_0 eV[a]								APR 1976
		Fragments of additional singlet and triplet systems in the region 29000 - 33000 cm^{-1}.								(1)(4)(10)*
C ($^1\Sigma$)	39120.2	1081.5 (H^Q)	9.1	(1.308)	(0.010)		(1.495)	C→A,[b] R	29683.1 (H^Q)	(5)(10)*
B $^1\Sigma^+$	21253.94	1370.82 Z	7.746[c]	1.5758[de]	0.0154	8.41[f]	1.3623	B→A,[g] V	11961.78 Z	(10)
								B[h]↔X,[i] R	21196.70 Z	(1)(2)(7)(10)(13)*(16)
b $^3\Sigma^+$	j									
A $^1\Pi$	9405.61	1144.24 Z	8.415[k]	1.3661[ℓde]	0.01628[m]	7.79[n]	1.4631	A→X, R	9234.92 Z	(3)(6)* (9)(10)*
a $^3\Pi$	(8480)[o]									
X $^1\Sigma^+$	0	1487.32 Z	11.830[p]	1.6510[e]	0.0190	8.20[q]	1.3309	For computed ground state properties see (17)(24).		

BeI: [a]Λ-type doubling, $\Delta\nu_{fe}$ = +0.0938(J+½).
[b]Spin splitting constant γ_0 = +0.0459.

(1) Murty, Rao, CS 38, 187 (1969); PRIA A 72, 71 (1972).

(2) Carleer, Colin, to be published.

BeKr$^+$: [a]Reduced mass of Be$^+$ + Kr.

(1) Subbaram, Coxon, Jones, CJP 53, 2016 (1975).

BeO: [a]Thermochemical value (mass-spectrom.)(14); in good agreement with 4.52 eV derived from an <u>ab initio</u> calculation (24) of X $^1\Sigma^+$. Extrapolations of X and A to their common limit Be(^{1}S) + O(^{1}D) lead to 3.9 and 4.8_2 eV, respectively (12). A considerably higher thermochemical value of 5.5_1 eV was determined by (11).

[b]Franck-Condon factors (22).

[c]$\omega_e y_e = -0.00027$.

[d]Numerous perturbations between levels of A $^1\Pi$ and B $^1\Sigma$, A $^1\Pi$ and X $^1\Sigma$, as well as perturbations by unidentified levels [probably belonging to a $^3\Pi$ and b $^3\Sigma$, see (20)]. For an extensive treatment see (6)(7)(8)(10).

[e]RKRV potential curves (21).

[f]$\beta_e = -0.07 \times 10^{-6}$; $H_v = +[27 - 2(v+\frac{1}{2})] \times 10^{-12}$.

[g]Very weak system.

[h]Radiative lifetime $\tau(v=0) = 90$ ns (25); $f_{00} = 0.0335$. A much smaller value, $f_{00}(B-X) = 0.00194$, was estimated from shock tube measurements (19).

[i]Franck-Condon factors (15)(22); approximate electronic transition moments, band oscillator strengths (19).

[j]The theoretical calculations of (18) place the $^3\Sigma^+$ state at 4100 cm^{-1} below B $^1\Sigma^+$, in rough agreement with (23) who calculate its energy at 15400 cm^{-1} above X $^1\Sigma$.

[k]$\omega_e y_e = +0.0339$.

[ℓ]Λ-type doubling, $\Delta\nu_{fe} = +0.00055\,J(J+1)$.

[m]$\gamma_e = +0.000055$.

[n]$\beta_e = -0.044 \times 10^{-6}$.

[o]Theoretical calculations (20)(18) place the $^3\Pi$ state at 920 to 2600 cm^{-1} below A $^1\Pi$, in reasonable agreement with (26) who predict it at 5900 cm^{-1} above X $^1\Sigma^+$.

[p]$\omega_e y_e = +0.0224$.

[q]$\beta_e \approx -0.01$; $H_v = +[12.5 - 1.1(v+\frac{1}{2})] \times 10^{-12}$.

BeO (continued):

(1) Bengtsson, AMAF A <u>20</u>, No. 28 (1928).
(2) Rosenthal, Jenkins, PR <u>33</u>, 163 (1929).
(3) Herzberg, ZP <u>84</u>, 571 (1933).
(4) Ciccone, RiSc (VI) <u>2</u>, 3 (1935).
(5) Harvey, Bell, PPS <u>47</u>, 415 (1935).
(6) Lagerqvist, Westöö, AMAF A <u>31</u>, No. 21 (1945).
(7) Lagerqvist, Westöö, AMAF A <u>32</u>, No. 10 (1945).
(8) Lagerqvist, AMAF A <u>33</u>, No. 8 (1946).
(9) Lagerqvist, AMAF B <u>34</u>, No. 23 (1947).
(10) Lagerqvist, Dissertation (Stockholm, 1948).
(11) Drummond, Barrow, TFS <u>49</u>, 599 (1953).
(12) Lagerqvist, AF <u>7</u>, 473 (1954).
(13) Gatterer, Junkes, Salpeter, Rosen, METOX (1957).
(14) Chupka, Berkowitz, Giese, JCP <u>30</u>, 827 (1959).
(15) Nicholls, Fraser, Jarmain, McEachran, ApJ <u>131</u>, 399 (1960).
(16) Thrush, PCS (1960), p. 339.
(17) Yoshimine, JCP <u>40</u>, 2970 (1964); JPSJ <u>25</u>, 1100 (1968).
(18) Verhaegen, Richards, JCP <u>45</u>, 1828 (1966).
(19) Drake, Tyte, Nicholls, JQSRT <u>7</u>, 639 (1967).
(20) Huo, Freed, Klemperer, JCP <u>46</u>, 3556 (1967).
(21) Thakur, Singh, JSRBHU <u>18</u>, 253 (1968).
(22) Liszt, Smith, JQSRT <u>11</u>, 1043 (1971).
(23) O'Neil, Pearson, Schaefer, CPL <u>10</u>, 404 (1971).
(24) Schaefer, JCP <u>55</u>, 176 (1971).
(25) Capelle, Johnson, Broida, JCP <u>56</u>, 6264 (1972).
(26) Pearson, O'Neil, Schaefer, JCP <u>56</u>, 3938 (1972).

State	T_e	ω_e	$\omega_e x_e$	B_e	α_e	D_e $(10^{-6} cm^{-1})$	r_e (Å)	Observed Transitions Design.	ν_{00}	References
$^{9}Be^{32}S$		$\mu = 7.0304599$		$D_0^0 = 3.8$ eV[a]						APR 1976
b $(^{3}\Pi)$								(b← a), R	25961.8 H; 25941.8 H; 25924.5 H	(2)*
B $^{1}\Sigma^{+}$	25941.6	851.35 Z	4.85	0.72894	0.00604[b]	2.14	1.8136_8	B↔X, R	25868.61 Z	(1)(2)*
C $(^{1}\Delta)$	[13545.8][c]	[660.7][c]	5.5	[0.5963][c]	0.0069		[2.005][c]			
A $^{1}\Pi$	7960.1	762.46[d] Z	4.12	0.6590[def]	0.00605[g]	2.00	1.907_5	A→X, R	7842.9[d] Z	(2)
a $(^{3}\Pi)$										
X $^{1}\Sigma^{+}$	0[h]	997.94 Z	6.137	0.79059	0.00664[i]	2.00	1.7415_3			
$^{9}Be^{(132)}Xe^{+}$		$(\mu = 8.4353358)$[a]								APR 1976
A $^{2}\Pi$	22273; 21922	545 H	3.9					A→X, V	22360 H; 22009 H	(1)*
X $^{2}\Sigma^{+}$	0	371 H	3.7							
$^{11}B^{19}F$		$\mu = 6.9701832_5$		$D_0^0 = 7.8_1$ eV[a]		I.P. = 11.115 eV[b]				APR 1976
		Rydberg series (abs.) beginning with D, J, P, ... : $\left.\begin{matrix} \nu(1-0) = 91330 \\ \nu(0-0) = 89650 \end{matrix}\right\} - R/(n-0.52)^2 \left\{\begin{matrix} n \le 18. \\ n \le 15. \end{matrix}\right.$								(24)*
		Rydberg series (abs.) beginning with C, I, O, ... : $\left.\begin{matrix} \nu(1-0) = 91330 \\ \nu(0-0) = 89650 \end{matrix}\right\} - R/(n-0.66)^2$, $n \le 9.$								(24)
R $^{1}\Sigma^{+}(6s\sigma)$[c]	[85848]							R← X, V	85150 H	(24)
P $^{1}\Pi$ $(5p\pi)$	84077	[1673] H[Q]		$[1.651_1]$[d]		[6.4][d]	$[1.210_3]$	P← X, V	84215 H[Q]	(24)
O $^{1}\Sigma^{+}(5p\sigma)$	$83680._2$	1676 H	9.5	[1.6275]		[5.6]	[1.2190]	O← X, V	83817.71 Z	(24)
L $^{1}\Sigma^{+}(5s\sigma)$	[83348.32]			[1.5578]		$[-21._3]$	[1.2460]	L← X, V	82650.21 Z	(24)
J $^{1}\Pi$ $(4p\pi)$	80544	$[1673.0_9]$ Z		1.6516[e]	0.0162	[6.5]	1.2101	J← X, V	80681.98 Z	(24)
h $^{3}\Pi$	80230	[1679.2] H[Q]		$[1.6475]$[f]		[6.2]	$[1.211_6]$	h→b, V	19225.28 Z	(24)
I $^{1}\Sigma^{+}(4p\sigma)$	79631.3_9	1666.28 Z	12.57	1.6382	0.0174	[6.4]	1.2150	I← X, V	79763.28 Z	(24)
H $^{1}\Sigma^{+}(3d\sigma)$	$[79389.3_2]$			[1.6511]		$[13._0]$	[1.2103]	H← X, V	78691.2_1 Z	(24)

BeS: [a]From Gaydon (4).

[b]$\gamma_e = -0.00003$.

[c]Constants for the lowest observed level and interval, vibrational numbering unknown. From perturbations in $A\ ^1\Pi$.

[d]Vibrational numbering uncertain.

[e]Constants for the f component; $\Delta\nu_{fe} \approx +0.00015 \times J(J+1)$.

[f]The $A\ ^1\Pi$ state is perturbed by three states, one of them being $X\ ^1\Sigma^+$, another probably $^1\Delta$.

[g]$\gamma_e = +0.000025$.

[h]Theoretical calculations (3) support a $^1\Sigma$ ground state.

[i]$\gamma_e = -0.00002$.

(1) Gissane, Barrow, PPS 82, 1065 (1963).

(2) Cheetham, Gissane, Barrow, TFS 61, 1308 (1965).

(3) Verhaegen, Richards, PPS 90, 579 (1967).

(4) Gaydon, DISSEN (1968).

$BeXe^+$: [a]Reduced mass of Be^+ + Xe.

(1) Coxon, Jones, Subbaram, CJP 53, 2321 (1975).

BF: [a]Thermochemical value (mass-spectrom.)(15)(18). Extrapolation of $A\ ^1\Pi$ gives 8.0_2 eV (9).

[b]Extrapolation of Rydberg series; 11.06 eV by electron impact (26).

[c]Approximate description of the Rydberg electron (24); see also (17).

[d]Computed from the data for $^{10}B^{19}F$: $B_0 = 1.7551$, $D_0 = 7.2 \times 10^{-6}$, $\Delta\nu_{ef} = +0.0146 \times J(J+1)$.

[e]$\Delta\nu_{ef} = +0.005_6 \times J(J+1) - \ldots$.

[f]$\Delta\nu_{ef} = +0.0024 \times N(N+1)$.

(continued p. 87)

State	T_e	ω_e	$\omega_e x_e$	B_e	α_e	D_e ($10^{-6}cm^{-1}$)	r_e (Å)	Observed Transitions Design.	ν_{00}	References
$^{11}B^{19}F$ (continued)										
g $^3\Sigma^+$	[78771.9]			$[1.676_5]$		$[5._1]$	$[1.201_1]$	g→b,^{g} V	16927.59 Z	(24)
F $^1\Pi$ (3dπ)	(77406)	$(1670)^{h}$		$[1.672_3]^{ij}$		$[8._9]$	$[1.202_6]$	F→A, V	26454.1_3	(12)*
								F←X, V	77542.7_9 Z	(24)
f $^3\Pi$	77405	[1678.1] Z		$[1.641_7]^{k}$		[5.0]	$[1.213_8]$	f→c, V	10428.15 Z	(24)*
								f→b, V	16400.10 Z	(24)
G $^1\Sigma^+$(4sσ)	76952	[1685.63] Z		1.6054	0.0147	$[1.2]^{\ell}$	1.2274	G→B, R	11596.99	(12)*
								G↔X, V	77096.41 Z	(10)(24)*
E $^1\Delta$ (3dδ)	(76292)	$(1581)^{h}$		$[1.6209]^{m}$		[6.4]	[1.2215]	E→A, V	25295.69 Z	(12)* (24)*
e $^3\Sigma^+$	75916	[1654.29] Z		1.6447	0.0151	[6.3]	1.2126	e→b,^{g}	14899.56 Z	(12)* (22) (24)*
D $^1\Pi$ (3pπ)	72144.4_2	[1661.96] Z	11.7 H^{Q}	1.6282^{n}	0.0170	[6.3]	1.2188	D↔X, V	72286.06 Z	(4)(5)(10) (24)*
d $^3\Pi$	70710.4	1696.71 Z	11.01	1.6517^{op}	0.0176^{q}	[6.5]	1.2101	d→b,	9711.65 Z	(24)*
C $^1\Sigma^+$(3pσ)	69030.3_8	1613.1_0 Z	14.5_0	1.6238	0.0194	[7.3]	1.2204	C→A, V	18046.53	(12)*
								C↔X, V	69135.19 Z	(4)(5)* (10) (24)
c $^3\Sigma^+$	67045	[1541.3] Z		$[1.603_0]^{r}$		$[5._5]^{r}$	$[1.228_3]$	c→a, V	38011.97 Z	(2)(8)(16)
B $^1\Sigma^+$(3sσ)	65353.9_3	1693.5_1 Z	12.6_1	1.6590	0.0178	[7.6]	1.2074	B→A, V	14410.76	(12)*
								B↔X, V	65499.42 Z	(4)(5)* (10) (24)
b $^3\Sigma^+$	61035.3	1629.28 Z	22.25_5	1.6385	0.0200^{s}	[7.0]	1.2149	b→a,^{t} V	32040.42 Z	(1)(2)(3)* (6)(8)(16)
A $^1\Pi$	51157.45	1264.9_6 Z	$12.5_3{}^{u}$	1.4227^{v}	0.0180^{w}	[7.3]	1.3038	A^{x}↔X,^{y} R	51088.66 Z	(4)(5)* (7) (11)* (24)
a $^3\Pi_r$	29144.3^{z}	1323.8_6 Z	$9.20^{a'}$	$1.413_5{}^{b'}$	0.0158	[6.3]	1.308_1	a→X, R	29105.8 Z	(28)
X $^1\Sigma^+$	0	1402.1_3 Z	$11.8_4{}^{c'}$	[1.507235]	0.0198	[7.6]	1.2625_9	Microwave sp.$^{d'}$		(27)

BF (continued):

[g]Headless band.

[h]Estimated from observed isotope shifts.

[i]$\Delta v_{ef} = + 0.0423 \times J(J+1) - \dots$.

[j]Lines with $J' \lesssim 6$ are weak or absent, both in emission and in absorption.

[k]$\Delta v_{ef} = - 0.0289 \times N(N+1) + \dots$.

[ℓ]$D_1 = 22._8 \times 10^{-6}$.

[m]$\Delta v_{ef} \approx - 4 \times 10^{-7} J^2(J+1)^2$.

[n]$\Delta v_{ef} = + 0.0009 \times J(J+1)$.

[o]$\Delta v_{ef} = + 0.0011 \times N(N+1)$.

[p]The emission from v=4 consists of Q branch lines only.

[q]$\gamma_e = + 0.0002$.

[r]From (24). Barrow et al. (8) give $B_0 = 1.605_2$, $D_0 = 8.2 \times 10^{-6}$.

[s]$\gamma_e = - 0.0012$.

[t]Franck-Condon factors (21).

[u]$\omega_e y_e = - 0.24_2$. This state may have a potential hump of $\sim 0.2_2$ eV.

[v]$\Delta v_{ef} \approx - 0.0002 \times J(J+1)$.

[w]$\gamma_e = - 0.00040$.

[x]Radiative lifetime $\tau(v=0,1,2) = 2.8$ ns; $f_{00} = 0.40$ (20)(23).

[y]Franck-Condon factors (25).

[z]$A = + 24.2_5$.

[a']$\omega_e y_e = + 0.047$.

[b']Λ-type doubling; see (8).

[c']$\omega_e y_e = + 0.05_6$.

[d']$\mu_{e\ell}(v=0) = 0.5$ D. Values of $eqQ(^{10,11}B)$ in (27). For computed ground state properties see (13)(14)(19).

(1) Dull, PR 47, 458 (1935).
(2) Strong, Knauss, PR 49, 740 (1936).
(3) Paul, Knauss, PR 54, 1072 (1938).
(4) Chrétien, Miescher, Nature 163, 996 (1949); HPA 22, 588 (1949).
(5) Chrétien, HPA 23, 259 (1950).
(6) Dodsworth, Barrow, PPS A 68, 824 (1955).
(7) Onaka, JCP 27, 374 (1957).
(8) Barrow, Premaswarup, Winternitz, Zeeman, PPS 71, 61 (1958).
(9) Barrow, TFS 56, 952 (1960).
(10) Mal'tsev, OS(Engl. Transl.) 9, 225 (1960).
(11) Verma, CJP 39, 1377 (1961); 40, 1852 (1962)[Erratum].
(12) Robinson, JMS 11, 275 (1963).
(13) Nesbet, JCP 40, 3619 (1964); 43, 4403 (1965).
(14) Huo, JCP 43, 624 (1965).
(15) Hildenbrand, Murad, JCP 43, 1400 (1965).
(16) Krishnamachari, Singh, CS 34, 655 (1965).
(17) Lefebvre-Brion, Moser, JMS 15, 211 (1965).
(18) Murad, Hildenbrand, Main, JCP 45, 263 (1966).
(19) Hegstrom, Lipscomb, JCP 45, 2378 (1966); 48, 809 (1968).
(20) Hesser, Dressler, JCP 45, 3149 (1966).
(21) Pathak, Maheshwari, IJPAP 5, 138 (1967).
(22) Czarny, Felenbok, CPL 2, 533 (1968).
(23) Hesser, JCP 48, 2518 (1968).
(24) Caton, Douglas, CJP 48, 432 (1970).
(25) Wentink, Spindler, JQSRT 10, 609 (1970).
(26) Hildenbrand, IJMSIP 7, 255 (1971).
(27) Lovas, Johnson, JCP 55, 41 (1971).
(28) Lebreton, Ferran, Marsigny, JP B 8, L465 (1975).

State	T_e	ω_e	$\omega_e x_e$	B_e	α_e	D_e ($10^{-3}cm^{-1}$)	r_e (Å)	Observed Transitions Design.	ν_{00}	References
$^{11}B^{1}H$		μ = 0.92330324		D_0^0 = 3.42 eV[a]		I.P. = 9.77 eV[b]				APR 1976 A
(5d)		Fragments in the region 74420 - 74523 cm^{-1}.						5d← X,		(7)
K $^1\Pi$ (4d)		Fragments only.						K← X,	71840	(7)
J $^1\Sigma^+$ (4p)		" " .						J← X,	70040	(7)
I $^1\Sigma^+$ 4s				[11.99][c]			[1.234]	I← X,[d]	67395.8 Z	(7)(22)
H $^1\Delta$ } 3d								H← A,	43345.7 Z	(7)* (22)
G $^1\Pi$ } 3d				$[12.25_5]$[c]			$[1.220_6]$	H← X,	66419.7 Z	
								G← X,	66399.3 Z	
F $^1\Sigma^+$ } 3d								F← X,	66079.5 Z	
E $^1\Sigma^+$ } 3p								E← X,	61872.3 Z	(7)*
D $^1\Pi$ } 3p				$[12.32_1]$		[1.3]	$[1.217_3]$	D← X,	61105.4 Z	
C $^1\Sigma^+$	55281.1	2474.7_2 Z	54.42_4[e]	12.410[f]	0.432	$[1.24_7]$[g]	1.2129	C→ A, V	32259.7_8 Z	(4)(13)
								C← X,[d]	55333.6_9 Z	(7)
B $^1\Sigma^+$ 3s	52335.8	2399.9_1 Z	69.51_9[h]	12.339[i]	0.485[j]	1.26[k]	1.2164	B→ A, V	29272.7_3 Z	(4)
								B← X,[d]	52346.6_9 Z	(7)* (13)
c $^3\Sigma^+$	ℓ									
C' $^1\Delta$	45981.0	2610.0_2 Z	46.6_2	12.757	0.390	1.219[m]	1.1963	C'↔ A, V	23029.2_1 Z	(13)*
b $^3\Sigma^{(-)}$				[12.126]		[1.28]	[1.2271]	b→ a, R	27060.8 Z	(1)(2)
A $^1\Pi$	23135.8	2250.9_9 Z	56.66_5[n]	12.295_2[opq]	0.834_6[r]	[1.451][g]	1.2186	A[s]↔ X, R_V	23073.9_6 Z	(1)* (2)(5) (13)*
a $^3\Pi$	t			[12.667][u]		[1.22]	[1.2006]			
X $^1\Sigma^+$	0	2366.9_0 Z	49.39_5[v]	12.021[w]	0.412	1.242[x]	1.2324	Summary of theoretical calculations: see ref. in (12)(14)(20)(23).		

B^1H: [a]From the predissociation by rotation in $A\ ^1\Pi$ (13); see [p]. The estimated height (see [n]) of the potential hump in $A\ ^1\Pi$ was subtracted from the extrapolated energy of the potential maximum. Good agreement with theory [(9), additional results summarized in (23)].

[b]From the observation of Rydberg states in the absorption spectrum (7); theoretical computations (21) give 9.53 eV.

[c]The rotational constants for the 4s state and the 3d complex were re-determined by Johns and Lepard (22) using a model which gives proper consideration to the effects of s∼d mixing. Ginter (10) obtained effective constants for the four states involved.

[d]Nearly undegraded, headless band.

[e]$\omega_e y_e = +\ 0.22_8$.

[f]Predissociation in v=2 for $J > 8$. v=3 perturbed at low J.

[g]Additional D_v and higher order constants in (13).

[h]$\omega_e y_e = -\ 3.92_7$.

[i]Theoretical calculations (16)(18) predict a double minimum potential curve.

[j]$\gamma_e = -\ 0.042_4$.

[k]$\beta_e = +\ 0.10 \times 10^{-3}$; $H_0 = +\ 0.56 \times 10^{-9}$.

[ℓ]The lowest $^3\Sigma^+$ state, $T_e \approx 51700$ (11)(18), is predicted (18) to be Rydberg-like (3sσ) and quasi-bound at small internuclear distances, but valence-like and repulsive (dissociating into ground state atoms) for separations larger than 1.45 Å.

[m]$\beta_e = -\ 0.020 \times 10^{-3}$; $H_e \approx 1.00 \times 10^{-7}$.

[n]$\omega_e y_e = -\ 15.83_0$. This state has a potential hump (3)(13) of approximately 0.155 eV (6)(19).

[o]Λ-type doubling, $\Delta\nu_{ef} = +\ [0.0389 - 0.0027(v+\frac{1}{2})]J(J+1)$.

[p]Predissociation by rotation. J' of the first missing or first diffuse lines:

	v'=0	1	2	3
em./abs.:	27/30	21/24	14/17	-/6

[q]$\mu_{e\ell}(v=0) = 0.58$ D (15).

[r]$+\ 0.132_2(v+\frac{1}{2})^2 - 0.05188(v+\frac{1}{2})^3$.

[s]Radiative lifetime $\tau = 159$ ns; $f_{00} = 0.035$ (17).

[t]$A_0 = +\ 5.95$. Theoretical estimates of T_e range from 2800 to 9700 cm^{-1}; see the summary in (14), also (19).

[u]Λ-type doubling; see (2).

[v]$\omega_e y_e = +\ 0.36_4$.

[w]$\mu_{e\ell}(v=0) = 1.2_7$ D (15). Theory (8) predicts a large negative rotational magnetic moment of $-\ 8.27\ \mu_N$.

[x]$\beta_e = -\ 0.026 \times 10^{-3}$. Higher order constants in (13).

(1) Lochte-Holtgreven, van der Vleugel, ZP 70, 188 (1931).
(2) Almy, Horsfall, PR 51, 491 (1937).
(3) Herzberg, Mundie, JCP 8, 263 (1940).
(4) Douglas, CJR A 19, 27 (1941).
(5) Thrush, Nature 186, 1044 (1960).
(6) Hurley, PRS A 261, 237 (1961).
(7) Bauer, Herzberg, Johns, JMS 13, 256 (1964).
(8) Stevens, Lipscomb, JCP 42, 3666 (1965); ref. (19) of BF.
(9) Ellison, JCP 43, 3654 (1965).
(10) Ginter, JCP 44, 950 (1966).
(11) Grimaldi, Lecourt, Lefebvre-Brion, Moser, JMS 20, 341 (1966).
(12) Cade, Huo, JCP 47, 614 (1967).
(13) Johns, Grimm, Porter, JMS 22, 435 (1967).
(14) Harrison, Allen, JMS 29, 432 (1969).
(15) Thomson, Dalby, CJP 47, 1155 (1969).
(16) Browne, Greenawalt, CPL 7, 363 (1970).
(17) Smith, JCP 54, 1384 (1971).
(18) Pearson, Bender, Schaefer, JCP 55, 5235 (1971).
(19) Blint, Goddard, CP 3, 297 (1974).
(20) Banyard, Taylor, JP B 8, L137 (1975).
(21) Griffing, Simons, JCP 62, 535 (1975).
(22) Johns, Lepard, JMS 55, 374 (1975).
(23) Meyer, Rosmus, JCP 63, 2356 (1975).

State	T_e	ω_e	$\omega_e x_e$	B_e	α_e	D_e $(10^{-3}cm^{-1})$	r_e (Å)	Observed Transitions Design.	ν_{00}	References
$^{11}B^2H$		μ = 1.70261633		D_0^0 = 3.46 eV[a]						APR 1976
H $^1\Delta$ } 3d								H← X,	66448.0 Z	} (2)*
G $^1\Pi$ } 3d				[6.63$_5$]			[1.221$_6$]	G← X,	66362.8 Z	
F $^1\Sigma^+$ } 3d								F← X,	66068.6 Z	
E $^1\Sigma^+$ } 3p				[6.73$_6$]		[0.4]	[1.212$_4$]	E← X,	61852.8 Z	} (2)*
D $^1\Pi$ } 3p								D← X,	61110.2 Z	
B $^1\Sigma^+$ 3s	52348[b]	[1700.3$_4$] Z	46.$_6$	6.70$_5$	0.19$_5$[c]	0.4[d]	1.215$_2$	B← X,[e]	52360.2$_1$ Z	(2)*
A $^1\Pi$	23142	[1594.0$_8$] Z	(43)	6.648[f]	0.280	0.403[g]	1.220$_4$	A→X, V_R	23098.75 Z	(1)(2)
X $^1\Sigma^+$	0	[1703.2$_6$] Z	(28)	6.54$_2$	0.17$_1$	0.4$_2$[h]	1.230$_2$			
$^{11}B^1H^+$				D_0^0 = 1.95 eV[a]						APR 1976
A $^2\Pi_r$	b			[11.565][c]		[1.24]	[1.2565]	A→X, R	26376.2 Z	(1)
X $^2\Sigma^+$	0			[12.374]		[1.25][d]	[1.2147]			
$^{(11)}B^{127}I$?		(μ = 10.1304606)								APR 1976
		Absorption bands in the region 35870 - 37590 cm^{-1} have been attributed (1) to BI, but are almost certainly due to BCℓ (A← X).								

B^2H: [a]From the value for B^1H.

[b]Large electronic isotope shift; see (3).

[c]$\gamma_e = -0.014$.

[d]$H_0 = +0.3_5 \times 10^{-9}$.

[e]Headless bands.

[f]Thunberg's (1) value of 6.653 was derived from R and P branches only; $\Delta v_{ef} = +0.0102 \times J(J+1)$.

[g]$\beta_e = +0.034 \times 10^{-3}$.

[h]$\beta_e = -0.0_5 \times 10^{-3}$.

(1) Thunberg, ZP 100, 471 (1936).

(2) See ref. (7) of B^1H.

(3) Bunker, JMS 28, 422 (1968).

B^1H^+: [a]$D_0^0(B^1H) + I.P.(B) - I.P.(B^1H)$.

[b]$A = +14.0$.

[c]Λ-type doubling, $\Delta v_{ef} \approx +0.0164 \times N(N+1) - \ldots$.

[d]$H_0 = +6.9 \times 10^{-8}$.

(1) See ref. (2) of B^1H.

BI: (1) Briggs, Piercy, SA A 29, 851 (1973).

State[a]	T_e	ω_e	$\omega_e x_e$	B_e	α_e	D_e ($10^{-9}cm^{-1}$)	r_e (Å)	Observed Transitions Design.	Observed Transitions ν_{00}	References
$^{209}Bi_2$		μ = 104.490201		$D_0^0 = 2.0_4$ eV[b]						MAY 1976 A
y	x + 15746.3	94.7 H	5.2					y→x, R	15741.4 H	(9)
x	x	103.2 H	2.45							
		Fragments of several absorption systems in the region 44800 - 52900 cm^{-1}.								(1)(2)(10)
D	(42228)[c]	129 H	9.7					D←(A)[c], R	24485 H	(1)*
		Three diffuse absorption bands at 40115, 40275, 40467 cm^{-1}.								(1)
C	36456	$155._2$[d] H	2.3[e]					C←X, R	36447 H	(1)* (2)
I	33216.7	156.4 H	6.1					I→A, V	15487.9 H	(8)*
H	[32657.1]	Only v'=0.						H→A, V	14851.6 H	(8)*
		Continuous absorption with maximum at 32000 cm^{-1}.								(1)
G	(29609.0)[f]	107.0[f] H	0.2					G→(A)[f], R	11857.0 H	(8)*
E	(26504.7)	$(63._5)$[g]	(8.5)	(0.01425)[g]	(0.00015)			E→B,	21470.9[h] (Z)	(11)
A 0_u^+	17739.3	132.49 H	0.302[i]	0.01968[jk]	0.000053	ℓ	2.863	A↔X, R	17719.2 H	(1)* (2)(4) (8)* (11)
A'	(8000)[m]	141.2	0.37[n]							
B	(5000)[m]	127.0_5	0.29[o]	(0.01790)	(0.000046)					
X $^1\Sigma_g^+$	0[m]	172.71 H	0.341[p]	[0.022781][k]	0.000055[q]	[1.50]	2.6596			
X'	(-1500)[m]	154.3	0.42							

Bi_2: [a]The state designations adopted in the Bi_2 table agree with those of (11). They are compared below with designations used elsewhere:

This table, and (11):	X'	X	B	A'	A	E	G	H	I	C	D
(1):		A			B					C	D
(3):		X			B					D	E
(4):		X			A						
(7):		X			A					C	D
(8):		X			A		G	H	I	D	E

Bi_2 (continued):

[b] Thermochemical value (mass-spectrom.)(5)(6), calculated for a $^1\Sigma$ ground state and disregarding other low-lying states.

[c] Contrary to the conclusion of Almy and Sparks (1) that the violet system involves low vibrational levels of A, Gerber and Broida (11) consider it more probable that the absorption originates in high vibrational levels ($v'' \approx 20$) of X', thereby reducing T_e to approximately 26000 cm^{-1}. Additional unassigned diffuse absorption features in the region 22000 - 24000 cm^{-1}.

[d] (2) give $\omega_e = 146.0$, $\omega_e x_e = 0.50$. All bands except those with $v' = 1$ are diffuse. Observed to $v' = 4$. At shorter wavelengths are additional features probably belonging to $C \leftarrow X$ but not assigned by (1).

[e] The value of $\omega_e x_e$ given by (1) is clearly $2\omega_e x_e$. The constants listed above take account of this correction.

[f] Gerber et al. (11) have found very similar constants ($\omega_e = 105.6_8$, $\omega_e x_e = 0.63$) for the lower state of an unidentified transition in the laser photoluminescence spectrum and have tentatively identified this lower state with the upper state of Reddy and Ali's (8) $G \rightarrow A$ system. They suggest that the emission from G involves high vibrational levels ($v'' \approx 50$) of X, rather than $v=0,\ldots,4$ of A, and they estimate $T_e \approx 20000$.

[g] Constants derived from intensity data; see (11).

[h] Recalculated from data in (11).

[i] $\omega_e y_e = - 0.0021$, $\omega_e z_e = + 0.000055$ (8). Slightly different constants in (1)(11).

[j] Extrapolated from B_8, B_9, B_{11} (4).

[k] RKR potential curves (7a)(11).

[ℓ] $D_8 = 1.71 \times 10^{-9}$.

[m] All four states give rise to long lower state progressions in the laser photoluminescence spectrum of Bi_2 (11). Their relative energies were estimated (11) from the temperature dependence of the photoluminescence intensities. The upper levels of the transitions, except those belonging to A and E, could not be identified.

[n] $\omega_e y_e = - 0.001_9$.

[o] $\omega_e y_e = - 0.001$.

[p] $\omega_e y_e = - 0.0018$, $\omega_e z_e = + 0.00001_0$ (8). Slightly different constants in (1)(11).

[q] From the laser photoluminescence spectrum (11), adjusted for best overall fit of observed with calculated intensities.

(1) Almy, Sparks, PR 44, 365 (1933).
(2) Nakamura, Shidei, JJP 10, 11 (1935).
(3) Herzberg, MOLSPEC 1 (1950).
(4) Åslund, Barrow, Richards, Travis, AF 30, 171 (1965).
(5) Kohl, Uy, Carlson, JCP 47, 2667 (1967).
(6) Rovner, Drowart, Drowart, TFS 63, 2906 (1967).
(7) DONNSPEC (1970).
(7a) Rao, Lakshman, IJPAP 8, 785 (1970).
(8) Reddy, Ali, JMS 35, 285 (1970).
(9) Singh, Nair, Rai, SpL 4, 313 (1971).
(10) Topouzkhanian, Sibaï, d'Incan, ZN 29 a, 436 (1974).
(11) Gerber, Sakurai, Broida, JCP 64, 3410 (1976); Gerber, Broida, JCP 64, 3423 (1976).

State	T_e	ω_e	$\omega_e x_e$	B_e	α_e	D_e ($10^{-8}cm^{-1}$)	r_e (Å)	Observed Transitions Design.	Observed Transitions ν_{00}	References
$^{209}Bi^{79}Br$		$\mu = 57.285367_3$								MAY 1976
		Fragments of three systems of V shaded absorption bands in the regions 40800 - 45000 and 33700 - 36900 cm^{-1}.								(2)
B	24710.9	265.3_4 H	1.95_6					B← X, V?	24738.4 H	(1)*
A 0^+	20532.0	135.9_1 H	0.53_4[a]	(0.0364)	(0.0002_5)		(2.84)	A↔X, R	20495.2 H	(1)* (3)(4) (5)
X 0^+	0	209.5_0[b] H	0.45	0.04321526	0.000132695[c]	0.734_7	2.60950_3	Microwave sp.		(6)
$^{209}Bi^{35}Cl$		$\mu = 29.9562508$		$D_0^0 = 3.0_8$ eV[a]						MAY 1976 A
B	25492.7	403.5_0 H	3.76_8[b]					B↔X, V?	25539.5 H	(1)* (2)*
A'	23054.5	217.5 H	2.95	(0.0739)[c]			(2.76_0)	A'↔X, R	23008.6 H	(5)(6)* (8) (9)*
A 0^+	21801.8	220.3 H	2.47[d]	(0.07927)[c]		(3.9)	(2.66_4)	A↔X, R	21757.4 H	(1)* (2)* (3)(10)(11)
X 0^+	0	308.4[e] H	0.96	(0.0921)[c]		(3.1)	(2.47_2)			
$^{209}Bi^{19}F$		$\mu = 17.4151900$								MAY 1976 A
C	44222.0	615.0 H	2.50	a				C↔X, V	44274 H	(3)(4)(5)* (14)*
		Two systems[a] (emission only) with $\omega_e' = {611.8 \atop 614.4}$, $\omega_e'' = {542.7 \atop 535.0}$ (P, Q heads).						V	36979 H[Q]	(3)(5)*(12)*
								V	32179 H	(3)(5)*(13)*
B	25986.4	602.0[b] H	3.50					B↔X,[b]	26033 H	(11)
A 0^+	22959.7	381.0 H	3.00[c]	d				A↔X, R	22894.6 H	(1)* (2)(6)* (7)*(8)*(9)*
		R shaded emission bands in the region 14350 - 16050 cm^{-1}; $\omega' \approx 399$, $\omega'' \approx 535$.								(11)(15)
X 0^+	0	510.7 H	2.05	d						

BiBr: [a]$\omega_e y_e = -0.103$; convergence limit at 22120 cm^{-1}.
[b]From the A−X system. Similar constants from B−X.
[c]$\gamma_e = +7.68 \times 10^{-8}$.

(1) Morgan, PR 49, 41 (1936).
(2) Sur, Majumdar, PNISI 20, 235 (1954).
(3) Sankaranarayanan, Patel, Narayanan, PIAS A 56, 171 (1962).
(4) Singh, IJPAP 6, 299 (1968).
(5) Lal, Mohamed, Khanna, IJPAP 13, 53 (1975).
(6) Kuijpers, Dymanus, CPL 39, 217 (1976).

BiCl: [a]Thermochemical value (7), calculated for a $^3\Sigma$ ground state.
[b]$\omega_e y_e = +0.0016$.
[c]Rotational analyses by (8)(10)(11); different results in (4)(5)(6). Uncertain.
[d]$\omega_e y_e = -0.02$.
[e]From the A−X system; similar constants from B−X.

(1) See ref. (1) of BiBr.
(2) Ray, IJP 16, 35 (1942).
(3) Venkateswarlu, Khanna, PIAS A 51, 14 (1960).
(4) Khanna, JMS 6, 319 (1961).
(5) Rao, Rao, IJP 39, 65 (1965).
(6) Babu, Rao, IJPAP 5, 79 (1967).
(7) Cubicciotti, JPC 71, 3066 (1967).
(8) Mohanty, Nair, Upadhya, IJPAP 6, 494 (1968).
(9) Yamdagni, JMS 35, 149 (1970).
(10) Singh, Upadhya, IJP 45, 121 (1971).
(11) Rai, Upadhya, Ram, IJP 48, 554 (1974).

BiF: [a]Rotational analyses (10)(12)(13)(14).
[b]Complex system of V and R shaded bands. The constants above refer to the V shaded heads.
[c]$\omega_e y_e = +0.10$.
[d]Rotational analyses of several A−X bands in (6)(8) [$B_0' = 0.2097$, $B_0'' = 0.2307$] and (9) [$B_0' = 0.2090$, $B_0'' = 0.2295$].

(1) Howell, PRS A 155, 141 (1936).
(2) See ref. (1) of BiBr.
(3) Rochester, PR 51, 486 (1937).
(4) Joshi, PPS 78, 610 (1961).
(5) Rao, Rao, IJP 36, 85 (1962).
(6) Rao, Rao, CJP 40, 1077 (1962).
(7) Sankaranarayanan, Narayanan, Patel, PIAS A 59, 378 (1964).
(8) Rao, Rao, IJP 39, 572 (1965).
(9) Mohanty, Upadhya, CS 36, 478 (1967).
(10) Murty, Rao, CS 36, 661 (1967).
(11) Patel, Narayanan, IJPAP 5, 223 (1967).
(12) Mohanty, Rai, Upadhya, Singh, JP B 1, 523 (1968).
(13) Chaudhry, Upadhya, Rai, Mohanty, JP B 1, 1223 (1968).
(14) Chaudhry, Upadhya, Rai, JP B 2, 628 (1969).
(15) Murty, Rao, Reddy, Rao, SpL 8, 217 (1975).

State	T_e	ω_e	$\omega_e x_e$	B_e	α_e	D_e ($10^{-4}cm^{-1}$)	r_e (Å)	Observed Transitions		References
								Design.	ν_{00}	
$^{209}Bi^1H$		$\mu = 1.00298823$		$D_0^0 \leq 2.90$ eV^{a}						MAY 1976
E 0^+	32940.3	[1105.58] Z	$30._1$	3.5456^{b}	0.0525	$[1.401]^{c}$	2.1772	E← X, R	32674.08 Z	(5)(8)
D $^1\Sigma$				[3.88]			$[2.08_1]$	D→C, R	20647 Z	(1)(3)
C $^1\Sigma$		[1313.6] Z		4.37	0.11		1.96_1			
B 0^+	21263	$[1643.07]^{d}$ Z	(47)	5.3078^{d}	0.1861	$[2.010]^{e}$	1.7795	B→A, V	16341.72^{d} Z	(1)(3)(4)(6)*
								B→X, V	21278.35^{d} Z	(1)* (3)*
A 1 $^3\Sigma^-$	4917.1	$[1669.16]^{f}$ Z	35.4	5.2386^{fg}	0.1546	$[1.904]^{fh}$	1.7912^{f}			
X 0^+ $^3\Sigma^-$	0	[1635.73]	31.6	5.137^{i}	0.148	$[1.83]^{i}$	1.805			
$^{209}Bi^2H$		$\mu = 1.99487609$								MAY 1976
E 0^+	(32929)	j		$1.796^{k\ell}$	0.0232	m	2.169	E← X, R	n	(8)
B 0^+	21263.7	$[1185.10]^{d}$	25.0	2.6687^{d}	0.0644	$[0.507]^{o}$	1.7795	B→A, V	16341.10^{d} Z	(4)(6)*
								B→X, V	21276.25^{d} Z	(2)(3)
A 1 $^3\Sigma^-$	4916	$[1206.9]^{f}$	(19)	$[2.6084]^{fg}$	0.058	$[0.493]^{fp}$	1.790^{f}			
X 0^+ $^3\Sigma^-$	0	[1173.32]	16.1	2.592^{i}	0.054	$[0.506]^{i}$	1.804			

Bi^1H, Bi^2H:

[a]From the predissociation in E $0^+(v=2)$, assuming dissociation into $Bi(^2D_{3/2}) + H(^2S)$; see (8).

[b]Lines having v'=1, J' ≥ 16, and all lines with v'=2, are diffuse. The lack of emission from E 0^+, even from levels which appear sharp in absorption, was attributed (8) to weak predissociation by the $\Omega = 1$ component of the repulsive $^5\Sigma^-$ state which arises from ground state atoms.

[c]$D_1 = 1.563 \times 10^{-4}$.

[d]Constants from (8).

[e]$D_1 = 2.146 \times 10^{-4}$ (8).

[f]Constants for the f component (8). Ω-type doubling,

$$\left.\begin{aligned} Bi^1H:\ \Delta\nu_{ef}(v=0) &\approx +\,0.0200 \times J(J+1) - \ldots \\ Bi^2H:\ \Delta\nu_{ef}(v=0) &\approx +\,0.0049 \times J(J+1) - \ldots \end{aligned}\right\}\ (6)(8).$$

[g]Magnetic hyperfine structure (4)(6)(7).

[h]$D_1 = 1.915 \times 10^{-4}$; also higher order constants (8).

[i]Effective constants; see also (8).

[j]$\Delta G(3/2) = 768.75$. Using isotope relations: $\omega_e \approx 829._3$, $\omega_e x_e \approx 15._1$.

[k]Extrapolated from $B_1 = 1.7607$ and B_2.

[ℓ]Slight broadening of lines having v'=2, J' ≥ 18. v'=3 not observed.

[m]$D_1 = 0.378 \times 10^{-4}$, $D_2 = 0.423 \times 10^{-4}$.

[n]$\nu_0(1-0) = 33540.58$.

[o]$D_1 = 0.528 \times 10^{-4}$ (8).

[p]$D_1 = 0.489 \times 10^{-4}$; higher order constants in (8).

(1) Heimer, ZP 95, 328 (1935).
(2) Heimer, ZP 103, 621 (1936).
(3) Heimer, Dissertation (Stockholm, 1937).
(4) Hulthén, Neuhaus, PR 102, 1415 (1956).
(5) Khan, Khan, PPS 88, 211 (1966).
(6) Neuhaus, ZN 21 a, 2113 (1966).
(7) T. M. Dunn, in "Molecular Spectroscopy: Modern Research", edited by K. N. Rao and C. W. Mathews; p. 231. Academic Press (1972).
(8) Lindgren, Nilsson, JMS 55, 407 (1975).

State	T_e	ω_e	$\omega_e x_e$	B_e	α_e	D_e (10^{-8} cm^{-1})	r_e (Å)	Observed Transitions Design.	ν_{00}	References
$^{209}Bi^{127}I$		$\mu = 78.957256_0$								MAY 1976
C	40707.1	230.9 H	4.75					C← X,	40739.5 H	(3)
		Unclassified absorption bands in the region 24100 - 25000 cm^{-1}.								(5)
B	23389.1	198.0_9 H	1.44_4[a]					(B→a),[b]	R 17213.5 H	(2)(4)*
								B↔X,	R 23405.9 H	(1)(2)(6)*
A'	20318.7	126.8 H	0.90					A'← X,	R 20300.0 H	(5)
A	20006.0	145.0 H	2.50					A← X,	R 19996.0 H	(5)
a	(6190)	170.8 H	0.3							
X 0^+	0	163.8_8 H	0.28_0[c]	0.027222814	0.000069790_6[d]	0.29959[e]	2.80050_1	Microwave sp.		(7)
$^{209}Bi^{16}O$		$\mu = 14.8577355_7$		$D_0^0 = 3.47$ eV[a]						MAY 1976
F	(40941)	[748] H	Fragments only.					F← X_1,	40970 H	(2)*
E ($^2\Sigma$?)	38550	769.3 H	6.2					E← X_1,	38588 H	(2)*
D $(^2\Pi)_{1/2}$	(32805)	(343)		$[0.228_4]$[bc]		[45]	[2.229]	D← X_1,	R 32631.35 Z	(6)*
C $(^2\Delta)_{3/2}$	(30700)	(465)		[0.2548][de]		$[30._6]$	[2.1102]	C← X_1,[f]	R 30587.16[d] Z	(6)*
B $^4\Sigma^-_{1/2}$	$28738._2$	483	5	0.260[g]	0.0029	h	2.09	B↔X_1,[f]	R 28633.35 Z	(2)(5)* (6)*
A $^2\Pi_{1/2}$	14187.0	508.8 Z	2.78	0.24715[i]	0.00167	$23._3$	2.1426	A↔X_1,[f]	R 14095.6 Z	(1)(3)(4)* (6)*
X_2 $^2\Pi_{3/2}$	(8000)									
X_1 $^2\Pi_{1/2}$	0	692.4 Z	4.34	0.3034[jc]	0.0022	[22.1]	1.934			

BiI: [a]Also higher order constants.

[b]It is not certain that the upper state of this system ($\omega_e' = 200.6$, $\omega_e'x_e' = 1.4$) is identical with the B state observed in absorption and emission.

[c]$\omega_e y_e = -0.005$.

[d]$+1.72_1 \times 10^{-8}(v+\frac{1}{2})^2 - 9.9 \times 10^{-11}(v+\frac{1}{2})^3$.

[e]$\beta_e = +0.00035_4 \times 10^{-8}$.

(1) See ref. (1) of BiBr.

(2) Rao, IJP 23, 379 (1949).

(3) M. M. Joshi, Thesis (Allahabad University, 1958). Quoted in (5).

(4) Singh, IJPAP 6, 445 (1968).

(5) Yamdagni, SA A 26, 1071 (1970).

(6) Singh, Asthana, Singh, SpL 8, 101 (1975).

(7) Kuijpers, Törring, Dymanus, CP 12, 309 (1976).

BiO: [a]Thermochemical value (mass-spectrom.)(9).

[b]Both high and low rotational levels are predissociated; the e and f levels were only observed for $43.5 \leq J \leq 71.5$ and $57.5 \leq J \leq 81.5$, resp.. In this region, the Ω-type doubling [recalc. from (6)] is well represented by $\Delta\nu_{fe} = (+)[0.306 - 11.09 \times 10^{-6}(J+\frac{1}{2})^2](J+\frac{1}{2})$.

[c]Unresolved magnetic hyperfine structure; see (6)(7).

[d]Vibrational numbering uncertain. The single band reported by (6) agrees in position with the 1-0 band of a weak system observed by (2) at about 1480 cm^{-1} above the B − X bands.

[e]Very small Ω-type doubling.

[f]The lines have half-widths of ~ 0.25 cm^{-1} (independent of J), owing to the unresolved magnetic hyperfine splitting of the ground state levels.

BiO (continued):

[g]Large Ω-type doubling, $\Delta\nu_{fe}(v,J) = p_v(J+\frac{1}{2}) - \ldots$ where $p_v \approx +4B_v$ (6). The sign of the splitting can not be determined from the spectrum but is chosen here to give agreement with theoretical predictions [see (8)] for a $^4\Sigma^-_{\frac{1}{2}}$ state. This requires the parities of most rotational levels in figure 4 of (6) to be reversed and leads to negative values of $\Delta\nu_{fe}$ in the ground state of BiO, contrary to results for other group V oxides.

[h]D_v increases from $D_2 = 32._3 \times 10^{-8}$ to $D_5 = 80._3 \times 10^{-8}$ [see (6)].

[i]$\Delta\nu_{fe}(v=2,\ldots,9) \approx (+)0.025(J+\frac{1}{2}) - \ldots$; for more details see (6).

[j]$\Delta\nu_{fe}(v=0) = (-)0.187(J+\frac{1}{2})$.

(1) Sen Gupta, IJP 18, 182 (1944).

(2) Bridge, Howell, PPS A 67, 44 (1954).

(3) Scari, APH 6, 73 (1956).

(4) Gissane, Barrow, PPS 85, 1048 (1965); 86, 682 (1965)(Corrigendum).

(5) Babu, Rao, CJP 44, 705 (1966).

(6) Barrow, Gissane, Richards, PRS A 300, 469 (1967).

(7) Atkins, PRS A 300, 487 (1967).

(8) Kopp, Hougen, CJP 45, 2581 (1967).

(9) Uy, Drowart, TFS 65, 3221 (1969).

(10) Rao, Lakshman, CS 40, 316 (1971).

(11) Asthana, Kushawaha, Nair, APP A 42, 739 (1972).

(12) Singh, Shukla, JQSRT 12, 1249 (1972).

(13) Singh, JQSRT 12, 1343 (1972).

State	T_e	ω_e	$\omega_e x_e$	B_e	α_e	D_e ($10^{-8}cm^{-1}$)	r_e (Å)	Observed Transitions Design.	ν_{00}	References
$^{209}Bi^{32}S$		μ = 27.7296867		D_0^0 = 3.17 eV[a]						MAY 1976
		One of two systems of ultra-violet bands attributed (1) to BiS is, in fact, the B – X system of S_2 [see (4)]. The other, consisting of V(?) shaded bands in the region 43200 - 46900 cm^{-1}, appears to include bands which the same author (2) also assigned to SbS. In the latter case the bands were described as red-degraded.								
A $^2\Pi_{1/2}$	≤13343.9	≥303.74[b] H	1.159	≥0.09258[bc]	0.000416	3.5_5	≤2.563	A← X,[d]	R ≤13291.5[b] H	(4)
X $^2\Pi_{1/2}$	0	408.71 H	1.46	[0.112764][e]	(0.000486)	[3.34]	2.3194			
$^{209}Bi^{(80)}Se$		(μ = 57.809502_2)		D_0^0 = 2.80 eV[a]						MAY 1976
D	44425	316.0 H	2.0					D← X,	V 44450 H	(1)
C	35618	304.0 H	2.0					C← X,	V 35637 H	(1)
B	20411	169.4 H	0.8					B← X,	R 20363 H	(5)
A ½	≤13235.7	≥190.9[b] H	0.595[b]					A← X,[c]	R ≤13198.3 H	(3)
X ½	0	[264.8][b] H	0.4							
$^{209}Bi^{(130)}Te$		(μ = 80.108962_1)		D_0^0 = 2.4_0 eV[a]						MAY 1976
E		Bands in the region 43900 - 45300 cm^{-1}.						E← X,	V	(1)
D	43116	263.0 H	0.96					D← X,	V 43143 H	(1)
C	(42870)	(164.4) H	(0.4)					C← X,	(V)(42848) H	(1)
B		Bands in the region 40700 - 42000 cm^{-1}.						B← X,	V 41967 H	(1)
A		Bands in the region 34000 - 35500 cm^{-1}.						A← X,	V	(1)
X	0	208.5 H	0.52							

BiS: [a]Thermochemical value (mass-spectrom.)(5); (3).
[b]Vibrational numbering uncertain.
[c]$|\Delta\nu_{fe}| \approx 0.006(J+\frac{1}{2})$ [v=7,8].
[d]Line widths of 0.45 cm^{-1} result from unresolved nuclear magnetic hyperfine structure.
[e]Large Ω-type doubling, $|\Delta\nu_{fe}(v=0)| = 0.1135(J+\frac{1}{2})$.

(1) Sur, IJP 25, 65 (1951).
(2) Sur, PNASI A 20, 251 (1951).
(3) Cubicciotti, JPC 67, 118 (1963).
(4) Barrow, Stobart, Vaughan, PPS 90, 555 (1967).
(5) See ref. (9) of BiO.
(6) Singh, Pandey, IJPAP 7, 580 (1969).
(7) Asthana, APP A 42, 739 (1972).

BiSe: [a]Thermochemical value (mass-spectrom.)(2)(4).
[b]Constants for $Bi^{80}Se$.
[c]Broad lines on account of unresolved nuclear magnetic hyperfine structure.

(1) Sharma, PPS A 67, 935 (1954).
(2) Porter, Spencer, JCP 32, 943 (1960).
(3) See ref. (4) of BiS.
(4) See ref. (9) of BiO.
(5) Yamdagni, IJPAP 8, 51 (1970).

BiTe: [a]Thermochemical value (mass-spectrom.)(2)(4); (3).

(1) See ref. (1) of BiSe.
(2) See ref. (2) of BiSe.
(3) Boncheva-Mladenova, Pashinkin, Novoselova, IANNM 4, 291 (1968).
(4) See ref. (9) of BiO.

State	T_e	ω_e	$\omega_e x_e$	B_e	α_e	D_e ($10^{-6}cm^{-1}$)	r_e (Å)	Observed Transitions Design.	Observed Transitions ν_{00}	References
$^{11}B^{14}N$		$\mu = 6.1635127_6$								JUN 1976
		Incompletely analyzed singlet transitions (in emission).						R	34499 H	
								R	32817 H	(1)
								V	30963 H	
A $^3\Pi$	27875.0	1317.5 H	14.9	1.555	0.010	(8.7)	1.326	$A \leftrightarrow X$, R	27775.8 H	(1)* (2)(3) (5)
X $^3\Pi$ [a]	0	1514.6 H	12.3	1.666	0.025	(8.1)	1.281			
$^{11}B^{16}O$		$\mu = 6.5209400_9$		$D_0^0 = 8.28$ eV[a]						JUN 1976
C $^2\Pi_{(r)}$	55346.1[b]	1315.3 H	11.1	1.483	0.018	8	1.320	$C \rightarrow X$, R γ- bands	55061.5[c]	(7)* (10) (11)*
B $^2\Sigma^+$	43174.05	1281.69 Z	10.66	1.5171[de]	0.0210	8.5	1.3054	$B \rightarrow A$,[f] V	19225.9 19351.0	(1)
								$B \rightarrow X$,[g] R β- bands	42872.34 Z	(1)* (3)* (6)(8)*
A $^2\Pi_i$	23958.76[h] 23833.7	1260.70 Z	11.157[i]	[1.4018][je]	0.0196	[7.63]	1.3533	$A \leftrightarrow X$,[k] R α- bands	23646.43[ℓ] Z 23521.3[ℓ] Z	(1)* (2)* (4)* (5)* (14)(18)
X $^2\Sigma^+$	0	1885.69 Z	11.81	1.7820[me]	0.0166	6.32[n]	1.2045	ESR sp.[o]		
$^{11}B^{16}O^+$										JUN 1976
b $^1\Sigma$	(27941)	(1952)		[1.8202]		[6.33]	[1.1917]	$b \rightarrow a$, V	28023.99 Z	(16a)
a $^1\Sigma$	(0)	(1787)		[1.7799]		[7.06]	[1.2052]			
$^{(11)}B^{16}O^-$						I.P. = 3.1_2 eV[p]				JUN 1976 A
$^{(11)}B^{31}P$		($\mu = 8.1223128_4$)		$D_0^0 = 3.5_6$ eV[a]						JUN 1976

BN: [a]The observation of $A \leftarrow X$ in absorption in rare gas matrices (5) supports theoretical predictions (4)(6) of a $^3\Pi$ ground state.

(1) Douglas, Herzberg, CJR A 18, 179 (1940).
(2) Nicholls, Fraser, Jarmain, CF 3, 13 (1959).
(3) Thrush, Nature 186, 1044 (1960).
(4) Verhaegen, Richards, Moser, JCP 46, 160 (1967).
(5) Mosher, Frosch, JCP 52, 5781 (1970).
(6) Melrose, Russell, JCP 55, 470 (1971); 57, 2586 (1972).

BO, BO^+, BO^-:

[a]Thermochemical value (mass-spectrom.)(17)(19); in good agreement with 8.3 eV by flame photometry (15).
[b]$A = (+)46.4$.
[c]R_2 head at 55084.2 cm^{-1}.
[d]Spin splitting constant $\gamma \approx +0.025$ (8)(18).
[e]Potential curves (16).
[f]Franck-Condon factors (9).
[g]Franck-Condon factors (9)(13)(21); measured relative intensities (13).
[h]$A_0 = -122.26$ (slight J dependence)(18); $A_1 = -122.36$ (5).
[i]$\omega_e y_e = +0.049$.
[j]Λ-type doubling in $^2\Pi_{\frac{1}{2}}$, $\Delta v_{fe} \approx +0.025(J+\frac{1}{2}) - \ldots$ (5)(18); for $^2\Pi_{\frac{3}{2}}$ see (5).
[k]Franck-Condon factors (9)(12)(13)(21); measured relative (13) and estimated absolute (22) intensities.
[ℓ]$\{J'=0\}$ relative to $N''=0$.
[m]From (8); slightly different constants in (5). Spin splitting constant $\gamma(v=2) = +0.0065$ (18).
[n]$\beta_e = +0.02 \times 10^{-6}$.
[o]In rare gas matrices at 4 K (20).
[p]From the heat of formation for BO^- (21a).

BO, BO^+, BO^- (continued):

(1) Mulliken, PR 25, 259 (1925).
(2) Jenkins, PNASU 13, 496 (1927).
(3) Elliot, PKNAW 33, 644 (1930); 38, 736 (1935).
(4) Scheib, ZP 60, 74 (1930).
(5) Jenkins, McKellar, PR 42, 464 (1932).
(6) Funke, Simons, PKNAW 38, 142 (1935).
(7) Chrétien, HPA 23, 259 (1950).
(8) Lagerqvist, Nilsson, Wigartz, AF 13, 379 (1958).
(9) See ref. (2) of BN.
(10) Kuzyakov, Tatevskii, Tunitskii, OS(Engl. Transl.) 9, 84 (1960).
(11) Mal'tsev, Kataev, Tatevskii, OS(Engl. Transl.) 9, 376 (1960).
(12) Nicholls, Fraser, Jarmain, McEachran, ApJ 131, 399 (1960).
(13) Robinson, Nicholls, PPS 75, 817 (1960).
(14) See ref. (3) of BN.
(15) De Galan, Physica 31, 1286 (1965).
(16) Singh, Rai, JQSRT 5, 723 (1965).
(16a) Kataev, Mal'tsev, VMUK 22 (2), 23 (1967).
(17) Coppens, Smoes, Drowart, TFS 64, 630 (1968).
(18) Dunn, Hanson, CJP 47, 1657 (1969).
(19) Uy, Drowart, HTS 2, 293 (1970).
(20) Knight, Easley, Weltner, JCP 54, 1610 (1971).
(21) Liszt, Smith, JQSRT 11, 1043 (1971).
(21a) Srivastava, Uy, Farber, TFS 67, 2941 (1971).
(22) Kuz'menko, Kuznetsova, Kuzyakov, Chuev, JAS 20, 373 (1974).

BP: [a]Thermochemical value (mass-spectrom.)(1).

(1) Gingerich, JCP 56, 4239 (1972).

State	T_e	ω_e	$\omega_e x_e$	B_e	α_e	D_e $(10^{-8}cm^{-1})$	r_e (Å)	Observed Transitions Design.	Observed Transitions ν_{00}	References
$^{79}Br_2$		$\mu = 39.459166_0$ $D_0^0 = 1.9707_0$ eV[a] I.P. = 10.52 eV[b]								SEP 1976 A
		Fragments of additional Rydberg series converging to A $^2\Pi_u$ of Br_2^+.								(18)*
		Rydberg series converging to X_2 $^2\Pi_{g,\frac{1}{2}}$ of Br_2^+: $\nu = 88306^c - \{R/(n-2.416)^2, R/(n-2.446)^2, R/(n-2.591)^2, R/(n-2.629)^2\}$ n = 5,6,7.								(18)*
		Rydberg series converging to X_1 $^2\Pi_{g,\frac{3}{2}}$ of Br_2^+: $\nu = 85165^d - \{R/(n-1.843)^2$, n = 5,...,12.; $R/(n-1.938)^2$, n = 5,...,18.; $R/(n-2.225)^2, R/(n-2.422)^2, R/(n-2.593)^2\}$ n = 5,...,20.								(18)*
N	76537	230[e] H						N← X, R	76491 H	(18)*
M	74060	241[e] H	(0.3)					M← X, R	74019 H	(18)*
L	72727	218[e] H	3					L← X, R	72674 H	(18)*
		Several groups of diffuse emission bands in the region 23600 - 50000 cm^{-1} have been assigned (5) to transitions from four states at 47000, 55534, 61444, 66500 cm^{-1} to various repulsive states arising from $Br(^2P_{\frac{3}{2},\frac{1}{2}}) + Br(^2P_{\frac{3}{2},\frac{1}{2}})$.								(5)
(K)		Extensive system of absorption bands in the region 59000 - 67000 cm^{-1}; no analysis. This system may include transitions to the upper states of:						(K)← X,		(18)
		a) three emission systems, ℓ→X, M→X, L→X of (9), with $\omega' \approx$ 293[e], 426[e], 281[e];						R; V; R	62266 H; 60879 H; 59855 H	(9)
		b) a long resonance series (63817 - 53779 cm^{-1}).								(12)
H	(56820)	108.0[ef]	1.5					H→ B,	(40890)[f]	(10)
G	56337	(255)[e] H						G→ X,[g] R	56303 H	(9)
F	52191	(120)[e] H						F→ X,[h] R	52090 H	(9)
E	51634.0	150.9[e]	0.49_5[i]					E↔ B[j],	35724.0[e]	(7)(11)(28)

Br_2: [a]From (32); corresponding values for $^{79,81}Br_2$ and $^{81}Br_2$ are 1.9708_2 and 1.9709_5 eV (short extrapol. of B 0_u^+).

[b]From photoionization (21); supported by measurements at different temperatures. In good agreement with 10.51 eV obtained by photoelectron spectroscopy (15)(23)(26). A slightly higher value, 10.56 eV, was derived (18) from the Rydberg series in the VUV. It is probable that this value refers to v'=1.

[c]The interval of 3141 cm^{-1} between $X_1\ ^2\Pi_{3/2}$ and $X_2\ ^2\Pi_{1/2}$ of Br_2^+ derived from the Rydberg series does not agree with the value 2820 cm^{-1} from the photoelectron spectrum (23). The discrepancy may be accounted for by assuming that instead of v'=0 as suggested in (18) the $^2\Pi_{1/2}$ series listed here have v'=2 while the $^2\Pi_{3/2}$ series have v'=1 (see [d]).

[d]According to the photoionization and photoelectron value of the ionization potential (see [b]) the five $^2\Pi_{3/2}$ Rydberg series in the table refer to v'=1. Vibrational structure; see (18).

[e]Normal isotopic mixture.

[f]The vibrational analysis is doubtful since only v"=21-32 were observed. ν_{00} (extrapolated) and T_e are different from (10) to allow for the new data on the B state (32).

[g]System J – X of (9), not observed in absorption.

[h]System H – X of (9), not observed in absorption.

[i]$\omega_e z_e = +\ 0.00006_5$; vibrational constants from the reanalysis (28) of the emission data of (7) and the absorption data of (11). See [j].

[j]It is not entirely certain that the lower state is B $^3\Pi_{u,0^+}$ and not A $^3\Pi_{u,1}$.

(continued p. 106)

State	T_e	ω_e	$\omega_e x_e$	B_e	α_e	D_e ($10^{-8}cm^{-1}$)	r_e (Å)	Observed Transitions Design.	ν_{00}	References
$^{79}Br_2$ (continued)										
D	48499	162.0[e]	0.29					D→B,	32595	(8)
C $^1\Pi_u$ 1_u	(24000)	Several absorption continua beyond 19580 cm^{-1} corresponding to a number of electronic transitions including that to C $^1\Pi_u$ with maximum at 24000 cm^{-1}.						C[k]← X,	(24000)	(1)(2)(3)(4)(6)(30a)
B $^3\Pi_u$ 0_u^+	15902.47	167.607 Z	1.6361[ℓ]	0.059589[mno]	0.00048910[p]	3.013[q]	2.6775_7	B[kr]↔X,[o] R	15823.47 Z	(16)* (24)* (20)(32)(35)
A $^3\Pi_u$ 1_u	13905	153[e] H	2.7[s]	0.0588[tu]	(0.0008)		2.69_5	A[kr]↔X,[u] R	13818[v]	(13)(14)* (29)
X $^1\Sigma_g^+$	0	325.321 Z	1.0774[w]	0.082107[x]	0.0003187[y]	2.092[q]	2.2810_5	Raman sp.[z]		

Br_2 (continued):

[k]Also observed in magnetic circular dichroism (38) and photofragment (40) spectra. The latter authors confirm Mulliken's (3) prediction that C $^1\Pi_u$ dissociates into $^2P_{\frac{3}{2}} + {}^2P_{\frac{3}{2}}$ and observe evidence for several excited g states by two-photon photofragment studies near 28000 and 38000 cm^{-1}.

[ℓ]$\omega_e y_e = -0.009369$ (for $v \leq 8$). Vibrational levels observed to v=55, dissociation limit ($^2P_{\frac{3}{2}} + {}^2P_{\frac{1}{2}}$) at 19579.76 cm^{-1} above X $^1\Sigma_g^+$(v=0,J=0). See [o]. Absorption in the B 0_u^+ continuum (43).

[m]Hfs observed in v=12 ($^{81}Br_2$) and v=17 ($^{79}Br_2$); see (33).

[n]Predissociation was observed (41)(39) for v=42, J=33 by the laser-molecular beam technique. B→X emitted in the recombination of Br($^2P_{\frac{3}{2}}$) atoms shows strong enhancement of bands with $5 < v' < 10$ presumably on account of inverse predissociation (27). See also [r].

[o]RKR potential function and Franck-Condon factors (25)(32). For the behaviour of the potential function near the dissociation limit[ℓ] see (30)(31)(34).

[p]$\gamma_e = -6.637 \times 10^{-6}$ (valid for $v \leq 8$).

[q]D_v and higher order constants in (32).

[r]Estimated radiative lifetimes for A and B range from 1000 to 2000 and 12 to 70 μs, respectively (29)(30a)(43). For the B state (22) find total lifetimes of the order of 1 μs; minima (~0.2 μs) occur for v=1 and 14 probably on account of predissociation. For lifetimes near the dissociation limit[ℓ] of B see (42).

[s]Convergence limit for $^{79}Br_2$ at 15894.6 cm^{-1} above X $^1\Sigma_g^+$(v=0,J=0), corresponding to $^2P_{\frac{3}{2}} + {}^2P_{\frac{3}{2}}$. A weak continuous spectrum joins onto the limit and overlaps the main absorption system B← X; see (17).

[t]Extrapolated from v=7; constants for v=0...6 have not been

determined. B_v, D_v, H_v, and Λ-type doubling constants for v=7...24 in (29).

[u]RKR potential function and Franck-Condon factors (29).

[v]Based on $\Delta G'(v=0-7)$ from low-resolution emission spectra (14) of normal Br_2 and the origin of the 7-0 $^{79}Br_2$ band at 14739.14 cm^{-1} derived from (29) and (32).

[w]$\omega_e y_e = -0.002298$.

[x]RKR potential curve (25)(32). Hfs observed (33) in v=4 ($^{81}Br_2$) and v=7 ($^{79}Br_2$).

[y]$\gamma_e = -1.04_5 \times 10^{-6}$.

[z]Resonance Raman spectra in the gas (19)(37), in solid argon (35); pure rotational Raman spectrum (36).

(1) Cordes, Sponer, ZP <u>63</u>, 334 (1930).
(2) Aickin, Bayliss, TFS <u>34</u>, 1371 (1938).
(3) Mulliken, PR <u>57</u>, 500 (1940).
(4) Rees, PPS <u>59</u>, 1008 (1947).
(5) Venkateswarlu, PIAS A <u>25</u>, 138 (1947).
(6) Bayliss, Sullivan, JCP <u>22</u>, 1615 (1954).
(7) Venkateswarlu, Verma, PIAS A <u>46</u>, 251 (1957).
(8) Venkateswarlu, Verma, PIAS A <u>46</u>, 416 (1957).
(9) Haranath, Rao, JMS <u>2</u>, 428 (1958).
(10) Verma, PIAS A <u>47</u>, 196 (1958).
(11) Briggs, Norrish, PRS A <u>276</u>, 51 (1963).
(12) Rao, Venkateswarlu, JMS <u>13</u>, 288 (1964).
(13) Horsley, JMS <u>22</u>, 469 (1967).
(14) Clyne, Coxon, JMS <u>23</u>, 258 (1967).
(15) Frost, McDowell, Vroom, JCP <u>46</u>, 4255 (1967).
(16) Horsley, Barrow, TFS <u>63</u>, 32 (1967).
(17) Sulzmann, Bien, Penner, JQSRT <u>7</u>, 969 (1967).
(18) Venkateswarlu, CJP <u>47</u>, 2525 (1969).
(19) Holzer, Murphy, Bernstein, JCP <u>52</u>, 399 (1970).
(20) Holzer, Murphy, Bernstein, JCP <u>52</u>, 469 (1970).
(21) Dibeler, Walker, McCulloh, JCP <u>53</u>, 4715 (1970).
(22) Capelle, Sakurai, Broida, JCP <u>54</u>, 1728 (1971).
(23) Cornford, Frost, McDowell, Ragle, Stenhouse, JCP <u>54</u>, 2651 (1971).
(24) Coxon, JMS <u>37</u>, 39 (1971).
(25) Coxon, JQSRT <u>11</u>, 443 (1971); <u>12</u>, 639 (1972).
(26) Potts, Price, TFS <u>67</u>, 1242 (1971).
(27) Clyne, Coxon, Woon-Fat, TFS <u>67</u>, 3155 (1971).
(28) Wieland, Tellinghuisen, Nobs, JMS <u>41</u>, 69 (1972).
(29) Coxon, JMS <u>41</u>, 548, 566 (1972).
(30) Goscinski, MP <u>24</u>, 655 (1972).
(30a) Coxon, in "Molecular Spectroscopy", Vol. <u>1</u>, p. 177. The Chemical Society (1973).
(31) Yee, Stone, MP <u>26</u>, 1169 (1973).
(32) Barrow, Clark, Coxon, Yee, JMS <u>51</u>, 428 (1974).
(33) Eng, LaTourrette, JMS <u>52</u>, 269 (1974).
(34) Le Roy, CJP <u>52</u>, 246 (1974).
(35) Ault, Howard, Andrews, JMS <u>55</u>, 217 (1975).
(36) Baierl, Hochenbleicher, Kiefer, AS <u>29</u>, 356 (1975).
(37) Baierl, Kiefer, JCP <u>62</u>, 306 (1975).
(38) Brith, Rowe, Schnepp, Stephens, CP <u>9</u>, 57 (1975).
(39) Lum, Hozack, JMS <u>58</u>, 325 (1975).
(40) Oldman, Sander, Wilson, JCP <u>63</u>, 4252 (1975).
(41) Lum, McAfee, JCP <u>63</u>, 5029 (1975).
(42) McAfee, Hozack, JCP <u>64</u>, 2491 (1976).
(43) Bondybey, Bearder, Fletcher, JCP <u>64</u>, 5243 (1976).

State	T_e	ω_e	$\omega_e x_e$	B_e	α_e	D_e ($10^{-7}cm^{-1}$)	r_e (Å)	Observed Transitions Design.	ν_{00}	References
$^{(79,81)}Br_2^+$		(μ = 39.952275)		D_0^0 = 3.26 eV[a]						JUN 1976
		Highly excited X-ray states obtained in heavy ion collisions (Br^+ + Br), discussed in (4).								
($^2\Sigma_g^+$)	(30700)	Third state observed in the photoelectron spectrum (2)(3).								
A_2 ($^2\Pi_u$) $\frac{1}{2}$	21602[b]	152.0	0.35					$A_2 \rightarrow X_2$,	18670	(1)*
A_1 ($^2\Pi_u$) $\frac{3}{2}$	19290[b]	190.0	1.0					$A_1 \rightarrow X_1$,	19197	(1)*
X_2 $^2\Pi_g$ $\frac{1}{2}$	2820[c]	376.0	1.25							
X_1 $^2\Pi_g$ $\frac{3}{2}$	0									
$^{(79,81)}Br_2^-$		(μ = 39.952549)		D_0^0 = 1.15 eV[a]		I.P. = 2.55 eV[b]				JUN 1976
($^2\Pi_g$) $\frac{1}{2}$	[77500]	363[c]								(4)
($^2\Pi_g$) $\frac{3}{2}$	[74700]									
X ($^2\Sigma_u^+$)	[0]									
$^{79}Br^{35}Cl$		μ = 24.2317306		D_0^0 = 2.23_3 eV[a]		I.P. = 11.1 eV[b]				JUN 1976 A
		Additional absorption bands above 65000 cm^{-1}; no analysis.								(1)* (8)*
		Six continuous emission bands between 27800 and 39200 cm^{-1}.								(5)
D	61570	[504][c] H						D↔X, V	61600 H	(1)* (5)
C	59325	519[c] H	2.9					C↔X, V	59362 H	(1)* (5)
B $^3\Pi_{0^+}$	16879.91[d]	222.68[d]	2.884[d]	0.107704[ef]		(1.0_1)	2.5415	B↔X,[g] R	16735.9[h]	(7)* (12) (13)(14)
A $^3\Pi_1$		Several bands of this system observed but not analysed.						A←X, R		(12)
X $^1\Sigma^+$	0	444.27_6 Z	1.843[i]	0.152469_5[f]	0.000769_7[j]	0.7183[k]	2.13606_5	Infrared sp.[ℓ]		(3)(4)(7)
								Raman sp.[m]		(9)
								Microwave sp.[n]		(2)

Br_2^+: [a]$D_0^0(Br_2)$ + I.P.(Br)[= 11.8139 eV (1a)] - I.P.(Br_2).

[b]The energies of the two $^2\Pi_u$ components fit moderately well with two partially resolved peaks at ~12.8 and 13.1 eV in the photoelectron spectrum (2)(3). The rather large difference between the ω_e values in the two components raises some doubt in the correctness of the analysis.

[c]From the photoelectron spectrum (2)(3).

(1) Haranath, Rao, IJP 29, 205 (1955).
(1a) Tech, JRNBS A 67, 505 (1963).
(2) Cornford, Frost, McDowell, Ragle, Stenhouse, JCP 54, 2651 (1971).
(3) Potts, Price, TFS 67, 1242 (1971).
(4) Soff, Müller, LNC 9, 557 (1974).

Br_2^-: [a]$D_0^0(Br_2)$ + I.P.(Br_2^-) - I.P.(Br^-).

[b]From endoergic charge transfer (2)(3)(3a)(5). Dissociative electron attachment gives the slightly higher value of 2.8_7 eV (1).

[c]From two progressions of resonances in the scattering of electrons by Br_2 indicating the existence of a preionized doubly excited state of Br_2^- with the X $^2\Pi_g$ state of Br_2^+ as "grandparent" (4).

(1) DeCorpo, Franklin, JCP 54, 1885 (1971).
(2) Chupka, Berkowitz, Gutman, JCP 55, 2724 (1971).
(3) Baede, Physica 59, 541 (1972).
(3a) Hughes, Lifshitz, Tiernan, JCP 59, 3162 (1973).
(4) Spence, PR A 10, 1045 (1974).
(5) Tang, Leffert, Rothe, Reck, JCP 62, 132 (1975).

BrCℓ: [a]From the heats of formation of BrCℓ and Br_2 (10) and the dissociation energies of Br_2 and $Cℓ_2$.

[b]Electron impact appearance potential (6).

[c]Normal isotopic mixture.

[d]Deperturbed constants (12) allowing for the perturbation

BrCℓ (continued):

by the intersecting 0^+ state arising from normal atoms (interaction matrix element ≈ 360 cm^{-1}); $\omega_e y_e = -0.0673$. The observed vibrational levels (12) are pushed down from their calculated positions by shifts ranging from 42 (for v=2) to 135 (for v=8) cm^{-1}.

[e]Deperturbed value (12); α_e not given. Experimental B_v, D_v, H_v values for $2 \le v \le 8$ in (12).

[f]RKR potential curve (12).

[g]Franck-Condon factors (12).

[h]Extrapolated from the lowest observed level (v'=2); see (12). (7) give 16695 cm^{-1} based on low-dispersion spectra.

[i]$\omega_e y_e = -0.0040_0$.

[j]$\gamma_e = -26 \times 10^{-7}$.

[k]$\beta_e = +0.0035 \times 10^{-7}$; $H_e = -1.53 \times 10^{-14}$.

[ℓ]Integrated abs. coefficients, dipole moment derivative (4).

[m]See also (11)[liquid BrCℓ] and (14)[BrCℓ in rare gas matrices].

[n]$\mu_{eℓ} \approx 0.57$ D and values of eqQ($^{35,37}Cℓ$, $^{79,81}Br$).

(1) Cordes, Sponer, ZP 79, 170 (1932).
(2) Smith, Tidwell, Williams, PR 79, 1007 (1950).
(3) Mattraw, Pachucki, Hawkins, JCP 22, 1117 (1954).
(4) Brooks, Crawford, JCP 23, 363 (1955).
(5) Haranath, Rao, IJP 31, 368 (1957).
(6) Irsa, Friedman, JINC 6, 77 (1958).
(7) Clyne, Coxon, PRS A 298, 424 (1967).
(8) Donovan, Husain, TFS 64, 2325 (1968).
(9) Holzer, Murphy, Bernstein, JCP 52, 399 (1970).
(10) JANAF (1971).
(11) Wallart, CJS 17, 128 (1972).
(12) Coxon, JMS 50, 142 (1974).
(13) Hadley, Bina, Brabson, JPC 78, 1833 (1974).
(14) Wight, Ault, Andrews, JMS 56, 239 (1975).

State	T_e	ω_e	$\omega_e x_e$	B_e	α_e	D_e $(10^{-6}cm^{-1})$	r_e (Å)	Observed Transitions: Design.	Observed Transitions: ν_{00}	References
$^{79}Br^{19}F$		$\mu = 15.3122179$		$D_0^0 = 2.54_8$ eV[a]		I.P. = 11.78 eV[b]				JUN 1976 A
		Fragments of three emission systems between 57000 and 64000 cm^{-1}.								(4)
B $^3\Pi_{0^+}$	18272.0	372.2 H	3.49^c					B↔X, R	18122.8 H	(2)(3)(8)*
A $^3\Pi_1$	(17385)	$(378)^d$ H	(16)					A←X, R	$(17235)^d$	(3)
X $^1\Sigma^+$	0	670.75 H	4.054	0.35584_3^e	0.00261_2	(0.401)	1.75894	Microwave sp.[f]		(1)
$^{(79)}Br^{19}F^+$				$D_0^0 = 2.61$ eV[g]						JUN 1976
X_2 $^2\Pi_{1/2}$	2600^h									
X_1 $^2\Pi_{3/2}$	0	750^h								
$^{79}Br^{16}O$		$\mu = 13.2994292_5$		$D_0^0 = 2.39_7$ eV[a]						JUN 1976
A $(^2\Pi_{3/2})$	27871	$485._9{}^b$ H	$5.4_0{}^c$	d				A↔X_1, R	27725^b H	(1)(2)*
X_2 $^2\Pi_{1/2}$	$(900)^e$									
X_1 $^2\Pi_{3/2}$	0	$778._7{}^b$ H	6.8_2	0.429598^f	0.003639	(0.523)	$1.717_2{}^g$	Microwave sp.[h] EPR sp.		(4)(7) (5)(8)
$^{11}B^{32}S$		$\mu = 8.1893677_4$		$D_0^0 = 6.01$ eV[a]						JUN 1976 A
G $^2\Sigma$				$[0.6148]^b$		[0.89]	$[1.829_8]$	G→F, V	19506.94 Z 19829.62 Z	(3)*
F $^2\Pi_i$	c			$[0.5782]^d$ [0.5760]		[0.81]	$[1.888_6]$			
D $^2\Delta_i$	$(48078)^e$ (47724)	$(676)^f$		[0.6032] [0.6005]		[2.5]	$[1.849_4]$	D→A, R	31830.74 Z 31810.39 Z	(3)*
C $^2\Pi_r$	39041.2 38925.8	892.6_4 H	6.74	[0.7052] $[0.6998]^g$	h		$[1.711_8]$	C→X, R γ - bands	38897.28 Z 38781.93 Z	(1)*
B $^2\Sigma^+$	$36223._4$	770 H	4.0	$[0.6311]^i$	h	[1.53]	$[1.806_0]$	B→A, V β - bands	20022.84 Z 20354.99 Z	(1)* (2)(3)*

BrF, BrF^+:

[a]From the heats of formation (7) of BrF and Br_2 and the dissociation energies of Br_2 and F_2. A higher value, 2.71 eV, was suggested (11) on the basis of an assumed 20 % failure for the linear Birge-Sponer extrapolation of the ground state.

[b]From the photoelectron spectrum (9); in reasonable agreement with an electron impact appearance potential of 11.8 eV (5).

[c]$\omega_e y_e = -0.22$.

[d]Fragmentary observations; the constants are very uncertain. See (11).

[e]Rotational constants recalculated by (6) from (1).

[f]$\mu_{e\ell} = 1.29$ D; also values for eqQ($^{79,81}Br$). Zeeman effect (10).

[g]D_0^0(BrF) + I.P.(Br) - I.P.(BrF).

[h]From the photoelectron spectrum (9).

(1) Smith, Tidwell, Williams, PR 77, 420 (1950).
(2) Durie, PRS A 207, 388 (1951).
(3) Brodersen, Sicre, ZP 141, 515 (1955).
(4) Brodersen, Mayo, ZP 143, 477 (1955).
(5) See ref. (6) of BrCℓ.
(6) Calder, Ruedenberg, JCP 49, 5399 (1968).
(7) See ref. (10) of BrCℓ.
(8) Clyne, Coxon, Townsend, JCS FT II 68, 2134 (1972).
(9) DeKock, Higginson, Lloyd, Breeze, Cruickshank, Armstrong, MP 24, 1059 (1972).
(10) Ewing, Tigelaar, Flygare, JCP 56, 1957 (1972).
(11) Coxon, CPL 33, 136 (1975).

BrO: [a]From the near-convergence of the absorption bands $A \leftarrow X_1$ assuming dissociation of A into $C\ell(^2P_{3/2}) + O(^1D_2)$; see (2).

[b]Normal isotopic mixture. The vibrational numbering in neither A nor X_1 is quite certain. In order to fit the absorption and the emission bands into the same vibrational scheme it was assumed (2) that the longest-wavelength absorption band was 1-0 and that the v" numbering of the emission bands (1) had to be raised by four units.

[c]$\omega_e y_e = -0.07_4$.

[d]All $A \leftarrow X_1$ bands are diffuse on account of predissociation, though a few bands show evidence of rotational fine structure.

[e]$A_0 = -815$ from EPR sp. (5); (8) estimate $A_0 = -980$.

[f]Effective constants.

[g]From the "true" $B_e = 0.4299$ (7).

[h]$\mu_{e\ell}(v=0) = 1.76_5$ D (7); 1.61 D from Stark effect in the gas-phase EPR sp. (3)(6). For eqQ($^{79,81}Br$) and magnetic hyperfine parameters for both isotopes see (4)(5)(7)(8).

(1) Coleman, Gaydon, DFS 2, 166 (1947).
(2) Durie, Ramsay, CJP 36, 35 (1958).
(3) Carrington, Levy, Miller, JCP 47, 3801 (1967).
(4) Powell, Johnson, JCP 50, 4596 (1969).
(5) Carrington, Dyer, Levy, JCP 52, 309 (1970).
(6) Byfleet, Carrington, Russell, MP 20, 271 (1971).
(7) Amano, Yoshinaga, Hirota, JMS 44, 594 (1972).
(8) Brown, Byfleet, Howard, Russell, MP 23, 457 (1972).

BS: [a]Thermochemical value (mass-spectrom.)(5). Different values suggested by (4).

[b]Spin-splitting constant $\gamma_0 = +0.0245$.

[c]$A_0 = -321.53$.

[d]$|\Delta v_{fe}| = 0.0197(J+\frac{1}{2})$.

[e]$A_0 = -175.05$.

[f]Estimated from the observed isotope shift for the 0-0 band.

[g]$\Delta v_{fe} \approx +0.01(J+\frac{1}{2})$.

(continued p. 113)

State	T_e	ω_e	$\omega_e x_e$	B_e	α_e	D_e ($10^{-6}cm^{-1}$)	r_e (Å)	Observed Transitions: Design.	Observed Transitions: ν_{00}	References
$^{11}B^{32}S$ (continued)										
A $^2\Pi_i$	16209.7[j] 15876.0	753.61 H	4.67	[0.6209][k] [0.6185]	0.0059[ℓ]	[1.69]	1.818_2	A→X, R α - bands	16002.2 H 15996.8 H 15668.5 H 15663.1 H	(1)* (6)
X $^2\Sigma^+$	0	1180.17 H	6.31[m]	0.7948_9[n]	0.0060_5[ℓ]	[1.40]	1.6092	ESR sp.[o].	*Ab initio* calc. (8)	
$^{(11)}B^{(80)}Se$		($\mu = 9.6762979_4$)		$D_0^0 = 4.7_5$ eV[a]						JUN 1976
$^{(11)}B^{(28)}Si$		($\mu = 7.9003924_8$)		$D_0^0 = 2.9_5$ eV[a]						JUN 1976
$^{12}C_2$		$\mu = 6.00000000$		$D_0^0 = 6.21$ eV[a]		I.P. = 12.15 eV[b]				JUL 1976 A
		Theoretical work and potential functions (16)(29)(35)(49).								
F $^1\Pi_u$	[75456.9]	[1557.5] Z		1.645	0.019	6	1.307	F← X, R	74532.9 Z	(51)*
g $^3\Delta_g$	[73183.6][c]	[1458.06] Z		1.5238	0.0170	6.6	1.3579	g← a, R	71649.6 Z	(51)*
f $^3\Sigma_g^-$	71045.8	1360.5 Z	14.8	1.448[d]	0.040[e]	10	1.393	f← a, R	70188.4 Z	(51)*
E $^1\Sigma_g^+$	55034.7	1671.5_0 Z	40.02[f]	1.7897	0.0387[g]	8.3[h]	1.2529	E→A,[i] V	46668.3 Z	(10)*
D $^1\Sigma_u^+$	43239.4_4	1829.57 Z	13.94	1.8332[j]	0.0196	7.32[j]	1.2380	D↔X,[k] Mulliken b.	43226.7_4[j] Z	(2)* (11) (46)
e $^3\Pi_g$	40796.65[ℓ]	1106.56 Z	39.260[m]	1.1922	0.0242	6.3[n]	1.5351	e→a,[o] R Fox-Herzberg b.	39806.46 Z	(7)*
C' $^1\Pi_g$		Preliminary constants from perturbations in C $^1\Pi_g$; see (40).								
C $^1\Pi_g$	34261.3	1809.1 Z	15.81[p]	1.7834	0.0180[p]	6.8	1.2552	C→A,[q] V_R Deslandres-d'Azambuja b.	25969.19 Z	(1)(3)(8) (50)
d $^3\Pi_g$	20022.50[r]	1788.22 Z	16.440[s]	1.7527[t]	0.01608[u]	6.7_4[v]	1.2661	d↔a,[w] V_R Swan b. (d← X)[x]	19378.44 Z	(6)* (25) (42)* (48)
c $^3\Sigma_u^+$	$13312._1$	$1961._6$[y]	$13._7$	1.87[y]			1.23			

BS (continued):

[h]v=1 of C $^2\Pi_r$ interacts with v=5 of B $^2\Sigma^+$ and with an unidentified state; see (3).

[i]Spin splitting constant $\gamma_0 = -0.0901$.

[j]$A_0 = -330.91$ (from B→A).

[k]$\Delta v_{fe} = +0.0176(J+\frac{1}{2})$.

[ℓ]Perturbations between higher levels of A $^2\Pi_i$ and of X $^2\Sigma^+$; see (3).

[m]$\omega_e y_e = -0.004$ (3).

[n]Spin splitting constant $\gamma = +0.013$ (7).

[o]In inert matrices (Ne, Ar) at 4 K (7).

(1) Zeeman, CJP 29, 336 (1951).
(2) Koryazhkin, Maltsev, VMUK No. 4, 92 (1968).
(3) McDonald, Innes, JMS 29, 251 (1969).
(4) Gingerich, CC (1970), p. 580.
(5) Uy, Drowart, HTS 2, 293 (1970).
(6) Singh, Tewari, Mohan, IJPAP 9, 269 (1971).
(7) Brom, Weltner, JCP 57, 3379 (1972).
(8) Ball, Thomson, CPL 36, 6 (1975).

BSe: [a]Thermochemical value (mass-spectrom.)(1).

(1) See ref. (5) of BS.

BSi: [a]Thermochemical value (mass-spectrom.)(1).

(1) Verhaegen, Stafford, Drowart, JCP 40, 1622 (1964).

C_2: [a]Average of thermochemical values (15)(55). Smaller values of 6.07 and 6.11 eV have been derived on the basis of a somewhat doubtful extrapolation of the vibrational levels in d $^3\Pi_g$ (39) and of an extrapolated limiting curve of the C $^1\Pi_g$ state (41). See also (13).

[b]From photoionization and heat of formation of C_2N_2 (38).

[c]$A_0 = -8.8 + 0.020\,J$; $A_1 = -7.4$.

[d]Spin splitting constant $\lambda \approx 0.41$.

[e]$\gamma_e = +0.006$.

C_2 (continued):

[f]$\omega_e y_e = +0.248$.

[g]$\gamma_e = -0.0005_5$. Rotational constants re-evaluated; (10) gives $B_e = 1.7930$, $\alpha_e = 0.0421$.

[h]$\beta_e = +0.6 \times 10^{-6}$.

[i]Franck-Condon factors (30); electronic trans. moment (57).

[j]From (46); slightly different constants in (2). $\beta_e \approx 3 \times 10^{-8}$.

[k]Radiative lifetime $\tau = 14._6$ ns, $f_{00} = 0.055$ (53). The corresponding electronic transition moment is in only moderate agreement with (57). Franck-Condon factors (14)(30).

[ℓ]A not determined, but much smaller than for a $^3\Pi_u$.

[m]$\omega_e y_e = +2.8050$, $\omega_e z_e = -0.1271$.

[n]$\beta_e = +0.29 \times 10^{-6}$.

[o]Franck-Condon factors (14)(20)(30); el. trans. moment (57).

[p]The ΔG and B_v curves are irregular (9)(40) and cannot be represented by the constants given without the use of higher order terms. The perturbation is strongest near v=5. Breaking-off at high J observed (41) in v=0,1,2.

[q]Franck-Condon factors (14)(30); el. trans. moment (57).

[r]$A = -16.9$.

[s]$\omega_e y_e = -0.5067$. (52) have observed bands up to v'=14.

[t]Numerous small perturbations by higher levels of b $^3\Sigma_g^-$ (19) and X $^1\Sigma_g^+$ (47) and by unidentified states.

Swan bands emitted in low-pressure oxy-acetylene flames show a distinct intensity alternation ascribed to excitation by collisions with c $^3\Sigma_u^+$ carbon molecules (24)(34).

Under certain conditions in discharges through CO the v'=6 progression of the Swan system appears almost exclusively and was at one time considered as a separate band system, the so-called high-pressure bands of carbon (5)(36) (45)*. Isotope studies (44) leave no doubt that the high-pressure bands are the v'=6 progression of the Swan system.

(continued p. 114)

State	T_e	ω_e		$\omega_e x_e$	B_e	α_e	D_e ($10^{-6} cm^{-1}$)	r_e (Å)	Observed Transitions Design.	ν_{00}	References
$^{12}C_2$ (continued)											
A $^1\Pi_u$	8391.00	1608.35	Z	12.07_8[z]	1.6163_4[a']	0.0168_6[z]	6.44[z]	1.3184_3	A↔X,[b'] R Phillips b.	8268.16 Z	(18)
b $^3\Sigma_g^-$	6434.2_7	1470.4_5	Z	11.1_9[c']	1.4985_2[d']	0.01634[e']	6.22	1.3692_8	b→a,[f'] R Ballik-Ramsay b.	5632.7 Z	(17)*
a $^3\Pi_u$	716.2_4[g']	1641.35	Z	11.67	1.6324_6[h']	0.01661	6.44	1.3119_0			
X $^1\Sigma_g^+$	0	1854.71	Z	13.34_0[i']	1.8198_4	0.0176_5[i']	6.92[i']	1.2425_3			

C_2 (continued):

[u] γ_e = - 0.001274.

[v] β_e = + 0.103 x 10^{-6}.

[w] Radiative lifetime τ = 170 ns (53), in reasonable agreement with (37) but much shorter than (22). f values obtained from the lifetime measurements as well as by other methods (28)(31)(32)(43) have been reviewed in (56). The latter authors' expression for the r dependence of the electronic transition moment was placed on an absolute scale using an average f_{00} = 0.020. For a more recent measurement of the electronic transition moment see (57). Franck-Condon factors (14)(20)(21)(27)(30).

[x] In solid matrices; tentative assignment by (23) who also report the observation of d← a. See, however, (33).

[y] From perturbations in A $^1\Pi_u$ (18).

[z] $\omega_e y_e$ = - 0.01_0; γ_e = - 0.00005_4; β_e = + 0.036 x 10^{-6}. } Very slightly revised constants in (54) based on the same data.

[a'] Λ-type doubling, $\Delta\nu_{ef}$ = - 0.00023 x J(J+1). Perturbations by c $^3\Sigma_u^+$.

[b'] f_{00} = 0.0025; see (59) for a comparison with other absolute measurements. Reasonably consistent with the electronic transition moment obtained by (57). Franck-Condon factors (14)(21)(27)(30).

[c'] $\omega_e y_e$ = + 0.028 (19).

[d'] Spin-splitting parameters λ_0 = 0.11, γ_0 = - 0.0036_5 (58). Small perturbations by levels of X $^1\Sigma_g^+$ (17).

[e'] γ_e = - 0.000087 (19).

[f'] Franck-Condon factors (30); electronic transition moment (57).

[g'] A = - 15.25; (58) gives additional spin-coupling constants.

[h'] Λ-type doubling; see (17) and (58).

[i'] $\omega_e y_e$ = - 0.172; γ_e = - 0.00023; β_e = + 0.081 x 10^{-6}. } From (18); very slightly revised constants in (54) based on the same data.

C_2 (continued):

(1) Dieke, Lochte-Holtgreven, ZP 62, 767 (1930).
(2) Landsverk, PR 56, 769 (1939).
(3) Herzberg, Sutton, CJR A 18, 74 (1940).
(4) Gerö, Schmid, PR 62, 82 (1942).
(5) Herzberg, PR 70, 762 (1946).
(6) Phillips, ApJ 108, 434 (1948).
(7) Phillips, ApJ 110, 73 (1949).
(8) Herzberg, MOLSPEC 1 (1950).
(9) Phillips, ApJ 112, 131 (1950).
(10) Freymark, AP(Leipzig) 8, 221 (1951).
(11) Norrish, Porter, Thrush, Nature 169, 582 (1952).
(12) Nicholls, PPS A 69, 741 (1956).
(13) Drowart, Burns, DeMaria, Inghram, JCP 31, 1131 (1959).
(14) Nicholls, Fraser, Jarmain, CF 3, 13 (1959).
(15) Brewer, Hicks, Krikorian, JCP 36, 182 (1962).
(16) Read, Vanderslice, JCP 36, 2366 (1962).
(17) Ballik, Ramsay, ApJ 137, 61 (1963).
(18) Ballik, Ramsay, ApJ 137, 84 (1963).
(19) Callomon, Gilby, CJP 41, 995 (1963).
(20) Jain, JQSRT 4, 427 (1964).
(21) Ortenberg, OS(Engl. Transl.) 16, 398 (1964).
(22) Jeunehomme, Schwenker, JCP 42, 2406 (1965).
(23) Barger, Broida, JCP 43, 2371 (1965).
(24) Bleekrode, Nieuwpoort, JCP 43, 3680 (1965); Nieuwpoort, Bleekrode, JCP 51, 2051 (1969).
(25) Bugrim, Lyutyi, Rossikhin, Tsikora, OS(Engl. Transl.) 19, 292 (1965).
(26) Mentall, Nicholls, PPS 86, 873 (1965).
(27) Spindler, JQSRT 5, 165 (1965).
(28) Fairbairn, JQSRT 6, 325 (1966).
(29) Fougere, Nesbet, JCP 44, 285 (1966).
(30) Halmann, Laulicht, ApJ(Suppl.) 12, 307 (1966); JCP 44, 2398 (1966).
(31) Harrington, Modica, Libby, JCP 44, 3380 (1966).
(32) Sviridov, Sobolev, Novgorodov, Arutyunova, JQSRT 6, 337, 875 (1966).
(33) Weltner, McLeod, JCP 45, 3096 (1966).
(34) Bleekrode, PRR(Suppl.) No. 7 (1967).
(35) Verhaegen, Richards, Moser, JCP 46, 160 (1967).
(36) Kunz, Harteck, Dondes, JCP 46, 4157 (1967).
(37) Fink, Welge, JCP 46, 4315 (1967).
(38) Dibeler, Liston, JCP 47, 4548 (1967).
(39) Messerle, Krauss, ZN 22 a, 1744 (1967).
(40) Messerle, Krauss, ZN 22 a, 2015 (1967).
(41) Messerle, Krauss, ZN 22 a, 2023 (1967).
(42) Tyte, Innanen, Nicholls, IAMS 5 (1967).
(43) Arnold, JQSRT 8, 1781 (1968).
(44) Dhumwad, Narasimham, CJP 46, 1254 (1968).
(45) Meinel, Messerle, ApJ 154, 381 (1968).
(46) Messerle, ZN 23 a, 470 (1968).
(47) Phillips, JMS 28, 233 (1968).
(48) Phillips, Davis, BAMS 2 (1968)
(49) Verhaegen, JCP 49, 4696 (1968).
(50) Cisak, Dabrowska, Rytel, APP 36, 497 (1969).
(51) Herzberg, Lagerqvist, Malmberg, CJP 47, 2735 (1969).
(52) Kini, Savadatti, JP B 2, 307 (1969).
(53) Smith, ApJ 156, 791 (1969).
(54) Marenin, Johnson, JQSRT 10, 305 (1970).
(55) Kordis, Gingerich, JCP 58, 5058 (1973).
(56) Danylewych, Nicholls, PRS A 339, 197, 213 (1974).
(57) Cooper, Nicholls, JQSRT 15, 139 (1975).
(58) Veseth, CJP 53, 299 (1975).
(59) Roux, Cerny, d'Incan, ApJ 204, 940 (1976).

State	T_e	ω_e	$\omega_e x_e$	B_e	α_e	D_e $(10^{-7}cm^{-1})$	r_e (Å)	Observed Transitions Design.	Observed Transitions ν_{00}	References
$^{12}C_2^+$		μ = 5.99986286		$D_0^0 = 5.3_2$ eV[a]						JUL 1976
A $^2\Sigma_g^-$	(40143)	(1340)		[1.648][b]		[100]	[1.306]	A← X,[c]	40137.8 Z	(2)*
(X) $^2\Pi_u$ [d]	0[e]	(1350)		[1.659]		[100]	[1.301]			
$^{12}C_2^-$		μ = 6.00013715		$D_0^0 = 8.4_8$ eV[a]		I.P. = 3.54 eV[b]				JUL 1976 A
a $^4\Sigma_u^+$	(19448)	(1074)[c]	(25)	(1.135)[c]	(0.0035)		(1.573)			
B $^2\Sigma_u^+$	18390.88	1968.73 Z	14.43_3[d]	1.8774_5[e]	0.0177_6[f]	68.4[g]	1.2233_0	B↔ X,[h] V	18483.98 Z	(1)*
A $^2\Pi_u$		Fragments of absorption bands.						A← X,		(7)
X $^2\Sigma_g^+$	0	1781.04 Z	11.58_5[i]	1.7468_5	0.0167	66.9	1.2682_1			
$^{40}Ca_2$		μ = 19.9812961		$D_0^0 = 0.12_9$ eV[a]						JUL 1976 A
		A band system in the region 16100 - 16900 cm^{-1} and assigned to Ca_2 by (3) is in all probability the same as that usually ascribed to CaO.[b]								
		Emission continuum from 19500 to 25000 cm^{-1}.								(1)
A $^1\Sigma_u^+$	≤18963.7	≥136.66[c] Z	0.716_6[d]	≥0.058247[c]	0.0003104[e]	0.459[f]	≤3.8058	A← X,[g] V	18999.7[h] Z	(2)(4)*
X $^1\Sigma_g^+$	0	64.92_8 Z	1.065_1[i]	0.046113	0.000702_8[j]	0.95_2[k]	4.2773			
$^{40}Ca^{79}Br$		μ = 26.5289081		$D_0^0 = 3.2_8$ eV[a]						JUL 1976 A
H	36798.7	343.4 H	1.0					H→ A, V	20840.2 H 20904.6 H	(5)
E $^2\Sigma^+$	33942.2	318.6[b] H	1.2					E→ X, V	33958.8 H	(4)
D $^2\Sigma^+$	31190.8	326.6[c] H	1.02					D↔ X, V	31211.4 H	(2)(4)

C_2^+: [a]$D_0^0(C_2)$ + I.P.(C) - I.P.(C_2).
[b]Perturbations.
[c]It is not entirely certain that the absorption spectrum obtained in flash discharges through C_2H_2/He mixtures is due to C_2^+.
[d]Not certain that this is the ground state. Theoretical calculations (1) suggest that $^2\Pi_u$ lies at 0.7 eV above the predicted $^4\Sigma_g^-$ ground state.
[e]$A_0 = - 8._0$ cm^{-1}.

(1) Verhaegen, JCP 49, 4696 (1968).
(2) Meinel, CJP 50, 158 (1972).

C_2^-: [a]From $D_0^0(C_2)$ and the electron affinities of C_2 and of C(= 1.268 eV).
[b]Photodetachment threshold (3).
[c]Constants derived (11) from the perturbations (1) in B $^2\Sigma_u^+$. Ab initio calculations by (9) and (10) independently predict the existence of a $^4\Sigma_u^+$ state slightly above B $^2\Sigma_u^+$ with r_e and ω_e values near those derived from the perturbations.
[d]$\omega_e y_e = - 0.324$.
[e]Small perturbations by a $^4\Sigma_u^+$.
[f]$\gamma_e = - 0.00037$
[g]$+ [1.0(v+\frac{1}{2}) + 0.08(v+\frac{1}{2})^2] \times 10^{-7}$.
[h]The spectrum was observed in flash discharges in CH_4 (1), behind reflected shock waves [$f_{e\ell} \approx 0.017$ (8)], and in rare gas matrices (2)(4)(6)(11)(12). Conclusive proof that it belongs to C_2^- was supplied by two-photon photodetachment spectroscopy (7). Franck-Condon factors (5).
[i]$\omega_e y_e = - 0.027$.

(1) Herzberg, Lagerqvist, CJP 46, 2363 (1968).
(2) Milligan, Jacox, JCP 51, 1952 (1969).
(3) Feldmann, ZN 25 a, 621 (1970).
(4) Frosch, JCP 54, 2660 (1971).

C_2^- (continued):

(5) Singh, Maheshwari, IJPAP 9, 296 (1971).
(6) Bondybey, Nibler, JCP 56, 4719 (1972).
(7) Lineberger, Patterson, CPL 13, 40 (1972).
(8) Cathro, Mackie, JCS FT II 69, 237 (1973).
(9) Barsuhn, JP B 7, 155 (1974).
(10) Thulstrup, Thulstrup, CPL 26, 144 (1974).
(11) Bondybey, Brus, JCP 63, 2223 (1975).
(12) Brus, Bondybey, JCP 63, 3123 (1975).

Ca_2: [a]Extrapolation of the vibrational levels in X $^1\Sigma_g^+$.
[b]The B_v values quoted by (3) are of the same order as those for CaO.
[c]Vibrational numbering unknown.
[d]$- 0.0037_4(v+\frac{1}{2})^3 - 0.000080(v+\frac{1}{2})^4$.
[e]$- [2.15(v+\frac{1}{2})^2 + 0.097(v+\frac{1}{2})^3] \times 10^{-6}$.
[f]$H_e = + 1.9 \times 10^{-14}$.
[g]RKR curves and Franck-Condon factors (4).
[h]Energy of the lowest observed vibrational level relative to X $^1\Sigma_g^+(v=0)$.
[i]$\omega_e y_e = + 0.0025_9$.
[j]$\gamma_e = - 0.0000073_5$.
[k]$+ [0.043_1(v+\frac{1}{2}) + 0.0010_7(v+\frac{1}{2})^2] \times 10^{-7}$; $H_e = - 1.4 \times 10^{-13}$, and higher order constants (4).

(1) Hamada, PM 12, 50 (1931).
(2) Weniger, Proc. I. A. U. Colloquium on Late-type Stars, Trieste (1966), edited by M. Hack; p. 25.
(3) Kovalenok, Sokolov, ISOANK No. 4, 118 (1967); IVUZF 11(3), 27 (1968).
(4) Balfour, Whitlock, CJP 53, 472 (1975).

CaBr: [a]Flame photometric value (7)(8).
[b]Normal isotopic mixture.
[c]Slightly different constants in (6).
(continued p. 119)

State	T_e	ω_e	$\omega_e x_e$	B_e	α_e	D_e ($10^{-7}cm^{-1}$)	r_e (Å)	Observed Transitions Design.	Observed Transitions ν_{00}	References
$^{40}Ca^{79}Br$ (continued)										
C $^2\Pi$	25537.5 25314.0	265.2 H	0.97					$C^d \leftrightarrow X$,[e] R	25527.4 H 25303.9 H	(2)(10)
B $^2\Sigma^+$	16380.0	284.6 H	0.92					$B^f \leftarrow X$,[g] V	16379.6 H	(2)
A $^2\Pi$	15985.8 15922.5	288.1 H	0.92					$A^h \leftrightarrow X$,[i] V	15987.2 H 15923.9 H	(1)(2)
X $^2\Sigma^+$	0	285.3 H	0.86							
$^{40}Ca^{35}Cl$		μ = 18.6496606		$D_0^0 = 4.0_9$ eV[a]						JUL 1976 A
G ($^2\Delta$)	36712 36708	434.0 H^Q	1.1					G → A, V	20580 H^Q 20645 H^Q	(6)
F ($^2\Pi$)	35700 35676	432.5 H	0.8[b]					F → B, V	18884 H 18859 H	(6)
								F → A, V	19567 H 19612 H	(6)
E ($^2\Sigma$)	34266.4	413.3	1.68					E → B, V	17439.8 H	(11)
								E ← X, V	34288.1	(5)
D ($^2\Sigma$)	31107.8	423.4	1.61					D ← X, V	31134.5	(5)
C $^2\Pi_r$	26574.6 26498.9	[333.86] Z 336.0 H	1.4 H	0.14305 [0.14216][c]	0.000747	1.02	2.516_0	$C^d \leftrightarrow X$, R	26557.80 Z 26481.82 Z	(1)(3)(7)
B $^2\Sigma^+$	16849.4	366.7[e] H	1.4_3	f				$B^g \leftrightarrow X$,[h] V	16847.6 H	(1)(2)(3)(14)*
A $^2\Pi$	16163.2 16093.7	372.3 H	1.2[i]					$A^j \leftrightarrow X$, V	16164.3 H^Q 16094.8 H^Q	(1)(2)(3)
X $^2\Sigma^+$	0	[367.53] Z	1.31 H	0.15195	0.00078_3	1.0	2.4390			
$^{40}Ca^{19}F$		μ = 12.8767412		$D_0^0 = 5.4_8$ eV[a]						JUL 1976
F $^2\Pi$	37547.8	681.7 H	3.55					F ↔ X, V	37595.1 H	(4)(8)
E $^2\Sigma^+$	34134.6	646.3 H	3.24					E ↔ X, V	34164.4 H	(4)* (8)
D $^2\Sigma^+$	30771.9	650.7 H	2.89					D ← X, V	30803.9 H	(4)*
C $^2\Pi$	30284.4 30255.1	481.7 H	2.02					C ↔ X, R	30232.1 H 30202.8 H	(1)(4)*

State	T_e	ω_e	$\omega_e x_e$	B_e	α_e	D_e ($10^{-7}cm^{-1}$)	r_e (Å)	Observed Transitions Design.	ν_{00}	References
$^{40}Ca^{19}F$ (continued)										
B $^2\Sigma^+$	18844.5	566.1^b	2.80	$[(0.336_1)]^c$				$B^d \leftrightarrow X$,[e] R	18834.2 (Z)	(1)(2)*
A $^2\Pi_r$	16562.3^f 16489.8	593.4 H^Q [586.8] Z	3.113^g 3.427^h H^Q	0.343_6[ij]	0.0028	(4.6)	1.952	$A^k \leftrightarrow X$, V_R	16565.6 Z 16493.1 Z	(1)(2)* (9) (12)
X $^2\Sigma^+$	0	[581.1] Z	2.74 H	0.338_5	0.0026	(4.5)	1.967	ESR sp.[ℓ]		

CaBr (continued):

[d]Radiative lifetime $\tau(v=0,1) = 32._5$ ns (9).
[e]Four heads.
[f]Radiative lifetime $\tau(v=0) = 42._9$ ns (9).
[g]Double heads on account of large spin doubling in the upper state. Constants refer to the short-wavelength head.
[h]Radiative lifetime $\tau(v=1) = 34._0$ ns (9).
[i]Four heads. For a reproduction of the spectrum see Fig. 1e of (3) who wrongly attributed the spectrum to MnBr.

(1) Hedfeld, ZP 68, 610 (1931).
(2) Harrington, Dissertation (U. of California, 1942).
(3) Hayes, Nevin, PPS A 68, 665 (1955).
(4) Reddy, Rao, IJPAP 6, 181 (1968).
(5) Reddy, Reddy, Rao, CS 39, 485 (1970).
(6) Shah, IJPAP 8, 118 (1970).
(7) Gurvich, Ryabova, Khitrov, FSCS No. 8, 83 (1973).
(8) Khitrov, Ryabova, Gurvich, HT(USSR) 11, 1005 (1973).
(9) Dagdigian, Cruse, Zare, JCP 60, 2330 (1974).
(10) Joshi, Gopal, Pramāṇa 4, 276 (1975).

CaCℓ: [a]Thermochemical value (mass-spectrom.)(8)(9); in good agreement with flame photometric results (10)(12).
[b]$\omega_e y_e = +0.06$.

CaCℓ (continued):

[c]Λ-type doubling (7).
[d]Radiative lifetime $\tau(^2\Pi_{\frac{3}{2}}, v=0) = 25._0$ ns (13).
[e]Recalculated from the heads of the 0-0 sequence of B − X, using for the ground state the constants given in the Table. In good agreement with constants derived from E − B.
[f]The predissociation above v=15 reported by (4) is not compatible with a dissociation energy of more than 4 eV.
[g]Radiative lifetime $\tau(v=0) = 38._2$ ns (13).
[h]Double heads on account of ϱ-type doubling in the upper state. The constants refer to the short-wavelength head.
[i]$\omega_e y_e = -0.05$. Average of the constants from F − A and G − A. Considerably different constants are obtained from A − X owing to large head-origin separations.
[j]Radiative lifetime $\tau(v=2) = 28._9$ ns (13).

(1) Walters, Barratt, PRS A 118, 120 (1928).
(2) Asundi, PIAS A 1, 830 (1935).
(3) Parker, PR 47, 349 (1935).
(4) Hellwege, ZP 100, 644 (1936).
(5) See ref. (2) of CaBr.

(continued p. 121)

CaF: See p. 121.

State	T_e	ω_e	$\omega_e x_e$	B_e	α_e	D_e $(10^{-4}cm^{-1})$	r_e (Å)	Observed Transitions Design.	ν_{00}	References
$^{40}Ca^1H$		μ = 0.98303388		$D_0^0 \le 1.70$ eV[a]		I.P. = 5.86 eV[b]				JUL 1976 A
I $(^2\Sigma)$				[(4.60)]			[(1.93)]	I ← X,	(39477)	(14)
H $(^2\Sigma)$				[(4.41)]			[(1.97)]	H ← X,	(38798)	(14)
F $^2\Sigma^+$	36705	1487[c]	28	[4.6867]		[2.017]	[1.9128]	F ← X, V	36797.05 Z	(11)(12)* (14)
M $^2\Delta$ } (4d)				[4.89] [d]			[1.87]	M ← X,	35481.8 [d]	(22)*
J $^2\Pi$ } (4d)	[e]			$[4.89_8]$[d]		[2.8] [d]	$[1.87_1]$	J ← X,	35068.5_3[d]	(22)*
G $^2\Sigma^+$ } (4d)	(34735)	[1458][f]		[4.765][d]		$[1.1_7]$[d]	[1.897]	G ← X,	34819.5 [d]	(22)*
L $^2\Pi$ } (5p)	(32680)[g]	[1407.6][h]		$[4.620_4]$[h]	0.113_9	$[1.92_0]$[hi]	1.914_8	L ← X,	32739.3_7[h]	(21)*
K $^2\Sigma^+$ } (5p)	(32640)	[(1391)] [h]		[4.601] [hj]	(0.085)	[1.93] [hk]	1.922	K ← X,	32691.1 [h]	(21)*
C $^2\Sigma^+$	28276	1445 H	25	[4.58][ℓm]		[1.7][ℓ]	[1.94]	C ↔ X, V	$28348._0$[ℓ]	(1)* (4)* (7)
D $^2\Sigma^+$	22602	1150 Z	$33._0$	2.50[n]	0.01		2.62	D → X, R	22524 Z	(4)(6)
E $^2\Pi$	20418[o]	$1248._6$ H[Q]	$21._8$	[4.284]		[2.2]	[2.001]	E → X,[p] V	20392 Z	(9)

Ca^1H: [a]From the predissociation in C $^2\Sigma^+$, assuming dissociation into Ca(^3P) + H(^2S).

[b]From the observation of Rydberg states in the absorption spectrum (22).

[c]From isotope relations (12).

[d]Deperturbed constants for the three interacting states which form the d complex. The $^2\Delta_{5/2}$ component was not observed. Most of the lines are somewhat diffuse. The $^2\Pi - {}^2\Sigma$ band whose low J lines are sharp had been misinterpreted by (20) in terms of two overlapping $^2\Sigma - {}^2\Sigma$ transitions. Small local perturbations.

[e]$A_0 = +\ 5.0$.

[f]From the proposed assignment of the 1-0 band by (14). (22) observe a diffuse single branch only.

[g]$A_0 = +\ 18.8$, $A_1 = +\ 19.3$.

[h]Deperturbed constants for the two interacting states which form the p complex. Small local perturbations.

[i]$D_1 = 2.06_4 \times 10^{-4}$; $H_0 = +\ 4.7 \times 10^{-9}$.

[j]F_1 levels not observed, presumably on account of predissociation. In v=1 both spin components seem to be predissociated.

[k]$H_0 = +\ 2.5 \times 10^{-9}$.

Ca^1H (continued):

[ℓ]Constants derived from high N values. At low N, v=0 is perturbed by a state of smaller B value. The observed origin of the 0-0 band is at 28352.5 cm^{-1}.

[m]At low pressure predissociation above v=0, N=10.

[n]Strong perturbations for higher N values.

[o]A ≈ 9.3.

[p]Two Q heads.

(continued p. 123)

CaCℓ (continued):

(6) Schütte, ZN 9 a, 891 (1954).
(7) Morgan, Barrow, Nature 185, 754 (1960).
(8) Zmbov, CPL 4, 191 (1969).
(9) Hildenbrand, JCP 52, 5751 (1970).
(10) See ref. (7) of CaBr.
(11) Khanna, Dubey, IJPAP 11, 510 (1973); 13, 603 (1975).
(12) Ryabova, Khitrov, Gurvich, HT(USSR) 10, 669 (1973).
(13) See ref. (9) of CaBr.
(14) Darji, Shah, IJPAP 13, 187 (1975).

CaF: [a]Thermochemical value (mass-spectrom.)(5)(7); 5.8_5 eV by flame photometry (6).

[b]Recalculated from the heads of the 0-0 sequence using for the ground state the constants given in the table.

[c]Using data from (2) and (12).

[d]Radiative lifetime $\tau(v=0) = 25._1$ ns (11).

[e]Double heads on account of large spin-doubling in the upper state and high N values of the heads.

[f]$A_0 = +73.4$.

[g]$\omega_e y_e = +0.0051$.

CaF (continued):

[h]$\omega_e y_e = +0.0619$.

[i]Λ-type doubling in $^2\Pi_{\frac{1}{2}}$, $\Delta v_{fe} = -0.045(J+\frac{1}{2})$.

[j]The possibility of predissociation above v=16, as suggested by (3), is now ruled out by the new value for D_0^0.

[k]Radiative lifetime $\tau(v=0) = 20._2$ ns (11).

[ℓ]In rare gas matrices at 4 K (10).

(1) Johnson, PRS A 122, 161 (1926).
(2) Harvey, PRS A 133, 336 (1931).
(3) See ref. (4) of CaCℓ.
(4) Fowler, PR 59, 645 (1941).
(5) Blue, Green, Bautista, Margrave, JPC 67, 877 (1963).
(6) Ryabova, Gurvich, HT(USSR) 2, 749 (1964).
(7) Hildenbrand, Murad, JCP 44, 1524 (1966).
(8) Prasad, Narayan, IJP 43, 205 (1969).
(9) Subbaram, Rao, IJP 43, 312 (1969).
(10) Knight, Easley, Weltner, Wilson, JCP 54, 322 (1971).
(11) See ref. (9) of CaBr.
(12) Field, Harris, Tanaka, JMS 57, 107 (1975).

State	T_e	ω_e	$\omega_e x_e$	B_e	α_e	D_e $(10^{-4}cm^{-1})$	r_e (Å)	Observed Transitions Design.	Observed Transitions ν_{00}	References
$^{40}Ca^{1}H$ (continued)										
B $^2\Sigma^+$ (4p)	15762	1285	20	$[4.3410]^q$	0.116	$[2.020]^{qr}$	1.974_4	B↔X, V	$15754.9_6{}^q$	(2)(3)(8)* (15)(23)(24)
A $^2\Pi_r$ (4p)	14413^s	1333	20	$[4.3476_9]^q$	0.106	$[1.883]^{qt}$	1.974_0	A↔X, V	14430.39^q	(2)(3)(5) (15)(23)(24)
X $^2\Sigma^+$	0	1298.34 Z	19.10	4.2766^u	0.0970	1.837^v	2.0025	ESR sp.[w]		
$^{40}Ca^{2}H$		μ = 1.91746270		$D_0^0 \leq 1.72$ eV[a]						JUL 1976 A
F $^2\Sigma^+$	(36698)	[1036.1] Z	14^b	2.434	0.051	$[0.49]^c$	1.901	F←X, V	36771.40 Z	(3)(4)(5)
M $^2\Delta$ (4d)				d						
J $^2\Pi$ (4d)				d				J←X,	35034.16^e	(8)
G $^2\Sigma^+$ (4d)				d				G←X,	(34750)	(8)(4)
L $^2\Pi$ (5p)	$(32647)^f$	$[1020.5_7]^g$	$(15._9)$	$[2.3830]^g$	0.0412	$[0.5018]^{gh}$	1.9125	L←X,	32713.61^g	(8)
K $^2\Sigma^+$ (5p)	(32615)	$[(1012)]^g$		$[2.375_1]^{gi}$	(0.019)	$[0.503]^g$	1.920	K←X,	32665.5^g	(8)
C $^2\Sigma^+$				$[(2.35)]^j$		[(0.44)]	[(1.93)]	C→X, V	28322 Z	(1)* (2)*
B $^2\Sigma^+$				2.282^k	0.045	$[0.54]^\ell$	1.963	B→X, V	15748.8 Z	(2)
X $^2\Sigma^+$	0	$[910._4]$ Z		$[2.1769]^m$	0.035	$[0.479]^n$	2.001_6			
CaH^+	The spectrum of CaH^+ reported in the previous edition (MOLSPEC 1) is in fact due to MgH.									

Ca^1H (continued):

[q]Deperturbed constants for the strongly interacting states A $^2\Pi$ and B $^2\Sigma^+$; see (23). For an extensive discussion of the rotational structure of the two states (including spin and Λ-type doubling) see also (17)(18)(19).

[r]$H_0 = +4.3_1 \times 10^{-9}$.

[s]$A_0 = +79.01$; slight J dependence (23). See also (17).

[t]$H_0 = +5.08 \times 10^{-9}$.

[u]Spin-splitting constant $\gamma_v = +0.042_9 - 0.0010(v+\frac{1}{2})$.

[v]$H_e = +5.51 \times 10^{-9}$.

[w]In Ar matrix at 4 K (16).

(1) Mulliken, PR 25, 509 (1925).
(2) Hulthén, PR 29, 97 (1927).
(3) Watson, Bender, PR 35, 1513 (1930).
(4) Grundström, ZP 69, 235 (1931).
(5) Mulliken, Christy, PR 38, 87 (1931).
(6) Grundström, ZP 75, 302 (1932).
(7) Grundström, ZP 95, 574 (1935).
(8) Watson, PR 47, 27 (1935).
(9) Watson, Weber, PR 48, 732 (1935).
(10) Grundström, Dissertation (Stockholm, 1936).
(11) Khan, PPS 80, 593 (1962).
(12) Edvinsson, Kopp, Lindgren, Åslund, AF 25, 95 (1963).
(13) Khan, PPS 87, 569 (1966).
(14) Khan, Afridi, JP B 1, 260 (1968).
(15) Liberale, Weniger, Physica 41, 47 (1969).
(16) Knight, Weltner, JCP 54, 3875 (1971).
(17) Veseth, JMS 38, 228 (1971).
(18) Veseth, MP 20, 1057 (1971).
(19) Veseth, MP 21, 287 (1971).
(20) Khan, Hasnain, NC B 18, 384 (1973).
(21) Kaving, Lindgren, Ramsay, PS 10, 73 (1974).

Ca^1H (continued):

(22) Kaving, Lindgren, PS 10, 81 (1974).
(23) Berg, Klynning, PS 10, 331 (1974).
(24) Berg, Klynning, AA(Suppl.) 13, 325 (1974).

Ca^2H: [a]See [a] of Ca^1H.

[b]From isotope relations.

[c]$D_1 = 0.30 \times 10^{-4}$.

[d]Term values, but no constants, determined (8) for v=0 of J $^2\Pi$ and G $^2\Sigma$. M $^2\Delta$ not observed with certainty. G $^2\Sigma$(v=0) is perturbed by L $^2\Pi$(v=2). Smaller local perturbations in J $^2\Pi$.

[e]$Q_{11}(\frac{1}{2})$ line.

[f]$A_0 = +19.4_8$, $A_1 = +19.3$.

[g]See [h] of Ca^1H.

[h]$D_1 = 0.495 \times 10^{-4}$; $H_1 = -1.2_5 \times 10^{-9}$.

[i]See [j] of Ca^1H.

[j]Strong perturbations. Breaking off above v'=0, N'=18.

[k]Effective value. Spin-splitting constant $\gamma \approx -0.364$, but see (6).

[ℓ]$D_1 = 0.65 \times 10^{-4}$.

[m]Spin-splitting constant $\gamma_0 = +0.021_6$.

[n]$D_1 = 0.61 \times 10^{-4}$.

(1) Grundström, ZP 97, 171 (1935).
(2) See ref. (8) of Ca^1H.
(3) See ref. (12) of Ca^1H.
(4) See ref. (13) of Ca^1H.
(5) See ref. (14) of Ca^1H.
(6) See ref. (18) of Ca^1H.
(7) See ref. (20) of Ca^1H.
(8) Kaving, Lindgren, PS 13, 39 (1976).

State	T_e	ω_e	$\omega_e x_e$	B_e	α_e	D_e (10^{-7}cm^{-1})	r_e (Å)	Observed Transitions: Design.	ν_{00}	References
$^{40}Ca^{127}I$		μ = 30.3920472		$D_0^0 \geq 2.77$ eV[a]						JUL 1976
E ($^2\Delta$)	36715.7 36713.3	287.2 H^Q	0.80					E→A, V	21092.8 H^Q 21149.3 H^Q	(12)
D ($^2\Sigma$)	31011.4	256.0 H	0.8					D↔X, V	31019.9 H	(1)(8)
C $^2\Pi$	23743.96 23315.51	224.72 H 229.75 H	0.530[b] 0.633					C↔X, R	23736.98 H 23311.03 H	(1)(3)(5)* (6)*
B $^2\Sigma^+$	15715.2	239.9_5 H	0.62					B[c]↔X, V	15715.8[d] 15712.4[d] H	(1)(2)(6)* (10)
A $^2\Pi$	15645.6 15586.2	241.6_9 H 242.6_5 H	0.83 0.62					A[e]↔X, V	15647.2 H 15588.4 H	(1)(2)(6)* (10)
X $^2\Sigma^+$	0	238.70 H	0.628							
$^{40}Ca^{16}O$		μ = 11.4229224_5		$D_0^0 \geq 4.7_6 \pm 0.15$ eV[a]						SEP 1976 A
		Progression of absorption bands ($\Delta G \approx 850$) in Kr and Xe matrices, 20000 - 26000 cm^{-1}; observed for both $Ca^{16}O$ and $Ca^{18}O$, but not definitely assigned.								(15)
		Green system of strong emission bands, 17900 - 18300 cm^{-1}; only one band (18276 cm^{-1}) analyzed (18).						(B→A')		(1)(5)(8)* (18)
		Orange system of strong emission bands, 15700 - 16700 cm^{-1}; no analysis.								(1)(5)(7)* (8)*
C $^1\Sigma^+$	28857.8	560.9 Z	$4._0$	0.3731[b]	0.0032	(7)	1.989	C↔X,[c] R	28772.4 Z	(4)* (11)
B $^1\Pi$	25991	[574.4] Z		0.388_2[db]	0.005_5	(7)	1.950	B↔X,[e] R	25913.0 Z	(4)(11)
A $^1\Sigma^+$	11554.8	718.9 Z	2.11	0.4059_2[f]	0.00137	$5._4$	1.9067	A→X, R	11548.8_4 Z	(3)* (6)* (12)(17)
A' $^1\Pi$	8433	545.7 H	2.54	0.337	0.002_1		2.09_3	A'→X,[g] R	8340	(20)*
a $^3\Pi_i$	(8313)[h]	556	$3._3$	0.335	0.001_5		2.09_9			
X $^1\Sigma^+$	0	732.1_1 Z	4.8_1[i]	0.4445_2	0.00338	6.58[j]	1.8221	k		

CaI: [a]Lower bound from a crossed molecular beam experiment (7); in agreement with the theoretical lower limit of (4).

[b]$\omega_e y_e = -0.083$.

[c]Radiative lifetime $\tau(v=4) = 50._9$ ns (11).

[d]Double heads on account of large spin doubling in $B\,^2\Sigma$.

[e]Radiative lifetime $\tau(v=3,5) = 41._7$ ns (11).

(1) Walters, Barratt, PRS A 118, 120 (1928).
(2) Hedfeld, ZP 68, 610 (1931).
(3) Mesnage, AP(Paris) 12, 5 (1939).
(4) Krasnov, Karaseva, OS(Engl. Transl.) 19, 14 (1965).
(5) Murty, Reddy, Rao, JP B 3, 425 (1970).
(6) Maheshwari, Shukla, Singh, IJPAP 9, 327 (1971).
(7) Mims, Lin, Herm, JCP 57, 3099 (1972).
(8) Khanna, Dubey, IJPAP 11, 286 (1973).
(9) Khanna, Dubey, IJPAP 11, 375 (1973).
(10) Darji, Vaidya, IJPAP 11, 923 (1973).
(11) Dagdigian, Cruse, Zare, JCP 60, 2330 (1974).
(12) Kamalasanan, Shah, CS 44, 805 (1975).

CaO: [a]From the $Ca + C\ell O_2$ chemiluminescence spectrum (21). By contrast, the most recent mass-spectrometric (9)(10) and flame-photometric (16) determinations - corrected, where necessary, for a $^1\Sigma$ ground state - lead to D_0^0 = 4.04 eV.

[b]Perturbations.

[c]ζ system of (2).

[d]Value of B^Q; $B^{PR} - B^Q = +0.0005$.

[e]ε system of (2).

[f]Many rotational perturbations by $A'\,^1\Pi$ and $a\,^3\Pi_i$ (3)(17).

[g]Directly observed for $v' > 9$ only; vibrational numbering determined from the perturbations in $A\,^1\Sigma^+$ (17).

[h]$A = -58$, from perturbations in $A\,^1\Sigma^+$ as are the other constants for this state; see (17).

[i]The vibrational constants in the Table are derived from levels with $v \leq 4$ (3)(4). From band heads in the A−X system ($v'' \leq 13$), and accounting for head-origin separations, (12) have derived $\omega_e = 733.4$, $\omega_e x_e = 5.28$, $\omega_e y_e = +0.044$.

[j]$\beta_e = +0.03 \times 10^{-7}$.

[k]The question whether this is the ground state was for long in doubt but the observation by (19) of a fundamental frequency of 707 cm^{-1} in a nitrogen matrix seems to settle it; see also (17) and the theoretical work of (14). For computed $^1\Sigma$ ground state properties see (13).

(1) Lejeune, BSRSL 14, 318 (1945).
(2) Lejeune, Rosen, BSRSL 14, 322 (1945).
(3) Hultin, Lagerqvist, AF 2, 471 (1951).
(4) Lagerqvist, AF 8, 83 (1954).
(5) Gaydon, PRS A 231, 437 (1955).
(6) Gatterer, Junkes, Salpeter, Rosen, METOX (1957).
(7) Rosen, Weniger, CR 248, 1645 (1959).
(8) Pearse, Gaydon, IDSPEC (1963).
(9) Colin, Goldfinger, Jeunehomme, TFS 60, 306 (1964).
(10) Drowart, Exsteen, Verhaegen, TFS 60, 1920 (1964).
(11) Veits, Gurvich, DC 173, 377 (1967).
(12) Brewer, Hauge, JMS 25, 330 (1968).
(13) Yoshimine, JPSJ 25, 1100 (1968).
(14) Carlson, Kaiser, Moser, Wahl, JCP 52, 4678 (1970).
(15) Brewer, Wang, JCP 56, 4305 (1972).
(16) Kalff, Alkemade, JCP 59, 2572 (1973).
(17) Field, JCP 60, 2400 (1974).
(18) Volnyets, Kovalenok, Sokolov, OS(Engl. Transl.) 36, 609 (1974).
(19) Ault, Andrews, JCP 62, 2320 (1975).
(20) Field, Capelle, Jones, JMS 54, 156 (1975).
(21) Engelke, Sander, Zare, JCP 65, 1146 (1976).

State	T_e	ω_e	$\omega_e x_e$	B_e	α_e	D_e $(10^{-6}cm^{-1})$	r_e (Å)	Observed Transitions Design.	ν_{00}	References
$^{40}Ca^{32}S$		μ = 17.7617690		$D_0^0 = 3.4_6$ eV[a]						JUL 1976
		Continuous absorption from 41860 cm^{-1} to higher wave numbers, observed and ascribed to CaS by (1), but questioned by (4).								
A $^1\Sigma^+$	15220.79	409.04 Z	0.818	0.16666	0.00060_5	0.109	2.3864	A← X, R	15194.44 Z	(4)
X $^1\Sigma^+$	0	462.23 Z	1.78	0.17667	0.00083_7	0.102	2.3178			
$^{12}C^{(79)}Br$		(μ = 10.41616100)		$D_0^0 \le 4.11$ eV[a]						JUL 1976
A_2 $^2\Delta_{5/2}$	b	c		[0.4956]		[1.2] }	[1.815][e]	$A_2 \leftarrow X_2$, V	32753.10 Z	(1)(2)*
A_1 $^2\Delta_{3/2}$		c		d				$(A_2 \leftarrow X_1)$	(33218.4) H	
								$(A_1 \leftarrow X_2)$	(32699.1) H	
								$A_1 \leftarrow X_1$, f	33163.5	
X_2 $^2\Pi_{3/2}$	g	c		[0.4877]		[0.6] }	[1.823][h]			
X_1 $^2\Pi_{1/2}$	0	c								
$^{12}C^{35}Cl$		μ = 8.93413850		D_0^0 a						JUL 1976
A $^2\Delta_r$	b	$[848._1]$ H		$[0.7062_0]$		[1.84]	[1.6346]	A↔X, V	35870.28[c] Z 36003.92[c] Z	(3)(4)(5)* (7)
X $^2\Pi_{3/2}$	d	[865.48] Z		0.7009_9	0.00678	[1.89][e]	1.6450	Theoretical work on low-lying valence states; see (9).		
X $^2\Pi_{1/2}$	0	[866.72] Z	6.2 H	0.6936[f]	0.00672	[1.9][g]				
$^{12}C^{(35)}Cl^+$										JUL 1976
A $(^1\Pi)$	42350	922.5 H	21.5					A→X, R	42220 H	(1)(2)
X $(^1\Sigma)$	0	1175.0 H	5.0							

CaS: [a]Thermochemical value (mass-spectrom.)(3), corrected for a $^1\Sigma$ ground state. Good agreement with (2).

(1) Mathur, PRS A <u>162</u>, 83 (1937).

(2) Marquart, Berkowitz, JCP <u>39</u>, 283 (1963).

(3) See ref. (9) of CaO.

(4) Blues, Barrow, TFS <u>65</u>, 646 (1969).

CBr: [a]From the predissociation in $A_1\ ^2\Delta_{3/2}$(v=0). Study of flame reactions (3) suggests 2.8_6 eV.

[b]$A_0 = + 28.$

[c]$\Delta G''(\frac{1}{2}) - \Delta G'(\frac{1}{2}) = + 93.7$ (F_1) and + 82.8 (F_2).

[d]Both v=0 and v=1 of A_1 are predissociated.

[e]From the "true" $B_0 = 0.4912$.

[f]Undegraded diffuse band.

[g]$A_0 = + 466.$

[h]From the "true" $B_0 = 0.4872$.

(1) Simons, Yarwood, TFS <u>57</u>, 2167 (1961).

(2) Dixon, Kroto, TFS <u>59</u>, 1484 (1963).

(3) Miller, Palmer, JCP <u>40</u>, 3701 (1964).

CCℓ, $CC\ell^+$:

[a]Study of flame reactions suggests $D_0^0 = 3.3_4$ eV (6).

[b]$A_0 = + 7.04$; see (8).

[c]J'=3/2 (average of F_1 and $\{F_2\}$) relative to {J"=0}.

[d]$A_0 = + 134.92.$

[e]$D_1 = 2.30 \times 10^{-6}.$

[f]Λ-type doubling, $\Delta v = 0.0137(J+\frac{1}{2})$.

[g]$D_1 = 2.0 \times 10^{-6}.$

(1) Barrow, Drummond, Walker, PPS A <u>67</u>, 186 (1954).

(2) Kuzyakov, Tatevskii, NDVSK No. 2, 237 (1959).

(3) Gordon, King, CJP <u>39</u>, 252 (1961).

(4) See ref. (1) of CBr.

(5) Verma, Mulliken, JMS <u>6</u>, 419 (1961).

(6) See ref. (3) of CBr.

(7) Ovcharenko, Kuzyakov, Tatevskii, OS(Engl. Transl.) <u>19</u>, 294 (1965).

(8) Merer, Travis, Watson, CJP <u>44</u>, 447 (1966).

(9) Bialski, Grein, JMS <u>61</u>, 321 (1976).

State	T_e	ω_e	$\omega_e x_e$	B_e	α_e	D_e $(10^{-}\ cm^{-1})$	r_e (Å)	Observed Transitions Design.	Observed Transitions ν_{00}	References
$^{(114,112)}Cd_2$		$(\mu = 56.447101_2)$ $D_0^0 = 0.08_0$ eV[a]								NOV 1974 A
		Large number of continua and diffuse bands. See (2), also (3)(4).								
$^{(114)}Cd^{(79)}Br$		$(\mu = 46.618527_8)$ $0.9\ eV \leq D_0^0 \leq 1.6$ eV[a]								NOV 1974
C $(^2\Pi_{3/2})$	31463	253.8[b] H	0.73					C↔X,	V 31474 H	(1)(5)(8)*
C $(^2\Pi_{1/2})$	(30300)	c							30310 H	(5)(7)(8)*
B $(^2\Sigma)$		Unclassified emission bands, 12300 - 30300 cm^{-1}.[d]						B→X,	R	(1)(3)(4)(5)(6)
X $(^2\Sigma)$	0	230.5[b] H	0.50							
$^{(114)}Cd^{35}Cl$		$(\mu = 26.7549590)$ $D_0^0 = 2.1_2$ eV[a]								NOV 1974
E $(^2\Sigma)$	45398.4	$264._0$ H	7.5[b]					E→X,	R $45363._3$ H	(5)* (7)*
C $(^2\Pi_r)$	{32502, 31485}	399.0 H^Q	1.5 0-0 sequence only.[c]					C← X,	V {32536 H^Q, 31519 H^Q}	(1)(4)(6)(11)
B $(^2\Sigma)$		Unclassified emission bands, 11500 - 30000 cm^{-1}.[d]						B→X,	R	(2)(8)
X $(^2\Sigma)$	0	$331._0$ H	1.0							
$^{(114)}Cd^{133}Cs$?		$(\mu = 61.336453_4)$								NOV 1974
		Diffuse V shaded absorption bands at 18810 and 19120 cm^{-1}.								(1)
$^{(114)}Cd^{19}F$		$(\mu = 16.282568_9)$ $D_0^0 = (3.2)$ eV[a]								NOV 1974
		"Bands" found in emission in the same region (1)(3) have been shown by (4) to be peculiar Cd lines.								
		Narrow continuum at 35400 cm^{-1}, and unclassified bands in absorption at 35855, 35877, and 35897 cm^{-1}.								(2)
E $(^2\Sigma)$	(34200)	(535) H						E←X,[b]	R	(2)
X $(^2\Sigma)$	0	(535) H								

Cd_2: [a]From temperature dependence of diffuse molecular absorption; average of the values obtained by (1) and (5).

(1) Kuhn, Arrhenius, ZP 82, 716 (1933).
(2) Finkelnburg, "Kontinuierliche Spektren", Springer, Berlin (1938).
(3) Garton, PPS A 64, 430 (1951).
(4) Freedhoff, PPS 92, 505 (1967).
(5) Bruner, Corbett, unpubl., quoted in DISSEN.

CdBr: [a]Na D line chemiluminescence in Na/$CdBr_2$ flames [(2), revised].
[b]Average of constants given in (1) (em.) and (8) (abs.).
[c]No agreement between constants derived from the analysis of the emission spectrum by (7) (ω_e = 237.0, $\omega_e x_e$ = 0.50) and of the absorption spectrum by (8) (ω_e = 253.0, $\omega_e x_e$ = 2.30).
[d]Constants suggested by (6) are T_e = 24823, ω_e = 105.4, $\omega_e x_e$ = 1.70.

(1) Wieland, HPA 2, 46, 77 (1929).
(2) Horn, Polanyi, Sattler, ZPC B 17, 220 (1932).
(3) Oeser, ZP 95, 699 (1935).
(4) Wieland, in "Contribution à l'Etude de la Structure Moléculaire", Vol. comm. Victor Henri, Ed. Desoer, Liège (1948), p. 229.
(5) Ramasastry, IJP 23, 453 (1949).
(6) Patel, Patel, IJP 41, 155 (1967).
(7) Darji, IJPAP 8, 240 (1970).
(8) Gosavi, Greig, Young, Strausz, JCP 54, 983 (1971).

CdCℓ: [a]From temperature dependence of absorption spectrum (10). Na D line chemiluminescence in Na/$CdC\ell_2$ flames gives 1.3 eV $\leq D_0^0 \leq$ 2.1 eV [(3), revised].
[b]Predissociation suggested for v > 1.
[c]Revised analysis (11).

CdCℓ (continued):
[d]Constants suggested by (9) are T_e = 26010, ω_e = 153.5, $\omega_e x_e$ = 3.75.

(1) Walter, Barratt, PRS A 122, 201 (1929).
(2) See ref. (1) of CdBr.
(3) See ref. (2) of CdBr.
(4) See ref. (3) of CdBr.
(5) Cornell, PR 54, 341 (1938).
(6) Howell, PRS A 182, 95 (1943).
(7) Ramasastry, IJP 21, 267 (1947)
(8) See ref. (4) of CdBr.
(9) Patel, Patel, IJPAP 4, 388 (1966).
(10) See ref. (5) of Cd_2.
(11) Wieland, unpubl., quoted in DONNSPEC.

CdCs: (1) Barratt, TFS 25, 758 (1929).

CdF: [a]Estimated thermochemical value (5).
[b]Rather diffuse heads.

(1) Pearse, Gaydon, PPS 50, 711 (1938).
(2) Fowler, PR 62, 141 (1942).
(3) See ref. (6) of CdCℓ.
(4) Pearse, Feast, Nature 163, 686 (1949).
(5) Besenbruch, Kana'an, Margrave, JPC 69, 3174 (1965).

State		T_e	ω_e	$\omega_e x_e$	B_e	α_e	D_e ($10^{-4}cm^{-1}$)	r_e (Å)	Observed Transitions Design.	ν_{00}	References
$^{(114)}Cd^{1}H$			(μ = 0.99898612)		D_0^0 = 0.678 eV[a]						NOV 1974 A
D									D←X, V	44136 H	(9)
C	$^2\Sigma^+$	(40202)	[1567] H	(50)	[6.00][b]			[1.68]	C←X, V	40314.9 Z	(7)* (9)*
B	$^2\Sigma^+$	(24961)	(1000)[c]	(17)	[2.95][c]			[2.39]	B→X, R	$24749._0$ Z	(4)
A	$^2\Pi$ 3/2	23116[d]	1757.8 Z	38.6	6.143	0.205	2.9	1.657_4	A→X, V	23277.2 Z	(1)* (2)(4)
	1/2	(22117)	[1676.9] Z		6.061[e]	0.193	2.7	1.668_6		22276.5 Z[f]	(5)(6)
X	$^2\Sigma^+$	0	[1337.1_4] Z	g	[5.323][h]	i	[3.14][j]	[1.780_5]	ESR sp.[k]		
$^{(114)}Cd^{2}H$			(μ = 1.97910657)								NOV 1974 A
D			[1056] H						D←X, V	44117 H	(9)
C	$^2\Sigma^+$		[1149] H	(37)					C←X, V	40260 H	(7)(9)*
A	$^2\Pi$ 3/2	ℓ	[1209.7] Z		3.086	0.070	0.73	1.661_4	A→X, V	23236.4 Z	(4)(5)
	1/2				[3.025]		[0.57]	[1.678_0]		22230.1 Z[f]	
X	$^2\Sigma^+$	0			[2.704][m]	m	[0.76][m]	[1.774_8]			
$^{(114)}Cd^{1}H^+$					D_0^0 = 2.1 eV[a]						NOV 1974
A	$^1\Sigma^+$	42934.1	$1252._0$ Z	8.6	4.851	0.082	2.9[b]	1.865_1	A→X, R	42680.6 Z	(1)
X	$^1\Sigma^+$	0	1772.5 Z	35.40[c]	6.071	0.190	2.9[bd]	1.667_2			
$^{(114)}Cd^{2}H^+$											NOV 1974
A	$^1\Sigma^+$	42930.6	887.2 Z	3.4_4	2.452	0.028	0.72[e]	1.864	A→X, R	42746.8 Z	(2)
X	$^1\Sigma^+$	0	1262.5 Z	19.0_1	3.075	0.068_2	0.48[f]	1.664_3			
$^{(114)}Cd^{(202)}Hg$			(μ = 72.830101_6) Continuous emission at 21300 cm^{-1}.								MAY 1976 (1)

Cd^1H, Cd^2H:

[a]Short extrapolation of the ground state.

[b]All observed levels diffuse due to predissociation.

[c]Observed up to v=13. More accurate constants will depend on the analysis of numerous perturbations by A $^2\Pi$.

[d]$A_0 = +\ 1012.8$, $A_1 = +\ 1016.2$.

[e]Λ-type doubling $\Delta v_{fe}(v=0) = +\ 0.62(J+\frac{1}{2}) - \ldots$ (3).

[f]$J'=\frac{1}{2}$ above $N''=0$. Svensson (4) uses a different definition.

[g]$\Delta G(3/2, \ldots, 9/2) = 1213.1_0,\ 1065.8_4,\ 881.1_0,\ 635.2$.
Estimated zero point energy ≈ 708.

[h]Spin doubling $\Delta v_{12}(v=0) = +\ 0.60(N+\frac{1}{2}) - \ldots$ (3)(4)(8).
γ decreases rapidly with increasing v.

[i]$B_1, \ldots, B_5 = 5.065,\ 4.758,\ 4.388,\ 3.893,\ 3.114$.

[j]$D_1, \ldots, D_5\ (10^{-4} cm^{-1}) = 3.38,\ 3.63,\ 5.00,\ 7.00,\ 16.00$.

[k]In Ar matrix at 4 K (10).

[ℓ]$A_0 = +\ 1012.4$.

[m]$B_1 = 2.536$, $D_1 = 1.78 \times 10^{-4}$. Spin doubling $\Delta v_{12}(v=0) \approx +\ 0.31(N+\frac{1}{2}) - \ldots$ (4)(8).

(1) Bender, PR 36, 1543 (1930).
(2) Watson, PR 36, 1134 (1930).
(3) Mulliken, Christy, PR 38, 87 (1931).
(4) Svensson, Dissertation (Stockholm, 1935).
(5) Deile, ZP 106, 405 (1937).
(6) Stenvinkel, Svensson, Olsson, AMAF A 26, No. 10 (1939).
(7) Khan, PPS 80, 1264 (1962).
(8) Veseth, JP B 3, 1677 (1970).
(9) Breckenridge, Callear, TFS 67, 2009 (1971).
(10) Knight, Weltner, JCP 55, 2061 (1971).

Cd^1H^+, Cd^2H^+:

[a]Value given by Gaydon in DISSEN.

[b]Values for H_v also reported in (1).

[c]$\omega_e y_e = -\ 0.371$.

[d]$\beta_e = +\ 0.1 \times 10^{-4}$.

[e]$\beta_e = +\ 0.08 \times 10^{-4}$.

[f]$\beta_e = +\ 0.12 \times 10^{-4}$.

(1) Svensson, Tyrén, ZP 85, 257 (1933).
(2) Zumstein, Gabel, McKay, PR 51, 238 (1937).

CdHg: (1) McGeoch, Fournier, Ewart, JP B 9, L121 (1976).

State	T_e	ω_e	$\omega_e x_e$	B_e	α_e	D_e (10^{-} cm^{-1})	r_e (Å)	Observed Transitions Design.	Observed Transitions ν_{00}	References
$^{(114)}Cd^{127}I$		(μ = 60.026479$_5$) 0.4 eV ≤ D_0^0 ≤ 0.9 eV[a]								NOV 1974
E ($^2\Sigma^+$)	(41912)	108.$_5$[b] H	1.0					E→X, R	(41877)[b] H	(1)(5)(6)
D ($^2\Pi_{3/2}$)	29531.$_7$	196.8[c] H	0.8$_5$					D↔X, V	29540.$_7$ H	(1)(3)(6)* (7)(11)*
C ($^2\Pi_{1/2}$)	28230.$_5$	188.$_2$[d] H	0.8$_4$					C↔X, (V)	28235.$_2$ H	(5)(6)* (9)* (11)*
B ($^2\Sigma$)		Large number of emission bands from 15000 to 28000 cm^{-1}.[e]						B→X, R		(1)(3)(4)(8) (10)
X ($^2\Sigma$)	0	178.7[f] H	0.7$_6$							
$^{(114)}Cd^{(115)}In$		(μ = 57.200716$_7$)								NOV 1974
		Diffuse bands in emission at 22160 and 24370 cm^{-1}, extending to longer wavelengths from the In lines at 4511 and 4102 Å.								(2)
(F)	e + 17549.3	76.2[a] H						(F)→(E), R	17535.2[a] H	(2)
(E)	e	104.5[a] H								
D ($^2\Sigma$)	c + 17336.7[b]	103.8 H	0.3					D→C,[c] V	17354.9 H	(1)* (2)*
C ($^2\Sigma$)	c	67.3 H	0.1							
B ($^2\Pi$)	a + 18017.0	210.4 H						B→A,[c] V	18038.6 H	(1)* (2)*
	a + 18008.8	212.2 H	0.15						18031.3 H	
A ($^2\Sigma$)	a	167.5 H	0.70							
$^{(114)}Cd^{(39)}K$?		(μ = 29.0323970)								NOV 1974
		Diffuse V shaded absorption bands at 23850 and 23960 cm^{-1}.								(1)
$^{(114)}Cd^{23}Na$?		(μ = 19.128878$_9$)								NOV 1974
		Diffuse V shaded absorption bands at 24990 and 25160 cm^{-1}.								(1)

CdI: [a]Na D line chemiluminescence in Na/CdI_2 flames [(2), revised].

[b]Vibrational numbering uncertain; compare (5) and (6).

[c]Average of constants obtained by (1) (emission) and (11) (absorption).

[d]Average of constants obtained by (9) (emission) and (11) (absorption).

[e]It was suggested by (10) that these bands form two separate systems and that for the system at shorter wavelengths $\nu_e = 23868.4$, $\omega_e' = 74.0$, $\omega_e'x_e' = 2.0$.

[f]Average of constants obtained by (1)(6)(9)(11).

(1) Wieland, HPA 2, 46, 77 (1929).
(2) Horn, Polanyi, Sattler, ZPC B 17, 220 (1932).
(3) Oeser, ZP 95, 699 (1935).
(4) Subbaraya, Rao, Rao, PIAS A 5, 372 (1937).
(5) Howell, PRS A 182, 95 (1943).
(6) Ramasastry, Rao, IJP 20, 100 (1946).
(7) Wieland, Herczog, HCA 29, 1702 (1946).
(8) Wieland, in "Contribution à l'Etude de la Structure Moléculaire", Vol. comm. Victor Henri, Ed. Desoer, Liège (1948), p. 229.
(9) Patel, Patel, Darji, IJPAP 5, 526 (1967).
(10) Patel, Patel, Darji, IJPAP 6, 342 (1968).
(11) Gosavi, Greig, Young, Strausz, JCP 54, 983 (1971).

$CdIn$: [a]Fragments of a band system [system E of (2)] overlapping $D \rightarrow C$.

[b]$\nu_e = 17326.7$ in (2) seems erroneous.

[c]Wrongly attributed to HgIn by (1). More recently, it was suggested that the $D - C$ system is, in fact, due to $CdIn_2$; see (3).

(1) Purbrick, PR 81, 89 (1951).
(2) Santaram, Winans, JMS 16, 309 (1965).
(3) Santaram, Vaidyan, Winans, JP B 4, 133 (1971).

CdK, $CdNa$:

(1) Barratt, TFS 25, 758 (1929).

State	T_e	ω_e	$\omega_e x_e$	B_e	α_e	D_e	r_e (Å)	Observed Transitions Design.	ν_{00}	References
$^{(114)}Cd^{16}O$	(μ = 14.0253946$_9$) $D_0^0 \leq 3.8_2$ eV[a]									DEC 1974
	Bands originally ascribed by (1) to CdO have later been shown by (2) to be due to Bi_2.									
$^{(114)}Cd^{(85)}Rb$?	(μ = 48.646891$_4$)									DEC 1974
	Diffuse V shaded absorption bands at 22280 and 22600 cm^{-1}. Other unclassified bands in the region 22500 - 24000 cm^{-1}.									(1)
$^{(114)}Cd^{(32)}S$	(μ = 24.9646325) $D_0^0 \leq 2.0_4$ eV[a]									DEC 1974
	The assignment to CdS of two absorption continua with long wavelength limits at 31700 and 42500 cm^{-1} (1) appears doubtful in the light of mass-spectroscopic evidence (2)(3)(4).									
$^{(114)}Cd^{(80)}Se$	(μ = 46.965051$_9$) $D_0^0 \leq 1.9_9$ eV[a]									DEC 1974
	The assignment to CdSe of two absorption continua with long wavelength limits at 25500 and 43850 cm^{-1} (1) appears doubtful in the light of mass-spectroscopic evidence (2)(3).									
$^{(114)}Cd^{(205)}Tl$	(μ = 73.217007$_4$)									DEC 1974
	Continuous emission extending to longer wavelengths from the Tℓ line at 3775 Å. Intensity maxima at 26480, 25590, 24870 cm^{-1}.									(1)
	Broad "continuous" band in emission at 20520 cm^{-1}, accompanied by V shaded bands at 20929, 21018, 21109, 21199 cm^{-1}.									(1)
	Continuous emission extending to longer wavelengths from the Tℓ line at 5350 Å. Intensity maxima at 18690, 17790, 17460 cm^{-1}.									(1)
$^{(140)}Ce_2$	(μ = 69.952742) $D_0^0 = 2.5$ eV[a]									DEC 1974

CdO: [a]Thermochemical value (3).

(1) Walter, Barratt, PRS A 122, 201 (1929).
(2) Barratt, Bonar, PM 9, 519 (1930).
(3) Brewer, Mastick, JCP 19, 834 (1951).

CdRb: (1) See ref. (1) of CdCs.

CdS: [a]Thermochemical value (mass-spectrom.)(2)(3)(4).

(1) Sen Gupta, PRS A 143, 438 (1933).
(2) Colin, ICB 26, 1129 (1961).
(3) Goldfinger, Jeunehomme, TFS 59, 2851 (1963).
(4) Marquart, Berkowitz, JCP 39, 283 (1963).

CdSe: [a]Thermochemical value (mass-spectrom.)(2).

(1) Mathur, IJP 11, 177 (1937).
(2) See ref. (3) of CdS.
(3) Berkowitz, Chupka, JCP 45, 4289 (1966).

CdTℓ: (1) Santaram, Vaidyan, Winans, JP B 4, 133 (1971).

Ce_2: [a]Average of several thermochemical values (mass-spectrom.)(1)(2)(3).

(1) Balducci, De Maria, Guido, JCP 50, 5424 (1969).
(2) Gingerich, CC (1969), 9.
(3) Gingerich, Finkbeiner, JCP 54, 2621 (1971).

State	T_e	ω_e	$\omega_e x_e$	B_e	α_e	D_e (10^{-7}cm^{-1})	r_e (Å)	Observed Transitions Design.	Observed Transitions ν_{00}	References
$^{(140)}Ce^{(11)}B$		(μ = 10.2061713)		$D_0^0 = 3.1_2$ eV[a]						DEC 1974
$^{(140)}Ce^{12}C$		(μ = 11.05204212)		$D_0^0 = 4.7$ eV[a]						DEC 1974
$^{(140)}Ce^{(193)}Ir$		(μ = 81.102841)		$D_0^0 = 6.0_2$ eV[a]						DEC 1974
$^{(140)}Ce^{14}N$		(μ = 12.7290315$_8$)		$D_0^0 = 5.3_2$ eV[a]						DEC 1974
$^{(140)}Ce^{16}O$		(μ = 14.3538845$_4$)		$D_0^0 = 8.1_8$ eV[a]						DEC 1974 A
		Large number of bands in the region 11000 - 29000 cm^{-1}(2)*. Tentative assignments by (4)(7), revising earlier assignments in (1)(2):								
D_1								$D_1 \leftrightarrow X_1$, R	21379.1 H^Q	(4)(7)
F_2		(772)		[0.35290]		[2.9_5]	[1.8242_6]	$F_2 \leftrightarrow X_2$, R	20834.21 Z	(4)(7)(1)
D_3								$D_3 \leftrightarrow X_3$, R	20516.1 H^Q	(7)(1)
C_1		(798)		[0.34984]		[2.6_9]	[1.8322_2]	$C_1 \leftrightarrow X_1$, R	20273.84 Z	(4)(7)
A_4								$A_4 \leftrightarrow X_4$, R	19871.7 H^Q	(7)
C_3								$C_3 \leftrightarrow X_3$, R	16356.9 H^R	(7)
G $^1\Pi$								$G \leftrightarrow Y_1$, R	15035.4$_0$ H^Q	(7)
B_2		(771)		[0.34705]		[2.8_1]	[1.8395_8]	$B_2 \leftrightarrow X_2$, R	13804.01 Z	(4)(7)(1)
A_2								$A_2 \leftrightarrow X_2$, R	12687.9$_1$ H^Q	(7)(1)
A_1		(749)		[0.34672]		[2.9_7]	[1.8404_5]	$A_1 \leftrightarrow X_1$, R	12595.75 Z	(4)(7)
A_3								$A_3 \leftrightarrow X_3$, V	12162.2$_0$ H^Q	(7)
Y_4 ($^3\Delta_3$)[b]		(866)		[0.3506_5]		[2.3]	[1.8301]			
Y_3 ($^3\Delta_2$)[b]		(509)		[0.3614]		[7.3]	[1.802_7]			
Y_2 ($^3\Delta_1$)[b]		(844)		[0.35214]		[2.45]	[1.8262_3]			
Y_1 $^1\Sigma^{(+)}$[b]		(850)		[0.35790]		[2.54]	[1.8114_8]			

State	T_e	ω_e	$\omega_e x_e$	B_e	α_e	D_e ($10^{-7}cm^{-1}$)	r_e (Å)	Observed Transitions Design.	ν_{00}	References
$^{(140)}Ce^{16}O$ (continued)										
X_4 $(^1\Phi_3)^b$	$x_2 + (2065)^c$	(818)		0.35710	0.0010_8	[2.71]	1.8135_0			
X_3 $^3\Phi_4{}^b$		(932)		0.35377	0.00114	[2.03]	1.8220_2			
X_2 $^3\Phi_3{}^b$	x_2	[822.76] (Z)	(3.09)	0.35779	0.00179	[2.69]	1.8117_5			
X_1 $^3\Phi_2{}^b$		(862)		[0.35452]		[2.40]	[1.8200_9]			

CeB: [a]Thermochemical value (mass-spectrom.)(1).
(1) Gingerich, JCP 53, 746 (1970).

CeC: [a]Thermochemical value (mass-spectrom.)(1).
(1) Gingerich, JCP 50, 2255 (1969).

CeIr: [a]Thermochemical value (mass-spectrom.)(1).
(1) Gingerich, JCS FT II 70, 471 (1974).

CeN: [a]Thermochemical value (mass-spectrom.)(1).
(1) Gingerich, JCP 54, 3720 (1971).

CeO: [a]Thermochemical value (mass-spectrom.)(3)(5)(6)(8).
[b]These states give rise to absorption bands at a temperature of 1900 °C (7). No details given for transitions from Y_2, Y_3, Y_4.
[c]$X_4(v=0)$ is at 2060.25 cm^{-1} above $X_2(v=0)$. All other separations are unknown [(7), no details].

(1) Watson, PR 53, 639 (1938).
(2) Gatterer, Junkes, Salpeter, Rosen, METOX (1957).
(3) Walsh, Dever, White, JPC 65, 1410 (1961).
(4) Ames, Barrow, PPS 90, 869 (1967).
(5) Ames, Walsh, White, JPC 71, 2707 (1967).
(6) Coppens, Smoes, Drowart, TFS 63, 2140 (1967).
(7) Barrow, in DONNSPEC (1970).
(8) Ackermann, Rauh, JCP 60, 2266 (1974).

State	T_e	ω_e	$\omega_e x_e$	B_e	α_e	D_e $(10^{-6}cm^{-1})$	r_e (Å)	Observed Transitions Design.	Observed Transitions ν_{00}	References
$^{(140)}Ce^{(106)}Pd$		(μ = 60.276395)		$D_0^0 = 3.3_0$ eV[a]						DEC 1974
$^{(140)}Ce^{(195)}Pt$		(μ = 81.454361)		$D_0^0 = 5.7_1$ eV[a]						DEC 1974
$^{(140)}Ce^{103}Rh$		(μ = 59.293219)		$D_0^0 = 5.6_5$ eV[a]						DEC 1974
$^{(140)}Ce^{(32)}S$		(μ = 26.0247376)		$D_0^0 = 5.8_6$ eV[a]						DEC 1974
$^{12}C^{19}F$		$\mu = 7.3545996_4$		$D_0^0 = 5.67$ eV[a]		I.P. = 9.20 eV[b]				AUG 1976
		Fragments of additional systems in the absorption spectrum above 50000 cm^{-1}.								(15)
C' $^2\Sigma^+$				[1.5327][c]		[3.3]	[1.223]	C'← X, V	53597.0	(15)
D $^2\Pi$	52272.5[d]	1803.9 Z	13.0	1.7301	0.0193	6.5	1.1510	D← X, V	52519.9[e]	(15)
B $^2\Delta_r$	49399.6[f]	[1153.3_4] Z	19.4[g] H	1.320_6[h]	0.022_8	4.0	1.317_4	B[i]↔ X, R	49340.1[j] Z	(1)* (2) (13)*
A $^2\Sigma^+$	42692.9	1780.4_5 Z	30.7_3	1.7228[k]	0.0189[ℓ]	[6.80][m]	1.1535	A[n]↔ X, V	42924.17[o] Z	(1)* (2)(3) (4)(5)(6) (15)
a $^4\Sigma^-$	(22000)[p]	(1324)[p]	(11)[p]	(1.302)[p]	(0.013)[p]					
X $^2\Pi_r$	0[q]	1308.1 Z	11.10[r]	1.4172[s]	0.0184_0[t]	6.5	1.2718	ESR sp.($^2\Pi_{\frac{3}{2}}$)[u]		(12)
CF^+, CF^-		Theoretical calculations (14).								AUG 1976

CePd: [a]Thermochemical value (mass-spectrom.)(1).
(1) Cocke, Gingerich, JPC 76, 2332 (1972).

CePt, CeRh: [a]Thermochemical value (mass-spectrom.)(1).
(1) Gingerich, JCS FT II 70, 471 (1974).

CeS: [a]Thermochemical value (mass-spectrom.)(1), revised (2).

(1) See ref. (6) of CeO.

(2) Bergman, Coppens, Drowart, Smoes, TFS 66, 800 (1970).

CF, CF^+, CF^-:

[a]Thermochemical value (mass-spectrom.)(17). From the predissociation in A $^2\Sigma^+$: $D_0^0 \leq 5.74$ eV.

[b]Photoionization (11) and electron impact (17) appearance potential measurements; supported by theoretical calculations (14).

[c]Line width increases with N. Vibrational numbering uncertain.

[d]$A \approx + 0.2$ or $+ 6.5$.

[e]By extrapolation from the 1-0,2-0, and 3-0 bands; the 0-0 band was not observed.

[f]$A_0 = + 0.76$. The alternative value, $A_0 = + 4.48$, leads to disagreement between observed and calculated relative line intensities (15).

[g]$\omega_e y_e = - 0.4$.

[h]Strong perturbations in v=2 (15).

[i]Radiative lifetime $\tau(v'=0) = 18._8$ ns (8)(10); $f_{00}(B \leftarrow X) = 0.022$ (recalculated for a $^2\Delta$ upper state). See also (9).

[j]$J'=3/2$ (average of $\{F_1\}$ and F_2) relative to $J''=1/2$ (average of F_1 and $\{F_2\}$). The band centres in (13) must all be reduced by 1.4 cm^{-1} (15) owing to an error resulting from the incorrect application of the vacuum correction.

[k]Predissociation above v=1, due to a curve crossing with the a $^4\Sigma^-$ state at an internuclear distance smaller than r_e; see (15)(16).

[l]$\gamma_e = - 0.0028$ (B_0 and B_1 from (6), B_2 from (15)).

[m]$D_1 = 7.1_0 \times 10^{-6}$ (6), $D_2 = 9.0_0 \times 10^{-6}$ (15).

[n]Radiative lifetime $\tau(v'=1) = 19._0$ ns (8)(10); $f_{00}(A \leftarrow X) \approx 0.0027$, $f_{el} \approx 0.026$ (7)(10). See also (9).

[o]$N'=0$ relative to $J''=\frac{1}{2}$ (average of F_1'' and $\{F_2''\}$).

CF, CF^+, CF^- (continued):

[p]Preliminary results of theoretical calculations (15)(16).

[q]$A_v = + 77.12 - 0.655v + 0.0057v^2$.

[r]$\omega_e y_e = + 0.093$, $\omega_e z_e = - 0.0011$

[s]Λ-type doubling; see (6).

[t]$\gamma_e = + 0.00011$.

[u]$\mu_{el}(C^-F^+) = 0.65$ D; polarity predicted by the theoretical work of (14) and (16).

(1) Andrews, Barrow, Nature 165, 890 (1950); PPS A 64, 481 (1951).

(2) Margrave, Wieland, JCP 21, 1552 (1953).

(3) Mann, Broida, Squires, JCP 22, 348 (1954).

(4) Kuzyakov, Tatevskii, OS 5, 699 (1958).

(5) Thrush, Zwolenik, TFS 59, 582 (1963).

(6) Porter, Mann, Acquista, JMS 16, 228 (1965).

(7) Harrington, Modica, Libby, JCP 44, 3380 (1966); 45, 2720 (1966)(Erratum).

(8) Hesser, Dressler, JCP 45, 3149 (1966).

(9) Wentink, Isaacson, JCP 46, 603 (1967).

(10) Hesser, JCP 48, 2518 (1968).

(11) Walter, Lifshitz, Chupka, Berkowitz, JCP 51, 3531 (1969).

(12) Carrington, Howard, MP 18, 225 (1970).

(13) Carroll, Grennan, JP B 3, 865 (1970).

(14) O'Hare, Wahl, JCP 55, 666 (1971).

(15) W. P. White, Dissertation (Ohio State Univ., 1971), and C. W. Mathews, private communication.

(16) Hall, Richards, MP 23, 331 (1972).

(17) Hildenbrand, CPL 32, 523 (1975).

State		T_e	ω_e	$\omega_e x_e$	B_e	α_e	D_e $(10^{-4}cm^{-1})$	r_e (Å)	Observed Transitions Design.	ν_{00}	References
$^{12}C^{1}H$		$\mu = 0.92974056$			$D_0^0 = 3.46_5$ eV^{a}		I.P. = 10.64 eV^{b}				AUG 1976 A
		Several unassigned absorption bands^{c} above 80000 cm^{-1}.									(26)
		Rydberg series joining on to G, $\nu = 85850 - R/(n-0.09)^2$; n = 3,4,5,6									(26)*
G	^{d}	[74373]							G ← X,	72960	(26)*
F	$^2\Sigma^+$	[65945]			$[12.17]^{e}$			[1.221]	F ← X, R	64531.5^{f} Z	(26)*
E	$^2\Pi$	[65625]			$[12.6]^{g}$			[1.20]	E ← X, R	64211.7^{f} Z	(26)*
D	$^2\Pi_i$	$[60394]^{h}$			$[13.7]^{g}$			[1.15]	D ← B, V	33282.8^{f} Z	(26)*
									D ← X,	58981.0^{f} Z	(26)*
C	$^2\Sigma^+$	31801.5	2840.2 Z	$125.9_6{}^{i}$	14.603^{jk}	0.7185^{ℓ}	$[15.55]^{m}$	1.1143	$C^{n} \leftrightarrow X,^{o}$ V_R	31778.1^{f} Z	(1)(3)(10)* (26)
B	$^2\Sigma^-$	(26044)	$[1794.9]^{p}$ Z		$[12.645]^{qk}$	^{r}	$[22.2]^{s}$	[1.1975]	$B^{t} \leftrightarrow X,^{o}$ R	25698.2^{f} Z	(2)* (3) (10)* (26)*
A	$^2\Delta$	23189.8^{u}	2930.7 Z	96.65	14.934^{vk}	0.697	15.4^{w}	1.1019	$A^{x} \leftrightarrow X,^{o}$ V	23217.5^{f} Z	(3)(6)(10)* (12)*
a	$^4\Sigma^-$	(5844)	(3145)	(72)	(15.4)	(0.55)		(1.08_5)		5985^{y}	(48)
X	$^2\Pi_r$	0^{z}	2858.5 Z	63.0_2	$14.457^{a'b'k}$	0.534	14.5	1.1199	Λ doubling sp.$^{a'}$		(39)
									Ab initio calculations: ref. in (44)		

C^1H: ^{a}From the predissociation in the B state (26) as modified by (41) and (49). Confirmed thermochemically by (13).

^{b}From Rydberg series (26). Theoretical photoionization cross section (36).

^{c}According to theoretical work (35) they represent the nf series.

^{d}3d complex consisting of $^2\Sigma$, $^2\Pi$, $^2\Delta$.

^{e}Spin splitting constant $\gamma_0 = +0.6_2$. Heterogeneous predissociation.

^{f}The band origins refer to the zero points of the Hill-Van Vleck formulae for the ground and excited ($\Lambda \neq 0$) states. See also (11).

^{g}Homogeneous predissociation.

^{h}A = - 28.5.

^{i}$\omega_e y_e = +13.55$, $\omega_e z_e = -3.957$; from CD using isotope relations.

^{j}Spin splitting constant $\gamma \approx +0.05$ (26)(37). Predissociation, see ^{n}.

^{k}Slightly different sets of constants from the same data are given by (37) for X, A, B, C and (40) for X, A. Precise values for the rotational energy levels of X $^2\Pi(v=0)$ in (15); tables of term values for levels of X, A, B, C in (37).

[l] $+ 0.0258(v+\frac{1}{2})^2 - 0.023(v+\frac{1}{2})^3$; see [i].

[m] $D_1 = 16.7 \times 10^{-4}$, $D_2 = 20 \times 10^{-4}$.

[n] Lifetime of C $^2\Sigma^+$ anomalously small on account of predissociation (24)(26)(28)(38). According to (49) the lifetime in v=0 varies from 10 ns at N=1 to 25 ns at N=24 and is even shorter in v=1. The previous observation of different lifetimes for F_1 and F_2 levels is not confirmed. The lifetime for CD is more than twice that for CH (28) indicating that predissociation is weaker (26). Ab initio theory (46) gives a radiative lifetime $\tau(v=0)$ = 89 ns suggesting that the radiationless and the radiative transition probabilities are approximately in the ratio 8 : 1 (49). From absorption experiments (21) an f value of 0.006 was derived for the C – X transition.

[o] Wavenumber and wavelength tables and comparisons with the solar spectrum published by (7). ^{13}CH lines for A – X measured by (23) and used to determine $^{13}C/^{12}C$ ratio in the sun. Franck-Condon factors (17)(33). The laboratory absorption spectrum was first observed by (5) in the acetylene combustion initiated by the flash photolysis of NO_2; in flames by (9); and more recently by (14)(19)(26).

[p] Very shallow potential curve; $\omega_e \approx 2251$, $\omega_e x_e \approx 230$ (from CD). A small potential maximum has been established in this state (26); confirmed by theoretical calculations (32).

[q] Spin splitting constant $\gamma_0 = -0.0285$ (3), $\gamma_1 = -0.020$ (26). Breaking off in emission above v'=0, N'=15 and v'=1, N'=6 and broadening in absorption at higher N' due to predissociation; see also [t]. Selective excitation of v'=1 in hydrogen flames and suppression of breaking off (4).

[r] $B_1 = 11.160$.

[s] $D_1 = 32.8 \times 10^{-4}$.

[t] Radiative lifetime $\tau = 0.38$ µs (20)(21)(28) corresponding to f ≈ 0.0029. (41) and (49) have measured lifetimes of individual rotational lines finding a regular increase from 0.32 to 0.38 µs from N'=2 to 14 (v'=0); sudden drop to 0.12 µs for N'=15 due to predissociation; see, however, (45). In v'=1 the lifetime is 0.40 µs until breaking off occurs between N'=6 and 7.

[u] $A = -1.0_2$ (26), see also (6)(37).

[v] Stark effect, $\mu_{el}(v=0)$ = 0.89 D (42).

[w] $\beta_e = +0.4 \times 10^{-4}$.

[x] Lifetime $\tau(v=0) = 0.53_4$ µs (49) corresponding to f = 0.0053; superseding earlier less precise values by (8)(20)(28)(31)(43). (34) from shock tube absorption measurements obtain $f_e = 0.0019$. A theoretical $f_{00} = 0.0068$ is given in (46). The observation of a reduced lifetime in v=1 for N > 11 (49) suggests a weak predissociation probably caused by X $^2\Pi$.

[y] T_0, from laser photoelectron spectrometry of CH^- (48). Theoretical calculations (32) give 5395 cm^{-1}. The vibrational and rotational constants given are theoretical values; see also (29).

[z] A = + 27.95. Slightly higher values in (16) and (37).

[a'] Λ-type doubling, $\Delta\nu \approx 0.038 \times N(N+1) - \ldots$. For the lowest J values more complicated formulae apply, see (15) and (16). The transition between the two Λ components of $J=\frac{1}{2}$ has been observed with its hyperfine structure in emission in interstellar clouds (39). The derived $J=\frac{1}{2}$ Λ-doublet separation is 3320.987 MHz, close to earlier predictions by (15)(30) and a subsequent theoretical calculation by (39a). Predicted Λ-doubling and hyperfine splittings for other J values in (47).

[b'] Stark effect, μ_{el} = 1.46 D (18).

(continued p. 143)

State	T_e	ω_e	$\omega_e x_e$	B_e	α_e	D_e ($10^{-4}cm^{-1}$)	r_e (Å)	Observed Transitions Design.	Observed Transitions ν_{00}	References
$^{12}C^2H$		μ = 1.72463610		$D_0^0 = 3.50_0$ eV[a]		I.P. = 10.64 eV[b]				AUG 1976 A
G c								G ← X,	72955	(5)*
F $^2\Sigma^+$	[65605]			[6.86][d]			[1.194]	F ← X,	R 64563.9[e] Z	(5)*
D $^2\Pi_i$	(59038)[f]	(2025)		[7.425]		[4.0]	[1.1474]	D ← B,	V 33212.0[e] Z	(5)*
C $^2\Sigma^+$	31818.1	2081.3 Z	66.79[g]	7.879[hi]	0.283[j]	[4.5][k]	1.1138	C[i] ↔ X,[ℓ]	V_R 31801.3[e] Z	(2)(5)
B $^2\Sigma^-$	26043	1652.5[m] Z	123.8	7.104[i]	0.341[n]	[6.36][o]	1.173_0	B[i] ↔ X,[ℓ]	R 25796.9[e] Z	(1)(2)(5)
A $^2\Delta$	23184.4	2203.3 Z	78.50	8.032	0.260	[4.5][p]	1.1032	A ↔ X,[ℓ]	V 23225.1[e] Z	(2)
X $^2\Pi_r$	0[q]	2099.7_5 Z	34.02	7.806[r]	0.208	4.2	1.1190			

C^2H: [a]From the predissociation in B $^2\Sigma^-$ of CD. The revised value of D_0^0(CH) would imply $D_0^0(CD) = 3.51_2$ eV.

[b]From I.P.(CH).

[c]See [d] of C^1H.

[d]Heterogeneous predissociation.

[e]See [f] of C^1H.

[f]A = - 27.7.

[g]$\omega_e y_e$ = + 5.364, $\omega_e z_e$ = - 1.15; these two constants are derived under the assumption that ΔG of X $^2\Pi$ is linear in v+½.

[h]Spin splitting constant γ_0 = + 0.06 (5).

[i]Lifetime and predissociation see (4)(5)(6). For B $^2\Sigma^-$, v=0, breaking off occurs above N'=24, for v=1 above N'=16. For C $^2\Sigma^+$ the predissociation is much weaker for CD than for CH; see [n] of C^1H.

[j]$+ 0.0075(v+\frac{1}{2})^2 - 0.005(v+\frac{1}{2})^3$.

[k]$D_1,\ldots,D_4$($10^{-4}cm^{-1}$) = 4.9, 5.2, 6.3, 9.

[ℓ]Franck-Condon factors (3)(7).

[m]The B state is too shallow for these constants to have much physical meaning.

[n]γ_e = - 0.095.

[o]$D_1 = 7.2 \times 10^{-4}$, $D_2 = 13.9 \times 10^{-4}$.

[p]$D_1 = 4.7 \times 10^{-4}$.

[q]A = + 27.95.

[r]Λ-type doubling, $\Delta\nu \approx 0.009 \times N(N+1)$ for higher N values. The splitting for J=½ is predicted (8) to be 1241 MHz.

(1) See ref. (2) of C^1H.

(2) Gerö, ZP 117, 709 (1941).

(3) See ref. (17) of C^1H.

(4) See ref. (24) of C^1H.

(5) See ref. (26) of C^1H.

(6) See ref. (28) of C^1H.

(7) See ref. (33) of C^1H.

(8) Hammersley, Richards, ApJ 194, L61 (1974).

C^1H (continued):

(1) Heimer, ZP 78, 771 (1932).
(2) Shidei, JJP 11, 23 (1936).
(3) Gerö, ZP 118, 27 (1941).
(4) Durie, PPS A 65, 125 (1952).
(5) Norrish, Porter, Thrush, PRS A 216, 165 (1953).
(6) Kiess, Broida, ApJ 123, 166 (1956).
(7) Moore, Broida, JRNBS A 63, 19 (1959).
(8) Bennett, Dalby, JCP 32, 1716 (1960).
(9) Gaydon, Spokes, van Suchtelen, PRS A 256, 323 (1960).
(10) Bass, Broida, NBSM No. 24 (1961).
(11) Garstang, PPS 82, 545 (1963).
(12) Pearse, Gaydon, IDSPEC (1963).
(13) Brewer, Kester, JCP 40, 812 (1964).
(14) Bleekrode, Nieuwpoort, JCP 43, 3680 (1965).
(15) Douglas, Elliot, CJP 43, 496 (1965).
(16) Goss, ApJ 145, 707 (1966).
(17) Halmann, Laulicht, ApJ(Suppl.) 12, 307 (1966).
(18) Phelps, Dalby, PRL 16, 3 (1966).
(19) Bleekrode, Thesis (Amsterdam, 1966); PRR(Suppl.) No. 7 (1967).
(20) Fink, Welge, JCP 46, 4315 (1967).
(21) Linevsky, JCP 47, 3485 (1967).
(22) Pathak, Singh, IJPAP 5, 139 (1967).
(23) Richter, Tonner, ZA 67, 155 (1967).
(24) Hesser, Lutz, PRL 20, 363 (1968).
(25) Sharma, Singh, Pathak, IJPAP 6, 443 (1968)
(26) Herzberg, Johns, ApJ 158, 399 (1969).

C^1H (continued):

(27) Le Calvé, Bourène, Schmidt, Clerc, JP(Paris) 30, 807 (1969).
(28) Hesser, Lutz, ApJ 159, 703 (1970).
(29) Liu, Verhaegen, JCP 53, 735 (1970).
(30) Baird, Bredohl, ApJ 169, L83 (1971).
(31) Smith, JCP 54, 1384 (1971).
(32) Lie, Hinze, Liu, JCP 57, 625 (1972); 59, 1872, 1887 (1973).
(33) Liszt, Smith, JQSRT 12, 947 (1972).
(34) Kuz'menko, Kuzyakov, Kuznetsova, Kudryumova, Chuev, HT(USSR) 9, 823 (1972).
(35) Walker, Kelly, JCP 57, 936 (1972).
(36) Walker, Kelly, CPL 16, 511 (1972).
(37) Botterud, Lofthus, Veseth, PS 8, 218 (1973).
(38) Elander, Smith, ApJ 184, 663 (1973).
(39) Rydbeck, Elldér, Irvine, Nature 246, 466 (1973); Rydbeck, Elldér, Irvine, Sume, Hjalmarson, AA 34, 479 (1974).
(39a) Hammersley, Richards, Nature 251, 597 (1974).
(40) Krupp, ApJ 189, 389 (1974).
(41) Brooks, Smith, ApJ 194, 513 (1974).
(42) Scarl, Dalby, CJP 52, 1429 (1974).
(43) Jørgensen, Sørensen, JCP 62, 2550 (1975).
(44) Meyer, Rosmus, JCP 63, 2356 (1975).
(45) Anderson, Peacher, Wilcox, JCP 63, 5287 (1975).
(46) Hinze, Lie, Liu, ApJ 196, 621 (1975).
(47) Levy, Hinze, ApJ 200, 236 (1975).
(48) Kasdan, Herbst, Lineberger, CPL 31, 78 (1975).
(49) Brzozowski, Bunker, Elander, Erman, ApJ 207, 414 (1976).

State	T_e	ω_e	$\omega_e x_e$	B_e	α_e	D_e ($10^{-4}cm^{-1}$)	r_e (Å)	Observed Transitions Design.	ν_{00}	References
$^{12}C^{1}H^{+}$				$D_0^0 = 4.08_5$ eV[a]						AUG 1976 A
B $^1\Delta$	(52534)	2075.5 Z	76.3	11.937	0.620	13[b]	1.2325	$B^c \rightarrow A$, V	28537.9 Z	(5)(7)*
b $^3\Sigma^-$	(38200)	[1939]		11.705[d]	0.538	13.5	1.2446	$b^e \rightarrow a$, R	28580	(5)(7)*
A $^1\Pi$	(24111)	1865.35 Z	115.85[f]	11.898_8[ghi]	0.9414[j]	20	1.2344	$A^k \leftrightarrow X$,[ℓ] R	23596.9_4 Z	(1)* (2)*
a $^3\Pi$	(9200)[m]	(2814)		14.048[i]	0.603	14	1.1361			
X $^1\Sigma^+$	0	[2739.70] Z	(64)[n]	14.177_6[i]	0.4917	14	1.1309			
$^{12}C^{2}H^{+}$				$D_0^0 = 4.13$ eV[a]						AUG 1976 A
A $^1\Pi$	(24138)	[1248.5] (Z)						$A^b \rightarrow X$, R	23747.6 (Z)	(1)
X $^1\Sigma^+$	0	[2029.3] (Z)								
$^{12}C^{1}H^{-}$				$D_0^0 = 3.43_5$ eV[a]		I.P. = 1.238 eV[b]				AUG 1976
a $^1\Delta$	(6828)	(3000)					(1.10)[c]		6815[d]	(2)
X $^3\Sigma^-$	0	(3025)					(1.08_9)[c]			
$^{12}C^{127}I$		$\mu = 10.96331634$		a						AUG 1976

C^1H^+: [a] $D_0^0(C^1H) + I.P.(C) - I.P.(C^1H)$.

[b] $\beta_e = 3 \times 10^{-4}$.

[c] Lifetime 0.22 μs [average of (8)(19)(21)].

[d] Mass-spectrometric observations suggest a predissociation of this state for $v > 1$ into the $^3\Sigma^+$ state arising from $C^+(^2P) + H(^2S)$; see (10)(17).

[e] Lifetime 0.48 μs (19).

[f] $\omega_e y_e = +2.64$. These constants do not fit higher vibrational levels.

C^1H^+ (continued):

[g] Λ-type doubling, $\Delta\nu_{ef}(v=0) = +0.0398\,J(J+1)$. The splitting decreases with increasing v.

[h] Tables of term values, and somewhat different constants based on the same data, are given by (14).

[i] Theoretical potential functions for all states arising from $C^+(^2P) + H(^2S)$ are given by (11). A $^1\Pi$ RKR curve in (4).

[j] $\gamma_e = -0.0019$. See [f].

C^1H^+ (continued):

[k]There is considerable disagreement between lifetime measurements by different authors:

	v = 0	1	2	3	4	
Ref. (19):	τ = 408	495	530	465	525	ns,
Ref. (21):	τ = 250	270	290			ns.

Earlier determinations (9)(13) gave much lower values probably because of overlapping N_2 bands. Following are the averaged f values recommended by (21): $f_{00} = 0.013_6$, $f_{10} = 0.009_1$, $f_{20} = 0.003_6$. The theoretical values (18) are $f_{00} = 0.00645$, $f_{10} = 0.00431$, $f_{20} = 0.00173$.

[ℓ]Occurs in interstellar absorption (1)(6)(16). Franck-Condon factors (from *ab initio* potential energy curves) in (15); see also (3)(22).

[m]From a theoretical calculation (11). A ≈ 23 (7).

[n]Average of a theoretical value (20) and several estimates based on experimental results.

(1) Douglas, Herzberg, CJR A 20, 71 (1942).
(2) Douglas, Morton, ApJ 131, 1 (1960).
(3) Nicholls, Fraser, Jarmain, McEachran, ApJ 131, 399 (1960).
(4) Read, Vanderslice, Jenč, JCP 37, 205 (1962).
(5) Carré, Dufay, CR B 266, 1367 (1968).
(6) Herbig, ZA 68, 243 (1968).
(7) Carré, Physica 41, 63 (1969).
(8) See ref. (28) of C^1H.
(9) See ref. (31) of C^1H.
(10) Lorquet, Lorquet, Wankenne, Momigny, Lefebvre-Brion, JCP 55, 4053 (1971).
(11) Green, Bagus, Liu, McLean, Yoshimine, PR A 5, 1614 (1972).
(12) See ref. (33) of C^1H.
(13) Anderson, Wilcox, Sutherland, NIM 110, 167 (1973).

C^1H^+ (continued):

(14) See ref. (37) of C^1H.
(15) Green, Hornstein, Bender, ApJ 179, 671 (1973).
(16) Hobbs, ApJ 181, 79 (1973).
(17) Newton, Sciamanna, JCP 58, 1292 (1973).
(18) Yoshimine, Green, Thaddeus, ApJ 183, 899 (1973).
(19) Brzozowski, Elander, Erman, Lyyra, ApJ 193, 741 (1974).
(20) Banyard, Taylor, JP B 8, L137 (1975).
(21) Brooks, Smith, ApJ 196, 307 (1975).
(22) Rao, Murty, Rao, Rao, PL A 54, 177 (1975).

C^2H^+: [a]From the value for C^1H^+.

[b]Lifetime $\tau(v=0)$ = 64 ns (2); $f_{00} = 0.06_2$. This value seems doubtful when compared with more recent results for C^1H^+; see [k] of C^1H^+.

(1) Cisak, Rytel, APP A 39, 627 (1971).
(2) See ref. (31) of C^1H.

C^1H^-: [a]From $D_0^0(C^1H)$ and the electron affinities of C^1H [= I.P.(CH^-)] and C (1.268 eV).

[b]From laser photoelectron spectrometry of CH^- (2). A theoretical value by (1) is $1.6_1 \pm 0.3$ eV.

[c]Indirectly from a Franck-Condon analysis.

[d]T_0, from the photodetachment spectrum of CH^-.

(1) Cade, PPS 91, 842 (1967).
(2) See ref. (48) of C^1H.

CI: [a]Study of flame reactions (1) suggests $D_0^0 = 2.1_7$ eV.

(1) Miller, Palmer, JCP 40, 3701 (1964).

State	T_e	ω_e	$\omega_e x_e$	B_e	α_e	D_e ($10^{-7}cm^{-1}$)	r_e (Å)	Observed Transitions: Design.	Observed Transitions: ν_{00}	References
$^{35}Cl_2$		$\mu = 17.4844268_0$	$D_0^0 = 2.47936_7$ eV[a]	I.P. = 11.50 eV[b]						SEP 1976 A
		Rydberg series joining on to P: $\nu(1-0) = 93200 - R/(n-0.54)^2$, $n = 3 \ldots 7$; fragmentary vibrational structure.								(9)
		Fragments of additional band systems in absorption at $\nu > 65000$ cm^{-1}.								(8)(9)
		Emission continua in the ultraviolet with maxima at 32640, 33810, 34700, 35450, 36220, 36820; 38970, 41140, 42500, 43710, 45500, 46610, 47670; 50060, 51850, 53890 cm^{-1}.[c]								(3)
P	(74405)	(621)	(3)					P← X,	74436	(8)(9)
O				[0.184_0][d]				O→X,	R 74018.5[d] Z	(41)*
N				[0.119_3][d]				N→X,	R 73363.3[d] Z	(41)*
M	(72853)	(636)	(4)					M← X,	72891	(8)
K	(64024)	(460)						K→X,[e]	63975	(5)
J	(61638)	(520)	(3)					J← X,	61618	(8)
I	61438	262.3 H	0.812					I→B,	V 43632 H	(7)
H	(59432)	(510)						H← X,	59408	(8)
G	(58629)	(208)						G→X,[e]	R 58454	(5)
F	(58263)	(442)						F← X,	58205	(8)
E	57953	249.75 H	0.875					E↔B,	R 40140.0 H	(6)(13)* (34)
D	(53568)	(440)	(1.5)					D← X,	53508	(8)
		Continuous absorption above ~52600 cm^{-1} at high pressure.								(1)(8)
C $^1\Pi_u$		Continuous absorption with maximum at 30500 cm^{-1}.						C↔X,[f]		(2)(4)(15) (18)(19)
B $^3\Pi_{0^+u}$	17809	259.5[g] H	5.3[h]	0.1625_6[i]	0.0021_2[j]	2.365[k]	2.435_4	B[ℓ]↔X,[m]	R 17658[g] H	(14)* (16)* (23)*

$C\ell_2$: [a]From the convergence limit in B $^3\Pi_{0^+}$ (see [h]). From the same limit (36) gives $D_0^0 = 19997.14$ cm^{-1} or 2.47934_9 eV presumably by using a different value for the $^2P_{1/2} - {}^2P_{3/2}$ energy difference in $C\ell$ I. Here we used 882.36 cm^{-1} from (22).

[b]From the photoelectron spectrum; average of (25) and (29). Photoionization (12), in agreement with the Rydberg series, yields 11.48 eV.

[c]They have been interpreted (3) as being due to transitions from stable excited states at 58000 (possibly F), 67700 and 75000 cm^{-1} to the repulsive states arising from $^2P + {}^2P$. The upper states at 67700 and 75000 cm^{-1} are considered to be 1_g states and, therefore, are not observed in absorption from the ground state.

[d]Upper levels of four extensive resonance series (10)(41). The v' values are uncertain. The resonance fluorescence spectrum is excited by the $C\ell$ I lines at 73983 and 73344 cm^{-1} in a discharge through $C\ell_2$ and involves transitions to ground state levels with $v'' \leq 59$. The ground state dissociation energy derived from these resonance series (41) agrees now very well with the more accurate value from the B − X system.

[e]These systems [called J − X and H − X by (5)] have not been observed in absorption. For this reason (8) suggest that they may actually be due to $C\ell_2^+$.

[f]The angular distribution of photo-fragments confirms the assignment of the upper state of the continuum to $^1\Pi_u$ (21); see also (35) and (40). The B ← X transition, however, contributes to the weak low-frequency region of the continuum; for a discussion of quantitative data see (36a).

[g]Since high resolution data (14)(23) are available only for $v \geq 5$ the constants given here are from the low resolution emission work of (16) (band heads); they are valid only for $0 \leq v \leq 6$. For $6 < v < 22$, (11) give $\omega_e = 259.57$, $\omega_e x_e = 4.75_3$, $\omega_e y_e = -0.067_7$, $\omega_e z_e = +0.00212$. The band origin of the 6-0 band is at 18993.79 cm^{-1}.

[h]Convergence limit 20879.64 ± 0.14 cm^{-1} (36)(38). See (28)(30)(32)(37)(38) and the review in (36) for relation of high vibrational levels to long-range internuclear potential.

[i]These constants are based on bands with $5 \leq v' \leq 13$ (23). B_v values up to $v = 31$ have been determined (14)(23).

[j]$\gamma_e = -0.00009_1$.

[k]$+ 0.225 \times 10^{-7}(v+\frac{1}{2}) + 0.015 \times 10^{-7}(v+\frac{1}{2})^2$. See [i].

[ℓ]Estimated radiative lifetimes in (36a).

[m]Franck-Condon factors from RKR potentials (26). For a discussion of the repulsive part of the potential see (35).

(continued p. 148)

State	T_e	ω_e	$\omega_e x_e$	B_e	α_e	D_e $(10^{-7} cm^{-1})$	r_e (Å)	Observed Transitions		References
								Design.	ν_{00}	
$^{35}Cl_2$ (continued)										
A $(^3\Pi_{1u})$	(17440)	(265)	H (5)					$A^{\ell} \rightarrow X$,[n]		(36a)
A' $(^3\Pi_{2u})$	(17160)[o]	(280)[p]						$A' \rightarrow X$,[q]		(42)
X $^1\Sigma_g^+$	0	559.7_2[rs]	Z 2.67_5[t]	0.2439_9[r]	0.0014_9[u]	1.86	1.987_9	Pressure induced IR absorption at 549 cm^{-1}		(20)
								Raman sp.		(24)(27)(31)(33)(43)

$C\ell_2$ (continued):

[n]Two weak progressions, not belonging to B − X and tentatively assigned as 1-v" and 2-v" with v" = 8,9,..., were observed in the chlorine atom recombination spectrum and in the spectrum of the nitrogen trichloride decomposition flame; see references in (36a).

[o]Not observed in the gas phase (see [q]); in an Ar matrix this new state is located 650 cm^{-1} below the B $^3\Pi_{0^+u}$ state.

[p]Estimated from isotope shifts.

[q]Long-lived (~76 ms in Ar) emission in rare gas matrices from v=0 of a new low-lying state following excitation into the B or C state; see (42).

[r]These constants are based on the lowest six vibrational levels (23). The following Dunham coefficients have been derived by (41) from a detailed analysis at high resolution of the resonance series excited by the $C\ell$ I lines at 1351.7 and 1363.5 Å; they represent all levels up to v=40:

$Y_{10} = 559.7507$ $\quad$ $Y_{01} = 0.244153$

$Y_{20} = -2.694271$ $\quad$ $Y_{11} = -0.0015163$

$Y_{30} = -3.32527 \times 10^{-3}$ $\quad$ $Y_{21} = -3.9078 \times 10^{-6}$

$Y_{40} = -2.27337 \times 10^{-4}$ $\quad$ $Y_{31} = 7.0811 \times 10^{-8}$

$Y_{50} = 3.92041 \times 10^{-6}$ $\quad$ $Y_{41} = -5.5875 \times 10^{-9}$

$Y_{60} = -6.02984 \times 10^{-8}$ $\quad$ $Y_{02} = -1.9195 \times 10^{-7}$

$Y_{00} = -0.0351$ $\quad$ $Y_{32} = -3.1678 \times 10^{-12}$

The same authors give, in addition, G(v) and B_v values up to v=59 and have determined an accurate RKR potential function. The long-range portion agrees very well with that predicted from theory.

[s]550.8 in liquid $C\ell_2$ (33); 554.6 in solid argon (39)(42).

[t]$\omega_e y_e = -0.006_7$.

[u]$\gamma_e = -0.0000001_7$.

$C\ell_2$ (continued):

(1) Cordes, Sponer, ZP 63, 334 (1930).
(2) Gibson, Bayliss, PR 44, 188 (1933); Gibson, Rice, Bayliss, PR 44, 193 (1933).
(3) Asundi, Venkateswarlu, IJP 21, 101 (1947); Venkateswarlu, PIAS A 26, 22 (1947).
(4) Sulzer, Wieland, HPA 25, 653 (1952).
(5) Haranath, Rao, JMS 2, 428 (1958).
(6) Venkateswarlu, Khanna, PIAS A 49, 117 (1959).
(7) Khanna, PIAS A 49, 293 (1959).
(8) Lee, Walsh, TFS 55, 1281 (1959).
(9) Iczkowski, Margrave, Green, JCP 33, 1261 (1960).
(10) Rao, Venkateswarlu, JMS 9, 173 (1962).
(11) Richards, Barrow, PCS (1962), 297.
(12) Watanabe, Nakayama, Mottl, JQSRT 2, 369 (1962).
(13) Briggs, Norrish, PRS A 276, 51 (1963).
(14) Douglas, Møller, Stoicheff, CJP 41, 1174 (1963).
(15) Jacobs, Giedt, JQSRT 5, 457 (1965).
(16) Clyne, Coxon, PRS A 298, 424 (1967).
(17) Todd, Richards, Byrne, TFS 63, 2081 (1967).
(18) Clyne, Stedman, TFS 64, 1816 (1968).
(19) Palmer, Carabetta, JCP 49, 2466 (1968).
(20) Winkel, Hunt, Clouter, JCP 50, 1298 (1969).
(21) Busch, Mahoney, Morse, Wilson, JCP 51, 449 (1969).
(22) Radziemski, Kaufman, JOSA 59, 424 (1969).
(23) Clyne, Coxon, JMS 33, 381 (1970).
(24) Holzer, Murphy, Bernstein, JCP 52, 399 (1970).
(25) Cornford, Frost, McDowell, Ragle, Stenhouse, JCP 54, 2651 (1971).
(26) Coxon, JQSRT 11, 443, 1355 (1971).
(27) Hochenbleicher, Schrötter, AS 25, 360 (1971).
(28) Le Roy, Bernstein, JMS 37, 109 (1971).
(29) Potts, Price, TFS 67, 1242 (1971).
(30) Goscinski, MP 24, 655 (1972).
(31) Hendra, Vear, SA A 28, 1949 (1972).
(32) Le Roy, CJP 50, 953 (1972).
(33) Wallart, CJS 17, 128 (1972).
(34) Wieland, Tellinghuisen, Nobs, JMS 41, 69 (1972).
(35) Child, Bernstein, JCP 59, 5916 (1973).
(36) Le Roy, in "Molecular Spectroscopy", Vol. 1, p. 113. The Chemical Society (1973).
(36a) Coxon, in "Molecular Spectroscopy", Vol. 1, p. 177. The Chemical Society (1973).
(37) Yee, Stone, MP 26, 1169 (1973).
(38) Le Roy, CJP 52, 246 (1974).
(39) Ault, Howard, Andrews, JMS 55, 217 (1975).
(40) Brith, Rowe, Schnepp, Stephens, CP 9, 57 (1975).
(41) Douglas, Hoy, CJP 53, 1965 (1975).
(42) Bondybey, Fletcher, JCP 64, 3615 (1976).
(43) Edwards, Good, Long, JCS FT II 72, 927 (1976).

State	T_e	ω_e	$\omega_e x_e$	B_e	α_e	D_e $(10^{-7}cm^{-1})$	r_e (Å)	Observed Transitions: Design.	Observed Transitions: ν_{00}	References
$^{35}Cl_2^+$		$\mu = 17.4842896_5$		$D_0^0 = 3.95$ eV[a]						SEP 1976
B $(^2\Sigma_g^+)$	$(34400)^b$									
$A_2(^2\Pi_{\frac{1}{2}u})$	$(20000)^b$	c		$[0.1778]^d$	c		$[2.329]^d$	$A_2 \rightarrow X_2$, R	$22746.9_6{}^d$ Z	(4)*
$A_1(^2\Pi_{\frac{3}{2}u})$		c		$[0.1788]^d$	c	[4.9]	$[2.322]^d$	$A_1 \rightarrow X_1$, R	$22199.5_4{}^d$ Z	(4)*
$X_2\ {}^2\Pi_{\frac{1}{2}g}$	645^b	644.77 Z	2.988	0.2697	0.00167		1.890_7			
$X_1\ {}^2\Pi_{\frac{3}{2}g}$	0	645.61 Z	3.015^e	0.2695_0	0.00164	1.8	1.8915			
$^{(35)}Cl_2^-$		$(\mu = 17.4845639_4)$		$D_0^0 = 1.26$ eV[a]		I.P. = 2.39 eV[b]				SEP 1976 A
$(^2\Pi_g)$	[79400]	650^c								(6)
X $(^2\Sigma_u^+)$	[0]	For computed ground state characteristics see (3).								
$^{35}Cl^{19}F$		$\mu = 12.3102869$		$D_0^0 = 2.617_3$ eV[a]		I.P. = 12.66 eV[b]				SEP 1976 A
		For binding energies of F(1s) and Cℓ(2s) core electrons in CℓF see (14).								
B $^3\Pi_{0^+}$	$18826._4$	363.1 Z	$8.6_4{}^c$	0.3319^d	0.0047^e	(10)	2.031	B← X, R	$18614._3{}^f$ Z	(2)(5)
X $^1\Sigma^+$	0	786.15^g Z	$6.16_1{}^g$	0.516478_8	0.004357_7	8.7_7	1.628313	Infrared sp., Raman sp. (liquid), microwave and mol. beam el. reson. sp..[h]		(3) (1) (9)(15)
$^{(35)}Cl^{19}F^+$				$D_0^0 = 2.93$ eV[i]						SEP 1976
B $(^2\Sigma^+)$	$(41500)^j$									(7)(10)
A $(^2\Pi)$	$(29500)^j$									(7)(10)
X $^2\Pi_{\frac{1}{2}}$	630^j	870^k								(7)(10)
X $^2\Pi_{\frac{3}{2}}$	0									

$C\ell_2^+$: [a]From D_0^0 ($C\ell_2$), I.P.($C\ell_2$), and I.P.($C\ell$).

[b]From the photoelectron spectrum (5)(6).

[c]Several higher vibrational levels have been observed and analyzed (4) but they do not form regular progressions. The vibrational constants quoted in (1a) from (1) just as the later values of (2) must be considered as erroneous (4). The partial rotational analysis of (3) agrees essentially with the more extensive one of (4).

[d]Lowest observed vibrational level, not necessarily v=0; see [c].

[e]$\omega_e y_e$ = + 0.007.

(1) Elliott, Cameron, PRS A 158, 681 (1937); 164, 531, (1938).

(1a) Herzberg, MOLSPEC 1 (1950).

(2) Haranath, Rao, IJP 32, 401 (1958).

(3) Rao, Rao, CJP 36, 1557 (1958).

(4) Huberman, JMS 20, 29 (1966).

(5) See ref. (25) of $C\ell_2$.

(6) See ref. (29) of $C\ell_2$.

$C\ell_2^-$: [a]From D_0^0 ($C\ell_2$) and the electron affinities of $C\ell_2$ and $C\ell$ (3.613 eV).

[b]From endoergic charge transfer (2)(4)(5)(7); slightly higher value (2.5$_2$ eV) from dissociative electron attachment (1).

[c]Single progression of resonances in the scattering of electrons by $C\ell_2$ indicating the existence of a doubly excited (preionizing) state of $C\ell_2^-$ with the X $^2\Pi_g$ state of $C\ell_2^+$ as "grandparent" (6).

(1) DeCorpo, Franklin, JCP 54, 1885 (1971).

(2) Chupka, Berkowitz, Gutman, JCP 55, 2724 (1971).

(3) Gilbert, Wahl, JCP 55, 5247 (1971).

(4) Baede, Physica 59, 541 (1972).

(5) Hughes, Lifshitz, Tiernan, JCP 59, 3162 (1973).

(6) Spence, PR A 10, 1045 (1974).

(7) Tang, Leffert, Rothe, Reck, JCP 62, 132 (1975).

$C\ell F$, $C\ell F^+$:

[a]Of the two possible values derived from the convergence of the B−X system thermochemical data (16) strongly favour the higher value given here. But (17) considers the lower value (2.558$_1$ eV) equally likely; see also (6).

[b]From the photoelectron spectrum (7)(10); in good agreement with photoionization [12.65 eV (6)] and electron impact [12.7 eV (4)] appearance potentials.

[c]$\omega_e y_e = -0.12_4$. The vibrational and rotational constants have been recalculated from data in (5) for $3 \le v \le 8$. Lower levels are not observed, higher levels are affected by a perturbation. See also (13). Convergence of v'-0 absorption progression at 21514 cm^{-1}.

[d]Predissociation (diffuseness) of 11-0, 12-0, 13-0 bands at 21254 - 21399 cm^{-1}. See [c].

[e]$\gamma_e = -0.00047_7$; see [c].

[f]Extrapolated from the 3-0 band.

[g]Recalculated by (17) from the infrared data of (3). ν_0 of the fundamental, 773.83 cm^{-1}, agrees rather poorly with $\Delta G(\frac{1}{2}) = 773.46$ from the electronic absorption spectrum (5).

[h]$\mu_{e\ell}(v=0)$ = 0.8881 D (9); the sign of the dipole moment is +ClF- (11)(14)(18), see however (8)(12). eqQ($C\ell$) and other hyperfine structure constants in (9)(15). Zeeman effect (8), $g_J = -0.1089\ \mu_N$.

[i]From D_0^0($C\ell F$), I.P.($C\ell F$), and I.P.($C\ell$).

[j]From photoelectron spectra (7)(10).

[k]From (10); (7) give 912 ± 30 cm^{-1}.

(1) Jones, Parkinson, Burke, JCP 18, 235 (1950).

(2) Schumacher, Schmitz, Brodersen, AAQA 38, 98 (1950).

(3) Nielsen, Jones, JCP 19, 1117 (1951).

(4) Irsa, Friedman, JINC 6, 77 (1958).

(5) Stricker, Krauss, ZN 23 a, 1116 (1968).

(6) Dibeler, Walker, McCulloh, JCP 53, 4414 (1970).

(continued p. 153)

State	T_e	ω_e	$\omega_e x_e$	B_e	α_e	D_e ($10^{-6}cm^{-1}$)	r_e (Å)	Observed Transitions Design.	Observed Transitions ν_{00}	References
$^{35}Cl^{16}O$		$\mu = 10.9749309_5$		$D_0^0 = 2.7505$ eV[a]		I.P. = 11.0 eV[b]				SEP 1976 A
		Several unclassified absorption bands in the region 67000 - 79000 cm^{-1}.								(14)
H	(74125)	(1025) H						H← X, V	74212 H	(14)
G ($^2\Pi$)	73913 73878	1075 H	10					G← X, V	73705 H 73986 H	(14)*
F ($^2\Sigma$)	69109	[1001][c] H						F← X, V	68869 H 69181 H	(14)*
E ($^2\Sigma$)	67333	1070 H	4					E← X, V	67120 H 67445 H	(14)*
D ($^2\Sigma$)	64486	1050 H	2					D← X, V	64269 H 64582 H	(14)*
C ($^2\Sigma$)	58448	1062 H	3					C← X, V	58234 H 58554 H	(14)*
A $^2\Pi_i$	32169 31650	$519._5$ Z	7.2[d]	0.445_0[e]	0.006[f]	[1.31]	1.858	A↔X, R	$31682._9$[g] Z $31482._3$[g] Z	(1)(2)(3)* (4)* (16)* (17)*
X $^2\Pi_i$	318[h] 0	$853._8$[i] Z	5.5[j]	0.62344_8[k]	0.0058[ℓ]	[1.33]	1.56963	Matrix IR and Raman sp.		(12)(15)
								Microwave sp.[m]		(7)(8)
								EPR sp.[m]		(5)(6)(9)(11)
$^{(35)}Cl^{16}O^-$						I.P. = 2.5 eV[n]				SEP 1976

ClO, ClO^-:

[a]From the convergence limit of the A← X, $^2\Pi_{\frac{3}{2}} \leftarrow {}^2\Pi_{\frac{3}{2}}$ subbands (16) assuming dissociation into $^2P_{\frac{3}{2}} + {}^1D$.

[b]Mass-spectrometric studies and theoretical calculations; see (10).

[c]$\Delta G(3/2) = 950$, $\Delta G(5/2) = 980$.

[d]$\omega_e y_e = -0.11$; see [e].

[e]In absorption observed to v'=25 of $^2\Pi_{\frac{3}{2}}$ [B_v, D_v values in (16)(17)], close to the dissociation limit at 38052 cm^{-1} above X $^2\Pi_{\frac{3}{2}}$(v=0). The constants are for levels with $v \leq 9$. All absorption and emission bands are diffuse on account of predissociation. Linewidths in different bands vary from 0.3 to 3.1 cm^{-1} and are >5 cm^{-1} in v'=6 (16).

$C\ell O$, $C\ell O^-$ (continued):

[f] $\gamma_e = + 0.00002_4$.

[g] Note that there have been several changes of vibrational numbering [see, e.g., (13)]; here we have used that of (16) and (17).

[h] From the VUV absorption spectrum (14); confirmed by EPR spectrum (5)(11).

[i] The effective ω_e values for $^2\Pi_{\frac{1}{2}}$ and $^2\Pi_{\frac{3}{2}}$ are $853._0$ and 854.9 cm^{-1}, resp. (17).

[j] $\omega_e y_e = - 0.02$.

[k] From combined microwave and ultraviolet data; the microwave values of (8) are $B_0(^2\Pi_{\frac{3}{2}}) = 0.619773$ and $B_0(^2\Pi_{\frac{1}{2}}) = 0.621231$. Λ-type doubling in the $^2\Pi_{\frac{1}{2}}$ component, $\Delta v_{fe} = (+)0.02249(J+\frac{1}{2})$.

[ℓ] $\gamma_e = + 0.000016$.

[m] Dipole moment from Stark effect of microwave spectrum $\mu_{e\ell}(v=0) = 1.23_9$ D (8), from Stark effect of EPR spectrum 1.1_8 D (6). eqQ and other hyperfine structure constants in (8)(11).

[n] Indirectly from thermochemical data and mass-spectrometric studies; also theoretical calculations; see (10).

(1) Pannetier, Gaydon, Nature 161, 242 (1948).
(2) Porter, DFS 9, 60 (1950).
(3) Porter, Wright, ZE 56, 782 (1952).
(4) Durie, Ramsay, CJP 36, 35 (1958).
(5) Carrington, Dyer, Levy, JCP 47, 1756 (1967).
(6) Carrington, Levy, Miller, JCP 47, 3801 (1967).
(7) Amano, Hirota, Morino, JMS 27, 257 (1968).
(8) Amano, Saito, Hirota, Morino, Johnson, Powell, JMS 30, 275 (1969).
(9) Uehara, Morino, JMS 36, 158 (1970).
(10) O'Hare, Wahl, JCP 54, 3770 (1971).
(11) Uehara, Tanimoto, Morino, MP 22, 799 (1971).
(12) Andrews, Raymond, JCP 55, 3087 (1971).
(13) Briggs, Nature PS 239, 13 (1972).
(14) Basco, Morse, JMS 45, 35 (1973).
(15) Chi, Andrews, JPC 77, 3062 (1973).
(16) Coxon, Ramsay, CJP 54, 1034 (1976).
(17) Coxon, Jones, Skolnik, CJP 54, 1043 (1976).

$C\ell F$, $C\ell F^+$ (continued):

(7) Anderson, Mamantov, Bull, Grimm, Carver, Carlson, CPL 12, 137 (1971).
(8) Ewing, Tigelaar, Flygare, JCP 56, 1957 (1972).
(9) Davis, Muenter, JCP 57, 2836 (1972).
(10) DeKock, Higginson, Lloyd, Breeze, Cruickshank, Armstrong, MP 24, 1059 (1972).
(11) Green, JCP 58, 3117 (1973).
(12) McGurk, Norris, Tigelaar, Flygare, JCP 58, 3118 (1973).
(13) See ref. (35) of $C\ell_2$.
(14) Carroll, Thomas, JCP 60, 2186 (1974).
(15) Lovas, Tiemann, JPCRD 3, 609 (1974).
(16) Nordine, JCP 61, 224 (1974).
(17) Coxon, CPL 33, 136 (1975).
(18) Janda, Klemperer, Novick, JCP 64, 2698 (1976).

State	T_e	ω_e	$\omega_e x_e$	B_e	α_e	D_e $(10^{-6}cm^{-1})$	r_e (Å)	Observed Transitions Design.	Observed Transitions ν_{00}	References
$^{12}C^{14}N$		$\mu = 6.46219329$	$D_0^0 = 7.7_6$ eV[a]		I.P. = 14.1_7 eV[b]					SEP 1976 A
		Theoretical work (24)(46)(48)(50)(63). RKR potential curves (15)(44)(53). Franck-Condon factors (21a)(53), and ref. in (27)(44).								
J $^2\Delta_i$	65258.19[c]	1121.76 Z	14.203[d]	1.305_2	0.020_8	5.8	1.413_7	J→A, R	55667.14[e] Z	(8)*
H $^2\Pi_{(r)}$	[61969.7][f]			[1.520]			[1.310]	H→B, R	35140.8_4[e] Z	(7)
G $^2\Pi$	[61655.0][gh]			[1.085][g]			[1.551]	G→B, R	34826.1_0[g] Z	(7)(43)
F $^2\Delta_r$	60095.6_4[i]	1239.5_0 Z	12.7_5	1.3834	0.0187	7	1.3732	F→A, R	50563.8_0[e] Z	(7)* (22) (47)
E $^2\Sigma^+$	59151.1_8	1681.43 Z	3.60[j]	1.4871	0.00643[k]	5.0	1.3245	E→A, R	49842.4_7	(8)* (47)
								E↔X, R	58959.85 Z	(8)* (43)*
D $^2\Pi_i$	54486.3[l]	1004.7_1[m] Z	8.7_8	1.162[m]	0.013	7	1.498	D→A, R	44838.0_8[e] Z	(7)*
								D→X, R	53955.4_6	(7)*
a $^4\Sigma^{(+)}$	(32400)[n]									
B $^2\Sigma^+$	25752.0	2163.9[o] Z	20.2[e']	1.973[o]	0.023[e']	[6.6]	1.150	B→A, V	16680.4_6	(31)(54)*
		[For a comprehensive review of molecular data on the B−X system including a bibliography of references prior to 1972 see (44)].						B[p]↔X,[q] V_R	25797.84 Z	(1)* (2)* (4a)(5)(7) (9)(32)(64)
A $^2\Pi_i$	9245.28[r]	1812.5_6[s] Z	12.60_9[t]	1.7151[s]	0.01708[u]	5.93[sv]	1.2333	A[w]↔X,[x] R	9117.38[e] Z	(3)* (3a) (4b)(7)(9) (16)* (28) (32)(44)*
		[For a useful compilation of molecular data on the A−X system and a bibliography of references prior to 1971 see (44).]								
X $^2\Sigma^+$	0	2068.59 Z	13.087[y]	1.8997_4[z]	0.01736_9[a']	6.40[b']	1.1718_2	IR fundamental b.		(62)(70)
								Microwave sp.[c']		(66)(71)
								ESR sp.[d']		(41)(73)

CN: [a]Based on ΔH^o_{f0}(CN) = 103.2 ± 2.5 kcal/mole recommended in (46a). It represents an average value obtained from mass-spectrometric (14) and shock tube studies (10)(13)(39)(56) and from the photo-ionization (25)(33)(37) and -dissociation (35) of CN containing molecules and their decomposition in collisions with Ar metastable atoms (34). The most recent quantitative absorption measurements by (69) together with lifetime measurements in the B – X system (see [p]) tend to support the slightly higher value, ΔH^o_{f0} = 105.5 ± 2 kcal/mole, of (37), giving $D^0_0 = 7.6_6$ eV. Gaydon (30) recommends 7.7_5 eV.

[b]Indirectly from a combination of photo-ionization (25)(33) and -dissociation (35) threshold energies for HCN and CℓCN; in good agreement with an electron impact appearance potential of 14.2 eV (14). Similar photoionization data for HCN (37) suggest, however, 14.03 eV. Theory predicts 14.10 eV (69a).

[c]$A_0 = -26.9_0$, $A_1 = -25.6_7$, $A_2 = -25.4_0$, $A_3 = -24.9_7$; recalculated from (8) with an improved A_0 for A $^2\Pi_i$.

[d]$\omega_e y_e$ = + 0.180.

[e]Refers to the zero point of the Hill-Van Vleck formula for $\Lambda \neq 0$ states.

[f]$A_0 = (+)49.5_7$.

[g]Vibrational numbering unknown; only the v'→1 band at 32702.6_0 cm^{-1} was identified. See also (48).

[h]|A| even smaller than in the D state.

[i]A_0 = + 28.77, A_1 = + 27.86, A_2 = + 27.06.

[j]$\omega_e y_e$ = - 1.02.

[k]γ_e = - 0.00077.

[ℓ]A_0 = - 3.03.

[m]Vibrational numbering uncertain.

[n]Predicted by ab initio calculations (46)(48). This state is probably responsible for the small perturbation in v=11, N=20 of B $^2\Sigma^+$ (67); see also (58).

from (7); much improved $\Delta G(v+\frac{1}{2})$, B_v, D_v values for v ≤ 7 in (64), but no equilibrium constants given. Spin-rotation interaction constant γ_0 = + 0.01565 [from the magnetic resonance spectrum (59)], for higher v see (64). Perturbations by A $^2\Pi$ (in v=0 at N=4, 7, 11, 15; see [s]), and by a $^4\Sigma^{(+)}$ (see [n]). From Stark effect of the P(1) and R(0) lines in the B – X, 0-0 band (36) determine for the B state $\mu_{e\ell}(^-CN^+)$ = 1.15 D.

[p]Radiative lifetime τ(v=0) = $65._6$ ns for the unperturbed levels and 72 ns for the pertubed N=4 level [tunable laser excitation (65)]. Phase shift measurements (42) give 59 ns for v=0-2, in good agreement with the direct decay measurements, $60._8$ ns, of (57). From the zero-electric-field-limit linewidth of the anticrossing spectrum (51) derive 39 ns. The electronic f value for the B – X system at 3860 Å, f = 0.033, is primarily based on shock tube emission and absorption measurements; for a review see (56).

[q]For 0-0, 0-1, 0-2 bands of $^{13}C^{14}N$ see (4).

[r]$A_v = -52.64 + 0.036_5 v + 0.0086 v^2$, v ≤ 12 (21).

[s]The vibrational and rotational constants given here are those of (21) based on the measurements of (16); see also (7). More elaborate evaluations, primarily intended to assist in the identification of CN lines in the solar spectrum, have been published by (43a) and (45); note, however, that the constants tabulated in (43a) are not the usual Dunham coefficients. An extension of the system to bands with v' ≤ 25 (32) requires the use of slightly different vibrational constants.

Λ-type doubling, for details see (3)(21). The rotational perturbations by B $^2\Sigma^+$ in v=10 have been the subject of many experimental and theoretical investigations (12)(17): microwave-optical double resonance (11)(18)(19)(29)(38)(40)(52), level anticrossing spectroscopy (49)(51)(60), magnetic (continued p. 156)

CN (continued):

resonance (59). Hyperfine structure constants in (52), $\mu_{e\ell}(^+CN^-) = 0.56$ D (60), $\Pi \sim \Sigma$ interaction parameters (59).

[t] $\omega_e y_e = -0.0118$.

[u] $\gamma_e = -0.000036_4$.

[v] $\beta_e = +0.042_5 \times 10^{-6}$.

[w] Radiative lifetime $\tau = 0.68$ μs (average for $1 \leq v \leq 9$), corresponding to $f_{00} = 0.0034$ (20), in good agreement with (70). A shock tube emission study (55) gives $f_{e\ell} = 0.0045$ at 10970 Å. See also (26)(68). A considerably shorter life-time for v=10, $\tau = 0.14$ μs, was derived by (51) from the anticrossing and microwave linewidths. Calculated relative absorption coefficients at three different temperatures in (61).

[x] Rotational analyses of the 0-0, 1-0, 2-0, 2-1, 3-1 bands of $^{13}C^{14}N$ in (23)(72).

[y] $\omega_e y_e = -0.00909_3$. Vibrational constants from (7), slightly different constants in (21).

[z] Spin-splitting constant $\gamma_0 = +0.00725$, from the microwave spectrum (66)(71); see also (21)(64). $\mu_{e\ell}(^+CN^-) = 1.45$ D from Stark effect in the B−X, 0-0 band (36).

[a'] $\gamma_e = -0.0000310_7$. Rotational constants from (64) who gives $\gamma_e = -0.0003107$ which appears to be a misprint; see also (7)(21).

[b'] $\beta_e = +0.01_2 \times 10^{-6}$ (21)(64).

[c'] In emission from interstellar sources; eqQ and other hyper-fine structure constants.

[d'] In rare gas matrices at 4K.

[e'] The ΔG and B_v curves are strongly non-linear.

(1) Jevons, PRS A 112, 407 (1926).
(2) Jenkins, PR 31, 539 (1928).
(3) Jenkins, Roots, Mulliken, PR 39, 16 (1932).
(3a) Parker, PR 41, 274 (1932).
(4) Jenkins, Wooldridge, PR 53, 137 (1938).
(4a) White, JCP 8, 79, 459 (1940).
(4b) Herzberg, Phillips, ApJ 108, 163 (1948).
(5) Feast, PPS A 62, 121 (1949).
(6) Kiess, ApJ 109, 551 (1949).
(7) Douglas, Routly, ApJ(Suppl.) 1, 295 (1955).
(8) Carroll, CJP 34, 83 (1956).
(9) Kiess, Broida, JMS 7, 194 (1961).
(10) Knight, Rink, JCP 35, 199 (1961).
(11) Barger, Broida, Estin, Radford, PRL 9, 345 (1962).
(12) Radford, Broida, PR 128, 231 (1962).
(13) Tsang, Bauer, Cowperthwaite, JCP 36, 1768 (1962).
(14) Berkowitz, JCP 36, 2533 (1962).
(15) Fallon, Vanderslice, Cloney, JCP 37, 1097 (1962).
(16) Davis, Phillips, BAMS 1 (1963).
(17) Radford, Broida, JCP 38, 644, 3031(Erratum) (1963).
(18) Evenson, Dunn, Broida, PR A 136, 1566 (1964).
(19) Radford, PR A 136, 1571 (1964).
(20) Jeunehomme, JCP 42, 4086 (1965).
(21) Poletto, Rigutti, NC 39, 519 (1965).
(21a) Halmann, Laulicht, ApJ(Suppl.) 12, 307 (1966).
(22) Jha, Rao, PIAS A 63, 316 (1966).

CN (continued):

(23) Wyller, ApJ 143, 828 (1966).
(24) Purcell, JCP 47, 1198 (1967); 48, 5735 (1968).
(25) Dibeler, Liston, JCP 47, 4548 (1967).
(26) Gippius, Kudryavtsev, Pechenov, Sobolev, HT(USSR) 5, 27 (1967).
(27) Ortenberg, Antropov, SPU 9, 717 (1967).
(28) Weinberg, Fishburne, Rao, JMS 22, 406 (1967).
(29) Evenson, APL 12, 253 (1968).
(30) Gaydon, DISSEN (1968).
(31) LeBlanc, JCP 48, 1841 (1968).
(32) LeBlanc, JCP 48, 1980 (1968).
(33) Dibeler, Liston, JCP 48, 4765 (1968).
(34) Setser, Stedman, JCP 49, 467 (1968).
(35) Davis, Okabe, JCP 49, 5526 (1968).
(36) Thomson, Dalby, CJP 46, 2815 (1968).
(37) Berkowitz, Chupka, Walter, JCP 50, 1497 (1969).
(38) Evenson, PR 178, 1 (1969).
(39) Levitt, Parsons, TFS 65, 1199 (1969).
(40) Pratt, Broida, JCP 50, 2181 (1969).
(41) Easley, Weltner, JCP 52, 197 (1970).
(42) Liszt, Hesser, ApJ 159, 1101 (1970).
(43) Lutz, CJP 48, 1192 (1970).
(43a) Swensson, Benedict, Delbouille, Roland, MSRSL* No. 5 (1970).
(44) Brocklehurst, Hébert, Innanen, Seel, Nicholls, IAMS 8 (1971); IAMS 9 (1972).
(45) Fay, Marenin, van Citters, JQSRT 11, 1203 (1971).
(46) Heil, Schaefer, ApJ 163, 425 (1971).
(46a) JANAF (1971).
(47) Lutz, ApJ 164, 213 (1971).
(48) Schaefer, Heil, JCP 54, 2573 (1971).
(49) Levy, JCP 56, 5493 (1972).
(50) Green, JCP 57, 4694 (1972).
(51) Cook, Levy, JCP 57, 5059 (1972).
(52) Meakin, Harris, JMS 44, 219 (1972).
(53) Rao, Lakshman, JQSRT 12, 1063 (1972).
(54) Schoonveld, JQSRT 12, 1139 (1972); JCP 58, 403 (1973).
(55) Arnold, Nicholls, JQSRT 12, 1435 (1972).
(56) Arnold, Nicholls, JQSRT 13, 115 (1973).
(57) Luk, Bersohn, JCP 58, 2153 (1973).
(58) Coxon, Setser, Duewer, JCP 58, 2244 (1973).
(59) Cook, Levy, JCP 58, 3547 (1973).
(60) Cook, Levy, JCP 59, 2387 (1973).
(61) Phillips, Leung, ApJ 180, 607 (1973).
(62) Phillips, ApJ 180, 617 (1973).
(63) Das, Janis, Wahl, JCP 61, 1274 (1974).
(64) Engleman, JMS 49, 106 (1974).
(65) Jackson, JCP 61, 4177 (1974).
(66) Penzias, Wilson, Jefferts, PRL 32, 701 (1974).
(67) Coxon, Ramsay, Setser, CJP 53, 1587 (1975).
(68) Jain, JQSRT 15, 571 (1975).
(69) Engleman, Rouse, JQSRT 15, 831 (1975).
(69a) Moffat, JMSt 25, 303 (1975).
(70) Treffers, ApJ 196, 883 (1975).
(71) Turner, Gammon, ApJ 198, 71 (1975).
(72) Hosinsky, Lindgren, AA(Suppl.) 25, 1 (1976).
(73) Adrian, Bowers, CPL 41, 517 (1976).

State	T_e	ω_e		$\omega_e x_e$	B_e	α_e	D_e (10^{-6}cm^{-1})	r_e (Å)	Observed Transitions Design.		ν_{00}		References
$^{12}C^{14}N^+$					$D_0^0 = 4.8_5$ eV[a]								SEP 1976
d $^1\Pi$	a + (46253)	[890.76]	Z	[b]	[1.6018]	[c]	[19.5][c]	[1.2762]	d[d] → b,	R	37544.49	Z	(1)* (4)*
f $^1\Sigma$	a + 45533.6	2670.5	Z	46.9	1.903[e]	0.032	[4.7]	1.171	f → b,	V	37703.14	Z	(4)*
									f → a,	[f]	45844.65	Z	(1)*
c $^1\Sigma$	a + (31771)	(1265)[g]		(11)	(1.403)[g]	(0.002)	[13]	(1.364)	c → a,	R	31381.60[g]	Z	(1)*
b $^1\Pi$	a + 8313.6	1688.35	Z	15.12	1.6767	0.0191[h]	[6.84]	1.2473					
a $^1\Sigma$	a	2033.05	Z	16.14	1.8964	0.0188	[7.0]	1.1729					
$^{12}C^{14}N^-$					$D_0^0 = 10.3_1$ eV[a]		I.P. = 3.82 eV[b]						SEP 1976 A

$^{12}C^{16}O$

$\mu = 6.85620871$ $D_0^0 = 11.09_2$ eV[a] I.P. = 14.0139 eV[b] — OCT 1976 A

Absorption in the 100 - 20 Å region. K absorption of C and O.[c] — (93)

Absorption cross sections from 650 - 180 Å. — (115)(149)(157)

Several weak and fragmentary progressions in absorption in the region 620 - 530 Å (161000 - 186000 cm^{-1}), probably corresponding to excitation of two electrons and tentatively assigned as first members of Rydberg series converging to higher electronic states of CO^+ derived from the photoelectron spectrum [revised assignments (125)].[d] — (130)

Rydberg series[e] converging to B $^2\Sigma^+$(v=0) of CO^+ (also series or fragments of series with v'=1, 2)[f]:

	Series	Formula	References
(ndσ,π)	Ogawa and Ogawa's series IV (joining on to R)	$\nu = 158664 - R/(n-0.19)^2$; n = 3,4,...,10.	(101)*
(npσ,π)	Tanaka's diffuse series (joining on to D_2, D_3)	$\nu = 158664 - R/(n-0.55)^2$; n = 4,5,...,9.	(17)* (101)*
(npσ,π)	Ogawa and Ogawa's series V (joining on to U)	$\nu = 158664 - R/(n-0.61)^2$; n = 5,6,7.	(101)*
(npσ,π)	Tanaka's sharp series (joining on to S_1, S_2)	$\nu = 158664 - R/(n-0.650-0.084/n-0.13/n^2)^2$; n = 3,4,...,13.	(17)* (101)*
(nsσ)	Ogawa and Ogawa's series III (joining on to T[g])	$\nu = 158664 - R/(n-0.902-0.232/n)^2$; n = 4,5,...,18.	(101)*

CN^+: [a]From D_0^0(CN) and the ionization potentials of CN and C.

[b]$\Delta G(3/2) = 548.54$, $\Delta G(5/2) = 542.18$, $\Delta G(7/2) = 550.36$. Vibrational numbering uncertain.

[c]$B_1 = 1.4319$, $D_1 = 47.3 \times 10^{-6}$;
$B_2 = 1.3023$, $D_2 = 17.9 \times 10^{-6}$;
$B_3 = 1.2598$, $D_3 = 19.4 \times 10^{-6}$;
$B_4 = 1.2196$, $D_4 = 9 \times 10^{-6}$.
Local perturbations in nearly all vibrational levels. Vibrational numbering uncertain.

[d]$\tau = 24$ ns (3).

[e]Perturbations; see p. 251 of (2a).

[f]0-0 sequence consisting of four headless bands.

[g]Vibrational numbering uncertain. Homogeneous perturbations by an unidentified state.

[h]$\gamma_e = -0.0005$.

(1) Douglas, Routly, ApJ 119, 303 (1954).
(2) See ref. (8) of CN.
(2a) Mulliken, JCP 33, 247 (1960).
(3) Smith, JCP 51, 3410 (1969).
(4) Lutz, ApJ 163, 131 (1971).
(5) See ref. (69a) of CN.

CN^-: [a]From D_0^0(CN) and the electron affinities of CN and C (1.268 eV).

[b]Photoionization of HCN (1). Earlier measurements are reviewed in this paper. Theory (2) predicts 3.69 eV.

(1) See ref. (37) of CN.
(2) Griffing, Simons, JCP 64, 3610 (1976).

CO: [a]From the predissociation in the B $^1\Sigma^+$ state (see [n] p.163). The uncertainty of ±0.017 eV corresponds to the uncertainty as to which combination of ^3P component states arises at the dissociation limit.

[b]From Rydberg series (101). For the second and third I.P. (1π and 4σ orbitals) see the higher Rydberg limits in the Table. The fourth, fifth, and sixth I.P. (3σ, 2σ, 1σ) have been determined from X-ray photoelectron spectroscopy to be 38.9, 296.24 (K limit of C), and 542.57 eV (K limit of O), respectively (62)(78)(162). See also [f].

[c]Preceding the two K limits (see [b]) are strong Rydberg series of absorption bands. The longest-wavelength absorptions correspond to excitation to the 2π orbital yielding a weak $^3\Pi \leftarrow X^1\Sigma^+$ and a strong $^1\Pi \leftarrow X^1\Sigma^+$ peak at 283 and 285 eV for the transitions from $1s_C$ and at 529 and 532 eV for the transitions from $1s_O$ (93). The transitions to the $^1\Pi$ states have also been observed in electron impact experiments at 287.7 and 534.4 eV (79).

[d]Dissociation produced by absorption in these bands and subsequent atomic fluorescence (153); predissociation into $C^+ + O^-$ (151).

[e]Absorption and photoionization coefficients from 1000 to 600 Å (33).

[f]Calculated Franck-Condon factors for ionization $X\ ^2\Sigma^+ \leftarrow X\ ^1\Sigma^+$, $A\ ^2\Pi \leftarrow X\ ^1\Sigma^+$, and $B\ ^2\Sigma^+ \leftarrow X\ ^1\Sigma^+$, see (51) and (53). Observed Franck-Condon factors from photoelectron spectrum (44)(82)(132); absolute ionization cross-sections (113)(161).

[g]The progression P_5 of (17) [called T in MOLSPEC 1 and (40)] must be reclassified as representing the members n=6, 7, and 9 of series III; see the spectrograms of (17) and (101).

State	T_e	ω_e	$\omega_e x_e$	B_e	α_e	D_e $(10^{-6}cm^{-1})$	r_e (Å)	Observed Transitions: Design.	Observed Transitions: ν_{00}	References
$^{12}C^{16}O$ (continued)										
D_3	(153271)	(1705)	(18.5)					$D_3 \leftarrow X$, [a]	153037	(17)* (101)*
U	(153199)	[1676]						$U \leftarrow X$,	152955	(101)*
D_2	(149294)	(1730)	(30)					$D_2 \leftarrow X$, [a]	149070	(17)* (101)*
S_2	(148929)	(1750)	(30)					$S_2 \leftarrow X$,	148715	(17)* (101)*
T	(147065)	(1658)	(11)	e				$T \leftarrow X$,	146810	(101)*
R	(144939)	(1735)	(28)	e				$R \leftarrow X$,	144718	(101)*
S_1	(138038)	(1771)	(29)	e				$S_1 \leftarrow X$,	137835	(101)*
Rydberg series[b] converging to A $^2\Pi(v=0)$ of CO^+ [also series with v'=1...8 (Ogawa a. Ogawa) and v'=1,2,3 (Tanaka)][c];										
(nsσ)		Ogawa and Ogawa's series[d] (joining on to W [n=3, see (135)] and O_1, O_2)		$\nu = 133484(A^2\Pi_{\frac{1}{2}}) - R/(n-1.077+...)^2$; n = 4,5,...,12.						(101)*
(npσ,π)		Tanaka's α series (joining on to P)		$\nu = 133380(A^2\Pi_{\frac{3}{2}}) - R/(n-0.67)^2$; n = 4,5,...,8.						(17)* (101)
Q	129043	1558	10.6					$Q \leftarrow X$, [b] R	128738	(17)
$O_2(^1\Pi)$	126729	1560	13.3	e				$O_2 \leftarrow X$, R	126424	(17)* (101)*
P	123656	[1521]						$P \leftarrow X$, R	123335	(17)* (101)*
$O_1(^1\Pi)$	121137	1570	13.4	e				$O_1 \leftarrow X$, R	120837	(4)(101)*
N	(119882)	(1600)		e				$N \leftarrow X$, [b]	(119600)	(4)
Rydberg series[b] converging to X $^2\Sigma^+(v=0)$ of CO^+ [also series with v'=1][c];										
npπ		Ogawa and Ogawa's series[d] (joining on to E, L)		$\nu_\infty = 113029$; formula not given, merging into npσ above n=8.						(101)*
npσ		Ogawa and Ogawa's series[d] (joining on to C, K)		$\nu = 113029 - R/(n-0.615-0.263/n-0.165/n^2)^2$; n = 3,4,...,32.						(101)*
nsσ		Lindholm's series (joining on to B, J, I')		$\nu_\infty = 113029$; formula not given, n = 3,4,...,10.						(52)
I' (5sσ)								$I' \leftarrow X$,	106383[f]	(31)* (52)
Z $^1\Sigma^+$				[(1.9)]	(v' unknown)		[(1.1_4)]	$Z \leftrightarrow X$, [b] R	105724[g] (Z)	(140)
$H^h(^1\Pi)$	(105811)	(1097)	(47)					$H \leftarrow X$, [bi] R Hopfield-Birge b.	105266	(2)

State	T_e	ω_e	$\omega_e x_e$	B_e	α_e	D_e ($10^{-6}cm^{-1}$)	r_e (Å)	Observed Transitions Design.	ν_{00}	References
$^{12}C^{16}O$ (continued)										
H' $^1\Pi$				[1.415]			[1.318]	H'←X, R	$104119._6$ Z	(135)
L $^1\Pi$ 4pπ	103251	[2181]	(15)	[1.981_2]		[14]	[1.114_0]	L←X, V	$103271._5$ Z	(101)* (135)
L'($^1\Pi$)								L'←X, R	103215 H^Q	(135)
K $^1\Sigma^+$ 4pσ				[1.918_9]		[66]	[1.132_0]	K←X, R	$103054._3$ Z	(101)* (135)
W $^1\Pi$ [j]				[1.558_5]		[65]	[1.256_0]	W↔X, R	$102807._1$ Z	(135)* (140)
W' $^1\Pi$				[1.536]			[1.265]	W'←X, R	$102310._6$ Z	(135)*
J $^1\Sigma^+$ 4sσ	(101409)	[$2235._3$] Z	(15)	[1.9203][k]		[5.8][k]	[1.1315][k]	J←X, R	101456 (Z)	(135)*
G $^1\Pi$ (3dπ)				[1.9625]		[7.0]	[1.1193]	G←X, V	$101031._0$ Z	(135)*
G' $^1\Pi$				[(1.59)]	(v' unknown)		[(1.24)]	G'←X, R	100651	(135)*
h ($^3\Sigma^+$, 4sσ)								h←X, R	100392 (Z)	(135)
Y $^1\Sigma^+$				[(1.83)]	(v' unknown)		[(1.16)]	Y→X, R	99963 (Z)	(140)
F $^1\Sigma^+$(3dσ)	(99803)	[2034.4] Z	[ℓ]	[1.86]	[m]	[80][m]	[1.15_0]	F←X,[b] R Hopfield-Birge b.	$99739._0$ Z	(135)*
V $^1\Pi$				[1.70]	(v' unknown)		[1.20]	V↔X, R	$98919._0$ H^Q	(135)(140)

CO: [a]Diffuse looking bands.

[b]See [e] p.159.

[c]See [f] p.159.

[d]These series and progressions from the high resolution work of (101) agree only partially with the early work (17)(18) at lower resolution.

[e]Preionization observed in electro-ionization of CO (114).

[f]This strong absorption band is clearly present but not assigned on the spectrogram of (101).

[g]An absorption band at this wavenumber is visible but not identified on the published spectrogram of (101).

[h]Previously called G [see (40)]. The present G $^1\Pi$ is from (135).

[i]This is the strongest system of (2). It is clearly present on the reproduction of (101), but these authors consider the first band at 950 Å as due to v'=1 of K(4pσ)←X, do not assign the second (strongest) band at 941 Å and consider the third band at 933 Å as n=5, v'=0 of the Rydberg series which starts with C(3pσ) and E(3pπ).

[j]cf. Ogawa and Ogawa's Rydberg series converging to A $^2\Pi_{\frac{1}{2}}$.

[k]v=0 diffuse by predissociation, v=1 sharp. The rotational constants are for v=1 (135).

[ℓ](3a) gives ω_e = 2112, $\omega_e x_e$ = 198, presumably from private communication by Hopfield-Birge.

[m]B_1 = 1.837, D_1 = 3×10^{-6}.

State	T_e	ω_e	$\omega_e x_e$	B_e	α_e	D_e ($10^{-6}cm^{-1}$)	r_e (Å)	Observed Transitions Design.	Observed Transitions ν_{00}	References
$^{12}C^{16}O$ (continued)										
		A $^1\Pi$ state at 98836 cm^{-1} reported by (15) was shown (140) to be due to N_2.								
g $^3\Sigma^+$								g←X, R	$98129._1$ (Z)	(135)
E $^1\Pi$ 3pπ	(92903)	[2153.8] Z	(42)	1.9771[a]	0.0254	[6.5]	1.1152	E→A, V	28185.1_8[b] Z	(86)* (154)
								E←X,[c] V Hopfield-Birge b.	92930.0_3 Z	(37)* (135)
c $^3\Pi$ 3pπ	[93158.5]			[1.935][d]		[-131][e]	[1.127]	c→a, V "3A" bands	43603.7	(57)*
								c←X, V	92076.9 Z	(64)*
C $^1\Sigma^+$ 3pσ	91916.5	2175.9_2 Z	14.7_6[f]	1.9533[g]	0.0196	6.2	1.1219	C[h]→A,[i] V Herzberg b.	27174.4_0[b] Z	(7)
								C[h]↔X,[j] V Hopfield-Birge b.	91919.1_5 Z	(55)(66)
j ($^3\Sigma^+$, 3pσ)	90975	[2166] Z	(15)	[1.8785][k]	(0.020)	[7.9]	[1.1441]	j←X, R	90988.0_4 Z [ℓ]	(55)
		The E_0 $^1\Sigma^+$ state at 90866 cm^{-1} reported by (15) was shown (140) to be due to N_2.								
k	[90972]	Single 0-v" progression.						k→a, V Kaplan b.	41417 H	(3)
B $^1\Sigma^+$ 3sσ	86945.2	2112.7_0 Z	15.2_2[m]	1.9612[n]	0.0261	7.1	1.1197	B[o]→A,[p] V Ångstrom b.	22171.3_5[b] Z	(7)
								B[o]↔X,[q] V Hopfield-Birge b.	86916.16 Z	(55)* (66)
b $^3\Sigma^+$ 3sσ	(83814)	$[2199._3]$ Z		1.986[rs]	0.042		1.113	b[t]→a,[u] V 3rd positive gr.[v]	$35358._5$ Z	(5)(9)(11)
								b←X,[u] V Hopfield-Birge b.	$83831._7$ H	(2)
(f $^3\Sigma^+$)		This state, first suggested by (10) and listed by (40) at T_0 = 83744, is in all probability not a separate state but represents v = 31 and 35 of a' $^3\Sigma^+$.[w]								

CO: [a]Clear case of accidental predissociation for J=31 (e level) at 94872 cm^{-1} above v=0, J=0 of X $^1\Sigma^+$ (136).

[b]The v_{00} values for B→A, C→A, E→A are not deperturbed and, therefore, do not add up with the deperturbed v_{00} for A−X (see also [m] p.165) to the v_{00} values listed for B−X, C−X, and E−X.

[c]Oscillator strength f_{00} = 0.094 (87).

[d]Λ-type doubling, Δv = 0.011 x N(N+1). Triplet splitting unobservably small as for most Rydberg states.

[e]H_0 = − 1.9 x 10^{-7}. The rotational constants represent average values for the two Λ-doubling components.

[f]Only v=0 and 1 observed in absorption, only v=0 in emission. ω_e, $\omega_e x_e$ derived with the aid of isotope ($^{12,13}CO$) data, see (55).

[g]$\mu_{e\ell}$ = 4.5 D, from Stark effect observations on the Herzberg bands (155).

[h]Lifetime τ(v=0) = 1.5 ns (47)(120); electronic branching ratios (120).

[i](60) and (67) have studied the bands of the isotopic molecules $^{13}C^{16}O$ and $^{12}C^{18}O$.

[j]Oscillator strength f_{00} = 0.16_3 (87).

[k]Rotational lines are diffuse because of predissociation.

[ℓ]In the electron energy loss spectrum (144) find a peak at 90858 cm^{-1} which, according to them, cannot be identified with the j $^3\Sigma^+$ state.

[m]Only two vibrational levels observed, ΔG(½) = 2082.26. ω_e, $\omega_e x_e$ derived with the aid of isotope data (55).

[n]A partial breaking off of the rotational structure in the Ångstrom bands occurs above J=37 in v'=0 and above J=17 in v'=1 leading to a dissociation limit at 89595 ± 30 cm^{-1} (21). RKR potential (40). $\mu_{e\ell}$ = 1.6_0 D from Stark effect measurements on the Ångstrom bands (155).

[o]Lifetime τ(v=0) = 21.8 ns (100), good agreement with (47) (73)(120). τ(v=1) = 15.5 ns (100); (73) give 23.8 ns. Electronic branching ratios (120).

[p]Franck-Condon factors (40). (75) have studied the rotational structure in the Ångstrom bands of the isotopic molecules $^{12}C^{18}O$, $^{13}C^{16}O$, $^{13}C^{18}O$ and the perturbations in these isotopes as well as in $^{12}C^{16}O$; see also (21)(59).

[q]Oscillator strength f_{00} = 0.015_3 (87). Discussion of the r-dependence of the transition moment (100).

[r]This state is strongly perturbed by the higher vibrational levels of the a' $^3\Sigma^+$ state (near its dissociation limit). (5) derived B_0 = 1.89 from lines with N values between 7 and 15. (6) from deperturbed term values derived B_0 = 2.058; (8) gave α = 0.033. The listed values of B_e and α_e are from a revised deperturbation by (16) who also gives the deperturbed constants ΔG(½) = 2188 and T_0 = 83816.

[s]Only two vibrational levels, v=0 and 1, have been observed. Breaking off on account of predissociation in v=0 above N=55, in v=1 above N=42 (6)(9). The absence of v=2 is puzzling since it is expected to lie below the dissociation limit (22).

[t]Lifetimes τ(v=0) = 53.6 ns, τ(v=1) = 69.1 ns (74)(118).

[u]Franck-Condon factors (40).

[v]The b→a bands with v'=1 were previously called "5B" bands (1).

[w]This interpretation was first suggested by (12). It is in agreement with the data of (109) on the a'−X system. The occurrence of these particular vibrational levels of a' $^3\Sigma^+$ in the emission spectrum is due to strong interaction with b $^3\Sigma^+$. Indeed, the levels mentioned were observed as "extra" bands accompanying the b $^3\Sigma^+$ → a $^3\Pi$ (third positive) bands.

State	T_e	ω_e	$\omega_e x_e$	B_e	α_e	D_e $(10^{-6}cm^{-1})$	r_e (Å)	Observed Transitions Design.	Observed Transitions ν_{00}	References
$^{12}C^{16}O$ (continued)										
D $^1\Delta$	65928	1094.0	10.20	1.257	0.017		1.399	D ← X, R	65391[a]	(43)* (96) (109)
I $^1\Sigma^-$	65084.4_0	1092.22 Z	10.70_4[b]	1.2705[c]	0.01848[d]	$D_2 = 9.0$	1.3911	I[e] ← X, R	64546.2_6 Z	(39)* (96) (109)*
A $^1\Pi$	65075.7_7	1518.2_4 Z	19.40[f]	1.6115[g]	0.0232_5[h]	7.33[i]	1.2353	A[j] ↔ X,[kℓ] R 4th positive gr.	64748.48[m] Z	(63)* (109)*
e $^3\Sigma^-$	64230.2_4	1117.7_2 Z	10.686[n]	1.2836[o]	0.01753[p]	6.77[q]	1.3840	e → a,[r] R Herman b.	15231.6	(19)* (28)
								e ← X, R	63704.8_5 Z	(23)* (96) (109)*
d $^3\Delta_i$	61120.1[s]	1171.94 Z	10.635[t]	1.3108[u]	0.01782[v]	6.59[q]	1.3696	d[w] → a,[r] R Triplet b.	12148.7	(13)(29)(76)
								d ← X, R	60621.9[t] Z	(71)(109)*
a' $^3\Sigma^+$	55825.4_9	1228.60 Z	10.468[x]	1.3446[y]	0.0189_2[z]	6.41[q]	1.3523	a'[α] → a,[β] R Asundi b.	6882.4	(1)(11)(14)
								a' ← X,[β] R Hopfield-Birge b.	55355.6 Z	(23)* (96) (109)*

CO: [a]Extrapolated, only v'=1, 6, 21 observed.

[b] $+ 0.055_4(v+\frac{1}{2})^3 - \ldots$; for higher order coefficients see (109).

[c]RKR potential (109).

[d] $+ 0.00029_1(v+\frac{1}{2})^2 - + \ldots$; for higher order coefficients see (109). Revised coefficients from deperturbed B_v values in (98a).

[e]Lifetime of this state (and/or D $^1\Delta$) 97 ± 15 μs (124).

[f] $+ 0.76_6(v+\frac{1}{2})^3 - + \ldots$; higher order coefficients in (109). Because of numerous perturbations (see [g]) these constants do not accurately represent the observed ($v \leq 23$) vibrational levels. Revised coefficients from deperturbed T_v values [see (105)] in (98a).

[g]Numerous perturbations produced by e $^3\Sigma^-$, d $^3\Delta$, a' $^3\Sigma^+$, D $^1\Delta$, I $^1\Sigma^-$, discussed by many investigators and summarized in (63). Deperturbed T_v and B_v values are given by (105), see also (98a). RKR potential (109); the potential function has a maximum, the last observed level lies above the dissociation limit.

[h] $+ 0.00159(v+\frac{1}{2})^2 - + \ldots$; higher order coefficients in (109), revised coefficients from deperturbed B_v values in (98a).

[i]Calculated value, $\beta_e = + 0.10 \times 10^{-6}$; see (105).

CO: [j] Lifetimes for v=0,...,6 are 10.7, 10.4, 9.4, 9.0, 9.7, 9.8, 10.5 ns, respectively (47)(89)(99). Values that are about 50% larger were given by (81).

[k] Oscillator strength $f_{e\ell} = 0.195$, $f_{00} = 0.020$ (87). f values from lifetime measurements (47) are approximately a factor of 2 smaller. See also (91)[r-dependence of electronic transition moment, $f_{e\ell} \approx 0.15$] and (94)(121). Franck-Condon factors (38)(40)(160).

[ℓ] See (27)(116) for spectroscopic data on ^{13}CO and $C^{18}O$.

[m] This is a nominal, rotationally deperturbed value. The lowest observed levels (v=0, J=1) lie at 64747.90(-) and at 64748.09(+) which would correspond to a J=0 level at 64744.8 cm^{-1}.

[n] $+ 0.117_4(v+\frac{1}{2})^3 - + \ldots$; from (109), see also (98a).

[o] Spin-splitting constant $\lambda_0 = + 0.51$; for its dependence on v see (98a)(131). RKR potential (109).

[p] $+ 7._1 \times 10^{-6}(v+\frac{1}{2})^2 + - \ldots$; the coefficients are from (109), but considerably different values for the higher order terms were obtained by (98a) from deperturbed rotational constants.

[q] Calculated value, see (98a).

[r] Intensity distribution, relative electronic transition moments: e→a (156), d→a (108). Franck-Condon factors: d→a (40). Rotational intensity distribution in the "triplet" bands (36).

[s] $A_v = - 16.00_5 - 0.113(v+\frac{1}{2}) - 0.0035_7(v+\frac{1}{2})^2$; from (98a).

[t] $+ 0.0785(v+\frac{1}{2})^3 - 0.00163_4(v+\frac{1}{2})^4$. The constants refer to the $^3\Delta_2$ component (109); see also (98a).

[u] RKR potential (109).

[v] $+ 0.000113(v+\frac{1}{2})^2$. The constants refer to the $^3\Delta_2$ component (109). From properly averaged term values of v=3, 4, 7, 9 (29) gives $B_e = 1.3099$, $\alpha_e = 0.0167_7$, in good agreement with the first two of the expansion coefficients determined by (98a).

[w] Lifetime strongly dependent on J and Ω because of perturbations by A $^1\Pi$ (117).

[x] $+ 0.0091(v+\frac{1}{2})^3 + 0.00259(v+\frac{1}{2})^4 - + \ldots$ (109). Revised coefficients from deperturbed T_v values in (98a).

[y] Spin-splitting constants $\lambda_0 = - 1.23$, $\gamma_0 \approx - 0.007$; dependence on v (98a)(131). See also (143). Dipole moment $\mu_{e\ell} = 1.0_6$ D ($^-CO^+$), from the radiofrequency spectrum of a $^3\Pi$; see (111)(147). RKR potential (109).

[z] $+ 0.00034_5(v+\frac{1}{2})^2 - + \ldots$ (109); revised coefficients from deperturbed B_v values in (98a).

[α] Lifetime 3.7 to 2.9 μs for v=5 to 8 (49).

[β] Franck-Condon factors (38)(40).

State	T_e	ω_e	$\omega_e x_e$	B_e	α_e	D_e ($10^{-6}cm^{-1}$)	r_e (Å)	Observed Transitions Design.	Observed Transitions ν_{00}	References
$^{12}C^{16}O$ (continued)										
a $^3\Pi_r$	48686.70[a]	1743.4_1	14.3_6[b]	1.69124[c]	0.01904[d]	6.36[e]	1.20574	a[f] ⟷ X,[g] R Cameron b.	48473.22[b]	(20)* (104) (109)*
X $^1\Sigma^+$	0	2169.81358 Z	13.28831[h]	1.93128087[i]	0.01750441[j]	6.1214_7[k]	1.128323[ℓ]	Rot.-vibr. sp.[mn]:		
								3-0		(45)(48)(129)
								2-0[o]		(139)
								1-0[p]		(26)(41)
								Rotation sp.:		
								Far IR sp.[q]		(25)
								Microwave sp.		(24)(70)(134)
								Mol. beam el. reson.[r]		(141)
								Mol. beam magn. reson.[s]		(46)(54)

CO: [a] $A_v = +41.53 - 0.14(v+\frac{1}{2}) - 0.009(v+\frac{1}{2})^2$; $A_J(v=0) = -0.000206$.

[b] $-0.04_5(v+\frac{1}{2})^3 + 0.002_5(v+\frac{1}{2})^4$; all vibrational and rotational constants for this state are from deperturbed levels (104)(105).

[c] Very precise values for the Λ-type doubling in $^3\Pi_1$ and $^3\Pi_2$, v=0-7, J=1-8, have been obtained (77)(83)(111)(147) from the study of the radiofrequency spectrum in a molecular beam electric resonance apparatus. While these doublings are small and increase rapidly with J the Λ-doubling for $^3\Pi_0$ [from combination defects (5)(34)] is fairly large at low J (~1.7 cm^{-1}) and decreases with J. Hyperfine structure in $^{13}C^{16}O$ (84). Dipole moment ($^+CO^-$) from molecular beam electric resonance spectrum $\mu_{e\ell}(v=0) = 1.374$ D (111); dipole moment function and radiative lifetimes for vibrational transitions in $a^3\Pi$ (150)(152).

[d] $-0.000041(v+\frac{1}{2})^2$; see [b].

[e] Calculated value, $\beta_e = +0.04 \times 10^{-6}$ (104)(105).

[f] Lifetime from time of flight studies ~9.5 ms (112); from afterglow decay $7._5$ ms (88)(110); theoretical values (92).

[g] Franck-Condon factors (38)(40).

[h] $+0.010511(v+\frac{1}{2})^3 + 5.74 \times 10^{-5}(v+\frac{1}{2})^4 + 9.8_3 \times 10^{-7}(v+\frac{1}{2})^5 - 3.16_6 \times 10^{-8}(v+\frac{1}{2})^6$; $v \leq 37$ (142).

[i] RKR potential functions (97)(103)(106).

[j] $+5.487 \times 10^{-7}(v+\frac{1}{2})^2 + 2.5_4 \times 10^{-8}(v+\frac{1}{2})^3$ (142).

[k] $-1.15_3 \times 10^{-9}(v+\frac{1}{2}) + 1.80_5 \times 10^{-10}(v+\frac{1}{2})^2$; $H_v = [5.83 - 0.173_8(v+\frac{1}{2})] \times 10^{-12}$ (142).

[ℓ] From the effective B_e value; the "true" $B_e = 1.93160$ found by (69) after introducing adiabatic and non-adiabatic corrections (and using older data) leads to $r_e = 1.12823$ Å. See also (102)(122).

[m] For data on $^{13}C^{16}O$, $^{12}C^{18}O$, $^{13}C^{18}O$ see (137)(158).

CO: [n]Intensities in 1-0, 2-0, 3-0 rotation-vibration bands and dipole moment function (42)(65)(90)(107)(127)(128)(146)(159); for $\Delta v=1$ transitions with v=4-10 (123). Pressure shift and pressure broadening (50)(58)(68)(129)(145)(148).

[o]$\Delta v=2$ sequence up to 33-31 in chemiluminescence (61) and flames (139).

[p]$\Delta v=1$ sequence up to 37-36 in the CO laser (72)(80)(138); 1-0 band in resonance fluorescence (30)(32).

[q]Line widths and intensities (56)(95). High pressure gas and liquid far IR absorption spectra in Ar (119); the quadrupole moment derived from this and other experimental and theoretical work [see (126)] is -2.0×10^{-26} esu cm^2.

[r]$\mu_{e\ell}(v=0,J=0) = 0.10980$ D ($^-CO^+$); with the dipole moment function of (65) (see [n]) this gives 0.1222 D at r_e (141).

[s]$g_J = -0.2689_0\ \mu_N$ for $^{12}C^{16}O$ (46);
$g_J = -0.2569_1\ \mu_N$ for $^{13}C^{16}O$ (54).

CO: (1) Asundi, PRS A 124, 277 (1929).
(2) Hopfield, Birge, PR 29, 922 (1929).
(3) Kaplan, PR 35, 1298 (1930).
(3a) Jevons, "Band Spectra of Diatomic Molecules" (Physical Society, London 1932).
(4) Henning, AP(Leipzig) 13, 599 (1932).
(5) Dieke, Mauchly, PR 43, 12 (1933).
(6) Gerö, ZP 95, 747 (1935).
(7) Schmid, Gerö, ZP 93, 656 (1935).
(8) Schmid, Gerö, ZP 96, 198 (1935).
(9) Gerö, ZP 101, 311 (1936).
(10) Schmid, Gerö, Nature 140, 508 (1937).
(11) Beer, ZP 107, 73 (1937).
(12) Gerö, ZP 109, 216 (1938).
(13) Gerö, Szabo, AP(Leipzig) 35, 597 (1939).
(14) Gerö, Lörinczi, ZP 113, 449 (1939).
(15) Tschulanowsky, JPUSSR 1, 341 (1939).
(16) Stepanov, JPUSSR 2, 197 (1940).
(17) Tanaka, Sci. Pap. IPCR (Tokyo) 39, 447 (1942).
(18) Takamine, Tanaka, Iwata, Sci. Pap. IPCR (Tokyo) 40, 371 (1943).
(19) Herman, Herman, JPR 9, 160 (1948).
(20) Rao, ApJ 110, 304 (1949).
(21) Douglas, Møller, CJP 33, 125 (1955).
(22) Barrow, Gratzer, Malherbe, PPS A 69, 574 (1956).
(23) Herzberg, Hugo, CJP 33, 757 (1955).
(24) Rosenblum, Nethercot, Townes, PR 109, 400 (1958).
(25) Loewenstein, JOSA 50, 1163 (1960).
(26) Rank, Skorinko, Eastman, Rao, Wiggins, JMS 4, 518 (1960).
(27) Shvangiradze, Oganezov, Chikhladze, OS(Engl. Transl.) 8, 239 (1960).
(28) Barrow, Nature 189, 480 (1961).
(29) Carroll, JCP 36, 2861 (1962).
(30) Millikan, JCP 38, 2855 (1963).
(31) Huffman, Larrabee, Tanaka, JCP 40, 2261 (1964)
(32) McCaa, Williams, JOSA 54, 326 (1964).
(33) Cook, Metzger, Ogawa, CJP 43, 1706 (1965).
(34) Freund, Klemperer, JCP 43, 2422 (1965).
(35) Kovács, APH 18, 107 (1965).
(36) Kovács, Törös, APH 18, 101 (1965).
(37) Tilford, Vanderslice, Wilkinson, CJP 43, 450 (1965).
(38) Halmann, Laulicht, ApJ(Suppl.) 12, 307 (1966).
(39) Herzberg, Simmons, Bass, Tilford, CJP 44, 3039 (1966).
(40) Krupenie, "The Band Spectrum of Carbon Monoxide", NSRDS-NBS 5 (1966).

(continued p. 168)

CO (continued):

(41) Rao, Humphreys, Rank, "Wavelength Standards in the Infrared". Academic Press (1966).
(42) Young, Eachus, JCP 44, 4195 (1966).
(43) Simmons, Tilford, JCP 45, 2965 (1966).
(44) Turner, May, JCP 45, 471 (1966).
(45) Bouanich, Lévy, Haeusler, CR B 264, 944 (1967).
(46) Ozier, Yi, Khoshla, Ramsey, JCP 46, 1530 (1967).
(47) Hesser, JCP 48, 2518 (1968).
(48) Bouanich, Lévy, Haeusler, JP(Paris) 29, 641 (1968).
(49) Hartfuss, Schmillen, ZN 23 a, 722 (1968).
(50) Hunt, Toth, Plyler, JCP 49, 3909 (1968).
(51) Krupenie, Benesch, JRNBS A 72, 495 (1968).
(52) Lindholm, AF 40, 103 (1969).
(53) Nicholls, JP B 1, 1192 (1968).
(54) Ozier, Crapo, Ramsey, JCP 49, 2314 (1968).
(55) Tilford, Vanderslice, JMS 26, 419 (1968).
(56) Dowling, JQSRT 9, 1613 (1969).
(57) Ginter, Tilford, JMS 31, 292 (1969).
(58) Hoover, Williams, JOSA 59, 28 (1969).
(59) Janjić, Pešić, Janković, GHDB 34, 301 (1969).
(60) Kepa, APP A 36, 1109 (1969).
(61) Schwartz, Thrush, JMS 32, 343 (1969).
(62) Siegbahn, Nordling, Johannson, Hedman, Hedén, Hamrin, Gelius, Bergmark, Werme, Manne, Baer, "ESCA Applied to Free Molecules". North-Holland, Amsterdam (1969).
(63) Simmons, Bass, Tilford, ApJ 155, 345 (1969).
(64) Tilford, JCP 50, 3126 (1969).
(65) Toth, Hunt, Plyler, JMS 32, 85 (1969).
(66) Aarts, de Heer, JCP 52, 5354 (1970).
(67) Asundi, Dhumwad, Patwardhan, JMS 34, 528 (1970).
(68) Bouanich, Larvor, Haeusler, CR B 269, 1238; 270, 396, 1220 (1970).
(69) Bunker, JMS 35, 306 (1970); 37, 197 (1971) (erratum).
(70) Helminger, De Lucia, Gordy, PRL 25, 1397 (1970).
(71) Herzberg, Hugo, Tilford, Simmons, CJP 48, 3004 (1970).
(72) Mantz, Nichols, Alpert, Rao, JMS 35, 325 (1970).
(73) Rogers, Anderson, JOSA 60, 278 (1970).
(74) Rogers, Anderson, JQSRT 10, 515 (1970).
(75) Rytel, et al., APP A 37, 559, 585; 38, 299 (1970); 39, 29 (1971); 41, 377, 757 (1972).
(76) Slanger, Black, CPL 4, 558 (1970).
(77) Stern, Gammon, Lesk, Freund, Klemperer, JCP 52, 3467 (1970).
(78) Thomas, JCP 53, 1744 (1970).
(79) Van der Wiel, El-Sherbini, Brion, CPL 7, 161 (1970).
(80) Yardley, JMS 35, 314 (1970).
(81) Chervenak, Anderson, JOSA 61, 952 (1971).
(82) Comes, Speier, ZN 26 a, 1998 (1971).
(83) Gammon, Stern, Klemperer, JCP 54, 2151 (1971).
(84) Gammon, Stern, Lesk, Wicke, Klemperer, JCP 54, 2136 (1971).
(85) Hasson, Nicholls, JP B 4, 681 (1971).
(86) Kepa, Rytel, APP A 39, 629 (1971).
(87) Lassettre, Skerbele, JCP 54, 1597 (1971).
(88) Lawrence, CPL 9, 575 (1971).
(89) Imhof, Read, CPL 11, 326 (1971).
(90) Moskalenko, Mirumyants, SPJ (1973), 721.
(91) Mumma, Stone, Zipf, JCP 54, 2627 (1971).
(92) James, JCP 55, 4118 (1971).
(93) Nakamura, Morioka, Hayaishi, Ishiguro, Sasanuma, 3rd International Conference on Vacuum Ultraviolet Radiation Physics (Tokyo, August 30 - September 2, 1971), paper 1pA1-6.

CO (continued):

(94) Pilling, Bass, Braun, JQSRT 11, 1593 (1971).
(95) Sanderson, Scott, White, JMS 38, 252 (1971).
(96) Simmons, Tilford, JRNBS A 75, 455 (1971).
(97) Mantz, Watson, Rao, Albritton, Schmeltekopf, Zare, JMS 39, 180 (1971).
(98) James, JMS 40, 545 (1971).
(98a) R. W. Field, Thesis (Harvard University, 1971).
(99) Burnham, Isler, Wells, PR A 6, 1327 (1972).
(100) Imhof, Read, Beckett, JP B 5, 896 (1972).
(101) Ogawa, Ogawa, JMS 41, 393 (1972).
(102) Bunker, JMS 42, 478 (1972).
(103) Dickinson, JMS 44, 183 (1972).
(104) Field, Tilford, Howard, Simmons, JMS 44, 347 (1972).
(105) Field, Wicke, Simmons, Tilford, JMS 44, 383 (1972).
(106) Fleming, Rao, JMS 44, 189 (1972).
(107) Roux, Effantin, d'Incan, JQSRT 12, 97 (1972).
(108) Slanger, Black, JP B 5, 1988 (1972).
(109) Tilford, Simmons, JPCRD 1, 147 (1972).
(110) Wauchop, Broida, JCP 56, 330 (1972).
(111) Wicke, Field, Klemperer, JCP 56, 5758 (1972).
(112) Johnson, JCP 57, 576 (1972).
(113) Judge, Lee, JCP 57, 455 (1972).
(114) Carbonneau, Marmet, CJP 51, 2202 (1973).
(115) Lee, Carlson, Judge, Ogawa, JQSRT 13, 1023 (1973).
(116) Rytel, Siwiec, APP A 44, 67 (1973).
(117) Slanger, Black, JCP 58, 194 (1973).
(118) Smith, Imhof, Read, JP B 6, 1333 (1973).
(119) Buontempo, Cunsolo, Jacucci, JCP 59, 3750 (1973).
(120) Dotchin, Chupp, Pegg, JCP 59, 3960 (1973).
(121) Vargin, Pasynkova, Trekhov, JAS 13, 1340 (1973).
(122) Watson, JMS 45, 99; 48, 479 (1973).
(123) Weisbach, Chackerian, JCP 59, 4272 (1973).
(124) Wells, Borst, Zipf, PR A 8, 2463 (1973).
(125) Åsbrink, Fridh, Lindholm, Codling, PS 10, 183 (1974).
(126) Billingsley, Krauss, JCP 60, 2767 (1974).
(127) Billingsley, Krauss, JCP 60, 4130 (1974).
(128) Bouanich, Brodbeck, JQSRT 14, 1199 (1974).
(129) Bouanich, Brodbeck, RPA 9, 475 (1974).
(130) Codling, Potts, JP B 7, 163, 314 (1974).
(131) Field, Lefebvre-Brion, APH 35, 51 (1974).
(132) Gardner, Samson, JCP 60, 3711 (1974).
(133) Klump, Lassettre, JCP 60, 4830 (1974).
(134) Lovas, Krupenie, JPCRD 3, 245 (1974).
(135) Ogawa, Ogawa, JMS 49, 454 (1974).
(136) Simmons, Tilford, JMS 49, 167 (1974).
(137) Johns, McKellar, Weitz, JMS 51, 539 (1974).
(138) Kildal, Eng, Ross, JMS 53, 479 (1974).
(139) Mantz, Maillard, JMS 53, 466 (1974).
(140) Tilford, Simmons, JMS 53, 436 (1974).
(141) Muenter, JMS 55, 490 (1975).
(142) Mantz, Maillard, Roh, Rao, JMS 57, 155 (1975).
(143) Sink, Lefebvre-Brion, Hall, JCP 62, 1802 (1975).
(144) Swanson, Celotta, Kuyatt, Cooper, JCP 62, 4880 (1975).
(145) Varanasi, JQSRT 15, 191 (1975).
(146) Varanasi, Sarangi, JQSRT 15, 473 (1975).
(147) Wicke, Klemperer, Field, JCP 62, 3544 (1975).
(148) Moskalenko, OS(Engl. Transl.) 38, 382 (1975).
(149) Watson, Stewart, Gardner, Lynch, PSS 23, 384 (1975).
(150) Wicke, Klemperer, MP 30, 1021 (1975).
(151) Locht, Dürer, CPL 34, 508 (1975).
(152) Wicke, Klemperer, JCP 63, 3756 (1975).
(153) Lee, Carlson, Judge, Ogawa, JCP 63, 3987 (1975).
(154) Kepa, Knot-Wisniewska, Rytel, APP A 48, 819 (1975).
(continued p. 171)

State	T_e	ω_e	$\omega_e x_e$	B_e	α_e	D_e ($10^{-6}cm^{-1}$)	r_e (Å)	Observed Transitions Design.	Observed Transitions ν_{00}	References
$^{12}C^{16}O^+$		$D_0^0 = 8.33_8$ eV[a]			I.P. = 26.8 eV[b]					OCT 1976 A
K' $^2\Sigma^+$	[528.69 eV]	Upper state of oxygen K radiation in CO (see [b] p.159).							528.56 eV	(40)
K $^2\Sigma^+$	[282.34 eV]	Upper state of carbon K radiation in CO (see [b] p.159).[c]							282.21 eV	(34)(40)
H $^2\Sigma^+$	(~25 eV)	This highly excited state (21a)(22b) is probably unstable and dissociating into C^+ + O(3s) as suggested by the photoelectron and photoion spectra of (36).								
F ($^2\Sigma$)	(105690)	(1780)[d]	(30)						(105470)[d]	(32)
E ($^2\Sigma$)	(87140)	(1420)[d]							86750[d]	(32)
G ($^2\Sigma$)	(73190)	(1400)[d]							72790[d]	(32)
D ($^2\Pi$)	(65230)	(1350)[d]							(64800)[d]	(32)
C ($^2\Delta_r$)	63012[e]	1144[f] (Z)	$33._3$	1.35_7[f]	0.024		1.34_6	C→A,	R 42168.5[f] H 41950.9[f] H	(21)
B $^2\Sigma^+$	45876.7	1734.1_8 Z	27.92_7[g]	1.7999_2[hi]	0.0302_5	7.7_5[j]	1.1687_7	B[k]→A,[ℓm] Baldet-Johnson b.	V 25226.0	(1)(4)(22)*
								B[k]↔X,[ℓmn] 1st negative b.	R 45633.5_2 Z	(2)* (16)
A $^2\Pi_i$	20733.3[o]	1562.0_6[p] Z	13.53_2[q]	1.5894_0[i]	0.0194_2	6.6	1.2437_7	A[r]→X,[ℓm] Comet-tail b.	R 20407.6[s] Z	(3)* (22)*
X $^2\Sigma^+$	0	2214.2_4 Z	15.16_4[t]	1.9772_0[ui]	0.0189_6	6.3_5	1.1151_4	Microwave sp.		(35)
$^{12}C^{16}O^{++}$										OCT 1976
(A $^3\Pi$)	(40000)[v]	Only one vibrational level is predicted to exist in the local minimum.								(24)
X ($^3\Pi$)	0[w]	(2400)[x]								(24)

CO^+, CO^{++}:

[a] $D_0^0(CO) + I.P.(C) - I.P.(CO)$.

[b] From the electron impact appearance potential of CO^{++} in CO (8)(22a), reduced by 0.9 eV following (24); see also [w].

[c] Several higher lying states corresponding to excitation of a second electron have been observed by (34) ("1s shake-up spectrum").

[d] Fairly long progressions in the photoelectron spectrum; vibrational numbering uncertain. Predissociation into C^+ + O (38)

[e] $A \approx + 100$.

[f] Vibrational numbering uncertain; rotational structure not fully resolved.

[g] $\omega_e y_e = + 0.328_3$.

[h] Spin-splitting constant $\gamma \approx + 0.018$.

[i] RKR potential functions (10)(14)(15).

[j] $\beta_e = + 0.2_2 \times 10^{-6}$.

[k] Lifetime $\tau(v=0) = 55 \pm 4$ ns (17)(20)(31)(37), slight increase with v (17)(31). Electronic branching ratio $A(B \to A) = 0.08/\tau(B)$, $A(B \to X) = 0.92/\tau(B)$ (29); $f_{00}(B-A) \approx 0.0007_1$, $f_{00}(B-X) \approx 0.0065$.

[l] Fluorescence by photoexcitation from the ground state X $^1\Sigma^+$ of CO (29)(33).

[m] Franck-Condon factors (9)(11)(14)(19)(23). Independence of electronic transition moment on r (29)(30); see, however, (6)(13)(28).

[n] Isotopic bands (7)(34a).

[o] $A = - 117._5$.

[p] Vibrational numbering confirmed by isotope studies (22).

[q] $\omega_e y_e = + 0.013_1$.

[r] Lifetimes $\tau(v=1) = 3.4_9$ μs, $\tau(v=3) = 2.7_8$ μs, $\tau(v=4) = 2.6_3$ μs, $\tau(v=6) = 2.4_1$ μs (27). (26) give values that are 10 to 50 % higher; earlier values (5)(18) are 20 % lower.

[s] Refers to zero point of Hill - Van Vleck formula.

[t] $\omega_e y_e = - 0.000_7$.

[u] From the microwave spectrum $B_0 = 1.967465$, spin-splitting constant $\gamma_0 = + 0.009105$.

[v] Rough estimate based on appearance potentials of CO^{++} from CO (8)(22a), and taking into account tentative assignments of the carbon and oxygen Auger spectra of CO (21a)(24).

[w] The potential minimum of the ground state of CO^{++} lies about 4.7 eV above the dissociation limit $C^+(^2P_{\frac{1}{2}}) + O^+(^4S)$ (which lies 35.97_0 eV above the ground state of CO) and is separated by a maximum of ~1.5 eV from this limit (22a)(24).

[x] Derived by (24) from the carbon Auger spectrum of CO (21a). At least six vibrational levels are observed.

References on page 173.

CO (continued):

(155) Fisher, Dalby, CJP 54, 258 (1976).
(156) Slanger, Black, JCP 64, 219 (1976).
(157) Wight, Van der Wiel, Brion, JP B 9, 675 (1976)
(158) Chen, Rao, McDowell, JMS 61, 71 (1976)
(159) Tipping, JMS 61, 272 (1976).
(160) Shimauchi, SL 25, 1 (1976).
(161) Samson, Gardner, JESRP 8, 35 (1976).
(162) Smith, Thomas, JESRP 8, 45 (1976).

State	T_e	ω_e	$\omega_e x_e$	B_e	α_e	D_e	r_e (Å)	Observed Transitions Design.	Observed Transitions ν_{00}	References
$^{12}C^{16}O^-$				$D_0^0 = 8.1_3$ eV[a]		I.P. = - 1.5 eV				OCT 1976 A
F		1730	40					b	13.95 eV[c]	(4)(6)
E								d	12.2 eV[c]	(3)(4)(6)
D								d	11.3 eV[c]	(3)(4)(6)
C $\{^2\Pi$									10.7 eV[c]	(2)(3)(4)(6)
									10.42 eV[c]	(2)(4)(6)
B $^2\Sigma^+$		[1940]						e	10.04 eV[c]	(2)(3)(4)(6)
A		Resonances coincident with the vibrational levels v=0, 1, 2 of $a^3\Pi$ of CO, ascribed to a state of CO^- with the same potential function as $a^3\Pi$.							6.0 eV[c]	(7)
X $^2\Pi$		Very broad resonance, vibrational structure barely detectable.							1.5 eV[c]	(1)

CO^-: [a]From D_0^0(CO) and the electron affinities of CO (-1.5 eV) and O (+1.465 eV).

[b]Progression of six resonances, observed in electron transmission (4) and electroionization (5) experiments; vibrational numbering uncertain. Its "grandparent" is assumed to be the A $^2\Pi$ state of CO^+, with the additional two electrons in a Rydberg orbital.

[c]Energies in eV above X $^1\Sigma^+$(v=0) of CO, all obtained as resonances in electron scattering experiments.

[d]Resonances in the excitation functions of CO, B $^1\Sigma^+$ and b $^3\Sigma^+$ (3), and in the electron transmission current (4).

[e]Sharp resonance (width ~0.045 eV); v'=0 and 1. The parent states are b $^3\Sigma^+$ and B $^1\Sigma^+$ of CO. Decay into CO($X^1\Sigma$, $a^3\Pi$, $a'^3\Sigma$, $A^1\Pi$) + e^-, C + O^-, and C^- + O; see also (8).

(1) Boness, Hasted, Larkin, PRS A 305, 493 (1968).
(2) Comer, Read, JP B 4, 1678 (1971).
(3) Mazeau, Gresteau, Joyez, Reinhardt, Hall, JP B 5, 1890 (1972).
(4) Sanche, Schulz, PR A 6, 69 (1972).
(5) See ref. (114) of CO.
(6) Schulz, RMP 45, 423 (1973).
(7) Wong, Schulz, PRL 33, 134 (1974).
(8) See ref. (144) of CO.

CO^+, CO^{++}:

(1) Bulthuis, Physica 1, 873 (1934).
(2) Rao, ApJ 111, 50 (1950).
(3) Rao, ApJ 111, 306 (1950).
(4) Rao, Sarma, MSRSL (4) 13, 141 (1953).
(5) Bennett, Dalby, JCP 32, 1111 (1960).
(6) Robinson, Nicholls, PPS 75, 817 (1960).
(7) See ref. (27) of CO.
(8) Dorman, Morrison, JCP 35, 575 (1961).
(9) Nicholls, CJP 40, 1772 (1962).
(10) Krupenie, Weissman, JCP 43, 1529 (1965).
(11) See ref. (38) of CO.
(12) Hesser, Dressler, JCP 45, 3149 (1966).
(13) Joshi, Sastri, Parthasarathi, JQSRT 6, 215 (1966).
(14) See ref. (40) of CO.
(15) Singh, Rai, JMS 19, 424 (1966).
(16) Herzberg, CPAS 2 (15), 1 (1968).
(17) See ref. (47) of CO.
(18) Fink, Welge, ZN 23 a, 358 (1968).
(19) See ref. (51) of CO.
(20) Fowler, Skwerski, Anderson, Copeland, Holzberlein, JCP 50, 4133 (1969).
(21) Marchand, d'Incan, Janin, SA A 25, 605 (1969).
(21a) See ref. (62) of CO.
(22) See ref. (67) of CO.
(22a) Newton, Sciamanna, JCP 53, 132 (1970).
(22b) See ref. (78) of CO.
(23) See ref. (82) of CO.
(24) Hurley, JCP 54, 3656 (1971).
(25) See ref. (93) of CO.
(26) Anderson, Sutherland, Frey, JOSA 62, 1127 (1972).
(27) Holland, Maier, JCP 56, 5229 (1972).
(28) Jain, JP B 5, 199 (1972).
(29) See ref. (113) of CO.
(30) Maier, Holland, JP B 5, L118 (1972).
(31) See ref. (120) of CO.
(32) See ref. (125) of CO.
(33) Lee, Carlson, Judge, Ogawa, JGR 79, 5286 (1974).
(34) Gelius, JESRP 5, 985 (1974).
(34a) Pešić, Janjić, Marković, Rytel, Siwiec, GHDB 39, 249 (1974); Pešić, Marković, Janković, Fizika 7, 83 (1975).
(35) Dixon, Woods, PRL 34, 61 (1975).
(36) Gardner, Samson, JCP 62, 1447 (1975).
(37) Jørgensen, Sørensen, JCP 62, 2550 (1975).
(38) See ref. (151) of CO.
(39) Lee, Carlson, Judge, JP B 9, 855 (1976).
(40) See ref. (162) of CO.

System	ν_e	ω'_e	$\omega'_e x'_e$	ω''_e	$\omega''_e x''_e$	Description	Degrad.	ν_{00}	References
$^{59}Co_2$		μ = 29.466594$_0$		D_0^0 = 1.6$_9$ eV^a					DEC 1974
$^{59}Co^{(79)}Br$		(μ = 33.738539$_1$) Emission bands in the visible (1)(2) and photographic infrared (4) regions. Tentative assignments:							DEC 1974
System A:	23049.$_8$	269.$_1$	0.4	271.$_6$	- 0.0$_5$	Single heads of line-like app.		23048.$_4$	(3)
" B:	22252.$_6$	304.$_3$	0.7$_5$	318.$_3$	2.1	Single heads	R	22245.$_9$	(3)
" C:	22132.$_1$	306.$_8$	1.6$_5$	324.$_0$	1.5	" "	R	22123.$_5$	(3)
" D:	21731.$_1$	300.$_2$ H^Q	- 0.7$_2$	322.$_8$	0.1$_5$	Double heads	R	21720.$_0$ H^Q	(3)
" E:	21111.5	306.3 H^Q	1.6	332.7	2.2	" "	R	21098.4 H^Q	(5)
" F:	20712.3	309.0	2.8	332.7	1.95	Single heads	R	20700.4	(5)
" G:	18354.5	317.0 H^Q	1.6	329.5	2.55	Double heads	R	18348.5 H^Q	(5)
" H:	17927.8	329.3	0.3	337.5	0.2	Single heads	R	17923.7	(5)
" I:	17863.7	326.5	0.8	336.0	0.25	" "	R	17858.8	(5)
" N_1:	13831.4	331.3	1.35	316.4	0.88	" "	R	13838.7	(4)
" N_2:	13740.5	332.2 H^Q	1.40	313.9	0.50	Double heads	R	13749.4 H^Q	(4)
$^{59}Co^{(35)}Cl$		(μ = 21.9465518)		D_0^0 = 4.0 eV^a					DEC 1974
		The following classification (1)(5) of emission bands [see also (2)] should be considered as tentative only:							
System A:	22967.3^b	420.0	1.66	421.8	1.34	Single heads	R	22966.3	(1)
" B:	22404.3^b	416.6	0.82	419.4	0.28	" "	R	22402.8	(1)
" C:	22190	401.$_0$		416.$_2$		" "		22182.5	(4)
" D:	22075	410.$_2$		416.$_4$		" "		22072.3	(4)
" E:	22014.6^b	420.0	1.14	421.2	0.74	" "	R	22013.9	(1)
" F:	21641	408.$_0$		409.0	- 0.45	" "	R	21640.4	(5)
" G:	21343.0	391.2	2.15	407.1	2.55	" "	R	21335.1	(5)
" H:	21266.1	413.2	2.00	416.8	0.38	" "	R	21263.9	(5)

System	ν_e	ω'_e	$\omega'_e x'_e$	ω''_e	$\omega''_e x''_e$	Description	Degrad.	ν_{00}	References
$^{59}Co^{(35)}Cl$ (continued)									
System I:	20971	405.0	0.80	$409._0$		Single heads	R	20969.1	(5)
" J:	19957	408.6 H^Q	- 1.00	$431._0$		Double heads	R	19945.8 H^Q	(5)
" K:	19878	408.4 H^Q	- 0.50	$428._0$		" "	R	19868.1 H^Q	(5)
" L:		0-0 sequence only.				" "	R	19826.3 H^Q	(5)
" M:	17641 17558 17486 17413 17330	$412._3$		$416._0$		Multiple heads		17638.8 17555.9 17484.1 17411.0 17328.0	(3)
" N_1:	14054.0	462.5 H^Q	2.65	482.6	3.65	Double heads	R	14044.2 H^Q	(5)
" N_2:	13946.1	467.6 H^Q	1.25	484.3	0.90	" "	R	13937.7 H^Q	(5)
" O:	12693.4	478.8	1.38	498.4	- 1.05	" "	R	12683.0	(5)

Co_2: [a]Thermochemical value (mass-spectrom.)(1).

(1) Kant, Strauss, JCP 41, 3806 (1964).

CoBr:
(1) Mesnage, CR 204, 1929 (1937).
(2) Mesnage, AP(Paris) 12, 5 (1939).
(3) Rao, Rao, IJP 36, 609 (1962).
(4) Rao, Rao, CS 37, 608 (1968).
(5) Rao, Reddy, Rao, IJPAP 10, 389 (1972).

CoCℓ: [a]From the heat of formation of CoCℓ(g) reported by (6).
[b]These three band systems may be the three components of a triplet transition.

(1) More, PR 54, 122 (1938).
(2) See ref. (2) of CoBr.
(3) Krishnamurty, CS 20, 323 (1951); IJP 26, 177 (1952).
(4) Krishnamurty, quoted in (5).
(5) Rao, Rao, IJP 35, 556 (1961).
(6) Kulkarni, Dadape, HTS 3, 277 (1971).

State	T_e	ω_e	$\omega_e x_e$	B_e	α_e	D_e ($10^{-4}cm^{-1}$)	r_e (Å)	Observed Transitions Design.	Observed Transitions ν_{00}	References
$^{59}Co^{(63)}Cu$		(μ = 30.432929_6) D_0^0 = 1.6_2 eV[a]								DEC 1974
$^{59}Co^{(74)}Ge$		(μ = 32.790873_4) D_0^0 = 2.4_3 eV[a]								DEC 1974
$^{59}Co^1H$		μ = 0.99088003								DEC 1974
A_2								$A_2 \leftarrow X_2$, R	21982 H	(6)
A_1 Ω = 4[a]		[1527.8] Z[b]		6.701[c]	0.305	[6.20][d]	1.593_4	$A_1 \rightarrow X_1$, R	22243.3[b] Z	(1)* (2)*
X_2										
X_1 Ω = 4[a]	0			[7.151]		[4.05]	[1.542_4]			
$^{59}Co^2H$		μ = 1.94754294 D_0^0 = (3.2) eV								DEC 1974
A_2 Ω = 3[a]		[1110.39] Z[b]		3.4312[e]	0.1059	[1.270][f]	1.5883_0	$A_2 \leftrightarrow X_2$, R	21929.19[b] Z	(5)* (6)
A_1 Ω = 4[a]	22367.4	1180.9 Z[b]	32.7	3.394[g]	0.106	1.0[h]	1.597_0	$A_1 \leftrightarrow X_1$, R	22267.5[b] Z	(3)(4)* (6)*
X_2 Ω = 3[a]	x_2[i]			[3.7559][j]		[1.223]	[1.5181]			
X_1 Ω = 4[a]	0	1373.22 Z[b]	17.59	3.7571	0.0747[k]	[1.125][ℓ]	1.5178_5			
$^{59}Co^{16}O$		μ = 12.5804777_6 D_0^0 = 3.8_1 eV[a]								DEC 1974
		R shaded bands in emission from 10900 to 15900 cm^{-1}, suggesting ω_e'' = 850, $\omega_e''x_e''$ = 6.								(1)(2)
$^{59}Co^{(32)}S$		(μ = 20.7272512) D_0^0 = 3.3_9 eV[a]								DEC 1974
$^{59}Co^{(28)}Si$		(μ = 18.9709743) D_0^0 = 2.8_1 eV[a]								DEC 1974

CoCu: [a]Thermochemical value (mass-spectrom.)(1).

(1) Kant, Strauss, Lin, JCP 52, 2384 (1970).

CoGe: [a]Thermochemical value (mass-spectrom.)(1).

(1) Kant, Strauss, JCP 49, 3579 (1968).

Co^1H, Co^2H:

[a]Possibly $^3\Phi_4$ and $^3\Phi_3$; $^3\Phi_2$ not identified.

[b]Contrary to the definitions adopted in these tables the band origins here do not include the J-independent terms $-\Omega^2 B_v$ of the rotational energy expressions for the upper and lower states; the zero lines are obtained by adding $-\Omega^2(B_v' - B_v'')$ to the values given in the table. Heimer (1)(2) uses a different definition.

[c]$v=0$ perturbed from $J \approx 10$ to $J \approx 20$.

[d]$D_1 = 5.90 \times 10^{-4}$.

[e]Ω-type doubling $\Delta v \begin{Bmatrix} (v=0) = 0.36 \\ (v=1) = 0.45 \end{Bmatrix} \times 10^{-8} J^6$.

[f]$D_1 = 1.326 \times 10^{-4}$.

[g]Perturbations in all three observed vibrational levels (4).

[h]$\beta_e = + 0.2 \times 10^{-4}$.

[i]$x_2 \approx 800$ cm^{-1}, estimated from the difference $B_0(\Omega=3) - B_0(\Omega=4)$.

[j]Constants derived from the upper Ω-type doubling component; $\Delta v = 3.42 \times 10^{-8} J^6 + \ldots$

[k]$\gamma_e = - 0.0013$.

[ℓ]$H_0 = + 2.5_6 \times 10^{-9}$; $D_1 = 1.15_7 \times 10^{-4}$, $H_1 = + 5._9 \times 10^{-9}$; $D_2 = 1.09 \times 10^{-4}$.

Co^1H, Co^2H (continued):

(1) Heimer, Dissertation (Stockholm, 1937).
(2) Heimer, ZP 104, 448 (1937).
(3) Klynning, Neuhaus, ZN 18 a, 1142 (1963).
(4) Klynning, Kronekvist, PS 6, 61 (1972).
(5) Klynning, Kronekvist, PS 7, 72 (1973).
(6) Smith, PRS A 332, 113 (1973).

CoO: [a]Thermochemical value (mass-spectrom.)[(3), recalculated (4)].

(1) Malet, Rosen, BSRSL 14, 382 (1945).
(2) Rosen, Nature 156, 570 (1945).
(3) Grimley, Burns, Inghram, JCP 45, 4158 (1966).
(4) Smoes, Mandy, Vander Auwera-Mahieu, Drowart, BSCB 81, 45 (1972).

CoS: [a]Thermochemical value (mass-spectrom.)(1).

(1) Drowart, Pattoret, Smoes, PBCS No.8, 67 (1967).

CoSi: [a]Thermochemical value (mass-spectrom.)(1).

(1) Vander Auwera-Mahieu, McIntyre, Drowart, CPL 4, 198 (1969).

State	T_e	ω_e	$\omega_e x_e$	B_e	α_e	D_e $(10^{-6}cm^{-1})$	r_e (Å)	Observed Transitions Design.	Observed Transitions ν_{00}	References
$^{12}C^{31}P$		$\mu = 8.64911823$		$D_0^0 = 5.28$ eV[a]						OCT 1976 A
B $^2\Sigma^+$	29100.4	836.32 H	5.917	0.6829[bc]	0.00628	(1.82)	1.689_4	B→A, R	21934.3 H 22092.6 H	(1)*
								B→X,[d] R	28898.9 H	(1)*
A $^2\Pi_i$	7053.2 6894.9	1061.99 H[Q]	6.035	0.713_5[e]	0.0058[e]	(1.3)	1.65_3			
X $^2\Sigma^+$	0	1239.67 H	6.86	0.7986[fc]	0.00597	(1.33)	1.562_2			
$^{(52)}Cr_2$		$(\mu = 25.970255_1)$		$D_0^0 = 1.5_6$ eV[a]						DEC 1974
		A rotationally resolved absorption band at 21751 cm^{-1} was observed in the photolysis of $Cr(CO)_6$ and could possibly be due to Cr_2.								(2)

CP: [a]Thermochemical value (mass-spectrom.) (5)(6)(7).

[b]Spin-splitting constant $\gamma = (-)0.006$.

[c]RKR potential curves (2).

[d]Morse-potential Franck-Condon factors (4).

[e]Recalculated from head-head separations in the B – A system. Similar calculations by (1) gave $B_e = 0.698$, $\alpha_e = 0.0077$. The rotational analysis of the B – A, 0-0 band (3) yields $B_0 = 0.7101$.

[f]Spin-splitting constant $\gamma = (+)0.017$.

(1) Bärwald, Herzberg, Herzberg, AP(Leipzig) 20, 569 (1934).
(2) Thakur, Singh, JSRBHU 18, 253 (1967).
(3) Chaudhry, Upadhya, IJP 43, 83 (1969).
(4) Wentink, Spindler, JQSRT 10, 609 (1970).
(5) Gingerich, TA 2, 233 (1971).
(6) Smoes, Myers, Drowart, CPL 8, 10 (1971).
(7) Kordis, Gingerich, JCP 58, 5058 (1973).

Cr_2: [a]Thermochemical value (mass-spectrom.) (1).

(1) Kant, Strauss, JCP 45, 3161 (1966).
(2) Efremov, Samoilova, Gurvich, OS(Engl. Transl.) 36, 381 (1974).

State	T_e	ω_e	$\omega_e x_e$	B_e	α_e	D_e ($10^{-4} cm^{-1}$)	r_e (Å)	Observed Transitions Design.	Observed Transitions ν_{00}	References
$^{(52)}Cr^{(79)}Br$		(μ = 31.324275$_5$) D_0^0 = 3.3$_6$ eV[a]								DEC 1974
		Group of sixteen emission bands in the region 15800 - 16200 cm^{-1}. No analysis.								(1)
$^{(52)}Cr^{(35)}Cl$		(μ = 20.8987849) D_0^0 = 3.7$_5$ eV[a]								DEC 1974
		Five complex groups of line-like bands in the region 15400 - 16900 cm^{-1}. Tentative analysis suggesting ω' = 342, ω'' = 292. Uncertain.						$(^6\Pi)\rightarrow(^6\Sigma)$		(1)
$^{(52)}Cr^{(63)}Cu$		(μ = 28.454706$_7$) D_0^0 = 1.6$_2$ eV[a]								DEC 1974
$^{(52)}Cr^{19}F$		(μ = 13.9103739) D_0^0 = 4.5$_7$ eV[a]								DEC 1974
System II:		R shaded bands in the region 25000 - 26000 cm^{-1}. Tentative analysis suggesting ω' = 500, ω'' = 536. Uncertain.						$(^6\Sigma)\rightarrow(^6\Sigma)$		(1)
System I:		R shaded bands in the region 22400 - 23800 cm^{-1}. Tentative analysis. Complex structure.[b]						$(^6\Pi)\rightarrow(^6\Sigma)$		(1)
$^{(52)}Cr^{(74)}Ge$		(μ = 30.505738$_2$) D_0^0 = 1.7$_2$ eV[a]								DEC 1974
$^{(52)}Cr^1H$		(μ = 0.98864215)								DEC 1974
		Complex band[a] in emission and absorption with multiple heads between 30075 and 30480 cm^{-1}.							(30386)	(2)(6)
B $(^6\Pi)$		Complex V shaded band in the region 26975 - 27930 cm^{-1}.[b]						B$\leftrightarrow$(X),	(27181)	(1)(2)(3) (5)(6)
A $^6\Sigma^{(+)}$	(11611)	[1479.4] Z	21.$_5$ H	5.337[c]	0.13$_1$	[2.47][d]	1.787	A$\rightarrow$X, R	11552.29 Z	(2)(3)* (4)*
X $^6\Sigma^{(+)}$	0	[1581.$_2$] H	(32)	6.220[e]	0.179	[3.47][f]	1.655$_7$			

CrBr: [a]Thermochemical value (flame photometry)(2).

(1) Rao, CS 18, 338 (1949).
(2) Bulewicz, Phillips, Sugden, TFS 57, 921 (1961).

CrCℓ: [a]Thermochemical value (flame photometry)(2).

(1) Rao, Rao, IJP 23, 508 (1949).
(2) See ref. (2) of CrBr.

CrCu: [a]Thermochemical value (mass-spectrom.)(1).

(1) Kant, Strauss, Lin, JCP 52, 2384 (1970).

CrF: [a]Thermochemical value (mass-spectrom.)(2).

[b]Barrow (see DONNSPEC) points out the existence of many coincidences between band heads of system I and band heads of the $A\,^2\Sigma - X\,^2\Pi$ system of SiF.

(1) Durgavathi, Rao, IJP 28, 525 (1954).
(2) Kent, Margrave, JACS 87, 3582 (1965).

CrGe: [a]Thermochemical value (mass-spectrom.)(1).

(1) Kant, Strauss, JCP 49, 3579 (1968).

Cr^1H, Cr^2H:

[a]Described as rather diffuse by (6).

[b]The weakness of the emission at low pressures, and the stronger appearance of the corresponding deuteride system suggest predissociation of the upper state through a potential hill (5).

[c]Spin splitting constants $\gamma_0 = 1.20$, $\lambda_0 = 0.14$ (3)(4). Perturbations in both v=0 and v=1 may be caused by a $^8\Sigma$ state (3).

[d]$H_0 = +0.7 \times 10^{-8}$; $D_1 = 2.3_6 \times 10^{-4}$.

[e]Spin splitting constants $\gamma_0 \approx 0.1$, $\lambda_0 = 0.03$ (3)(4).

[f]$H_0 = +1.0 \times 10^{-8}$; $D_1 = 3.43 \times 10^{-4}$.

(continued p. 183)

State	T_e	ω_e	$\omega_e x_e$	B_e	α_e	D_e (10^{-4}cm^{-1})	r_e (Å)	Observed Transitions Design.	ν_{00}	References
$^{(52)}Cr^{2}H$		(μ = 1.93891666)								DEC 1974
B ($^6\Pi$)		Complex band structure in the region 27020 - 27540 cm^{-1}.						B$\leftrightarrow$(X)		(5)(6)
A $^6\Sigma^{(+)}$	(11609)	1089 H	13	$[2.737]^{g}$		$[0.677]^{h}$	[1.782]	A$\rightarrow$X, R	11559.65 Z	(4)*
X $^6\Sigma^{(+)}$	0	(1182)		$[3.142]^{i}$		$[0.888]^{j}$	[1.663_5]			
$^{(52)}Cr^{127}I$		(μ = 36.855845$_7$)		D_0^0 = 2.9$_4$ eV^a						DEC 1974
$^{(52)}Cr^{14}N$		(μ = 11.0295312$_7$)		D_0^0 = 3.8$_7$ eV^a						DEC 1974
$^{(52)}Cr^{16}O$		(μ = 12.2290254$_9$)		D_0^0 = 4.4 eV^a						DEC 1974
B $^5\Pi$	16586^b	750.$_5$ H	9.4	0.4874 0.4801 0.4751 0.4735 0.4675	0.0044 0.0048 0.0057 0.0070 0.0050		1.703	B$\rightarrow$X, R	16487.6 H 16501.2 H 16511.3 H 16519.4 H 16515.2 H	(1)(2)(3)* (5)* (7)
		Five R shaded emission bands with multiple heads between 11800 and 11990 cm^{-1}, tentatively assigned as 1-0 sequence of a new transition by (4).							(11117)	(5)*
X $^5\Pi$	0^c	898.4 H	6.7$_5$	0.5410 0.5348 0.5284 0.5233 0.5231	0.0049 0.0049 0.0050 0.0036 0.0070		1.615			
$^{(52)}Cr^{(32)}S$		(μ = 19.7901878)		D_0^0 = 3.3$_7$ eV^a						DEC 1974
B	23448	510 H	2					B$\rightarrow$A, R	23393 H	(2)
A	0	621 H	4							

Cr^1H, Cr^2H (continued):

[g]Spin splitting constants $\gamma_0 = 0.64$, $\lambda_0 = 0.28$. Several perturbations.
[h]$H_0 = + 0.1_0 \times 10^{-8}$.
[i]Spin splitting constants $\gamma_0 = 0.02_3$, $\lambda_0 = 0.03$.
[j]$H_0 = + 0.07 \times 10^{-8}$.

(1) Gaydon, Pearse, Nature <u>140</u>, 110 (1937).
(2) Kleman, Liljeqvist, AF <u>9</u>, 345 (1955).
(3) Kleman, Uhler, CJP <u>37</u>, 537 (1959).
(4) O'Connor, PRIA A <u>65</u>, 95 (1967).
(5) O'Connor, JP B <u>2</u>, 541 (1969).
(6) Smith, PRS A <u>332</u>, 113 (1973).

CrI: [a]Thermochemical value (flame photometry)(1).

(1) See ref. (2) of CrBr.

CrN: [a]Thermochemical value (mass-spectrom.)(1).

(1) Srivastava, Farber, HTS <u>5</u>, 489 (1973).

CrO: [a]Thermochemical value (mass-spectrom.)(6).
[b]$|A| \approx 100$.
[c]$|A| \approx 110$. Not certain that this is the ground state.

(1) Ferguson, JRNBS <u>8</u>, 381 (1932).
(2) Ghosh, ZP <u>78</u>, 521 (1932).
(3) Ninomiya, JPSJ <u>10</u>, 829 (1955).
(4) Gaspard, Rosen, quoted in ref. (5).
(5) Gatterer, Junkes, Salpeter, Rosen, METOX.
(6) Grimley, Burns, Inghram, JCP <u>34</u>, 664 (1961).
(7) Murthy, Nagaraj, PPS <u>84</u>, 827 (1964).

CrS: [a]Thermochemical value (mass-spectrom.)(1).

(1) Drowart, Pattoret, Smoes, PBCS No. 8, <u>67</u>, (1967).
(2) Monjazeb, Mohan, SpL <u>6</u>, 143 (1973).

State	T_e	ω_e	$\omega_e x_e$	B_e	α_e	D_e (10^{-6}cm^{-1})	r_e (Å)	Observed Transitions		References
								Design.	ν_{00}	
$^{12}C^{32}S$		μ = 8.72519418 $D_0^0 = 7.35_5$ eV[a] I.P. = 11.33_5 eV[b]								OCT 1976 A
		Fragments of further band systems and Rydberg series.								(12)*
G	(81373)	[1229]						G← X,	81347	(12)*
F								F← X,	77537	(12)*
		Continuous absorption to a repulsive state; 74600 - 76300 cm^{-1}.								(12)*
E	(71890)	[1459] H						E← X, (V)	71979 H	(12)*
c ($^3\Sigma^+$)								c← X,[c]	71803 H	(12)*
C ($^1\Sigma^+$)	(71255)	[1425] H						C← X,[d]	71327 H	(12)*
B ($^1\Sigma^+$)	(64868)	[1332] H						B← X, (V)	64893 H	(12)*
A' $^1\Sigma^+$	56505	462.4 H	7.46[e]	0.511_4	0.0109[f]	(2.5)	1.944	A'→ X, R	56093 H	(15)*
x, y		Fragments of two perturbing states (B_x< 0.61, B_y< 0.77) near 39170 and 39950 cm^{-1}.								(3)
		A new band at 39138 cm^{-1}, originally (14) attributed to a $^3\Delta$ state, is now believed to be due to v=11 of a $^3\Pi$ (24).								
A $^1\Pi$	38904.4	$1073._4$[g] Z	$10._1$	0.7800[ghi]	0.0063[j]	(1.65)	1.573_9	A[k]↔ X,[ℓ] R	38797.6 Z	(1a)* (3)*
e $^3\Sigma^-$	38683	752[m]	$4._7$	0.619_4[m]	0.004_0	(1.6_8)	1.766	e← X, R	38417[m]	(3)*
d $^3\Delta_i$[n]	35675.0	795.6[o]	4.91	0.6367[o]	0.0061	(1.6_3)	1.742_0	d← X, R	35430.6[o]	(3)* (14)(27)
a' $^3\Sigma^+$	31331.4	830.7[o]	5.04	0.6489[op]	0.0060	(1.5_8)	1.725_5	a'← X, R	31104.6[o]	(3)* (14)(27)
		Unclassified emission bands, probably due to triplet - triplet transitions, in the region 13300 - 22200 cm^{-1}.								(22)*
a $^3\Pi_r$	27661.0[q]	1135.1[o]	7.73	0.7851[o]	0.0072	[1.9_4][r]	1.568_7	a→ X,[s] R	27585.7[o]	(8)(22)* (27)
X $^1\Sigma^+$	0	1285.08 Z	6.46	0.8200462[i]	0.0059224	1.43	1.534941	Microwave sp.[t]		(1)(4)(23a)

CS, CS^+:

[a]From a short extrapolation of the vibrational levels in A' $^1\Sigma^+$ (15), assuming that the atomic products arising at the dissociation limit are C, $^3P_2$ + S, $^3P_2$. The latest thermochemical (mass-spectrometric) value is 7.21 eV (18). Both values agree with an upper limit (< 7.7 eV) derived from infrared chemiluminescence studies (14a), and the corresponding heats of formation, ΔH^o_{f0} = 66.11 or 69.5 kcal/mole, respectively, are supported by photodissociation (20a) and photoionization (7a) results for CS_2. See, however, (26)(29) who suggest ΔH^o_{f0} = 33 kcal/mole, implying $D^0_0 = 8.7_9$ eV.

[b]From the photoelectron spectrum (16)(17)(20).

[c]Single weak absorption band.

[d]Bands described as diffuse.

[e]$\omega_e y_e = -0.108$, $\omega_e z_e = -0.0377$.

[f]$+0.00112(v+\frac{1}{2})^2 - 0.000208(v+\frac{1}{2})^3$.

[g]Deperturbed constants (3); all observed vibrational levels of this state are strongly perturbed by interactions with a $^3\Pi$, a' $^3\Sigma^+$, d $^3\Delta$, e $^3\Sigma^-$ (1a)(3)(14). The following rather different set of deperturbed parameters is given by (27): $T_e = 38895.7$, $\omega_e = 1077.3$, $\omega_e x_e = 10.66$, $B_e = 0.7881$, $\alpha_e = 0.0092$.

[h]Λ-doubling intervals in v=0 (J=1-9) range from 0.00045 to 0.05961 cm^{-1} (13)(14). The variation with J of the Stark effect was observed in optical-rf double resonance and was analyzed (14) to give $\mu_{el}(v=0) = 0.63$ D ($^+CS^-$); see also (13).

[i]RKR potential functions (6).

[j]$\gamma_e = -0.0004$.

[k]Lifetimes from Hanle effect observations $\tau(v=0) = 176$ ns (corrected for lengthening by triplet mixing), $\tau(v=2)$ = 203 ns (21). The phase shift method (10) gives $\tau(v=0)$ = 255, $\tau(v=1) = 339$, $\tau(v \geq 2) = 292$ ns; $f_{00} = 0.0059$.

[l]Morse-potential Franck-Condon factors (5); compare with experimental values from the fluorescence spectrum excited in the VUV photolysis of CS_2 and OCS (25). $C^{34}S$ isotopic bands (7)(11).

[m]Only v=1 and 2 observed (3); the vibrational numbering is from isotope studies (14). The following set of deperturbed parameters is given by (27): $T_e = 38681.9$, $\omega_e = 752.8$, $\omega_e x_e = 4.95$, $B_e = 0.6227$, $\alpha_e = 0.0062$. Spin-splitting in v=1: $\lambda + \frac{1}{2}\gamma = 1.75$ (3).

[n]This state, originally (3)(14) considered to be $^3\Pi$ and labelled k, is now believed to be $^3\Delta$ (19)(24). The name has been changed to d $^3\Delta$ in order to emphasize the similarity to CO. $A \approx -50$.

[o]Deperturbed constants (27).

[p]Spin-splitting constant $\lambda(v=10) = -1.28$ (14).

[q]$A \approx 95$ cm^{-1}.

[r]From (22).

[s]Two subbands corresponding to $^3\Pi_1 - {}^3\Sigma^+$ and $^3\Pi_0 - {}^3\Sigma^+$ have been observed (22).

[t]Dipole moments [$^-CS^+$, see (23)] from Stark effect $\mu_{el}(v=0)$ = 1.958 D, $\mu_{el}(v=1) = 1.93_6$ D (9). Zeeman effect (23), molecular g factor -0.2702. $^{32}S/^{33}S$ and $^{32}S/^{34}S$ mass ratios from microwave spectra (2).

(continued p. 186)

State		T_e	ω_e	$\omega_e x_e$	B_e	α_e	D_e ($10^{-6}cm^{-1}$)	r_e (Å)	Observed Transitions Design.	ν_{00}	References
$^{12}C^{32}S^+$					$D_0^0 = 6.38_0$ eV[u]						OCT 1976 A
C	$^2\Sigma^+$	54120	1055[v]							53960[v]	(20)
B	$^2\Sigma^+$	36470	868[v]							36210[v]	(20)
A	$^2\Pi_i$	11990[w]	1012.8 Z	6.52	0.71776	0.00622	1.82[x]	1.6407	A→X, R	11806.3 Z	(28)
X	$^2\Sigma^+$	0	1384[y]		[0.8640_0][z]		[1.26]	[1.4954]			

CS, CS^+ (continued):

[u] D_0^0(CS) + I.P.(S) - I.P.(CS).
[v] From the photoelectron spectrum (20); see also (16).
[w] A = - 298.46 (28).
[x] $\beta_e = + 0.15 \times 10^{-6}$.
[y] From the photoelectron spectrum (20); (16) and (17) give 1330 and 1290 cm^{-1}, respectively. Only one level has been found in the optical spectrum (28).
[z] Spin-splitting constant γ_0 = + 0.0201.

CS, CS^+:

(1) Mockler, Bird, PR 98, 1837 (1955).

(1a) Lagerqvist, Westerlund, Wright, Barrow, AF 14, 387 (1958).

(2) Rosenblum, Townes, Geschwind, RMP 30, 409 (1958).

(3) Barrow, Dixon, Lagerqvist, Wright, AF 18, 543 (1960).

(4) Kewley, Sastry, Winnewisser, Gordy, JCP 39, 2856 (1963).

(5) Felenbok, PPS 86, 676 (1965).

(6) Nair, Singh, Rai, JCP 43, 3570 (1965).

(7) Narasimham, Gopal, CS 35, 485 (1966).

(7a) Dibeler, Walker, JOSA 57, 1007 (1967).

(8) Tewarson, Palmer, JMS 27, 246 (1968).

(9) Winnewisser, Cook, JMS 28, 266 (1968).

(10) Smith, JQSRT 9, 1191 (1969).

(11) Chaudhury, Upadhya, Thakur, IJP 44, 375 (1970).

(12) Donovan, Husain, Stevenson, TFS 66, 1 (1970).

(13) Silvers, Bergeman, Klemperer, JCP 52, 4385 (1970).

(14) Field, Bergeman, JCP 54, 2936 (1971).

(14a) Hancock, Morley, Smith, CPL 12, 193 (1971); Hancock, Ridley, Smith, JCS FT II 68, 2117 (1972).

(15) Bell, Ng, Suggitt, JMS 44, 267 (1972).

(16) Jonathan, Morris, Okuda, Smith, Ross, CPL 13, 334 (1972); FDCS No. 54, 48 (1972).

(17) King, Kroto, Suffolk, CPL 13, 457 (1972).

(18) Hildenbrand, CPL 15, 379 (1972).

(19) Robbe, Schamps, CPL 15, 596 (1972).

(20) Frost, Lee, McDowell, CPL 17, 153 (1972).

(20a) Okabe, JCP 56, 4381 (1972).

(21) Silvers, Chiu, JCP 56, 5663 (1972).

(22) Taylor, Setser, Coxon, JMS 44, 108 (1972).

(23) McGurk, Tigelaar, Rock, Norris, Flygare, JCP 58, 1420 (1973).

(23a) Lovas, Krupenie, JPCRD 3, 245 (1974).

(24) Bruna, Kammer, Vasuderan, CP 9, 91 (1975).

(25) Lee, Judge, JCP 63, 2782 (1975).

(26) Hubin-Franskin, Locht, Katihabwa, CPL 37, 488 (1976).

(27) Cossart, Horani, Rostas, quoted by Cossart, Bergeman, JCP 65, 5462 (1976).

(28) M. Horani, unpublished.

(29) Hubin-Franskin, Katihabwa, Collin, IJMSIP 20, 285 (1976).

State	T_e	ω_e	$\omega_e x_e$	B_e	α_e	D_e $(10^{-9}cm^{-1})$	r_e (Å)	Observed Transitions Design.	ν_{00}	References
$^{133}Cs_2$		μ = 66.452718		D_0^0 = 0.394 eV[a]		3.592 < I.P.(eV)[b] < 3.821				OCT 1976 A
		Unidentified structure in the absorption spectrum from 31800 to 34900, 36700 to 41700 cm^{-1}.								(9)
		Fragments of other systems, as well as diffuse bands near atomic lines.[c]								(1)(4)
(E)		Strong system in the region 19140 - 21700, maximum at 20800 cm^{-1}.						$E^d \leftrightarrow X$,		(1)* (7)(14)
D		Weak system in the region 16500 - 18000 cm^{-1}.						$D \leftarrow X$,		(1)(7)
C $(^1\Pi_u)$ 1_u	15948.60	29.703[e] (Z)	0.0576[e]	0.01347[f]	0.0000785	12.10[g]	4.34_0	$C^d \leftrightarrow X$,	V_R 15942.45 (Z)	(1)* (2)(10)
B $(^1\Pi_u)$ 1_u	13043.88	34.329[h] H	0.0800[h]					$B \leftarrow X$,	R 13040.03 H	(1)* (12)
A $(^1\Sigma_u^+)$		Extended system in the region 8800 - 11500 cm^{-1}; partial analysis.						$A \leftrightarrow X$,	R	(1)(3)(8)(15)
b $(^3\Pi_u)$		Unresolved system from 8000 to 8600 cm^{-1}, maximum at 8370 cm^{-1}.						$b \leftrightarrow X$,		(8)(15)
a $^3\Sigma_u^+$	(3140)	Repulsive curve with small van der Waals minimum.								
X $^1\Sigma_g^+$	0	42.022[i]	0.0823[i]	0.0127_1[f]	0.0000264	4.64[f]	4.47	Mol. beam magn. reson.[j]		
$^{133}Cs_2^+$				D_0^0 = 0.61 eV[k]						OCT 1976
X $^2\Sigma_g^+$							(4.44)[ℓ]			

Cs_2, Cs_2^+:

[a]Short extrapolation of ground state vibrational levels (12).

[b]Associative ionization of cesium vapour (16); see also (18). (11) give an appearance potential of 3.8_0 eV, (16) recommend 3.68 eV as the most likely value.

[c]Satellites and self-broadening of Cs principal lines (17) (21). Resonances in the two-photon ionization spectrum of cesium corresponding to the fundamental and diffuse series in absorption result from reactions initiated by molecular transitions to intermediate dissociative states; see (20).

[d]Intermediates in the two-photon ionization of Cs_2 (19)(22).

[e]ω_e, ..., $\omega_e z_e$ obtained (10) from V shaded P heads [also observed in the magnetic rotation spectrum (12)] and calculated head-origin separations; see [f].

[f]The rotational structure of the C − X bands is not resolved. The rotational constants for both states are based on estimated values for B_0'', D_0'' and have been adjusted to reproduce the observed V as well as R shaded heads in the P, Q, and R branches. The red-degraded heads occur through the D terms in the rotational energy expressions. See (10).

[g]$\beta_e = -0.193 \times 10^{-9}$; see [f].

[h]ω_e, ..., $\omega_e z_e$ from band heads in the magnetic rotation spectrum (12). Vibrational levels up to v'=83 observed in absorption. The B state extrapolates to a potential maximum at ~250 cm^{-1} above $6^2P_{3/2} + 6^2S_{1/2}$ (12).

[i]Average of the constants from B − X and C − X (10)(12). Also higher order constants. Vibrational levels observed up to v"=100.

[j]$eqQ(^{133}Cs) = +0.23$ MHz (5); $g_J = 0.00547\ \mu_N$ (6).

[k]$D_0^0(Cs_2)$ + I.P.(Cs) − I.P.(Cs_2); see [b].

[ℓ]Rough estimate from the analysis of charge-exchange cross sections (13). Theoretical values tend to be larger (20a).

(1) Loomis, Kusch, PR 46, 292 (1934).
(2) Kusch, Loomis, PR 49, 217 (1936).
(3) Finkelnburg, Hahn, PZ 39, 98 (1938).
(4) Tsi-Ze, Shang-Yi, JPR 9, 169 (1938).
(5) Logan, Coté, Kusch, PR 86, 280 (1952).
(6) Brooks, Anderson, Ramsey, PRL 10, 441 (1963); PR A 136, 62 (1964).
(7) Lapp, Harris, JQSRT 6, 169 (1966).
(8) Bayley, Eberlin, Simpson, JCP 49, 2863 (1968).
(9) Creek, Marr, JQSRT 8, 1431 (1968).
(10) Kusch, Hessel, JMS 25, 205 (1968).
(11) Foster, Leckenby, Robbins, JP B 2, 478 (1969).
(12) Kusch, Hessel, JMS 32, 181 (1969).
(13) Olson, PR 187, 153 (1969).
(14) Baumgartner, Demtröder, Stock, ZP 232, 462 (1970).
(15) Sorokin, Lankard, JCP 55, 3810 (1971).
(16) Marr, Wherrett, JP B 5, 1735 (1972).
(17) Niemax, PL A 38, 141 (1972).
(18) Popescu, Pascu, Collins, Johnson, Popescu, PR A 8, 1666 (1973).
(19) Collins, Johnson, Popescu, Musa, Pascu, Popescu, PR A 8, 2197 (1973).
(20) Collins, Johnson, Mirza, Popescu, Popescu, PR A 10, 813 (1974).
(20a) Bellomonte, Cavaliere, Ferrante, JCP 61, 3225 (1974).
(21) Niemax, Pichler, JP B 7, 1204 (1974); 8, 2718 (1975).
(22) Granneman, Klewer, Nygaard, Van der Wiel, JP B 9, 865 (1976).

State	T_e	ω_e	$\omega_e x_e$	B_e	α_e	D_e $(10^{-9}cm^{-1})$	r_e (Å)	Observed Transitions Design.	ν_{00}	References
$^{133}Cs^{40}Ar$		$\mu = 30.7241573$								OCT 1976 A
D		Strong emission at 17800 and weaker bands at 18500 cm^{-1}, just to the red of the forbidden 7s-6s Cs line at 18536 cm^{-1}.[a]						D→X,		(4)
		Weak emission bands on either side of the forbidden 5d-6s Cs lines at 14499 and 14597 cm^{-1}.								(4)
B $^2\Sigma^+$		Unstable.[b]						B↔X,[b]		(2)(3)
A $^2\Pi_{3/2}$		Weakly bound state, well depth $D_e \approx 300\ cm^{-1}$.[b]						A↔X,[b]		(2)(3)
$^2\Pi_{1/2}$		Weakly bound state, well depth $D_e \approx 200\ cm^{-1}$.[b]								
X $^2\Sigma^+$		Unstable.[b]								
$^{133}Cs^{79}Br$		$\mu = 49.516045_4$ $D_0^0 = 4.17$ eV[a] I.P. = 7.72 eV[b]								NOV 1976
		Strong autoionization peaks[c] in the photoionization spectrum at and above 12.5 eV (13); also observed by (12) in the energy loss spectrum of 25 keV electrons.								(12)(13)
		Absorption continua with maxima at 46900, 39300, 36300 cm^{-1} (1)(3)(10), followed by diffuse absorption (fluctuation) bands[d] in the region 34200 - 27900 cm^{-1} (1)(2)(3). The chemiluminescence spectrum[d] (19) consists of a long ground-state vibrational progression which extends from 28000 to 17000 cm^{-1}.								(1)(2)(3) (10)(19)
X $^1\Sigma^+$	0	149.66[e]	0.374[e]	0.03606925	0.00012401_2[f]	8.380_1[g]	3.07225_1	Microwave sp.[h]		(4)(8)(17) (18)
$^{133}Cs^{(79)}Br^+$		$D_0^0 = 0.34$ eV[i]								NOV 1976 A
A $(\frac{1}{2})$	3200[j]									
X $(\frac{3}{2}, \frac{1}{2})$	0[j]									

CsAr: [a]The bands are believed to arise from transitions between a bound Cs(7sσ)-Ar excited molecular state and the unstable 6sσ ground state.

[b]The far-wing emission profiles of the collision-broadened Cs resonance lines at 11178 and 11732 cm^{-1} have been analyzed (2) to give the potential curves for the states X, A, B in the interval 3.5 - 6 Å; see also the theoretical calculations of (1). Absorption coefficients in the wings of these lines have been measured (3); barely detectable maxima in the red wings may be due to vibrational structure in the A $^2\Pi$ components.

(1) Baylis, JCP 51, 2665 (1969).
(2) Hedges, Drummond, Gallagher, PR A 6, 1519 (1972).
(3) Chen, Phelps, PR A 7, 470 (1973).
(4) Tam, Moe, Park, Happer, PRL 35, 85 (1975).

CsBr, $CsBr^+$:

[a]From the threshold for dissociative photoionization (13), in good agreement with the thermochemical value of (6); see also (7)(9). The value quoted here corresponds to dissociation into normal atoms; see, however, (11) who demonstrated that under some conditions (shock waves) the primary dissociation products are ions ($Cs^+ + Br^-$) rather than normal atoms in violation of the non-crossing rule.

[b]Photoionization mass-spectrometry (13). Average values (photoelectron spectroscopy) for the vertical first and second ionization potentials: 8.12 and 8.51 eV (15)(20).

[c]Associated with excitation of an electron from the metal 5p shell.

[d]The absorption and the emission bands arise from transitions between a shallow homopolar upper state (from normal atomic products) and the low- and high-lying vibrational levels, resp., of the ionic ground state.

[e]Derived from the rotational constants (18). See also (5).

CsBr, $CsBr^+$ (continued):

[f] $+ 1.02 \times 10^{-7}(v+\frac{1}{2})^2 + 3._2 \times 10^{-10}(v+\frac{1}{2})^3$ (18).

[g] $\beta_e = +0.0064 \times 10^{-9}$; also higher order constants (18).

[h] $\mu_{el} = 10.8_2$ D (molecular beam electric deflection) (21); $eqQ(^{79}Br) = -[6.79 - 0.73(v+\frac{1}{2})]$ MHz (14); $g_J(v=0) = (-)0.009_9\ \mu_N$ (16).

[i]From I.P.(CsBr) and the threshold energy for dissociative ionization (Cs^+ from CsBr), see (13). (20) give 0.24 eV.

[j]From the photoelectron spectrum (15)(20). The two observed broad peaks (see [b]) correspond to removal of an electron from the halogen 4p shell.

(1) Müller, AP(Leipzig) 82, 39 (1927).
(2) Sommermeyer, ZP 56, 548 (1929).
(3) Barrow, Caunt, PRS A 219, 120 (1953).
(4) Honig, Stitch, Mandel, PR 92, 901 (1953); Honig, Mandel, Stitch, Townes, PR 96, 629 (1954).
(5) Rice, Klemperer, JCP 27, 573 (1957).
(6) Brewer, Brackett, CRev 61, 425 (1961).
(7) Bulewicz, Phillips, Sugden, TFS 57, 921 (1961).
(8) Rusk, Gordy, PR 127, 817 (1962).
(9) Scheer, Fine, JCP 36, 1647 (1962).
(10) Davidovits, Brodhead, JCP 46, 2968 (1967).
(11) Berry, Cernoch, Coplan, Ewing, JCP 49, 127 (1968).
(12) Geiger, Pfeiffer, ZP 208, 105 (1968).
(13) Berkowitz, JCP 50, 3503 (1969).
(14) Hoeft, Tiemann, Törring, ZN 27 a, 702 (1972).
(15) Berkowitz, Dehmer, Walker, JCP 59, 3645 (1973).
(16) Honerjäger, Tischer, ZN 28 a, 458 (1973).
(17) Miller, Finney, Inman, AD 5, 1 (1973).
(18) Honerjäger, Tischer, ZN 29 a, 819 (1974).
(19) Oldenborg, Gole, Zare, JCP 60, 4032 (1974).
(20) Potts, Williams, Price, PRS A 341, 147 (1974).
(21) Story, Hebert, JCP 64, 855 (1976).

State	T_e	ω_e	$\omega_e x_e$	B_e	α_e	D_e $(10^{-7}cm^{-1})$	r_e (Å)	Observed Transitions Design.	ν_{00}	References
$^{133}Cs^{35}Cl$		μ = 27.6847083		D_0^0 = 4.58 eV[a]		I.P. = 8.3_2 eV[b]				NOV 1976 A
		Strong autoionization peaks[c] at and above 12.4 eV in photoionization (15) and electron energy loss spectra (13).								(13)(15)
		Absorption continua with maxima at 51500 and 40500 cm^{-1} (1)(3)(12), followed by diffuse absorption bands (fluctuation b.) from 40850 to 29840 cm^{-1} (3). The chemiluminescence spectrum (21) consists of a long lower-state vibrational progression and extends from 29400 to 20800 cm^{-1}. See [d] of CsBr.								(1)(3)(12) (21)
X $^1\Sigma^+$	0	214.17[d]	0.731[d]	0.07209149	0.00033756[e]	0.32675[f]	2.90627_2	IR sp.		(6)
								Microwave sp.[g]		(4)(10)(19) (20)
								Mol. beam el. reson.[h]		(2)(5)(14)
								Mol. beam magn. reson.		(11)
$^{12}C^{80}Se$		μ = 10.43336114		D_0^0 = 5.9_8 eV[a]						NOV 1976
A $^1\Pi$	(35243)	[817.7][b]		[0.487][c]			$[1.82_1]$	A↔X, R	35135.25[d]	(1)* (2)(3)* (6)*
a $^3\Pi_2$		(not observed)								
a $^3\Pi_1$	(24466)			[0.544]			$[1.72_3]$	a→X, R	24396.6 Z	(4)
a $^3\Pi_{0^+}$	(24150)	[893.2] H		[0.544]					24080.0 Z	
X $^1\Sigma^+$	0	1035.36 Z	4.86	0.5750[e]	0.00379	7.1[f]	1.67647[g]	Microwave sp.[h]		(5)

CsCl: [a]Thermochemical value (7), confirmed by the photoionization data of (15). See also (8)(9).

[b]Onset of a broad band in the photoelectron spectrum, maximum (vertical I.P.) at 8.75 eV (22). The photoelectron spectrum was also investigated by (17) who find 7.84 and 8.54 eV, respectively.

CsCl (continued):

[c]Interpretation analogous to CsBr (footnote [c]).

[d]Calculated from the rotational constants (20). Good agreement with the less precise values from the infrared spectrum (6).

[e] $+ 3.42 \times 10^{-7}(v+\frac{1}{2})^2 + 1.8 \times 10^{-9}(v+\frac{1}{2})^3$ (20).

CsCℓ (continued):

[f] β_e = + 0.00038 x 10^{-7}; also higher order constants (20).
[g] eqQ(^{35}Cℓ) = + [1.830 - 0.118(v+½)] MHz,
|eqQ(^{133}Cs)| ≤ 1.1 MHz (16).
g_J = - [0.02815 - 0.00031(v+½)] μ_N (18); see, however, (11) who give g_J = (-)0.021$_2$.
[h] $\mu_{e\ell}$ = [10.358 + 0.058(v+½)] D (14).

(1) Schmidt-Ott, ZP 69, 724 (1931).
(2) Luce, Trischka, PR 82, 323; 83, 851 (1951); JCP 21, 105 (1953).
(3) See ref. (3) of CsBr.
(4) See ref. (4) of CsBr.
(5) Trischka, JCP 25, 784 (1956).
(6) See ref. (5) of CsBr.
(7) See ref. (6) of CsBr.
(8) See ref. (7) of CsBr.
(9) See ref. (9) of CsBr.
(10) Clouser, Gordy, PR A 134, 863 (1964).
(11) Mehran, Brooks, Ramsey, PR 141, 93 (1966).
(12) See ref. (10) of CsBr.
(13) See ref. (12) of CsBr.
(14) Hebert, Lovas, Melendres, Hollowell, Story, Street, JCP 48, 2824 (1968).
(15) See ref. (13) of CsBr.
(16) Hoeft, Tiemann, Törring, ZN 27 a, 1516 (1972).
(17) See ref. (15) of CsBr.
(18) See ref. (16) of CsBr.
(19) See ref. (17) of CsBr.
(20) See ref. (18) of CsBr.
(21) See ref. (19) of CsBr.
(22) See ref. (20) of CsBr.

CSe: [a] Thermochemical value (mass-spectrom.)(3a).
[b] $\Delta G(3/2)$ = 833.7; from the Q(1) lines of the 0-1, 1-1, 2-1 bands (6). This state is strongly perturbed and no meaningful vibrational and rotational constants can be given. The absorption spectrogram of (3) shows in addition to $^1\Pi - {}^1\Sigma$ some absorption bands probably due to some of the perturbing levels, in particular one progression which may be $a'^3\Sigma^+ \leftarrow X^1\Sigma^+$ in analogy to CO and CS.
[c] B_1 = 0.497, B_2 = 0.482, B_3 = 0.447; see [b].
[d] Q(1) line for the 0-0 band (6).
[e] B_0 = 0.573155 from the microwave spectrum (5).
[f] β_e = - 0.2$_5$ x 10^{-7}.
[g] From the B_0 values for various isotopes obtained from the microwave spectrum (5).
[h] Dipole moment from Stark effect 1.99 D (5); molecular g factor from Zeeman effect -0.2431.

(1) Barrow, PPS 51, 989 (1939).
(2) Laird, Barrow, PPS 66, 836 (1953).
(3) Callear, Tyerman, TFS 61, 2395 (1965).
(3a) D. Detry (unpublished), quoted in DONNSPEC (1970).
(4) Lebreton, Bosser, Marsigny, JP B 6, L226 (1973).
(5) McGurk, Tigelaar, Rock, Norris, Flygare, JCP 58, 1420 (1973).
(6) Stringat, Bacci, Pischedda, CJP 52, 813 (1974).

State	T_e	ω_e	$\omega_e x_e$	B_e	α_e	D_e ($10^{-6}cm^{-1}$)	r_e (Å)	Observed Transitions: Design.	Observed Transitions: ν_{00}	References
$^{133}Cs^{19}F$		μ = 16.6223003 D_0^0 = 5.1$_5$ eV[a] I.P. = 8.8$_0$ eV[b]								NOV 1976
		Several autoionizing states[c] at and above 12.1 eV (16), also observed in the electron energy loss spectra of (14).								(14)(16)
		Continuous absorption with maximum at 47700 cm^{-1}, preceded by diffuse absorption bands (fluctuation b.) in the region 43700 - 36900 cm^{-1}. See [d] of CsBr.								(2)
X $^1\Sigma^+$	0	352.56[d]	1.61$_5$[d]	0.18436969	0.0011756$_2$[e]	0.20168[f]	2.345351	IR sp.		(12)
								Microwave sp.		(3)(10)(20)(21)
								Mol. beam el. reson.[g]		(1)(4)(9)(13)(15)(17)
								Mol. beam magn. reson.		(11)
$^{133}Cs^{1}H$		μ = 1.00024037 D_0^0 = 1.81 eV[a]								NOV 1976
a ($^3\Delta$)	[28534][b]	[276][b]		[1.51][b]						(4)
B $^1\Sigma^+$	[28350.5][c]	[82.1][c] Z		[0.70][c]			(4.63)	B← X, R	27908.2[c] Z	(4)
A $^1\Sigma^+$	17845.$_8$	165.7 Z	-7.7$_7$[d]	1.075	-0.021$_9$[e]	98[f]	3.96$_0$	A↔X, R	17488.$_3$ Z	(1)(2)(5)
X $^1\Sigma^+$	0	891.0 Z	12.9$_3$[g]	2.7099	0.0579	[113]	2.4938			
$^{133}Cs^{2}H$		μ = 1.98403535								NOV 1976
A $^1\Sigma^+$	(17833)	(123.9)	(-3.08)[h]	i				A← X, R	j	(2)(3)
X $^1\Sigma^+$	0	[619.1] H		[1.354]		[20]	[2.505]			
$^{133}Cs^{4}He$		μ = 3.8855843$_3$								NOV 1976 A
		Potential curves for X $^2\Sigma$, A $^2\Pi$, and B $^2\Sigma$ in the interval r = 3.5 to 5.5 Å have been derived from the analysis of the extreme-wing emission profiles of the Cs resonance lines at 11178 and 11732 cm^{-1}. Only A $^2\Pi_{3/2}$ has a small potential minimum, $D_e \approx$ 170 cm^{-1}.								(1)

CsF: [a]Thermochemical value (5); see also (6)(7)(18).

[b]Adiabatic ionization potential from the photoelectron spectrum (19); vertical I.P. at 9.6_0 eV.

[c]Interpretation analogous to CsBr (footnote [c]).

[d]Derived from the rotational constants (21); good agreement with infrared results (12). See also (8).

[e]$+ 1.18 \times 10^{-6}(v+\frac{1}{2})^2 + 1.7_7 \times 10^{-8}(v+\frac{1}{2})^3$ (21).

[f]$\beta_e = + 0.00031 \times 10^{-6}$; also higher order constants (21).

[g]$\mu_{e\ell} = [7.8478 + 0.07026(v+\frac{1}{2}) + 0.000195(v+\frac{1}{2})^2]$ D (15)(17); $eqQ(^{133}Cs) = + [1245.2 - 16.2(v+\frac{1}{2}) + 0.31(v+\frac{1}{2})^2]$ kHz (13)(17); $g_J(v=0) = (-)0.06420\ \mu_N$, for $v \neq 0$ see (9)(20); see also (11).

(1) Trischka, PR 74, 718 (1948); 76, 1365 (1949).
(2) See ref. (3) of CsBr.
(3) See ref. (4) of CsBr.
(4) See ref. (5) of CsCℓ.
(5) See ref. (6) of CsBr.
(6) See ref. (7) of CsBr.
(7) See ref. (9) of CsBr.
(8) Ritchie. Lew, CJP 42, 43 (1964).
(9) Gräff, Runolfsson, ZP 187, 140 (1965).
(10) Veazey, Gordy, PR A 138, 1303 (1965).
(11) See ref. (11) of CsCℓ.
(12) Baikov, Vasilevskii, OS(Engl. Transl.) 22, 198 (1967).
(13) English, Zorn, JCP 47, 3896 (1967).
(14) See ref. (12) of CsBr.
(15) See ref. (14) of CsCℓ.
(16) See ref. (13) of CsBr.
(17) Bennewitz, Haerten, Klais, Müller, CPL 9, 19 (1971); ZP 249, 168 (1971).
(18) JANAF (1971).
(19) See ref. (15) of CsBr.
(20) See ref. (16) of CsBr.
(21) See ref. (18) of CsBr.

Cs^1H, Cs^2H:

[a]Short extrapolation of vibrational levels in B $^1\Sigma^+$ to the limit $^2D + ^2S$ (4). A longer extrapolation of the ground state levels (5) gives 2.08 eV.

[b]Constants for the lowest of three levels observed in small perturbations of B $^1\Sigma^+$; $A \approx 14$. Vibrational numbering unknown, $\Omega = 3$ not observed.

[c]Constants for the lowest of eleven observed levels, v' = 11 ± 3. ΔG increases to 93.7 for v=18; $B_v(v \leq 18) = 0.787 - 0.0076(v+\frac{1}{2})$. For v > 18 both ΔG and B_v decrease rapidly.

[d]$- 0.29_7(v+\frac{1}{2})^3 + 0.0040(v+\frac{1}{2})^4$ (valid for $v \leq 15$).

[e]$- 0.00131(v+\frac{1}{2})^2$ (valid for $v \leq 13$).

[f]$\beta_e = - 2.0 \times 10^{-6}$.

[g]$\omega_e y_e = + 0.10_5$, $\omega_e z_e = - 0.0018$. Observed up to v"=14 in laser-excited fluorescence (5).

[h]$\omega_e y_e = - (0.058)$; from band heads, $8 \leq v' \leq 18$.

[i]Only the 10-0 and 11-0 bands have been rotationally analyzed (2); $B_{10} = 0.587$, $B_{11} = 0.586$.

[j]Band origins for the 10-0 and 11-0 bands at 19091.2 and 19261.8 cm^{-1}.

(1) Almy, Rassweiler, PR 53, 890 (1938).
(2) Bartky, JMS 21, 25 (1966).
(3) Császár, Koczkás, APH 23, 211 (1967).
(4) Ringström, JMS 36, 232 (1970).
(5) Tam, Happer, JCP 64, 2456 (1976).

CsHe: (1) See ref. (2) of CsAr.

State	T_e	ω_e	$\omega_e x_e$	B_e	α_e	D_e ($10^{-9}cm^{-1}$)	r_e (Å)	Observed Transitions: Design.	Observed Transitions: ν_{00}	References
$^{133}Cs^{127}I$		$\mu = 64.917826_1$ $D_0^0 = 3.56$ eV[a] I.P. = 7.25 eV[b]								NOV 1976 A
		Autoionizing levels at and above 12.6 eV (12); these states are also seen in the electron energy loss spectrum of (11). Interpretation analogous to CsBr.								(11)(12)
		Continuous absorption with maxima at 54050, 50250, 46500, 41400, 38800, 30900 cm^{-1} (2)(3)(10), followed by diffuse absorption bands (fluctuation b.) in the region 29140 - 22900 cm^{-1} (1)(3). The chemiluminescence spectrum (17) consisting of a long lower-state vibrational progression extends from 25000 - 18000 cm^{-1}. See [d] of CsBr.								(1)(2)(3)(10)(17)
X $^1\Sigma^+$	0	119.178[c]	0.2505[c]	0.02362735_7	0.000068263[d]	3.7146[e]	3.31519_2	Microwave sp.[f]		(4)(8)(16)
$^{133}Cs^{127}I^+$		$D_0^0 = 0.21$ eV[g]								NOV 1976 A
	80000 - 100000	Two or perhaps three peaks in the 17 - 20 eV region of the photoelectron spectrum are believed to arise from the removal of an electron from the metal 5p shell of Cs^+I^-.								(14)
A $(\frac{1}{2})$ X $(\frac{3}{2}, \frac{1}{2})$	(7400) 0	From the photoelectron spectrum; removal of an electron from the halogen 5p shell of Cs^+I^-. The $\frac{3}{2}$ and $\frac{1}{2}$ components of the stronger peak are not resolved.								(14)(18)
$^{133}Cs^{(84)}Kr$		$(\mu = 51.436456_6)$								NOV 1976 A
		Strong emission at 17700 and weaker bands near 18500 cm^{-1}, attributed to transitions from a stable Cs(7sσ)-Kr upper state to the unstable 6sσ ground state.								(2)
		Weak emission bands on either side of the forbidden 5d-6s lines at 14499 and 14597 cm^{-1}.								(2)
		The analysis of the far-wing emission profiles of the Cs resonance lines (11178, 11732 cm^{-1}) indicates that only A $^2\Pi_{\frac{3}{2},\frac{1}{2}}$ are stable states with well depths of ~350 cm^{-1}. Atomic scattering data [see ref. in (1)] predict a small ground state minimum.								(1)

State	T_e	ω_e	$\omega_e x_e$	B_e	α_e	D_e	r_e (Å)	Observed Transitions		References
								Design.	ν_{00}	
$^{133}Cs^{(20)}Ne$	(μ = 17.3782925)									NOV 1976 A
	Strong emission at 18020 and a narrow band at 18570 cm^{-1}, as well as weaker bands in the 14000 - 15000 cm^{-1} region; see the analogous transitions of CsAr and CsKr.									(2)
	The molecular states associated with the Cs resonance lines at 11178 and 11732 cm^{-1} (see Cs-Ar, He, Kr) are unstable except A $^2\Pi_{3/2}$ which has a potential minimum of $\sim$130 cm^{-1}.									(1)

CsI, CsI^+:

[a]Average of a thermochemical value (6) and a value obtained from photodissociative ionization of CsI (12). See also (7)(9).

[b]From photoionization mass-spectrometry (12). The adiabatic ionization potential from photoelectron spectroscopy (14)(18) is 7.10 eV; the first and second vertical potentials are 7.54 and 8.46 eV (average values).

[c]Derived from the rotational constants (16). See also (3)(5).

[d]$+ 4.89 \times 10^{-8}(v+\frac{1}{2})^2 + 1.1_4 \times 10^{-10}(v+\frac{1}{2})^3$ (16).

[e]$\beta_e = +0.0023_0 \times 10^{-9}$; also higher order constants (16).

[f]$\mu_{e\ell} = 11.6_9$ D (molecular beam electric deflection)(19); $eqQ(^{127}I) = -[14.28 + 2.10(v+\frac{1}{2})]$ MHz, $|eqQ(^{133}Cs)| \leq 1$ MHz (13); $|g_J| < 0.0036\ \mu_N$ (15).

[g]From the photoionization data of (12). Using 7.10 instead of 7.25 eV for I.P.(CsI) (see [b]) would increase this value to 0.36 eV.

CsI, CsI^+ (continued):

(1) See ref. (2) of CsBr.
(2) See ref. (1) of CsCℓ.
(3) ⋮ (10) See ref. (3)...(10) of CsBr, respectively.
(11) See ref. (12) of CsBr.
(12) See ref. (13) of CsBr.
(13) Hoeft, Tiemann, Törring, ZN 27 a, 1017 (1972).
(14) See ref. (15) of CsBr.
(15) See ref. (16) of CsBr.
(16) See ref. (18) of CsBr.
(17) See ref. (19) of CsBr.
(18) See ref. (20) of CsBr.
(19) See ref. (21) of CsBr.

CsKr, CsNe:

(1) See ref. (2) of CsAr.
(2) See ref. (4) of CsAr.

State	T_e	ω_e	$\omega_e x_e$	B_e	α_e	D_e ($10^{-8}cm^{-1}$)	r_e (Å)	Observed Transitions Design.	Observed Transitions ν_{00}	References
$^{133}Cs^{16}O$		$\mu = 14.2767370_3$								NOV 1976
X $^2\Sigma^+$	0	Ground state symmetry from matrix ESR spectroscopy (2) and magnetic and electric deflection analysis of Cs + NO_2 scattering products (1).								(1)(2)
$^{133}Cs^{(132)}Xe$		($\mu = 66.201451_7$)								NOV 1976 A
		Absorption bands arising from molecular upper states correlated with various excited levels of atomic Cs, 22100 - 24700 cm^{-1}.								(4)
		Strong emission at 17470, accompanied by much weaker features at 18500 cm^{-1} (3); also observed in absorption (4). The blue wing of the main peak shows undulations which may be due to vibrational levels in the upper state. See the analogous transitions of Cs-Ar, Kr, Ne.								(3)(4)(5)
		Complex structure in emission and absorption from 13500 to 15900 cm^{-1}; see Cs-Ar, Kr, Ne.								(3)(4)(5)
		The analysis of the far-wing emission profiles of the Cs resonance lines at 11178 and 11732 cm^{-1}[(2); see also (1)] shows that B $^2\Sigma$ is unstable, A $^2\Pi_{\frac{1}{2}}$ and $^2\Pi_{\frac{3}{2}}$ have potential minima of 430 and 500 cm^{-1}, resp., and X $^2\Sigma$ is very weakly bound.								(2)
$^{63}Cu_2$		$\mu = 31.464794_9$ $D_0^0 = 2.03$ eV[a] I.P. = 7.3_7 eV[b]								DEC 1974 A
		Additional band systems attributed to Cu_2 in the regions 34500 - 37000, 39000 - 40100, and 40600 - 42900 cm^{-1}. No analysis.								(5)
B $^1\Sigma_u^+$	21758.3_5	[242.15] Z	2.0[c] H	0.09889_0	0.000606	6.30	2.3276	B↔X, R	21747.88 Z	(1)* (2)(7)(8)(10)
A ($^1\Pi_u$)	20433.2	191.9 H	0.348[d]	(0.08185)	(0.00062)	(3.8)	(2.558_4)	A↔X, R	20396.0 H	(1)* (2)(7)(9)
X $^1\Sigma_g^+$	0	[264.55] Z	1.025[e] H	0.10874_3	0.000614	7.16	2.2197			

CsO: (1) Herm, Herschbach, JCP 52, 5783 (1970).
(2) Lindsay, Herschbach, Kwiram, JCP 60, 315 (1974).

CsXe: (1) Herman, Herman, JQSRT 4, 487 (1964).
(2) Hedges, Drummond, Gallagher, PR A 6, 1519 (1972).
(3) Tam, Moe, Park, Happer, PRL 35, 85 (1975).
(4) Happer, Moe, Tam, PL A 54, 405 (1975).
(5) Sayer, Ferray, Lozingot, Berlande, JP B 9, L293 (1976).

Cu_2: [a]Thermochemical value (mass-spectrom.)(3)(4)(6), recalculated (12).
[b]Experimental value by (11), quoted in an ab initio study of Cu_2 and Cu_2^+ by (13). No details.
[c]$\omega_e y_e = -0.02$.
[d]$\omega_e y_e = -0.018$.
[e]$\omega_e y_e = +0.0008$.

(1) Kleman, Lindkvist, AF 8, 333 (1954).
(2) Ruamps, CR 238, 1489 (1954).
(3) Drowart, Honig, JCP 25, 581 (1956); JPC 61, 980 (1957).
(4) Schissel, JCP 26, 1276 (1957).
(5) Ruamps, AP(Paris) 4, 1111 (1959).
(6) Ackerman, Stafford, Drowart, JCP 33, 1784 (1960).
(7) Åslund, Barrow, Richards, Travis, AF 30, 171 (1965).
(8) Pesič, Weniger, CR B 272, 46 (1971).
(9) Pesič, Weniger, CR B 273, 602 (1971).
(10) Rao, Lakshman, JQSRT 11, 1157 (1971).
(11) Cabaud, Ph.D. Thesis (U. Claude Bernard, Lyon, 1972).
(12) Smoes, Mandy, Vander Auwera-Mahieu, Drowart, BSCB 81, 45 (1972).
(13) Joyes, Leleyter, JP B 6, 150 (1973).

State	T_e	ω_e	$\omega_e x_e$	B_e	α_e	D_e ($10^{-8}cm^{-1}$)	r_e (Å)	Observed Transitions Design.	ν_{00}	References
$^{(63)}Cu^{27}Al$		(μ = 18.8846172)		$D_0^0 = 2.2_1$ eV[a]						DEC 1974
$^{(63)}Cu^{75}As$		($\mu = 34.201994_0$)								DEC 1974
		Sequences of mostly V shaded bands in thermal emission at 13529, 13531, 14134, 14633, 14684, 14811 cm^{-1}. $\omega' < \omega''$.								(1)
$^{63}Cu^{209}Bi$		$\mu = 48.365456_8$								DEC 1974
B	x_2 + 19708.6	197.4 H	0.62	a				$B \rightarrow X_2$, R	19708.0 H	(1)* (2)
A	15922.0	196.9 H	0.85	a				$A \rightarrow X_1$, V_R	15920.6 H	(1)(2)
X_2 b	x_2	198.6 H	0.55							
X_1	0	199.6 H	0.66	a						
$^{63}Cu^{79}Br$		$\mu = 35.011427_7$		$D_0^0 = 3.4_3$ eV[a]						NOV 1975
D ($^1\Sigma^+$)	25538.6	281.9 H	1.35					$D \rightarrow X$, R	25522.1 H	(3)*
C $^1\Sigma^+$ (0^+)	23460.9	294.7[b] Z	1.06	0.0942[b]	0.0004_1	3	2.26_1	$C \leftrightarrow X$, R	23450.9 Z	(1)(4)*
B ($^1\Pi$)	23044.7	284.22 H	1.32					$B \leftrightarrow X$,[c] R	23029.3 H	(1)
A ($^1\Pi$)	20498.5	296.13 H	1.01					$A \leftrightarrow X$,[c] R	20489.2 H	(1)
X $^1\Sigma^+$	0	314.8[b] Z	0.96	0.10192625	0.00045214[d]	4.2737[e]	2.17344_1[f]	Microwave sp.		(5)

CuAℓ: [a]Thermochemical value (mass-spectrom.)(1)(2).

(1) Blue, Gingerich, 16th Annual Conference on Mass Spectrometry and Allied Topics, Pittsburgh (May 1968); paper 129.

(2) Uy, Drowart, TFS 67, 1293 (1971).

CuAs: (1) Lefebvre, Houdart, CR B 274, 178 (1972).

CuBi: [a]For rotational constants derived from incompletely resolved rotational structure in the 0-0 bands see (2). Uncertain.

[b]It is not known whether X_1 is the ground state, nor is it clear whether or not X_1 and X_2 are identical.

(1) Lefebvre, Houdart, CR B 270, 1485 (1970); 272, 1301 (1971).

(2) Lefebvre, Bocquet, Houdart, RPA 8, 149 (1973).

CuBr: [a]Thermochemical value (2).

[b]Vibrational and rotational constants derived from data for the C − X system of $^{63}Cu^{81}Br$ [Tables II and III of (4)] using $\mu_i/\mu = 1.01108$.

[c]These bands appear to have R and Q heads (3).

[d]$+ 6.78 \times 10^{-7}(v+\frac{1}{2})^2 - 1.0 \times 10^{-9}(v+\frac{1}{2})^3$.

[e]$\beta_e = -0.00187 \times 10^{-8}$; $H_e = -5.074 \times 10^{-15}$.

[f]From the corrected $B_e = 0.1019274$.

(1) Ritschl, ZP 42, 172 (1927).

(2) Brewer, Lofgren, JACS 72, 3038 (1950).

(3) Rao, Apparao, PIAS A 60, 57 (1964).

(4) Rao, Apparao, CJP 45, 2805 (1967).

(5) Manson, De Lucia, Gordy, JCP 63, 2724 (1975).

State	T_e	ω_e	$\omega_e x_e$	B_e	α_e	D_e ($10^{-7}cm^{-1}$)	r_e (Å)	Observed Transitions Design.	Observed Transitions ν_{00}	References
$^{65}Cu^{35}Cl$		μ = 22.7279945		D_0^0 = 3.93 eV[a]						DEC 1975
F $^1\Pi$ (1)	25285.30	384.94 Z	1.65	0.1607	0.00091	1.2	2.148	F↔X, R	25270.11 Z	(2)(4)(5)* (7)* (11)
E $^1\Sigma^+(0^+)$	23074.24	403.30 Z	1.62[b]	0.1663	0.00108	1.0	2.112	E↔X, R	23068.23 Z	(1)(3)(4)(5) (9)* (11)
D $^1\Pi$ (1)	22969.74	392.89 Z	1.74_5	0.1677_7[c]	0.0009_8		2.102_6	D↔X, R	22958.50 Z	(1)(4)(5) (9)* (11) (15)
C $^1\Sigma^+(0^+)$	20630.94	396.93 Z	1.48[d]	0.1691[e]	0.00089	1.2	2.094	C↔X, R	20621.78 Z	(1)(4)(5)(6) (8)*
B $^1\Pi$ (1)	20484.08	399.29 Z	1.61[d]	0.1684[fe]	0.00092	1.2	2.099	B↔X, R	20476.07 Z	(1)(4)(5)(6) (8)*
A $^1\Pi$ (1)	19001.4	407.0 H	1.70					A↔X, R	18997.2 H	(1)(5)*
A'($^1\Sigma^+$)	(13440)	(510)[g]						(A'→X), V	13479.5 H	(13)
X $^1\Sigma^+$	0	415.29[e] Z	1.58	0.17628802[h]	0.00099647[i]	1.2706_0[j]	2.051183[k]	Microwave sp.		(14)
$^{63}Cu^{19}F$		μ = 14.5928363		D_0^0 = 4.4_2 eV[a]						AUG 1975
C $^1\Pi$	20258.4_7	645.0_7 H^Q	4.19	[0.3746][b]		(5.1)	[1.756_1]	C↔X, R	20269.62 Z	(1)(2)*
B $^1\Sigma^+$	19717.5	657.0 c	3.9_2	0.3716	0.0032[d]	(4.8)	1.763_2	B↔X, R	19734.66 Z	(1)(2)*
A $^1\Pi$	17543.4	649.2[e] H^Q	4.0_0	[0.3675][f]			[1.773_0]	A↔X , R	17556.7 H^Q	(1)(2)
X $^1\Sigma^+$	0	622.6_5 H^Q	3.95	0.3794029	0.0032298[g]	5.63	1.744930	Microwave sp.[h]		(5)(6)
$^{(63)}Cu^{(69)}Ga$		(μ = 32.895626_8)								DEC 1974
A	15276.7	151.1 H	2.28					A→X, R	15240.8 H	(1)
X	0	222.0 H	0.55							
$^{(63)}Cu^{(74)}Ge$		(μ = 33.991986_3)		D_0^0 = 2.0_9 eV[a]						DEC 1974

CuCℓ: [a]Thermochemical value (mass-spectrom.)(10)(12).

[b]$\omega_e y_e = -0.009_3$ (recalculated for $v \leq 4$).

[c]Λ-type doubling $\Delta\nu_{ef}(v=0) = -0.0011_1\, J(J+1)$.

[d]From the value for the $^{63}Cu^{35}C\ell$ isotope [see (6)].

[e]Consistent with constants for $^{63}Cu^{35}C\ell$ in (6).

[f]Λ-type doubling $\Delta\nu_{ef}(v=0) = -0.0008_7\, J(J+1)$.

[g]Preliminary data.

[h]For constants of $^{63}Cu^{35}C\ell$, $^{63}Cu^{37}C\ell$, $^{65}Cu^{37}C\ell$ see (14).

[i]$+ 1.96_4 \times 10^{-6}(v+\frac{1}{2})^2 - 2 \times 10^{-9}(v+\frac{1}{2})^3$.

[j]$\beta_e = -0.0007_6 \times 10^{-7}$; $H_e = -2.007 \times 10^{-14}$.

[k]From the corrected $B_e = 0.1762895\ cm^{-1}$ [see (14)].

(1) Ritschl, ZP 42, 172 (1927).
(2) Bloomenthal, PR 54, 497 (1938).
(3) Sinha, CS 17, 208 (1948).
(4) Asundi, Rao, Brody, Nature 192, 444 (1961).
(5) Rao, Brody, JCP 35, 776 (1961).
(6) Lagerqvist, Lazarova-Girsamof, Naturw. 48, 68 (1961); AF 20, 543 (1962).
(7) Rao, Asundi, Brody, CJP 40, 412 (1962).
(8) Rao, Asundi, Brody, CJP 40, 423 (1962).
(9) Rao, Brody, Asundi, CJP 40, 1443 (1962).
(10) Hildenbrand, JCP 52, 5751 (1970).
(11) Puri, Mohan, IJPAP 8, 759 (1970); CS 43, 340 (1974).
(12) Guido, Gigli, Balducci, JCP 57, 3731 (1972).
(13) Rao, Rao, SpL 7, 463 (1974).
(14) Manson, De Lucia, Gordy, JCP 62, 1040 (1975).
(15) Ahmed, Barrow, JP B 8, L362 (1975).

CuF: [a]Thermochemical value (mass-spectrom.)(4). See also (3).

[b]Λ-type doubling $\Delta\nu_{ef} = +0.0010\, J(J+1)$.

[c]From bandheads, taking into account head-origin separations.

[d]Slight modification of the analysis of the 1-1 band in (2).

[e]0-0 sequence only. Constants recalculated assuming that the lower state is X $^1\Sigma^+$.

[f]From a partial rotational analysis, assuming that the lower state is X $^1\Sigma^+$.

[g]$\gamma_e = +0.0000123$.

[h]$\mu_{e\ell}(v=0) = 5.7_7$ D. Values for eqQ(Cu) in (5), magnetic constants from Zeeman effect measurements in (6).

(1) See ref. (1) of CuCℓ.
(2) Woods, PR 64, 259 (1943).
(3) Kent, McDonald, Margrave, JPC 70, 874 (1966).
(4) Hildenbrand, JCP 48, 2457 (1968).
(5) Hoeft, Lovas, Tiemann, Törring, ZN 25 a, 35 (1970).
(6) Honerjäger, Tischer, ZN 29 a, 1919 (1974).

CuGa: (1) Biron, CR B 271, 1096 (1970).

CuGe: [a]Thermochemical value (mass-spectrom.)(1).

(1) Kant, Strauss, JCP 49, 3579 (1968).

State	T_e	ω_e	$\omega_e x_e$	B_e	α_e	D_e ($10^{-4}cm^{-1}$)	r_e (Å)	Observed Transitions		References
								Design.	ν_{00}	
$^{63}Cu^1H$		$\mu = 0.99193919$		$D_0^0 = 2.7_3$ eV^a						JAN 1975
D $^1\Pi$	(44669)	[1804.0] Z	(55)	7.72^b	0.31	4.7_6^c	1.484	D← X,	R 44651.2 Z	(2)*
d $^3\Pi_r$	$(41000)^{de}$	$[1760]^e$		$[7.8]^e$		$[10]^e$	[1.48]		40920^{fe}	(10)
E $^1\Sigma^+$	39299	574^g Z	-3.6_6^h	3.09_3^g	-0.036_6^h	3.8^h	2.34_4	E← X,	R 38626^g	(10)
b Δ_2	(28470)	$[(1475)]^i$		$[(6.7)]^i$			$[(1.5_9)]$	b← X,	R $(28250)^i$	(9)
e $(^3\Pi_2)$ 2				$[>6.2_2]^j$			[<1.65]		$>27958^{fj}$	(9)
c 1	(28161)	[1388.7] Z	(85)	6.43^k	0.42		1.62_6	c↔ X,	R 27957.5 Z	(3)(4)(9)
C 1	27270.4	1627.3 Z	86.0^{ℓ}	6.553^m	0.352^n	$[4.76]^n$	1.610_4	C↔ X,	R 27101.3 Z	(3)(4)(9)*
a $(^3\Sigma^+)$				$[(5.9)]^o$					$(26365)^{fo}$	(4)(9)
B $^3\Pi_{0^+}$	26420.9	1669.7^p Z	51.2	6.582	0.290^p	$[4.05]^p$	1.606_9	B↔ X,	R 26281.7 Z	(1)(4)(9)
A $^1\Sigma^+$	23434.2	1698.4 Z	44.0^q	6.874^r	0.263^q	4.35^{sq}	1.572_4	A↔ X,	R 23311.1 Z	(1)(4)(8)(9)
X $^1\Sigma^+$	0	1941.26 Z	37.51^t	7.9441	0.2563^t	5.20^t	1.46263			
$^{63}Cu^2H$		$\mu = 1.95163876$		$D_0^0 = 2.7_6$ eV^a						JAN 1975
c 1		b		$[3.1614]^c$	b	$[0.73]^{db}$	$[1.6529_5]$	c← X,	R 27989.11 Z	(3)
C 1	27271.6	$1143._5$ Z	$29._9$	$[3.2589]^e$	0.131	$[1.211]^e$	1.611_9	C← X,	R 27148.59^e Z	(3)
a $^3\Sigma^+$	(26490)	$[(1090)]^f$		$[3.20]^f$	0.10		1.63_0	a← X,	R $(26346)^f$	(3)
B $^3\Pi_{0^+}$	26381.8	1222.0 Z	35.7	3.317^g	0.086	$[1.2]^g$	1.613_7	B← X,	R 26296.5 Z	(3)
A $^1\Sigma^+$	23412.9	1210.9 Z	19.50^h	3.521	0.096^i	$[1.16]^j$	1.566_3	A↔ X,	R 23326.1 Z	(1)(2)(3)
X $^1\Sigma^+$	0	1384.14 Z	18.97	4.0381	0.0917	$[1.362]^k$	1.4625_5			

Cu^1H: [a]Extrapol. of A $^1\Sigma^+$ to the limit $Cu(^2D_{5/2}) + H(^2S)$. Predissociation in A $^1\Sigma^+$ gives <2.89 eV, flame photom. (7) 2.86 eV.
[b]Λ-type doubling $\Delta\nu_{ef} = +0.047\,J(J+1)$, for $J < 15$.
[c]$\beta_e = +0.6_8 \times 10^{-4}$.
[d]$A \approx +117$.
[e]From perturbations in E $^1\Sigma^+$; vibr. numbering uncertain.
[f]T_0, referring to X $^1\Sigma^+(v=0)$.

Cu^1H (continued):
[g]Lowest observed level is v=3. Numerous perturbations in v=4...11 by three levels of d $^3\Pi_r$.
[h]$\omega_e y_e = -0.27$; $\gamma_e = -0.0048$; $\beta_e = -0.08 \times 10^{-4}$.
[i]v=0,1 interact with v=1,2, resp., of C 1. Deperturbed constants for v=1 (e levels) are $B = 6.36_4$, $D = 7.6 \times 10^{-4}$.
[j]From a perturbation in v=0 of c 1.

Cu^1H (continued):

[k] Deperturbed constants; v=0 interacts with Ω=2(v=0), v=1 with B $^3\Pi_0$+(v=2). Ω-type doubling Δv_{ef}(v=0) ≈ + 0.055 x J(J+1).

[ℓ] $\Delta G(5/2) \approx 1100$.

[m] Ω-type doubling $\Delta v_{ef} \begin{Bmatrix} (v=0) \approx +0.022 \\ (v=1) \approx +0.060 \\ (v=2) \approx +0.095 (\text{depert.}) \end{Bmatrix} \times J(J+1) + \ldots$

[n] Deperturbed constants; B_2 = 5.47, B_3 < 4.7. D_0 refers to e levels, $D_1 \approx 3.2 \times 10^{-4}$, $D_2 = 6.8 \times 10^{-4}$. Perturbations by levels of b Δ_2.

[o] From a perturbation in B $^3\Pi_0$+(v=0). Very uncertain.

[p] Deperturbed constants; B_2 = 5.83; $D_1 = 4.95 \times 10^{-4}$, $D_2 \approx 7.8 \times 10^{-4}$. Perturbations by levels of A $^1\Sigma^+$, a ($^3\Sigma^+$), c 1.

[q] v=2, 3, 4 perturbed by B $^3\Pi_0$+(v=0, 1, 2). (Deperturbed) constants for v=2, ..., 6 are

B_v	= 6.199,	5.885,	5.540,	5.144,	4.576.
$D_v(10^{-4}cm^{-1})$	= 4.92,	5.22,	6.3,	7.39,	8.8.
$\Delta G(v+\frac{1}{2})$	=	1427.0,	1319.7,	1166.8,	1020.5.

[r] Predissociation above v=0, J=0 (5) caused by an unstable $^3\Sigma^+$ state from Cu, ^2S + H, ^2S. See also (6).

[s] $\beta_e = + 0.22 \times 10^{-4}$.

[t] From a combination of Ringström's data for v-0, 1, 2 and Heimer's data for v=3, 4. $\omega_e y_e = + 0.06_7$, $\gamma_e = + 0.0015$, $\beta_e = - 0.07_3 \times 10^{-4}$.

(1) Heimer, Heimer, ZP 84, 222 (1933)

(2) Grundström, ZP 98, 128 (1935).

(3) Heimer, ZP 95, 321 (1935).

(4) Heimer, Dissertation (Stockholm, 1937).

(5) Herzberg, Mundie, JCP 8, 263 (1940).

(6) Kleman, AF 6, 17 (1953).

(7) Bulewicz, Sugden, TFS 52, 1475 (1956).

(8) Loginov, OS(Engl. Transl.) 16, 220 (1964).

(9) Ringström, AF 32, 211 (1966).

(10) Ringström, CJP 46, 2291 (1968).

Cu^2H: [a] From the value for Cu^1H.

[b] A second level is observed at 2240.6 cm^{-1} above v=0, but could equally well be C 1(v=3). It is perturbed, probably by b Δ_2(v=2). Approximate constants for the e levels are B = 2.85, D = 1.0×10^{-4}.

[c] Ω-type doubling Δv_{ef} = + 0.0166 x J(J+1) - ...

[d] $H_0 = - 19.6 \times 10^{-9}$ (average of e and f components).

[e] Constants for e component. f component (B_0 = 3.2558, $D_0 = 1.270 \times 10^{-4}$, v_{00} = 27152.83) appears to be perturbed by a lower lying 0^- level of smaller B value. v=1 and 2 are perturbed by levels of a $^3\Sigma^+$ and b Δ_2.

[f] The only observed transition is to v=1 (deperturbed constants B = 3.102, D = 1.36×10^{-4}, v(1-0) = 27435.7). Evidence for additional levels comes from the observation of perturbations in B $^3\Pi_0$+ and C 1.

[g] Deperturbed constants; $D_1 = 1.07 \times 10^{-4}$, $D_2 = 1.1 \times 10^{-4}$. Perturbations by levels of a $^3\Sigma^+$.

[h] $\omega_e y_e = - 0.570$.

[i] B_4 = 3.0785, B_5 = 2.9633, B_6(depert.) = 2.83.

[j] D_1, ..., $D_5(10^{-4}cm^{-1})$ = 1.16, 1.18, 1.28, 1.35, 1.44.

[k] $H_0 = + 2.8 \times 10^{-9}$; $D_1 = 1.354 \times 10^{-4}$, $H_1 = 3.3 \times 10^{-9}$; $D_2 = 1.31 \times 10^{-4}$; $D_3 = 1.29 \times 10^{-4}$.

(1) Jeppesen, PR 50, 445 (1936).

(2) See ref. (4) of Cu^1H.

(3) See ref. (9) of Cu^1H.

State	T_e	ω_e	$\omega_e x_e$	B_e	α_e	D_e ($10^{-8}cm^{-1}$)	r_e (Å)	Observed Transitions Design.	Observed Transitions ν_{00}	References
$^{(63)}Cu^{1}H^+$?		Fragments of a band system in the region 34500 - 45500 cm^{-1}. Lower state constants were reported in (1), but information is incomplete.								JAN 1975 (1)
$^{63}Cu^{127}I$		$\mu = 42.068564_3$		$D_0^0 < 3.27$ eV[a]						AUG 1975 A
E $^1\Sigma^+$ (0^+)	24001.4	229.4_0 Z	0.95	0.0658[b]	0.0003_6	(2.2)	2.46_8	E↔X, R	23983.8_0 Z	(1)(2)(5) (6)* (8)*
D $^1\Pi$ (1)	22957.5	212.8 H	0.92	b				D↔X, R	22931.6 H	(1)(2)(8)*
C $^1\Sigma^+$ (0^+)	21867.3	229.7 H	0.53	[0.0681][b]			[2.42_6]	C↔X, R	21849.9_4 Z	(1)(2)(6) (8)*
c										
A $^1\Pi$ (1)	19734.2	213.3 H	2.22	0.0676_1[d]	0.0004_2	(2.7)	2.435	A↔X, R	19708.2 H	(1)(2)(10)*
X $^1\Sigma^+$	0	264.5 Z	0.60	0.07328742	0.00028390_3[e]	2.2439[f]	2.338324	Microwave sp.[g]		(11)
$^{(63)}Cu^{(7)}Li$		($\mu = 6.3122532$)		$D_0^0 = 1.98$ eV[a]						APR 1975
$^{(63)}Cu^{23}Na$		($\mu = 16.838309_9$)		$D_0^0 = 1.7_9$ eV[a]						JAN 1975
$^{(63)}Cu^{(58)}Ni$		($\mu = 30.164639_6$)		$D_0^0 = 2.1_0$ eV[a]						JAN 1975

CuH^+: (1) Mahanti, Nature 127, 557 (1931).

CuI: [a]Thermochemical value (3).

[b]Rotational analyses of C − X and E − X by (6); analyses of C − X, D − X, E − X by (8) and of D − X by (7) are erroneous or doubtful. See also (9).

[c]Bands previously assigned to a B − X system are now known to be due to Cu_2 (4).

[d]Recalculated from (10).

[e]$+ 3.12 \times 10^{-7}(v+\frac{1}{2})^2 - 1.3 \times 10^{-9}(v+\frac{1}{2})^3$.

[f]$H_e = - 2.288 \times 10^{-15}$.

[g]Also value for $eqQ(^{127}I)$.

(1) Mulliken, PR 26, 1 (1925).
(2) See ref. (1) of CuBr.
(3) Brewer, Lofgren, JACS 72, 3038 (1950).
(4) See ref. (1) of Cu_2.
(5) Nair, Upadhya, Nature 211, 1170 (1966).
(6) Nair, Upadhya, CJP 44, 1267 (1966).
(7) Nair, Upadhya, CS 35, 593 (1966).
(8) Rao, Apparao, CJP 44, 2241, 2247 (1966).
(9) Nair, Rai, CJP 45, 2810 (1967).
(10) Pandey, Upadhya, Mohanty, IJP 42, 154 (1968).
(11) Manson, De Lucia, Gordy, JCP 62, 4796 (1975).
(12) Wu, Dows, JMS 58, 384 (1975)

CuLi: [a]Thermochemical value (mass-spectrom.)(1).

(1) Neubert, Zmbov, JCS FT I 70, 2219 (1974).

CuNa: [a]Thermochemical value (mass-spectrom.)(1).

(1) Piacente, Gingerich, ZN 28 a, 316 (1973).

CuNi: [a]Thermochemical value (mass-spectrom.)(1).

(1) Kant, Strauss, Lin, JCP 52, 2384 (1970).

State	T_e	ω_e	$\omega_e x_e$	B_e	α_e	D_e ($10^{-7}cm^{-1}$)	r_e (Å)	Observed Transitions Design.	Observed Transitions ν_{00}	References
$^{63}Cu^{16}O$		μ = 12.7533703$_7$		D_0^0 = 2.7$_9$ eV[a]						AUG 1975 A
		Unidentified transitions in matrix absorption (F← X?) and fluorescence.								(5)(9)
P $^2\Pi_{3/2}$		[574][b]		[0.384][b]	0.005		[1.85$_5$][b]	P→X_1, R	25194[b]	(12)
M $^2\Pi_{3/2}$				[0.419][c]			[1.77$_6$][c]	M↔X_1, R	23898[c] Z	(3)(8)(10)
I $^2\Pi_{3/2}$		[608][d]		[0.416][d]	0.004$_6$		[1.78$_3$][d]	I→X_1, R	22449[d]	(12)
H $^2\Pi_{3/2}$		[557][d]		[0.4176][d]	0.005$_6$		[1.779$_1$][d]	H→X_1, R	22326[d]	(12)
G $^2\Sigma^{(-)}$ ($\frac{1}{2}$)	21618.$_6$	[582.74] Z	(4.0)	0.41481[e]	0.00370	7.24	1.7850$_9$	G→X_2,[f] X_1,[f] R	21316.94[g] Z 21593.98[g] Z	(11)*
F $^2\Pi_i$	21237[h]	[600.8] Z	(4.4)	0.4121[i]	0.0038	8	1.791$_0$	F→X, R	21082.8[j] Z	(4)* (8) (10)* (11)
E $^2\Delta_{5/2}$	21058.$_0$	733 H^Q	5.5	0.4445[k]	0.0036		1.724$_4$	E→X_1,	21104.1 H^Q	(12)
		Bands in the green region, partially analysed (13) in terms of a $^2\Pi_i \rightarrow X\ ^2\Pi_i$ transition.								(4)* (13)*
A $^2\Sigma^{(+)}$	16491.$_3$	[631.02] Z	(6.0)	0.43387[ℓ]	0.00475	7.93	1.7454$_3$	A↔X_2, X_1, R	16215.33[g] Z 16492.37[g] Z	(4)* (7)* (8)(10)*
X_2 $^2\Pi_{1/2}$	279.02[m]	636.18 Z	4.36	0.44415[n]	0.00449	8.4	1.7251$_3$	o		
X_1 $^2\Pi_{3/2}$	0	640.17 Z	4.43	0.44454	0.00456	8.5	1.7243$_7$			

CuO: [a]Thermochemical value (mass-spectrom.)(1)(6).

[b]Lowest observed level and ΔG, vibrational numbering unknown. The observed transitions are v-3 at 23327 cm^{-1}, v-4 and (v+1)-4.

[c]One level only. Vibrational numbering uncertain; the authors in (4) suggest v=1. Perturbations.

[d]These are values of $\Delta G(\frac{3}{2})$, B_1, r_1, ν(1-0). v=0 not observed.

[e]Spin doubling $\Delta\nu_{12}\begin{Bmatrix}(v=0) = +\ 0.1674 \\ (v=1) = +\ 0.1743\end{Bmatrix} \times N(N+1) + \ldots$

[f]$G \rightarrow X_1$ is considerably weaker than $G \rightarrow X_2$. Relative branch intensities in both sub-bands are unusual.

[g]N'=0 relative to {J"=0}.

[h]$A_0 = -\ 6.24$, $A_1 = -\ 31.87$. Also J-dependent terms (10).

[i]v=0 perturbed. Λ-type doubling in v=1:

$^2\Pi_{1/2}$, $\Delta\nu_{fe} = (-)[0.319_0(J+\frac{1}{2}) - \ldots]$;

$^2\Pi_{3/2}$, $\Delta\nu_{fe} = (-)[2.74 \times 10^{-5}(J-\frac{1}{2})(J+\frac{1}{2})(J+3/2) - \ldots]$.

[j]J'=½ relative to J"=½ (average of $\{F_1\}$ and F_2).

[k]v=0 strongly perturbed.

[ℓ]Spin-doubling constants $\gamma_0 = -\ 0.1952$, $\gamma_1 = -\ 0.1908$.

[m]$A_0 = -\ 276.11$, $A_1 = -\ 272.28$, $A_2 = -\ 268.69$; also J-dependent terms (10).

[n]Λ-type doubling in $^2\Pi_{1/2}$, $\Delta\nu_{fe} = (-)[0.014_8(J+\frac{1}{2}) - \ldots]$ (average of v=0, 1, 2).

[o]The absence of an ESR spectrum attributable to matrix isolated CuO is compatible with a $^2\Pi$ ground state (9).

CuO (continued):

(1) R. P. Burns, quoted in (2).
(2) Cheetham, Barrow, AdHTC 1, 7 (1967).
(3) Lagerqvist, Uhler, ZN 22 b, 551 (1967).
(4) Antić-Jovanović, Pěsić, Gaydon, PRS A 307, 399 (1968).
(5) Shirk, Bass, JCP 52, 1894 (1970).
(6) Smoes, Mandy, Vander Auwera-Mahieu, Drowart, BSCB 81, 45 (1972).
(7) Antić-Jovanović, Pěsić, JP B 6, 2473 (1973).
(8) Appelblad, Lagerqvist, JMS 48, 607 (1973).
(9) Thompson, Easley, Knight, JPC 77, 49 (1973).
(10) Appelblad, Lagerqvist, PS 10, 307 (1974).
(11) Appelblad, Lagerqvist, CJP 53, 2221 (1975).
(12) O. Appelblad, A. Lagerqvist, USIP Annual Report 1974 (May-75), and private communication.
(13) Lefebvre, Pinchemel, Bacis, CJP 54, 735 (1976).

State	T_e	ω_e	$\omega_e x_e$	B_e	α_e	D_e ($10^{-8}cm^{-1}$)	r_e (Å)	Observed Transitions Design.	ν_{00}	References
$^{(63)}Cu^{(32)}S$		(μ = 21.2007816)		$D_0^0 = 2.8_0$ eV[a]						SEP 1976 A
A $^2\Sigma^{(+)}$	17946.1	375.2[b] H^Q	3.6_6	[0.1806][c]		[18]	[2.099]	A→X, R	17493.1 H^Q 17925.7 H^Q	(1)(4)
X $^2\Pi_{(i)}$	433.4 0	413.4 H^Q 415.0 H^Q	1.6_5 1.7_5	[0.1891]		[18]	[2.051]			
$^{63}Cu^{121}Sb$		$\mu = 41.387622_9$								JAN 1975
		Additional sequences of R shaded bands in thermal emission at 13014, 13843, 14476, 15423, 15825, 16482 cm^{-1}.								(1)*
A	18511.7	222.7_1 H	0.86					A→X, R	18505.6 H	(1)
X[a]	0	234.8_3 H	0.73							
$^{63}Cu^{80}Se$		$\mu = 35.206516_8$		$D_0^0 = 2.5_5$ eV[a]						SEP 1976
A $^2\Sigma^{(+)}$	17960.4	253.0 H	2.74	[0.10169][b]		$[7._0]$	$[2.169_9]$	A→X, R	16344.37 Z 17935.32 Z	(1)* (3)*
X $^2\Pi_i$	1590.9 0	302.3_7 H	0.99	[0.10775][c] [0.10774]		[5.9]	[2.1081]			
$^{(63)}Cu^{(120)}Sn$		($\mu = 41.269608_7$)		$D_0^0 = 1.8_0$ eV[a]						JAN 1975
$^{63}Cu^{130}Te$		$\mu = 42.393295_1$		$D_0^0 = 2.3_5$ eV[a]						OCT 1975
		Fragments of an additional system in the region 17200 - 18200 cm^{-1}.						(B→X)		(3)
A [b]	15991.92	200.58 H	2.009	0.0673_0	0.0005_2	3.41[c]	2.431	A→(X), R	15965.55 H	(1)* (3)*
(X) [b]	0	252.67 H	0.701	0.0720_5	0.0002_8	2.65[d]	2.349			

CuS: [a]Thermochemical value (mass-spectrom.)(2)(3)(5).
[b]Average of Biron's (1) constants for the upper states of his systems A and B.
[c]Spin-splitting constant $\gamma_0 = -0.0360$.

(1) Biron, CR 258, 4228 (1964).
(2) Drowart, Pattoret, Smoes, PBCS No. 8, 67 (1967).
(3) Uy, Drowart, TFS 67, 1293 (1971).
(4) Biron, CR B 274, 978 (1972); 281, 401 (1975).
(5) See ref. (6) of CuO.

CuSb: [a]Not certain that this is the ground state.

(1) Lefebvre, Houdart, CR B 273, 662 (1971).

CuSe: [a]Thermochemical value (mass-spectrom.)(2).
[b]Spin-splitting constant $\gamma_0 = -0.0022$.
[c]Λ-type doubling, $\Delta v_{fe}(v=0) = (-)[0.0284(J+\frac{1}{2}) - ...]$.

(1) Joshi, JMS 8, 79 (1962).
(2) See ref. (6) of CuO.
(3) Lefebvre, Bocquet, CJP 54, 1664 (1976).

CuSn: [a]Thermochemical value (mass-spectrom.)(1).

(1) Ackerman, Drowart, Stafford, Verhaegen, JCP 36, 1557 (1962).

CuTe: [a]Thermochemical value (mass-spectrom.)(2).
[b]$\Omega' = \Omega''$.
[c]$\beta_e = +0.14 \times 10^{-8}$.
[d]$\beta_e = +0.01 \times 10^{-8}$.

(1) Maheshwari, Sharma, PPS 81, 898 (1963).
(2) See ref. (6) of CuO.
(3) Lefebvre, Bocquet, JP B 8, 1322 (1975).

State	T_e	ω_e	$\omega_e x_e$	B_e	α_e	D_e (10^{-} cm^{-1})	r_e (Å)	Observed Transitions Design.	ν_{00}	References
$^{(164)}Dy^{19}F$		(μ = 17.0252779) D_0^0 = 5.4$_6$ eV[a]								JAN 1975
$^{(164)}Dy^{16}O$		(μ = 14.5729973$_4$) D_0^0 = 6.2$_5$ eV[a] Large number of mostly R shaded emission bands, 15900 - 19600 and 20800 - 23500 cm^{-1}. Tentative vibrational analysis by (1); uncertain.								JAN 1975 A (1)(2)* (3)
$^{(166)}Er^{19}F$		(μ = 17.0466290) D_0^0 = 5.8$_3$ eV[a]								JAN 1975
$^{(166)}Er^{16}O$		(μ = 14.5886378$_7$) D_0^0 = 6.3$_0$ eV[a] V and R shaded emission bands in the regions 17200 - 18500 and 19200 - 20400 cm^{-1}.								JAN 1975 A (1)* (2)
$^{(153)}Eu^{19}F$		(μ = 16.8989393) D_0^0 = 5.4$_2$ eV[a] Emission in the region 14000 - 33000 cm^{-1}, maximum at 26000 cm^{-1}.								OCT 1975 (2)
$^{(153)}Eu^{16}O$		(μ = 14.4803335$_6$) D_0^0 = 4.8$_0$ eV[a] Emission spectrum from 14000 to 25000 cm^{-1}.								OCT 1975 A (1)* (5)(6)
X	0	[668.0][b]						IR abs. sp.[b]		(4)
$^{(153)}Eu^{(32)}S$		(μ = 26.4434070) D_0^0 = 3.7$_1$ eV[a]								JAN 1975
$^{(153)}Eu^{(80)}Se$		(μ = 52.486909$_4$) D_0^0 = 3.0$_8$ eV[a]								JAN 1975
$^{(153)}Eu^{(130)}Te$		(μ = 70.238662) D_0^0 = 2.4$_5$ eV[a]								JAN 1975

DyF: [a]Thermochemical value (mass-spectrom.)(1).

(1) Zmbov, Margrave, JPC 70, 3379 (1966).

DyO: [a]Thermochemical value (mass-spectrom.)(4), recalc.(5).

(1) Piccardi, SA 1, 533 (1941).

(2) Gatterer, Junkes, Salpeter, Rosen, METOX (1957).

(3) Mavrodineanu, Boiteux, "Flame Spectroscopy", Wiley (1965).

(4) Ames, Walsh, White, JPC 71, 2707 (1967).

(5) Smoes, Coppens, Bergman, Drowart, TFS 65, 682 (1969).

ErF: [a]Thermochemical value (mass-spectrom.)(1).

(1) See ref. (1) of DyF.

ErO: [a]Thermochemical value (mass-spectrom.)(3), recalc.(4).

(1) See ref. (2) of DyO.

(2) Herrmann, Alkemade, "Chemical Analysis by Flame Photometry", Wiley-Interscience (1963).

(3) See ref. (4) of DyO.

(4) See ref. (5) of DyO.

EuF: [a]Thermochemical value (mass-spectrom.)(1). From the short-wavelength cutoff of the chemiluminescence spectrum (2) obtain $D_0^0 \geq 5.62$ eV.

(1) Zmbov, Margrave, JINC 29, 59 (1967).

(2) Dickson, Zare, CP 7, 361 (1975).

EuO: [a]Thermochemical value (mass-spectrom.)(7). Considerably higher values (5.7_2 eV) in (2)(3)(6).
[b]In Ar matrix at 15 K. For $Eu^{18}O$ at 633.7 cm^{-1}. Derived constants are $\omega_e = 671._8$, $\omega_e x_e = 1.9$ (4).

(1) See ref. (2) of DyO.

(2) See ref. (4) of DyO.

(3) See ref. (5) of DyO.

(4) Gabelnick, Reedy, Chasanov, JCP 60, 1167 (1974).

(5) Edelstein, Eckstrom, Perry, Benson, JCP 61, 4932 (1974).

(6) See ref. (2) of EuF.

(7) Hildenbrand, Murad, ZN 30 a, 1087 (1975).

EuS: [a]Thermochemical value (mass-spectrom.)(1), recalc.(2).

(1) See ref. (5) of DyO.

(2) Bergman, Coppens, Drowart, Smoes, TFS 66, 800 (1970).

EuSe: [a]Thermochemical value (mass-spectrom.)(2). Based on $D_0^0(Se_2) = 3.411$ eV; (1) prefer $D_0^0(Se_2) = 3.164$ eV.

(1) Barrow, Chandler, Meyer, PTRSL A 260, 395 (1966).

(2) See ref. (2) of EuS.

EuTe: [a]Thermochemical value (mass-spectrom.)(1).

(1) See ref. (2) of EuS.

State	T_e	ω_e	$\omega_e x_e$	B_e	α_e	D_e (10^{-6} cm^{-1})	r_e (Å)	Observed Transitions: Design.	Observed Transitions: ν_{00}	References
$^{19}F_2$		μ = 9.4992023		D_0^0 = 1.602 eV[a]		I.P. = 15.686 eV[b]				JUL 1976 A
		Many strong absorption bands observed and partially analyzed up to 126000 cm^{-1}, heavily perturbed and not assigned.[c]								(19)*
K $^1\Pi_u$				[1.040][d]			[1.306]	K← X,	V 116855.72 Z	(16)(19)*
J $^1\Pi_u$ (4pσ)	116409	[1032.6] Z		[1.041]	e		[1.306]	J← X,	V 116469.4 Z	(19)*
I $^1\Sigma_u^+$ (4pπ)	113841	[1108.92] Z	f	[0.8009]	g	[1.8][h]	[1.488_6]	I→ f,	R 17081.6[i] H	(7)* (8)(19)
								I→ F,	R 20732[j] H	(7)* (8)(19)
								I←→ X,[k]	R 113940.24 Z	(14)(16)(19)*
H $^1\Pi_u$ (3pσ)	105520.14	1088.19 Z	9.875	1.021[ℓ]	0.014		1.318	H← X,	V 105606.27 Z	(16)(19)
h $^3\Pi_{1u}$ (3pσ)	(104904)	(1100)[m]		1.022[n]	0.016		1.318	h← X,	V 104998.7 Z	(19)
G $^1\Sigma_u^+$ (3pπ)	(104300)	Inferred from strong perturbations of the higher vibrational levels of C $^1\Sigma_u^+$.								(19)
E($^1\Sigma_u^+$)	≤100912	[196.3][o] Z	(0.96)	[0.194][o]			[3.02][o]	E← X,	R $100555._5$[o] Z	(19)*
D($^1\Sigma_u^+$)	≤98756	[221.6][o] Z	(1.22)	[0.207][o]			[2.93][o]	D← X,	R $98411._9$[o] Z	(19)*
f($^3\Pi_{1g}$)(3sσ)	[97314][p]	p		[1.005][p]			[1.329][p]			
C $^1\Sigma_u^+$	≤93499	[493.2][q] Z		[0.484][q]			[1.91_5][q]	C←→ X,	R 93290.4[q] Z	(19)*
F $^1\Pi_g$ (3sσ)	93099	1133.34[r] H	9.17_3	1.047[r]	0.012		1.302			
A $^1\Pi_u$		Continuous absorption with maximum at 35000 cm^{-1}.						A← X,		(4)(5)
a $^3\Pi_{0^+u}$		Continuous absorption with maximum at 25500 cm^{-1}.						a← X,[s]		(4)(5)
X $^1\Sigma_g^+$	0	916.64 Z	11.236[t]	0.89019	0.013847[u]	3.3	1.41193	Raman sp.		(2)(9)(18) (20)
								Mol. beam magn. reson.[v]		(6)
								Ab initio calc.		(13)

F_2: [a]From the observed vibrational levels of the ground state (19); the highest observed level (presumably the last stable level) is at 12830.38 cm^{-1}. Shock tube experiments of (15) give the same value within ±0.05 eV; see also the earlier work of (3)(5a)(10).

[b]From photoionization (12). Photoelectron spectra (11) give 15.70 eV.

[c]The assignments to two Rydberg series by (16) are questioned by (19).

[d]B_Q; B_{PR} = 1.034.

[e]B_2 = 0.9916.

[f]$\Delta G(3/2) = 754.06$, $\Delta G(5/2) = 553.78$, $\Delta G(7/2) = 733.01$; strongly perturbed, not certain whether one or two electronic states are involved.

[g]$B_1 = 0.8129$, $B_2 = 0.8980$, $B_3 = 0.8946$, $B_4 = 0.891$; see [f].

[h]D_v(v=1...4, $10^{-6} cm^{-1}$) = 4.0, 14.0, -2.6, -50.0.

[i]The lower state vibrational numbering of this band is unknown, see [p].

[j]v" uncertain, see [r].

[k]Four of the strongest absorption bands; in emission only bands with v=0 and 1 and very weakly v=2.

[ℓ]From Q branches, $B_Q - B_{PR} \approx +0.004$. v'=0,1,2,3 analyzed, a weak and highly perturbed band at 911 Å (109770 cm^{-1}) may be 4-0.

[m]Estimated from the 0-0 and 2-0 (ν_0 = 107069.4) bands.

[n]From Q branches; $B_Q - B_{PR} = +0.005$. v=0 strongly perturbed.

[o](Deperturbed) constants determined from the lowest observed levels; vibrational numbering unknown. Only those levels which interact strongly with C $^1\Sigma_u^+$ have been found in absorption. It is possible that D and E are not two independent states.

[p]From I→f (7); vibrational numbering unknown. The assignments of (8) suggest the existence of additional levels at 1058 below and 1044 cm^{-1} above the single level reported by (7).

[q]Constants for the lowest observed level, vibrational numbering unknown (v=n). In absorption levels up to v = n+30 have been observed. Numerous perturbations by levels of D and E (interaction matrix elements ≈ 10 cm^{-1}); a much stronger interaction with G affects levels having v ≥ n+25. For details see Table 5 of (19).

[r]Vibrational numbering uncertain. Extensive perturbations. Strong predissociation leading to line broadening in I→F bands having v" = 1,2,4,5,6,8; bands with v" = 3,7 are sharp.

[s]The existence of the a←X absorption becomes clear only after subtraction of the much stronger effect of the A←X absorption from the observed absorption intensities (5). It is generally assumed that the $^3\Pi_{0^+u}$ state has a minimum but no discrete absorption or emission has been observed, see (1). (17) predict a dissociation energy D_e = 3300 cm^{-1} and r_e = 1.9 Å for this state.

[t]$\omega_e y_e$ = - 0.113. These constants represent only the lowest nine levels (i.e. v ≤ 8) (19). Levels have been observed up to v=22; this last level lies only 90 cm^{-1} below the extrapolated limit.

[u]$+ 0.0001179(v+\frac{1}{2})^2 - 0.000020_3(v+\frac{1}{2})^3$, representing $B_0 \ldots B_{12}$.

[v]Rotational gyromagnetic ratio $g_J = -0.120_8\ \mu_N$, (nuclear) spin-rotation and spin-spin interaction constants c = - 157.3 kHz, d = 8.0 kHz.

References on page 217 .

State	T_e	ω_e	$\omega_e x_e$	B_e	α_e	D_e (10^{-} cm^{-1})	r_e (Å)	Observed Transitions Design.	ν_{00}	References
$^{19}F_2^+$		$D_0^0 = 3.33_9$ eV[a]								JUL 1976 A
		Ab initio calculations (3) predict several further stable states. Ground state calc. (7).								
B ($^2\Sigma_g^+$)	(43000)	From the photoelectron spectrum; uncertain observation.								(4)
A $^2\Pi_{u,i}$	(22755) 22484.8[b]	520.4[b]	7.3_0[b]					A→X,[c] R	22139 22208.8[b] H	(2)
X $^2\Pi_{g,i}$	340[d] 0	1073.3[b] H	9.13[b]	1.015[e]	0.010		1.322			
$^{19}F_2^-$		$D_0^0 = 1.28$ eV[a]				I.P. = 3.08 eV[b]				JUL 1976
F	(120490)	1210[c]	60							(4)
E ($^2\Pi_g$)	(115300)	1050[c]								(4)
X $^2\Sigma_u^+$	0	(510)[d]		(0.50)[d]			(1.8_8)[d]			

F_2^+: [a]From D_0^0 (F_2), I.P.(F_2), and I.P.(F).

[b]From the R heads in the 3/2 – 3/2 sub-system. The (arbitrary) lower state vibrational numbering of (2) has been reduced by 5 leading to better agreement with the excitation energy of 2.70 eV (21800 cm^{-1}) and the ground state vibrational frequency found by photoelectron spectroscopy (4)(5); see also (6).

[c]Bands attributed to this system by (1) are probably all due to F_2 (I→F), see (2).

[d]From the photoelectron spectrum (4). At slightly lower resolution (5) find 240 cm^{-1}.

[e]Revised vibrational numbering, see [b].

F_2^+ (continued):

(1) Stricker, ZN 21 a, 1518 (1966).

(2) See ref. (7) of F_2.

(3) Balint-Kurti, MP 22, 681 (1971).

(4) See ref. (11) of F_2.

(5) Potts, Price, TFS 67, 1242 (1971).

(6) See ref. (16) of F_2.

(7) Ellis, Banyard, Tait, Dixon, JP B 6, L233 (1973).

F_2^-: [a]From $D_0^0(F_2)$ and the electron affinities of F_2 (3.08 eV) and F (3.399 eV).

[b]From endoergic charge transfer reactions (1).

[c]Two progressions of resonances in the electron transmission current starting at 11.25 and 11.90 eV. They indicate the existence of doubly excited (preionizing) states of F_2^- having the F $^1\Pi_g$ and f $^3\Pi_g$ Rydberg states of F_2 as parent states and X $^2\Pi_g$ of F_2^+ as grandparent.

[d]Theoretical values from MO-SCF calculations (2), D_e = 1.66 eV. Multiconfiguration valence bond calculations (3) give ω_e = 293, $\omega_e x_e$ = 1, B_e = 0.425, r_e = 2.04 Å, D_e = 1.06 eV. See also (3a)(matrix Raman sp.).

F_2^- (continued):

(1) Chupka, Berkowitz, Gutman, JCP 55, 2724 (1971).
(2) Gilbert, Wahl, JCP 55, 5247 (1971).
(3) Copsey, Murrell, Stamper, MP 21, 193 (1971).
(3a) Howard, Andrews, JACS 95, 3045 (1973).
(4) Spence, PR A 10, 1045 (1974).

F_2 (continued):

(1) Nathans, JCP 18, 1122 (1950).
(2) Andrychuk, CJP 29, 151 (1951).
(3) Barrow, Caunt, PRS A 219, 120 (1953).
(4) Steunenberg, Vogel, JACS 78, 901 (1956).
(5) Rees, JCP 26, 1567; 27, 1424 (erratum) (1957).
(5a) Stamper, Barrow, TFS 54, 1592 (1958).
(6) Ozier, Crapo, Cederberg, Ramsey, PRL 13, 482 (1964).
(7) Porter, JCP 48, 2071 (1968).
(8) Stricker, Krauss, ZN 23 a, 486 (1968).
(9) Claassen, Selig, Shamir, AS 23, 8 (1969).
(10) DeCorpo, Steiger, Franklin, Margrave, JCP 53, 936 (1970).
(11) Cornford, Frost, McDowell, Ragle, Stenhouse, JCP 54, 2651 (1971).
(12) Berkowitz, Chupka, Guyon, Holloway, Spohr, JCP 54, 5165 (1971).
(13) Das, Wahl, JCP 56, 3532 (1972).
(14) Di Lonardo, Douglas, JCP 56, 5185 (1972).
(15) Blauer, Solomon, JCP 57, 3587 (1972).
(16) Gole, Margrave, JMS 43, 65 (1972).
(17) Child, Bernstein, JCP 59, 5916 (1973).
(18) Stricker, Hochenbleicher, ZN 28 a, 27 (1973).
(19) Colbourn, Dagenais, Douglas, Raymonda, CJP 54, 1343 (1976).
(20) Edwards, Good, Long, JCS FT II 72, 984 (1976).

State	T_e	ω_e	$\omega_e x_e$	B_e	α_e	D_e $(10^{-}\ cm^{-1})$	r_e (Å)	Observed Transitions Design.	Observed Transitions ν_{00}	References
$^{(56)}Fe_2$		$(\mu = 27.967467_0)$ $D_0^0 = 1.0_6$ eV[a]								MAR 1976
C		Continuous absorption with maximum at 24120 cm^{-1}.						C← X,[b]		(2)
B		(218)[c]						B← X,[b]	(21095)[d]	(2)
A		(194)[c]						A← X,[b]	(18355)[d]	(2)
X	0									
$^{(56)}Fe^{(79)}Br$		$(\mu = 32.734036_2)$								MAR 1976
		Very diffuse absorption feature at 40890 cm^{-1}.								(5)
		Narrow group of absorption bands[a] in the region 28100 - 28400 cm^{-1}.								(5)
		Bands in the region 26600 - 27400 cm^{-1}; ω' ≈ 315, ω" ≈ 304, possibly						$^4\Pi \leftrightarrow {}^4\Sigma$. V	27064.9 H 27007.8 H 26935.1 H 26832.1 H	(1)(2)(4)(5)
		V and R shaded emission bands in the region 15400 - 17500 cm^{-1}; tentative assignments to four systems.								(2)(3)
$^{(56)}Fe^{35}Cl$		$(\mu = 21.5170409)$								MAR 1976
		Diffuse absorption bands in the region 41700 - 42700 cm^{-1}.								(10)
C ($^4\Pi$)		Fragments of a complex system; $\Delta G'(\frac{1}{2}) = 416$, $\Delta G''(\frac{1}{2}) = 427$.						$(C \leftrightarrow X_1)$, R	31255.9 H 31243.2 H 31221.2 H 31180.8 H	(1)(2)(4)(6) (9)
B ($^6\Pi$)	x_2 + 29255.9 29184 29120 29065 29021 28986	434.8 H	2.1					$B \leftrightarrow X_2$, V	29269.5 H 29197.2 H 29133.5 H 29079.1 H 29034.6 H 28999.7 H	(2)(3)(7)*
A ($^4\Pi$)	x_1 + 28016 27958 27922 27894	(427.8)[a] H	(1.2)					$A \leftrightarrow X_1$, V	28024.7 H 27967.3 H 27931.4 H 27903.2 H	(4)(9)

State	T_e	ω_e	$\omega_e x_e$	B_e	α_e	D_e (10^{-} cm^{-1})	r_e (Å)	Observed Transitions Design.	ν_{00}	References
$^{(56)}Fe^{35}Cl$ (continued)										
A' ($^4\Pi$)		Complex system in the region 20000 - 21000 cm^{-1}; ω' = 370, ω" = 395.						(A'→X_1), R	20549.3 H 20484.5 H 20429.0 H 20378.0 H	(5)
		Fragments of an emission system; ω' ≈ 405, ω" ≈ 394.						V	17379	(5)
		Tentative assignments of emission bands in the regions 16480 - 16720						V	(16486)	(3)(5)
		and 15250 - 15750 cm^{-1}.						V_R	(15443)	(3)(5)(8)
X_2 ($^6\Sigma$)	x_2[b]	407.2 H	1.4							
X_1 ($^4\Sigma$)	x_1[b]	(409.9)[a] H	(2.2)							

Fe_2: [a]Thermochemical value (mass-spectrom.)(1).
[b]In Ar matrices at 10 K.
[c]Long upper state progression.
[d]Lowest observed level, vibrational numbering unknown.

(1) Lin, Kant, JPC 73, 2450 (1969).
(2) De Vore, Ewing, Franzen, Calder, CPL 35, 78 (1975).

FeBr: [a]Resembling the B − X_2 system of FeCℓ.

(1) Miescher, HPA 11, 463 (1938).
(2) Mesnage, AP(Paris) 12, 5 (1939).
(3) Reddy, JSIR B 18, 188 (1959).
(4) Rao, Rao, JP B 3, 725 (1970).
(5) See ref. (3) of FeF.

FeCℓ: [a]Vibrational constants from (9). An earlier analysis (4) of A − X_1 gave ω_e' = 431.0, $\omega_e'x_e'$ = 2.9, ω_e'' = 397.0, $\omega_e''x_e''$ = 1.6.
[b]It is not certain whether X_1 or X_2 is the ground state.

(1) Mesnage, CR 201, 389 (1935).
(2) See ref. (1) of FeBr.
(3) See ref. (2) of FeBr.
(4) Müller, HPA 16, 3 (1943).
(5) Reddy, Rao, JMS 4, 16 (1960).
(6) Rao, Rao, CS 38, 87 (1969).
(7) Rao, Rao, JP B 3, 878 (1970).
(8) Rao, Rao, Rao, CS 39, 392 (1970).
(9) Rao, Rao, IJPAP 9, 102 (1971).
(10) See ref. (3) of FeF.

State	T_e	ω_e	$\omega_e x_e$	B_e	α_e	D_e (10^{-6} cm^{-1})	r_e (Å)	Observed Transitions Design.	ν_{00}	References
$^{(56)}Fe^{19}F$		(μ = 14.1815983)								FEB 1975
C	x + (42743)	(694) H	(7)					C ↔ (X), V	42751 H	(3)*
		Single sequence of bands near 32060 cm^{-1}, in emission only.						(R)		(1)
		Narrow group of bands[a], 30900 - 31200 cm^{-1}, in em. and abs..						(V)		(2)(3)*
		Narrow group of bands[b] near 30240 cm^{-1}, in em. and abs..						(V)		(2)(3)*
(X)	x	(678) H	(7)							
$^{(56)}Fe^{(74)}Ge$		(μ = 31.841213$_8$)		D_0^0 = 2.1$_4$ eV[a]						JAN 1975
$^{(56)}Fe^{1}H$		(μ = 0.98998780)								AUG 1975 A
$^{(56)}Fe^{2}H$		(μ = 1.94409920)								
		Complex abs. and em. spectra, 18500 - 19200 and 19900 - 21300 cm^{-1}. No analysis.								(2)* (3)
		Additional band at 10100 cm^{-1}.								(4)
		Theoretical calculations (1)(5) predict a $^6\Sigma^+$ or $^6\Delta$ ground state.								
$^{(56)}Fe^{16}O$		(μ = 12.4381536$_7$)		D_0^0 = 4.2$_0$ eV[a]		I.P. = 8.7$_1$ eV[b]				JUN 1977
		Single band, in emission and absorption.							R 41480 H	(8)* (9)(12)*
c	26441	[545] H						c → A[c],	R 22326 H	(1)(3)(19)
c'	23569	[535] H	Fragments of an emission system, 18500 - 20000 cm^{-1}.					c' → A[c],	R 19449 H	(19)
b	21962 (21865)	[667] H [(661)] H		[0.471$_7$][d]			[1.695]	b[e] ↔ A[c],	R 17908 H 17808 H	(1)(2)* (7)* (8)(9)(14) (19)
a	21245	820 H	1	[0.497][f]			[1.65$_1$]	a[g] ↔ A[c],	R 17267 H	(1)(2)* (7)* (8)(9)(19)
B ($^5\Pi$)	14404	650[h] H	5	Large number of emission bands, 7100 - 15000 cm^{-1}; tentative vibrational analyses.				{ B → A[c], B → X,	R 10340 H R 14245 H	(5)(19) (3)(19)(21)
A $^5\Sigma^+$ [i]	3948	880.53[d] Z	4.63	0.5127$_1$[d]	0.0037$_6$	[0.6$_6$]	1.6259		T_0 = 3905[j]	
X $^5\Delta$ [i]	0	965[h] H					(1.57)[k]	ℓ		
$^{(56)}Fe^{16}O^-$				D_0^0 = 4.2$_3$ eV[m]		I.P. = 1.49$_2$ eV[n]				JUN 1977
X	0	740[n]					(1.63)[k]			

FeF: [a]Similar in appearance to the $B - X_2$ system of FeCℓ.
[b]Similar in appearance to the $A - X_1$ system of FeCℓ. Additional bands at 29580 cm^{-1} (2) suggest $\Delta G''(\frac{1}{2}) \approx 660$.

(1) Barrow, Carroll, unpublished, quoted in DONNSPEC.
(2) Senior, Barrow, unpublished, quoted in DONNSPEC.
(3) Brinton, Callear, JCS FT II 70, 203 (1974).

FeGe: [a]Thermochemical value (mass-spectrom.)(1).

(1) Kant, Strauss, JCP 49, 3579 (1968).

Fe^1H, Fe^2H:

(1) Walker, Walker, Kelly, JCP 57, 2094 (1972).
(2) Carroll, McCormack, ApJ (Part 2) 177, L33 (1972).
(3) Smith, PRS A 332, 113 (1973).
(4) Klynning, Lindgren, USIP Report 73-20 (1973).
(5) Scott, Richards, JCP 63, 1690 (1975).

FeO, FeO^-:

[a]Thermochemical value (mass-spectrom.)(10)(15)(20). See also (4)(6).
[b]Electron impact appearance potential (20).
[c]The similarity of the vibrational constants suggests a common lower state for the five systems. Uncertain.
[d]From the rotational analysis (14) of four bands of the short-wavelength component of $b - A$. They appear to be the $\Omega' = \Omega'' = 0$ subbands of a quintet or septet $\Sigma - \Sigma$ transition. An earlier analysis by (11) is incorrect.
[e]Lifetime $\tau \approx 500$ ns (19).
[f]Rotational analysis using tunable-laser excited fluorescence spectroscopy (19).
[g]Lifetime $\tau \approx 450$ ns (19).
[h]From a re-analysis of $B \rightarrow X$ by (21). The FeO^- photoelectron sp. (21) gives a ground state frequency of 970 ± 60 cm^{-1}.
[i]Symmetries assigned (21) on the basis of *ab initio* calculations (18).

FeO, FeO^- (continued):

[j]From 0-0 band assignments for $B \rightarrow X$ (21) and $B \rightarrow A$ (5)(19). From the FeO^- photoelectron sp. (21) obtain 3990 ± 100 cm^{-1}.
[k]Estimated (21) from relative vibrational intensities in the photoelectron spectrum of FeO^-.
[ℓ]The IR transition strength measured at 880 cm^{-1} (16)(17) requires re-interpretation in view of the recent re-assignment of the low-lying states (21).
[m]From D_0^0(FeO) and the electron affinities of FeO and O.
[n]From the laser photoelectron spectrum of FeO^- (21).

(1) Rosen, Nature 156, 570 (1945).
(2) Delsemme, Rosen, BSRSL 14, 70 (1945).
(3) Malet, Rosen, BSRSL 14, 377 (1945).
(4) Brewer, Mastik, JCP 19, 834 (1951).
(5) Bass, Benedict, ApJ 116, 652 (1952).
(6) Lagerqvist, Huldt, ZN 8 a, 493 (1953).
(7) Gatterer, Junkes, Salpeter, Rosen, METOX (1957).
(8) Callear, Norrish, PRS A 259, 304 (1960).
(9) Bass, Kuebler, Nelson, JCP 40, 3121 (1964).
(10) Burns (1966), quoted in (13).
(11) Dhumwad, Narasimham, PIAS A 64, 283 (1966).
(12) Callear, Oldman, TFS 63, 2888 (1967).
(13) Cheetham, Barrow, AdHTC 1, 7 (1967).
(14) Barrow, Senior, Nature 223, 1359 (1969).
(15) Balducci, De Maria, Guido, Piacente, JCP 55, 2596 (1971).
(16) Von Rosenberg, Wray, JQSRT 12, 531 (1972).
(17) Fissan, Sulzmann, JQSRT 12, 979 (1972).
(18) Bagus, Preston, JCP 59, 2986 (1973).
(19) West, Broida, JCP 62, 2566 (1975).
(20) Hildenbrand, CPL 34, 352 (1975).
(21) Engelking, Lineberger, JCP 66, 5054 (1977).

State	T_e	ω_e	$\omega_e x_e$	B_e	α_e	D_e (10^{-7} cm^{-1})	r_e (Å)	Observed Transitions Design.	ν_{00}	References
$^{(56)}Fe^{(32)}S$		(μ = 20.3437228) $D_0^0 = 3.3_1$ eV[a]								JAN 1975
$^{(56)}Fe^{(28)}Si$		(μ = 18.6491827) $D_0^0 = 3.0_4$ eV[a]								JAN 1975
$^{19}F^{16}O$		$\mu = 8.6838822_5$ $D_0^0 = 2.2_3$ eV[a] I.P. = 12.7_9 eV[b]								JUL 1976
X $^2\Pi$	0	[1028.7][c]	(5.15)[d]	(1.104)[d]	(0.0097)[d]		(1.32_6)[d]			
$^{(69,71)}Ga_2$		($\mu = 34.955426_7$) $D_0^0 = 1.4_0$ eV[a]								JUL 1976
		R shaded emission bands in the region 18200 - 21700 cm^{-1}; $\omega \approx 165$.								(3)
GaAs, GaBi		See p. 226.								
$^{69}Ga^{81}Br$		$\mu = 37.220586_4$ $D_0^0 = 4.3_1$ eV[a]								JUL 1976
C $^1\Pi$	(36000)	Diffuse absorption bands (fluctuation b.) indicating a shallow upper state potential curve.						C← X,	37310[b]	(1)
B $^3\Pi_1$	28532.0	271.6[c] H	2.50					B↔X,	V 28535.9 H	(1)(5)*
A $^3\Pi_{0^+}$	28161.8	272.2[c] H	2.53					A↔X,	V 28166.0 H	(1)(5)*
X $^1\Sigma^+$	0	263.0[c] H	0.81	0.081839_3	0.0003207	(0.32)[d]	2.35248	Microwave sp.		(2)
$^{69}Ga^{35}Cl$		μ = 23.1990149 D_0^0 = 4.92 eV[a]								JUL 1976
		Continuous absorption at 41200 and near 47600 cm^{-1}.								(1)(2)
C $^1\Pi$	(40261)	[120][b] H						C← X,	R 40139 H	(1)(2)*
B $^3\Pi_1$	29859.4	395.3 H	2.5	$[0.1568_6]$[c]			$[2.152_3]$	B↔X,	V 29874.0_3[c] Z	(1)(2)*
A $^3\Pi_{0^+}$	29527.2	395.3 H	2.3[d]	$[0.1577_9]$[e]			$[2.146_0]$	A↔X,	V 29541.9 (Z)	(1)(2)*
X $^1\Sigma^+$	0	365.3[f] H	1.2	0.1499045	0.0007936[g]	(1.01)[h]	2.201690	Microwave sp.[i]		(3)(7)

FeS: [a]Thermochemical value (mass-spectrom.)(1)(2).

(1) Marquart, Berkowitz, JCP 39, 283 (1963).
(2) Drowart, Pattoret, Smoes, PBCS No. 8, 67 (1967).

FeSi: [a]Thermochemical value (mass-spectrom.)(1).

(1) Vander Auwera-Mahieu, McIntyre, Drowart, CPL 4, 198 (1969).

FO: [a]Indirectly from the difference between the electron impact appearance potentials of FO^+ from FO and F_2O (4) and the known heat of atomization of F_2O; see also (2)(5).
[b]Photoionization mass-spectrometry of F_2O (7). See also (2).
[c]From matrix IR absorption and Raman spectra in Ar (1)(3) (6). (2) suggest a corrected gas phase frequency of 1050.
[d]Theoretical calculations (2).

(1) Arkell, Reinhard, Larson, JACS 87, 1016 (1965).
(2) O'Hare, Wahl, JCP 53, 2469 (1970).
(3) Andrews, Raymond, JCP 55, 3078 (1971).
(4) Clyne, Watson, CPL 12, 344 (1971).
(5) Levy, JCP 56, 1415 (1972).
(6) Andrews, JCP 57, 51 (1972).
(7) Berkowitz, Dehmer, Chupka, JCP 59, 925 (1973).

Ga_2: [a]Thermochemical value (mass-spectrom.)(1)(2)(4).

(1) Drowart, Honig, BSCB 66, 411; JPC 61, 980 (1957).
(2) Chupka, Berkowitz, Giese, Inghram, JPC 62, 611 (1958).
(3) Ginter, Ginter, Innes, JPC 69, 2480 (1965).
(4) Gingerich, Blue, 18th Annual Conference on Mass-Spectrometry and Allied Topics, San Francisco (June 1970), paper F2.

GaBr: [a]Limit from the analysis of the fluctuation bands, and extrapolation of vibrational levels in $^3\Pi_{0^+}$ and $^3\Pi_1$; for correlation with atomic products see (3). Flame photometry (4) gives 4.4_5 eV.
[b]Vertical transition.
[c]Analysis of (1). (5) revised some of the vibrational assignments and gave constants for $^{69}Ga^{79}Br$.
[d]Calculated from $4B_e^3/\omega_e^2$.

(1) Miescher, Wehrli, HPA 6, 458 (1933); 7, 331 (1934).
(2) Barrett, Mandel, PR 109, 1572 (1958).
(3) Barrow, TFS 56, 952 (1960).
(4) Bulewicz, Phillips, Sugden, TFS 57, 921 (1961).
(5) Savithry, Rao, Murty, Rao, Physica 75, 386 (1974).

GaCℓ: [a]From spectroscopic evidence concerning C $^1\Pi$, see (4). Good agreement with the flame photometric value of (6).
[b]Bands with v'=1 are diffuse on account of predissociation.
[c]Recalculated by (5) from data in (2).
[d]$\omega_e y_e = + 0.015$.
[e]$B_0' - B_0''$ redetermined from the differences between consecutive lines in the R and P branches measured by (2).
[f]$\Delta G(\frac{1}{2}) = 363.3_2$ from band origins (5).
[g]$\gamma_e = + 1.8_0 \times 10^{-6}$.
[h]See [d] of GaBr.
[i]$eqQ(^{69}Ga) = - [92.4_0 - 0.6_8(v+\frac{1}{2})]$ MHz,
$eqQ(^{35}C\ell) = - [13.2_0 + 0.2_0(v+\frac{1}{2})]$ MHz (7).

(1) See ref. (1) of GaBr.
(2) Levin, Winans, PR 84, 431 (1951).
(3) See ref. (2) of GaBr.
(4) See ref. (3) of GaBr.
(5) Bartky, JMS 5, 206 (1960); 6, 275 (erratum) (1961).
(6) See ref. (4) of GaBr.
(7) Tiemann, Grasshoff, Hoeft, ZN 27 a, 753 (1972).

State	T_e	ω_e	$\omega_e x_e$	B_e	α_e	D_e (10^{-6}cm^{-1})	r_e (Å)	Observed Transitions Design.	ν_{00}	References
$^{69}Ga^{19}F$		μ = 14.8932747		$D_0^0 = 5.9_8$ eV^a						JUL 1976
C $^1\Pi$	47365.7	542.3_5 (Z)	$9.5_5{}^b$	(0.358)	$(0.0053_5)^c$		(1.778)	C↔X, R_V	47324.1 (Z)	(1)* (2)(3)
B $^3\Pi_1$	33427.8	662.1 Z	1.45^d	$0.3719_8{}^{ef}$	0.0030_2		1.7444	B↔X, V	33448.1_2 Z	(1)* (2)(3)
A $^3\Pi_{0^+}$	33105.5	663.0_2 Z	2.18^g	$0.3709_8{}^e$	(0.0030_2)		1.7467	A↔X, V	33126.1_7 Z	(1)* (2)(3)
X $^1\Sigma^+$	0	622.2 Z	3.2	0.3595161	0.0028642^h	0.50	1.774369	Microwave sp.^{i}		(6)(7)
$^{(69)}Ga^1H$		(μ = 0.99330124)		$D_0^0 < 2.84$ eV^a						JUL 1976
		Open-structure absorption bands in the region 41650 - 46300 cm^{-1}, provisionally ascribed to Ga^1H.								(1)
A $^1\Pi$				$[5.1]^{bc}$			$[1.8_2]$	A↔X, R	23714 H^Q	(2)(3)(6)
a $^3\Pi_2$				[6.811]		$[620]^d$	$[1.578_5]$	a →X, V	17909.43 Z	(3)*
a $^3\Pi_1$	17622.0_1	1631.1_7 Z	$58.2_2{}^e$	6.692^f	0.326^g	489^h	1.592_5	a ↔X, V_R	17626.84 Z	(2)(3)*
a $^3\Pi_{0^+}$	17337.08	1640.54 Z	62.72^i	6.394^f	0.276^j	262^k	1.629_2	a ↔X R_V	17345.78 Z	(2)(3)*
a $^3\Pi_{0^-}$	17333	[1492.5] Z	ℓ	6.358	0.220^m	$[243]^n$	1.633_8	a →X,^{o} R_V	17340.41 Z	(5)
X $^1\Sigma^+$	0	1604.52 Z	28.77^p	6.137^f	0.181^q	342^r	1.663_0			
$^{(69)}Ga^2H$		(μ = 1.95691832)		$D_0^0 < 2.86$ eV^a						JUL 1976
A $^1\Pi$				$[2.61]^c$		[100]	[1.82]	A←X, R	23860.2^s Z	(6)
a $^3\Pi_1$				[3.339]		[113]	$[1.606_2]$	a←X, V	17634.36 Z	(2)
X $^1\Sigma^+$	0			[3.083]	0.06_3	[84]	1.663_1			

GaF:

[a]Thermochemical value (mass-spectrom.)(5).

[b]$\omega_e y_e = -0.6_0$. This state may have a small potential maximum of ~0.26 eV (5).

[c]$\gamma_e = -(0.0003)$. Rotational constants estimated from observed R - Q or P - Q head separations and B".

[d]$\omega_e y_e = -0.43_3$.

[e]Recalculated from (3) with the more accurate ground state constants of (6).

[f]Λ-type doubling, $B_{RP} - B_Q \approx +6 \times 10^{-5}$.

[g]$\omega_e y_e = -0.3_1$.

[h]$\gamma_e = +0.0000102$.

[i]$\mu_{e\ell}(v=0) = 2.45$ D (6); $eqQ(^{69}Ga) = -[107.07 - 1.09(v+\frac{1}{2})]$ MHz (6)(7), $g_J(v=0) = -0.0601_2\ \mu_N$ (7).

(1) Welti, Barrow, Nature 168, 161 (1951); PPS A 65, 629 (1952).
(2) Barrow, Jacquest, Thompson, PPS A 67, 528 (1954).
(3) Barrow, Dodsworth, Zeeman, PPS A 70, 34 (1957).
(4) See ref. (3) of BaBr.
(5) Murad, Hildenbrand, Main, JCP 45, 263 (1966).
(6) Hoeft, Lovas, Tiemann, Törring, ZN 25 a, 1029 (1970).
(7) Honerjäger, Tischer, ZN 29 a, 1919 (1974).

Ga^1H, Ga^2H:

[a]From the predissociation of A $^1\Pi(v=0)$, see [c].

[b]From the value for Ga^2H.

[c]Deuteride bands involving this level have diffuse rotational lines whose width increases with J. Corresponding lines of the hydride are even broader, only diffuse Q and R heads can be recognized. The diffuseness results from tunnelling through a potential maximum which separates the shallow potential well from the dissociation limit $^2P_{3/2} + {}^2S$.

Ga^1H, Ga^2H (continued):

[d]$D_{RP} - D_Q = +12 \times 10^{-6}$.

[e]$\omega_e y_e = -7.47$.

[f]RKR potential functions (4).

[g]$\gamma_e = -0.0315$. Rotational constants derived from R and P branches. Slightly different B_v and D_v values from the Q branches; see (3).

[h]$-124 \times 10^{-6}(v+\frac{1}{2}) + 83.5 \times 10^{-6}(v+\frac{1}{2})^2$; see [g].

[i]$\omega_e y_e = -6.195$.

[j]$\gamma_e = -0.0400$.

[k]$+75 \times 10^{-6}(v+\frac{1}{2}) + 14.5 \times 10^{-6}(v+\frac{1}{2})^2$.

[ℓ]$\Delta G(3/2) = 1313.6$.

[m]$\gamma_e = -0.058$.

[n]$D_1 = 413 \times 10^{-6}$, $D_2 = 46_3 \times 10^{-6}$.

[o]Each band consists of a single Q branch.

[p]$\omega_e y_e = +0.360$.

[q]$\gamma_e = -0.0005$.

[r]$+1 \times 10^{-6}(v+\frac{1}{2})^2$.

[s]Calculated from $\nu_0(0-1) = 22745.8$ using ground state vibrational constants derived from Ga^1H. The 0-0 band was not observed owing to a strong overlapping impurity.

(1) Garton, PPS A 64, 509 (1951).
(2) Neuhaus, Nature 180, 433 (1957); AF 14, 551 (1959).
(3) Ginter, Innes, JMS 7, 64 (1961).
(4) Ginter, Battino, JCP 42, 3222 (1965).
(5) Poynor, Innes, Ginter, JMS 23, 237 (1967).
(6) Kronekvist, Lagerqvist, Neuhaus, JMS 39, 516 (1971).

State	T_e	ω_e	$\omega_e x_e$	B_e	α_e	D_e ($10^{-7}cm^{-1}$)	r_e (Å)	Observed Transitions Design.	ν_{00}	References
$^{69}Ga^{127}I$		$\mu = 44.666098_4$		$D_0^0 = 3.4_7$ eV[a]						SEP 1976
C $^1\Pi$		Continuous absorption with maximum at 32600 cm^{-1}.						C← X,		(1)
B $^3\Pi_1$	25900.6	185.0 H^Q	2.7					B↔X, [b]	25884.3 H^Q	(1)*
A $^3\Pi_{0^+}$	25571.0	193.2 H	2.4					A↔X, [b]	25559.0 H	(1)*
X $^1\Sigma^+$	0	216.6 H^Q	0.5	0.056934_7	0.000189	(0.157)	2.57467	Microwave sp.[c]		(2)
$^{(69)}Ga^{16}O$		($\mu = 12.9822464_3$)		$D_0^0 = 3.9_1$ eV[a]						SEP 1976
		Additional unclassified emission bands in the region 20000 - 23000 cm^{-1}.								(1)
B $^2\Sigma$	25706.9	762.9[b] H	3.44	[(0.4013)]		[(2.9)]	[(1.798_8)]	B↔X, R	25705.3 H	(1)* (2)(4)(5)
X $^2\Sigma$	0	767.5[b] H	6.24	[(0.4271)]		[(3.7)]	[(1.743_6)]			
$^{(69)}Ga^{31}P$		($\mu = 21.3703567$)		$D_0^0 = 2.3_8$ eV[a]						SEP 1976
$^{(69)}Ga^{(130)}Te$		($\mu = 45.032342_9$)		$D_0^0 = 2.7_3$ eV[a]						SEP 1976
$^{(69)}Ga^{75}As$		($\mu = 35.899311_5$)		$D_0^0 = 2.1_8$ eV[a]						SEP 1976
$^{(69)}Ga^{209}Bi$		($\mu = 51.830821_4$)		$D_0^0 = 1.6_0$ eV[a]						SEP 1976

GaI: [a]Extrapolation of vibrational intervals in A and B; see the discussion in (3).
[b]Both directions of shading occur, even in one and the same band.
[c]$eqQ(^{69}Ga) = -66$ MHz,
$eqQ(^{127}I) = -549$ MHz.

(1) See ref. (1) of GaBr.
(2) See ref. (2) of GaBr.
(3) See ref. (3) of GaBr.

GaO: [a]Thermochemical value (mass-spectrom.)(6). For flame photometric results see (3)(4).
[b]Constants from (5) who attribute the double heads observed for most bands to isotope splitting; (1) assumes spin doubling. (1), (2), and (4) give all slightly different constants.

(1) Guernsey, PR 46, 114 (1934).
(2) Sen, IJP 10, 429 (1936).
(3) Gurvich, Veits, BASPS 22, 670 (1958).
(4) Gurvich, Novikov, Ryabova, OS(Engl. Transl.) 18, 68 (1965).
(5) Raziunas, Macur, Katz, JCP 39, 1161 (1963); 42, 2634 (1965).
(6) Burns, JCP 44, 3307 (1966).

GaP: [a]Thermochemical value (mass-spectrom.)(1)(2).

(1) Gingerich, Piacente, JCP 54, 2498 (1971).
(2) Piacente, Gingerich, HTS 3, 219 (1971).

GaTe: [a]Thermochemical value (mass-spectrom.)(1), adjusted to the new value of $D_0^0(Te_2)$; see DONNSPEC (1970).

(1) Uy, Muenow, Ficalora, Margrave, TFS 64, 2998 (1968).

GaAs: [a]Thermochemical value (mass-spectrom.)(1).

(1) De Maria, Malaspina, Piacente, JCP 52, 1019 (1970).

GaBi: [a]Thermochemical value (mass-spectrom.)(1).

(1) Piacente, Desideri, JCP 57, 2213 (1972).

State	T_e	ω_e	$\omega_e x_e$	B_e	α_e	D_e (10^{-} cm^{-1})	r_e (Å)	Observed Transitions Design.	ν_{00}	References
$^{(158)}Gd^{19}F$		(μ = 16.9583061) $D_0^0 = (6.0_8)$ eV[a]								JAN 1975
$^{(158)}Gd^{16}O$		(μ = 14.5239012$_1$) $D_0^0 = 7.3_7$ eV[a]								JAN 1975 A
		Large number of mostly R shaded bands in emission from 13300 to 22500 cm^{-1}.								(2)* (4)*
		Two systems of bands with multiple heads were identified by (1) and confirmed and extended by (4)(7):								
System I:		$\nu_H = 21700.5 + 748.0(v'+\frac{1}{2}) - 3.0(v'+\frac{1}{2})^2 - 830.1(v''+\frac{1}{2}) + 2.3(v''+\frac{1}{2})^2$						R	21659.3 H	(1)(4)(7)
System II:		$\nu_H = 20470._5 + 767._0(v'+\frac{1}{2}) - 4.9(v'+\frac{1}{2})^2 - 836._0(v''+\frac{1}{2}) + 2.8(v''+\frac{1}{2})^2$						R	20435.$_5$ H	
		Assignments at longer wavelengths are less certain, and considerable disagreement exists between analyses by (1)(3)(7) except in the region 15700 - 16200 cm^{-1} where three long sequences have been identified:						R R R	16170 H 16122 H 16094 H	(1)(3)(7)
$^{(158)}Gd^{(32)}S$		(μ = 26.5890611) $D_0^0 = 5.3_7$ eV[a]								JAN 1975
$^{(158)}Gd^{(80)}Se$		(μ = 53.063878$_1$) $D_0^0 = 4.4_1$ eV[a]								JAN 1975
$^{(158)}Gd^{(130)}Te$		(μ = 71.275762) $D_0^0 = 3.4_9$ eV[a]								JAN 1975
$^{(72,74)}Ge_2$		(μ = 36.4539647) $D_0^0 = 2.8_2$ eV[a]								SEP 1976

GdF: [a]Thermochemical value (mass-spectrom.), quoted from (1). Re-evaluation using the reaction enthalpies and auxiliary data in (1) results, however, in 6.9_5 eV.

(1) Zmbov, Margrave, JINC 29, 59 (1967).

GdO: [a]Thermochemical value (mass-spectrom.)(5)(6)(8).

(1) Piccardi, GCI 63, 887 (1933).
(2) Gatterer, RS 1, 153 (1942).
(3) Lemaître, Rosen, quoted in (4).
(4) Gatterer, Junkes, Salpeter, Rosen, METOX (1957).
(5) Ames, Walsh, White, JPC 71, 2707 (1967).
(6) Smoes, Coppens, Bergman, Drowart, TFS 65, 682 (1969).
(7) Suárez, Grinfeld, JCP 53, 1110 (1970).
(8) Drowart, Myers, Szwarc, Vander Auwera-Mahieu, Uy, to be published.

GdS: [a]Thermochemical value (mass-spectrom.)(1), recalc. (2).

(1) See ref. (6) of GdO.
(2) Bergman, Coppens, Drowart, Smoes, TFS 66, 800 (1970).

GdSe, GdTe:

[a]Thermochemical value (mass-spectrom.)(1).

(1) See ref. (2) of GdS.

Ge_2: [a]Thermochemical value (mass-spectrom.)(1).

(1) Kant, JCP 44, 2450 (1966).

State	T_e	ω_e	$\omega_e x_e$	B_e	α_e	D_e (10^{-} cm^{-1})	r_e (Å)	Observed Transitions Design.	Observed Transitions ν_{00}	References
$^{(74)}Ge^{(79)}Br$		(μ = 38.169031_6)		D_0^0 = (3.5) eV^a						SEP 1976
		Additional unassigned absorption bands above 43000 cm^{-1}.								(6)
G	47544	358 H	1.5					G← X, V	47577 H	(6)*
F								F← X, V	45946 H	(6)*
E ($^2\Sigma^+$)	44805	366 H	5	Very diffuse bands.[b]				E← X, V	43689 H 44840 H	(6)*
D	c							D← X, V	43296 H 44426 H	(6)*
C ($^2\Pi$)	(41156) 41046	359 H	3	Only v'=2 progression observed. Predissociation begins between v'=1 and 2.				C← X, V	(40037) 41077 H	(6)*
B ($^2\Sigma^+$)	33413	$383._7$ H	0.7					B↔X, V	32307 H 33457 H	(1)* (6)
A'($^2\Delta$)	27252 27156	$197._3$ H $190._1$ H	$7._2$ $7._3$					A'→X, R	26051.5 H 27102.1 H	(2)*
A ($^2\Sigma$)		Bands in the region 17000 - 23000 cm^{-1}. Different vibrational analyses, $\omega_e' \approx 190$.						A→X, R		(3)(4)(5)
X $^2\Pi_r$	1150 0	$295._4$ H	0.7_2							
$^{(74)}Ge^{12}C$		(μ = 10.32404534)		D_0^0 = 4.7_3 eV^a						SEP 1976
$^{74}Ge^{35}Cl$		μ = 23.7389853		D_0^0 = (4.4) eV^a						SEP 1976
(D)		Tentative assignments of bands above 46000 cm^{-1}.						(D← X),	46640 H 47586 H	(7)
C ($^2\Pi$)	45845 45824	493.4 H	2.0					C← X, V	44914 H 45867 H	(7)*
C'	42181	506.9 H	3.7					C'← X, V	42230 H	(7)*
B ($^2\Sigma$)	$33992._2$	526.6[b] H	(0.3)	c				B↔X, V	$33078._0$ H $34052._0$ H	(1)* (2)(4) (7)*

State	T_e	ω_e	$\omega_e x_e$	B_e	α_e	D_e (10^{-} cm^{-1})	r_e (Å)	Observed Transitions Design.	ν_{00}	References
$^{74}Ge^{35}Cl$ (continued)										
$A'(^2\Delta_r)$	29561.7 29499.8	335.5 H 342.1 H	4.58 4.65					$A' \rightarrow X$,[d] R	28550.9 H 29466.2 H	(2)*
$A\ (^2\Sigma^+)$		Bands in the region 16000 - 25000 cm^{-1}. Different vibrational analyses, $\omega_e' \approx 250$.						$A \rightarrow X$, R		(3)(5)
$X\ ^2\Pi_r$	975.0 0	405.5 H 407.6 H	1.23 1.36	c						

GeBr: [a]Extrapolation of vibrational levels in A'; atomic products at the limit uncertain.
[b]It is possible that more than one transition is involved.
[c]Doublet separation ≈ 20 cm^{-1}.

(1) Jevons, Bahford, Briscoe, PPS 49, 532 (1937).
(2) Andrews, Barrow, PPS A 63, 957 (1950).
(3) Kuznetsova, Kuzyakov, IVUZK 12 (9), 1183 (1969).
(4) Rao, Haranath, JP B 2, 1385 (1969).
(5) Chatalic, Deschamps, Pannetier, JCPPB 67, 335 (1970).
(6) Oldershaw, Robinson, TFS 67, 2499 (1971).

GeC: [a]Thermochemical value (mass-spectrom.)(1).

(1) Drowart, De Maria, Boerboom, Inghram, JCP 30, 308 (1959).

GeCℓ: [a]See [a] of GeBr.
[b](4) and (6) give slightly different constants.
[c]Partial rotational analysis of two subbands (6).
[d]The bands have at least two heads of comparable intensity.

(1) See ref. (1) of GeBr.
(2) Barrow, Lagerqvist, AF 1, 221 (1949).
(3) Deschamps, Robert, Pannetier, JCPPB 65, 1084 (1968).
(4) Filippova, Kuzyakov, VMUK No. 3, 25 (1968).
(5) Rao, Haranath, JP B 2, 1080 (1969).
(6) Mishra, Khanna, IJPAP 8, 825 (1970).
(7) Oldershaw, Robinson, TFS 66, 532 (1970).

State	T_e	ω_e	$\omega_e x_e$	B_e	α_e	D_e $(10^{-7}cm^{-1})$	r_e (Å)	Observed Transitions Design.	ν_{00}	References
$^{74}Ge^{19}F$		μ = 15.1139772		D_0^0 = 5.0_0 eV[a]		I.P. = 7.4_6 eV[b]				SEP 1976
G $^2\Delta_r(4d\delta)$	49412.89[c]	[710.37] Z	2.82 H	0.38408	0.00261	[4.43][d]	1.7041_1	G→X, V	48523.4 H^Q 49415.6 H^Q	(2)(7)*
D $^2\Sigma^+(6s\sigma)$	48581.26	833.12 Z[e]	6.52 H	0.39972[e]	0.00214	[3.7_3][f]	1.6704	D→A, V	25473.3_0 Z	(7)*
								D→X, V	47726.6 H^Q 48662.6 H^Q	(7)*
D' $^2\Pi$ $(4d\pi)$	47920.73[g]	803.96[h] Z	3.38 H	0.40068[hi]	0.00259	3.6_2[h]	1.6684	D'→A, V	24798.9_8[h] Z	(7)*
								D'→X,[j] V	47043.0 H 47976.3 H	(7)*
E $^2\Sigma^+(5p\sigma)$	46645.41	760.08 Z	2.967	0.39845[k]	0.00290	4.33	1.6731_0	E→B,[ℓ]	11616.26 Z	(7)*
C $^2\Delta$	43977.49[m]	[684.00] Z	9.31	0.38835	0.00421	[4.97][n]	1.6947_2	C→X, V	43059.27[o] Z 43994.43[o] Z	(2)(4)(7)*
C' $^2\Pi$ $(5p\pi)$	43369.61[p]	796.88 Z	3.415	0.39957[q]	0.00258	4.13	1.6707_5	C'→A,[r] R	20244.31	(7)
								C'→X,[s] V	42547.8 H 43377.9 H	(4)(7)*
a $^4\Sigma^-$	35194.68[t]	[628.31] Z	6.66[u]	0.36676[t]	0.00369	[4.94][v]	1.7439	a→X,	35181.77 Z	(6)*
B $^2\Sigma^+(5s\sigma)$	35010.85	796.99 Z	3.613[w]	0.39440[x]	0.00255	3.88	1.6816_7	B→A, V	11885.56 Z	(2)
								B→X,[y] V	34141.23 Z 35076.39 Z	(1)* (6)*
A $^2\Sigma^+$	23316.65	413.03 Z	1.124[z]	0.32039[a']	0.00307[b']	7.78[c']	1.8658_2	A→X,[y] R	22255.67 Z 23190.83 Z	(1)* (6)*
X $^2\Pi_{3/2}$	934.33	667.33 Z	3.150[d']	0.36660	0.00267_5[f']	4.50	1.7452			
$^2\Pi_{1/2}$	0	665.67 Z		0.36578[e']		4.47				

GeF: [a]Thermochemical value (mass-spectrom.)(3); see also (5).

[b]Approximate limit of the ns (n = 5, 6, ...) Rydberg series (7). (5) give 7.2 eV from electron impact mass-spectrometry.

[c]$A_0 = +22.20$, $A_1 = +21.5$; small J dependence.

[d]$D_1 = 4.52 \times 10^{-7}$.

[e]Rotational analysis of v=2, and tentative results for v=0.

[f]$D_2 = 3.3_3 \times 10^{-7}$.

[g]$A_2 = (+)4.23$, $A_3 = (+)4.35$.

[h]Extrapolation from the rotationally analyzed levels v'=2 and 3; $\beta_e = +0.2_1 \times 10^{-7}$.

[i]Large Λ-type doubling, also spin-rotation interaction; see (7).

[j]The 0-1 and 0-2 bands were previously considered by (2) as 0-0 bands of their transitions F − X and E − X, resp..

[k]Spin-doubling constant $\gamma(v=0) = -0.0358$.

[ℓ]$\Delta v=0$ sequence of slightly V shaded bands, provisionally assigned in (2) to a $E(^2\Pi) - B(^2\Sigma)$ transition.

[m]$A_0 = 13.88$, $A_1 = 14.25$.

[n]$D_1 = 5.15 \times 10^{-7}$.

[o]Referring in the upper state to the zero point of the Hill - Van Vleck expression.

[p]$A_0 = 105.63$, $A_1 = 105.88$, $A_2 = 105.96$; small J dependence.

[q]Λ-type doubling in $^2\Pi_{\frac{1}{2}}$, $\Delta\nu_{fe}(v=0) = -0.0373(J+\frac{1}{2})$.

[r]Extremely weak system consisting of a long 0-v" progression; not analyzed.

[s]The 3-1 and 1-0 bands of $C'\,^2\Pi_{\frac{1}{2}} - X\,^2\Pi_{\frac{1}{2}}$ correspond to the 0-0 subbands of a system D − X proposed earlier by (2). The 2-1 band of the 3/2 − 3/2 system was assigned by (2) as 0-0 band of $C' - X\,^2\Pi_{\frac{1}{2}}$.

[t]Spin-splitting constants for v=0: $\lambda = +8.086$ [$4\lambda = T_0(^4\Sigma_{3/2}) - T_0(^4\Sigma_{1/2})$], $\gamma_1 \approx \gamma_2 = 0.0119$. See also [x].

[u]Based on the interpretation of a perturbation in B $^2\Sigma^+$, v=4.

[v]$D_1 = 4.90 \times 10^{-7}$.

[w]$\omega_e y_e = +0.0124$.

[x]Spin-doubling constant $\gamma(v=0) = +0.00100$. Extensive perturbations between B $^2\Sigma^+$, v=0 and a $^4\Sigma^-$, v=0.

[y]Franck-Condon factors (8).

[z]$\omega_e y_e = -0.0129$.

[a']Spin-doubling constant $\gamma(v) = -[0.03662 + 0.00014(v+\frac{1}{2})]$.

[b']$\gamma_e = +0.000038$.

[c']$-0.22 \times 10^{-7}(v+\frac{1}{2}) + \ldots$

[d']$\omega_e y_e = +0.0068$.

[e']Λ-type doubling, $\Delta\nu_{fe} = -[0.0218_4 + 0.00019_5(v+\frac{1}{2})](J+\frac{1}{2})$.

[f']$\gamma_e = +0.000008_3$.

(1) Andrews, Barrow, PPS A 63, 185 (1950).
(2) Barrow, Butler, Johns, Powell, PPS 73, 317 (1959).
(3) Ehlert, Margrave, JCP 41, 1066 (1964).
(4) Uzikov, Kuzyakov, VMUK No. 5, 30 (1969).
(5) Harland, Cradock, Thynne, INCL 9, 53 (1973).
(6) Martin, Merer, CJP 51, 125 (1973).
(7) Martin, Merer, CJP 52, 1458 (1974).
(8) Singh, IJPAP 13, 204 (1975).

State	T_e	ω_e	$\omega_e x_e$	B_e	α_e	D_e $(10^{-4}cm^{-1})$	r_e (Å)	Observed Transitions Design.	ν_{00}	References
$^{72}Ge^{1}H$		μ = 0.99389799		$D_0^0 \le 3.3$ eV[a]						SEP 1976
B $(^2\Sigma)$	[41074]	Incompletely resolved bands.						B← X, R	39686 H 40573 H	(3)
A $^2\Delta$	25454[b]	[1185.15] Z	(127)[c]	6.535[d]	0.619_6	[5.71][e]	1.611	A↔X, R	25197.0 Z	(1)* (6)* (9)
a $^4\Sigma^{(-)}$	[16747][f]			[6.7654][f]		[4.60]	[1.5834]	a→X, V	15802.8 Z	(2)(5)
X $^2\Pi_r$	0[g]	[1833.77] Z	(37)[c]	6.725_9[h]	0.191_6	[3.26][i]	1.5880			
$^{72}Ge^{2}H$		μ = 1.95923588		$D_0^0 \le 3.3$ eV[j]						SEP 1976
A $^2\Delta$	25460[k]	1027.82 Z	65.73	3.286_0[ℓ]	0.167_0[m]	[1.40][n]	1.6182	A↔X, R	25283.0 Z	(6)* (9)
X $^2\Pi_r$	0[o]	[1320.09] Z	(19)[c]	3.414_5[p]	0.070_2	[0.832][q]	1.5874			
$^{74}Ge^{127}I$		$\mu = 46.711803_0$								SEP 1976
		Additional absorption bands above 40000 cm^{-1}; tentative assignments.								(3)
G								$G \leftarrow X_1$,[a] V	43959	(4)*
F	42769	292 H	2					$F \leftarrow X_2$,[a] $F \leftarrow X_1$, V	41376 42792 H	(3)(4)*
E								$E \leftarrow X_1$,	42639	(4)*
D								$D \leftarrow X_1$,[a] V	41781	(3)(4)*
C								$C \leftarrow X_1$,[a]	41404	(3)(4)*
B $(^2\Sigma)$	32650.1	305.9[b] H	0.67					$B \leftarrow X_2$, fragments only.		(2)
								$B \leftarrow X_1$,[a] V	32679.9 H	(1)* (2)
A $(^2\Sigma)$	18663	155.2 H	0.58					$A \rightarrow X_1$, R	18617 H	(2)
X_2 $^2\Pi_{3/2}$	(1413)									
X_1 $^2\Pi_{1/2}$	0	246.3 H	0.75							

Ge^1H, Ge^2H:

[a]From the predissociation in A $^2\Delta$ (6); thermochemical data give the same value (4).

[b]$A_0 = 10.3$, $A_1 = 6.1$; moderate J dependence. For a more detailed theoretical discussion of this $^2\Delta$ state see (8).

[c]Estimated from isotope shifts (6).

[d]Spin-doubling constants $\gamma = 0.473$ (v=0) and 0.362 (v=1). Broadening of absorption lines due to predissociation above 27000 cm^{-1}.

[e]$D_1 = 8.01 \times 10^{-4}$; also higher order constants.

[f]Spin-splitting constants $\lambda = 6.52$, $\gamma_1 = 0.037$, $\gamma_2 = 0.048$. Note, that the definitions of λ, γ_1, and γ_2 used in (5) are those of Hougen (4a) and differ from those used by Martin and Merer [see ref. (6) of GeF]. The B_0 value in the abstract of (5) and quoted in DONNSPEC is clearly wrong.

[g]$A_0 = 892.52$, $A_1 = 896.12$ from (6); small J dependence. See also (9).

[h]Λ-type doubling; in $^2\Pi_{\frac{1}{2}}(v=0)$, $\Delta v = 0.496(J+\frac{1}{2}) - \ldots$

[i]$D_1 = 3.40 \times 10^{-4}$.

[j]From the value for Ge^1H.

[k]$A_0 = 14.4$, $A_1 = 9.9$, $A_2 = 7.2$; moderate J dependence.

[l]Spin-doubling constants $\gamma = 0.300$ (v=0), 0.281 (v=1), 0.22 (v=2). Predissociation, see [d].

[m]$\gamma_e = -0.021_9$.

[n]$D_1 = 1.65 \times 10^{-4}$, $D_2 = 1.7 \times 10^{-4}$; also higher order constants.

Ge^1H, Ge^2H (continued):

[o]$A_0 = 891.86$, $A_1 = 894.35$ from (6); small J dependence. See also (9).

[p]Λ-type doubling; in $^2\Pi_{\frac{1}{2}}(v=0)$, $\Delta v = 0.2506(J+\frac{1}{2}) - \ldots$

[q]$D_1 = 0.824 \times 10^{-4}$.

(1) Kleman, Werhagen, AF 6, 359 (1953).
(2) Kleman, Werhagen, AF 6, 399 (1953).
(3) Barrow, Drummond, Garton, PPS A 66, 191 (1953).
(4) Barrow, Deutsch, PCS (1960), 122.
(4a) Hougen, CJP 40, 598 (1962).
(5) Klynning, AF 32, 563 (1966).
(6) Klynning, Lindgren, AF 32, 575 (1966).
(7) Kovács, Pacher, JP B 4, 1633 (1971).
(8) Veseth, Physica 56, 286 (1971).
(9) Veseth, JMS 48, 283 (1973).

GeI: [a]The systems B − X, C − X, D − X, F − X, and G − X of (1)(4) correspond to C − X, D − X, E − X, F − X, and H − X of (2)(3).

[b]Vibrational numbering of (2), based on observed isotope shifts and confirmed by (4).

(1) Oldershaw, Robinson, TFS 64, 2256 (1968).
(2) Chatalic, Deschamps, Pannetier, JCPPB 67, 1567 (1970).
(3) Chatalic, Iacocca, Pannetier, CR C 274, 1784 (1972).
(4) Oldershaw, Robinson, JMS 44, 602 (1972).

State	T_e	ω_e	$\omega_e x_e$	B_e	α_e	D_e (10^{-7} cm^{-1})	r_e (Å)	Observed Transitions: Design.	Observed Transitions: ν_{00}	References
$^{74}Ge^{16}O$		$\mu = 13.1496256_4$		$D_0^0 = 6.78$ eV[a]		I.P. = 11.1_0 eV[b]				SEP 1976 A
		Additional unassigned R shaded absorption bands between 66800 and 69900 cm^{-1}.								(3)
F	$67474._5$	809.0 H	5.66					F← X, R	67390 H	(3)*
E	$49637._3$	504.3 H	4.8					E↔X, R	49397 H	(3)* (16)*
A $^1\Pi$	37766.9 H	650.4 H	4.21	0.413_3[cd]	0.0033	8.2	1.761	A↔X,[d] R	37595.4 Z	(1)* (2)* (8)* (13) (16)(18)*
a ($^3\Pi_1$)	32132	$734._9$ H	5.3	(0.438)			(1.71_1)	a→X,[d] R	32007 H	(7)(25)*
a'($^3\Sigma^+$)	27733	$633._3$ H	2.7	(0.389)			(1.81_5)	a'→X,[de] R	27553 H	(21)(24)* (25)*
(B)	(21117)	($580._0$) H	($3._5$)	Observed in thermal emission.[f]				(B→X), R	(20917)	(15)
X $^1\Sigma^+$	0	985.5[g] H	4.29[g]	0.4856962[d]	0.0030756	4.72[h]	1.624648	Microwave sp. Mol. beam el. reson.[ij]		(6)(9)(17) (10)(20)
$^{74}Ge^{32}S$		$\mu = 22.3188284$		$D_0^0 = 5.67$ eV[a]						SEP 1976
E	38884.8	310.3 H	1.43[b]					E← X, R	38752.1 H	(1)(3)*
A $^1\Pi$	32889.5	375.0 H	1.51					A↔X, R	32789.2 H	(1)(2)* (3)*
a ($^3\Pi$)								a→X,	(22400)[c]	(8)
X $^1\Sigma^+$	0	575.8 H	1.80	0.18656575_7	0.00074910_3[d]	0.7883[e]	2.012086[f]	IR sp.[g] Microwave sp.[h]		(9) (5)(10)

GeO: [a]Thermochemical value (mass-spectrom.)(4).
[b]Vertical I.P. from electron impact mass spectrometry (12).
[c]Perturbation between A $^1\Pi$(v=0) and a $^3\Pi$(v=8), see (25).
[d]RKR potential curves (5)(22), Franck-Condon factors (19) (23)(25); variation of A – X electronic transition moment (26).

GeO (continued):
[e]Also observed in phosphorescence in various solid matrices at low temperature (14); lifetime in matrices (11).
[f]The vibrational analysis seems uncertain since the intensity distribution does not agree with that expected for such a large change of ω_e.

GeO (continued):

[g]From band origins (13) obtain ω_e = 986.84, $\omega_e x_e$ = 4.47.
[h]Calculated D_e.
[i]$\mu_{el}(^{+}GeO^{-})$ = [3.2720 + 0.0208(v+½)] D (10).
[j]g_J(v=0) = − 0.1411 (17)(20); also magnetic susceptibility anisotropy and molecular quadrupole moment.

(1) Jevons, Bashford, Briscoe, PPS 49, 543 (1937).
(2) Drummond, Barrow, PPS A 65, 277 (1952).
(3) Barrow, Rowlinson, PRS A 224, 374 (1954).
(4) Drowart, Degrève, Verhaegen, Colin, TFS 61, 1072 (1965).
(5) Nair, Singh, Rai, JCP 43, 3570 (1965); IJPAP 9, 130 (1971).
(6) Törring, ZN 21 a, 287 (1966).
(7) Sharma, Padur, PPS 90, 269 (1967).
(8) Majumdar, Mohan, IJPAP 6, 183 (1968).
(9) Hoeft, Lovas, Tiemann, Tischer, Törring, ZN 24 a, 1217 (1969).
(10) Raymonda, Muenter, Klemperer, JCP 52, 3458 (1970).
(11) Meyer, Smith, Spitzer, JCP 53, 3616 (1970).
(12) Hildenbrand, IJMSIP 7, 255 (1971).
(13) Korzh, Kuznetsova, OS(Engl. Transl.) 31, 286 (1971).
(14) Meyer, Jones, Smith, Spitzer, JMS 37, 100 (1971).
(15) Tewari, Mohan, JMS 39, 290 (1971).
(16) Murty, Reddy, Rao, IJPAP 10, 834 (1972).
(17) Honerjäger, Tischer, ZN 28 a, 1374 (1973).
(18) Murty, Rao, Rao, PRIA A 73, 213 (1973).
(19) Sinha, Chatterjee, IJPAP 11, 57 (1973).
(20) Davis, Muenter, JCP 61, 2940 (1974).
(21) Hager, Wilson, Hadley, CPL 27, 439 (1974).
(22) Savithry, Rao, Rao, CS 43, 329 (1974).
(23) Singh, IJPAP 12, 528 (1974).
(24) Hager, Harris, Hadley, JCP 63, 2810 (1975).
(25) Capelle, Brom, JCP 63, 5168 (1975).
(26) Rao, Rao, Rao, JQSRT 16, 467 (1976).

GeS: [a]Thermochemical value (mass-spectrom.)(4). The convergence limit in E − X, assuming dissociation into $^3P + ^3P$, gives $5.6_6 \pm 0.13$ eV, the error being due to the uncertainty with regard to the particular triplet components involved.
[b]Extended progression (v"=0) converging to 46715 cm^{-1}. The vibrational constants given represent this progression up to v'=17 only.
[c]Extrapolated from observed phosphorescence spectra in various solid matrices at low temperature (8). Lifetime in these matrices (7).
[d]$-4.4 \times 10^{-8}(v+\frac{1}{2})^2 - 2.8 \times 10^{-9}(v+\frac{1}{2})^3$.
[e]$\beta_e = +0.0010 \times 10^{-7}$.
[f]From the effective B_e; 2.012043 Å at the minimum of the Born-Oppenheimer potential curve, see (10).
[g]In low-temperature argon and nitrogen matrices.
[h]μ_{el}(v=0) = 2.00 D from Stark effect of microwave spectrum (5). Quadrupole hyperfine structure for isotopic molecules with nuclear spins I ≥ 1 (6).

(1) Shapiro, Gibbs, Laubengayer, PR 40, 354 (1932).
(2) Barrow, PPS 53, 116 (1941).
(3) See ref. (2) of GeO.
(4) Coppens, Smoes, Drowart, TFS 63, 2140 (1967).
(5) See ref. (9) of GeO.
(6) Hoeft, Lovas, Tiemann, Törring, JCP 53, 2736 (1970).
(7) See ref. (11) of GeO.
(8) See ref. (14) of GeO.
(9) Marino, Guérin, Nixon, JMS 51, 160 (1974).
(10) Stieda, Tiemann, Törring, Hoeft, ZN 31 a, 374 (1976).

State	T_e	ω_e	$\omega_e x_e$	B_e	α_e	D_e $(10^{-8}cm^{-1})$	r_e (Å)	Observed Transitions: Design.	Observed Transitions: ν_{00}	References
$^{74}Ge^{80}Se$		$\mu = 38.401013_5$		$D_0^0 = 4.9_8$ eV[a]						SEP 1976
E	35462.6	217.7 H	1.02[b]					E← X, R	35367 H	(2)*
A $^1\Pi$	30845.7	269.4 H	0.89					A↔X, R	30776.2 H	(1)* (2)*
X $^1\Sigma^+$	0	408.7 H	1.36	0.096340508	0.000289040[c]	2.207_1[d]	2.134629[e]	IR sp.[f]		(6)
								Microwave sp.[g]		(3)(7)
$^{(74)}Ge^{(28)}Si$		$(\mu = 20.2956423)$		$D_0^0 = 3.0_8$ eV[a]						SEP 1976
$^{74}Ge^{130}Te$		$\mu = 47.112514_2$		$D_0^0 = 4.2_4$ eV[a]						SEP 1976
E	(31470)	(170) H	(1.2)					E←X, R	31458 H 31401 H [b]	(2)
A $^1\Pi$	27750.8	221.0 H	0.89					A↔X, R	27699.3 H	(1)* (2)
X $^1\Sigma^+$	0	323.9 H	0.75	0.06533821	0.00017246[c]	1.1_8	2.340165	IR sp.[d]		(8)
								Microwave sp.[e]		(4)

GeSe: [a]From the convergence of the E − X, v"=0 progression assuming dissociation into $^3P + {}^3P$. Since the particular triplet components involved are uncertain the possible error is rather large, viz. ± 0.25 eV.

[b]Long progression of absorption bands converging at 42360 cm^{-1}. The vibrational constants represent the levels only to v'=14.

[c]$- 3.4 \times 10^{-8}(v+\frac{1}{2})^2 - 1.1 \times 10^{-9}(v+\frac{1}{2})^3$.

[d]$\beta_e = + 0.0019 \times 10^{-8}$.

[e]From the effective B_e. According to (7) the minimum of the Born-Oppenheimer potential curve is at 2.134603 Å.

[f]In low-temperature argon and nitrogen matrices.

[g]$\mu_{el}(v=0) = 1.64_8$ D from Stark effect of microwave spectrum (4). Hyperfine structure for odd isotopes (5).

(1) Barrow, Jevons, PPS 52, 534 (1940).
(2) See ref. (2) of GeO.
(3) Hoeft, ZN 21 a, 1240 (1966).
(4) Hoeft, Lovas, Tiemann, Törring, ZN 25 a, 539 (1970).
(5) See ref. (6) of GeS.
(6) See ref. (9) of GeS.
(7) See ref. (10) of GeS.

GeSi: [a]Thermochemical value (mass-spectrom.)(1).

(1) See ref. (1) of GeC.

GeTe: [a]Thermochemical value (mass-spectrom.)(3), corrected for the new value of D_0^0 (Te_2). Extrapolation of the vibrational levels of the E state (2) gives 4.1 ± 0.4 eV.

[b]There seem to be two components of this band system, possibly owing to strong perturbations.

[c]$\gamma_e = - 5.0 \times 10^{-8}$.

[d]In low-temperature nitrogen matrix.

[e]$\mu_{el}(v=0) = 1.06$ D from Stark effect of microwave spectrum (5); hyperfine structure of ^{73}Ge (7). See also (6).

(1) See ref. (1) of GeSe.
(2) See ref. (2) of GeO.
(3) Colin, Drowart, JPC 68, 428 (1964).
(4) Hoeft, Nolting, ZN 22 a, 1121 (1967).
(5) See ref. (4) of GeSe.
(6) See ref. (6) of GeS.
(7) Tiemann, Hoeft, Törring, ZN 26 a, 1930 (1971).
(8) See ref. (9) of GeS.

State	T_e	ω_e	$\omega_e x_e$	B_e	α_e	D_e (10^{-2}cm^{-1})	r_e (Å)	Observed Transitions Design.	Observed Transitions ν_{00}	References
$^1H_2$		μ = 0.50391261		D_0^0 = 4.4781$_3$ eV[a]		I.P. = 15.4258$_9$ eV[b]				NOV 1976 A

WAVELENGTH TABLES of the H_2 spectrum from 2800 to 29000 Å with assignments of many of the lines (109). The TABLES OF ENERGY LEVELS (24) are also very useful as long as it is realized that the absolute values of the energy levels (n ≥ 2) relative to the ground state need correction. Graphs and tables of POTENTIAL ENERGY CURVES for all known states of H_2, H_2^+, and H_2^- (107).

See v p. 241

Fragments of three other triplet systems.[c]

State	T_e	ω_e	$\omega_e x_e$	B_e	α_e	D_e	r_e	Design.	ν_{00}	References
u $^3\Pi_u$ 6pπ	[123488.$_0$]	Only v=0 observed.		[29.3]		[2.3]	[1.06$_9$]	u→a, δ bands	26232.3[f]	(1)(24)
t[d] $^3\Sigma_u^+$ 5fσ	(121292)	(2661.4)	(121.9)	e				t→a,	(25342)	(4)
q[d]($^3\Sigma_g^+$) 5dσ	(121295)	[2172.6]		e				q→c,	(25325)[f]	(3)
n $^3\Pi_u$ 5pπ	120952.$_9$	2321.4	62.8$_6$	29.9$_5$	1.24[g]	[2.3]	1.057	n→a, γ bands	24847.3[f]	(1)(24)
m[h] $^3\Sigma_u^+$ 4fσ	(119317)	[2457.1]		e				m→a,	23295.1[i]	(4)
s $^3\Delta_g$ 4dδ	118875.$_2$	2291.7[j]	62.4$_4$[j]	k				s→c,	22949.3[ℓ]	(1)(18)(24)
r $^3\Pi_g$ 4dπ	118613.$_7$	2280.3[m]	57.9$_6$[m]	k				r→c,	22683.2[m]	(1)(18)(24)
p $^3\Sigma_g^+$ 4dσ	118509.$_8$	2303.1	76.9$_0$	e				p−k,[n]		(154)
								p→c,	22586.0[f]	(1)(18)(24)
v ($^3\Pi_g$) [o]	(118330)	(2340)	(57)	([29.1])			([1.07$_2$])	v→c,	(22430)	(3)
k $^3\Pi_u$ 4pπ	118366.$_2$[p]	2344.37	67.2$_9$[q]	30.07$_4$	1.46$_2$[r]	[1.8$_5$]	1.054$_7$	k→a, β bands	22271.0[f]	(1)(15a)(24)
f $^3\Sigma_u^+$ 4pσ	(116705)	[2143.6][s]		[27.0][s]			[1.11]	f→a,	20526.0[s]	(1)(24)
o[t] $^3\Sigma_u^+$	(114234)	2399.1	91.0	[35]			[0.98]	o→a,	(18160)	(4)
ℓ[u] $^3\Pi_u$	113825	2596.8	106.0	[36]			[0.96]	ℓ→a,	17846[f]	(4)

$^1H_2$: [a]This is an upper limit (36118.3 ± 0.5 cm^{-1}), the lower limit being 4.4779 eV. According to (95) the true value is probably close to the upper limit; see also (101) who gives D_0^0 = 36118.6 cm^{-1} on the basis of a reassignment of the last vibrational levels of the B state. The most recent theoretical value of (70) - including a small non-adiabatic correction of (161) - is 36117.9 cm^{-1}. An earlier independent calculation (54) (not including the non-adiabatic correction) gave 36118.1 cm^{-1}.

[b]From the limit of the npσ, $^1\Sigma_u^+$ Rydberg series (124417.2 cm^{-1}) taking account of perturbations and pressure shift of high n lines (114). The earlier value of (98) was higher by 1.2 cm^{-1} because it was not corrected for pressure shift. The latest theoretical (<u>ab initio</u>) value (79) including relativistic, Lamb shift, and non-adiabatic corrections is 15.4259_0 eV; see (114).

[c] $^3B \rightarrow c$, $^3C \rightarrow c$, $7p\pi \rightarrow a$ (1)(3).

[d]t and q are designated ^3F and ^3G, resp., in (3)(4).

[e]The states g $^3\Sigma_g^+$(3dσ), p $^3\Sigma_g^+$(4dσ), q $^3\Sigma_g^+$(5dσ), m $^3\Sigma_u^+$(4fσ) and t $^3\Sigma_u^+$(5fσ) are strongly affected by ℓ-uncoupling. The N=1 levels lie below N=0 for v=0 and 1; meaningful B values cannot be given until the whole d and f complexes have been fully analysed, see (58).

[f]Referred to the (non-existent) N=0 level in $^3\Pi$ states; the N=1 levels of c $^3\Pi$ (+and -) lie 60.7 cm^{-1} above N=0.

[g]Represents B_0 and B_1 of $^3\Pi^-$ only; (4) gives B_2 = 26.26, B_3 = 24.54.

[h] ^3E of (4).

[i]Refers to N'=0 which lies above N'=4 because of strong ℓ-uncoupling.

[j]Constants refer to N=2; from v=0, 1, 2.

[k]Because of strong ℓ-uncoupling no meaningful B values can be given; see [e].

[ℓ]Refers to the N=2 level of s $^3\Delta_g^-$ above the hypothetical level N=0 of c $^3\Pi_u$; see [f].

[m]The constants refer to N=1 of r $^3\Pi_g^-$; ν_{00} is the energy above the hypothetical level N=0 of c(v=0), see [f].

[n]Anticrossings and microwave transitions. The energy difference between k $^3\Pi_u$(v=1, N=3) and p $^3\Sigma_g^+$(v=1, N=5) is +0.2785 cm^{-1}. Fine structure parameters.

[o] ^3A of (3); probably a doubly excited state. The possibility (1sσ)(4fπ) mentioned by (3) and quoted in MOLSPEC <u>1</u> can be ruled out since it does not give rise to an even state.

[p]A_0(ortho) = - 0.00937, A_0(para) = - 0.0071_0 cm^{-1} (123)(133) (154); also hyperfine structure investigated by these authors.

[q]$\omega_e y_e$ = + 0.99 from (15a).

[r]From B_0 and B_1 of Π^- only (1).

[s]Calculated from the data in (1) and (24). $\Delta G(\frac{1}{2})$ and ν_{00} refer to actual N=0 level which is strongly perturbed.

[t] ^3D of (4). Probably a doubly excited state: (2pσ)(3dσ).

[u] ^3Y of (4). Probably a doubly excited state: (2pσ)(3dπ).

[v]The T_e values for the upper states of the <u>triplet</u> transitions are based on T_e'' for the lower state (a or c) and have been calculated assuming $Y_{00}' \approx Y_{00}''$.

State	T_e See [v] p.241	ω_e	$\omega_e x_e$	B_e	α_e	D_e $(10^{-2}cm^{-1})$	r_e (Å)	Observed Transitions Design.	ν_{00}	References
$^1H_2$ (continued)										
j $^3\Delta_g$ 3dδ	(113533)	2345.2_6[a]	66.5_6[b]	30.08_5[a]	1.69_2	1.9_0	1.0545	j ↔ c,[c] R	17633.0[p]	(1)(24)
i $^3\Pi_g$ 3dπ	(113132)	2253.5_5[a]	67.0_5[e]	29.22_1[a]	1.50_6	1.7_6	1.0700	i – d,[f]		(132)
								i → e, R	5384.81[g]	(47)
								i ↔ c,[c] R	17185.8[p]	(1)(24)
h $^3\Sigma_g^+$ 3sσ	(112913)	$[2268.7_3]$[h]		$[30.6_2]$[h]			[1.045]	h → c,	16990.8[d]	(2)(24)
g $^3\Sigma_g^+$ 3dσ	$112854._4$	2290.8_6	105.4_3[i]	j				g → e, R	5116.6	(47)
								g ↔ c,[c]	16917.6[d]	(1)(2)(24)
d $^3\Pi_u$ 3pπ	$112700._3$[k]	2371.58[ℓ]	66.27[m]	30.364[ℓn]	1.545	[1.91]	1.0496	d^o → a, R Fulcher (α) b.	16619.0[d]	(5)(24)
e $^3\Sigma_u^+$ 3pσ	$107774._7$	2196.13	65.80[q]	27.30	1.515		1.107	e → a, R	11605.6	(1)(6)
a $^3\Sigma_g^+$ 2sσ	$95936._1$[r]	2664.83	71.65[s]	34.216	1.671	[2.16]	0.9887_9	a^t → b,[u]		
								(a – X)	95076.4[r]	
c $^3\Pi_u$ 2pπ	$95838._5$[v]	2466.8_9	63.51[w]	31.07[xy]	1.42_5	$[1.9_5]$	1.037_6	(c – X)	$94881._0$[z]	
b $^3\Sigma_u^+$ 2pσ		Unstable; lower state of the continuous spectrum of H_2 (a → b). Pot. function (43).								

$^1H_2$: [a]These constants [from (58)] refer to the $^3\Pi^-$ and $^3\Delta^-$ components and are based on "Approximation 2" of (53) for the evaluation of the ℓ-uncoupling. The observed levels are given by (24).

[b]$\omega_e y_e = +\ 0.74_5$.

[c]Observed in absorption in flash discharges (60).

[d]See [f] p. 241 .

[e]$\omega_e y_e = -\ 1.27_2$. Ab initio calculations (41)(45) give a pronounced potential maximum near 2.5 Å for this state.

[f]Anticrossings and microwave transitions; i $^3\Pi_g$(v=3, N=2) is 1.9244 cm^{-1} above d $^3\Pi_g$(v=3, N=1).

[g]Refers to Π^-(N=1). Π^+(N=1) is at 5471.70 cm^{-1} above e $^3\Sigma_u^+$(v=0, N=0). The rotational levels are very irregular, only partly on account of ℓ-uncoupling.

[h]From (24). (2) give ω_e = 2395.2, $\omega_e x_e$ = 64.2, B_o = 30.0. According to (24) the v=0 levels may be spurious. If so, only v=1 remains with $B_1 = 28.7_2$.

[i]$\omega_e y_e = +\ 2.40_3$; calculated from the N=0 levels of (24).

$^1H_2$ (continued):

[j]See [e] p. 241 .

[k]The fine structure in the N=1 levels of both ortho- and para-H_2 has been observed in microwave-optical double resonance by (127) who give A_e = 0.0281 as well as spin-spin coupling constants. For para-H_2, v=0, N=1 the three component levels J=1, 2, and 0 are at -0.01241, -0.00695, and +0.07197 cm^{-1}, resp.. For ortho-H_2 the hyperfine structure has also been studied.

[ℓ]Constants refer to $^3\Pi^-$. $^3\Pi^+$ is strongly perturbed, i.e. the Λ-type doubling is fairly large and irregular (7).

[m]$\omega_e y_e$ = + 0.88.

[n]Breaking-off of P and R branches ($^3\Pi^+$) above v'=3 on account of predissociation. Breaking-off of Q branches ($^3\Pi^-$) for v'=7, 8 above N=1 on account of preionization (9).

[o]Lifetime 63 ns (81); see, however, (118) who give 31 ns.

[p]Lower component of N'=1 ($i\,^3\Pi$) or 2 ($j\,^3\Delta$) relative to the (non-existent) N"=0 level of $c\,^3\Pi$.

[q]$\omega_e y_e$ = - 0.433.

[r]The T_0 (v_{00}) value is derived from singlet-triplet anti-crossings in a magnetic field (134)(140) and corresponds to v=0, N=0. It agrees fairly well with $95073._2$ obtained from the energy of a $^3\Sigma_u^+$(v=0, N=0) below the ionization limit, 29344 ± 2 cm^{-1} (9), combined with the new value of I.P.(H_2). (24) gives T_0 = 95226 without explanation; the most recent theoretical value is 95077.3 (150). The T_e value in the table takes account of Y_{00} in both upper (Y'_{00} = 4.9_2) and lower state.

[s]$\omega_e y_e$ = + 0.92. Precise <u>ab initio</u> potential function (incl. diagonal corrections) and predicted vibrational levels in (67). Except for a constant shift, the latter agree well with the observed levels (24).

[t]Lifetime τ(v=0,1) = 10.4_5 ns (111)(149).

[u]Reproduction in MOLSPEC <u>1</u>, Fig. 12.

[v]A = - 0.1249 cm^{-1} (135)(138). T_e takes account of Y_{00} in both upper (Y'_{00} = 4.1_8) and lower state.

[w]$\omega_e y_e$ = + 0.552.

[x]The Λ-type doubling is quite small ($\sim$0.5 cm^{-1} for N=6); the constants refer to the average. The triplet splitting in N=2 of para-H_2 has been fully resolved in molecular beam experiments of (28) yielding $\Delta\nu$(J=2-1) = 0.16438, $\Delta\nu$(J=2-3) = 0.19674 cm^{-1} with J=2 at the top. The hyperfine structure in N=1,J=2 of ortho-H_2 is $\Delta\nu$(F=3-2) = 0.0236, $\Delta\nu$(F=2-1) = 0.0154 cm^{-1} as quoted by (32). (18) give spin splittings for N = 1,2,3,4,5 without resolving J=N+1 from J=N-1.

[y]The levels of $c\,^3\Pi_u^+$ are strongly predissociated by the $b\,^3\Sigma_u^+$ state (60); the levels of $c\,^3\Pi_u^-$ are either very weakly affected by a forbidden predissociation to $b\,^3\Sigma_u^+$ (33) (36) or decay radiatively (by magnetic dipole radiation) to the $b\,^3\Sigma_u^+$ state as suggested by the lifetime measurements of (117), τ(v=0) = 1.02 ms independent of spin component and isotope. (139) observed quenching of $c\,^3\Pi_u^-$ in an electric field. The Stark effect is large ($\sim 10^4$ times greater than for the ground state) and has been studied experimentally by (160) and compared with the theoretical values of (145).

[z]This number, obtained from v_{00}(a-X) + v_{00}(e-a) + v_{00}(g-e) - v_{00}(g-c), is 87 cm^{-1} higher than given in MOLSPEC <u>1</u>, a change made necessary by the work of (47). See also [r].

State	T_e	ω_e	$\omega_e x_e$	B_e	α_e	D_e ($10^{-2}cm^{-1}$)	r_e (Å)	Observed Transitions Design.	Observed Transitions ν_{00}	References
$^1H_2$ (continued)										
		Several excited states above the ionization limit, established by electron impact studies and leading to two excited atoms or $H + H^+$.								(164)(165)
		Continuous absorption above ~130000 cm^{-1} [a].								(38)(49)
v'=0 Rydberg series of rotational levels observed in low temperature absorption from X $^1\Sigma_g^+$, v"=0, J"=0 and 1 and converging to										
N=2 of H_2^+:		J=1 levels of npπ $^1\Pi_u^+$ (n = 6,...,32, joining on to C, D, D', D")[b]; $\nu = 124591.5^c - R/(n+0.082)^2$. Similar series with v' = 1,...,6[d].						R(0) lines (para-H_2)		(85)(98) (114)*
N=1 of H_2^+:		J=1 levels of npπ $^1\Pi_u^-$ (n = 6,...,43, joining on to C, D, D', D")[e]; $\nu = 124476.0^c - R/(n+0.082)^2$. Similar series with v' = 1,...,5.						Q(1) lines (ortho-H_2)		(85)(98) (114)*
		J=0 levels of npσ $^1\Sigma_u^+$ (n = 5,...,19, joining on to B, B', B")[e]; $\nu = 124476.0^c - R/(n-0.203)^2$. Similar series with v' = 1, 2, 3.						P(1) lines (ortho-H_2)		(98)(114)*
N=0 of H_2^+:		J=1 levels of npσ $^1\Sigma_u^+$ (n = 5,...,40, joining on to B, B', B")[b]; $\nu = 124417.2^c - R/(n-0.203)^2$. Similar series with v' = 1,...,6[d].						R(0) lines (para-H_2)		(85)(98) (114)*
$\bar{B}$ $^1\Sigma_u^+$		State causing ion-pair formation after excitation of higher Rydberg states; also responsible for perturbations in B' $^1\Sigma_u^+$. Correlates at small r with B" $^1\Sigma_u^+$, forming a double-minimum state								(129)(155) (163).
D" $^1\Pi_u$ 5pπ	121211.0[f]	2319.92[f]	63.041	30.7_6[gh]	1.45[g]	(3)	1.043	D"←X, R	120176.0[f]	(46)(73)
D' $^1\Pi_u$ 4pπ	118865.3[f]	2329.97[f]	63.140	29.8_9[ih]	1.11[i]	[2.5][i]	1.058	D'←X, R	117835.2[f]	(40)(46)(73)
S[j] $^1\Delta_g$ 4dδ	[(119893)]	Only v=0 observed.		[(28.8)][k]			[(1.07_8)]	S→B, V	(27510)[ℓ]	(1)(24)
O[m] $^1\Sigma_g^+$ 4sσ	[(119870)]	Only v=0 observed.		[(32)]			[(1.0_2)]	O→B, V	(27487)[n]	(1)
R[o] $^1\Pi_g$ 4dπ	(118688)	[2142][p]		[(30)][k]			[(1.0_6)]	(R→C) R→B, V	(18488) (27376)[q]	(1)(24)
P[r] $^1\Sigma_g^+$ 4dσ	[119531]	Only v=0 observed.		[(30)][k]			[(1.0_6)]	(P→C) P→B, V	18260 27148[s]	(1)(24)
T[t] $^1\Sigma_g^+$	[119512.6]	Only v=0 observed.		[(25.4)]			[(1.14_8)]	T→B, V	27130.1	(1)(24)

$^1H_2$: [a]Theoretical and experimental values for the ionization probability into the various vibrational levels of H_2^+ are given by (50)(71)(75)(147) and (68)(76), resp.. The ionization cross section near the ionization limit has been studied at high resolution by (62)(63). See also (168).

[b]For high n there is strong ℓ-uncoupling and the two series of $^1\Sigma_u^+$ and $^1\Pi_u^+$ levels of para-H_2 should be called np0 and np2, resp., corresponding to the fact that the first converges to N=0 , the second to N=2 of H_2^+. There are strong systematic perturbations between the J=1 levels of these two series (because of ℓ-uncoupling) so that the formulae as given do not represent the series very well. An accurate representation can be obtained by Fano's quantum defect theory; see (114). Levels of npπ, $^1\Pi_u^+$ above N=0 of H_2^+ are preionized resulting in asymmetrically broadened absorption lines with apparent emission wings.

[c]Limits of Rydberg series above v"=0, J"=0.

[d](155) have observed Rydberg levels with v = 9,10,11 in the study of ion-pair formation.

[e]These two series of ortho levels are essentially unperturbed.

[f]Average of Π^+ and Π^-. v_{00} referred to $\{N'=0\}$.

[g]Refers to Π^-; Π^+ is perturbed, $B_0(\Pi^+) = 30.178$, $B_1(\Pi^+) = 31.370$.

[h]RKR potential function in (72).

[i]Refers to Π^-; $\gamma_e = -0.53$. Π^+ is perturbed, $B_0(\Pi^+) = 31.09_5$, $B_1(\Pi^+) = 29.165$.

[j]4F of (24), $4^1\chi$ of (1).

[k]The states P, R, S form a d complex with strong uncoupling. As a result the constants given have only limited meaning.

[ℓ]The two J=2 levels are observed at 27631.3 and 27732.9 cm^{-1} above J=0, v=0 of B $^1\Sigma_u^+$ (24). The v_{00} value given is an extrapolated average for J=0 and, because of the uncoupling, is rather uncertain.

[m]4^1O of (1), not given by (24).

[n]From R(0) and P(1) according to the data of (1).

[o]4^1B of (1), 4E of (24).

[p]Refers to $^1\Pi^-$.

[q]The two J=1 levels are observed at 27385.8 and 27487.1 cm^{-1} above J=0, v=0 of B $^1\Sigma_u^+$ (24). The v_{00} value given is an extrapolated average for J=0 and, because of the uncoupling, is rather uncertain.

[r]4^1C of (1), 4D of (24).

[s]The J=1 level is observed at 27207.62 cm^{-1} above J=0, v=0 of B $^1\Sigma_u^+$. The value given for J=0 is extrapolated and, because of the uncoupling, is rather uncertain.

[t]4^1K of (1), doubly excited state.

State	T_e	ω_e	$\omega_e x_e$	B_e	α_e	D_e $(10^{-2}cm^{-1})$	r_e (Å)	Observed Transitions		References
								Design.	ν_{00}	
$^1H_2$ (continued)										
B" $^1\Sigma_u^+$ 4pσ	117984.5	2197.50	68.136	26.68[ab]	1.19[a]	[3.4]	1.119_8	B"← X, R	116886.9[c]	(40)(46)(73) (106)*
N $^1\Sigma_g^+$ [d]	(116287)	[1983.3]		[(18.4)]			[(1.35)]	N → B, R	24896.4	(1)(24)
U ($^1\Sigma_g^+$) [d]	[116707.7]	Only v=0.		[(18.8)]			[(1.33)]	U → B, [e] R	24325.1	(1)(24)
M $^1\Sigma_g^+$ [d]	(114485)	[2176.0]		[(13)]			[(1.6_0)]	M → B, R	23190.0[f]	(1)(24)
L $^1\Sigma_g^+$ [d]	(114520)	[(1835)]		[(9.7)]			[(1.8_6)]	L → B, R	23054.8[f]	(1)(24)
H[g] $^1\Sigma_g^+$ 3sσ	113899	2538	124	[(29.5)]			[(1.06_5)]	H → C, R H → B, V	13866.6[h] 22754.1[h]	(1)(24)
D $^1\Pi_u$ 3pπ	113888.7	2359.91	68.81_6[i]	30.29_6[jkb]	1.42[j]	2.01[ℓ]	1.0508	D → E, R D ⟷ X,[m] R	$13709._7$[n] 112872.3[n]	(11)(24) (40)* (46) (73)(106)*
J $^1\Delta_g$ 3dδ	(113550)	2341.1_5[o]	63.2_3[o]	30.08_1[o]	1.71_8[o]	1.8_9[o]	1.0546	J → C, R J → B, [q] V	13435.6[p] 22322.5[p]	(1)(24)
I[r] $^1\Pi_g$ 3dπ	(113142)	2259.1_5[o]	78.4_1[or]	29.25_9[os]	1.58_4[o]	1.8_0[o]	1.0693	I[s] → C, R I → B, V	12982.5[t] 21869.5[t]	(1)(10)(24)
G[u] $^1\Sigma_g^+$ 3dσ	112834	2343.9	55.9[v]	[(28.4)][w]			[(1.08_5)]	G[w] → C,[x] R G → B, V	12722.2[y] 21609.2[y]	(1)(24)
K[z]($^1\Sigma_g^+$)	(112669)	[2232.59]	30	[10.8]			[1.76]	K → C, R K → B, R	12538.6 21425.4	(1)(24)

$^1H_2$: [a]Representing only B_0 and B_1. The B_v curve has a positive curvature for low v and a strong negative curvature for high v. (46) gives $B_v = 27.1_3 - 2.35(v+\frac{1}{2}) + 0.665(v+\frac{1}{2})^2 - 0.0729(v+\frac{1}{2})^3$.

[b]RKR potential function (72). Ab initio pot. function (163).

[c]Deperturbed value from (40). The observed value for J=0 [perturbed by B'(v=4)] is 116885.6 according to (40) and 116885.3 according to (46), while in the more recent paper (73) gives 116882.00.

[d]All these states are considered as doubly excited states by (24). They may well form one or two double-minimum states (similar to E, F) together with H $^1\Sigma_g^+$.

$^1H_2$ (continued):

[e] This is the λ4142.8 progression of (1) as revised by (24).

[f] These values agree with (24); (1) gives 23057.22 and 23191.66 for L and M, respectively.

[g] 3^1O of (1).

[h] From R(0) of the 0-0 band and F(1)-F(0) as given by (1). The basis for 22751.6 in (1) is not clear.

[i] $+ 1.027_4(v+\frac{1}{2})^3 - 0.0420_2(v+\frac{1}{2})^4$; the vibrational constants (73) refer to the average of Π^+ and Π^-. See also [k].

[j] $\gamma_e = -0.025$; the rotational constants (40) represent only the levels v=0, 1, 2 of Π^-. The Π^+ levels are strongly perturbed by the B' state which also causes the predissociation of $^1\Pi^+$ for $v' \geq 3$; see [k]. (46) gives for the deperturbed values

$$B_v(\Pi^+) = 32.5_1 - 2.00(v+\tfrac{1}{2}) + 0.071(v+\tfrac{1}{2})^2 - 0.0040(v+\tfrac{1}{2})^3,$$
$$B_v(\Pi^-) = 30.8_1 - 1.96(v+\tfrac{1}{2}) + 0.102(v+\tfrac{1}{2})^2 - 0.0053(v+\tfrac{1}{2})^3.$$

[k] Strong predissociation for $v' \geq 3$; no bands with $v' \geq 3$ have ever been observed in emission. In absorption strongly broadened lines with apparent emission wings (Beutler-Fano shapes) in $D^1\Pi_u^+ \leftarrow X^1\Sigma_g^+$ (106); line widths of 4 and 11.5 cm^{-1} for J=1 and 2, resp., have been observed (103) and accounted for by interaction with the continuum of $B'^1\Sigma_u^+$ (105)(108)(112). Widths for $D^1\Pi_u^- \leftarrow X^1\Sigma_g^+$ (Q) lines are much smaller. Ly_α fluorescence as a result of prediss. (63)(82)(97). Electric field induced component of prediss. (92).

[ℓ] From (40); (46) gives $D_v(\Pi^+) = 0.033 + 0.0010(v+\frac{1}{2})$, $D_v(\Pi^-) = 0.0283 - 0.0012(v+\frac{1}{2})$.

[m] RKR Franck-Condon factors (89). Absorption coefficients of $D \leftarrow X$ bands (38). Oscillator strengths $f_{00} = 0.0061_4$, $f_{20} = 0.010_9$ (142).

[n] Average of Π^+ and Π^- extrapolated to J=0. The Λ-type doubling for v=0, J=1 is 4.2 cm^{-1} with Π^+ above Π^-.

[o] These constants (58) refer to Π^- and Δ^- and take into account the effects of ℓ-uncoupling in the d complex according to the formulae of (53). They cannot be used to derive energy levels without the use of these formulae. The observed levels are given in (24).

[p] Refers to J=2 of Δ^- at 10.8 cm^{-1} below J=2 of Δ^+.

[q] The forbidden $^1\Delta_g \rightarrow {}^1\Sigma_u^+$ transition occurs because of strong uncoupling in the upper state. Only Q branches are observed in these bands.

[r] 3^1B of (1), 3E of (24). (39) and (41) predict a fairly high (0.4 eV) maximum in the potential function of this state.

[s] Zeeman effect studies (20) yield $g(v=0,J=1) = 0.49_8$, $g(v=0,J=2) = 0.412$, etc.; lifetime $\tau(v=0,J=2) = 38$ ns (119), see [w].

[t] Referred to J'=1 of $I^1\Pi^-$; J=1 of $I^1\Pi^+$ is 62.32 cm^{-1} higher.

[u] 3^1C of (1), 3D of (24).

[v] No levels higher than v=3 have been observed which suggests that the dissociation limit is $1^2S + 2^2S, {}^2P$ at 118377.6 cm^{-1}. The constants represent only v=0,1,2.

[w] This value (1) does not represent the low rotational levels because of ℓ-uncoupling, e.g. the J=1 level is below J=0. The actual levels are given in (24). Hyperfine structure for v=1,J=1; A = 1.0 ± 0.17 MHz (136). Large Zeeman splittings corresponding to the strong ℓ-uncoupling (20), $g(v=0,J=1) = 0.901$, $g(v=0,J=2) = 0.571$, etc.; see also (113). Lifetimes from Hanle effect observations (119); $\tau(v=0,J=1) = 27$ ns, $\tau(v=0,J=2,3) = 39$ ns.

[x] The $G \rightarrow B$ system gives rise to the strongest lines in the visible region.

[y] Referred to J'=0 which, because of ℓ-uncoupling, has an anomalous position.

[z] 3^1K of (1), probably due to $(2s\sigma)^2$.

State	T_e	ω_e	$\omega_e x_e$	B_e	α_e	D_e ($10^{-2}cm^{-1}$)	r_e (Å)	Observed Transitions		References
								Design.	ν_{00}	
$^1H_2$ (continued)										
Q ($^1\Pi_g$) [a]	(113163)	[742]		[(16.3)]			[(1.43)]	Q→B, R	$21151._1$	(4)(24)
B' $^1\Sigma_u^+$ 3pσ	$111642._8$ [b]	2039.52	83.406[c]	26.70_5 [e]	2.781[d]	[1.2][f]	1.119_2	B'→E,F	11311.5[g]	(34)
								B'←X,[h] R	110478.2	(40)(44)(46)
F[i] $^1\Sigma_g^+$ $2p\sigma^2$	100911[j]	[1199][i]		$B_4 = 6.24$[k]			$r_4 = 2.31_5$[k]	F→B,[ℓ] R	ν_{40}=13635.1	(14)(34)
E[i] $^1\Sigma_g^+$ 2sσ	100082.3[m]	2588.9[m]	130.5[m]	32.68[m]	1.818[m]	[2.28][m]	1.011_8	E→B,[ℓ] V	8961.23	(8)(22)(24)(34)
C $^1\Pi_u$ 2pπ	100089.8[n]	2443.77	69.524[op]	31.362_9 [p]	1.664_7[q]	2.23[r]	1.0327_9	C[s]↔X,[t] R Werner b.	99120.1_7[n]	(12)(37)(44)(129)

$^1H_2$: [a]Fragmentary, possibly (2pσ)(2pπ).

[b]Takes account of Y_{00} in both upper and lower state. Y'_{00} = 15.3 cm^{-1} is rather uncertain and depends strongly on the number of levels included. See [d].

[c] $+ 3.533(v+\frac{1}{2})^3 - 0.93750(v+\frac{1}{2})^4$; these are the constants of (40)[except T_e which is taken from (129)], they apply only to v=0,...,4. (73) gives a very different set of constants based on seven levels v=0,...,6. The ΔG curve (in H_2, HD, and D_2) has a characteristic tail which makes representation of the higher vibrational levels by a conventional formula meaningless (40)(129).

[d] $+ 0.540(v+\frac{1}{2})^2 - 0.091_7(v+\frac{1}{2})^3$; these constants (40) represent only the first five (deperturbed) B_v values. If only three levels are used $B_v = 26.371 - 1.9000(v+\frac{1}{2}) - 0.0050(v+\frac{1}{2})^2$ leading to a very different Y_{00} value (3.6) from the one used here (see [b]).

[e]RKR potential functions (44)(72). A very slight maximum of the potential function at 2.9 Å has been predicted by (156) but not confirmed in the calculations of (163); see also (151). The experimental data, while suggesting an anomalous form of the potential function, do not indicate a maximum (129).

[f]The higher D_v values are quite irregular.

[g]From the 0-1 band of (34); from $T_0(B')-T_0(E)$ one obtains 11313.6_2.

[h]RKR Franck-Condon factors (89). Oscillator strengths f_{10} = 0.0028, f_{30} = 0.0048 (142).

[i]Because of strong interaction the two states E [2^1X of (1), 2A of (24)] and F, in zero approximation 1sσ2sσ and $(2p\sigma)^2$, form a single state with two minima as first recognized by (30). The most detailed calculation of the potential function and the energy levels is that of (86) whose numbering and ΔG(½) value for the $F^1\Sigma_g^+$ component has been adopted in

the table. According to (86) v_{00}(F-B) would be at 9146.8 cm^{-1} but v=0,1,2,3 of F have not been observed. The observed v=4 level lies just below the potential maximum.

[j]From the observed v_{40} and the energy of v=4 above the (outer) minimum as calculated by (86).

[k]Vibrational numbering of (86). See [i].

[ℓ]Franck-Condon factors (137). Electronic trans. moment (88).

[m]These numbers represent only the lower vibrational levels near the inner minimum. Owing to the interaction of E and F (see [i]) higher $\Delta G(v+\frac{1}{2})$, B_v, D_v values are irregular.

[n]The T_e values for B and C include the effects of Y_{00} on the zero point energies in both upper and lower states; $Y'_{00}(B) = 8.7$, $Y'_{00}(C) = 5.0$ cm^{-1}. On the other hand, the T_e value of C $^1\Pi_u$ and v_{00}(C-X) exclude the term $-B\Lambda^2$ in the energy formula, a term that is usually included to form part of the effective potential energy. With this inclusion and disregarding Y_{00} (44) gives T_e = 100063.42 and v_{00} = 99090.35 on the basis of older data for v=0...4 and his own precise data for v=5...13.

[o]$+ 0.7312(v+\frac{1}{2})^3 - 0.0415(v+\frac{1}{2})^4$. These constants refer to the (unperturbed) Π^- component and are based on an 8-level fit to the data of (129) [v=0-4] and (37) [v=5-7]. Somewhat different constants are given by (44). Note, that the T_e values in (24) are too low by 8.4 cm^{-1} (44). The constants of (73) are affected by not recognizing this error.

[p]Theoretical work (13)(29)(43)(52)(61) has predicted, and the analysis of the spectrum (40)(129) has confirmed, that the potential curve of C $^1\Pi_u$ has a van der Waals maximum of ~105 cm^{-1} above the asymptote near r = 4.8 Å. Ab initio potential function (without diagonal corrections) and predicted vibrational levels (67). RKR potential functions (44) (72); see, however, (124).

[q]$+ 0.0296(v+\frac{1}{2})^2 - 0.00296(v+\frac{1}{2})^3$. These constants refer to the Π^- component (Π^+ is strongly perturbed by B $^1\Sigma_u^+$) and are from an 8-level least-squares fit of the data of (129) [v=0-4] and (37) [v=5-7]. Somewhat discordant B_v values for both Π^- and Π^+ (the latter after deperturbation) are given by (37)(46)(129). The Λ-type doubling for v=0, J=1 is 1.17 cm^{-1}; for other v, J as well as theoretical values see (124)(131).

[r]$\beta_e = -0.0007_4$.

[s]Lifetime $\tau(v=0,1,2,3)$ = 0.6 ns (66).

[t]RKR Franck-Condon factors calculated by (51)(89) and "measured" by (83)(128)(130) who have also determined the dependence of the transition moment on r. Ab initio calculation of the latter by (88). Theoretical transition probabilities and f values (88)(90)(91)(93), experimental values (66) (130)(142): $f_{10} = 0.059$, $f_{20} = 0.060$, $f_{30} = 0.044$, ... Calculated transitions to the continuum of X $^1\Sigma_g^+$ (120). Selective enhancements of v=0 and 2 of C $^1\Pi_u$ in Ar-H_2 mixtures have been studied by (55); similar enhancements have also been observed in Kr-H_2 mixtures. For stimulated emission in the Q(1) and P(3) lines of the 1-4, 2-5, 2-6, 3-7 Werner bands see (116)(121).

State	T_e	ω_e	$\omega_e x_e$	B_e	α_e	D_e ($10^{-2}cm^{-1}$)	r_e (Å)	Observed Transitions Design.	ν_{00}	References
$^1H_2$ (continued)										
B $^1\Sigma_u^+$ 2pσ	91700.0[a]	1358.09	20.888[b]	20.015_4[c]	1.1845[d]	1.625[e]	1.2928_2	B[f]$\leftrightarrow$X,[gh] Lyman b.	R 90203.3_5	(25)(77)(129)
X $^1\Sigma_g^+$ 1sσ^2	0	4401.21_3	121.33_6[i]	60.853_0[k]	3.062_2[j]	4.71[ℓ]	0.74144	Quadrupole[m] and field-induced sp..[n]		(15)(48) (26)(56)(74)
								Raman sp.[o]		(23)(56)
								Rotational[p] and nuclear rf magn. reson.		(17a)(21) (17)(19)

$^1H_2$: [a]See [n] p. 249.

[b]$+ 0.7196(v+\frac{1}{2})^3 - 0.0598(v+\frac{1}{2})^4 + 0.00216(v+\frac{1}{2})^5$, $Y_{00} = 8.7$; from a least squares fit (129) to the first eight levels as given by (25). (77) gives slightly different constants based on the first five levels only. (73) and (37) have observed levels up to v=35 and 37, resp., very close to the dissociation limit at $118377._6$ cm^{-1} (95). The dissociation energy of the B $^1\Sigma_u^+$ state is 28174.2 cm^{-1}.

[c]RKR potential functions (31)(44)(72)(89); see also (126). Precise <u>ab initio</u> potential function (incl. diagonal corrections) and predicted vibrational levels (67)(152).

[d]$+ 0.1214(v+\frac{1}{2})^2 - 0.0117(v+\frac{1}{2})^3 + 0.0004_6(v+\frac{1}{2})^4$, from a least squares fit (129) to the first eight levels. (77) gives slightly different constants based on the first five levels only. For $v \geq 8$ there are strong rotational perturbations caused by interaction with C $^1\Pi_u$. Only after deperturbation can meaningful B_v values for these levels be obtained [see (129)]. For a theoretical discussion of the intensities in the perturbed region see (131).

[e]$- 2.16_5 \times 10^{-3}(v+\frac{1}{2}) + 2.28_9 \times 10^{-4}(v+\frac{1}{2})^2 - 1.18_5 \times 10^{-5}(v+\frac{1}{2})^3$. For individual B_v and D_v values see (25)(37)(129).

[f]Lifetime $\tau(v=3...7) = 0.8$ ns (66); $\tau(v=8...11) = 1.0$ ns (111).

[g]Franck-Condon factors from RKR potentials (51)(89); from <u>ab initio</u> potential functions (64)(90)(91), including theoretical oscillator strengths; see also (167). J dependence of Franck-Condon factors and transition probabilities (87)(88)(102). Experimental Franck-Condon factors and oscillator strengths (57)(65)(69)(83)(130)(142)(157); $\sum f_{v'0} = 0.29$. Variation of transition moment with r (69)(83)(157) and, <u>ab initio</u>, (64)(88). Selective enhancements of v=3 and 10 of B $^1\Sigma_u^+$ in an Ar-H_2 mixture, first observed by Lyman,

$^{1}H_2$ (continued):

have recently been studied by (55); similar enhancements were also observed in Kr-H_2 mixtures. Stimulated emission in the P branches of the 3-10, 4-11, 5-12, 6-13, 7-13 Lyman bands (96)(99).

[h] A continuous spectrum corresponding to transitions to the continuum of $X\ ^1\Sigma_g^+$ has been observed (94) and the intensity distribution found to be in agreement with calculations. (93)(120) have calculated transition probabilities and the fractions that go to the continuum for v'=0...36. (80) calculated the continuous spectrum corresponding to absorption from the ground state to the continuum of $B^1\Sigma_u^+$.

[i] $+ 0.812_9(v+\frac{1}{2})^3$; these constants (56) represent only the levels v=0,1,2,3. (25) has less accurate constants representing higher G(v) values. The "true" ω_e (including Dunham corrections) is $4403._2$ (48). The zero-point energy ($Y_{00} = 8.9_3$ included) is 2179.2_7 cm^{-1} (27).

[j] $+ 0.057_7(v+\frac{1}{2})^2 - 0.005_1(v+\frac{1}{2})^3$; these constants (56) represent only $B_{0...3}$ which are the best known B_v values. (74) from the field-induced spectrum give a very slightly different B_0 (59.334_3 versus 59.336_2); see also (104). The formula

$$B_v = 60.863_5 - 3.0763_8(v+\tfrac{1}{2}) + 0.0601_7(v+\tfrac{1}{2})^2 - 0.0048_1(v+\tfrac{1}{2})^3$$

of (25) holds up to v=8. Higher B_v values (25) require higher and higher terms in the formula. All the constants given are $Y_{01}...Y_{31}$ values; (48) have introduced Dunham corrections and give the "true" $B_e = 60.867_9$. According to (16) the hyperfine levels F=1 and 2 for J=1,v=0 are 1.823×10^{-5} and 2.005×10^{-5} cm^{-1} below the F=0 component.

[k] RKR potential functions (31)(35)(42), see also (100); ab initio potential functions (141)(153). Rotational and vibrational levels calculated from the latter are given in (153); see also (59)(70). (59) include some of the quasi-bound levels above the dissociation limit [see also (78)]; for their experimental observation see (25)(162). Recent comparisons between ab initio calculated and observed energy levels (115)(144)(158)(159).

[ℓ] $- 0.0027_4(v+\frac{1}{2}) + 0.0004_0(v+\frac{1}{2})^2$; $H_v = [4.9 - 0.5(v+\frac{1}{2})] \times 10^{-5}$; from (48), see also (56).

[m] (48) give absolute intensity measurements of the quadrupole rotation-vibration spectrum (1-0, 2-0, 3-0) as well as corrections for pressure shifts; see also (125)(143)(169). Dependence of quadrupole moment on r (43). Predicted intensities in the rotation-vibration spectrum (84), in the rotation spectrum (110). Predicted lifetimes of rotation-vibration levels (166), e.g. $\tau(v=1,J=1) = 1.17 \times 10^6$ s.

[n] The rotation and rotation-vibration spectrum has been observed in pressure-induced absorption, see the review by (122).

[o] Raman cross sections (148).

[p] Rotational g factor $g_J = 0.88291$.

$^1H_2$ (continued):

(1) Richardson, "Molecular Hydrogen and Its Spectrum", Yale University Press (1934).
(2) Richardson, Rymer, PRS A 147, 24 (1934).
(3) Richardson, Rymer, PRS A 147, 251 (1934).
(4) Richardson, Rymer, PRS A 147, 272 (1934).
(5) Dieke, Blue, PR 47, 261 (1935).
(6) Dieke, PR 48, 606 (1935).
(7) Dieke, PR 48, 610 (1935).
(8) Dieke, PR 50, 797 (1936).
(9) Beutler, Jünger, ZP 101, 285 (1936).
(10) Dieke, Lewis, PR 52, 100 (1937).
(11) Richardson, PRS A 160, 487 (1937); 164, 316 (1938).
(12) Dieke, PR 54, 439 (1938).
(13) King, Van Vleck, PR 55, 1165 (1939).
(14) Dieke, PR 76, 50 (1949).
(15) Herzberg, CJR A 28, 144 (1950).
(15a) Cunningham, Dieke, Report No. NYO-692, Johns Hopkins Univ., Dept. of Physics (1950).
(16) Ramsey, PR 85, 60 (1952).
(17) Kolsky, Phipps, Ramsey, Silsbee, PR 87, 395 (1952).
(17a) Harrick, Ramsey, PR 88, 228 (1952).
(18) Foster, Richardson, PRS A 217, 433 (1953).
(19) Harrick, Barnes, Bray, Ramsey, PR 90, 260 (1953).
(20) Dieke, Cunningham, Byrne, PR 92, 81 (1953).
(21) Barnes, Bray, Ramsey, PR 94, 893 (1954).
(22) Porto, Dieke, JOSA 45, 447 (1955).
(23) Stoicheff, CJP 35, 730 (1957).
(24) Dieke, JMS 2, 494 (1958).
(25) Herzberg, Howe, CJP 37, 636 (1959).
(26) Terhune, Peters, JMS 3, 138 (1959).
(27) Herzberg, Monfils, JMS 5, 482 (1960).
(28) Lichten, PR 120, 848 (1960); 126, 1020 (1962).
(29) Mulliken, PR 120, 1674 (1960).
(30) Davidson, JCP 35, 1189 (1961).
(31) Tobias, Vanderslice, JCP 35, 1852 (1961).
(32) Frey, Mizushima, PR 128, 2683 (1962).
(33) Lichten, BAPS 7, 43 (1962).
(34) Porto, Jannuzzi, JMS 11, 379 (1963).
(35) Weissman, Vanderslice, Battino, JCP 39, 2226 (1963).
(36) Chiu, JCP 40, 2276 (1964).
(37) Namioka, JCP 40, 3154 (1964).
(38) Cook, Metzger, JOSA 54, 968 (1964).
(39) Mulliken, PR A 136, 962 (1964).
(40) Namioka, JCP 41, 2141 (1964).
(41) Browne, PR A 138, 9 (1965).
(42) Ginter, Battino, JCP 42, 3222 (1965).
(43) Kolos, Wolniewicz, JCP 43, 2429 (1965).
(44) Namioka, JCP 43, 1636 (1965).
(45) Wright, Davidson, JCP 43, 840 (1965).
(46) Monfils, JMS 15, 265 (1965).
(47) Gloersen, Dieke, JMS 16, 191 (1965).
(48) Fink, Wiggins, Rank, JMS 18, 384 (1965).
(49) Samson, Cairns, JOSA 55, 1035 (1965).
(50) Dunn, JCP 44, 2592 (1966).
(51) Halmann, Laulicht, JCP 44, 2398 (1966); 46, 2684 (1967).
(52) Rothenberg, Davidson, JCP 44, 730 (1966).
(53) Ginter, JCP 45, 248 (1966).
(54) Hunter, JCP 45, 3022 (1966).
(55) Takezawa, Innes, Tanaka, JCP 45, 2000 (1966).
(56) Foltz, Rank, Wiggins, JMS 21, 203 (1966).
(57) Geiger, Topschowsky, ZN 21 a, 626 (1966).
(58) Ginter, JCP 46, 3687 (1967).

$^{1}H_2$ (continued):

(59) Waech, Bernstein, JCP 46, 4905 (1967).
(60) Herzberg, SL 16, 14 (1967).
(61) Kolos, IJQC 1, 169 (1967).
(62) Chupka, Berkowitz, JCP 48, 5726 (1968); 51, 4244 (1969).
(63) Comes, Wellern, ZN 23 a, 881 (1968).
(64) Dalgarno, Allison, ApJ 154, L95 (1968).
(65) Haddad, Lokan, Farmer, Carver, JQSRT 8, 1193 (1968).
(66) Hesser, JCP 48, 2518 (1968).
(67) Kolos, Wolniewicz, JCP 48, 3672 (1968).
(68) Villarejo, JCP 48, 4014 (1968).
(69) Hesser, Brooks, Lawrence, JCP 49, 5388 (1968).
(70) Kolos, Wolniewicz, JCP 49, 404 (1968).
(71) Villarejo, JCP 49, 2523 (1968).
(72) Monfils, BCSARB (5) 54, 44 (1968).
(73) Monfils, JMS 25, 513 (1968).
(74) Brannon, Church, Peters, JMS 27, 44 (1968).
(75) Nicholls, JP B 1, 1192 (1968).
(76) Turner, PRS A 307, 15 (1968).
(77) Wilkinson, CJP 46, 1225 (1968).
(78) Allison, CPL 3, 371 (1969).
(79) Jeziorski, Kolos, CPL 3, 677 (1969).
(80) Allison, Dalgarno, AD 1, 91 (1969).
(81) Cahill, JOSA 59, 875 (1969).
(82) Comes, Wenning, ZN 24 a, 587 (1969).
(83) Geiger, Schmoranzer, JMS 32, 39 (1969).
(84) James, JMS 32, 512 (1969).
(85) Herzberg, PRL 23, 1081 (1969).
(86) Kolos, Wolniewicz, JCP 50, 3228 (1969).
(87) Villarejo, Stockbauer, Inghram, JCP 50, 1754 (1969).
(88) Wolniewicz, JCP 51, 5002 (1969).
(89) Spindler, JQSRT 9, 597, 627, 1041 (1969).
(90) Allison, Dalgarno, AD 1, 289 (1970).
(91) Allison, Dalgarno, MP 19, 567 (1970).
(92) Comes, Wenning, ZN 25 a, 406 (1970).
(93) Dalgarno, Stephens, ApJ 160, L107 (1970).
(94) Dalgarno, Herzberg, Stephens, ApJ 162, L49 (1970).
(95) Herzberg, JMS 33, 147 (1970).
(96) Hodgson, PRL 25, 494 (1970).
(97) Mentall, Gentieu, JCP 52, 5641 (1970).
(98) Takezawa, JCP 52, 2575, 5793 (1970).
(99) Waynant, Shipman, Elton, Ali, APL 17, 383 (1970).
(100) Zhirnov, Vasilevskii, OS(Engl. Transl.) 29, 352 (1970).
(101) Stwalley, CPL 6, 241 (1970).
(102) Becker, Fink, ZN 26 a, 319 (1971).
(103) Comes, Schumpe, ZN 26 a, 538 (1971).
(104) Buijs, Gush, CJP 49, 2366 (1971).
(105) Fiquet-Fayard, Gallais, MP 20, 527 (1971).
(106) Herzberg, in "Topics in Modern Physics" (Condon Vol.), p. 191. Colorado Associated University Press (1971).
(107) Sharp, AD 2, 119 (1971).
(108) Julienne, CPL 8, 27 (1971).
(109) Crosswhite, "The Hydrogen Molecule Wavelength Tables of Gerhard Heinrich Dieke". Wiley-Interscience (1972).
(110) Dalgarno, Wright, ApJ 174, L49 (1972).
(111) Smith, Chevalier, ApJ 177, 835 (1972).
(112) Fiquet-Fayard, Gallais, CPL 16, 18 (1972).
(113) Freund, Miller, JCP 56, 2211 (1972).
(114) Herzberg, Jungen, JMS 41, 425 (1972).
(115) Bunker, JMS 42, 478 (1972).
(116) Hodgson, Dreyfus, PRL 28, 536 (1972); PR A 9, 2635 (1974).
(117) Johnson, PR A 5, 1026 (1972).

(continued p. 255)

State	T_e	ω_e	$\omega_e x_e$	B_e	α_e	D_e $(10^{-2}cm^{-1})$	r_e (Å)	Observed Transitions Design.	ν_{00}	References
$^1H^2H$		μ = 0.67171137		D_0^0 = 4.5138_3 eV[a]		I.P. = 15.4446_6 eV[b]				NOV 1976 A
k $^3\Pi_u$ 4pπ	(118384.2)	2030.56	50.36[c]	22.548	0.951	0.92	1.0550	k→a, R	22295.24	(7)
d $^3\Pi_u$ 3pπ	$(112717._4)$	2054.59[d]	49.74[e]	22.810[df]	1.020	[1.16]	1.0489	d→a, R Fulcher (α)	16640.6 b.	(2)
e $^3\Sigma_u^+$ 3pσ	$(107776._6)$	1905.17	51.70[g]	20.766	1.010	[0.89]	1.0993	e→a, R	11624.6	(1)
a $^3\Sigma_g^+$ 2sσ	(95947.1)[h]	2308.44	53.77[i]	25.685	1.099	[1.28]	0.9885	(a−X)	(95201.5)[h]	
c $^3\Pi_u$ 2pπ		No constants of this state have yet been determined.[j]								
b $^3\Sigma_u^+$ 2pσ		Repulsive, lower state of hydrogen continuum.						a→b		

$^1H^2H$: [a]36406.2 cm^{-1}, from (29). From <u>ab initio</u> calculations (21) obtain 36405.5 cm^{-1}, including a very small non-adiabatic correction by (51).

[b]From the Rydberg series of (36) and corrected for pressure shift, see (34).

[c]$\omega_e y_e$ = + 0.6968.

[d]Refers to $^3\Pi^-$; $^3\Pi^+$ is strongly perturbed.

[e]$\omega_e y_e$ = + 0.58.

[f]The Λ-type doubling is large and irregular (1). Breaking-off of P and R branches for v' > 3 on account of predissociation (2).

[g]$\omega_e y_e$ = + 0.522, $\omega_e z_e$ = + 0.091.

[h]The energy of none of the triplet states above X $^1\Sigma_g^+$(v=0, J=0) has yet been experimentally established. The T_e value in the table is the average of those of H_2 and D_2; the electronic isotope shift is fairly large. T_0 is calculated from this T_e value taking account of Y'_{00} = 3.8_5. The theoretical T_e = 95950 cm^{-1} is based on the observed dissociation limit and D_e from (17).

[i]$\omega_e y_e$ = + 0.60.

[j]See footnote [y] on p. 243 ($^1H_2$).

$^1H_2$ (continued):

(118) Marechal, Jost, Lombardi, PR A 5, 732 (1972).
(119) Van der Linde, Dalby, CJP 50, 287 (1972).
(120) Stephens, Dalgarno, JQSRT 12, 569 (1972).
(121) Waynant, PRL 28, 533 (1972).
(122) Welsh, in MTP Review of Science, Phys. Chem. Ser. 1, Vol. 3, p. 33 (1972); PC 30, 84 (1974).
(123) Freund, Miller, Zegarski, CPL 23, 120 (1973).
(124) Julienne, JMS 48, 508 (1973).
(125) Margolis, JMS 48, 409 (1973).
(126) Stwalley, JCP 58, 536 (1973).
(127) Freund, Miller, JCP 58, 2345, 3565; 59, 4093, 5770 (1973).
(128) Schmoranzer, Geiger, JCP 59, 6153 (1973).
(129) Dabrowski, Herzberg, CJP 52, 1110 (1974).
(130) Fabian, Lewis, JQSRT 14, 523 (1974).
(131) Ford, JMS 53, 364 (1974).
(132) Freund, Miller, JCP 60, 4900 (1974).
(133) Miller, Freund, Zegarski, JCP 60, 3195 (1974).
(134) Miller, Freund, JCP 61, 2160 (1974); 63, 256 (1975).
(135) Jette, JCP 61, 816 (1974).
(136) Melieres-Marechal, Lombardi, JCP 61, 2600 (1974).
(137) Lin, JCP 60, 4660 (1974).
(138) Jette, Miller, CPL 29, 547 (1974).
(139) Johnson, PR A 9, 576 (1974).
(140) Jost, Lombardi, PRL 33, 53 (1974).
(141) Kolos, Wolniewicz, CPL 24, 457 (1974).
(142) Lewis, JQSRT 14, 537 (1974).
(143) McKellar, Icarus 22, 212 (1974).
(144) Orlikowski, Wolniewicz, CPL 24, 461 (1974).
(145) English, Albritton, JP B 8, 2123 (1975).
(146) Ford, JMS 56, 251 (1975).
(147) Ford, Docken, Dalgarno, ApJ 195, 819; 200, 788 (1975).
(148) Harney, Randolph, Milanovich, ApJ 200, L179 (1975).
(149) King, Read, Imhof, JP B 8, 665 (1975).
(150) Kolos, CPL 31, 43 (1975).
(151) Wolniewicz, CPL 31, 248 (1975).
(152) Kolos, Wolniewicz, CJP 53, 2189 (1975).
(153) Kolos, Wolniewicz, JMS 54, 303 (1975).
(154) Miller, Freund, JCP 62, 2240 (1975).
(155) Chupka, Dehmer, Jivery, JCP 63, 3929 (1975).
(156) Ford, Browne, Shipsey, DeVries, JCP 63, 362 (1975).
(157) Schmoranzer, JP B 8, 1139 (1975).
(158) Dabrowski, Herzberg, CJP 54, 525 (1976).
(159) Bishop, Shih, JCP 64, 162 (1976).
(160) Kagann, English, PR A 13, 1451 (1976).
(161) Bunker, unpublished.
(162) Herzberg, McKenzie, to be published.
(163) Kolos, JMS 62, 429 (1976).
(164) Crowe, McConkey, PRL 31, 192 (1973).
(165) Misakian, Zorn, PR A 6, 2180 (1972).
(166) Black, Dalgarno, ApJ 203, 132 (1976).
(167) Lin, CJP 53, 310 (1975).
(168) Backx, Wight, Van der Wiel, JP B 9, 315 (1976).
(169) Chackerian, Giver, JMS 58, 339 (1975).

State	T_e	ω_e	$\omega_e x_e$	B_e	α_e	D_e ($10^{-2}cm^{-1}$)	r_e (Å)	Observed Transitions Design.	ν_{00}	References
$^1H^2H$ (continued)		Ionization continua joining on to Rydberg series.[a]								
Rydberg series of rotational levels observed in low temperature absorption from $X^1\Sigma_g^+(v=0)$ and converging to										
N=2 (v=1) of HD^+:		J=1 (v=1) levels of npπ $^1\Pi_u^+$ (n = 36...46)[bc]; $\nu = 126606.4$[d] $- R_{HD}/(n+0.082)^2$.						R(0) lines		(36)
N=1 (v=0) of HD^+:		J=1 (v=0) levels of npπ $^1\Pi_u^-$ (n = 6...23, joining on to C, D, D', D"); $\nu = 124613.3$[d] $- R_{HD}/(n+0.082)^2$. Similar series with v' = 1, 2, 3.						Q(1) lines		(36)
N=0 (v=0) of HD^+:		J=1 (v=0) levels of npσ $^1\Sigma_u^+$ (n = 5...48, joining on to B, B', B")[c]; $\nu = 124568.6$[d] $- R_{HD}/(n-0.203)^2$. Similar series with v' = 1.								
$\bar{B}$ $^1\Sigma_u^+$		See $^1H_2$.								(42)(46)
D" $^1\Pi_u$ 5pπ	121231.2[e] 121216.3[e]	2006.17[f]	45.801[f]	[22.865] [22.144] [g]	h	[2.1] [1.3]	[1.0477] [1.0646]	D"← X, R	120332.6[e] 120317.7[e]	(13)(18)
D' $^1\Pi_u$ 4pπ	118879.2[f]	2014.9_1[f]	47.018[i]	22.3_5[jg]	1.25[k]	[2.2][j]	1.060	D'← X, R	117984.7	(13)(18)
B" $^1\Sigma_u^+$ 4pσ	117980.4	1896.60	48.924	20.34[ℓ]	0.398[ℓ]	[2.5]	1.111	B"← X, R	117026.2	(13)(18)
M $^1\Sigma_g^+$	[115073]	v=0 (?) only.		[10.4]			[1.55]	M→B, R	22782.5	(4)
D $^1\Pi_u$ 3pπ	113901.7[m]	2039.13[f]	48.91_7[n]	22.9_1[og]	0.97[p]	(1.2)[q]	1.047	D← X,[r] R	113018.8_4[s]	(13)(18)
J $^1\Delta_g$ 3dδ	(113536)	[1832.8][t]						J→B, V	22162.3[t]	(4)
I $^1\Pi_g$ 3dπ	(113110)	1962.14[o]	58.21[u]	22.36[o]	1.21	[0.7]	1.059	I→B, V	21786.2[v]	(4)
G $^1\Sigma_g^+$ 3dσ	(112843)	[1879.9], vibrational perturbations for v > 0.[u]						G→B, R	21492.4	(4)
K ($^1\Sigma_g^+$)	(112663)	[(1981)], only fragmentary data.						K→B, R	(21363.2)[w]	(4)
B' $^1\Sigma_u^+$ 3pσ	$111649._7$[x]	$1775._2$	67.6_6[yu]	20.0_0[zg]	1.2_8[z]	[0.16][z]	1.120	B'← X,[r] R	110632.5_8	(13)(18)(48)

$^1H^2H$: [a]Cross sections for photoionization into the various vibrational levels of HD^+ and the adjoining continuum (dissociative photoionization) calculated by (39)(44), observed by (37). Photoionization near I.P. studied by (12).

[b]Except for n=2...5 (i.e. C...D") only the diffuse (preionized) members above n=35 have been observed. The corresponding v'=0 series has not been found for $n > 5$.

[c]There are strong perturbations between $np\sigma\ ^1\Sigma_u^+$ and $np\pi\ ^1\Pi_u^+$ similar to those in H_2, but in HD they have not yet been studied in detail.

[d]The Rydberg limits are from (36) but corrected for pressure shift; see (34). The quantum defects, given only for the Q(1) series by (36), are taken from the corresponding series in H_2 (34).

[e]Large J=0 splitting (18), Π^+ above Π^-.

[f]Average of Π^+ and Π^- which differ for HD much more than for H_2 and D_2.

[g]RKR potential functions (19).

[h]$B_1(\Pi^+) = 22.618$, $B_1(\Pi^-) = 22.310$.

[i]$\omega_e y_e = + 0.126_6$.

[j]The rotational constants refer to Π^-; Π^+ is perturbed (see [c]); $B_0(\Pi^+) = 22.289$, $B_1(\Pi^+) = 21.901$.

[k]$\gamma_e = - 0.05$.

[ℓ]The rotational constants represent B_0 and B_1 only; strongly non-linear B_v curve.

[m]Y'_{00} not included. (18) gives 113900.7_5.

[n]$\omega_e y_e = + 0.2171$.

[o]Refers to Π^-.

[p]$\gamma_e = - 0.028$.

[q]The D_v values show considerable scatter. The H_v values (13) are hardly significant.

[r]Franck-Condon factors from electron energy loss spectra in (26).

[s]Average of Π^+ and Π^-, extrapolated to J=0. The Λ-type doubling for J=1, v=0 is 3.5_2 cm^{-1} with Π^+ above Π^-.

[t]Referred to J=2 of $^1\Delta^-$.

[u]There are two dissociation limits with adjoining continua at 118665.9 and 118687.4 cm^{-1} corresponding to H(n=2) + D(n=1) and H(n=1) + D(n=2), respectively (29). It appears (48) that the first limit corresponds to B $^1\Sigma_u^+$, E,F $^1\Sigma_g^+$, and C $^1\Pi_u$, while the second corresponds to B' $^1\Sigma_u^+$, G $^1\Sigma_g^+$, and I $^1\Pi_u$; see also (32). The C $^1\Pi_u$ state, unlike H_2 or D_2, apparently does not have a potential maximum.

[v]Refers to J=1 of Π^-; J=1 of Π^+ lies $28._0$ cm^{-1} higher.

[w]Refers to the J=1 level.

[x]Takes account of Y_{00} in the upper as well as in the lower state. For the states B, C, and B' $Y'_{00} = 7._1$, $3._7$, and $2._0$, respectively.

[y]$\omega_e y_e = + 3.6_6$, $\omega_e z_e = - 0.6_5$; five-level fit. All levels up to the last (v=11) have been observed.

[z]Five-level fit. The deviations for $v > 2$ are large and irregular because of numerous perturbations. The B_v values of (13) deviate by up to 1 cm^{-1} from those of (48) used here. The latter are effective, non-deperturbed values. Higher D_v values show considerable irregularities because of local perturbations.

State	T_e	ω_e	$\omega_e x_e$	B_e	α_e	D_e $(10^{-2}cm^{-1})$	r_e (Å)	Observed Transitions: Design.	Observed Transitions: ν_{00}	References
$^1H^2H$ (continued)										
F[a] $^1\Sigma_g^+$ $2p\sigma^2$	100927.$_5$[b]	(1087.9)[c]	(21.6)[cd]	B_1 = 4.50			r_1 = 2.36	E→B,	V 8901.72	(3)(48)
E[a] $^1\Sigma_g^+$ $2s\sigma$	100120.$_4$[b]	2204.4[e]	81.6[e]	24.568[e]	1.288[e]	[1.23]	1.0107	E,F← X,[f]	R 99301.5$_9$[g]	(48)*
C $^1\Pi_u$ $2p\pi$	100092.9[h]	2119.6$_5$	53.31[i]	23.522[j]	1.096[k]	1.4$_9$[ℓ]	1.0329	C↔X,[my] Werner b.	R 99252.8$_6$	(13)(18)(48)*
B $^1\Sigma_u^+$ $2p\sigma$	91698.3[h]	1177.16	15.59[n]	15.071[j]	0.820[o]	0.882[p]	1.2904	B↔X,[qy] Lyman b.	R 90399.8$_6$	(23)* (48)*
X $^1\Sigma_g^+$ $1s\sigma^2$	0	3813.1$_5$	91.65[r]	45.655[s]	1.986[t]	2.60$_5$[u]	0.74142	Rotation-vibration sp.[v]		(10)(49)
								Pure rotation sp.[w]		(22)
								Raman sp.		(9)
								Field- and		(20)
								collision-induced sp.		(45)
								Rf magn. reson. sp.[x]		(5)(6)

$^1H^2H$: [a]The states E and F may be considered as forming one double-minimum state. The potential maximum is at 104480 cm^{-1} above X(v=0,J=0) (27). See also [i] on p. 248 ($^1H_2$).

[b]Derived by extrapolation of differences between observed vibrational levels (48) and those calculated from the double-minimum potential function of (27).

[c]From the theoretical energy levels assuming an independent (outer) potential minimum (48). The lowest observed level is v=1 and the observed intervals are $\Delta G(3/2, 5/2, 7/2)$ = 1002.6, 956.0, 916.7, resp.. Higher levels show the effects of interaction with E $^1\Sigma_g^+$. See [a].

[d]See [u] p. 257 .

[e]These constants (3) are from the lowest vibrational levels ($v \leq 2$ of the inner minimum) neglecting the interaction with the F state; see [a].

[f]This transition, forbidden in H_2 and D_2, is weakly allowed in HD since the g,u symmetry is no longer rigorous.

[g]The 0-0 band has not been observed in VUV absorption but is obtained by adding $\nu_{00}(E-B)$ (3) to $\nu_{00}(B-X)$. The first observed VUV absorption band is at 100618.50 cm^{-1} and corresponds to the transition to the second lowest level in the outer minimum (1-0).

$^1H^2H$ (continued):

[h] See [x] p. 257. Note, that the T_e value for C $^1\Pi_u$ and $\nu_{00}(C-X)$ exclude the term $-B\Lambda^2$ of the rotational energy expression.

[i] $\omega_e y_e = +0.656$, $\omega_e z_e = -0.03_3$; eight-level fit. Only Π^- levels have been included in the fit since many of the Π^+ levels are strongly perturbed by B $^1\Sigma_u^+$. After deperturbation the Π^+ levels agree fairly well with corresponding Π^- levels. Levels up to v=15 have been observed; this level is within 42 cm^{-1} of the lower of the two dissociation limits, see [u] p. 257.

[j] RKR potential functions (19).

[k] $+0.037(v+\frac{1}{2})^2 - 0.005_3(v+\frac{1}{2})^3$, eight-level fit of B_v values of Π^-.

[ℓ] $\beta_e = -0.0007_7$.

[m] Selective enhancement of v'=0 of C $^1\Pi$ in Ar-H_2 mixtures studied by (15). Franck-Condon factors from electron energy loss spectra (26).

[n] $+0.42_7(v+\frac{1}{2})^3 - 0.02_9(v+\frac{1}{2})^4 + 0.0008(v+\frac{1}{2})^5$, fit of first eight levels; all levels up to v=43 have been observed (48).

[o] $+0.116(v+\frac{1}{2})^2 - 0.021_6(v+\frac{1}{2})^3 + 0.0024(v+\frac{1}{2})^4 - 0.00011(v+\frac{1}{2})^5$; eight-level fit, see [n].

[p] $\beta_e = -0.00050_5$.

[q] Selective enhancement of v'=3 and 5 of B $^1\Sigma_u^+$ in Ar-H_2 mixtures studied by (15). Franck-Condon factors from electron energy loss spectra (26), from fluorescence spectra (25); large vibration-rotation interaction effects (25)(28)(30).

[r] $+0.723(v+\frac{1}{2})^3 - 0.013_3(v+\frac{1}{2})^4 + 0.0016_5(v+\frac{1}{2})^5$; ten-level fit. The zero-point energy ($Y_{00} = 6.5_1$ included) is 1890.2_6. All levels up to the last, v=17, have been observed. This level lies 5.1 cm^{-1} below the dissociation limit.

[s] Theoretical values for all bound and quasibound levels in the ground state of HD are given by (31).

[t] $+0.031_5(v+\frac{1}{2})^2 - 0.0022_1(v+\frac{1}{2})^3$; ten-level fit (48). Somewhat more accurate B_v values than used by (48) for v=0...6 have been derived from the rotation-vibration spectrum by (49).

[u] $\beta_e = -0.00054$; $H_v = 2.2 \times 10^{-5}$ has been assumed.

[v] The transition moments for the 1-0, 2-0, 3-0, 4-0 and 5-0 vibration bands are observed to be 5.0, 1.9, 0.80, 0.42 and 0.21×10^{-5} D, respectively (40)(41)(43)(49); for theoretical discussions see (38)(47)(50). In addition to the electric dipole infrared spectrum one line of the quadrupole component of the fundamental, S(0), has been observed (43).

[w] From the rotation spectrum (22) have obtained a dipole moment in the lowest vibrational level of 5.8×10^{-4} D, or, after a small correction for rotation (16), 5.5×10^{-4} D; see also (38). Predicted IR emissivities in the pure rotation lines (33).

[x] Rotational magnetic moment for J=1 0.662 μ_N (6)(8).

[y] Theoretical band oscillator strengths, transition probabilities and photodissociation cross sections in (24).

References on page 261.

State	T_e	ω_e	$\omega_e x_e$	B_e	α_e	D_e ($10^{-2}cm^{-1}$)	r_e (Å)	Observed Transitions Design.	Observed Transitions ν_{00}	References
$^1H^3H$		μ = 0.75540394		D_0^0 = (4.5269$_4$) eV[a]		I.P. = (15.4514$_6$) eV[b]				NOV 1976
k $^3\Pi_u$ 4pπ	(118384.8)	1915.05	44.67[c]	20.106	0.8638[d]	0.86	1.0535	k→a, R	22303.7$_8$	(2)
d $^3\Pi_u$ 3pπ	(112717.9)	1936.93	43.439[e]	20.219[f]	0.823[f]	0.812[g]	1.0506	d→a, R	16648.1$_0$	(1)
e $^3\Sigma_u^+$ 3pσ	(107772.7)	1796.42	45.69[h]	18.3167	0.819[i]	0.744[j]	1.1037$_9$	e→a, R	11631.98	(3)
a $^3\Sigma_g^+$ 2sσ	(95950.$_8$)[k]	2177.01	47.84[ℓ]	22.819	0.9182[m]	0.97[n]	0.9889$_2$	(a − X)	(95243.5)[o]	
C $^1\Pi_u$ 2pπ	(100094.$_7$)[k]							(C − X)	(99301.$_5$)[p]	
B $^1\Sigma_u^+$ 2pσ	(91698.$_1$)[k]							(B − X)	(90472.$_2$)[p]	
X $^1\Sigma_g^+$ 1sσ^2	0	3597.0$_5$[q]	81.67$_8$[q]	40.595	1.6640[r]		(0.74142)			

$^1H^3H$: [a]36511.9 cm^{-1}, from ab initio potential function (7); non-adiabatic corrections which are certainly less than +0.35 cm^{-1} and Lamb shift corrections ($\approx$ - 0.2 cm^{-1}) are not included. No observed value is available yet.

[b]From theoretical D_0^0(HT) and D_0^0(HT^+) values and I.P.(H).

[c]$\omega_e y_e$ = + 0.527.

[d]γ_e = + 0.0123.

[e]$\omega_e y_e$ = + 0.459, $\omega_e z_e$ = - 0.036.

[f]γ_e = + 0.0080; the rotational constants refer to $^3\Pi^-$ since $^3\Pi^+$ is perturbed; the Λ-type doubling is somewhat irregular and fairly large.

[g]β_e = - 0.00008.

[h]$\omega_e y_e$ = + 0.34, $\omega_e z_e$ = - 0.060.

[i]γ_e = - 0.0039.

[j]β_e = - 0.00041.

[k]From the T_e values of H_2 and D_2 assuming that the electronic isotope shift is proportional to (1 - μ_{H_2}/μ_{HT}).

[ℓ]$\omega_e y_e$ = + 0.502, $\omega_e z_e$ = - 0.015.

[m]γ_e = + 0.0123.

[n]β_e = - 0.00038.

[o]From T_e assuming Y'_{00} = 0, but taking account of Y''_{00} (see [q]).

[p]From T_e and the zero-point energy calculated by (6).

[q]All constants calculated by (5) from the potential function of (4) and based on v=0,1,2,3 only. Experimental values are not available. $\omega_e y_e$ = + 0.575; Y_{00} = 5.7.

[r]$+ 0.0238(v+\frac{1}{2})^2 - 0.0015(v+\frac{1}{2})^3$. See also [q].

$^{1}H^{3}H$ (continued):

(1) Dieke, Tomkins, PR 76, 283 (1949).
(2) See ref. (15a) of $^{1}H_2$.
(3) Dieke, Tomkins, PR 82, 796 (1951)
(4) See ref. (43) of $^{1}H_2$.
(5) Cashion, JCP 45, 1037 (1966).
(6) See ref. (67) of $^{1}H_2$.
(7) See ref. (70) of $^{1}H_2$.

$^{1}H^{2}H$ (continued):

(1) See ref. (6)(7) of $^{1}H_2$.
(2) See ref. (5) of $^{1}H_2$.
(3) See ref. (8) of $^{1}H_2$.
(4) See ref. (10) of $^{1}H_2$.
(5) Kellogg, Rabi, Ramsey, Zacharias, PR 57, 677 (1940).
(6) Ramsey, PR 58, 226 (1940).
(7) See ref. (15a) of $^{1}H_2$.
(8) Ramsey, "Molecular Beams", p. 239. Clarendon Press, Oxford (1956).
(9) See ref. (23) of $^{1}H_2$.
(10) Durie, Herzberg, CJP 38, 806 (1960).
(11) Blinder, JCP 35, 974 (1961).
(12) Dibeler, Reese, Krauss, JCP 42, 2045 (1965).
(13) See ref. (46) of $^{1}H_2$.
(14) Kolos, Wolniewicz, JCP 45, 944 (1966).
(15) Takezawa, Innes, Tanaka, JCP 46, 4555 (1967).
(16) Karl, CJP 46, 1973 (1968).
(17) See ref. (67) of $^{1}H_2$.
(18) See ref. (73) of $^{1}H_2$.
(19) See ref. (72) of $^{1}H_2$.
(20) See ref. (74) of $^{1}H_2$.
(21) See ref. (70) of $^{1}H_2$.
(22) Trefler, Gush, PRL 20, 703 (1968); CJP 47, 2115 (1969).
(23) See ref. (77) of $^{1}H_2$.
(24) See ref. (80)(90)(91) of $^{1}H_2$.
(25) Fink, Akins, Moore, CPL 4, 283 (1969).
(26) See ref. (83) of $^{1}H_2$.
(27) See ref. (86) of $^{1}H_2$.
(28) Allison, JCP 52, 4909 (1970).
(29) See ref. (95) of $^{1}H_2$.
(30) See ref. (102) of $^{1}H_2$.
(31) Le Roy, JCP 54, 5433 (1971).
(32) Thorson, JMS 37, 199 (1971).
(33) See ref. (110) of $^{1}H_2$.
(34) See ref. (114) of $^{1}H_2$.
(35) See ref. (117) of $^{1}H_2$.
(36) Takezawa, Tanaka, JCP 56, 6125 (1972).
(37) Berkowitz, Spohr, JESRP 2, 143 (1973).
(38) Bunker, JMS 46, 119 (1973).
(39) Itikawa, JESRP 2, 125 (1973).
(40) McKellar, CJP 51, 389 (1973).
(41) Bejar, Gush, CJP 52, 1669 (1974).
(42) See ref. (129) of $^{1}H_2$.
(43) McKellar, CJP 52, 1144 (1974).
(44) See ref. (147) of $^{1}H_2$.
(45) Prasad, Reddy, JCP 62, 3582 (1975); 65, 83 (1976).
(46) See ref. (155) of $^{1}H_2$.
(47) Bunker, JMS 61, 319 (1976).
(48) See ref. (158) of $^{1}H_2$.
(49) McKellar, Goetz, Ramsay, ApJ 207, 663 (1976).
(50) Wolniewicz, CJP 53, 1207 (1975); 54, 672 (1976).
(51) Bunker, unpublished.

State	T_e	ω_e	$\omega_e x_e$	B_e	α_e	D_e $(10^{-2}cm^{-1})$	r_e (Å)	Observed Transitions Design.	Observed Transitions ν_{00}	References
$^2H_2$		μ = 1.00705111		D_0^0 = 4.55632 eV[a]		I.P. = 15.46660 eV[b]				NOV 1976 A
u $^3\Pi_u$ 6pπ	122365.6	1649.03	35.13[c]	15.03_6	0.587[d]	0.53	1.055_1	u→a,	26286.75	(6)
w ($^3\Pi_g$)5dπ		Fragment						w→c,		(6)
q ($^3\Sigma_g^+$)5dσ		Fragment						q→c,		(6)
n $^3\Pi_u$ 5pπ	120976.9	1652.73	34.25[e]	15.04_0[f]	0.560[g]	0.53	1.055_0	n→a,	24900.14	(6)
r $^3\Pi_g$ 4dπ	[(119380)]			hi				r→c, R	(22650)[h]	(6)
p $^3\Sigma_g^+$ 4dσ	[119242]			h				p→c, R	$22509._9$[h]	(6)
k $^3\Pi_u$ 4pπ	118396.7	1658.85	33.88[j]	15.07_5[k]	0.566[ℓ]	0.46	1.053_8	k→a,	22323.06	(6)
f $^3\Sigma_u^+$ 4pσ	116640	1618	32.8	14.66	0.62		1.069	f→a, R	20546.0	(6)
j $^3\Delta_g$ 3dδ	[$114194._1$]			h				j→c, V	17462.3[h]	(6)
i $^3\Pi_g$ 3dπ	(113093)	[1541.9]		hm				i→e, R	5320.0[n]	(15)*
								i→c, R	17131.9[n]	(6)
g $^3\Sigma_g^+$ 3dσ	(112856)	[1511.3]		h				g→e, R	5067.8	(15)*
								g→c, R	16879.8	(6)
d $^3\Pi_u$ 3pπ	112729.8[o]	1678.22[p]	32.94[q]	15.200[p]	0.5520	[0.49]	1.0494	d→a, R Fulcher b.	16666.0	(1)
e $^3\Sigma_u^+$ 3pσ	107774.0	1556.64	34.51[r]	13.856	0.451	[0.4]	1.0991	e→a, R	11649.1	(2)
a $^3\Sigma_g^+$ 2sσ	95958.0_8[s]	1885.84	35.96[s]	17.109	0.606	[0.55]	0.9891	a[t]→b,		
								(a−X)	95348.1_8[u]	(1)
c $^3\Pi_u$ 2pπ	[$96731._8$]			[15.305][v]		[0.51_4][v]	[1.0458]	(c[w]−X)	$95185._3$[x]	(6)
b $^3\Sigma_u^+$ 2pσ		Lower state of continuous spectrum of D_2 (a→b).								

$^{2}H_2$: [a]36748.9 cm^{-1}, from the dissociation limit (beginning of continuum) in the B' − X system (34). The same value has been derived by (53) from the last observed levels in the ground state by relations involving the long-range behaviour of the potential function. 36748.2 cm^{-1} from <u>ab initio</u> calculations (52).

[b]From the Rydberg limits of (54) after correction for pressure shift (41).

[c]$\omega_e y_e$ = + 0.627, uncertain.

[d]γ_e = + 0.008, uncertain.

[e]$\omega_e y_e$ = + 0.627, uncertain.

[f]Λ-type doubling constant q(v=0) = 0.25 cm^{-1} (6).

[g]γ_e = + 0.035.

[h]Strongly affected by ℓ-uncoupling, no constants given by (6); ν_{00} roughly evaluated from their wave number data. See also [e] and [k] p. 241 ($^{1}H_2$).

[i]Anti-crossings of r $^{3}\Pi_g$(v=0,N=2) with G $^{1}\Sigma_g^+$(v=4,N=2) yielding orbital g factors and hyperfine structure (57).

[j]$\omega_e y_e$ = + 0.508, $\omega_e z_e$ = + 0.0345.

[k]Λ-type doubling constant q(v=0) = 0.29 cm^{-1} (6).

[ℓ]γ_e = + 0.008, uncertain.

[m]Anti-crossings between i $^{3}\Pi_g$(v=1,N=1) and I $^{1}\Pi_g$(v=1,N=1) observed by (56).

[n]Refers to Π^-(v=0,N=1); Π^+(v=0, N=1) is at 5348.9 cm^{-1} above e $^{3}\Sigma_u^+$(v=0,N=0). The rotational levels are very irregular.

[o]Microwave optical magnetic resonance induced by electrons (45) gives the following triplet splittings for v=0, N=1 of para-D_2: $\Delta\nu_{10}$ = 0.04301, $\Delta\nu_{02}$ = 0.00656, and of ortho-D_2: $\Delta\nu_{02}$ = 0.08286, $\Delta\nu_{21}$ = 0.00402 cm^{-1}; similar splittings for v=1...5. (47) derive A_e = − 0.02809 cm^{-1}.

[p]Refers to the $^{3}\Pi^-$ component; $^{3}\Pi^+$ is strongly perturbed. The Λ-type doubling is large and irregular (3); for v=0, N=1 it is 0.13 cm^{-1} (47). Breaking-off of P and R branches for v' > 4 on account of predissociation (1); see also (45).

[q]$\omega_e y_e$ = + 0.24.

[r]$\omega_e y_e$ = + 0.287, $\omega_e z_e$ = − 0.04.

[s]$\omega_e y_e$ = + 0.34. T_e takes account of Y_{00} in both upper (Y'_{00} = 2.6$_7$) and lower state.

[t]Lifetime τ(v=0,1) = 12.$_5$ ns (42).

[u]From singlet-triplet anti-crossings (56)(57).

[v]From the assignments of (6) in the g − c, i − c, j − c, ... bands by evaluating combination differences.

[w]Lifetime τ(v=0) = 1.02 ms (42a), refers to the non-predissociating component c $^{3}\Pi_u^-$ and corresponds to radiative (magnetic dipole) transitions to b $^{3}\Sigma_u^+$; see [y] p. 243 ($^{1}H_2$).

[x]From T_0 of a $^{3}\Sigma_g^+$ and the ν_{00} values for the transitions e − a, g − e, and g − c.

State	T_e	ω_e	$\omega_e x_e$	B_e	α_e	D_e ($10^{-2}cm^{-1}$)	r_e (Å)	Observed Transitions Design.	ν_{00}	References
$^2H_2$ (continued)										
		Ionization continua joining on to Rydberg series.[a]								
v'=0 Rydberg series of rotational levels observed in low temperature absorption from $X^1\Sigma_g^+$(v=0) and converging to										
N=2 of D_2^+:		J=1 levels of npπ $^1\Pi_u^+$ (n = 6...9, joining on to C, D, D', D")[b]; $\nu = 124833 - R_{D_2}/(n+0.082)^2$.						R(0) lines (ortho-D_2)		(54)
N=1 of D_2^+:		J=1 levels of npπ $^1\Pi_u^-$ (n = 6...24, joining on to C, D, D', D")[c]; $\nu = 124775.0 - R_{D_2}/(n+0.082)^2$. Similar series with v' = 1, 2.						Q(1) lines (para-D_2)		(54)
N=0 of D_2^+:		J=1 levels of npσ $^1\Sigma_u^+$ (n = 5...25,36...45, joining on to B, B', B")[b]; $\nu = 124745.5_5 - R_{D_2}/(n-0.203)^2$. Similar series with v' = 1, 2.						R(0) lines (ortho-D_2)		(54)
$\overline{B}$ $^1\Sigma_u^+$		See $^1H_2$.								(48)(50)(60)
D" $^1\Pi_u$ 5pπ	121227.5[d]	1648.6_8[d]	33.63_8[de]	15.13_3[fg]	0.652_1[f]	[0.70_4][f]	1.051_7	D"← X, R	120497.0[d]	(14)(22)(54)
D' $^1\Pi_u$ 4pπ	118887.9[d]	1653.1_5[d]	33.35[dh]	15.04_1[ig]	0.550_8[i]	[0.32_3][i]	1.055_0	D'← X, R	118159.7[d]	(14)(22)(54)
B" $^1\Sigma_u^+$ 4pσ	117970.7	1563.0_2	35.41_6[j]	13.68_5[g]	0.384_2[k]	[0.02_4]	1.106_0	B"← X, R	117196.9	(14)(22)(54)
M $^1\Sigma_g^+$	[$114504._5$]	v=0 (?) only, fragmentary.		[4.0]			[2.0_6]	M → B, R	22324.2	(5)
D $^1\Pi_u$ 3pπ	113914.0	1667.60	33.34_3[ℓ]	15.1_1[mg]	0.54[n]	0.5 [n]	1.05_3	D← X,[o] R	113193.0[p]	(14)(22)
I $^1\Pi_g$ 3dπ	113081.5	1600.14	39.42	14.739[qr]	0.526[q]	[0.25][q]	1.0657	I → B, V	21691.4[s]	(5)
G $^1\Sigma_g^+$ 3dσ	(112893)	[1440.8]		(perturbed)[tu]				G → B, R	21433.2	(5)
K ($^1\Sigma_g^+$)	(112610)	[1660]		[6.6]			[1.5_9]	K → B,	(21260)[v]	(5)
B' $^1\Sigma_u^+$ 3pσ	$111642._2$[w]	1451.98	45.679[x]	13.60_5[g]	0.920[y]	[0.41_5][z]	1.109_2	B'← X, R	110815.65	(14)(22)(48)*

$^2H_2$: [a]Cross sections for photoionization into the various vibrational levels of D_2^+ and the adjoining continuum (dissociative photoionization) observed by (24)(43) and calculated by (18)(26)(46)(51).

[b]See the remarks in [b] p.245 concerning the corresponding series of $^1H_2$. Note, however, that an accurate representation using Fano's quantum defect theory has not yet been attempted for D_2.

[c]This series of levels is obtained (54) from a Rydberg series of Q(1) lines whose limit is at 124715.2 cm^{-1}. A similar series of Q(2) lines with a limit at 124654 cm^{-1} converges to the N=2 level of D_2^+; also observed for v=1 and 2. These series, unlike J=1 of npσ, $^1\Sigma_u^+$ and npπ, $^1\Pi_u^+$, are essentially unperturbed.

[d]Average of Π^+ and Π^-.

[e]$\omega_e y_e = + 0.303_4$; v = 0-5.

[f]$\gamma_e = + 0.0132_9$; constants refer to Π^-. For Π^+ (54) give $B_e = 16.19_8$, $\alpha_e = 0.618_8$, $\gamma_e = -0.0413_2$; $D_0 = 0.0078_5$.

[g]RKR potential functions (21).

[h]$\omega_e y_e = + 0.226$; a very small quartic term differs in sign for Π^+ and Π^- (54); v = 0-7.

[i]$\gamma_e = + 0.0050_3$; constants refer to Π^-. Π^+ is perturbed by B" $^1\Sigma_u^+$, particularly for v=3 and 7. After deperturbation, and excluding v=2,4,8, (54) obtain $B_v(\Pi^+, v=0\text{-}10) = 15.33_6 - 0.496_6(v+\frac{1}{2}) - 0.0048_9(v+\frac{1}{2})^2$; $D_0(\Pi^+) = 0.0075_6$.

[j]$\omega_e y_e = + 0.084_3$, $\omega_e z_e = - 0.0136_4$; (22) gives slightly different constants. There are strong perturbations which make vibrational constants somewhat ambiguous.

[k]$\gamma_e = - 0.0173_6$. Strong rotational perturbations in v=4, weaker ones in v=3, 5, and 9 caused by D' $^1\Pi_u^+$ (54).

[l]$+ 0.1698(v+\frac{1}{2})^3 + 0.00296(v+\frac{1}{2})^4 - 0.000307(v+\frac{1}{2})^5$; the vibrational constants refer to the average of Π^+ and Π^- (22).

[m]Strong predissociation for $v \geq 4$, not yet studied in detail but (36) observe line widths of 3.5 cm^{-1} for J=2, v=4...7 of Π^+. (31) observe Ly_α of D in fluorescence as a consequence of predissociation and find a noticeable increase of predissociation when an electric field is applied (field-induced predissociation). Theoretical discussion (40).

[n]$\gamma_e = - 0.00_2$, D_v irregular; the rotational constants refer to Π^-.

[o]Franck-Condon factors from electron energy loss spectra (28).

[p]Average of Π^+ and Π^- extrapolated to J=0. The Λ-type doubling for v=0,J=1 is 2.1_4 cm^{-1} with Π^+ above Π^-.

[q]Effective constants for Π^-, strongly affected by ℓ-uncoupling. See also I $^1\Pi_g$ of $^1H_2$.

[r]See [m] p. 263.

[s]Refers to the J=1 level of $^1\Pi^-$; the J=1 level of $^1\Pi^+$ lies $18._7$ cm^{-1} higher.

[t]Zeeman effect in 0-0 band (9).

[u]See [i] p. 263.

[v]Refers to J'=1.

[w]Takes account of Y_{00} in both upper and lower state. The Y'_{00} values for B, C, B' are $4._2$, $2._2$, $5._1$ cm^{-1}, resp., but Y_{00}(B') is uncertain; see comments regarding $^1H_2$.

[x]$\omega_e y_e = + 2.096$, $\omega_e z_e = - 0.294$, seven-level fit (48). (22) gives rather different constants based on a nine-level fit; his ninth level (v=8) disagrees strongly with that of (48).

[y]$+ 0.102(v+\frac{1}{2})^2 - 0.0134(v+\frac{1}{2})^3$, seven-level fit (48).

[z]$D_1 = 0.0037_1$, higher D_v values are irregular.

State	T_e	ω_e	$\omega_e x_e$	B_e	α_e	D_e $(10^{-2}cm^{-1})$	r_e (Å)	Observed Transitions Design.	ν_{00}	References
$^2H_2$ (continued)										
F $^1\Sigma_g^+$ $2p\sigma^2$	(100931.2)[a]	[859.1][b]		$B_6 = 3.5$[c]			$r_6 = 2.2$	F→B, R	$\nu_{60} = 13912.70$	(16)
E $^1\Sigma_g^+$ $2s\sigma$	100128.1	1784.42	48.10_5	16.369_6	0.6764	[0.54]	1.0112_4	E→B, V	8827.99	(16)
C $^1\Pi_u$ $2p\pi$	100097.2[d]	1729.92	34.917[e]	15.673_1[f]	0.5679[g]	0.532[h]	1.0334_6	C↔X,[i] R Werner b.	99409.18[j]	(44)* (48)*
B $^1\Sigma_u^+$ $2p\sigma$	91697.2[d]	963.08	11.038[k]	10.068_0[f]	0.4198[ℓ]	0.40_3[m]	1.2894_4	B↔X,[no] R Lyman b.	90633.79	(27)* (44)* (48)*
X $^1\Sigma_g^+$ $1s\sigma^2$	0	3115.50	61.82[p]	30.443_6[q]	1.0786[r]	1.141[s]	0.74152	Field- and		(23)
								pressure-induced sp.[t]		(17)(38)(49)
								Raman sp.		(11)
								Rf magn. reson. sp.[u]		(7)(10)(33) (35)

$^2H_2$: [a]From the observed ν_{60} and the energy of v=6 above the (outer) minimum as calculated by (29); see $^1H_2$.

[b]Calculated $\Delta G(\frac{1}{2})$ value of the outer minimum of the double-minimum state (29); see $^1H_2$. According to (29) the lowest level of the outer minimum is 9190.1 cm^{-1} above B $^1\Sigma_u^+(v=0)$, but the v=0...5 levels have not yet been observed. The v=6 level lies just below the potential maximum.

[c]Vibrational numbering of (29). The D_6 value is large and negative. Higher vibrational levels lie above the potential maximum and have larger B_v values (e.g. B_{12} = 5.688) corresponding to the fact that for these levels the vibrational motion covers both minima of the E,F state. A few rotational levels of v=4 have been observed.

[d]See [w] p. 265 . T_e of C $^1\Pi_u$ and ν_{00}(C-X) both exclude $-B\Lambda^2$.

[e]$\omega_e y_e = +0.2612$, $\omega_e z_e = -0.00946$; the zero-point energy ($Y_{00} = 2._2$ included) is 858.4_6 cm^{-1}. The eight-level fit refers to Π^- (48). All vibrational levels up to v=19 have been observed. The last level lies 50 cm^{-1} above the dissociation limit confirming the theoretical prediction (13) of a maximum in the potential function.

[f]RKR potential functions (21).

$^2H_2$ (continued):

[g] $+ 0.00419(v+\frac{1}{2})^2 - 0.00010_1(v+\frac{1}{2})^3$, eight-level fit referring to $^1\Pi_u^-$ (48). Several of the $^1\Pi_u^+$ levels are strongly perturbed by B $^1\Sigma_u^+$.

[h] $- 0.000216(v+\frac{1}{2}) + 0.000011(v+\frac{1}{2})^2$.

[i] Franck-Condon factors from electron energy loss spectra (28). Theoretical band oscillator strengths, transition probabilities and photodissociation cross sections (30).

[j] Excludes $- B\Lambda^2$.

[k] $+ 0.4109(v+\frac{1}{2})^3 - 0.0370(v+\frac{1}{2})^4 + 0.00154(v+\frac{1}{2})^5$, the zero-point energy ($Y_{00} = 4._2$ included) is 483.0_3 cm^{-1}; eight-level fit (48). All vibrational levels up to v=51 have been observed.

[ℓ] $+ 0.0296(v+\frac{1}{2})^2 - 0.0015(v+\frac{1}{2})^3$; eight-level fit (48).

[m] $- 0.000320(v+\frac{1}{2}) + 0.000013(v+\frac{1}{2})^2$.

[n] Selective enhancements of v'=7 and 9 in Ar-D_2 mixtures studied by (20). Experimental Franck-Condon factors (28), calculated (18a). Theoretical band oscillator strengths, transition probabilities, and photodissociation cross sections (30).

[o] Continuous component of B-X (corresponding to the continuum of X $^1\Sigma_g^+$) observed by (32).

[p] $\omega_e y_e = + 0.562$, $\omega_e z_e = - 0.0228_6$; the zero-point energy ($Y_{00} = 4.1_3$ included) is 1546.4_9 cm^{-1}. Data from the Raman measurements of (11) and the field-induced spectrum of (23) have been combined with the somewhat less accurate VUV results in the least-squares solution (10-level fit) for the vibrational and rotational constants (44). All vibrational levels have been observed, the last one, v=21, being only 2.1 cm^{-1} below the dissociation limit (44). Theoretical values for all bound and quasi-bound levels are given by (37); see also (52). For a discussion of the small differences obs.-calc. see (39)(44)(55).

[q] According to (8) the hyperfine levels F=1 and 2 for v=0,J=1 (para-D_2) are 0.6609×10^{-5} and 0.4669×10^{-5}cm^{-1} below the F=0 component.

[r] $+ 0.01265(v+\frac{1}{2})^2 - 0.00069(v+\frac{1}{2})^3$; see [p]. As for $^1H_2$ the B_v curve has a slightly negative curvature at low v.

[s] $- 0.00022_4(v+\frac{1}{2}) - + \ldots$, from the data of (11)(23)(44).

[t] 1-0 and 2-0 bands.

[u] Nuclear spectrum (7); the rotational spectrum gives the rotational magnetic moment for J=1: 0.44288 μ_N (10). (35) determine spin-rotation and quadrupole interaction constants for J=1,2 and derive the quadrupole moment of D. Polarizability anisotropy $\alpha_\parallel - \alpha_\perp = 0.289_7$ Å^3 (33).

(1) See ref. (5) of $^1H_2$.
(2) See ref. (6) of $^1H_2$.
(3) See ref. (7) of $^1H_2$.
(4) Jeppesen, PR 49, 797 (1936).
(5) See ref. (10) of $^1H_2$.
(6) See ref. (15a) of $^1H_2$.
(7) See ref. (17) of $^1H_2$.
(8) See ref. (16) of $^1H_2$.
(9) Dieke, JPR 15, 393 (1954).
(10) Ramsey, "Molecular Beams", p. 238. Clarendon Press, Oxford (1956).
(11) See ref. (23) of $^1H_2$.
(12) See ref. (12) of $^1H^2H$.
(13) See ref. (43) of $^1H_2$.
(14) See ref. (46) of $^1H_2$.

(continued p. 269)

State	T_e	ω_e	$\omega_e x_e$	B_e	α_e	D_e (10^{-2} cm^{-1})	r_e (Å)	Observed Transitions		References
								Design.	ν_{00}	
$^2H^3H$		μ = 1.20764393		$D_0^0 = (4.5727_1)$ eV[a]		I.P. = (15.4749_6) eV[b]				NOV 1976
a $^3\Sigma_g^+$ 2sσ	$(95961._8)$[c]							(a – X)	$(95404._6)$[d]	
C $^1\Pi_u$ 2pπ	$(100098._4)$[c]							(C – X)	$(99470._0)$[d]	
B $^1\Sigma_u^+$ 2pσ	$(91696._7)$[c]							(B – X)	$(90724._9)$[d]	
X $^1\Sigma_g^+$ 1sσ^2	0	2845.5_2[e]	51.38_6[e]	25.395[e]	0.8221[e]	0.809[f]	(0.74142)			(1)(3)

$^2H^3H$: [a]36881.1 cm^{-1}, calculated from ab initio potential function (5); non-adiabatic corrections which are certainly less than + 0.2 cm^{-1} and Lamb shift corrections ($\approx$ - 0.2 cm^{-1}) are not included. No observed value is available yet.

[b]From the theoretical D_0^0 (DT) and D_0^0 (DT^+) values and I.P.(D).

[c]From the T_e values of H_2 and D_2 assuming that the electronic isotope shift is proportional to $(1 - \mu_{H_2}/\mu_{DT})$.

[d]From T_e and the zero-point energy calculated by (4).

[e]Calculated by (3) from the potential function of (2) and based on v=0...3 only; $\omega_e y_e = +\ 0.336$, $\gamma_e = +\ 0.0087$, $Y_{00} = 3.4$. Slightly different numbers were obtained by (1) from the constants of H_2 by using isotope relations.

[f]Calculated by (1) from the constants of H_2 using isotope relations; $\beta_e = -\ 0.000114$.

(1) Jones, JCP 17, 1062 (1949).

(2) See ref. (43) of $^1H_2$.

(3) See ref. (5) of $^1H^3H$.

(4) See ref. (67) of $^1H_2$.

(5) See ref. (70) of $^1H_2$.

$^{2}H_2$ (continued):

(15) See ref. (47) of $^{1}H_2$.
(16) Dieke, Cunningham, JMS 18, 288 (1965).
(17) Watanabe, Welsh, CJP 43, 818 (1965).
(18) See ref. (50) of $^{1}H_2$.
(18a) See ref. (51) of $^{1}H_2$.
(19) Fowler, Holzberlein, JCP 45, 1123 (1966).
(20) See ref. (15) of $^{1}H^{2}H$.
(21) See ref. (72) of $^{1}H_2$.
(22) See ref. (73) of $^{1}H_2$.
(23) See ref. (74) of $^{1}H_2$.
(24) See ref. (68) of $^{1}H_2$.
(25) See ref. (70) of $^{1}H_2$.
(26) See ref. (71) of $^{1}H_2$.
(27) See ref. (77) of $^{1}H_2$.
(28) See ref. (83) of $^{1}H_2$.
(29) See ref. (86) of $^{1}H_2$.
(30) See ref. (80)(90)(91) of $^{1}H_2$.
(31) See ref. (92) of $^{1}H_2$.
(32) See ref. (94) of $^{1}H_2$.
(33) English, MacAdam, PRL 24, 555 (1970).
(34) See ref. (95) of $^{1}H_2$.
(35) Code, Ramsey, PR A 4, 1945 (1971).
(36) See ref. (103) of $^{1}H_2$.
(37) See ref. (31) of $^{1}H^{2}H$.
(38) Reddy, Kuo, JMS 37, 327 (1971).
(39) See ref. (115) of $^{1}H_2$.
(40) See ref. (112) of $^{1}H_2$.
(41) Herzberg, Science 177, 123 (1972).
(42) See ref. (111) of $^{1}H_2$.
(42a) See ref. (117) of $^{1}H_2$.
(43) See ref. (37) of $^{1}H^{2}H$.
(44) Bredohl, Herzberg, CJP 51, 867 (1973).
(45) Freund, Miller, JCP 59, 4073 (1973).
(46) See ref. (39) of $^{1}H^{2}H$.
(47) See ref. (127) of $^{1}H_2$.
(48) See ref. (129) of $^{1}H_2$.
(49) Russell, Reddy, Cho, JMS 52, 72 (1974).
(50) See ref. (155) of $^{1}H_2$.
(51) See ref. (147) of $^{1}H_2$.
(52) See ref. (153) of $^{1}H_2$.
(53) Le Roy, Barwell, CJP 53, 1983 (1975).
(54) Takezawa, Tanaka, JMS 54, 379 (1975).
(55) See ref. (158) of $^{1}H_2$.
(56) Jost, Lombardi, Derouard, Freund, Miller, Zegarski, CPL 37, 507 (1976).
(57) Miller, Freund, Zegarski, JCP 64, 1842 (1976).
(58) Bunker, unpublished.
(59) Herzberg, unpublished.
(60) Kolos, JMS 62, 429 (1976).

State	T_e	ω_e	$\omega_e x_e$	B_e	α_e	D_e $(10^{-2}cm^{-1})$	r_e (Å)	Observed Transitions Design.	ν_{00}	References
$^3H_2$		μ = 1.50802486		D_0^0 = (4.5909_7) eV[a]		I.P. = (15.4867_0) eV[b]				NOV 1976
n $^3\Pi_u$ 5pπ	$(120984._3)$	1348.89	22.52	10.021	0.294	0.22	1.0562	n→a, R	24923.03	(2)
k $^3\Pi_u$ 4pπ	$(118403._2)$	1355.39	22.026[c]	10.053	0.296	0.22	1.0545	k→a, R	22345.34	(2)
f $^3\Sigma_u^+$ 4pσ	(116653)	[1278]		9.90	0.30		1.063	f→a, R	20561.9	(2)
d $^3\Pi_u$ 3pπ	$(112736._0)$	1372.11	22.135[d]	10.150[e]	0.3050[e]	0.217[f]	1.0494	d→a, R Fulcher b.	16686.4_4	(1)
e $^3\Sigma_u^+$ 3pσ	$(107770._8)$	1272.28	23.03[g]	9.2056	0.2803[h]	0.1887[i]	1.1019_7	e→a, R	11671.06	(3)
a $^3\Sigma_g^+$ 2sσ	$(95965._4)$[j]	1541.57	24.47[k]	11.4374	0.3258[ℓ]		0.98862	(a−X)	(95464.4)[m]	
b $^3\Sigma_u^+$ 2pσ		Repulsive state, lower state of T_2 continuum.						a→b		
F } $^1\Sigma_g^+$ { $2p\sigma^2$	$(100935._9)$[n]	[706.0][o]		$B_8 = 2.5_0$[p]			r_8 = 2.11	F→B, R	ν_{80} = 14302.50[q]	(4)
E } $^1\Sigma_g^+$ { 2sσ	$(100136._7)$	1454.18	30.52	10.9306	0.3659	0.2403	1.0112_8	E→B, V	8765.40	(4)
C $^1\Pi_u$ 2pπ	$(100099._7)$[j]							(C−X)	$(99536._9)$[r]	
B $^1\Sigma_u^+$ 2pσ	$(91696._3)$[j]	787.28[s]	7.013[s]	6.716[t]	0.207_6[t]	0.17_3[t]	1.290_1	(B−X)	$(90825._0)$[u]	
X $^1\Sigma_g^+$ $1s\sigma^2$	0	2546.4_7[v]	41.23[v]	20.335[v]	0.5887[w]		(0.74142)			(6)

$^3H_2$: [a]37028.4 cm^{-1}, calculated from ab initio potential function (8); non-adiabatic corrections which are certainly less than +0.2 cm^{-1} and Lamb shift corrections (≈ −0.2 cm^{-1}) are not included. No observed value is available yet.

[b]From the theoretical values of $D_0^0(T_2)$ and $D_0^0(T_2^+)$, and I.P.(T).

[c]$\omega_e y_e$ = + 0.133.

[d]$\omega_e y_e$ = + 0.159, $\omega_e z_e$ = − 0.002.

[e]γ_e = + 0.0038. The rotational constants refer to $^3\Pi^-$; $^3\Pi^+$ is perturbed. The Λ-type doubling is somewhat irregular.

[f]β_e = − 0.000065.

[g]$\omega_e y_e$ = + 0.170, $\omega_e z_e$ = − 0.0204.

[h]γ_e = − 0.00156.

$^3H_2$ (continued):

[i] $\beta_e = -0.000045$.

[j] From the T_e values of H_2 and D_2 assuming that the electronic isotope shift is proportional to $(1 - \mu_{H_2}/\mu_{T_2})$.

[k] $\omega_e y_e = +0.312$, $\omega_e z_e = -0.016$.

[ℓ] $\gamma_e = +0.00273$.

[m] From T_e assuming $Y_{00}(a\,^3\Sigma_g^+) = 0$ but taking account of Y_{00} in the ground state $X\,^1\Sigma_g^+$ (see [v]).

[n] From the observed v_{80} and the energy of v=8 above the outer minimum of the E,F double-minimum state as calculated by (9).

[o] Calculated $\Delta G(\frac{1}{2})$ value for the outer minimum (9).

[p] Vibrational numbering of (9). The D_8 value is large and negative. For higher vibrational levels (above the potential maximum) B_v is larger, e.g. (4) give $B_{13} = 3.892$, and $D_{13} = 0.00109$ has the normal sign, in agreement with the fact that for these levels the vibrational motion covers both minima of the E,F state. v=0...7 levels not yet observed.

[q] According to (9) the lowest level of the outer minimum is expected at 9204.5 cm^{-1} above $B\,^1\Sigma_u^+(v=0)$. The v=8 level lies just below the potential maximum.

[r] From T_e and the zero-point energy calculated by (7).

[s] From the R(0) lines in the E,F − B system (4); $\omega_e y_e = +0.0912$. The zero-point energy ($Y_{00} = 2.6$ included) is 394.4_6 cm^{-1}.

[t] From combination differences formed from the data on E,F − B (4); $\gamma_e = +0.007_2$, $\beta_e = -0.0008$.

[u] From the calculated T_e and the zero-point energies as given in [s] and [v].

[v] Calculated by (6) from the potential function of (5) and based on v=0...3 only; experimental values are not available. $\omega_e y_e = +0.258$; the zero-point energy ($Y_{00} = 2.8$ included) is 1265.7_4 cm^{-1}.

[w] $+0.0053(v+\frac{1}{2})^2 - 0.00018(v+\frac{1}{2})^3$; see [v].

(1) See ref. (1) of $^1H^3H$.
(2) See ref. (15a) of $^1H_2$.
(3) See ref. (3) of $^1H^3H$.
(4) See ref. (16) of $^2H_2$.
(5) See ref. (43) of $^1H_2$.
(6) See ref. (5) of $^1H^3H$.
(7) See ref. (67) of $^1H_2$.
(8) See ref. (70) of $^1H_2$.
(9) See ref. (86) of $^1H_2$.
(10) See ref. (152) of $^1H_2$.

State	T_e	ω_e	$\omega_e x_e$	B_e	α_e	D_e $(10^{-2}cm^{-1})$	r_e (Å)	Observed Transitions Design.	ν_{00}	References
$^1H_2^+$		μ = 0.50377547		$D_0^0 = 2.6507_8$ eV[a]						NOV 1976 A
		Several other predicted excited states, mostly repulsive.[b]								
C $^2\Pi_u$ 2pπ	$(102696._7)$[c]	(266.0_1)	(6.45_4)	(1.899)	(0.075_8)	(0.039)	(4.19_8)	(C − B)	(8806.3)[c]	(27)
B $^2\Sigma_g^+$ 3dσ	$(93804._5)$[c]	(437.1_5)	(5.24_7)	(1.530)	(0.031_2)	(0.0075)	(4.67_7)	(B − X)	$(92877._4)$	(27)
A $^2\Sigma_u^+$ 2pσ		Repulsive state.[d]						A ← X		(11)
X $^2\Sigma_g^+$ 1sσ	0	$2321._7$[e]	66.2[e]	30.2_1[efg]	1.68_5[e]	e	1.052[h]	Spin reorientation sp.[i]		(13)
$^1H^2H^+$		μ = 0.67155901		$D_0^0 = 2.6677_1$ eV[a]						NOV 1976 A
X $^2\Sigma_g^+$ 1sσ	0	[1913.1][b]		(22.45_2)[c]	(1.00_1)[c]	(1.1)[c]	(1.057)	Rotation-vibration sp.[d]		(5)
$^1H^3H^+$		μ = 0.75523263		$D_0^0 = (2.67401_7)$ eV[e]						NOV 1976 A
X $^2\Sigma_g^+$ 1sσ	0	[(1809.234)][e]								

$^1H_2^+$: [a]Experimental value derived from $D_0^0(H_2)$ and I.P.(H_2) (21). The latest theoretical value [non-adiabatic calculation (25), and including relativistic and Lamb shift corrections] is 2.65073 eV; see also (7)(14).

[b]Electronic wavefunctions and energies are given by (2) where references to earlier literature may be found.

[c]Data for these two states are entirely theoretical [adiabatic approximation (27); see also (19)]. ν_{00}(C–B) refers to N=1 of C $^2\Pi_u$ and N=0 of B $^2\Sigma_g^+$; see Table VI of (27). Beckel (lecture at Columbus 1976) gives 8803.9_0.

[d]Observed in the photodissociation of H_2^+ (11)(13) and in the photoelectron spectrum of H_2 (22); the direct transition from the $X^1\Sigma_g^+$ ground state causes a shoulder in the absorption spectrum of H_2 near 380 Å. Of recent <u>ab initio</u> calculations of the potential function the most detailed seems to be that of (4). This state has a small van der Waals minimum (D_0 = 3.4 cm^{-1}) at 6.64 Å (16).

[e]From Rydberg series limits of H_2 (21); $\Delta G(\frac{1}{2})$ = 2191.2, $\omega_e y_e$ = + 0.6, γ_e = + 0.041_2. Similar constants (and D_e = 0.018) were obtained (1) by extrapolation from low members of the series $np^3\Pi$ (n=2-5). Higher vibrational levels and their B_v values have been derived from photoelectron spectra (8)(17). See also [i].

[f]For two recent <u>ab initio</u> calculations of the potential function s. (4)(26); relativistic corrections (15), Lamb shift corrections (12)(14), non-adiabatic corrections (25). Rotational and vibrational levels up to v=18 from <u>ab initio</u> calculations in (5)(7)(18)(26), incl. quasi-bound levels.

[g]Franck-Condon factors for photoionization of H_2 (3)(6) (10). Effect of r-dependence of the transition moment (24). Experimental cross sections (9)(23).

[h]The most recent theoretical value (20) is 1.0569 Å.

[i]Observed for N=1 and N=2 of v= 4...8. For v=4,N=2 the spin splitting is 0.0027059 cm^{-1}; for odd N the hyperfine structure due to the proton spins is superposed which gives a much larger splitting. The strongest of the observed five transitions for v=4,N=1 occurs at 0.0423810 cm^{-1}; the extrapolated wavenumber for v=0,N=1 is 0.046842.

(1) See ref. (15a) of $^1H_2$.

(2) Bates, Ledsham, Stewart, PTRSL A 246, 215 (1953).

(3) Halmann, Laulicht, JCP 43, 1503 (1965).

(4) Peek, JCP 43, 3004 (1965); see also Sandia Corp. Rep. No. SC-RR-65-77 (1965) and supplement.

(5) Wind, JCP 43, 2956 (1965).

(6) See ref. (50) of $^1H_2$.

(7) Hunter, Pritchard, JCP 46, 2153 (1967).

(8) Siegbahn, Nordling, Fahlman, Nordberg, Hamrin, Hedman, Johansson, Bergmark, Karlsson, Lindgren, Lindberg, "ESCA - Atomic, Molecular and Solid State Structure Studied by Means of Electron Spectroscopy", p. 208f.; NARSSU (IV) 20 (Uppsala, 1967).

(9) Spohr, von Puttkamer, ZN 22 a, 705 (1967).

(10) See ref. (75) of $^1H_2$.

(11) Richardson, Jefferts, Dehmelt, PR 165, 80 (1968).

(12) Gersten, JCP 51, 3181 (1969).

(13) Jefferts, PRL 23, 1476 (1969).

(14) See ref. (79) of $^1H_2$.

(15) Luke, Hunter, McEachran, Cohen, JCP 50, 1644 (1969)

(16) Peek, JCP 50, 4595 (1969).

(17) Åsbrink, CPL 7, 549 (1970).

(18) Beckel, Hansen, Peek, JCP 53, 3681 (1970).

(19) Kroll, JMS 35, 436 (1970).

(20) Shaad, Hicks, JCP 53, 851 (1970).

(21) See ref. (114) of $^1H_2$.

(22) Samson, CPL 12, 625 (1972).

(23) See ref. (37) of $^1H^2H$.

(24) See ref. (39) of $^1H^2H$.

(25) Bishop, MP 28, 1397 (1974).

(26) Hunter, Yau, Pritchard, ADNDT 14, 11 (1974).

(27) Bishop, Shih, Beckel, Wu, Peek, JCP 63, 4836 (1975).

[a]From D_0^0(HD), I.P.(HD), and I.P.(H). A second dissociation limit corresponding to $H + D^+$ is only 0.00370 eV higher. The theoretical value for $D_0^0(HD^+)$ is 2.66767_9 eV (3)(6).

[b]$\Delta G(3/2) = 1816.7$, $\Delta G(5/2) = 1723.7$; from Rydberg series of HD (1). Ab initio calculations (6) give $\Delta G(1/2) = 1913.005$.

[c]All bound rotational and vibrational levels derived from the ab initio potential function in the adiabatic approximation are listed by (4); see also (2). The rotational constants in the table are based on these calculated levels (4) using v=0 and 1 only.

[d]Several lines of the 1-0, 2-1, 3-2 infrared bands have been observed and measured with high accuracy by (5) but there are not enough lines to obtain "observed" values for the rotational constants. The lines show a splitting of 0.0010 cm^{-1} corresponding to the difference in spin splitting in the upper and lower state.

[e]Theoretical value (6).

(1) See ref. (36) of $^1H^2H$.

(2) Bishop, Wetmore, JMS 46, 502 (1973).

(3) See ref. (25) of $^1H_2^+$.

(4) See ref. (26) of $^1H_2^+$.

(5) Wing, Ruff, Lamb, Spezeski, PRL 36, 1488 (1976).

(6) Bishop, PRL 37, 484 (1976).

State	T_e	ω_e	$\omega_e x_e$	B_e	α_e	D_e $(10^{-2}cm^{-1})$	r_e (Å)	Observed Transitions Design.	Observed Transitions ν_{00}	References
$^{2}H_2^+$		$\mu = 1.00691397$		$D_0^0 = 2.6919_6$ eV[a]						NOV 1976 A
C $^2\Pi_u$ 2pπ	b	(188.0)	(3.14)[c]	(0.950)	(0.026)[d]		(4.19_8)	(C − B)	(8836)	(7)
B $^2\Sigma_g^+$ 3dσ	b	(309.4)	(2.62)[e]	(0.766)	(0.011)		(4.67_5)			
X $^2\Sigma_g^+$ 1sσ	0	[1577.3][f]		15.016[g]	0.560[g]	0.53[g]	1.0559			
$^{2}H^{3}H^+$		$\mu = 1.20750134$		$D_0^0 = (2.69999_3)$ eV[h]						NOV 1976 A
X $^2\Sigma_g^+$ 1sσ	0	[(1445.410)][h]								
$^{3}H_2^+$		$\mu = 1.50788772$		$D_0^0 = (2.7077_4)$ eV[i]						NOV 1976
X $^2\Sigma_g^+$	0	1336[j]	22[j]	10.014[j]	0.294[j]	0.23[j]	1.0566			
$^{1}H_2^-$		$\mu = 0.50404975$								NOV 1976 A
		A comprehensive review of resonances in electron impact on H_2, HD, and D_2 has been published by (3) and contains references to earlier literature.								
G		(2260)						a	15.07 eV[b]	(3)(5)
F		2420	40					a	13.63 eV[b]	(3)(5)
D $^2\Sigma_g^+$ $1s\sigma 2p\pi^2$		2540	53	$[34._5]$[c]			0.97[d]	e	11.32 eV[b]	(2)(3)(4)(6)
C $(^2\Delta_g$ $1s\sigma 2p\pi^2)$[f]		2540	42				1.03[d]	e	11.19 eV[b]	(2)(3)(4)(6)
B $^2\Sigma_g^+(1s\sigma 2p\sigma^2)$[g]		(1530)	40				1.18[d]	e	10.93 eV[b]	(2)(3)
A $^2\Sigma_g^+$		Repulsive curve, having the unstable b $^3\Sigma_u^+$ state of H_2 as its parent state. Observed in dissociative attachment experiments.							~10 eV[b]	(3)
X $^2\Sigma_u^+$ $1s\sigma^2 2p\sigma$		Slightly bound state ($D_0^0 \approx$ 1-2 eV).[h]								(3)

$^{2}H_2^+$, $^{2}H^{3}H^+$, $^{3}H_2^+$:

[a]From $D_0^0(D_2)$ and I.P.(D_2); theoretical value 2.69192 eV (8).
[b]Data for these two states are entirely theoretical (7).
[c]$\omega_e y_e = -0.027$.
[d]$\gamma_e = -0.0003$.
[e]$\omega_e y_e = +0.011$.
[f]$\Delta G(3/2) = 1512.1$; from Rydberg series of D_2 (10). Corresponding theoretical values 1577.15, 1512.47 [adiabatic approximation (9)]. Higher vibrational levels [observed in photoionization (6) and photoelectron (5) spectra] and rotational levels, incl. quasi-bound levels, calculated by (4)(9). Franck-Condon factors for photoionization of D_2 from X $^1\Sigma_g^+$ calculated by (2)(3) and measured by (5)(6).
[g]Extrapolated (1) from low members of the Rydberg series $np^3\Pi$ (n=2...6) of D_2. See also [f].
[h]Theoretical value (11).
[i]Calculated from the Born-Oppenheimer pot. function and the adiabatic corrections (scaled down for T_2^+) given by (12).
[j]Extrapolated (1) from low members of the Rydberg series $np^3\Pi$ (n=2...5) of T_2.

(1) See ref. (15a) of $^{1}H_2$.
(2) See ref. (3) of $^{1}H_2^+$.
(3) See ref. (50) of $^{1}H_2$.
(4) See ref. (7) of $^{1}H_2^+$.
(5) See ref. (9) of $^{1}H_2^+$.
(6) See ref. (68) of $^{1}H_2$.
(7) See ref. (19) of $^{1}H_2^+$.
(8) See ref. (25) of $^{1}H_2^+$.
(9) See ref. (26) of $^{1}H_2^+$.
(10) See ref. (54) of $^{2}H_2$.
(11) See ref. (6) of $^{1}H^{2}H^+$, $^{1}H^{3}H^+$.
(12) Bishop, Wetmore, MP 26, 145 (1973).

$^{1}H_2^-$:

[a]Two progressions of resonances (series "f" and "g") proceeding across the ionization threshold of H_2^+. Higher members correlate with structure previously observed in photoionization of H_2.
[b]Energies in eV above X $^1\Sigma_g^+$(v=0) of H_2, all obtained as resonances in electron scattering experiments. The value for B $^2\Sigma_g^+$ is extrapolated from v=2.
[c]From resolved rotational structure of the resonances (2).
[d]From a fit of calculated to observed excitation cross sections (2); for D $^2\Sigma_g^+$ in good agreement with the value obtained from the observed rotational structure.
[e]The progressions of resonances attributed to B, C, D are normally referred to as series "b", "c", "a", resp..
[f]Assignment according to (6); (2) favour $1s\sigma 2s\sigma 2p\pi$, $^2\Pi_u$.
[g]Possible parent state B $^1\Sigma_u^+$.
[h]The potential function is well defined only outside the potential well of H_2 [see, however, (1)]; its asymptote lies 0.75421 eV below that of H_2 $X^1\Sigma_g^+$. The lifetime against preionization is estimated to be 10^{-15} s corresponding to a width of 1 eV. The ground state of H_2^- causes a broad resonance in the scattering of low energy electrons by H_2 and shows up both in the formation of H^- by dissociative attachment and in the vibrational excitation of H_2.

(1) Eliezer, Taylor, Williams, JCP 47, 2165 (1967).
(2) Comer, Read, JP B 4, 368 (1971); Joyez, Comer, Read, JP B 6, 2427 (1973).
(3) Schulz, RMP 45, 423 (1973).
(4) Spence, JP B 7, L87 (1974).
(5) Schowengerdt, Golden, PR A 11, 160 (1975).
(6) Chang, PR A 12, 2399 (1975).

State	T_e	ω_e	$\omega_e x_e$	B_e	α_e	D_e $(10^{-4}cm^{-1})$	r_e (Å)	Observed Transitions Design.	ν_{00}	References
$^1H^{81}Br$		$\mu = 0.99542702$		$D_0^0 = 3.758$ eV[a]		I.P. = 11.67 eV[b]				DEC 1976 A
		Numerous absorption bands above 114000 cm^{-1}, tentatively assigned to higher members of two Rydberg series starting with L and M and converging to A $^2\Sigma^+$ of HBr^+; I.P.[$A\,^2\Sigma^+, v=0$] = 123373 cm^{-1} (15.296$_4$ eV).								(43)
M ($^1\Sigma^+$)	(109473)	[1308]	v=0...4 observed. Assigned as $4p\sigma\, 4p\pi^4\, 6s\sigma$.[c]					M← X,	108814	(43)*
L ($^1\Sigma^+, ^1\Pi$)	(104201)	[1262]	v=0...3 observed. Assigned as $4p\sigma\, 4p\pi^4\, 5p\sigma$ and/or $5p\pi$.[c]					L← X,	103519	(43)*
		Further absorption bands of doubtful assignment between 75200 and 83600 cm^{-1}.								(13)(15)
K[d] 1	(83902)	(2518)[e]		[8.195]		[22.0]	[1.437$_5$]	K← X, R	83847.9[f] Z	(15)
J[d] 1	(81243)	(2502)[e]		[8.02$_7$][g]		[3.6$_1$]	[1.453]	J← X, R	81180.7[h] Z	(13)*
I[d] 1	(80436)	(2525)[e]		[8.16$_9$][i]		[10.$_4$]	[1.440]	I← X, R	80385.6[j] Z	(13)*
g ($^3\Sigma^-$)0^+ t	[79253.2]			[7.6$_3$][k]		[-17]	[1.49]	g← X, R	77940.$_0$ Z	(13)* (36)*
F $^1\Delta$ t	[78322.3]			[8.20]			[1.43$_7$]	F← X, R	77009.1 Z	(36)*
f_1 $^3\Delta_1$ t	(76814)	[2299.7] Z		8.02$_7$	0.21$_3$		1.453	f_1← X, R	76650.9 Z	(13)(36)*
D $^1\Pi$ u	(76310)	[2405.5] Z		8.12$_5$	0.21		1.444	D← X, R	76199.4 Z	(13)* (36)*
d_0 $^3\Pi_0$ u	(76193)	[2418.5] Z		[7.624][ℓ]	(0.32)		[1.490$_4$]	d_0← X, R	76088.8 Z	(13)(36)*
E ($^1\Sigma^+$)0^+ t	[76691]			[7.34][m]			[1.51$_9$]	E← X, R	75378	(36)
V $^1\Sigma^+$ n	(75800)	(790)	Bands in emission above 46500 cm^{-1}, in absorption above 75700. Incomplete analysis.					V↔X,[o] R	(74900)	(14)(36)*
f_2 $^3\Delta_2$ t	[75533.8]			[8.67$_5$][p]		[16.$_5$][p]	[1.397]	f_2← X, R	74220.6 Z	(13)(36)*
f_3 $^3\Delta_3$ t	[75403.1]	Weak transition.		[7.41]		[-7.6]	[1.51$_2$]	f_3← X, R	74089.9 Z	(36)*
e $^3\Sigma^+$ t	[75053]	Very diffuse, unresolved band.						e← X, R	73740	(36)
d_1 $^3\Pi_1$ u	[74855]	Diffuse band, rotational structure unresolved.						d_1← X, R	73542	(13)(36)
d_2 $^3\Pi_2$ u	[74753]	Diffuse band, rotational structure unresolved.						d_2← X, R	73440	(13)(36)

State	T_e	ω_e	$\omega_e x_e$	B_e	α_e	D_e ($10^{-4}cm^{-1}$)	r_e (Å)	Observed Transitions Design.	ν_{00}	References
$^1H^{81}Br$ (continued)										
C $^1\Pi$	v 70578	2552 Z	52	7.89	0.30		1.46_5	$C\leftarrow X$,[q] R	70527.6 Z	(13)(32)*
b_0 $^3\Pi_0$ 0^+	v (68998)	[2452]		$[7.99_6]^r$			[1.455]	$b_0\leftarrow X$, R	68911.2 Z	(13)(32)*
b_0 $^3\Pi_0$ 0^-	v (68991)							R	68904^s H	
b_1 $^3\Pi_1$	v (67180)	[2444.2] Z		$8.14_8{}^r$	0.29_2		1.442	$b_1\leftarrow X$, R	67088.4 Z	(13)* (32)*
b_2 $^3\Pi_2$	v [67663.0]			$[7.80_5]^r$			[1.473]	$b_2\leftarrow X$, R	66349.8 Z	(13)(32)*

^{1}HBr: [a]From $D_0^0(H_2)$, $D_0^0(Br_2)$, and ΔH_{f0}(HBr; from gaseous H_2, Br_2).

[b]Average value from photoionization (10) and photoelectron spectra (23)(29); refers to X $^2\Pi_{3/2}$ of the ion. A more recent paper (39) gives 11.64_5 eV.

[c]Strongly broadened by preionization; estimated lifetime against preionization 9.5×10^{-15} s (46).

[d]I, J, K correspond to absorption bands with clear analogues in DBr.

[e]From the observed HBr-DBr isotope shift assuming that the observed bands are 0-0 bands.

[f]Band [37] of (15).

[g]Ω-type doubling, $\Delta\nu_{ef} = + 0.14_2 \times J(J+1) - \ldots$; B and D represent average values.

[h]Band [28] of (13). Sharp P, Q, R branches; the Q levels appear to be predissociated for $J \geq 14$.

[i]From R, P branches. $\Delta\nu_{ef} = - 0.041 \times J(J+1)$.

[j]Band [26] of (13).

[k]Perturbed at high J.

[ℓ]Slightly diffuse lines.

[m]Perturbed.

[n]Derived from $H^+ + Br^-$; configuration $\ldots\sigma\pi^4\sigma^*$.

[o]Heavily perturbed extensive band system. Absorption lines above 75923 cm^{-1} are diffuse. B' varies irregularly between 3.4 and 4.5 cm^{-1}.

[p]Average values for the two Ω-type doubling components.

[q]Very strong absorption, lines are diffuse.

[r]Diffuse rotational structure.

[s]Diffuse Q head.

[t]Configuration $\ldots\sigma^2\pi^3$ $5p\pi$.

[u]Configuration $\ldots\sigma^2\pi^3$ $5p\sigma$.

[v]Configuration $\ldots\sigma^2\pi^3$ $5s\sigma$.

State	T_e	ω_e	$\omega_e x_e$	B_e	α_e	D_e ($10^{-4}cm^{-1}$)	r_e (Å)	Observed Transitions: Design.	Observed Transitions: ν_{00}	References
$^1H^{81}Br$ (continued)										
A ($^1\Pi$) [w]		Continuous absorption starting at ~35000 with maximum at 56400 cm^{-1}.						A←X		(1)(2)(4)(5)(28)
X $^1\Sigma^+$	0	2648.975^{x} Z	45.217_5^{y}	8.46488_4^{x}	0.23328^{z}	$3.457_5^{a'}$	1.41443_5	Rot.-vibr. sp.[b'c']		(21)
								Rotation spectrum[d'c']		(8)(17)(31)
								Raman sp.[e']		(45)
								Mol. beam el. reson.[f']		(42)

^{1}HBr (continued):

[w] Configuration $...\sigma^2\pi^3\sigma^*$.

[x] These are Y_{10} and Y_{01} values; applying Dunham corrections (21) obtain $\omega_e = 2649.21_5$, $B_e = 8.46506_5$. Additional corrections (adiabatic, non-adiabatic) are discussed by (38). The microwave B_0 value of (17) was included in the evaluation of B_e. See also [b'f'].

[y] $\omega_e y_e = -0.0029$.

[z] $+0.000873_5(v+\frac{1}{2})^2 - 0.000120(v+\frac{1}{2})^3$.

[a'] $-0.039_7 \times 10^{-4}(v+\frac{1}{2}) + 0.0038(v+\frac{1}{2})^2$; $H_v = 7.63 \times 10^{-9} - 0.55 \times 10^{-9}(v+\frac{1}{2})$.

[b'] In absorption the 1-0, 2-0, 3-0, 3-1, 4-0, 5-0, 6-0 bands have been studied (6)(7)(12)(21)(41); in emission 1-0, 2-1, 3-2, 4-3 (11)(20). The constants in the table are from (21), those of (20)(41) are very similar and of comparable accuracy. See also (47). Absolute intensities have been measured (16)(18)(30)(33) and the dipole moment function has been calculated; (40) give for $H^{79}Br$ [D,Å]: $\mu_{e\ell}(r) = +0.788 + 0.315(r-r_e) + 0.575(r-r_e)^2$; see also (24)(34)(37).

[c'] For observations and measurements of pressure-induced bands and pure rotation lines ($\Delta J=2$) see (22)(27). The pressure broadening of the lines has been studied by (16)(25).

[d'] Absolute intensities have been measured by (19).

[e'] Raman cross sections in gaseous HBr.

[f'] The following constants (as well as corresponding values for $H^{79}Br$) are given in (42):

- $\mu_{e\ell}(v=0,J=1) = 0.8265$ D [in a later paper (44) derive 0.8282 D from Stark effect of rotation spectrum];
- quadrupole and other hyperfine coupling constants;
- $g_J = 0.3712$.

These constants supersede earlier values of (9)(17)(26)(31)(35).

^{1}HBr (continued):

(1) Bates, Halford, Anderson, JCP 3, 531 (1935).
(2) Goodeve, Taylor, PRS A 152, 221 (1935).
(3) Price, PRS A 167, 216 (1938).
(4) Datta, Chakravarty, PNISI 7, 297 (1941).
(5) Romand, AP(Paris) (12) 4, 527 (1949).
(6) Naudé, Verleger, PPS 63, 470 (1950).
(7) Thompson, Williams, Callomon, SA 5, 315 (1952).
(8) Hansler, Oetjen, JCP 21, 1340 (1953).
(9) Schurin, Rollefson, JCP 26, 1089 (1957).
(10) Watanabe, JCP 26, 542 (1957).
(11) Mould, Price, Wilkinson, SA 16, 479 (1960).
(12) Plyler, JRNBS A 64, 377 (1960).
(13) Barrow, Stamper, PRS A 263, 259, 277 (1961).
(14) Stamper, Barrow, JPC 65, 250 (1961).
(15) Stamper, CJP 40, 1279 (1962).
(16) Babrov, JCP 40, 831 (1964).
(17) Jones, Gordy, PR A 136, 1229 (1964).
(18) Babrov, Shabott, Rao, JCP 42, 4124 (1965).
(19) Chamberlain, Gebbie, Nature 208, 480 (1965).
(20) James, Thibault, JCP 42, 1450 (1965).
(21) Rank, Fink, Wiggins, JMS 18, 170 (1965).
(22) Atwood, Vu, Vodar, SA A 23, 553 (1967).
(23) Frost, McDowell, Vroom, JCP 46, 4255 (1967).
(24) Jacobi, JMS 22, 76 (1967).
(25) Pourcin, Bachet, Coulon, CR B 264, 975 (1967).
(26) Tokuhiro, JCP 47, 109 (1967).
(27) Weiss, Cole, JCP 46, 644 (1967).
(28) Huebert, Martin, JPC 72, 3046 (1968).
(29) Lempka, Passmore, Price, PRS A 304, 53 (1968).
(30) Rao, Lindquist, CJP 46, 2739 (1968).
(31) Van Dijk, Dymanus, CPL 4, 170 (1969).
(32) Ginter, Tilford, JMS 34, 206 (1970).
(33) Gustafson, Rao, CJP 48, 330 (1970).
(34) Tipping, Herman, JMS 36, 404 (1970).
(35) Van Dijk, Dymanus, CPL 5, 387 (1970).
(36) Ginter, Tilford, JMS 37, 159 (1971).
(37) Rao, JP B 4, 791 (1971).
(38) Bunker, JMS 42, 478 (1972).
(39) Delwiche, Natalis, Momigny, Collin, JESRP 1, 219 (1972).
(40) Urquhart, Clark, Rao, ZN 27 a, 1563 (1972).
(41) Bernage, Niay, Bocquet, Houdart, RPA 8, 333 (1973); CR B 278, 235 (1974).
(42) Dabbousi, Meerts, de Leeuw, Dymanus, CP 2, 473 (1973).
(43) Terwilliger, Smith, JMS 50, 30 (1974).
(44) Van Dijk, Dymanus, CP 6, 474 (1974).
(45) Cherlow, Hyatt, Porto, JCP 63, 3996 (1975).
(46) Terwilliger, Smith, JCP 63, 1008 (1975).
(47) Ogilvie, Koo, JMS 61, 332 (1976).

State	T_e	ω_e	$\omega_e x_e$	B_e	α_e	D_e $(10^{-4}cm^{-1})$	r_e (Å)	Observed Transitions Design.	Observed Transitions ν_{00}	References
$^2H^{81}Br$		μ = 1.96518641 D_0^0 = 3.804 eV[a] I.P. = 11.67_3 eV[b]								DEC 1976 A
		Numerous absorption bands above 114000 cm^{-1}; see ^1HBr.								(13)
M $(^1\Sigma^+)$	(109374)	[1000] v=0...5 observed. See ^1HBr.						M← X,	108937	(13)*
L $(^1\Sigma^+, ^1\Pi)$	(104141)	[971] v=0...4 observed. See ^1HBr.						L← X,	(103690)	(13)*
		Further absorption bands of doubtful assignment between 72000 and 83800 cm^{-1}.								(6)
K^c 1	(83908)	(1792)[d]		[4.121][e]		[0.80]	[1.442_8]	K← X,	R 83867.1[f] Z	(6)
J^c 1	(81248)	(1781)[d]		[4.083][g]		[1.36]	[1.449_5]	J← X,	R 81202.2[f] Z	(6)*
I^c 1	(80442)	(1797)[d]		[4.107]		[1.38]	[1.445_2]	I← X,	R 80403.8[f] Z	(6)
g $(^3\Sigma^-)0^+$	(78125)	[$1440._6$] Z		[3.79]	(0.40)[h]	[14]	[1.50_4]	g← X,	R 77908.8 Z	(6)(10)
F $^1\Delta$	[77989.0]			[4.095]			[1.447_3]	F← X,	R 77052.3 Z	(6)(10)
f_1 $^3\Delta_1$	(76787)	[1657.3] Z		4.06_4	0.07_6		1.453	f_1← X,	R 76679.2 Z	(6)(10)
D $^1\Pi$	(76280)	[1732.6] Z		4.128[i]	0.076		1.441_5	D← X,	R 76209.1 Z	(6)(10)*
d_0 $^3\Pi_0$	[77027.4]			[3.99_3][j]			[1.466]	d_0← X,	R 76090.7 Z	(10)*
E $(^1\Sigma^+)0^+$	(75682)	[1365] Z		[3.953]	(0.78)[h]		[1.472_9]	E← X,	R 75428.1 Z	(6)(10)
V $^1\Sigma^+$		Extensive absorption system above 75700 cm^{-1}, incompletely analysed because of strong perturbations. B_v values range from 1.68 to 2.12.						V← X,	R	(10)
f_2 $^3\Delta_2$	(74370)	[1660.3] Z		4.29_4	0.097	[2.4]	1.413	f_2← X,	R 74263.1 Z	(6)(10)*
f_3 $^3\Delta_3$	[75088.6]	Weak transition.		[3.83]		[-2.1]	[1.49_7]	f_3← X,	R 74151.9 Z	(10)*
e $^3\Sigma^+$	[74677]	Very diffuse, unresolved band.						e← X,	73740	(10)
d_1 $^3\Pi_1$	[74499]	Diffuse band, rotational structure unresolved.						d_1← X,	R 73562	(6)(10)
d_2 $^3\Pi_2$	[74391.5]			[3.906][k]		[0.36]	[1.482]	d_2← X,	R 73454.8 Z	(6)(10)*
C $^1\Pi$	70563	1811.2 Z	25.5	4.02[ℓ]	0.10_5		1.46_1	C← X,	R (70501) H^Q	(6)(8)*

State	T_e	ω_e	$\omega_e x_e$	B_e	α_e	D_e ($10^{-4}cm^{-1}$)	r_e (Å)	Observed Transitions Design.	ν_{00}	References
$^2H^{81}Br$ (continued)										
$b_0\ ^3\Pi_0\ 0^+$	[69837.1]			$[4.08_2]^m$			[1.450]	$b_0 \leftarrow X$,	R 68900.4 Z	(6)* (8)*
0^-	[69832]	Unresolved Q branch, assignment uncertain.							R (68895)	
$b_1\ ^3\Pi_1$	$67113._9$	1814 Z	27	4.10^n	0.09		1.44_6	$b_1 \leftarrow X$,	R 67077.5 Z	(6)(8)
$b_2\ ^3\Pi_2$	[67290.7]			$[4.01_8]^m$			[1.461]	$b_2 \leftarrow X$,	R 66354.0 Z	(6)(8)
A ($^1\Pi$)		Continuous absorption beginning at $\sim$39000 cm^{-1}.						$A \leftarrow X$,		(1)
X $^1\Sigma^+$	0	1884.75 Z	$22.71_8{}^o$	$[4.245596]^p$	0.083_8	$[0.8832]^p$	1.4145	Rot.-vibr. sp.[q]		(2)(5)
								Rotation spectrum[r]		(3)(4)(9)(12)

^{2}HBr: [a]From $D_0^0(^1HBr)$.

[b]From the photoelectron sp. (11) of the isotopic mixture.

[c]See [d] of ^{1}HBr.

[d]See [e] of ^{1}HBr.

[e]Ω-type doubling, $\Delta\nu_{ef} \approx + 0.0371\,J(J+1)$.

[f]Bands {35}, {27}, {26} of (6).

[g]Ω-type doubling, $\Delta\nu_{ef} = + 0.0314\,J(J+1)$.

[h]v=0 and 1 only; v=1 is perturbed.

[i]From P, R branches. $B_0(Q) = 4.109$.

[j]From P, R branches. $B_0(Q) = 3.92_6$.

[k]Branches are slightly diffuse.

[ℓ]Diffuse rotational structure; v=0...4.

[m]Slightly diffuse lines.

[n]v=2, 3 diffuse. Slightly diffuse lines for v=0, 1.

[o]$\omega_e y_e = - 0.010_6$.

[p]Microwave value (9)(12).

[q]1-0, 2-0, 3-0 bands in absorption (2), Δv=1 sequence in emission (5).

[r]$\mu_{e\ell}(v=0) = 0.8233$ D from Stark effect of 1-0 line (12). Quadrupole and other hyperfine coupling const. (7)(9)(12).

(1) See ref. (1) of ^{1}HBr.
(2) Keller, Nielsen, JCP 22, 294 (1954).
(3) Palik, JCP 23, 217 (1955).
(4) Cowan, Gordy, PR 111, 209 (1958)
(5) See ref. (11) of ^{1}HBr.
(6) See ref. (15) of ^{1}HBr.
(7) See ref. (26) of ^{1}HBr.
(8) See ref. (32) of ^{1}HBr.
(9) De Lucia, Helminger, Gordy, PR A 3, 1849 (1971).
(10) See ref. (36) of ^{1}HBr.
(11) See ref. (39) of ^{1}HBr.
(12) See ref. (44) of ^{1}HBr.
(13) See ref. (43) of ^{1}HBr.

State	T_e	ω_e	$\omega_e x_e$	B_e	α_e	D_e ($10^{-4} cm^{-1}$)	r_e (Å)	Observed Transitions Design.	ν_{00}	References
$^3H^{81}Br$		$\mu = 2.9076700_9$		$D_0^0 = 3.825$ eV[a]						DEC 1976 A
X $^1\Sigma^+$	0	[1519.26][b] Z	(15.4_8)	[2.874466][c]	0.0459[b]	0.45[b]	1.4145_6	Rotation-vibration b.		(2)
								Rotation sp.[d]		(1)
$^1HBr^+$				$D_0^0 = 3.894$ eV[a]						DEC 1976 A
A $^2\Sigma^+$	28421.0	1404.0[b]	37.7_5[b]	5.970_2[cd]	0.247_6	4.30_0[e]	1.6842	A→X,[f]	R 27904.6[g] Z	(1)(2)(3) (6)*
X $^2\Pi_i$	0[h]	2441.5_2[i] Z	47.4_0[j]	8.072_1[k]	0.236_3	[3.48]	1.4484			
$^2HBr^+$				$D_0^0 = 3.937$ eV[ℓ]						DEC 1976 A
A $^2\Sigma^+$	28421.2	999.2_5 Z	19.1_2	3.024_1[mn]	0.089_4	1.1_0	1.6842	A→X,	R 28054.4[g] Z	(3)(6)
X $^2\Pi_i$	0[o]	1734.3_6[i] Z	22.1_8[p]	4.087_7[q]	0.084_8	0.87_0	1.4486			
$^1HBr^-$										DEC 1976
		Resonances due to inverse preionization have been found in the transmission of electrons through HBr in the energy range 8.0 - 10.8 and 11.5 - 13.2 eV. They correspond to excited states of HBr^- in which two electrons are in Rydberg orbitals $(ns\sigma)^2$, $ns\sigma\, np\sigma$, $(np\sigma)^2$, ... while the core is in the X $^2\Pi_{\frac{3}{2},\frac{1}{2}}$ or A $^2\Sigma^+$ state of HBr^+.								(1)

^3HBr: [a]From $D_0^0(^1HBr)$.

[b]From the rotation-vibration spectrum of the isotopic mixture ($^{79,81}Br$).

[c]From the microwave spectrum (1) assuming the infrared D value.

[d]Quadrupole coupling constants (3).

(1) Burrus, Gordy, Benjamin, Livingston, PR 97, 1661 (1955).

(2) Jones, Robinson, JCP 24, 1246 (1956).

(3) Tokuhiro, JCP 47, 109 (1967).

$^1HBr^+$, $^2HBr^+$:

[a]From the predissociation in the A $^2\Sigma^+$ state (4); see [d]. $D_0^0(^1HBr) + I.P.(Br) - I.P.(^1HBr) = 3.90_2$ eV.

[b]From isotope relations and the constants for DBr^+ (6). (2) derive $\Delta G(\frac{1}{2}) = 1328.7$ from band origins.

[c]Spin splitting constant $\gamma(v=0) = 2.10_5$; small variation with N.

[d]Only v'=0 and 1 observed in emission. (4) have observed a sharp breaking-off at N'=12 for v'=1 and a less distinct breaking-off at N'=21 for v'=0 yielding a dissociation limit at 31407 ± 20 cm^{-1}. In the photoelectron spectrum of HBr (5) the bands with v'=2 and 3 are as narrow as those with v'=0 and 1 but bands with v' ≥ 4 are strongly broadened indicating strong predissociation (lifetime $\sim 2 \times 10^{-14}$s).

[e]$\beta_e = -0.01_4 \times 10^{-4}$.

[f]Transition moment variation with r from observed vibrational intensity distribution and calculated Franck-Condon factors (7).

$^1HBr^+$, $^2HBr^+$ (continued):

[g]Referring in the lower state to the zero point of the Hill-Van Vleck formula.

[h]$A_v = -2652.8_0 + 2.81(v+\frac{1}{2})$.

[i]Value for the heavier isotope (^{81}Br).

[j]$\omega_e y_e = +0.40$.

[k]Λ-type doubling constants ($p \approx 2.05$) in (2)(6).

[ℓ]From $D_0^0(^1HBr^+)$.

[m]Spin splitting constant $\gamma(v=0) = 1.06_8$.

[n]Only v'=0,1,2 observed in emission; breaking-off for N' > 12 of v'=2 (4). In the photoelectron spectrum v' = 0...8 have been observed (5) but the bands with v'=6,7, 8 are very broad indicating strong predissociation beginning at the same energy as in $^1HBr^+$.

[o]$A_0 = -2651.4_5$, $A_1 = -2649.5_4$, ...

[p]$\omega_e y_e = -0.1$.

[q]Λ-type doubling constants ($p \approx 1.05$) in (6).

(1) Norling, ZP 95, 179 (1935).

(2) Barrow, Caunt, PPS 66, 617 (1953).

(3) Marsigny, Lebreton, Ferran, Lagrange, CR C 266, 176, 1731 (1968); Marsigny, Lebreton, Petit, CR C 270, 1632 (1970).

(4) Haugh, Bayes, JPC 75, 1472 (1971).

(5) Delwiche, Natalis, Momigny, Collin, JESRP 1, 219 (1973).

(6) Lebreton, JCPPB 70, 1188 (1973).

(7) Haugh, Schneider, Smith, JMS 51, 123 (1974).

$^1HBr^-$: (1) Spence, Noguchi, JCP 63, 505 (1975).

State	T_e	ω_e	$\omega_e x_e$	B_e	α_e	D_e $(10^{-4}cm^{-1})$	r_e (Å)	Observed Transitions Design.	Observed Transitions ν_{00}	References
$^1H^{35}Cl$		$\mu = 0.97959272$		$D_0^0 = 4.433_6$ eV[a]		I.P. = 12.74_8 eV[b]				DEC 1976 A
		Rydberg series corresponding to excitation of a 2p electron.						200-210 eV		(50)(60)
		Numerous absorption bands above 123000 cm^{-1}, tentatively assigned to higher members of the Rydberg series starting with L and M and converging to A $^2\Sigma^+$ of HCl^+.								(55)
M ($^1\Sigma^+$)	(117811)	[1529]	v=0...5 observed. Assigned as $3p\sigma\, 3p\pi^4\, 5s\sigma$.[c]					M← X,	117093	(55)*
L ($^1\Sigma^+$, $^1\Pi$)	111280	1531	52	Assigned as $3p\sigma\, 3p\pi^4\, 4p\sigma/\pi$.[c]				L← X,	110555	(55)*
		Many other absorption bands in the region 83000 - 93000 cm^{-1} corresponding to Rydberg states strongly perturbed by the V $^1\Sigma^+$ state which itself gives rise to many perturbed bands.								(62)
K $^1\Pi$	(89861)	[2604.6] Z		$[9.230]^d$		$[-12._6]^d$	[1.3654]	K← X, R	89680.5 Z	(62)
H $^1\Sigma^+$	(89120)	[2093.8] Z		[8.4410]		[8.93]	[1.4278]	H← X, R	88684.5 Z	(62)
E $^1\Sigma^+$	(84193)	[2138.6] Z		[6.6423]		[36.2]	[1.6096]	E← X, R	83780.2 Z	(62)
g ($^3\Sigma^-$)1	^{e} [84329.7]			$[10.36]^f$		$[17]^f$	$[1.28_9]$	g← X,	82847.4 Z	(48)*
f_1 $^3\Delta_1$	^{e} [84006.1]			$[10.27_0]^g$		$[-13]^g$	[1.294]	f_1← X,	82523.8 Z	(48)*
D $^1\Pi$	^{h} [83972.0]			$[9.79_4]^i$		$[20._5]^i$	[1.326]	D← X, R	82489.7 Z	(48)*
d_0 $^3\Pi_0$	^{h} [83753.6]			$[9.40_4]^j$		$[-2._2]^j$	[1.353]	d_0← X, R	82271.3 Z	(48)*
f_2 $^3\Delta_2$	^{e} [83497.7]			$[10.85_1]^k$		$[29._5]^k$	[1.259]	f_2← X, V	82015.4 Z	(48)*
f_3 $^3\Delta_3$	^{e} [83308.2]			$[9.45]^g$		$[-1._3]^g$	$[1.34_9]$	f_3← X, R	81825.9 Z	(48)*
d_1 $^3\Pi_1$	^{h} [83255.6]			$[9.76_8]^\ell$		$[8]^\ell$	[1.327]	d_1← X, R	81773.3 Z	(48)*
d_2 $^3\Pi_2$	^{h} [83083.0]			$[8.63_2]^m$		$[-14]^m$	[1.412]	d_2← X, R	81600.7 Z	(48)*
C $^1\Pi$	^{n} 77575	[2684.0] Z	o	$[9.33_3]$	o		[1.358]	C← X,[p] R	77485.3 Z	(1)* (44)
V $^1\Sigma^+$	^{q} $77293._0$	877.16 Z	16.04^r	2.727	-0.026	1.02^r	2.512	V↔ X,[s] R	76245.3 Z	(8)(9)(48)*
		Continuous emission spectrum with maximum at 38900 cm^{-1}.						V→A		(9)

State	T_e	ω_e	$\omega_e x_e$	B_e	α_e	D_e ($10^{-4} cm^{-1}$)	r_e (Å)	Observed Transitions Design.	ν_{00}	References
$^1H^{35}Cl$ (continued)										
$b_0\ ^3\Pi_0$	[n] (75617)	[2712]		[10.36][t]			[1.289]	$b_0 \leftarrow X$,[u]	75490.4 Z	(1)(44)*
$b_1\ ^3\Pi_1$	[n] (75195)	(2900)	(79)	[9.87][t]			[1.320]	$b_1 \leftarrow X$, R	$75142._6$ Z	(1)(44)*
$b_2\ ^3\Pi_2$	[n] [$76322._2$]			[9.18][t]			[1.369]	$b_2 \leftarrow X$,[u] R	$74839._9$ Z	(1)(44)*
A ($^1\Pi$)	[v]	Continuous absorption starting at 44000 cm^{-1}, maximum[w] at 65500 cm^{-1}						$A \leftarrow X$		(2)(3)

$^1H^{35}C\ell$: [a]From $D_0^0(H_2)$, $D_0^0(C\ell_2)$, and $\Delta H_{f0}^o(HC\ell)$.

[b]From the photoelectron spectrum (33)(38)(46); photoionization measurements give similar results (20)(27). A somewhat smaller I.P.(12.73_0 eV) may be derived from the second band system in the photoelectron spectrum at 16.25_4 eV corresponding to A $^2\Sigma^+$ of $HC\ell^+$. Higher ionization potentials at 207.1 and 208.7 eV correspond to the removal of a 2p electron (50).

[c]Strongly broadened by preionization (lifetime 1.1 x 10^{-14}s)(55).

[d]Average B, D values; B(R,P) - B(Q) = + 0.385.

[e]Configuration $...\sigma^2\pi^3 4p\pi$.

[f]Average B, D values; $B(1^+) - B(1^-) = - 0.06_0$.

[g]Refers to Δ^+; Q branch not resolved.

[h]Configuration $...\sigma^2\pi^3 4p\sigma$.

[i]Average B, D values; $B(\Pi^+) - B(\Pi^-) = + 0.063$.

[j]Average B, D values; $B(\Pi^+) - B(\Pi^-) = - 0.04_0$.

[k]Average B, D values; $B(\Delta^+) - B(\Delta^-) = - 0.03_0$.

[ℓ]Average B, D values; $B(\Pi^+) - B(\Pi^-) = - 0.16_0$.

[m]Average B, D values; $B(\Pi^+) - B(\Pi^-) = - 0.06_7$.

[n]Configuration $...\sigma^2\pi^3 4s\sigma$.

[o]v=1,2,3 are increasingly diffuse; $B_1 = 9.29_6$. (44) give $\omega_e = 2817.5$, $\omega_e x_e = 66.0$, $B_e = 9.44$, $\alpha_e = 0.15$.

[p]Also observed in inert matrices (57).

[q]Typical "V" state with configuration $...\sigma\pi^4\sigma^*$.

[r]$\omega_e y_e = 0.113_3$; $\beta_e = 0.2_0 \times 10^{-4}$.

[s]Very extended progression in absorption, not yet analyzed in detail. The higher vibrational levels are strongly perturbed by Rydberg states (48)(62). The vibrational and rotational constants given were obtained from the emission spectrum with $v \leq 3$ (8)(9) but because of the perturbations have only very limited meaning.

[t]Diffuse rotational structure; 1-0 and 2-0 are increasingly diffuse.

[u]The $b_2 \leftarrow X$ and $b_0 \leftarrow X$ components have only 1/50 of the intensity of $b_1 \leftarrow X$.

[v]Configuration $...\sigma^2\pi^3\sigma^*$.

[w]Absorption coefficient k=40.

1.2745 x10^{-10} m

State	T_e	ω_e	$\omega_e x_e$	B_e	α_e	D_e (10^{-4} cm^{-1})	r_e (Å)	Observed Transitions Design.	ν_{00}	References
$^1H^{35}Cl$ (continued)										
X $^1\Sigma^+$	0	2990.946_3[x]	52.8186[y]	10.59341_6[xz]	0.30718_1[a']	5.3194[zb']	1.27455_2[c']	Rot.-vibr. bands[d'e']		(19)(28)
								Rotation spectrum[f'e']		(4)(23)(52)
								Raman cross sections		(58)
								Mol. beam electric[g'] and		(41)(64)
								magnetic[h'] reson.		(37)

$^1H^{35}Cl$ (continued):

[x] Applying the Dunham corrections (28) obtain ω_e = 2991.0904 and B_e = 10.593553. Additional corrections (adiabatic, non-adiabatic) discussed by (49). Vibrational levels up to v=5 have been observed in infrared absorption (12)(19)(28) and emission (10), higher levels in the V→X bands (8)(9). Dunham potential coefficients (61). Most recent ab initio values of the ground state molecular constants (59); charge distribution (40).

[y] $\omega_e y_e = +\ 0.2243_7$, $\omega_e z_e = -\ 0.0121_8$ (28).

[z] Slightly different constants in (11)(26)(31). These papers and (39) give also constants for $H^{37}Cl$.

[a'] $+\ 0.001772_4(v+\frac{1}{2})^2 - 0.0001201(v+\frac{1}{2})^3$.

[b'] $-\ 7.51_0 \times 10^{-6}(v+\frac{1}{2}) + 4.0_0 \times 10^{-7}(v+\frac{1}{2})^2$; higher order terms in (28). See also (30).

[c'] Uncorrected value from the $B_e(\equiv Y_{01})$ given in the table. The internuclear distance at the minimum of the Born-Oppenheimer curve is r_e = 1.2746149 Å (49)(63).

[d'] Absolute intensities (cm^{-2}atm^{-1}) of the

1-0 band:	130	(5)
2-0 band:	2.9	(5)
	3.70	(17)(45)
3-0 band:	0.023	(5)

[e'] Pressure-induced shifts (by foreign gases) of rotation-vibration and rotation lines (13)(14)(21)(22)(24). For discussions of pressure-induced bands and pure rotation lines (ΔJ=2) see (32)(36). Self and foreign-gas line broadening (5)(7)(16)(17)(18)(29)(43)(45)(47)(52). Infrared absorption in liquid and solid phases (42)(51).

[f'] Absolute intensity measurements (25)(34).

[g'] μ_{el}(v=0,1,2) = 1.1085, 1.1390, 1.1685 D, resp. (41). Dipole moment function (41)(54); see also (53)(56). g_J = 0.4594, also quadrupole and other hyperfine coupling constants (41)(64); see also (35)(53).

[h'] Proton spin - rotation interaction constant (15)(37).

$^{1}HC\ell$: (1) Price, PRS A 167, 216 (1938).
(2) Datta, Banerjee, PNISI 7, 305 (1941).
(3) Romand, AP(Paris) (12) 4, 527 (1949).
(4) Hansler, Oetjen, JCP 21, 1340 (1953).
(5) Benedict, Herman, Moore, Silverman, CJP 34, 850 (1956); JCP 26, 1671 (1957).
(6) Legay, CaP 11, 383 (1957); RO 37, 11 (1958).
(7) Babrov, Ameer, Benesch, JMS 3, 185 (1959).
(8) Jacques, D.Phil. Thesis (Oxford, 1959).
(9) Jacques, Barrow, PPS 73, 538 (1959).
(10) Mould, Price, Wilkinson, SA 16, 479 (1960).
(11) Plyler, Tidwell, ZE 64, 717 (1960).
(12) Rank, Birtley, Eastman, Rao, Wiggins, JOSA 50, 1275 (1960).
(13) Rank, Eastman, Birtley, Wiggins, JCP 33, 323 (1960).
(14) Ben-Reuven, Kimel, Hirshfeld, Jaffe, JCP 35, 955 (1961).
(15) Leavitt, Baker, Nelson, Ramsey, PR 124, 1482 (1961).
(16) Goldring, Benesch, CJP 40, 1801 (1962).
(17) Jaffe, Kimel, Hirshfeld, CJP 40, 113 (1962).
(18) Plyler, Thibault, JRNBS A 66, 435 (1962).
(19) Rank, Eastman, Rao, Wiggins, JOSA 52, 1 (1962).
(20) Watanabe, Nakayama, Mottl, JQSRT 2, 369 (1962).
(21) Gebbie, Stone, PPS 82, 309 (1963).
(22) Jaffe, Friedmann, Hirshfeld, Ben-Reuven, JCP 39, 1447 (1963).
(23) Jones, Gordy, PR A 135, 295; 136, 1229 (1964).
(24) Jaffe, Hirshfeld, Ben-Reuven, JCP 40, 1705 (1964).
(25) Chamberlain, Gebbie, Nature 208, 480 (1965).
(26) Lévy, Rossi, Joffrin, Thanh, JCPPB 62, 600 (1965).
(27) Nicholson, JCP 43, 1171 (1965).
(28) Rank, Rao, Wiggins, JMS 17, 122 (1965).
(29) Alamichel, Legay, JP(Paris) 27, 233 (1966).
(30) Herman, Asgharian, JCP 45, 2433 (1966).
(31) Lévy, Rossi, Haeusler, JP(Paris) 27, 526 (1966).
(32) Atwood, Vu, Vodar, SA A 23, 553 (1967).
(33) Frost, McDowell, Vroom, JCP 46, 4255 (1967).
(34) Sanderson, AO 6, 1527 (1967).
(35) Tokuhiro, JCP 47, 109 (1967).
(36) Weiss, Cole, JCP 46, 644 (1967).
(37) Code, Khosla, Ozier, Ramsey, Yi, JCP 49, 1895 (1968).
(38) Lempka, Passmore, Price, PRS A 304, 53 (1968).
(39) Webb, Rao, JMS 28, 121 (1968).
(40) Cade, Bader, Henneker, Keaveny, JCP 50, 5313 (1969).
(41) Kaiser, JCP 53, 1686 (1970).
(42) Katz, Ron, CPL 7, 357 (1970).
(43) Lévy, Mariel-Piollet, Bouanich, Haeusler, JQSRT 10, 203 (1970).
(44) Tilford, Ginter, Vanderslice, JMS 33, 505 (1970).
(45) Toth, Hunt, Plyler, JMS 35, 110 (1970).
(46) Weiss, Lawrence, Young, JCP 52, 2867 (1970).
(47) Rich, Welsh, CPL 11, 292 (1971).
(48) Tilford, Ginter, JMS 40, 568 (1971).
(49) Bunker, JMS 42, 478 (1972).
(50) Hayes, Brown, PR A 6, 21 (1972).
(51) Khatibi, Vu, JCPPB 69, 654, 662, 674 (1972).
(52) Rosenberg, Lightman, Ben-Reuven, JQSRT 12, 219 (1972).
(53) Bunker, JMS 45, 151 (1973).
(54) Smith, JQSRT 13, 717 (1973).
(55) Terwilliger, Smith, JMS 45, 366 (1973); JCP 63, 1008 (1975).
(56) Kaiser, JQSRT 14, 317 (1974).
(57) Boursey, JCP 62, 3353 (1975).
(58) Cherlow, Hyatt, Porto, JCP 63, 3996 (1975).
(59) Meyer, Rosmus, JCP 63, 2356 (1975).
(60) Schwarz, CP 11, 217 (1975).
(61) Ogilvie, Koo, JMS 61, 332 (1976).
(62) Douglas, Greening, unpublished.
(63) Watson, JMS 45, 99 (1973).
(64) de Leeuw, Dymanus, JMS 48, 427 (1973).

State	T_e	ω_e	$\omega_e x_e$	B_e	α_e	D_e (10^{-4} cm^{-1})	r_e (Å)	Observed Transitions Design.	Observed Transitions ν_{00}	References
$^2H^{35}Cl$		μ = 1.90441364		D_0^0 = 4.485$_2$ eV[a]		I.P. = 12.75$_6$ eV[b]				DEC 1976 A
		Numerous absorption bands above 123000 cm^{-1}; see ^{1}HCℓ.								(25)
M ($^1\Sigma^+$)	(117670)	[1220]	v=0...8 observed.	See ^{1}HCℓ.				M← X,	117214	(25)
L ($^1\Sigma^+$, $^1\Pi$)	111268	1126	37	See ^{1}HCℓ.				L← X,	110748	(25)
		Many other absorption bands in the region 83000 - 93000 cm^{-1}; see ^{1}HCℓ.								(27)
K $^1\Pi$	(89945)	[1858.8] Z		[4.9306][c]			[1.3399]	K← X,	R 89708.9 Z	(27)
H $^1\Sigma^+$	(88944)	[1631.8] Z		[4.6578]		[0.14$_7$]	[1.3786]	H← X,	R 88694.0 Z	(27)
E $^1\Sigma^+$	(84417)	[1186.9] Z		[3.2960]		[-2.48]	[1.6388]	E← X,	R 83944.9 Z	(27)
f_1 $^3\Delta_1$	[83626.2]			[5.21$_0$]		[0.96]	[1.303]	f_1← X,	R 82560.4 Z	(22)
D $^1\Pi$	(82632)	[1918.8] Z		5.14$_2$[d]	0.152	[2.$_0$][d]	1.312	D← X,	R 82525.9 Z	(22)
d_0 $^3\Pi_0$	[83350.3]			[5.01$_6$]		[1.3$_5$]	[1.328]	d_0← X,	R 82284.5 Z	(22)
f_2 $^3\Delta_2$	[83140.0]			[5.32$_8$]		[3.9]	[1.289]	f_2← X,	R 82074.2 Z	(22)
d_1 $^3\Pi_1$	[82855.0]			[5.13$_7$][e]		[1.9][e]	[1.313]	d_1← X,	R 81789.2 Z	(22)
d_2 $^3\Pi_2$	[82695.6]			[4.74$_7$][f]		[-7][f]	[1.366]	d_2← X,	R 81629.8 Z	(22)
C $^1\Pi$	77558.5	2027.1 Z	34.98[g]	4.962[h]	0.120	[1.16]	1.336	C← X,	R 77497.6 Z	(9)(18)*
b_0 $^3\Pi_0$	[76548.0]			[5.21$_8$]			[1.302]	b_0← X,	R 75482.2 Z	(18)*
b_1 $^3\Pi_1$	75199.3	2015.4 Z	29.1	[5.10$_0$][i]	0.12$_9$[j]	[3.$_5$]	1.309	b_1← X,	R 75133.9 Z	(9)(18)*
b_2 $^3\Pi_2$	[75912.7]			[4.90$_5$]			[1.343]	b_2← X,	R 74846.9 Z	(18)*
X $^1\Sigma^+$	0	2145.163 Z	27.1825[k]	5.448794[ℓ]	0.113291[k]	1.39[m]	1.274581	Rot.-vibr. bands[nop]		(2)(3)(7)(8)
								Rotation sp.[p]		(1)(5)(21)(23)
								Mol. beam el. reson.[q]		(16)

$^2HC\ell$: [a]From $D_0^0(^1HC\ell)$.

[b]From the photoelectron spectrum (19); via the A $^2\Sigma^+$ state of $DC\ell^+$ one finds 12.73_2 eV.

[c]Average B value; B(R,P) - B(Q) = + 0.1028.

[d]Average B, D values; $B(\Pi^+) - B(\Pi^-)$ = + 0.022 in v=0 and - 0.006 in v=1.

[e]Average B, D values; $B(\Pi^+) - B(\Pi^-)$ = - 0.022.

[f]Average B, D values; $B(\Pi^+) - B(\Pi^-)$ = - 0.05_4.

[g]$\omega_e y_e$ = + 0.39.

[h]$B_0(\Pi^+) - B_0(\Pi^-)$ = + 0.0034 (9). Lines are slightly diffuse but much sharper than in the corresponding $^1HC\ell$ bands.

[i]$B(\Pi^+) - B(\Pi^-)$ = + 0.01_0 (9).

[j]v=1 and 2 are increasingly diffuse.

[k]$\omega_e y_e$ = + 0.08649, $\omega_e z_e$ = - 0.00355, γ_e = + 0.0004589, $\delta_e(Y_{31})$ = - 0.00002103 from $^1HC\ell$ by isotope relations (8). Constants for $^2H^{37}C\ell$ in (2)(3)(5)(7)(15).

[ℓ]The rotational constants, derived from infrared data (8), agree with the more recent microwave data of (21).

[m]β_e = - $0.013_3 \times 10^{-4}$; H_e = $2.3_1 \times 10^{-9}$ (7).

[n]1-0, 2-0, 3-0 bands in absorption (2)(3)(8), Δv=1 sequence in emission (7); P(7), P(8), P(9) lines of the 2-1 band observed in D_2-$C\ell_2$ explosion laser (14). IR absorption in solid argon and other low-temperature matrices, hindered rotation (20).

[o]Absolute intensities [$cm^{-2}atm^{-1}$] of the

1-0 band:	66.6	(4)
2-0 band:	1.06	(4)
3-0 band:	0.0059	(4)

[p]Pressure broadening in the 1-0 band (both self-broadening and broadening by foreign gases) (10)(12), in the rotation spectrum (23); pressure shifts (11).

[q]$\mu_{e\ell}$(v=0) = 1.1033 D, $\mu_{e\ell}$(v=1) = 1.1256 D (16). Dipole moment function (16)(24); see also (26). Quadrupole and other hyperfine coupling constants (16); see also (13).

(1) See ref. (4) of $^1HC\ell$.
(2) Pickworth, Thompson, PRS A 218, 37 (1953).
(3) Van Horne, Hause, JCP 25, 56 (1956).
(4) See ref. (5) of $^1HC\ell$.
(5) Cowan, Gordy, PR 111, 209 (1958).
(6) Burrus, JCP 31, 1270 (1959).
(7) See ref. (10) of $^1HC\ell$.
(8) See ref. (19) of $^1HC\ell$.
(9) Stamper, CJP 40, 1274 (1962).
(10) Crane-Robinson, Thompson, PRS A 272, 441, 453 (1963).
(11) See ref. (24) of $^1HC\ell$.
(12) James, Thibault, JCP 40, 534 (1964).
(13) See ref. (35) of $^1HC\ell$.
(14) Corneil, Pimentel, JCP 49, 1379 (1968).
(15) See ref. (39) of $^1HC\ell$.
(16) See ref. (41) of $^1HC\ell$.
(17) See ref. (42) of $^1HC\ell$.
(18) See ref. (44) of $^1HC\ell$.
(19) See ref. (46) of $^1HC\ell$.
(20) Davies, Hallam, TFS 67, 3176 (1971).
(21) De Lucia, Helminger, Gordy, PR A 3, 1849 (1971).
(22) See ref. (48) of $^1HC\ell$.
(23) See ref. (52) of $^1HC\ell$.
(24) See ref. (54) of $^1HC\ell$.
(25) See ref. (55) of $^1HC\ell$.
(26) See ref. (56) of $^1HC\ell$.
(27) See ref. (62) of $^1HC\ell$.

State	T_e	ω_e	$\omega_e x_e$	B_e	α_e	D_e $(10^{-4}cm^{-1})$	r_e (Å)	Observed Transitions Design.	Observed Transitions ν_{00}	References
$^3H^{35}Cl$		$\mu = 2.7765715_{3}$		$D_0^0 = 4.507_{8}$ eV[a]						DEC 1976
X $^1\Sigma^+$	0	[1739.10] Z	(18.64)	[3.705089][b]	0.0611[b]	0.77[b]	1.2749	Rotation-vibration sp.		(2)
								Rotation sp.[c]		(1)
$^1H^{35}Cl^+$				$D_0^0 = 4.65_{3}$ eV[a]						DEC 1976 A
A $^2\Sigma^+$	28626.5	1606.4_{7}[b] Z	40.3_{1}[b]	7.505_{4}[bcd]	0.331_{3}[b]	[6.43_{5}][e]	1.5142	A→X,[f]	R 28095.97[g] Z	(1)(3)(5)
X $^2\Pi_i$	0[h]	2673.69[i] Z	52.537[i]	9.9566_{1}[jk]	0.3271_{6}[j]	[5.47_{7}][ℓ]	1.31468			
$^2H^{35}Cl^+$				$D_0^0 = 4.69_{7}$ eV[m]						DEC 1976
A $^2\Sigma^+$	28629.2	1152.0_{9}[b] Z	20.4_{5}[b]	3.859_{5}[bn]	0.122_{2}[b]	[1.717][o]	1.5144	A→X,[f]	R 28247.65[g] Z	(1)(3)(5)
X $^2\Pi_i$	0[p]	1918.56[i] Z	27.263[i]	5.1216_{4}[qk]	0.1203_{7}[q]	[1.459][r]	1.31466			
$^1HCl^-$										DEC 1976 A
		Resonances due to inverse preionization have been found in the transmission of electrons through HCℓ in the energy range 9.1 - 11.0 and 12.5 - 13.9 eV. Interpretation analogous to HBr^-.								(2)
X $^2\Sigma^+$	a									

$^{3}HC\ell$: [a]From $D_0^0(^1HC\ell)$.

[b]B_0 from the microwave spectrum (1), α_e and D_e from the infrared spectrum (2). (1)(2) give also data for $T^{37}C\ell$.

[c]Nuclear quadrupole coupling constants (3).

(1) Burrus, Gordy, Benjamin, Livingston, PR 97, 1661 (1955).

(2) Jones, Robinson, JCP 24, 1246 (1956).

(3) Tokuhiro, JCP 47, 109 (1967).

$^{1}HC\ell^+$, $^{2}HC\ell^+$:

[a]Refers to $^2\Pi_{3/2}$; from $D_0^0(HC\ell)$ + I.P.($C\ell$) - I.P.($HC\ell$).

[b]Equilibrium values of (3); G(v) and B_v are far from linear but higher terms in $(v+\frac{1}{2})$ have not been given. (3)(5) give T_v, B_v values up to v=6 ($HC\ell^+$) and 9 ($DC\ell^+$); their T_v values differ systematically by ~ 2.34 cm^{-1}.

[c]Spin doubling constants for v=0...6 in (3)(5);
$\gamma(v=0;\ N) = +\ 0.5942 - 0.000034\,N(N+1)$.

[d]Higher vibrational levels are predicted (2) to be predissociated by the $^4\Pi$ state from $H(^2S) + C\ell^+(^3P)$.

[e]$H_0 = 2.3_0 \times 10^{-8}$. Additional D_v, H_v values in (3)(5).

[f]Variation of transition moment with r derived from observed intensities by (4) who also give Franck-Condon factors.

[g]From (5); (3) give 28098.35 ($HC\ell^+$) and 28250.08 ($DC\ell^+$); see [b].

[h]Spin-orbit coupling constants for v=0,1,2 in (5);
$A(v=0;\ J) = -\ 648.13_1 + 0.00209(J+\frac{1}{2})^2$.

[i]From $\Delta G(1/2)$ and $\Delta G(3/2)$ of (5).

[j]From B_0, B_1, B_2 of (5); $\gamma_e = +\ 0.0026_6$.

[k]Λ-type doubl. const. [$p \approx 0.60(HC\ell^+)$, $0.31(DC\ell^+)$] in (5).

[ℓ]$H_0 = 1.5_2 \times 10^{-8}$; additional D_v, H_v values in (5).

$^{1}HC\ell^+$, $^{2}HC\ell^+$ (continued):

[m]Refers to $^2\Pi_{3/2}$; from $D_0^0(DC\ell)$ + I.P.($C\ell$) - I.P.($DC\ell$).

[n]Spin-doubling constants for v=0...9 in (3)(5);
$\gamma(v=0;\ N) = +\ 0.3060 - 8.6 \times 10^{-6} N(N+1)$.

[o]$H_0 = 0.40 \times 10^{-8}$; additional D_v, H_v values in (3)(5).

[p]Spin-orbit coupling constants for v=0,1,2 in (5);
$A(v=0;\ J) = -\ 648.17_0 + 0.00095(J+\frac{1}{2})^2$.

[q]From B_0, B_1, B_2 of (5); $\gamma_e = +\ 0.00061$.

[r]$H_0 = +\ 0.29 \times 10^{-8}$; additional D_v, H_v values in (5).

(1) Norling, ZP 104, 638; 106, 177 (1937).

(2) Raftery, Richards, JP B 6, 1301 (1973).

(3) Sheasley, Mathews, JMS 47, 420 (1973).

(4) Haugh, Schneider, Smith, JMS 51, 123 (1974).

(5) Saenger, Zare, Mathews, JMS 61, 216 (1976).

$^{1}HC\ell^-$: [a]The ground state X $^2\Sigma^+$ of $HC\ell^-$ lies presumably above the ground state X $^1\Sigma^+$ of $HC\ell$; see (1).

(1) Fiquet-Fayard, JP B 7, 810 (1974).

(2) Spence, Noguchi, JCP 63, 505 (1975).

State	T_e	ω_e	$\omega_e x_e$	B_e	α_e	D_e $(10^{-4} cm^{-1})$	r_e (Å)	Observed Transitions Design.	Observed Transitions ν_{00}	References
$^4He_2$		$\mu = 2.0013016_3$		$D_e^0 = 0.00090$ eV^{a}		I.P. = 22.223 eV^{c}				NOV 1976 A
				$D_0(a^3\Sigma_u^+) = 1.850$ eV^{b}						
				$D_0(A^1\Sigma_u^+) = 2.355$ eV^{b}						
		Rydberg series of npπ $^3\Pi_g$ levels (b, e, i, ℓ, p, r, s, t, u, ...)		} $\nu^d = 34302.20 - R/(n - 0.071)^2$, n = 2...17.						(23)(39)
		Rydberg series of npσ $^3\Sigma_g^+$ levels (c, g, k', n, p', r', s', t', ...)		} $\nu^d = 34302.20 - R/(n - 0.777)^2$, n = 3...12.						(23)(39)
u $^3\Pi_g$ 10pπ	177291	$[1628.6_9]^{e}$ Z	(35.2_5)	7.21_2^{e}	0.22_2^{e}	$[5.03]^{e}$	1.081	u→a, R	33189.16 Z	(23)
t' $^3\Sigma_g^+$ 10pσ	[177969]			$[8.27]^{f}$				t'→a, V	33026.6 Z	(23)
t $^3\Pi_g$ 9pπ	177027	$[1629.1_5]^{g}$ Z	(35.2_5)	7.21_2^{g}	0.23_0^{g}	$[5.07]^{g}$	1.081	t→a, R	32925.96 Z	(23)
s' $^3\Sigma_g^+$ 9pσ	[177636]			$[8.04]^{f}$				s'→a, V	32693.2 Z	(23)
s $^3\Pi_g$ 8pπ	176658	$[1629.30]^{h}$ Z	(35.2_5)	7.213^{h}	0.223_4^{h}	5.1^{h}	1.080_6	s→a, R	32556.66 Z	(23)
r' $^3\Sigma_g^+$ 8pσ	[177154]			$[7.78]^{f}$				r'→a, V	32211.7 Z	(23)
r $^3\Pi_g$ 7pπ	176117	$(1700.5_6)^{i}$	(35.2_5)	$[7.104_4]^{i}$	i	$[5.08]^{i}$	[1.0889]	r→d, R	11625.3 Z	(39)
								r→a, R	32016.56 Z	(23)
p' $^3\Sigma_g^+$ 7pσ	[176421]			$[7.543]^{f}$				p'→a,	31478.6 Z	(23)
q { $^3\Delta_u$ 6dδ	[176195]	}								
q { $^3\Pi_u$ 6dπ	[(176169)]	}		$[7.092]^{j}$		$[5.0]^{j}$	$[1.089_8]$	q→c, { V / R / R	20365 / (20330) / 20288	(1)(39)
q { $^3\Sigma_u^+$ 6dσ	[176120]	}						q→b, { V / R / R	26483 / (26466) / 26409	(1)(39)
p $^3\Pi_g$ 6pπ	175281	1701.18 Z	35.35	7.220^{k}	0.224	5.14	1.080_1	p→d, R	10788.6_8 Z	(39)
								p→a, R	31179.93 Z	(23)
o $^3\Sigma_u^+$ 6sσ	[176001]			[7.109]		[5.1]	$[1.088_5]$	o→c, V	20168.8 Z	(2)(39)
								o→b, R	26290.3 Z	(2)(39)

State	T_e	ω_e	$\omega_e x_e$	B_e	α_e	D_e ($10^{-4}cm^{-1}$)	r_e (Å)	Observed Transitions Design.	ν_{00}	References
$^4He_2$ (continued)										
n $^3\Sigma_g^+$ 6pσ	174389	[1619.52] Z	$(36._5)$	7.475_4 [ℓ]	0.249_0	$[7.2_1]$ [m]	1.0615	n→a,	30283.26 Z	(3)(67)
m $^3\Delta_u$ 5dδ	[174863]							m→c,	V (19039) R (18944) R 18899	(39)
m $^3\Pi_u$ 5dπ	[(174778)]			[7.09] [n] [7.07] [o]			$[1.09_1]$			
m $^3\Sigma_u^+$ 5dσ	[174730]							m→b,	V 25152 R (25070) R 25019	(39)
ℓ $^3\Pi_g$ 5pπ	173884	$[1633.9_6]$ [p] Z	$(35.2_5$	7.226 [p]	0.222 [p]	[5.12] [p]	1.079_7	ℓ→d,	R 9393.9 Z	(39)
								ℓ→a,	R 29785.31 Z	(23)

He_2: [a]Average of two independent values for the well depth obtained from measurements of the total (44) and differential (45)(66) elastic scattering cross sections (0.88_8 and 0.90_5 meV, resp.). Ab initio values range from 0.78 to 1.04 meV (33)(43)(51)(53)(56)(64). Both experiment and theory agree that no bound vibrational level exists in the potential well, i.e. $D_0^0 = 0$; see (27)(63). A somewhat higher D_e value (0.99 meV) was derived (61) from the temperature dependence of the relaxation time in dilute ^3He. For measurements of the short-range potential (0.49 < r(Å) < 1.56) see (57); the long-range potential is discussed by (28)(29).

[b]Based on $D_0^0(He_2^+)$. From a detailed interpretation of the Hopfield continuum and the 600 Å absorption and emission bands (49) derives $D_e(A^1\Sigma_u^+) = 2.50$ eV.

[c]Relative to $He(^1S) + He(^1S)$, i.e. $I.P.(He_2) = I.P.(He) - D_0^0(He_2^+)$. The I.P. for the lowest stable state (a $^3\Sigma_u^+$) is 4.25297 eV.

[d]Giving the v=0,N=0 levels (real or hypothetical) of npσ $^3\Sigma_g^+$ and npπ $^3\Pi_g^-$ relative to 2sσ $^3\Sigma_u^+$, v=0, N=0.

[e]Refers to Π^-. $B_0(^3\Pi^+) \approx 5.6$, strongly affected by ℓ-uncoupling.

[f]Effective value, strongly affected by ℓ-uncoupling.

[g]Refers to Π^-. $B_0(^3\Pi^+) = 5.87$, strongly affected by ℓ-uncoupling.

[h]Refers to Π^-. $B_0(^3\Pi^+) = 6.13$, strongly affected by ℓ-uncoupling.

[i]Constants refer to Π^-; $B_0(^3\Pi^+) = 6.37_5$ affected by ℓ-uncoupling. v=1 is perturbed; approximate deperturbed constants for Π^-: $B_1 = 6.88_6$, $\Delta G(\frac{1}{2}) = 1629.7$.

[j]Strong ℓ-uncoupling. The rotational constants (1) refer to the average of Π^- and Δ^-.

[k]Refers to Π^-. $B_0(^3\Pi^+) = 6.630$ affected by ℓ-uncoupling.

[ℓ]Several small accidental perturbations.

[m]$D_1 = 6.8_2 \times 10^{-4}$, $H_0 = 14._2 \times 10^{-8}$, $H_1 \approx -26 \times 10^{-8}$.

[n]Average of Π^- and Δ^- as given by (39).

[o]Average of Σ^+, Π^+, Δ^+ as given by (39).

[p]Refers to Π^-.

State	T_e	ω_e	$\omega_e x_e$	B_e	α_e	D_e ($10^{-4}cm^{-1}$)	r_e (Å)	Observed Transitions		References
								Design.	ν_{00}	
$^{4}He_2$ (continued)										
k $^{3}\Sigma_u^+$ 5sσ	(173698)	[1635.3] Z		7.23_{2}	0.23		1.079	k→c, V	18683.5 Z	(39)
								k→b, R	$24804._{8}$ Z	(39)
k' $^{3}\Sigma_g^+$ 5pσ	172236	1686.90 Z	38.10	7.379[a]	0.349[b]	[5.8]	1.068_{4}	k'→a, R	28127.58 Z	(67)
j $^{3}\Delta_u$ 4dδ	171573	1702.2_{4}[c] Z	35.0_{7}	7.208_{8}[c]	0.224_{8}[d]	5.2[c]	1.0810	j→c, V	16583.18[c] Z	(50)
j $^{3}\Pi_u$ 4dπ	171402	1680.9_{4}[c] Z	40.8_{1}	7.186_{0}[c]	0.229_{6}[e]	5.4[c]	1.0827	R	16400.69[c] Z	
								R	16316.54[f]	
j $^{3}\Sigma_u^+$ 4dσ	171323	1669.7_{9}[f]	39.0_{9}	Strongly perturbed by ℓ-uncoupling and by interaction with the h state.[h]				j→b, V	22704.5[c] Z	(50)
									22522.0[c] Z	
								R	22437.8[f]	
i $^{3}\Pi_g$ 4pπ	171290	[1637.94][g] Z	(35.2_{5})	7.242[g]	0.223[g]	$[5.1_{4}]$[g]	1.078_{5}	i→a, R	27193.01 Z	(23)(39)
h $^{3}\Sigma_u^+$ 4sσ	(170884)	[1637.9] Z		7.26_{4}[h]	0.23	(5.2_{4})	1.077	h→c, V	15870.7 Z	(2)(39)(50)
								h→b, R	21992.2 Z	(2)(39)(50)*
g $^{3}\Sigma_g^+$ 4pσ	167714	[1589.92] Z	(41)	7.220_{7}	0.247_{8}	[5.38][i]	1.0801	g→a, R	23597.00 Z	(67)
f $^{3}\Delta_u$ 3dδ	166303	1706.82 Z	35.10	7.230[j]	0.227[k]	[5.26][ℓ]	1.079_{4}	f→c, V	11316.06 Z	(17)
f $^{3}\Pi_u$ 3dπ	165877 [m]	1661.48 Z	44.79	7.136[j]	0.235[n]	[5.34][o]	1.086_{5}		10864.53 Z	
								R	10659.33 Z	
f $^{3}\Sigma_u^+$ 3dσ	165685	1635.77 Z	44.41	7.071[j]	0.246[p]	[5.31][q]	1.091_{4}	f→b, V	17437.3 Z	(13)*
									16985.8 Z	(13)(39)
								R	16780.6 Z	
e $^{3}\Pi_g$ 3pπ	165598	1721.22 Z	34.97_{0}[s]	7.283_{8}[r]	0.2215[t]	5.22	1.0754	e↔a,[u] R	21507.26 Z	(36)
d $^{3}\Sigma_u^+$ 3sσ	164479	1728.01 Z	36.13[v]	7.341_{2}	0.224_{4}[w]	5.32	1.0712	d→c, V	9502.7 Z	(13)
								d→b, R_V	15623.1 Z	(13)*
c $^{3}\Sigma_g^+$ 3pσ	155053	1583.85 Z	52.7_{4}[x]	7.004_{8}	0.310_{5}[y]	[5.56][z]	1.0966	c→a, R	10889.48 Z	(11)
b $^{3}\Pi_g$ 2pπ	148835	1769.07 Z	35.02[a']	7.447_{3}[b']	0.219_{6}[c']	[5.30][d']	1.0635	b→a, R	4768.2 Z	(5)(14)
a $^{3}\Sigma_u^+$ 2sσ	144048	1808.5_{6} Z	38.2_{1}[e'f']	7.703_{6}[g']	0.228_{1}[h']	5.56[i']	1.0457	(a − X)	144935[j']	

He_2: [a]Several small accidental perturbations.

[b]$\gamma_e = +0.030$.

[c]The vibrational and rotational constants refer to Π^- and Δ^- which are less affected by ℓ-uncoupling.

[d]$\gamma_e = -0.003_8$.

[e]$\gamma_e = -0.004_6$.

[f]Constants refer to $N' = 1$.

[g]Constants refer to $^3\Pi^-$.

[h](50) give average effective constants for the four interacting components $j(^3\Delta_u^+, {}^3\Pi_u^+, {}^3\Sigma_u^+)$ and $h\ {}^3\Sigma_u^+$.

[i]$H_0 = 2._0 \times 10^{-8}$.

[j]These constants are corrected for ℓ-uncoupling effects.

[k]$\gamma_e = -0.004_6$.

[ℓ]$D_1 = 5.28 \times 10^{-4}$, $D_2 = 5.5_0 \times 10^{-4}$, $H_0 = 2.5 \times 10^{-8}$.

[m]*Ab initio* calculations of $f\ {}^3\Pi_u$ and $F\ {}^1\Pi_u$ (10)(26) yield excellent agreement with the observed constants and confirm the presence of substantial potential maxima; see also (9).

[n]$\gamma_e = -0.0069$.

[o]$D_1 = 5.34 \times 10^{-4}$, $D_2 = 5.7_4 \times 10^{-4}$, $H_0 = 2.9 \times 10^{-8}$.

[p]$\gamma_e = -0.0095$.

[q]$D_1 = 5.34 \times 10^{-4}$, $D_2 = 5.45 \times 10^{-4}$, $H_0 = 0.9_1 \times 10^{-8}$.

[r]The rotational constants refer to Π^- (36); $B(^3\Pi^-) - B(^3\Pi^+) \approx +0.072$. Slightly different constants were given by (4) who also derived constants for $^3He_2$.

[s]$\omega_e y_e = -0.03_8$. See also [r].

[t]$\gamma_e = -0.0013_8$.

[u]Observed in absorption in a pulsed discharge (19).

[v]$\omega_e y_e = -0.126_7$.

[w]$\gamma_e = -0.0027_3$.

[x]$\omega_e y_e = -1.256_6$, $\omega_e z_e = -0.487_5$.

[y]$+ 0.162_9(v+\frac{1}{2})^2 - 0.065_5(v+\frac{1}{2})^3$.

[z]$D_1 = 5.76 \times 10^{-4}$, $D_2 = 6.11 \times 10^{-4}$, ...; $H_0 = 1.3_2 \times 10^{-8}$, $H_1 = -0.4_0 \times 10^{-8}$, $H_2 = -5.3 \times 10^{-8}$, ...

[a']$\omega_e y_e = -0.048_3$.

[b']The constants refer to $^3\Pi^-$ and were derived by (13) from the d,f→b bands; $B(^3\Pi^-) - B(^3\Pi^+) \approx +0.026$. The triplet splitting is partially resolved in the d→b bands (13).

[c']$\gamma_e = -0.0017_0$.

[d']$D_1 = 5.34 \times 10^{-4}$, ...; $H_0 = 2.8_0 \times 10^{-8}$, $H_1 = 3.6_0 \times 10^{-8}$,...

[e']$\omega_e y_e = -0.3_8$ (36).

[f']There is good experimental (20)(30) and theoretical (7) (21)(48) evidence for a potential maximum in this state. (20) place the maximum at 0.067 eV above the asymptote; the net dissociation energy is 1.850 eV.

[g']From molecular beam magnetic resonance experiments (58) (65) have determined the triplet splitting for N=1 and 3. The splitting constants (extrapolated to N=0) are $\lambda = -0.03666$, $\gamma = -0.000080_8\ cm^{-1}$. An *ab initio* calculation (55) gives $\lambda = -0.04089$.

[h']$\gamma_e = -0.0046_2$ (36).

[i']$\beta_e = +0.07 \times 10^{-4}$; $H_e \approx 2.7 \times 10^{-8}$ (36).

[j']Energy of a $^3\Sigma_u^+$, v=0, N=0 above $He(^1S) + He(^1S)$, based on $D_0^0(He_2^+) = 19073\ cm^{-1}$. See also [c'] p. 297 .

State		T_e	ω_e	$\omega_e x_e$	B_e	α_e	D_e ($10^{-4} cm^{-1}$)	r_e (Å)	Observed Transitions Design.	ν_{00}	References
$^4He_2$ (continued)											
S	$^1\Pi_g$ 8pπ	[177515]			(7.21)	(0.22)		(1.08_1)	S→A, R	$30228._6$ Z	(39)
R	$^1\Pi_g$ 7pπ	[176983]			(7.22)	(0.22)		(1.08_0)	R→A, R	$29696._4$ Z	(39)
P	$^1\Pi_g$ 6pπ	[176160]			(7.22)	(0.22)		(1.08_0)	P→A, R	$28873._9$ Z	(39)
M	$^1\Delta_u$ 5dδ	[(174838)]									
	$^1\Pi_u$ 5dπ	[(174788)]			[7.09][a] [7.07][b]			[1.09_1]	M→B,	(24050) (24000) (23960)	(39)
	$^1\Sigma_u^+$ 5dσ	[(174748)]									
L	$^1\Pi_g$ 5pπ	[174794]			(7.23)	(0.22_2)		(1.07_9)	L→A, R	$27507._8$ Z	(39)
J	$^1\Delta_u$ 4dδ	[172416]			[7.097][c]		[5.0]	[1.089_4]	J→C, V V R	14183.90[c] Z 14058.37[c] Z 13990.32[d]	(50)
	$^1\Pi_u$ 4dπ	[172290]			[7.080][c]		[5.4]	[1.090_8]			
	$^1\Sigma_u^+$ 4dσ	[172222][d]	Strongly perturbed by ℓ-uncoupling and interaction with H $^1\Sigma_u^+$.[e]						J→B, V V R	21627.9[c] Z 21502.4[c] Z 21434.3[d]	(50)
I	$^1\Pi_g$ 4pπ	[172266]			(7.24_2)	(0.22_3)		(1.078)	I→A, R	24979.6 Z	(39)
H	$^1\Sigma_u^+$ 4sσ	[171951]			(7.26)[e]	(0.23)		(1.07_7)	H→C, V	13719.5 Z	(39)(50)
									H→B, R	21163.5 Z	(39)(50)
F	$^1\Delta_u$ 3dδ	166304	1706.59 Z	35.06	7.230[g]	0.225[h]	[5.20][i]	1.079_4	F→B, V V R	16360.9 Z 16008.3 Z 15837.5 Z	(12)(17)
	$^1\Pi_u$ 3dπ	165971 [f]	1670.57 Z	40.03	7.156[g]	0.235	[5.24][j]	1.084_9			
	$^1\Sigma_u^+$ 3dσ	(165813)	[1564.25] Z	(40)	7.098[g]	0.246	[5.21][k]	1.089_4			
E	$^1\Pi_g$ 3pπ	165911	1721.1_9 Z	34.7_6[ℓ]	7.270_5[m]	0.215_6[n]	5.20	1.0764	E→A, R	19476.61 Z	(36)
D	$^1\Sigma_u^+$ 3sσ	165085	1746.43 Z	35.54	7.365	0.218_0[o]	5.24[p]	1.069_4	D→B,[q] R_V	15161.81 Z	(12)
									D→X,[r]	continuum	
C	$^1\Sigma_g^+$ 3pσ	157415	1653.43 Z	41.0_4[s]	7.052	0.215[t]	5.08[u]	1.092_9	C→A, R	10945.50 Z	(11)(39)
B	$^1\Pi_g$ 2pπ	149914	1765.76 Z	34.39[v]	7.403[w]	0.216[w]	5.02[w]	1.066_7	(B − A)	3501.5[x]	
A	$^1\Sigma_u^+$ 2sσ	146365 [y]	1861.3_3 Z	35.2_8[z]	7.778_9	0.216_6[a']	5.44	1.0406	A↔X,[b']	147279[c']	(8)(18)(22)
									Hopfield continuum		(38)

State	T_e	ω_e	$\omega_e x_e$	B_e	α_e	D_e (10^{-4} cm^{-1})	r_e (Å)	Observed Transitions Design.	v_{00}	References
$^4He_2$ (continued)										
X $^1\Sigma_g^+$	0	Repulsive potential with very small well (D_e = 0.90 meV).					2.97[d']			

He_2: [a]Average of Π^- and Δ^- as given by (39).

[b]Average of Σ^+, Π^+, Δ^+ as given by (39).

[c]These constants refer to Π^- and Δ^- which are less affected by ℓ-uncoupling.

[d]Refers to N=1.

[e](50) give average effective constants for the four interacting components J($^1\Delta_u^+$, $^1\Pi_u^+$, $^1\Sigma_u^+$) and H $^1\Sigma_u^+$.

[f]See [m] p. 295.

[g]These constants are corrected for ℓ-uncoupling effects.

[h]γ_e = - 0.0045.

[i]D_1 = 5.26 x 10^{-4}, H_0 = 2.1$_8$ x 10^{-8}.

[j]D_1 = 5.26 x 10^{-4}, H_0 = 1.9$_7$ x 10^{-8}.

[k]D_1 = 5.29 x 10^{-4}.

[ℓ]$\omega_e y_e$ = - 0.02$_3$.

[m]The rotational constants refer to Π^- (36); B(Π^-) - B(Π^+) ≈ + 0.044.

[n]γ_e = - 0.0022$_7$.

[o]γ_e = - 0.005$_9$.

[p]β_e = + 0.1$_9$ x 10^{-4}.

[q]Franck-Condon factors (25).

[r]The weak maximum near 676 Å in the Hopfield continuum is ascribed by (37) to the transition D→X.

[s]$\omega_e y_e$ = + 0.354$_9$, $\omega_e z_e$ = - 0.131$_5$. Calculations of (47) give a potential hump of 0.22 eV at 2.09 Å; see also (59) and (60).

[t]γ_e = - 0.0111.

[u]β_e = + 0.1$_0$ x 10^{-4}; H_0 = 1.7$_2$ x 10^{-8}, ...

[v]$\omega_e y_e$ = - 0.026$_7$ (12).

[w]B_e refers to Π^-; B(Π^-) - B(Π^+) = + 0.019; γ_e = - 0.001$_5$, β_e = + 0.05 x 10^{-4} (12).

[x]From (39).

[y]RKR potential curve (34)[see also (39)]; ab initio potential (42). The latter gives vibrational and rotational levels in good agreement with the experimental values. (8) have established, from the absorption and emission bands near 600 Å, a potential maximum of 0.059 eV in the A $^1\Sigma_u^+$ state. Theoretical work by (15)(16)(47) gives maxima of 0.084, 0.153, 0.061 eV, respectively; see also (59)(60).

[z]$\omega_e y_e$ = - 0.13$_6$ (36).

[a']γ_e = - 0.0027$_3$ (36).

[b']Transitions from the low vibrational levels of A $^1\Sigma_u^+$ to X $^1\Sigma_g^+$ give rise to the Hopfield continuum; see MOLSPEC 1, 404. Transitions from the high vibrational levels as well as the continuous range of energy levels of A $^1\Sigma_u^+$ to X $^1\Sigma_g^+$ give rise to diffuse bands near 600 Å observed in emission (18)(22) and absorption (8)(38) with quite different intensity distribution. See also (6)(37) and (31)(52)(54). Observed absorption coefficients near 600 Å (38) agree fairly well with those predicted by (41).

[c']Energy of the v=0,N=0 level of A $^1\Sigma_u^+$ relative to He(^{1}S) + He(^{1}S), calculated from the corresponding value for a $^3\Sigma_u^+$ by (continued p. 299)

State	T_e	ω_e	$\omega_e x_e$	B_e	α_e	D_e $(10^{-4} cm^{-1})$	r_e (Å)	Observed Transitions Design.	ν_{00}	References
$^4He_2^+$		$\mu = 2.0011645$		$D_0^0 = 2.365$ eV[a]						NOV 1976 A
A $^2\Sigma_g^+$		Repulsive state arising from $He(1^1S) + He^+(1^2S)$.[b]								(4)(6)(8)
X $^2\Sigma_u^+$	0	$1698._5$[c]	$35._3$	7.211[c]	0.224	5.1	1.080_8			(9)(10)(14)
$^4He_2^{++}$		$\mu = 2.0010273$								NOV 1976 A
		Calculations of excited states by (3)(5).								
X $^1\Sigma_g^+$		(3295)[d]					(0.704)[e]			(1)(2)(7)

He_2^+, He_2^{++}:

[a]From the theoretical $D_e = 2.469 \pm 0.006$ eV (12). From He-He$^+$ differential scattering cross sections (6) and (13) obtain $D_e = 2.34$ and 2.55 eV, respectively.

[b]The potential functions of several excited states have been calculated by (4)(11). $A^2\Sigma_g^+$, the lowest $^2\Sigma_g^+$ state, has a very small van der Waals minimum at 5.3 Å and a hump near 0.8 Å caused by an avoided crossing with the second lowest $^2\Sigma_g^+$ state; see (8). The states $^4\Sigma_u^+$, $^2\Sigma_g^+$, $^2\Sigma_u^+$ arising from $He(2^3S) + He^+(1^2S)$ are found to have minima with $D_e \approx 1.1$, 0.57, 0.46 eV, respectively; several other states are purely repulsive.

[c]Constants obtained by extrapolation from the $np\pi\ ^3\Pi_g$ Rydberg series of He_2; they agree with values derived from the theoretical potential function of (4) and (8).

[d]Theoretical value (2).

[e]The theoretical potential function according to (2)(7) shows a local minimum at 0.704 Å and 69600 cm^{-1} above $He^+ + He^+$, separated from this limit by a maximum at 1.151 Å and 82200 cm^{-1}. He_2^{++} has not yet been found mass-spectrometrically, nor has its spectrum been observed.

(1) Kolos, Roothaan, RMP 32, 219 (1960).
(2) Fraga, Ransil, JCP 37, 1112 (1962).
(3) Browne, JCP 42, 1428 (1965).
(4) Browne, JCP 45, 2707 (1966).
(5) Jennings, JCP 46, 2442 (1967).
(6) Olson, Mueller, JCP 46, 3810 (1967).
(7) Conroy, Bruner, JCP 47, 921 (1967).
(8) Gupta, Matsen, JCP 47, 4860 (1967).
(9) See ref. (23) of He_2.
(10) See ref. (30) of He_2.
(11) Bardsley, PR A 3, 1317 (1971).
(12) Liu, PRL 27, 1251 (1971).
(13) Weise, Mittmann, Ding, Henglein, ZN 26 a, 1122 (1971).
(14) See ref. (46) of He_2.

He_2 (continued):

adding the energy difference $\Delta v = 2343.91 \pm 0.05$ cm^{-1} as determined by (62) from singlet-triplet anticrossings. Optical measurements give $\Delta v = 2344.1$ cm^{-1} (30). The relative position of the levels is much more accurately known than their absolute values.

[d'] From differential elastic scattering measurements (45)(66).

(1) Dieke, ZP 57, 71 (1929).
(2) Dieke, Imanishi, Takamine, ZP 57, 305 (1929).
(3) Jevons, "Band Spectra of Diatomic Molecules", Physical Society, London (1932).
(4) Dieke, Robinson, PR 80, 1 (1950).
(5) Hepner, Herman, CR 243, 1504 (1956).
(6) Tanaka, Jursa, LeBlanc, JOSA 48, 304 (1958).
(7) Poshusta, Matsen, PR 132, 307 (1963).
(8) Tanaka, Yoshino, JCP 39, 3081 (1963); 50, 3087 (1969).
(9) Mulliken, PR A 136, 962 (1964).
(10) Browne, PR A 138, 9 (1965).
(11) Ginter, JCP 42, 561 (1965).
(12) Ginter, JMS 17, 224 (1965).
(13) Ginter, JMS 18, 321 (1965).
(14) Gloersen, Dieke, JMS 16, 191 (1965).
(15) Allison, Browne, Dalgarno, PPS 89, 41 (1966).
(16) Scott, Greenawalt, Browne, Matsen, JCP 44, 2981 (1966).
(17) Ginter, JCP 45, 248 (1966).
(18) Mies, Smith, JCP 45, 994 (1966).
(19) Callear, Hedges, Nature 215, 1267 (1967).
(20) Ludlum, Larson, Caffrey, JCP 46, 127 (1967).
(21) Klein, Greenawalt, Matsen, JCP 47, 4820 (1967).
(22) Smith, JCP 47, 1561 (1967); 49, 4817 (1968).
(23) Ginter, Ginter, JCP 48, 2284 (1968).
(24) Murrell, Shaw, MP 15, 325 (1968).
(25) Zhirnov, Shadrin, OS(Engl. Transl.) 24, 478 (1968).
(26) Gupta, Matsen, JCP 50, 3797 (1969).
(27) Murrell, MP 16, 601 (1969).
(28) Alexander, JCP 52, 3354 (1970).
(29) Bruch, McGee, JCP 52, 5884 (1970).
(30) Ginter, Battino, JCP 52, 4469 (1970).
(31) Michaelson, Smith, CPL 6, 1 (1970).
(32) DONNSPEC (1970).
(33) Schaefer, McLaughlin, Harris, Alder, PRL 25, 988 (1970).
(34) Smith, Chow, JCP 52, 1010 (1970).
(35) Bennewitz, Busse, Dohmann, CPL 8, 235 (1971).
(36) Brown, Ginter, JMS 40, 302 (1971).
(37) Chow, Smith, JCP 54, 1556 (1971).
(38) Chow, Smith, Waggoner, JCP 55, 4208 (1971).
(39) Ginter, in (32).
(40) Liu, PRL 27, 1251 (1971).
(41) Sando, Dalgarno, MP 20, 103 (1971).
(42) Mukamel, Kaldor, MP 22, 1107 (1971).
(43) McLaughlin, Schaefer, CPL 12, 244 (1971).
(44) Bennewitz, Busse, Dohmann, Oates, Schrader, ZP 253, 435 (1972); PRL 29, 533 (1972).
(45) Farrar, Lee, JCP 56, 5801 (1972).
(46) Ginter, Brown, JCP 56, 672 (1972).
(47) Guberman, Goddard, CPL 14, 460 (1972).
(48) Gupta, MP 23, 75 (1972).
(49) Sando, quoted by (46).
(50) Brown, Ginter, JMS 46, 256 (1973).
(51) Liu, McLean, JCP 59, 4557 (1973).

(continued p. 301)

State	T_e	ω_e	$\omega_e x_e$	B_e	α_e	D_e $(10^{-2}cm^{-1})$	r_e (Å)	Observed Transitions Design.	ν_{00}	References
$^4He^{40}Ar$		$\mu = 3.6382034_6$ $D_e^0 = 0.0024$ eV[a]								SEP 1976 A
X $^1\Sigma^+$	0	Only v=0 is bound; see (6)					3.51^a	Translational sp.[b]		(1)
$^4He^{40}Ar^+$		$D_e(A^2\Sigma^+) = 0.19$ eV[a]								SEP 1976
A $^2\Sigma^+$		Two broad groups of bands with partially resolved rotational structure. No vibrational assignments.						$A \rightarrow X_2$,[b] V $A \rightarrow X_1$, V	68630^c 70130^c	(2)*
X_2 $(\frac{1}{2})$	1500	Arising from He + $Ar^+(^2P$ ground state splitting 1432 cm^{-1}).								
X_1 $(\frac{1}{2},\frac{3}{2})$	0									
$^4He^1H$		$\mu = 0.80510570$ $D_e^0 = (0.003)$ eV[a]								SEP 1976 A
		Excited states derived from $He(1s^2) + H(2s,2p)$ and from $He(1s2s,1s2p) + H(1s)$ have been calculated by (1)(4) and (3), respectively. Several of these states have dissociation energies of the order of 2 eV.								
X $^2\Sigma^+$		Calculated potential function (3); comparison with H-He scattering results (2).						Translational sp.[b]		
$^4He^1H^+$		$D_0^0 = (1.8450)$ eV[a]								SEP 1976 A
		Several other excited $^1\Sigma$ states calculated by (8), lowest $^{1,3}\Pi$ states by (2); see also (1).								
A $^1\Sigma^+$	(104707)	$[(157._5)]^b$					(2.9)	$(A-X)^c$	(103244)	(12)
a $^3\Sigma^+$	(104273)	$[(297._6)]$	Theoretical data				(2.3_6)	$(a-X)$	(102847)	(12)
X $^1\Sigma^+$	0	(3228.4)	$(157.71)^d$	$(34.887)^e$	$(2.636_2)^d$	(1.61)	(0.7743)			(11)

HeAr: [a]From molecular beam scattering measurements; average of values given by (4) and (5).
[b]Dipole moment function from *ab initio* calculations (2) and from translational absorption spectra (3).

(1) Bosomworth, Gush, CJP 43, 751 (1965)
(2) Matcha, Nesbet, PR 160, 72 (1967).
(3) Bar-Ziv, Weiss, CPL 19, 148 (1973).
(4) Chen, Siska, Lee, JCP 59, 601 (1973).
(5) Smith, Rulis, Scoles, Aziz, Duquette, JCP 63, 2250 (1975).
(6) Bobetic, Barker, JCP 64, 2367 (1976).

$HeAr^+$: [a]From He^+-off-Ar elastic scattering data (1).
[b]$A \rightarrow X_1$ is stronger than $A \rightarrow X_2$.
[c]The transition energy is close to the difference of $He + Ar^+$ and $He^+ + Ar$ which is 71201 cm^{-1}. This agreement, together with the characteristic splitting of the lower states, makes the explanation by (2) of the observed bands as a charge transfer spectrum very convincing.

(1) Weise, BBPC 77, 578 (1973).
(2) Tanaka, Yoshino, Freeman, JCP 62, 4484 (1975).

He^1H: [a]Theoretical value (3); 0.0007_3 eV (5) seems less likely to be correct considering that $D_e^0(He_2)$ = 0.00090 eV.
[b]Theoretical intensity distrib.(5); no experimental data.

(1) Michels, Harris, JCP 39, 1464 (1963).
(2) Fischer, Kemmey, JCP 53, 50 (1970).
(3) Miller, Schaefer, JCP 53, 1421 (1970).
(4) Slocomb, Miller, Schaefer, JCP 55, 926 (1971).
(5) Ulrich, Ford, Browne, JCP 57, 2906 (1972).

He^1H^+: [a]Theoretical value (11) from the potential function of (9) (10); D_e^0 = 2.0402 eV. Proton scattering by He gives D_e^0 = 2.0_0 eV (4).
[b]$\Delta G(3/2) = 78.8$, zero-point energy 112.1, derived by (12)

He^1H^+ (continued):

from the potential function of (9); see also (8).
[c]Franck-Condon factors for bound-continuum transitions (3).
[d]$\omega_e y_e = + 0.454$, $\gamma_e = - 0.0305$. A number of quasi-bound levels (below the centrifugal barrier) have been observed (5) in the momentum distribution of protons formed in the predissociation of HeH^+; they have been calculated by (6) (7)(10)(11).
[e]$\mu_{e\ell}$(referred to the center of mass) = 1.66 D (11).

(1) Michels, JCP 44, 3834 (1966).
(2) Hoyland, JCP 47, 49 (1967).
(3) Sizun, Durup, MP 22, 459 (1971).
(4) Weise, Mittmann, Ding, Henglein, ZN 26 a, 1122 (1971).
(5) Schopman, Fournier, Los, Physica 63, 518 (1973).
(6) Peek, Physica 64, 93 (1973).
(7) Bernstein, CPL 25, 1 (1974).
(8) Green, Michels, Browne, Madsen, JCP 61, 5186 (1974).
(9) Kolos, IJQC 10, 217 (1976).
(10) Kolos, Peek, CP 12, 381 (1976).
(11) Dabrowski, Herzberg, N.Y. Acad. Sci. (II) 38, 14 (1977).
(12) Dabrowski, Herzberg, unpublished.

He_2 (continued):

(52) Mukamel, Kaldor, MP 26, 291 (1973).
(53) Bertoncini, Wahl, PRL 25, 991 (1970); JCP 58, 1259 (1973).
(54) Peatman, Wu, CP 2, 335 (1973).
(55) Beck, Nicolaides, Musher, PR A 10, 1522 (1974).
(56) Kleinman, Wolfsberg, JCP 61, 4366 (1974).
(57) Foreman, Rol, Coffin, JCP 61, 1658 (1974).
(58) Lichten, McCusker, Vierima, JCP 61, 2200 (1974).
(59) Andresen, Kuppermann, MP 30, 997 (1975).
(60) Guberman, Goddard, PR A 12, 1203 (1975).
(61) Chapman, PR A 12, 2333 (1975).
(62) Miller, Freund, Zegarski, Jost, Lombardi, Derouard, JCP 63, 4042 (1975).
(63) Poulat, Larsen, Novaro, MP 30, 645 (1975).
(64) Snook, Spurling, JCS FT II 71, 852 (1975).
(65) Vierima, JCP 62, 2925 (1975).
(66) Burgmans, Farrar, Lee, JCP 64, 1345 (1976).
(67) Orth, Ginter, JMS 61, 282 (1976).

State	T_e	ω_e	$\omega_e x_e$	B_e	α_e	D_e (10^{-} cm^{-1})	r_e (Å)	Observed Transitions: Design.	Observed Transitions: ν_{00}	References
$^4He^1H^{++}$		$D_e(A^2\Sigma^+) = (0.849_5)$ eV[a]								SEP 1976 A
		Potential functions of several states calculated by (1)(2). Only $A\,^2\Sigma^+$ [from H(1s) + He^{++}] has a minimum. Satisfactory agreement with charge exchange experiments [$He^{++} + H \rightarrow He^+ + H^+$; (2)].								
A $^2\Sigma^+$	(322378)[a]	(950)[a]	(39)[a]				(2.05_9)[a]			
X $^2\Sigma^+$		Repulsive state (corresponding to $He^+ + H^+$).								
$^4He^{(84)}Kr$		(μ = 3.8203705) $D_e^0 = 0.0021_3$ eV[a]								SEP 1976 A
X $^1\Sigma^+$		Doubtful whether any level other than v=0 is bound (2).					3.75[a]			
$^4He^{(84)}Kr^+$		$D_e(A^2\Sigma^+) = 0.22$ eV[a]								SEP 1976
A $^2\Sigma^+$		Two broad groups of partially resolved bands, no vibr. assignments.						$A \rightarrow X_2$, V_R $A \rightarrow X_1$,	78370 83820[b]	(2)*
X_2 ($\frac{1}{2}$)	5450	Arising from He + $Kr^+(^2P$ ground state splitting 5371 cm^{-1}).								
X_1 ($\frac{1}{2}, \frac{3}{2}$)	0									
$^4He^{(20)}Ne$		(μ = 3.3349306_8) $D_e^0 = 0.0012_3$ eV[a]; $D_0^0 = (0.0002_3)$ eV, see (4).								SEP 1976 A
		Similar but unclassified bands near other strong Ne lines.								(2)
F		[120][b] Progression of five bands converging to $3d'[\frac{3}{2}]1_u$ of Ne.						$F \leftarrow X$,[c]	162050	(2)
E		[40][b] Progression of six bands converging to $3d[\frac{1}{2}]1_u$ of Ne.						$E \leftarrow X$,[c]	161340	(2)
D		[30][b] Progression of four bands converging over a maximum to $4s'[\frac{1}{2}]1_u$ of Ne.						$D \leftarrow X$,[c]	159590	(2)
C		[100][b] Progression of six bands converging to $4s'[\frac{1}{2}]1_u$ of Ne.						$C \leftarrow X$,[d] V	159050 H	(2)
B		[40][b] Progression of four bands converging over a maximum to $4s[\frac{3}{2}]1_u$ of Ne.						$B \leftarrow X$,[c]	158860	(2)
A		[150][b] Progression of three bands converging to $4s[\frac{3}{2}]1_u$ of Ne.						$A \leftarrow X$,[d] V	158440 H	(2)
X $^1\Sigma^+$	0	Only v=0 is bound; see (4).					3.21[a]	e		

State	T_e	ω_e	$\omega_e x_e$	B_e	α_e	D_e ($10^{-4}cm^{-1}$)	r_e (Å)	Observed Transitions: Design.	Observed Transitions: ν_{00}	References
$^4He^{20}Ne^+$		$D_0^0 = 0.69_2$ eV[a]								APR 1978
		RKR potential curves (7), theoretical potential functions (4).								
B $^2\Sigma^+$	(30233)	152.649 Z	21.718[b]	0.72116[c]	0.08978[d]	[0.855][e]	2.6475_2	B→X, R	v(0-6) = 24970.73[f] Z	(1)* (2) (6)* (7)*
A_2 $^2\Pi_{1/2}$	(6649)	[129.69] Z		[0.86571][g]	h	[1.592][e]	2.3186	B→A_2, R	23590.23 Z	
A_1 $^2\Pi_{3/2}$		Not yet observed.								
X $^2\Sigma^+$	0	(1308.3)[f]	(84.8)[i]	(2.991)[j]	(0.191)[k]	$D_6 = 2.098$[e]	(1.300)			

He^1H^{++}: [a]Theoretical values (1); T_e is relative to $He^+ + H^+$.

(1) Bates, Carson, PRS A 234, 207 (1956).
(2) Piacentini, Salin, JP B 7, 1666 (1974).

HeKr: [a]From molecular beam scattering measurements (1).

(1) See ref. (4) of HeAr.
(2) See ref. (6) of HeAr.

$HeKr^+$: [a]From He^+-off-Kr elastic scattering data (1).
[b]The transition energy is close to the difference of $He^+ + Kr$ and $He + Kr^+$ which is 85396. See [c] of $HeAr^+$.

(1) See ref. (1) of $HeAr^+$.
(2) See ref. (2) of $HeAr^+$.

HeNe: [a]From molecular beam scattering measurements (3).
[b]First observed ΔG, not necessarily ΔG(½).
[c]Very diffuse features.
[d]Diffuse band heads.
[e]Theoretical dipole moment function (1).

(1)(3)(4) See ref. (2)(4)(6), resp., of HeAr.
(2) Tanaka, Yoshino, JCP 57, 2964 (1972).

$HeNe^+$: [a]Extrapol. to v=0 from the lowest observed ground state level (v=6). The more accurately known $D_6^0 = 882.5$ cm^{-1} is based on a short extrapolation of the vibrational levels in B $^2\Sigma^+$.

$HeNe^+$ (continued):
[b]$\omega_e y_e = +1.005$. $D_0 = 292.8$, $D_e = 364.0$ cm^{-1}(7) confirming the less precise value from elastic scattering data [$He^+ + Ne$, see (5)].
[c]No spin doubling observed. A hfs splitting of ~ 0.30 cm^{-1} occurs in the spectrum of $^3HeNe^+$ (7).
[d]$\gamma_e = -0.00348$.
[e]For other D_v and higher order constants see (7).
[f]Only v"=6...9 observed because of Franck-Condon factors; ΔG"(6½,...) = 341.59, 233.42, 146.27. Vibrational numbering established by the study of the spectrum of $^3HeNe^+$ (7); see also (3).
[g]Very large Ω-type doubling. Rotational structure similar to that of a $^2\Sigma^-$ state with $\gamma_0 = +0.3047_0$.
[h]$B_1 = 0.71658$.
[i]$\omega_e y_e = +(1.5)$.
[j]See [f]; $B_6 \ldots B_9 = 1.58983, 1.35755, 1.09037, 0.8431$. Large spin splitting, $\gamma_6 = +0.82985$. For other γ_v and higher order const. see (7).
[k]$\gamma_e = -(0.0038)$.

(1) Oskam, Jongerius, Physica 24, 1092 (1958).
(2) Fache, CR 256, 2145 (1963).
(3) Henderson, Matsen, Robertson, JCP 43, 1290 (1965).
(4) Sidis, Lefebvre-Brion, JP B 4, 1040 (1971).
(5)(6) See ref. (1)(2), resp., of $HeAr^+$.
(7) Dabrowski, Herzberg, to be published.

State	T_e	ω_e	$\omega_e x_e$	B_e	α_e	D_e $(10^{-4} cm^{-1})$	r_e (Å)	Observed Transitions Design.	ν_{00}	References
$^4He^{(132)}Xe$		(μ = 3.8847222)		$D_e^0 = 0.0021_7$ eV[a] $D_0^0 = (0.0010_3)$ eV[b]						SEP 1976 A
X $^1\Sigma^+$		[(8.2)][b]					4.15[a]	Translational sp.		(1)
$^4He^{(132)}Xe^+$				$D_e(A\,^2\Sigma^+)$ = 0.28 eV[a]						SEP 1976
A $^2\Sigma^+$		Two broad and only partially resolved groups of indistinct band heads; no vibrational assignments.						$A \rightarrow X_2$, $A \rightarrow X_1$,	87800 98520 [b]	(2)
X_2 (½)	10720	Arising from He + Xe^+(^2P ground state splitting 10537 cm^{-1}).								
X_1 (½, 3/2)	0									
$^1H^{19}F$		μ = 0.95705545		D_0^0 = 5.869 eV[a]		I.P. = 16.039 eV[b]				JAN 1977 A
		Rydberg levels converging to the ground state of HF^+ have been observed in the electron energy loss spectrum.								(44)
D $^1\Sigma^+$		Two strong bands between 104000 and 116000 cm^{-1}, not yet analysed.						D←X,		(30)
C $^1\Pi$	(105820)	[2636]		[16.0]			[1.04_9]	C←X,	R 105090.8	(25)(30)
b $^3\Pi$		Absorption bands above 100000 cm^{-1}, not yet analysed.						b←X,		(25)
B $^1\Sigma^+$	84776.65	1159.18 Z	18.005[c]	4.0291[d]	0.0177[e]	1.932[f]	2.0908_6	B↔X,[g]	R 83304.96 Z	(3)(30)*
A		Continuous absorption starting at 60600 cm^{-1}.[h]								(2)
X $^1\Sigma^+$	0	4138.32[i] Z	89.88[j]	20.9557[ikℓ]	0.798[m]	21.51[n]	0.91680_8	Rot.-vibr. sp.[opq]		(6)(7)(10)(19)
								Rotation sp.[rq]		(9)(11)(16)
								Mol. beam el. reson.[s]		(8)(20)(26)(31)
								Mol. beam magn. reson.[t]		(5)

HeXe: [a]From molecular beam scattering measurements (2).

[b]Theoretical values from (3).

(1) Marteau, Vu, Vodar, CR B 266, 1068 (1968).

(2) See ref. (4) of HeAr.

(3) See ref. (6) of HeAr.

$HeXe^+$: [a]From He^+-off-Xe scattering data (1).

[b]The transition energy is close to the difference of He + Xe^+ and He^+ + Xe which is 100477 cm^{-1}. See [c] of $HeAr^+$.

(1) See ref. (1) of $HeAr^+$.

(2) See ref. (2) of $HeAr^+$.

^1HF: [a]From the limiting curve of dissociation for the ground state (30); see [k].

[b]From photoelectron spectra (32)(41). Earlier photoionization studies yielded 16.00_7 eV (23), a value strongly affected by the presence of autoionizing Rydberg levels in the threshold region (41). The second ionization potential (removal of a 3σ electron) from the photoelectron spectrum (21)(41) is 19.118 eV in agreement with the value derived from the spectrum of HF^+. The third and fourth ionization potentials (removal of a 2σ and 1σ electron, respectively) are 39.61 (38)(39) and 694.25 eV (39); these are vertical potentials from X-ray photoelectron spectra.

[c]$\omega_e y_e$ = + 0.184; the vibrational and rotational constants were obtained from a fit to the seven lowest vibrational levels (30). See [g].

[d]RKR potential curves (4)(30).

[e] $- 0.000950(v+\frac{1}{2})^2 + 0.0000601(v+\frac{1}{2})^3$; see [c].

[f] $+ 0.182 \times 10^{-4}(v+\frac{1}{2}) + 0.00551 \times 10^{-4}(v+\frac{1}{2})^2$; see [c].

[g]Very extensive band system (also called V − X) extending in absorption from 96000 to 117000, in emission from

^1HF (continued):

36000 to 70000 cm^{-1}. Strong perturbations above v'=27, but bands have been identified to v'=73. The B (or V) state was also observed in the electron energy loss spectrum (44).

[h]HF is quite transparent to 1650 Å (2). Theoretical potential curves for three repulsive states ($^3\Pi$, $^1\Pi$, $^3\Sigma^+$) arising from ground state atomic products were given by (42).

[i]Introduction of the Dunham correction (19) gives ω_e = 4138.767 and B_e = 20.9561.

[j] $+ 0.90(v+\frac{1}{2})^3 - 0.0110(v+\frac{1}{2})^4 - 0.00067(v+\frac{1}{2})^5$, $v \leq 9$ (19). A different formula for higher vibrational levels ($v \leq 19$) was derived by (3). All levels up to the last (v=19) are tabulated in (30).

[k]For v=14...19 the rotational levels break off at decreasing J on account of predissociation by rotation. A few broadened lines near these limits have been observed (30). From the limiting curve the dissociation energy 47333 ± 60 cm^{-1} has been determined (30).

[ℓ]RKR potential curves (4)(30), Dunham potential coefficients (19)(43). Ab initio calculations of molecular constants (27)(34)(37)(42).

[m] $+ 0.0127(v+\frac{1}{2})^2 - 0.00044(v+\frac{1}{2})^3$, from (19).

[n] $- 0.68 \times 10^{-4}(v+\frac{1}{2}) + 0.029 \times 10^{-4}(v+\frac{1}{2})^2$, $H_e = 1.59 \times 10^{-7}$; from (19), see also (6).

[o]1-0, 2-0 bands studied in absorption under high resolution by (7)(19), 3-0, 4-0, 5-0 in the photographic infrared by (1)(10). In emission, rotation-vibration bands have been studied by (6) and (40), the latter extending the chemical laser emission, first observed in the 2-1 band by (15), to v=9 and Δv=6. Electric discharge induced laser emission in the 3-2, 2-1, 1-0 bands, see (12)(24).

(continued on p. 307)

State	T_e	ω_e	$\omega_e x_e$	B_e	α_e	D_e $(10^{-4} cm^{-1})$	r_e (Å)	Observed Transitions Design.	ν_{00}	References
$^2H^{19}F$		$\mu = 1.82104540$		$D_0^0 = 5.938$ eV[a]		I.P. = 16.058 eV[b]				JAN 1977
B $^1\Sigma^+$	84824.0[c]	839.4 Z	8.90	2.1210	0.0071_2	0.5543[d]	2.0891	B→X, R	83753.8 Z	(2)
X $^1\Sigma^+$	0	2998.19_2[e] Z	45.76_1[e]	11.0102[e]	0.3017[e]	5.94[e]	0.91694	Rot.-vibr. sp.[fgh]		(1)(4)
								Rotation sp.[h]		(5)(7)(11)
								Mol. beam el. reson.[i]		(9)
								Mol. beam magn. reson.		(3)

^{2}HF: [a]From $D_0^0(^1HF)$.

[b]From photoelectron spectra (6)(8)(14). Photoionization studies yield 16.03 eV (10).

[c]Large electronic isotope shift.

[d]$\beta_e = +0.0464 \times 10^{-4}$.

[e]Based on the 1-0 and 2-0 rotation-vibration bands (4); $\gamma_e = +0.0027_5$, $\beta_e = -0.12 \times 10^{-4}$. Using the older measurements of (1) for v=0,1,2 and their own measurements of the B–X system at high v" (2) obtain:

$$G(v) = 3001.00_8(v+\tfrac{1}{2}) - 47.96_9(v+\tfrac{1}{2})^2 + 0.58504(v+\tfrac{1}{2})^3 - 0.028102(v+\tfrac{1}{2})^4 + 9.9959 \times 10^{-4}(v+\tfrac{1}{2})^5 - 2.0290 \times 10^{-5}(v+\tfrac{1}{2})^6 \quad (v \leq 24);$$

$$B_v = 11.0037_5 - 0.3036_2(v+\tfrac{1}{2}) + 0.003849_5(v+\tfrac{1}{2})^2 - 1.759_3 \times 10^{-4}(v+\tfrac{1}{2})^3 + 9.168_7 \times 10^{-6}(v+\tfrac{1}{2})^4 - 3.104_4 \times 10^{-7}(v+\tfrac{1}{2})^5;$$

also higher terms in the expression for D_v. (11) obtain from the submillimeter microwave spectrum $B_0 = 10.86034_6$.

[f]Laser emission in the 4-3, 3-2, 2-1, 1-0 bands in transverse discharges through $D_2 + SF_6$ (12); in a chemical laser source the emission extends to v=12 and Δv=6 (17).

[g]The radiative lifetime of v=1 is 0.032 s; see (15)(16).

[h]Mid- and far-infrared spectra in rare-gas matrices (13).

[i]$\mu_{el}(v=0, J=1) = 1.81881$ D; also nuclear quadrupole and other hyperfine coupling constants.

(1) Talley, Kaylor, Nielsen, PR 77, 529 (1950).
(2) See ref. (3) of ^{1}HF.
(3) Nelson, Leavitt, Baker, Ramsey, PR 122, 856 (1961).
(4) Spanbauer, Rao, Jones, JMS 16, 100 (1965).
(5) See ref. (11) of ^{1}HF.
(6) See ref. (18) of ^{1}HF.
(7) Perkins, SA A 24, 285 (1968).
(8) Brundle, CPL 7, 317 (1970).
(9) See ref. (20) of ^{1}HF.
(10) See ref. (23) of ^{1}HF.
(11) De Lucia, Helminger, Gordy, PR A 3, 1849 (1971).
(12) See ref. (24) of ^{1}HF.
(13) See ref. (22) of ^{1}HF.
(14) See ref. (32) of ^{1}HF.
(15) See ref. (33) of ^{1}HF.
(16) Bonczyk, PR A 11, 1522 (1975).
(17) See ref. (40) of ^{1}HF.

^{1}HF (continued):

[p]Line strengths, collision-broadened widths, dipole moment function (28)(29)(35)(36)(40)(45). The radiative lifetime of v=1 [P(4) line] is 6.16 ms (33). (40) give a vibrational dipole moment matrix for $v \leq 9$ based on intensity measurements in chemical laser emission.

[q]Rotation and rotation-vibration spectra in rare-gas matrices (22).

[r]Laser emission in the pure rotation spectrum (13).

[s]$\mu_{e\ell}(v=0, J=1) = 1.82618$ D (20)(26); $g_J = 0.7410$, quadrupole moment $\Theta = 2.36 \times 10^{-26}$ esu cm^2 (31); also nuclear spin - rotation and other hyperfine structure constants.

[t]Nuclear reorientation spectrum.

(1) Naudé, Verleger, PPS A 63, 470 (1950).
(2) Safary, AP(Paris) 9, 203 (1954).
(3) Johns, Barrow, PRS A 251, 504 (1959).
(4) Fallon, Vanderslice, Mason, JCP 32, 698; 33, 944 (1960).
(5) Baker, Nelson, Leavitt, Ramsey, PR 121, 807 (1961).
(6) Mann, Thrush, Lide, Ball, Acquista, JCP 34, 420 (1961).
(7) Herget, Deeds, Gailar, Lovell, Nielsen, JOSA 52, 1113 (1962).
(8) Weiss, PR 131, 659 (1963).
(9) Rothschild, JOSA 54, 20 (1964).
(10) Fishburne, Rao, JMS 19, 290 (1966).
(11) Revich, Stankevich, DANS 170, 1376 (1966); engl. transl. DPC 170, 699 (1966).
(12) Deutsch, APL 10, 234 (1967).
(13) Deutsch, APL 11, 18 (1967).
(14) Frost, McDowell, Vroom, JCP 46, 4255 (1967).
(15) Kompa, Pimentel, JCP 47, 857 (1967).
(16) Mason, Nielsen, JOSA 57, 1464 (1967).
(17) Lempka, Passmore, Price, PRS A 304, 53 (1968).
(18) Lempka, Price, JCP 48, 1875 (1968).
(19) Webb, Rao, JMS 28, 121 (1968).
(20) Muenter, Klemperer, JCP 52, 6033 (1970).
(21) Berkowitz, CPL 11, 21 (1971).
(22) Mason, Von Holle, Robinson, JCP 54, 3491 (1971).
(23) Berkowitz, Chupka, Guyon, Holloway, Spohr, JCP 54, 5165 (1971).
(24) Goldhar, Osgood, Javan, APL 18, 167 (1971).
(25) Di Lonardo, Douglas, JCP 56, 5185 (1972).
(26) Muenter, JCP 56, 5409 (1972).
(27) Bondybey, Pearson, Schaefer, JCP 57, 1123 (1972).
(28) Meredith, JQSRT 12, 485 (1972); Meredith, Smith, JQSRT 13, 89 (1973).
(29) Spellicy, Meredith, Smith, JCP 57, 5119 (1972).
(30) Di Lonardo, Douglas, CJP 51, 434 (1973).
(31) de Leeuw, Dymanus, JMS 48, 427 (1973).
(32) Walker, Dehmer, Berkowitz, JCP 59, 4292 (1973).
(33) Hinchen, JOSA 64, 1162 (1974).
(34) Krauss, Neumann, MP 27, 917 (1974).
(35) Lie, JCP 60, 2991 (1974).
(36) Rimpel, ZN 29 a, 588 (1974).
(37) Meyer, Rosmus, JCP 63, 2356 (1975).
(38) Banna, Shirley, JCP 63, 4759 (1975).
(39) Shaw, Thomas, PR A 11, 1491 (1975).
(40) Sileo, Cool, JCP 65, 117 (1976).
(41) Guyon, Spohr, Chupka, Berkowitz, JCP 65, 1650 (1976).
(42) Dunning, JCP 65, 3854 (1976).
(43) Ogilvie, Koo, JMS 61, 332 (1976).
(44) Salama, Hasted, JP B 9, L333 (1976).
(45) Yardley, Balint-Kurti, MP 31, 921 (1976).

State	T_e	ω_e	$\omega_e x_e$	B_e	α_e	D_e $(10^{-4}cm^{-1})$	r_e (Å)	Observed Transitions Design.	ν_{00}	References
$^3H^{19}F$		$\mu = 2.6028413_9$		$D_0^0 = 5.968$ eV[a]						JAN 1977
X $^1\Sigma^+$	0	2508.5_4[b] Z	32.5_4[b]	7.692	0.176[c]	[2.6]	0.9176	Rotation-vibration b.		(1)
$^1H^{19}F^+$				$D_0^0 = 3.423$ eV[a]		I.P.; see [b].				JAN 1977
		Highly excited states of HF^+ with configuration $1\sigma\, 2\sigma^2\, 3\sigma^2\, 1\pi^4$						c	678.2 eV	(5)
		and $1\sigma^2\, 2\sigma\, 3\sigma^2\, 1\pi^4$.						c	23.6 eV	(5)
A $^2\Sigma^+$	25449.82[d]	1496.07 Z	88.42_3[e]	11.753_6[fg]	1.0261[h]	$[28.8_9]$[i]	1.2242	A→X, R	24648.91[j]	(4)
X $^2\Pi_i$	0[k]	3090.48 Z	88.99_6[ℓ]	17.577_1[mn]	0.8863[o]	$[22.0_5]$[p]	1.0011			
$^2H^{19}F^+$				$D_0^0 = 3.482$ eV[a]						JAN 1977
A $^2\Sigma^+$	25490	1080[b]	55					(A − X)	24900[b]	
X $^2\Pi_i$	0[c]	2250[b]	45.8							
$^1H^{19}F^-$										JAN 1977
A $^2\Pi$		[2860] Progression of four resonances (due to inverse preionization) in the electron transmission current. The $^2\Pi$ state is derived from its "grandparent" X $^2\Pi_i$ of HF^+ by the addition of a pair of $3s\sigma$ Rydberg electrons.							103440[a]	(2)
X $^2\Sigma^+$		*Ab initio* calculations (1) predict this state to be entirely repulsive.								

^{3}HF: [a]From $D_0^0(^1HF)$.

[b]From the 1-0 and 2-0 rotation-vibration bands using calculated values of $\omega_e y_e$ and $\omega_e z_e$ (1).

[c]$\gamma_e = +0.001_9$.

(1) Jones, Goldblatt, JMS 1, 43 (1957).

$^1HF^+$: [a]Average of two values, 27650 and 27562 cm^{-1}, the former from $D_0^0(HF) + I.P.(H) - I.P.(HF)$, the latter from the predissociation in A $^2\Sigma^+$ (see [g]).

[b]Several excited states of HF^{++} were observed in the Auger electron spectrum of HF (5). Tentatively, the $...\sigma^2\pi^2\ ^1\Delta$ and $^1\Sigma$ states may be identified at 33.9 and 35.9 eV above X $^2\Pi_{3/2}(v=0)$ of HF^+; the $^3\Sigma$ state remains undetected.

[c]From X-ray photoelectron spectra; satellite ("shake-up") peaks corresponding to 25 - 35 eV higher excitation energies than the main 1σ peak were observed by (7) and attributed to simultaneous removal of a F 1s electron and excitation of a 3σ or 1π electron into a higher orbital.

[d]Includes the Y_{00} corrections in the upper and lower states; $Y''_{00} - Y'_{00} = 2.93$ cm^{-1}.

[e]$\omega_e y_e = - 7.328$.

[f]Spin splitting constants $\gamma(v=0...3) = 0.533_9,\ 0.540_0,\ 0.566_5,\ 0.644$.

[g]Breaking off of rotational levels at N=3 for v=3 and N=10 for v=2 because of predissociation by rotation. The predissociation limit is 27966 ± 50 cm^{-1} above $J=\frac{3}{2}$, v=0 of $^2\Pi_{3/2}$ and corresponds to dissociation into $H^+ + F(^2P_{1/2})$; see, however, the discussion in (4).

[h]$\gamma_e = - 0.0933$; from v=0,1,2 only. $B_3 = 6.741$.

[i]$D_1 = 31.65 \times 10^{-4}$, $D_2 = 39.3 \times 10^{-4}$;
$H_0 = -3.2 \times 10^{-7}$, $H_1 = -11.3 \times 10^{-7}$, $H_2 = -34 \times 10^{-7}$.

[j]Refers to the zero-point of the Hill-Van Vleck formula in the lower state.

[k]$A(v) = - 292.85 + 0.58(v+\frac{1}{2})$; see [l].

$^1HF^+$ (continued):

[l]Only v=0,1,2 have been observed with high accuracy in the optical spectrum, v=3...11 in the photoelectron sp.(6).

[m]For Λ-type doubling constants ($p \approx 0.60$) see (4).

[n]_Ab initio_ calculations (1)(2)(3) give molecular constants in agreement with the experimental results quoted here.

[o]$\gamma_e = + 0.0142$; see [l].

[p]$D_1 = 21.41 \times 10^{-4}$, $D_2 = 21.2_4 \times 10^{-4}$.

(1) Julienne, Krauss, Wahl, CPL 11, 16 (1971).
(2) See ref. (27) of ^1HF.
(3) Raftery, Richards, JP B 5, 425 (1972).
(4) Gewurtz, Lew, Flainek, CJP 53, 1097 (1975).
(5) See ref. (39) of ^1HF.
(6) See ref. (41) of ^1HF.
(7) Martin, Mills, Shirley, JCP 64, 3690 (1976).

$^2HF^+$: [a]From $D_0^0(DF) + I.P.(D) - I.P.(DF)$; from $D_0^0(HF^+)$ one obtains 3.473 eV.

[b]From the photoelectron spectrum of DF (1)(2)(3). The optical spectrum of DF^+ has not yet been analysed.

[c]$A = - 266$ cm^{-1}, from incompletely resolved photoelectron peaks (3).

(1) See ref. (21) of ^1HF.
(2) See ref. (32) of ^1HF.
(3) See ref. (41) of ^1HF.

$^1HF^-$: [a]Energy relative to HF, X $^1\Sigma^+(v=0)$.

(1) See ref. (27) of ^1HF.
(2) Spence, Noguchi, JCP 63, 505 (1975).

State	T_e	ω_e	$\omega_e x_e$	B_e	α_e	D_e ($10^{-7}cm^{-1}$)	r_e (Å)	Observed Transitions Design.	Observed Transitions ν_{00}	References
$^{(180)}Hf^{(79)}Br$		(μ = 54.859052_6)								JAN 1975
		Three sequences of R shaded bands in emission at 15860, 16110 (0-0), and (16345) cm^{-1}.								(1)
$^{(180)}Hf^{127}I$		(μ = 74.420556)								JAN 1975
		Complex groups of emission bands in the regions 19700 - 20300 and 20450 - 20750 cm^{-1}.								(1)
		Long 0-0 sequence of R shaded bands in emission.						R	14559.7 H	(2)
$^{180}Hf^{16}O$		$\mu = 14.6892328_7$		D_0^0 = 8.19 eV[a]		I.P. = 7.5_5 eV[b]				JAN 1975 A
J				[0.3719][cd]			[1.756_7]	$J \rightarrow x_3$,[c] R	22655.82 Z	(1)* (3)(6)
H				[0.3699][cd]			[1.761_4]	$H \rightarrow x_2$,[c] R	21239.87 Z	(1)* (6)
C	x_1 + 19719.3	$852._5$ H^Q	4.1	[0.3696][c]		[3.0]	[1.762_1]	$C \rightarrow x_1$,[c] R	19682.94 Z	(1)* (6)
x_3		(945)[e]	(5)	[0.3788][c]		[2.6]	[1.740_6]			
x_2				[0.3781][c]		[2.6]	[1.742_2]			
x_1	x_1	$925._0$ H^Q	3.5	[0.3776][c]		[2.8]	[1.743_3]			
G $^1\Sigma^{(+)}$	30090.0	[852.29] Z	(3.7_1)	0.370106	0.002071	[2.764][f]	1.76090_4	$G \leftrightarrow X$,[g] R	30032.71 Z	(1)* (3)(6)
F $^1\Sigma^{(+)}$	27413.59	849.40 Z	3.67	0.36563	0.00188	2.702[h]	1.7716_5	$F \leftrightarrow X$,[g] R	27351.14 Z	(1)* (3)(6)*
E $^1\Pi$	25230.94	866.93 Z[i]	3.68	0.36928[j] ℓ 0.36868[k]	0.00198 0.00197	2.69 2.67	1.7636	$E \leftrightarrow X$,[g] R	25177.25[i] Z	(1)* (3)(6) (7)
D $^1\Pi$	23554.41	872.60 Z[i]	3.31	0.36912[k] ℓ 0.36837[j]	0.00182 0.001800	2.624[m] 2.631[n]	1.7642	$D \leftrightarrow X$,[g] R	23503.65[i] Z	(1)* (3)(6)* (7)
B $^1\Pi^o$	17562.22	907.01 Z[i]	3.38	0.378060[k] 0.377537[j]	0.001852 0.001848	2.63 2.618[p]	1.7429	$B \leftrightarrow X$,[g] R	17528.65[i] Z	(1)* (3)(4) (6)(7)
A $^1\Sigma^{(+)o}$	16616.92	914.24 Z	3.38	0.377985	0.001827	2.587[q]	1.74245_4	$A \leftrightarrow X$,[g] R	16586.96 Z	(1)* (3)(6) (7)
X $^1\Sigma^{(+)}$	0	974.09 Z	3.228	0.386537	0.001724	2.438[r]	1.72307_1	s		

HfBr: (1) Savithry, Rao, Rao, Physica 67, 400 (1973).

HfI: (1) Savithry, Rao, Rao, CS 40, 516 (1971).
(2) Savithry, Rao, Rao, CS 42, 533 (1973).

HfO: [a]Thermochemical value (mass-spectrom.)(9). See also (2).
[b]Electron impact appearance potential (8).
[c]P, Q, and R branches. No Λ-type doubling observed. The three systems are probably subsystems of a triplet - triplet transition, possibly $^3\Phi - {}^3\Delta$ [see (5)].
[d]v=0 perturbed by state of smaller B value.
[e]From the fluorescence spectrum of HfO in a Ne matrix (3).
[f]$D_1 = 3.73 \times 10^{-7}$, perturbed.
[g]Absorption spectra of HfO in solid inert gas matrices; see (3).
[h]$\beta_e = + 0.023 \times 10^{-7}$.
[i]These band origins do not conform to the usual convention adopted in these tables. Subtract B_v' in order to obtain zero lines.
[j]f levels.
[k]e levels.
[l]Perturbations between D and E.
[m]$\beta_e = + 0.032 \times 10^{-7}$.
[n]$\beta_e = + 0.019 \times 10^{-7}$.
[o]Possibly $^3\Pi_{0^+}$ and $^3\Pi_1$; see (5).
[p]$\beta_e = + 0.009 \times 10^{-7}$.
[q]$\beta_e = + 0.016 \times 10^{-7}$.
[r]$\beta_e = + 0.006 \times 10^{-7}$.
[s]IR spectrum of HfO trapped in Ne matrix (3).

HfO (continued):

(1) Gatterer, Junkes, Salpeter, Rosen, METOX.
(2) Panish, Reif, JCP 38, 253 (1963).
(3) Weltner, McLeod, JPC 69, 3488 (1965).
(4) Edvinsson, Naturw. 53, 177 (1966).
(5) Edvinsson, Dissertation (Stockholm, 1971). USIP Report 71-09.
(6) Edvinsson, Nylén, PS 3, 261 (1971).
(7) Wentink, Spindler, JQSRT 12, 1569 (1972).
(8) Rauh, Ackermann, JCP 60, 1396 (1974).
(9) Ackermann, Rauh, JCP 60, 2266 (1974).

State	T_e	ω_e	$\omega_e x_e$	B_e	α_e	D_e (10^{-8}cm^{-1})	r_e (Å)	Observed Transitions Design.	ν_{00}	References
$^{(200,202)}Hg_2$		(μ = 100.482247) $0.06_5 < D_0^0 < 0.09_1$ eV[a] $9.39 <$ I.P. < 9.61 eV[b]								JAN 1975 A
		Numerous bands and continua in absorption and emission from 19000 to 71000 cm^{-1}. Literature before 1938 reviewed in (6); details concerning the 2540 Å (39353 cm^{-1}) band in (7).[c] - See also HgO.								(6) (7)
		Large number of additional emission features reported by (8) whose tentative vibrational assignments in the regions 21000 - 22800 and 18800 - 20100 cm^{-1} suggest $\omega_e' = 122$, $\omega_e' x_e' = 0.6$; $\omega_e'' = 145$, $\omega_e'' x_e'' = 0.5$,							22152	(8)
		and $\omega_e' = 121$, $\omega_e'' = 141$, $\omega_e'' x_e'' = 0.5$,							19615	(8)
X $^1\Sigma_g^+$	0	(36)					(3.3)			(2)(3)(4) (10)
$^{(200,202)}Hg_2^+$?		$0.91 < D_0^0 < 1.11$ eV[a]								JAN 1975
		Group of bands near 40300 cm^{-1}.								(1)(2)
$^{(202)}Hg^{40}Ar$		(μ = 33.3614163)								JAN 1975 A
		Molecular features associated with the Hg resonance line at 2537 Å (39412 cm^{-1}), observed in emission and absorption by (1)(2)(3)(4)(10). See also the discussions by (7)(8)(9). Calculated constants for the ground state in (5)(6)(7).								
$^{202}Hg^{40}Ar^+$		D_0^0 = (0.20) eV								JAN 1975
A $(J_a = \frac{5}{2})_{\frac{1}{2}}$	35444.5[a]	104.5 H	1.5	0.05789[b]	0.00076	6.95[c]	2.954_4	A→X,	R_V 35447.2 H	(1)* (2)(3)
X $^2\Sigma^+$	0	99.0 H	1.5	[0.06142]		[9.7_8]	[2.868_3]			
$^{(202)}Hg^{81}Br$		(μ = 57.771192) D_0^0 = 0.71 eV								JAN 1975 A
E	(40720)	(166)	(1.1)					E→X,	a (40710)	(7)
D ($^2\Pi_{3/2}$)	38574.4	228.5[b] H	0.950					D↔X,	V 38595.5 H	(1)(5)*(13)*

Hg_2: [a]Thermochemical value (1)(2)(4), disregarding corrections for rotation suggested by (9) which in our view do not apply.
[b]From (5).
[c]For radiative lifetimes of the upper states of the 3350 Å (29840 cm^{-1}) and 4850 Å (20610 cm^{-1}) emissions see (11)(12)(13).

(1) Koernicke, ZP 33, 219 (1925).
(2) Kuhn, Freudenberg, ZP 76, 38 (1932).
(3) Ekstein, Magat, CR 199, 264 (1934).
(4) Kuhn, PRS A 158, 230 (1937).
(5) Arnot, M'Ewen, PRS A 165, 133 (1938).
(6) Finkelnburg, "Kontinuierliche Spektren", Springer (Berlin, 1938).
(7) Mrozowski, PR 76, 1714 (1949).
(8) Takeyama, JSHU A 15, 235 (1952).
(9) Winans, Heitz, ZP 133, 291 (1952); 135, 406 (1953).
(10) Epstein, Powers, JPC 57, 336 (1953).
(11) Ladd, Freeman, McEwan, Claridge, Phillips, JCS FT II 69, 849 (1973).
(12) Phaneuf, Skonieczny, Krause, PR A 8, 2980 (1973).
(13) Skonieczny, Krause, PR A 9, 1612 (1974).

Hg_2^+: [a]From $(I.P.)_{Hg}$ and $(I.P. - D_0^0)_{Hg_2}$ [see ref. (5) of Hg_2].

(1) Winans, PR 42, 800 (1932).
(2) See ref. (7) of Hg_2.

HgAr:

(1) Oldenberg, ZP 47, 184 (1928); 55, 1 (1929).
(2) Kuhn, Oldenberg, PR 41, 72 (1932).
(3) Kuhn, PRS A 158, 212 (1937).
(4) Preston, PR 51, 298 (1937).
(5) Heller, JCP 9, 154 (1941).
(6) Legowski, BAPS(MAP) 6, 127 (1958).
(7) Michels, De Kluiver, Ten Seldam, Physica 25, 1321 (1959).
(8) Fiutak, Frackowiak, BAPS(MAP) 11, 175 (1963).
(9) Behmenburg, ZN 27 a, 31 (1972).
(10) Kielkopf, Miller, JCP 61, 3304 (1974).

$HgAr^+$: [a]The observed isotope shifts closely parallel those of the corresponding transition in Hg^+, i.e. $5d^9 6s^2\ ^2D_{5/2} \rightarrow 5d^{10} 6s\ ^2S_{1/2}$.
[b]Large Ω-type doubling, $\Delta\nu_{fe} = +p_v(J+\frac{1}{2}) - \ldots$ where $p_v = +5.30\,B_v$. Theory (3) predicts $p \approx +6B$.
[c]$\beta_e = +0.23 \times 10^{-8}$.

(1) Santaram, Winans, CJP 44, 1517 (1966).
(2) Bridge, CC (1970), p. 358; JMS 42, 370 (1972).
(3) Hougen, JMS 42, 381 (1972).

HgBr: [a]Headless diffuse bands.
[b]Constants for $^{202}Hg^{81}Br$.
(continued p. 315)

State	T_e	ω_e	$\omega_e x_e$	B_e	α_e	D_e	r_e (Å)	Observed Transitions Design.	ν_{00}	References
$^{(202)}Hg^{81}Br$ (continued)										
C $(^2\Pi_{1/2})$	$34722._0$	278.6^c H	1.82					C↔X, V	$34767._5{}^c$ H	(1)(2)(6)(8) (12)*(13)*
B $^2\Sigma^+$	23485.0	135.075 H	0.275					B→X, R	23459.5 H	(3)(9)(11)*
X $^2\Sigma^+$	0	186.47 H	0.9665^d							(11)
$^{(202)}Hg^{35}Cl$		(μ = 29.8079550)		$D_0^0 = 1.0_4$ eV						JAN 1975 A
D $(^2\Pi_{3/2})$	39703.5	341.8 H^Q	1.87					D↔X, V	39727.7 H^Q	(1)(3)* (9)* (13)*
C $(^2\Pi_{1/2})$	(35782)	$(383)^a$						C↔X, (V)	35828	(1)(2)(3)* (6)* (7)(8) (10)(12)(13)*
B $^2\Sigma^+$	23421.0	192.0^b H	0.50^b					B→X, R_V	23371.0 H	(1)(2)(4)* (7)* (11)
X $^2\Sigma^+$	0	292.61 H	1.6025^c				$[(2.23)]^d$			
$^{(202)}Hg^{133}Cs$		(μ = 80.157999)		D_e = 0.050 eV^a						JAN 1975 A
		Diffuse V shaded absorption bands at 18710, 19140, 19560, 20270, 20750 cm^{-1}. Bands showing fine structure at 20030, 20060, 20090 cm^{-1}. Tentatively attributed to HgCs.								(1)
X $^2\Sigma^+$	0						$5.0_9{}^a$			(5)
$^{(202)}Hg^{19}F$		(μ = 17.3649660)		D_0^0 = (1.8) eV						JAN 1975
E	$46770._6$	587.6 H	10.1					E→X,[a] R	$46818._4$ H	(2)*
D $(^2\Pi_{3/2})$	42999.6	469.4^b H^Q	10.05^b					D→X, V	42987.4 H^Q	(1)*
C $(^2\Pi_{1/2})$	(39044)	(506) H						C→X, V	39053 H	(1)*
X $^2\Sigma^+$	0	490.8^c H^Q	4.05							

HgBr (continued):

[c]From (12)(13). Earlier analyses of C – X by (4)(5) and (10) as well as the suggested existence of an additional system in the region 36100 - 37000 cm^{-1} (10)(12) are not convincing.

[d]$\omega_e y_e$ = - 0.0090.

(1) Wieland, HPA 2, 46, 77 (1929).
(2) Wieland, ZP 77, 157 (1932).
(3) Wieland, HPA 12, 295 (1939).
(4) Sastry, CS 10, 197 (1941).
(5) Sastry, PNISI 7, 359 (1941).
(6) Howell, PRS A 182, 95 (1943).
(7) Rao, Rao, IJP 18, 281 (1944).
(8) Wieland, JCPPB 45, 3 (1948).
(9) Wieland, in "Contribution a l'Etude de la Structure Moléculaire", Vol. comm. Victor Henri, Ed. Desoer, Liège (1948), p. 229.
(10) Krishnamurthy, ZP 152, 242 (1958).
(11) Wieland, ZE 64, 761 (1960).
(12) Patel, Darji, IJP 42, 110 (1968).
(13) Greig, Gunning, Strausz, JCP 52, 3684 (1970).

HgCℓ: [a]From the absorption spectrum (13).

[b]For v ≥ 30: ω_e = 186.2, $\omega_e x_e$ = 0.40.

[c]$\omega_e y_e$ = - 0.01493, $\omega_e z_e$ = - 0.000033. All constants refer to R heads of B – X. From Q heads of D – X ω_e = 293.4, $\omega_e x_e$ = 1.82.

[d]From electron diffraction data (5); according to (7) more likely corresponding to the Hg-Cℓ separation in $HgC\ell_2$ than in HgCℓ.

(1) See ref. (1) of HgBr.
(2) See ref. (2) of HgBr.
(3) Cornell, PR 54, 341 (1938).
(4) Wieland, ZPC B 42, 422 (1939).

HgCℓ (continued):

(5) Maxwell, Mosley, PR 57, 21 (1940).
(6) Sastry, PNISI 7, 351 (1941).
(7) Wieland, HPA 14, 420 (1941).
(8) See ref. (6) of HgBr.
(9) Wieland, HCA 26, 1939 (1943).
(10) See ref. (8) of HgBr.
(11) See ref. (9) of HgBr.
(12) Krishnamurthy, ZP 150, 287 (1958).
(13) Horne, Gosavi, Strausz, JCP 48, 4758 (1968).

HgCs: [a]From Cs – Hg scattering data (5); see also (2)(3)(4).

(1) Barratt, TFS 25, 758 (1929).
(2) Morse, Bernstein, Hostettler, JCP 36, 1947 (1962).
(3) Morse, Bernstein, JCP 37, 2019 (1962).
(4) Neumann, Pauly, JCP 52, 2548 (1970).
(5) Buck, Kick, Pauly, JCP 56, 3391 (1972).

HgF: [a]Close double heads.

[b]Above v=4: ω_e = 410, $\omega_e x_e$ = 1.5.

[c]From D – X.

(1) See ref. (6) of HgBr.
(2) Babu, Rao, Reddy, IJPAP 4, 467 (1966).

State	T_e	ω_e	$\omega_e x_e$	B_e	α_e	D_e ($10^{-4}cm^{-1}$)	r_e (Å)	Observed Transitions		References
								Design.	ν_{00}	
$^{(202)}Hg^1H$		(μ = 1.00282118)		D_0^0 = 0.3744 eV[a]						JAN 1976 A
D $^2\Sigma^+$		Fragments only.		[(4.7)]			$[(1.8_9)]$	D → X, R	(37040)	(2)(8)
C $^2\Sigma^+$				[4.519][b]		[34.8][c]	$[1.928_7]$	C → X, R	35587.4 Z	(2)(8)(12)
B $^2\Sigma^+$				[4.028][d]		[16.27][e]	$[2.042_9]$	B → X,[f] R	33876.47 Z	(1)(2)(8)(12)
A_2 $^2\Pi_{3/2}$	28274	2068.24[g] Z	43.04[h]	6.741_1[gi]	0.2295[j]	[2.818][gk]	1.5791	A_2 → X,[ℓ] V	28616.23[m] Z	(2)(4)(5)(6)(8)
A_1 $^2\Pi_{1/2}$		[1939.18][g] Z	n	[6.5609][gop]	q	[2.850][gr]	[1.6007]	A_1 → X,[ℓ] V_R	24933.10[m] Z	(2)(4)(5)*(6)(7)(8)(9)* (15)
X $^2\Sigma^+$	0	[1203.24][g] Z	s	[5.3888][gptu]	v	[3.953][gw]	$[1.7662_0]$	ESR sp.[x]	Potential curve (14)(15).	
$^{(202)}Hg^2H$		(μ = 1.99421540)		D_0^0 = 0.3976 eV[a]						JAN 1976
A $^2\Pi_r$				[3.3342][bc]		[0.724][b]	[1.5923]	A → X,[d] V	26400.4[b]	(1)(2)(3)(6)(7)
X $^2\Sigma^+$	0	[896.12] Z	e	$[2.738_5]$[cfgh]	i	[0.91][fj]	[1.7569]	ESR sp.[k]		
$^{(202)}Hg^3H$		(μ = 2.9716734_0)		D_0^0 = 0.4086 eV[a]						JAN 1976
A_1 $^2\Pi_{1/2}$		[1150.8] H		ℓ				A_1 → X, V	24776.1 H	(3)
X $^2\Sigma^+$	0	[748.72] Z	m	[1.8464][ℓno]	p	[0.4111][q]	[1.7528]			

Hg^1H: [a]From the predissociation in X $^2\Sigma^+$ of the hydride and deuteride (15). See also Fig. 189 of MOLSPEC, Vol. 1.

[b]Spin doubling $\Delta\nu_{12} = +[2.71 - 0.0093 \times N(N+1)](N+\frac{1}{2})$, increasing rapidly above N=17. See also (13).

[c]$H_0 = +3.42 \times 10^{-6}$.

[d]Spin doubling $\Delta\nu_{12} = +[1.11 - 0.00374 \times N(N+1)](N+\frac{1}{2})$, increasing very rapidly above N=19. See also (13).

[e]$H_0 = +2.04 \times 10^{-6}$.

Hg^1H (continued):

[f]Q branches weak in 0-0, strong in 0-1 and 0-2, not observed in 0-3.

[g]These are effective constants; "true" vibronic energies and rotational constants calculated by (12).

[h]$-0.052_0(v+\frac{1}{2})^4$ [v = 0, ..., 6].

[i]Perturbations.

Hg^1H (continued):

[j] $+ 0.0083_9(v+\frac{1}{2})^2 - 0.0013_0(v+\frac{1}{2})^3$ [v = 0, ..., 6].

[k] $D_1, \ldots, D_3(10^{-4}cm^{-1})$ = 2.880, 2.867, 3.182; $H_0, \ldots, H_3(10^{-8}cm^{-1})$ = + 0.09, + 0.07, - 5.4, + 2.46.

[ℓ] Hg "nuclear" isotope shifts; see (6).

[m] Constants in (8) are slightly different since they refer to J'=1/2 or 3/2 instead of {J'=0} relative to N"=0.

[n] $\Delta G(3/2, 5/2)$ = 1809.66, 1509.13.

[o] Λ-type doubling $\Delta\nu_{fe}(v=0) = + 3.360(J+\frac{1}{2}) - \ldots$ (8); decreasing with increasing v, but very anomalous in v=3. See also (13).

[p] Magnetic hyperfine structure in X $^2\Sigma^+$ and A_1 $^2\Pi_{1/2}$ of $^{199}Hg^1H$ and $^{201}Hg^1H$, see (7)(9).

[q] B_1 = 6.3239, B_2 = 6.0271 (perturbation by v=0 of A_2), B_3 = 5.104 (very anomalous, see [r]).

[r] $D_1, \ldots, D_3(10^{-4}cm^{-1})$ = 2.952, 3.462, - 18.42; $H_0, \ldots, H_3(10^{-8}cm^{-1})$ = + 0.082, - 1.21, - 18.4, - 31.

[s] $\Delta G(3/2, \ldots, 7/2)$ = 965.34, 632.53, 167.82. G(0) = 681.2 (15).

[t] Spin doubling $\Delta\nu_{12}(v=0) = + 2.14_4(N+\frac{1}{2}) - \ldots$; decreasing rapidly with increasing v (4)(9).

[u] Predissociation by rotation. Lines with N" > 29, 22, 15 in v"=0, 1, 2, resp., are broad. Lines with N" > 30, 24, 18, 8, 6 in v"=0, ..., 4, resp., are absent (3)(5)* (9).

[v] $B_1, \ldots, B_4$ = 4.9512, 4.3473, 3.2510, 1.451.

[w] $D_1, \ldots, D_4(10^{-4}cm^{-1})$ = 5.016, 8.08, 33.87, 40.7; $H_0, \ldots, H_3(10^{-8}cm^{-1})$ = - 0.06, - 4.99, - 21.4, - 1490; higher order constants in (8).

[x] In solid Ar matrix at 4 K (11).

(1) Hulthén, ZP 50, 319 (1928).
(2) Rydberg, ZP 73, 74 (1932).
(3) Rydberg, ZP 80, 514 (1933).
(4) Fujioka, Tanaka, Sci. Pap. IPCR (Tokyo) 34, 713 (1938).
(5) Porter, JOSA 52, 1201 (1962).
(6) Porter, Davis, JOSA 52, 1206 (1962).

Hg^1H (continued):

(7) Porter, Davis, JOSA 53, 338 (1963).
(8) Phillips, Davis, BAMS Vol. 2 (1968).
(9) Eakin, Davis, JMS 35, 27 (1970).
(10) Veseth, JP B 3, 1677 (1970).
(11) Knight, Weltner, JCP 55, 2061 (1971).
(12) Veseth, JMS 44, 251 (1972).
(13) Veseth, JP B 6, 1484 (1973).
(14) Kosman, Hinze, JMS 56, 93 (1975).
(15) Stwalley, JCP 63, 3062 (1975).

Hg^2H, Hg^3H:

[a] See [a] of Hg^1H.

[b] "True" constants of (6) who gives doublet splitting and Λ-type doubling parameters. The heads of the 0-0 subbands are at 24810.6 and 28493.6 cm^{-1}.

[c] Magnetic hfs in X $^2\Sigma^+$(b = 0.2010 in v=1) and A $^2\Pi_{1/2}$ of $^{199}Hg^2H$; see (3).

[d] Hg isotope effects (1).

[e] $\Delta G(3/2 \ldots 11/2)$ = 790.48, 660.35, 497.80, 264.92, 58.35. G(0) = 488.8 (7).

[f] Effective constants; "true" values for v=0 in (6).

[g] Spin doubling $\Delta\nu_{12}(v=0) = +1.085(N+\frac{1}{2}) - \ldots$; decreasing rapidly with increasing v.

[h] Predissociation by rotation. Highest observed levels in v=0...6 are N=43, 35, 31, 23, 15, 10, 8 (3)(7).

[i] $B_{1\ldots6}$ = 2.595_4, 2.422_5, 2.196_1, 1.845, 1.1729, 0.732_5.

[j] $D_{1\ldots6}(10^{-4}cm^{-1})$ = 1.14, 1.41, 2.02, 1.7, 16.27, $(5._8)$. $H_{0\ldots5}(10^{-8}cm^{-1})$ = -0.9, -0.6_1, $-1._7$, $-5._0$, -110, -300.

[k] In solid argon at 4 K (5).

[ℓ] Magnetic hfs in X $^2\Sigma^+$ and A_1 $^2\Pi_{1/2}$ of $^{199}Hg^3H$ (3).

[m] $\Delta G(3/2 \ldots 11/2)$ = 680.34, 601.75, 509, 390, 240. G(0) = 402.6 (7).

(continued p. 319)

State	T_e	ω_e	$\omega_e x_e$	B_e	α_e	D_e (10^{-4}cm^{-1})	r_e (Å)	Observed Transitions Design.	ν_{00}	References
$^{(202)}Hg^1H^+$		$D_0^0 = (2.9_9)$ eV								FEB 1975
A $^1\Sigma^+$	44316.6	1623.6[a] Z	45.1$_1$[b]	5.867[a]	0.201[c]	3.1[d]	1.692$_7$	A→X, R	44112.6 Z	(1)* (4)(5)
X $^1\Sigma^+$	0	2027.7[a] Z	40.9	6.613	0.206	2.8$_5$	1.594$_4$			
$^{(202)}Hg^2H^+$		$D_0^0 = (3.0_1)$ eV								FEB 1975
A $^1\Sigma^+$	44306.3	1151.2[a] Z	22.4$_5$[e]	2.953	0.074[f]	0.77[g]	1.691$_9$	A→X, R	44161.9 Z	(2)(3)(4)(5)
X $^1\Sigma^+$	0	1438.5[a] Z	20.7	3.328	0.073$_6$	0.72	1.593$_8$			
$^{(202)}Hg^4He$		(μ = 3.9248221$_3$) See ref. (1)(4)(5) of HgAr.								FEB 1975
$^{(202)}Hg^{127}I$		(μ = 77.935292$_4$)		$D_0^0 = 0.35_4$ eV[a]						FEB 1975 A
H	47110	97.1 H	1.7	Predissociation above v=2.				H→X, R	47096 H	(2)(8)*
G	45542	88.4 H	0.2					G→X, R	45524 H	(2)(8)*
F_3	(44531)	(85.5) H	(0.8)					F_3→X, R	(44510) H	(7)(9)
F_2		Unclassified bands in the region 40800 - 42200 cm^{-1}, converging near 40800 cm^{-1}.						F_2→X, R		(7)(13)*
F_1	40152	90.8 H	0.93					F_1→X, R	40135 H	(7)(9)
E		Bands in the region 37500 - 39500 cm^{-1}; tentative analyses (4)(14).						E→X, R		(3)(4)(14)
D ($^2\Pi_{3/2}$)	36269	178.0 H	1.14					D→X, V	36295 H	(3)(5)(6)
C ($^2\Pi_{1/2}$)	32730.$_0$	235.6 H	2.21					C↔X, V	32785.$_0$ H	(1)(3)(5)(6) (10)(12)(16)
B $^2\Sigma^+$	24187.1	110.45 H	0.15					B↔X, R	24180.0 H	(11)(15)*
X $^2\Sigma^+$	0	125.0 H	1.0[b]							

Hg^2H, Hg^3H (continued):

[n]Spin doubling $\Delta\nu_{12}(v=0) = +0.744(N+\frac{1}{2}) - \dots$; decreasing with increasing v.

[o]Predissociation by rotation. The last observed levels in v=0,1,2 are N=51, 44, 39.

[p]B_1, B_2, B_3 = 1.7691, 1.6806, 1.5757.

[q]$D_1, D_2, D_3(10^{-4}cm^{-1})$ = 0.4795, 0.5586, 0.7957; $H_0 \dots H_3(10^{-8}cm^{-1})$ = -0.272, +0.057, -0.247, +1.66.

(1) Mrozowski, ZP 99, 236 (1936).
(2) See ref. (4) of Hg^1H.
(3) See ref. (9) of Hg^1H.
(4) See ref. (10) of Hg^1H.
(5) See ref. (11) of Hg^1H.
(6) See ref. (12) of Hg^1H.
(7) See ref. (15) of Hg^1H.

Hg^1H^+, Hg^2H^+:

[a]Re-evaluated from the data of (4).

[b]$\omega_e y_e = -7.2_0$, calculated from the value for Hg^2H^+.

[c]$\gamma_e = -0.045_5$.

[d]$\beta_e = +0.4 \times 10^{-4}$.

[e]$\omega_e y_e = -2.5_7$.

[f]$\gamma_e = -0.0105$.

[g]$\beta_e = +0.10 \times 10^{-4}$.

(1) Hori, ZP 61, 481 (1930).
(2) Mrozowski, APP 4, 405 (1935).
(3) Hori, Huruiti, ZP 101, 279 (1936).
(4) Mrozowski, Szulc, APP 6, 44 (1937).
(5) Mrozowski, PR 58, 332 (1940).

HgI: [a]Extrapolation for X $^2\Sigma^+$ (15). Good agreement with an earlier thermochemical value (12).

[b]Above v=7: $\omega_e x_e \approx 1.5$.

(1) See ref. (1) of HgBr.
(2) Prileshajewa, PZS 1, 189 (1932).
(3) Wieland, ZP 76, 801 (1932).
(4) Sastry, PNISI 8, 289 (1942).
(5) See ref. (6) of HgBr.
(6) Rao, Sastry, Krishnamurti, IJP 18, 323 (1944).
(7) Rao, Rao, IJP 20, 148 (1946).
(8) Ramasastry, Rao, IJP 21, 143 (1947).
(9) Ramasastry, IJP 22, 95 (1948).
(10) See ref. (8) of HgBr.
(11) See ref. (9) of HgBr.
(12) Wieland, Herczog, HCA 32, 889 (1949).
(13) Ramasastry, PNISI 18, 487 (1952).
(14) Krishnamurthi, ZP 160, 438 (1960).
(15) See ref. (11) of HgBr.
(16) Greig, Gunning, Strausz, JCP 52, 4569 (1970).

State	T_e	ω_e	$\omega_e x_e$	B_e	α_e	D_e	r_e (Å)	Observed Transitions Design.	ν_{00}	References
$^{(202)}Hg^{(115)}In$		(μ = 73.237853)								FEB 1975
		Two diffuse emission bands extending to longer wavelengths from the In lines at 4511 Å (22160 cm^{-1}) and 4102 Å (24373 cm^{-1}).								(1)
		Fragments of a weak system of V shaded emission bands at 20020 cm^{-1}.								(1)
B	a + 19106	198 H						B→A,[a] V	19130 H	(1)* (2)*
A	a	151 H								
$^{(202)}Hg^{(39)}K$		(μ = 32.6625294) D_e = 0.052 eV[a]								FEB 1975 A
		Diffuse, V shaded absorption bands at 16160, 16260, 24310, 25070 cm^{-1}; tentatively assigned to HgK.								(1)
X $^2\Sigma^+$	0						4.9_1[a]			(5)
$^{(202)}Hg^{(84)}Kr$		(μ = 59.281983_1)								FEB 1975 A
		See ref. (1)(2)(5)(7)(10) of HgAr.								
$^{(202)}Hg^{(7)}Li$		(μ = 6.7804667) D_e = 0.105 eV[a]								FEB 1975
X $^2\Sigma^+$	0						3.0_0[a]			(2)(4)
$^{(202)}Hg^{23}Na$		(μ = 20.640336_8) D_e = 0.055 eV[a]								SEP 1976 A
		Diffuse, V shaded absorption bands at 15380, 15490, 21510, 22590, 22690 cm^{-1}; tentatively assigned to HgNa (1). Theoretical calculation of the lowest excited states (4).								
X $^2\Sigma^+$	0						4.7_2[a]			(2)(3)
$^{(202)}Hg^{(20)}Ne$		(μ = 18.1917011)								FEB 1975
		See ref. (1)(5) of HgAr.								

HgIn: [a]Double heads.

(1) Purbrick, PR 81, 89 (1951).
(2) Santaram, Winans, JMS 16, 309 (1965).

HgK: [a]From K – Hg scattering data (5); see also (2)(3)(4).

(1)...(5) See ref. (1)...(5), resp., of HgCs.

HgLi: [a]From Li – Hg scattering data (2)(4); see also (1)(3).

(1) Groblicki, Bernstein, JCP 42, 2295 (1965).
(2) Olson, JCP 49, 4499 (1968).
(3) See ref. (4) of HgCs.
(4) Buck, Hoppe, Huisken, Pauly, JCP 60, 4925 (1974).

HgNa: [a]From Na – Hg scattering data (2)(3).

(1) See ref. (1) of HgCs.
(2) See ref. (4) of HgCs.
(3) Buck, Pauly, JCP 54, 1929 (1971).
(4) Düren, CPL 39, 481 (1976).

State	T_e	ω_e	$\omega_e x_e$	B_e	α_e	D_e	r_e (Å)	Observed Transitions Design.	ν_{00}	References
$^{(202)}Hg^{16}O$?		(μ = 14.8211639$_5$)								FEB 1975
		Absorption bands in the region 33900 - 36500 cm^{-1}, usually ascribed to Hg_2 ("wing" bands), were tentatively assigned to HgO by (1).								(1)
$^{(202)}Hg^{(85)}Rb$		(μ = 59.779505$_9$) D_e = 0.049 eV[a]								FEB 1975 A
		Diffuse, V shaded absorption bands at 15710, 15780, 20490 cm^{-1} and other unclassified bands in the region 22700 - 23800 cm^{-1}; tentatively assigned to HgRb.								(1)
X $^2\Sigma^+$	0						5.1[a]			(2)
$^{(202)}Hg^{(32)}S$		(μ = 27.6025696) $D_0^0 \leq 2.1_7$ eV[a]								FEB 1975
$^{(202)}Hg^{(80)}Se$		(μ = 57.259761$_4$) $D_0^0 \leq 1.6_9$ eV[a]								FEB 1975
$^{(202)}Hg^{(205)}Tl$		(μ = 101.730727)								FEB 1975 A
		Four band systems in the red (15300 cm^{-1}, em. and abs.), green (19200 cm^{-1}), blue (22000 cm^{-1}, em.), and violet regions of the spectrum (26200 cm^{-1}, em. and abs.). Different vibrational analyses in (1) and (2), no details.								(1)(2)
$^{(202)}Hg^{(132)}Xe$		(μ = 79.792686$_9$)								FEB 1975 A
		See ref. (1)(5)(10) of HgAr.								

HgO: (1) Walter, Barratt, PRS A 122, 201 (1929).

HgRb: [a]From Rb – Hg scattering data (2).

(1) See ref. (1) of HgCs.
(2) See ref. (4) of HgCs.

HgS, HgSe:

[a]Thermochemical value (mass-spectrom.)(3)(4)(5). Based on mass-spectroscopic evidence it is unlikely that three absorption continua with long-wavelength limits at 22500, 32200, 44400 cm^{-1}, observed by (1), and two absorption continua with long-wavelength limits at 22200, 38630 cm^{-1}, observed by (2), are due to HgS and HgSe, resp..

(1) Sen-Gupta, PRS A 143, 438 (1934).
(2) Mathur, IJP 11, 177 (1937).
(3) Colin, ICB 26, 1129 (1961).
(4) Goldfinger, Jeunehomme, TFS 59, 2851 (1963)
(5) Marquart, Berkowitz, JCP 39, 283 (1963).

HgTℓ: (1) Winans, Pearce, PR 74, 1262 (1948).
(2) Winans, Santaram, Pearce, Proc. Int. Conf. Spectrosc., Bombay (1967), Vol. I, p. 149.

State	T_e	ω_e	$\omega_e x_e$	B_e	α_e	D_e ($10^{-4}cm^{-1}$)	r_e (Å)	Observed Transitions Design.	ν_{00}	References
$^1H^{127}I$		μ = 0.99988453		D_0^0 = 3.054_1 eV[a]		I.P. = 10.38 eV[b]				SEP 1976 A
L	[100640]	Broad absorption peak (width ~4500 cm^{-1}).[c]						L← X,	99500	(31)
H (1)	[75435]	First member of a Rydberg series converging to X $^2\Pi_{\frac{1}{2}}$(v=0) of HI^+ (I.P. = 11.05 eV); fragments of additional series.[d]						H← X,	74290	(3)* (27)*
F $^1\Delta$	[71372.8]			$[6.33_5]$[e]		$[2._3]$[e]	[1.631]	F← X, R	70228.2 Z	(30)*
f_1 $^3\Delta_1$	[70831.5]			$[6.01_5]$[f]			[1.674]	f_1← X, R	69686.9 Z	(30)*
D $^1\Pi$	[70389.0]			[6.198][g]		[2.1][g]	[1.649]	D← X, R	69244.4 Z	(30)*
d_0 $^3\Pi_0$	[70302.4]			[6.117][h]		[2.1][h]	[1.660]	d_0← X, R	69157.8[h] Z	(30)*
		Additional unclassified absorption bands between 68100 and 69000 cm^{-1}.								(30)
G 1	[70136.4]			[6.406][i]		[3.2][i]	[1.622]	G← X, (R)	68991.8 Z	(3)* (30)*
V $^1\Sigma^+$				[2.84][j]		[2.0][j]	[2.44][j]	V← X, R	68004.4[j] Z	(30)*
E $^1\Sigma^+$	(66326)	[1681.8] Z		[6.110]	k	[2.5][k]	$[1.661_1]$	E← X, R	66022.6 Z	(3)* (30)*
f_2 $^3\Delta_2$	[65838.6]			[6.757][ℓ]		$[12._3]$[ℓ]	[1.580]	f_2← X, V	64694.0 Z	(30)*
f_3 $^3\Delta_3$	[65717.5]			[5.706][m]		[-8.3][m]	[1.719]	f_3← X, R	64572.9 Z	(30)*
e $^3\Sigma^+$	[65345]	Very diffuse feature.						e← X,	64200	(30)*
d_1 $^3\Pi_1$	[65028]	Diffuse feature.						d_1← X,	63883 H^Q	(30)*
d_2 $^3\Pi_2$	(63922)	[2154.4] Z		[6.065]	n	[1.7]	$[1.667_3]$	d_2← X, R	63854.9 Z	(30)*
C $^1\Pi$	(62378)	[2183] H^Q	Diffuse, no rotational structure.					C← X,	62325 H^Q	(3)(20)
b_0 $^3\Pi_0$ 0^+	60858.7	2314.7° Z	54.3°	6.493°	0.118°		1.611_4	b_0← X,	60857.9 Z	(20)* (30)*
b_0 $^3\Pi_0$ 0^-	(60840)	Diffuse Q head only.							60839 H^Q	(20)*
b_1 $^3\Pi_1$	(56783)	[2200]		[6.427]	(v=1 diffuse Q head)		$[1.619_6]$	b_1← X,	56738.3 Z	(3)(20)
b_2 $^3\Pi_2$	(55874)	[2207.4] Z		6.436	0.175		1.618_5	b_2← X,	55833.1 Z	(20)*

State	T_e	ω_e	$\omega_e x_e$	B_e	α_e	D_e ($10^{-4}cm^{-1}$)	r_e (Å)	Observed Transitions Design.	ν_{00}	References
$^1H^{127}I$ (continued)										
A ($^1\Pi$) a ($^3\Pi_{0^+}$, $^3\Pi_1$)	Continuous absorption starting at ~28000 with maximum at ~46000 cm^{-1}.							A, a ← X,[p]		(2)(4)(5) (17)(24)
X $^1\Sigma^+$	0	2309.01_4 Z	39.643_5[q]	[6.4263650][rs]	0.1688_6[t]	[2.069][r]	1.60916	Rot.-vibr. sp.[u]		(6)(12)(23) (26)
								Rotation sp.[vw]		(1)(7)(8)(22)
								Raman sp.[x]		(29)

[1]HI: [a]From $D_0^0(H_2)$, $D_0^0(I_2)$, and ΔH_{f0}(HI, from gaseous H_2, I_2).

[b]From photoionization studies by (10); refers to $X^2\Pi_{3/2}$ of HI^+. (18) give the same value, (15) give 10.42 eV from the photoelectron spectrum.

[c]Diffuse on account of predissociation and preionization; presumably first member of a Rydberg series converging to A $^2\Sigma^+$ of HI^+ (31).

[d]Above the first ionization limit (X $^2\Pi_{3/2}$) the members of the series are subject to preionization and are seen as photoionization peaks (27).

[e]Average B and D, $B(2^+) - B(2^-) = -0.05$.

[f]Average B value, $B(1^+) - B(1^-) = +0.240$.

[g]Refers to the 1^+ component; $B(1^-) \approx 6.25$.

[h]Constants refer to the 0^+ component; for 0^- $B_0 = 6.091$, $\nu_{00} = 69149.5$.

[i]Average B and D, $B(1^+) - B(1^-) = +0.107$.

[j]Vibrational numbering uncertain; the numbers given refer to the lowest level observed in absorption for which v is probably fairly high. Several higher vibrational levels have been found; strong perturbations.

[k]$B_1 = 5.62$, $D_1 = 28 \times 10^{-4}$, perturbed at high J.

[ℓ]Average B and D, $B(2^+) - B(2^-) = -0.040$.

[m]Average B and D, $B(3^+) - B(3^-) = +0.018$.

[n]$B_1(^3\Pi_2) = 5.92_3$.

[o]From v=0,1,3 only; $\gamma_e = -0.031_7$.

[p]Photofragment spectroscopy at 37550 cm^{-1} (28) shows that the continuum is of composite nature; 36% of the absorption is due to $^3\Pi_{0^+}$ yielding H + I($^2P_{\frac{1}{2}}$). (28) have analysed the continuum in terms of three overlapping transitions $^1\Pi$, $^3\Pi_{0^+}$, $^3\Pi_1$ ← X. A very weak continuum with maximum at 23500 cm^{-1} was reported by (4).

[q]$\omega_e y_e = -0.0200$, $\omega_e z_e = +0.01621$, from the 1-0,...,4-0 vibr.-rot. bands (23); very slightly different constants are given by (26) who have measured the 5-0 and 6-0 bands.

[r]Microwave value (22).

[s]Dunham potential coefficients (32).

[t]$\gamma_e = -0.00095$; from (23), see [q].

[u]The 1-0,...,6-0 bands have been observed in absorption. Ab-

(continued on p. 327)

State	T_e	ω_e	$\omega_e x_e$	B_e	α_e	D_e ($10^{-4} cm^{-1}$)	r_e (Å)	Observed Transitions: Design.	Observed Transitions: ν_{00}	References
$^2H^{127}I$		μ = 1.98263579		$D_0^0 = 3.094_9$ eV[a]						SEP 1976 A
F $^1\Delta$	[71069.9]			[3.161]		[0.51]	[1.640]	F← X, R	70255.0 Z	(9)
		Further unclassified absorption bands between 68100 and 69800 cm^{-1}.								(9)
f_1 $^3\Delta_1$	[70513.3]			[3.089][b]		[0.4_5][b]	[1.659]	f_1← X, R	69698.4 Z	(9)*
D $^1\Pi$	[70065.5]			[3.155][c]		[0.55][c]	[1.642]	D← X, R	69250.6 Z	(9)*
d_0 $^3\Pi_0$	[69972.1]			[3.131][d]		[0.47][d]	[1.648]	d_0← X, R	69157.2[d] Z	(9)*
V $^1\Sigma^+$				[1.30][e]			[2.56][e]	V← X, R	67893.6[e] Z	(9)
E $^1\Sigma^+$	66163.5	1455.9 Z	75.9_5	[3.092]	f	[0.60]	[1.658]	E← X, R	66057.6 Z	(9)*
f_2 $^3\Delta_2$	[65533.6]			[3.296][g]		[1.6_4][g]	[1.606]	f_2← X,	64718.7 Z	(9)
f_3 $^3\Delta_3$	[65440.7]			[3.00_2][g]		[0.22][g]	[1.68_3]	f_3← X, R	64625.8 Z	(9)
e $^3\Sigma^+$	[65095]	Very diffuse feature.						e← X,	64280	(9)
d_1 $^3\Pi_1$	(63958)	[(1586)][h]		[3.135][g]		[0.39][g]	[1.647]	d_1← X, R	63936.2 Z	(9)
d_2 $^3\Pi_2$	[64681.5]			[3.111]		[0.39]	[1.653]	d_2← X, R	63866.6 Z	(9)
C $^1\Pi$	62321	1664	30	Diffuse Q heads only.				C← X,	62331 H[Q]	(5)
b_0 $^3\Pi_0$ 0^+	[61667.4]			[3.255]			[1.616]	b_0← X,	60852.5 Z	(5)*
0^-	[61647]			Diffuse, unresolved Q branch.					60832 H	
b_1 $^3\Pi_1$	[57547.6]			[3.245][i]			[1.619]	b_1← X,	56732.8 Z	(5)*
b_2 $^3\Pi_2$	(55862)	[1585.2] Z		3.265_1	0.063_2		1.613_7	b_2← X,	55840.1 Z	(5)*
A ($^1\Pi$), a ($^3\Pi$)		Continuous absorption starting at 33000 with maximum at 45000 cm^{-1}.						A, a ← X,[j]		(8)
X $^1\Sigma^+$	0	1639.65_5 Z	19.87_3[k]	[3.2534872][ℓ]	0.06082[k]	[0.5264][ℓ]	1.60909	Rot.-vibr. sp.		(2)(7)
								Rotation sp.[m]		(1)(3)(4)(6)

^{2}HI: [a]From D_0^0(^{1}HI).

[b]Average B and D, $B(1^+) - B(1^-) = + 0.064$.

[c]Average B and D, $B(1^+) - B(1^-) = - 0.014$.

[d]The constants refer to the 0^+ component; the numbering of the Q branch (0^- component, $\nu_{00} \approx 69147.0$) is uncertain.

[e]See j of ^{1}HI.

[f]$B_1 = 2.97_4$ (perturbed at intermediate J), $B_2 = 2.80_2$.

[g]Constants refer to the Ω^+ component.

[h]The 1-0 band is quite diffuse and its assignment uncertain.

[i]Broad P and R lines, diffuse and unresolved Q branch.

^{2}HI (continued):

[j]At 37550 cm^{-1} 26% of the absorption is due to $^3\Pi_{0^+}$; see [p] of ^{1}HI.

[k]$\omega_e y_e = -0.045_9$, $\gamma_e = -0.00017_6$ from the IR sp. (7).

[ℓ]Microwave value (6).

[m]$\mu_{e\ell}(v=0) = 0.31_6$ D (3). Iodine hfs constants in (6).

(1) See ref. (7) of ^{1}HI.
(2) Jones, JMS 1, 179 (1957).
(3) Burrus, JCP 28, 427 (1958).
(4) Cowan, Gordy, PR 111, 209 (1958).
(5) See ref. (20) of ^{1}HI.
(6) See ref. (22) of ^{1}HI.
(7) See ref. (23) of ^{1}HI.
(8) See ref. (24) of ^{1}HI.
(9) See ref. (30) of ^{1}HI.

^{1}HI (continued):

solute intensities, dipole moment function (9)(11)(14)(16)(25). The R branch of the fundamental is much stronger than the P branch on account of rotation-vibration interaction; for the overtones this effect is very small (11)(14). The overall intensities decrease rather slowly in the series 1-0, 2-0, 3-0 (11)(14). Line width, pressure broadening studied by (9)(14).

[v]From the hfs of the microwave spectrum (19) derive nuclear quadrupole (I) and other hyperfine coupling constants; see also (22). From the Stark effect in the hfs of the 1-0 transition (21) obtain $\mu_{e\ell}(v=0) = 0.4477$ D.

[w]Absolute intensities (13).

[x]Vibrational Raman cross sections.

(1) Czerny, ZP 44, 235 (1927).
(2) Goodeve, Taylor, PRS A 154, 181 (1936).
(3) Price, PRS A 167, 216 (1938).
(4) Datta, Kundu, PNISI 7, 311 (1941).
(5) Romand, AP(Paris) (12) 4, 529 (1949).
(6) Boyd, Thompson, SA 5, 308 (1952).
(7) Palik, JCP 23, 217 (1955).
(8) Cowan, Gordy, PR 104, 551 (1956).
(9) Ameer, Benesch, JCP 37, 2699 (1962).
(10) Watanabe, Nakayama, Mottl, JQSRT 2, 369 (1962).
(11) Benesch, JCP 39, 1048 (1963); 40, 422 (1964).
(12) Haeusler. Meyer, Barchewitz, JP(Paris) 25, 961 (1964).
(13) Chamberlain, Gebbie, Nature 208, 480 (1965).
(14) Meyer, Haeusler, Barchewitz, JP(Paris) 26, 305 (1965).
(15) Frost, McDowell, Vroom, JCP 46, 4255 (1967).
(16) Jacobi, JMS 22, 76 (1967).
(17) Huebert, Martin, JPC 72, 3046 (1968).
(18) Lempka, Passmore, Price, PRS A 304, 53 (1968).
(19) Van Dijk, Dymanus, CPL 2, 235 (1968).
(20) Tilford, Ginter, Bass, JMS 34, 327 (1970).
(21) Van Dijk, Dymanus, CPL 5, 387 (1970).
(22) De Lucia, Helminger, Gordy, PR A 3, 1849 (1971).
(23) Hurlock, Alexander, Rao, Dreska, JMS 37, 373 (1971).
(24) Ogilvie, TFS 67, 2205 (1971).
(25) Tipping, Forbes, JMS 39, 65 (1971).
(26) Bernage, Niay, Houdart, CR B 278, 235 (1974).
(27) Tsai, Baer, JCP 61, 2047 (1974).
(28) Clear, Riley, Wilson, JCP 63, 1340 (1975).
(29) Cherlow, Hyatt, Porto, JCP 63, 3996 (1975).
(30) Ginter, Tilford, Bass, JMS 57, 271 (1975).
(31) Terwilliger, Smith, JCP 63, 1008 (1975).
(32) Ogilvie, Koo, JMS 61, 332 (1976).

State	T_e	ω_e	$\omega_e x_e$	B_e	α_e	D_e $(10^{-5}cm^{-1})$	r_e (Å)	Observed Transitions Design.	Observed Transitions ν_{00}	References
$^3H^{127}I$		$\mu = 2.9460334_1$								SEP 1976
X $^1\Sigma^+$	0			[2.193261]		[2.385][a]	[1.615230]	Microwave sp.[b]		(1)(2)
$^1H^{127}I^+$				$D_0^0 = 3.12_5$ eV[a]		I.P. = $19._6$ eV[b]				SEP 1976
A $^2\Sigma^+$	(28000)[c]	d								
X $^2\Pi_{1/2}$	(5400)[c]									
$^2\Pi_{3/2}$	0	[(2170)][e]					(1.62)[e]			
$^1H^{127}I^-$										SEP 1976
		Resonances due to inverse preionization have been found in the electron transmission current in the range 6.7 - 10.0 eV. They correspond to excited states of HI^- in which two electrons are in Rydberg orbitals $ns\sigma^2$, $np\sigma^2$, or $nd\lambda^2$ while the core is in the X $^2\Pi_{3/2}$ or $^2\Pi_{1/2}$ state of HI^+.								(1)
$^{165}Ho_2$		$\mu = 82.465179$		$D_0^0 = 0.8_2$ eV[a]						FEB 1975
$^{165}Ho^{19}F$		$\mu = 17.0360178$		$D_0^0 = 5.5_7$ eV[a]						FEB 1975
B	21240.9	502.2 H	2.9					B←X,[b] R	21184.2 H	(2)
A	19152.77	539.45 Z	4.3_9	0.24591[c]	0.00217	[0.0173]	2.0060	A←X,[b] R	19114.41 Z	(2)*
X d	0	615.28 Z	2.60_3	0.26295	0.00145	[0.0178]	1.9399			
$^{165}Ho^{16}O$		$\mu = 14.5808654_3$		$D_0^0 = 6.3_9$ eV[a]						FEB 1975 A
		R shaded emission bands from 21700 to 23000 cm^{-1}.								(3)
		Two complicated groups of R and V shaded emission bands in the regions 16900 - 18000 and 18500 - 19800 cm^{-1}.							(18795) (17881)	(1)* (2)*

State	T_e	ω_e	$\omega_e x_e$	B_e	α_e	D_e (10^{-} cm^{-1})	r_e (Å)	Observed Transitions		References
								Design.	ν_{00}	
$^{165}Ho^{(32)}S$		(μ = 26.7806009)		$D_0^0 = 4.3_8$ eV[a]						FEB 1975
$^{165}Ho^{(80)}Se$		($\mu = 53.832260_1$)		$D_0^0 = 3.4_1$ eV[a]						FEB 1975

^3HI: [a]Calculated by isotope relations from HI and DI (2).
[b]Iodine hyperfine structure constants.

(1) Rosenblum, Nethercot, PR 97, 84 (1955).
(2) De Lucia, Helminger, Gordy, PR A 3, 1849 (1971).

$^1HI^+$: [a]From D_0^0(HI) + I.P.(I) - I.P.(HI).
[b]From the electron impact appearance potential of HI^{++} (1).
[c]From the photoelectron spectrum (2)(3).
[d]Unlike those of HF, HCℓ, HBr, the $A^2\Sigma$ photoelectron peak of HI is very broad and without vibrational structure, clearly suggesting that the $A^2\Sigma^+$ state of HI^+ is strongly predissociated. (3) estimate $\Delta G(\frac{1}{2}) \approx 1040$, $r_e \approx 1.90$.
[e]Estimated by (3) from the constants for HF^+, $HC\ell^+$, HBr^+.

(1) Dorman, Morrison, JCP 35, 575 (1961).
(2) Frost, McDowell, Vroom, JCP 46, 4255 (1967).
(3) Lempka, Passmore, Price, PRS A 304, 53 (1968).

$^1HI^-$: (1) Spence, Noguchi, JCP 63, 505 (1975).

Ho_2: [a]Thermochemical value (mass-spectrom.)(1).

(1) Cocke, Gingerich, JPC 75, 3264 (1971).

HoF: [a]Thermochemical value (1).
[b]P, Q, and R branches.
[c]Perturbations.
[d]Possibly lowest component of an inverted triplet or quintet state with large spin-orbit splitting.

(1) Zmbov, Margrave, JPC 70, 3379 (1966).
(2) Robbins, Barrow, JP B 7, L234 (1974).

HoO: [a]Thermochemical value (mass-spectrom.)(4), recalc. (5).

(1) Gatterer, RS 1, 139 (1942).
(2) Gatterer, Junkes, Salpeter, Rosen, METOX (1957).
(3) Mavrodineanu, Boiteux, "Flame Spectroscopy", Wiley (1965).
(4) Ames, Walsh, White, JPC 71, 2707 (1967).
(5) Smoes, Coppens, Bergman, Drowart, TFS 65, 682 (1969).

HoS: [a]Thermochemical value (mass-spectrom.)(1), recalc. (2).

(1) See ref. (5) of HoO.
(2) Bergman, Coppens, Drowart, Smoes, TFS 66, 800 (1970).

HoSe: [a]Thermochemical value (mass-spectrom.)(1).

(1) See ref. (2) of HoS.

State	T_e	ω_e	$\omega_e x_e$	B_e	α_e	D_e $(10^{-}\ cm^{-1})$	r_e (Å)	Observed Transitions Design.	ν_{00}	References
$^{127}I_2$		$\mu = 63.452237_8$ $D_0^0 = 1.54238$ eV[a] I.P. = (9.311) eV[b]								JAN 1977 A
		The absorption spectrum from 450000 to 870000 cm^{-1} (55.8 to 107.9 eV) at low resolution has been described by (55). It corresponds to excitation from the 4d shell to various unfilled orbitals.								(55)
		The absorption spectrum in the VUV region at high resolution has most recently been photographed by (27) who gives an extensive table of observed features in the region 56500 - 75800 cm^{-1}. Most of the bands are assigned to extended Rydberg series converging to a common limit at 75814 cm^{-1} (9.400 eV), a smaller number to fragments of series converging to 80895 cm^{-1} (10.03 eV). The limits are assumed to correspond to v=0 of X $^2\Pi_g$, $\frac{3}{2}$ and $\frac{1}{2}$, respectively, of I_2^+; see, however, [b]. Several of the progressions observed in absorption (27) appear to correspond to emission bands recorded by (11) under medium resolution in the region 56000 - 68000 cm^{-1} and classified by them as belonging to twelve systems. See also (6)(25)*.								(11)(27)*
I	(51973)[c]	112.4 H	0.70_5[d]					I→(B),[c] R	36197 H	(9)(12)(32)
(H)	(46063)	103.7 H	0.09_5					(H→B),[e] R	30283 H	(12)(32)
F ($^1\Sigma_u^+$) [f]	47217.8	95.95_5 H	0.3623				(3.6)[g]	F↔X,[h] R	47158.6 H	(12)(32)(47)*
F'	45230	93.4 H	0.6					F'→X,[i] R	45169 H	(11)
G'($^3\Pi_{1g}$) [j]	(42300)	Suggested upper state of high temperature absorption "continuum" shortward of 3263 Å (30640 cm^{-1}).						G'←A,		(5)(32)
E $^3\Pi_{0^+g}$ [j]	41411.4	101.59 H	0.238_0				(3.65)	E→B,[k] R	25630.5 H	(13)(32)(47)*
D $^1\Sigma_u^+$ [ℓ]	(40679)	104.41 H	0.243[m]					D↔X,[n] R	(40624)	(25)* (32) (36)
G $^3\Pi_{2g}$ [j]	(40300)	Suggested upper state of high temperature absorption "continuum" shortward of 3427 Å (29170 cm^{-1}).						G←A',[o]		(5)(32)
C $^3\Sigma_{1u}^+$ [ℓ]		Repulsive state from $^2P_{3/2} + {}^2P_{1/2}$ responsible for a weak but broad absorption continuum with maximum at 2700 Å (37000 cm^{-1}).[p]						C←X		(7)(10)(32)

I_2: [a]From the convergence of the vibrational levels in the B $^3\Pi_{0^+u}$ state (49)(50).

[b]From the photoelectron spectrum (33)(56); adiabatic potential established by temperature variation. The same method yields 9.953 eV for the ionization potential to $^2\Pi_{\frac{1}{2}g}(v=0)$. Neither result agrees with the values obtained by (27) from Rydberg series, i.e. 9.400 and 10.03 eV. The discrepancy could be understood if the Rydberg series were to correspond to v'=3, but the absence of series with v'=0,1,2 would still be puzzling.

[c]Weak emission bands in the presence of foreign gases, 2785 - 2731 Å. T_e is based on the assumption that the lower state is the B state, but (36) has suggested that instead it may be the D state leading to $T_e \approx 76872$ cm^{-1}.

[d]$\omega_e y_e = + 0.004$.

[e]Strong emission bands in the presence of foreign gases, 3460 - 3015 Å. It is by no means certain that the lower state has been correctly identified as the B state. (35) suggests that the bands arise from the transition G→A'.

[f]Configuration $...\sigma_g^2 \pi_u^3 \pi_g^3 \sigma_u^2$.

[g]From the intensity distribution and Franck-Condon principle (32)(47).

[h]In emission in electric discharges in the presence of foreign gases, 2740 - 2490 Å. Also observed for $^{129}I_2$, confirming the vibrational numbering. Emission bands in the region 2240 - 1950 Å are assigned by (11) to a separate system (called H→X) with $\nu_{00} = 48072$ and $\omega \approx 79$. It seems, however, possible that these bands belong to F→X.

[i]The analysis of this fairly extensive system [2400 - 2240 Å, called E→X by (11)] is not yet supported by isotope studies, nor is it seen in absorption.

[j]Configuration $...\sigma_g \pi_u^4 \pi_g^3 \sigma_u^2$.

[k]Emission bands in the presence of foreign gases, 4400 - 4000 Å. Also observed for $^{129}I_2$, confirming the vibrational numbering. The E→B fluorescence spectrum following two-photon absorption (77) consists of transitions both to the discrete and to the continuous part of B, the latter giving rise to diffuse bands ("structured" continuum)(83). From a comparison of the calculated with the observed intensity distribution (83) obtains the potential function of E as well as the variation of the transition moment with r. The lifetime of E→B is 27 ns (82) confirming that this is an allowed transition and that the E state is $^3\Pi_{0^+g}$.

[l]Configuration $...\sigma_g \pi_u^4 \pi_g^4 \sigma_u$.

[m]$\omega_e y_e = + 0.00045$; the v' numbering is uncertain and, therefore, the vibrational constants are subject to change.

[n]The system includes the absorption bands of (2)(3)(6), remeasured by (20). It also includes the resonance series of (14) in the region 1830 - 2370 Å which arise from very high vibrational levels (v' ≈ 195) of the D state excited by the 1830 Å atomic line of iodine. The system further includes the diffuse emission bands in the region 2500 - 5000 Å with a characteristic group near 3250 Å [McLennan bands (1)]. The diffuse bands have been recognized by (32) to correspond to transitions from D to the continuum of X [Condon diffraction bands, see also (72)]. Earlier summaries (10)(11)(27) gave an electronic state at $T_e = 51427.9$ with $\omega_e = 169.41$, $\omega_e x_e = 0.941$, $\omega_e y_e = +0.0022$ which was to represent the Cordes absorption bands from 1950 to 1795 Å (6). Following (32)(36) we consider these bands as part of D←X.

[o]The G→A' transition has been observed to lase strongly when mixtures of HI or CF_3I or CH_3I with argon (1000 - 4000 torr) are excited by a pulsed high current electron beam (89). See also H→B, footnote [e].

[p]Nature of the upper state (1_u) and of the dissociation products confirmed by photofragment spectroscopy (54).

State	T_e	ω_e	$\omega_e x_e$	B_e	α_e	D_e $(10^{-9}cm^{-1})$	r_e (Å)	Observed Transitions Design.	ν_{00}	References
$^{127}I_2$ (continued)										
B" $^1\Pi_{1u}$ [q]		Repulsive state from $^2P_{\frac{3}{2}} + {}^2P_{\frac{3}{2}}$, responsible for absorption continuum with maximum at 20050 cm^{-1} and for the predissociation of B $^3\Pi_{0^+u}$.						$B'' \leftrightarrow X$,[r]		(31)(60)(73)
B' $^3\Pi_{0^-u}$ [q]		Repulsive state from $^2P_{\frac{3}{2}} + {}^2P_{\frac{3}{2}}$. The previous assignment of B' as the state responsible for the magnetic field induced predissociation of B is now in doubt; see [u].								
B $^3\Pi_{0^+u}$ [q]	15769.01	125.69_7	Z 0.764_2[st]	0.02903_9[uv]	0.000158_2[wt]	5.4_3[x]	3.024_7	$B \leftrightarrow X$,[ryz]	R 15724.57[a'] Z	(50)(59)* (70)
A $^3\Pi_{1u}$ [q]	(11888)	(44.0)[b']	H (1.0)	[b']				$A \leftarrow X$,[c']	R (11803)[b']	(4)(60)
A' $^3\Pi_{2u}$ [q]	(10100)	Suggested as lower state of high temperature absorption bands near 3427 Å (32)(35).								
X $^1\Sigma_g^+$	0	214.50_2[d']	Z 0.614_7[d']	0.03737_2[d'g']	0.0001113_8[d']	4.2_5[e']	2.666_3	Raman sp.[f']		(48)(71)

I_2 (continued):

[q]Configuration $\ldots\sigma_g^2\pi_u^4\pi_g^3\sigma_u$.

[r]f values based on magnetic circular dichroism spectra have been estimated as 0.0018 ($B'' \leftarrow X$) and 0.009 ($B \leftarrow X$) and have been compared with earlier results (78). For a comparison of theoretical and observed intensities in the $B \rightarrow X$ resonance series see (17).

[s]$- 0.00178(v+\frac{1}{2})^3 - 0.0000738(v+\frac{1}{2})^4 + 0.00000103(v+\frac{1}{2})^5$, from levels with $4 \leq v \leq 50$ (50).

[t]Somewhat different constants, valid for $4 \leq v \leq 77$, are given by (70): $T_e = 15768.32$, $\omega_e = 126.165$, $\omega_e x_e = 0.8673$, ..., $B_e = 0.028939$, $\alpha_e = 0.0001204$, ... (using calculated D_v values); see also (51)(90). RKR potential curve (50). For a discussion of the long-range potential and ΔG, B_v values near the dissociation limit see (29)(39)(50)(58)(62)(66).

[u]Collision induced predissociation of the B state (21); magnetic field induced predissociation (22)(42)(43); spontaneous predissociation (46); hyperfine predissociation (86). The purely radiative lifetime (37)(46)(86) increases smoothly from 0.91 μs at v=7 to approximately 10 μs at the highest observed levels. The measured lifetimes (37)(41) (52)(57)(76)(80) are considerably reduced by spontaneous predissociation due to rotational and hyperfine mixing with B" $^1\Pi_{1u}$, the latter leading to differences in lifetime between ortho and para levels (86). Only near v=12 and above v≈50 are the actual lifetimes close to the purely radiative ones. The magnetic field induced predissociation of B $^3\Pi_{0^+u}$ was previously assumed to be caused by B' $^3\Pi_{0^-u}$, and a potential function for this latter state was derived

I_2 (continued):

from magnetic quenching data (43)(53). The recent observation, however, of a quantum interference effect between magnetic and spontaneous predissociations (64)(68)(85) has established that the magnetic predissociation, too, is produced by the B" $^1\Pi_{1u}$ state.

[v] g_J varies from -0.059 at low v to -5.4_5 μ_N near the dissociation limit; from Hanle effect observations (38)(79)(87)(88). See also (69).

[w] $-3.3_6 \times 10^{-7}(v+\frac{1}{2})^2 - 4.7_8 \times 10^{-8}(v+\frac{1}{2})^3 + 3.2_6 \times 10^{-10}(v+\frac{1}{2})^4$, from levels with $4 \leq v \leq 77$ (50).

[x] $\beta_e \approx +0.3_0 \times 10^{-9}$ for $v \leq 10$ (50). D_v increases rapidly above v=20; for more details see (51)(70).

[y] The continuum joining onto the discrete bands is overlapped by the B"← X continuum. A resolution of these two continua and the A← X continuum was given by (60). See also (61).

[z] The hyperfine structure of several lines has been observed by various high resolution laser techiques; electric quadrupole, magnetic octupole, and other magnetic hfs constants have been evaluated (23)(24)(28)(30)(34)(40)(45)(63)((65)(74)(81); similar analyses for $^{129}I_2$ and $^{127,129}I_2$ (84).

[a'] Extrapolated from data with $v' \geq 4$. The vibrational numbering, changed (19) by 1 from the previous table in MOLSPEC 1, has been confirmed by isotope studies (18).

[b'] (60) suggests that the v' numbering of (4) may have to be raised substantially. Preliminary results of a rotational analysis (92) of nine bands in the A← X, v"=5 progression and of three bands in the v"=4 progression indicate that $\omega' \approx 57.5$, $\omega'x' \approx 1.85$, B'(for the lowest analyzed level) = 0.02375, $\alpha' \approx 0.0005$.

[c'] The continuum joining onto the discrete bands has been studied by many investigators, most recently by (60) who derives an f value of 0.00062; see also (78).

[d'] These constants (50) represent the levels v=0-5; (70), for v=0-6, give $\omega_e = 214.582$, $\omega_e x_e = 0.6243$, $B_e = 0.037363$, $\alpha_e = 0.0001145$ using calculated D_v values. On the basis of the resonance series of (8)(16) and (14), (26) has given polynomial formulae for G(v), B_v, and D_v valid up to v=82: $\omega_e = 214.548$, $\omega_e x_e = 0.6163$, ..., $B_e = 0.037395$, $\alpha_e = 0.0001244$, ..., $D_e = 4.54 \times 10^{-9}$, $\beta_e = 0.017 \times 10^{-9}$, ... The most accurate constants for v=0 were derived (91) from the analysis by means of Fourier transform spectroscopy of the B← X, 30-0 band: $B_0 = 0.037311_5$, $D_0 = 4.5_5 \times 10^{-9}$, $H_0 = -0.76 \times 10^{-15}$.

The vibrational levels of the ground state have been observed up to v=84 [D→ X resonance series (14)], i.e. to within 400 cm^{-1} of the dissociation limit. The levels v" = 98...115 originally reported by (14) were found to be due to an NO impurity (67). As a consequence the RKR potential function of (14) must be corrected at high v. The RKR curves of (17) and (26) extend only to v=82 and are unaffected by this correction.

[e'] $\beta_e = +0.32 \times 10^{-9}$; see [d'].

[f'] High resolution resonance Raman spectra of I_2 vapour up to the eleventh overtone (12-0). Raman spectra in rare gas matrices (75).

[g'] $g_J(v=0, J=12,14) = 9.13 \times 10^{-4}$ μ_N from non-linear level crossing (44).

References on p.335 and 337.

State	T_e	ω_e	$\omega_e x_e$	B_e	α_e	D_e (10^{-} cm^{-1})	r_e (Å)	Observed Transitions Design.	ν_{00}	References
$^{127}I_2{}^+$		μ = 63.452101		D_0^0 = (2.683) eV[a]						JAN 1977
B ($^2\Sigma_g^+$)	(27900)[b]									
A $^2\Pi_{\frac{1}{2}u}$	(18950)[b]									
A $^2\Pi_{\frac{3}{2}u}$	(12420)[b]									
X $^2\Pi_{\frac{1}{2}g}$	5180[c]	220[d]								
X $^2\Pi_{\frac{3}{2}g}$	0	240[d]								
$^{127}I_2{}^-$		μ = 63.452375		D_0^0 = 1.04 eV[a]		I.P. = 2.56 eV[b]				JAN 1977
E ($^2\Pi_{\frac{1}{2}g}$)	(72100)	Resonances (due to inverse preionization) in the electron transmission current at 5.78 and 6.38 eV.[c]								(3)
E ($^2\Pi_{\frac{3}{2}g}$)	(67300)									
X ($^2\Sigma_u^+$)	0									

$I_2{}^+$: [a]From $D_0^0(I_2)$ + I.P.(I) - I.P.(I_2); the uncertainty in the ionization potential of I_2 (see [b] of I_2) makes $D_0^0(I_2{}^+)$ equally uncertain.
[b]From photoelectron spectra (1)(4).
[c]From the photoelectron spectrum (5). (1) obtain 5080.
[d]Average vibrational spacings in the photoelectron spectrum of (5). Good agreement for the $\frac{3}{2}$ component with ω = 238 from the resonance Raman spectrum in solution (2).

(1) Frost, McDowell, Vroom, JCP 46, 4255 (1967).
(2) Gillespie, Morton, JMS 30, 178 (1969).
(3) See ref. (27) of I_2.
(4) Cornford, Frost, McDowell, Ragle, Stenhouse, JCP 54, 2651 (1971).
(5) See ref. (56) of I_2.

$I_2{}^-$: [a]From $D_0^0(I_2)$ and the electron affinities of I and I_2.
[b]From endoergic charge transfer, weighted average of the values in (1) and (2).
[c]The $^2\Pi_{ig}$ excited state is derived from the "grandparent" X $^2\Pi_{ig}$ of $I_2{}^+$ by the addition of a pair of nsσ Rydberg electrons. Weaker resonances at 6.85 and 7.15 eV may be associated with the configuration ...nsσ npσ/π.

(1) Chupka, Berkowitz, Gutman, JCP 55, 2724 (1971).
(2) Baede, Physica 59, 541 (1972).
(3) Spence, PR A 10, 1045 (1974).

I_2 (continued):

(1) McLennan, PRS A 88, 289 (1913); 91, 23 (1914).
(2) Pringsheim, Rosen, ZP 50, 1 (1928).
(3) Kimura, Miyanishi, Sci. Pap. IPCR (Tokyo) 10, 33 (1929).
(4) Brown, PR 38, 1187 (1931).
(5) Skorko, Nature 131, 366 (1933); APP 3, 191 (1934).
(6) Cordes, ZP 97, 603 (1935).
(7) Kortüm, Friedheim, ZN 2 a, 20 (1947).
(8) Rank, Baldwin, JCP 19, 1210 (1951).
(9) Venkateswarlu, PR 81, 821 (1951).
(10) Mathieson, Rees, JCP 25, 753 (1956).
(11) Haranath, Rao, JMS 2, 428 (1958).
(12) Verma, PIAS A 48, 197 (1958).
(13) Haranath, Rao, IJP 34, 123 (1960).
(14) Verma, JCP 32, 738 (1960).
(15) Weissman, Vanderslice, Battino, JCP 39, 2226 (1963).
(16) Rank, Rao, JMS 13, 34 (1964).
(17) Zare, JCP 40, 1934 (1964).
(18) Brown, James, JCP 42, 33 (1965).
(19) Steinfeld, Zare, Jones, Lesk, Klemperer, JCP 42, 25 (1965).
(20) Nobs, Wieland, HPA 39, 564 (1966).
(21) Steinfeld, JCP 44, 2740 (1966).
(22) Degenkolb, Steinfeld, Wasserman, Klemperer, JCP 51, 615 (1969).
(23) Hanes, Dahlstrom, APL 14, 362 (1969).
(24) Kroll, PRL 23, 631 (1969).
(25) Myer, Samson, JCP 52, 716 (1970).
(26) Le Roy, JCP 52, 2683 (1970).
(27) Venkateswarlu, CJP 48, 1055 (1970).
(28) Hänsch, Levenson, Schawlow, PRL 26, 946 (1971).
(29) Le Roy, Bernstein, JMS 37, 109 (1971).
(30) Hanes, Lapierre, Bunker, Shotton, JMS 39, 506 (1971).
(31) Oldman, Sander, Wilson, JCP 54, 4127 (1971).
(32) Mulliken, JCP 55, 288 (1971).
(33) Potts, Price, TFS 67, 1242 (1971).
(34) Sorem, Levenson, Schawlow, PL A 37, 33 (1971).
(35) Tellinghuisen, quoted in (32).
(36) Wieland, quoted in (32).
(37) Brewer, Tellinghuisen, JCP 56, 3929 (1972).
(38) Broyer, Lehmann, PL A 40, 43 (1972).
(39) Le Roy, CJP 50, 953 (1972).
(40) Levenson, Schawlow, PR A 6, 10 (1972).
(41) Shotton, Chapman, JCP 56, 1012 (1972).
(42) Capelle, Broida, JCP 57, 5027 (1972).
(43) Chapman, Bunker, JCP 57, 2951 (1972).
(44) Solarz, Levy, CPL 17, 35 (1972).
(45) Sorem, Hänsch, Schawlow, CPL 17, 300 (1972).
(46) Tellinghuisen, JCP 57, 2397 (1972).
(47) Wieland, Tellinghuisen, Nobs, JMS 41, 69 (1972).
(48) Kiefer, Bernstein, JMS 43, 366 (1972).
(49) Barrow, Broyd, Pederson, Yee, CPL 18, 357 (1973).
(50) Barrow, Yee, JCS FT II 69, 684 (1973).
(51) Brown, Burns, Le Roy, CJP 51, 1664 (1973).
(52) Capelle, Broida, JCP 58, 4212 (1972).
(53) Child, JMS 45, 293 (1973).
(54) Clear, Wilson, JMS 47, 39 (1973).
(55) Comes, Nielsen, Schwarz, JCP 58, 2230 (1973).
(56) Higginson, Lloyd, Roberts, CPL 19, 480 (1973).
(57) Keller, Broyer, Lehmann, CR B 277, 369 (1973).
(58) Le Roy, in "Molecular Spectroscopy", Vol. 1, p. 113. The Chemical Society (1973).

(Continued on p. 337)

State	T_e	ω_e	$\omega_e x_e$	B_e	α_e	D_e (10^{-} cm^{-1})	r_e (Å)	Observed Transitions Design.	Observed Transitions ν_{00}	References
$^{127}I^{79}Br$		$\mu = 48.658793_7$ $D_0^0 = 1.817_6$ eV[a] I.P. = 9.85 eV[b]								JAN 1977 A
	62000 - 77000	Fragments of several Rydberg series converging to X $^2\Pi_{\frac{3}{2}}$ and $^2\Pi_{\frac{1}{2}}$ of IBr^+. Tentative assignments. The three strongest series join on to E, H, J, resp..								(18)
J	65793	[267] H						J← X,	65792 H	(18)
H	(64092)	(290) H						H← X, (V)	(64103)[c] H	(14)*
G	(60877)	(280) H						G← X, (V)	(60883)[c] H	(14)*
F	56349	[310] H						F↔X,[d] V	56370 H	(3)(5)(14)*
E	51677	[314] H						E↔X,[e] V	51700 H	(3)(5)(14)*
D	(38849)	90.2 H	0.15					D→A,[f] R	(26476) H	(5)(6)
		Diffuse emission bands from 19000 to 29000 and 36000 to 43000 cm^{-1}.								(4)
		Fragments of three emission band systems near 18600, 22300, and 24500 cm^{-1}.								(6)
		Weak continuous absorption with maximum at 37000 cm^{-1}.						(B'← X)		(12)(14)
		Strong continuous absorption with maximum at 20200 cm^{-1}.[g]						(B← X)		(12)(14)
B' 0^+		(60) Potential well resulting from an avoided intersection of a repulsive 0^+ state with B $^3\Pi_{0^+}$; see (2)(10)(19)(23).[h]						B'← X, R		(2)(10)*

IBr: [a]From the convergence of A $^3\Pi_1$; see ℓ.

[b]Vertical potential from the photoelectron spectrum (17); vibrational structure not resolved. From the temperature dependence of the photo-ion yield curve in the threshold region (17a) conclude that the adiabatic potential is probably 9.79 eV.

[c](18) give 63573 and 60624 as 0-0 bands of these systems [called F - X, and E - X, resp., by (14)]; the band intervals seem to fit better with the choice made here.

[d]Formerly called D - X.

[e]Formerly called C - X. An extended resonance series excited by the 1849 Å Hg line (1) may originate from a high level (v'=8?) of E.

[f]The analysis of this system was based on the old (2) numbering of the A - X system. The shift in the numbering by one unit established by (9) means either that the v"=0 column has to be omitted from the D - A Deslandres table of (6), and thus ν_{00} = 26340, or that the ΔG" values of (6) are systematically higher than those of (2) and (9). Our choice of ν_{00} [from (6)] corresponds to the latter alternative.

IBr (continued):

[g]Absorption in the continuum at $\nu \lesssim 33000$ cm^{-1} produces predominantly excited Br $^2P_{\frac{1}{2}}$ atoms [diabatic dissociation, see (14)]. By contrast, absorption at 18830 cm^{-1}, i.e. closer to the region of the avoided crossing of the two 0^+ curves, leads mainly to ground state Br $^2P_{\frac{3}{2}}$ atoms [adiabatic dissociation (15)]. See also the discussion in (19a).

[h]Vibrational levels of this state, some sharp, others diffuse, have been observed (2) from 17215 cm^{-1} (relative to X $^1\Sigma$, v=0) to 18315 cm^{-1}, i.e. to within 30 cm^{-1} of the limit I($^2P_{\frac{3}{2}}$) + Br($^2P_{\frac{1}{2}}$). The vibrational numbering is still undecided; (10), following (2), assigns v=8 to the lowest observed level, (19) suggest v=2, while (22) prefer v=5. Narrow J regions of a number of levels have been analyzed by (10); B_v varies from 0.03226 (I^{81}Br) for v=8 in the numbering of (10) to 0.0229 (I^{79}Br) for v=27. The levels v=25 of I^{81}Br and v=26 of I^{79}Br have also been observed in the magnetic rotation spectrum (7).

(references on p. 339)

I_2 (continued):

(59) Singh, Tellinghuisen, JMS 47, 409 (1973).
(60) Tellinghuisen, JCP 58, 2821 (1973).
(61) Tellinghuisen, JCP 59, 849 (1973).
(62) Yee, CPL 21, 334 (1973).
(63) Youmans, Hackel, Ezekiel, JAP 44, 2319 (1973).
(64) Broyer, Vigué, Lehmann, CPL 22, 313 (1973).
(65) Ruben, Kukolich, Hackel, Youmans, Ezekiel, CPL 22, 326 (1973).
(66) Le Roy, CJP 52, 246 (1974).
(67) Verma, Le Roy, JCP 61, 438 (1974).
(68) Vigué, Broyer, Lehmann, JP B 7, L158 (1974).
(69) Wallenstein, Paisner, Schawlow, PRL 32, 1333 (1974).
(70) Wei, Tellinghuisen, JMS 50, 317 (1974).
(71) Williams, Rousseau, Dworetsky, PRL 32, 196 (1974).
(72) Tellinghuisen, CPL 29, 359 (1974).
(73) Brown, Burns, CJP 52, 1862 (1974).
(74) Bunker, Hanes, CPL 28, 377 (1974).
(75) Howard, Andrews, JRS 2, 447 (1974).
(76) Paisner, Wallenstein, JCP 61, 4317 (1974).
(77) Rousseau, Williams, PRL 33, 1368 (1974).
(78) Brith, Rowe, Schnepp, Stephens, CP 9, 57 (1975).
(79) Broyer, Lehmann, Vigué, JP(Paris) 36, 235 (1975).
(80) Broyer, Vigué, Lehmann, JCP 63, 5428 (1975).
(81) Hackel, Casleton, Kukolich, Ezekiel, PRL 35, 568 (1975).
(82) Rousseau, JMS 58, 481 (1975).
(83) Tellinghuisen, PRL 34, 1137 (1975).
(84) Tesic, Pao, JMS 57, 75 (1975).
(85) Vigué, Broyer, Lehmann, JCP 62, 4941 (1975).
(86) Broyer, Vigué, Lehmann, JCP 64, 4793 (1976).
(87) Gouedard, Broyer, Vigué, Lehmann, PRL 36, 906, 996 (1976).
(88) Gouedard, Broyer, Vigué, Lehmann, CPL 43, 118 (1976).
(89) Hays, Hoffman, Tisone, CPL 39, 353 (1976).
(90) Tellinghuisen, JMS 62, 294 (1976).
(91) Gerstenkorn, Luc, Perrin, JMS 64, 56 (1977).
(92) Ashby, private communication (1975).

State	T_e	ω_e	$\omega_e x_e$	B_e	α_e	D_e $(10^{-8} cm^{-1})$	r_e (Å)	Observed Transitions Design.	ν_{00}	References
$^{127}I\,^{79}Br$ (continued)										
B $^3\Pi_{0^+}$	16168	$142._5$ Z	2.57^i	0.0432^{jk}	0.0005_3	$D_2 = 2.2$	2.83	B← X, R	16104^j Z	(2)(11)
A $^3\Pi_1$	12350	138	1.7^ℓ	$B_9 = 0.03803^m$		$D_9 = 3.0^m$		A↔X,[n] R	12285°	(2)(9)* (13)*
X $^1\Sigma^+$	0	$268.64_0{}^p$ Z	$0.814_0{}^q$	$0.0568325_2{}^r$	$0.0001969_0{}^s$	(1.02)	2.46898_9	Raman sp.[t] Microwave sp.[u]		(16) (20)
$^{127}I^{(79)}Br^+$		$D_0^0 = 2.42$ eV[a]								JAN 1977
B $^2\Sigma^+$	(31050)	From the photoelectron spectrum (band maxima) (1).								
A $^2\Pi_{1/2}$	(20410)									
A $^2\Pi_{3/2}$	(17260)									
X $^2\Pi_{1/2}$	4600									
X $^2\Pi_{3/2}$	0									
$^{127}I^{(79)}Br^-$		$D_0^0 = 1.1_2$ eV[b]			I.P. = 2.6_7 eV[c]					JAN 1977

IBr (continued):

[i] $\omega_e y_e = -0.11$; from v=2,3,4, including data for $I^{81}Br$.

[j] Extrapolated from data with v'=2,3,4.

[k] Bands with v'=5 are diffuse, presumably because of predissociation into the intersecting 0^+ state from $^2P_{3/2} + {}^2P_{3/2}$. Higher levels are not observed; see, however, B' 0^+.

[ℓ] $\omega_e y_e = -0.02$; the constants represent the levels v' = 5...16 (9)(13), vibrational numbering confirmed by the observed isotope shifts (9). Levels have been identified up to v=43, convergence at 14660 cm^{-1}.

[m] B_v and D_v values from v=9 to 30 are listed by (9). The Λ-type doubling constant q [≈B(R,P)-B(Q)] for v=14 is $+10\times10^{-6}$ and increases to 42×10^{-6} for v=29.

[n] An extended magnetic rotation spectrum has been observed by (7).

[o] Extrapolated from data with v' ≥ 5.

IBr (continued):

[p]These vibrational constants from laser-excited fluorescence data (24) differ only very slightly from the earlier constants of (9).

[q]$\omega_e y_e = -0.0017_7$.

[r]Microwave value of (20), in good agreement with the less accurate values from the electronic spectrum (9)(24). An earlier microwave value by (8) was clearly erroneous.

[s]$\gamma_e = -4.7 \times 10^{-7}$.

[t]Resonance Raman spectra in argon matrices (21).

[u]Iodine and bromine eqQ values for v=0,1,2.

(1) Loomis, Allen, PR 33, 639 (1929).
(2) Brown, PR 42, 355 (1932).
(3) Cordes, Sponer, ZP 79, 170 (1932).
(4) Asundi, Venkateswarlu, IJP 21, 77 (1947).
(5) Haranath, Rao, IJP 31, 368 (1957).
(6) Venkateswarlu, Verma, PIAS A 47, 150, 161 (1958).
(7) Eberhardt, Cheng, Renner, JMS 3, 664 (1959).
(8) Jaseja, JMS 5, 445 (1960).
(9) Selin, AF 21, 479 (1962).
(10) Selin, AF 21, 529 (1962).
(11) Selin, Söderborg, AF 21, 515 (1962).
(12) Seery, Britton, JPC 68, 2263 (1964).
(13) Clyne, Coxon, JMS 23, 258 (1967).
(14) Donovan, Husain, TFS 64, 2325 (1968).
(15) Busch, Mahoney, Morse, Wilson, JCP 51, 837 (1969).
(16) Holzer, Murphy, Bernstein, JCP 52, 399 (1970).
(17) Potts, Price, TFS 67, 1242 (1971).
(17a) Dibeler, Walker, McCulloh, Rosenstock, IJMSIP 7, 209 (1971).
(18) Donovan, Robertson, SpL 5, 361 (1972).

IBr (continued):

(19) Child, Bernstein, JCP 59, 5916 (1973).
(19a) Coxon, in "Molecular Spectroscopy", Vol. 1, p. 177. The Chemical Society (1973).
(20) Tiemann, Möller, ZN 30 a, 986 (1975).
(21) Wight, Ault, Andrews, JMS 56, 239 (1975).
(22) Couillaud, Ducasse, Garrido, Joly, JP B 9, 2091 (1976).
(23) Faist, Bernstein, JCP 64, 2971 (1976).
(24) Weinstock, JMS 61, 395 (1976).

IBr^+, IBr^-:

[a]From D_0^0(IBr) + I.P.(I) - I.P.(IBr).

[b]From D_0^0(IBr) and the electron affinities of Br and IBr.

[c]From endoergic charge transfer, average of three values (2)(3).

(1) See ref. (17) of IBr.
(2) Chupka, Berkowitz, Gutman, JCP 55, 2724 (1971).
(3) Baede, Physica 59, 541 (1972).

State	T_e	ω_e	$\omega_e x_e$	B_e	α_e	D_e ($10^{-8}cm^{-1}$)	r_e (Å)	Observed Transitions Design.	ν_{00}	References
$^{127}I^{35}Cl$		μ = 27.4146708 D_0^0 = 2.1531 eV[a] I.P. = 10.08 eV[b]								JAN 1977 A
		Fragments of many Rydberg series in the absorption spectrum above 60000 cm^{-1} have been observed and tabulated by (23); see also (19). The intensities in these series are most irregular, possibly because of perturbations by valence states in this region. The Rydberg series are believed to converge to two limits at 81362 and 85996 cm^{-1} corresponding to the $^2\Pi_{3/2}$ and $^2\Pi_{1/2}$, v=0 components of the ground state of ICl^+.								(23)*
L	71006	[420] H	Strong progression, F ← X of (14), n_5 of (23).					L ← X, V	71024	(14)* (23)*
G	66484	[421] H	Strong progression, E ← X of (14), g_5 of (23).					G ← X, V	66503 H	(14)* (23)*
		Extensive system of absorption bands in the region 60250 - 63300 cm^{-1}, not yet analyzed.								(23)
F	58167	[445] H	Extensive band system, formerly D − X, b_6 of (23).					F ↔ X, V	58198 H	(2)(7)(14)* (23)*
E	53477	[434] H	Extensive band system, formerly C − X, a_6 of (23).					E ↔ X, V	53502 H	(2)(7)(14)* (23)*
		Continuous absorption with maximum at 41600 cm^{-1}.								(11)
D	37585	173.2 H	1.1	Extensive band system.				D → A,[c] R	23824 H	(7)*
		Diffuse emission bands between 18700 and 27100 cm^{-1}.								(4)
		Continuous absorption with maximum at 21000 cm^{-1}.[d]								(11)
B' 0^+	(18157)	[32]	Potential well arising from the avoided crossing of the "original" B $^3\Pi_{0^+}$ curve going to $I(^2P_{\frac{3}{2}}) + Cl(^2P_{\frac{1}{2}})$ and the repulsive 0^+ curve coming from unexcited atoms; see also (20b).[e]					B' ← X, R	17981 H	(1)
B $^3\Pi_{0^+}$	$17363._1$	221.1[f] Z	9.62	0.0872[fg]	0.0017	(10)	2.66	B ← X,[h] R	$17279._5$[f] Z	(1)(10)
A $^3\Pi_1$	13742	212.3[i] H	2.39[j]	0.084832[k]		(5.4)	2.6923	A[ℓ] ↔ X,[m] R	13656[i] H	(3)* (8)(10) (12)*
X $^1\Sigma^+$	0	384.29_3 Z	1.501	0.1141587[n]	0.0005354	(4.03)	2.32087_8	Infrared sp.[o]		(6)
								Raman sp.[p]		(15)
								Microwave sp.[o]		(5)(20)

ICℓ: [a]From the convergence limit of A $^3\Pi_1$, see [j].

[b]Average value obtained by photoelectron spectroscopy (17)(18) and photoion mass-spectrometry (18a).

[c]Constants of (7) for the lower state (ω_e'' = 209,7, $\omega_e''x_e''$ = 1.9) are not in agreement with the work of (12) and (21) concerning the A $^3\Pi_1$ state; see (12). This system was also observed in fluorescence following two-step laser excitation D← A← X (22). (Note that ν_e given by (7) does not match their observed ν_{00}.)

[d]At least partly due to B← X; see the discussion in (20a).

[e]There are fragments of fairly sharp branches at places corresponding to extrapolated levels of B $^3\Pi_{0^+}$.

[f]Extrapolated from data with v' = 1,2,3.

[g]v'=4 is strongly predissociated and gives rise to broad, diffuse absorption bands; no higher levels are observed. At high resolution (25) even v'=3 shows appreciable broadening of the individual lines which rapidly increases above J'=37 and is presumably caused by predissociation into the 0^+ state arising from $^2P_{\frac{3}{2}} + {}^2P_{\frac{3}{2}}$; see B' 0^+. The non-radiative lifetime varies from ~ 0.2 ns at low J to 0.02 ns at J=42 (25).

[h]The f value is estimated to be 0.0026 corresponding to a radiative lifetime of 1 μs (25).

[i]Vibrational constants from (12) whose measurements of the I + Cℓ chemiluminescence spectrum include bands with v' = 1...18. (21), extrapolating from the data (zero lines) of (10) for $3 \le v' \le 19$, obtain ν_{00} = 13655.23. Earlier, from band origins with $7 \le v' \le 19$, (10) derived ω_e = 209.111, $\omega_e x_e$ = 1.886, $\omega_e y_e$ = - 0.03558, ν_{00} = 13660.29. See [j].

[j]$\omega_e y_e$ = - 0.012 (12). T_v values for 38 levels ($v \ge 3$) are listed in (10). There is a strong vibrational perturbation above v=34, but the last few levels are again regular and converge to 17366.0 cm^{-1}.

[k]B_e, as well as B_0 = 0.084583, B_1 = 0.083995, B_2 = 0.083308, have been extrapolated by (21) from the B_v values of (10) for $3 \le v \le 19$. B_v values up to v=40 are tabulated in (10). The Λ-type doubling constant q [≈ B(R,P)-B(Q)] increases from $+1.32 \times 10^{-5}$ for v=10 to $+11.0 \times 10^{-5}$ for v=27, but the doubling becomes irregular at higher v values (10). The dipole moment in A $^3\Pi_1$ varies from 2.0_0 D at v=7 to 0.6 D at v=27 (21); sign opposite to that in X $^1\Sigma^+$. RKR potential curve (20a).

[ℓ]Fluorescence lifetimes of 110 and 76 μs have been measured (16) following laser excitation at 6068 and 5922 Å, resp..

[m]Magnetic rotation spectrum (9); its intensity drops sharply above v'=28. See also (13). Laser induced fluorescence series in argon matrix (24).

[n]From the microwave spectrum (20); in excellent agreement with the much earlier values from the electronic spectrum (3).

[o]$\mu_{e\ell}$(v=0) = 1.24 D (20); dipole derivative 2.1 D/Å (6). Electric quadrupole coupling constants for I and Cℓ in (20), also iodine spin-rotation constant.

[p]Resonance Raman spectrum in argon matrix (24).

(1) Brown, Gibson, PR 40, 529 (1932).
(2) Cordes, Sponer, ZP 79, 170 (1932).
(3) Curtis, Patkowski, PTRSL A 232, 395 (1934).
(4) Asundi, Venkateswarlu, IJP 21, 76 (1947).
(5) Townes, Merritt, Wright, PR 73, 1334 (1948).
(6) Brooks, Crawford, JCP 23, 363 (1955).
(7) Haranath, Rao, IJP 31, 156 (1957).
(8) Hulthén, Johansson, Pilsäter, AF 14, 31 (1958).
(9) Eberhardt, Cheng, Renner, JMS 3, 664 (1959).
(10) Hulthén, Järlsäter, Koffman, AF 18, 479 (1960).

(continued on p. 343)

State	T_e	ω_e	$\omega_e x_e$	B_e	α_e	D_e $(10^{-6} cm^{-1})$	r_e (Å)	Observed Transitions: Design.	Observed Transitions: ν_{00}	References
$^{127}I^{(35)}Cl^+$		$D_0^0 = 2.52$ eV[a]								JAN 1977
B $(^2\Sigma)$	(33550)	From the photoelectron spectrum (band maxima) (1).								
A $(^2\Pi)$	(22420)									
X $^2\Pi_{1/2}$	4680									
X $^2\Pi_{3/2}$	0	[390]								
$^{127}I^{19}F$		$\mu = 16.5245715$ $D_0^0 = 2.87_9$ eV[a] I.P. = (10.5) eV[b]								JAN 1977 A
B $^3\Pi_{0^+}$	19052.2_4	411.34[c] Z	2.825[d]	0.2272_1[e]	0.00139_8[f]	0.2_8	2.118_9	B→X,[g] R	18952.8_6 Z	(2)* (3)* (7)*
A $^3\Pi_1$	15706	380.5 H	3.8					A→X, R	15591 H	(7)*
X $^1\Sigma^+$	0	610.24[c] Z	3.123[h]	0.279710_8	0.001873_4[i]	0.237	1.90975_9	Microwave sp.[j]		(5)(6)
$^{(115)}In_2$		$(\mu = 57.451938)$ $D_0^0 = 1.0_1$ eV[a]								APR 1977
		R and V shaded bands in emission from 16800 to 20000 cm^{-1} ($\omega' \approx 115$, $\omega'' \approx 142$) and in absorption[b] from 26000 to 28000 cm^{-1}.								(3)

ICℓ (continued):

(11) Seery, Britton, JPC 68, 2263 (1964).
(12) Clyne, Coxon, PRS A 298, 424 (1967).
(13) Stalder, Eberhardt, JCP 47, 1445 (1967).
(14) Donovan, Husain, TFS 64, 2325 (1968).
(15) Holzer, Murphy, Bernstein, JCP 52, 399 (1970).
(16) Holleman, Steinfeld, CPL 12, 431 (1971).
(17) Potts, Price, TFS 67, 1242 (1971).
(18) Cornford, Thesis (Univ. of British Columbia, 1971).
(18a) Dibeler, Walker, McCulloh, Rosenstock, IJMSIP 7, 209 (1971).
(19) Donovan, Robertson, SpL 5, 281 (1972).
(20) Herbst, Steinmetz, JCP 56, 5342 (1972).
(20a) Coxon, in "Molecular Spectroscopy", Vol. 1, p. 177. The Chemical Society (1973).
(20b) Child, Bernstein, JCP 59, 5916 (1973).
(21) Cummings, Klemperer, JCP 60, 2035 (1974).
(22) Barnes, Moeller, Kircher, Verber, APL 24, 610 (1974).
(23) Venkateswarlu, CJP 53, 812 (1975).
(24) Wight, Ault, Andrews, JMS 56, 239 (1975).
(25) Olson, Innes, JCP 64, 2405 (1976).

$ICℓ^+$: [a]From $D_0^0(ICℓ) + I.P.(I) - I.P.(ICℓ)$.

(1) See ref. (17) of ICℓ.

IF: [a]This is the value of (7) and (8) based on a discussion of the vibrational levels in A $^3\Pi_1$ and B $^3\Pi_{0^+}$; an upper limit of 2.894 eV follows from the observed predissociation in B. (4) suggest $D_0^0 = 2.54$ eV on the basis of comparisons with other interhalogens.

[b]Electron impact appearance potential (1).

[c]The vibrational constants for both B $^3\Pi_{0^+}$ and X $^1\Sigma^+$ have

IF (continued):

been recalculated from the data of (2). Similar revised constants for the ground state may be found in (8). Additional higher order constants are needed to represent the B state levels above v=6.

[d]$\omega_e y_e = -0.0744$; see [c].

[e]Onset of predissociation at J=45 of v=11 (23341 cm^{-1}).

[f]$\gamma_e = -0.00008_2$.

[g]RKR Franck-Condon factors, variation of transition moment with r (7).

[h]$\omega_e y_e = -0.00347$.

[i]$\gamma_e = -0.0000027$.

[j]Iodine hyperfine coupling constants (6).

(1) Irsa, Friedman, JINC 6, 77 (1958).
(2) Durie, CJP 44, 337 (1966).
(3) Clyne, Coxon, Townsend, JCS FT II 68, 2134 (1972).
(4) See ref. (20b) of ICℓ.
(5) McGurk, Flygare, JCP 59, 5742 (1973).
(6) Tiemann, Hoeft, Törring, ZN 28 a, 1405 (1973).
(7) Birks, Gabelnick, Johnston, JMS 57, 23 (1975).
(8) Coxon, CPL 33, 136 (1975).

In_2: [a]Thermochemical value (mass-spectrom.)(2)(4).

[b]See also the earlier work of (1).

(1) Wajnkranc, ZP 104, 122 (1936/7).
(2) De Maria, Drowart, Inghram, JCP 31, 1076 (1959).
(3) Ginter, Ginter, Innes, JPC 69, 2480 (1965).
(4) Gingerich, Blue, 18th Annual Conference on Mass Spectrometry and Allied Topics. San Francisco (June 1970).

State	T_e	ω_e	$\omega_e x_e$	B_e	α_e	D_e (10^{-8} cm^{-1})	r_e (Å)	Observed Transitions: Design.	Observed Transitions: ν_{00}	References
$^{115}In^{81}Br$		μ = 47.480276 D_0^0 = 3.9$_9$ eV[a] I.P. = 9.09 eV[b]								JAN 1977 A
C $^1\Pi$	(34000)	Fluctuation bands with various v"; shallow upper potential curve.						C← X,	35053[c]	(1)*
B $^3\Pi_1$	27382.2	218.0[d] H	1.60					B↔X, V_R	27380.5 H	(1)* (5)*
A $^3\Pi_{0^+}$	26596.0	227.4[d] H	1.58					A↔X, V	26599.0 H	(1)* (5)*
X $^1\Sigma^+$	0	221.0 H	0.65	0.0548944[e]	0.0001862[e]	(1.35)	2.54318	Microwave sp.[f]		(2)
$^{115}In^{(79)}Br^+$		D_0^0 = 0.6$_9$ eV[g]								JAN 1977
B ($^2\Sigma$)	(26500)	From the photoelectron spectrum (adiabatic potentials)(6).[h]								
A ($^2\Pi$)	(4300)									
X ($^2\Sigma$)	0									
$^{115}In^{35}Cl$		μ = 26.8097927 D_0^0 = 4.44 eV[a] I.P. = 9.51 eV[b]								JAN 1977
		Continuous absorption, ~47600 cm^{-1}.								(1)(3)
D		Narrow continuum, maximum at 38260 cm^{-1}.						D← X,		(1)(2)* (3)
C $^1\Pi$	37483.6	177.3 H[Q]	12.6	[c]				C↔X, R	37410.7 H[Q]	(1)(2)* (3)*
B $^3\Pi_1$	28560.2	339.4[d] H	2.1	[0.1152]		[2.$_4$]	[2.336]	B↔X, V	28570.9 H	(2)* (7) (14)*
A $^3\Pi_{0^+}$	27764.7	340.3 H	2.0	0.1155	0.000654	6.$_5$	2.333	A↔X V	27775.9 H	(2)* (7)* (14)*
X $^1\Sigma^+$	0	317.4 H[Q]	1.01	0.1090580	0.0005178[e]	(5.1$_4$)	2.401169	Microwave sp.[f]		(5)(9)(10)
$^{115}In^{(35)}Cl^+$		D_0^0 = 0.72 eV[g]								JAN 1977
B ($^2\Sigma$)	(26700)	From the photoelectron spectrum (adiabatic potentials)(12).[h]								
A ($^2\Pi$)	(5300)									
X ($^2\Sigma$)	0									

$InBr$, $InBr^+$:

[a]From thermochemical data and spectroscopic evidence about the $^1\Pi$ and $^3\Pi$ excited states; see (3). In good agreement with 4.0_2 eV obtained by flame photometry (4).

[b]Adiabatic potential from the photoelectron spectrum (6); vertical potential 9.41 eV.

[c]Vertical transition from v"=0.

[d]Constants for $^{115}In^{79}Br$ are given by (5) who propose small changes in the vibrational numbering of the B − X bands observed by (1).

[e]Calculated from the rotational constants for the $^{115}In^{79}Br$ isotope (2).

[f]Electric quadrupole coupling constants.

[g]From $D_0^0(InBr) + I.P.(In) - I.P.(InBr)$.

[h]The $^2\Pi$ photoelectron peak is very broad with a maximum at 10.20 eV ($^2\Pi_{\frac{1}{2}}$?) and a shoulder at ~9.8 eV ($^2\Pi_{\frac{3}{2}}$?). These vertical potentials are close to the dissociation limits $In^+(^1S) + Br(^2P_{\frac{1}{2}}, {}^2P_{\frac{3}{2}})$ at 10.2_4 and 9.7_8 eV, respectively, and seem compatible with the predicted instability or near instability of A $^2\Pi$.

(1) Wehrli, Miescher, HPA 6, 457 (1933); 7, 298 (1934).
(2) Barrett, Mandel, PR 109, 1572 (1958).
(3) Barrow, TFS 56, 952 (1960).
(4) Bulewicz, Phillips, Sugden, TFS 57, 921 (1961).
(5) Lakshminarayana, Haranath, IJP 44, 504 (1970).
(6) Berkowitz, Dehmer, JCP 57, 3194 (1972).

$InC\ell$, $InC\ell^+$:

[a]From thermochemical data and spectroscopic evidence about the $^1\Pi$ excited state (6); 4.5_1 eV by flame photometry (8).

$InC\ell$, $InC\ell^+$ (continued):

[b]Adiabatic potential from the photoelectron spectrum (12); vertical potential 9.75 eV.

[c]The rotational analysis of v=0 by (3) is probably not correct [see (4), p.967-8]. Predissociation for v ≥ 1; the diffuseness is stronger for v=2 than for v=1.

[d]Constants from (2), slightly different constants in (14).

[e]$\gamma_e = + 8.4 \times 10^{-7}$.

[f]$\mu_{e\ell}(v=0) = 3.79$ D (13). Quadrupole coupling constants (11).

[g]From $D_0^0(InC\ell) + I.P.(In) - I.P.(InC\ell)$.

[h]The maximum (vertical potential) of the very broad $^2\Pi$ peak is at 10.85 eV, i.e. the $^2\Pi$ potential curve in the Franck-Condon region has risen ~0.6 eV above its dissociation limit $In^+(^1S) + C\ell(^2P)$.

(1) Miescher, Wehrli, HPA 6, 256 (1933).
(2) See ref. (1) of InBr, InBr+.
(3) Froslie, Winans, PR 72, 481 (1947).
(4) Barrow, Glaser, Zeeman, PPS A 68, 962 (1955).
(5) See ref. (2) of InBr, InBr+.
(6) See ref. (3) of InBr, InBr+.
(7) Youngner, Winans, JMS 4, 23 (1960).
(8) See ref. (4) of InBr, InBr+.
(9) Hoeft, ZP 163, 262 (1961).
(10) Delvigne, de Wijn, JCP 45, 3318 (1966).
(11) Schenk, Tiemann, Hoeft, ZN 25 a, 1827 (1970).
(12) See ref. (6) of InBr, InBr+.
(13) Tiemann, Hoeft, Törring, ZN 27 a, 869 (1972).
(14) Ashrafunnisa, Rao, Murthy, Rao, Physica 73, 421 (1974).

State	T_e	ω_e	$\omega_e x_e$	B_e	α_e	D_e (10^{-5}cm^{-1})	r_e (Å)	Observed Transitions: Design.	Observed Transitions: ν_{00}	References
$^{115}In^{19}F$		μ = 16.3028614		D_0^0 = 5.2_5 eV[a]						JAN 1977
		Unidentified weak system in emission consisting of two sequences of R shaded bands; $\Delta G'(\frac{1}{2})$ = 550, $\Delta G''(\frac{1}{2})$ = 610. R and Q heads.							45709.1 H^Q	(2)
D	[47803]	Weak system.						D← X, R	47536 H	(2)
C $^1\Pi$	42809.2	463.9 H^Q	7.35[b]	0.2674	0.0047_2	0.036	1.966	C↔X, R_V	42772.3 Z	(1)(3)(10)*
B $^3\Pi_1$	31255.7_4	572.2_5 H^Q	2.63[c]	0.2736_2[d]	0.0020_4	(0.025)	1.944_0	B↔X, V	31274.1_8 Z	(1)(2)(3)
A $^3\Pi_{0^+}$	30445.8_6	575.2 (Z)	3.67_6	0.2732_0[d]	0.0020_2	(0.025)	1.945_5	A↔X, V	30465.5_2 Z	(1)(2)(3)
X $^1\Sigma^+$	0	535.3_5 Z	2.6_4	0.2623241_1	0.0018797_7[e]	0.0252	1.985396	Microwave sp.[f]		(7)(8)
								Mol. beam el. reson.[g]		(9)
$^{115}In^1H$		μ = 0.99906241		D_0^0 = 2.48 eV[a]						JAN 1977
		Open band structure in absorption from 40000 to 42900 cm^{-1}, no prominent heads.								(2)
A $^1\Pi$	(22655)	[141.7] Z	b	[3.850][c]	d	[389]	[2.093]	A← X, R	22016.9 Z	(4)(5)
a $^3\Pi$ 2	(17800)	[1300.9][e] Z		5.489[ef]	0.332	[34.6]	1.753	a: →X, V_R	17780.9 Z	(6)
a $^3\Pi$ 1	16941.6_1	1415.1_1[e] Z	43.5_5[g]	5.399_6[ehijk]	0.235_6[h]	[$32._1$][h]	1.7678	↔X, R_V	16904.9_8 Z	(1)(4)(6)
a $^3\Pi$ 0^+	16278.1_5	1458.5_7[e] Z	61.0_3[ℓ]	5.329[eik]	0.246_8[m]	[27.7][n]	1.779_4	↔X, R_V	16259.6_8 Z	(1)(6)
a $^3\Pi$ 0^-	(16230)	[1303.4_2][e] Z		5.349[e]	0.326	[28.1][o]	1.776	→X, V_R	16211.5_3 Z	(10)
a' $^3\Sigma^+$		Not observed, but suggested (12) as the state responsible for anomalies in the Λ-type doubling of $^3\Pi_1$ and for the predissociation of all $^3\Pi$ components.								
X $^1\Sigma^+$	0	1476.0_4 Z	25.61[p]	4.994_5[k]	0.142_8[q]	22.3[r]	1.8380			
$^{115}In^2H$		μ = 1.97940604		D_0^0 = 2.51 eV[s]						JAN 1977
A $^1\Pi$	(22600)	[178.2] Z	t	[2.096][u]	v	[42.6]	[2.016]	A← X, R	22167.1 Z	(5)
a $^3\Pi$ 1	16933	[950.03] Z	w	2.724[j]	0.078[x]	[8.2]	1.768_2	a: ←X, V_R	16922.40 Z	(5)
a $^3\Pi$ 0^+	(16270)	[967.79] Z		2.722	0.112	[7.9]	1.769	←X, V	16273.99 Z	(5)
X $^1\Sigma^+$	0	1048.2_4 Z	12.3_7	2.523	0.051	[5.8]	1.837_3			

InF: [a]Thermochemical value (mass-spectrom.)(6). Flame photometry gives 5.4 eV (5).

[b]$\omega_e y_e$ = - 0.5. This state may have a potential maximum of ~0.21 eV (4)(6).

[c]$\omega_e y_e$ = - 0.095.

[d]From the B' - B" values of (3) combined with B" from the microwave spectrum.

[e]γ_e = + 0.0000049$_7$.

[f]μ_{el}(v=0) = 3.40 D (8).

[g]Hyperfine interaction constants.

(1) Welti, Barrow, Nature 168, 161 (1951); PPS A 65, 629 (1952).
(2) Barrow, Jacquest, Thompson, PPS A 67, 528 (1954).
(3) Barrow, Glaser, Zeeman, PPS A 68, 962 (1955).
(4) Barrow, TFS 56, 952 (1960).
(5) Bulewicz, Phillips, Sugden, TFS 57, 921 (1961).
(6) Murad, Hildenbrand, Main, JCP 45, 263 (1966).
(7) Lovas, Törring, ZN 24 a, 634 (1969).
(8) Hoeft, Lovas, Tiemann, Törring, ZN 25 a, 1029 (1970).
(9) Hammerle, Van Ausdal, Zorn, JCP 57, 4068 (1972).
(10) Nampoori, Kamalasanan, Patel, JP B 8, 2841 (1975).

In^{1}H, In^{2}H:

[a]From the predissociations in A $^1\Pi$ and a $^3\Pi$.

[b]$\Delta G(3/2)$ = 80.8.

[c]Λ-type doubling, Δv_{ef}(v=0) = + 0.0047 J(J+1). Breaking off due to predissociation above J' = 9, 7, 4 in v' = 0, 1, 2, respectively; the limiting curve intersects the ordinate axis near 22250 cm^{-1} above X $^1\Sigma$(v=0,J=0). A few additional diffuse lines have been observed as well as fragments of an "extra" band; see [d].

[d]B_1 = 1.915, B_2 = 1.363. D_v values are also given (4), but

In^{1}H, In^{2}H (continued):

these constants are not sufficient to reproduce the levels of this perturbed state. Zeeman effect studies (7).

[e]Effective constants determined by (6)(10). For a more rigorous treatment of the fine structure of a$^3\Pi$ see (12).

[f]Predissociation in v=0 above J=26 ($^3\Pi^-$) and 27 ($^3\Pi^+$), in v=1 above J=17 ($^3\Pi^+$). The break-off points are in agreement with the limiting curve of dissociation for $^3\Pi_1^+$ (6).

[g]$\omega_e y_e$ = - 13.14.

[h]Average constants for $^3\Pi_1^+$ and $^3\Pi_1^-$; γ_e = - 0.056$_7$. The Λ-type doubling is irregular, see (12), and may be caused by the unobserved a' $^3\Sigma^+$ state. Zeeman effect studies (7).

[i]The $^3\Pi_1$ and $^3\Pi_{0^+}$ components have nearly identical limiting curves of predissociation; the derived dissociation limit at 20125 cm^{-1} above X $^1\Sigma$, v=0, J=0, appears to correspond to a potential maximum at ~3.3 Å (6). Possible correlations of the low-lying states of InH with those of the separated atoms have been discussed by (8) and, more recently, by (12).

[j]The hyperfine structure of J=1 has been investigated both experimentally (4)(5)(11) and theoretically (9)(13).

[k]RKR potential curves (8).

[l]$\omega_e y_e$ = - 6.83.

[m]γ_e = - 0.039$_0$.

[n]H_0 = - 3.76 x 10^{-8}. $|D_v|$ and $|H_v|$ increase rapidly with v.

[o]H_0 = - 2.9 x 10^{-8}.

[p]$\omega_e y_e$ = + 0.30$_8$.

[q]γ_e = + 0.0012$_3$.

[r]H_0 = + 0.46$_3$ x 10^{-8}.

[s]From the value for In^{1}H and the predissociation in A $^1\Pi$.

[t]$\Delta G(3/2)$ = 62.8, $\Delta G(5/2)$ = 51.1.

(continued on p. 349)

State	T_e	ω_e	$\omega_e x_e$	B_e	α_e	D_e $(10^{-8}cm^{-1})$	r_e (Å)	Observed Transitions Design.	ν_{00}	References
$^{115}In^{127}I$		$\mu = 60.303194$		$D_0^0 = 3.4_3$ eV[a]		I.P. = 8.50 eV[b]				JAN 1977
C $^1\Pi$		Continuum with maximum at 31400 cm^{-1}.						C← X,		(1)
B $^3\Pi_1$	25050.5	146.7 H	2.3					B↔X, R_V	25034.8 H[Q]	(1)*
A $^3\Pi_{0^+}$	24401.6	158.5 H	1.7					A↔X, V_R	24392.0 H	(1)* (2)*
X $^1\Sigma^+$	0	177.1 H	0.4	0.0368670	0.0001041_1[c]	0.76	2.75365	Microwave sp.[d]		(3)(6)
$^{115}In^{127}I^+$				$D_0^0 = 0.7_2$ eV[e]						JAN 1977
B $(^2\Sigma)$	(27300)	From the photoelectron spectrum (7).								
A $(^2\Pi_{1/2})$	(7700)									
$(^2\Pi_{3/2})$	(2300)									
X $(^2\Sigma)$	0									
$^{(115)}In^{16}O$		$(\mu = 14.0404485)$		2.9_4 eV < $D_0^0 \leq 3.2_5$ eV[a]						JAN 1977
A	23595.1 23033.1	626.66 H	3.40	b				A→X, R_V	23557.0 H 22995.0 H	(1)*
(X $^2\Sigma$)	0	703.09 H	3.71[c]	b						
$^{(115)}In^{(32)}S$		$(\mu = 25.0123666)$		$D_0^0 = 2.9_4$ eV[a]						JAN 1977
$^{(115)}In^{(121)}Sb$		$(\mu = 58.913758_3)$		$D_0^0 = 1.5_4$ eV[a]						JAN 1977
$^{(115)}In^{(80)}Se$		$(\mu = 47.134275_6)$		$D_0^0 = 2.5_0$ eV[a]						JAN 1977
$^{(115)}In^{(130)}Te$		$(\mu = 60.972685)$		$D_0^0 = 2.1_9$ eV[a]						JAN 1977

In^1H, In^2H (continued):

[u]Λ-type doubling, $\Delta v_{ef}(v=0) = + 0.0012\,J(J+1)$. Breaking-off due to predissociation above $J'=13(v'=0,1)$, $10(v'=2)$, $7(v'=3)$.

[v]$B_1 = 1.098$, $B_2 = 0.981$, $B_3 = 0.751$. D_v values have been determined (5) but their meaning is limited in view of the strong perturbations affecting this state. Fragments of three "extra" bands with $v''=0,1,2$.

[w]$\Delta G(3/2) = 857.42$.

[x]$\gamma_e = - 0.0185$.

(1) Grundström, ZP 113, 721 (1939).
(2) Garton, PPS A 64, 509 (1951).
(3) Kleman, Dissertation (Stockholm, 1953).
(4) Neuhaus, ZP 150, 4 (1958).
(5) Neuhaus, ZP 152, 402 (1958).
(6) Ginter, JMS 11, 301 (1963).
(7) Larsson, Neuhaus, AF 27, 275 (1964).
(8) Ginter, Battino, JCP 42, 3222 (1965).
(9) Freed, JCP 45, 1714 (1966).
(10) Ginter, JMS 20, 240 (1966).
(11) Larsson, Neuhaus, Åslund, AF 37, 141 (1968).
(12) Veseth, Lofthus, JMS 49, 414 (1974).
(13) Veseth, JMS 59, 51 (1976).

InI, InI^+:

[a]Based on thermochemical data and spectroscopic evidence about the $^1\Pi$ and $^3\Pi$ excited states; see (4). Flame photometry gives 3.3_8 eV (5).

[b]Adiabatic potential from the photoelectron spectrum (7); vertical potential 8.82 eV.

[c]$\gamma_e = + 8._0 \times 10^{-8}$.

InI, InI^+ (continued):

[d]Indium and iodine quadrupole coupling constants (6).

[e]From $D_0^0(\text{InI}) + \text{I.P.(In)} - \text{I.P.(InI)}$.

(1) Wehrli, Miescher, HPA 6, 457 (1933); 7, 298 (1934).
(2) Wehrli, HPA 7, 611, 676 (1934); 9, 587 (1936).
(3) Barrett, Mandel, PR 109, 1572 (1958).
(4) Barrow, TFS 56, 952 (1960).
(5) Bulewicz, Phillips, Sugden, TFS 57, 921 (1961).
(6) Schenk, Tiemann, Hoeft, ZN 25 a, 1827 (1970).
(7) Berkowitz, Dehmer, JCP 57, 3194 (1972).

InO: [a]The lower limit is D_0^0(InS); the upper limit, also based on mass-spectrometric evidence, was estimated by (3)(4).

[b]A satisfactory interpretation of the observed bands is still lacking; see (1)(2). Preliminary rotational analysis of two bands by (5).

[c]$\omega_e y_e = - 0.285$.

(1) Watson, Shambon, PR 50, 607 (1936).
(2) Howell, PPS 57, 32 (1945).
(3) Burns, De Maria, Drowart, Inghram, JCP 38, 1035 (1963).
(4) Colin, Drowart, TFS 64, 2611 (1968).
(5) Jacquinot, Lavendy, CR B 281, 397 (1975).

InS, InSe, InTe:

[a]Thermochemical value (mass-spectrom.)(1).

(1) See ref. (4) of InO.

InSb: [a]Thermochemical value (mass-spectrom.)(1).

(1) De Maria, Drowart, Inghram, JCP 31, 1076 (1959).

State	T_e	ω_e	$\omega_e x_e$	B_e	α_e	D_e ($10^{-7}cm^{-1}$)	r_e (Å)	Observed Transitions		References
								Design.	v_{00}	
$^{127}I^{16}O$		$\mu = 14.2045833_3$		$D_0^0 = 1.8$ eV[a]						JAN 1977
A $^2\Pi_{3/2}$	21557.8_1	514.5_7 Z	5.5_2	0.2763_5[b]	0.0027_3	3.2	2.072_3	$A \leftrightarrow X_1$,[c]	R 21474.0_5 Z	(1)* (2)* (3)*
X_2 $^2\Pi_{1/2}$	(2330)[d]									
X_1 $^2\Pi_{3/2}$	0	681.4_7 Z	4.2_9[e]	0.34026[f]	0.00269_6[g]	3.6	1.8676	Microwave sp.[h]		(8)
								EPR sp.		(5)(6)

IO: [a]Based on an extrapolation of the vibrational levels of A $^2\Pi_{3/2}$ and on the assumption that A dissociates into $^2P_{3/2} + {}^1D$. Flame photometry (4) gives a value of 2.4 eV which seems less likely since it is as high as the value for BrO. The observed predissociation in A indicates $D_0^0 \leq 2.72$ eV. Extrapolation of the ground state gives 1.9_4 eV (7).

[b]The rotational lines of absorption bands with v'=0 and 2 are sharp; the lines of the 3-0 absorption band are distinctly diffuse, and the 1-0 and 4-0 bands are completely diffuse owing to predissociation (2).

[c]Franck-Condon factors (9).

[d]Estimated by (6) from spin-orbit coupling of component atoms.

[e]$\omega_e y_e = -0.01_3$.

[f]From the microwave spectrum (8) obtained $B_0(^2\Pi_{3/2}) = 0.338804$ which agrees well with the value from the electronic spectrum.

[g]$\gamma_e = -0.000009_7$.

[h]Magnetic hyperfine interaction, nuclear quadrupole coupling constant.

(1) Coleman, Gaydon, Vaidya, Nature 162, 108 (1948).

(2) Durie, Ramsay, CJP 36, 35 (1958).

(3) Durie, Legay, Ramsay, CJP 38, 444 (1960).

(4) Phillips, Sugden, TFS 57, 914 (1961).

(5) Carrington, Dyer, Levy, JCP 52, 309 (1970).

(6) Brown, Byfleet, Howard, Russell, MP 23, 457 (1972).

(7) Trivedi, Gohel, JP B 5, L38 (1972).

(8) Saito, JMS 48, 530 (1973).

(9) Rao, Rao, Rao, PL A 50, 341 (1974).

State	T_e	ω_e	$\omega_e x_e$	B_e	α_e	D_e ($10^{-7}cm^{-1}$)	r_e (Å)	Observed Transitions Design.	Observed Transitions ν_{00}	References
$^{(193)}Ir^{(11)}B$		(μ = 10.4150834)		$D_0^0 = 5.2_7$ eV[a]						FEB 1975
$^{(193)}Ir^{12}C$		(μ = 11.29743405)		D_0^0 = 6.45 eV[a]						FEB 1975
		Two unidentified bands at 15846 and 16504 cm^{-1}.								(3)
L $^2\Phi_{7/2}$	(20940)	(810)		[0.4812]		[6.8]	[1.760_9]	L↔X_1,	R 20816.52[b] Z	(2)(3)
K $^2\Pi_{3/2}$	19349	[832.7] Z		0.5053	0.0051	7.4	1.718_4	K↔X_1,	R 19236.64[b] Z	(2)* (3)
E_2 $^2\Delta_{3/2}$	x_2 + (12180)	(960)		0.5148	0.0043	[5.9][c]	1.702_5	E_2→X_2,	R 12145.3[d] Z	(3)
E_1 $^2\Delta_{5/2}$	15149.2	963.9 Z	5.4_4	0.5132	0.0040	6.2	1.705_2	E_1↔X_1,	R 15100.89[d] Z	(3)*
D $^2\Phi_{7/2}$	14413.5	935.7 Z	5.9_6	0.5053	0.0038	5.9	1.718_4	D↔X_1,	R 14350.95[d] Z	(3)*
X_2 $^2\Delta_{3/2}$	x_2[e]	(1030)		0.5272	0.0035	5.5	1.682_4			
X_1 $^2\Delta_{5/2}$	0	1060.1 Z	4.5_3	0.5268	0.0032	5.2	1.683_0			
$^{193}Ir^{16}O$		μ = 14.7705663		$D_0^0 = (3.6_4)$ eV[a]						FEB 1975
		Additional bands at 14338, 14490 cm^{-1} (2), and 17070 cm^{-1} (2)(4).								(2)(4)
H' (3/2)		d		[0.3378][b]	d	[$4._0$][b]	[1.838_1]	H'→A',	R 22005.8[cb] Z	(4)
H" 5/2				[0.3371]		[$4._0$]	[1.840_0]	H"→A",	R 21869.7[c] Z	(4)
D (3/2)	a' + 17720	[672.5] Z		0.3534[e]	0.0026	[3.2]	1.797_1	D→A',	R 17602.72[c] Z	(4)*
A" 5/2[f]		903.3 Z	4.7	0.3867[g]	0.0024	[2.9]	1.718_0			
A' (3/2)[f]	a'	909.4 Z	4.7	0.3847	0.0025	3.0	1.722_4			
$^{(193)}Ir^{(28)}Si$		($\mu = 24.434297_5$)		$D_0^0 = 4.7_6$ eV[a]						FEB 1975
$^{(193)}Ir^{232}Th$		(μ = 105.352106)		D_0^0 = 5.9 eV[a]						FEB 1975

IrB: [a]Thermochemical value (mass-spectrom.)(1).

(1) Vander Auwera-Mahieu, Peeters, McIntyre, Drowart, TFS 66, 809 (1970).

IrC: [a]Thermochemical value (mass-spectrom.)(1)(4).

[b]J'=0 relative to J"=0, calculated from the data in (3).

[c]$D_1 = 5.4 \times 10^{-7}$.

[d]J'=0 relative to J"=0. A different definition of the band origins is used in (3).

[e]$x_2 \approx 3200\ cm^{-1}$; see (3).

(1) McIntyre, Vander Auwera-Mahieu, Drowart, TFS 64, 3006 (1968).
(2) Jansson, Scullman, Yttermo, CPL 4, 188 (1969).
(3) Jansson, Scullman, JMS 36, 248 (1970).
(4) Gingerich, CPL 23, 270 (1973).

IrO: [a]Thermochemical value, estimated (3) from mass-spectrometric results of (1).

[b]Only one perturbed vibrational level of H' analyzed; vibrational numbering uncertain.

[c]J'=0 relative to J"=0.

[d]Three additional levels, numbered I, II, and III by (4), have been identified in transitions to A" and may represent v=1, 2, and 3 of H". With this interpretation $B_1, \ldots, B_3$ = 0.3307, 0.326, 0.3205, and $\Delta G(1/2, \ldots, 3/2)$ = 605.2 Z, 564.6, 544.3.

[e]Perturbation in v=0 at J=59½.

[f]A' and A" may be the two components of a $^2\Delta$ state.

[g]Perturbations.

(1) Norman, Staley, Bell, JCP 42, 1123 (1965).
(2) Raziunas, Macur, Katz, JCP 43, 1010 (1965).
(3) Drowart, Goldfinger, AC (Int. Ed. Engl.) 6, 581 (1967).
(4) Jansson, Scullman, JMS 43, 208 (1972).

IrSi: [a]Thermochemical value (mass-spectrom.)(1).

(1) See ref. (1) of IrB.

IrTh: [a]Thermochemical value (mass-spectrom.)(1).

(1) See ref. (4) of IrC.

State	T_e	ω_e	$\omega_e x_e$	B_e	α_e	D_e $(10^{-8}cm^{-1})$	r_e (Å)	Observed Transitions Design.	ν_{00}	References
$^{39}K_2$		$\mu = 19.4818545$ $D_0^0 = 0.514$ eV[a] I.P. = 4.0 eV[b]								JAN 1977 A
		Unidentified diffuse emission bands, 17460 - 17840 cm^{-1}.								(24)
		Diffuse bands close to lines of the principal series of K; fragments of additional systems.								(5)(7)(8)
G	28091	64.9_0[c] H	0.55					G← X,	R 28077 H	(10)
F	27571	62.2_9[c] H	0.2_4					F← X,	R 27556 H	(10)
E	$26494._0$	$61._8$[d] H	0.2_8					E← X,	R $26478._9$ H	(10)(13)
D	24627.7	61.60 H	0.90[e]					D← X,	R 24612.3 H	(12)
C $^1\Pi_u$	22969.4_3	61.48_5 Z	0.13_3	0.04404	0.00011[f]		4.433	C← X,	R 22954.2_0 Z	(1)(12)(15)
B $^1\Pi_u$	15376.7_4	75.00	0.3876[g]	0.04876_3[h]	0.00024	8.2_5	4.212_5	B[i]↔X,[j]	R 15368.2_0 Z	(4)(6)(20)
A $^1\Sigma_u^+$	11681.9	69.09 H	0.153	[k]				A↔X,	R 11670.5 H	(2)(25)
a $^3\Sigma_u^+$		Not observed; scattering calculations predict a very shallow potential minimum at ~8.7 Å.								(27)
X $^1\Sigma_g^+$	0	92.021 Z	0.2829[ℓ]	0.056743	0.000165[m]	8.63	3.9051	Mol. beam magn. reson.[n]		(11)(14)(16)
$^{(39)}K_2^+$		$D_0^0 = 0.8_5$ eV[o]								JAN 1977
		No spectrum observed. The formation of K(4p) in (K^+,K) collisions has been studied by (28) who have interpreted the magnitude and velocity dependence of the cross section in terms of properties of the potential curves for the lowest excited states of K_2^+.								
X $^2\Sigma_g^+$	0	(67)[p]					(4.11)[p]			
$^{39}K^{40}Ar$		$\mu = 19.7283641$ $D_e^0 = 0.0053$ eV[a]								JAN 1977
		Predicted potential energy curves for higher excited states (6).								
C $^2\Sigma$		Continuous emission with maximum at 19672 cm^{-1}, shifted by 1355 cm^{-1} from the forbidden K 5s-4s line at 21027 cm^{-1}.						C → X,		(10)
A $^2\Pi$, B $^2\Sigma$; X $^2\Sigma^+$	0	Theoretical interatomic potentials (3).					5.2[a]	Mol. beam magn. reson.[b]		(4)(5)(9)

K_2, K_2^+:

[a]From the convergence limit of B $^1\Pi_u$. Thermochemical value [(3), recalc. (6)] 0.56 eV.

[b]Photoionization of K_2 (21); in agreement with limits (3.57 < I.P. < 4.11 eV) obtained from chemi-ionization thresholds (17)(19). (15) estimate 4.09 eV.

[c]Analysis by (10); not confirmed by (13).

[d]Average of the constants obtained by (10)(13).

[e]$\omega_e y_e$ = + 0.0010, $\omega_e z_e$ = - 0.00030; convergence at 25590.

[f]γ_e = + 3 x 10^{-6}.

[g]$\omega_e y_e$ = + 0.004366, $\omega_e z_e$ = - 0.0001830; vibrational constants for v' < 25 from low-resolution magnetic rotation spectra. Higher vibrational levels converge rapidly at 17160 cm^{-1} above X $^1\Sigma_g^+$, v=0. T_e is from the analysis of high-resolution laser-induced fluorescence spectra (20).

[h]Recalculated by (20) from the data of (4).

[i]Radiative lifetime τ(v=6,7,8) = 12.4 ns (23); (22) measured 9.7 ns.

[j]Absorption cross sections (18).

[k]A complex magnetic rotation spectrum has been observed which may be due to perturbations of A $^1\Sigma_u^+$ by an unidentified $^3\Pi$ state (9).

[ℓ]$\omega_e y_e$ = - 0.002055. Vibrational and rotational constants from the laser-induced B→X fluorescence spectrum (20).

[m]$- 7.2 \times 10^{-6}(v+\frac{1}{2})^2 + 1.5 \times 10^{-7}(v+\frac{1}{2})^3$. See [ℓ].

[n]g_J = 0.02163 μ_N (16). For NMR spectrum and potassium eqQ see (11)(14).

[o]From $D_0^0(K_2)$ + I.P.(K) - I.P.(K_2). The experimentally observed limits (photoionization of potassium vapour) are 0.74 eV < D_0^0 < 1.27 eV (19). Theoretical calculations (26) give 1 eV.

[p]Theoretical calculations (26).

(1) Yamamoto (1929), revised by (2).

K_2, K_2^+ (continued):

(2) Crane, Christy, PR 36, 421 (1930).
(3) Lewis, ZP 69, 786 (1931).
(4) Loomis, PR 38, 2153 (1931).
(5) Kuhn, ZP 76, 782 (1932).
(6) Loomis, Nusbaum, PR 39, 89 (1932).
(7) Chakraborti, IJP 10, 155 (1935).
(8) Okuda, Nature 138, 168 (1936).
(9) Carroll, PR 52, 822 (1937).
(10) Yoshinaga, PPMSJ 19, 847, 1073 (1937).
(11) Kusch, Millman, Rabi, PR 55, 1176 (1939).
(12) Sinha, PPS 60, 436 (1948).
(13) Sinha, PPS A 63, 952 (1950).
(14) Logan, Cote, Kusch, PR 86, 280 (1952).
(15) Robertson, Barrow, PCS (1961), 329.
(16) Brooks, Anderson, Ramsey, PRL 10, 441 (1963); PR A 136, 62 (1964).
(17) Lee, Mahan, JCP 42, 2893 (1965).
(18) Lapp, Harris, JQSRT 6, 169 (1966).
(19) Williams, JCP 47, 4281 (1967).
(20) Tango, Link, Zare, JCP 49, 4264 (1968).
(21) Foster, Leckenby, Robbins, JP B 2, 478 (1969).
(22) Baumgartner, Demtröder, Stock, ZP 232, 462 (1970).
(23) Tango, Zare, JCP 53, 3094 (1970).
(24) Rebbeck, Vaughan, JP B 4, 258 (1971).
(25) Sorokin, Lankard, JCP 55, 3810 (1971).
(26) Bellomonte, Cavaliere, Ferrante, JCP 61, 3225 (1974).
(27) Geittner, ZP A 272, 359 (1975).
(28) Aquilanti, Casavecchia, JCP 64, 751 (1976).

KAr, KAr^+: see p. 357 .

State	T_e	ω_e	$\omega_e x_e$	B_e	α_e	D_e $(10^{-8}cm^{-1})$	r_e (Å)	Observed Transitions Design.	ν_{00}	References
$^{(39)}K^{40}Ar^+$		$D_e^0 = 0.12_2$ eV[c]								JAN 1977 A
X $^1\Sigma^+$							2.8_8[c]			
$^{39}K^{79}Br$		$\mu = 26.0849820$ $D_0^0 = 3.91$ eV[a] I.P. $= 7.8_5$ eV[b]								JAN 1977
		Two peaks in the electron energy loss spectrum at 19.7 and 23.7 eV (13) may be associated with excitation of an electron from the metal 3p shell of K^+Br^-.								
		Continuous absorption above 31000 cm^{-1}, maxima at 36000, 39300, 47600 cm^{-1}.[c]								(1)(4)
A		Diffuse absorption bands, 25000 - 31000 cm^{-1}.[d] The chemiluminescence spectrum (gas - beam, or flame) consists of a long ground-state vibrational progression stretching from 17500 to 31500 cm^{-1}, the onsets and cutoffs depending on the prevailing conditions of the experiment.[d]						A↔X		(2)(3)(4)(5)(15)(16)
X $^1\Sigma^+$	0	213[e]	(0.80)	0.08122109	0.00040481[f]	4.461_9[g]	2.82078	Rotation-vibration sp.		(7)
								Rotation sp.		(6)(10)
								Mol. beam rf el. reson.[h]		(12)(14)
$^{(39)}K^{(79)}Br^+$		$D_0^0 = 0.4_0$ eV[i]								JAN 1977
D $(\frac{1}{2})$	(137300)[j]	Removal of an electron from the metal 3p shell.								
C $(\frac{3}{2},\frac{1}{2})$	134700[j]									
A $(\frac{1}{2})$	3900[j]	Removal of an electron from the halogen 4p shell.								
X $(\frac{3}{2},\frac{1}{2})$	0[j]									

KAr, KAr^+:

[a]Based on atomic scattering data (1)(2)(8). The ground state potential supports nine vibrational levels (9).

[b]The spin-rotation interaction [$\gamma(v=0) = 8.0 \times 10^{-6}$] (4)(9) as well as the hyperfine interaction (5) have been investigated.

[c]From K^+-off-Ar total scattering cross sections (7); references to earlier experimental and theoretical work may be found in this paper.

(1) Buck, Pauly, ZP 208, 390 (1968).
(2) Düren, Raabe, Schlier, ZP 214, 410 (1968).
(3) Baylis, JCP 51, 2665 (1969).
(4) Mattison, Pritchard, Kleppner, PRL 32, 507 (1974).
(5) Freeman, Mattison, Pritchard, Kleppner, PRL 33, 397 (1974).
(6) Pascale, Vandeplanque, JCP 60, 2278 (1974).
(7) Budenholzer, Galante, Gislason, Jorgensen, CPL 33, 245 (1975).
(8) Fluendy, Kerr, Lawley, McCall, JP B 8, L190 (1975).
(9) Freeman, Mattison, Pritchard, Kleppner, JCP 64, 1194 (1976).
(10) Tam, Moe, Bulos, Happer, OC 16, 376 (1976).

KBr, KBr^+:

[a]Thermochemical value (8)(9); from the analysis of the fluctuation bands in the UV absorption spectrum (4) and (5) estimate 3.96 and ≤ 4.10 eV, respectively.

[b]Adiabatic potential from the photoelectron spectrum (17), not corrected for thermal population of the ground state vibrational levels; vertical potential 8.34 eV.

[c]Absorption cross sections (11). The electron energy loss spectrum (13) has peaks at 37000, 52000, 63000 cm^{-1}.

KBr, KBr^+ (continued):

[d]See [d] of CsBr. From the emission data (16) have constructed a tentative upper-state potential energy curve which is, however, dependent on the model potential used for the ground state.

[e]From the IR spectrum. The Dunham theory applied to the microwave results gives $\omega_e = 219.17$, $\omega_e x_e = 0.758$ (10).

[f]$\gamma_e = +7.7 \times 10^{-7}$.

[g]$\beta_e = -0.0002 \times 10^{-8}$.

[h]$\mu_{e\ell}[D] = 10.60256 + 0.0504_5(v+\frac{1}{2}) + 0.00015(v+\frac{1}{2})^2$ (12)(14); nuclear quadrupole and spin-rotation constants in (14), see also (6).

[i]From $D_0^0(KBr) + I.P.(K) - I.P.(KBr)$; (17) give 0.33 eV.

[j]From the maxima of the photoelectron spectrum (17)(18).

(1) Müller, AP(Leipzig) 82, 39 (1927).
(2) Beutler, Josephy, ZP 53, 747 (1929).
(3) Sommermeyer, ZP 56, 548 (1929).
(4) Levi, Dissertation (Berlin, 1934).
(5) Barrow, Caunt, PRS A 219, 120 (1953).
(6) Fabricand, Carlson, Lee, Rabi, PR 91, 1403 (1953).
(7) Rice, Klemperer, JCP 27, 573 (1957).
(8) Brewer, Brackett, CRev 61, 425 (1961).
(9) Bulewicz, Phillips, Sugden, TFS 57, 921 (1961).
(10) Rusk, Gordy, PR 127, 817 (1962).
(11) Davidovits, Brodhead, JCP 46, 2968 (1967).
(12) van Wachem, de Leeuw, Dymanus, JCP 47, 2256 (1967).
(13) Geiger, Pfeiffer, ZP 208, 105 (1968).
(14) de Leeuw, van Wachem, Dymanus, JCP 50, 1393 (1969).
(15) Oldenborg, Gole, Zare, JCP 60, 4032 (1974).
(16) Kaufmann, Kinsey, Palmer, Tewarson, JCP 61, 1865 (1974).
(17) Potts, Williams, Price, PRS A 341, 147 (1974).
(18) Potts, Williams, JCS FT II 72, 1892 (1976).

State	T_e	ω_e	$\omega_e x_e$	B_e	α_e	D_e $(10^{-7} cm^{-1})$	r_e (Å)	Observed Transitions: Design.	Observed Transitions: ν_{00}	References
$^{39}K^{35}Cl$		μ = 18.4291764 D_0^0 = 4.34 eV[a] I.P. = 8.4_4 eV[b]								JAN 1977
		Several peaks in the electron energy loss spectrum above 19 eV (13) may be associated with excitation of an electron from the metal 3p shell of K^+Cl^-.								
		Continuous absorption above 36000 cm^{-1}, maxima at 40100, 41100 cm^{-1}.[c]								(1)(3)(11)
A		Diffuse fluctuation bands in absorption from 30000 to 36000 cm^{-1}.[d] The chemiluminescence spectrum consists of a long lower-state vibrational progression extending from 17000 to 24000 cm^{-1} in a beam-gas arrangement (15) and from 20000 to 36000 in flames (2)(3)(16).[d]						A↔X		(2)(3)(15) (16)
X $^1\Sigma^+$	0	281[e]	(1.30)	0.1286347_6	0.0007899[f]	1.087_4[g]	2.66665	Rot.-vibr. sp.[h]		(7)
								Rotation sp.		(5)(6)(10)
								Mol. beam rf. el. reson.[i]		(12)(14)
$^{(39)}K^{(35)}Cl^+$		D_0^0 = 0.2_4 eV[j]								JAN 1977
D (½)	(131500)[k]	Removal of an electron from the metal 3p shell.								
C (3/2, ½)	129500[k]									
A (½)	0[k]	Removal of an electron from the halogen 3p shell.								
X (3/2, ½)										
$^{(39)}K^{133}Cs$		(μ = 30.1304153) D_0^0 = (0.47) eV[a]								JAN 1977
		Diffuse absorption band at 18558 cm^{-1}.								(1)
X $^1\Sigma^+$	0	(66.2)[a]					(4.28)[a]			
$^{39}K^{19}F$		μ = 12.7712442 D_0^0 = 5.07 eV[a]								JAN 1977
		The electron energy loss spectrum (16) has peaks at 6.8, 8.5, 9.6 and 18.4, 22.5, 23.8, 24.9 eV.								

$KC\ell$, $KC\ell^+$:

[a]Thermochemical value (8)(9); the analysis of the fluctuation bands gave $D_0^0 = 4.39$ eV (and ω" = 280) according to (3) or $D_0^0 \leq 4.66$ eV (and ω" = 305) according to (4).

[b]Adiabatic potential from the photoelectron spectrum of (17), not corrected for thermal population of ground state vibrational levels; the band maximum (vertical potential) is at 8.92 eV.

[c]UV absorption cross sections (11). In the electron energy loss spectrum (13) there are peaks at 42000, 53000, and 68000 cm^{-1}.

[d]See [d] of CsBr. From the emission data (16) have derived a tentative potential energy curve for the upper state (see [d] of KBr); $D_e' \approx 450$ cm^{-1}.

[e]From the IR spectrum. The Dunham theory applied to the microwave results gives $\omega_e = 279.8_0$, $\omega_e x_e = 1.167$.

[f]$\gamma_e = +1.6_3 \times 10^{-6}$.

[g]$\beta_e = -8._3 \times 10^{-11}$.

[h]IR study of matrix isolated $KC\ell$ (18).

[i]$\mu_{e\ell}[D] = 10.239_1 + 0.0596_6(v+\frac{1}{2}) + 0.00019(v+\frac{1}{2})^2$ (12)(14); nuclear electric quadrupole and spin-rotation coupling constants (12).

[j]From $D_0^0(KC\ell)$ + I.P.(K) - I.P.($KC\ell$); (17) give 0.19 eV.

[k]From the maxima in the photoelectron spectrum (17)(19). The two peaks corresponding to the $^2P_{3/2}$ and $^2P_{1/2}$ components of normal $C\ell$ are not resolved.

(1) Müller, AP(Leipzig) 82, 39 (1927).
(2) Beutler, Josephy, ZP 53, 747 (1929).
(3) Levi, Dissertation (Berlin, 1934).
(4) Barrow, Caunt, PRS A 219, 120 (1953).
(5) Lee, Fabricand, Carlson, Rabi, PR 91, 1395 (1953).

$KC\ell$, $KC\ell^+$ (continued):

(6) Tate, Strandberg, JCP 22, 1380 (1954).
(7) Rice, Klemperer, JCP 27, 573 (1957).
(8) Brewer, Brackett, CRev 61, 425 (1961).
(9) Bulewicz, Phillips, Sugden, TFS 57, 921 (1961).
(10) Clouser, Gordy, PR A 134, 863 (1964).
(11) Davidovits, Brodhead, JCP 46, 2968 (1967).
(12) van Wachem, Dymanus, JCP 46, 3749 (1967).
(13) Geiger, Pfeiffer, ZP 208, 105 (1968).
(14) Hebert, Lovas, Melendres, Hollowell, Story, Street, JCP 48,2824 (1968).
(15) Oldenborg, Gole, Zare, JCP 60, 4032 (1974).
(16) Kaufmann, Kinsey, Palmer, Tewarson, JCP 61, 1865 (1974).
(17) Potts, Williams, Price, PRS A 341, 147 (1974).
(18) Ismail, Hauge, Margrave, JMS 54, 402 (1975).
(19) Potts, Williams, JCS FT II 72, 1892 (1976).

KCs: [a]Interpolated values based on the constants for K_2 and Cs_2 (2).

(1) Walter, Barratt, PRS A 119, 257 (1928).
(2) Cavaliere, Ferrante, Lo Cascio, JCP 62, 4753 (1975).

KF: [a]Thermochemical value (6)(7); an upper limit, $D_0^0 \leq 5.28$ eV, was derived (3) from the analysis of the fluctuation bands.

(continued on p. 361)

State	T_e	ω_e	$\omega_e x_e$	B_e	α_e	D_e $(10^{-5}cm^{-1})$	r_e (Å)	Observed Transitions Design.	ν_{00}	References
$^{39}K^{19}F$ (continued)										
A		Fluctuation bands, 34300 - 46700 cm^{-1} (absorption).						A← X		(3)
X $^1\Sigma^+$	0	428^b	(2.4)	0.27993741_3	2.33503_8[c] x 10^{-3}	0.04834[d]	2.17145_7	Rot.-vibr. sp.[e]		(9)(12)
								Rotation sp.		(2)(5)(10)(17)
								Mol. beam rf electric[f] and magn. reson.[g]		(13)(14)(15)(11)
$^{39}K^1H$		μ = 0.98241434 D_0^0 = (1.8_6) eV [extrapol. of A $^1\Sigma$]								JAN 1977
		Theoretical potential energy curves, coupling between molecular states (8).								
		Fluctuating continuum in emission, 25000 - 33000 cm^{-1}.								(3)
A $^1\Sigma^+$	$19052._8$	228.2 Z	-5.7_5[a]	1.26_9	-0.037_5[b]	$9._5$	3.68	A[c]↔X, R	$18680._1$ Z	(1)(2)(5)(6)
X $^1\Sigma^+$	0	983.6 Z	14.3_2	3.412_3	0.081_7	$15._3$	2.242_5			
$^{39}K^2H$		μ = 1.91510699 D_0^0 = (1.8_9) eV								JAN 1977
A $^1\Sigma^+$	19060	161 Z	-3.3[d]	0.65_9	-0.012_7[e]	$2._7$	3.65	A↔X, R	18790 Z	(4)(6)
X $^1\Sigma^+$	0	$706._6$ Z	7.7_3	1.753_9	0.0318	$4._7$	2.240_3			
$^{39}K^{127}I$		μ = 29.8108348 D_0^0 = 3.31 eV[a] I.P. = 7.2_1 eV[b]								JAN 1977
		Features in the electron energy loss spectrum (12)(22) at ~3.8, 4.7, 5.1 eV correspond to maxima in the absorption spectrum. Peaks at higher energies could indicate that KI absorbs strongly in the VUV region; structure above 19 eV may arise from excitation of a metal 3p electron.								
		Continuous absorption[c] above 26500 cm^{-1}, maxima at 30800, 38400, 41700[d] cm^{-1}.								(1)(4)(11)

KF (continued):

[b]From the IR spectrum (9)(12). (10), applying the Dunham theory to the microwave results, calculate $\omega_e = 426.0_4$ and $\omega_e x_e = 2.43$.

[c]$\gamma_e = + 6.96_3 \times 10^{-6}$.

[d]$\beta_e = - 2 \times 10^{-10}$.

[e]IR spectrum of matrix isolated KF (18).

[f]$\mu_{e\ell}[D] = 8.5583 + 0.06841(v+\frac{1}{2}) + 0.000256(v+\frac{1}{2})^2$ (8)(15), electric quadrupole and other hyperfine coupling constants (4)(13)(14); see also (2)(5).

[g]$g_J = (-)0.036_4\ \mu_N$ (11).

(1) Müller, AP(Leipzig) 82, 39 (1927).
(2) Grabner, Hughes, PR 79, 819 (1950).
(3) Barrow, Caunt, PRS A 219, 120 (1953).
(4) Schlier, ZP 147, 600 (1957).
(5) Green, Lew, CJP 38, 482 (1960).
(6) Brewer, Brackett, CRev 61, 425 (1961).
(7) Bulewicz, Phillips, Sugden, TFS 57, 921 (1961).
(8) Gräff, Runolfsson, ZP 176, 90 (1963).
(9) Ritchie. Lew, CJP 42, 43 (1964).
(10) Veazey, Gordy, PR A 138, 1303 (1965).
(11) Mehran, Brooks, Ramsey, PR 141, 93 (1966).
(12) Baikov, Vasilevskii, OS(Engl. Transl.) 22, 198 (1967).
(13) Bonczyk, Hughes, PR 161, 15 (1967).
(14) van Wachem, Dymanus, JCP 46, 3749 (1967).
(15) van Wachem, de Leeuw, Dymanus, JCP 47, 2256 (1967).
(16) Geiger, Pfeiffer, ZP 208, 105 (1968).
(17) Dijkerman, Flegel, Gräff, Mönter, ZN 27 a, 100 (1972).
(18) Ismail, Hauge, Margrave, JINC 35, 3201 (1973).

K^1H, K^2H:

[a]$\omega_e y_e = - 0.169$. The $\Delta G(v+\frac{1}{2})$ curve has a maximum at $v \approx 11$.

[b]$\gamma_e = - 0.0023_2$. The B_v curve has a maximum at $v \approx 7$.

[c]Radiative lifetime $\tau < 10$ ns (7).

[d]$\omega_e y_e = - 0.07_2$. The $\Delta G(v+\frac{1}{2})$ curve has a maximum at $v \approx 16$.

[e]$\gamma_e = - 0.0005_7$. The B_v curve has a maximum at $v \approx 11$.

(1) Almy, Hause, PR 42, 242 (1932).
(2) Hori, Mem. Ryojun Coll. Engng. 6, 1 (1933).
(3) Hori, Mem. Ryojun Coll. Engng. 6, 115 (1933).
(4) Imanishi, Sci. Pap. IPCR (Tokyo) 39, 45 (1941).
(5) Almy, Beiler, PR 61, 476 (1942).
(6) Bartky, JMS 20, 299 (1966).
(7) Cruse, Zare, JCP 60, 1182 (1974).
(8) Numrich, Truhlar, JPC 79, 2745 (1975).

KI: [a]Thermochemical value (8); flame photometry (9) gives 3.4_9 eV. Earlier spectroscopic estimates were 3.31 eV (4) and ≤ 3.47 eV (6).

[b]Adiabatic potential from the photoelectron spectrum (19), not corrected for thermal population of ground state vibrational levels; band maximum (vertical potential) at 7.68 eV.

[c]Absorption cross sections (11). A preliminary measurement of the photodissociation product anisotropy by photofragment spectroscopy at 28800 cm^{-1} (18) indicates strong contributions from perpendicular transitions.

[d]Photodissociation produces K(4p ^2P); similarly, photodissociation at 52600 cm^{-1} leads to K(5p ^2P) (17).

(continued on p. 363)

State	T_e	ω_e	$\omega_e x_e$	B_e	α_e	D_e (10^{-8} cm^{-1})	r_e (Å)	Observed Transitions Design.	ν_{00}	References
$^{39}K^{127}I$ (continued)										
A (0^+)	(26620)	This is one of five very shallow states ($0^+, 0^-, 1, 1, 2$)[e] arising from normal atomic products $K(^2S_{1/2}) + I(^2P_{3/2})$. The analysis of K-off-I differential elastic scattering measurements (14) suggests that only 0^+ ($D_e \approx 150$ cm^{-1}, $r_e \approx 3.85$ Å) is favorably situated for transitions from the ground state. Accordingly, (15) have analyzed the fluctuation bands observed in absorption from 19600 to 27000 cm^{-1} (3)(4) and in chemiluminescence[f] from 16300 to 26200 cm^{-1} (4)(15)(16) in terms of the A↔X transition, assigning vibrational quantum numbers v" from 2 to 64, and constructing an accurate potential curve for the excited state.								
X $^1\Sigma^+$	0	186.53[g]	0.574	0.06087473	0.00026776[h]	2.593_4[i]	3.04784_4	Rotation spectrum[j]		(7)(10)
$^{(39)}K^{127}I^+$		$D_0^0 = 0.4_4$ eV[k]								JAN 1977
D ($\frac{1}{2}$)	(143700)[ℓ]	Removal of an electron from the metal 3p shell.								
C ($\frac{3}{2}, \frac{1}{2}$)	140800[ℓ]									
A ($\frac{1}{2}$)	7900[ℓ]	Removal of an electron from the halogen 5p shell.								
X ($\frac{3}{2}, \frac{1}{2}$)	0[ℓ]									
$^{(39)}K^{(84)}Kr$		(μ = 26.6083236) $D_e^0 = 0.0089$ eV[a]								JAN 1977
C $^2\Sigma$		Continuous emission with maximum at 19530 cm^{-1}, shifted by 1497 cm^{-1} from the forbidden K 5s-4s line at 21027 cm^{-1}.						C→X		(4)
B $^2\Sigma$		Theoretical interatomic potentials (3).								
A $^2\Pi$										
X $^2\Sigma^+$	0						5.1[a]			

KI, KI^+ (continued):

[e]A differential scattering experiment investigating the collisional depolarization of polarized K atoms (20) leads to the conclusion that J,J coupling would provide the most adequate description of these states.

[f]The actual onset and cutoff wavelengths depend on the experimental conditions: a beam-gas arrangement (16) or flames (4)(15).

[g]From the microwave results by use of Dunham's theory.

[h]$\gamma_e = + 3.8_8 \times 10^{-7}$.

[i]$\beta_e = + 4 \times 10^{-12}$.

[j]Quadrupole hyperfine structure (13). The dipole moment $\mu_{el} = 10.8_2$ D was measured by the electric deflection method (23); see also (5).

[k]From D_0^0(KI) + I.P.(K) - I.P.(KI); (19) give 0.37 eV.

[ℓ]From band maxima of the photoelectron spectrum (19)(21).

(1) Müller, AP(Leipzig) 82, 39 (1927).
(2) Beutler, Josephy, ZP 53, 747 (1929).
(3) Sommermeyer, ZP 56, 548 (1929).
(4) Levi, Dissertation (Berlin, 1934).
(5) Rodebush, Murray, Bixler, JCP 4, 372 (1936).
(6) Barrow, Caunt, PRS A 219, 120 (1953).
(7) Honig, Mandel, Stitch, Townes, PR 96, 629 (1954).
(8) Brewer, Brackett, CRev 61, 425 (1961).
(9) Bulewicz, Phillips, Sugden, TFS 57, 921 (1961).
(10) Rusk, Gordy, PR 127, 817 (1962).

KI, KI^+ (continued):

(11) Davidovits, Brodhead, JCP 46, 2968 (1967).
(12) Geiger, Pfeiffer, ZP 208, 105 (1968).
(13) Tiemann, El Ali, Hoeft, Törring, ZN 28 a, 1058 (1973).
(14) Kaufmann, Lawter, Kinsey, JCP 60, 4016 (1974).
(15) Kaufmann, Kinsey, Palmer, Tewarson, JCP 60, 4023 (1974).
(16) Oldenborg, Gole, Zare, JCP 60, 4032 (1974).
(17) Earl, Herm, JCP 60, 4568 (1974).
(18) Ormerod, Powers, Rose, JCP 60, 5109 (1974).
(19) Potts, Williams, Price, PRS A 341, 147 (1974).
(20) Carter, Pritchard, JCP 62, 927 (1975).
(21) Potts, Williams, JCS FT II 72, 1892 (1976).
(22) Rudge, Trajmar, Williams, PR A 13, 2074 (1976).
(23) Story, Hebert, JCP 64, 855 (1976).

KKr: [a]From atomic scattering data (1)(2)

(1) Buck, Pauly, ZP 208, 390 (1968).
(2) Düren, Raabe, Schlier, ZP 214, 410 (1968).
(3) Baylis, JCP 51, 2665 (1969).
(4) Tam, Moe, Bulos, Happer, OC 16, 376 (1976).

State	T_e	ω_e	$\omega_e x_e$	B_e	α_e	D_e $(10^{-}\ cm^{-1})$	r_e (Å)	Observed Transitions Design.	ν_{00}	References
$^{(39)}K^{16}O$		$(\mu = 11.3398256_4)$								JAN 1977
A $^2\Pi$	$(347)^a$	$(442)^a$					$(2.33)^a$			
X $^2\Sigma^+$	$0\ ^a$	$(384)^b$					$(2.22)^a$			
$^{(39)}K^{16}O^+$		$D_0^0(K^+ - O) = 0.2$ eV[c]								JAN 1977
$^{(84)}Kr_2$		$(\mu = 41.955752_7)$ $D_0^0 = 0.0157$ eV[a] I.P. = 12.87 eV[b]								APR 1977 A
		Several emission continua in the near infrared, visible, and ultra-violet; interpretation doubtful.								(2)(11)
		Four unclassified band systems, 92530 - 94200 cm^{-1}; the upper states are probably derived from the configuration $4p^6\ {}^1S + 4p^5({}^2P_{\frac{3}{2}})5p$.								(7)
D (1_u)	(86000)	Probably a repulsive state; several diffuse bands shortward of the second resonance line (upper state $5s'[\frac{1}{2}]1_u$) at 1165 Å (85848 cm^{-1}).						D↔X,		(7)*
C (0_u^+)	85522.0	43.3_1 H	1.74	Band system converging to $^1S + 5s'[\frac{1}{2}]1_u$; v" = 0...9, v' = 0...5.				C←X,	V 85531.5 H	(7)*
		Additional unclassified absorption bands shortward of the first resonance line (upper state $5s[\frac{3}{2}]1_u$) at 1236 Å (80918 cm^{-1}); 80927 - 81001 cm^{-1}.								(7)
B (0_u^+)	(80006)	$[80.8]^c$ H		Band system converging to $^1S + 5s[\frac{3}{2}]1_u$; v" = 0...9, v' = (0...12)[c].				B←X,	V 80035[c] H	(7)*
A (1_u)	(79613)	$[67.3]^c$ H		Band system converging to $^1S + 5s[\frac{3}{2}]2_u$; v" = 1...7, v' = (0...5)[c].				A←X,[d]	V 79635[c] H	(7)*
		Continuous emission with two maxima usually referred to as the "first continuum" closer to the first resonance line at 1236 Å (80918 cm^{-1}) and the "second continuum" at 1470 Å (68000 cm^{-1}).						A,B→X,[e]		(1)*
X $^1\Sigma_g^+$	0	24.1_8 H	1.34^f				$4.0_3{}^g$			
$^{(84)}Kr_2^+$		$(\mu = 41.955615_5)$ $D_0^0 = 1.15$ eV[h]								APR 1977 A
		Some excited states have been qualitatively discussed by (10a).								
X $(^2\Sigma_u^+)$	0						(2.6) [estimated value (9a)]			

KO, KO^+:

[a]*Ab initio* calculation by (4). The relative position of the two states is in accordance with results of a magnetic deflection analysis of $M + NO_2$ scattering products (M = alkali metal) by (1).

[b]Fundamental in solid nitrogen matrix (2); the *ab initio* value is $\omega_e = 467\ cm^{-1}$ (4).

[c]From a merging beam study of the reaction $CO^+(K,C)KO^+$ (3).

(1) Herm, Herschbach, JCP 52, 5783 (1970).
(2) Spiker, Andrews, JCP 58, 713 (1973).
(3) Rol, Entemann, Wendell, JCP 61, 2050 (1974).
(4) So, Richards, CPL 32, 227 (1975).

Kr_2, Kr_2^+:

[a]From the absorption spectrum (7). Various other methods, reviewed in (7), agree well with the spectroscopic value; see also [g].

[b]From photoionization studies (14); see also the earlier work of (3)(4).

[c]The v' numbering assumes that the lowest observed level has v'=0 which may, however, not be the case.

[d]A fairly strong diffuse R shaded band (called "spectral demarcation") at 79923 cm^{-1} is considered to belong to the same electronic transition (7).

[e]The emission is attributed to transitions from the lowest excited states A $^3\Sigma_u^+(1_u)$ and B $^1\Sigma_u^+(0_u^+)$ to the repulsive part of the ground state potential; for a detailed discussion of the analogous spectrum of Xe_2 see (4a). The "second continuum" was recently observed in emission from high-pressure krypton excited by high-current short-duration electron bursts from a Febetron source (12) and was found to be characterized by three radiative decay con-

Kr_2, Kr_2^+ (continued):

stants of 9, 32, 350 ns. Similar excitation (10) produces "long-lived" (τ = 353 ns) molecular species which absorb at 10100, 10160, 10250, 10350 cm^{-1}. The 1470 Å emission was also observed from Kr_2 in a neon matrix excited by α particles (6).

[f]$\omega_e y_e = + 0.022$; vibrational levels observed to v=9.

[g]From viscosity data, virial coefficients (9) and collision cross sections (5). More recently, (8) have derived r_e = 4.007 Å and D_e = 0.0174 eV from a combination of all available data; see also (13).

[h]$D_0^0(Kr_2) + I.P.(Kr) - I.P.(Kr_2)$. From elastic scattering of Kr^+ by Kr (9a) obtain D_e = 1.21 eV.

(1) Tanaka, JOSA 45, 710 (1955).
(2) Herman, Herman, Nature 195, 1086 (1962).
(3) Huffman, Katayama, JCP 45, 138 (1966).
(4) Samson, Cairns, JOSA 56, 1140 (1966).
(4a) Mulliken, JCP 52, 5170 (1970).
(5) Buck, Dondi, Valbusa, Klein, Scoles, PR A 8, 2409 (1973).
(6) Gedanken, Raz, Jortner, JCP 59, 1630 (1973).
(7) Tanaka, Yoshino, Freeman, JCP 59, 5160 (1973).
(8) Barker, Watts, Lee, Schafer, Lee, JCP 61, 3081 (1974).
(9) Gough, Smith, Maitland, MP 27, 867 (1974).
(9a) Mittmann, Weise, ZN 29 a, 400 (1974).
(10) Oka, Rao, Redpath, Firestone, JCP 61, 4740 (1974).
(10a) Barr, Dee, Gilmore, JQSRT 15, 625 (1975).
(11) Birot, Brunet, Galy, Millet, Teyssier, JCP 63, 1469 (1975).
(12) Koehler, Ferderber, Redhead, Ebert, PR A 12, 968 (1975).
(13) Nain, Aziz, Jain, Saxena, JCP 65, 3242 (1976).
(14) Ng, Trevor, Mahan, Lee, JCP 66, 446 (1977).

State	T_e	ω_e	$\omega_e x_e$	B_e	α_e	D_e (10^{-} cm^{-1})	r_e (Å)	Observed Transitions Design.	ν_{00}	References
$^{(39)}K^{(85)}Rb$		(μ = 26.7080933) D_0^0 = (0.50) eV[a]								JAN 1977
		Diffuse absorption band at 20160 cm^{-1}.								(1)
X $^1\Sigma^+$		(75.5)[a]					(4.07)[a]			
$^{(84)}Kr^{(79)}Br$		(μ = 40.669180_4)								APR 1977
B	(48130)	(166) Broad diffuse peaks, principal peak at 48440 cm^{-1}.[a]						B→X,		(1)
X	0	Repulsive state with rather flat potential function.								
$^{(84)}Kr^{(35)}Cl$		(μ = 24.6827076)								APR 1977
D								D→X,	50250	(2)
B	(44890)	(210) Broad diffuse peaks, principal peak at 45350 cm^{-1}.[a]						B→X,	45040	(1)(2)
X	0	Repulsive state with very flat potential function.								
$^{(84)}Kr^{19}F$		(μ = 15.4910711)								APR 1977
E ($^2\Sigma$)	(71500)[a]	(608)[a] Rydberg state arising[b] from Kr(^3P) + F(^2P).					(1.83)[a]			
D ($^2\Pi$) $\frac{1}{2}$	(48000)[a]	(356)[a] Ionic state arising[b] from $Kr^+(^2P_{\frac{1}{2}})$ + F^-.					(2.47)[a]	D→X,[c]	45340	(8)(12)
C ($^2\Pi$) $\frac{3}{2}$	(42800)[a]	(349)[a] } Ionic states arising[b] from $Kr^+(^2P_{\frac{3}{2}})$ + F^-.[e]					(2.44)[a]			
B ($^2\Sigma$) $\frac{1}{2}$	40840 [d]	310 [d] }					(2.27)[f]	B→X,[g]	40230	(4)(5)(6)(10)
A ($^2\Pi$)		Repulsive state (11).								
X ($^2\Sigma$)	0	Repulsive state with rather flat potential function (3)(10)(11).						ESR spectrum[h]		(1)
$^{(84)}Kr^{19}F^+$		$D_0^0 \geq$ 1.58 eV[i]								APR 1977
X $^1\Sigma^+$		(621)[j]	(8.3)[j]	(0.355)[j]	(0.0044)[j]		(1.752)[j]			

KRb: [a]Interpolated values based on the constants for K_2 and Rb_2 (2).

(1) Walter, Barratt, PRS A 119, 257 (1928).
(2) Cavaliere, Ferrante, Lo Cascio, JCP 62, 4753 (1975).

KrBr, KrCℓ:

[a]In emission from low-pressure (0.5 - 5 torr) reactions of metastable Kr atoms with Br_2 or CH_2Br_2, $Cℓ_2$ or $CCℓ_4$ (1). The KrCℓ bands were also observed in electron-beam-excited mixtures of Ar and Kr + $Cℓ_2$ (2). The upper states are ionic states arising from $Kr^+ + Br^-$ or $Cℓ^-$ although the adiabatic dissociation products are $Kr(^3P_2)$ + Br or Cℓ; the dissociation energies can be estimated as 3.93 eV for KrBr and 4.33 eV for KrCℓ and are very similar to those of the ground states of RbBr and RbCℓ.

(1) Golde, JMS 58, 261 (1975).
(2) Murray, Powell, APL 29, 252 (1976).

KrF, KrF^+:

[a]Theoretical calculations of (11). T_e relative to the $Kr(^1S) + F(^2P_{3/2})$ limit. For the D state a vibrational frequency of ~336 was observed in argon matrices (9).

[b]In the diabatic sense.

[c]In reactions of metastable Kr atoms with small fluorine containing molecules (8) and from electron-beam-excited mixtures of Ar and Kr + NF_3 (12). Also in absorption of matrix isolated KrF (7)(9).

[d]From the analysis through trial-and-error theoretical simulations of the structured continuum observed at moderate to high pressures (10). An earlier interpretation (5) of the much broader peaks observed at low pressure

KrF, KrF^+ (continued):

lead to $\omega_e \approx 280$ cm^{-1}. The theoretical calculations of (11) give 341 cm^{-1}. In neon and argon matrices $\omega_e \approx 340$ and 315, respectively (9).

[e]The B state dissociates adiabatically to $Kr(^3P_2)$ + F; $D_0' \approx$ 4.85 eV is very similar to the ground state dissociation energy of RbF.

[f]Estimated value used by (10) in the analysis of the structured continuum. The theoretical value (11) is 2.51 Å.

[g]In emission from electron-beam-excited mixtures of Ar and Kr + F_2 at moderate to high pressures. The low-pressure spectrum observed in reactions of metastable Kr atoms with F_2 or N_2F_4 (5) may also contain contributions from C → X. Also observed in matrix absorption (7)(9).

[h]Observed in γ-irradiated KrF_4 at low temperature (77 K) confirming that the ground state is $^2\Sigma$. KrF seems to be stabilized in the solid.

[i]From mass-spectrometric studies (2); ab initio calculations (3) yield D_0^0 = 1.90 eV.

[j]Theoretical values (3); no spectroscopic data available.

(1) Falconer, Morton, Streng, JCP 41, 902 (1964).
(2) Berkowitz, Chupka, CPL 7, 447 (1970).
(3) Liu, Schaefer, JCP 55, 2369 (1971).
(4) Brau, Ewing, JCP 63, 4640 (1975).
(5) Golde, JMS 58, 261 (1975).
(6) Tisone, Hays, Hoffman, OC 15, 188 (1975).
(7) Ault, Andrews, JCP 64, 3075 (1976).
(8) Velazco, Kolts, Setser, JCP 65, 3468 (1976).
(9) Ault, Andrews, JCP 65, 4192 (1976).
(10) Tellinghuisen, Hays, Hoffman, Tisone, JCP 65, 4473 (1976).
(11) Dunning, Hay, APL 28, 649 (1976).
(12) Murray, Powell, APL 29, 252 (1976).

State	T_e	ω_e	$\omega_e x_e$	B_e	α_e	D_e (10^{-} cm^{-1})	r_e (Å)	Observed Transitions Design.	Observed Transitions ν_{00}	References
$^{(84)}Kr^1H^+$		$D_0^0 = 4.35$ eV[a]								APR 1977 A
X $^1\Sigma^+$							1.50[b]			
$^{(84)}Kr^{16}O$		($\mu = 13.4341455_9$)								APR 1977 A
		Emission band near 5577 Å (17925 cm^{-1}), shaded to the violet and assigned to a KrO van der Waals molecule with O in the $^1S_0$ or $^1D_2$ metastable states.								(1)* (2)
$^{(84)}Kr^{(132)}Xe$		($\mu = 51.285788_3$) $D_e^0 = 0.0197$ eV[a] $D_0^0 = 0.0184$ eV[b]								APR 1977 A
		Emission continuum with maximum at 1540 Å (64900 cm^{-1}), presumably corresponding to transitions from the lowest excited state(s) to the repulsive part of the ground state potential.								(5)
X $^1\Sigma^+$		[20.3][b]					4.18[a]			
$^{(84)}Kr^{(132)}Xe^+$										APR 1977
		Violet shaded emission system extending from 20300 to 21700 cm^{-1} with several diffuse maxima between 20400 and 20800 cm^{-1}, observed in the luminescence resulting from bombardment of Kr/Xe mixtures by fast electrons, and assigned by (4) as a charge-transfer spectrum of $KrXe^+$ similar to the spectra of $HeNe^+$ etc..								(1)(2)*

State	T_e	ω_e	$\omega_e x_e$	B_e	α_e	D_e (10^{-} cm^{-1})	r_e (Å)	Observed Transitions Design.	ν_{00}	References
$^{(39)}K^{(132)}Xe$	(μ = 30.0786526) $D_e^0 = 0.013_8$ eV[a]									JAN 1977
C $^2\Sigma$	Continuous emission with maximum at 19207 cm^{-1}, shifted by 1820 cm^{-1} from the forbidden K 5s - 4s line at 21027 cm^{-1}.							C→X,		(3)
B $^2\Sigma$	Interatomic potential calculations (2).									
A $^2\Pi$										
X $^2\Sigma^+$							5.2[a]			

KrH^+: [a]From a study of the proton transfer reaction $H_3^+ + Kr = KrH^+ + H_2$, and based on $D_0(H_2 - H^+)$ (3). From elastic scattering of protons by Kr (1) and (2) have derived D_e = 4.60 and 4.45 eV, resp..
[b]Average value obtained from proton scattering (1)(2).

(1) Rich, Bobbio, Champion, Doverspike, PR A 4, 2253 (1971).
(2) Weise, Mittman, Ding, Henglein, ZN 26 a, 1122 (1971).
(3) Payzant, Schiff, Bohme, JCP 63, 149 (1975).

KrO:

(1) Cooper, Cobb, Tolnas, JMS 7, 223 (1961).
(2) Kugler, AP(Leipzig) (7) 14, 137 (1964).

KrXe, $KrXe^+$:
[a]From the "excess properties" of Kr - Xe mixtures (3).
[b]Theoretical values (6), calculated from the potential of (3).

(1) Friedl, ZN 14 a, 848 (1959); 15 a, 398 (1960).
(2) See ref. (2) of KrO.
(3) Lee, Henderson, Barker, MP 29, 429 (1975).
(4) Tanaka, Yoshino, Freeman, JCP 62, 4484 (1975).
(5) Verkhovtseva, Ovechkin, Fogel, CPL 30, 120 (1975).
(6) Bobetic, Barker, JCP 64, 2367 (1976).

KXe: [a]From atomic scattering data (1).

(1) Buck, Pauly, ZP 208, 390 (1968).
(2) Baylis, JCP 51, 2665 (1969).
(3) Tam, Moe, Bulos, Happer, OC 16, 376 (1976).

State	T_e	ω_e	$\omega_e x_e$	B_e	α_e	D_e $(10^{-7} cm^{-1})$	r_e (Å)	Observed Transitions: Design.	Observed Transitions: ν_{00}	References
$^{139}La_2$		$\mu = 69.453202$ $D_0^0 = 2.5_0$ eV[a]								FEB 1975
		Narrow V shaded bands in emission from a King furnace, tentatively attributed to La_2 and assigned to three sequences in the region 16400 - 16600 cm^{-1}.								(2)
$^{139}La^{19}F$		$\mu = 16.7126010$								FEB 1975
		Emission spectra attributed to LaF:								
		three systems of V shaded bands (P and Q heads)								
		$\nu_e = 30188.9$, $\omega_e' = 674.7$, $\omega_e' x_e' = 2.35$, $\omega_e'' = 603.8$, $\omega_e'' x_e'' = 1.96$,								(3)
		30076.4 682.4 2.28 607.1 1.88,								(1)
		28468 657 1.1 607 1.8;								(1)
		one system of R shaded bands in the region 27500 - 28200 cm^{-1}.								(1)
		Absorption spectra:								
e $^3\Phi$, Ω = 4 3 2				[0.2302] [0.2294] [0.2277]		[1.7] [2.0] [1.7]	[2.098]	e ← a, R	22340.20 Z 22149.20 Z 21834.46 Z	(4)
d $^3\Phi$, Ω = 4 3 2		[483] [473.2] [484.6]		[0.2336] [0.2321] [0.2322]		[2.2] [2.3] [1.9]	[2.082]	d ← a, R	18809.40 Z 18519.90 Z 18295.80 Z	(4)(5)
c_2 $^3\Delta_2$				[0.2191]		[1.9]	[2.146]	$c_2 \leftarrow a_2$, R	13597.27 Z	(2)(4)
c_1 $^3\Delta_1$				[0.2189]		[2.0]	[2.147]	$c_1 \leftarrow a_1$, R	13567.00 Z	(2)(4)
b_2 $^3\Delta_2$				[0.2216]		[2.9]	[2.133]	$b_2 \leftarrow a_2$, R	13316.3_5 Z	(4)
a_3 a_2 $^3\Delta$, Ω = 3 2 a_1 1	a	[537] $[537.6_5]$ $[537.1_4]$		[0.23875] 0.23902 0.23831	 0.00121 0.00120	[1.87] [1.88] [1.79]	2.054_8			
E $^1\Sigma^+$		[(421)]		[0.229][b]			[2.10]	E ← X, R	22574.22 H	(4)
D $^1\Sigma^+$				[0.242][b]			[2.04]	D ← X, R	22485.35 H	(4)
C $^1\Pi$		[549]		$[0.2374_5]$[c]		$[1.6_5]$	$[2.061_1]$	C ← X, R	20959.83 Z	(4)(5)
B $^1\Pi$		[474]		$[0.2293_6]$[c]		[2.7]	$[2.097_1]$	B ← X, R	16184.00 Z	(4)(5)
A $^1\Sigma^+$		[489.4] Z		0.2278	0.0011	[2.1]	2.104	A ← X, R	11661.9 Z	(2)(4)
X $^1\Sigma^+$	0 a	[570]		$[0.2456_2]$		$[1.7_8]$	$[2.026_5]$			

State	T_e	ω_e	$\omega_e x_e$	B_e	α_e	D_e ($10^{-4}cm^{-1}$)	r_e (Å)	Observed Transitions Design.	ν_{00}	References
$^{139}La^1H$		μ = 1.00056569								SEP 1976
b $^3\Phi_4$				[3.8939][a]	0.0892	[1.28]	[2.0801]	b↔a,	16346.74 Z	(1)*
$^3\Phi_3$				[3.8774][a]	0.1071	[1.40]	[2.0845]		16006.73 Z	
$^3\Phi_2$				-	-	-	-		-	
a $^3\Delta_3$				[3.8477]	0.0764	[1.18]	[2.0925]			
$^3\Delta_2$				[3.8431]	0.0773	[1.20]	[2.0938]			
$^3\Delta_1$				-	-	-	-			
C $^1\Sigma$				[3.8162][a]		[0.50]	[2.1012]	C→A,	18595.10 Z	(1)*
B $^1\Delta$				[3.8232][a]		[2.0]	[2.0992]	B↔A,	15618.76 Z	(1)*
A $^1\Pi$				[3.7837][b]		[1.12]	[2.1102]			
$^{139}La^2H$		μ = 1.98531572								SEP 1976
b $^3\Phi_4$				[1.9743][a]	0.0331	[0.35]	[2.0738]	b→a,	16359.92 Z	(1)*
$^3\Phi_3$				[1.9691][a]	0.0403	[0.42]	[2.0766]		16010.97 Z	
$^3\Phi_2$				-	-	-	-		-	
a $^3\Delta_3$				[1.9523]	0.0274	[0.31]	[2.0855]			
$^3\Delta_2$				[1.9510]	0.0282	[0.34]	[2.0862]			
$^3\Delta_1$				-	-	-	-			
B $^1\Delta$				[1.9261][a]		[0.57]	[2.0996]	B→A,	15611.92 Z	(1)*
A $^1\Pi$				[1.9346][c]		[0.35]	[2.0950]			

La_2: [a]Thermochemical value (mass-spectrom.)(1).

(1) Verhaegen, Smoes, Drowart, JCP 40, 239 (1964).

(2) Carette, Blondeau, CR B 269, 16 (1969).

La^1H, La^2H:

[a]Emission breaks off at relatively low J' values.

[b]Λ-type doubling, $\Delta\nu = 0.0023\,J(J+1)$.

[c]Λ-type doubling, $\Delta\nu = 0.00115\,J(J+1)$.

(1) Bernard, Bacis, CJP 54, 1509 (1976).

LaF: [a]Not certain whether X $^1\Sigma$ or a $^3\Delta$ is the ground state.

[b]Strong interaction between the two $^1\Sigma$ states.

[c]Λ-type doubling.

(1) Shenyavskaya, Gurvich, Mal'tsev, VMUK 20(4), 10 (1965).

(2) Barrow, Bastin, Moore, Pott, Nature 215, 1072 (1967).

(3) Shenyavskaya, Gurvich, Mal'tsev, OS(Engl. Transl.) 24, 556 (1968).

(4) Barrow, Lee, Partridge, unpubl. } quoted by Barrow in DONNSPEC (1970).

(5) Hauge, unpubl.

State	T_e	ω_e	ω_ex_e	B_e	α_e	D_e ($10^{-7}cm^{-1}$)	r_e (Å)	Observed Transitions Design.	ν_{00}	References
$^{139}La^{16}O$		μ = 14.3433002		D_0^0 = 8.23 eV[a]		I.P. = 4.9_5 eV[b]				FEB 1975 A
F $(^2\Sigma)$	28015	858 H	3.7	c				F→X, V	28035 H	(6)* (7)(20)
D $(^2\Sigma)$	(27749)	(790)[d] H	3	c				D→X, V	(27735)[d] H	(6)* (7)(20)
C $^2\Pi_r$	22849.0 e $22631._2$	798.4[f] Z $792._5$[g]	2.2_0[h] (2.2_1)[h]	0.3520[i] 0.3503[i]	0.0017 0.0016	[2.8][j]	1.829_5	C→A',[k] V	14671.19[ℓ] Z 15150.00[ℓ] Z	(6)*(7)(25)* (27)(32)
								C↔X, R	22839.61[ℓ] Z 22618.87[ℓ] Z	(1)(2)* (3)* (4)* (6)* (19)(26)* (27)(32)
B $^2\Sigma^+$	17879.1	[730.4] Z	2.04[m]	0.3413[n]	0.0017	[2.5]	1.855_7	B↔X, R	17837.8 Z	(1)(2)(3)* (4)* (6)* (7)(9)(10)* (11)(13)(19) (21)(28)(30) (32) (22)[$La^{18}O$]
A $^2\Pi_r$	13525.6 o 12663.3	[757.17] Z 761.8[p]	2.0_1[p] 2.1_2	[0.3463][q]	r	[3.1]	[1.842_2]	A↔X, R	13497.63[ℓ] Z 12635.65[ℓ] Z	(1)(2)* (6)* (7)(10)* (11)(19)(32)
A' $^2\Delta_r$	8191.2 7493.4	771.6[f] Z $768._0$[s]	2.3_0[h] (2.1_7)[h]	0.3444 0.3435	0.0017 0.0016	[2.8][t]	1.848_5			
X $^2\Sigma^+$ u	0	[812.7_5] Z	2.22[mh]	0.3526[v]	0.0014	2.6	1.825_7	ESR sp.[w]		

LaO: [a]Thermochemical value (mass-spectrom.)(5)(8)(15)(16)(17) (18)(29); photoionization [Parr and Inghram, quoted in (23)].

[b]Electron impact appearance potential (31).

[c]Tentative rotational analysis (24).

LaO (continued):

[d](20) favour a slightly different vibrational analysis, increasing v' by 1 and choosing ν_{00} = 26949.

[e]A_0 = +221.44.

[f] From band origins (25) as well as band heads (7) in C – A', the latter corrected for head-origin separations.

[g] From band heads in the 0-0 sequence of C – X (6) after correcting for head-origin separations.

[h] Slightly different constants in (27).

[i] Large Λ-type doubling in $^2\Pi_{\frac{1}{2}}$: $\Delta\nu_{ef}(v=0) = +\ 0.121(J+\frac{1}{2}) - \ldots$; in $^2\Pi_{3/2}$: $\Delta\nu_{ef}(v=0) = +\ 0.80 \times 10^{-6}(J-\frac{1}{2})(J+\frac{1}{2})(J+3/2)$.

[j] $D_1 = 2.5 \times 10^{-7}$.

[k] Systems G→? and E→? of (7).

[ℓ] J'=0 relative to N"=0 or J"=0.

[m] From band heads and calculated head-origin separations in B – X.

[n] Large spin doubling in B $^2\Sigma^+$: $\Delta\nu_{12} = (-)0.256(N+\frac{1}{2})$; small hyperfine structure [see (28)].

[o] $A_0 = +\ 862.6$.

[p] From band heads and calculated head-origin separations in A – X.

[q] Large Λ-type doubling in $^2\Pi_{\frac{1}{2}}$: $\Delta\nu_{ef} = +\ 0.267(J+\frac{1}{2})$.

[r] $B_1(^2\Pi_{3/2}) = 0.3449$.

[s] From Q as well as P heads, the latter corrected for head-origin separations, in the 0-0 sequence of C $^2\Pi_{\frac{1}{2}}$ – A' $^2\Delta_{3/2}$.

[t] $D_1 = 2.5 \times 10^{-7}$.

[u] Matrix studies at 4 K (14)(19) confirm that X $^2\Sigma^+$ is the ground state. See also (12)(13).

[v] Small spin doubling in X $^2\Sigma^+$: $\Delta\nu_{12} = +\ 0.0025_7(N+\frac{1}{2})$ (28); large hyperfine structure, 4b = 0.494 (10)(26)(28).

[w] In Ar matrix at 4 K (14)(19).

(1) Jevons, PPS 41, 520 (1929).
(2) Meggers, Wheeler, JRNBS 6, 239 (1931); 9, 268 (1932).
(3) Piccardi, GCI 63, 127 (1933).
(4) Gatterer, RS 1, 153 (1942).
(5) Chupka, Inghram, Porter, JCP 24, 792 (1956).
(6) Gatterer, Junkes, Salpeter, Rosen, METOX (1957).
(7) Hautecler, Rosen, BCSARB 45, 790 (1959).
(8) Goldstein, Walsh, White, JPC 65, 1400 (1961).
(9) Tawde, Chandratreya, CS 30, 137 (1961).
(10) Åkerlind, AF 22, 65 (1962).
(11) Ortenberg, Glasko, Dimitriev, SAAJ 8, 258 (1964).
(12) Berg, Wharton, Klemperer, Büchler, Stauffer, JCP 43, 2416 (1965).
(13) Brewer, Walsh, JCP 42, 4055 (1965).
(14) Kasai, Weltner, JCP 43, 2553 (1965).
(15) Smoes, Drowart, Verhaegen, JCP 43, 732 (1965).
(16) Ames, Walsh, White, JPC 71, 2707 (1967).
(17) Coppens, Smoes, Drowart, TFS 63, 2140 (1967).
(18) Drowart, Pattoret, Smoes, PBCS, No. 8, 67 (1967).
(19) Weltner, McLeod, Kasai, JCP 46, 3172 (1967).
(20) Carette, Houdart, CR B 271, 110 (1970).
(21) Murthy, Murthy, JP B 3, L15 (1970); 5, 714 (1972).
(22) Suárez, JP B 3, 729 (1970); CPL 16, 515 (1972).
(23) Uy, Drowart, HTS 2, 293 (1970).
(24) Carette, Houdart, CR B 272, 595 (1971).
(25) Green, JMS 38, 155 (1971).
(26) Green, CJP 49, 2552 (1971).
(27) Green, JMS 40, 501 (1971).
(28) Bacis, Collomb, Bessis, PR A 8, 2255 (1973).
(29) Ackermann, Rauh, JCP 60, 2266 (1974).
(30) Marañon, Suárez, SpL 7, 303 (1974).
(31) Rauh, Ackermann, JCP 60, 1396 (1974).
(32) Schoonveld, Sundaram, ApJ(Suppl.) 27(246), 307 (1974).

State	T_e	ω_e	$\omega_e x_e$	B_e	α_e	D_e (10^{-6} cm^{-1})	r_e (Å)	Observed Transitions Design.	ν_{00}	References
^{139}La^{103}Rh		μ = 59.113028		$D_0^0 = 5.4_4$ eV[a]						FEB 1975
^{139}La^{32}S		μ = 25.9899651		$D_0^0 = 5.9_1$ eV[a]						FEB 1975
		Strong absorption bands in the region 22200 - 23800 cm^{-1}. Complex structure.								(5)
B $^{2}\Sigma^{+}$	13790.1$_7$	410.07$_5$ H	0.94	[0.11099][b]	(0.00034)	[0.031]	[2.4174]	B$\leftrightarrow$X, R	13766.86 Z	(5)
X $^{2}\Sigma^{+}$	0	456.7$_0$ H	0.96$_5$	[0.11693][c]	(0.00034)	[0.030]	[2.3552]			
^{139}La$^{(80)}$Se		(μ = 50.730136$_7$)		$D_0^0 = 4.9_3$ eV[a]						FEB 1975
^{139}La$^{(130)}$Te		(μ = 67.127825)		$D_0^0 = 3.9_1$ eV[a]						FEB 1975
^{139}La^{89}Y		μ = 54.209522		$D_0^0 = 2.0_4$ eV[a]						FEB 1975
^{7}Li$_2$		μ = 3.5080024		$D_0^0 = 1.04_6$ eV[a]		I.P. = 5.0$_0$ eV[b]				MAR 1977 A
		Fragments of other absorption band systems in the ultraviolet.								(5)
D $^{1}\Pi_u$	(34518)[c]	[201.68][c] Z		0.4628[d]	0.0073	11.4	3.222	D$\leftarrow$X, R	34443.58[c] Z	(11)(17)
C $^{1}\Pi_u$	30550.6	237.9 Z	3.35[e]	0.5075 [i]	0.0097$_3$[f]	9	3.077	C$\leftarrow$X, R	30493.6 Z	(5)(9)(11)(12)(16)
B $^{1}\Pi_u$	20436.01	270.12 Z	2.67$_3$[g]	0.5577[hi]	0.0085[j]	9.45[k]	2.935	B$\leftrightarrow$X,[ℓ] R	20395.32 Z	(1)* (3)(4)(18)(19)(20)(21)
A $^{1}\Sigma_u^{+}$	14068.35	255.47 Z	1.58$_2$[m]	0.4975$_0$[i]	0.0054$_0$	7.54[n]	3.107$_9$	A$\leftarrow$X,[ℓ] R	14020.63 Z	(6)(8)*
a $^{3}\Sigma_u^{+}$		Not observed; predicted potential minimum of ~300 cm^{-1} at 4.3 Å (22).								
X $^{1}\Sigma_g^{+}$	0	351.43 Z	2.610[o]	0.67264[i]	0.00704[p]	9.87[q]	2.6729	Mol. beam magn. reson.[r]		(10)(13)

LaRh: [a]Thermochemical value (mass-spectrom.)(1).

(1) Cocke, Gingerich, Kordis, HTS 5, 474 (1973).

LaS: [a]Thermochemical value (mass-spectrom.)(1)(2)(3)(4).
[b]Large spin-splittings, $\gamma_0 = (-)0.0962_7$.
[c]Large hyperfine splittings, $4b = 0.47\ cm^{-1}$.

(1) Cater, Lee, Johnson, Rauh, Eick, JPC 69, 2684 (1965).
(2) Coppens, Smoes, Drowart, TFS 63, 2140 (1967).
(3) Cater, Steiger, JPC 72, 2231 (1968).
(4) Bergman, Coppens, Drowart, Smoes, TFS 66, 800 (1970).
(5) Marcano, Barrow, JP B 3, L121 (1970).

LaSe: [a]Thermochemical value (mass-spectrom.)(1)(2).

(1) See ref. (4) of LaS.
(2) Ni, Wahlbeck, HTS 4, 326 (1972).

LaTe: [a]Thermochemical value (mass-spectrom.)(1).

(1) See ref. (4) of LaS.

LaY: [a]Thermochemical value (mass-spectrom.)(1).

(1) Verhaegen, Smoes, Drowart, JCP 40, 239 (1964).

Li_2, Li_2^+, Li_2^-:

[a]Value recommended by (33). Extrapolation of the ground state vibrational levels, taking into account the contribution due to long-range forces, gives 8473 cm^{-1} [1.051 eV, see (33)]. The latest thermochemical value [mass-spectrom.(32)] is 1.11 eV; see also (2) who obtained 1.03 eV using a molecular beam method.

[b]Average of a photoionization (15) and electron impact (32) appearance potential (5.15 and 4.86 eV, resp.). The Rydberg series B,C,D... extrapolates to 4.99 eV (26).

Li_2, Li_2^+, Li_2^- (continued):

[c]Vibrational numbering uncertain.

[d]Rotational constants from Q branches only; see also [c].

[e]$\omega_e y_e = +0.064$.

[f]$\gamma_e = +0.00015_4$. Rotational constants from Q branches only; (12) gives $B(R,P) - B(Q) = +0.0024$.

[g]$\omega_e y_e = -0.082_5$. Vibrational constants recalculated from band origins (v'=0...4) of (20). Slightly different constants in (3) from a low-resolution magnetic rotation spectrum including much higher vibrational levels. The B $^1\Pi_u$ state has a potential maximum of ~0.09 eV (7)(18)(29); the theoretical calculations of (35) predict a barrier of 0.0724 eV at 5.61 Å.

[h]Slightly different constants in (25); Λ-type doubling $\Delta\nu_{ef} \approx +0.00016\,J(J+1)$.

[i]RKR potential curves (14).

[j]$\gamma_e \approx -0.00014$.

[k]$\beta_e = +0.14 \times 10^{-6}$.

[ℓ]The B−X and A−X systems of $^6Li_2$ have been analysed by (24) and (23), respectively.

[m]$\omega_e y_e = +0.002_5$.

[n]$H_e = +1.23 \times 10^{-10}$.

[o]$\omega_e y_e = +0.00295$, $\omega_e z_e = -0.000636$; from the B→X fluorescence spectrum (18), including levels from v=0 to 18.

[p]$\gamma_e = -0.00004$ (v=0...4) (8). For v > 10 B_v and D_v are better represented by $B_v = B_{10} - 0.0077(v-10) - 0.00016(v-10)^2$ and $D_v = D_{10} + 0.22 \times 10^{-6}(v-10)$ (18).

[q]$H_e = +1.54 \times 10^{-10}$. See also [p].

[r]$g_J = 0.1079_7\ \mu_N$ (13). Li nuclear electric quadrupole coupling constant (10).

(continued on p. 376)

State	T_e	ω_e	$\omega_e x_e$	B_e	α_e	D_e $(10^{-6} cm^{-1})$	r_e (Å)	Observed Transitions Design.	ν_{00}	References
$^7Li_2^+$		$D_0^0 = 1.4_4$ eV[s]								MAR 1977
		No spectrum observed. For theoretical calculations see (27)(30)(31) and references given by these authors.								
$^7Li_2^-$		$D_0^0 = (0.88)$ eV[t] I.P. = (0.45) eV[u]								MAR 1977
$^{(7)}Li^{40}Ar$		$(\mu = 5.9681970)$ $D_e^0 = 0.0052_8$ eV[a]								JAN 1977
B $^2\Sigma$		Unstable except for a small van der Waals minimum.[b]								
A $^2\Pi$	(14150)	Attractive potential, $D_e = 800\ cm^{-1}$.[c]					3.0_8[c]	A→X,		(5)
X $^2\Sigma'$	0	Unstable except for a small van der Waals minimum.[d]					4.78[a]			
$^{(7)}Li^{40}Ar^+$		$D_e^0 = 0.303$ eV[e]								JAN 1977
$^7Li^{79}Br$		$\mu = 6.4431916$ $D_0^0 = 4.3_3$ eV[a] I.P. = (10.0) eV[b]								JAN 1977
		Peaks in the electron energy loss spectrum at 6.9 and 8.5 eV (15).								
		Continuous absorption above 33000 cm^{-1}, first maximum at ~39000 cm^{-1}.[c]								
A		Diffuse absorption bands at 31560, 31018, 30467, 29879, (29442) cm^{-1}.						A←X		(5)
X $^1\Sigma^+$	0	563.1_6	3.5_3[d]	0.555399_0	0.005644_2[e]	2.159	2.170427	Rot.-vibr. sp.[f]		(4)(7)
								Rotation sp.		(3)(10)(12)
								Mol. beam rf electric[g]		(12)(16)(17)
								and magn. reson.[h]		(13)

Li_2, Li_2^+, Li_2^- (continued): [s] $D_0^0(Li_2)$ + I.P.(Li) - I.P.(Li_2).

[t] From $D_0^0(Li_2)$ and the electron affinities of Li_2 and Li.

[u] Theoretical electron affinity of Li_2 from an <u>ab initio</u> calculation by (34). A value of 0.90 eV was computed earlier by (28), but see (34).

Li_2, Li_2^+, Li_2^- (continued):

(1) Harvey, Jenkins, PR 35, 789 (1930).
(2) Lewis, ZP 69, 786 (1931).
(3) Loomis, Nusbaum, PR 38, 1447 (1931).
(4) McKellar, PR 44, 155 (1933).
(5) Vance, Huffman, PR 47, 215 (1935).
(6) Almy, Irwin, PR 49, 72 (1936).
(7) King, Van Vleck, PR 55, 1165 (1939).
(8) McKellar, Jenkins, PDAO 7, 155 (1939).
(9) Sinha, PPS 60, 443 (1948).
(10) Logan, Coté, Kusch, PR 86, 280 (1952).
(11) Barrow, Travis, Wright, Nature 187, 141 (1960).
(12) Velasco, ARSEFQ A 56, 175 (1960).
(13) Brooks, Anderson, Ramsey, PRL 10, 441 (1963); PR A 136, 62 (1964).
(14) Krupenie, Mason, Vanderslice, JCP 39, 2399 (1963).
(15) Foster, Leckenby, Robbins, JP B 2, 478 (1969).
(16) Rico, OPA 2 (1), 33 (1969).
(17) Mercier, Rico, Velasco, OPA 2 (2), 96 (1969).
(18) Velasco, Ottinger, Zare, JCP 51, 5522 (1969).
(19) Ottinger, Velasco, Zare, JCP 52, 1636 (1970).
(20) Velasco, Ruano, Rico, OPA 3 (3), 159 (1970).
(21) Ottinger, Poppe, CPL 8, 513 (1971).
(22) Kutzelnigg, Staemmler, Gélus, CPL 13, 496 (1972).
(23) Velasco, Rivero, OPA 5 (2), 76 (1972).
(24) Velasco, Morales, Publ. Inst. Optica "Daza de Valdes" Madrid, No. 36 (1972); OPA 6 (1), 52 (1973).
(25) Velasco, Ennen, Ottinger, OPA 6 (2), 11 (1973).
(26) Velasco, OPA 6 (2), 16 (1973).
(27) Bottcher, Dalgarno, CPL 36, 137 (1975).
(28) Andersen, Simons, JCP 64, 4548 (1976).
(29) Olson, Konowalow, CPL 39, 281 (1976).
(30) Müller, Jungen, CPL 40, 199 (1976).

Li_2, Li_2^+, Li_2^- (continued):

(31) Kirby-Docken, Cerjan, Dalgarno, CPL 40, 205 (1976).
(32) Wu, JCP 65, 2040, 3181 (1976).
(33) Stwalley, JCP 65, 2038 (1976).
(34) Dixon, Gole, Jordan, JCP 66, 567 (1977).
(35) Kahn, Dunning, Winter, Goddard, JCP 66, 1135 (1977).

LiAr, $LiAr^+$:

[a]From differential (3) and total (2) scattering cross sections.
[b]See the semiempirical calculations of (1).
[c]Morse potential parameters (average of two suggested values) from the analysis of the far-wing fluorescence spectrum longward of the Li resonance line at 14904 cm^{-1} (5). See also (6).
[d]See the potential energy curves in (2)(3)(5).
[e]From Li^+-off-Ar differential scattering measurements (4).

(1) Baylis, JCP 51, 2665 (1969).
(2) Ury, Wharton, JCP 56, 5832 (1972).
(3) Klingbeil, JCP 59, 797 (1973).
(4) Böttner, Dimpfl, Ross, Toennies, CPL 32, 197 (1975).
(5) Scheps, Ottinger, York, Gallagher, JCP 63, 2581 (1975).
(6) Scheps, Gallagher, JCP 65, 859 (1976).

LiBr: [a]Thermochemical value (8)(9).
[b]Maximum of a very broad photoelectron peak with two additional ill-defined peaks at 10.6 and 11.6 eV, the latter possibly due to the dimer $(LiBr)_2$ (19).
[c]Absorption cross sections (14).
[d]$\omega_e y_e$ = + 0.02; vibrational constants from the infrared spectrum of the natural isotopic mixture.
[e]γ_e = + 0.0000244; rotational constants evaluated (12) from the microwave results for $^6Li^{79}Br$.
[f]For IR spectrum in inert gas matrices see (11).

(continued on p. 379)

State	T_e	ω_e	$\omega_e x_e$	B_e	α_e	D_e $(10^{-6}cm^{-1})$	r_e (Å)	Observed Transitions Design.	ν_{00}	References
$^7Li^{35}Cl$		$\mu = 5.8435744$		$D_0^0 = 4.8_4$ eV[a]						JAN 1977
K $3p\pi$ $^2\Pi$	513500	1090	First members of two Rydberg series converging to the Li 1s ionization limit of LiCℓ at ~66 eV (532300 cm^{-1}); vibrational numbering not established.					K← X,	513700	(28)
J 3σ $^2\Sigma$	505900	1030						J← X,	506100	(28)
I $2p\pi$ $^2\Pi$	479200	850						I← X,	479300	(28)
H 2σ $^2\Sigma$	463900	950						H← X,	464000	(28)
		The energy loss spectrum of 25 keV electrons has peaks at 5.3, 7.2, and 8.9 eV.								(19)
		Continuous absorption above 40000 cm^{-1}, first maximum[b] at 42800 cm^{-1}.[c]								(1)(18)
A		Diffuse absorption bands at 35642, 35032, 34482 cm^{-1}.						A← X		(6)
X $^1\Sigma^+$	0	643.31[d]	4.50_1[d]	0.7065222_4	0.0080096_1[e]	3.4087[f]	2.020673[g]	Rot.-vibr. sp.[h]		(5)(9)
								Rotation sp.		(14)(21)
								Mol. beam rf electric[i]		(4)(20)(22)(24)
								and magn. reson.[j]		(17)
$^{(7)}Li^{(35)}Cl^-$				$D_0^0 = 1.8_4$ eV[k]		I.P. = 0.61 eV[ℓ]				JAN 1977
X $^2\Sigma^+$	0	(480)[ℓ]					2.18[m]			

LiCℓ, LiCℓ⁻:

[a]Thermochemical value (10)(11)(15). A slightly higher value was suggested by (7).

[b]Also observed in the electron energy loss spectrum.

[c]Absorption cross sections (18).

[d]Calculated (21) from the rotational constants by use of Dunham's theory. From the infrared spectrum of the isotopic mixture (9) obtain $\omega_e = 641._1$, $\omega_e x_e = 4.2$. For $^6Li^{35}C\ell$ (12) find $\omega_e \approx 705$ by the molecular beam electric reson. method.

[e]$\gamma_e = +\ 0.00003966$.

[f]$\beta_e = -\ 0.0190 \times 10^{-6}$.

[g]From the effective B_e. Using the data of (21) for the four LiCℓ isotopes (23) has determined r_e at the minimum of the Born-Oppenheimer potential as 2.020700 Å.

[h]IR spectrum of matrix isolated LiCℓ (13)(16).

[i]Electric dipole moment of $^6Li^{35}C\ell$: $\mu_{e\ell}[D] = 7.085_3 + 0.0868(v+\frac{1}{2}) + 0.0005_6(v+\frac{1}{2})^2$ (20), see also (4)(14). For electric quadrupole and other hyperfine coupling constants see (4)(22). The Zeeman spectrum was also studied by the

LiCℓ, $LiCℓ^-$ (continued):

molecular beam electric reson. method (24); $g_J(^7Li^{35}Cℓ)$ = +0.10042 and +0.10064 μ_N for v=0 and 1, resp., superseding an earlier value by the magnetic resonance method (17).

[j]Nuclear reorientation spectrum of Li (2)(3)(8).

[k]From D_0^0(LiCℓ) and the electron affinities of LiCℓ and Cℓ.

[ℓ]From the photoelectron spectrum of $LiCℓ^-$ (25); I.P. is reasonably close to the calculated electron affinity of LiCℓ [0.54 eV (26)]; see also (27).

[m]The relative intensities of the photoelectron peaks have been compared (25) with calculated Franck-Condon factors.

(1) Müller, AP(Leipzig) 82, 39 (1927).
(2) Kusch, PR 75, 887 (1949).
(3) Logan, Coté, Kusch, PR 86, 280 (1952).
(4) Marple, Trischka, PR 103, 597 (1956).
(5) Klemperer, Rice, JCP 26, 618 (1957).
(6) Berry, Klemperer, JCP 26, 724 (1957).
(7) Gurvich, Veits, BASPS 22, 670 (1958).
(8) Kusch, JCP 30, 52 (1959).
(9) Klemperer, Norris, Büchler, Emslie, JCP 33, 1534 (1960).
(10) Brewer, Brackett, CRev 61, 425 (1961).
(11) Bulewicz, Phillips, Sugden, TFS 57, 921 (1961).
(12) Moran, Trischka, JCP 34, 923 (1961).
(13) Snelson, Pitzer, JPC 67, 882 (1963).
(14) Lide, Cahill, Gold, JCP 40, 156 (1964).
(15) Hildenbrand, Hall, Ju, Potter, JCP 40, 2882 (1964).
(16) Schlick, Schnepp, JCP 41, 463 (1964).
(17) Mehran, Brooks, Ramsey, PR 141, 93 (1966).
(18) Davidovits, Brodhead, JCP 46, 2968 (1967).
(19) Geiger, Pfeiffer, ZP 208, 105 (1968).
(20) Hebert, Lovas, Melendres, Hollowell, Story, Street, JCP 48, 2824 (1968).
(21) Pearson, Gordy, PR 177, 52 (1969).
(22) Gallagher, Hilborn, Ramsey, JCP 56, 5972 (1972).
(23) Watson, JMS 45, 99 (1973).
(24) Freeman, Johnson, Ramsey, JCP 61, 3471 (1974).
(25) Carlsten, Peterson, Lineberger, CPL 37, 5 (1976).
(26) Jordan, Luken, JCP 64, 2760 (1976).
(27) Jordan, JCP 65, 1214 (1976).
(28) Radler, Sonntag, Chang, Schwarz, CP 13, 363 (1976).

LiBr (continued):

[g]Dipole moment of $^6Li^{79}Br$: $\mu_{eℓ}[D] = 7.226_2 + 0.083_2(v+\frac{1}{2}) + 0.0005_7(v+\frac{1}{2})^2$ (12). For electric quadrupole and other hyperfine coupling constants of the various isotopes see (12)(16)(17). The Zeeman spectrum was studied by the electric resonance method (18); $g_J(^7Li^{79}Br)$ = 0.11206 superseding an earlier value by the magnetic resonance method (13); also $^{79,81}Br$ nuclear magnetic moments.

[h]Li nuclear reorientation spectrum (1)(2)(6).

(1)(2) See ref. (2)(3), resp., of LiCℓ.
(3) Honig, Mandel, Stitch, Townes, PR 96, 629 (1954).
(4)...(9) See ref. (5)(6)(8)...(11), resp., of LiCℓ.
(10) Rusk, Gordy, PR 127, 817 (1962).
(11) See ref. (16) of LiCℓ.
(12) Hebert, Breivogel, Street, JCP 41, 2368 (1964).
(13)...(15) See ref. (17)...(19), resp., of LiCℓ.
(16) Hebert, Street, PR 178, 205 (1969).
(17) Hilborn, Gallagher, Ramsey, JCP 56, 855 (1972).
(18) Cecchi, Ramsey, JCP 60, 53 (1974).
(19) Goodman, Allen, Cusachs, Schweitzer, JESRP 3, 289 (1974).

State	T_e	ω_e	$\omega_e x_e$	B_e	α_e	D_e (10^{-5}cm^{-1})	r_e (Å)	Observed Transitions: Design.	Observed Transitions: ν_{00}	References
$^{(7)}Li^{133}Cs$		(μ = 6.6642051)		D_e^0 = (0.72) eV[a]						JAN 1977
B	(16477)	(77) H	Fragment					B← X,	16432 H	(1)(2)
X $^1\Sigma^+$	0	(167) H					(3.5_4)[b]			
$^7Li^{19}F$		μ = 5.1238103		D_0^0 = 5.91 eV[a]						JAN 1977
K 3pπ $^2\Pi$	510700	1420	First members of two Rydberg series converging to the Li 1s ionization limit of LiF at $65._5$ eV (528300 cm^{-1}); vibrational numbering not established.					K← X,	510900	(31)
J 3σ $^2\Sigma$	502200	1400						J← X,	502500	(31)
I 2pπ $^2\Pi$	477500	1240						I← X,	477600	(31)
H 2σ $^2\Sigma$	458600	(1000)						H← X,	458600	(31)
		Peaks in the electron energy loss spectrum at 6.6, 8.7, 10.9, 62.0 eV.								(22)
		Ab initio studies of the lowest $^1\Sigma$ states (including the ground state), curve crossings (28)(29)(32).								
X $^1\Sigma^+$	0	910.34[b] Z	7.929[b]	1.3452576	0.0202868[c]	1.1754[d]	1.563864	Rotation-vibr. sp.[e]		(8)(9)(14)
								Rotation sp.		(16)(19)(24)(25)
								Mol. beam rf electric[f]		(16)(23)(26)(30)
								and magn. reson.[g]		(20)

LiCs: [a]From Li - Cs total scattering cross sections (3), dependent on the assumed value for r_e.
[b]Theoretical value, quoted in (3).

(1) Walter, Barratt, PRS A 119, 257 (1928).
(2) Weizel, Kulp, AP(Leipzig) 4, 971 (1930).
(3) Kanes, Pauly, Vietzke, ZN 26 a, 689 (1971).

LiF: [a]Thermochemical value (5)(10)(11)(17).
[b]From the infrared spectrum [constants corresponding to the J numbering "M_{orig} - 2" in table III of (9)]. In good agreement with constants calculated from the microwave results: $\omega_e = 910.2_5$, $\omega_e x_e = 8.10$.
[c]$+ 0.0001558(v+\frac{1}{2})^2 - 3.5 \times 10^{-7}(v+\frac{1}{2})^3$.

LiF (continued):

[d] $\beta_e = -0.0124 \times 10^{-5}$.

[e] For IR frequencies in inert gas matrices see (12)(15)(18)(21). The lifetime of the lowest vibrationally excited level of ^{6}LiF, $\tau(v=1) = 14._3$ ms, was determined by (27) using the molecular beam electric resonance method.

[f] Dipole moment of ^{7}LiF: $\mu_{el}[D] = 6.2839 + 0.08153(v+\frac{1}{2}) + 0.000445(v+\frac{1}{2})^2$, v=0,1,2 (30); see also (16)(23)(26). For electric quadrupole and other hyperfine coupling constants see (25)(30). Earlier electric resonance work in (2)(3)(4)(13) and (6) who found $g_J(^7\text{LiF}) = +0.0642\ \mu_N$ from the Zeeman splitting of the hyperfine structure; see also [g].

[g] $g_J(^7\text{LiF}) = (+)0.0737\ \mu_N$ by the magnetic resonance method (20); see also (6). Li NMR spectrum (1)(7).

(1) Kusch, PR 75, 887 (1949).
(2) Swartz, Trischka, PR 88, 1085 (1952).
(3) Braunstein, Trischka, PR 98, 1092 (1955).
(4) Kastner, Russell, Trischka, JCP 23, 1730 (1955).
(5) Pugh, Barrow, TFS 54, 671 (1958).
(6) Russell, PR 111, 1558 (1958)
(7) Kusch, JCP 30, 52 (1959).
(8) Klemperer, Norris, Büchler, Emslie, JCP 33, 1534 (1960).
(9) Vidale, JPC 64, 314 (1960).
(10) Brewer, Brackett, CRev 61, 425 (1961).
(11) Bulewicz, Phillips, Sugden, TFS 57, 921 (1961).
(12) Linevsky, JCP 34, 587 (1961); 38, 658 (1963).
(13) Moran, Trischka, JCP 34, 923 (1961).
(14) Vasilevskii, Baikov, OS(Engl. Transl.) 11, 21 (1961).
(15) Snelson, Pitzer, JPC 67, 882 (1963).
(16) Wharton, Klemperer, Gold, Strauch, Gallagher, Derr, JCP 38,1203 (1963).
(17) Hildenbrand, Hall, Ju, Potter, JCP 40, 2882 (1964).
(18) Schlick, Schnepp, JCP 41, 463 (1964).
(19) Veazey, Gordy, PR A 138, 1303 (1965).
(20) Mehran, Brooks, Ramsey, PR 141, 93 (1966).
(21) Snelson, JCP 46, 3652 (1967).
(22) Geiger, Pfeiffer, ZP 208, 105 (1968).
(23) Hebert, Lovas, Melendres, Hollowell, Story, Street, JCP 48, 2824 (1968).
(24) Pearson, Gordy, PR 177, 52 (1969).
(25) Cupp, Smith, Contini, Woods, Gallagher, PL A 44, 305 (1973).
(26) Mariella, Herschbach, Klemperer, JCP 58, 3785 (1973).
(27) Bedding, Moran, PR A 9, 2324 (1974).
(28) Kahn, Hay, Shavitt, JCP 61, 3530 (1974).
(29) Botter, Kooter, Mulder, CPL 33, 532 (1975).
(30) Hebert, Hollowell, JCP 65, 4327 (1976).
(31) Radler, Sonntag, Chang, Schwarz, CP 13, 363 (1976).
(32) Yardley, Balint-Kurti, MP 31, 921 (1976).

State	T_e	ω_e	$\omega_e x_e$	B_e	α_e	D_e $(10^{-3}cm^{-1})$	r_e (Å)	Observed Transitions: Design.	Observed Transitions: ν_{00}	References
$^7Li^1H$		μ = 0.88123833		D_0^0 = 2.42871 eV[a]						JAN 1977 A
		For ab initio calculations of X $^1\Sigma$, A $^1\Sigma$, B $^1\Pi$, a $^3\Sigma$, b $^3\Pi$ (lowest stable triplet state at ~1700 cm^{-1} below B $^1\Pi$) see (13). Excitation energies and oscillator strengths for higher-lying states have been computed by (19). The most recent ground state studies are those of (17) and (22), the latter including other low-lying $^1\Sigma^+$ states.								
B $^1\Pi$	34912	[130.73] Z	[b]	3.383[c] [d]	0.986[e]	[2.6][f]	2.378	B← X, R	34312.26 Z	(3)*
A $^1\Sigma^+$	26516	[280.96][g] Z		[2.8536][gd]		[1.187][g]	2.60_5	A[h]↔ X,[i] R	25943.13 Z	(1)
X $^1\Sigma^+$	0	1405.65 Z	23.20[j]	7.5131 [d]	0.2132[k]	0.8617[ℓ]	1.5957	Rot.-vibr. sp.		(2)(4)(5)
								Mol. beam electric[m] and magn. reson.		(7)(12)(18) (8)
$^7Li^2H$		μ = 1.56487081		D_0^0 = 2.45090 eV[a]		I.P. = 7.7 eV[b]				JAN 1977
B $^1\Pi$	$34908._8$	178.70 Z	29.13[c]	1.904[d]	0.425[e]	0.44[f]	2.379	B← X, R	34466.27 Z	(2)(9)
A $^1\Sigma^+$	26513	[205.6][g]		[1.6125][hi]		$[0.350_8]$[h]	2.59_0	A↔ X, R	26082.8 Z	(1)(11)
X $^1\Sigma^+$	0	1054.80_3 Z	12.93_5[j]	4.239_4[k]	0.100_0[ℓ]	0.2756[m]	1.5941	Rotation spectrum		(7)
								Mol. beam electric[n] and magn. reson.		(5)(8)(13) (6)

Li^1H: [a]From the predissociation in B $^1\Pi$; the evaluation by (14) takes into account the long-range potential of this state.

[b]$\Delta G(3/2)$ = 45.9.

[c]Predissociation by rotation; breaking off above J'=8,5,2 in v'=0,1,2, respectively; see also (14). Dissociation limit at 34492.5 cm^{-1} above X $^1\Sigma$, v"=0, J"=0.

[d]RKR potential curves (6); (9)[A state]; (14)[B state, combination with long-range tail and exponential inner wall].

[e]γ_e = - 0.045.

[f]$D_1 = 4.8 \times 10^{-3}$; $H_0 = -1.7 \times 10^{-5}$, $H_1 = -5.6 \times 10^{-5}$.

[g]$\Delta G(v+\frac{1}{2})$, B_v, D_v, H_v have been determined up to v=14. The ΔG and B_v curve have maxima for v=9 and 3, resp.; $\omega_e \approx 235$, $\omega_e x_e \approx -28$, $\omega_e y_e \approx -4$; $B_e \approx 2.819$, $\alpha_e \approx -0.078$, $\gamma_e \approx -0.026$, and higher order constants.

Li^1H (continued):

[h]Radiative lifetimes $\tau(v',J')$: $\tau(2,3) = 29._4$ ns, $\tau(5,3) = 30._5$ ns, $\tau(7,12) = 36._9$ ns (20); $\tau(5,5\text{-}15) = 31$ ns (21).

[i]Intensity distribution in the v'-0 bands (11); RKR Franck-Condon factors (10). The A–X system of $^6Li^1H$ was analyzed by (15).

[j]$\omega_e y_e = + 0.163$.

[k]$\gamma_e = + 0.0007_5$; all rotational constants are from v=0,1,2 only.

[ℓ]$\beta_e = - 0.0160 \times 10^{-3}$, see [k]. $H_v = 11.4 \times 10^{-8} - \ldots$

[m]$\mu_{e\ell}(v=0,1,2) = 5.8820,\ 5.9905,\ 6.098$ D (7)(12). Hyperfine structure constants (7)(12)(18). Zeeman spectrum (18), $g_J(v=0,J=1) = -0.6584_2$ in agreement with an earlier less precise value obtained by (8) using the magnetic resonance method. For a combination of both theoretical (13) and experimental results see (16).

(1) Crawford, Jorgensen, PR 47, 932 (1935); 49, 745 (1936).
(2) Klemperer, JCP 23, 2452 (1955)
(3) Velasco, CJP 35, 1204 (1957); OPA 7 (1), 14 (1974).
(4) Norris, Klemperer, JCP 28, 749 (1958).
(5) James, Norris, Klemperer, JCP 32, 728 (1960).
(6) Fallon, Vanderslice, Mason, JCP 32, 1453 (1960); 33, 944 (1960) [Erratum].
(7) Wharton, Gold, Klemperer, JCP 33, 1255 (1960); 37, 2149 (1962).
(8) Lawrence, Anderson, Ramsey, PR 130, 1865 (1963).
(9) Singh, Jain, PPS 79, 274, 753 (1963).
(10) Halmann, Laulicht, JCP 46, 2684 (1967).
(11) Fernandez-Florez, Velasco, OPA 2 (3), 123 (1969).
(12) Rothstein, JCP 50, 1899 (1969).
(13) Docken, Hinze, JCP 57, 4928, 4936 (1972).

Li^1H (continued):

(14) Way, Stwalley, JCP 59, 5298 (1973).
(15) Velasco, Rivero, OPA 7 (1), 45 (1974).
(16) Docken, Freeman, JCP 61, 4217 (1974).
(17) Meyer, Rosmus, JCP 63, 2356 (1975).
(18) Freeman, Jacobson, Johnson, Ramsey, JCP 63, 2597 (1975).
(19) Stewart, Watson, Dalgarno, JCP 63, 3222 (1975).
(20) Dagdigian, JCP 64, 2609 (1976).
(21) Wine, Melton, JCP 64, 2692 (1976).
(22) Yardley, Balint-Kurti, MP 31, 921 (1976).

Li^2H: [a]From the predissociation in B $^1\Pi$ (9). A thermochemical value (mass-spectrom.) of 2.49 eV was determined by (12).

[b]Electron impact appearance potential (12).

[c]$\omega_e y_e = - 3.51$ (v=0...3) (9).

[d]All four observed levels are predissociated by rotation; breaking off above J'=12,9,6,2 in v'= 0,1,2,3. Dissociation limit at 34671.5 cm^{-1} above X $^1\Sigma$, v"=0, J"=0.

[e]$- 0.005(v+\frac{1}{2})^2 - 0.001(v+\frac{1}{2})^3$, v=0...3 (9).

[f]$\beta_e = +0.63 \times 10^{-3}$; $H_0 = - 2.5 \times 10^{-6}$, $H_1 = - 5.3 \times 10^{-6}$, $H_2 = - 16.5 \times 10^{-6}$.

[g]Extrapolated from the observed $\Delta G(v+\frac{1}{2})$ values for v=1...18. The ΔG curve has a maximum at v=13. $\omega_e \approx 181._9$ $\omega_e x_e \approx - 13.4$, $\omega_e y_e \approx - 1.0_3$, ...

[h]Extrapolated from observations for v=1...17; H_v values were also determined. The B_v curve goes through a maximum at v=4. $B_e \approx 1.6054$, $\alpha_e \approx - 0.0152$, $\gamma_e \approx - 0.0021_1$, ...

[i]RKR potential curves (3)(4).

[j]$+ 0.0392_3(v+\frac{1}{2})^3 + 0.00325_3(v+\frac{1}{2})^4 - 0.000148_9(v+\frac{1}{2})^5$; the vibrational and rotational constants (except D_v, H_v) are

(continued on p. 385)

State	T_e	ω_e	$\omega_e x_e$	B_e	α_e	D_e $(10^{-6}cm^{-1})$	r_e (Å)	Observed Transitions Design.	ν_{00}	References
$^7Li^4He$		$\mu = 2.5486235_2$								JAN 1977
		$X\ ^2\Sigma$ (4)(5)(6) and $B\ ^2\Sigma$ (6) are unstable except for very small van der Waals minima; $A\ ^2\Pi$ (3)(6) is attractive with an estimated well depth of 500 to 1500 cm^{-1}. Additional excited state potential curves computed by (1)(5). Theoretical coefficients for collision-induced absorption by Li - He (ν = 0 - 300 cm^{-1}) calculated by (4a). The noble-gas broadening of the Li resonance line was studied by (7).								
$^7Li^4He^+$										JAN 1977 A
		Theoretical calculations of the ground state potential (2)(3)(9) predict D_e and r_e values varying from 0.064 eV and 1.98 Å to 0.075 eV and 1.92 Å, respectively. See also (1)(8)								
$^7Li^{127}I$		$\mu = 6.6484410$ $\quad D_0^0 = 3.5_4$ eV[a]								JAN 1977
		Peaks in the electron energy loss spectrum at 7.3 and 9.7 eV.								(16)
		Continuous absorption above 28600 cm^{-1} with maxima at 33900 and 45000 cm^{-1}.[b]								(15)
A		Diffuse absorption bands from 29146 to 24507 cm^{-1}.[c]						A← X		(7)(19)
X $^1\Sigma^+$	0	498.1_6 (Z)	3.3_9[d]	0.443182_0	0.004090[e]	1.448	2.391924	Rot.-vibr. sp.		(6)(9)
								Rotation spectrum		(4)(12)(13)
								Mol. beam rf electric[f]		(13)(18)
								and magn. reson.[g]		(14)
$^{(7)}Li^{(39)}K$		(μ = 5.9454387)								JAN 1977 A
B	(17578)	(130) H	Fragment					B← X, (R)	17539 H	(1)* (2)
X $^1\Sigma^+$	0	(207) H		(0.265)[a]			(3.27)[a]	Mol. beam rf el. reson.[b]		(3)

Li^2H (continued):

derived from A→X fluorescence series extending to v" = 20, i.e. 82% of the dissociation energy (11); the more accurate data for low v" from (1) are also included, but not the microwave data of (7).

[k] RKR potential curve (11).

[ℓ] $+ 0.0028_7(v+\frac{1}{2})^2 - 0.000109_6(v+\frac{1}{2})^3$; see [j]. From the millimeter wave spectrum (7) obtain $Y_{01} = 4.233107$ and $Y_{11} = - 0.091550$ (v=0,1 only).

[m] $- 0.0066_3 \times 10^{-3}(v+\frac{1}{2}) + 0.00057_5 \times 10^{-3}(v+\frac{1}{2})^2$; $H_v = [2.118 - 0.105(v+\frac{1}{2})] \times 10^{-8}$ (1).

[n] $\mu_{e\ell}(v=0) = 5.8677$ D (5)(8). Hfs constants (5)(8)(13). $g_J(v=0,J=1) = -0.2767_4\ \mu_N$ from the Zeeman sp. (13); a less precise value was earlier obtained (6) by the magnetic resonance method. See also (10).

(1) Crawford, Jorgensen, PR 47, 358 (1935); 49, 745 (1936).
(2) Velasco, CJP 35, 1204 (1957)
(3) Fallon, Vanderslice, Mason, JCP 32, 1453 (1960); 33, 944 (1960) [Erratum].
(4) Singh, Jain, CJP 40, 520 (1962).
(5) Wharton, Gold, Klemperer, JCP 37, 2149 (1962).
(6) Lawrence, Anderson, Ramsey, PR 130, 1865 (1963).
(7) Pearson, Gordy, PR 177, 59 (1969).
(8) Rothstein, JCP 50, 1899 (1969).
(9) Stwalley, Way, Velasco, JCP 60, 3611 (1974).
(10) Docken, Freeman, JCP 61, 4217 (1974).
(11) Ennen, Ottinger, CPL 36, 16 (1975).
(12) Ihle, Wu, JCP 63, 1605 (1975).
(13) Freeman, Jacobson, Johnson, Ramsey, JCP 63, 2597 (1975).

$LiHe, LiHe^+$:

(1) Schneiderman, Michels, JCP 42, 3706 (1965).
(2) Catlow, McDowell, Kaufman, Sachs, Chang, JP B 3, 833 (1970).
(3) Krauss, Maldonado, Wahl, JCP 54, 4944 (1971).
(4) Dehmer, Wharton, JCP 57, 4821 (1972).
(4a) Bottcher, Dalgarno, Wright, PR A 7, 1606 (1973).
(5) Pascale, Vandeplanque, JCP 60, 2278 (1974).
(6) Bottcher, Cravens, Dalgarno, PRS A 346, 157 (1975).
(7) Gallagher, PR A 12, 133 (1975).
(8) Morrison, Akridge, Ellis, Pai, McDaniel, JCP 63, 2238 (1975).
(9) Hariharan, Staemmler, CP 15, 409 (1976).

LiI: [a] Close agreement between several thermochemical values [(10); mass-spectrom.(5); flame-photom. (11)] and the spectroscopic value of (1).

[b] Absorption cross sections (15). UV absorption in inert matrices (17) shows banded structure in the region 34500 - 40000 cm^{-1} indicating the existence of a stable upper state (possibly lowest excited 0^+) with $\omega_e \approx 365$.

[c] A broad single band appears at 28560 cm^{-1} in matrix absorption spectra (17).

[d] $\omega_e y_e = + 0.08$; vibrational constants from the IR sp.(9).

[e] $\gamma_e = + 0.000015_3$. For constants of ^6LiI see (13).

[f] Dipole moments for ^6LiI: $\mu_{e\ell}(v=0,1) = 7.428_5, 7.512_0$ D (13). Hfs constants (13)(18); see also (4) and (2)(3)(8).

[g] $g_J(^7LiI) = (+)0.107\ \mu_N$. Li NMR spectrum (2)(3)(8).

References on p. 387 .

LiK: [a] Estimated constants for $^7Li^{39}K$ (3). See also (4).

[b] $\mu_{e\ell}(^7Li^{39}K;\ v=0,J=1) = 3.4_5$ D; also values of eqQ.

References on p. 387 .

State	T_e	ω_e	$\omega_e x_e$	B_e	α_e	D_e (10^{-} cm^{-1})	r_e (Å)	Observed Transitions Design.	Observed Transitions ν_{00}	References
$^{(7)}Li^{(84)}Kr$		(μ = 6.4746469) D_e^0 = 0.0084 eV[a]								JAN 1977
B $^2\Sigma^+$		Unstable except for a small van der Waals minimum.[b]								
A $^2\Pi$	(13790)	Attractive potential, $D_e \approx 1185$ cm^{-1}.[c]					3.1_8[c]			
X $^2\Sigma^+$	0	Unstable except for a small van der Waals minimum.[d]					4.78[a]			
$^{(7)}Li^{(84)}Kr^+$		D_e^0 = 0.391 eV[e]								JAN 1977
$^7Li^{14}N$		μ = 4.6741171 Predicted spectroscopic constants in (1)(2).								JAN 1977
$^7Li^{23}Na$		μ = 5.3755098								JAN 1977 A
B $^1\Pi$				[0.307][ab]			[3.20][b]	B→X, R	20226.01[b] Z	(3)
X $^1\Sigma$	0	256.8 Z	1.612[c]	0.396	0.0036		2.81	Mol. beam rf electric[d] and magn. reson.		(5) (4)
$^{(7)}Li^{23}Na^+$		D_e^0 = 0.99_4 eV[e]								JAN 1977

LiKr, $LiKr^+$:

[a]From high-resolution differential scattering cross sections (3); **see also** (2).

[b]Semiempirical calculations of (1).

[c]Morse potential parameters (average of two suggested values) from the analysis of the far-wing fluorescence spectrum longward of the Li resonance line at 14904 cm^{-1} (5). See also (6).

[d]See the potential energy curves in (2)(3)(5).

[e]From Li^+-off-Kr differential scattering cross sections (4).

(1) Baylis, JCP 51, 2665 (1969).
(2) Dehmer, Wharton, JCP 57, 4821 (1972).
(3) Auerbach, JCP 60, 4116 (1974).
(4) Böttner, Dimpfl, Ross, Toennies, CPL 32, 197 (1975).
(5) Scheps, Ottinger, York, Gallagher, JCP 63, 2581 (1975).
(6) Scheps, Gallagher, JCP 65, 859 (1976).

LiN: (1) Dykstra, Pearson, Schaefer, JACS 97, 2321 (1975).
(2) Jordan, JCP 65, 1214 (1976).

$LiNa$, $LiNa^+$:

[a]Λ-type doubling, $\Delta\nu_{ef} = +\ 1.9_5 \times 10^{-4} J(J+1)$.
[b]v' numbering unknown.
[c]$\omega_e y_e = -\ 0.0075$.
[d]$\mu_{e\ell}(v=0) = 0.463$ D; quadrupole coupling constants (5), for ^{7}Li in disagreement with the value from the magnetic resonance spectrum (4). See also (2)[molecular beam electric deflection] and (9)[*ab initio* calculation, dipole moment function].
[e]Theoretical calculation of (8), confirmed by the elastic Na^+-on-Li scattering data of (7). Slightly lower values computed by (6) and (1) who give also results for other spectroscopic constants.

$LiNa$, $LiNa^+$ (continued):

(1) Bertoncini, Das, Wahl, JCP 52, 5112 (1970).
(2) Dagdigian, Graff, Wharton, JCP 55, 4980 (1971).
(3) Hessel, PRL 26, 215 (1971).
(4) Brooks, Anderson, Ramsey, JCP 56, 5193 (1972).
(5) Graff, Dagdigian, Wharton, JCP 57, 710 (1972).
(6) Oppenheimer, Bottcher, Dalgarno, CPL 15, 24 (1972).
(7) von Busch, Hormes, Liesen, CPL 34, 244 (1975).
(8) Habitz, Schwarz, CPL 34, 248 (1975).
(9) Rosmus, Meyer, JCP 65, 492 (1976).

LiI: (1) Beutler, Levi, ZPC B 24, 263 (1934).
(2) Kusch, PR 75, 887 (1949).
(3) Logan, Coté, Kusch, PR 86, 280 (1952).
(4) Honig, Mandel, Stitch, Townes, PR 96, 629 (1954).
(5) Friedman, JCP 23, 477 (1955).
(6) Klemperer, Rice, JCP 26, 618 (1957).
(7) Berry, Klemperer, JCP 26, 724 (1957).
(8) Kusch, JCP 30, 52 (1959).
(9) Klemperer, Norris, Büchler, Emslie, JCP 33, 1534 (1960).
(10) Brewer, Brackett, CRev 61, 425 (1961).
(11) Bulewicz, Phillips, Sugden, TFS 57, 921 (1961).
(12) Rusk, Gordy, PR 127, 817 (1962).
(13) Breivogel, Hebert, Street, JCP 42, 1555 (1965).
(14) Mehran, Brooks, Ramsey, PR 141, 93 (1966).
(15) Davidovits, Brodhead, JCP 46, 2968 (1967).
(16) Geiger, Pfeiffer, ZP 208, 105 (1968).
(17) Oppenheimer, Berry, JCP 54, 5058 (1971).
(18) Jacobson, Ramsey, JCP 65, 1211 (1976).
(19) Levi, Dissertation (Berlin, 1934).

LiK: (1) Walter, Barratt, PRS A 119, 257 (1928).
(2) Weizel, Kulp, AP(Leipzig) 4, 971 (1930).
(3) Dagdigian, Wharton, JCP 57, 1487 (1972).
(4) Cavaliere, Ferrante, Lo Cascio, JCP 62, 4753 (1975).

State	T_e	ω_e	$\omega_e x_e$	B_e	α_e	D_e (10^- cm^{-1})	r_e (Å)	Observed Transitions Design.	Observed Transitions ν_{00}	References
$^{(7)}Li^{(20)}Ne$		(μ = 5.1934518) D_e^0 = 0.0011 eV[a]								JAN 1977
B $^2\Sigma$		Semiempirical interaction potential calculations (1). Broadening of								
A $^2\Pi$		the Li resonance line at 6708 Å (14904 cm^{-1}) by neon (4).								
X $^2\Sigma$		Model potential calculation (3); see also (1).					5.05[a]			
$^7Li^{16}O$		μ = 4.8768325 D_0^0 = 3.49 eV[a] I.P. = 8.4_5 eV[b]								JAN 1977
A $^2\Sigma^+$	(2330)[c]	(866.7)[c]	(6.7)[c]	(1.349)[c]	(0.0199)[c]		(1.599)[c]			
X $^2\Pi_i$	0 [d]	(851.5)[ce]	(12.5)[c]	(1.202)[c]	(0.0151)[c]		(1.695)[c]	IR spectrum[f]		(2)(9)
								Mol. beam rf el. reson.[g]		(5)
$^{(7)}Li^{(85)}Rb$		(μ = 6.4805376)								JAN 1977
B	(17552)	(81) H	Fragment[a]					B← X,	17500 H	(1)(2)
X $^1\Sigma^+$	0	(185) H						Mol. beam rf el. reson.[b]		(3)
$^{(7)}Li^{(132)}Xe$		(μ = 6.6616695) D_e^0 = 0.0127 eV[a]								JAN 1977
B $^2\Sigma$		Repulsive state except for a small van der Waals minimum.[b]								
A $^2\Pi$	(13380)	Attractive potential, $D_e \approx$ 1630 cm^{-1}.[c]					3.0_6[c]			
X $^2\Sigma$	0	Repulsive state except for a small van der Waals minimum.[d]					4.80[a]			
$^{(7)}Li^{(132)}Xe^+$		D_e^0 = 0.51 eV[e]								JAN 1977

LiNe: [a]Approximate constants from total scattering cross sections (2).

(1) Baylis, JCP 51, 2665 (1969).
(2) Dehmer, Wharton, JCP 57, 4821 (1972).
(3) Bottcher, Dalgarno, Wright, PR A 7, 1606 (1973).
(4) Gallagher, PR A 12, 133 (1975).

LiO: [a]Thermochemical value [mass-spectrom.(1)(2)(7), flame photom.(4)].
[b]Electron impact appearance potential (7).
[c]CI calculation of (6); see also (8).
[d]A_0 = - 112.0, A_1 = - 108; from the radio frequency spectrum (5). The $^2\Pi$ symmetry of the ground state was originally established by molecular beam electric deflection (3).
[e]The fundamental frequencies in krypton (2) and nitrogen (9) matrices are 745 and 700 cm^{-1}, respectively.
[f]In inert gas matrices.
[g]$\mu_{e\ell}(v=0)$ = 6.84 D. The hfs parameters have been re-evaluated by (10).

(1) Berkowitz, Chupka, Blue, Margrave, JPC 63, 644 (1959).
(2) White, Seshadri, Dever, Mann, Linevsky, JCP 39, 2463 (1963).
(3) Berg, Wharton, Klemperer, JCP 43, 2416 (1965).
(4) Dougherty, Dunn, McEwan, Phillips, CPL 11, 124 (1971).
(5) Freund, Herbst, Mariella, Klemperer, JCP 56, 1467 (1972).
(6) Yoshimine, JCP 57, 1108 (1972); Yoshimine, McLean, Liu, JCP 58, 4412 (1973).
(7) Hildenbrand, JCP 57, 4556 (1972).

LiO (continued):

(8) Marchetti, Julienne, Krauss, JRNBS A 76, 665 (1972).
(9) Spiker, Andrews, JCP 58, 702 (1973).
(10) Veseth, JMS 59, 51 (1976).

LiRb: [a]Only five bands, four of which have $\Delta v=0$.
[b]Dipole moment of $^7Li^{85}Rb$: $\mu_{e\ell}(v=0)$ = 4.0_0 D, calculated (3) with B_0 = 0.218 estimated from Badger's rule [see, however, (4)]. Quadrupole coupling constants (3).

(1) Walter, Barratt, PRS A 119, 257 (1928).
(2) Weizel, Kulp, AP(Leipzig) 4, 971 (1930).
(3) Dagdigian, Wharton, JCP 57, 1487 (1972).
(4) Cavaliere, Ferrante, Lo Cascio, JCP 62, 4753 (1975).

LiXe, $LiXe^+$:

[a]From high-resolution differential scattering cross sections (3). See also (2).
[b]Semiempirical calculation of (1).
[c]Morse potential parameters (average of two suggested values) derived from the far-wing fluorescence spectrum longward of the Li resonance line at 14904 cm^{-1} (5). See also (6).
[d]See the potential curves in (2)(5).
[e]From Li^+-on-Xe differential scattering cross sections (4).

(1) See ref. (1) of LiNe.
(2) See ref. (2) of LiNe.
(3) Auerbach, JCP 60, 4116 (1974).
(4) Böttner, Dimpfl, Ross, Toennies, CPL 32, 197 (1975).
(5) Scheps, Ottinger, York, Gallagher, JCP 63, 2581 (1975).
(6) Scheps, Gallagher, JCP 65, 859 (1976).

State	T_e	ω_e	$\omega_e x_e$	B_e	α_e	D_e $(10^{-6} cm^{-1})$	r_e (Å)	Observed Transitions Design.	Observed Transitions ν_{00}	References
$^{175}Lu^{19}F$		μ = 17.1373091		D_0^0 [a]						MAR 1976 A
G $^1\Sigma$	33225.9	599.1 H	2.6					G → X,	33220.1 H	(2)
F $^1\Sigma$	25831.8	[555.59] Z	2.6 H	0.25815	0.00169	0.210	1.9520	F → X, R	25806.28 Z	(2)(3)
E $^1\Pi$	24474.10	543.42 Z	2.28	0.25647[b]	0.00161	0.227	1.9584	E → X, R	24439.98 Z	(3)
D $^1\Pi$	20047.8	569.7 H	2.5	[0.2592][c]		[0.22]	[1.948]	D → X, R	20027.33 Z	(2)(3)
C	(18894.1)	605.5 H	2.5					C → (X),	18891.5 H	(2)
B $^1\Pi$	16799.9	[576.08] Z	2.5 H	[0.26241][d]	0.0016	[0.203]	[1.9361]	B → X, R	16784.52 Z	(2)(3)*
A $^1\Sigma$	16164.6_6	587.9_5 Z	2.58	0.26356	0.00162	0.208	1.9319	A → X, R	16152.73 Z	(3)*
X $^1\Sigma^{(+)}$	0	611.79 Z	2.54	0.26764	0.00156	0.204	1.9171			
$^{175}Lu^{1}H$		μ = 1.00205245								MAR 1975
H $^1\Pi$		a		[4.5548] b [4.5365] c	(0.131)[d] 0.135_4	[179] [177] [e]	[1.923_8]	H → X,	23525.00 Z	(1)* (2)
G $^1\Sigma$		f		4.546_0	0.120_6[g]	178	1.9237	G → X, R	19767.00 Z	(1)* (2)
F $^1\Pi$				[4.7025][h] c [4.649_6][h] b		[213] [171] [i]	[1.897]	F → X, V	18921.49 Z	(2)
E $^1\Sigma$				[4.627_5][h]		[200]	[1.9067]	E → X,	17732.92 Z	(2)
D $^1\Pi$		j		4.534[k] c 4.518 b	0.131 0.137	230[k] [217][ℓ]	1.928_0	D → X,[m] R	17050.1 Z	(2)
C $^1\Pi$		n		[4.230_6][o] c [4.196_5] b	 0.102_7	[136][o] [118][p]	[1.998]	C → X, R	16721.9 Z	(2)
B $^1\Sigma$		q		4.572_3	0.117_1	195[r]	1.9182	B → X,	15270.00 Z	(2)
A $^1\Pi$		s		[4.862_3] b [4.467_4] c	 0.114_9	[690][t] [195][u]	[1.90]	A → X,	12988.63 Z	(2)
X $^1\Sigma^{(+)}$	0	(1520)	(22)	4.6021	0.099_0	169[v]	1.9119			

LuF: [a]The dissociation energy was estimated (1) at 5.9 eV.

[b]Λ-type doubling, $\Delta\nu_{fe} = (+)0.00030 \times J(J+1)$.

[c]Λ-type doubling, $\Delta\nu_{fe} = (-)0.0032 \times J(J+1)$.

[d]Λ-type doubling, $\Delta\nu_{fe}(v=0) = (-)0.00060 \times J(J+1)$.

(1) Zmbov, Margrave, AdC No. 72, 267 (1968).

(2) D'Incan, Effantin, Bacis, JP B 5, L189 (1972).

(3) Effantin, Wannous, D'Incan, Athénour, CJP 54, 279 (1976).

Lu^1H: [a]$\nu_0(1-1) = 23473$, $\nu_0(2-2) = 23389$.

[b]e levels.

[c]f levels.

[d]Perturbation in v=1.

[e]$D_1 = 183 \times 10^{-6}$; $H_0 = +0.2_5 \times 10^{-8}$.

[f]$\nu_0(1-1) = 19696$.

[g]v=1 perturbed.

[h]Perturbations in v=0 and 1; constants for v=1 could not be determined.

[i]$H_0 = -3 \times 10^{-8}$.

[j]$\nu_0(1-1) = 16918.6$.

[k]Perturbation in v=0. $H_0 = +4 \times 10^{-8}$, $H_1 = +7._6 \times 10^{-8}$.

[ℓ]$D_1 = 200 \times 10^{-6}$; $H_0 = +1.4 \times 10^{-8}$.

[m]Very weak system, $J' < 7$ not observed.

[n]$\nu_0(1-1) = 16591.0$.

[o]Perturbations in v=0 and 1. $H_0 = -0.9 \times 10^{-8}$. Constants for v=1 could not be determined.

[p]$D_1 = 126 \times 10^{-6}$.

[q]$\nu_0(1-1) = 15173.17$.

[r]$H_0 = +0.9 \times 10^{-8}$.

[s]$\nu_0(1-1) = 12904.32$, $\nu_0(2-2) = 12810.00$.

[t]$H_0 = +5._0 \times 10^{-7}$.

[u]$D_1 = 194 \times 10^{-6}$; $H_0 = -0.5 \times 10^{-8}$, $H_1 = -2.4 \times 10^{-8}$.

[v]$H_e = +0.5 \times 10^{-8}$.

(1) D'Incan, Effantin, Bacis, CJP 50, 1810 (1972).

(2) Effantin, D'Incan, CJP 51, 1394 (1973).

State	T_e	ω_e	$\omega_e x_e$	B_e	α_e	D_e ($10^{-6}cm^{-1}$)	r_e (Å)	Observed Transitions Design.	Observed Transitions ν_{00}	References
$^{175}Lu^{2}H$		$\mu = 1.99117769$		$D_0^0 = (3.4)$ eV						MAR 1975
H $^1\Pi$	23545.9	1058.6 Z	17.4	2.323_6[a]	0.048_4	44.4[b]	1.908_8	H→X,	23535.73 Z	(1)(2)
G $^1\Sigma$	19815.5	1023.6 Z	11.3[c]	2.272_5	0.0362_2[c]	42.6[c]	1.930_2	G→X, R	19789.28 Z	(1)(2)
F $^1\Pi$	(18922)	$[1072.5_3]$ Z		2.373[d]	0.036	[110][d]	1.889	F→X, V	18923.26 Z	(1)(2)
E $^1\Sigma$	17744	$[1033.2_6]$ Z	$(15._5)$[e]	2.369_0	0.049_3[e]	$46._4$[f]	1.890_4	E→X,	17737.05 Z	(1)(2)
D $(^1\Pi)$								D→X,[g]	17129.1 Z	(1)
C $^1\Pi$	16778.2	997.2 Z	16.2[h]	2.205[i]	0.038[h]	46.3[h]	1.959	C→X, R	16737.64 Z	(1)(2)
B $^1\Sigma$	$15306._3$	$[988.5_4]$ Z	$(12._1)$	2.268_1	0.040_0	$44._4$[j]	1.932_0	B→X, R	15274.59 Z	(1)(2)
A $^1\Pi$								A→X,	13039.9 Z	(1)
X $^1\Sigma^{(+)}$	0	1075.3_5 Z	10.3[k]	2.320_3	0.0356	43.3[k]	1.910_2			
$^{175}Lu^{16}O$		$\mu = 14.6550016_8$		$D_0^0 = 7.1_9$ eV[a]						MAR 1975 A
		Two short continua in the red.								(3)
C $^2\Sigma^+$	24440	(770) H	(5)	[0.34411][b]		[0.297]	$[1.8283_4]$	C→X, R	24402.90 Z	(3)(8)(10)*
B $(^2\Pi_{3/2})$	21470	$793._0$[c] H	4.0	[0.3528][d]			$[1.805_7]$	B→X, R	21445 H	(1)(2)*(4)* (7)(9)
A $(^2\Pi_{1/2})$	(19392)[e]	(800) H	(5)					A→X, R	(19370)[e] H	(2)* (3) (4)* (9)
X $^2\Sigma^+$	0	$842._5$[c] H	3.1_0	[0.35806][f]	0.0016	[0.255]	1.7904			
$^{(175)}Lu^{(195)}Pt$		($\mu = 92.205411$)		$D_0^0 = 4.1$ eV[a]						MAR 1975
$^{(175)}Lu^{(32)}S$		($\mu = 27.0317641$)		$D_0^0 = 5.1_9$ eV[a]						MAR 1975
$^{(175)}Lu^{(80)}Se$		($\mu = 54.856813_5$)		$D_0^0 = 4.2_9$ eV[a]						MAR 1975
$^{(175)}Lu^{(130)}Te$		($\mu = 74.548536$)		$D_0^0 = 3.3_3$ eV[a]						MAR 1975

Lu^2H: [a]Λ-type doubling $\Delta\nu_{ef} = + 0.0052 \times J(J+1) - \ldots$.
[b]$\beta_e = + 1.4 \times 10^{-6}$; $H_0 = + 3.0 \times 10^{-10}$.
[c]$\omega_e y_e = - 0.51$; $\gamma_e = - 0.0018$; $\beta_e = + 1.3 \times 10^{-6}$; $H_0 \approx + 2 \times 10^{-10}$. Perturbations by E $^1\Sigma$; constants in the Table are effective values.
[d]Λ-type doubling $\Delta\nu_{fe}(v=1) = + 0.022 \times J(J+1) - \ldots$. $D_1 = 45 \times 10^{-6}$. Strong perturbations.
[e]$v \geq 2$ perturbed by G $^1\Sigma$. Deperturbed constants for v=2 in (2).
[f]$\beta_e = + 1._6 \times 10^{-6}$.
[g]Identification uncertain.
[h]$\omega_e y_e = + 0.21$, $\gamma_e \approx - 0.0011$, $\beta_e = + 0.6_3 \times 10^{-6}$; $H_0 = + 7.2 \times 10^{-10}$.
[i]Λ-type doubling $\Delta\nu_{fe} = + 0.0100 \times J(J+1) - \ldots$.
[j]$\beta_e = + 0.4 \times 10^{-6}$.
[k]$\omega_e y_e = + 0.03_4$, $\beta_e = - 0.6 \times 10^{-6}$; $H_e = + 6.0 \times 10^{-10}$.

(1) D'Incan, Effantin, Bacis, JP B 5, L187 (1972).
(2) Effantin, d'Incan, CJP 52, 523 (1974).

LuO: [a]Thermochemical value (mass-spectrom.)(5), recalc. (6).
[b]Large spin doubling, $\gamma_0 = (-)0.4940$. v=0 perturbed for $N > 30$.
[c]Recalculated using bandheads of the B – X 0-0 and 1-0 sequences only.
[d]B_v appears to depend irregularly on v; see (9).
[e]From (9). The vibrational numbering in (3) differs by -1 in both upper and lower state leading to $\nu_{00} \approx 19332$.
[f]Large hyperfine splitting, $4b = 0.663$ cm^{-1} (10).

LuO (continued):

(1) Watson, Meggers, JRNBS 20, 125 (1938).
(2) Gatterer, RS 1, 153 (1942).
(3) Gatterer, Krishnamurty, PPS A 65, 151 (1952).
(4) Gatterer, Junkes, Salpeter, Rosen, METOX (1957).
(5) Ames, Walsh, White, JPC 71, 2707 (1967).
(6) Smoes, Coppens, Bergman, Drowart, TFS 65, 682 (1969).
(7) Suárez, JP B 3, 1389 (1970).
(8) Bacis, Bernard, d'Incan, CR B 273, 272 (1971).
(9) Effantin, Bacis, d'Incan, CR B 273, 605 (1971).
(10) Bacis, Bernard, CJP 51, 648 (1973).

LuPt: [a]Thermochemical value (mass-spectrom.)(1).

(1) Gingerich, HTS 3, 415 (1971).

LuS: [a]Thermochemical value (mass-spectrom.)(1); recalculated (2).

(1) See ref. (6) of LuO.
(2) Bergman, Coppens, Drowart, Smoes, TFS 66, 800 (1970).

LuSe, LuTe:

[a]Thermochemical value (mass-spectrom.)(1).

(1) See ref. (2) of LuS.

State	T_e	ω_e	$\omega_e x_e$	B_e	α_e	D_e (10^{-6} cm^{-1})	r_e (Å)	Observed Transitions: Design.	Observed Transitions: ν_{00}	References
$^{24}Mg_2$		$\mu = 11.9925222$		$D_0^0 = 0.0501_0$ eV[a]						MAR 1977 A
C $(^1\Sigma_u^+)$				$[0.135_4]$[b]		$[0.4]$[b]	$[3.22_2]$[b]	C← X,	V $[38048.8]$[b] Z	(4)
B $(^1\Pi_u)$		Continuous absorption and diffuse bands, 35050 - 37600 cm^{-1}, tentatively assigned to the $^1\Pi_u$ state arising from $^1S + {}^1P$.						B← X,		(2)(3)(4)
A $^1\Sigma_u^+$	26068.76	190.61_5 Z	1.1456[c]	0.14799_9	0.001316_4[d]	0.334[e]	3.0818_6	A↔X,[fg]	V 26138.63 Z	(1)(2)* (4)*
X $^1\Sigma_g^+$	0	51.12_1 Z	1.645[h]	0.09287	0.00377_6[i]	1.21_7[j]	3.890_5	kℓ		
$^{24}Mg^{(79)}Br$		$(\mu = 18.3945343)$		$D_0^0 \leq 3.35$ eV[a]						MAR 1977
		R shaded bands in the region 36800 - 39400 cm^{-1} have been variously attributed to a $^2\Pi \leftarrow X^2\Sigma$ transition (3) or to a $^2\Sigma \rightarrow X^2\Sigma$ system (5); $\omega_e' \approx 223$ or 272, respectively.								
A $^2\Pi$	25877 25766.9	393.9 H	2.04	$[(0.1685)]$[bc]			$[(2.332)]$	A↔X,	V 25887 H 25776.8 H	(1)(2)(3)(6)
X $^2\Sigma^+$	0	373.8 H	1.34	$[(0.1645)]$[b]			$[(2.360)]$			
$^{24}Mg^{35}Cl$		$\mu = 14.2268710$		$D_0^0 = 3.29$ eV[a]						MAR 1977
		Fragments of four band systems at 12700 (1), 25900 (2), 37060 and 40850 cm^{-1} [(4), abs.].								
B $^2\Sigma$	47630	$[552.0]$[b] H						B→A,	V 21136.8 H 21191.9[b] H	(6)
A $^2\Pi$ (1/2)	26523.4	491.6 H	2.54[c]	$[0.25116]$	(0.0018)	0.22_5	$[2.1720]$	A↔X,[d]	V 26535.89 Z	(3)(5)
(3/2)	$26469._4$			$[0.24914]$[e]	(0.0018)	0.22_5	$[2.1808]$		26481.95 Z	
X $^2\Sigma^+$	0	$[462.12]$ Z	2.10	0.24502	0.0015_8	0.25_2	2.1991			
$^{(24)}Mg^{133}Cs$?		$(\mu = 20.318267_7)$								APR 1978
		One diffuse and one V shaded band at 17520 and 20659 cm^{-1}, respectively.								(1)

Mg_2: [a]From the convergence of the vibrational levels v=0...12 of the ground state (4) including the slight extension (v=13) and correction by (9).

[b]Lowest observed level, vibrational numbering uncertain.

[c]$\omega_e y_e = +0.001772$. Vibrational numbering established by the observed $^{24}Mg_2 - ^{26}Mg_2$ isotope shifts (4).

[d]$\gamma_e = +0.147_4 \times 10^{-5}$.

[e]$H_e = +0.23 \times 10^{-11}$.

[f]Franck-Condon factors and their dependence on rotation (8).

[g]Observed in Kr and Xe matrices by (6) and in Ne, Ar, N_2 matrices by (11). The same authors also observe other absorptions probably related to B − X and C − X in the gas.

[h]$\omega_e y_e = +0.0162_4$, $\omega_e z_e = -0.00061_1$.

[i]$+0.000106_8(v+\frac{1}{2})^2 - 0.0000096_8(v+\frac{1}{2})^3$.

[j]$-0.031 \times 10^{-6}(v+\frac{1}{2}) + 0.02_1 \times 10^{-6}(v+\frac{1}{2})^2$;
$H_v = -0.26 \times 10^{-10} - 0.08_5 \times 10^{-10}(v+\frac{1}{2})$.

[k]Term values of ro-vibrational levels, RKR potential functions (4)(9). For other potential functions see (7)(10).

[ℓ]Long-range potential studied by (5)(9).

(1) Hamada, PM (7) 12, 50 (1931).
(2) Weniger, JP(Paris) 25, 946 (1964).
(3) Edelhoff, Kusch, Lochte-Holtgreven, RRP 13, 125 (1968).
(4) Balfour, Douglas, CJP 48, 901 (1970).
(5) Stwalley, JCP 54, 4517 (1971).
(6) Brewer, Wang, JMS 40, 95 (1971).
(7) Brett, Chan, CJP 50, 1587 (1972).
(8) Balfour, Whitlock, CJP 50, 1648 (1972).
(9) Li, Stwalley, JCP 59, 4423 (1973).
(10) Muhlhausen, Konowalow, CP 7, 143 (1975).
(11) Knight, Ebener, JMS 61, 412 (1976).

MgBr: [a]From the predissociation in A $^2\Pi$.

[b]Partial rotational analysis of the 0-0 band (4).

[c]Predissociation above v=3.

(1) Olmsted, ZWP 4, 293 (1906).
(2) Morgan, PR 50, 603 (1936).
(3) Harrington, Dissertation (U. of California, 1942).
(4) Patel, Patel, JP B 2, 515 (1969).
(5) Reddy, Rao, CS 39 (22), 509 (1970).
(6) Puri, Mohan, CS 42, 442 (1974).

MgCℓ: [a]Thermochemical value (mass-spectrom.); average of three reactions (8)(9).

[b]No clear identification of the various isotopic band heads was possible.

[c]$\omega_e y_e = -0.025$.

[d]The assignment of the emission band heads by (2) is not correct.

[e]From (5). Somewhat different results in (7).

(1) Querbach, ZP 60, 109 (1930).
(2) Parker, PR 47, 349 (1935).
(3) Morgan, PR 50, 603 (1936).
(4) Harrington, Dissertation (U. of California, 1942).
(5) Morgan, Barrow, Nature 192, 1182 (1961).
(6) Rao, Rao, IJP 37, 640 (1963).
(7) Patel, Patel, IJP 42, 254 (1968).
(8) Hildenbrand, JCP 52, 5751 (1970).
(9) Farber, Srivastava, CPL 42, 567 (1976).

MgCs: (1) Barratt, PRS A 109, 194 (1925).

State	T_e	ω_e	$\omega_e x_e$	B_e	α_e	D_e ($10^{-4}cm^{-1}$)	r_e (Å)	Observed Transitions: Design.	Observed Transitions: ν_{00}	References
$^{24}Mg^{19}F$		μ = 10.6012334		D_0^0 = 4.75 eV[a]						MAR 1977
		(V shaded emission bands in the region 1800 - 1900 Å, attributed to MgF^+ by (15), are due to the B→X transition of AℓF.)								
G ($^2\Sigma$)	74304.2	800.0 H	6.18	Four 0-0 sequences; no details.				G↔X, (V)	74343	(15)
F ($^2\Sigma$)	57067.2	756.6 H	4.06					F↔X, (V)	57085	(15)
E ($^2\Pi$)	55694.9	775.9 H	4.00					E↔X, (V)	55722	(15)
D ($^2\Pi$)	54263.6	792.3 H	4.38					D↔X, (V)	54299	(15)
C $^2\Sigma^+$	42538.9	[813.10] Z	(5.04)	0.55102	0.00449	[0.00991][b]	1.6988	C←X,[c] V	42589.64 Z	(3)* (6)
B $^2\Sigma^+$	37167.3	[750.94] Z	(5.60)	0.53844	0.00510	[0.01083][d]	1.7185	B↔X,[e] V	37187.45 Z	(1)(2)* (3)* (6)
A $^2\Pi_r$	27851.$_2$ [f] 27816.1	746.0 H [740.12] Z	3.97	0.52105[g]	0.00327	[0.0038][h]	1.7469	A↔X,[i] V	27863.$_7$ Z 27829.60 Z	(2)* (3)* (6)
X $^2\Sigma^+$	0	[711.69] Z	(4.94)	0.51922	0.00470	[0.01080][j]	1.7500	ESR sp.[k]		(14)
$^{24}Mg^1H$		μ = 0.96718516		D_0^0 = 1.34 eV[a]						MAR 1977
		Unidentified line structure at 42570 cm^{-1} [em. and abs., (2)(3)(8)(10)] and 43180 - 43520 cm^{-1} [abs., (12)].								
I ($^2\Pi$)				[5.96]		[2]	[1.71$_0$]	I←X, V	47997.$_6$ Z	(12)
H ($^2\Sigma^+$)		R shaded band with head at 2100 Å.						H←X, R	47543 (Z)	(12)
G ($^2\Sigma^+$)		V shaded band with head at 2172 Å.						G←X, V	46082 (Z)	(12)
D $^2\Sigma^-$	(42065)	(1630)[b]		[6.296][c]		[3.8]	[1.664]	D→A, V	22861.$_5$[d]	(6)(10)
C $^2\Pi_{(r)}$	(41164)[e]	[1623.4] Z		6.161[f]	0.144	3.0	1.682	C→A,	21956.$_5$[d]	(1)(2)(7)(10)
								C↔X, V	41235.$_9$ Z	(1)* (3)(4) (7)(8)(10) (17)*
E $^2\Sigma^+$	(35568)	[1444.8] Z	[g]	6.23	0.30[h]	[3.3][i]	1.67$_3$	E←X, V_R	35550.6 Z	(11)(13)* (20)
B' $^2\Sigma^+$	22410.$_5$	828.$_0$ Z	11.6[j]	[2.596]	[k]	1.2[ℓ]	2.597	B'→X, R	22081.$_9$ Z	(20)(22)

MgF: [a]Thermochemical value (mass-spectrom.)(7); similar results can be obtained from (4) and (5) with the use of appropriately revised auxiliary data.

[b]$D_1 = 0.00984 \times 10^{-4}$.

[c]Franck-Condon factors (8)(11).

[d]$D_1 = 0.01085 \times 10^{-4}$.

[e]Franck-Condon factors (8)(12).

[f]$A = +37$ (6)(13). The regular character of this state is suggested by the theoretical calculations of (9) (13) and supported by the analysis of the ESR sp.(14).

[g]For Λ-type doubling constants see (6) and (13).

[h]$D_1 = 0.0078 \times 10^{-4}$.

[i]Morse potential Franck-Condon factors (10).

[j]$D_1 = 0.01075 \times 10^{-4}$.

[k]In Ne and Ar matrices at 4K.

(1) Jevons, PRS A 122, 211 (1929).
(2) Jenkins, Grinfeld, PR 45, 229 (1934).
(3) Fowler, PR 59, 645 (1941).
(4) Ehlert, Blue, Green, Margrave, JCP 41, 2250 (1964).
(5) Murad, Hildenbrand, Main, JCP 45, 263 (1966).
(6) Barrow, Beale, PPS 91, 483 (1967).
(7) Hildenbrand, JCP 48, 3657 (1968).
(8) Maheshwari, Singh, Shukla, JP B 1, 993 (1968).
(9) Walker, Richards, JP B 1, 1061 (1968).
(10) Singh, Shukla, Maheshwari, JQSRT 9, 533 (1969).
(11) Rao, Lakshman, JQSRT 10, 945 (1970).
(12) Rao, Lakshman, Physica 46, 609 (1970).
(13) Walker, Richards, JP B 3, 271 (1970).
(14) Knight, Easley, Weltner, Wilson, JCP 54, 322 (1971).
(15) Novikov, Gurvich, JAS 14, 820 (1971).

Mg^1H: [a]From extrapolations of the vibrational levels in X, A, and B'. Close agreement with the theoretical value of (21), $D_0 = 1.25$ eV.

[b]Estimated from $\omega^2 = 4B^3/D$ and from the observed isotope shift of the 0-0 band.

[c]Spin doubling, $|\gamma| = 0.01$.

[d]Average of the two subband origins (10).

[e]$A_0 = (+)3.7$; (17) gives -3.7 but his arguments are not conclusive.

[f]The P and R branches of the $C \rightarrow X$, 0-0 band break off above N'=10; bands with v' > 0 have Q branches only. The predissociation is caused by B' $^2\Sigma^+$; see (20).

[g]$\Delta G(3/2) = 1490$. The v=2 level was formerly believed to be v=0 of a new $^2\Sigma$ state (B $^2\Sigma^+$), see (4)(5)(9)(10). Similarly, (11) assigned the 1-0 band as 0-0 band of a new system $F \leftarrow X$.

[h]B_2(see [g]) = 5.44_8. Transitions to v=1 are discrete but those to v=2 are diffuse indicating that the state causing the predissociation [B' $^2\Sigma^+$, see (20)] crosses the E state potential curve between v=1 and 2. Emission from v=2 has been observed at high pressure.

[i]$D_1 = 4.0 \times 10^{-4}$; $D_2 = 2.7_1 \times 10^{-4}$, $H_2 = +8._6 \times 10^{-8}$ (see [g]).

[j]The equilibrium constants in (22) do not reproduce the observed intervals, probably owing to an error in $\omega_e z_e$. $D_e \approx 10900$ cm^{-1} (22).

[k]$B_v(v=1,2,3,4...9) = 2.605, 2.618, 2.608, 2.592 ... 2.419$; $B_e = 2.585$, but the higher order constants of (22) do not reproduce their data. Small perturbations by A $^2\Pi$. RKR potential curve (22). See also (24).

[l]$\beta_e \approx +0.13 \times 10^{-4}$.

(continued on p. 398)

State	T_e	ω_e	$\omega_e x_e$	B_e	α_e	D_e $(10^{-4} cm^{-1})$	r_e (Å)	Observed Transitions Design.	Observed Transitions ν_{00}	References
$^{24}Mg^1H$ (continued)										
A $^2\Pi_r$	$19226._8$[m]	1598.17 Z	31.08_5[n]	6.1913	0.1931[o]	3.60[p]	1.6778	A ↔ X,[qr] V	$19278._4$[d]	(6)(10)(16) (17)(23)
X $^2\Sigma^+$	0	1495.20 Z	31.889[s]	5.8257	0.1859[t]	3.44[p]	1.7297	ESR sp.[u]		(19)
$^{24}Mg^2H$		μ = 1.85807372		D_0^0 = 1.36 eV[a]						MAR 1977
G $^2\Sigma^+$				[3.26][b]		[0.8]	$[1.66_8]$	G → X, V	46071.4 Z	(6)
D $^2\Sigma^-$	(42080)	(1180)[c]		$[3.27_3]$		[1.0]	[1.665]	D → A, V	$22860._0$ Z	(4)
C $^2\Pi_{(r)}$	(41180)	[1140.1] Z		3.256[de]	0.073	0.92	1.669	C → A, V	21940[f]	(4)
								C ↔ X, V	41214.2[g] Z	(1)(3)(4)(8)
E $^2\Sigma^+$		h		[3.20]	h		$[1.68_4]$	E ← X, V_R	35549.0 Z	(5)*
B' $^2\Sigma^+$	$22410._3$	$597._5$ Z	6.0_5[i]	[1.350]	j	0.25[k]	2.596	B' → X,[ℓ] R	$22172._7$ Z	(10)
A $^2\Pi_r$	$19235._5$[m]	1154.75 Z	16.67_5[n]	3.2190	0.0679_5[o]	0.964[p]	1.6788	A ↔ X, V	19273.7_1[g] Z	(2)* (8)(9)
X $^2\Sigma^+$	0	1077.8_9 Z	16.1_2[q]	3.0306	0.06289[r]	0.92_1[s]	1.7302			

Mg^1H (continued):

[m]A_0 = + 35.3 (17).

[n]$\omega_e y_e$ = - 0.633. These constants (for v ≤ 3) have been recalculated from the three ΔG values obtained by (23); the equilibrium constants determined by (23) do not reproduce their data. $D_e \approx 14200$ cm^{-1} (23).

[o]γ_e = + 0.0005. Small perturbations by B' $^2\Sigma^+$.

[p]D_v increases rapidly with v.

[q]Oscillator strength f_{00} = 0.257 [see (23) and references given there], much larger than the earlier experimental value of (14).

[r](16) has measured the 0-0 and 0-1 bands of ^{25}MgH and ^{26}MgH; see also (15)(18).

[s]$\omega_e y_e$ = + 0.384, $\omega_e z_e$ = - 0.113 (23); $D_e^0 \approx 11700$ cm^{-1} (23).

[t]$+ 0.0039_3(v+\frac{1}{2})^2 - 0.0012_3(v+\frac{1}{2})^3$. The last term was not reported by (23) but must be included for a satisfactory fit to their data (v=0...6).

[u]In Ar matrices at 4K.

(1) Pearse, PRS A 122, 442 (1929).

(2) Guntsch, ZP 87, 312 (1934).

Mg^1H (continued):

(3) Guntsch, ZP 93, 534 (1935).
(4) Grundström, Dissertation (Stockholm, 1936).
(5) Grundström, Nature 137, 108 (1936).
(6) Guntsch, ZP 104, 584 (1937).
(7) Guntsch, ZP 107, 420 (1937).
(8) Turner, Harris, PR 52, 626 (1937).
(9) Guntsch, ZP 110, 549 (1938).
(10) Guntsch, Dissertation (Stockholm, 1939).
(11) Khan, PPS 77, 1133 (1961).
(12) Khan, PPS 80, 209 (1962).
(13) Khan, PPS 82, 572 (1963).
(14) Main, Carlson, DuPuis, JQSRT 7, 805 (1967).
(15) Branch, ApJ 159, 39 (1970).
(16) Balfour, ApJ 162, 1031 (1970).
(17) Balfour, JP B 3, 1749 (1970).
(18) Boyer, AA 12, 464 (1971).
(19) Knight, Weltner, JCP 54, 3875 (1971).
(20) Balfour, Cartwright, CPL 32, 82 (1975).
(21) Meyer, Rosmus, JCP 63, 2356 (1975).
(22) Balfour, Cartwright, CJP 54, 1898 (1976).
(23) Balfour, Cartwright, AA(Suppl.) 26, 389 (1976).
(24) Sink, Bandrauk, Henneker, Lefebvre-Brion, Raseev, CPL 39, 505 (1976).

Mg^2H: [a]From the value for Mg^1H, in good agreement with an extrapolation of the ground state vibrational levels.

[b]R branch breaks off at N'=12.

[c]See [b] of Mg^1H.

[d]$B(R,P) - B(Q) \approx + 0.008$.

[e](4) reports breaking-off of P and R branches in the $C \rightarrow X$, 0-0 band above N'=14. No line broadening is observed in the absorption spectrum (8).

Mg^2H (continued):

[f]See [d] of Mg^1H.

[g]Refers in the upper state to the zero point of the Hill-Van Vleck equation.

[h]Fragments of what is probably the $E \leftarrow X$, 1-0 band have been found (7) between 36690 and 36930 cm^{-1}. Two additional levels, formerly attributed to a separate $^2\Sigma^+$ state ($B\ ^2\Sigma$), have been identified at 39593.8 cm^{-1} [in emission (4) and absorption (7), $B_v = 2.862$] and at 38330.3 cm^{-1} [in absorption (7), $B_{v-1} = 2.915$].

[i]$D_e \approx 11200\ cm^{-1}$ (10).

[j]$B_v(v=1,2,4,5,6) = 1.355, 1.359, 1.358, 1.354, 1.346$; $B_e = 1.346$. No emission from v'=3 has been found. Small perturbations by $A\ ^2\Pi$.

[k]$\beta_e \approx + 0.04_3 \times 10^{-4}$.

[l]RKR Franck-Condon factors (10).

[m]$A_0 = + 35.0_4$, $A_1 = 36._1$, $A_2 = 36._7$.

[n]$\omega_e y_e = - 0.183_3$; $D_e \approx 15500\ cm^{-1}$ (9).

[o]$\gamma_e = - 0.00122_5$. Small perturbations by $B'\ ^2\Sigma$.

[p]$\beta_e = + 0.008_9 \times 10^{-4}$.

[q]$\omega_e y_e = + 0.08_5$, $\omega_e z_e = - 0.029_3$; 4th-order fit to the ΔG values of (9) whose equilibrium constants do not satisfactorily reproduce their data. $D_e^0 \approx 11500\ cm^{-1}$ (9).

[r]$\gamma_e = - 0.00144$.

[s]$+ 0.012 \times 10^{-4}(v+\frac{1}{2}) + \ldots$

(1) See ref. (3) of Mg^1H.
(2) Fujioka, Tanaka, Sci. Pap. IPCR (Tokyo) 30, 121 (1936).
(3) See ref. (8) of Mg^1H.
(4)...(7) See ref. (10)...(13) of Mg^1H, respectively.
(8) See ref. (17) of Mg^1H.
(9) Balfour, Cartwright, CJP 53, 1477 (1975).
(10) See ref. (22) of Mg^1H.

State	T_e	ω_e	$\omega_e x_e$	B_e	α_e	D_e (10^{-4}cm^{-1})	r_e (Å)	Observed Transitions: Design.	Observed Transitions: ν_{00}	References
$^{24}Mg^{1}H^{+}$				D_0^0 = (2.08) eV[a]						MAR 1977
B $^1\Pi$	50476	[527.3] Z		[3.380][b]	0.283	[4.7][c]	[2.271]	B→X,[d] R	49898.6 Z	(3)(5)(8)
A $^1\Sigma^+$	35904.52	1135.8 Z	8.18_1[e]	4.3295	0.06803_4[f]	2.510[g]	2.0064	A→X,[dh] R	35628.81 Z	(1)(2)(5)(7)(8)*
X $^1\Sigma^+$	0	1699.1 Z	31.93_5[i]	6.3870	0.1819_4[j]	3.664[k]	1.6519			
$^{(24)}Mg^{2}H^{+}$				D_0^0 = (2.1_1) eV[ℓ]						MAR 1977
A $^1\Sigma^+$	35902.6	817.0 Z	3.47[m]	2.252	0.024_3[n]	0.64	2.007	A→X, R	35701.0 Z	(4)(6)
X $^1\Sigma^+$	0	1226.6 Z	16.30[o]	3.324	0.066_7[p]	0.96	1.652			
$^{(24)}Mg^{127}I$		(μ = 20.1724379)		D_0^0 = (2.9_2) eV[a]						MAR 1977
D	26680	295 H	0.5					D→X,	26670 H	(3)
C	26099	270 H	0.72					C→X,	26076 H	(3)
B	25612	295 H						B↔X R	25602 H	(1)(3)
A	24319	323 H	1.0					A↔X, V	24322 H	(1)(3)
X ($^2\Sigma$)	0	316 H	0.50							
$^{(24)}Mg^{(39)}K$		(μ = 14.8461445)		One R shaded and two V shaded absorption bands at 15264 and 19411, 21678 cm^{-1}.						APR 1978 (1)*
$^{(24)}Mg^{23}Na$		(μ = 11.7384318)		V shaded absorption band at 18895 cm^{-1}.						
$^{(24)}Mg^{(85)}Rb$		(μ = 18.7022250)		V shaded absorption band at 21144 cm^{-1}.						

Mg^1H^+, Mg^2H^+:

[a]Extrapolation of vibrational levels in X $^1\Sigma$ and A $^1\Sigma$.

[b]$B(R,P) - B(Q) = + 0.004$.

[c]$H_0 = - 8 \times 10^{-8}$.

[d]The spectra of the isotopes $^{25}Mg^1H^+$ and $^{26}Mg^1H^+$ have also been analyzed (8).

[e]$\omega_e y_e = - 0.147_9$, $\omega_e z_e = - 0.0078_8$; $D_e \approx 18000$ cm^{-1}.

[f]$+ 0.0000039_8(v+\frac{1}{2})^2 - 0.0001208(v+\frac{1}{2})^3$. RKR potential curve (8).

[g]$H_e = + 1.82 \times 10^{-8}$; higher order constants in (8).

[h]RKR Franck-Condon factors (8).

[i]$\omega_e y_e = - 0.188_2$, $\omega_e z_e = - 0.0204_4$; $D_e^0 \approx 17000$ cm^{-1}.

[j]$+ 0.00020_2(v+\frac{1}{2})^2 - 0.00062_9(v+\frac{1}{2})^3 + 0.000045_7(v+\frac{1}{2})^4 - 0.0000023_5(v+\frac{1}{2})^5$. RKR potential curve (8).

[k]$H_e = + 2.07 \times 10^{-8}$; higher order terms in (8).

[ℓ]From the value for Mg^1H^+.

[m]$\omega_e y_e = - 0.11_7$.

[n]$\gamma_e = - 0.0003$; from the average B_v values of (4)(6).

[o]$\omega_e y_e = - 0.16_7$.

[p]$\gamma_e = - 0.0009$; see [n].

(1) Pearse, PRS A 125, 157 (1929).
(2) Guntsch, ZP 87, 312 (1934).
(3) Guntsch, ZP 107, 420 (1937).
(4) Juraszynska, Szulc, APP 7, 49 (1938).
(5) Guntsch, Dissertation (Stockholm, 1939).
(6) Guntsch, AMAF A 31, No. 22 (1945).
(7) Pillow, PPS A 62, 237 (1949).
(8) Balfour, CJP 50, 1082 (1972).

MgI: [a]From an ill-defined threshold for dissociative photoionization, and from the heat of atomization of MgI_2, using an assumed value for the ionization potential of MgI (2).

(1) Morgan, PR 50, 603 (1936).
(2) Berkowitz, Chupka, JCP 45, 1287 (1966).
(3) Puri, Mohan, Pramāṇa 4, 171 (1975).

MgK, MgNa, MgRb:

(1) Barratt, PRS A 109, 194 (1925).

State	T_e	ω_e	$\omega_e x_e$	B_e	α_e	D_e ($10^{-6}cm^{-1}$)	r_e (Å)	Observed Transitions Design.	ν_{00}	References
$^{24}Mg^{16}O$		$\mu = 9.5957762_8$		$D_0^0 = (3.5_3)$ eV^a						MAR 1977 A
G $^1\Pi$	$[40259._8]$			$[0.522_4]^b$		$[2.2_7]$	$[1.834]$	G→A, V	$36365._4$ Z	(22)*
								G→X, R	$39868._6$ Z	(22)*
F $^1\Pi$	(37922)	[696] H		$[0.5590]^c$		$[1.42_4]$	$[1.772_8]$	F→X, R	37879.1 Z	(21)*
E $^1\Sigma^+$	(37722)	$[705]^d$ H		$[0.524_9]^b$		$[1.1_4]$	$[1.829]$	E→A, V	34180 H^Q	(27)*
								E→X, R	37683.5 Z	(18)* (25)*
C $^1\Sigma^-$	30080.6	632.4 Z	5.2	0.5008	0.0048	$[1.27]^e$	1.872_9	C→A,^{f} R	26500.94 Z	(4)* (10)
e $^3\Sigma^-$		Very weakly bound or repulsive state; see (19). Continuous absorption above 31250 cm^{-1} by shock-heated MgO in Ar/O_2 mixtures.						(e← a)		(23)
D $^1\Delta$	29851.6	632.5 Z	5.3	0.5014	0.0048	$[1.26]^g$	1.871_8	D→A,^{fh} R	26272.04 Z	(10)
d $^3\Delta_i$	$(29300)^i$	(650)		(0.50)			(1.8_7)	d↔a, (V)	26867 H^Q	(1b)(23)(26)
c $^3\Sigma^+$	(28300)	The assignment of the bands to MgO is still uncertain. See also (19).						(c← a)	25900	(23)
B $^1\Sigma^+$	19984.0	824.0_8 Z	$4.7_6{}^j$	0.5822^k	0.0045	1.14^ℓ	1.737_1	B→A,^{m} V	16500.2_9 Z	(1a)*
								B↔X,^{n} V	20003.5_7 Z	(1)(1a)(3)*
A $^1\Pi$	3563.3	664.4_4 Z	3.9_1	0.5056^o	0.0046	1.18	1.864_0			
a $^3\Pi_i$	$(2400)^p$	(650)		(0.50)			(1.8_7)			
X $^1\Sigma^+$	0	785.0_6 Z	5.1_8	0.5743^k	0.0050	1.22^q	1.749_0	r		
$^{24}Mg^{32}S$		$\mu = 13.7042725$		$D_0^0 \leq 2.4$ eV^a						MAR 1977
B $^1\Sigma^+$	23052.59	497.34 Z	2.333	0.25518	0.00155	0.269	2.1956	B↔X, R	23036.98 Z	(1)(4)*
X $^1\Sigma^+$	0	528.74 Z	2.704	0.26797	0.00176	0.276	2.1425			

For MgRb see p. 400.

MgO: [a]The dissociation energy (to Mg ^1S + O ^3P) is quite uncertain. The value given is the thermochemical value of (6) as corrected (19) for the presence of the low-lying $a^3\Pi$ state. From flame photometry (15) obtain 4.1_6 eV assuming a $^3\Sigma$ ground state, while (2) also from flame photometry but assuming a $^1\Sigma$ ground state obtain 4.3_4 eV.

[b]Several rotational perturbations.

[c]Small Λ-type doubling; intensity perturbations.

[d]Data for $Mg^{18}O$ are given by (25).

[e]$D_1 = 1.30 \times 10^{-6}$; $H_0 = 0.48 \times 10^{-11}$.

[f]Franck-Condon factors (12)(17)(24).

[g]$D_1 = 1.29 \times 10^{-6}$.

[h]For $^{18}O - ^{16}O$ isotope shifts see (10).

[i]$A \approx -25$.

[j]Vibrational isotope shifts for $^{24}Mg^{18}O$ and $^{26}Mg^{16}O$ (8).

[k]RKR potential functions (13).

[ℓ]$\beta_e = +0.02_5 \times 10^{-6}$.

[m]Franck-Condon factors (5); oscillator strength (11)(16).

[n]Observed in absorption in shock-heated Ar/O_2 mixtures containing MgO (23). Franck-Condon factors (5)(7)(9); oscillator strength (11)(16); dependence of transition moment on r from measured intensities (20).

[o]Small Λ-type doubling.

[p]$A \approx -50$; T_e is the theoretical value of (19). An experimental value of 3200 ± 1000 is given by (23).

[q]$\beta_e = +0.02 \times 10^{-6}$.

[r]Theoretical ground state properties (14); more recent calculations of ground and excited states (19).

(1) Lagerqvist, AMAF A 29, No. 25 (1943).
(1a) Lagerqvist, Uhler, AF 1, 459 (1949).
(1b) Brewer, Porter, JCP 22, 1867 (1954).
(2) Veits, Gurvich, OS 1, 22 (1956); ZFK 31, 2306 (1957).

MgO (continued):

(3) Gatterer, Junkes, Salpeter, Rosen, METOX (1957).
(4) Brewer, Trajmar, Berg, ApJ 135, 955 (1962).
(5) Nicholls, JRNBS A 66, 227 (1962).
(6) Drowart, Exsteen, Verhaegen, TFS 60, 1920 (1964).
(7) Ortenberg, Glasko, Dmitriev, SAAJ 8, 258 (1964).
(8) Pešić, PPS 83, 885 (1964).
(9) Prasad, PPS 85, 810 (1965).
(10) Trajmar, Ewing, ApJ 142, 77 (1965).
(11) Main, Carlson, DuPuis, JQSRT 7, 805 (1967).
(12) Srivastava, Maheshwari, PPS 90, 1177 (1967).
(13) Thakur, Singh, JSRBHU 18, 253 (1967).
(14) Yoshimine, JPSJ 25, 1100 (1968).
(15) Cotton, Jenkins, TFS 65, 376 (1969).
(16) Main, Schadee, JQSRT 9, 713 (1969).
(17) Gandara, Schamps, Bécart, CR B 270, 1213 (1970).
(18) Singh, JP B 4, 565 (1971).
(19) Schamps, Lefebvre-Brion, JCP 56, 573 (1972).
(20) Dube, IJPAP 11, 445 (1973).
(21) Singh, JP B 6, 1339 (1973).
(22) Singh, JP B 6, 1917 (1973).
(23) Evans, Mackie, CP 5, 277 (1974).
(24) Shadrin, Zhirnov, OS(Engl. Transl.) 38, 367 (1975).
(25) Antić-Jovanović, Pešić, Bojović, JMS 60, 416 (1976).
(26) Schamps, Gandara, JMS 62, 80 (1976).
(27) Antić-Jovanović, Bojović, Pešić, JP B 9, L575 (1976).

MgS: [a]From mass-spectrometric studies (2)(3).

(1) Wilhelm, Iowa State Coll. J. Sci. 6, 475 (1932).
(2) Colin, ICB 26, No. 9 (1961).
(3) Colin, Goldfinger, Jeunehomme, TFS 60, 306 (1964).
(4) Marcano, Barrow, TFS 66, 2936 (1970).

State	T_e	ω_e	$\omega_e x_e$	B_e	α_e	D_e	r_e (Å)	Observed Transitions Design.	Observed Transitions ν_{00}	References
$^{55}Mn_2$		$\mu = 27.469023_2$ $D_0^0 = 0.2_3$ eV^a								MAR 1976
		Absorption in the region 14400 - 16500 cm^{-1}; in Ar matrices at 10 K. $\omega' \approx 111$.								(2)
$^{55}Mn^{79}Br$		$\mu = 32.390081_4$ $D_0^0 = 3.2_2$ eV^a								MAR 1975
		Bands in the region 15800 - 16400 cm^{-1}, attributed to MnBr by (6), are probably due to CaBr.								(10)
		Emission bands in the visible and near infrared:								
		$\nu_e = 20024.6$; $\omega_e' = 286.6$, $\omega_e'x_e' = 1.4$; $\omega_e'' = 295.9$, $\omega_e''x_e'' = 1.6$;						b	R 20020.0 H^{Q_5}	(8)*
								c	R 19705.9 19669.8 H^Q	(8)*
		$\nu_e = 10682.1$; $\omega_e' = 290.4$, $\omega_e'x_e' = 0.13$; $\omega_e'' = 291.3$, $\omega_e''x_e'' = 0.01$;						d	R 10681.6 H^Q	(7)*
		$\nu_e = 10667.5$; $\omega_e' = 295.6$, $\omega_e'x_e' = 0.62$; $\omega_e'' = 298.8$, $\omega_e''x_e'' = 0.61$;						e	R 10665.9 H	(7)*
A ($^7\Pi$)	26303.7^f	302.3 H^{Q_4}	0.6					$A \leftrightarrow X$,	V 26311.6 H^{Q_4}	(1)(2)(3)* (4)(5)
X ($^7\Sigma$)	0	286.7 H^{Q_4}	0.8							
$^{55}Mn^{(35)}Cl$		($\mu = 21.3678872$) $D_0^0 = 3.7_0$ eV^a								MAR 1975
		Emission bands in the visible and near infrared:								
		$\nu_e = 20115$; $\omega_e' = 378$,^{b} $\omega_e'' = 386$,^{b}						c	R 20111 H^Q	(8)*
		$\nu_e = (19938)$; $\omega_e' = (385)$, $\omega_e'' = (410)$,						d	R (19925)	(8)*
		$\nu_e = 11420$; $\omega_e' = 385$,^{b} $\omega_e'' = 398$,^{b}						e	R 11414 H^Q	(7)*
B ($^7\Sigma$)	40807	[320] H						$B \leftarrow X$,	V 40776 H	(4)
A ($^7\Pi$)	27005.$_0$^{f}	[407.9] H	^{g}					$A \leftrightarrow X$,	V 27017.8 H^{Q_4}	(1)(2)(3) (4)* (5)(6)
X ($^7\Sigma$)	0	[382.4] H	^{g}							

Mn_2: [a]Thermochemical value (mass-spectrom.)(1).

(1) Kant, Lin, Strauss, JCP 49, 1983 (1968).

(2) De Vore, Ewing, Franzen, Calder, CPL 35, 78 (1975).

MnBr: [a]Thermochemical value (flame photometry)(9).

[b]Q, R, S heads; possibly quintet system.

[c]Two sequences of double headed bands.

[d]Q, R heads; possibly additional branches.

[e]Single heads.

[f]$|A| \approx 58$.

(1) Mesnage, AP(Paris) 12, 5 (1939).

(2) Müller, HPA 16, 3 (1943).

(3) Bacher, HPA 21, 379 (1948).

(4) Rao, IJP 23, 517 (1949).

(5) Hayes, Nevin, PRIA A 57, 15 (1955).

(6) Hayes, Nevin, PPS A 68, 665 (1955).

(7) Hayes, PPS A 68, 670 (1955).

(8) Hayes, PPS A 68, 1097 (1955).

(9) Bulewicz, Phillips, Sugden, TFS 57, 921 (1961).

(10) Barrow, in DONNSPEC (1970).

MnCℓ: [a]Thermochemical value (flame photometry)(9).

[b]The band head analyses by (7) and (8) lead to negative values of $\omega_e x_e$ (≈ -0.5).

[c]Q, R, S heads; possibly quintet system.

[d]Strong 0-0 sequence, multiple heads.

[e]Multiple heads.

[f]$|A| \approx 41$.

[g]Different values for $\omega_e x_e$ suggested by (3) and (6).

(1) See ref. (1) of MnBr.

(2) Miescher, Müller, HPA 15, 319 (1942).

(3) See ref. (2) of MnBr.

(4) See ref. (3) of MnBr.

(5) See ref. (4) of MnBr.

(6) See ref. (5) of MnBr.

(7) Hayes, Nevin, NC(Suppl.) 2, 734 (1955).

(8) See ref. (8) of MnBr.

(9) See ref. (9) of MnBr.

State	T_e	ω_e	$\omega_e x_e$	B_e	α_e	D_e (10^{-4} cm^{-1})	r_e (Å)	Observed Transitions: Design.	Observed Transitions: ν_{00}	References
$^{55}Mn^{19}F$		$\mu = 14.1166531$ $D_0^0 = 4.3_5$ eV[a]								MAR 1975
		Emission bands in the visible and near infrared:								
		$\nu_e = 20298._2$; $\omega_e' = 637.1$, $\omega_e'x_e' = 1.9$; $\omega_e'' = 649.1$, $\omega_e''x_e'' = 1.5$;						b R	20292.$_1$ H^Q	(6)*
								c R	20023.3 H^R 19971.6 H^Q	(6)*
		$\nu_e = 14527._7$; $\omega_e' = 595.4$, $\omega_e'x_e' = 3.15$; $\omega_e'' = 645.4$, $\omega_e''x_e'' = 3.2$;						d R	14502.$_7$ H^{QR}	(5)*
								e V_R	12179.6 H^Q 12153.6 H^P	(4)*
B ($^7\Sigma$)	41231.5	637.2 H	4.46					B← X, V	41240.3 H	(1)*
A ($^7\Pi$)	28465.$_0$[f]	673[g] H	4					A↔X, V_R	28491.$_9$[h] H	(1)(2)(3)* (7)*
X ($^7\Sigma$)	0	618.8[i] H	3.01							
$^{55}Mn^1H$		$\mu = 0.98966996$ $D_0^0 = (2.5)$ eV								OCT 1975
		Complex rotational structure from 22600 to 24000 (max. 23360 - 23920) cm^{-1}; in emission.								(7)*
		Open but strongly perturbed rotational structure in the region 21270 - 22580 cm^{-1}; in emission and absorption. Probably quintet transition.								(2)* (7)* (12)
		Extremely complicated R shaded band with very dense structure centered around 20900 cm^{-1}; in emission and absorption. Probably quintet system.								(1)(2)* (7)* (12)
		Complex rotational structure from 11200 to 14500 cm^{-1}; in emission.								(7)*
A $^7\Pi$ a	17597[b]	[1623][c]	(33)	6.425[de]	0.187	[3.62][f]	1.628	A↔X, V	17666[c]	(1)(2)* (3)* (4)(5)* (7) (8)(10)(14)
X $^7\Sigma$ a	0	1548.$_0$ Z	28.8	5.6841[d]	0.1570[g]	[3.039][h]	1.7311			
$^{55}Mn^2H$		$\mu = 1.94287387$ $D_0^0 = (2.6)$ eV								OCT 1975
		Complex rotational structure in the region 23250 - 23800 cm^{-1}; in emission.								(7)*
		Complex R shaded rotational structure with heads at 22355 and 22564 cm^{-1}; in emission and absorption.								(7)* (12)

MnF:

[a]Thermochemical value (mass-spectrom.)(8).

[b]Complex bands; R, Q heads.

[c]Two sequences of single heads, interpreted as R and Q heads.

[d]Four heads.

[e]Single sequence of bands having apparently both Q and P (or R) heads.

[f]$|A| \approx 25$.

[g]Average of (3) and (7).

[h]Strongest head of F_4 component.

[i]From B − X system.

(1) Rochester, Olsson, ZP 114, 495 (1939).

(2) Bacher, Miescher, HPA 20, 245 (1947).

(3) See ref. (3) of MnBr.

(4) See ref. (7) of MnCℓ.

(5) See ref. (6) of MnBr.

(6) See ref. (8) of MnBr.

(7) Rao, Reddy, Rao, PPS 79, 741 (1962).

(8) Kent, Ehlert, Margrave, JACS 86, 5090 (1964).

Mn^1H, Mn^2H:

[a]For an *ab initio* calculation of these two states see (11).

[b]$A = 40.6$ [see (14)].

[c]From approximate origins for the F_4 component.

[d]Approximate ("effective") constants. Improved ("true") constants for Mn^1H (v=0) together with spin-orbit, spin-spin, spin-rotation interaction and ℓ-uncoupling constants in (14). See also (9) (13).

[e]Perturbations.

[f]$D_1 = 3.64 \times 10^{-4}$; $H_0 = 8.2 \times 10^{-9}$, $H_1 = 7.4 \times 10^{-9}$.

[g]$\gamma_e = -0.0019$.

[h]$D_1 = 3.053 \times 10^{-4}$, $D_2 = 3.08 \times 10^{-4}$; $H_0 = 9.3_3 \times 10^{-9}$, $H_1 = 9.7_9 \times 10^{-9}$, $H_2 = 10.2 \times 10^{-9}$. H_v values in (13) differ by an order of magnitude.

(continued p. 409)

State	T_e	ω_e	$\omega_e x_e$	B_e	α_e	D_e $(10^{-5}cm^{-1})$	r_e (Å)	Observed Transitions Design.	Observed Transitions ν_{00}	References
$^{55}Mn^{2}H$ (continued)										
		Complex R shaded rotational structure between 20000 and 21270 cm^{-1}; in emission and absorption. Perturbations. Probably quintet system.								(7)* (12)
		Complex structure in emission from 11330 to 14530 cm^{-1}.								(7)*
A $^7\Pi$	(17602)	(1205)[i]	(17)	[3.244][de]		[9.51][j]	[1.635]	A↔X, V	17652[c]	(6)(7)* (8) (12)
X $^7\Sigma$	0	$1102._5$ Z	13.9	2.8957[d]	0.051_3[k]	[7.95][ℓ]	1.7310			
$^{55}Mn^{127}I$		$\mu = 38.340229_2$		$D_0^0 = 2.8_9$ eV[a]						MAR 1975
A $(^7\Pi)$								A←X,[b] V	(25000)	(1)(2)*
X $(^7\Sigma)$	0	(240) H	(1.5)							
$^{55}Mn^{16}O$		$\mu = 12.3881671_6$		$D_0^0 = 3.7_0$ eV[a]						MAR 1975 A
B								B←(X),[b]	(38950)	(6)* (8)
A $^6\Sigma$	17949	762.8 H	9.6[c]	(0.390)[d]			(1.86_8)[d]	A↔X, R	17909[c] H	(1)(3)* (4) (6)(7)(8) (11)(13)
X $^6\Sigma$ e	0	839.6 H	4.7_9	(0.435)[d]			(1.76_9)[d]			
$^{55}Mn^{(32)}S$		$(\mu = 20.2103419)$		$D_0^0 = 2.8_5$ eV[a]						MAR 1975
		Fragments of an additional system with $\omega' \approx 461$, $\omega'' \approx 480$;						[b] R	22320 H	(4)
B	a + 18917.4	371.5 H	1.30					B→A,[b] R	18858.0 H	(4)(5)
A	a	490.5 H	1.78							
$^{(96,98)}Mo_2$?		$(\mu = 48.447357_7$								MAR 1975
		Sequence of unclassified emission bands, extending from 5184 Å (19285 cm^{-1}) to longer wavelengths. Tentatively attributed to Mo_2.								(1)

Mn^1H, Mn^2H (continued):

[i]From the values for Mn^1H.

[j]$H_0 = 1.2 \times 10^{-9}$.

[k]$\gamma_e = -0.002_0$.

[ℓ]$D_1 = 8.22 \times 10^{-5}$, $D_2 = 8.36 \times 10^{-5}$; $H_0 = 1.4_1 \times 10^{-9}$, $H_1 = 2.1_7 \times 10^{-9}$, $H_2 = 2.9_3 \times 10^{-9}$.

(1) Heimer, Naturw. 24, 521 (1936).
(2) Pearse, Gaydon, PPS 50, 201 (1938).
(3) Nevin, PRIA A 48, 1 (1942).
(4) Nevin, PRIA A 50, 123 (1945).
(5) Nevin, Doyle, PRIA A 52, 35 (1948).
(6) Nevin, Conway, Cranley, PPS A 65, 115 (1952).
(7) Nevin, Stephens, PRIA A 55, 109 (1953).
(8) Hayes, McCarvill, Nevin, PPS A 70, 904 (1957).
(9) Kovács, PRIA A 60, 15 (1959).
(10) Kovács, Scari, APH 9, 423 (1959).
(11) Bagus, Schaeffer, JCP 58, 1844 (1973).
(12) Smith, PRS A 332, 113 (1973).
(13) Pacher, APH 35, 73 (1974).
(14) Kovács, Pacher, JP B 8, 796 (1974).

MnI: [a]Thermochemical value (flame photometry)(3).
[b]Complex system, incomplete analysis.

(1) Bacher, Miescher, HPA 20, 245 (1947).
(2) Bacher, HPA 21, 379 (1948).
(3) Bulewicz, Phillips, Sugden, TFS 57, 921 (1961).

MnO: [a]Thermochemical value (mass-spectrom.)[Burns, quoted in (9)(10)]. Flame photometric values are approximately 4.1 eV (2)(5).

[b]Seemingly continuous absorption, strongest in the region 38800 - 39100 cm^{-1}; some indication of band heads. It is not certain that this system is due to MnO.

MnO (continued):

[c]$\omega_e y_e = +0.06$. Heads having v'=0 are ill defined.

[d]Partial rotational analysis of the 1-0 band (13); spin-spin interaction constants $\lambda' \approx 0$, $\lambda'' = +0.66$ cm^{-1}. Hyperfine broadening of all rotational lines; many bands are highly perturbed (12).

[e]Matrix studies at 4 K (11) support the view that the lower state of A—X is the ground state.

(1) Sen Gupta, ZP 91, 471 (1934).
(2) Huldt, Lagerqvist, AF 3, 525 (1952).
(3) Gatterer, Junkes, Salpeter, Rosen, METOX (1957).
(4) Das Sarma, ZP 157, 98 (1959).
(5) Padley, Sugden, TFS 55, 2054 (1959).
(6) Callear, Norrish, PRS A 259, 304 (1960).
(7) Joshi, SA 18, 625 (1962).
(8) Garrett, Lee, Kay, JCP 45, 2698 (1966).
(9) Cheetham, Barrow, AdHTC 1, 7 (1967).
(10) Coppens, Smoes, Drowart, TFS 63, 2140 (1967).
(11) Thompson, Easley, Knight, JPC 77, 49 (1973).
(12) J. G. Kay, private communication (1974).
(13) Pinchemel, Schamps, CJP 53, 431 (1975).

MnS: [a]Thermochemical value (mass-spectrom.)(1)(2)(3).
[b]Thermal emission, attributed to MnS.

(1) Colin, Goldfinger, Jeunehomme, Nature 194, 282 (1962).
(2) Wiedemeyer, Gilles, JCP 42, 2765 (1965).
(3) Drowart, Pattoret, Smoes, PBCS No. 8, 67 (1967).
(4) Monjazeb, Mohan, SpL 6, 143 (1973).
(5) Biron, Boulet, Ruamps, CR B 278, 835 (1974).

Mo_2: (1) Becker, Schürgers, ZN 26 a, 2072 (1971).

State	T_e	ω_e	$\omega_e x_e$	B_e	α_e	D_e	r_e (Å)	Observed Transitions Design.	ν_{00}	References
$^{(98)}Mo^{19}F$		(μ = 15.9109144) D_0^0 = 4.78 eV[a]								SEP 1976
$^{(98)}Mo^{14}N$?		(μ = 12.25087398) 14 R shaded emission bands in the region 15840 - 16690 cm^{-1}, most of them previously believed to be due to MoO (1), have been attributed (3) to a nitride, Mo_xN, of molybdenum on the basis of observed $^{14}N/^{15}N$ isotope shifts.								MAR 1975 (1)(3)
$^{(98)}Mo^{16}O$		(μ = 13.74876422) D_0^0 = 5.0 eV[a] Emission bands at 15358, 15511 cm^{-1} (open structure), 11631 cm^{-1} (V), 11549, 11606, 11626 cm^{-1} (R) are believed to be due to an oxide of molybdenum. Bands in the region 15930 - 16690 cm^{-1}, previously attributed to MoO, are due to Mo_xN.								MAR 1975 (1)(3)

MoF: [a]Thermochemical value (mass-spectrom.)(1).

(1) Hildenbrand, JCP 65, 614 (1976).

MoN, MoO:

[a]Thermochemical value (mass-spectrom.)(2).

(1) Gatterer, Junkes, Salpeter, Rosen, METOX (1957).

(2) De Maria, Burns, Drowart, Inghram, JCP 32, 1373 (1960).

(3) Howard, Conway, JCP 43, 3055 (1965).

State	T_e	ω_e	$\omega_e x_e$	B_e	α_e	D_e (10^{-} cm^{-1})	r_e (Å)	Observed Transitions Design.	ν_{00}	References
$^{14}N_2$		$\mu = 7.00153720$		$D_0^0 = 9.759_4$ eV[a]		I.P.($3\sigma_g$) = 15.5808 eV[b]				FEB 1977 A
						($1\pi_u$) = 16.6986 eV[bc]				
						($2\sigma_u$) = 18.7507 eV[b]				
						($2\sigma_g$) = 37.9 eV[d]				
						($1s_N$) = 409.9$_8$ eV[e]				

For a very detailed and critical review of the spectrum of molecular nitrogen and its ions see the recent publication by Lofthus and Krupenie (196). Atlas of the VUV absorption spectrum 1060 - 1520 Å and table of absorption lines (87). Tables of band head wavelengths (39)(48)(196). Photoionization and absorption cross sections (45)(57)(147). Potential functions (60)(68)(182)(196).

State	T_e	ω_e	$\omega_e x_e$	B_e	α_e	D_e	r_e	Design.	ν_{00}	References
		Several Rydberg series (excitation of $1s_N$) with limit (K edge) at 409.5 eV.								(114)(166) (173)(174)
x" ($^1\Sigma_u^+$)		X-ray absorption ($3s\sigma \leftarrow 1s_N$).						x"← X,	405.5$_9$ eV	
x' ($^1\Pi_u$)		X-ray absorption and emission ($1\pi_g \leftrightarrow 1s_N$), broad peak.[f]						x'↔ X,	400.8$_4$ eV	
Photoionization and dissociative photoionization processes corresponding to various excited states of N_2^+.[g]										(177)(194)
v $^1\Pi_g$		From high-energy electron impact spectroscopy.						v← X,	253000	(180)
		Fragments of other progressions and unassigned bands, 198000 - 207000 cm^{-1}.								(76)(163)
u_5	(183640)	(2100)	(15)	First two members of Codling's Rydberg series.[h]				u_5← X,	(183510)	
u_4	(178565)	(2070)	(15)					u_4← X,	(178420)	
		Hopfield's Rydberg series converging to B $^2\Sigma_u^+$(v=0) of N_2^+:								
$\ldots 2\sigma_u 1\pi_u^4 3\sigma_g^2 \left\{ \begin{matrix} ns\sigma \\ nd\sigma \end{matrix} \right\}$		$\nu = 151233 - \left\{ \begin{matrix} R/(m+0.141-0.199/m)^2,\ m = 3\ldots 11 \text{ (apparent emission series)}^{i} \\ R/(m-0.070-0.041/m)^2,\ m = 3\ldots 20 \text{ (absorption series)}^{jk\ell} \end{matrix} \right.$								(1)(4)* (37)*
		Worley's ("third") Rydberg series joining on to o_3, o_4, o_5 and converging to A $^2\Pi_{u\frac{1}{2}}$(v=0) of N_2^+:								
$\ldots 2\sigma_u^2 1\pi_u^3 3\sigma_g^2 ns\sigma$		$\nu = 134721 - R/(n-1.06)^2$, n = 3...16.[mnop]								(8)(37)* (190)*
		Ogawa and Tanaka's Rydberg series joining on to O_4, O_5 and converging to A $^2\Pi_{u\frac{3}{2}}$(v=0) of N_2^+:								
$\ldots 2\sigma_u^2 1\pi_u^3 3\sigma_g^2 ns\sigma$ [q]		$\nu = 134644 - R/n^{*2}$, n* = 2.84, 3.85, 4.86, ..., 14.91.[mnop]								(37)* (190)*
		Several dissociation continua in the region 100000 - 160000 cm^{-1}.								(105)(156)
		Several unidentified bands in the region 126100 - 131550 cm^{-1}.								(51)

State	T_e	ω_e	$\omega_e x_e$	B_e	α_e	D_e (10^{-} cm^{-1})	r_e (Å)	Observed Transitions Design.	Observed Transitions ν_{00}	References
$^{14}N_2$ (continued)										
s^r	133355 133316	1885 H	(12)	Ogawa's "new progression 4" o				$s \leftarrow X$,	133119 H 133080 H	(51)
r^r	132878	1903 H	(15)	Ogawa's "new progression 3" os				$r \leftarrow X$, R	132650 H	(51)
q^r	132136	1900 H	(18)	Ogawa's "new progression 2" os				$q \leftarrow X$, t	131906 H	(51)
p^r	129136	1869 H	(10)	Ogawa's "new progression 1" os				$p \leftarrow X$, R	128892 H	(51)

N_2: [a]From the predissociation in C $^3\Pi_u$ assuming dissociation into $^4S_{3/2} + {^2D_{5/2}}$. The latest *ab initio* calculation of the ground state gives D_e = 8.58 eV (188).

[b]From the Rydberg series.

[c]Average of the two limits corresponding to $^2\Pi_{\frac{3}{2}}$ and $^2\Pi_{\frac{1}{2}}$.

[d]From the data on N_2^+.

[e]From $x \rightarrow X$ of N_2^+ and I.P.(N_2); the extrapolated K limit is 409.5 eV (114).

[f]Confirmed by electron-energy-loss measurements (130). Preionization to X $^2\Sigma_g^+$ and A $^2\Pi_u$ of N_2^+ observed by Auger electrons of 384.7 and 383.8 eV (119).

[g]Absorption cross sections 140000-500000 cm^{-1} (163)(165).

[h]This series, of which only two members have been observed, probably converges to C $^2\Sigma_u^+$ of N_2^+. In u_4 v=1...13, in u_5 v=3...8 have been observed, but the vibrational numbering in both states is uncertain. Evidence of preionization.

[i]Longward of the apparent emission "lines" are close-lying fairly sharp absorption "lines" (37). Both features belong to the same Fano shape produced by the interaction of the series of Rydberg levels with the continuum joining on to the A $^2\Pi$ (or, less likely, X $^2\Sigma$) limit.

[j]Similar series with v'=1.

[k]The first three members at 138330, 144090, 146690 cm^{-1} are very broad (presumably because of preionization), the higher members are sharper and shaded to the red. No rotational structure has been resolved. Preionization also observed by electron spectroscopy (162)(195).

[ℓ]Oscillator strengths from absorption coefficients: f(m=3,4,...) = 0.0131, 0.0053, ... (126).

[m]Similar series for v'=1 [observed to n=45 ($^2\Pi_{\frac{1}{2}}$) and n*=40.2 ($^2\Pi_{\frac{3}{2}}$)] and for v'=2...7 (190).

[n]Corresponding series in $^{15}N_2$ (51).

[o]Preionization observed in photoionization studies (58)(105)(161).

[p]Preionization in absorption series having v"=1...4 was observed in active nitrogen using the photoionization technique (146).

[q]Interpreted as ...ndσ by (112).

[r]These designations should not be confused with the older designations of component states of b $^1\Pi_u$ and b' $^1\Sigma_u^+$.

[s]Preionization also observed by electron spectroscopy (162)(195).

[t]Line-like, not shaded.

State	T_e	ω_e	$\omega_e x_e$	B_e	α_e	D_e ($10^{-6}cm^{-1}$)	r_e (Å)	Observed Transitions Design.	ν_{00}	References
$^{14}N_2$ (continued)										
$o_5\ ^1\Pi_u$	(127868)	(1935)[a]	(19)					$o_5 \leftarrow X$,	R 127655 H^Q	(37)* (190)*
$O_5(^3\Pi_u)$	127445	1925 H^Q	18.4					$O_5 \leftarrow X$,	R 127227 H^Q	(37)* (190)*
		Rydberg series converging to X $^2\Sigma_g^+$(v=0) of N_2^+:								
$c'_n\ ^1\Sigma_u^+ \ldots 1\pi_u^4\, 3\sigma_g\, np\sigma$ [b]		– Carroll and Yoshino's series joining on to c'_4, c'_5.						$c'_n \leftarrow X$		(88)(144)* (186)*
$c_n\ ^1\Pi_u \ldots 1\pi_u^4\, 3\sigma_g\, np\pi$ [b]		– Ledbetter's series c_4, c_5, c_6.						$c_n \leftarrow a''$		(151)*
		– Worley and Jenkins' series joining on to c_3, c_4: $\nu = 125666.8^c - R/(m + 0.3697 - 0.3459/m + 0.532/m^2 - 0.960/m^4)^2$, $m(\equiv n-1) = 2 \ldots 31$.						$c_n \leftarrow X$,[def]		(5)(37)* (144)*(186)*
$o_4\ ^1\Pi_u$	122419	[1824.$_1$] H		[1.7338][g]		[4]	[1.1784]	$o_4 \leftarrow X$,	R 122155.4 Z	(37)(184)
$O_4(^3\Pi_u)$	121263	1982 H	27.$_0$					$O_4 \leftarrow X$,	R 121071.1 H	(37)
$c'_5\ ^1\Sigma_u^+$	(115876)	[2221.8] Z	[h]	[1.345][i]				$c'_5 \leftarrow X$,	R 115849.8 Z	(118)(144)*
$c_4\ ^1\Pi_u$	115635.$_9$	2220.3 Z	19.4	[1.926$_1$][j]	0.01$_5$	[6.3][j]	1.116	$c_4 \leftarrow a''$,	16725.1$_2$ Z	(151)*
		[A progression of six bands (v"=1-6) arises from c_4(v'=0)→a, (11)]						$c_4 \leftarrow X$,	R 115565.5$_3$ Z	(144)*(151)*
$z\ ^1\Delta_g$	(115435)	(1700)		1.761	0.0153		1.169	z→w,	V 43411.2[k] Z	(25)
$y\ ^1\Pi_g$	114305.2[ℓ]	1906.43[ℓ]	37.51[ℓ]	1.739[ℓm]	0.017[ℓ]	(5.8)	1.177	y→w, Kaplan II s.	V 42467.5[n] Z	(26)* (175)
								y→a', Kaplan I s.	V 46426.7[n] Z	(26)* (175)
$k\ ^1\Pi_g$	(113808)[o]	[2182.32][ℓ]		1.959[ℓ]	0.031[ℓ]	(5.9)	1.109	k→w, Carroll-Subbaram II s.	V 41932.4[n] Z	(175)*
								k→a', Carroll-Subbaram I s.	V 45891.7[n] Z	(175)*
$x\ ^1\Sigma_g^-$	113438.0	1910.0 Z	20.7	1.750[p]	0.0225	(6)	1.173	x→a', 5th pos. gr.	V 45472.8 Z	(9)(23)* (172)
$d'\ \{^1\Sigma_u^-$ or $^1\Delta_u\}$	[112500]	Only v'=0 observed.						d'→a,	42373 H	(11)(12)(106)

State	T_e	ω_e	$\omega_e x_e$	B_e	α_e	D_e $(10^{-6}cm^{-1})$	r_e (Å)	Observed Transitions Design.	ν_{00}	References
$^{14}N_2$ (continued)										
$o_3\ ^1\Pi_u$	105869[q]	1987.4[q]	16.3[q]	1.7339[q]	0.0088[q]	(5.3)	1.1784	$o_3 \rightarrow a$,	V 36731	(13)(15)
								$o_3 \leftarrow X$,[r]	R 105683	(8)(184)
H $^3\Phi_u$	(105720)[s]	924.21 Z	12.29[t]	1.0873	0.0191	$[7._0]$[u]	1.4881	$(H \rightarrow ?)$[v]	12407.2 H	(17)(18)
								$H \rightarrow G$,[w]	V 17897.08[x] Z	(10)(11)(19)
								Gaydon-Herman green s.		(145)*

N_2: [a]Vibrational intervals decrease irregularly; v=0 was shown [by electron spectroscopy (195)] to be preionized.

[b]For higher n values c_n and c'_{n+1} lie close together and interact strongly (ℓ-uncoupling). Band structures for n = 5...12 have been discussed (144)(154)(186).

[c]The limit according to Yoshino [see (196)] lies at 125667.5 cm^{-1} but is estimated (196) to have an uncertainty of 5 cm^{-1}.

[d]Similar series with v'=1.

[e]Corresponding series in $^{15}N_2$ (51).

[f]See [p] on p. 413.

[g]Only v=0 [perturbed by c'_6(v=1)] is sharp; bands with v'=1, 2,3 are diffuse owing to predissociation or preionization (levels with v ≳ 2 are above the first ionization potential).

[h]$\Delta G(3/2)$ = 2119.7, see [i].

[i]Strong homogeneous perturbations with the higher vibrational levels (v ≳ 18) of b' $^1\Sigma_u^+$ (144). The B_0 value is an effective value at low J; B_{eff}(v=1,2) = 1.28_5, 1.17_3. In addition, there are heterogeneous interactions with the close-lying levels of $c_4\ ^1\Pi_u$. For deperturbed constants see (152)(152a).

[j]Constants for Π^- (151); $B_0(\Pi^+)$ = 1.906. α_e from (144).

[k]Re-evaluated from the origin of the 0-2 band. (25) gives 43667.0 which was undoubtedly calculated with the constants of the a rather than of the w state.

[ℓ]Strong homogeneous interaction between k $^1\Pi_g$ and y $^1\Pi_g$. The constants given are the deperturbed values from (175) and refer to Π^-, the only component observed in k $^1\Pi_g$.

[m]Predissociation of the Π^+ component above J=10 of v=0. Λ-type doubling and predissociation discussed in (189).

[n]Not deperturbed.

[o]From the deperturbed T_{00} = 113723.58.

[p]Only v'=0,1,2 observed. Predissociation (weakening of emission) at v'=2, J'≈15 corresponding to the limit $^2D + ^2D$ (33); actual breaking-off occurs at J'=25. See also (189).

[q]Deperturbed constants (184). Homogeneous interactions with levels (v ≥ 6) of b $^1\Pi_u$ and heterogeneous perturbations by b' $^1\Sigma_u^+$.

[r]Oscillator strenghts (147).

[s]$A_0 \ldots A_3$ = -12.073 ... -12.094 (145); see also (164).

[t]$\omega_e y_e$ = - 0.173.

[u]$D_{1,2,3}(10^{-6}cm^{-1})$ = 4.5, 6.0, 5.0 (145). See, however, (164).

[v]Fragment of near infrared spectrum, $\Delta G''(\frac{1}{2}) \approx 712$.

[w]Franck-Condon factors and r-centroids (171).

[x]From a more detailed theoretical treatment (164) derives ν_{00} = 17902.400.

State	T_e	ω_e	$\omega_e x_e$	B_e	α_e	D_e (10^{-6} cm^{-1})	r_e (Å)	Observed Transitions		References
								Design.	ν_{00}	
$^{14}N_2$ (continued)										
c_4' $^1\Sigma_u^+$	104519[a]	2201.78[a]	25.199[a]	1.9612[a]	0.0436[a]		1.1080	$c_4' \rightarrow a$, V_R Gaydon-Herman b.	35371.2 Z	(12)(25)
								$c_4' \leftrightarrow X$,[b] R	104323.3[c] Z	(55)(88)(104)(106)(118)
c_3 $^1\Pi_u$	104476[d]	2192.20[d]	14.70[d]	1.9320[d]	0.0395[d]		1.1163	$c_3 \rightarrow a$, R_V	35187.0 (Z)	(15)
								$c_3 \leftrightarrow X$,[e] R	104138.2[f] Z	(104)(106)
b' $^1\Sigma_u^+$	104498[g]	760.08[g]	4.418[g]	1.1549[gh]	0.007387[g]		1.4439	$b' \rightarrow a$, R	[i]	(25)
								$b' \leftrightarrow X$, R Birge-Hopfield b.	103673.8[j] Z	(22)(118)
D $^3\Sigma_u^+$	[104746.$_6$]	Only v=0 is observed		[1.961]		[20]	[1.108$_0$]	D[k]$\rightarrow B$, V 4th pos. gr.	44264.$_1$[ℓ] Z	(6)
b $^1\Pi_u$	(101675)	[634.8][m]		[1.448$_3$][no]		[29][p]	1.284$_1$[q]	$b \rightarrow a$, R Gaydon-Herman b.	31865.7 (Z)	(9)(11)(25)(172)
								$b \leftrightarrow X$,[e] R	100816.9 Z	(104)(184)
a″ $^1\Sigma_g^+$	[100016.$_0$]			[1.9133][r]		[6.2][r]	[1.1218]	$a'' \leftarrow X$,[s]	98840.30[r] Z	(113)
C′ $^3\Pi_u$	98351[t]	791	33.5	[1.0496]	u	[10.9][v]	[1.514$_6$]	$C' \leftrightarrow B$, R Goldstein-Kaplan b.	38255.5[w] Z	(44)* (191)* (198)*

N_2: [a]Deperturbed constants (152)(152a); $\omega_e y_e = +0.7874$. Strong perturbations produced by interaction with b′ $^1\Sigma_u^+$; before these perturbations were recognized (106)(106a) the vibrational levels were attributed to independent states called p′, r′, k, s′, h, h′, h″, h‴. The observed vibrational intervals (from band origins) and rotational constants for v=0,1,2,3... are:

$\Delta G(v+\frac{1}{2})$ = 2046.2, 2175.5, 2112.2, 2111.7 ...

B_v = 1.929, 1.711, 1.436, 1.594 ...

[b]Radiative lifetime $\tau(v'=0)$ = 0.9 ns (80)(99); oscillator strengths (98)(147).

[c]Observed ν_{00}, not deperturbed.

[d]Deperturbed constants (152)(152a). Strong perturbations produced by interaction with b $^1\Pi_u$; the observed vibrational intervals (from band heads) and rotational constants for v=0,1,2,... are:

$\Delta G(v+\frac{1}{2})$ = 2401, 2146, 2103, 2042;

$B_v(\Pi^-)$ = 1.516, 1.755, 1.813 [see (104)(152a)].

N_2 (continued):

[e]Oscillator strengths (98)(147).

[f]Quoted from (196); not deperturbed. (106) gives 104139.

[g]Deperturbed constants (152)(152a); $\omega_e y_e$ = + 0.1093, ... γ_e = - 0.0000750, ... Strong perturbations on account of interactions with c_4', c_5' $^1\Sigma_u^+$. Before these perturbations were recognized (104)(106)(106a)(118) several of the vibrational levels were assumed to be independent states called b', g, f, r, s, t, u by (8). The observed vibrational intervals (from band origins) and rotational constants for v=0, 1, 2, 3, ... are:

$\Delta G(v+\frac{1}{2})$ = 744.9, 732.9, 717.6, 777.7, ... } highest obs'd
B_v = 1.151_5, 1.15, 1.142, 1.152, ... } level v=28.

Intensity perturb. in the electron energy loss sp. (108).

[h]The b' ← X absorption bands show diffuseness indicating predissociation for v'=20, 21, 22 (118). Emission bands have only been observed to v'=9. For v'=5 and above J'=12 an intensity anomaly suggesting inverse predissociation has been observed in emission (56); it corresponds to the limit $^4S + {}^2P$. Selective emission from v'=0, 2, 7 in discharges in Ar and Kr with traces of N_2 (94).

[i]Only the 7-0 band was observed at v_0 = 40000.7 (25).

[j]Observed band origin, not deperturbed.

[k]Lifetime $\tau(v=0) = 14._1$ ns (159). Franck-Condon factors (110).

[ℓ]Extrapolated from $Q_2(3)$ of the 0-1 band.

[m]$\Delta G(\frac{3}{2},\frac{5}{2},...)$ = 700.0, 711.9, 685.2, 1151.4, 646.2, ... [(104), from band origins]. (152)(152a) give the deperturbed constants ω_e = 461.01, $\omega_e x_e$ = - 132.257, $\omega_e y_e$ = - 35.005, $\omega_e z_e$ = + 5.822, ...; see [n].

[n]B_v(v=1, 2, 3...) = 1.408_6, 1.387_2, 1.381_5, 1.421_3, ... (104); (152)(152a) give the deperturbed constants B_e = 1.4601, α_e = 0.02624, γ_e = - 0.00362, ... Strong perturbations on account of interaction with the c_3 $^1\Pi_u$ and o_3 $^1\Pi_u$ Rydberg states. Before these perturbations were recognized (104)(106) several of the vibr. levels were assumed to be independent electronic states called i, j, b, ℓ, m, p, q (8). Intensity perturb. in the electron energy loss sp. (71)(108).

[o]The lines of absorption bands with v'=0,2,3,4 are broadened on account of predissociation (especially v'=3); corresponding emission bands have not been observed. The state causing the predissociation is probably C' $^3\Pi_u$ (104); see, however, (137) who find that an additional diffuse level, very likely the still missing o_3 $^3\Pi_u$(v=0) level at ~103000 cm^{-1}, is required to explain the broadening of v'=3.

[p]Effective (perturbed) D_0 value.

[q]From the deperturbed B_e (see [n]).

[r]From Rydberg series having a" as lower state (151).

[s]First thought to be observed as quadrupole absorption (90), later recognized as pressure-induced dipole transition represented by a broad diffuse absorption band at ~99005 cm^{-1} (113); note the large pressure shift of +165 cm^{-1}. The electron energy loss spectrum (81) shows a peak at 12.25 eV. Another $^1\Sigma_g^+$ state, non-Rydberg in character, is predicted to intersect a" not far from its minimum (129).

[t]$A_0 = 2._{10}$, recalculated by (191) from the data of (44) who obtained 1.1_5; A_1 = 2.73, deperturbed value (191).

[u]B_1(obs.) = 1.2056, B_1(depert.) = 1.026. Strong mixing of C'(v=1) with C(v=5). Deperturbed constants and RKR potential functions are given by (191) who have analyzed in detail the C'(v=1) ← B(v=5) band for $^{14}N_2$, $^{14}N^{15}N$ and $^{15}N_2$. The perturbing level C(v=5) was recently observed by (198) in absorption from B(v=6).

[v]$H_0 = 8.3 \times 10^{-10}$.

[w]v_{00} = 38296.75 in (44) refers to the F_1 component.

State	T_e	ω_e	$\omega_e x_e$	B_e	α_e	D_e ($10^{-6}cm^{-1}$)	r_e (Å)	Observed Transitions: Design.	ν_{00}	References
$^{14}N_2$ (continued)										
E $^3\Sigma_g^+$	(95858)	[2185] H		[1.927_3]		[6.0]	[1.117_7]	E→B, V E→A, V Herman-Kaplan b. E←X,[a]	36467.9 (Z) 46019.7_2 Z	(107) (169)*
C" $^5\Pi_u$	(93500)	Arising from $^4S+{}^2D$; according to (72) responsible for the main predissociation in C $^3\Pi_u$.								
C $^3\Pi_u$	89136.88[b]	2047.17_8 Z	28.445_0[c]	1.8247_3[d]	0.01868[e]		1.1486_9	C[f]→B,[g] V 2nd pos. gr. C←X,[h] R Tanaka b.	29671.0 Z 88977.89 Z	(1a)(29)(3)* (15a)* (53)* (21)(54) (67)*
G $^3\Delta_g$	(87900)[i]	[742.49] Z	11.85 (H)	0.9280	0.0161	[$5._0$]	1.6107			
A' $^5\Sigma_g^+$	(78800)[j]	(650)[j]	[For more recent results of an <u>ab initio</u> calculation see (199).]				(1.55)[j]			
w $^1\Delta_u$	72097.4	1559.26[k] H	11.63	1.498[ℓ]	0.0166		1.268	w→a,[m] R McFarlane IR b. w←X,[hn] R Tanaka b.	2747.29 Z 71698.4[o] Z	(84) (54)(155)
a $^1\Pi_g$	69283.06	1694.20_8 Z	13.949_1[p]	1.6169[q]	0.01793[r]	(5.89)	1.2203	a→a',[m] V McFarlane IR b. a[s]↔X,[t] R Lyman-Birge-Hopfield b.	1212.28 Z 68951.20 Z	(62)(83) (64)* (121)
a' $^1\Sigma_u^-$	68152.66	1530.25_4 Z	12.0747[u]	1.4799	0.01657[v]	(5.55)	1.2755	a'↔X,[w] R Ogawa-Tanaka-Wilkinson-Mulliken b.	67739.31 Z	(30)* (65)* (103)(183) (155)

N_2: [a]Resonance-like electron impact excitation function centered at 12.2 eV with a half-width of 0.4 eV (143). Lifetime of the E state 190 μs (132).

[b]A = 39.2 (2a).

[c]$\omega_e y_e = +\ 2.0383_3$, $\omega_e z_e = -\ 0.5350$.

[d]Breaking-off on account of predissociation (3) in v'=1, 2, 3, 4 above N'=65, 55, 43, 28, resp., yielding an accurate dissociation limit at 97938 cm^{-1} ($^4S+{}^2D$). A second prediss. in high-pressure discharges (when the first prediss. disappears) has been found in v'=2 and 3 above N'=80 and 67, respectively (7). According to (72) the first predissociation is caused by C" $^5\Pi_u$, the second by C' $^3\Pi_u$. Predissociation in $^{15}N_2$ (50). Intensity perturbations (1a)(3a)(3b).

N_2 (continued):

[e] $- 0.00228(v+\frac{1}{2})^2 + 0.000733(v+\frac{1}{2})^3 - 0.00015(v+\frac{1}{2})^4$.

[f] Lifetimes for v=0,1,2 vary between 35 and 41 ns (128)(136)(160)(178)(192). For f values of C − B see (47)(52).

[g] The head of the 0-0 band produces laser oscillation; high resolution measurements of the laser lines (101), see also (91). An anomalous intensity alternation has been observed by (100), see also (93). C(v=5)←B(v=6) band in absorption (198). $^{14}N^{15}N$ and $^{15}N_2$ isotope shifts (34). RKR Franck-Condon factors (73)(77), dependence on rotation (117). Integrated band intensities (38). Dependence of the electronic transition moment on r (139)(150); absolute transition probabilities (139).

[h] RKR Franck-Condon factors (73)(75)(196).

[i] $A_0 = -0.21$, $A_1 = -0.25$. All constants for this state are from H→G (145); for a more detailed theoretical treatment and somewhat different constants see (164).

[j] From the predissociations in a and B (42); see also (27)(43). The dissociation energy of this state is estimated to be between 850 and 1100 cm^{-1}. According to (31) the $^5\Sigma_g^+$ state plays an important role in the mechanism of the Lewis-Rayleigh afterglow of nitrogen.

[k] Vibrational constants from the absorption spectrum (54), good agreement with band origin data for k→w (175).

[ℓ] Rotational constants from y→w (26).

[m] Appears in stimulated emission.

[n] The w← X Tanaka bands appear diffuse even under high resolution (197) indicating that this is a pressure-induced transition which has apparently no measurable spontaneous transition probability (187). Observed in solid N_2 by (92).

[o] From $v_{00}(a-X) + v_{00}(w-a)$ (84). The value from the w← X absorption spectrum is 71740.3 (head) indicating a pressure shift of ~ +40 cm^{-1}; compare with a"← X.

[p] $\omega_e y_e = +0.00793_5$, $\omega_e z_e = +0.000291$, from (64); (196) give very slightly different numbers.

[q] Small Λ-type doubling, $|q_0| = 0.00010$ (84). Breaking-off at low pressure above v=6, J=13 for both Λ components because of predissociation (16). The state causing the predissociation is $^5\Sigma_g^+$ from $^4S + {}^4S$.

[r] $\gamma_e = -0.000029_3$.

[s] The lifetime is about 100 μs but depends strongly on v. Non-exponential decay because of radiative interactions with a' $^1\Sigma_u^-$ and w $^1\Delta_u$ (148). See also (89)(140) who give f values.

[t] This transition has both a magnetic dipole and an electric quadrupole component (28)(69), see also (140). Observed in absorption in solid N_2 by (92). RKR Franck-Condon factors (73)(75)(196). From intensity measurements and the Franck-Condon factors of (73) it is concluded by (82) that the electronic transition moment can be considered as constant for most bands of this system. Comparison with intensities in the electron energy loss spectrum (61).

[u] $\omega_e y_e = +0.04129$, $\omega_e z_e = -0.000290$, from (65); (196) give very slightly different numbers.

[v] $\gamma_e = +0.0000241$.

[w] Franck-Condon factors (75)(196); $f_{00}(a'-X) = 8.4 \times 10^{-11}$, corresponding to a lifetime of 0.013 s (187).

State	T_e	ω_e		$\omega_e x_e$	B_e	α_e	D_e ($10^{-6}cm^{-1}$)	r_e (Å)	Observed Transitions Design.	ν_{00}	References
$^{14}N_2$ (continued)											
B' $^3\Sigma_u^-$	66272.4$_7$	1516.88	Z	12.18$_1$[a]	1.473$_3$[b]	0.0166$_6$[c]	(5.56)	1.278$_4$	B'→B, R "Y" bands,[d]	6545.5 (Z)	(32)(36) (182)
									B'↔X,[e] R Ogawa-Tanaka-Wilkinson b.	65852.35 Z	(30)* (35) (66)* (149) (155)
W $^3\Delta_u$	59808	1501.$_4$	(Z)	11.6					W↔B, R,V Wu-Benesch b.	73	(102)(124) (131)(157)
									W←X,[f] R Saum-Benesch b.	59380	(123)* (155)
B $^3\Pi_g$	59619.3$_5$[g]	1733.39	Z	14.122[h]	1.6374$_5$[i]	0.0179$_1$[j]	[5.9]	1.2126$_0$	B[k]↔A,[ℓ] V 1st pos. gr.	9552.0$_3$ Z	(29)(196)
									B←X,[m] R Wilkinson b.	59306.81 Z	(40)
A $^3\Sigma_u^+$	50203.6$_3$	1460.64	Z	13.87$_2$[n]	1.4546[o]	0.0180[p]	[6.1$_5$]	1.2866	A[q]↔X,[m] R Vegard-Kaplan b.	49754.78 Z	(29)(70)(85)
X $^1\Sigma_g^+$	0	2358.57	Z	14.324[r]	1.99824$_1$[s]	0.017318[s]	[5.76]	1.09768$_5$	Rot.-vibr.[t] and rot. sp.:		
									- pressure induced		(14)(59)(63) (86)(135) (141)(181)
									- el. field induced		(185)
									Raman spectra[u]		(20)(134) (167)
									Mol. beam magn. reson.[v]		

N_2: [a]$\omega_e y_e$ = + 0.0418$_6$, $\omega_e z_e$ = - 0.000732 (196).
[b]Spin splitting constants (v=5): λ = +0.66, γ = -0.0030 (66).
[c]γ_e = + 0.000009 (196).
[d]Also referred to as "infrared afterglow bands".
[e]RKR Franck-Condon factors (75)(196). Rotational intensity distribution (120).
[f]Franck-Condon factors (123).
[g]A_e = 42.24 (133).

N_2 (continued):

[h] $\omega_e y_e = -0.0569$, $\omega_e z_e = +0.00361$ (196); slightly different constants are given by (74)(133).

[i] Predissociation above v=12, N=33 (2)(153). The state causing the predissociation is probably A' $^5\Sigma_g^+$. Inverse predissociation A' $^5\Sigma_g^+ \rightarrow$ B $^3\Pi_g$ seems to be responsible for some of the phenomena in active nitrogen [see (142) and references quoted there]. The levels v'=12, 11, 10 are preferably excited in the Lewis-Rayleigh afterglow; for excitation of other levels in the afterglow see (193).

[j] $\gamma_e = -0.00007_7$ (196); slightly different constants in (133).

[k] Lifetime τ(v=0...12) = 5.0 μs (176); see also (79). For B – A absorption f values (≈ 0.0025) see (41)(78)(97).

[ℓ] Stimulated emission for some of the lines of the 4-2, 3-1, 2-0, 2-1, 1-0, 0-0, 0-1 bands has been observed (91). RKR Franck-Condon factors (73)(77), dependence on rotation (117). Dependence of the electronic transition moment on r (139)(150); absolute transition probabilities (111)(139).

[m] Franck-Condon factors (75)(196). Rotational intensity distribution in the Vegard-Kaplan bands (122).

[n] $\omega_e y_e = +0.010_3$, $\omega_e z_e = -0.0019_7$ (196); slightly different constants in (74)(133).

[o] Spin splitting constants (v=0): $\lambda = -1.33$, $\gamma = -0.003$ (70); see also (133). The radio-frequency spectrum of this state was studied by the molecular beam magnetic resonance method (127)(158); magnetic hyperfine and electric quadrupole coupling constants.

[p] $\gamma_e = -0.00008_8$ (196).

[q] Lifetime 1.3 s for the F_2 levels, 2.5 s for the F_1 and F_3 levels (115)(116)(138); electronic transition moment, variation with r (96)(109)(115).

[r] $\omega_e y_e = -0.00226$, $\omega_e z_e = -0.00024$ (196). (167) gives $\Delta G(\frac{1}{2}) = 2329.916_8$ and similar data for $^{14}N^{15}N$ and $^{15}N_2$.

[s] From B_0 and B_1 of (167) and using $\gamma_e = -0.00003_3$ from (196).

[t] Predicted transition moments for quadrupole vibration spectrum (170). The quadrupole moment in the v=0 level is measured to be -1.4×10^{-26} e.s.u. cm^2 (46)(95).

[u] Raman spectra of $^{14}N^{15}N$ and $^{15}N_2$ (167)(168).

[v] $g_J(^{15}N_2) = 0.2593$, sign not determined (49). For magnetic resonance spectra of metastable N_2 in the A $^3\Sigma_u^+$ state see (127)(158).

(1) Hopfield, PR 35, 1133; 36, 789 (1930).
(1a) Coster, Brons, van der Ziel, ZP 84, 304 (1933).
(2) van der Ziel, Physica 1, 353 (1934).
(2a) Budó, ZP 96, 219 (1935).
(3) Büttenbender, Herzberg, AP(Leipzig) (5) 21, 577 (1935).
(3a) Gerö, ZP 96, 669 (1935).
(3b) Coster, Brons, ZP 97, 570 (1935).
(4) Takamine, Suga, Tanaka, Sci. Pap. IPCR (Tokyo) 34, 854 (1938).
(5) Worley, Jenkins, PR 54, 305 (1938).
(6) Gerö, Schmid, ZP 116, 598 (1940).
(7) Hori, Endo, PPMSJ 23, 834 (1941).
(8) Worley, PR 64, 207 (1943); 89, 863 (1953).
(9) Gaydon, PRS A 182, 286 (1944).
(10) Gaydon, PPS 56, 85 (1944).
(11) Herman, AP(Paris) (11) 20, 241 (1945).
(12) Gaydon, Herman, PPS 58, 292 (1946).
(13) Janin, Crozet, CR 223, 1114 (1946).
(14) Crawford, Welsh, Locke, PR 75, 1607 (1949).
(15) Janin, JRCNRS 3 (12), 156 (1950).
(15a) Herzberg, MOLSPEC 1 (1950).
(16) Douglas, Herzberg, CJP 29, 294 (1951).

N_2 (continued):

(17) Herman, CR 233, 738 (1951).
(18) Carroll, Sayers, PPS A 66, 1138 (1953).
(19) Grün, ZN 9 a, 1017 (1954).
(20) Stoicheff, CJP 32, 630 (1954).
(21) Tanaka, JOSA 45, 663 (1955).
(22) Wilkinson, Houk, JCP 24, 528 (1956).
(23) Lofthus, JCP 25, 494 (1956).
(24) Hepner, Herman, AGEP 13, 242 (1957).
(25) Lofthus, CJP 35, 216 (1957).
(26) Lofthus, Mulliken, JCP 26, 1010 (1957).
(27) Oldenberg, in "The Threshold of Space", p. 180, ed. Zelikoff (1957).
(28) Wilkinson, Mulliken, ApJ 126, 10 (1957).
(29) Dieke, Heath, Johns Hopkins Spectroscopic Report No. 17 (1959).
(30) Ogawa, Tanaka, JCP 30, 1354 (1959); 32, 754 (1960).
(31) Bayes, Kistiakowsky, JCP 32, 992 (1960).
(32) Carroll, Rubalcava, PPS 76, 337 (1960).
(33) Lofthus, Nature 186, 302 (1960).
(34) Shvangiradze, Oganezov, Chikhladze, OS(Engl. Transl.) 8, 239 (1960).
(35) Wilkinson, JCP 32, 1061 (1960).
(36) Dieke, Heath, JCP 33, 432 (1960).
(37) Ogawa, Tanaka, CJP 40, 1593 (1962).
(38) Tyte, PPS 80, 1347, 1354 (1962).
(39) Wallace, ApJ(Suppl.) 6, 445 (1962).
(40) Wilkinson, JQSRT 2, 343 (1962).
(41) Wurster, JCP 36, 2111 (1962).
(42) Carroll, JCP 37, 805 (1962).
(43) Mulliken, JCP 37, 809 (1962).
(44) Carroll, PRS A 272, 270 (1963).
(45) Huffman, Tanaka, Larrabee, JCP 39, 910 (1963).
(46) Ketelaar, Rettschnick, MP 7, 191 (1963).
(47) Nicholls, JATP 25, 218 (1963).
(48) Pearse, Gaydon, IDSPEC (1963).
(49) Chan, Baker, Ramsey, PR A 136, 1224 (1964).
(50) Frackowiak, BAPS(MAP) 12, 361 (1964).
(51) Ogawa, CJP 42, 1087 (1964).
(52) Reis, JQSRT 4, 783 (1964).
(53) Tyte, Nicholls, IAMS 2 (1964).
(54) Tanaka, Ogawa, Jursa, JCP 40, 3690 (1964).
(55) Tilford, Wilkinson, JMS 12, 231 (1964).
(56) Tilford, Wilkinson, JMS 12, 347 (1964).
(57) Cook, Metzger, JCP 41, 321 (1964).
(58) Cook, Ogawa, CJP 43, 256 (1965).
(59) Bosomworth, Gush, CJP 43, 751 (1965).
(60) Gilmore, JQSRT 5, 369 (1965).
(61) Lassettre, Meyer, Longmire, JCP 42, 807 (1965).
(62) McFarlane, PR A 140, 1070 (1965).
(63) Reddy, Cho, CJP 43, 2331 (1965).
(64) Vanderslice, Tilford, Wilkinson, ApJ 141, 395 (1965).
(65) Tilford, Wilkinson, Vanderslice, ApJ 141, 427 (1965).
(66) Tilford, Vanderslice, Wilkinson, ApJ 141, 1226 (1965).
(67) Tilford, Vanderslice, Wilkinson, ApJ 142, 1203 (1965).
(68) Benesch, Vanderslice, Tilford, Wilkinson, ApJ 142, 1227 (1965).
(69) Vanderslice, Wilkinson, Tilford, JCP 42, 2681 (1965).
(70) Miller, JCP 43, 1695 (1965).
(71) Meyer, Skerbele, Lassettre, JCP 43, 3769 (1965).
(72) Carroll, Mulliken, JCP 43, 2170 (1965).
(73) Zare, Larsson, Berg, JMS 15, 117 (1965).
(74) Artym, OS(Engl. Transl.) Suppl. 2, 2 (1966).

N_2 (continued):

(75) Benesch, Vanderslice, Tilford, Wilkinson, ApJ 143, 236 (1966).
(76) Codling, ApJ 143, 552 (1966).
(77) Benesch, Vanderslice, Tilford, Wilkinson, ApJ 144, 408 (1966).
(78) Dronov, Sobolev, Faizullov, OS(Engl.Transl.) 21, 301 (1966).
(79) Jeunehomme, JCP 45, 1805 (1966).
(80) Hesser, Dressler, JCP 45, 3149 (1966).
(81) Lassettre, Skerbele, Meyer, JCP 45, 3214 (1966).
(82) McEwan, Nicholls, Nature 209, 902 (1966).
(83) McFarlane, JQE 2, 229 (1966).
(84) McFarlane, PR 146, 37 (1966).
(85) Miller, JMS 19, 185 (1966).
(86) Shapiro, Gush, CJP 44, 949 (1966).
(87) Tilford, Wilkinson, Franklin, Naber, Benesch, Vanderslice, ApJ(Suppl.) 13, 31 (1966).
(88) Carroll, Yoshino, JCP 47, 3073 (1967).
(89) Ching, Cook, Becker, JQSRT 7, 323 (1967).
(90) Dressler, Lutz, PRL 19, 1219 (1967).
(91) Kasuya, Lide, AO 6, 69 (1967).
(92) Roncin, Damany, Romand, JMS 22, 154 (1967).
(93) Fishburne, Lazdinis, Seibert, JMS 23, 100 (1967).
(94) Tanaka, Nakamura, SL 16, 73 (1967).
(95) Buckingham, Disch, Dunmur, JACS 90, 3104 (1968).
(96) Chandraiah, Shepherd, CJP 46, 221 (1968).
(97) Cunio, Jansson, JQSRT 8, 1763 (1968).
(98) Lawrence, Mickey, Dressler, JCP 48, 1989 (1968).
(99) Hesser, JCP 48, 2518 (1968).
(100) Bleekrode, JCP 49, 951 (1968); Physica 44, 24 (1969).
(101) Parks, Rao, Javan, APL 13, 142 (1969).
(102) Wu, Benesch, PR 172, 31 (1968).
(103) Campbell, Thrush, TFS 65, 32 (1969).
(104) Carroll, Collins, CJP 47, 563 (1969).
(105) Comes, Weber, ZN 24 a, 1941 (1969).
(106) Dressler, CJP 47, 547 (1969).
(106a) Lefebvre-Brion, CJP 47, 541 (1969).
(107) Freund, JCP 50, 3734 (1969).
(108) Geiger, Schröder, JCP 50, 7 (1969).
(109) Broadfoot, Maran, JCP 51, 678 (1969).
(110) Hébert, Nicholls, JP B 2, 626 (1969).
(111) Kupriyanova, Kolesnikov, Sobolev, JQSRT 9, 1025 (1969).
(112) Lindholm, AF 40, 111 (1969).
(113) Lutz, JCP 51, 706 (1969).
(114) Nakamura, Sasanuma, Sato, Watanabe, Yamashita, Iguchi, Ejiri, Nakai, Yamaguchi, Sagawa, Nakai, Oshio, PR 178, 80 (1969).
(115) Shemansky, JCP 51, 689 (1969).
(116) Shemansky, Carleton, JCP 51, 682 (1969).
(117) Shumaker, JQSRT 9, 153 (1969).
(118) Carroll, Collins, Yoshino, JP B 3, L127 (1970).
(119) Carlson, Moddeman, Pullen, Krause, CPL 5, 390 (1970).
(120) Kovács, OS(Engl. Transl.) 28, 239 (1970).
(121) Miller, JOSA 60, 171 (1970).
(122) Miller, PR A 1, 590 (1970).
(123) Saum, Benesch, PR A 2, 1655 (1970).
(124) Saum, Benesch, AO 9, 195 (1970).
(125) Vaidyan, Santaram, JCP 52, 3068 (1970).
(126) Cook, Ogawa, JCP 53, 1292 (1970).
(127) Freund, Miller, De Santis, Lurio, JCP 53, 2290 (1970).
(128) Johnson, Fowler, JCP 53, 65 (1970).

N_2 (continued):

(129) Michels, JCP 53, 841 (1970).
(130) Van der Wiel, El-Sherbini, Brion, CPL 7, 161 (1970).
(131) Benesch, Saum, JP B 4, 732 (1971).
(132) Borst, Zipf, PR A 3, 979 (1971).
(133) Bullock, Hause, JMS 39, 519 (1971).
(134) Butcher, Willets, Jones, PRS A 324, 231 (1971).
(135) De Remigis, Welsh, Bruno, Taylor, CJP 49, 3201 (1971).
(136) Imhof, Read, JP B 4, 1063 (1971).
(137) Leoni, Dressler, ZAMP 22, 794 (1971).
(138) Meyer, Klosterboer, Setser, JCP 55, 2084 (1971).
(139) Shemansky, Broadfoot, JQSRT 11, 1385 (1971).
(140) Pilling, Bass, Braun, JQSRT 11, 1593 (1971).
(141) Sheng, Ewing, JCP 55, 5425 (1971).
(142) Becker, Fink, Groth, Jud, Kley, DFS No. 53, 35 (1972).
(143) Borst, Wells, Zipf, PR A 5, 648, 1744 (1972).
(144) Carroll, Yoshino, JP B 5, 1614 (1972).
(145) Carroll, Collins, Murnaghan, JP B 5, 1634 (1972).
(146) Cook, McNeal, JCP 56, 1388 (1972).
(147) Carter, JCP 56, 4195 (1972).
(148) Freund, JCP 56, 4344 (1972).
(149) Golde, Thrush, PRS A 330, 121 (1972).
(150) Jain, JQSRT 12, 759 (1972).
(151) Ledbetter, JMS 42, 100 (1972).
(152) Leoni, Dressler, HPA 45, 959 (1972).
(152a) Leoni, Dissertation (ETH Zürich, 1972).
(153) Polak, Slovetskii, Sokolov, OS(Engl. Transl.) 32, 247 (1972).
(154) Carroll, JCP 58, 3597 (1973).
(155) Chutjian, Cartwright, Trajmar, PRL 30, 195 (1973).
(156) Cook, Ogawa, Carlson, JGR 78, 1663 (1973).
(157) Covey, Saum, Benesch, JOSA 63, 592 (1973).
(158) De Santis, Lurio, Miller, Freund, JCP 58, 4625 (1973).
(159) Kurzweg, Egbert, Burns, JCP 59, 2641 (1973).
(160) Dotchin, Chupp, Pegg, JCP 59, 3960 (1973).
(161) Carter, Berkowitz, JCP 59, 4573 (1973).
(162) Hicks, Comer, Read, JP B 6, L65 (1973).
(163) Lee, Carlson, Judge, Ogawa, JQSRT 13, 1023 (1973).
(164) Veseth, MP 26, 101 (1973).
(165) Watson, Lang, Stewart, JP B 6, L148 (1973).
(166) Werme, Grennberg, Nordgren, Nordling, Siegbahn, Nature 242, 453 (1973).
(167) Bendtsen, JRS 2, 133 (1974).
(168) Butcher, Jones, JCS FT II 70, 560 (1974).
(169) Carroll, Doheny, JMS 50, 257 (1974).
(170) Cartwright, Dunning, JP B 7, 1776 (1974).
(171) Mohamed, Khanna, IJPAP 12, 77 (1974).
(172) Rajan, PRIA A 74, 17 (1974).
(173) Vinogradov, Shlarbaum, Zimkina, OS(Engl. Transl.) 36, 383 (1974).
(174) Vinogradov, Zimkina, Akimov, Shlarbaum, IANSF 38, 508 (1974). Engl. Transl. in BASPS 38, 69 (1974).
(175) Carroll, Subbaram, CJP 53, 2198 (1975).
(176) Chen, Anderson, PR A 12, 468 (1975).
(177) Gardner, Samson, JCP 62, 1447 (1975).
(178) Chen, Anderson, JCP 63, 1250 (1975).
(179) Wong, Lee, Wellenstein, Bonham, JCP 63, 1538 (1975).
(180) Lee, Wong, Bonham, JCP 63, 1643 (1975).
(181) Buontempo, Cunsolo, Jacucci, Weiss, JCP 63, 2570 (1975).
(182) Gartner, Thrush, PRS A 346, 103 (1975).
(183) Golde, CPL 31, 348 (1975).
(184) Yoshino, Tanaka, Carroll, Mitchell, JMS 54, 87 (1975).

N_2 (continued):

(185) Courtois, Jouvé, JMS 55, 18 (1975).

(186) Johns, Lepard, JMS 55, 374 (1975).

(187) Tilford, Benesch, JCP 64, 3370 (1976).

(188) Dunning, Cartwright, Hunt, Hay, Bobrowicz, JCP 64, 4755 (1976).

(189) Mulliken, JMS 61, 92 (1976).

(190) Yoshino, Ogawa, Tanaka, JMS 61, 403 (1976).

(191) Ledbetter, Dressler, JMS 63, 370 (1976).

(192) Osherovich, Gorshkov, OS(Engl. Transl.) 41, 92 (1976).

(193) Ung, JCP 65, 2987 (1976).

(194) Wight, Van der Wiel, Brion, JP B 9, 675 (1976).

(195) Wilden, Hicks, Comer, JP B 9, 1959 (1976).

(196) Lofthus, Krupenie, JPCRD 6, 113 (1977).

(197) Tilford, Vanderslice, Tanaka, unpubl., quoted by (196).

(198) Ledbetter, JCP 67, 3400 (1977).

(199) Krauss, Neumann, MP 32, 101 (1976).

State	T_e	ω_e	$\omega_e x_e$	B_e	α_e	D_e $(10^{-6}cm^{-1})$	r_e (Å)	Observed Transitions Design.	ν_{00}	References
$^{14}N_2^+$	**)	μ = 7.00140006		D_0^0 = 8.712$_8$ eV[a]		I.P. = 27.1 eV[b]				MAR 1977 A
		Several other much weaker X-ray emissions corresponding to higher levels of N_2^+.								
x		K state of N_2 (removal of one K electron)						x→C,	384.5 eV[c]	
								x→B,	391.33 eV	(47)(48)
								x→A,[d]	393.34 eV	
								x→X,	394.40 eV[e]	
G $^2\Sigma_g^+$	(195000)	Repulsive state corresponding to removal of a 2σ_g electron and dissociating into N(^{4}S) + N^+(2s2p^3 ^{3}P), observed in photoelectron and photoion spectra with maximum at 39.8 and limit at 37.8$_6$ eV.								(58)
C $^2\Sigma_u^+$	64608.$_1$[f]	2071.5[f] Z	9.2$_9$[g]	[1.509$_8$]	hi	4.0	[1.262$_8$]	C[j]→X, R 2nd neg. gr.	64542.0 Z	(9)(10)* (11)* (72)
D $^2\Pi_{gi}$	52318.2[k]	907.7$_1$ Z	11.91[ℓ]	1.113	0.020	5	1.471	D→A, R Janin-d'Incan b.	42654.0 Z	(13)* (16)
(a $^4\Sigma_u^+$)	(25467)	(2398)	(14)	(2.071)	(0.014)			(a→X)[m]	(25563.26)	
B $^2\Sigma_u^+$	25461.4$_6$	2419.84 Z	23.18$_9$[n] b'	[2.07456][o] p	0.024[q]	[6.17]	1.074$_2$	B[r]↔X,[s] V_R 1st neg. gr.	25566.04 Z	(1)* (1a)(6) (20)* (38) (73)
A $^2\Pi_{ui}$	9166.9$_5$[t]	1903.7$_0$[u] Z	15.02 b'	1.7444[v] p	0.0188$_3$[v]	5.6[v]	1.1749	A[w]→X,[x] R Meinel b.	9015.5$_7$ Z	(8)(73)
X $^2\Sigma_g^+$	0	2207.00 Z	16.10[y] b'	1.93176[z] p	0.01881	[6.10][a']	1.11642			

N_2^+: [a]From $D_0^0(N_2)$ + I.P.(N) - I.P.(N_2).

[b]From the electron impact appearance potential of N_2^{++} (15) and I.P.(N_2). See also (39) and ref. given there.

[c]Vertical transition.

[d]Consisting of a v" progression (v"=0...5).

[e]Observed X-ray emission line leading to X $^2\Sigma_g^+$(v=0); the value predicted from the K limit and I.P.(N_2) is 393.9 eV. The X-ray line has a shoulder at 394.1$_7$ eV probably corresponding to v"=1.

[f]Different constants are derived by (23) who assumes that the levels v=0,1,2 are perturbed and therefore does not include them in the evaluation; he gives T_e = 64562.93 cm^{-1}.

N_2^+ (continued):

[g] $\omega_e y_e = -0.43$. Vibrational numbering confirmed by $^{14}N-^{15}N$ isotope shifts (7)(17)(24). Additional vibrational levels (to v'=17) observed by (49) in the photoelectron spectrum.

[h] $B_v(v=1...6) = 1.509_6, 1.503_5, 1.497_0, 1.489_8, 1.480_4, 1.470_0$. The shape of the B_v curve is anomalous; meaningful equilibrium constants have not been determined. Spin doubling partially resolved for high N values (11).

[i] In a discharge through $He + N_2$ the v'=3 bands are strongly enhanced. It has been suggested (6) that this enhancement is due to inverse predissociation, i.e. $N(^4S) + N^+(^3P) \rightarrow N_2^+(C^2\Sigma_u^+)$; see also (11). Direct predissociation, i.e. the production of N^+ ions, has been observed mass-spectrometrically (35)(66); see also (30)(34). According to (35) the "metastable" ion $N_2^+(C^2\Sigma_u^+)$ has a lifetime of 0.8 μs before radiationless decomposition; see, however, (55) and [j]. The question which state causes the predissociation has been the subject of much discussion (54)(61)(62) following the discovery of a large isotope effect on this predissociation (43)(53)(55)(59). See also (36). Another predissociation limit at 26.70 eV above $N_2(X^1\Sigma_g^+, v=0)$ and corresponding to $^2D^o + ^3P$ has been observed by (49).

[j] Radiative lifetime of the non-predissociated levels $\tau(v \leq 2) = 0.09$ μs (34).

[k] $A = -16.5$.

[ℓ] $\omega_e y_e = +0.016$. Vibrational numbering confirmed by $^{14}N-^{15}N$ isotope shifts (17).

[m] The identification of this quartet - doublet transition by (68) has been questioned by (70) who considers it as the 1st negative system of $^{14}N^{15}N^+$.

[n] $\omega_e y_e = -0.537_5$, $\omega_e z_e = -0.0495$; these constants, derived from ω_0, $\omega_0 x_0$, ... of (1a), represent the unperturbed vibrational levels only to v = 10; for v > 10 the ΔG curve shows a positive curvature. Levels observed up to v = 29 (6).

[o] Spin splitting constant $\gamma_0 = +0.0229$ (73); (67) give $\gamma' - \gamma'' = 0.015$. Rotational intensity distribution (67); see also (57).

[p] RKR potential functions (25), see also (22)(60).

[q] Valid only to v=3. Rotational perturbations by A $^2\Pi_u$ in v = 1, 3, 5, 8, 9, 13; see (1)(1a)(2)(3)(4)(38).

[r] Lifetime $\tau(v=0) = 60.5$ ns, see (42)(65) and earlier work quoted by them. Perturbed rotational levels (e.g. v=1, N= 13, 14) have lifetimes up to 95 ns (52). (18) gives $f_{00} = 0.019$; the electronic f value at the wavelength of the 0-1 band is 0.034 (21).

[s] Observed in absorption in flash discharges (28); laser-induced fluorescence (63). From intensity measurements (45) conclude that the electronic transition moment is nearly independent of r; see, however, (18)(27)(32). Franck-Condon factors (31)(33)(72).

[t] $A(v=2,3) = -74.62$, small J dependence (73).

[u] From (16); (73) give $\Delta G(5/2) = 1813.32_7$. Levels observed in nitrogen ion beams to v=30 (46). Vibrational numbering confirmed by $^{15}N_2^+$ data (12).

[v] From the rotational constants for v=2,3 (73); also Λ-type doubling constants. For higher but less accurate B_v values see (16). $\beta_e = +0.18 \times 10^{-6}$ (73).

[w] Lifetime varies from 13.9 μs for v=1 to 7.3 μs for v=8; see (40), who give references to earlier work, and (37)(46). Transition probabilities, f values (41); $f_{00} = 0.00168$.

[x] The bands were observed in the aurora (5) before they were discovered in the laboratory (8). Dependence of the

(continued on p. 428)

N_2^+ (continued):

electronic transition moment on r (29)(41)(46)(69)(71).

[y] $\omega_e y_e$ = - 0.040. These constants (38) are only very slightly different from those implied by (6) who gives ω_0, $\omega_0 x_0$ ... (73) give $\Delta G(\frac{1}{2})$ = 2174.72_8.

[z] From B_0 and B_1 of (73). Spin splitting constant γ(v=0,1) = + 0.0083 (73); see also (67).

[a'] $D_1 = 6.38 \times 10^{-6}$(73); higher, less accurate D_v values in (38).

[b'] Experimental Franck-Condon factors for photoionization into these states from the ground state of N_2 (26)(33)(50); theoretical Franck-Condon factors (72)(74); see also (44)(51).

(1) Herzberg, AP(Leipzig) (4) 86, 189 (1928).
(1a) Coster, Brons, ZP 70, 492 (1931); 73, 747 (1932).
(2) Childs, PRS A 137, 641 (1932).
(3) Parker, PR 44, 90, 914 (1933).
(4) Brons, Physica 1, 739 (1934); PKNAW 38, 271 (1935).
(5) Meinel, ApJ 112, 562 (1950); 113, 583 (1951); 114, 431 (1951).
(6) Douglas, CJP 30, 302 (1952).
(7) Baer, Miescher, HPA 26, 91 (1953).
(8) Douglas, ApJ 117, 380 (1953).
(9) Tanaka, JCP 21, 1402 (1953).
(10) Wilkinson, CJP 34, 250 (1956).
(11) Carroll, CJP 37, 880 (1959).
(12) Liu, ApJ 129, 516 (1959).
(13) Tanaka, Namioka, Jursa, CJP 39, 1138 (1961).
(14) Wallace, ApJ(Suppl.) 6, 445 (1962).
(15) Dorman, Morrison, JCP 39, 1906 (1963).
(16) Janin, d'Incan, Stringat, Magnaval, RO 42, 120 (1963).
(17) Namioka, Yoshino, Tanaka, JCP 39, 2629 (1963).
(18) Nicholls, JATP 25, 218 (1963).
(19) Gilmore, JQSRT 5, 369 (1965).
(20) Tyte, Nicholls, IAMS 3 (1965).
(21) Wray, Connolly, JQSRT 5, 633 (1965).
(22) Joshi, JQSRT 6, 211 (1966).
(23) Joshi, PPS 87, 285 (1966).
(24) Joshi, PPS 87, 561 (1966).
(25) Singh, Rai, JMS 19, 424 (1966).
(26) Spohr, von Puttkamer, ZN 22 a, 705 (1967).
(27) Kiselyovskii, Shimanovich, OS(Engl. Transl.) 24, 266 (1968).
(28) Herzberg, CPAS 2, No. 15 (1968).
(29) Koppe, Koval, Grytsyna, Fogel, OS(Engl. Transl.) 24, 440 (1968).
(30) Fournier, Ozenne, Durup, JCP 53, 4095 (1970).
(31) Grandmontagne, Jorus, Vincent, JP(Paris) 31, 749 (1970).
(32) Brown, Landshoff, JQSRT 11, 1143 (1971).
(33) Comes, Speier, ZN 26 a, 1998 (1971).
(34) Fournier, van de Runstraat, Govers, Schopman, de Heer, Los, CPL 9, 426 (1971).
(35) Wankenne, Momigny, IJMSIP 7, 227 (1971).
(36) Fournier, Govers, van de Runstraat, Schopman, Los, JP(Paris) 33, 755 (1972).
(37) Gray, Roberts, Morack, JCP 57, 4190 (1972).
(38) Klynning, Pagès, PS 6, 195 (1972).
(39) Appell, Durup, Fehsenfeld, Fournier, JP B 6, 197 (1973).
(40) Peterson, Moseley, JCP 58, 172 (1973).
(41) Cartwright, JCP 58, 178 (1973).
(42) Dotchin, Chupp, Pegg, JCP 59, 3960 (1973).

N_2^+ (continued):

(43) Govers, van de Runstraat, de Heer, JP B 6, L73 (1973).

(44) Kosinov, Skovorodko, OS(Engl. Transl.) 35, 344 (1973).

(45) Lee, Judge, JP B 6, L121 (1973).

(46) Maier, Holland, JCP 59, 4501 (1973).

(47) Werme, Grennberg, Nordgren, Nordling, Siegbahn, Nature 242, 453 (1973).

(48) Werme, Grennberg, Nordgren, Nordling, Siegbahn, PRL 30, 523 (1973).

(49) Åsbrink, Fridh, PS 9, 338 (1974).

(50) Gardner, Samson, JCP 60, 3711 (1974).

(51) Lee, Rabalais, JCP 61, 2747 (1974).

(52) Dufayard,Negre, Nedelec, JCP 61, 3614 (1974).

(53) Govers, Fehsenfeld, Albritton, Fournier, Fournier, CPL 26, 134 (1974).

(54) Lorquet, Lorquet, CPL 26, 138 (1974).

(55) van de Runstraat, de Heer, Govers, CP 3, 431 (1974).

(56) Cartwright, Dunning, JP B 8, L100 (1975).

(57) Désesquelles, Do Cao, Vaissade, JP(Paris) 36, 795 (1975).

(58) Gardner, Samson, JCP 62, 1447 (1975).

(59) Govers, van de Runstraat, de Heer, CP 9, 285 (1975).

(60) Hajj, PL A 50, 427 (1975).

(61) Tellinghuisen, Albritton, CPL 31, 91 (1975).

(62) Roche, Lefebvre-Brion, CPL 32, 155 (1975).

(63) Engelking, Smith, CPL 36, 21 (1975).

(64) Thulstrup, Andersen, JP B 8, 965 (1975).

(65) Smith, Read, Imhof, JP B 8, 2869 (1975).

(66) Wankenne, Bolduc, Marmet, CJP 53, 770 (1975).

(67) Bouchoux, Bacis, Goure, Lambert, JQSRT 16, 451 (1976).

(68) d'Incan, Topouzkhanian, JCP 63, 2683 (1975); 64, 3494 (1976).

(69) Wu, Shemansky, JCP 64, 1134 (1976).

(70) Dressler, JCP 64, 3493 (1976).

(71) Mandelbaum, Feldman, JCP 65, 672 (1976).

(72) Lofthus, Krupenie, JPCRD 6, 113 (1977).

(73) Colbourn, Douglas, JMS 65, 332 (1977).

(74) Albritton, Schmeltekopf, Zare, "Diatomic Intensity Factors", Wiley, to be published.

**) Potential functions of the observed states (19). Predicted electronic states and potential functions (56)(64). Table of band head wavelengths (14).

State	T_e	ω_e	$\omega_e x_e$	B_e	α_e	D_e ($10^{-6}cm^{-1}$)	r_e (Å)	Observed Transitions Design.	ν_{00}	References
$^{14}N_2^{++}$		$\mu = 7.00126291$								MAY 1976 A
		Following the early work of (3), (6) have recently given the results of detailed calculations of many low-lying singlet and triplet states. On the basis of this theoretical work the assignment of the observed transition to N_2^{++} seems very probable but not conclusive.								
D $^1\Sigma_u^+$	(62928)	(1910)[a]		[1.8644]		[7.1]	[1.1364]	D→X,	62903.1_8 Z	(1)* (2)
		A few intermediate states are suggested by Auger electron (4) and double charge transfer spectra (5).								
X $^1\Sigma_g^+$ [b]	0	(1960)[a]		[1.8801]		[6.9]	[1.1317]			
$^{14}N_2^{-}$		$\mu = 7.00167435$ $D_0^0 = 7.93$ eV[a] I.P. = -1.90_1 eV[b]								MAY 1976 A
		Theoretical calculations of N_2^- states (10).								
		Additional resonances in electron scattering at higher energies, several of them associated with the lowest N_2 Rydberg states which derive from A $^2\Pi_u$ of N_2^+.								(6a)(7)(8)
E $^2\Sigma_g^+$	(77150)[c]	(2180)[c]	(16)[c]				1.11_5[c]			(4)
		Two long progressions of shape resonances in electron scattering at 8.2 - 9.6 and 9.0 - 11.2 eV above X $^1\Sigma_g^+(v=0)$ of N_2. They correspond to negative ion formation with subsequent decay to A $^3\Sigma_u^+$ and B $^3\Pi_g$, respectively, of neutral N_2.								(5)(6)(7)
X $^2\Pi_g$	0[d]	(1968)[ef]	(10)[e]				1.19_3[e]	Vibr. Raman sp.[f] ESR sp.[g]		(2) (1)

N_2^{++}: [a]Estimated from $(4B^3/D)^{\frac{1}{2}}$.

[b]Not certain that this is the ground state. As for C_2, a low-lying $^3\Pi_u$ state has been predicted (6).

(1) Carroll, CJP 36, 1585 (1958).
(2) Carroll, Hurley, JCP 35, 2247 (1961).
(3) Hurley, JMS 9, 18 (1962).
(4) Stalherm, Cleff, Hillig, Mehlhorn, ZN 24 a, 1728 (1969).
(5) Appell, Durup, Fehsenfeld, Fournier, JP B 6, 197 (1973).
(6) Thulstrup, Anderson, JP B 8, 965 (1975).

N_2^-: [a]From $D_0^0(N_2)$ and the electron affinities of N_2 and N, assuming for the latter -0.07 eV as given by (9).

[b]From the energy difference (corrected for zero-point energy) of the potential minima given by (3); see [e].

[c]Molecular constants estimated by (4) from a short progression of three very sharp resonances in electron scattering at 11.48, 11.75, 12.02 eV above X $^1\Sigma_g^+(v=0)$ of N_2. The parent is the E $^3\Sigma_g^+$ Rydberg state of N_2, the grandparent X $^2\Sigma_g^+$ of N_2^+.

[d]The minimum of the ground state potential function lies 1.92_5 eV above that of N_2 (3); see [e].

[e]Ground state constants derived by (3) from a comparison of experimental with calculated cross sections for the vibrational excitation of N_2 by resonance scattering of electrons in the 2 - 4 eV region. The vibrational levels

N_2^- (continued):

have a considerable width owing to preionization, but the width parameter Γ decreases rapidly with increasing r from 0.57 eV at $r_e(N_2)$.

[f](2) observe $\Delta G(\frac{1}{2}) \approx 1840$ for N_2^- in potassium halide single crystals.

[g]Spectrum of N_2^- in potassium halides, confirming the $^2\Pi$ nature of the ground state.

(1) Brailsford, Morton, Vannotti, JCP 50, 1051 (1969).
(2) Holzer, Murphy, Bernstein, JMS 32, 13 (1969).
(3) Birtwistle, Herzenberg, JP B 4, 53 (1971).
(4) Comer, Read, JP B 4, 1055 (1971).
(5) Snache, Schulz, PR A 6, 69 (1972).
(6) Mazeau, Gresteau, Hall, Joyez, Reinhardt, JP B 6, 862 (1973).
(6a) Mazeau, Hall, Joyez, Landau, Reinhardt, JP B 6, 873 (1973).
(7) Schulz, RMP 45, 423 (1973).
(8) Golden, Burns, Sutcliffe, PR A 10, 2123 (1974).
(9) Hotop, Lineberger, JPCRD 4 (3), 539 (1975).
(10) Thulstrup, Andersen, JP B 8, 965 (1975).

State	T_e	ω_e	$\omega_e x_e$	B_e	α_e	D_e $(10^{-7}cm^{-1})$	r_e (Å)	Observed Transitions		References
								Design.	ν_{00}	
$^{23}Na_2$		$\mu = 11.4948852$		$D_0^0 = 0.720$ eV[a]		I.P. = 4.90 eV[b]				JUN 1977 A
		Diffuse bands of Na_2 van der Waals molecules close to the lines of the principal series of Na.								(5)(7)
		Several fragments of other UV emission and absorption band systems.[c]								(4)(10)(13)
E $(^1\Pi_u)$	35557.0	106.2 H	0.6_5					E← X, R	35530.6 H	(15)(19)
D $^1\Pi_u$	33486.8	111.3[d] H	0.4_8	e				D↔X, R	33462.9 H	(13)(15)(17)(19)
$^1\Sigma_g^+$	(33000)	Fragment observed in two-photon excited Na_2 fluorescence.								(37)
C $^1\Pi_u$	29382	119.33[f] H	0.53	e				C↔X, R	29362 H	(12)(14)(17)
B $^1\Pi_u$	20320.02	124.090 Z	0.6999[g]	0.12527_7[h]	0.000723_7[i]	3.24_8[j]	3.4228	B[k]↔X,[ℓ] R	20302.49[m] Z	(1)(6)(21)(30)
A $^1\Sigma_u^+$	14680.58	117.323 Z	0.3576[n]	0.11078_4[op]	0.000548_8[q]	3.88_2[r]	3.6384	A[s]↔X, R	14659.80 Z	(2)(8)(31)(36)(40)
a $^3\Pi$	< 14680	(145)[t]		(0.140)[t]						
X $^1\Sigma_g^+$	0	159.124_5 Z	0.7254_7[u]	0.154707[h]	0.0008736[v]	5.81_1[w]	3.0788_7	Mol. beam magn. reson.[x]		

Na_2: [a]From $D_e^0 = 5890 \pm 70$ cm^{-1} based on the RKR potential curve for the ground state (21)(30). The thermochemical value of (3), obtained by a molecular beam technique, is 0.73_2 eV.

[b]From photoionization (20)(23). A similar value is obtained by extrapolation of the Rydberg series B,C,D,E (17)(19)(26).

[c]Molecular absorption cross sections 27000 - 62500 cm^{-1} (20).

[d]Vibrational constants from (15).

[e](17) report the following rotational constants for

D: $B_e = 0.1185$, $\alpha_e = 0.001$;

C: $B_e = 0.1281_5$, $\alpha_e = 0.0008_4$.

Considerably different constants, however, are quoted by Richards in (25):

D: $B_e = 0.1152$, $\alpha_e = 0.00110$;

C: $B_e = 0.1185$, $\alpha_e = 0.00096$.

[f]Vibrational constants from (14) (except T_e which has been recalculated). (17), without details, give $T_e = 29393$, $\omega_e = 117.3$, $\omega_e x_e = 0.5_5$, while Richards (25) quotes 29384.8, 119.53, and 0.782, respectively.

[g] $- 0.00495(v+\frac{1}{2})^3 - 0.000153(v+\frac{1}{2})^4 + 7.01 \times 10^{-6}(v+\frac{1}{2})^5 - 1.804 \times 10^{-7}(v+\frac{1}{2})^6$; from the laser-induced fluorescence spectrum, including levels with $v' \leq 29$ (30). This state has a potential hump of ~ 550 cm^{-1} (0.069 eV); see (30), also (27). The non-appearance of levels with $v' > 26$ in the magnetic rota-

tion spectrum may be due to weak predissociation; see (11).

[h]RKR potential functions (30).

[i] $- 3.15_9 \times 10^{-5}(v+\frac{1}{2})^2 + 1.040 \times 10^{-6}(v+\frac{1}{2})^3 - 2.92_0 \times 10^{-8}(v+\frac{1}{2})^4$. The constants are for P and R lines, $B(R,P) - B(Q) = +0.0000128$ (30).

[j] $\beta_e = + 4.75 \times 10^{-8}$, ...; $H_e = - 2.14_5 \times 10^{-11}$, and higher order constants (30).

[k]Radiative lifetimes of 24 different levels (v',J') have been measured with an accuracy of ~1% by means of a delayed coincidence single-photon counting technique (34); the observed lifetimes [see also (22)(24)] vary from 7.0 to 7.5 ns and have been used to determine the electronic transition moment and its variation with r. The results are in good agreement with (28) and with the ab initio calculations of (39). See, however, (27)(29).

[ℓ]Franck-Condon factors, dependence on rotation (34).

[m]The band origin given here does not include $-B\Lambda^2$.

[n] $+ 5.167 \times 10^{-6}(v+\frac{1}{2})^3 + 9.277 \times 10^{-6}(v+\frac{1}{2})^4 - 1.456 \times 10^{-7}(v+\frac{1}{2})^5$; from (40), see also (31). The observations extend to $v' = 44$ (36)(40).

[o]Rotational perturbations in v=0 and 1 are caused by levels belonging to the three components of the lower-lying a $^3\Pi$ state (31). Similar perturbations affect the higher vibrational levels and are responsible for the appearance of an A−X magnetic rotation spectrum (8)(9).

[p]RKR potential function (40).

[q] $+ 1.625 \times 10^{-8}(v+\frac{1}{2})^2 + 3.165 \times 10^{-8}(v+\frac{1}{2})^3 - 9.205 \times 10^{-10}(v+\frac{1}{2})^4$; from (40), see also (31).

[r] $H_e = + 1.129 \times 10^{-12}$ (40), see also (31).

[s]Radiative lifetimes (35) are nearly constant for $1 \leq v \leq 25$, $\tau = 12.5 \pm 0.5$ ns, in very good agreement with theory (39).

[t]Constants estimated from the perturbations in A $^1\Sigma_u^+$ (9), see [o]. (33) predict $T_e = 13500$, $\omega_e = 160$, $B_e = 0.154$.

[u] $- 0.001095(v+\frac{1}{2})^3 - 4.72 \times 10^{-5}(v+\frac{1}{2})^4 + 3.2_1 \times 10^{-7}(v+\frac{1}{2})^5 - 7.5_3 \times 10^{-9}(v+\frac{1}{2})^6$; the analysis of the B→X system includes levels with $v'' \leq 46$ (30).

[v] $- 3.14_6 \times 10^{-6}(v+\frac{1}{2})^2 - 2.40_0 \times 10^{-7}(v+\frac{1}{2})^3 + 4.8_4 \times 10^{-9}(v+\frac{1}{2})^4 - 8.7_3 \times 10^{-11}(v+\frac{1}{2})^5$, from (30).

[w] $\beta_e = + 3.59 \times 10^{-9}$; $H_e = + 1.92 \times 10^{-12}$, also higher order constants (30).

[x] $g_J = (+)0.0389_2\ \mu_N$ (18). From the nuclear resonance spectrum (16) determined eqQ, but much improved hyperfine coupling constants eqQ and c [spin-rot. const., see also (38)] in the v=0, J=28 level have recently been obtained by (32) using a laser-fluorescence molecular-beam-resonance technique.

(1) Loomis, Wood, PR 32, 223 (1928).
(2) Fredrickson, PR 34, 207 (1929).
(3) Lewis, ZP 69, 786 (1931).
(4) Kimura, Uchida, Sci. Pap. IPCR (Tokyo) 18, 109 (1932).
(5) Kuhn, ZP 76, 782 (1932).
(6) Loomis, Nusbaum, PR 40, 380 (1932).
(7) Wurm, ZP 79, 736 (1932).
(8) Fredrickson, Stannard, PR 44, 632 (1933).
(9) Carroll, PR 52, 822 (1937).
(10) Yoshinaga, PPMSJ 19, 847, 1073 (1937).
(11) King, Van Vleck, PR 55, 1165 (1939).
(12) Pearse, Sinha, Nature 160, 159 (1947).
(13) Sinha, PPS 59, 610 (1947).
(14) Sinha, PPS A 62, 124 (1949).

(continued on p. 435)

State	T_e	ω_e	$\omega_e x_e$	B_e	α_e	D_e (10^{-} cm^{-1})	r_e (Å)	Observed Transitions Design.	ν_{00}	References
$^{23}Na_2{}^+$		μ = 11.4947480 D_0^0 = 0.96 eV[a]								MAR 1977
		Theoretical calculations of excited state potential curves (1)(2)(3), oscillator strengths (4).								
X $^2\Sigma_g^+$	0	(126)[b]		(0.117)[b]			(3.54)[b]			
$^{23}Na_2{}^-$		μ = 11.4950223 D_0^0 = (0.44) eV[c] I.P. = (0.27) eV[d]								MAR 1977
$^{23}Na^{40}Ar$		μ = 14.5940364 D_e^0 = 0.0052 eV[a]								MAR 1977 A
C $^2\Sigma^+$		Bound state; continuous emission with maximum at ~23580 cm^{-1}, shifted to the red by 2160 cm^{-1} from the forbidden Na 4s-3s line at 25740 cm^{-1}.						C→X,		(5)
B $^2\Sigma^+$		Repulsive state except for a very small van der Waals minimum (2). Responsible for the far-blue-wing fluorescence spectrum of the Na D lines (16956 and 16973 cm^{-1}) in the presence of Ar gas.						B→X,		(4)
A $^2\Pi$		Attractive potential, D_e = 550 cm^{-1}.[b]					3.17[b]	A→X,		(4)
X $^2\Sigma^+$	0	Unstable except for a small van der Waals minimum (2)(4)					4.8_1[a]			
$^{23}Na^{79}Br$		μ = 17.803435_5 D_0^0 = 3.74 eV[a] I.P. = 8.3_1 eV[b]								MAR 1977
		The electron energy loss spectrum has peaks at 5.2, 7.1, 8.8, (20.0), 31.0 eV.								(13)
		Continuous absorption above ~30000 cm^{-1} with maxima near 35500 and 40000 cm^{-1}.[cd]								
A		Diffuse absorption bands from 28800 to 22600 cm^{-1}; in emission continua and diffuse bands from 31000 to 17000 cm^{-1}.						A↔X		(1)(2)(3)

Na_2 (continued):

(15) Chang, CJPS 7, 377 (1950).
(16) Logan, Coté, Kusch, PR 86, 280 (1952).
(17) Barrow, Travis, Wright, Nature 187, 141 (1960).
(18) Brooks, Anderson, Ramsey, PRL 10, 441 (1963); PR A 136, 62 (1964).
(19) Morales, ARSEFQ A 59, 3 (1963).
(20) Hudson, JCP 43, 1790 (1965).
(21) Demtröder, McClintock, Zare, JCP 51, 5495 (1969).
(22) McClintock, Demtröder, Zare, JCP 51, 5509 (1969).
(23) Foster, Leckenby, Robbins, JP B 2, 478 (1969).
(24) Baumgartner, Demtröder, Stock, ZP 232, 462 (1970).
(25) DONNSPEC (1970).
(26) Velasco, OPA 6 (2), 16 (1973).
(27) Callender, Gersten, Leigh, Yang, PRL 32, 917 (1974); PR A 14, 1672 (1976).
(28) Hessel, Smith, Drullinger, PRL 33, 1251 (1974).
(29) Williams, Rousseau, PRL 33, 1516 (1974).
(30) Demtröder, Stock, JMS 55, 476 (1975).
(31) Kusch, Hessel, JCP 63, 4087 (1975).
(32) Rosner, Holt, Gaily, PRL 35, 785 (1975).
(33) Bardsley, Junker, Norcross, CPL 37, 502 (1976).
(34) Demtröder, Stetzenbach, Stock, Witt, JMS 61, 382 (1976).
(35) Ducas, Littman, Zimmerman, Kleppner, JCP 65, 842 (1976).
(36) Kaminsky, Hawkins, Kowalski, Schawlow, PRL 36, 671 (1976).
(37) Woerdman, CPL 43, 279 (1976).
(38) König, Weber, CPL 44, 293 (1976).
(39) Stevens, Hessel, Bertoncini, Wahl, JCP 66, 1477 (1977).
(40) Kaminsky, JCP 66, 4951 (1977).

Na_2^+, Na_2^-:

[a]From $D_0^0(Na_2)$ + I.P.(Na) - I.P.(Na_2); theoretical values are D_e^0 = 1.02 eV (2) and 0.980 eV (3).
[b]Theoretical calculations of (2); (3) predict r_e = 3.3 Å and ω_e = 109.2, $\omega_e x_e$ = 0.21 (4).
[c]From $D_0^0(Na_2)$ and the electron affinities of Na_2 and Na.
[d]Estimated on theoretical grounds (2).

(1) Bottcher, Allison, Dalgarno, CPL 11, 307 (1971).
(2) See ref. (33) of Na_2.
(3) Cerjan, Docken, Dalgarno, CPL 38, 401 (1976).
(4) Kirby-Docken, Cerjan, Dalgarno, CPL 40, 205 (1976).

NaAr: [a]From atomic scattering data (1); see also (3).
[b]Morse potential parameters from the analysis of the extreme-red-wing fluorescence spectrum of the sodium D lines (16956, 16973 cm^{-1}) in the presence of Ar gas (4); $\omega_e \approx 25\ cm^{-1}$. See also (2).

(1) Düren, Raabe, Schlier, ZP 214, 410 (1968).
(2) Baylis, JCP 51, 2665 (1969).
(3) Nikiforov, Shcherba, OS(Engl. Transl.) 32, 567 (1972).
(4) York, Scheps, Gallagher, JCP 63, 1052 (1975); Scheps, Gallagher, JCP 65, 859 (1976).
(5) Tam, Moe, Bulos, Happer, OC 16, 376 (1976).

NaBr: [a]Thermochemical value [(9)(16); flame photom.(10)].
[b]Onset of a broad photoelectron peak with maximum at 8.80 eV, not corrected for thermal population of ground state levels (21). See also (20).
[c]UV absorption cross sections (12). Absorption at higher energies produces fluorescence from Na(3p ^2P), maximum efficiency at 50500 cm^{-1} (19).

(continued on p.437)

State	T_e	ω_e	$\omega_e x_e$	B_e	α_e	D_e $(10^{-7}cm^{-1})$	r_e (Å)	Observed Transitions Design.	ν_{00}	References
$^{23}Na^{79}Br$ (continued)										
X $^1\Sigma^+$	0	$302._1$[e] (Z)	(1.5_0)	0.1512533_1	0.0009409_5[f]	1.553_5[g]	2.50203_8	Rot.-vibr. sp. Rotation sp. Mol. beam electric[h] and magnetic reson.		(8) (7)(11) (14) (4)(5)(6)
$^{23}Na^{(79)}Br^+$		$D_0^0 = 0.5_7$ eV[i]								MAR 1977
A $(\frac{1}{2})$	5240[j]	Removal of an electron from the halogen 4p shell of Na^+Br^-.								
X $(\frac{3}{2},\frac{1}{2})$	0[j]									
$^{23}Na^{35}Cl$		$\mu = 13.8706867$ $D_0^0 = 4.23$ eV[a] I.P. = 8.9_3 eV[b]								MAR 1977
		Continuous absorption above $\sim$34000 cm^{-1} with maxima at (38500) and 42000[c] cm^{-1}.[d]								
A		Diffuse bands are seen (2) in absorption from 33000 to 27000 cm^{-1}; the emission from 33000 to 18000 cm^{-1} appears to be continuous.						A↔X		(2)
X $^1\Sigma^+$	0	366[e] (Z)	(2.0_5)	0.2180630_9	0.0016248_2[f]	3.120_2[g]	2.36079_5	Rot.-vibr. sp.[h] Rotation sp. Mol. beam electric[i] and magnetic reson.		(11) (6)(9)(15) (18)(19)(20) (3)(4)(5)(7)(8)
$^{23}Na^{(35)}Cl^+$		$D_0^0 = 0.4_4$ eV[j]								MAR 1977
A $(\frac{1}{2})$	3710[k]	Removal of an electron from the halogen 3p shell of $Na^+C\ell^-$.								
X $(\frac{3}{2},\frac{1}{2})$	0[k]									
$^{23}Na^{(35)}Cl^-$		$D_0^0 = (1.29)$ eV[ℓ] I.P. = (0.67) eV[m]								MAR 1977

NaBr (cont'd), $NaBr^+$:

[d]The absorption spectrum of NaBr trapped in inert gas matrices shows band structure at $v > 32300$ cm^{-1}; under certain conditions as many as 16 peaks have been observed with spacings of approximately 453 and 523 cm^{-1} in Ar, 529 in Kr, and 472 in N_2 (17).

[e]From the IR spectrum (8); $\omega_e = 298.4_9$ and $\omega_e x_e = 1.16$ have been calculated by (11) from the microwave results using Dunham's theory.

[f]$\gamma_e = +0.00000243$.

[g]$\beta_e = -0.0050 \times 10^{-7}$.

[h]$\mu_{el}[D] = 9.091_8 + 0.0531(v+\frac{1}{2})$, $v \leq 2$ (14). (15), quoted in (18), gives eqQ values and their dependence on v.

[i]From D_0^0(NaBr) + I.P.(Na) - I.P.(NaBr). (21) give 0.49 eV using a corrected value of I.P.(NaBr).

[j]From band maxima in the photoelectron spectrum (20)(21).

(1) Müller, AP(Leipzig) 82, 39 (1927).
(2) Beutler, Josephy, ZP 53, 747 (1929).
(3) Levi, Dissertation (Berlin, 1934).
(4) Nierenberg, Ramsey, PR 72, 1075 (1947).
(5) Logan, Coté, Kusch, PR 86, 280 (1952).
(6) Coté, Kusch, PR 90, 103 (1953).
(7) Honig, Mandel, Stitch, Townes, PR 96, 629 (1954).
(8) Rice, Klemperer, JCP 27, 573 (1957).
(9) Brewer, Brackett, CRev 61, 425 (1961).
(10) Bulewicz, Phillips, Sugden, TFS 57, 921 (1961).
(11) Rusk, Gordy, PR 127, 817 (1962).
(12) Davidovits, Brodhead, JCP 46, 2968 (1967).
(13) Geiger, Pfeiffer, ZP 208, 105 (1968).
(14) Hebert, Lovas, Melendres, Hollowell, Story, Street, JCP 48, 2824 (1968).
(15) Melendres, Thesis (U. of California, 1968).

NaBr, $NaBr^+$ (continued):

(16) JANAF (1971).
(17) Oppenheimer, Berry, JCP 54, 5058 (1971).
(18) Miller, Finney, Inman, AD 5, 1 (1973).
(19) Earl, Herm, JCP 60, 4568 (1974).
(20) Goodman, Allen, Cusachs, Schweitzer, JESRP 3, 289 (1974).
(21) Potts, Williams, Price, PRS A 341, 147 (1974).

NaCl, $NaCl^+$, $NaCl^-$:

[a]Thermochemical value [(13)(21), flame photom.(10)(12)(14)].

[b]Adiabatic potential from the onset of a broad photoelectron peak with maximum at 9.34 eV (23); not corrected for thermal population of ground state levels.

[c]Also observed in the electron energy loss spectrum (17) which has additional peaks at 7.1, 9.6, (20.5), 31.5 eV.

[d]UV absorption cross sections (16).

[e]From the IR spectrum (11). The Dunham relations applied to the microwave results of (15) give $\omega_e = 364.6_0$, $\omega_e x_e = 1.76$. In argon matrix $\Delta G(\frac{1}{2}) = 335$ (24).

[f]$\gamma_e = +0.0000051_4$.

[g]$\beta_e = -0.008_3 \times 10^{-7}$; $H_e = -3.40 \times 10^{-14}$.

[h]For IR spectrum of matrix isolated sodium chloride see (24).

[i]$\mu_{el}[D] = 8.97141 + 0.05940(v+\frac{1}{2}) + 0.00025(v+\frac{1}{2})^2$, $v \leq 3$ (18) (20). For electric quadrupole coupling constants and their dependence on v see (19)(20); earlier results by the magnetic resonance method.

[j]From D_0^0(NaCl) + I.P.(Na) - I.P.(NaCl); (23) suggest 0.33 eV using a corrected value for I.P.(NaCl).

[k]From band maxima in the photoelectron spectrum (22)(23).

[l]From D_0^0(NaCl) and the electron affinities of NaCl and Cl.

[m]Estimated electron affinity (25).

References on p. 439.

State	T_e	ω_e	$\omega_e x_e$	B_e	α_e	D_e (10^{-6} cm^{-1})	r_e (Å)	Observed Transitions Design.	ν_{00}	References
$^{23}Na^{133}Cs$		$\mu = 19.599483_0$		$D_e^0 = (0.42)$ eV[a]						MAR 1977
		Fragment of another system overlapping C – X.								
C	(24270)	(62) H						C← X, (R)	24250 H	(1)* (2)
B	(18250)	(65) H						B← X, R	18233 H	(1)* (2)
X $^1\Sigma^+$	0	(98) H					(4.0)[a]	Mol. beam rf el. reson.[b]		
$^{23}Na^{19}F$		$\mu = 10.4021896$		$D_0^0 = (5.3_3)$ eV[a]						MAR 1977
		The electron energy loss spectrum has peaks at 5.7, 7.7, (18.5), 32.5 eV.								(15)
		Continuous absorption above 41000 cm^{-1}.								(1)
A		Fluctuation bands in absorption from 39350 to 36600 cm^{-1}.						A← X		(4)
X $^1\Sigma^+$	0	536[bc] (Z)	3.4[b]	0.4369012_7	0.0045586_9[d]	1.16_1	1.92594_7	Rot.-vibr. sp.[c]		(11)(14)
								Rotation sp.		(8)(10)(13)
								Mol. beam rf electric[e]		(10)(12)
								and magnetic reson.		(2)(3)(5)

NaCs: [a]Potential data for the $^1\Sigma^+$ ground state from total scattering cross-section measurements (3); see also (4). (3) give also potential parameters for the $^3\Sigma^+$ excited state ($D_e = 0.031$ eV, $r_e = 6.2$ Å) which arises from the same atomic products $^2S + ^2S$.

[b]From the measured Stark coefficient $\mu_0^2/B_0 = 1.81 \times 10^{-18}$ cm^3 and assuming $B_0 = 0.0631$ cm^{-1} (5) obtain $\mu_{e\ell}(v=0) = 4.7_5$ D.

(1) Walter, Barratt, PRS A 119, 257 (1928).
(2) Weizel, Kulp, AP(Leipzig) 4, 971 (1930).
(3) Neumann, Pauly, PRL 20, 357 (1968); JCP 52, 2548 (1970).

NaCs (continued):

(4) Schlier, Vietzke, CPL 3, 250 (1969).
(5) Dagdigian, Wharton, JCP 57, 1487 (1972).

NaF: [a]Value recommended by (17) and based on the highest Na levels observed in chemiluminescent emission from the reaction $Na_2 + F \rightarrow NaF + Na^*$. It is in agreement with the flame-photometric value (5.2_5 eV) of (7), but substantially higher than the thermochemical value (4.9_3 eV) given by (6) or derived from the data in (16).

[b]From the IR spectrum (11)(14).

NaF (continued):

[c]For IR frequencies in low-temperature rare gas matrices see (9).

[d]γ_e = + 0.00002335.

[e]$\mu_{e\ell}[D] = 8.123_5 + 0.0644(v+\frac{1}{2}) + 0.0003_7(v+\frac{1}{2})^2$, $v \leq 2$ (10)(12); see also (8). Na quadrupole coupling constant, dependence on v (8)(10)(12); earlier, less accurate values by the magnetic resonance method (2)(3)(5).

(1) Müller, AP(Leipzig) 82, 39 (1927).
(2) Zeiger, Bolef, PR 85, 788 (1952).
(3) Logan, Coté, Kusch, PR 86, 280 (1952).
(4) Barrow, Caunt, PRS A 219, 120 (1953).
(5) Coté, Kusch, PR 90, 103 (1953).
(6) Brewer, Brackett, CRev 61, 425 (1961).
(7) Bulewicz, Phillips, Sugden, TFS 57, 921 (1961).
(8) Bauer, Lew, CJP 41, 1461 (1963); 42, 830 (1964).
(9) Snelson, Pitzer, JPC 67, 882 (1963).
(10) Hollowell, Hebert, Street, JCP 41, 3540 (1964).
(11) Ritchie, Lew, CJP 42, 43 (1964).
(12) Gräff, Werth, ZP 183, 223 (1965).
(13) Veazey, Gordy, PR A 138, 1303 (1965).
(14) Baikov, Vasilevskii, OS(Engl. Transl.) 22, 198 (1967).
(15) Geiger, Pfeiffer, ZP 208, 105 (1968).
(16) JANAF (1971).
(17) Ham, JCP 60, 1802 (1974).

$NaC\ell$, $NaC\ell^+$, $NaC\ell^-$ (continued):

(1) See ref. (1) of NaF.
(2) Levi, Dissertation (Berlin, 1934).
(3) Nierenberg, Ramsey, PR 72, 1075 (1947).
(4) See ref. (2) of NaF.
(5) See ref. (3) of NaF.
(6) Stitch, Honig, Townes, PR 86, 813 (1952).
(7) See ref. (5) of NaF.
(8) Ochs, Coté, Kusch, JCP 21, 459 (1953).
(9) Honig, Mandel, Stitch, Townes, PR 96, 629 (1954).
(10) Gurvich, Veits, DPC 116, 639 (1957).
(11) Rice, Klemperer, JCP 27, 573 (1957).
(12) Gurvich, Veits, BASPS 22, 670 (1958).
(13) See ref. (6) of NaF.
(14) See ref. (7) of NaF.
(15) Clouser, Gordy, PR A 134, 863 (1964).
(16) Davidovits, Brodhead, JCP 46, 2968 (1967).
(17) See ref. (15) of NaF.
(18) Hebert, Lovas, Melendres, Hollowell, Story, Street, JCP 48, 2824 (1968).
(19) Cederberg, Miller, JCP 50, 3547 (1969).
(20) de Leeuw, van Wachem, Dymanus, JCP 53, 981 (1970).
(21) JANAF (1971).
(22) Goodman, Allen, Cusachs, Schweitzer, JESRP 3, 289 (1974).
(23) Potts, Williams, Price, PRS A 341, 147 (1974).
(24) Ismail, Hauge, Margrave, JMS 54, 402 (1975).
(25) Jordan, JCP 65, 1214 (1976).

State	T_e	ω_e	$\omega_e x_e$	B_e	α_e	D_e (10^{-4}cm^{-1})	r_e (Å)	Observed Transitions Design.	ν_{00}	References
$^{23}Na^{1}H$		μ = 0.96549966 D_0^0 = (1.88) eV[a]								MAR 1977
b $^3\Pi$	(30940)[b]	(419)[b]	(50)[b]	(3.53)[b]	(0.85)[b]		(2.22)[b]			
A $^1\Sigma^+$	22719	310.$_6$[c] Z		1.696[d]		2.27[e]	3.209	A[f]↔X, R	22294.$_5$ Z	(1)* (2)* (3)(4)
X $^1\Sigma^+$	0	1172.2 Z	19.72[g]	4.9012	0.1353[h]	3.32[h]	1.8874			

Extensive theoretical calculations have been made of X $^1\Sigma^+$ (7)(9)(10)[see also (6)(8)] and of A $^1\Sigma^+$, B $^1\Pi$, a $^3\Sigma^+$, b $^3\Pi$, c $^3\Sigma^+$ (9); B, a, and c are predicted to be unstable. The calculations of (9) include transition moments, band strengths, line strengths, and the far-wing broadening of the sodium D lines; see also (14). Additional higher excited states have been computed by (11).

State	T_e	ω_e	$\omega_e x_e$	B_e	α_e	D_e	r_e	Design.	ν_{00}	References
$^{23}Na^{2}H$		μ = 1.85186304								MAR 1977
A $^1\Sigma^+$		i		j		k		A←X, R	v(7-0)= 24106.2 Z	(3)
X $^1\Sigma^+$	0	[826.1$_0$] Z		2.5575	0.0520	0.915[ℓ]	1.8866			

$^{23}Na^{4}He$
$^{23}Na^{4}He^+$

μ = 3.4090714$_3$ — MAR 1977

Interaction potentials for X $^2\Sigma^+$, A $^2\Pi$, B $^2\Sigma^+$ have been calculated by (1)(2)(5)(6), for X $^1\Sigma^+$ by (2); see also (3). The coefficients of collision-induced absorption by Na - He in the region 0 - 300 cm^{-1} have been evaluated by (4).

$^{23}Na^{127}I$

μ = 19.463754$_1$ D_0^0 = 3.00 eV[a] I.P. = 7.64 eV[b] — MAR 1977 A

Peaks in the electron energy loss spectrum at 3.3, 5.7, 8.1, (20.0), 31.0 eV. (15)

A — Dense, discrete rotational structure in absorption from 18500 to 33950 cm^{-1}(2)(8)(18), maximum of absorption at 30800.[c] At higher energies continuous absorption with maxima at ~38900

Na^1H, Na^2H:

[a]Theoretical value as modified by (10); the computational results are $D_e^0 = 1.92$ (10) and 1.88 eV (9). Extrapolation of the ground state vibrational levels suggests 2.1 eV.

[b]Theoretical predictions of (9) who feel, however, that the true potential of this state is considerably deeper than the calculated potential curve.

[c]Anomalous potential curve [an RKR curve has been constructed by (5)]; $\Delta G(3/2...39/2)$ = 329.9, 337.2, 343.7, 349.6, 353.9, 357.4, 359.2, 360.3, 360.1, 359.5, 357.7, 354.3, 352.3, 348.4, 343.9, 340.0, 333.7, 330.0, 323.7 [from band origins (4)].

[d]$B_1...B_7$ = 1.823, 1.875, 1.908, 1.930, 1.938, 1.941, 1.936 (4).

[e]$H_e \approx +5.7 \times 10^{-8}$ (4); both D_v and H_v decrease with increasing v.

[f]Radiative lifetimes $\tau(v'=3;J'=8) = 24._0$ ns, $\tau(4;11) = 28._3$ ns, $\tau(5;16) = 27._1$ ns (12); $\tau(8;3) = 22._7$ ns (13).

[g]$\omega_e y_e = +0.160$, $\omega_e z_e = -0.005$ (4).

[h]Rotational constants from (3), $\beta_e = -0.03 \times 10^{-4}$. (4) gives $B_e = 4.886$, $\alpha_e = 0.129$, $D_e \approx 3.15 \times 10^{-4}$, $H_e \approx +1.7 \times 10^{-8}$ but includes only levels with $v'' \geq 3$.

[i]$\Delta G(15/2...33/2) = 255.0_5$, 256.92, 258.48, 259.39, 260.04, 260.22, 260.08, 259.61, 258.85, 257.6 (from b. origins).

[j]$B_7...B_{17}$ = 1.010, 1.012, 1.012, 1.010, 1.007, 1.003, 0.998, 0.991, 0.984, 0.975, 0.95.

[k]$D_7...D_{17} = 0.53 \times 10^{-4}...0.44 \times 10^{-4}$.

[l]$\beta_e = -0.009 \times 10^{-4}$.

(1) Hori, ZP 62, 352 (1930).
(2) Hori, ZP 71, 478 (1931).
(3) Olsson, ZP 93, 206 (1935).
(4) Pankhurst, PPS A 62, 191 (1949).

Na^1H, Na^2H (continued):

(5) Jain, Sah, JCP 38, 1553 (1963).
(6) Varshni, Shukla, RMP 35, 130 (1963).
(7) Cade, Huo, JCP 47, 649 (1967).
(8) Cade, Bader, Henneker, Keaveny, JCP 50, 5313 (1969).
(9) Sachs, Hinze, Sabelli, JCP 62, 3367, 3377, 3384, 3389, 3393 (1975).
(10) Meyer, Rosmus, JCP 63, 2356 (1975).
(11) Numrich, Truhlar, JPC 79, 2745 (1975).
(12) Baltayan, Jourdan, Nedelec, PL A 58, 443 (1976).
(13) Dagdigian, JCP 64, 2609 (1976).
(14) Watson, Stewart, Dalgarno, MP 32, 1661 (1976).

NaHe, $NaHe^+$:

(1) Baylis, JCP 51, 2665 (1969).
(2) Krauss, Maldonado, Wahl, JCP 54, 4944 (1971).
(3) Bottcher, CPL 18, 457 (1973).
(4) Bottcher, Dalgarno, Wright, PR A 7, 1606 (1973).
(5) Pascale, Vandeplanque, JCP 60, 2278 (1974).
(6) Bottcher, Cravens, Dalgarno, PRS A 346, 157 (1975).

NaI: [a]Dissociative photoionization of NaI (12). Thermochemical values (9) tend to be slightly higher, atomic fluorescence gives slightly lower values (neglecting, however, the thermal population of the ground state vibrational levels); for references see (12). Flame photometry suggests 3.1_3 eV (10).

[b]From photoionization (12); in good agreement with the photoelectron spectrum (21).

[c]In inert gas matrices absorption bands from 25000 to 26000 cm^{-1} (19); the vibrational interval in Ar is ~166.

(continued on p. 442)

State	T_e	ω_e	$\omega_e x_e$	B_e	α_e	D_e ($10^{-7}cm^{-1}$)	r_e (Å)	Observed Transitions Design.	ν_{00}	References
$^{23}Na^{127}I$ (continued)										
		and 46300 cm^{-1}(1)(2).[d] Bands in emission from 24000 to 20000 cm^{-1}(2).								
X $^1\Sigma^+$	0	258[e] (Z)	(1.0_8)	0.1178056_1	0.0006477_7[f]	0.973_4[g]	2.71145_2	Rot.-vibr. sp.		(7)
								Rotation sp.		(6)(11)
								Mol. beam rf electric[h]		(16)(17)
								and magn. reson.[i]		(3)(4)(5) (13)
$^{23}Na^{127}I^+$		$D_0^0 = 0.50$ eV[j]								MAR 1977
A ($^2\Sigma$)	9520[k]	Removal of an electron from the halogen 5p shell of Na^+I^-.								
X ($^2\Pi_{1/2}$)	1770[k]									
X ($^2\Pi_{3/2}$)	0[k]									
$^{23}Na^{39}K$		$\mu = 14.4586992$		$D_0^0 = 0.62$ eV[a]		I.P. = 4.5 eV[b]				MAR 1977 A
E	25228[c]	95.8_5 H	0.94					E← X, R	25214 H	(2)
D ($^1\Pi$)	20090.6	82.17 H	0.350[d]					D← X, R	20070.0 H	(1)(2)*
C $^1\Pi$	16993.8	71.3_3 Z	1.23_6[e]	0.0574_4 [i]	0.00007_8[f]	$32._1$[g]	4.50_5	C← X, R	16967.2 Z	(1)(2)* (8)
A ($^1\Sigma^+$)	12139.7	79.85_2 H	0.0872[h]					A← X, R	12118.0 H	(1)
X $^1\Sigma^+$	0	124.134 Z	0.511[j]	0.0905	0.000458_4[k]	8.1_9[g]	3.58_9	Mol. beam rf electric[ℓ]		(5)
								and magn. reson.[m]		(4)

Note: It now appears that the ground state constants of (6) are unreliable; the uncertainty also affects the new results for the C state (8) [Arimondo, private comm. (1978)].

NaI (continued), NaI^+:

[d] UV absorption cross sections (14).

[e] From the IR spectrum (7). Application of the Dunham relations to the microwave results of (11) gives $\omega_e = 259.2_0$, $\omega_e x_e = 0.96_4$.

[f] $\gamma_e = +0.00000143$.

[g] $\beta_e = -0.0005 \times 10^{-7}$.

[h] $\mu_{e\ell}[D] = 9.210_3 + 0.0507(v+\frac{1}{2})$, $v \leq 2$ (16). Electric quadrupole coupling constants, dependence on v (17); earlier

NaI, NaI$^+$ (continued):

results by the magnetic resonance method (3)(4)(5).

[i] $g_J = (+)0.027\ \mu_N$ (13).

[j] From $D_0^0(NaI) + I.P.(Na) - I.P.(NaI)$; (21) give 0.60 eV.

[k] From the maxima in the photoelectron spectrum (20)(21).

(1) Schmidt-Ott, ZP 69, 724 (1931).
(2) Levi, Dissertation (Berlin, 1934).
(3) Nierenberg, Ramsey, PR 72, 1075 (1947).
(4) Logan, Coté, Kusch, PR 86, 280 (1952).
(5) Cóte, Kusch, PR 90, 103 (1953).
(6) Honig, Mandel, Stitch, Townes, PR 96, 629 (1954).
(7) Rice, Klemperer, JCP 27, 573 (1957).
(8) Berry, JCP 27, 1288 (1957).
(9) Brewer, Brackett, CRev 61, 425 (1961).
(10) Bulewicz, Phillips, Sugden, TFS 57, 921 (1961).
(11) Rusk, Gordy, PR 127, 817 (1962).
(12) Berkowitz, Chupka, JCP 45, 1287 (1966).
(13) Mehran, Brooks, Ramsey, PR 141, 93 (1966).
(14) Davidovits, Brodhead, JCP 46, 2968 (1967).
(15) Geiger, Pfeiffer, ZP 208, 105 (1968).
(16) Hebert, Lovas, Melendres, Hollowell, Story, Street, JCP 48, 2824 (1968).
(17) Miller, Zorn, JCP 50, 3748 (1969).
(18) Berg, Skewes, JCP 51, 5430 (1969).
(19) Oppenheimer, Berry, JCP 54, 5058 (1971).
(20) Goodman, Allen, Cusachs, Schweitzer, JESRP 3, 289 (1974).
(21) Potts, Williams, Price, PRS A 341, 147 (1974).

NaK: (8) Allegrini, Moi, Arimondo, CPL 45, 245 (1977).

NaK: [a] Extrapolation of the vibrational levels in C $^1\Pi$ assuming that this state has no potential maximum.

[b] Photoionization appearance potential (3).

[c] Recalculated; (2) gives 25201 which must be erroneous.

[d] $\omega_e y_e = -0.00814$.

[e] $\omega_e y_e = +0.008_8$, $\omega_e z_e = -0.00022_3$. All constants obtained from a study of the C state vibrational-rotational structure by laser excitation and collision-induced dissociation (8). Similar vibrational constants have been derived by (1) from absorption and magnetic rotation spectra; convergence limit at 18025 ± 250 cm^{-1} above X $^1\Sigma^+(v=0)$, corresponding to $Na(^2S) + K(4p\ ^2P)$.

[f] $\gamma_e = -0.000002_3$.

[g] Centrifugal distortion constants as given by (8) and (6); there is no agreement with D_e values calculated from $4B_e^3/\omega_e^2$.

[h] $\omega_e y_e = -0.00389$.

[i] Perturbations (possibly by a $^3\Pi$ state) are very likely the cause for the appearance of a weak magnetic rotation sp..

[j] $-7.9 \times 10^{-5}(v+\frac{1}{2})^3 - 2.4_7 \times 10^{-7}(v+\frac{1}{2})^4 - 2.5_0 \times 10^{-7}(v+\frac{1}{2})^5$ (6).

[k] $-9.4 \times 10^{-7}(v+\frac{1}{2})^2 - 9.0_8 \times 10^{-8}(v+\frac{1}{2})^3$ (6).

[ℓ] $\mu_{e\ell}(v=0) = 2.66_7$ D [recalculated from $\mu_0^2/B_0 = 3.966 \times 10^{-19}$ cm^3 (5) with the new value for B_0]. Electric quadrupole coupling constants (5); see also (4).

[m] $g_J = 0.0253\ \mu_N$ (4). For an NMR experiment using optical pumping see (7).

(1) Loomis, Arvin, PR 46, 286 (1934).
(2) Sinha, PPS 60, 447 (1948).
(3) Foster, Leckenby, Robbins, JP B 2, 478 (1969).
(4) Brooks, Anderson, Ramsey, JCP 56, 5193 (1972).
(5) Dagdigian, Wharton, JCP 57, 1487 (1972).
(6) Toueg, Thesis (Columbia U., 1974), quoted by (8).
(7) König, Weber, CPL 44, 293 (1976).

State	T_e	ω_e	$\omega_e x_e$	B_e	α_e	D_e (10^{-} cm^{-1})	r_e (Å)	Observed Transitions Design.	Observed Transitions ν_{00}	References
$^{23}Na^{(84)}Kr$		(μ = 18.045680$_2$)		D_e^0 = 0.0085 eV[a]						MAR 1977
C $^2\Sigma$		Attractive potential; emission continuum with maximum at 23260 cm^{-1}, shifted to the red by 2480 cm^{-1} from the forbidden Na 4s-3s line at 25740 cm^{-1}.						C→X,		(5)
B $^2\Sigma$		Unstable except for a very small van der Waals minimum (2); responsible for the far-blue-wing fluorescence spectrum of the Na D lines (16956 and 16973 cm^{-1}) in a Kr atmosphere.						B→X,		(4)
A $^2\Pi$		Attractive potential; $D_e \approx 730$ cm^{-1}.[b]					3.25[b]	A→X,		(4)
X $^2\Sigma^+$	0	Repulsive except for a small van der Waals minimum (2)(4).					4.7$_3$[a]			
$^{23}Na^{(20)}Ne$		(μ = 10.6932986)								MAR 1977 A
B $^2\Sigma$				D_e = (0.0050) eV[a]			(5.3)[a]			
A $^2\Pi$				D_e = 0.014$_9$ eV[a]			4.2$_3$[a]			
X $^2\Sigma^+$	0			D_e^0 = 0.0014 eV[a]			4.8$_2$[a]			
$^{23}Na^{16}O$		μ = 9.4324071		D_0^0 = 2.6$_0$ eV[a]						MAR 1977
A ($^2\Sigma$)	(1500)[b]	(548)[b]		(0.470)[b]			(1.95)[b]			
X ($^2\Pi$)	0 [b]	(526)[b]		(0.425)[b]			(2.05)[b]			
$^{23}Na^{16}O^+$				D_0^0 = 0.8 eV[c]						
$^{23}Na^{(85)}Rb$		(μ = 18.091514$_0$)								MAR 1977
		A second system (18400 - 18800 cm^{-1}) has not been analyzed.								(1)
A ($^1\Pi$)	16421.8	61.49 H	0.945					A←X,	(R) 16399.1 H	(2)
X ($^1\Sigma$)	0	106.64 H	0.455					Mol. beam el. deflection[a]		(3)

State	T_e	ω_e	$\omega_e x_e$	B_e	α_e	D_e (10^{-} cm^{-1})	r_e (Å)	Observed Transitions Design.	ν_{00}	References
$^{23}Na^{(132)}Xe$		($\mu = 19.577567_2$) $D_0^0 = 0.0124$ eV[a]								MAR 1977
C $^2\Sigma$		Attractive potential; continuous emission with maximum at 22690 cm^{-1}, shifted to the red by 3050 cm^{-1} from the forbidden Na 4s-3s line at 25740 cm^{-1}.						C→X		(5)
B $^2\Sigma$		Unstable except for a small van der Waals minimum (2); responsible for the far-blue-wing fluorescence spectrum of the Na D lines (16956 and 16973 cm^{-1}) in a Xe atmosphere.						B→X,		(4)
A $^2\Pi$		Attractive potential; $D_e = 1130$ cm^{-1}.[b]					3.3_1[b]	A→X,		(4)
X $^2\Sigma^+$	0	Repulsive except for a small van der Waals minimum (2)(4).					4.9_1[a]			

NaKr, NaXe:

[a]From atomic scattering data (1); see also (2)(3).

[b]Morse potential parameters from the analysis (4) of the extreme-red-wing fluorescence spectrum of the Na D lines in the presence of Kr or Xe gas; $\omega_e \approx 74$ or 83, respectively. See also (2).

(1) Düren, Raabe, Schlier, ZP 214, 410 (1968).

(2) Baylis, JCP 51, 2665 (1969).

(3) Nikiforov, Shcherba, OS(Engl. Transl.) 32, 567 (1972).

(4) York, Scheps, Gallagher, JCP 63, 1052 (1975); Scheps, Gallagher, JCP 65, 859 (1976).

(5) Tam, Moe, Bulos, Happer, OC 16, 376 (1976).

NaNe: [a]All constants derived from high-resolution measurements of differential cross-sections for scattering of Na in the ^2S ground and ^2P excited states from Ne (3). There is serious disagreement with the pseudo-potential calculations of (1)(2).

NaNe (continued):

(1) See ref. (2) of NaKr, NaXe.

(2) Pascale, Vandeplanque, JCP 60, 2278 (1974).

(3) Carter, Pritchard, Kaplan, Ducas, PRL 35, 1144 (1975).

NaO, NaO^+:

[a]Thermochemical value [mass-spectrom.(2), calorim.(3)].

[b]Theoretical results, average of two calculations (4)(6). The relative term values of the two states are in qualitative agreement with deductions made from a magnetic deflection analysis of the reactive scattering of Na + NO_2, see (1).

[c]From a merging-beam study of the reaction $CO^+(Na,C)NaO^+$ (5). See also (4).

References on p. 447 .

NaRb: See p. 447 .

NaXe: See NaKr.

State	T_e	ω_e	$\omega_e x_e$	B_e	α_e	D_e ($10^{-7}cm^{-1}$)	r_e (Å)	Observed Transitions Design.	ν_{00}	References
$^{93}Nb_2$		$\mu = 46.453190_2$								MAR 1975
A		$(280)^a$						$A \leftarrow X,^a$	(18704)	(1)
X	0									
$^{93}Nb^{14}N$		$\mu = 12.1689420_2$								OCT 1975
A $^3\Phi$		0-0 sequence only.		$[0.4951]^a$		$[3.8]^a$	$[1.672_7]$	$^3\Phi_4 \rightarrow {}^3\Delta_3$, R $^3\Phi_3 \rightarrow {}^3\Delta_2$, R $^3\Phi_2 \rightarrow {}^3\Delta_1$, R	16859.91 H^Q 16542.89 H^Q 16144.57 H^Q	(1)
(X) $^3\Delta$		$[(1002.5)]^b$		$[0.5010]^a$		$[3.5]^a$	$[1.662_8]$			
$^{93}Nb^{16}O$		$\mu = 13.6456564_1$		$D_0^0 = 7.8$ eV^a						OCT 1975
G $^4\Sigma^-$ b	21385.3	850.5 Z	3.3_7	0.4001^b	0.0019	2.7	1.757_2	$G \leftrightarrow X$, R	21316.2 Z	(1)(3)* (4)* (5)* (9)(11)
F	(20740)	$(908)^c$	(4)					$F \leftarrow X,^c$	(20704)	(11)
D	(16640)	$[(932)]^d$						$D \leftarrow X,^d$	(16618)	(11)
C	(15580)	$(904)^d$	(2)					$C \leftarrow X,^d$	(15544)	(10)(11)
B	(15300)	$(891)^d$	(4)					$B \leftarrow X,^d$	(15256)	(11)
E	(15270)	$(847)^d$						$E \leftarrow X,^d$	(15206)	(11)
A'								$A' \leftarrow X,^d$	(14467)	(11)
A	(13720)	$[(899)]^d$						$A \leftarrow X,^d$	(13676)	(11)
X $^4\Sigma^-$ b	0	989.0 Z	3.8_3	0.4321^b	0.0021	2.2	1.690_9	IR and ESR sp.^{e}		(10)(11)
NbO^+		The emitter of a spectrum tentatively attributed to NbO^+ (1) has been shown by isotopic substitution to be NbN [T. M. Dunn, quoted in (2)]. For references see NbN.								

Nb_2: [a]Absorption in Ar matrix at 14 K.

(1) Green, Gruen, JCP 57, 4462 (1972).

NbN, NbO^+:

[a]Constants for the F_2 component. Large nuclear hyperfine structure in $^3\Delta_1$ and $^3\Delta_3$, $b \approx 0.196\ cm^{-1}$ (3). A smaller effect was also observed in the excited $^3\phi$ state (3).
[b]IR spectrum of $Nb^{14}N$ in Ar matrix at 14 K (2). Identity with the lower state of the gas phase sp. not certain.

(1) Dunn, Rao, Nature 222, 266 (1969).
(2) Green, Korfmacher, Gruen, JCP 58, 404 (1973).
(3) Féménias, Athénour, Dunn, JCP 63, 2861 (1975); see also Féménias, Athénour, Stringat, CJP 52, 361 (1974); 53, 542, 2353(erratum) (1975).

NbO: [a]Thermochemical value (mass-spectrom.)(6).
[b]The rotational analysis by (3) of seven bands in the G – X system assumed that they represent a $^2\Delta - {}^2\Delta$ transition. More recently, additional branches were found (7)(8) having very wide nuclear hyperfine structure [$b \approx 0.19\ cm^{-1}$ (12), as compared to $0.165\ cm^{-1}$ from the matrix ESR spectrum (11)]. The four branches of (3) probably represent the F_2 and F_3 components of the $^4\Sigma^- - {}^4\Sigma^-$ transition. The rotational analysis by (4) appears to be in error.

NbO (continued):

[c]Only observed in a Ne matrix where the F state interacts with the nearby G $^4\Sigma^-$ state.
[d]Absorption in rare gas matrices. Data are for Ne except for E← X which is only observed in Ar. So far, no satisfactory analysis was given for the complex systems of R shaded emission bands in the gas phase spectrum from 13300 to 18200 cm^{-1} (1)(2)(5)*; see, however, (8)(12).
[e]In rare gas matrices.

(1) Rao, IJP 24, 35 (1950).
(2) Rao, Premaswarup, IJP 27, 399 (1953).
(3) Uhler, AF 8, 265 (1954).
(4) Rao, Nature 173, 1240 (1954); PNISI A 21, 188, 219 (1955).
(5) Gatterer, Junkes, Salpeter, Rosen, METOX (1957).
(6) Shchukarev, Semenov, Frantseva, ZNK 11, 233 (1966). Engl. transl. in RJIC 11, 129 (1966).
(7) D. Richards, Thesis (Oxford, 1969).
(8) Dunn, in "Molecular Spectroscopy: Modern Research" (ed. Rao and Mathews), p. 231. Academic Press (1972).
(9) Singh, Shukla, JQSRT 12, 1249 (1972).
(10) See ref. (2) of NbN.
(11) Brom, Durham, Weltner, JCP 61, 970 (1974).
(12) See ref. (3) of NbN.

NaO, NaO^+:

(1) Herm, Herschbach, JCP 52, 5783 (1970).
(2) Hildenbrand, Murad, JCP 53, 3403 (1970).
(3) O'Hare, JCP 56, 4513 (1972).
(4) O'Hare, Wahl, JCP 56, 4516 (1972).
(5) Rol, Entemann, Wendell, JCP 61, 2050 (1974).
(6) So, Richards, CPL 32, 227 (1975).

NaRb: [a]$\mu^2/B = 6.6 \times 10^{-19}\ cm^3$; assuming $B = 0.0734\ cm^{-1}$ (3) derive an electric dipole moment of 3.1 D.

(1) Walter, Barratt, PRS A 119, 257 (1928).
(2) Kusch, PR 49, 218 (1936).
(3) Dagdigian, Wharton, JCP 57, 1487 (1972).

State		T_e	ω_e		$\omega_e x_e$	B_e	α_e	D_e ($10^{-6}cm^{-1}$)	r_e (Å)	Observed Transitions Design.	ν_{00}		References
$^{14}N^{79}Br$			$\mu = 11.8928384_5$			$D_0^0 = 2.9_0$ eV[a]							MAY 1977
b	$^1\Sigma^+$	14787.3	785.5	H	4.363	0.4733[b]	0.0152		1.730_6	b→X,[c]	V 14834.3	H	(1)(2)*
X	$^3\Sigma^-(0^+)$	0[d]	691.75	H	4.720	0.444[db]	0.0040[d]		1.78_7	Vibration sp.[e]			(3)
$^{14}N^{35}Cl$			$\mu = 9.9990234_9$										MAY 1977
b	$^1\Sigma^+$	14984.6	935.6	H[Q]	5.4	[0.6828_4]		[1.65]	[1.5713]	b→X,	V 15038.94	Z	(3)*
X	$^3\Sigma^-$	0	827.0	H[Q]	5.1	[0.6468_5][a]		[1.78]	[1.6144]	Vibration sp.[b]			(1)(2)
$^{(142)}Nd^{19}F$			(μ = 16.7552378)			$D_0^0 = 5.8_7$ eV[a]							MAR 1975
$^{(142)}Nd^{16}O$			($\mu = 14.3746936_0$)			$D_0^0 = 7.3_3$ eV[a]							MAR 1975 A
			Additional bands at higher energies (< 24000 cm^{-1}).										(1)(3)
			Numerous R shaded and a number of V shaded bands in emission from 11460 to 15950 cm^{-1}.										(2)*
$^{(142)}Nd^{(32)}S$			(μ = 26.0932231)			$D_0^0 = 4.8_5$ eV[a]							MAR 1975
$^{(142)}Nd^{(80)}Se$			($\mu = 51.125039_1$)			$D_0^0 = 3.9_3$ eV[a]							MAR 1975
$^{(142)}Nd^{(130)}Te$			(μ = 67.821022)			$D_0^0 = 3.1_2$ eV[a]							MAR 1975

NBr: [a]Estimated from the highest v' value observed in b→X emission from the reaction $N(^4S) + Br(^2P_{\frac{3}{2}}) \rightarrow NBr^* \rightarrow NBr + h\nu$ (2).

[b]Potential functions (4).

[c]Franck-Condon factors (5).

[d]Only levels with N = J+1, i.e. $F_1(N)$ corresponding in case "c" to the 0^+ component, have been observed at high resolution. For this reason, and since only B''_3 and B''_7 have been determined (assuming $\lambda \approx 8.7$, $\gamma \approx -0.7$) the values of B_e and α_e are not very precise.

[e]In argon and nitrogen matrices, $\Delta G(\frac{1}{2}) = 691$.

(1) Elliott, PRS A 169, 469 (1939).
(2) Milton, Dunford, Douglas, JCP 35, 1202 (1961).
(3) Milligan, Jacox, JCP 40, 2461 (1964).
(4) Singh, Rai, IJPAP 4, 102 (1966).
(5) Itagi, Shamkuwar, Itagi, IJP 45, 385 (1971).

NCℓ: [a]Spin coupling constants $\lambda_0 = 1.77_6$, $\gamma_0 = -0.0071_5$.

[b]In argon and nitrogen matrices, $\Delta G(\frac{1}{2}) = 825$.

(1) Milligan, JCP 35, 372 (1961).
(2) See ref. (3) of NBr.
(3) Colin, Jones, CJP 45, 301 (1967).

NdF: [a]Thermochemical value (mass-spectrom.)(1).

(1) Zmbov, Margrave, JCP 45, 3167 (1966).

NdO: [a]Thermochemical value (mass-spectrom.)(4), recalc. (5).

(1) Piccardi, AANL (Ser. 6) 21, 584 (1935).
(2) Gatterer, Junkes, Salpeter, Rosen, METOX (1957).
(3) Herrmann, Alkemade, "Chemical Analysis by Flame Photometry", Interscience (1963).
(4) Ames, Walsh, White, JPC 71, 2707 (1967).
(5) Smoes, Coppens, Bergman, Drowart, TFS 65, 682 (1969).

NdS: [a]Thermochemical value (mass-spectrom.)(1), recalc. (2).

(1) See ref. (5) of NdO.
(2) Bergman, Coppens, Drowart, Smoes, TFS 66, 800 (1970).

NdSe, NdTe:

[a]Thermochemical value (mass-spectrom.)(1).

(1) See ref. (2) of NdS.

State	T_e	ω_e	$\omega_e x_e$	B_e	α_e	D_e $(10^{-}\ cm^{-1})$	r_e (Å)	Observed Transitions Design.	ν_{00}	References
$^{(20)}Ne_2$		(μ = 9.9962203)		D_0^0 = 0.00202 eV[a]						MAR 1977
M (0_u^+)	(166580)	[30][b] H	(10)	Progr. of three b.[c] converging to $^1S + 5s'[\frac{1}{2}]_1$.				M← X,	166590[b] H	(4)
L (0_u^-)	(166362)	[70][b] H	(7.5)	Progr. of four b.[c] converging to $^1S + 5s'[\frac{1}{2}]_0$.				L← X, (V)	166390[b] H	(4)
K (0_u^+)	(165750)	[50][b] H	(10)	Progr. of five b. converging to $^1S + 5s[\frac{3}{2}]_1$.				K← X, V	165770[b] H	(4)
J (1_u)	(164220)	[190][b]		Progr. of nine b.[d] converging to $^1S + 5s[\frac{3}{2}]_2$.				J← X,	164310[b]	(4)
		Diffuse unclassified bands 162940 - 164040 cm^{-1}, correlated to ^1S + states derived from $2p^54p$.								(4)
I	(161950)	[79][b] H		Progr. of five b.[e] converging to $^1S + 3d'[\frac{3}{2}]_1$.				I← X,	161980[b] H	(4)
H (0_u^+)	(160322)	[201][b] H	(7)	Progr. of eight b.[f] converging to $^1S + 3d[\frac{1}{2}]_1$.[g]				H← X, V	160415[b] H	(4)*
G (0_u^+)	(160233)	[213][b] H	(9.3)	Progr. of eight b.[f] converging to $^1S + 3d[\frac{3}{2}]_1$.				G← X, V	160334[b] H	(4)*
F (0_u^+)	(159347)	[60][b] H	(7.5)	Progr. of four b. converging to $^1S + 4s'[\frac{1}{2}]_1$.				F← X, V	159370[b] H	(4)*
E (0_u^-)	(159135)	[80][b] H	(10)	Progr. of four b. converging to $^1S + 4s'[\frac{1}{2}]_0$.				E← X, V	159170[b] H	(4)*
D (0_u^+)	(158635)	[60][b] H	(10)	Progr. of three b. converging to $^1S + 4s[\frac{3}{2}]_1$.				D← X, V	158660[b] H	(4)*
C (1_u)	(156480)	[290][b]	(10)	Progr. of ten b.[h] converging to $^1S + 4s[\frac{3}{2}]_2$.				C← X, (V)	156620[b]	(4)*
		Diffuse unclassified bands 150400 - 153600 cm^{-1}, correlated to ^1S + states derived from $2p^53p$.								(4)
Theoretical potential functions, vibrational levels, and lifetimes for states derived from $^1S + 2p^53s\ ^{1,3}P$ are given by (9).										
(1_u)	(135900)	Unstable; responsible for broadening to shorter wavelengths of the resonance line at 735.9 Å.								(4)*
B (0_u^+)	(135761)	[58][b] H		Only two levels[f] observed, correlated to $^1S + 3s'[\frac{1}{2}]_1$.[g]				B← X, V	135778[b] H	(4)*
(1_u)	(134500)	Unstable; responsible for broadening to shorter wavelengths of the resonance line at 743.7 Å.								(4)*
A $^1\Sigma_u^+(0_u^+)$	(133800)	[176][b] H	[i]	Progr. of four b.[f] converging to $^1S + 3s[\frac{3}{2}]_1$.[g]				A← X, V	133899[b] H	(4)*
		Continuous emission with maxima at 134400 ("first continuum") and 121200[j] ("second continuum")								(1)*

State	T_e	ω_e	$\omega_e x_e$	B_e	α_e	D_e (10^{-} cm^{-1})	r_e (Å)	Observed Transitions Design.	ν_{00}	References
$^{(20)}Ne_2$ (continued)										
a $^3\Sigma_u^+$		Not observed in absorption from the ground state, but tentatively assigned as lower state of an absorption at 12270 cm^{-1} in Ne (200 - 1060 torr) excited by high-current short-duration electron bursts from a Febetron source (10); lifetime $\tau \approx 6.6_2$ μs.								
X $^1\Sigma_g^+$	0	$[13.7]^k$ H		[0.17]	(0.06)		$[3.1_5]$			
$^{(20)}Ne_2^+$		$D_e^0 = 1.30$ eV^a								MAR 1977
		Theoretical potential functions for the ground state and excited states arising from $^1S + {}^2P_{\frac{3}{2},\frac{1}{2}}$ (4).								
X $^2\Sigma_u^+$	0	$(510)^b$		$(0.554)^b$			$(1.75)^b$			

Footnotes and references on p. 453.

Ne_2: [a]From $\Delta G''(\frac{1}{2})$ observed in the VUV absorption spectrum (4) and the binding energy of v=1 (6) as corrected by (7). (7) recommend $D_e^0 = 28._6$ cm^{-1}(0.0035_5 eV), a value derived from solid-state data (5a) and in fair agreement with the Ne-Ne scattering data of (2) and (5) [$31._9$ and $29._7$ cm^{-1}, resp.] and the ab initio calculations of (3) and (8) [$28._2$ and $27._2$ cm^{-1}, resp.]; see also (11) who give $30._2$.

[b]Lowest observed level and interval $\Delta G(v+\frac{1}{2})$; the first observed band may not have v' = 0.

[c]It is not certain that these two progressions belong to two separate electronic states.

[d]All bands are very diffuse.

[e]Irregular intervals and intensities; tentative classification.

[f]Rotational structure partially resolved.

[g]Also weaker progression with v" = 1.

[h]Diffuse bands.

[i]$\Delta G(v+\frac{3}{2}) = 93$, $\Delta G(v+\frac{5}{2}) = 40$.

[j]Attributed to A→X (4), possibly also contributions from a→X.

[k]Zero-point energy 12.3 cm^{-1}[see (7)].

(1) Tanaka, Jursa, LeBlanc, JOSA 48, 304 (1958).
(2) Siska, Parson, Schafer, Lee, JCP 55, 5762 (1971).
(3) Gordon, Kim, JCP 56, 3122 (1972).
(4) Tanaka, Yoshino, JCP 57, 2964 (1972).
(5) Farrar, Lee, Goldman, Klein, CPL 19, 359 (1973).
(5a) Goldman, Klein, JLTP 12, 101 (1973).
(6) Tanaka, Yoshino, Freeman, JCP 59, 564, 5748 (erratum) (1973).
(7) Le Roy, Klein, McGee, MP 28, 587 (1974)
(8) Stevens, Wahl, Gardner, Karo, JCP 60, 2195 (1974).
(9) Cohen, Schneider, JCP 61, 3230, 3240 (1974).
(10) Oka, Rao, Redpath, Firestone, JCP 61, 4740 (1974).
(11) Nain, Aziz, Jain, Saxena, JCP 65, 3242 (1976).

State	T_e	ω_e	$\omega_e x_e$	B_e	α_e	D_e (10^{-} cm^{-1})	r_e (Å)	Observed Transitions Design.	ν_{00}	References
$^{(20)}Ne^{40}Ar$		(μ = 13.3257930$_9$)		D_e^0 = 0.0062 eV[a]						MAR 1977
X $^1\Sigma^+$	0	[(20.9)][b]		D_0^0 = (0.0045) eV[b]			3.43[a]	Translational sp.[c]		(1)
$^{(20)}Ne^{40}Ar^+$										MAR 1977
A_3 ($\frac{1}{2}$)	(46960)[a]	Five band groups [called A,B,C,D,E by (1)] with some vibrational structure which has, however, not been assigned.						$A_3 \rightarrow X_2$	45380	(1)
A_2 ($\frac{3}{2}$)	(46180)[a]							$A_3 \rightarrow X_1$	46960	
A_1 ($\frac{1}{2}$)	(45920)[a]							$A_2 \rightarrow X_2$	44590	
X_2 ($\frac{1}{2}$)	(1585)[a]							$A_2 \rightarrow X_1$	46180[b]	
X_1 ($\frac{1}{2}$,$\frac{3}{2}$)	0 [a]							$A_1 \rightarrow X_1$	45920	
$^{(20)}Ne^{19}F$		(μ = 9.7413757$_7$)								MAR 1977 A
		No experimental results. Theoretical potential energy curves for X $^2\Sigma^+$ and A $^2\Pi$ (1).								
$^{(20)}Ne^{1}H$		(μ = 0.95945861)								MAR 1977
		No spectra observed, all data are theoretical (5):								
C $^2\Sigma^+$	(78170)[a]	(2737)[b]		(14.8)			(1.089)			
B $^2\Pi$	(70200)[a]	(2913)		(18.0)			(0.988)			
A $^2\Sigma^+$	(70000)[a]	(2801)		(17.7)			(0.996)			
X $^2\Sigma^+$	0 [a]	Repulsive st. with van der Waals minimum.								
$^{(20)}Ne^{1}H^+$				D_0^0 = 2.08 eV[c]						MAR 1977
		Several excited states briefly discussed in (6).								
X $^1\Sigma^+$		(2917)[d]		(18.0)[d]			(0.98$_9$)[d]			

Ne_2^+: [a]From elastic scattering of Ne^+ by Ne (3). A similar value ($D_0^0 = 1.3_5$ eV) from spectral line profiles in a Ne afterglow (dissociative recombination radiation of neon) is given by (1)(2). From <u>ab initio</u> theory (4) derive $D_e = 1.17$ eV.
[b]Theoretical values (4).

(1) Connor, Biondi, PR A <u>140</u>, 778 (1965).
(2) Frommhold, Biondi, PR <u>185</u>, 244 (1969).
(3) Mittmann, Weise, ZN <u>29</u> a, 400 (1974).
(4) Cohen, Schneider, JCP <u>61</u>, 3230, 3240 (1974).

NeAr: [a]From crossed-beam differential scattering measurements (4).
[b]Theoretical values (5); the ground state potential supports four bound vibrational levels.
[c]Dipole moment function from <u>ab initio</u> calculations (2), from translational absorption spectra (3).

(1) Bosomworth, Gush, CJP <u>43</u>, 751 (1965).
(2) Matcha, Nesbet, PR <u>160</u>, 72 (1967).
(3) Bar-Ziv, Weiss, CPL <u>19</u>, 148 (1973).
(4) Ng, Lee, Barker, JCP <u>61</u>, 1996 (1974).
(5) Bobetic, Barker, JCP <u>64</u>, 2367 (1976).

$NeAr^+$: [a]The splittings between X_1 and X_2 and between A_2 and A_3 are very similar to those of the ^2P ground states of Ar^+ and Ne^+ (1431 and 780 cm^{-1}, resp.). The splitting between A_1 and A_2 is molecular in nature; both states arise from $^2P_{3/2}$ of Ne^+ ($\Omega = 1/2$ and $3/2$). The corresponding splitting of X_1 is not resolved.

$NeAr^+$ (continued):

[b]The transition energy $A_2 - X_1$ is close to the difference of $Ne^+ + Ar$ and $Ne + Ar^+$ which is 46820 cm^{-1}. This agreement together with that mentioned in [a] makes the interpretation by (1) of the observed spectrum as a charge transfer spectrum very convincing.

(1) Tanaka, Yoshino, Freeman, JCP <u>62</u>, 4484 (1975).

NeF: (1) Gardner, Karo, Wahl, JCP <u>65</u>, 1222 (1976).

NeH, NeH^+:

[a]Energies relative to $Ne(^1S) + H(1s\,^2S)$. A, B, C have calculated dissociation energies D_e of 1.53, 1.50, 0.51 eV (5).
[b]Predicted potential hump of 0.87 eV at 2.1 Å (5).
[c]From a comparative study of the reactions $H_2^+ + He \rightarrow HeH^+ + H$ and $H_2^+ + Ne \rightarrow NeH^+ + H$ by photoionization mass-spectrometry (2); $D_0(NeH^+) - D_0(HeH^+) = +0.24$ eV. Theoretical values for D_e vary between 2.1 and 2.3 eV (1)(5)(6); similar values are obtained from proton scattering on Ne (3)(4).
[d]Theoretical values (5).

(1) Peyerimhoff, JCP <u>43</u>, 998 (1965).
(2) Chupka, Russell, JCP <u>49</u>, 5426 (1968).
(3) Rich, Bobbio, Champion, Doverspike, PR A <u>4</u>, 2253 (1971).
(4) Weise, Mittmann, Ding, Henglein, ZN <u>26</u> a, 1122 (1971).
(5) Bondybey, Pearson, Schaefer, JCP <u>57</u>, 1123 (1972).
(6) Vasudevan, MP <u>30</u>, 437 (1975).

State	T_e	ω_e	$\omega_e x_e$	B_e	α_e	D_e ($10^{-6}cm^{-1}$)	r_e (Å)	Observed Transitions Design.	ν_{00}	References
$^{(20)}Ne^{(84)}Kr$		(μ = 16.1456407)		D_e^0 = 0.0064 eV[a]						MAR 1977
X $^1\Sigma^+$	0	[(18.7)][b]		D_0^0 = (0.0048) eV[b]			3.58[a]			
$^{(20)}Ne^{(84)}Kr^+$										MAR 1977
A_3 ($\frac{1}{2}$)	(60790)[c]	Five band groups [called A,B,C,D,E by (3)] with some vibrational structure which has, however, not been assigned.						$A_3 \rightarrow X_2$	55310	(3)
A_2 ($\frac{3}{2}$)	(60020)[c]							$A_3 \rightarrow X_1$	60790	
A_1 ($\frac{1}{2}$)	(59770)[c]							$A_2 \rightarrow X_2$	54560	
X_2 ($\frac{1}{2}$)	(5470)[c]							$A_2 \rightarrow X_1$	60020[d]	
X_1 ($\frac{1}{2}$,$\frac{3}{2}$)	0[c]							$A_1 \rightarrow X_1$	59770	
$^{(20)}Ne^{(132)}Xe$		(μ = 17.3610604)		D_e^0 = 0.0065 eV[a]						MAR 1977 A
X $^1\Sigma^+$	0	[(17.5)][b]		D_0^0 = (0.0048) eV[b]			3.74_5[a]	Translational sp.		(1)
$^{(20)}Ne^{(132)}Xe^+$										MAR 1977
A_3 ($\frac{1}{2}$)	(75300)[c]	Five band groups [called A,B,C,D,E by (3)] with some vibrational structure which has, however, not been assigned.						$A_3 \rightarrow X_2$	64730	(3)
A_2 ($\frac{3}{2}$)	(74520)[c]							$A_3 \rightarrow X_1$	75300	
A_1 ($\frac{1}{2}$)	(74460)[c]							$A_2 \rightarrow X_2$	63940	
X_2 ($\frac{1}{2}$)	(10575)[c]							$A_2 \rightarrow X_1$	74520[d]	
X_1 ($\frac{1}{2}$,$\frac{3}{2}$)	0[c]							$A_1 \rightarrow X_1$	(74460)	
$^{14}N^{19}F$		μ = 8.0613378_9		D_0^0 = 3.5 eV[a]						MAY 1977
		Theoretical potential functions and spectroscopic constants for the ground and several excited states (5)(6)(8).								
b $^1\Sigma^+$	18877.05	1197.49 Z	8.64	1.2377_0	0.01448	5.28	1.2998_3	$b^b \rightarrow X$,[c] V	18905.20 Z	(2)*
a $^1\Delta$	[12003.60]			[1.2225][d]		[4.5]	[1.3079]	$a \rightarrow X$, V	11435.16 Z	(3)*
X $^3\Sigma^-$	0	1141.37 Z	8.99	1.2056_8[e]	0.01492	5.39	1.3169_8	Vibration sp.[f]		(1)

State	T_e	ω_e	$\omega_e x_e$	B_e	α_e	D_e (10^{-} cm^{-1})	r_e (Å)	Observed Transitions Design.	ν_{00}	References
NF^+, NF^-	No experimental data; ab initio calculations (5)(6)(7a).									MAY 1977

NeKr, $NeKr^+$;
NeXe, $NeXe^+$:

[a]From crossed-beam differential scattering measurements (2).
[b]Theoretical values (4); the ground state potentials support five or possibly six (NeXe) bound vibrational levels.
[c]The splittings between X_1 and X_2 and between A_2 and A_3 are very similar to those of the ^2P ground states of Kr^+ (5371 cm^{-1}) or Xe^+ (10537 cm^{-1}) and Ne^+ (780 cm^{-1}). The splitting between A_1 and A_2 is molecular in nature; both states arise from $^2P_{3/2}$ of Ne^+ (Ω = 1/2 and 3/2). The corresponding splitting in the ground state is not resolved.
[d]The transition energy $A_2 - X_1$ is close to the difference between the ionization potentials of Ne and Kr or Xe, i.e. 61015 or 76096 cm^{-1}. This agreement together with that mentioned in [c] makes the interpretation by (3) of the observed spectra as charge transfer spectra very convincing.

(1) Marteau, Granier, Vu, Vodar, CR B 265, 685 (1967).
(2) Ng, Lee, Barker, JCP 61, 1996 (1974).
(3) Tanaka. Yoshino, Freeman, JCP 62, 4484 (1975).
(4) Bobetic, Barker, JCP 64, 2367 (1976).

NF, NF^+, NF^-:

[a]Estimated by (5) on the basis of spectroscopic and thermochemical data; the ab initio computed value is 3.6 eV.
[b]Lifetime 0.1_6 s, estimated from the rate of attenuation of the green emission in the reaction $N + NF_2$ along a flow tube (4).
[c]Franck-Condon factors (9).
[d]The EPR spectrum of $NF(^1\Delta)$ (7) yields $\mu_{e\ell}$ = 0.37 D as well as hyperfine coupling constants.
[e]Spin coupling constants $\lambda_0 = 1.21_5$, $\gamma_0 = -0.0048$.
[f]Observed in argon and nitrogen matrices, $\Delta G(\frac{1}{2}) = 1115$.

(1) Milligan, Jacox, JCP 40, 2461 (1964).
(2) Douglas, Jones, CJP 44, 2251 (1966).
(3) Jones, CJP 45, 21 (1967).
(4) Clyne, White, CPL 6, 465 (1970).
(5) O'Hare, Wahl, JCP 54, 4563 (1971); O'Hare, JCP 59, 3842 (1973).
(6) Andersen, Öhrn, JMS 45, 358 (1973).
(7) Curran, MacDonald, Stone, Thrush, PRS A 332, 355 (1973).
(7a) Ellis, Banyard, Tait, Dixon, JP B 6, L233 (1973).
(8) Ellis, Banyard, JP B 7, 2021 (1974).
(9) Mohamed, Khanna, Lal, IJPAP 12, 243 (1974).

State	T_e	ω_e		$\omega_e x_e$	B_e	α_e	D_e ($10^{-4}cm^{-1}$)	r_e (Å)	Observed Transitions: Design.		Observed Transitions: ν_{00}		References
$^{14}N^{1}H$		μ = 0.94016028			$D_0^0 \leq 3.47$ eV[a]		I.P. = (13.63) eV[b]						MAR 1977 A
Theoretical potential functions and spectroscopic constants for the ground and excited states: (21)(44)(46); (32)(36)(51).													
d $^1\Sigma^+$	83160	2672.6	Z	71.2	14.390[c]	0.621	16.0[d]	1.1163	$d^e \rightarrow c$,[m]	V	39512.2_6	Z	(5)* (19) (23)(27) (55)*
									$d^e \rightarrow b$,[m]	R	61619.6_0	Z	(29)(55)*
c $^1\Pi$	(43744)	[2122.6_4]	Z	f	14.537[ghi]	0.593[j]	[22.0][k]	1.110_6	$c^\ell \rightarrow b$,[m]	R	22106.6_2	Z	(4)* (26)*
									$c^\ell \rightarrow a$,[m]	R	30755.5_4	Z	(1)(2)* (6) (14)*
A $^3\Pi_i$	$29807._4$[n]	3231.2	Z	98.6	16.6745[opi]	0.7454	[17.80][q]	1.0369_8	$A^r \leftrightarrow X$,[m]	s	29776.76	Z	(3)(8)* (18) (35)
b $^1\Sigma^+$	21202	3352.4	Z	74.2_4[t]	16.705[u]	0.591	16.0[v]	1.0360	$b^w \rightarrow X$,	s	21238	Z	(45)
a $^1\Delta$	(12566)[x]	[3188]	Z	(68)[y]	[16.439][i]	0.66	[16.2]	1.034_1	(a − X)		12589[z]		
X $^3\Sigma^-$	0	3282.2_7	Z	78.3_5	16.6993[a']	0.6490	[17.097][b']	1.0362_1	Rotation sp.[c']				(48)(54)
									Fundamental b. in matrices (45a)(55a)				

N^1H: [a]From the limiting curve of dissociation in c $^1\Pi$ (55), see [h] and [f]. Theoretical calculations by (33)(39)(43) suggest D_0^0 = 3.43, 3.31, 3.17 eV, resp.; the most recent prediction on theoretical and empirical grounds (46) is D_0^0 = 3.40 eV. From the electron impact appearance potential of N_2^+ from HN_3 (7) follows $D_0^0 = 3.5_9$ eV; a shock-tube measurement (20) gives 3.2_1 eV. Both results are compatible with limits derived from the study of reactions of rare gas metastables with small NH-containing molecules (31), the upper limit being closer to the semi-empirical calculations of (10)(12), the lower limit being in better agreement with the thermochemical measurements of (38).

[b]Theoretical value (33); (16) give an electron impact appearance potential of 13.1 eV.

[c](27) report a breaking-off in the d→c 1-1 band above J'=15 which they attribute to predissociation in the upper state. Intensity anomalies are confirmed (55) for the d→c 1-1 and 1-0 bands, but higher rotational levels (except J'=16) do emit in transitions to the b state (55). Similar intensity perturbations are seen in other d→b and d→c bands.

[d]$H_e = +10 \times 10^{-8}$.

[e]Radiative lifetime $\tau(v=0)$ = 18 ns (28).

[f]$\Delta G(3/2) = 1694.0_8$. The theoretical calculations of (32) predict a potential maximum resulting from the avoided

crossing of the two $^1\Pi$ states arising from $N(^2D) + H(^2S)$ and $N(^2P) + H(^2S)$.

[g] Λ-type doubling $\Delta\nu_{ef}(v=0) = +\ 0.0165\,J(J+1)$.

[h] Predissociation by rotation in v=0 above J=22 and in v=1 above J=15 (27)(55). A much weaker predissociation, not seen on the photographic plates but detected by high-resolution lifetime measurements [(50), see ℓ], affects the lower J levels of both v=0 and 1 and may be caused by interaction with the unstable $^5\Sigma^-$ state arising from ground state atomic products. All rotational structure in v=2 is diffuse except for J=1.

[i] Electric dipole moments of a $^1\Delta$, A $^3\Pi$, c $^1\Pi$ = 1.49, 1.31, 1.70 D, respectively (15); see also (25).

[j] $\gamma_e = -\ 0.347$ (55).

[k] $D_1 = 26.6 \times 10^{-4}$, $D_2 = 51._0 \times 10^{-4}$; $H_0 = -26 \times 10^{-8}$, $H_1 = -115 \times 10^{-8}$ (55).

[ℓ] Lifetime $\tau(v=0,J=2) = 411$ ns (50), decreasing with increasing rotation to 226 ns for J=17 owing to weak predissociation; in v=1 the longest lifetime (57.1 ns) was observed (50) for J=4. See also (13)(28) whose low-resolution measurements gave $\tau(v=0) \approx 460$ ns. Relative transition probabilities (41).

[m] Franck-Condon factors (37)(42).

[n] $A(v=0,1) = -34.79$; from (40) who gives also spin-spin and spin-rotation interaction constants. See also (22)(30).

[o] "True" rotational constants of (40); see also (22)(30). Λ-type doubling parameters may be found in (22)(40). The effective constants (18) are $B_e = 16.6901$, $\alpha_e = 0.7440$.

[p] Weak predissociation in v=0 and 1 above N=25 and 15, resp., observed by high-resolution lifetime measurements (50) and attributed to interaction with the unstable $^5\Sigma^-$ state arising from $^4S + {}^2S$.

[q] $+\ 9.9 \times 10^{-8} J^3(J+1)^3 - 13 \times 10^{-12} J^4(J+1)^4$; $D_1 = 17.85 \times 10^{-4}$, $H_1 = +6.0 \times 10^{-8}$ (40).

[r] Lifetime $\tau(v=0,N=4) = 404$ ns (50), first increasing to 453 ns at N=25, then decreasing rapidly to 96 ns at N=31 owing to predissociation. Similar results for v=1, see p. Absorption oscillator strength 0.0076 (11)(13)(28); f values in (17) refer to emission and should be multiplied by 2 for comparison. Relative transition probabilities (41).

[s] Undegraded 0-0 band.

[t] $\omega_e y_e = +\ 0.70$, $\omega_e z_e = -\ 0.035_3$ $(v \leq 9)$.

[u] Rotational constants from the analysis of the d→b bands [(55), $1 \leq v'' \leq 9$]. (26) gives $B_e = 16.7326$ and $\alpha_e = 0.6049$ from the c→b 0-0 and 0-1 bands.

[v] $H_e = 11 \times 10^{-8}$.

[w] Radiative lifetime $\tau = 18$ ms (47)(49).

[x] Earlier theoretical predictions by (9) and (24) gave 14200 and 13100 cm^{-1}, respectively.

[y] From a comparison with $\Delta G(\frac{1}{2})$ of ND.

[z] $\nu_{00}(b-X) + \nu_{00}(c-b) - \nu_{00}(c-a)$; in very good agreement with the singlet-triplet separation of 12500 cm^{-1} derived by (34) from the difference of the threshold energies for the production of $NH(c^1\Pi) + CO(X^1\Sigma)$ or $NH(X^3\Sigma) + CO(a^3\Pi)$ by photodissociation of HNCO. From the photoelectron spectrum of NH^- (53) find 12735 ± 137 cm^{-1}.

[a'] "True" rotational constants of (40); spin-splitting constants $\gamma = -0.011_7$, $\lambda \approx +\ 0.82$. The evaluation of the constants by (40) takes fully into account the $^3\Pi \sim {}^3\Sigma^-$ interaction and thus leads to results which differ considerably from the effective constants of earlier investigators (8)(18)(30)(35). The latter are in good agreement with more precise constants obtained recently from the laser-magnetic-

(continued on p. 459)

State	T_e	ω_e	$\omega_e x_e$	B_e	α_e	D_e ($10^{-4}cm^{-1}$)	r_e (Å)	Observed Transitions Design.	Observed Transitions ν_{00}	References
$^{14}N^2H$		μ = 1.76083613		$D_0^0 \leq 3.54$ eV[a]						MAR 1977 A
d $^1\Sigma^+$	83168	1953.7 Z	38.2	7.693	0.257[b]	[4.81][b]	1.1156	d→c, V	39484.22 Z	(8)* (16)
								d→b, R	61721.5 (Z)	(10)
c $^1\Pi$	43786	1756.5 Z	50.9_5[c]	7.833[de]	0.379[f]	[6.07][g]	1.105_5	c→b, R	22237.3_1 Z	(9)*
								c↔a,[h] R	30849.0_6 Z	(1)(2)(3)* (4)*
A $^3\Pi_i$	29820[i]	2361[j]	53[j]	[8.7575][k]	0.282_9[ℓ]	[5.08_7][k]	1.0372	A↔X, [m]	29798.7_5 Z	(3)* (5)(6) (7)(11)*
b $^1\Sigma^+$	21198	[2371.8_1] Z	$38._8$[n]	8.947_2	0.238_3	4.6_4	1.0344	b→X, [m]	21225 Z	(14)
a $^1\Delta$	(12596)	[2356.1_7] Z	(36)[o]	8.954_2	0.242_7	[4.83_8][p]	1.0340	(a−X)	12613[q]	
X $^3\Sigma^-$	0	2398[j]	42[j]	[8.7913][r]	0.2531[s]	[4.904][r]	1.0361_2	Rotation sp.[t]		(15)

N^2H: [a]From the limiting curve of dissociation in c $^1\Pi$ (16); see [e]. See also [f] of NH.

[b]$\gamma_e = +0.0033$ (v=0,1,2), $B_3 = 6.864$; $D_{1,2,3}(10^{-4}cm^{-1})$ = 4.50, 4.51, 12; $H_0 = +2.49 \times 10^{-8}$ (16).

[c]$\omega_e y_e = -10.3_7$.

[d]Λ-type doubling $\Delta\nu_{ef}(v=0) = +0.00436\,J(J+1)$, and increasing with v (8)(16).

[e]Predissociation by rotation in v=0...3 above J=30, 25, 18, 5, respectively (16).

[f]$\gamma_e = -0.047$ ($v \leq 2$); $B_3 = 5.80_3$ (16).

[g]$D_{1,2,3}(10^{-4}cm^{-1})$ = 6.49, 8.06, 27; $H_{0,1,2}(10^{-8}cm^{-1})$ = -4.0, -13.2, -38.5.

[h]Observed in absorption by (11) in the flash photolysis of DN_3.

[i]$A_0 = -34.58$; from (13) who gives additional multiplet splitting constants. See also (12).

[j]Calculated from the constants for N^1H. The origins of the A↔X 1-1 and 2-2 bands are at 29738.4_2 and 29658.1_9 cm^{-1}.

[k]"True" constants calculated by (13) from the data of (11), $H_0 = +1.57 \times 10^{-8}$. The Λ-type doubling parameters were also evaluated by (13). B_0(effective) = 8.7610 cm^{-1} (6); for effective D_v, H_v values (v=0,1,2) see (6)(7).

[ℓ]$\gamma_e = -0.003_3$. α_e and γ_e refer to the effective rotational constants of (6)(7).

[m]Undegraded 0-0 band.

[n]Average of two values obtained a) from $\omega_e x_e$ of N^1H and b) from bandhead measurements and calculated head-origin separations in the d→b system.

[o]From a comparison with $\Delta G(\frac{1}{2})$ of N^1H.

[p]$D_1 = 4.71 \times 10^{-4}$, $H_0 = +2.09 \times 10^{-8}$.

[q]ν_{00}(b-X) + ν_{00}(c-b) - ν_{00}(c-a).

[r]"True" B_0 and D_0 from (13), $H_0 = +1.82 \times 10^{-8}$; multiplet (continued on p. 461)

N^1H (continued):

resonance spectrum (54) in v=0 and 1: $B_e = 16.6668$, $\alpha_e = 0.6470$; $\gamma(v=0) = -0.0546_6$, $\gamma(v=1) = -0.0517$; $\lambda(v=0,1) = +0.9198$. See also (52).

b' $D_1 = 16.88 \times 10^{-4}$, $H_0 = +10.48 \times 10^{-8}$, $H_1 = +10.2 \times 10^{-8}$ (40).

c' By the laser-magnetic-resonance method; fine (see a') and hyperfine structure constants.

(1) Dieke, Blue, PR 45, 395 (1934).
(2) Pearse, PRS A 143, 112 (1934).
(3) Funke, ZP 96, 787 (1935); 101, 104 (1936).
(4) Lunt, Pearse, Smith, PRS A 151, 602 (1935).
(5) Lunt, Pearse, Smith, PRS A 155, 173 (1936).
(6) Florent, Leach, JPR 13, 377 (1952).
(7) Franklin, Dibeler, Reese, Krauss, JACS 80, 298 (1958).
(8) Dixon, CJP 37, 1171 (1959).
(9) Hurley, PRS A 249, 402 (1959).
(10) Companion, Ellison, JCP 32, 1132 (1960).
(11) Bennett, Dalby, JCP 32, 1716 (1960).
(12) Jordan, Longuet-Higgins, MP 5, 121 (1962).
(13) Fink, Welge, ZN 19 a, 1193 (1964).
(14) Shimauchi, SL 13, 53 (1964).
(15) Irwin, Dalby, CJP 43, 1766 (1965).
(16) Foner, Hudson, JCP 45, 40 (1966).
(17) Harrington, Modica, Libby, JQSRT 6, 799 (1966).
(18) Murai, Shimauchi, SL 15, 48, 165(erratum) (1966).
(19) Narasimham, Krishnamurty, PIAS A 64, 97 (1966).
(20) Seal, Gaydon, PPS 89, 459 (1966).
(21) Cade, Huo, JCP 47, 614 (1967).
(22) Horani, Rostas, Lefebvre-Brion, CJP 45, 3319 (1967).
(23) Whittaker, PPS 90, 535 (1967).
(24) Cade, CJP 46, 1989 (1968).
(25) Huo, JCP 49, 1482 (1968).
(26) Whittaker, JP B 1, 977 (1968).
(27) Krishnamurty, Narasimham, JMS 29, 410 (1969).
(28) Smith, JCP 51, 520 (1969).
(29) Whittaker, CJP 47, 1291 (1969).
(30) Bollmark, Kopp, Rydh, JMS 34, 487 (1970).
(31) Stedman, JCP 52, 3966 (1970).
(32) Kouba, Öhrn, JCP 52, 5387 (1970).
(33) Liu, Verhaegen, JCP 53, 735 (1970).
(34) Okabe, JCP 53, 3507 (1970).
(35) Malicet, Brion, Guénébaut, JCPPB 67, 25 (1970).
(36) O'Neil, Schaefer, JCP 55, 394 (1971).
(37) Smith, Liszt, JQSRT 11, 45 (1971).
(38) Kaskan, Nadler, JCP 56, 2220 (1972).
(39) Liu, Legentil, Verhaegen, in "Selected Topics in Molecular Physics"(ed. Clementi), p.19. Chemie GmbH (1972).
(40) Veseth, JP B 5, 229 (1972).
(41) Lents, JQSRT 13, 297 (1973).
(42) Rao, Lakshman, IJPAP 11, 539 (1973).
(43) Stevens, JCP 58, 1264 (1973).
(44) Das, Wahl, Stevens, JCP 61, 433 (1974).
(45) Gilles, Masanet, Vermeil, CPL 25, 346 (1974); JPhoC 3, 417 (1974/75).
(45a) Milligan, Jacox, JCP 41, 2838 (1964).
(46) Meyer, Rosmus, JCP 63, 2356 (1975).
(47) Zetzsch, Stuhl, CPL 33, 375 (1975).
(48) Radford, Litvak, CPL 34, 561 (1975).
(49) Gelernt, Filseth, Carrington, CPL 36, 238 (1975).
(50) Smith, Brzozowski, Erman, JCP 64, 4628 (1976).
(51) Hay, Dunning, JCP 64, 5077 (1976).
(52) Palmiere, Sink, JCP 65, 3641 (1976).
(53) Engelking, Lineberger, JCP 65, 4323 (1976).
(54) Wayne, Radford, MP 32, 1407 (1976).
(55) Graham, Lew, CJP 56, 85 (1978).
(55a) Rosengren, Pimentel, JCP 43, 507 (1965).

State	T_e	ω_e	$\omega_e x_e$	B_e	α_e	D_e ($10^{-4}cm^{-1}$)	r_e (Å)	Observed Transitions Design.	ν_{00}	References
$^{14}N^{1}H^{+}$		D_0^0 = (3.39) eV[a]								MAR 1977
		Ab initio calculated potential energy curves for the ground and excited states (3).								
C $^2\Sigma^+$	(35000)	2150.5_6 Z	73.0_7	13.265_2[b]	0.789_1[c]	20.0[d]	1.1626	$C^t \rightarrow X$, R	34561.27[e]	(1)* (2)(5)*
B $^2\Delta_i$	(23300)[f]	(2280)		[13.516]		[19]	[1.1518]	$B^t \rightarrow X$, R	22960.46[e]	(2)*
A $^2\Sigma^-$	(22200)	[1585.49] Z	(61)[g]	11.4553[h]	0.6897	[20.2][i]	1.2511	$A^t \rightarrow X$, R	21567.67[e]	(2)*
a $^4\Sigma^-$	(500)	[2520][j]		[14.69][j]		(17)	[1.105]	(a – X)	354[ej]	
X $^2\Pi_r$	0[k]	[2922][ℓ]		[15.35][ℓm]	(0.64)[ℓ]	[17][ℓ]	1.070			
$^{14}N^{2}H^{+}$										MAR 1977
B $^2\Delta_i$	(23300)[n]	(1672)		[7.2715]		[5.5][o]	[1.1474]	B → X, R	23063.83[e]	(2)
A $^2\Sigma^-$	(22200)	[1182.40] Z	(32)[g]	6.1206[p]	0.2752	[5.8][q]	1.2507	A → X, R	21750.59[e]	(2)
X $^2\Pi_r$	0[r]	[2143.04] Z		[8.244]	s	[5.3]	$[1.077_6]$			
$^{14}N^{1}H^{-}$						I.P. = 0.38_1 eV[a]				MAR 1977
X $^2\Pi_i$	0[b]						1.047[c]			
$^{14}N^{127}I$		$\mu = 12.6114804_6$								MAY 1977
X $^3\Sigma^-$	0	[590][a]						Vibrational sp.[a]		(1)

N^1H^+, N^2H^+:

[a]Theoretical value (3), corresponding to $N(^4S) + H^+$. The $^2\Pi$ ground state must dissociate into $N^+(^3P) + H(^2S)$.

[b]Spin-splitting constants $\gamma_0 \ldots \gamma_2$ = +0.105, +0.108, +0.111.

[c]$\gamma_e = + 0.008_9$.

[d]$\beta_e \approx -0.8 \times 10^{-4}$, $H_e \approx +11 \times 10^{-8}$.

[e]Energy of lowest rotational level relative to the lowest level $[F_{1e}(\frac{1}{2})]$ of the ground state.

[f]$A_0 = - 3.6$.

N^1H^+, N^2H^+ (continued):

[g]Estimated from the $\Delta G(\frac{1}{2})$ values for NH^+ and ND^+.
[h]Spin-splitting constants $\gamma_0 = -0.097$, $\gamma_1 = -0.100$.
[i]$D_1 = 18.6 \times 10^{-4}$, $H_0 = +18.5 \times 10^{-8}$.
[j]From the analysis of perturbations in X $^2\Pi(v=0,1)$ (2).
[k]$A_0 = +77.8$.
[ℓ]v=0,1 strongly perturbed by v=0,1, resp., of $a^4\Sigma^-$ (2)(5).
[m]Λ-doubling in $^2\Pi_{\frac{1}{2}}(v=0,J=\frac{1}{2})$: $\Delta\nu_{fe} = +0.45_1 \pm 0.010$ cm^{-1}.
[n]$A_0 = -3.5$.
[o]$H_0 = +1.5 \times 10^{-8}$.
[p]Spin-splitting constants $\gamma_0 = -0.052$, $\gamma_1 = -0.053$.
[q]$D_1 = 5.5 \times 10^{-4}$.
[r]$A_0 = +78.4$.
[s]v=1 strongly perturbed by v=1 of $a^4\Sigma^-$.
[t]Lifetimes of A, B, C: 1.0_9, 0.9_8, 0.40 μs, resp. (4).

(1) Feast, ApJ 114, 344 (1951).
(2) Colin, Douglas, CJP 46, 61 (1968).
(3) Liu, Verhaegen, JCP 53, 735 (1970).
(4) Brzozowski, Elander, Erman, Lyyra, PS 10, 241 (1974).
(5) Krishnamurthy, Saraswathy, Pramāṇa 6, 235 (1976).

NH^-: [a]From the laser photodetachment spectrum (1)(2).
[b]$A = -63$ cm^{-1} [estimated (2)].
[c]From a Franck-Condon factor analysis of the observed $^3\Sigma^- \leftarrow {}^2\Pi$ photoelectron spectrum (2).

(1) Celotta, Bennett, Hall, JCP 60, 1740 (1974).
(2) Engelking, Lineberger, JCP 65, 4323 (1976).

NI: [a]Observed in the condensate at 10 K of the products of a discharge in an N_2/I_2 mixture. The observation of a corresponding band at 573 cm^{-1} for ^{15}NI supports the assignment.

(1) Minkwitz, Froben, CPL 39, 473 (1976).

N^2H (continued):

splitting parameters $\lambda_0 \approx +0.89_5$, $\gamma_0 = -0.0061$. The evaluation of these constants (13) takes fully into account the $^3\Pi \sim {}^3\Sigma^-$ interaction and thus gives results that differ considerably from the effective constants (5)(6)(11). The latter are in good agreement with the more precise constants from the laser-magnetic-resonance sp. (15): $B_0 = 8.7815$, $\lambda_0 = +0.9184$, $\gamma_0 = -0.0294$. Effective D_v, H_v values in (6)(7).
[s]From the effective B_v values (5)(6)(7).
[t]Laser magnetic resonance; hyperfine structure constants.

(1) Dieke, Blue, PR 45, 395 (1934).
(2) Florent, Leach, JPR 13, 377 (1952).
(3) Shimauchi, SL 13, 53 (1964).
(4) Hanson, Kopp, Kronekvist, Åslund, AF 30, 1 (1965).
(5) Kopp. Kronekvist, Åslund, AF 30, 9 (1965).
(6) Shimauchi, SL 15, 161 (1966).
(7) Shimauchi, SL 16, 185 (1967).
(8) Whittaker, PPS 90, 535 (1967).
(9) Whittaker, JP B 1, 977 (1968).
(10) Whittaker, CJP 47, 1291 (1969).
(11) Bollmark, Kopp, Rydh, JMS 34, 487 (1970).
(12) Kovács, Korwar, APH 29, 85 (1970).
(13) Veseth, JP B 5, 229 (1972).
(14) Gilles, Masanet, Vermeil, CPL 25, 346 (1974); JPhoC 3, 417 (1974/75).
(15) Wayne, Radford, MP 32, 1407 (1976).
(16) Graham, Lew, CJP 56, 85 (1978).

System	ν_e	ω'_e	$\omega'_e x'_e$	ω''_e	$\omega''_e x''_e$	Remarks Description	Degrad.	ν_{00}	References
$^{(58)}Ni_2$		(μ = 28.967667$_9$) D_0^0 = 2.3$_6$ eV[a]							MAR 1976 A
		Absorption in the region 21790 - 23900 cm^{-1}; in Ar matrices at 10 K. $\omega' \approx 192\ cm^{-1}$.							(2)
$^{(58)}Ni^{(79)}Br$		(μ = 33.409116$_0$) D_0^0 = 3.6$_9$ eV[a]							MAR 1975
		Large number of R shaded emission bands in the region 20000 - 25600 cm^{-1}. Below are preliminary assignments by (3):							
System α_1:	24326.2	297.0 H^R	3.35	311.6	1.7	R, Q heads. b	R	24318.5[c] H^R	(3)* (5)
" α_2:	23920.3	262 H		293.2		b	R	23904.7 H	(3)
" α_3:	23791.1	305.4 H^R	0.2	315.0	0.5	R, Q(?) heads. d	R	23786.4 H^R	(3)
" β_1:	22975.5	293.8 H	2.5	322.8	1.0	b	R	22960.6 H	(3)* (5)
" β_2:	22435.9	257 H^Q		274		d	R	22427.4 H^Q	(3)*
" γ :	21792.6	297.5 H^Q	1.1	323.0	1.2	b	R	21779.9 H^Q	(3)
$^{(58)}Ni^{(35)}Cl$		(μ = 21.8066837) D_0^0 = 3.8$_2$ eV[a]							MAR 1975
		Large number of R shaded emission bands in the regions 11000 - 15000 and 18000 - 26000 cm^{-1}. The ground state has not been identified and all assignments must be considered as preliminary.							
System A_1:	24623.7	394.5 H^Q	0.35	415.5	1.3	b	R	24613.4 H^Q	(1)(4)*
" A_2:		0-0 sequence only.				g	R	24411.6 H^Q	(1)(2)(4)*
" A_3:	24138.3	380 H		397		c	R	24129.8 H	(1)(4)*
" B_1:	23347.1	375 H		402		d	R	23333.6 H	(4)
" B_2:	23233.1	406.6 H	2.75	426.3	1.9	e	R	23223.0 H	(1)(2)(4)*
" C:	22749.0	397.8 H^Q	0.75	418.2	0.70	f	R	22738.8 H^Q	(1)(2)(4)*
" D:	21914.7	398 H^Q		417		g	R	21905.2 H^Q	(4)*
" E_1:	21762.2	405.2 H	1.3	435.3	1.85	g	R	21747.3 H	(4)*
" E_2:	21649.4	383.1 H^Q		404.0	1.65	g	R	21639.4 H^Q	(4)*
" I:	20545.3	409.6 H^Q	0.80	431.1	0.20		R	20534.4 H^Q	(7)
" II:	20284.2	382.8 H^Q		423.9	1.35		R	20264.0 H^Q	(7)
" III:	19980.1	403.6 H^Q	0.05	430.0	0.30		R	19967.0 H^Q	(7)

System	ν_e	ω'_e	$\omega'_e x'_e$	ω''_e	$\omega''_e x''_e$	Remarks Description	Degrad.	ν_{00}	References
$^{(58)}Ni^{(35)}Cl$ (continued)									
System IV:	18689.4	412.6 H	1.60	433.2	1.65		R	18679.1 H	(7)
" F:	14441.2	407.2 H	2.5	420		h	R	14434.2 H	(6)
" G:	12976.5	395.0 H	1.55	424		h	R	12961.6 H	(6)
" H:	12276.8	416.0 H^Q		428			R	12270.8 H^Q	(6)
" I:	11913.1	395.0 H	1.40	402			R	11909.3 H	(6)
" J:	11511.3	398.7 H^Q	1.45	405.2	1.15		R	11508.0 H^Q	(6)

Ni_2: [a]Thermochemical value (mass-spectrom.)(1).

(1) Kant, JCP 41, 1872 (1964).

(2) De Vore, Ewing, Franzen, Calder, CPL 35, 78 (1975).

NiBr: [a]Thermochemical value (flame photometry)(4).

[b]Alternative assignments in (1)(2).

[c]Q head at 24311.9 cm^{-1}.

[d]Alternative assignments in (1)(2)(5).

(1) Mesnage, AP(Paris) 12, 5 (1939).

(2) Krishnamurty, IJP 26, 429 (1952).

(3) Reddy, Rao, PPS 75, 275 (1960).

(4) Bulewicz, Phillips, Sugden, TFS 57, 921 (1961).

(5) Sundarachary,PNASI A 32, 311 (1962).

NiCℓ: [a]Thermochemical value (flame photometry)(5).

[b]System 4 of (1). Alternative assignments in (3).

[c]System 3 of (1). Alternative assignments in (2)(3).

[d]Alternative assignments in (2)(3).

[e]System 2 of (1). Alternative assignments in (3).

[f]System 1 of (1). Alternative assignments in (3).

[g]Alternative assignments in (3).

[h]Results of a rotational analysis in (7). Uncertain.

(1) More, PR 54, 122 (1938).

(2) See ref. (1) of NiBr.

(3) Krishnamurty, IJP 26, 207 (1952).

(4) See ref. (3) of NiBr.

(5) See ref. (4) of NiBr.

(6) Rao, Reddy, Rao, ZP 166, 261 (1962).

(7) Rao, Rao, CS 38, 589 (1969).

State	T_e	ω_e	$\omega_e x_e$	B_e	α_e	D_e ($10^{-4}cm^{-1}$)	r_e (Å)	Observed Transitions Design.	ν_{00}	References
$^{(58)}Ni^{19}F$		(μ = 14.3068431)								APR 1975
A		Fragments of a band system.						$A \rightarrow (X)$,	22144 H	(1)
(X)	0	(740)								
$^{(58)}Ni^{(74)}Ge$		($\mu = 32.479611_1$)		$D_0^0 = 2.8_8$ eV^{a}						APR 1975
$^{(58)}Ni^{1}H$		(μ = 0.99059317)		$D_0^0 \leq 3.07$ eV^{a}						APR 1975
		Diffuse R shaded head in absorption at 34073 cm^{-1}.								(6)*
$C_2\ ^2\Delta_{3/2}$	e			$[6.31_1]^{bc}$		[7.6]	[1.642]	$C_2 \leftarrow X_2$, R $C_2 \leftarrow X_1$, R	23100.80^{d} Z $24081.2_3{}^{d}$ Z	(4)(5)
$C_1\ ^2\Delta_{5/2}$				$[6.156]^{bf}$		[6.1]	$[1.662_7]$	$C_1 \leftrightarrow X_1$, R	23760.7^{d} Z	(2)* (4)(5)
		R shaded absorption bands with heads at 19681 and 19890 cm^{-1}. Perturbations.								(6)*
$B\ ^2\Delta_{5/2}$	16193	1570.9 Z	34.55^{g}	h				$B \leftrightarrow X_1$, R	15977.3^{d} Z	(1)* (2)(3) (6)
$A\ ^2\Delta_{5/2}$				$[6.283]^{i}$		$[4.7_5]$	$[1.645_8]$	$A \leftrightarrow X_1$, R	15520.1^{d} Z	(1)* (2)(3) (6)
$X_2\ ^2\Delta_{3/2}$	(980)			$[7.781]^{b}$		[5.9]	[1.4789]			
$X_1\ ^2\Delta_{5/2}$	0 j	[1926.6] Z	$(38)^{k}$	$[7.700]^{b}$	$0.23_1{}^{\ell}$	$[4.81]^{\ell}$	1.475_6			
$^{(58)}Ni^{2}H$		(μ = 1.94643507)		$D_0^0 \leq 3.10$ eV^{a}						APR 1975
		Diffuse R shaded head in absorption at 34073 cm^{-1}.								(6)*
$C_2\ ^2\Delta_{3/2}$		$[728.9_7]^{m}$ Z		[3.271]	mn	$[1.62]^{m}$	$[1.627_2]$	$C_2 \leftarrow X_2$, R	$23171.6_9{}^{d}$ Z	(4)(5)
$C_1\ ^2\Delta_{5/2}$		$[918.3_0]$ Z		[3.240]	0.188^{n}	$[1.52]^{o}$	1.611_7	$C_1 \leftarrow X_1$, R	$23830.4_7{}^{d}$ Z	(4)(5)
		R shaded absorption bands with heads at 19707 and 19845 cm^{-1}. Perturbations.								(6)*
$B\ ^2\Delta_{5/2}$	16172	1130.0 Z	23.4	p				$B \leftarrow X_1$, R	16021.8^{d} Z	(6)*
$X_2\ ^2\Delta_{3/2}$	(980)			[3.996]		[1.42]	$[1.472_2]$			
$X_1\ ^2\Delta_{5/2}$	0	$[1390.0_9]$ Z	$(19)^{k}$	[3.992]	0.092	$[1.30]^{q}$	1.464_5			

NiF: (1) Krishnamurty, IJP 27, 354 (1953).

NiGe: [a]Thermochemical value (mass-spectrom.)(1).

(1) Kant, JCP 44, 2450 (1966).

Ni^1H, Ni^2H:

[a]From the predissociation in C_1.
[b]For Λ-type doubling see (5).
[c]Lines with $J' \gtrsim 11\frac{1}{2}$ are diffuse.
[d]These band origins are corrected for J-independent terms not usually taken into account in these tables.
[e]$A_0 = -160.2 + \ldots$
[f]Lines with $J' \gtrsim 12\frac{1}{2}$ are diffuse, both in absorption and in thermal emission.
[g]$\omega_e y_e = -2.87$; from v=0, ..., 3 only. See [h].
[h]All observed levels perturbed. Heimer (2)(3) obtained the following B_v values by extrapolation to J=0:

v = 0,	B_v = 5.113	(increasing with J),
1	5.350	(" " J),
2	5.480	(" " J),
3	5.900	(decreasing with increasing J).

A reanalysis of the red band systems of NiH is considered necessary by (5).

Ni^1H, Ni^2H (continued):

[i]Λ-type doubling increases rapidly above $J=9\frac{1}{2}$. See (2).
[j]$A_0 = -490.2 + \ldots$
[k]From isotope relations.
[ℓ]$D_1 = 7.8 \times 10^{-4}$. Heimer (2)(3) finds $\alpha_e = 0.248$, $D_1 = 5.1 \times 10^{-4}$, but his interpretation of a perturbation in the 0-1 band of $C_1 - X_1$ is questioned by (5).
[m]It is not certain that the level assigned as v=1 ($B_1 = 3.410$, $D_1 = 3.6 \times 10^{-4}$) belongs to the same electronic state.
[n]Lines of the 1-0 band with $J' \geq 9\frac{1}{2}$ are broad.
[o]$D_1 = 2.0 \times 10^{-4}$.
[p]All observed levels perturbed. Extrapolation to J=0 gives $B_0 = 2.930$, $B_1 = 2.710$, $B_2 = 2.463$.
[q]$D_1 = 1.4_0 \times 10^{-4}$.

(1) Gaydon, Pearse, PRS A 148, 312 (1935).
(2) A. Heimer, Dissertation (Stockholm, 1937).
(3) Heimer, ZP 105, 56 (1937).
(4) Andersén, Lagerqvist, Neuhaus, Åslund, PPS 82, 637 (1963).
(5) Åslund, Neuhaus, Lagerqvist, Andersén, AF 28, 271 (1964).
(6) Smith, PRS A 332, 113 (1973).

State	T_e	ω_e	$\omega_e x_e$	B_e	α_e	D_e	r_e (Å)	Observed Transitions Design.	Observed Transitions ν_{00}	References
$^{(58)}Ni^{127}I$		(μ = 39.776352$_0$) D_0^0 = 2.9$_9$ eV[a]								APR 1975
$^{(58)}Ni^{16}O$		(μ = 12.5343923$_8$) D_0^0 = 3.8$_7$ eV[a]								APR 1975
		Three additional systems in the same spectral region.								(1)(2)
b	a + (21262)	(570) H						b→a,	R (21135) H	(1)(2)
a	a	(825) H								
B	(16447)	(560) H						B→X,	R (16420) H	(1)(2)
A	(12725)	(475) H						A→X,	R (12655) H	(1)(2)
(X)	0	(615) H								
$^{(58)}Ni^{(32)}S$		(μ = 20.6024487) D_0^0 = 3.5$_3$ eV[a]								APR 1975
$^{(58)}Ni^{(28)}Si$		(μ = 18.8663722) D_0^0 = 3.2$_6$ eV[a]								APR 1975
$^{14}N^{16}O$		μ = 7.46643323 D_0^0 = 6.496$_8$ eV[a] I.P. = 9.26436 eV[b]								MAR 1977 A

For a detailed discussion of the electronic spectrum with particular emphasis on Rydberg~Rydberg and Rydberg~non-Rydberg interactions see (192a); this review contains references to spectra of four isotopes as well as a short summary of theoretical calculations.

State		References
($^2\Sigma^-$, $^2\Delta$, $^2\Sigma^+$)	Broad unresolved peak in the oxygen K shell electron energy loss spectrum at 532.7 eV.[c]	(182a)
	Rydberg states converging to the nitrogen K edge at 410.2 ($^3\Pi$) and 411.7 ($^1\Pi$) eV, observed in X-ray absorption and electron energy loss spectra at 406.3, 407.3, 408.6 eV, ...	(178)(182a)
($^2\Sigma^-$) } ($^2\Delta$, $^2\Sigma^+$, $^2\Sigma^-$) }	Two very weak bands in the X-ray absorption spectrum at 402.3 and 403.9 eV.[d]	(178)
($^2\Sigma^-$, $^2\Delta$, $^2\Sigma^+$)	Strong unresolved peak in X-ray absorption and electron energy loss spectra at 399.8 eV.[e]	(178)(182a)

NiI: [a]Thermochemical value (flame photometry)(1).

(1) Bulewicz, Phillips, Sugden, TFS 57, 921 (1961).

NiO: [a]Thermochemical value (mass-spectrom.)(5), recalc. (6). See also (3)(4).

(1) Rosen, Nature 156, 570 (1945).
(2) Malet, Rosen, BSRSL 14, 382 (1945).
(3) Brewer, Mastick, JCP 19, 834 (1951).
(4) Huldt, Lagerqvist, ZN 9 a, 358 (1954).
(5) Grimley, Burns, Inghram, JCP 35, 551 (1961).
(6) Smoes, Mandy, Vander Auwera-Mahieu, Drowart, BSCB 81, 45 (1972).

NiS: [a]Thermochemical value (mass-spectrom.)(1).

(1) Drowart, Pattoret, Smoes, PBCS No. 8, 67 (1967).

NiSi: [a]Thermochemical value (mass-spectrom.)(1).

(1) Vander Auwera-Mahieu, McIntyre, Drowart, CPL 4, 198 (1969).

NO: [a]From the breaking-off below $N'=4$ in the $C \rightarrow A$ 0-0 band emitted during radiative recombination of N and O atoms via inverse predissociation (187), see also (169); in good agreement with (123). A very slightly higher value, i.e. $6.4977 \leq D_0^0 \leq 6.5007$ eV, is suggested (179) by the failure to detect $F \rightarrow C$ laser transitions ending on the lowest C level observed in the N + O recombination spectrum.

[b]Extrapolation of selected rotational lines in the $nf \leftarrow X$ Rydberg series (180)(204), based on the fine structure analysis of the 4f and 5f complexes (119).

[c]Excitation of a $1s_O$ electron to the 2π orbital.

[d]Tentatively interpreted as arising from two-electron excitation from the $1s_N$ and 1π to the 2π orbital (178). Only one peak (404.7 eV) is observed in the electron energy loss spectrum (182a).

[e]Excitation of a $1s_N$ electron to the 2π orbital.

State	T_e	ω_e	$\omega_e x_e$	B_e	α_e	D_e $(10^{-}\ cm^{-1})$	r_e (Å)	Observed Transitions: Design.	Observed Transitions: ν_{00}	References
$^{14}N^{16}O$ (continued)										
$...4\sigma 5\sigma^2 1\pi^4 2\pi \begin{cases} nd\lambda \\ np\lambda \end{cases}$	Narayana and Price's absorption Rydberg series converging to c $^3\Pi$ of NO^+: $\nu = 175220 - \begin{cases} R/(n+0.05)^2, & n = 3...7. \\ R/(n-0.70)^2, & n = 3...6. \end{cases}$ All bands diffuse.[a] Fragments of an additional series (181). Absorption cross sections 700 - 180 Å (143000 - 556000 cm^{-1}) (164)(168).									(157)* (181)
$...5\sigma 1\pi^4 2\pi \cdot np\lambda$	Tanaka's absorption Rydberg series converging to A $^1\Pi(v=0)$ and b $^3\Pi(v=0)$ of NO^+: $\nu \approx \begin{cases} 147805 - R/(n-0.78)^2; & \gamma \text{ series, } n = 3...11.^b \\ 133570 - R/(n-0.70)^2; & \beta \text{ series, } n = 3...15.^b \end{cases}$ Also fragments of weak series with v'=1.									(5)(51)* (126)
$...1\pi^3 2\pi\ ns\sigma$	Rydberg series converging to a $^3\Sigma^+$, w $^3\Delta$, b' $^3\Sigma^-$, A' $^1\Sigma^-$, W $^1\Delta$ of NO^+. Only the first two or three members of each absorption series have been identified; long upper state progressions. Tables of absorption features 950 - 650 Å (105000 - 154000 cm^{-1}) (5)(51)* (97)*. Absorption coefficients, photoionization and photodissociation yields (85)(97)(103)(146)(176).									(126)
	The band structure of the absorption spectrum from 80000 to 105000 cm^{-1} has not yet been analyzed. Absorption coefficients, photoionization efficiency curves (58)(85)(103); the data of (103) are conveniently plotted in Figure 6 of (165) and Figure 4 of (153). Autoionization processes have been studied by photoelectron spectroscopy (137)(165)(199); partial cross-sections for the formation of vibrationally excited NO^+ (153).									
	Atlas of the absorption spectrum 1420 - 1250 Å (70400 - 80000 cm^{-1}) (192); for a photographic reproduction of the spectrum at longer wavelengths (1920 - 1400 Å) see (42). A useful quantitative low-resolution plot of the absorption from 2300 to 1100 Å may be found in Figure 2.1 of (192a), adapted from (13). Absorption coefficients, photoionization efficiency curves (17)(103)(166) and (195) whose supersonic molecular beam technique made it possible to resolve the autoionization structure superimposed on the first four vibrational steps due to direct ionization.									

State	T_e	ω_e	$\omega_e x_e$	B_e	α_e	D_e (10^{-} cm^{-1})	r_e (Å)	Observed Transitions: Design.	Observed Transitions: ν_{00}	References
$^{14}N^{16}O$ (continued)										
		Rydberg series converging to v=0...4 of X $^1\Sigma^+$ of NO^+ and fragments of series with v'=5:								
nf series		n = 4...15. Sharp rotational structure.						nf← X		(192)*(204)*
ndδ series		Joining on to $F^2\Delta$(n=3), $N^2\Delta$(n=4), $U^2\Delta$(n=5), and incompletely observed to n=8. Perturbations by stable and unstable states.						ndδ ← X		(192)*(192a) (204)*
npπ,σ series		Joining on to $C^2\Pi$,$D^2\Sigma^+$(n=3), $K^2\Pi$,$M^2\Sigma^+$(n=4), $Q^2\Pi$,$R^2\Sigma^+$(n=5), and $W^2\Pi$,$Y^2\Sigma^+$(n=6); bands of varying diffuseness have been observed to n=11. The influence of the unstable A' $^2\Sigma^+$ state is briefly discussed in (204).						npπ,σ ← X		(192)*(192a) (204)*
nsσ series		Joining on to $A^2\Sigma^+$(n=3), $E^2\Sigma^+$(n=4), $S^2\Sigma^+$(n=5), $T^2\Sigma^+$(n=6), $Z^2\Sigma^+$(n=7). The B_e values decrease from 1.997 (n=3) to 1.713 for the highest observed state (n=11) as a consequence of nsσ∼(n-1)dσ interactions, see [f] on p. 471 . Sharp rotational structure.						nsσ← X		(192)*(192a) (204)*
		Several unassigned non-Rydberg levels, mixed with Rydberg levels, near the dissociation limit $^2D + {}^3P$ at 71627 cm^{-1}.								(74)(192)*
6f	71427	2376 H	16.2					6f← X,	71543 71662.6 H	(91)(192)*
6dδ	(71342)	(2397) H	(23)	[1.86]	v=1,2,3 diffuse.			6dδ ← X, V	71467 71586[c] Z	(91)(192)* (204)
Z $^2\Sigma^+$ 7sσ	71224	2377 H	16.4	[1.938]				Z← X, V	71340 71460[c] Z	(91)(192)*

NO: [a]Photoionization yields (NO^+, N^+, O^+) in the region of these Rydberg series (175).

[b]The Rydberg formulae do not accurately reproduce the observed bands owing to the slow variation of the quantum defect with n.

[c]These band origins refer to N'=0 (non-existent for $\Lambda \neq 0$) in the excited state and to the hypothetical level J"=0 of the X $^2\Pi_{\frac{1}{2}}$ ground state, in accordance with definitions adopted in these tables. The corresponding numbers for the X $^2\Pi_{\frac{3}{2}}$ component are obtained by subtracting 119.7 cm^{-1}.

State	T_e	ω_e	$\omega_e x_e$	B_e	α_e	D_e (10^{-} cm^{-1})	r_e (Å)	Observed Transitions Design.	Observed Transitions ν_{00}	References
$^{14}N^{16}O$ (continued)										
Y $^2\Sigma^+$ 6pσ	70614	2370	$15._0$	[2.11] v=1,2,3 diffuse				Y← X, V	70728[a] 70847[a] Z	(74)(91) (192)*
W $^2\Pi$ 6pπ	70512	2375	$15._6$	v=0 perturbed by non-Rydberg level; v=1,2,3 very diffuse.				W← X,	70627 70747	(74)(91) (192)*
5f	70079	2377 H	16.5	[1.988][b]				5f← X,	70195[c] 70315[c]	(91)(119) (192)*
U $^2\Delta$ 5dδ	(69977)	2371	$16._4$	{ Partial rot. analyses for v=0,1 ($^{15}N^{18}O$). Perturbations by non-Rydberg levels. v=2,3,4 diffuse to varying degrees.				U← X,	(70090) (70210)	(91)(192)* (204)
T $^2\Sigma^+$ 6sσ	(69728)	2372	15.7	v=0 coincides with I(v=6) and E(v=4), strong perturbation. B_1 = 1.92.				T← X, V	(69841) (69961)	(91)(192)*
R $^2\Sigma^+$ 5pσ	(68598)			[2.04][d] v=1,2 diffuse				R← X, V	68710.9[a] 68830.7[a] Z	(74)(192)*
Q $^2\Pi$ 5pπ		v=0,1 mixed with non-Rydberg levels, v=2,3,4 diffuse.						Q← X,	68526 68646	(74)(192)*
O' $^2\Pi$ 4dπ	(67762)[e]	(2371)	(16)	[2.022][f]				0,0'→D,	14702.2 14697.9	(64)(121)*
O $^2\Sigma^+$ 4dσ	(67757)			[1.990][f]				0,0'→C,	15623 15619	(64)(121)*
								0,0'← X,	67874.8 67870.5[a] 67994.5[a] Z 67990.3[a] Z	(91)(121)*
4f	67596	2381 H	18.5	[1.988][b]			$[1.065_7]$	4f← X,	67713[c] 67833[c]	(91)(119)*
N $^2\Delta$ 4dδ	67374	2375[g]	15[g]	1.969[g]	0.026[g]			N→C,[h]	15238[g]	(64)(135)*
								N↔X, V	67489 67609[g]	(74)(76)* (88)*
S $^2\Sigma^+$ 5sσ	66900	2378 Z	16.5	1.980	0.020			S← X, V	67016[a] 67136[a] Z	(52)(74)(90)

State	T_e	ω_e	$\omega_e x_e$	B_e	α_e	D_e (10^{-} cm^{-1})	r_e (Å)	Observed Transitions: Design.	Observed Transitions: ν_{00}	References
$^{14}N^{16}O$ (continued)										
M $^2\Sigma^+$ 4pσ	64437	2352 Z	$19._5$	2.022[i]	0.018			M← X, V	64540[a] 64660[a] Z	(36)(52)(74) (90)* (192)*
K $^2\Pi$ 4pπ		v=0...3 observed.		[1.895][jk]				K← X, V	64167 64287[aℓ] Z	(52)(74) (90)*
I $^2\Sigma^+$	(63500)	Five levels (v=4...8?) have been observed for various isotopes in the region 67800 - 72000 cm^{-1}. Erratic behaviour with regard to diffuseness and isotope shifts on account of interactions with the unstable A' $^2\Sigma^+$ state and with npσ Rydberg states. See also (160).						I← X, R		(74)(204)

NO: [a]See [c] on p.469 .

[b]B value of the NO^+ core. For details of the analysis and derived core parameters (polarizability, quadrupole moment) see (119).

[c]Energy of the Δ (or $\mathcal{L}=2$) component relative to the hypothetical level J"=0, calculated using results from the analysis of the $^{14}N^{18}O$ spectrum; see (119).

[d]The interaction between $R^2\Sigma^+(v=0)$, $I^2\Sigma^+(v=5?)$, and the continuous $A'^2\Sigma^+$ state has been observed in the spectra of four isotopes; see Figure 2.5 of (192a).

[e]A slight mixing of the ground state into the ndπ components is responsible for the larger than expected spin-orbit coupling in H' (A = +0.96, ξ = +0.92) and O' (A = +0.36, ξ = +0.34); see (129), also (57)(121)(143).

[f]Strong ℓ-uncoupling, $\eta(v=0)$ = 1.92 and 1.88 for 3d and 4d, resp. (143). The magnitude of η was interpreted in terms of s∼d mixing (129); the interaction matrix elements are 910 (4sσ∼3dσ) and 430 cm^{-1} (5sσ∼4dσ). The non-negligible spin-orbit coupling in ndπ (see [e]) gives rise to small perturbations between e levels of the $^2\Pi$ F_1 and F_2 components (56)(57)(121)(143). Additional perturbations in H,H' by Rydberg and non-Rydberg levels (143). For H,H'(v=3) only Π^- has been observed. The Π^+ and Σ^+ components of 0,0'(v=0) are weakly predissociated for all N, Π^- above N=16 (121).

[g]Approximate deperturbed constants, see [p] on p. 473; v=3 at 74580 cm^{-1} is very diffuse (204).

[h]The N→C 0-0 band is strongly mixed with B'→C 7-0; see [q] on p. 473 .

[i]Heterogeneous perturbations by levels of B $^2\Pi$ (112). Levels having v ≥ 1 are diffuse to varying degrees.

[j]A small perturbation by L $^2\Pi$(v=2?) affects the first few rotational levels in v=0; higher vibrational levels (v = 1,2,3) are strongly mixed with non-Rydberg states (B $^2\Pi$ and L $^2\Pi$).

[k]Λ-type doubling, $\Delta\nu_{fe}(F_1)$ = +0.034 N(N+1).

[ℓ]Deperturbed.

$^{14}N^{16}O$ (continued)

State	T_e	ω_e	$\omega_e x_e$	B_e	α_e	D_e $(10^{-6}cm^{-1})$	r_e (Å)	Observed Transitions Design.	Observed Transitions ν_{00}	References
G $^2\Sigma^-$	62913.0	1085.54 Z	11.083[a]	1.2523[b]	0.0204		1.3427	G← X,[c] R	62384.7 62504.4[d] Z	(69)* (192)* (204)
L $^2\Pi_i$	(62500)[e]	Fragments of several levels (vibr. numbering not established) in perturbations with levels of B, C, K. Constants comparable to $B^2\Pi$.						L← X		(74)(90)*
H' $^2\Pi$ 3d π	62485.4[f]	2371.3 Z	16.17	2.015[g]	0.021		1.058_5	H,H'→D,[h]	9426.0 9414.2	(55)(64)*
H $^2\Sigma^+$ 3d σ	62473.4	[2339.4] Z		2.003[g]	0.018		1.061_7	H,H'→C,[h]	10348 10336	(55)(64) (121)*
								H,H'→A,[h]	18518.2 18506.4	(55)(64)*
								H,H'← X,[i]	62598.6 62586.8 62718.4[d] Z 62706.6[d] Z	(56)(91) (143)*
F $^2\Delta$ 3dδ	61800	2394[j]	20[j]	1.982[j]	0.023[j]		1.067	F→C,[kh]	9670[j]	(64)(135)*
								F↔X,[i] V	61924 62044[j]	(52)(74) (76)* (88)*
E $^2\Sigma^+$ 4sσ	60628.8	2375.3 Z	16.4_3	1.9863[ℓ]	0.0182	5.6		E→D,[hm]	7571.5	(9)(45)
								E→A,[h] "6000 Å" b.	16663.63 Z	(9)* (143)*
								E← X,[n] V	60744.1 60863.8[d] Z	(11)(26) (42)* (74)
B' $^2\Delta_i$	60364.2[o]	1217.4[p]	15.61[p]	1.332[p]	0.021[p]		1.302	B'→C,[q]		(64)(135)*
								B'[r]→B,[n] V	14508.6 14538.7	(12)(14)(31) (64)
								B'[r]↔X,[sn] R β' bands	59900.7 60020.4[d] Z	(10)(12)(15) (16)(26)(22)* (74)(76)*(88)*
D $^2\Sigma^+$ 3pσ	53084.7	2323.90 Z	22.885[t]	2.0026[u]	0.02175	[5.8]	1.061_8	D[v]→A,[h] "11000 Å" b.	9092.17 Z	(9)(64) (112)*(143)*
								D[v]↔X,[nw] V ε bands	53172.7 53292.4[d] Z	(1)* (6)* (7) (12)(16)(24) (37)(42)* (90)(112)* (144)

NO: [a]$\omega_e y_e$ = - 0.1439. The levels v=10, 11, 12 are diffuse and lie above the limit $^2D + {}^3P$; v=13 not observed. See also (160).

[b]Small perturbations in isotope spectra.

[c]Absorption in rare gas matrices (101)(197), in high pressure argon (190).

[d]See [c] on p. 469.

[e]A ≈ - 80.

[f]See [e] on p. 471.

[g]See [f] on p. 471.

[h]For experimental and theoretical f values see (122)(140) and (138), respectively.

[i]Also observed by non-resonant multiphoton ionization spectroscopy (189).

[j]Approximate deperturbed constants; see [p].

[k]Lines of the perturbed F→C 1-1 band are prominent in the NO laser spectrum (179); see [p,q].

[ℓ]v=3,4 somewhat diffuse, v=5 sharp. Emission observed from v ≤ 2; (64) reports an abrupt breaking-off in the E→A 2-2 band for an upper state energy of 68100 cm^{-1}.

[m]E→C not observed, in agreement with theoretical predictions (138) regarding the dipole transition strengths of E→C and E→D.

[n]For references to Franck-Condon factor calculations see the review by (99).

[o]A_0 = -2.2, A_1 = -2.4, ..., A_9 = -4.9 (88).

[p]Deperturbed constants; $B'^2\Delta$ interacts strongly with $F^2\Delta$ (matrix element $H_e \approx 450\ cm^{-1}$) and $N^2\Delta$ ($H_e \approx 400$) (84)(88); see also [s]. Perturbations by $B^2\Pi$ are unobservably small because of unfavourable Franck-Condon factors (191b).

[q]Fragments of two bands, 4-1 at 9800 cm^{-1} and 7-0 at 15300 cm^{-1}, both appearing on account of configuration interaction, in the upper state with $F^2\Delta$ and $N^2\Delta$, respectively, in the lower state with $B^2\Pi$. Lines of the 4-1 band, together with F→C 1-1, are seen in the NO laser spectrum (65)(89)(162)(179).

[r]Lifetime $\tau(v=1)$ = 75 ns (174).

[s]An experimentally deperturbed spectrum of B'–X is observed in matrix absorption (101)(116)(197). A gradual deperturbation in the gas phase is induced by increasingly high foreign gas pressures (190).

[t]$\omega_e y_e$ = + 0.75, $\omega_e z_e$ = - 0.22, from v ≤ 4 (36), not including v=5 and 6 (90). The vibrational constants clearly differ from those of other Rydberg $^2\Sigma$ states or of the NO^+ ground state. It has been suggested [see e.g. (143), also (160)] that there is an avoided crossing of the potential curves of $D^2\Sigma^+$ and $A'^2\Sigma^+$ (unstable, arising from $^4S + {}^3P$).

[u]From (36). Heterogeneous perturbations by $B^2\Pi$; for details see (112). According to (64) the rotational structure of D→A 1-1, 2-2, 3-3 breaks off abruptly at D state energies of 59270 cm^{-1} in v=1,2 and 60100 cm^{-1} in v=3.

[v]Lifetimes $\tau(v=0)$ = $18._4$ ns (106), $19._0$ ns (183), $25._7$ ns (174); $\tau(v=1)$ = $26._4$ ns (174).

[w]f_{00} = 0.0025, f_{10} = 0.0046, f_{20} = 0.0033; from integrated absorption intensities (43). See also (62)(124).

State	T_e	ω_e	$\omega_e x_e$	B_e	α_e	D_e $(10^{-6}cm^{-1})$	r_e (Å)	Observed Transitions Design.	Observed Transitions ν_{00}	References
$^{14}N^{16}O$ (continued)										
C $^2\Pi_r$ 3pπ	52126[a]	2395[b]	15[b]	2.000[bc]	0.030[b]		1.062	$C^c \rightarrow A$,[d]	8172	(45)(117)* (187)*
								$C^c \leftrightarrow X$,[ef] V δ bands	52251 52371	(1)* (3)(4)* (6)(30)(42)* (90)(117)* (144)
b ($^4\Sigma^-$)	(48680)	1206[g] H	15					b → a, V	10395 10375 10350 g H 10323 10300 10272	(18)*(19)*
B $^2\Pi_r$	45942.6[h] 45913.6	1039.8[i] Z 1037.2[i] Z	8.3_2[j] 7.7_0[k]	1.152[i] 1.092[iℓ]	0.012 0.012	4.9	1.416_7	$B^m \leftrightarrow X$,[nf] R β bands	45392.1[o] Z 45481.7[o] Z	(1)*(2)*(6)* (3)(15)(18)* (24)(31)(37) (42)* (73)* (52)(90)(127)
A $^2\Sigma^+$ 3sσ	43965.7	2374.31 Z	16.106[p]	1.9965[qrs]	0.01915[q]	5.4	1.0634	$A^t \leftrightarrow X$,[uvf] V γ bands	44080.5 44200.2[w] Z	(1)* (3)(8) (24)(37)(46) (127)
a ($^4\Pi_i$)	(38440)	1017 H	11					(a → X)[x] M bands	(38000)	(48)(66)

NO: [a]$A_0 = +3.0$ cm^{-1} (104).

[b]Approximate deperturbed constants; strong interaction with $B^2\Pi$, see [i]. Λ-type doubling, $\Delta\nu_{fe}(F_1) = +0.016\,N(N+1)$.

[c]Weak predissociation in v=0 above N=3 or 4 [see (179)(187) and [a] on p. 467]. The predissociation is assumed to occur via the continuum of the $a^4\Pi$ state and causes a reduction of the measured lifetimes in v=0 from 20 ns for N ≲ 4 to 3 ns for N ≳ 4 (183); τ(v=1) ≤ 0.3 ns. No emission has been observed from levels having v ≥ 1.

[d]See [h] on p. 473.

[e]$f_{00} = 0.0023$ (43)(177), higher value in (124); $f_{10} = 0.0058$, $f_{20} = 0.0027$ (43). See also (62).

[f]See [n] on p. 473. RKR Franck-Condon factors for the β bands (111)(128), for the γ bands (111)(134).

[g]A different vibrational numbering was suggested by (75).

[h]$A_v = +31.32 + 1.152(v+\frac{1}{2}) + 0.0448(v+\frac{1}{2})^2$. The expression represents the data of (127) for the first seven levels. A_v increases to +77 for v=25; see (90).

NO (continued):

[i]Effective constants for $v \leq 5$ (73). The re-evaluation of the constants by (127), based on new measurements of the β bands and using a modified Hill - Van Vleck expression, gave $G(v) = 1037.45(v+\frac{1}{2}) - 7.472(v+\frac{1}{2})^2 + 0.0725_3(v+\frac{1}{2})^3$,

$$B_v = 1.1250 - 0.01348(v+\tfrac{1}{2}) + 0.00012_5(v+\tfrac{1}{2})^2.$$

The highest level observed in emission is v'=7 [mixed with C(v=0)] (49)(117); vibrational levels as high as v'=29 have been identified in the absorption spectrum (74). They are strongly perturbed by interaction with the Rydberg states $C^2\Pi$ (matrix element $H_e \approx 1200$ cm^{-1}) and $K^2\Pi$ ($H_e \approx 800$); see (42)(84). A complete deperturbation, taking also into account the interaction with $L^2\Pi$, was attempted by (136); more recent results by Gallusser and Dressler ($\omega_e = 1025.0$, $\omega_e x_e = 4.52$, $\omega_e y_e = -0.0846$) are quoted by (185) who observed the deperturbed spectrum of $B^2\Pi$ in matrix absorption (101)(116). A similar deperturbation is induced by high pressure foreign gases (190). Heterogeneous interactions with levels of $D^2\Sigma^+$ and $M^2\Sigma^+$ are discussed by (112).

[j]$\omega_e y_e = +0.17$.

[k]$\omega_e y_e = +0.10$.

[ℓ]Λ-type doubling, $\Delta v_{fe}(v=0) = -0.0064(J+\frac{1}{2})$ (64).

[m]Radiative lifetimes $\tau(v=0,1,4) = 1.9_9, 1.7_8, 1.6_5$ μs, resp. (174). (68) give somewhat longer lifetimes.

[n]$f_{00} = 2.5 \times 10^{-8}$ (141); $f_{v'0}$ values increase to 4.6×10^{-5} for v'=6 (43)(141)(150). Above v'=7 the intensities are governed by the strong interactions with the 3p and 4p Rydberg states; see [i]. See also (59)(62)(70)(105).

[o]Referring to the hypothetical J=0 levels in both upper and lower state.

[p]$\omega_e y_e = -0.0465$ (127).

[q]Rotational constants re-evaluated from data in (36) and (127); the equilibrium constants of the latter appear unreliable.

[r]Spin splitting constant $\gamma(v=3) = -0.0027650$, $\mu_{e\ell}(v=3) = 1.10$ D. These constants, as well as eqQ and magnetic hf constants, have been recalculated by (194) from the optical-rf double resonance experiment of (172); see also (152) (191). Hanle effect (139)(151a)(159).

[s]According to (8)(41) the intensity of the emission bands drops sharply at N' = 74, 64, 52, 38 in v' = 0, 1, 2, 3, respectively; bands with $v' \geq 4$ have not been observed in emission.

[t]Radiative lifetimes $\tau(v=0) = 215$ ns, $\tau(v=1) = 203$ ns, $\tau(v=2) = 174$ ns (206); good agreement with (174) except for v=2 where these authors find $\tau = 195$ ns. See also (87)(159) (183).

[u]$f_{00} = 0.00038$, $f_{10} = 0.00081$, $f_{20} = 0.00069$, $f_{30} = 0.00030$; weighted average values from (38)(43)(132)(150)(151). Variation of transition moment with r (70)(83)(87)(93)(105) (111)(149); see also (144).

[v]Also observed in two-photon excitation (173)(186)(206) and magnetic rotation spectra (100). $^{15}N^{16}O$ band head measurements (125).

[w]See [c] on p.469.

[x]Assignment uncertain, only observed in rare gas matrices. Predicted lifetime 0.1 s (115). See also (170).

State	T_e	ω_e	$\omega_e x_e$	B_e	α_e	D_e $(10^{-6}cm^{-1})$	r_e (Å)	Observed Transitions Design.	ν_{00}	References
$^{14}N^{16}O$ (continued)										
X $^2\Pi_r$	119.82^a 0	1904.04_0 Z 1904.20_4 Z	14.100^b 14.075^b	$[1.72016]^c$ $[1.67195]^{cd}$	0.0182 0.0171	$[10.2_3]^c$ $[0.5_4]^c$	1.15077	$\frac{3}{2} \leftarrow \frac{1}{2}$, [e]	119.73^f Z	(148)
								Rotation-vibration sp.:		
								$4 \rightarrow 2$		(63)
								$4 \leftarrow 0$		(77)(78)
								$3 \rightarrow 1$		(63)
								$3 \leftarrow 0$, [gh]		(23)(54)(71)(72)(77)
								$2 \leftarrow 1$		(208)
								$2 \leftrightarrow 0$, [ghi]		(23)(63)(72)(77)
								$1 \leftrightarrow 0$, [ghijkℓmn]		(32)(34)(61)(67)(113)(205)(209)
								Rotation sp. [o]		(29)(33)(44)(86)(148)
								Raman sp. [p]		(120)(133)
								EPR sp.		(82)(145)(193)
								Hyperfine Λ-doubl. sp. [q]		(131)(154)(200)

NO: [a] $A_v = + 123.26 - 0.1906(v+\frac{1}{2}) - 0.0108(v+\frac{1}{2})^2$; from the analysis of β and γ bands having $v'' \leq 16$ (127). Much more precise constants for v=0 and 1 (A_{eff} = 123.1393 and 122.8935, respectively) and their J dependence have been determined from measurements on the vibration-rotation fundamental and on the pure rotation spectrum (205)(209). See also (148)(156).

[b] $\omega_e y_e = + 0.011_0$ ($^2\Pi_{3/2}$) and $+ 0.007_7$ ($^2\Pi_{1/2}$); these are effective vibrational constants obtained from rotation-vibration spectra (67)(77). (205)(209) have accurately evaluated $\Delta G(\frac{1}{2}) = 1875.972$; see [a]. (127), see [a], give the following expression, valid for $v \leq 16$: $G(v) = 1904.40_5(v+\frac{1}{2}) - 14.187_0(v+\frac{1}{2})^2 + 0.0240_0(v+\frac{1}{2})^3 - 0.00093(v+\frac{1}{2})^4$. The vibrational levels have been observed to v = 23 (18).

[c] Effective rotational constants from rotation (29)(44)(86) and rotation-vibration spectra (191a)(205). Precise B and D values for v=0 and 1 have been calculated by (209), see [a]: $B_0 = 1.69611_5$, $B_1 = 1.67854_4$; $D_0 = 5.3_4 \times 10^{-6}$, $D_1 = 5.3_7 \times 10^{-6}$; good agreement with (205). (127), see [a], give the following expression for $v \leq 16$: $B_v = 1.7042_7 - 0.01728(v+\frac{1}{2}) - 0.000037(v+\frac{1}{2})^2$.

[d] Λ-type doubling, $\Delta\nu_{fe} \approx (+)0.0117(J+\frac{1}{2})$. Precise Λ-doubling constants have been evaluated by (154)(200)(205)(209), the variation of p and q with v agrees with the measurements of (208).

[e] Observed in the electronic-rotational Raman spectrum (4a)(118)(130), and as magnetic dipole transition in the far IR (148). Laser Zeeman spectrum (156).

[f] See [o] on p. 475.

[g] Magnetic rotation (50)(81)(107)(114)(161).

[h] Integrated band intensities, dipole moment function (92)(142)(163)(167); (184)(198).

[i] 2-0 b. of $^{15}N^{18}O$ (96), 1-0 b. of $^{15}N^{16,18}O$ (39)(96)(113).

NO (continued):

[j] Λ-doubling, nuclear hfs, and Zeeman splittings (147)(158). $^2\Pi_{3/2}$ laser magnetic resonance spectra (171)(191a).

[k] Laser Stark spectrum (188); $\mu_{e\ell}(^2\Pi_{\frac{1}{2}}, v=0) = 0.157_4$ D (see also [q]), $\mu_{e\ell}(^2\Pi_{\frac{1}{2}}, v=1) = 0.1416$ D. For $^2\Pi_{\frac{3}{2}}$ the difference $\mu_{e\ell}(v=1) - \mu_{e\ell}(v=0) = -0.01735$ D was determined.

[ℓ] From pressure-broadened linewidths (196) derive a value of 2.4×10^{-26} esu cm^2 for the quadrupole moment of NO. Earlier results are reviewed in this paper.

[m] Absorption of CO laser radiation by NO (202)(203)(207).

[n] $\Delta v = 1$ sequence in emission (201). Several laser lines have been observed in the P branches of the 6-5, ..., 11-10 bands (83a).

[o] Zeeman effect (21), Stark effect (40), both in $^2\Pi_{1/2}$.

[p] See also references in [e].

[q] $\mu_{e\ell} = 0.15872$ D from Stark effect in $^2\Pi_{\frac{1}{2}}, v=0, J=\frac{1}{2}$ (131); eqQ and other hf coupling constants (154)(200). See also (152)(191).

(1) Leifson, ApJ 63, 73 (1926).
(2) Jenkins, Barton, Mulliken, PR 30, 150 (1927).
(3) Schmid, ZP 49, 428 (1928); 64, 84 (1930).
(4) Schmid, ZP 64, 279 (1930).
(4a) Rasetti, ZP 66, 646 (1930).
(5) Tanaka, Sci. Pap. IPCR (Tokyo) 39, 456 (1942).
(6) Gaydon, PPS 56, 95, 160 (1944).
(7) Gerö, Schmid, von Szily, Physica 11, 144 (1944).
(8) Gerö, Schmid, PPS 60, 533 (1948).
(9) Feast, CJR A 28, 488 (1950).
(10) Baer, Miescher, HPA 24, 331 (1951).
(11) Tanaka, Seya, Mori, JCP 19, 979 (1951).
(12) Baer, Miescher, Nature 169, 581 (1952); HPA 26, 91 (1953).
(13) Marmo, JOSA 43, 1186 (1953).
(14) Ogawa, SL 2, 87 (1953).
(15) Sutcliffe, Walsh, PPS A 66, 209 (1953).
(16) Tanaka, JCP 21, 788 (1953).
(17) Watanabe, Marmo, Inn, PR 91, 1155 (1953); Watanabe, JCP 22, 1564 (1954).
(18) Brook, Kaplan, PR 96, 1540 (1954).
(19) Ogawa, SL 3, 39 (1954).
(20) Tanaka, JCP 22, 2045 (1954).
(21) Mizushima, Cox, Gordy, PR 98, 1034 (1955).
(22) Miescher, CJP 33, 355 (1955); HPA 29, 401 (1956).
(23) Nichols, Hause, Noble, JCP 23, 57 (1955).
(24) Ogawa, SL 3, 90 (1955).
(25) Sun, Weissler, JCP 23, 1372 (1955).
(26) Ueda, SL 3, 143 (1955).
(27) Walker, Weissler, JCP 23, 1962 (1955).
(28) Granier, Astoin, CR 242, 1431 (1956).
(29) Gallagher, Johnson, PR 103, 1727 (1956).
(30) Herzberg, Lagerqvist, Miescher, CJP 34, 622 (1956).
(31) Ogawa, Shimauchi, SL 5, 147 (1956).
(32) Shaw, JCP 24, 399 (1956).
(33) Palik, Rao, JCP 25, 1174 (1956).
(34) Thompson, Green, SA 8, 129 (1956).
(35) Astoin, JRCNRS No. 38, 1 (1957).
(36) Barrow, Miescher, PPS A 70, 219 (1957).
(37) Deézsi, Mátrai, APH 7, 111 (1957).
(38) Weber, Penner, JCP 26, 860 (1957).
(39) Fletcher, Begun, JCP 27, 579 (1957).

NO: (40) Burrus, Graybeal, PR 109, 1553 (1958).
(41) Deézsi, APH 9, 125 (1958).
(42) Lagerqvist, Miescher, HPA 31, 221 (1958).
(43) Bethke, JCP 31, 662 (1959).
(44) Favero, Mirri, Gordy, PR 114, 1534 (1959).
(45) Heath, Los Alamos Report LA-2335 (1959).
(46) Koczkás, APH 10, 117 (1959).
(47) Maryott, Kryder, JCP 31, 617 (1959).
(48) Broida, Peyron, JCP 32, 1068 (1960).
(49) Deézsi, APH 11, 155 (1960).
(50) Mann, Hause, JCP 33, 1117 (1960).
(51) Huber, HPA 34, 929 (1961).
(52) Lagerqvist, Miescher, CJP 40, 352 (1962).
(53) Miescher, JQSRT 2, 421 (1962).
(54) Arcas, Haeusler, Joffrin, Meyer, van Thanh, Barchewitz, AO 2, 909 (1963).
(55) Huber, Huber, Miescher, PL 3, 315 (1963).
(56) Huber, Miescher, HPA 36, 257 (1963).
(57) Kovács, HPA 36, 699 (1963).
(58) Nicholson, JCP 39, 954 (1963).
(59) Antropov, Dronov, Sobolev, OS(Engl. Transl.) 17, 355 (1964).
(60) Callear, Smith, DFS 37, 96 (1964).
(61) James, JCP 40, 762 (1964).
(62) Ory, JCP 40, 562 (1964).
(63) Horn, Dickey, JCP 41, 1614 (1964).
(64) Huber, HPA 37, 329 (1964).
(65) Huber, PL 12, 102 (1964).
(66) Frosch, Robinson, JCP 41, 367 (1964).
(67) James, Thibault, JCP 41, 2806 (1964).
(68) Jeunehomme, Duncan, JCP 41, 1692 (1964).
(69) Lofthus, Miescher, CJP 42, 848 (1964).
(70) Marr, PPS 83, 293 (1964).
(71) Meyer, Haeusler, van Thanh, Barchewitz, JP(Paris) 25, 337 (1964).
(72) Olman, McNelis, Hause, JMS 14, 62 (1964); 21, 111 (1966) (erratum).
(73) Callear, Smith, TFS 61, 1303 (1965).
(74) Dressler, Miescher, ApJ 141, 1266 (1965).
(75) Gilmore, JQSRT 5, 369 (1965).
(76) Jungen, Miescher, ApJ 142, 1660 (1965).
(77) Meyer, Haeusler, Barchewitz, JP(Paris) 26, 799 (1965).
(78) Meyer, Haeusler, CR 260, 4182 (1965).
(79) Abels, Shaw, JMS 20, 11 (1966).
(80) Alamichel, JP(Paris) 27, 345 (1966).
(81) Aubel, Hause, JCP 44, 2659 (1966).
(82) Brown, Radford, PR 147, 6 (1966).
(83) Callear, Pilling, Smith, TFS 62, 2997 (1966).
(83a) Deutsch, APL 9, 295 (1966).
(84) Felenbok, Lefebvre-Brion, CJP 44, 1677 (1966).
(85) Reese, Rosenstock, JCP 44, 2007 (1966).
(86) Hall, Dowling, JCP 45, 1899 (1966).
(87) Jeunehomme, JCP 45, 4433 (1966).
(88) Jungen, CJP 44, 3197 (1966).
(89) Jungen, Miescher, Suter, PL 21, 36 (1966).
(90) Lagerqvist, Miescher, CJP 44, 1525 (1966).
(91) Miescher, JMS 20, 130 (1966).
(92) Schurin, Ellis, JCP 45, 2528 (1966).
(93) Antropov, Kolesnikov, Ostrovskaya, Sobolev, OS (Engl. Transl.) 22, 109 (1967).
(94) Crosley, Zare, PRL 18, 942 (1967). [Erroneous, s.(139)]
(95) Feinberg, Camal, JQSRT 7, 581 (1967).
(96) Griggs, Rao, Jones, Potter, JMS 22, 383 (1967).
(97) Metzger, Cook, Ogawa, CJP 45, 203 (1967).

NO: (98) Oppenheim, Yair Aviv, Goldman, AO 6, 1305 (1967).
(99) Ortenberg, Antropov, SPU 9, 717 (1967).
(100) Robinson, JCP 46, 4525 (1967); 50, 5018 (1969).
(101) Roncin, Damany, Romand, JMS 22, 154 (1967).
(102) Varanasi, Penner, JQSRT 7, 279 (1967).
(103) Watanabe, Matsunaga, Sakai, AO 6, 391 (1967).
(104) Ackermann, Miescher, CPL 2, 351 (1968).
(105) Antropov, Proc. (Trudy) P. N. Lebedev Phys. Inst. 35, 1 (1966) [Engl. Transl. Consultants Bureau, New York (1968)].
(106) Hesser, JCP 48, 2518 (1968).
(107) Buckingham, Segal, JCP 49, 1964 (1968).
(108) Callear, Pilling, Smith, TFS 64, 2296 (1968).
(109) Chardon, Théobald, CR B 266, 602 (1968).
(110) Crosley, Zare, JCP 49, 4231 (1968). [Erroneous, s. (139)]
(111) Jain, Sahni, TFS 64, 3169 (1968).
(112) Jungen, Miescher, CJP 46, 987 (1968).
(113) Keck, Hause, JMS 26, 163 (1968).
(114) Keck, Hause, JCP 49, 3458 (1968).
(115) Lefebvre-Brion, Guérin, JCP 49, 1446 (1968).
(116) Roncin, JMS 26, 105 (1968).
(117) Ackermann, Miescher, JMS 31, 400 (1969).
(118) Fast, Welsh, Lepard, CJP 47, 2879 (1969).
(119) Jungen, Miescher, CJP 47, 1769 (1969).
(120) Renschler, Hunt, McCubbin, Polo, JMS 32, 347 (1969).
(121) Suter, CJP 47, 881 (1969).
(122) Wray, JQSRT 9, 255 (1969).
(123) Callear, Pilling, TFS 66, 1618 (1970).
(124) Callear, Pilling, TFS 66, 1886 (1970).
(125) Cisak, Danielak, Rytel, APP A 37, 67 (1970).
(126) Edqvist, Lindholm, Selin, Sjögren, Åsbrink, AF 40, 439 (1970).
(127) Engleman, Rouse, Peek, Baiamonte, Los Alamos Sci. Lab. Report LA-4364 (1970); Engleman, Rouse, JMS 37, 240 (1971).
(128) Generosa, Harris, JCP 53, 3147 (1970).
(129) Jungen, JCP 53, 4168 (1970).
(130) Lepard, CJP 48, 1664 (1970).
(131) Neumann, ApJ 161, 779 (1970).
(132) Pery-Thorne, Banfield, JP B 3, 1011 (1970).
(133) Shotton, Jones, CJP 48, 632 (1970).
(134) Spindler, Isaacson, Wentink, JQSRT 10, 621 (1970).
(135) Ackermann, CJP 49, 76 (1971).
(136) Bartholdi, Leoni, Dressler, ZAMP 22, 797 (1971).
(137) Collin, Delwiche, Natalis, IJMSIP 7, 19 (1971).
(138) Gallusser, Dressler, ZAMP 22, 792 (1971).
(139) German, Zare, Crosley, JCP 54, 4039 (1971).
(140) Groth, Kley, Schurath, JQSRT 11, 1475 (1971).
(141) Hasson, Nicholls, JP B 4, 1769 (1971).
(142) Michels, JQSRT 11, 1735 (1971).
(143) Miescher, CJP 49, 2350 (1971).
(144) Poland, Broida, JCP 54, 4515 (1971); JQSRT 11, 1863 (1971).
(145) Ashford, Jarke, Solomon, JCP 57, 3867 (1972).
(146) Bahr, Blake, Carver, Gardner, Kumar, JQSRT 12, 59 (1972).
(147) Blum, Nill, Calawa, Harman, CPL 15, 144 (1972).
(148) Brown, Cole, Honey, MP 23, 287 (1972).
(149) Bubert, JCP 56, 1113 (1972).
(150) Farmer, Hasson, Nicholls, JQSRT 12, 627, 635 (1972).
(151) Hasson, Farmer, Nicholls, Anketell, JP B 5, 1248 (1972).
(151a) Gouédard, AP(Paris) 7, 159 (1972).
(152) Green, CPL 13, 552 (1972); 23, 115 (1973).
(153) Kleimenov, Chizhov, Vilesov, OS(Engl. Transl.) 32, 371 (1972).
(154) Meerts, Dymanus, JMS 44, 320 (1972).
(155) - s. (192a)
(156) Mizushima, Evenson, Wells, PR A 5, 2276 (1972).

NO: (157) Narayana, Price, JP B 5, 1784 (1972).
(158) Nill, Blum, Calawa, Harman, CPL 14, 234 (1972).
(159) Weinstock, Zare, Melton, JCP 56, 3456 (1972).
(160) Ben-Aryeh, JQSRT 13, 1441 (1973).
(161) Blum, Nill, Strauss, JCP 58, 4968 (1973).
(162) Broida, Miescher, JQE 9, 1029 (1973).
(163) Chandraiah, Cho, JMS 47, 134 (1973).
(164) Gardner, Lynch, Stewart, Watson, JP B 6, L262 (1973).
(165) Gardner, Samson, JESRP 2, 153 (1973).
(166) Killgoar, Leroi, Berkowitz, Chupka, JCP 58, 803 (1973).
(167) Konkov, Vorontsov, OS(Engl. Transl.) 34, 595 (1973).
(168) Lee, Carlson, Judge, Ogawa, JQSRT 13, 1023 (1973).
(169) Mandelman, Carrington, Young, JCP 58, 84 (1973).
(170) Zarur, Chiu, JCP 59, 82 (1973).
(171) Zeiger, Blum, Nill, JCP 59, 3968 (1973).
(172) Bergeman, Zare, JCP 61, 4500 (1974).
(173) Bray, Hochstrasser, Wessel, CPL 27, 167 (1974).
(174) Brzozowski, Elander, Erman, PS 9, 99 (1974).
(175) Hertz, Jochims, Schenk, Sroka, CPL 29, 572 (1974).
(176) Hertz, Jochims, Sroka, PL A 46, 365 (1974).
(177) Mandelman, Carrington, JQSRT 14, 509 (1974).
(178) Morioka, Nakamura, Ishiguro, Sasanuma, JCP 61, 1426 (1974).
(179) Miescher, JMS 53, 302 (1974).
(180) Miescher, in "Vacuum Ultraviolet Radiation Physics", p. 61 (ed. Koch, Haensel, Kunz). Pergamon-Vieweg, Braunschweig (1974).
(181) Sasanuma, Morioka, Ishiguro, Nakamura, JCP 60, 327 (1974).
(182) Takezawa, paper MG12, 29th Symposium on Molecular Structure and Spectroscopy, Columbus, Ohio (1974).
(182a) Wight, Brion, JESRP 4, 313 (1974).
(183) Benoist d'Azy, López-Delgado, Tramer, CP 9, 327 (1975).
(184) Billingsley, JCP 62, 864; 63, 2267 (1975).
(185) Boursey, Roncin, JMS 55, 31 (1975).
(186) Bray, Hochstrasser, Sung, CPL 33, 1 (1975).
(187) Dingle, Freedman, Gelernt, Jones, Smith, CP 8, 171 (1975).
(188) Hoy, Johns, McKellar, CJP 53, 2029 (1975).
(189) Johnson, Berman, Zakheim, JCP 62, 2500 (1975).
(190) Miladi, Le Falher, Roncin, Damany, JMS 55, 81 (1975); Miladi, Thèse (U. de Paris-Sud, Centre d'Orsay, 1976).
(191) Walch, Goddard, CPL 33, 18 (1975).
(191a) Hakuta, Uehara, JMS 58, 316 (1975).
(191b) Field, Gottscho, Miescher, JMS 58, 394 (1975).
(192) Miescher, Alberti, JPCRD 5, 309 (1976).
(192a) Miescher, Huber, International Review of Science, Physical Chemistry Series Two, Vol. 3, Spectroscopy (ed. Ramsay). Butterworths (1976).
(193) Jarke, Ashford, Solomon, JCP 64, 3097 (1976).
(194) Woods, Dixon, JCP 64, 5319 (1976).
(195) Ng, Mahan, Lee, JCP 65, 1956 (1976).
(196) Tejwani, Golden, Yeung, JCP 65, 5110 (1976).
(197) Boursey, JMS 61, 11 (1976).
(198) Billingsley, JMS 61, 53 (1976).
(199) Caprace, Delwiche, Natalis, Collin, CP 13, 43 (1976).
(200) Meerts, CP 14, 421 (1976).
(201) Mantz, Shafer, Rao, AO 15, 599 (1976).
(202) Richton, AO 15, 1686 (1976).
(203) Hanson, Monat, Kruger, JQSRT 16, 705 (1976).
(204) Miescher, CJP 54, 2074 (1976).
(205) Valentin, Boissy, Cardinet, Henry, Chen, Rao, CR B 283, 233 (1976).
(206) Zacharias, Halpern, Welge, CPL 43, 41 (1976).
(207) Garside, Ballik, Elsherbiny, Shewchun, AO 16, 398 (1977).
(208) Guerra, Sanchez, Javan, PRL 38, 482 (1977).
(209) Johns, Reid, Lepard, JMS 65, 155 (1977).

NO^+, NO^{++} (continued from p.483):

(12) Stair, Gauvin, in "Aurora and Airglow", p.365 (ed. McCormac). Reinhold, New York (1967).

(13) Hesser, JCP 48, 2518 (1968).

(14) Huber, CJP 46, 1691 (1968).

(15) Nicholls, JP B 1, 1192 (1968).

(16) Price, in "Molecular Spectroscopy", p.221 (ed. Hepple). The Institute of Petroleum, London (1968).

(17) Samson, PL A 28, 391 (1968).

(18) Sjögren, Szabo, AF 37, 551 (1968).

(19) Jungen, Miescher, CJP 47, 1769 (1969).

(20) Siegbahn, Nordling, Johansson, Hedman, Hedén, Hamrin, Gelius, Bergmark, Werme, Manne, Baer, "ESCA Applied to Free Molecules". North-Holland, Amsterdam (1969).

(21) Edqvist, Lindholm, Selin, Sjögren, Åsbrink, AF 40, 439 (1970).

(22) Jungen, Lefebvre-Brion, JMS 33, 520 (1970).

(23) Wentink, Spindler, JQSRT 10, 609 (1970).

(24) Aarts, de Heer, Physica 54, 609 (1971).

(25) Edqvist, Åsbrink, Lindholm, ZN 26 a, 1407 (1971).

(26) Lefebvre-Brion, CPL 9, 463 (1971).

(27) Maier, Holland, JCP 54, 2693 (1971).

(28) Moddeman, Carlson, Krause, Pullen, Bull, Schweitzer, JCP 55, 2317 (1971).

(29) Davis, Shirley, JCP 56, 669 (1972).

(30) Mentall, Morgan, JCP 56, 2271 (1972).

(31) Stone, Zipf, JCP 56, 2870 (1972).

(32) Thulstrup, Öhrn, JCP 57, 3716 (1972).

(33) Appell, Durup, Fehsenfeld, Fournier, JP B 6, 197 (1973).

(34) Billingsley, CPL 23, 160 (1973).

(35) Davis, Martin, Banna, Shirley, JCP 59, 4235 (1973).

(36) Field, JMS 47, 194 (1973).

(37) Bagus, Schrenk, Davis, Shirley, PR A 9, 1090 (1974).

(38) Billingsley, Krauss, JCP 60, 2767 (1974).

(39) Hertz, Jochims, Schenk, Sroka, CPL 29, 572 (1974).

(40) Thulstrup, Thulstrup, Andersen, Öhrn, JCP 60, 3975 (1974).

(41) Alberti, Douglas, CJP 53, 1179 (1975).

(42) Coxon, Clyne, Setser, CP 7, 255 (1975).

(43) Hillier, Kendrick, JCS FT II 71, 1654 (1975).

(44) Darko, Hillier, Kendrick, CPL 45, 188 (1977).

State	T_e	ω_e	$\omega_e x_e$	B_e	α_e	D_e ($10^{-6}cm^{-1}$)	r_e (Å)	Observed Transitions Design.	ν_{00}	References
$^{14}N^{16}O^{+}$		$D_0^0 = 10.850_6$ eV[a] I.P. = 30.3 eV[b]								MAR 1977 A
$^1\Pi$	534.4 eV	Removal of a $1\sigma(1s_O)$[c] electron; observed in X-ray photoelectron spectra (20)(29).								
$^3\Pi$	533.9 eV									
$^1\Pi$	402.3 eV	Removal of a $2\sigma(1s_N)$[c] electron; observed in X-ray photoelectron spectra (20)(29).								
$^3\Pi$	400.9 eV									
$^1\Pi$	34.5 eV	Removal of a 3σ electron; observed in X-ray photoelectron spectra (20)(29).								
$^3\Pi$	31.3 eV									
$(B'\ ^1\Sigma^+)$[d]	(109400)	(730)[d]	7							
c $^3\Pi$ [e]	(100780)	(1900)								
A $^1\Pi$ [f]	73471.72	1601.93 Z	20.207[gh]	1.5860[i]	0.0245	[5.6]	1.1931	A[j]→X,[k] R	73083.46 Z	(2)(3)(4)* (36)(41)
W $^1\Delta$	71450	1278[ℓ]	16.0 [h]	(1.361)[m]	(0.0192)		(1.288)			
A' $^1\Sigma^+$	69540	1283[ℓ]	13.5 [h]	(1.363)[m]	(0.0192)		(1.287)			
b' $^3\Sigma^-$	67720	1284[ℓ]	$10._7$ [h]	1.357[n]	(0.0192)		1.290	b'→X,[o]		(27)(36)
w $^3\Delta$	61880	1313[ℓ]	$10._6$ [h]	(1.377)[m]	(0.0192)		(1.280)			
b $^3\Pi$ [p]	59240	1710	(14)[m] [h]	(1.634)[m]	(0.0184)[m]		(1.175)	b→X,	58900	(27)
a $^3\Sigma^+$	52190	1293[ℓ]	$15._1$ [h]	(1.369)[m]	(0.0192)		(1.284)			
		Neither photoelectron spectra (9)(16)(17) nor charge exchange experiments (18) have confirmed the existence of a previously predicted (1)(6) electronic state of NO^+ at 39960 cm^{-1} above X $^1\Sigma^+$. See (14).								
X $^1\Sigma^+$	0	2376.42 Z	16.262 [h]	1.99727[q]	0.01889	5.64	1.06322	Vibr.-rot. sp.[r]		(12)
$^{14}N^{16}O^{++}$										MAR 1977
C	64000	Three states or unresolved groups of states observed by double charge transfer spectroscopy (33); reasonable agreement with theoretical predictions (40).								
B $(^2\Sigma^+)$	25000									
A $(^2\Pi)$, X $(^2\Sigma^+)$	0									

NO^+, NO^{++}:

[a] $D_0^0(NO)$ + I.P.(O) − I.P.(NO).

[b] Average of two values obtained by electron impact mass-spectrometry (5) and double charge transfer spectroscopy (33). A similar value (30.8 eV) can be derived from Auger electron spectra (28), but the assignments are highly tentative and can also be interpreted as indicating an ionization potential of only ~26 eV.

[c] The $1s_N$ and $1s_O$ $^1\Pi - ^3\Pi$ splittings are 1.41 and 0.53 eV, respectively (29)(35). The $1s_N$ photoelectron peaks have an anomalous intensity ratio of 3.43 (37); theoretical interpretation (43)(44). Predicted $1s_N$ satellite ("shake-up") peaks (43).

[d] Long vibrational progression in the photoelectron spectrum at ~23 eV (21)(25); neither the vibrational numbering nor the assignment to $B'^1\Sigma^+$ are certain. (26) suggests a mixed $^1\Pi$ state resulting from configuration interaction between states arising from $...4\sigma\, 5\sigma^2\, 1\pi^4\, 2\pi$ (like c $^3\Pi$) and $...4\sigma^2\, 5\sigma\, 1\pi^3\, 2\pi^2$. See also (32)(40).

[e] Limit of the Narayana-Price Rydberg series. Single strong peak in the photoelectron spectrum at 21.72 eV (16)(17)(21)(25). The high-energy wing is overlapped by weak unresolved structure ($\omega \approx 400$) tentatively attributed to the corresponding $^1\Pi$ state by (25); see, however, [d]. Predissociation into N^+ + O (39).

[f] Limit of Tanaka's γ Rydberg series. Short progression in the photoelectron spectrum (9)(16)(21)(25); see [h].

[g] $\omega_e y_e = -0.2683$.

[h] Franck-Condon factors for ionizing transitions from X $^2\Pi$ (7)(8)(11)(15)(36)(42).

[i] Perturbations by $b'^3\Sigma^-$ (4)(36)(41).

[j] Radiative lifetime $\tau(v=0)$ = 56 ns (10)(13); f_{00} = 0.00025. Variation of transition moment with r (24)(30)(31).

[k] Franck-Condon factors (15)(23)(42).

[ℓ] Long upper state progression in the photoelectron spectrum (21)(25); see [h].

[m] Interpolated using data for CO and N_2 (36).

[n] Adjusted to give agreement with B_5 = 1.2512 as obtained from a perturbation analysis (36).

[o] Fragments in emission from perturbed $A^1\Pi \sim b'^3\Sigma^-$ levels.

[p] Limit of Tanaka's β Rydberg series. Very short progression in the photoelectron spectrum (9)(16)(21)(25); see [h].

[q] Quadrupole moment of NO^+ = $+0.79 \times 10^{-26}$ esu cm^2 (19); see also (22)(38).

[r] Observed in the IR spectrum of hot air resulting from a high altitude nuclear detonation; the 1-0 and 2-0 bands have been identified. Theoretical intensities (34).

(1) Tanaka, Sci. Pap. IPCR (Tokyo) 39, 456 (1942).
(2) Baer, Miescher, HPA 26, 91 (1953).
(3) Tanaka, JCP 21, 562 (1953).
(4) Miescher, CJP 33, 355 (1955); HPA 29, 135 (1956).
(5) Dorman, Morrison, JCP 35, 575 (1961).
(6) Huber, HPA 34, 929 (1961).
(7) Wacks, JCP 41, 930 (1964).
(8) Halmann, Laulicht, JCP 43, 1503 (1965).
(9) Turner, May, JCP 45, 471 (1966).
(10) Hesser, Dressler, JCP 45, 3149 (1966).
(11) Spohr, von Puttkamer, ZN 22 a, 705 (1967).

(continued on p.481)

State	T_e	ω_e	$\omega_e x_e$	B_e	α_e	D_e $(10^{-6}cm^{-1})$	r_e (Å)	Observed Transitions Design.	ν_{00}	References
$^{14}N^{16}O^-$		$D_0^0 = 5.056$ eV[a] I.P. = 0.024 eV[b]								MAR 1977 A
		Evidence for additional compound states ("resonances" related to the "grandparents" $b^3\Pi$, $A^1\Pi$, and $c^3\Pi$ of NO^+) in the 12-18 eV region of the electron transmission (2)(7) and electroionization (9) spectra of NO.								
		The nature of the state (or states) involved in the production of $N(^2D) + O^-$ by dissociative electron attachment (7-12 eV) has been discussed by (11); see also (10).								
$(^3\Sigma^-)^c$	51700	2320	13	Short vibrational progressions of resonances in the electron transmission current (2); predicted widths range from 1 to 25 meV (14).						
$(^3\Sigma^+)^c$	43800	2380	12							
$(^3\Pi)^c$	43400	2370	12							
$(^1\Sigma^+)^c$	40400	2330	8							
b $^1\Sigma^+$	$(9300)^d$									
a $^1\Delta$	6050	1492^d	(8)				$1.26_2{}^d$			
X $^3\Sigma^-$	0	1363^d	8	$1.42_7{}^e$			$1.25_8{}^e$			
$^{(237)}Np^{16}O$		$(\mu = 14.9838726_3)$ [a] I.P. = 5.7 eV[b]								JUN 1975
$^{14}N^{32}S$		$\mu = 9.7380289_4$ $D_0^0 = 4.8$ eV[a] I.P. = 8.87 eV[b]								MAY 1977
		Theoretical potential curves for most of the observed valence states have been calculated by (28).								
F $^2\Delta$				$[0.8367]^c$		[1.05]	$[1.438_4]$	F→X, V	55959.8 Z $56181._3$ Z	(11)(29)*
J $^2\Sigma^+$				$[(0.825)]^d$			$[(1.44_9)]$	J→X, V	55562 55784	(29)
E $^2\Pi_r$	^{e}			$[0.821]^e$			$[1.45_2]$	E→X,	51154^e 51330^e	(11)(29)*
I $^2\Sigma^+$	(44400)	[1008]		$[0.6940]^f$		[5]	$[1.579_3]$	I→X, R	44050.1 Z 44271.6 Z	(7)* (11)(15) (23)* (29)

NO^-: [a]From D_0^0(NO) and the electron affinities of O and NO.

[b]From the photodetachment spectrum (5). Good agreement with (3) and (4).

[c]Symmetries assigned on the basis of theoretical calculations (6). The states consist of two Rydberg electrons temporarily bound to the NO^+ $X^1\Sigma^+$ core.

[d]From the analysis (13)(15) of electron scattering data (1)(7)(8)(12). For the ground state (5) estimate $\omega_e \approx 1470$ cm^{-1}, see [e].

[e]Franck-Condon factor analysis of the photodetachment spectrum (5). The analysis of electron scattering data (15) leads to $r_e = 1.26_7$ Å.

(1) Spence, Schulz, PR A 3, 1968 (1971).
(2) Sanche, Schulz, PRL 27, 1333 (1971); PR A 6, 69 (1972).
(3) McFarland, Dunkin, Fehsenfeld, Schmeltekopf, Ferguson, JCP 56, 2358 (1972).
(4) Parkes, Sugden, JCS FT II 68, 600 (1972).
(5) Siegel, Celotta, Hall, Levine, Bennett, PR A 6, 607 (1972).
(6) Lefebvre-Brion, CPL 19, 456 (1973).
(7) Schulz, RMP 45, 423 (1973).
(8) Burrow, CPL 26, 265 (1974).
(9) Carbonneau, Marmet, CJP 52, 1885 (1974).
(10) Thulstrup, Thulstrup, Andersen, Öhrn, JCP 60, 3975 (1974).
(11) Van Brunt, Kieffer, PR A 10, 1633 (1974).
(12) Zecca, Lazzizzera, Krauss, Kuyatt, JCP 61, 4560 (1974).
(13) Tronc, Huetz, Landau, Pichou, Reinhardt, JP B 8, 1160 (1975).

NO^- (continued):

(14) Pearson, Lefebvre-Brion, PR A 13, 2106 (1976).
(15) Teillet-Billy, Fiquet-Fayard, JP B 10, L111 (1977).

NpO: [a]Thermodynamic properties of NpO (1).

[b]Corrected electron impact appearance potential (2).

(1) Ackermann, Rauh, JCP 62, 108 (1975).
(2) Rauh, Ackermann, JCP 62, 1584 (1975).

NS: [a]Estimate based on a linear Birge-Sponer extrapolation for the ground state (19). *Ab initio* calculations (19) give 5.2 eV.

[b]From the photoelectron spectrum (30).

[c]Intensity perturbations and predissociations in both doublet components.

[d]Only fragments observed.

[e]Approximate deperturbed constants, $A \approx +45$. This level is strongly perturbed by v=11 of H $^2\Pi_i$ ($B_{11} = 0.525$, $A_{11} = -110$).

[f]Spin doubling constant $\gamma_0 \approx +0.1$ for $N \lesssim 20$. The spin splitting increases rapidly at higher N because of a perturbation by H $^2\Pi$(v=1). I $^2\Sigma^+$(v=1) ($B_1 \approx 0.695$) interacts with H $^2\Pi$(v=2) at somewhat lower N values. A strong homogeneous interaction with levels of C $^2\Sigma^+$ explains the larger than expected $\Delta G(\frac{1}{2})$ value of I $^2\Sigma^+$.

(References on p. 487)

State	T_e	ω_e	$\omega_e x_e$	B_e	α_e	D_e $(10^{-6}cm^{-1})$	r_e (Å)	Observed Transitions: Design.	Observed Transitions: ν_{00}	References
$^{14}N^{32}S$ (continued)										
H $^2\Pi_i$	(44049)[g] 43876	767.6	5.0	[0.5972] [0.5915] h	0.0059	[1.75]	1.702	H→X, R	43824.6 Z 43429.6 Z	(21)* (23) (29)
C $^2\Sigma^+$	43290	[1389]	i	[0.8275]	i	[1.2]	$[1.446_4]$	C[j]↔X,[k] V γ system	43165.9 Z 43387.4 Z	(1)(2)* (3) (6)* (15) (24)* (29)*
G $^2\Sigma^-$	43346	[879.8] Z		[0.6905][ℓ]	m	[2.5]	$[1.583_4]$	G→X, R	42956.8 Z 43178.3 Z	(23)*
A $^2\Delta_r$	(40046) 40005	[934.4] Z [943.9] Z	n 8.4 H	n [0.6850]	n n		$[1.589_7]$	A↔X,[k] R β system	39688.1 Z 39875.7 Z	(1)(4)(6)* (15)(23)*
B' $^2\Sigma^+$	(36255)	(1060)	(15)	(0.78)[o]			(1.49)	B'→X,	(35952)[o] (36174)[o]	(26)(27)
B $^2\Pi_r$	30384.1 30294.9	798.78 Z 797.31 Z	3.59 3.72	0.6013[p] 0.5962[p]	0.0046 0.0048	1.3 1.3	1.697	B→X, R	29953.6 Z 30085.1 Z	(8)(9)(12)* (13)(15)* (18)(23)*
X $^2\Pi_r$	221.5[q] 0	1218.7 Z	7.28	[0.775156] [0.769602] r	0.0063_5	1.2	1.4940_2	Microwave sp.[s] EPR sp.($^2\Pi_{3/2}$)		(16) (10)(14)(17)
$^{14}N^{32}S^+$		D_0^0 = 6.3 eV[t]								MAY 1977
		Bands previously (5) attributed to NS^+ have been reclassified by (23) as G→X system of NS.								
X $^1\Sigma^+$	0	1415[u]	15				1.440[v]			
NS^-		<u>Ab initio</u> calculations (22).								MAY 1977

NS (cont'd), NS^+, NS^-:

[g] A_0 = - 172.3.

[h] Perturbations by levels of E, I, C; see [e,f,i].

[i] $\Delta G(3/2) = 1414$ (29), $\Delta G(5/2) \approx 1372$, $\Delta G(7/2) \approx 1378$ (11). $B_1 \approx 0.7975$, $B_2 = 0.8150$ (29); (24) give $B_0 = 0.8287_6$ and $D_0 = 1.40_4 \times 10^{-6}$ as well as $\gamma_0 = +0.0055$. C $^2\Sigma^+$ interacts strongly with I $^2\Sigma^+$; in addition, v=1 is extensively perturbed by H $^2\Pi$(v=2) and G $^2\Sigma^-$(v=2).

[j] Estimated lifetime $\tau(v=0) \approx 6.5$ ns [Hanle effect measurements (25)].

NS, NS^+, NS^- (continued):

[k] Observed in absorption by (20) following the flash photolysis of OCS in the presence of excess N_2O.

[ℓ] Spin doubling constant $\gamma_0 = +0.034$.

[m] $B_1 = 0.6780$.

[n] The A state is strongly perturbed, particularly the $^2\Delta_{5/2}$ component [see (23)]; the $\Delta G(\frac{1}{2})$ values are from (4), $\omega_e x_e$ from (6), and B_0 from (23). Additional constants may be found in these references but, because of the perturbations, have limited meaning. (23) suggest that the perturbing state is $^2\Phi_{5/2}$.

[o] Only the 1-0 band has been observed and analyzed ($B_1 = 0.7716$, $\nu_0 = 37203.9$ and 36982.5). The v=0 level of $^{14}N^{32}S$ interacts with B $^2\Pi_{\frac{1}{2}}$(v=8) (27); weak emission from this level was observed for $^{15}N^{32}S$ (27).

[p] Several perturbations. Except for v=8 (see [o]), these interactions have not yet been analyzed.

[q] $A_0 = +\ 222.9_4$, $A_1 = +\ 223.0_9$, recalculated by (16) from the data of (2).

[r] Λ-type doubling, $\Delta\nu_{fe}(v=0) = +0.01325_3(J+\frac{1}{2})$ (16).

[s] $\mu_{e\ell}(^2\Pi_{\frac{1}{2}}, v=0) = 1.81$ D from Stark effect measurements in the rotation spectrum (16); (14), using the EPR method, obtained 1.3_5 D for $^2\Pi_{3/2}$. This value was recently revised to 1.86 D (31). See also (19). Hyperfine coupling constants (14)(16)(17).

[t] D_0^0(NS) + I.P.(S) - I.P.(NS).

[u] Vibrational constants from the photoelectron spectrum (30).

[v] Estimated value from a Franck-Condon factor analysis of the photoelectron peaks (30).

(1) Fowler, Bakker, PRS A 136, 28 (1932).
(2) Zeeman, CJP 29, 174 (1951).
(3) Barrow, Downie, Laird, PPS A 65, 70 (1952).
(4) Barrow, Drummond, Zeeman, PPS A 67, 365 (1954).
(5) Dressler, HPA 28, 563 (1955).
(6) Narasimham, Srikameswaran, PIAS A 56, 316 (1962).
(7) Narasimham, Srikameswaran, PIAS A 56, 325 (1962).
(8) Narasimham, Srikameswaran, PIAS A 59, 227 (1964); Nature 197, 370 (1963).
(9) Smith, Meyer, JMS 14, 160 (1964).
(10) Carrington, Levy, JCP 44, 1298 (1966).
(11) Joshi, ZP 191, 126 (1966).
(12) Peyron, Lam Thanh My, JCPPB 64, 129 (1967).
(13) Goudmand, Dessaux, JCPPB 64, 135 (1967).
(14) Carrington, Howard, Levy, Robertson, MP 15, 187 (1968).
(15) Narasimham, Subramanian, JMS 29, 294 (1969).
(16) Amano, Saito, Hirota, Morino, JMS 32, 97 (1969).
(17) Uehara, Morino, MP 17, 239 (1969).
(18) Vidal, Dessaux, Marteel, Goudmand, CR C 268, 2140 (1969).
(19) O'Hare, JCP 52, 2992 (1970).
(20) Donovan, Breckenridge, CPL 11, 520 (1971).
(21) Narasimham, Balasubramanian, JMS 40, 511 (1971).
(22) O'Hare, JCP 54, 4124 (1971).
(23) Jenouvrier, Pascat, CJP 51, 2143 (1973).
(24) Balasubramanian, Narasimham, JMS 53, 128 (1974).
(25) Sivers, Chiu, JCP 61, 1475 (1974); JMS 61, 316 (1976).
(26) Narasimham, Raghuveer, Balasubramanian, JMS 54, 160 (1975).
(27) Jenouvrier, Daumont, JMS 61, 313 (1976).
(28) Bialski, Grein, JMS 61, 321 (1976).
(29) Vervloet, Jenouvrier, CJP 54, 1909 (1976).
(30) Dyke, Morris, Trickle, JCS FT II 73, 147 (1977).
(31) Byfleet, Carrington, Russell, MP 20, 271 (1971).

State	T_e	ω_e	$\omega_e x_e$	B_e	α_e	D_e (10^{-6} cm^{-1})	r_e (Å)	Observed Transitions Design.	Observed Transitions ν_{00}	References
$^{14}N^{80}Se$		$\mu = 11.9152663_9$		$D_0^0 = (4.0)$ eV						MAY 1977
C $^2\Delta_{5/2}$	(34650)	$\Delta G(3/2) = 738.32$		$B_2 = 0.4410$ $B_1 = 0.4568$		2.5 (D_2) 0.01(D_1)	$r_1 = 1.760$	$C \rightarrow X_2$,[a] R	$\nu(1-0)=$ 34431.48 Z	(8)(12)* (13)*
B $^2\Sigma^{(-)}$	(34400)	b		[0.4503][c]		[2.8]	[1.772_5]	$B \rightarrow X_2$, R $B \rightarrow X_1$,[d] R	33431.21 Z 34322.03 Z	(12)(13)*
b ($^4\Sigma^-_{1/2}$)	(24840)	[766] H		[0.4407][e]			[1.791_7]	$b \rightarrow X_1$,[f] R	24744.3 H	(16)
A_2 $^2\Pi_{3/2}$	(24800)	[612.5] Z	g	[0.4173]	g	[0.67]	[1.841]	$A_2 \rightarrow X_2$,[h] R	23765.4 Z	(9)* (16)*
A_1 $^2\Pi_{1/2}$	(24350)	[658.9] Z	i	[0.4117]	i	[0.60]	[1.854]	$A_1 \rightarrow X_1$,[h] R	24204.0 Z	(9)* (16)*
a ($^4\Pi_i$)	(19700)[j]	(710.7)[j]	(10.3)[j]	(0.361)[j]	(0.002)[j]		(1.98_0)			
X_2 $^2\Pi_{3/2}$	891.8[k]	954.96 Z	5.64_8	0.5189	0.0040	0.65	1.6518			
X_1 $^2\Pi_{1/2}$	0	956.81 Z	5.64_3	0.5182[ℓ]	0.0040	0.65				

NSe: [a](12) originally assigned the v'=1 progression to a $^2\Pi_{\frac{1}{2}} \rightarrow X_1\,^2\Pi_{\frac{1}{2}}$ transition.

[b]A weak band at 2845.37 Å was assigned by (12) to the $B \rightarrow X_1$ 1-0 transition, giving $\Delta G'(\frac{1}{2}) = 804.6$.

[c]Spin splitting constant $\gamma_0 = -0.035$.

[d](8) assigned the v'=0 progression to a $^2\Pi_{\frac{1}{2}} \rightarrow X_1\,^2\Pi_{\frac{1}{2}}$ transition. These authors list a number of unclassified bands in the same wavelength region.

[e]Calculated from the corresponding value for $^{15}N^{80}Se$. The b state levels interact with $A_1\,^2\Pi_{\frac{1}{2}}$, see [i].

[f]Two very weak v" progressions.

[g]Strongly perturbed state; $\Delta G(3/2) = 803.8$, $\Delta G(5/2) = 654.4$, $\Delta G(7/2) = 637.7$; $B_1 = 0.4252$, $B_2 = 0.4164$, $B_3 = 0.4112$, $B_4 = 0.4069$. According to (16) the perturbing state may well be $^4\Sigma^-_{3/2}$.

[h]Earlier assignments of bands belonging to these two subsystems postulated the existence of two close lying and interacting $^2\Pi$ states called $A^2\Pi_{1/2}$, $A'^2\Pi_{1/2}$, and $A,A'^2\Pi_{3/2}$; see (14), also (1)(2)(3)(4)(5)(6)(7)(10)(11). (16) use A and A" in place of A_1 and A_2, respectively.

[i]Strongly perturbed state; $\Delta G(3/2) = 615.8$, $\Delta G(5/2) = 695.7$; $B_1 = 0.4096$, $B_2 = 0.4205$, $B_3 = 0.4093$, $B_5 = 0.3980$, $B_7 = 0.3920$, $B_8 = 0.3901$. Λ-type doubling $\Delta v(v=0) = 0.032(J+\frac{1}{2})$, the sign being opposite to that in $X_1\,^2\Pi_{\frac{1}{2}}$. For higher vibrational levels the Λ-type doubling is irregular owing to the strong interaction with $b\,^4\Sigma^-_{1/2}$.

[j]$A \approx -25$; all constants have been derived from the analysis of perturbations in the A_1, A_2 states of $^{15}N^{80}Se$ and $^{14}N^{80}Se$; see (16).

[k]$A_{eff} = 890.84$ as given in (13) corresponds to the difference between the two hypothetical J=0 levels.

[l]Λ-type doubling $\Delta v = 0.043(J+\frac{1}{2})$. The interaction (due to s-uncoupling) between the nearly degenerate levels v+1 and v of $^2\Pi_{1/2}$ and $^2\Pi_{3/2}$, respectively, leads for v = 4,5 to an anomalous Λ-type doubling in both components (15).

(1) Pannetier, Goudmand, Dessaux, Arditi, CR 260, 2155 (1965).

(2) Goudmand, Dessaux, JCPPB 64, 135 (1967).

(3) Dessaux, Goudmand, CR C 267, 1198 (1968).

(4) Pascat, Daumont, Jenouvrier, Guénébaut, CR B 269, 1309 (1969).

(5) Pascat, Daumont, Jenouvrier, Guénébaut, CR C 270, 20 (1970).

(6) Daumont, Jenouvrier, Pascat, Guénébaut, CR B 271, 120 (1970).

(7) Daumont, Jenouvrier, Pascat, CR C 271, 712 (1970).

(8) Jenouvrier, Daumont, Pascat, CR C 271, 1358 (1970).

(9) Subbaram, Rao, JMS 36, 163 (1970).

(10) Daumont, Jenouvrier, Pascat, Guénébaut, CR C 272, 1545 (1971).

(11) Jenouvrier, Daumont, Pascat, Guénébaut, CR C 272, 1627 (1971).

(12) Yee, Jones, JMS 37, 304 (1971).

(13) Harding, Jones, Yee, Jenouvrier, Daumont, Pascat, Guénébaut, CJP 49, 2033 (1971).

(14) Daumont, Jenouvrier, Pascat, Guénébaut, JCPPB 69, 218 (1972).

(15) Jenouvrier, Pascat, Lefebvre-Brion, JMS 45, 46 (1973).

(16) Daumont, Jenouvrier, Pascat, CJP 54, 1292 (1976).

State	T_e	ω_e	$\omega_e x_e$	B_e	α_e	D_e (10^{-} cm^{-1})	r_e (Å)	Observed Transitions Design.	Observed Transitions ν_{00}	References
$^{16}O_2$		$\mu = 7.9974575_1$		$D_0^0 = 5.115_6$ eV[a]		I.P.($1\pi_g$) = 12.071 eV[b] ($1\pi_u$) = 16.092 eV[c] ($3\sigma_g$) = 18.159 eV[c] ($2\sigma_u$) = 24.549 eV[c] ($2\sigma_g$) = 39.6 eV[d] ($1s_0$) = 543.1 eV[d]				MAR 1977 A
		A detailed review of the entire spectrum of molecular oxygen has been published by (141).								
		Potential energy diagrams (63)(128)(141)(190); predicted electronic states and potential functions (167)(176)(182).								
		Several Rydberg states converging to the oxygen K limits at 543.1($^4\Sigma^-$) and 544.2($^2\Sigma^-$) eV, in X-ray absorption and electron energy loss spectra.								(133)(166)(175)
Z ($^3\Pi_u$)		Strong X-ray absorption peak (excitation $1s_0 \rightarrow 1\pi_g$).						Z← X,	532 eV[e]	(133)(166)(175)
		Absorption cross sections and cross sections for the production of atomic fluorescence by photodissociation in the region 175 - 850 Å (570000 - 115000 cm^{-1}) (156)(158)(161)(164). Earlier results in (18)(23)(58).								
		Rydberg states with the outer electrons in $3s\sigma$, $3p\sigma$, $3d\sigma$ orbitals and the O_2^+ core in the highest $\ldots 1\pi_u^3\, 1\pi_g^2\ ^2\Pi_u$ state have been tentatively identified in the electroionization spectrum of O_2 at 20.73, 21.75, 22.28 eV, respectively.								(160)
		Codling and Madden's Rydberg series converging to c $^4\Sigma_u^-$(v=0) of O_2^+ : $\nu = 198125 - \begin{cases} R/(n-0.16)^2 & n = 3\,(\text{Y state}),4\ldots11.^{fg} \\ R/(n-0.95)^2 & n = 3\,(\text{W state}),4\ldots.8.^{fgh} \end{cases}$ Similar series with v'=1.								(65)*
Y [f]	(184440)	[1510]		[g]				Y← X,	184410	(65)*
W ($^3\Sigma_u^-$)[f]	(168290)	[1510]		[g]				W← X,	168260	(65)*
		Yoshino and Tanaka's weak Rydberg series converging to B $^2\Sigma_g^-$(v=0) of O_2^+ : $\nu = 163700 - R/(n-0.54)^2$ n = 6 (V state),7...12.[f] Similar series with v'=1,2,3.								(98)
V [f]	(160270)	(1100)						V ← X,	160031	(98)
		Tanaka and Takamine's strong Rydberg s. of R shaded dif. b. converging to B $^2\Sigma_g^-$(v=0) of O_2^+ : $\nu = 163702 - R/(n-0.70)^2$ n = 3 (U state),4...23.[fi] Similar series with v'=1,2,3.								(9)(86)*(98)*

State	T_e	ω_e	$\omega_e x_e$	B_e	α_e	D_e (10^{-} cm^{-1})	r_e (Å)	Observed Transitions Design.	ν_{00}	References
$^{16}O_2$ (continued)										
		Fragments of Rydberg series (155000 - 160000 cm^{-1}) converging to D $^2\Delta_g$ of O_2^+.								(93)
		Namioka, Ogawa and Tanaka's Rydberg s. of weak R shaded b. converging to b $^4\Sigma_g^-(v=0)$ of O_2^+: $\nu = 146560^j - R/(n-0.53)^2$ n = 4 (R state),5...16.[f] Similar series with v'=1.2.								(44)(98)*
		Tanaka and Takamine's Rydberg s. of strong R shaded b. converging to b $^4\Sigma_g^-(v=0)$ of O_2^+: $\nu = 146556^j - R/(n-0.68)^2$ n = 4 (Q state),5...30.[fk] Similar series with v'=1...4.								(9)(44)(98)*
U[f]	142548	1148 H	23	[i]				U← X,	R 142329 H	(9)(86)
R[f]	(137643)	(1152) H						R← X,	R 137432 H	(98)
Q[f]	136759	1207 H	18	[i]				Q← X,	R 136571 H	(6)(44)(98)

O_2: [a] = 41260 ± 15 cm^{-1}, from the convergence limit of the B← X bands (21).

[b]From the high-resolution photoelectron spectrum of (178), see also (60)(72)(97). Photoionization studies [(73)(75), additional references in (178)] give appearance potentials of ~ 12.067 eV.

[c]Calculated from the energy levels of O_2^+.

[d]From the X-ray photoelectron spectrum (112).

[e](166) obtain 530.8 eV from the electron energy loss spectrum.

[f]Possible upper state symmetries have been discussed on theoretical (81) and empirical (93) grounds. Several of these Rydberg levels have also been observed in the high-resolution electron energy loss spectrum (99).

[g]Strongly preionized.

[h]A weak satellite series approximately 50 cm^{-1} longward of the main bands has been observed by (65).

[i]Preionization observed by photoionization mass-spectrometry (170).

[j]The limits refer to band origins; the approximate head-origin separation has been subtracted from the observed heads.

[k]Both preionization (to $O_2^+ + e^-$) and predissociation (to $O^+ + O^-$, for n ≥ 5) have been established by photoionization mass-spectrometry (170).

State	T_e	ω_e	$\omega_e x_e$	B_e	α_e	D_e $(10^{-6} cm^{-1})$	r_e (Å)	Observed Transitions Design.	Observed Transitions ν_{00}	References

$^{16}O_2$ (continued)

Additional unclassified bands in the region 100000 - 135000 cm^{-1} (9). Absorption and photoionization cross sections of $O_2(X\ ^3\Sigma_g^-)$ 100000 - 170000 cm^{-1} (30)(53)(54)(82). Dissociation continua with maxima at 125000, 131000, 138000 cm^{-1} (153).

State	T_e	ω_e	$\omega_e x_e$	B_e	α_e	D_e $(10^{-6} cm^{-1})$	r_e (Å)	Design.	ν_{00}	References
p $^1\Phi_u$	118951	1071[a] H	8.3	1.116[b]	0.014	(4.5)	1.374	p← a,	R 110815 H	(79)* (108)* (169)*
I"	(118200)	(1050)[c]	(15)	d				I"← X,	(117900)	(9)(170)
I'	117750	1050[e]	$9._9$	df				I'← X,	117490	(6)* (9)* (53)*
I	116420	1070[e]	$14._5$	df				I← X,	116160	(6)* (9)* (53)*
H $(^3\Pi_u)$	99880	[1070][g]		d				H← X,	99630	(2)* (6)* (9)* (53)*

Additional discrete and diffuse absorption bands in the region 80000 - 100000 cm^{-1} (only partly assigned) may belong to various Rydberg series converging to the first ionization potential. Onset of the ionization continuum observed at 1027.6 Å (97314 cm^{-1}) by photoionization mass-spectrometry (170). Absorption cross sections of $O_2(X\ ^3\Sigma_g^-)$ 51000 - 100000 cm^{-1} (30)(68)(177). Absorption cross sections of $O_2(a\ ^1\Delta_g)$ have been measured (177) from 63000 to 92000 cm^{-1} [see also (118)], photoionization cross sections (114a) from 89400 to 96600 cm^{-1}.

References: (6)* (17) (146)(151) (173)

State	T_e	ω_e	$\omega_e x_e$	B_e	α_e	D_e	r_e	Design.	ν_{00}	References
4f complex	[91300]	Very complex spectrum 90400 - 90700 cm^{-1}.						4f← a,	82500	(152)
								4f← X,	90500	(151)
L $(^3\Pi_u)$	[90044] [89948] [89858] h			[1.588] [1.531] [1.486] h		[29] [20] [30] h	[1.152] [1.173] [1.191] h	L← X,	V 89257.3 h Z 89161.0 Z 89070.7 Z	(151)
k $(^1\Delta_u)$	[89066]			[1.451]		[20.8]	[1.205]	k← a,	V 80395.8 Z	(84)(146)
j $(^1\Sigma_u^+)$ i	(87209)	[1896]		[1.701]	j	[12][j]	[1.113]	j← X,	V 87370.2 Z	(111)* (151)
G $(^3\Sigma_u^+)$	(86998)	[1822]	k	$[1.69_8]$	0.026[k]		[1.114]	G← X,	V 87122[ℓ] Z	(151)(173)*

A Rydberg series (observed in absorption from a $^1\Delta_g$) joins on to e,e' and i,i' and converges to X $^2\Pi_g$ of O_2^+. (146)(152)

State	T_e	ω_e	$\omega_e x_e$	B_e	α_e	D_e ($10^{-6}cm^{-1}$)	r_e (Å)	Observed Transitions Design.	ν_{00}	References
$^{16}O_2$ (continued)										
i ($^1\Delta_{2u}$)	(86846)	[2062]		[1.688]	0.042	[10.5]	[1.117]	i } ← a, V	79208.0[m] Z	(84)(146)
i'($^3\Delta_{2u}$)	(86843)	[1699]		[1.791]		[140]	[1.085]	i' } ← a, V	79022.6[m] Z	(152)
h ($^1\Pi_u$)	(86750)	(2200)		[1.451][n]			[1.205][n]	h← a, V	81362.5[n] Z	(84)(146)
g ($^1\Pi_u$)	(86604)	[2048]		[1.615][o]		[6.0][o]	[1.142]	g← X, V	86841.4 Z	(151)

O_2: [a]The 0-0, 1-0, 2-0 bands are overlapped. Vibrational numbering confirmed by $^{18}O_2$ isotope shifts.

[b]Rotational analyses for v=3,5,7; v=4,6,8,9 are diffuse.

[c]Probably progression II of (9), extended and reassigned by Katayama, Huffman, Tanaka [unpubl., see Figure 1 of (170)].

[d]Preionization observed by photoionization mass-spectrometry (170). Several autoionizing levels have been studied by photoelectron spectroscopy (123)(155)(157). See also (47).

[e]These progressions have been reassigned and extended by Katayama, Huffman, Tanaka (see [c]) and include most of the bands of progressions I, N, I', P of (6). They occur in the region of the second member ($4s\sigma_g$) of the Rydberg series beginning with H [(93), see [g]]. Other Rydberg series going to $a^4\Pi_u$ or $A^2\Pi_u$ may also be present; higher members possibly account for many unassigned bands in the region 810 - 740 Å (123000 - 135000 cm^{-1}).

[f]That the diffuse nature of the bands is at least partly due to predissociation has been shown by the observation of O I lines in fluorescence; (161) gives cross sections for this reaction from 850 to 650 Å (117000 - 154000 cm^{-1}).

[g]Long but strongly perturbed v' progression composed of bands previously (6) assigned to four shorter progressions H, H', M, M'; first member ($3s\sigma_g$) of a Rydberg series converging to $a^4\Pi_u$ of O_2^+ (93)(195). The intensity distribution [(53)(82), see also (170)] closely resembles that of the $a^4\Pi_u$ progression in the photoelectron spectrum (195).

[h]Vibrational numbering uncertain.

[i](111) assumed this to be a $^3\Sigma_u^+$ state; reassigned by (151).

[j]B_1 = 1.698, D_1 = 42 x 10^{-6}.

[k]Partial rotational analyses of a weak and diffuse 0-0 band and of stronger 1-0 and 2-0 bands (151).

[ℓ]The $^{18}O_2$ isotope effect shows that this is a 0-0 band (173).

[m]The two components are assumed to correspond to the ground state splitting (A = 200) of O_2^+ (146)(152).

[n]Perturbed rotational structure. According to (146) these constants refer to the 1-0 band, the unresolved 0-0 band being at 79180 cm^{-1}.

[o]Constants for Π^+; $B_0(\Pi^-)$ = 1.611, $D_0(\Pi^-)$ = 14 x 10^{-6}. Constants for the diffuse v=1 level were also determined.

State	T_e	ω_e	$\omega_e x_e$	B_e	α_e	D_e (10^{-6} cm^{-1})	r_e (Å)	Observed Transitions: Design.	Observed Transitions: ν_{00}	References
$^{16}O_2$ (continued)										
F'	[87510]	Group of six line-like features similar to F← X.						F'← X,	86720	(17)(173)*
F $^3\Pi_u$	(85868)	[2008] H	v=1 diffuse	[1.434]		[11]	[1.212]	F← X,	86085.0 Z	(136)(151) (173)*
	(85780)	[2000] H		[1.398]		[6.0]	[1.228]		85992.6 [a] Z	
	(85689)	[2001] H		[1.352]		[5.3]	[1.249]		85902.3 Z	
E $^3\Sigma_u^-$	(79883)	[2547]	[b]	[b]				E← X, R	80369[b]	(6)* (17) (150)(173)*
f $^1\Sigma_u^+$ [c]	76091	1927	19.0	1.703[d]	0.020	[e]	1.113	f← b, V	63141.5 Z	(84)
								f← X, V	76262.4[f]	(84)(111)* (173)*
D ($^3\Sigma_u^+$) [g]	(75260)	1957	19.7	1.73[h]	0.025	[i]	1.10_4	D← X, V	(75450)	(84)(111)* (173)*
e ($^1\Delta_{2u}$)	(75254)	[1830] H		[1.682]	(diffuse lines)		[1.119]	e } ← a, V	67499.6[k] Z	(84)(118) (146)(152)
e' ($^3\Delta_{2u}$)	(74915)	[2052] H	[j]					e' }	67272 [k] H	
d ($^1\Pi_g$)	(69180)	[1860]	[ℓ]					(d← X)	69320[m]	(192)
C ($^3\Pi_g$)	(65530)	[1840]	[n]					(C← X)	65670[m]	(150)(171)
B $^3\Sigma_u^-$	49793.28	709.31[o] Z	10.65[o]	0.8190_2[opq]	0.01206[o]	4.55[r]	1.6042_6	B↔X,[st] R Schumann-Runge b.	49358.15 Z	(5)*(7)*(21)* (78)* (96)* (115)(168)

O_2: [a]The $^{18}O_2$ isotope shift shows that this is a 0-0 band. F $^3\Pi_u$ is a mixed state resulting from the avoided crossing of the unstable $^3\Pi_u$ state (arising from $^3P + {}^3P$) with the lowest $^3\Pi_u$ Rydberg state ($3p\sigma_u$); see (167)(194). Oscillator strengths (171).

[b]The three strongest bands in this region at 80369, 82916, 85345 cm^{-1} [called "longest band", "second band", "third band" by (17)] have long resisted attempts at identification. Recent _ab initio_ calculations (186)(194) have shown that very probably they correspond to the second $^3\Sigma_u^-$ state formed by the avoided crossing of B $^3\Sigma_u^-$ with the lowest $^3\Sigma_u^-$ Rydberg state ($3p\pi_u$). The predicted ω_e is of the order of 3000 cm^{-1}. All three bands are diffuse [O(^{1}D) atoms have been detected in the predissociation of E $^3\Sigma_u^-$ (193)] and show double peaks (two close double peaks for the "second band"). In $^{18}O_2$ the rotational structure of the "longest band" is resolved [$B' = 1.307_2$, $D' = 1._8 \times 10^{-6}$, $\lambda' = -3.3_7$, $\gamma' = +0.045$ (179)] and confirms that the upper state is indeed $^3\Sigma_u^-$ (173). On the basis of the observed isotope shift (173) prefer the assignment of the "longest band" as 1-0 b. [see

also (194)]. f values of 0.0102, 0.0080, 0.0015 for the three bands have been determined from electron energy loss measurements (171).

[c] α state of (84), progression II of (17).

[d] v=2 diffuse. Rotational constants for $^{18}O_2$ in (179).

[e] $D_2 = 25.8 \times 10^{-6}$, $D_3 = 7 \times 10^{-6}$, $D_4 = 10 \times 10^{-6}$.

[f] The 0-0 band is not observed since it is in the continuum which covers the 1300 Å region.

[g] β state of (84) who assumed it to be $^1\Sigma_u^+$; reassigned by (111). Progression I of (17).

[h] Levels other than v=2 and 3 are too diffuse for analysis, both in $^{16}O_2$ and $^{18}O_2$; for the latter see (179).

[i] $D_2 = 14.8 \times 10^{-6}$, $D_3 = 21.0 \times 10^{-6}$.

[j] $\Delta G(3/2) = 1698$, $\Delta G(5/2) = 1838$.

[k] See [m] on p. 493.

[ℓ] $\Delta G(3/2) = 1770$, $\Delta G(5/2) \approx 1800$.

[m] From electron energy loss spectra. C and d are considered to be the lowest Rydberg states ($3s\sigma_g$) of O_2. Apparent oscillator strengths, summed over the first four bands of the C − X progression, yield an f value of 0.00074 (171).

[n] $\Delta G(3/2) = 1960$, $\Delta G(5/2) = 1780$ [average of values given by (150) and (171)].

[o] $\omega_e y_e = -0.139$, $\gamma_e = -0.00055_6$, from a low order fit to $v \leq 4$; the representation of levels having $v \leq 13$ requires seven Y_{i0} and seven Y_{i1} coefficients (141)(190). Band origins (v"=0), B_v, D_v values for v'=0...21 (115)(168)(190); T_0 values of (115) (absorption) and (168) (emission) agree to better than 0.1 cm^{-1} [note, however, two typographical errors for G_0 and G_3 in Table 5 of (168)]. Convergence limit of the vibrational levels at $57127._5$ cm^{-1} (21). RKR potential (141)

[p] The spin splitting constants at low v are $\lambda = 1.5$, $-\gamma \approx 0.04$ cm^{-1} | (37)(64). They increase rapidly above v≈12 (21)(135).

[q] Predissociation above v=2 established by line width measurements in absorption (27)(31)(92)(115)(119)(139); maximum at v=4, subsidiary peaks at v = 7,11. Ab initio calculations (134)(174)(185) show that the repulsive $^5\Pi_u$ state from normal atoms is the main contributor to the predissociation with smaller contributions from $^1\Pi_u$, $^3\Pi_u$ [earlier investigators assumed this to be the only contributor (101)(110)(114)(125)] and $^3\Sigma_u^+$. Evidence for inverse predissociation has been found by (95); see also (122)(131).

[r] $\beta = +0.22 \times 10^{-6}$ for low v; D_v increases rapidly above $v \approx 4$.

[s] The B state levels have been observed in absorption from v'=0 to the convergence limit (see [o]) (21)(115). Absorption by vibrationally excited O_2 ($v'' \leq 5$) (74)(96); data for $^{17}O^{16}O$, $^{18}O^{16}O$, $^{18}O_2$ (52)(61); absorption in inert gas matrices (58a)(63a)(115a) and (197). The formation of $O(^1D)$ atoms by photoabsorption in the adjoining continuum has been verified by (193). Emission bands with low v' and high v" are observed in various electrical discharges (13)(39)(168).

[t] For intensity measurements in the discrete portion of the B − X system see (30a)(69)(90)(92)(113)(117)(171), and in the continuum (68)(69)(71)(171); at the absorption maximum near 1445 Å (69200 cm^{-1}) the absorption coefficient is 382 cm^{-1} ($\sigma = 1.42 \times 10^{-17}$ cm^2) (71). Absorption f values vary from 3.4×10^{-10} for the 0-0 band to 3.4×10^{-5} for the 14-0, 15-0 bands to 1.3×10^{-5} for the 20-0 band, yielding an oscillator strength sum of $\sim 32 \times 10^{-5}$ for the Schumann-Runge bands. The overall electronic absorption oscillator strength is 0.162 which represents an upper limit if, as suggested by (171) and recently confirmed by (196), the continuum contains contributions from other dissociative states; see

(continued on p. 497)

State	T_e	ω_e	$\omega_e x_e$	B_e	α_e	D_e ($10^{-6}cm^{-1}$)	r_e (Å)	Observed Transitions Design. ν_{00}	References
$^{16}O_2$ (continued)									
A $^3\Sigma_u^+$	35397.8	799.07 Z	12.16[a]	0.9106	0.0141_6[a]	4.7[b]	1.5215	(A→b)[c] (21886) (A→a)[c] (27125) A↔X,[de] R 35007.1_5 Z Herzberg I b.	 (16)* (22)* (89)*
A' $^3\Delta_u$	(34690)[f]	(850)[g]	(20)[g]	(0.96)[h]	(0.026_2)[h]		(1.48)	(A'→a)[c] (26440) A'←X,[ij] R (34320)[g] Herzberg III b.	 (19)*
c $^1\Sigma_u^-$	$33057._3$	794.2_9 Z	12.73_6[k]	0.915_5	0.0139_1[k]	[$7._4$]	1.517_4	c→a,[ℓ] (24782) c↔X,[m] R $32664._1$ Z Herzberg II b.	(188) (19)* (87)

O_2: [a]$\omega_e y_e = -0.55_0$, $\gamma_e = -0.0009_7$. The constants of (16) have been adjusted (80)(141) to the revised vibrational numbering (v' raised by one unit) of (22). The spin splitting constants for low v are $\lambda = -4.9_5$ and $\gamma \approx 0$; they decrease appreciably above $v \approx 7$. RKR potential (37)(89)(140)(141).

[b]D_v increases rapidly above $v \approx 4$.

[c]The tentative identification of the A→b transition in an oxygen afterglow by (22) was not confirmed by (26). Other unidentified features in the nightglow and in the oxygen afterglow have been variously attributed to the A→a and A'→a transitions by (189) and (28), respectively. A high resolution trace of one of these bands at 4007 Å can be seen in Figure 1 of (87).

[d]First observed in absorption at atmospheric pressure and a path of > 25 m (4a)(16). The bands occur in emission in the nightglow (24)(28) and in various afterglows (22)(26)(42)(88). According to (34)(58a) bands correlated with this system have also been observed in matrix isolation studies; these bands have recently been reassigned, see [j].

[e]For detailed intensity measurements in the discrete region and in the adjoining continuum see (43)(69)(104)(127)(129). The electronic absorption oscillator strength is $\sim 10^{-7}$; cross sections in the continuum vary from $\sim 0.5 \times 10^{-24}$ cm^2 at 2400 Å to $\sim 30 \times 10^{-24}$ cm^2 at 1920 Å where transitions to other dissociative states begin to make significant contributions to the observed intensity (129). Franck-Condon factors and Franck-Condon densities (80)(89)(140)(141).

[f]The separation of the F_3 and F_2 components in v=6, extrapolated to J=0, is 145.9 cm^{-1}.

O_2 (continued):

[g]The vibrational constants and ν_{00} have been estimated from measurements of the diffuse high-pressure bands (see [i]). The only accurately known vibrational interval is $\Delta G(5\frac{1}{2}) = 611.2$ for the F_3 component (19). The vibrational numbering is uncertain.

[h]Extrapolated from B_5 and B_6 assuming a linear B_v curve; the v numbering has been estimated (see [g]).

[i]Only two weak bands have been analyzed at low pressure and 800 m path length (19). At high pressure and in liquid O_2 a fairly strong progression of diffuse triplets has been studied by many investigators. This progression appears to be the analogue in $(O_2)_2$ of the $A' \leftarrow X$ bands (their intensity increases with the square of the pressure) (1)(4)(8)(19). For lack of other information the $A' \leftarrow X$ 0-0 band is assumed to be at the position of the first diffuse high-pressure band.

[j]Visible emission bands of oxygen in low temperature matrices (34) have recently been reinterpreted (188) as belonging to the $A' \rightarrow X$ system.

[k]$\omega_e y_e = -0.244_4$, $\omega_e z_e = +0.0005_5$; $\gamma_e = -0.00074_0$. The constants refer to the revised vibrational numbering suggested by (87); see [m].

[ℓ]This system was only observed in Xe matrices ($\nu_{00} = 24552$) by excitation with VUV light.

[m]In absorption the 6-0,...,11-0 bands [new v' numbering of (87), 1-0,...,6-0 in the old numbering of (19)] have been observed with path lengths of 800 m atm (19); in emission several bands with low v' are seen in the afterglow of an oxygen-argon mixture (70)(87). The v'=0 progression is the strongest feature of the Venus night airglow (191).

O_2 (continued from p. 495):

also (187). A rather different total f value of 0.040 is derived from shock-tube absorption and emission studies (33)(51)(103); the discrepancy is probably due to the r-dependence of the electronic transition moment (56)(77)(121) (187). Franck-Condon factors based on RKR and similar potentials (50)(77)(106)(141)(190); (77) give data for $^{18}O_2$. The spectral emissivity in the Schumann-Runge bands has been discussed by (85)(103). Franck-Condon densities (55).

State	T_e	ω_e	$\omega_e x_e$	B_e	α_e	D_e ($10^{-6}cm^{-1}$)	r_e (Å)	Observed Transitions: Design.	Observed Transitions: ν_{00}	References
$^{16}O_2$ (continued)										
b $^1\Sigma_g^+$	13195.1	1432.77[a]	Z 14.00[a]	1.40037[a]	0.01820[a]	5.351[b]	1.22688	$b \rightarrow a$,[c]	5238.5	(40)
								$b \leftrightarrow X$,[de] R Atmospheric oxygen b.	13120.91[f] Z	(12)*
a $^1\Delta_g$	7918.1	$[1483.5_0]$	Z $(12._9)$	1.4264	0.0171	[4.86]	1.2156_3	a[g] $\leftrightarrow X$,[he] R IR atmosph. oxygen b.	7882.39 Z	(10)*
X $^3\Sigma_g^-$	0	1580.19_3	Z 11.98_1[i]	[1.4376766][j] $B_e = 1.44563$	0.0159_3[kℓ]	[4.839][jℓ]	1.20752	Rot.-vibr. sp. (collision induced)		(12a)(75a) (142)
								Rotation sp.[mn]		(94)(105)
								Spin reorientation (fine structure) sp.[mo]		(20)(41)(76) (120)(159)
								Raman sp.[p]		(38)* (124)* (162)(183)*
								EPR sp.		(25)(138)(154)

O_2: [a]These constants have been re-evaluated [(148), see also (168)] from the measurements of the b – X system (12) using improved lower state constants; $\gamma_e = -0.00004_2$. RKR potential curve (148). Constants for $^{16}O^{18}O$, $^{16}O^{17}O$ in (12).

[b]$+ 0.0318(v+\frac{1}{2}) + 0.0012(v+\frac{1}{2})^2$. The D_v values have been calculated (148) using vibrational wavefunctions computed from the experimental potential curve; see (147).

[c]Q branch of the 0-0 band observed in a discharge through O_2 and He. Absolute transition probability $\sim 2.5 \times 10^{-3}\,s^{-1}$.

[d]In absorption observed in the solar spectrum; in the laboratory with more than 1 m path. In emission in the aurora and nightglow (14) as well as in various discharges (11)(15)(39)(40). Band intensities [in $cm^{-1}km^{-1}atm^{-1}$(STP)] for the 0-0, 1-0, 2-0 bands are 532, 40.8, 1.52, respectively (102); slightly smaller values in (137). The transition probability for the 0-0 band is 0.075 s^{-1} [average of values given by (102) and (137)]. (49) gives the band oscillator strengths $f_{00} = 2.5 \times 10^{-10}$, $f_{10} \approx 0.2 \times 10^{-10}$. RKR Franck-Condon factors (141)(190); rotational intensity distribution and pressure broadening (100)(102)(137).

[e]Pressure induced spectra $a \leftarrow X$, $b \leftarrow X$ as well as simultaneous transitions in two colliding molecules have been studied by many investigators. See recent papers by (116)(142) which refer to earlier work.

[f](148) give $\nu_{00} = 13122.235\ cm^{-1}$, differing by $+\frac{2}{3}\lambda$ (spin-spin interaction in X $^3\Sigma_g^-$) from the zero line of (12).

O_2 (continued):

[g]EPR spectra of $O_2(^1\Delta_g)$ (62)(126); for $^{17}O^{16}O$ see (132).

[h]Observed in absorption in the solar spectrum (10), in emission in a discharge (40) and in the day and twilight glow (29)(45)(91). Values given for the transition probability $A_{00}(s^{-1})$ are $2.5_8 \times 10^{-4}$ (67), 1.9×10^{-4} (29), 1.5×10^{-4} (51a). Franck-Condon factors (36)(141)(190); (107).

[i]$\omega_e y_e = + 0.0474_7$, $\omega_e z_e = - 0.00127_3$ (141)(190), see also (5). $\Delta G(\frac{1}{2}) = 1556.381$ (12)(148)(162), higher ΔG values are less accurately known. G(v) values for $v \leq 28$ are listed in (168). RKR potential curve (32)(141)(190).

[j]From a re-evaluation by (172) of all available microwave and photographic (electronic and Raman) data; these constants supersede earlier results of (144) and are in very good agreement with (180)(181). Spin splitting constants $\lambda_0 = + 1.9847511$, $\gamma_0 = - 0.00842536$; higher order (centrifugal distortion) constants in (172)(180)(181), see also (165). For v=1, $\lambda_1 = + 1.989586$, $\gamma_1 = - 0.0084468$ (159), s. also (154).

[k]$+ 0.00006_{41}(v+\frac{1}{2})^2 - 2._{85} \times 10^{-6}(v+\frac{1}{2})^3$ (141)(190).

$B_1 = 1.42192$ (148)(159)(168); see also (12).

[l]B_v and D_v values for $v \leq 28$ are listed in (168).

[m]For microwave data on $^{18}O_2$ see (149), on $^{16}O^{18}O$ (159)(180).

[n]Laser magnetic resonance spectra (143)(145)(181).

[o]The Stark effect of the 118 GHz fine structure transition (N=1, J=1← J=0) has been observed by (163) leading to a reliable value for the polarizability anisotropy $\alpha_{\parallel} - \alpha_{\perp} = 1.12$ Å^3.

[p]For Raman data on $^{16}O^{18}O$ and $^{18}O_2$ see (183)(184). The 2-1 hot band was recently resolved in the purely isotropic part of the scattered light (198). Spin structure (130).

(1) Wulf, PNASU 14, 609 (1928).
(2) Hopfield, ApJ 72, 133 (1930).
(3) Childs, Mecke, ZP 68, 344 (1931).
(4) Finkelnburg, Steiner, ZP 79, 69 (1932).
(4a) Herzberg, Naturw. 20, 577 (1932).
(5) Curry, Herzberg AP(Leipzig) 19, 800 (1934).
(6) Price, Collins, PR 48, 714 (1935).
(7) Knauss, Ballard, PR 48, 796 (1935).
(8) Herman, AP(Paris) 11, 548 (1939).
(9) Tanaka, Takamine, PR 59, 771 (1941); Sci. Pap. IPCR (Tokyo) 39, 437 (1942).
(10) Herzberg, Herzberg, ApJ 105, 353 (1947).
(11) Kaplan, Nature 159, 673 (1947).
(12) Babcock, Herzberg, ApJ 108, 167 (1948).
(12a) Crawford, Welsh, Locke, PR 75, 1607 (1949).
(13) Feast, PPS A 63, 549 (1950).
(14) Meinel, ApJ 112, 464 (1950); 113, 583 (1951).
(15) Kvifte, Nature 168, 741 (1951).
(16) Herzberg, CJP 30, 185 (1952).
(17) Tanaka, JCP 20, 1728 (1952).
(18) Weissler, Lee, JOSA 42, 200 (1952); Lee, JOSA 45, 703 (1955).
(19) Herzberg, CJP 31, 657 (1953).
(20) Miller, Townes, PR 90, 537 (1953).
(21) Brix, Herzberg, CJP 32, 110 (1954).
(22) Broida, Gaydon, PRS A 222, 181 (1954).
(23) Aboud, Curtis, Mercure, Rense, JOSA 45, 767 (1955).
(24) Chamberlain, ApJ 121, 277 (1955).
(25) Tinkham, Strandberg, PR 97, 951 (1955).
(26) Barth, Kaplan, JCP 26, 506 (1957); JMS 3, 583 (1959).
(27) Wilkinson, Mulliken, ApJ 125, 594 (1957).
(28) Chamberlain, ApJ 128, 713 (1958).
(29) Vallance Jones, Harrison, JATP 13, 45 (1958).

O_2: (30) Watanabe, AdGp 5, 153 (1958).
(30a) Bethke, JCP 31, 669 (1959).
(31) Carroll, ApJ 129, 794 (1959).
(32) Vanderslice, Mason, Maisch, JCP 32, 515 (1960).
(33) Treanor, Wurster, JCP 32, 758 (1960).
(34) Broida, Peyron, JCP 32, 1068 (1960); Schoen, Broida, JCP 32, 1184 (1960).
(35) Nicholls, CJP 38, 1705 (1960).
(36) Nicholls, Fraser, Jarmain, McEachran, ApJ 131, 399 (1960).
(37) Vanderslice, Mason, Maisch, Lippincott, JCP 33, 614 (1960).
(38) Weber, McGinnis, JMS 4, 195 (1960).
(39) Herman, Herman, Rakotoarijimy, JPR 22, 1 (1961).
(40) Noxon, CJP 39, 1110 (1961).
(41) Zimmerer, Mizushima, PR 121, 152 (1961).
(42) Barth, Patapoff, ApJ 136, 1144 (1962).
(43) Ditchburn, Young, JATP 24, 127 (1962).
(44) Namioka, Ogawa, Tanaka, Proc. Int. Symp. Mol. Structure and Spectroscopy, Tokyo (1962), p. B208-1.
(45) Noxon, Vallance Jones, Nature 196, 157 (1962).
(46) Singh, Jain, CJP 40, 520 (1962).
(47) Nicholson, JCP 39, 954 (1963).
(48) Dorman, Morrison, JCP 39, 1906 (1963).
(49) Dianov-Klokov, OS(Engl. Transl.) 16, 224 (1964).
(50) Jarmain, CJP 41, 1926 (1963).
(51) Krindach, Sobolev, Tunitskii, OS(Engl. Transl.) 15, 326 (1963).
(51a) Vallance Jones, Gattinger, PSS 11, 961 (1963).
(52) Halmann, JCS (1964), 3729.
(53) Huffman, Larrabee, Tanaka, JCP 40, 356 (1964).
(54) Cook, Metzger, JCP 41, 321 (1964).
(55) Jarmain, Nicholls. PPS 84, 417 (1964).
(56) Marr, CJP 42, 382 (1964).
(57) Ory, Gittleman, ApJ 139, 357 (1964).
(58) de Reilhac, Damany-Astoin, CR 258, 519 (1964).
(58a) Bass, Broida, JMS 12, 221 (1964).
(59) Wacks, JCP 41, 930 (1964).
(60) Al-Joboury, May, Turner, JCS (1965), 616.
(61) Halmann, Laulicht, JCP 42, 137 (1965).
(62) Falick, Mahan, Myers, JCP 42, 1837 (1965).
(63) Gilmore, JQSRT 5, 369 (1965).
(63a) Schnepp, Dressler, JCP 42, 2482 (1965).
(64) Ginter, Battino, JCP 42, 3222 (1965).
(65) Codling, Madden, JCP 42, 3935 (1965).
(66) Halmann, Laulicht, JCP 43, 1503 (1965).
(67) Badger, Wright, Whitlock, JCP 43, 4345 (1965).
(68) Kosinskaya, Startsev, OS(Engl. Transl.) 18, 416 (1965).
(69) Blake, Carver, Haddad, JQSRT 6, 451 (1966).
(70) Degen, Nicholls, JGR 71, 3781 (1966).
(71) Goldstein, Mastrup, JOSA 56, 765 (1966).
(72) Turner, May, JCP 45, 471 (1966).
(73) McNeal, Cook, JCP 45, 3469 (1966).
(74) Ogawa, SL 15, 97 (1966).
(75) Samson, Cairns, JOSA 56, 769 (1966).
(75a) Shapiro, Gush, CJP 44, 949 (1966).
(76) West, Mizushima, PR 143, 31 (1966).
(77) Halmann, Laulicht, JCP 46, 2684 (1967).
(78) Hébert, Innanen, Nicholls, IAMS 4 (1967).
(79) Huffman, Larrabee, Tanaka, JCP 46, 2213 (1967).
(80) Jarmain, Nicholls, PPS 90, 545 (1967).
(81) Leclercq, AAp 30, 93 (1967).
(82) Matsunaga, Watanabe, SL 16, 31 (1967).
(83) Oppenheim, Goldman, JCP 46, 3493 (1967).
(84) Alberti, Ashby, Douglas, CJP 46, 337 (1968).
(85) Ben-Aryeh, JOSA 58, 679 (1968).

O_2: (86) Ogawa, CJP 46, 312 (1968).
(87) Degen, CJP 46, 783, 2850 (erratum) (1968).
(88) Degen, Nicholls, JP B 1, 983 (1968).
(89) Degen, Innanen, Hébert, Nicholls, IAMS 6 (1968).
(90) Farmer, Fabian, Lewis, Lokan, Haddad, JQSRT 8, 1739 (1968).
(91) Gattinger, CJP 46, 1613 (1968).
(92) Hudson, Carter, JOSA 58, 1621 (1968).
(93) Lindholm, AF 40, 117 (1968).
(94) McKnight, Gordy, PRL 21, 1787 (1968).
(95) Myers, Bartle, JCP 48, 3935 (1968).
(96) Ogawa, Chang, SL 17, 45 (1968).
(97) Turner, PRS A 307, 15 (1968).
(98) Yoshino, Tanaka, JCP 48, 4859 (1968).
(99) Geiger, Schröder, JCP 49, 740 (1968).
(100) Burch, Gryvnak, AO 8, 1493 (1969).
(101) Riess, Ben-Aryeh, JQSRT 9, 1463 (1969).
(102) Miller, Boese, Giver, JQSRT 9, 1507 (1969); 14, 793 (1974); 16, 595 (1976).
(103) Buttrey, JQSRT 9, 1527 (1969).
(104) Degen, Nicholls, JP B 2, 1240 (1969).
(105) Gebbie, Burroughs, Bird, PRS A 310, 579 (1969).
(106) Harris, Blackledge, Generosa, JMS 30, 506 (1969).
(107) Haslett, Fehsenfeld, JGR 74, 1878 (1969).
(108) Huffman, Larrabee, Baisley, JCP 50, 4594 (1969).
(109) Albritton, Schmeltekopf, Zare, JCP 51, 1667 (1969).
(110) Murrell, Taylor, MP 16, 609 (1969).
(111) Ogawa, Yamawaki, CJP 47, 1805 (1969).
(112) Siegbahn, Nordling, Johansson, Hedman, Hedén, Hamrin, Gelius, Bergmark, Werme, Manne, Baer, "ESCA Applied to Free Molecules", North-Holland (1969).
(113) Ackerman, Biaumé, Kockarts, PSS 18, 1639 (1970).
(114) Child, JMS 33, 487 (1970).
(114a) Clark, Wayne, JGR 75, 699 (1970); MP 18, 523 (1970).
(115) Ackerman, Biaumé, JMS 35, 73 (1970).
(115a) Boursey, Roncin, Damany, CPL 5, 584 (1970).
(116) Findlay, CJP 48, 2107 (1970).
(117) Hasson, Hébert, Nicholls, JP B 3, 1188 (1970).
(118) Ogawa, JCP 53, 3754 (1970).
(119) Snopko, OS(Engl. Transl.) 29, 445 (1970).
(120) Wilheit, Barrett, PR A 1, 213 (1970).
(121) Allison, Dalgarno, Pasachoff, PSS 19, 1463 (1971).
(122) Wray, Fried, JQSRT 11, 1171 (1971).
(123) Bahr, Blake, Carver, Gardner, Kumar, JQSRT 11, 1853 (1971).
(124) Butcher, Willetts, Jones, PRS A 324, 231 (1971).
(125) Durmaz, Murrell, MP 21, 209 (1971).
(126) Miller, JCP 54, 330 (1971).
(127) Ogawa, JCP 54, 2550 (1971).
(128) Freund, JCP 54, 3125 (1971).
(129) Hasson, Nicholls, JP B 4, 1778, 1789 (1971).
(130) Rich, Lepard, JMS 38, 549 (1971).
(131) Sharma, Wray, JCP 54, 4578 (1971).
(132) Arrington, Falick, Myers, JCP 55, 909 (1971).
(133) Nakamura, Morioka, Hayaishi, Ishiguro, Sasanuma, 3rd International Conference on Vacuum Ultraviolet Radiation Physics, Tokyo (1971), p. 1pA1-6.
(134) Schaefer, Miller, JCP 55, 4107 (1971).
(135) Bergeman, Wofsy, CPL 15, 104 (1972).
(136) Chang, Ogawa, JMS 44, 405 (1972).
(137) Galkin, Zhukova, Mitrofanova, OS(Engl. Transl.) 33, 462 (1972).
(138) Gerber, HPA 45, 655 (1972).
(139) Hudson, Mahle, JGR 77, 2902 (1972).
(140) Jarmain, JQSRT 12, 603 (1972).

O_2: (141) Krupenie, JPCRD 1, 423 (1972).
(142) McKellar, Rich, Welsh, CJP 50, 1 (1972).
(143) Mizushima, Wells, Evenson, Welch, PRL 29, 831 (1972).
(144) Welch, Mizushima, PR A 5, 2692 (1972).
(145) Evenson, Mizushima, PR A 6, 2197 (1972).
(146) Yamawaki, Ogawa, Internal Technical Report University of Southern California Vac-UV-130 (1972).
(147) Albritton, Harrop, Schmeltekopf, Zare, JMS 46, 25 (1973).
(148) Albritton, Harrop, Schmeltekopf, Zare, JMS 46, 103 (1973).
(149) Steinbach, Gordy, PR A 8, 1753 (1973).
(150) Cartwright, Hunt, Williams, Trajmar, Goddard, PR A 8, 2436 (1973).
(151) Chang, Ogawa, Internal Technical Report University of Southern California Vac-UV-140 (1973).
(152) Collins, Husain, Donovan, JCS FT II 69, 145 (1973).
(153) Cook, Ogawa, Carlson, JGR 78, 1663 (1973).
(154) Cook, Zegarski, Breckenridge, Miller, JCP 58, 1548 (1973).
(155) Kinsinger, Taylor, IJMSIP 11, 461 (1973).
(156) Lee, Carlson, Judge, Ogawa, JQSRT 13, 1023 (1973).
(157) Tanaka, Tanaka, JCP 59, 5042 (1973).
(158) Watson, Lang, Stewart, PL A 44, 293 (1973).
(159) Amano, Hirota, JMS 53, 346 (1974).
(160) Carbonneau, Marmet, PR A 9, 1898 (1974).
(161) Carlson, JCP 60, 2350 (1974).
(162) Fletcher, Rayside, JRS 2, 3 (1974).
(163) Gustafson, Gordy, PL A 49, 161 (1974).
(164) Lee, Carlson, Judge, Ogawa, JCP 61, 3261 (1974).
(165) Veseth, Lofthus, MP 27, 511 (1974).
(166) Wight, Brion, JESRP 4, 313 (1974).
(167) Buenker, Peyerimhoff, CP 8, 324; CPL 34, 225 (1975).
(168) Creek, Nicholls, PRS A 341, 517 (1975).
(169) Katayama, Huffman, Tanaka, JCP 62, 2939 (1975).
(170) Dehmer, Chupka, JCP 62, 4525 (1975).
(171) Huebner, Celotta, Mielczarek, Kuyatt, JCP 63, 241 (1975).
(172) Johns, Lepard, JMS 55, 374 (1975).
(173) Ogawa, Yamawaki, Hashizume, Tanaka, JMS 55, 425 (1975).
(174) Julienne, Krauss, JMS 56, 270 (1975).
(175) LaVilla, JCP 63, 2733 (1975).
(176) Moss, Goddard, JCP 63, 3523 (1975).
(177) Ogawa, Ogawa, CJP 53, 1845 (1975).
(178) Samson, Gardner, CJP 53, 1948 (1975).
(179) Ogawa, CJP 53, 2703 (1975).
(180) Steinbach, Gordy, PR A 11, 729 (1975).
(181) Tomuta, Mizushima, Howard, Evenson, PR A 12, 974 (1975).
(182) Beebe, Thulstrup, Andersen, JCP 64, 2080 (1976).
(183) Edwards, Good, Long, JCS FT II 72, 865 (1976).
(184) Harney, Milanovich, CJS 21, 162 (1976).
(185) Julienne, JMS 63, 60 (1976).
(186) Yoshimine, Tanaka, Tatewaki, Obara, Sasaki, Ohno, JCP 64, 2254 (1976).
(187) Julienne, Neumann, Krauss, JCP 64, 2990 (1976).
(188) Richards, Johnson, JCP 65, 3948 (1976).
(189) Wraight, Nature 263, 310 (1976).
(190) Albritton, Schmeltekopf, Zare, "Diatomic Intensity Factors", Harper and Row (to be published).
(191) Lawrence, Barth, Argabright, Science 195, 573 (1977).
(192) Trajmar, Cartwright, Hall, JCP 65, 5275 (1976).
(193) Stone, Lawrence, Fairchild, JCP 65, 5083 (1976).
(194) Buenker, Peyerimhoff, Perić, CPL 42, 383 (1976).

O_2: (195) See ref. (29) of O_2^+.
(196) Cartwright, Fiamengo, Williams, Trajmar, JP B 9, L419 (1976).
(197) Fugol, Gimpelevich, Timchenko, OS(Engl. Transl.) 40, 159 (1976).
(198) Altmann, Klöckner, Strey, CPL 46, 461 (1977).

O_2^+ (continued from p. 505):

been re-evaluated from more precise measurements by (48) who also give improved Λ-type doubling constants.

c' $+ 0.03 \times 10^{-6}(v+\frac{1}{2}) + \ldots$; D_v computed from RKR potential (34).

(1) Bozóky, ZP 104, 275 (1937).
(2) Nevin, PTRSL A 237, 471 (1938); PRS A 174, 371 (1940).
(3) Nevin, Murphy, PRIA A 46, 169 (1941).
(4) Branscomb, PR 79, 619 (1950).
(5) Feast, PPS A 63, 557 (1950).
(6) Nicolet, Dogniaux, JGR 55, 21 (1950).
(7) Vegard, Nature 165, 1012 (1950); AGEP 6, 157 (1950).
(8) Dahlstrom, Hunten, PR 84, 378 (1951).
(9) Budó, Kovács, APH 4, 273 (1954).
(10) Herman, Ferguson, Nicholls, CJP 39, 476 (1961).
(10a) Weniger, JPR 23, 225 (1962).
(11) Kovács, Weniger, JPR 23, 377 (1962).
(12) LeBlanc, JCP 38, 487 (1963).
(13) Dorman, Morrison, JCP 39, 1906 (1963).
(14) Rao, PPS 81, 240 (1963).
(15) Rao, Nature 201, 1112 (1964).
(16) Halmann, Laulicht, JCP 43, 1503 (1965).
(17) Dufay, Druetta, Eidelsberg, CR 260, 1123 (1965).
(17a) Jeunehomme, JCP 44, 4253 (1966).
(18) Turner, May, JCP 45, 471 (1966).
(19) Spohr, von Puttkamer, ZN 22 a, 705 (1967).
(20) Bhale, Rao, PIAS A 67, 350 (1968).
(21) Doolittle, Schoen, Schubert, JCP 49, 5108 (1968).
(22) Fink, Welge, ZN 23 a, 358 (1968).
(23) Nishimura, JPSJ 24, 130 (1968).
(24) Dixon, Hull, CPL 3, 367 (1969).
(25) Asundi, Ramachandrarao, CPL 4, 89 (1969).
(26) Lindholm, AF 40, 117 (1969).
(27) See ref. (112) of O_2.
(27a) Albritton, Schmeltekopf, Zare, JCP 51, 1667 (1969).
(28) Borst, Zipf, PR A 1, 1410 (1970).
(29) Edqvist, Lindholm, Selin, Åsbrink, PS 1, 25 (1970).
(30) Jonathan, Morris, Ross, Smith, JCP 54, 4954 (1971).
(31) Bhale, JMS 43, 171 (1972).
(32) Krupenie, JPCRD 1, 423 (1972).
(33) Zare, in "Molecular Spectroscopy: Modern Research", ed. Rao and Mathews, Academic Press (1972), p. 207.
(34) Albritton, Harrop, Schmeltekopf, Zare, JMS 46, 89 (1973).
(35) Fairbairn, JCP 60, 521 (1974).
(36) Gardner, Samson, JCP 61, 5472 (1974).
(37) Jonathan, Morris, Okuda, Ross, Smith, JCS FT II 70, 1810 (1974).
(38) Schopman, Locht, CPL 26, 596 (1974).
(39) Stockdale, Deleanu, CPL 28, 588 (1974).
(40) Gardner, Samson, CPL 32, 315 (1975).
(41) Raftery, Richards, JCP 62, 3184 (1975).
(42) Gardner, Samson, JCP 62, 4460 (1975).
(43) LaVilla, JCP 63, 2733 (1975).
(44) Ogawa, Ogawa, JMS 55, 56 (1975).
(45) Rao, Kota, Rao, Rao, CS 44, 877 (1975).
(46) Veseth, PS 12, 125 (1975).
(47) Beebe, Thulstrup, Andersen, JCP 64, 2080 (1976).
(48) Colbourn, Douglas, JMS 65, 332 (1977).
(49) See ref. (190) of O_2.
(50) Bhale, Narasimham, Pramāṇa 7, 324 (1976).
(51) Tabché-Fouhaillé, Durup, Moseley, Ozenne, Pernot, Tadjeddine, CP 17, 81 (1976); PRL 37, 891 (1976).
(52) See ref. (3) of O_2^{++}.

State	T_e	ω_e	$\omega_e x_e$	B_e	α_e	D_e $(10^{-6}cm^{-1})$	r_e (Å)	Observed Transitions Design.	ν_{00}	References
$^{16}O_2^+$		$\mu = 7.9973203_7$ $D_0^0 = 6.663$ eV[a] I.P. = 24.2 eV[b]								MAR 1977 A
		A detailed review of O_2^+ and its spectrum may be found in (32). Predicted electronic states and potential functions (47); contains references to earlier theoretical work.								
x_2 $^2\Sigma^-$	532.1 eV	Removal of a $1s_0$ electron from the ground state of O_2.[c]						$x_2 \rightarrow A$, $x_1 \rightarrow a$,	526.4 eV[d]	(43)
x_1 $^4\Sigma^-$	531.0 eV							$(x_2 \rightarrow X)$,	531.8 eV[e]	(43)
Several additional states observed in ESCA studies (27) and tentatively assigned by (37).										
$(^2\Sigma_g^-)$	29.5 eV	Removal of a $2\sigma_g$ electron from the ground state of O_2.[f]								
$(^4\Sigma_g^-)$	27.5 eV									
$(^2\Sigma_u^-)$	15.8 eV	Removal of a $2\sigma_u$ electron from the ground state of O_2.[g]								
c $^4\Sigma_u^-$	(100914)	[1545][h]		[1.561][i]		[6.7]	[1.162_0]	$c \rightarrow b$, V Hopfield b.	51540.7 Z	(12)(44)*
$(^2\Pi_u)$	(90000)	Diffuse (predissociating) state observed in the photoelectron spectrum of O_2; vertical I.P. ≈ 24 eV (29)(42). Probably highest $^2\Pi_u$ state of configuration $...\pi_u^3\pi_g^2$ (24).								
B $^2\Sigma_g^-$	66719	1156[j]	22[j]				(1.29_8)[k]			
D $(^2\Delta_g)$	62730	920[l]	(12)				(1.33)[m]			
C $(^2\Phi_u)$	(53620)	(900)[n]								
b $^4\Sigma_g^-$	49552	1196.77 Z	17.09	1.2872_9[p]	0.02206	5.81[q]	1.2796_4	$b^r \rightarrow a$,[s] V 1st negative b.	16666.74 Z	(2)(3)(10a)
b' $(^4\Pi_g)$	(48000)	Weakly bound state arising from $^3P + {}^4S$ (47); in its unstable region observed by laser photofragment spectroscopy.						$b' \leftarrow a$,		(51)
A $^2\Pi_u$	$40669._3$[t]	898.2_5 Z	13.57_3	1.0617_0	0.0193_6[u]	5.94[v]	1.4090_5	$A^w \rightarrow X$,[x] R 2nd negative b.	$40068._1$ Z	(1)(5)(20)(32)(34)(48)(50)
a $^4\Pi_{ui}$	32964[y]	1035.69 Z	10.39	1.1046_6	0.01575[z]	4.88[z]	1.3813_8			
X $^2\Pi_g$	197.3[a'] 0	1904.7_7 Z	16.25_9	1.6913	0.01976[b']	5.32[c']	1.1164			

O_2^+: [a] $D_0^0(O_2)$ + I.P.(O) - I.P.(O_2).

[b] From the electron impact appearance potential of O_2^{++}, A.P. = 36.3 eV (13), and I.P.(O_2). (52) give A.P. = 37.2 eV.

[c] Highly excited states (K limits) observed in X-ray photoelectron (27) and emission (43) spectra.

[d] Unresolved vertical transitions.

[e] Predicted vertical transition; in the X-ray emission spectrum of (43) this transition is hidden by an artefact.

[f] Observed in the low-resolution X-ray photoelectron spectrum of (27). In the 304 Å photoelectron spectrum (40)(42) find a very broad maximum corresponding to $^4\Sigma_g^-$ and two sharp peaks (I.P. 40.33 and 40.40 eV) corresponding to $^2\Sigma_g^-$.

[g] Observed in the X-ray PE spectrum (27). A very weak broad maximum appears near 27.5 eV in the 304 Å PE spectrum of (29); not confirmed by (42). See also (37).

[h] Average of values obtained by PE spectroscopy (29) and from the limits of Codling and Madden's Rydberg series.

[i] Spin splitting constant ε = 0.44 cm^{-1}. Only bands having v'=0 occur in emission; predissociation (29)(38).

[j] From the limits of Tanaka and Takamine's Rydberg series; in good agreement with constants obtained from PE spectra (29). Predissociation (21)(38)(39).

[k] Franck-Condon factor analysis of PE band intensities (37).

[ℓ] Only observed in the PE spectrum of $O_2(^1\Delta_g)$ (30)(37); tentatively identified as convergence limit of a fragmentary Rydberg series (26). Predissociation (38).

[m] Franck-Condon factor analysis of the PE spectrum (37).

[n] Only observed in the PE spectrum of $O_2(^1\Delta_g)$ (37); vibrational numbering uncertain.

[o] RKR potential curves (32)(49). Calculated Franck-Condon factors for ionizing transitions from X $^3\Sigma_g^-$, a $^1\Delta_g$, b $^1\Sigma_g^+$ (16)(25)(32)(49), and for recombination transitions from X $^2\Pi_g$ and a $^4\Pi_u$ to B $^3\Sigma_u^-$ (32). (16) give also results for $^{16}O^{18}O$ and $^{18}O_2$; note, however, that their calculations for transitions to X $^2\Pi_g$ are based on the previously accepted but now abandoned vibrational numbering for the ground state of O_2^+ and lead to disagreement with observed photoelectron intensities (19). Experimental Franck-Condon factors from photoelectron spectra (29)(36)(42).

[p] Spin splitting constant ε = 0.1487 cm^{-1}.

[q] β_e = + $0.18_5 \times 10^{-6}$.

[r] Radiative lifetime τ = 1.1_5 μs (17a)(22)(28)(35).

[s] Observed in various discharges (14) and in aurorae (6)(7)(8)(15). Excitation by electron impact (23), by fast ions (10)(17). Franck-Condon factors (32)(49). Rotational line strengths (33).

[t] A_v increases from A_0 = -3.6 to A_{15} = +10.0 (31)(34)(48)(50). Theoretical interpretation (41).

[u] γ_e = - 0.00017_3; constants fitted to v' ≦ 8 (34). Additional B_v values up to v=15 are listed by (48) who also give Λ-type doubling constants.

[v] + 0.06(v+½) + 0.01_2(v+½)^2; the D_v values have been computed (34) from the experimental potential curve.

[w] Radiative lifetime τ = 0.69 μs (17a)(22).

[x] Excitation by electron impact; its effect on the rotational temperature (4). Franck-Condon factors (25)(27a)(32)(49).

[y] $A_0 \ldots A_6$ = -47.79 ... -48.01 (11); anomalous dependence on J of the multiplet splitting (9)(11). T_e calculated from the limit (b $^4\Sigma_g^-$) of Tanaka and Takamine's Rydberg series with I.P.(O_2), ν_{00}(b→a), and the constants for a, X.

[z] Constants representing v=0,1,2; β_e = - $0.09_5 \times 10^{-6}$ (2). For v=3...6 see (3)(10a). Slightly different constants in (46).

[a'] A_v decreases from A_0 = +200.33 (48) to A_{10} = $+192.0_5$ (34). See also (41).

[b'] Constants fitted to v ≦ 10 (34). Selected B_v values have (continued on p. 503)

State	T_e	ω_e	$\omega_e x_e$	B_e	α_e	D_e (10^{-} cm^{-1})	r_e (Å)	Observed Transitions: Design.	Observed Transitions: ν_{00}	References
$^{16}O_2^{++}$		$\mu = 7.9971832_2$								MAR 1977
		Ab initio predicted electronic states and potential curves (7); empirical calculations of spectroscopic constants (1).								
		Additional states observed by Auger electron spectroscopy (4).								
c	94000	Detected by double charge transfer spectroscopy [$H^+ + O_2 \rightarrow H^- + O_2^{++}$] (5).								
b	54000									
a ($^3\Sigma_u^+$)	32500	Observed in electron impact experiments (3) and in the Auger electron spectrum (4).								
X $^1\Sigma_g^+$	0	Locally stable, observed by electron impact mass-spectrometry (2)(3)(6).								
$^{16}O_2^-$		$\mu = 7.9975946_6$		$D_0^0 = 4.09_4$ eV[a]		I.P. = 0.440 eV[b]				MAR 1977
b	118540	1290		Short progression of resonances in electron transmission.[c]					14.27 eV[d]	(8)(14)
a ($^4\Pi_u$)	97800	1044	10	Long progression of resonances in electron transmission.[e]					11.68 eV[f]	(8)(14)
		Additional resonances in the electron transmission current at 8 - 11 eV[d].								(8)(14)
		Several bound excited states predicted by theoretical calculations (12).								
A ($^2\Pi_u$)	(25300)	(574.5)[g]	(7.1)[g]					A↔X,[h]	(25000)[i]	(1a)(11)
		Photodetachment cross sections 6700 - 4600 Å (14900 - 21800 cm^{-1}).								(15)
X $^2\Pi_{gi}$	0[j]	1090[k]	$8._1$[k]				1.35[ℓ]	Raman sp.[m]		(1b)(3)
								EPR sp.[n]		(1)(2a)

O_2^{++}:
(1) Hurley, JMS 9, 18 (1962).
(2) Dorman, Morrison, JCP 39, 1906 (1963).
(3) Daly, Powell, PPS 90, 629 (1967).
(4) Moddeman, Carlson, Krause, Pullen, Bull, Schweitzer, JCP 55, 2317 (1971).
(5) Appell, Durup, Fehsenfeld, Fournier, JP B 6, 197 (1973).
(6) Meyerson, Ihrig, IJMSIP 10, 497 (1973).
(7) Beebe, Thulstrup, Andersen, JCP 64, 2080 (1976).

O_2^-:
[a] From $D_0^0(O_2)$ and the electron affinities of O (1.462 eV) and O_2.
[b] From the O_2^- photodetachment spectrum (9); see also (2). From endothermic negative-ion charge-transfer reactions (6) obtain I.P. $\geq 0.45 \pm 0.1$ eV. The theoretical value is 0.42 eV (10).
[c] "Band b". Suggested "grandparent" state b $^4\Sigma_g^-$ of O_2^+.
[d] Energy relative to X $^3\Sigma_g^-(v=0)$ of neutral O_2.
[e] "Band a". The negative ion state results from the addition of two Rydberg electrons in the $3s\sigma_g$ orbital to the O_2^+ core in the a $^4\Pi_u$ state ("grandparent").
[f] Extrapolated energy of v=0 relative to X $^3\Sigma_g^-(v=0)$ of neutral O_2.
[g] Absorption in KBr, vibrational numbering uncertain (11).
[h] Observed in alkali halide crystals at 4.2 and 2 K.
[i] Estimated v_{00} for the free O_2^- ion, by extrapolation from data for various host crystals (3).
[j] $A = - 160$ cm^{-1} (13).
[k] From electron scattering cross sections for gaseous O_2 (4)(5); similar measurements by (7) suggest $\omega_e = 1140$, $\omega_e x_e = 12$. A direct measurement of $\Delta G''(\frac{1}{2})$ in the photodetachment spectrum (9) gives ~ 1090 cm^{-1}, in agreement with extrapolations from Raman frequencies in alkali

O_2^- (continued):

halide crystals (3). Anharmonicities derived from low-temperature fluorescence spectra (see [h]) are approximately 8.7 (11).
[ℓ] From a Franck-Condon factor analysis of the photodetachment spectrum (9) and a similar evaluation by (16) of the electron scattering data of (5).
[m] In alkali halide crystals and in solid KO_2 and NaO_2.
[n] In alkali halide crystals.

(1) Känzig, Cohen, PRL 3, 509 (1959).
(1a) Rolfe, JCP 40, 1664 (1964).
(1b) Creighton, Lippincott, JCP 40, 1779 (1964).
(2) Pack, Phelps, JCP 44, 1870 (1966).
(2a) Zeller, Känzig, HPA 40, 845 (1967).
(3) Rolfe, Holzer, Murphy, Bernstein, JCP 49, 963 (1968); JMS 26, 543 (1968).
(4) Boness, Schulz, PR A 2, 2182 (1970).
(5) Linder, Schmidt, ZN 26 a, 1617 (1971).
(6) Tiernan, Hughes, Lifshitz, JCP 55, 5692 (1971).
(7) Gray, Haselton, Krause, Soltysik, CPL 13, 51 (1972).
(8) Sanche, Schulz, PR A 6, 69 (1972).
(9) Celotta, Bennett, Hall, Siegel, Levine, PR A 6, 631 (1972).
(10) Zemke, Das, Wahl, CPL 14, 310 (1972).
(11) Ikezawa, Rolfe, JCP 58, 2024 (1973).
(12) Krauss, Neumann, Wahl, Das, Zemke, PR A 7, 69 (1973).
(13) Land, Raith, PRL 30, 193 (1973).
(14) Schulz, RMP 45, 423 (1973).
(15) Cosby, Ling, Peterson, Moseley, JCP 65, 5267 (1976).
(16) Parlant, Fiquet-Fayard, JP B 9, 1617 (1976).

State	T_e	ω_e	$\omega_e x_e$	B_e	α_e	D_e $(10^{-4}cm^{-1})$	r_e (Å)	Observed Transitions Design.	ν_{00}	References
$^{16}O^{1}H$		$\mu = 0.94808710$		$D_0^0 = 4.392$ eV[a]		I.P. $= 12.9_0$ eV[b]				MAY 1977 A
		Theoretical potential functions for 48 states (92); for X $^2\Pi$ and A $^2\Sigma^+$ see (112)(113)(118).								
C $^2\Sigma^+$	$89459._1$	1232.9 Z	19.1	4.247^{cd}	0.078	2	2.046_1	$C^e \rightarrow A,^m$ R	55820.7 Z	(20)(42)(64)*
		Strong many-line spectrum 1900 - 1700 Å, tentative identification.						$(C \rightarrow X)^e$	(88223)	(20)(46)*
D $^2\Sigma^-$	(82130)	(2954)		$[15.2179]^f$		[16.16]	$[1.0809_3]$	$D \leftarrow X,^g$ R	$81759.7_8{}^h$ Z	(106)
B $^2\Sigma^+$	69774	[660.0] Z	i	$[5.086]^{jd}$	k	$[9.29]^{\ell}$	$[1.869_8]$	$B \rightarrow A,^m$ R	35965.5 Z	(16)(20)(33) (42)* (58)* (64)* (73)
A $^2\Sigma^+$	32684.1	3178.8_6 Z	$92.91_7{}^n$	17.358^{opqd}	$0.786_8{}^r$	$[20.39]^s$	1.0121	$A^t \leftrightarrow X,^u$ R	$32402.3_9{}^h$ Z	(3)(57)(64)*
X $^2\Pi_i$	0^v	3737.76_1 Z	$84.881_3{}^w$	$18.910_8{}^{xd}$	0.7242^y	19.38^z	0.96966	$\frac{1}{2} \leftarrow \frac{3}{2}$ a'	126.23	(69a)(90)
								Rotation - vibr. b. [b'c']		
								1-0 sequence		(18)(27)(122)
								2-0 sequence		(8)(27)(29) (122)
								3-0 sequence		(36)(122)
								4-0 sequence		(10)(11)(14) (36)
								other sequences		(11)(14)(26) (36)
								Rotation sp.		(13)(107) (124)
								Hf Λ-doubling sp.[d'e'f']		(12)(41)(60) (105)(115) (117)(126)
								EPR sp.[g']		(35)(69)(72) (78)(93)(109)

O^1H: [a]Short extrapolation of the vibrational levels in A $^2\Sigma^+$, assuming that this state has no potential maximum; confirmed by the observed predissociation in B $^2\Sigma^+$ (64). $D_e = 4.621$ eV, in complete agreement with the most recent theoretical value (121).

[b]Photoionization mass-spectrometry of HOF (99); 13.01 eV from the photoelectron spectrum (125).

[c]Spin splitting constants $\gamma_0 \ldots \gamma_3 = +1.09 \ldots +0.67$ (20)(64).

[d]RKR potential functions (32)(66).

[e]Lifetime of the upper state of the C→A system ~6 ns; measurements in the 1700 - 1900 Å region give $\tau \approx 2$ ns which, in spite of the poor agreement, is not inconsistent with the assignment of at least part of these bands to C→X (120). Much longer lifetimes (~80 ns) have been reported by (83).

[f]Spin splitting constant $\gamma_0 = -0.293$.

[g]Theoretical oscillator strength $f_{00} = 0.0036$ (119).

[h]Energy of N'=0 relative to the zero-point of the Hill-Van Vleck expression for the ground state.

[i]Using isotope relations (16) estimates $\omega_e = 940$, $\omega_e x_e = 105$, $\omega_e y_e = -21._5$.

[j]Spin splitting const. $\gamma < 0.03$ (64). Prediss. by rotation in v=0 above N=15, in v=1 above N=9; diss. products $^1S + {}^2S$ (42)(64).

O^1H (continued):

[k] $B_1 = 4.119$. Constants for the B state are from (64).

[ℓ] $D_1 = 29.1 \times 10^{-4}$, $H_1 = -15 \times 10^{-5}$.

[m] Franck-Condon factors (42).

[n] $- 1.791_5(v+\frac{1}{2})^3 + 0.3236_2(v+\frac{1}{2})^4 - 0.03585(v+\frac{1}{2})^5$ (16). Energy levels and improved $\Delta G(v+\frac{1}{2})$ values are listed in (64).

[o] Spin splitting constants $\gamma_0 \ldots \gamma_3 = 0.201, 0.196, 0.192, 0.193$ (79); see also (3).

[p] Predissociation in v=0 above N=23, in v=1 above N=14, and of all levels in v=2; from the lifetime measurements of (89)(98)(116). A sharp decrease in the intensity of emission lines originating from the predissociated levels has been observed at or slightly above threshold in low pressure flames and discharges (5)(9)(55); (21) report corresponding intensity increases due to inverse predissociation in hydrogen flames; see also (63). The predissociation is noticeably stronger for the F_1 than for the F_2 levels; according to (59)(67)(75)(98)(101)(111)(116) it is caused by the $^4\Sigma^-$ state arising from $^3P + {}^2S$, but the possibility of predissociation by the $^2\Pi$ ground state has also been considered (74)(77). A much stronger predissociation of A $^2\Sigma^+$ leading to diffuseness in the $B \rightarrow A$ bands has been observed (20)(58)(64) for v=5,6,7 at all N levels and for v=8 above N=6; according to (73) it is produced by the $^4\Pi$ state arising from $^3P + {}^2S$.

[q] $\mu_{e\ell}(v=0) = 1.98$ D from high-field Stark effects on the $A \rightarrow X$ transition (84).

[r] $\gamma_e = -0.016$. The equilibrium constants were derived by (16) to fit B_v from v=0 to 4. Improved B_v values (v=0...3) have been obtained by (79) from the data of (3); additional B_v values for $v \leq 9$ are listed in (16)(20)(42)(64). For term values (v=0...3) see (79).

[s] Other D_v values in (79) and (16)(20)(42)(64); $H_0 = 8.7_1 \times 10^{-8}$ (79).

[t] Radiative lifetimes $\tau(v=0,N=1) = 693 \pm 10$ ns, $\tau(v=1,N=1) = 736 \pm 11$ ns (116). The increase from v=0 to v=1 and a similar variation with increasing rotation in v=0 [(116), see also (89)(98)] are explained by the dependence on r of the transition moment [see (116) and references given there]. (108) using a method very similar to (116) obtain 720 and 765 ns for the average lifetimes of several low-lying rotational levels in v=0 and 1, respectively. Earlier apparently less accurate determinations (89)(91)(98)(103)(104) gave somewhat higher τ values [see also (45)]; Hanle effect measurements (86)(100) are slightly lower. The decay time of the predissociated v=2 level (see [p]) is 203 ± 13 ns for N=0 and decreases rapidly at higher N (116); estimated non-radiative lifetime for N=1 ~ 270 ns. High-resolution line absorption measurements by (96) [see also (44)] give $f_{00} = 0.00095$ for the rotationless molecule in reasonable agreement with lifetime measurements but in sharp contrast to $f_{00} = 0.0014_8$ obtained by (50) using the hook method; f_{10} values are, respectively, 0.00024 and 0.0008_9.

[u] Observed in emission in all kinds of electric discharges (often as an impurity), in flames (34) and in the heads of comets; in absorption in H_2O vapour at high temperature (1) and in electric discharges (2), in flames (28), in the flash photolysis of H_2O, $O_3 + H_2O$, and other mixtures (31)(37)(53), and in stellar spectra, especially the solar spectrum (24). Emission and absorption in solid neon (61). Atlas of $A - X$ bands (7), new measurements of the 0-0 and 1-0 bands (87). Magnetic rotation spectrum (82). Franck-

(continued on p. 510)

O^1H (continued):

Condon factors (17)(25)(42). Vibrational intensity distribution (3)(114); rotational intensity distribution (39)(54); effect of variation of transition moment with r and dependence on J (51)(114).

[v] $A_v = -139.21 - 0.27_5 v$ (122).

[w] $+ 0.540_9(v+\frac{1}{2})^3 - 0.0213_4(v+\frac{1}{2})^4 - 0.0011_3(v+\frac{1}{2})^5$, representing the vibrational levels up to v=5 (122); see also (10)(11). $\Delta G(\frac{1}{2}) = 3569.64_0$ (122).

[x] Λ-type doubling parameters $p_v = 0.235 - 0.006\,v$, $q_v = -0.0391 + 0.0018\,v$, see (122) who give also centrifugal distortion terms. See [y].

[y] $+ 0.0070_6(v+\frac{1}{2})^2 - 0.00050(v+\frac{1}{2})^3$, representing $B_0 \ldots B_5$ (122); slightly different constants in (10)(79)(85)(90). Term values for $v \leq 3$ tabulated in (3)(79), for v=4,5,6 in (10), for v=7,8,9 in (36).

[z] $- 0.43_2 \times 10^{-4}(v+\frac{1}{2}) + 0.024 \times 10^{-4}(v+\frac{1}{2})^2$; $H_0 = 14.2 \times 10^{-8}$ (122). See [y].

[a'] The 79 μm electric dipole spectrum ($^2\Pi_{\frac{1}{2}}, J=\frac{1}{2} \leftarrow {}^2\Pi_{\frac{3}{2}}, J=\frac{3}{2}$) has been measured by the laser magnetic resonance method.

[b'] Observed in emission in the spectrum of the night sky (4)(11)(15)(22)(26), in the $H + O_3$ reaction (14)(19)(36)(81), in the $H + O_2$ reaction (27), and in oxyacetylene flames (8)(18)(29)(122). In absorption in rare gas matrices (62).

[c'] Radiative lifetimes derived from observed intensities: $\tau(v=1) = 24$ ms, $\tau(v=2) = 12$ ms (97); from the decay rate of the $9 \rightarrow 7$ radiation: $\tau(v=9) = 64$ ms (80). The dipole moment function has been studied by many authors, most recently by (38)(76)(81) and (97) from measured band strengths and transition probabilities. An extensive *ab initio* calculation of the dipole moment function is given by (112) [see also (113)]. (110) has used this *ab initio* function to predict absolute intensities of a large array of vibration-rotation transitions taking account of spin uncoupling and vibration-rotation interaction [see also the early work of (6) and (40)].

[d'] The 18 cm transition ($^2\Pi_{3/2}, J=3/2$) consists of four components (23)(47a)(88):

$F'=1 \leftarrow F''=2$	1612.2310_1 MHz
$F'=1 \leftarrow F''=1$	1665.4018_4 MHz
$F'=2 \leftarrow F''=2$	1667.3590_3 MHz
$F'=2 \leftarrow F''=1$	1720.5299_8 MHz

Einstein A coefficients for these transitions have been calculated by (49). Calculated frequencies for ^{17}OH (102).

[e'] Hfs and Λ-doubling constants. From Stark shifts of the hf Λ-doubling transitions (48)(95) determine $\mu_{e\ell}(v=0) = 1.6676$ D.

[f'] Also observed in interstellar clouds, see the reviews in (52)(56)(65). In some clouds there is strong evidence for maser action. In the laboratory population inversion between Λ-doublet states was recently observed by (123).

[g'] Observed in v=0...9. Hyperfine and Λ-doubling constants. EPR spectrum of ^{17}OH (68).

(1) Bonnhoeffer, Reichardt, ZPC A 139, 75 (1928).
(2) Oldenberg, JCP 3, 266 (1935).
(3) Dieke, Crosswhite, Bumblebee Series Report No. 87, Johns Hopkins University (1948); JQSRT 2, 97 (1962).
(4) Meinel, ApJ 111, 555 (1950).
(5) Gaydon, Wolfhard, PRS A 208, 63 (1951).
(6) Heaps, Herzberg, ZP 133, 48 (1952).
(7) Bass, Broida, NBS Circular 541 (1953).
(8) Benedict, Plyler, Humphreys, JCP 21, 398 (1953).
(9) Broida, Kane, PR 89, 1053 (1953).

O^1H: (10) Herman, Hornbeck, ApJ 118, 214 (1953).
(11) Chamberlain, Roesler, ApJ 121, 541 (1955).
(12) Dousmanis, Sanders, Townes, PR 100, 1735 (1955).
(13) Madden, Benedict, JCP 23, 408 (1955).
(14) McKinley, Garvin, Boudart, JCP 23, 784 (1955).
(15) Vallance Jones, Nature 175, 950 (1955).
(16) Barrow, AF 11, 281 (1956).
(17) Nicholls, PPS A 69, 741 (1956).
(18) Allen, Blaine, Plyler, SA 9, 126 (1957).
(19) Kraus, ZN 12 a, 479 (1957).
(20) Michel, ZN 12 a, 887 (1957).
(21) Charton, Gaydon, PRS A 245, 84 (1958).
(22) Connes, Gush, JPR 20, 915 (1959).
(23) Ehrenstein, Townes, Stevenson, PRL 3, 40 (1959).
(24) Moore, Broida, JRNBS A 63, 279 (1959).
(25) Nicholls, Fraser, Jarmain, CF 3, 13 (1959).
(26) Blackwell, Ingham, Rundle, ApJ 131, 15 (1960).
(27) Charters, Polanyi, CJC 38, 1742 (1960).
(28) Gaydon, Spokes, van Suchtelen, PRS A 256, 323 (1960).
(29) Rogge, Yarger, Dickey, JCP 33, 453 (1960).
(30) Wallace, ApJ 132, 894 (1960).
(31) Basco, Norrish, PRS A 260, 293 (1961).
(32) Fallon, Tobias, Vanderslice, JCP 34, 167 (1961).
(33) Herman, Felenbok, Herman, JPR 22, 83 (1961).
(34) Krishnamachari, Broida, JCP 34, 1709 (1961).
(35) Radford, PR 122, 114 (1961); 126, 1035 (1962).
(36) Bass, Garvin, JMS 9, 114 (1962).
(37) Black, Porter, PRS A 266, 185 (1962).
(38) Ferguson, Parkinson, PSS 11, 149 (1963).
(39) Learner, PRS A 269, 311 (1962).
(40) Cashion, JMS 10, 182 (1963).
(41) Ehrenstein, PR 130, 669 (1963).
(42) Felenbok, AAp 26, 393 (1963).
(43) Kayama, JCP 39, 1507 (1963).
(44) Golden, del Greco, Kaufman, JCP 39, 3034 (1963).
(45) Bennett, Dalby, JCP 40, 1414 (1964).
(46) Felenbok, Czarny, AAp 27, 244 (1964).
(47) Phelps, Dalby, CJP 43, 144 (1964).
(47a) Radford, PRL 13, 534 (1964).
(48) Powell, Lide, JCP 42, 4201 (1965).
(49) Turner, Nature 212, 184 (1966); 214, 379 (1967).
(50) Anketell, Pery-Thorne, PRS A 301, 343 (1967).
(51) Anketell, Learner, PRS A 301, 355 (1967).
(52) Barrett, Science 157, 881 (1967).
(53) Horne, Norrish, Nature 215, 1373 (1967).
(54) Meinel, ZN 22 a, 977 (1967).
(55) Naegeli, Palmer, JMS 23, 44 (1967); 28, 417 (1968).
(56) Robinson, McGee, ARAA 5, 183 (1967).
(57) Stoebner, Delbourgo, JCPPB 64, 1115 (1967).
(58) Czarny, Felenbok, AAp 31, 141 (1968).
(59) Palmer, Naegeli, JMS 28, 417 (1968).
(60) Poynter, Beaudet, PRL 21, 305 (1968).
(61) Tinti, JCP 48, 1459 (1968).
(62) Acquista, Schoen, Lide, JCP 48, 1534 (1968).
(63) Gutman, Lutz, Jacobs, Hardwidge, Schott, JCP 48, 5689 (1968).
(64) Carlone, Dalby, CJP 47, 1945 (1969).
(65) Cook, Physica 41, 1 (1969).
(66) Horsley, Richards, JCPPB 66, 41 (1969).
(67) Michels, Harris, CPL 3, 441 (1969).
(68) Carrington, Lucas, PRS A 314, 567 (1970).
(69) Churg, Levy, ApJ 162, L161 (1970).
(69a) Evenson, Wells, Radford, PRL 25, 199 (1970).
(70) Smith, JCP 53, 792 (1970).
(71) Veseth, JP B 3, 1677 (1970).

O^1H: (72) Clough, Curran, Thrush, PRS A 323, 541 (1971).
(73) Czarny, Felenbok, Lefebvre-Brion, JP B 4, 124 (1971).
(74) Durmaz, Murrell, TFS 67, 3395 (1971).
(75) Gaydon, Kopp, JP B 4, 752 (1971).
(76) d'Incan, Effantin, Roux, JQSRT 11, 1215 (1971); 12, 97 (1972).
(77) Julienne, Krauss, Donn, ApJ 170, 65 (1971).
(78) Lee, Tam, Larouche, Woonton, CJP 49, 2207 (1971).
(79) Moore, Richards, PS 3, 223 (1971).
(80) Potter, Coltharp, Worley, JCP 54, 992 (1971).
(81) Murphy, JCP 54, 4852 (1971).
(82) Nanes, Robinson, JCP 55, 963 (1971).
(83) Remy, SpL 4, 319 (1971).
(84) Scarl, Dalby, CJP 49, 2825 (1971).
(85) Veseth, JMS 38, 228 (1971).
(86) de Zafra, Marshall, Metcalf, PR A 3, 1557 (1971).
(87) Engleman, JQSRT 12, 1347 (1972).
(88) ter Meulen, Dymanus, ApJ 172, L21 (1972).
(89) Elmergreen, Smith, ApJ 178, 557 (1972).
(90) Mizushima, PR A 5, 143 (1972).
(91) Becker, Haaks, ZN 28 a, 249 (1973).
(92) Easson, Pryce, CJP 51, 518 (1973).
(93) Hinkley, Walker, Richards, PRS A 331, 553 (1973).
(94) Klein, JQSRT 13, 581 (1973).
(95) Meerts, Dymanus, CPL 23, 45 (1973).
(96) Rouse, Engleman, JQSRT 13, 1503 (1973).
(97) Roux, d'Incan, Cerny, ApJ 186, 1141 (1973).
(98) Sutherland, Anderson, JCP 58, 1226 (1973); 59, 6690 (1973) (erratum).
(99) Berkowitz, Appelman, Chupka, JCP 58, 1950 (1973).
(100) German, Bergeman, Weinstock, Zare, JCP 58, 4304 (1973).
(101) Palmer, Naegeli, JCP 59, 994 (1973).
(102) Valtz, Soglasnova, ApL 13, 23 (1973).
(103) Becker, Haaks, Tatarczyk, CPL 25, 564 (1974).
(104) Brophy, Silver, Kinsey, CPL 28, 418 (1974).
(105) Destombes, Marlière , Rohart, Burie, Journel, CR B 278, 275 (1974); 280, 809 (1975).
(106) Douglas, CJP 52, 318 (1974).
(107) Ducas, Javan, JCP 60, 1677 (1974).
(108) Hogan, Davis, CPL 29, 555 (1974).
(109) Lee, Tam, CP 4, 434 (1974).
(110) Mies, JMS 53, 150 (1974).
(111) Smith, Elmergreen, Brooks, JCP 61, 2793 (1974).
(112) Stevens, Das, Wahl, Krauss, Neumann, JCP 61, 3686 (1974).
(113) Chu, Yoshimine, Liu, JCP 61, 5389 (1974).
(114) Crosley, Lengel, JQSRT 15, 579 (1975).
(115) Destombes, Marlière , CPL 34, 532 (1975).
(116) German, JCP 62, 2584; 63, 5252 (1975).
(117) Meerts, Dymanus, CJP 53, 2123 (1975).
(118) Meyer, Rosmus, JCP 63, 2356 (1975).
(119) Ray, Kelly, ApJ 202, L57 (1975).
(120) Smith, Stella, JCP 63, 2395 (1975).
(121) Arnold, Whiting, Sharbaugh, JCP 64, 3251 (1976).
(122) Maillard, Chauville, Mantz, JMS 63, 120 (1976).
(123) ter Meulen, Meerts, van Mierlo, Dymanus, PRL 36, 1031 (1976).
(124) Downey, Robinson, Smith, JCP 66, 1685 (1977).
(125) Katsumata, Lloyd, CPL 45, 519 (1977).
(126) Meerts, CPL 46, 24 (1977).

O^2H (continued from p. 515):

(18) de Zafra, Marshall, Metcalf, PR A 3, 1557 (1971).

(19) Elmergreen, Smith, ApJ 178, 557 (1972).

(20) Becker, Haaks, ZN 28 a, 249 (1973).

(21) Clyne, Coxon, Woon Fat, JMS 46, 146 (1973).

(22) German, Bergeman, Weinstock, Zare, JCP 58, 4304 (1973).

(23) Meerts, Dymanus, ApJ 180, L93 (1973).

(24) Meerts, Dymanus, CPL 23, 45 (1973).

(25) Rouse, Engleman, JQSRT 13, 1503 (1973).

(26) Weinstock, Zare, JCP 58, 4319 (1973).

(27) Brophy, Silver, Kinsey, CPL 28, 418 (1974).

(28) Carlone, PR A 9, 606 (1974).

(29) Douglas, CJP 52, 318 (1974).

(30) Carlone, PR A 12, 2464 (1975).

(31) Coxon, JMS 58, 1 (1975).

(32) Coxon, Hammersley, JMS 58, 29 (1975).

(33) German, JCP 62, 2584; 63, 5252 (1975).

(34) Meerts, Dymanus, CJP 53, 2123 (1975).

(35) Smith, Stella, JCP 63, 2395 (1975).

(36) Wilcox, Anderson, Peacher, JOSA 65, 1368 (1975).

(37) German, JCP 64, 4192 (1976).

(38) Woods, Dixon, JCP 64, 5319 (1976).

(39) Crosley, Lengel, JQSRT 17, 59 (1977).

(40) Katsumata, Lloyd, CPL 45, 519 (1977).

State	T_e	ω_e	$\omega_e x_e$	B_e	α_e	D_e ($10^{-4}cm^{-1}$)	r_e (Å)	Observed Transitions: Design.	Observed Transitions: ν_{00}	References
$^{16}O^{2}H$		$\mu = 1.78884797$		$D_0^0 = 4.453$ eV[a]		I.P. = 12.9_1 eV[b]				MAY 1977
C $^2\Sigma^+$	(89470)	(898)	(10)	[2.235][c]			[2.053]	$C^d \rightarrow A$,[e] R	56090.3 Z	(7)(13)*
		Strong many-line spectrum 1900 - 1700 Å, tentative identification.						$(C \rightarrow X)^d$	(88568)	(8)*
D $^2\Sigma^-$	(82160)	(2074)		[8.2283][f]		[5.179]	[1.0701_8]	$D \leftarrow X$, R	81853.22[g] Z	(29)
B $^2\Sigma^+$	69775	[546.9] Z	[h]	[2.745][i]	[j]	[2.50][k]	[1.852_9]	$B \rightarrow A$,[ℓ] R	36275.7 Z	(5)(6)(7)* (9)(13)*
A $^2\Sigma^+$	32680.8_5[m]	2316.17 Z	50.433[n]	9.193_6[opqr]	0.318_1[s]	[5.763][t]	1.0124	$A^u \leftrightarrow X$,[v] R	32477.18[g] Z	(13)* (21) (31)
X $^2\Pi_i$	0[w]	2720.24 Z	44.055	10.020_9[xr]	0.275_7[y]	[5.374][z]	0.9697_5	Hf Λ-doubling sp.[a']		(4)(23)(34)
								EPR sp.[b']		(14)

O^2H: [a]From $D_0^0(O^1H)$ assuming zero electronic isotope shift.

[b]From I.P.(O^1H) and the zero-point energies of OH, OD, OH^+, OD^+. Photoelectron spectroscopy gives 13.01 eV (40).

[c]Spin splitting constant $\gamma_0 = +0.6$.

[d]Lifetimes of 6.1 and ~2 ns have been reported (35) for the upper state of the $C \rightarrow A$ bands and for the 1850 Å group, respectively. See [e] of O^1H.

[e]Only the 0-11 and 0-12 bands have been observed (13).

[f]Spin splitting constant $\gamma_0 = -0.156$.

[g]See [h] of O^1H.

[h]$\Delta G(3/2) = 357.9$ (13). Using isotope relations (5) derives $\omega_e = 684$, $\omega_e x_e = 55._7$, $\omega_e y_e = -8._3$.

[i]Spin splitting constant $\gamma < 0.05$.

[j]$B_1 = 2.445$, $B_2 = 1.947$ (13).

[k]$D_1 = 4.48 \times 10^{-4}$, $D_2 = 18.9 \times 10^{-4}$; $H_0 = -12 \times 10^{-7}$, $H_1 = -16 \times 10^{-7}$ (13).

[ℓ]Franck-Condon factors (7).

[m]T_e has been corrected for the effects of Y_{00} on the zero-point energy in both upper and lower state and for a small electronic term o_v^{Σ} due to interaction with the $^2\Pi$ state; see (31).

[n]$\omega_e y_e = -0.2350$; the constants represent only v=0...3 (31). Vibrational energy levels and ΔG values up to v=13 are listed in (13). Preliminary vibrational constants may be found in (5).

[o]Spin splitting constants (31): $\gamma_0 = +0.1201$ [good agreement with (29)], $\gamma_1 = +0.117_0$, $\gamma_2 = +0.111_4$.

[p]In low-pressure flames and discharges the intensity of emission lines originating in v'=0,1,2 decreases rapidly above N'=29,26,17, respectively (3), owing to predissociation by $^4\Sigma^-$ (see [p] of O^1H); substantially higher N' values in (10) correspond to the first lines of zero intensity. The lifetime of the v'=0 rotational levels drops sharply above N'≈34 (19)(36). A much stronger predissociation

O^2H (continued):

occurs for v=7...12 causing diffuseness in the B→A bands in low pressure discharges (13)(15); asymmetric line shapes in the 0-9 band have been studied by (28) and have been found (30) to contain a Q component because of mixing with the $^4\Pi$ state that causes the predissociation (15).

[q] $\mu_{e\ell}(v=0)$ = 2.16 D (17), see [q] of O^1H. A considerably lower value of 1.72 D was derived by (26) from Stark shifts in the location of high-field level crossings and appears to be favoured by ab initio calculations [for references see (26)]. Hfs constants from high-field level crossing experiments in (26) [eqQ corrected by (38)] and (37).

[r] RKR potential functions (31).

[s] $\gamma_e = -0.0011_9$. The equilibrium constants refer to the true mechanical B_v values derived by (31) from the effective constants (given in the same paper) for $v \leq 3$. Term values for $v \leq 3$ (21).

[t] $+1.65 \times 10^{-8} J^3(J+1)^3 - 6._0 \times 10^{-13} J^4(J+1)^4$ (31); good agreement with D_0 and H_0 of (29). (31) gives D_v, H_v for $v \leq 2$.

[u] Radiative lifetimes $\tau(v=0,N=1) = 691 \pm 9$ ns, $\tau(v=1,N=1) = 712 \pm 10$ ns, $\tau(v=2,N=1) = 736 \pm 13$ ns (33). (19)(20)(27)(36) give somewhat longer lifetimes for low rotational levels in v=0; Hanle effect measurements (18)(22) are slightly lower. Oscillator strengths from high-resolution line absorption: $f_{00} = 0.00096$ and $f_{10} = 0.00026$ for the rotationless molecule (25).

[v] Zeeman effect in the 0-0 band (12)(16), magnetic rotation spectrum (16). Absorption and emission in solid Ne (11). Franck-Condon factors (7)(31). Vibrational intensity distribution, variation of the transition moment with r, rotational dependence of transition probabilities (39).

[w] $A_0 = -139.23_0$, $A_1 = -139.44_0$, $A_2 = -139.64_4$ (31). Ab initio calculation of spin-orbit coupling parameters (32).

[x] Λ-type doubling parameters $p_0 = +0.1266$, $q_0 = -0.01093_4$; additional constants for $v \leq 3$ (31); ab initio calc. (32).

[y] $\gamma_e = +0.000_6$. Term values for $v \leq 3$ (21).

[z] $D_1 = 5.331 \times 10^{-4}$, $D_2 = 5.28 \times 10^{-4}$; $H_0 = +1.93_5 \times 10^{-8}$, $H_1 = +2.06_5 \times 10^{-8}$, $H_2 = +2.46 \times 10^{-8}$; $L_0 = -5.2 \times 10^{-13}$ (31). For v=0 good agreement with (29).

[a'] Λ-doubling and hfs coupling constants. Dipole moment $\mu_{e\ell}(v=0) = 1.6531_2$ D, from Stark shifts of hf Λ-doubling transitions (24); see also (17).

[b'] Spectrum of ^{17}OD; magnetic hf and electric quadrupole coupling constants.

(1) Sastry, IJP 15, 95, 455 (1941); 16, 27, 169, 343 (1942); Sastry, Rao, IJP 15, 27 (1941).
(2) Oura, JPSJ 6, 401 (1951); LTS 6, 41 (1951).
(3) Broida, Kane, PR 89, 1053 (1953).
(4) Dousmanis, Sanders, Townes, PR 100, 1735 (1955).
(5) Barrow, AF 11, 281 (1956).
(6) Herman, Felenbok, Herman, JPR 2, 83 (1961).
(7) Felenbok, AAp 26, 393 (1963).
(8) Felenbok, Czarny, AAp 27, 244 (1964).
(9) Czarny, Felenbok, AAp 31, 141 (1968).
(10) Palmer, Naegeli, JMS 28, 417 (1968).
(11) Tinti, JCP 48, 1459 (1968).
(12) Thakur, Rai, Singh, JCP 48, 3389 (1968).
(13) Carlone, Dalby, CJP 47, 1945 (1969).
(14) Carrington, Lucas, PRS A 314, 567 (1970).
(15) Czarny, Felenbok, Lefebvre-Brion, JP B 4, 124 (1971).
(16) Nanes, Robinson, JCP 55, 963 (1971).
(17) Scarl, Dalby, CJP 49, 2825 (1971).

(continued on p. 513)

State	T_e	ω_e	$\omega_e x_e$	B_e	α_e	D_e (10^{-4} cm^{-1})	r_e (Å)	Observed Transitions: Design.	Observed Transitions: ν_{00}	References
$^{16}O^1H^+$				$D_0^0 = 5.0_9$ eV[a]						MAY 1977
b $^1\Sigma^+$	(29050)	[2981][b]		[16.320][b]	(0.732)[b]	[19.2][b]	1.032	(b − X)	29058.8[b]	
A $^3\Pi_i$ {0^-, 0^+, 1, 2}	28438.55	2133.65 Z	79.55	13.791_6[cd]	0.8889[e]	[22.49_5][f]	1.1354	A[g] → X,[h] R	28034.04[i], 28028.31[i], 27948.43[i], 27864.31[i]	(2)* [n]
a $^1\Delta$	17660[j]									
X $^3\Sigma^-$	0	3113.37 Z	78.52	16.794_3[k]	0.7494[ℓ]	[19.17_4][m]	1.0289			
$^{16}O^2H^+$				$D_0^0 = 5.1_4$ eV[o]						MAY 1977
b $^1\Sigma^+$	(29050)	[2174][b]		[8.736_9][b]	(0.285)[b]	[5.44][b]	1.030	(b − X)	29051.2[b]	
A $^3\Pi_i$ {0^-, 0^+, 1, 2}	28452.75	1558.08 Z	44.438[p]	7.310_7[qr]	0.3398[s]	[6.38_8][t]	1.1354	A[u] → X,[h] R	28181.47[i], 28176.06[i], 28095.83[i], 28011.57[i]	(2)* [n]
a $^1\Delta$	17660[j]			$B_6 = 7.242_4$[v]		v				
X $^3\Sigma^-$	0	2271.80 Z	44.235[w]	8.911_6[x]	0.2896[y]	[5.44][z]	1.0283			
$^{16}O^1H^-$				$D_0^0 = 4.755$ eV[a]		I.P. = 1.825_4 eV[b]				MAY 1977
a ($^3\Pi$)	(28000)	Absorption ($\lambda < 3500$ Å) and long-lived emission ($\lambda_{max} \approx 4000$ Å) in aqueous solutions.								(3)
X $^1\Sigma^+$	0	(3700)[c]		(18.9)[c]			(0.970)[c]			
$^{16}O^2H^-$				$D_0^0 = 4.814$ eV[d]		I.P. = 1.823_0 eV[b]				
X $^1\Sigma^+$	0	(2700)[c]		(10.02)[c]			(0.970)[c]			
$^{(192)}Os^{16}O$?		($\mu = 14.7646702_7$) Mostly R shaded emission bands in the region 11400 - 16400 cm^{-1}. No analysis.								APR 1975 (1)* (2)

O^1H^+, O^2H^+:

[a] $D_0^0(O^1H) + I.P.(H) - I.P.(O^1H)$.
[b] Constants derived from the perturbations in $A\ ^3\Pi$. The b–X transition is not observed.
[c] Λ-doubling constants $p_0 = -0.251$, $q_0 = +0.0478$; for v=1,2 see (2).
[d] Perturbations by $b\ ^1\Sigma^+$.
[e] $\gamma_e = +\ 0.0173_0$.
[f] $H_0 = +\ 10.4 \times 10^{-8}$; other D_v, H_v values for $v \leq 2$ in (2).
[g] Lifetime $\tau(v=0) = 0.89$ μs (1); similar results for v=1,3.
[h] Vibrational intensity distribution (branching ratios), variation of the transition moment with r (3).
[i] Subband origins as defined by (2). From the data for the OD^+ $A^3\Pi(v=0)$ level (2) estimate the true spin-orbit and spin-spin interaction parameters $A = -83.83$ cm^{-1} and $2\alpha = -5.9_2$ cm^{-1}.
[j] From the photoelectron spectrum (4).
[k] Spin splitting constants for v=0: $\lambda_0 = 2.13_4$ cm^{-1} and $\gamma_0 = -\ 0.147_8$ cm^{-1}; similar results for v=1 and 2.
[ℓ] $\gamma_e = +\ 0.0109_7$.
[m] $H_0 = +\ 12.3_9 \times 10^{-8}$; values for D_1, D_2 in (2).
[n] A rather complete list and critical assessment of earlier references is given in (2).
[o] From $D_0^0(O^1H^+)$.
[p] $\omega_e y_e = +\ 0.368_3$.
[q] Λ-doubling constants $p_0 = -0.132$, $q_0 = +0.0122_2$ cm^{-1}; for v=1,2,3 see (2).
[r] Perturbations by $b\ ^1\Sigma^+$ and $a\ ^1\Delta$.
[s] $\gamma_e = +\ 0.00372$.
[t] $H_0 = +\ 2.0_2 \times 10^{-8}$; other D_v, H_v values for $v \leq 3$ in (2).
[u] Lifetime $\tau(v=0) = 1.06$ μs (1a).
[v] This level at 29434 cm^{-1} is only observed through perturbations in $A^3\Pi(v=1)$; $D_6 = 4.8 \times 10^{-4}$. The v=5 level produces a weak perturbation in $A^3\Pi(v=0)$.

O^1H^+, O^2H^+ (continued):

[w] $\omega_e y_e = +\ 0.426_7$.
[x] Spin splitting constants $\lambda_0 = +2.141$, $\gamma_0 = -0.079_0$ cm^{-1}; similar constants for v=1,2 (2).
[y] $\gamma_e = +\ 0.00300$.
[z] $H_0 = 2.1_7 \times 10^{-8}$; values for D_1, D_2, D_3 in (2).

(1) Brzozowski, Elander, Erman, Lyyra, PS 10, 241 (1974).
(1a) Brzozowski, Erman, Lew, CPL 34, 267 (1975).
(2) Merer, Malm, Martin, Horani, Rostas, CJP 53, 251 (1975).
(3) Gérard, Govers, van de Runstraat, Marx, CPL 44, 154 (1976).
(4) Katsumata, Lloyd, CPL 45, 519 (1977).

O^1H^-, O^2H^-:

[a] From $D_0^0(OH)$ and the electron affinities of OH and O.
[b] From high-resolution photodetachment studies of OH^- and OD^- (5); see also (1)(2)(4).
[c] Estimates based on the analysis of photodetachment data (1)(4)(5).
[d] Analogous to [a].

(1) Branscomb, PR 148, 11 (1966).
(2) Kay, Page, TFS 62, 3081 (1966).
(3) Merkel, Hamill, JCP 55, 2174 (1971).
(4) Celotta, Bennett, Hall, JCP 60, 1740 (1974).
(5) Hotop, Patterson, Lineberger, JCP 60, 1806 (1974).

OsO:

(1) Gatterer, Junkes, Salpeter, Rosen, METOX (1957).
(2) Raziunas, Macur, Katz, JCP 43, 1010 (1965).

State	T_e	ω_e	$\omega_e x_e$	B_e	α_e	D_e $(10^{-7}cm^{-1})$	r_e (Å)	Observed Transitions Design.	ν_{00}	References
$^{31}P_2$		$\mu = 15.4868817$		$D_0^0 = 5.033$ eV[a]		I.P. = 10.53 eV[b]				MAY 1977 A
		Carroll and Mitchell's Rydberg series V_n converging to F $^2\Sigma_u^+(v=0)$:								
$\ldots 4\sigma_u 5\sigma_g^2 2\pi_u^4 ns\sigma_g$ [c]		$\nu = 125225 - R/(n-1.86)^2$, $n = 5(V_5), 6 \ldots 16$. Similar series with v'=1.								(21)
$V_5(^1\Sigma_u^+)$	114085	[789]						$V_5 \leftarrow X$,	114090	(21)
		Carroll and Mitchell's Rydberg series G_n, H_n, P_n, R_n converging to A $^2\Sigma_g^+(v=0)$:								
$\ldots 4\sigma_u^2 5\sigma_g 2\pi_u^4$ $nf\pi_u$		$\nu = 87179 - R/(n-0.062-0.633/n^{*2})^2$, $n = 4(R_4), 5 \ldots 21$.								(21)*
$nf\sigma_u$		$R/(n-0.155+0.179/n^{*2})^2$, $n = 4(P_4), 5 \ldots 14$.								
$np\pi_u$		$R/(n-1.468-0.475/n^{*2})^2$, $n = 4(H), 5(H_5), 6 \ldots 28$.								
$np\sigma_u$		$R/(n-1.660-0.252/n^{*2})^2$, $n = 4(G), 5(N), 6(G_6), 7 \ldots 12$.								
		Carroll and Mitchell's Rydberg series E_n, K_n, M_n, L_n converging to X_2 $^2\Pi_{\frac{1}{2}}(v=1)$:								
$\ldots 4\sigma_u^2 5\sigma_g^2 2\pi_u^3$ $nd\delta_g$?		$\nu \approx 85895 - R/(n+0.04)^2$, $n = 3(L), 4$.				Fragments of series with v'=0 ($\nu_\infty = 85229$) and with v' > 1.				(21)*
$nd\pi_g$		$R/(n-0.03)^2$, $n = 3(M), 4(M_4), 5(S)$.								
$nd\sigma_g$		$R/(n-0.12)^2$, $n = 3(K), 4 \ldots 10$.								
$ns\sigma_g$		$R/(n-1.93)^2$, $n = 4(E), 5(E_5), 6$.								
		Unclassified bands 79700 - 84500 cm^{-1}.								(21)
U	83362	[551] H		Observed to v'=2.				$U \leftarrow X$,	R 83248 H	(21)*
$G_6(^1\Sigma_u^+)$	81327	[723]		Diffuse bands.				$G_6 \leftarrow X$,	R 81299 H	(9)(21)
T	80992	[713.1] H		Observed to v'=4.				$T \leftarrow X$,	R 80958.5 H	(9)(21)*
S $^1\Sigma_u^+$	(80840)			$B_1 = 0.2783$			$r_1 = 1.978$	$S \leftarrow X$,	R 81453.89[d] Z	(9)(21)
$R_4(^1\Pi_u)$	80115	[740]		Broad diffuse bands.				$R_4 \leftarrow X$,	80095	(21)
Q $(^1\Pi_u)$	(79860)	[617.6] H		0-0 and 1-1 bands only.				$Q \leftarrow X$,	R 79779.5_3 H	(9)(21)
$P_4(^1\Sigma_u^+)$	(79827)			Broad band.				$P_4 \leftarrow X$,	79804	(21)

State	T_e	ω_e	$\omega_e x_e$	B_e	α_e	D_e ($10^{-7}cm^{-1}$)	r_e (Å)	Observed Transitions Design.	ν_{00}	References
$^{31}P_2$ (continued)										
$M_4(^1\Sigma_u^+)$	(78271)	(705)	(5)	Only v'=1 and 3(diffuse) observed.				$M_4 \leftarrow X$, R	$(78233)^e$	(21)
$H_5(^1\Pi_u)$	(78256)			Very broad band.				$H_5 \leftarrow X$,	78233	(21)
$K_4(^1\Pi_u)$	(77968)	[732]		Diffuse bands.				$K_4 \leftarrow X$,	77944	(21)
O	77475	529 H	2					$O \leftarrow X$, R	77349 H	(21)*
N $^1\Sigma_u^+$	$77286._7$	[641.76] Z	(29.7)	0.2984^f	0.0050_4	[3.1]	1.909_9	$N \leftarrow X$, R	77240.2_3 Z	(9)(21)
L $(^1\Pi_u)$	73593	[672.2] H						$L \leftarrow X$, R	73539.3 H	(9)(21)
$E_5(^1\Pi_u)$	73385	685 H	$2._7$					$E_5 \leftarrow X$, R	73337	(21)
M $^1\Sigma_u^+$	73168.0	684.0^g H	2.9_7	0.2802^g	0.0016		1.971	$M \leftarrow X$, R	73119.6^g H	(9)(21)
K $^1\Pi_u$	72288	[701.7] H^Q	(5)	$[0.2704]^{hf}$			[2.006]	$K \leftarrow X$, R	72253.3 Z	(9)(21)
		Numerous diffuse irregularly spaced "lines" 69900 - 70400 cm^{-1}, no assignments.								(9)
H $(^1\Pi_u)$	(68960)	Very intense diffuse band, width ~80 cm^{-1}.						$H \leftarrow X$,	68940	(9)(21)

P_2: [a]From the predissociation limit in C $^1\Sigma_u^+$ which according to (1) is equal to the dissociation limit $^4S + ^2D$. The latest thermochemical (mass-spectrometric) value is 5.04 eV (17) eliminating the possibility that the predissociation limit corresponds to $^2D + ^2D$ as suggested by (16).

[b]From the photoelectron spectrum (25) which shows partially resolved peaks at 10.53 and 10.55 eV corresponding to $X_1\ ^2\Pi_{3/2}$ and $X_2\ ^2\Pi_{1/2}$, respectively. Subtraction of the estimated doublet splitting in the X $^2\Pi$ ground state [~ 260 cm^{-1}, see ref. (6) of P_2^+] from the Rydberg series limit at 85229 cm^{-1} (21) gives 10.53_5 eV.

[c]Core configuration as suggested by (25) on the basis of <u>ab initio</u> calculations; Carroll and Mitchell (21) proposed $\ldots 4\sigma_u^2 5\sigma_g\ 2\pi_u^3 2\pi_g$.

[d]Assigned as 0-0 band by (9), reassigned as 1-0 band by (21) [third member of the M_n(v'=1) Rydberg series].

[e]0-0 band obscured by $H_5 \leftarrow X$ 0-0.

[f]RKR potential functions (10)(13).

[g]Following (21) the v' numbering of (9) has been increased by one unit.

[h]All lines of this system are diffuse; the diffuseness is largest for v'=3 and 4 such that the rotational structure can no longer be recognized. The broadening is independent of J; the state causing the predissociation is probably $^1\Pi_u$.

State	T_e	ω_e	$\omega_e x_e$	B_e	α_e	D_e ($10^{-7}cm^{-1}$)	r_e (Å)	Observed Transitions Design.	Observed Transitions ν_{00}	References
$^{31}P_2$ (continued)										
I $^1\Pi_u$	[68849.3]			[0.2541]		[$2._5$]	[2.070]	I← X, R	68459.6_3 Z	(9)
G $^1\Sigma_u^+$	66313.37	694.12 Z	4.18_2	0.29730^f	0.00195	[2.25]	1.9135	G↔X, R	66269.71 Z	(6)(9)*
E $^1\Pi_u$	59446.21	700.66 H^Q	2.92	$[0.2807_2]^f$		[1.84]	$[1.969_2]$	E↔X,[i] R	59406.14 Z	(6)(9)*
B $^1\Pi_u$	50845.9	$[358.96]^j$ Z	k	$[0.2268]^j$	ℓ	[3.2]	[2.191]	B→A, R	16203.7_1[j] Z	(18)(23)*
c $^3\Pi_u$ 2 1 0	47176.8 47159.1 47139.2	393.67 Z	3.849^m	0.2190^n	0.0024	2.5	2.229	c→b,[o] R	18721.5 Z 18836.4 Z 18944.6 Z	(15)(22)* (26)(27)
C $^1\Sigma_u^+$	46941.26	473.93 Z	2.340^p	0.24211^{qf}	0.00175^r	2.57	2.1204	C↔X, R	46787.97 Z	(1)* (2)* (4) (9)(11)* (23)*
A $^1\Pi_g$	34515.25	618.95 Z	3.00	0.27524	0.00168	2.2	1.9887	A→X, R	34434.30 Z	(7)* (14) (18)(23)*
b $^3\Pi_g$ 2 1 0	28329.6 28197.0 28068.9	644.66 Z	3.213^s	0.2805	0.00178	2.0	1.970	b→a,[o] V	9574.8 Z 9442.2 Z 9314.1 Z	(12)(20) (22)* (26)
b' $^3\Sigma_u^-$	28503.4^t	604.4_8 Z	2.2	0.2584^u	0.0014	1.7	2.052	b'→X, R	28415.4 Z	(12)(19)*
a $^3\Sigma^+$	18794.5	565.1_7 H	2.7_5	$B_1 = 0.2503_3{}^v$	$D_1 = 3.3 \times 10^{-7}$		$r_1 = 2.085_3$	(a − X)	18686.7^w	
X $^1\Sigma_g^+$	0	780.77 Z	$2.83_5{}^x$	0.30362^f	0.00149^y	1.88	1.8934			

P_2 (continued):

[i]Franck-Condon factors (13).

[j]The assignment of v'=0 to the lowest observed level is arbitrary.

[k]$\Delta G(3/2) = 352.71$.

[ℓ]$B_1 = 0.2209$, $B_2 = 0.217_1$.

[m]$\omega_e y_e = -0.0915$ (26); slightly different constants in (27).

[n]Perturbations by C $^1\Sigma_u^+$. The observation of extra lines enables the relative position of singlet and triplet levels to be determined (26); very similar results have been obtained by (24)(27).

[o]Franck-Condon factors (22).

[p]$\omega_e y_e = +0.0066$. Constants from (9), but see (27) whose deperturbation of the C state levels (see [q]) leads to T_e =

P_2 (continued):

46945.5_7, $\omega_e = 472.81$, $\omega_e x_e = 2.24_6$, $\omega_e y_e = +\,0.00412_2$.

[q]Strong vibrational perturbations for v=1,2,3,5 (1) and many rotational perturbations (3)(4)(5)(9)(23) due to interaction with levels of c $^3\Pi_u$; see (27). Breaking off in emission (predissociation) above J=58 of v=10 and J=34 of v=11; the corresponding dissociation limit is $^4S + ^2D$ at 51959 ± 25 cm^{-1}. A second predissociation indicated by diffuseness in absorption begins at v=17 and reaches a maximum for low J values of v=19 (9).

[r]$\gamma_e = +\,3._3 \times 10^{-6}$.

[s]$\omega_e y_e < 0.001$ cm^{-1} (26); slightly different constants in (27).

[t]Recalculated; (19) give 28507.74 which probably refers to data obtained from band heads rather than band origins.

[u]Spin splitting constants $\lambda = +3.20$, $\gamma = -0.001$.

[v]Spin splitting constants $\lambda_1 = -3.2_1$, $\gamma_1 = -0.003$ (22).

[w]Not observed; indirectly derived (26) from perturbations in c $^3\Pi_u$; see [n].

[x]$\omega_e y_e = -\,0.00462$ (27); slightly different constants in (4) and (9).

[y]$\gamma_e = -\,2._7 \times 10^{-6}$. Constants recalculated from the data of (1)(7)(9)(11). Higher B_v values not included in the re-evaluation may be found in (1)(4).

(1) Herzberg, AP(Leipzig) (5) 15, 677 (1932).
(2) Ashley, PR 44, 919 (1933).
(3) Marais, PR 70, 499 (1946).
(4) Marais, Verleger, PR 80, 429 (1950).
(5) Naudé, Verleger, PR 80, 432 (1950).
(6) Dressler, HPA 28, 563 (1955).
(7) Douglas, Rao, CJP 36, 565 (1958).
(8) Gutbier, ZN 16 a, 268 (1961).
(9) Creutzberg, CJP 44, 1583 (1966).
(10) Singh, Rai, IJPAP 4, 102 (1966).
(11) Dixit, PIAS A 66, 325 (1967).
(12) Mrozowski, Santaram, JOSA 57, 522 (1967).
(13) Rao, Lakshman, IJPAP 8, 617 (1970).
(14) Verma, Broida, CJP 48, 2991 (1970).
(15) Brion, Da Paz, Mongin, Guenebaut, CR B 272, 999 (1971).
(16) Vaidyan, Santaram, IJPAP 9, 1022 (1971).
(17) Kordis, Gingerich, JCP 58, 5141 (1973).
(18) Malicet, Brion, Guenebaut, CR C 276, 991 (1973).
(19) Brion, Malicet, Guenebaut, CJP 52, 2143 (1974); 53, 201 (1975) (erratum).
(20) Brion, Malicet, JP B 8, L164 (1975).
(21) Carroll, Mitchell, PRS A 342, 93 (1975).
(22) Brion, Malicet, JP B 9, 2097 (1976).
(23) Brion, Malicet, Guenebaut, CJP 54, 362 (1976).
(24) Brion, Malicet, Merienne-Lafore, CR C 283, 171 (1976).
(25) Bulgin, Dyke, Morris, JCS FT II 72, 2225 (1976).
(26) Carroll, Nulty, JP B 9, L427 (1976).
(27) Brion, Malicet, Merienne-Lafore, CJP 55, 68 (1977).

State	T_e	ω_e	$\omega_e x_e$	B_e	α_e	D_e $(10^{-7}cm^{-1})$	r_e (Å)	Observed Transitions Design.	ν_{00}	References
$^{31}P_2^+$		μ = 15.4867445		D_0^0 = 4.99 eV[a]						MAY 1977
F $(^2\Sigma_u^+)$	(40180)[b]	[810][b]								
C_2 $^2\Pi_g$ 3/2	28870.6 c	441.47[d] Z	2.58	0.2162_9	0.00136	2.0	2.243_4	$C_2 \rightarrow X_1$, R	28754.2[d] Z	(1)* (2)(6)*
C_1 $^2\Pi_g$ 1/2	(28686.6)							$C_1 \rightarrow X_2$, R	28312.3[d] Z	
B $^2\Sigma_u^+$	(25566)	410.50[e] Z	3.23	0.2419[e]	0.00211	3.3	2.121	B→A, R	23224.8_9[e] Z	(1)*
D_2 $^2\Pi_g$ 1/2	(18832.5)	462.2	2.45	0.2196	0.00142	2.0	2.226	$D_2 \rightarrow X_2$, R	18467.6 Z	(3)(6)*
D_1 $^2\Pi_g$ 3/2	18740.7							$D_1 \rightarrow X_1$, R	18635.8 Z	
A $^2\Sigma_g^+$	(2179)[f]	[733.0][g] Z		0.3037[g]	0.0021		1.893			
X_2 $^2\Pi_u$ 1/2	(260)[h]	672.20[d] Z	2.74	0.2760_0	0.00151	2.0	1.985_9			
X_1 $^2\Pi_u$ 3/2	0									
$^{(208)}Pb_2$		(μ = 103.988329)		D_0^0 = 0.8_2 eV[a]						MAY 1977 A
		Emission bands in the region 12500 - 11100 cm^{-1}, observed in Ar and Ne matrices by excitation with argon/krypton ion laser lines.								(7)
		Absorption bands in the regions 33400 - 34600 cm^{-1} (preceded by continuous fluctuations from 31300 to 33300 cm^{-1}), 36800 - 38000, and 41000 - 42200 cm^{-1}.								(2)
B	19490.3	161.64[b] H	1.036[c]					B↔X,[d] R	19515.7 H	(1)* (2)* (4)
A	14465.6	162.4[b]	0.4					A↔X,[e] (R)	14491.5	(2)* (4)(5)
X	0	110.5[bf] H	0.35							
$^{(208)}Pb^{209}Bi$		(μ = 104.238661)		D_0^0 = 1.4_3 eV[a]						MAY 1977

P_2^+: [a]$D_0^0(P_2)$ + I.P.(P) - I.P.(P_2).

[b]From Rydberg series of P_2 (4) and observations by photoelectron spectroscopy (5).

[c]A_v strongly dependent on v, $A_{v+1} - A_v = -4.13$ cm^{-1} (6).

[d]Vibrational numbering of (6); the previous numbering of (1) has been increased by one unit in both upper (C) and lower (X) state following the recommendations of (4).

[e]Vibrational numbering uncertain; the B state has not been observed in the photoelectron spectrum (5).

[f]From Rydberg limits of P_2 (4) and the estimated spin splitting in the ground state of P_2^+; in reasonable agreement with the photoelectron-spectroscopic value of 2230 cm^{-1} (5).

[g]Vibrational numbering uncertain.

[h]The magnitude of the spin-orbit coupling in the $^2\Pi$ ground and excited states has been estimated (6) from the effective B values. The photoelectron spectrum (5) suggests for the ground state a spin splitting of $\sim$150 cm^{-1}.

(1) Narasimham, CJP 35, 1242 (1957).
(2) Brion, Malicet, Guenebaut, CR C 276, 471 (1973).
(3) Brion, Malicet, Guenebaut, CR C 276, 551 (1973).
(4) Carroll, Mitchell, PRS A 342, 93 (1975).
(5) Bulgin, Dyke, Morris, JCS FT II 72, 2225 (1976).
(6) Malicet, Brion, Guenebaut, CJP 54, 907 (1976).

Pb_2: [a]Thermochemical value (mass-spectrom.) (6).

[b]Constants obtained by (4) from a reclassification of bands in the blue-green and in the red system. The new analysis of the ground state seems to be confirmed by the observation of a lower state frequency of 111 cm^{-1} in laser-excited matrix emission spectra [see (7) who attribute the 111 cm^{-1} interval to a matrix-induced splitting of the ground state]. The revised lower state constants imply, however, a strong contradiction to the usual rule that if $\omega' > \omega''$ then $B' > B''$ since according to the published spectrograms of (2) the B − X bands are strongly shaded to the red, i.e. $B' < B''$.

[c]$\omega_e y_e = + 0.0055$.

[d]Previously called A − X. Also observed in inert gas matrices (3)(7).

[e]Previously believed to have B as its lower state.

[f](4) give 119.1 which seems to be a typographical or arithmetical error since it does not represent the band head measurements of (1).

(1) Shawhan, PR 48, 343 (1935).
(2) Weniger, JP(Paris) 28, 595 (1967).
(3) Brewer, Chang, JCP 56, 1728 (1972).
(4) Johnson, Cannell, Lunacek, Broida, JCP 56, 5723 (1972).
(5) Puri, Mohan, IJPAP 13, 206 (1975).
(6) Gingerich, Cocke, Miller, JCP 64, 4027 (1976).
(7) Teichman, Nixon, JMS 59, 299 (1976).

PbBi: [a]Thermochemical value (mass-spectrom.) (1).

(1) Rovner, Drowart, Drowart, TFS 63, 2906 (1967).

State	T_e	ω_e		$\omega_e x_e$	B_e	α_e	D_e (10^{-7} cm^{-1})	r_e (Å)	Observed Transitions Design.	ν_{00}	References
$^{(208)}Pb^{79}Br$		(μ = 57.209681_3)			D_0^0 = (2.5) eV[a]		I.P. = 7.8 eV[b]				JUL 1977
B ($^2\Sigma$)	$34523._7$	258.2	H	0.60	Diffuse bands (prediss.)				$B \leftarrow X_1$, V	$34549._0$ H	(2)*
A ($\frac{1}{2}$)	$20884._3$	152.5	H	0.40[c]	d				$A \leftrightarrow X_1$, R	$20856._8$ H	(1)(3)(6)
X_1 $^2\Pi_{1/2}$	0	207.5	H	0.50[e]	d						
$^{(208)}Pb^{35}Cl$		(μ = 29.9355409)			D_0^0 = (3.1) eV[a]		I.P. = 7.5_5 eV[b]				JUL 1977
B ($^2\Sigma$)	35199[c]	382.1[c]	H	1.05[c]	Diffuse bands (prediss.)				$B \leftarrow X_1$, V	35238[c] H	(3)*
A ($\frac{1}{2}$)	$21865._0$	228.7	H	0.78	d				$A \rightarrow X_2$, R	$13546._2$ H	(5)
X_2 $^2\Pi_{3/2}$	$8272._2$	321.6	H	0.3[e]					$A \leftrightarrow X_1$, R	$21827._4$ H	(1)(2)* (5)* (8)
X_1 $^2\Pi_{1/2}$	0	303.9	H	0.88	d						
$^{208}Pb^{19}F$		μ = 17.4081886			D_0^0 = 3.6_4 eV[a]		I.P. = 7.5 eV[b]				JUL 1977
		Strong absorption continuum with maximum at ~41000 cm^{-1}; emission continuum with maximum at ~32800 cm^{-1}.									(2)
F	(47866)	[628]	H						$F \leftarrow X_1$, (V)	47927 H	(2)
E	(45400)	(565)	H						$E \rightarrow X_2$, V		(5)
									$E \leftarrow X_1$, V	45430 H	(2)
D	(43818)	[597]	H						$D \leftarrow X_1$, (V)	43863 H	(2)
C	38046	594.0	H	2.50	Diffuse bands[c]				$C \leftarrow X_1$, V	38089 H	(2)
B $^2\Sigma^+$	35644.4	[605.75]	Z	3.42 H^Q	0.24810[de]	0.001479	1.63	1.9756	$B \leftrightarrow X_2$, V	27420.91 Z	(2)* (7)* (8)
									$B \leftrightarrow X_1$, V	35696.79 Z	(2)* (8)
A $\frac{1}{2}$ ($^2\Sigma^+$)	22556.5	[394.73]	Z	1.77 H	0.20762[f]	0.001430	2.22	2.1597	$A \rightarrow X_2$, R	14226.21	(4)
									$A \leftrightarrow X_1$, R	22502.09 Z	(1)(2)* (8)
X_2 $^2\Pi_{3/2}$	8263.5	[528.75]	Z	1.50 H^Q	0.23403	0.001450	1.78	2.0342	($X_2 - X_1$)	8275.88	
X_1 $^2\Pi_{1/2}$	0	[502.73]	Z	2.28[g]	0.22875[h]	0.001473	1.83	2.0575			

PbBr: [a]See (2) for a discussion of this value.

[b]Electron impact study of $PbBr_2$ (4); the same authors give 0.9 eV for the electron affinity of PbBr.

[c]$\omega_e y_e = -0.028$.

[d]Tentative rotational analysis (5).

[e](3) give $\omega_e x_e = 0.52$, $\omega_e y_e = +0.0023$. Similar constants in (6).

(1) Morgan, PR 49, 47 (1936).
(2) Wieland, Newburgh, HPA 25, 87 (1952).
(3) Pannetier, Deschamps, CR 261, 3109 (1965).
(4) Hastie, Bloom, Morrison, JCP 47, 1580 (1967).
(5) Lal, Khanna, CJP 46, 1991 (1968).
(6) Singh, IJPAP 6, 384 (1968).

PbCℓ: [a]See (3) for a discussion of this value.

[b]From an electron impact study of $PbC\ell_2$ by (6) who have also determined the electron affinity of PbCℓ, E.A. = 1.0 eV.

[c]The revised vibrational analysis of the $B \leftarrow X_1$ system by (5a) leads to the expression $v = 34937.5 + 386.3(v'+\frac{1}{2}) - 1.36(v'+\frac{1}{2})^2 - 300.8(v''+\frac{1}{2}) + 1.04(v''+\frac{1}{2})^2$ and $v_{00} = 34980.2$.

[d]Tentative rotational analyses (4)(7).

[e]$\omega_e y_e = -0.07$.

(1) See ref. (1) of PbBr.
(2) Rochester, PRS A 153, 407 (1936); 167, 567 (1938).
(3) See ref. (2) of PbBr.
(4) Rao, Rao, ZP 181, 58 (1964).
(5) Pannetier, Deschamps, BSCF (1965), 2933.

PbCℓ (continued):

(5a)Cordes, Gehrke, ZPC(Frankfurt a. M.) 51, 281 (1966).
(6) See ref. (4) of PbBr.
(7) Singh, Singh, CS 37, 282 (1968).
(8) Singh, IJPAP 8, 114 (1970).

PbF: [a]Thermochemical value (mass-spectrom.) (6). See also (3) who consider the possibilities $D_0^0 = 4.54$, 3.57, or 3.22 eV.

[b]Electron impact appearance potential (6).

[c](3) consider it possible that the observed C state levels (v=1...5) are actually higher vibrational levels of B $^2\Sigma^+$ (v=5...9).

[d]Spin splitting constant $\gamma = +0.0027$.

[e]Breaking off in emission above v'=1 (predissociation).

[f]Ω-type doubling $\Delta v_{fe}(v=0) = +0.6185(J+\frac{1}{2}) - 10.2 \times 10^{-7}(J+\frac{1}{2})^3$.

[g]Determined from head-origin calculations.

[h]Λ-type doubling $\Delta v_{fe}(v=0) = -0.1388(J+\frac{1}{2})$.

(1) See ref. (1) of PbBr.
(2) See ref. (2) of PbCℓ.
(3) See ref. (2) of PbBr.
(4) Barrow, Butler, Johns, Powell, PPS 73, 317 (1959).
(5) Singh, IJPAP 5, 292 (1967).
(6) Zmbov, Hastie, Margrave, TFS 64, 861 (1968).
(7) Singh, Singh, Singh, CJP 50, 2206 (1972).
(8) Lumley, Barrow, JP B 10, 1537 (1977).

State	T_e	ω_e		$\omega_e x_e$	B_e	α_e	D_e ($10^{-6}cm^{-1}$)	r_e (Å)	Observed Transitions: Design.	ν_{00}	References
$^{(208)}Pb^{1}H$		(μ = 1.00296500)			$D_0^0 \leq$ 1.59 eV[a]						MAY 1977
C ($^2\Delta$)		Single weak band, not yet analyzed.							C→X,	26205	(3)(5)
B [b]	(18030)	[478.8][c]	Z		[2.478][cd]			[2.604]	B→X,	R 17498.7[c] Z	(1)(3)
A	(17590)	(500)[e]		(10)[e]	(3.02_5)[e]	(0.05)[e]		(2.36)	(A − X)	(17060)[e]	
X ($^2\Pi_{1/2}$)[f]	0	1564.1	Z	29.75	4.971	0.144	[201]	1.838_8			
$^{(208)}Pb^{127}I$		(μ = 78.813543_3)			D_0^0 = (2.0) eV[a]						JUL 1977
B ($^2\Sigma$)	$33488._0$	198.7	H	0.35	Diffuse bands (prediss.)				B←X,	V $33507._1$ H	(1)*
A (½)	$20529._0$	142.0[b]	H	1.50					A↔X,	R $20519._4$[b] H	(1)*
X ($^2\Pi_{1/2}$)	0	160.5	H	0.25							
$^{208}Pb^{16}O$		μ = 14.8526392_3			D_0^0 = 3.83 eV[a]						JUL 1977
		Fragments of two further absorption systems in the region 54800 – 57500 cm^{-1}; not fully published.									(10)
G	51661	540.5	H	6					G←X,	R 51570 H	(10)
F	51153	558.5	H	3					F←X,	R 51072 H	(10)
E 0^+	34454	454	H	7	(0.239)[b]	(0.0014)		(2.18)	E↔X,	R 34320 H	(2)* (5)
D 1	30198.7	530.5	H	2.92	0.2711[cd]	0.0031	(0.28)	2.046	D↔X,[e]	R 30103.5 H	(1a)(2)* (5) (14)*
C' 1	24947	494	H	3.0	0.248[f]	0.0018	(0.25)	2.14	C'←X,	R 24833 H	(1a)(5)(10)
C 0^+	23820	532[g]	H	3.9	0.254	0.002	(0.25)	2.11	C←X,	R 23725 H	(1a)(5)(10)
B 1	22285	498.0[h]	H	2.20	0.2646[id]	0.0026	(0.30)	2.071	B[j]↔X,[k]	R 22173.4 H	(1a)(5)(14)*
A 0^+	$19862._6$	$444._3$[ℓ]	H	0.5_4[ℓ]	0.2586_9[d]	0.0013_8	(0.33)	2.094_6	A[j]↔X,[e]	R $19725._0$ H	(1)(3)* (5) (17)
b 0^- } ($^3\Sigma^+$)	(16454)	(441)							b→X,	R 16315 H	(15)(16)
a 1 } ($^3\Sigma^+$)	$16024._9$	$481._5$	H	2.4_5	(0.25_2)			(2.1_2)	a→X,	R $15905._4$ H	(15)* (16) (17)*
X $^1\Sigma^+$	0	721.0	H	3.54[m]	0.3073056[d]	0.0019148	(0.22_3)	1.921813	Microwave sp.[n]		(6)
									Matrix IR sp.		(12a)

PbH: [a]From the predissociation in B assuming dissociation at that limit into $^3P_1 + ^2S$.

[b](1)(3) assumed this to be a $^2\Sigma$ state; more recently, (5) suggested that the red system of PbH originates from a $^4\Sigma^-(\frac{1}{2},\frac{3}{2})$ upper state.

[c]Strong perturbations; $\Delta G(3/2...9/2)$ = 448.1, 438.5, 432.6, 403; $B_1...B_4$ (for low J values) = 2.660, 2.766, 2.770, 2.646. (2) has estimated the following "deperturbed" constants: $\omega_e = 535$, $\omega_e x_e = 15$, $B_e = 2.48$, $\alpha_e = 0.08$, $T_0 = 17520$.

[d]Breaking off (predissociation) at N'=30,24,20 for v'=3, 4,5, respectively (1). The v=5,N=20 level lies at about 20610 cm^{-1} above the lowest ground state level.

[e]All constants estimated from the perturbations in B (2).

[f]Originally believed to be $^2\Sigma$ (1)(3), reassigned by (4)(5). The $^2\Pi_{3/2}$ component is expected at $\sim$8000 cm^{-1} above $^2\Pi_{1/2}$; the corresponding spectrum is in the infrared and not yet observed.

(1) Watson, PR 54, 1068 (1938).
(2) Gerö, ZP 116, 379 (1940).
(3) Watson, Simon, PR 57, 708 (1940).
(4) Howell, PPS 57, 37 (1945).
(5) Kleman, Thesis (Stockholm, 1953).

PbI: [a]See the discussion in (1).

[b]Vibrational numbering uncertain.

(1) Wieland, Newburgh, HPA 25, 87 (1952).

PbO: [a]Thermochemical value (mass-spectrom.) (7). From the $Pb + O_3$ chemiluminescence spectrum under single-collision conditions (15) derive $D_0^0 \geq 3.74$ eV.

[b]Strong perturbations make the constants for this state somewhat uncertain. For ^{206}PbO $B_e = 0.2421$, $\alpha_e = 0.0026$ (10).

PbO (continued):

[c]Perturbations in v=0 (14).

[d]RKR potential functions (8).

[e]Franck-Condon factors (4).

[f](10) quotes $B_e = 0.2491$ (extrapol. from v=6,7) for ^{206}PbO.

[g]The vibrational numbering of (1a) has been increased by 2.

[h]Vibrational constants from (1a); (5) give $\omega_e = 493.5$, $\omega_e x_e = 2.26$. Irregular vibrational intervals.

[i]Rotational perturbations in v=1 (14).

[j]Lifetimes $\tau[B(v=0,1)] = 2.5_8$ μs, $\tau[A(v=2)] = 3.7_5$ μs (15).

[k]Relative intensities (11); transition probabilities (12).

[ℓ]Constants derived from band heads with $v' \leq 6$ (17), in good agreement with results quoted by (10) from an unpublished thesis by Travis (rot. anal. of v=0...3 of ^{206}PbO) but considerably smaller than earlier values ($\omega_e = 451.7$, $\omega_e x_e = 3.33$) proposed by (1).

[m]Ground state levels observed to v=15 (17).

[n]Stark effect (9), $\mu_{e\ell}(v=0) = 4.6_4$ D. Zeeman effect (13), $g_J(v=0) = -0.1623$.

(1) Bloomenthal, PR 35, 34 (1930).
(1a)Howell, PRS A 153, 683 (1936).
(2) Vago, Barrow, PPS 59, 449 (1947).
(3) Gatterer, Junkes, Salpeter, Rosen, METOX (1957).
(4) Nicholls, Fraser, Jarmain, CF 3, 13 (1959).
(5) Barrow, Deutsch, Travis, Nature 191, 374 (1961).
(6) Törring, ZN 19 a, 1426 (1964).
(7) Drowart, Colin, Exsteen, TFS 61, 1376 (1965).
(8) Nair, Singh, Rai, JCP 43, 3570 (1965).
(9) Hoeft, Lovas, Tiemann, Tischer, Törring, ZN 24 a, 1222 (1969).
(10) Barrow, in DONNSPEC (1970), p.320.
(11) Dube, Upadhya, Rai, JQSRT 10, 1191 (1970).

(continued p. 529)

State	T_e	ω_e	$\omega_e x_e$	B_e	α_e	D_e (10^{-8} cm^{-1})	r_e (Å)	Observed Transitions Design.	Observed Transitions ν_{00}	References
$^{31}P^{(79)}Br$		(μ = 22.2436175)								MAY 1977
b $^1\Sigma^+$	(11782)	[(482.4)] H						b→X,	V 11808.6 H	(1)*
X $^3\Sigma^-(0^+)$	0	[(428.4)] H								
$^{(208)}Pb^{32}S$		(μ = 27.7119401)		D_0^0 = 3.49 eV[a]						JUL 1977
F	47770	370 H	(7.8)					F←X,	R 47729.5[b] H	(2)*
E (0^+)	(34000)	Unclassified bands in the region 3100 - 2750 Å (32200 - 36400 cm^{-1}).						E←X,		(2)*
D 1	29653.2	297.8_3 Z	1.36_5	0.1016_0[c]	0.00064		2.447	D←X,[d]	R 29587.4 Z	(1)(3)
C'(1)	25024.4	283.95 H	1.171					C'←X,	R 24952.3 H	(1)
C (0^+)	23212.9	303.93 H	1.436					C←X,	R 23150.7 H	(1)
B 1	21847.4	282.1_7 H	0.85_6	0.0999_2[c]	0.00060_2		2.467	B←X,[d]	R 21774.5 H	(1)* (3)
A 0^+	18853.0	260.8_3 Z	0.36_3	0.0963_4[c]	0.00026_2		2.513	A←X,[d]	R 18768.9 Z	(1)* (3)
a 1	$14892._9$	285.9 H	(0.88)	0.0926_7	0.00037_4		2.562	a←X,[d]	R 14821.9 Z	(3)
X $^1\Sigma^+$	0	429.40 Z	1.30	0.11631868[ec]	0.00043509[f]	3.415[g]	2.286863	Vibration sp.[h]		(7a)
								Rotation sp.[i]		(5)(9)
$^{(208)}Pb^{(80)}Se$		(μ = 57.732425_8)		D_0^0 = 3.08 eV[a]						JUL 1977
F	45220.9	224.8 H	0.50					F←X,	R 45194.5 H	(3)(4)*
D	28418.0	190.4 H	0.53[b]					D↔X,	R 28374.4 H	(2)* (4)
C	23315.7	183.0 H	0.25					C←X,	R 23268.5 H	(1)
B	21005.8	184.8 H	0.43					B←X,	R 20959.4 H	(1)
A	18716.8	166.9 H	0.14					A←X,[c]	R 18661.5 H	(1)
X $^1\Sigma^+$	0	277.6[d] H	0.51[d]	0.05059953[e]	0.00012993[f]	0.70	2.402233	Vibration sp.[g]		(7a)
								Rotation sp.[h]		(5)

PbO (continued):

(12) Dube, CS 40, 32 (1971).
(12a) Ogden, Ricks, JCP 56, 1658 (1972).
(13) Honerjäger, Tischer, ZN 28 a, 1372 (1973).
(14) Ram, Singh, Upadhya, SpL 6, 515 (1973).
(15) Oldenborg, Dickson, Zare, JMS 58, 283 (1975).
(16) Kurylo, Braun, Abramovitz, Krauss, JRNBS A 80, 167 (1976).
(17) Linton, Broida, JMS 62, 396 (1976).

PBr:

(1) de Bie - Prévot, Thèse (U. Libre de Bruxelles, 1974).

PbS:

[a]Thermochemical value (mass-spectrom.) (2a), revised (6). There appears to be a convergence of the E← X bands near 2715 Å (2). If it is assumed that this limit corresponds to $^3P_1 + {}^3P_1$ (3) one finds $D_0^0 = 3.54$ eV.

[b]Observed value. T_e, ω_e, $\omega_e x_e$ represent only v'=1,2,...; the v'=0 bands are displaced by $\sim$12 cm^{-1} to lower energies owing to a perturbation.

[c]RKR potential functions (4).

[d]Also observed as laser-excited emission in Ne, Ar, Kr, SF_6 matrices (8). Lifetimes in solid Ar have been measured for a (260 μs), A (0.95 μs), B (1.8 μs) (8).

[e]Rotational constants for $^{208}Pb^{32}S$; B_e values for other isotopes and adiabatic corrections (9).

[f] $- 5.3_1 \times 10^{-7}(v+\frac{1}{2})^2 - 5._0 \times 10^{-9}(v+\frac{1}{2})^3$.

[g] $\beta_e = + 0.013 \times 10^{-8}$.

[h]In Ar matrix at 12 K. (8) have also observed the Raman spectrum of the PbS fundamental in solid argon.

[i]Dipole moment of $^{208}Pb^{32}S$, $\mu_{e\ell}(v=0) = 3.5_9$ D, from Stark effect of rotation spectrum (5). A somewhat different value, 4.0_2 D, is given by (7). $g_J = -0.06422$ [Honerjäger and Tischer, quoted in (9)].

(1) Rochester, Howell, PRS A 148, 157 (1935).

PbS (continued):

(2) Vago, Barrow, PPS 59, 449 (1947).
(2a) Colin, Drowart, JCP 37, 1120 (1962).
(3) Barrow, Fry, Le Bargy, PPS 81, 697 (1963).
(4) Nair, Singh, Rai, JCP 43, 3570 (1965).
(5) Hoeft, Lovas, Tiemann, Tischer, Törring, ZN 24 a, 1222 (1969).
(6) Uy, Drowart, TFS 65, 3221 (1969).
(7) Murty, Curl, JMS 30, 102 (1969).
(7a) Marino, Guerin, Nixon, JMS 51, 160 (1974).
(8) Teichman, Nixon, JMS 54, 78; 57, 14 (1975); 65, 258 (1977).
(9) Tiemann, Stieda, Törring, Hoeft, ZN 30 a, 1606 (1975).

PbSe:

[a]Thermochemical value (6) [based on $D_0^0(Se_2) = 3.41$ eV]. This value agrees well with $D_0^0 = 3.10$ eV derived from a dissociation limit at 2870 Å (4) if dissoc. into $^3P_1 + {}^3P_1$ is assumed.

[b] $\omega_e y_e = -0.004$.

[c]Also observed as laser-excited emission in Ne matrices (8).

[d]Average of (1) and (2).

[e]Rotational constants of $^{208}Pb^{80}Se$; data for fourteen other isotopic species (5).

[f] $\gamma_e = - 1.11 \times 10^{-7}$.

[g]In Ar matrix at 12 K.

[h]Dipole moment of $^{208}Pb^{80}Se$, $\mu_{e\ell}(v=0) = 3.2_8$ D, from Stark effect measurements on microwave transitions (7).

(1) Walker, Straley, Smith, PR 53, 140 (1938).
(2) Barrow, Vago, PPS 56, 76 (1944).
(3) Sharma, Nature 157, 663 (1946).
(4) See ref. (2) of PbS.
(5) Hoeft, Manns, ZN 21 a, 1884 (1966).
(6) See ref. (6) of PbS.
(7) Hoeft, Lovas, Tiemann, Törring, ZN 25 a, 539 (1970).
(7a) See ref. (7a) of PbS.
(8) Teichman, Nixon, JMS 57, 14 (1975).

State	T_e	ω_e	$\omega_e x_e$	B_e	α_e	D_e ($10^{-9}cm^{-1}$)	r_e (Å)	Observed Transitions Design.	ν_{00}	References
$^{(208)}Pb^{(130)}Te$		(μ = 79.961030_2)		$D_0^0 = 2.5_5$ eV[a]						JUL 1977
G	46541.7[b]	159.6 H	1.4					G← X, R	46515.3 H	(3)*
F	41658.8	176.4[c] H	1.0					F← X, R	41640.9 H	(3)*
D	27176.5	142.6 H	1.58					D← X, R	27141.5 H	(3)*
B	19737.8	144.9 H	0.45					B↔X, R	19704.3 H	(1)(5)*
A	18405.5	127.0_8 H	0.10_5					A↔X,[d] R	18363.1 H	(5)*
a	14925.5	146.64 H	0.24					a← X, R	14892.9 H	(7)
X $^1\Sigma^+$	0	211.9_6 H	0.43	0.03130774[e]	0.00006743_3[f]	2.7	2.59497_5	Microwave sp.[g]		(8)
$^{31}P^{(35)}Cl$		(μ = 16.4251442)								JUL 1977 A
B	41234	786 H	27	Diffuse bands[a]				B← X, V	41333 H	(1)
b $^1\Sigma^+$	12087	607 H	(3.5)					b→X, V	12102.1 H	(2)
X $^3\Sigma^-(0^+)$	0	577 H	3.5							
$^{(106,108)}Pd_2$		(μ = 53.447166)		$D_0^0 = 0.7_3$ eV[a]						APR 1975
$^{(106)}Pd^{27}Al$		(μ = 21.5030940)		$D_0^0 = 2.6_0$ eV[a]						MAY 1976
$^{(106)}Pd^{(11)}B$		(μ = 9.9725941)		$D_0^0 = 3.3_7$ eV[a]						APR 1975
$^{(106)}Pd^{(74)}Ge$		(μ = 43.534131_2)		$D_0^0 = 2.7_0$ eV[a]						APR 1975
$^{(106)}Pd^1H$		(μ = 0.99832471)								APR 1975
		Complex absorption spectrum 21300 - 24400 cm^{-1}; strong perturbations.								(1)
(A)		Strong band.						(A)← X,(R)	22167 H	(4)

State	T_e	ω_e	$\omega_e x_e$	B_e	α_e	D_e (10^{-} cm^{-1})	r_e (Å)	Observed Transitions Design.	ν_{00}	References
(106) Pd^1H (continued)										
X $^2\Sigma^+$	0							ESR sp.[a]		(3)

PbTe: [a]Thermochemical values of (4) and (6), corrected by (9) for change in value of $D_0^0(Te_2)$.

[b](2) gives T_e = 45918.0.

[c]Slightly different constants in (2).

[d]A different band system in the same region (T_e = 16362.3, ω_e' = 141.4, $\omega_e'x_e'$ = 0.22$_4$) was reported earlier by (1) in absorption but not found by (5).

[e]Rotational constants for $^{208}Pb^{130}Te$.

[f] $-0.56 \times 10^{-7}(v+\frac{1}{2})^2 - 0.3_7 \times 10^{-9}(v+\frac{1}{2})^3$.

[g]Dipole moment $\mu_{e\ell}(v=0) = 2.7_3$ D from Stark effect measurements on pure rotational transitions of $^{208}Pb^{130}Te$ (10). Zeeman effect (11), $g_J(v=0) = -0.0180_0$ μ_N for $^{208}Pb^{130}Te$.

(1) Walker, Straley, Smith, PR 53, 140 (1938).
(2) Sharma, Nature 157, 663 (1946).
(3) Vago, Barrow, PPS 59, 449 (1947).
(4) Pashinkin, Novoselova, RJIC 4, 1229 (1959).
(5) Grove, Ginsburg, SA 16, 730 (1960).
(6) Porter, JCP 34, 583 (1961).
(7) Le Bargy, Barrow, PPS 82, 332 (1963).
(8) Tiemann, Hoeft, Schenk, ZN 24 a, 787 (1969).
(9) Uy, Drowart, TFS 65, 3221 (1969).
(10) Hoeft, Lovas, Tiemann, Törring, ZN 25 a, 539 (1970).
(11) Honerjäger, Tischer, ZN 29 a, 1695 (1974).

PCℓ: [a]Observed in the flash photolysis of $PC\ell_3$.

(1) Basco, Yee, CC (1967), 1146.
(2) de Bie - Prévot, Thèse (U. Libre de Bruxelles, 1974).

Pd_2: [a]Thermochemical value (mass-spectrom.) (1)(2).

(1) Ackerman, Stafford, Verhaegen, JCP 36, 1560 (1962).
(2) Lin, Strauss, Kant, JCP 51, 2282 (1969).

PdAℓ: [a]Thermochemical value (mass-spectrom.) (1).

(1) Cocke, Gingerich, Chang, JCS FT I 72, 268 (1976).

PdB: [a]Thermochemical value (mass-spectrom.) (1).

(1) Vander Auwera-Mahieu, Peeters, McIntyre, Drowart, TFS 66, 809 (1970).

PdGe: [a]Thermochemical value (mass-spectrom.) (1).

(1) Peeters, Vander Auwera-Mahieu, Drowart, ZN 26 a, 327 (1971).

Pd^1H: [a]In rare gas matrices at 4 K.

(1), (3), (4) } See the corresponding ref. of Pd^2H (p. 533).

State	T_e	ω_e	$\omega_e x_e$	B_e	α_e	D_e $(10^{-6}cm^{-1})$	r_e (Å)	Observed Transitions Design.	ν_{00}	References
$^{108}Pd^2H$		$\mu = 1.97719645$								APR 1975
(E) $(^2\Sigma)$				$[2.94]^a$				(E)← X, R	24670.48 Z	(1)(2)*
(D) $(^2\Sigma)$				$[2.12]^a$				(D)← X, R	23866.87 Z	(2)
(C) $(^2\Sigma)$				$[2.57]^a$				(C)← X, R	23483.42 Z	(2)
(B) $(^2\Sigma)$				$[1.99]^a$				(B)← X, R	23073.43 Z	(2)
(A) $(^2\Sigma)$				$[3.09]^a$				(A)← X, R	22264.06 Z	(2)
X $^2\Sigma^+$	0	1446.02 Z	19.59	3.6489^b	0.0812	93.0^c	1.5285_9	ESR sp.[d]		(3)
$^{(106)}Pd^{16}O$		$(\mu = 13.8961401_2)$ $D_0^0 = 2.8_7$ eV[a]								APR 1975
$^{(106)}Pd^{(28)}Si$		$(\mu = 22.1306028)$ $D_0^0 = 3.2_1$ eV[a]								APR 1975
$^{31}P^{19}F$		$\mu = 11.7755965$								JUL 1977
		R shaded triple heads in the region 15600 - 17900 cm^{-1}, tentatively assigned by (4) to a triplet system of PF with $\omega'' \approx 1135$, probably belong to the D→B system of PO (0-0, 0-1, 0-2 bands). Several unassigned bands in the region 16200 - 25500 cm^{-1} (4).								
g $^1\Pi$	[52063.6]			[0.6186]		[0.85]	$[1.521_3]$	g→b, V	38277.74 Z	(1)
								g→a, V	44544.80 Z	(1)
d $^1\Pi$	36024	[413.19] Z	a	0.4848	(0.0062)	[2.8]	1.718	d→b, R	22444.95 Z	(1)*
								d→a, R	28712.14 Z	(1)
B $^3\Pi_2$	29827	[435.86] Z		0.4693	0.0038^e		1.752_2	B→X, R	29623.06 Z	(1)*
B $^3\Pi_1$	29686 b	[435.91] Z		0.4663					29481.80 Z	
B $^3\Pi_0$	29543	[436.06] Z	c	0.4632^d					29338.68 Z	
b $^1\Sigma^+$	13353.90	866.14 Z	4.51	0.5725	0.0045	[0.9]	1.581_3	b→X, V	13363.59 Z	(3)*
a $^1\Delta$	7090.43	858.79 Z	4.438^f	0.5699	0.00467	[1.0]	1.584_9			
X $^3\Sigma^-$	0	846.75 Z	4.489^g	0.5665^h	0.00456		1.589_7			

State	T_e	ω_e	$\omega_e x_e$	B_e	α_e	D_e ($10^{-6}cm^{-1}$)	r_e (Å)	Observed Transitions		References
								Design.	ν_{00}	
$^{31}P^{19}F^+$										JUL 1977
A $^2\Sigma$	35434.64	619.00 Z	4.62	0.5593[i]	0.0079	1.6	1.599_9	A→X,	R 35217.62 Z	(1)
X $^2\Pi_r$	0[j]	1053.25 Z	5.05	0.6360	0.0048	0.6	1.500_3			

Pd^2H: [a]Effective B values at N=0. All levels strongly perturbed. Neither the vibrational numbering nor the number of electronic states involved is known.

[b]Spin splitting constant $\gamma_v = -2.262 - 0.051(v+\frac{1}{2})$; slight N dependence.

[c]$\beta_e = -1.0 \times 10^{-6}$.

[d]In rare gas matrices at 4 K.

(1) Lagerqvist, Neuhaus, Scullman, PPS 83, 498 (1964).

(2) Malmberg, Scullman, Nylén, AF 39, 495 (1969).

(3) Knight, Weltner, JMS 40, 317 (1971).

(4) Scullman, Dissertation (Stockholm, 1971); see USIP Report 71-02.

PdO: [a]Thermochemical value (mass-spectrom.) (1).

(1) Norman, Staley, Bell, JPC 68, 662 (1964); 69, 1373 (1965).

PdSi: [a]Thermochemical value (mass-spectrom.) (1).

(1) Vander Auwera-Mahieu, Peeters, McIntyre, Drowart, TFS 66, 809 (1970).

PF, PF^+:

[a]$\Delta G(3/2...7/2) = 416.57, 418.96, 420.98$.

[b]$A_0 = +143.06$, $A_1 = +142.87$; see (2) who also accounts for spin-rotation interaction and for the combined effects of spin-spin interaction and perturbations by $^1\Pi$ states.

[c]$\Delta G(3/2) = 437.37$.

[d]Λ-type doubling independent of J, $\sim$0.07, 0.16, 0.25 cm^{-1} for v=0,1,2, respectively (1).

[e]Notice that Table VIII of (1) contains a number of misprints leading to disagreement with constants in Table XI.

[f]$\omega_e y_e = +0.0147$.

[g]$\omega_e y_e = +0.019$.

[h]Spin splitting constants $\lambda_0 = +2.9623$, $\gamma_0 = +0.0018$.

[i]Spin splitting constant $\gamma = 0.0073$.

[j]$A = +323.95$.

(1) Douglas, Frackowiak, CJP 40, 832 (1962).

(2) Kovács, CJP 42, 2180 (1964).

(3) Colin, Devillers, Prévot, JMS 44, 230 (1972).

(4) Skolnik, Goodfriend, JMS 50, 202 (1974).

State		T_e	ω_e		$\omega_e x_e$	B_e	α_e	D_e $(10^{-4}cm^{-1})$	r_e (Å)	Observed Transitions Design.		Observed Transitions ν_{00}		References
$^{31}P^{1}H$			$\mu = 0.97606596$			$D_0^0 = (3.02)$ eV^{a}								JUL 1977 A
d	$^1\Pi$					$[8.47_8]^{b}$		$[4.1_7]$	$[1.427_3]$	d← a,		62725.28	Z	(11)*
B	$^3\Pi_0$					[7.3]			$[1.5_4]$	B← X,	R	69587.8	Z	(11)
c	$^1\phi$					$[8.60_2]$	0.21	$[5.4_4]$	$[1.417_0]$	c← a,^{c}		61548.6_8	Z	(11)*
	0^+		[1833.78]	Z								29434.61^{j}		
A $^3\Pi_i$	0^-	29498^{d}	[1833.39]	Z	$(98._5)^{e}$	$[8.0222]^{f}$	g	$[5.683]^{h}$	$[1.4672_8]$	A^{i}↔ X,	R	29434.28^{j}		(1)*(3)*(4)* (9)(13)
	1		[1833.74]	Z								29316.81^{j}		
	2		[1834.38]	Z								29203.21^{j}		
b	$^1\Sigma^+$	$(15160)^{k}$												
a	$^1\Delta$	$(7660)^{k}$				[8.443]	0.12	$[4.1_8]$	[1.4302]	Rotation sp.$^{\ell}$				(14)
X	$^3\Sigma^-$	0	2365.2^{m}		44.5^{m}	8.5371^{n}	0.2514	4.36	1.4223_4	Rotation sp.$^{\ell}$				(14)
$^{31}P^{2}H$			$\mu = 1.89112949$			$D_0^0 = (3.06)$ eV^{o}								JUL 1977
	0^+		[1357.40]	Z								29495.66^{j}		
A $^3\Pi_i$	0^-	29505^{p}	[1357.14]	Z	$(50.8)^{e}$	$[4.1720]^{f}$	q	$[1.506]^{r}$	[1.4617]	A→X,	R	29495.37^{j}		(3)* (13)
	1		[1357.21]	Z								29377.90^{j}		
	2		[1357.54]	Z								29264.21^{j}		
X	$^3\Sigma^-$	0	1699.2^{m}		23.0^{m}	4.4081^{s}	0.0928	1.16	1.4220					
$^{31}P^{1}H^{+}$						$D_0^0 \leq 3.36$ eV^{a}								JUL 1977
A	$^2\Delta_r$	$(26221)^{b}$	[1398.76]	Z		$[6.983_3]^{c}$	d	[6.28]	[1.5726]	A→X,	R	25770.59^{e}	Z	(1)*
X	$^2\Pi_r$	0^{f}	[2299.60]	Z		$[8.385_1]^{g}$	h	[4.16]	[1.4352]					
$^{31}P^{2}H^{+}$														JUL 1977
A	$^2\Delta_r$	$(26259)^{i}$	[1017]	H		$[3.635]^{j}$		[1.71]	$[1.566_0]$	A→X,	R	25792 26077	H^{Q}	(2)*
X	$^2\Pi_r$	0^{k}	[1666]	H		$[4.350_5]^{\ell}$		[1.16]	[1.4314]					

P^1H, P^2H:

[a]Adjusted theoretical value recommended by (15); see also (6)(8)(12).

[b]Λ-type doubling $|\Delta\nu| = 0.1275\,J(J+1) - 0.54 \times 10^{-4} J^2(J+1)^2$.

[c]Sequence of nearly undegraded bands; the origins of the 1-1 and 2-2 bands are at 61554.5 and 61560.7 cm^{-1}, resp..

[d]$A_0 = -115.71$, $A_1 = -115.20$; see (13) who give also centrifugal distortion corrections A_D as well as estimated spin-spin and second-order spin-orbit parameters.

[e]Estimated using isotope relations.

[f]For Λ-doubling constants see (13).

[g]$B_1 = 7.549_2$. No emission has been observed from v=1 of P^1H, probably owing to weak predissociation by the repulsive $^5\Sigma^-$ state arising from ground state atoms (13).

[h]$D_1 = 6.54 \times 10^{-4}$, $H_0 = -1.6 \times 10^{-8}$.

[i]Lifetime 0.45 μs, corresponding to an absorption oscillator strength of 0.0078 (7).

[j]Subband origins as defined by (13).

[k]Theoretical predictions (6)(10)(12), for $a^1\Delta$ confirmed by laser photoelectron spectrometry of PH^-[see ref.(3) of PH^-].

[ℓ]$N = 4 \rightarrow 5$ rotational transitions observed by the laser magnetic resonance method.

[m]Constants deduced from isotope relations (13).

[n]Spin splitting constants $\lambda_0 = +2.21_2$, $\gamma_0 = -0.073_8$; $\lambda_1 = +2.20_7$, $\gamma_1 = -0.072_6$ (13).

[o]From the value for P^1H.

[p]$A_0 = -115.74$, $A_1 = -115.55$; see also [d].

[q]$B_1 = 4.0047$.

[r]$D_1 = 1.640 \times 10^{-4}$.

[s]Spin splitting constants $\lambda_0 = +2.211$, $\gamma_0 = -0.0385$; $\lambda_1 = +2.20_2$, $\gamma_1 = -0.038_1$ (13).

(1) Pearse, PRS A 129, 328 (1930).

P^1H, P^2H (continued):

(2) Ishaque, Pearse, PRS A 156, 221 (1936).
(3) Ishaque, Pearse, PRS A 173, 265 (1939).
(4) Legay, CJP 38, 797 (1960).
(5) Kovács, APH 13, 303 (1961).
(6) Jordan, JCP 41, 1442 (1964).
(7) Fink, Welge, ZN 19 a, 1193 (1964).
(8) Cade, Huo, JCP 47, 649 (1967).
(9) Horani, Rostas, Lefebvre-Brion, CJP 45, 3319 (1967).
(10) Cade, CJP 46, 1989 (1968).
(11) Balfour, Douglas, CJP 46, 2277 (1968).
(12) Liu, Legentil, Verhaegen, in "Selected Topics in Molecular Physics" (ed. Clementi), p.19. Chemie GmbH (1972).
(13) Rostas, Cossart, Bastien, CJP 52, 1274 (1974).
(14) Davies, Russell, Thrush, CPL 36, 280 (1975).
(15) Meyer, Rosmus, JCP 63, 2356 (1975).

P^1H^+, P^2H^+:

[a]From the predissociation in A $^2\Delta(v=0)$. A rough extrapolation of the A state to the dissociation limit $^1D + ^2S$ leads to $D_0^0 \approx 3.0_6$ eV (1).

[b]$A_0 = +1.38$, $A_1 = +0.82$.

[c]Spin-rotation interaction constant $\gamma = 0.175$. There is a sudden breaking off in v=0 above N=12.

[d]$B_1 = 6.558_8$. All lines originating from F_1 levels of v=1 are much weaker than those arising from F_2.

[e]Refers to the zero-point of the Hill-Van Vleck expression for both upper and lower state.

[f]$A_0 = +295.9_4$, $A_1 = +296.2$.

[g]Λ-type doubling constants for v=0: $|p| = 0.23$, $|q| = 0.011$.

(continued p. 537)

State	T_e	ω_e	$\omega_e x_e$	B_e	α_e	D_e ($10^{-6}cm^{-1}$)	r_e (Å)	Observed Transitions Design.	Observed Transitions ν_{00}	References
$^{31}P^1H^-$				$D_0^0 = (3.29)$ eV[a]		I.P. = 1.02_8 eV[b]				JUL 1977
X $^2\Pi_i$	0[c]	[2230][b]					[1.407][d]			
$^{31}P^{127}I$		$\mu = 24.8970930$								JUL 1977
b $^1\Sigma^+$	11528.7	407.8 H	2.9					b→X,	V 11541.7 H	(1)*
X $^3\Sigma^-$	0	381.7 H	2.9							
$^{31}P^{14}N$		$\mu = 9.64336165_5$		$D_0^0 = 6.3_6$ eV[a]		I.P. = 11.8_5 eV[b]				JUL 1977 A
A $^1\Pi$	39805.6_6	1103.09 Z	7.222	0.7307_1[c]	0.00663	1.29	1.5467	A[d]↔X,[e]	R 39688.5_2 Z	(1)* (2)
X $^1\Sigma^+$	0	1337.24 Z	6.983	0.7864854[c]	0.0055364[f]	1.091	1.490866	Microwave sp.		(8)(9)
								Mol. beam rf el. reson.[g]		(7)
$^{31}P^{14}N^+$				$D_0^0 = 5.0_0$ eV[h]						JUL 1977 A
		In the photoelectron spectrum there are indications of two additional peaks at 3.9_5 and 4.8_5 eV (31900 and 39100 cm^{-1}, resp.) above X $^2\Sigma^+$ (12); uncertain.								
A $^2\Pi$	(4030)	(1050)[b]								
X $^2\Sigma^+$	0	[(1200)][b]								

P^1H^+, P^2H^+ (continued):

[h] $B_1 = 8.145_0$.
[i] $A_0 = + 1.35$.
[j] Spin-rotation interaction constant $\gamma = 0.096$.
[k] $A_0 = + 295.83$.
[ℓ] Λ-type doubling constant $|p| = 0.08$.

(1) Narasimham, CJP 35, 901 (1957).
(2) Narasimham, Dixit, CS 36, 1 (1967).

P^1H^-: [a] From $D_0^0(P^1H)$ and the electron affinities of PH and H. Notice, however, that the ground $^2\Pi$ state of PH^- cannot dissociate into $P(^4S) + H^-(^1S)$ but must correlate with the slightly higher limit $P^-(^3P) + H(^2S)$ at 3.31 eV. The atomic electron affinities are taken from (2).
[b] From the photodetachment spectrum (3).
[c] $A = -212\ cm^{-1}$ [theoretical value (1)].
[d] Franck-Condon factor analysis of the photodetachment spectrum (3).

(1) Walker, Richards, JCP 52, 1311 (1970).
(2) Hotop. Lineberger, JPCRD 4(3), 539 (1975).
(3) Zittel, Lineberger, JCP 65, 1236 (1976).

PI: (1) de Bie - Prévot, Thèse (U. Libre de Bruxelles, 1974).

PN, PN^+:

[a] Latest thermochemical value (mass-spectrom.) (6). Previous thermochemical value 7.57 eV (5). The origin of the discrepancy is not clear. From ab initio calculations a value of 6.43 eV is obtained; see (6).
[b] From the photoelectron spectrum (12).
[c] Potential functions (3).
[d] Lifetime $\tau(v=0) = 0.23\ \mu s$ [Hanle effect measurement (11)].
[e] Relative transition probabilities from the fluorescence spectrum, Franck-Condon factors (10); see also (4).
[f] $- 6.4_0 \times 10^{-6}(v+\frac{1}{2})^2 - 2.5 \times 10^{-7}(v+\frac{1}{2})^3$ (9).
[g] $\mu_{e\ell}[D] = 2.7514 - 0.0086(v+\frac{1}{2})$, as corrected by (9) for improved B_v values; $(eqQ)_N[kHz] = -5172.8 + 60.7(v+\frac{1}{2})$. (7) give also magnetic hf coupling constants.
[h] From $D_0^0(PN)$, I.P.(PN), and I.P.(P).

(1) Curry, Herzberg, Herzberg, ZP 86, 348 (1933).
(2) Moureu, Rosen, Wetroff, CR 209, 207 (1939).
(3) Singh, Rai, IJPAP 4, 102 (1966).
(4) Smith, JP B 1, 89 (1968).
(5) Uy, Kohl, Carlson, JPC 72, 1611 (1968).
(6) Gingerich, JPC 73, 2734 (1969).
(7) Raymonda, Klemperer, JCP 55, 232 (1971).
(8) Hoeft, Tiemann, Törring, ZN 27 a, 703 (1972).
(9) Wyse, Manson, Gordy, JCP 57, 1106 (1972).
(10) Moeller, Silvers, CPL 19, 78 (1973).
(11) Moeller, McKeever, Silvers, CPL 31, 398 (1975).
(12) Wu, Fehlner, CPL 36, 114 (1975).

State	T_e	ω_e	$\omega_e x_e$	B_e	α_e	D_e $(10^{-6}cm^{-1})$	r_e (Å)	Observed Transitions Design.	Observed Transitions ν_{00}	References
$^{31}P^{16}O$		$\mu = 10.5479381_1$ $D_0^0 = 6.15$ eV[a] I.P. = (8.2_3) eV[b]								SEP 1977
		Theoretical calculations of the ground state (11)(21), of low-lying valence states (29)(30), and of Rydberg states (22).								
H $^2\Sigma^+$	(56017)[c]	(1391)	(7)	(0.780)	(0.005_4)		(1.431)	H→B, R_V	(25401)[d]	(20a)* (33)*
I $^2\Sigma^+$	55458.1	1390.2 Z	6.0	0.7798[e]	0.0048	0.8	1.4316	I→A,	15051.1 Z	(24)*
								I→B, V	24842.0	(20a)* (33)*
								I→X, V	55536.7	(33)*
E $^2\Delta$	53091	[1456.28] Z	(15.8)	0.758_3[f]	0.0074	0.8	1.451_8	E→X, V	53215.6 Z	(1)* (25)* (35)*
G $^2\Sigma^+$	(52412)[c]	(1382)	(13)	(0.780)	(0.005_4)		(1.431)	G→A,	(11999)[d]	(31)*
								G→B, R_V	(21790)[d]	(20a)* (33)*
								G→X, V	(52484)[d]	(33)*
F $^2\Sigma^+$	(49880)[c]	(850)	$(7._5)$	(0.608_2)	(0.0045)		(1.621)	F→A, R	(9202)[d]	(33)*
								F→B, R	(18993)[d]	(20a)* (24) (33)*
								F→X R	(49688)[d]	(33)*
D $^2\Pi_r$	48520[g]	[1358.1]	(7)	0.755[h]	0.007		1.455	D→B, R	17894.6[d]	(15)(18)* (34)*
								D↔X, R	48589.3[d]	(1)(15)(18)* (40)*
C $^2\Sigma^-$	44831.7_5	779.2_2[i] Z	5.14	0.590_3[i]	0.0056	(1.4)	1.645	C→X, R	44605.0_5 Z	(1)(20)(23)* (32)*
C' $^2\Delta$	43742.7_4	825.7_4 Z	6.93[j]	0.640_5[k]	0.0052	(1.5)	1.580	C'→X, R	43538.8_5 Z	(5)(6)(8)* (16)(20) (23)* (32)*
A $^2\Sigma^+$	40406.8_9	1390.9_4[ℓ] Z	6.91[ℓ]	0.7801[ℓmn]	0.0054_2[ℓ]	1.0	1.4313	A→B, V_R	9790.86 Z	(20a)* (24)* (31)*
								A↔X,[o] γ bands V	40485.60 Z	(2)* (4)(12) (16)* (36)*

State	T_e	ω_e	$\omega_e x_e$	B_e	α_e	D_e ($10^{-6}cm^{-1}$)	r_e (Å)	Observed Transitions Design.	v_{00}	References
$^{31}P^{16}O$ (continued)										
$B'\ ^2\Pi$	33120.7[p]	759.2[q]	3.8_5[q]	0.5420[q]	0.0049[q]	[1.0]	1.717_2	$B' \rightarrow B$, [r] R		(34)*
								$B' \leftrightarrow X$, R	32884.3[q]	(19)(20)(37) (40)*

PO: [a]Thermochemical value (mass-spectrom.) (27). The predissociation in the perturbed D(v=0) level (see [h]) gives the upper limit $D_0^0 \leq 6.161$ eV.

[b]Theoretical value (22). (27) report an electron impact appearance potential of $8._5$ eV.

[c]The G, H, I $^2\Sigma^+$ Rydberg states (as well as higher levels of A $^2\Sigma^+$) interact with the F $^2\Sigma^+$ non-Rydberg state. Most of the observed bands are strongly perturbed. The constants in the Table are deperturbed values taken from (33) who give results for $P^{16}O$ and $P^{18}O$; similar constants have earlier been reported by (20a). In addition to the strong homogeneous interactions a large number of local perturbations have been described for both isotopes (20a)(24)(33).

[d]The v_{00} values have been calculated from deperturbed constants and should not be expected to coincide precisely with observed transitions.

[e]Rotational perturbations in v=0,1,2 (24).

[f]Perturbations by C' $^2\Delta$.

[g]$A_0 = +26.2$, $A_1 = +22.8$. All constants given for D $^2\Pi$ (v=0,1) are deperturbed values from (40); see also (34). This state interacts strongly with high vibrational levels of B' $^2\Pi$. For v=2 and 3 only fragments have been observed in absorption (40) making the deperturbation results for these levels even less reliable.

[h]The highest level observed in emission from the interacting pair D(v=0)$\sim$B'(v=24?) is the N=34 level [predominantly D(v=0)] lying at 49647 cm^{-1} above X $^2\Pi_{\frac{1}{2}}$(v=0,J=½); see (15). Higher levels are predissociated and give rise to diffuse lines in the absorption spectrum. A summary of predissociation phenomena observed in B' and D levels is given by (40) who suggest that the predissociation is caused by the repulsive part of the $^4\Pi$ state arising from $^4S + {}^3P$.

[i]Vibrational numbering confirmed by $P^{16}O - P^{18}O$ isotope shifts (32).

[j](35) suggest that an additional term, $\omega_e y_e \approx +0.040$, is needed in order to account for higher vibrational levels (v=14...18) found in perturbations of E $^2\Delta$(v=0,1,2).

[k]Small spin doubling, $\gamma \approx 0.0085$ (32); local perturbations.

[ℓ]Average of values obtained by (2)(31)(36). Above v=3 the influence of the F $^2\Sigma^+$ state becomes noticeable.

[m]Spin splitting constant $\gamma = +0.0013$ (31). Many rotational perturbations (26)(31)(36).

[n]Potential curves (7).

[o]Franck-Condon factors (10).

[p]A_v increases from -13.3 for v=0 to about +30 for v=23...26 (19)(34)(40); theoretical explanation (29).

[q]These are the constants of (31) based on the rotational (continued p. 540)

State	T_e	ω_e	$\omega_e x_e$	B_e	α_e	D_e $(10^{-6}cm^{-1})$	r_e (Å)	Observed Transitions Design.	Observed Transitions ν_{00}	References
$^{31}P^{16}O$ (continued)										
b $^4\Sigma^-$	$(34837)^s$	$(889._0)^s$	$(6.6_2)^s$	$(0.644)^s$	$(0.006)^s$		(1.57_5)			
B $^2\Sigma^+$	30730.88	1164.51 Z	13.46^t	0.7463^{un}	0.0088^v	1.25	1.4634	$B \leftrightarrow X$, V_R β bands	30694.74 Z	(1)* (4)(9)* (14)(16)* (28)(37)* [w]
X $^2\Pi_r$	0^x	1233.34 Z	6.56	0.7337^{yn}	0.0055	1.3	1.4759			
$^{31}P^{16}O^+$				$D_0^0 = (8.4_1)$ eV[a]						JUL 1977
A	49930	1017 H	8					$A \rightarrow X$, R	49735 H	(1)
X $^1\Sigma^+$	0	1405 H	(5)							
$^{31}P^{16}O^-$				$D_0^0 = 5.78$ eV[a]		I.P. = 1.09_2 eV[b]				JUL 1977
		SCF calculations (1).								
a $^1\Delta$	4470	$[1020]^b$								
X $^3\Sigma^-$	0	$[1000]^b$					1.54_0[c]			
$^{210}Po_2$		$\mu = 104.991442$		$D_0^0 = (1.90)$ eV[a]						JUL 1977
A	25149.3	108.532 H	0.4417					$A \rightarrow X$, R	25125.7 H	(1)
X	0	155.715 H	0.3353^b							

PO (continued):

analysis of v=0 and 1 (19) and on the identification of several intermediate levels (v=6 and 12...22) in perturbations with B $^2\Sigma^+$ (37) and A $^2\Sigma^+$ (31). Higher vibrational levels (v=23...26, formerly D' $^2\Pi$) are observed in the region of strong interaction with the Rydberg D $^2\Pi$ state (34)(40). The vibrational numbering chosen by (19) and used in this Table is arbitrary and may have to be increased by 2 as suggested by (41) whose re-analysis of the B' − X system includes a large number of absorption bands previously attributed to OPCℓ (39) as well as a number of emission bands left unassigned by (20) and (37).

[r]The B' → B bands originate from highly excited levels with v' ≈ 24 which are strongly mixed with low vibrational levels of D $^2\Pi$.

PO (continued):

[s]Constants derived from perturbations in B $^2\Sigma^+$ and A $^2\Sigma^+$ (37), corrected for the new vibrational numbering of (36).

[t]$\omega_e y_e$ = - 0.05.

[u]Spin splitting constant γ = -0.0068. Extensive rotational perturbations in v=6,7 by levels of B' $^2\Pi$ and b $^4\Sigma^-$; see (37). The v=7 level appears to be preferentially populated in the P_4 + O glow (20).

[v]γ_e = -0.00006.

[w]Papers by (3)(13)(17) contain erroneous branch identifications; see (18)(28).

[x]$A_v = 224.03 + 0.18\,v - 0.013\,v^2$ ($v \leq 11$) (37); see also (38). Similar results in (32).

[y]Λ-type doubling $\Delta v_{ef}(^2\Pi_{\frac{1}{2}}) = -0.0070(J+\frac{1}{2})$.

(1) Dressler, HPA 28, 563 (1955).
(2) Rao, CJP 36, 1526 (1958).
(3) Singh, CJP 37, 136 (1959).
(4) Norrish, Oldershaw, PRS A 262, 10 (1961).
(5) Santaram, Rao, ZP 168, 553 (1962); IJP 37, 14 (1963).
(6) Narasimham, Dixit, Sethuraman, PIAS A 62, 314 (1965).
(7) Singh, Rai, JPC 69, 3461 (1965).
(8) Guenebaut, Couet, Coquart, JCPPB 63, 969 (1966).
(9) Meinel, Krauss, ZN 21 a, 1520, 1878 (1966).
(10) Sankaranarayanan, IJP 40, 678 (1966).
(11) Boyd, Lipscomb, JCP 46, 910 (1967).
(12) Coquart, Couet, Ngo, Guenebaut, JCPPB 64, 1197 (1967).
(13) Mohanty, Upadhya, Singh, Singh, JMS 24, 19 (1967).
(14) Couet, Ngo, Coquart, Guenebaut, JCPPB 65, 217 (1968).
(15) Couet, Coquart, Ngo, Guenebaut, JCPPB 65, 1241 (1968).
(16) Dixit, Narasimham, PIAS A 68, 1 (1968).
(17) Mohanty, Rai, Upadhya, PIAS A 68, 165 (1968).
(18) Verma, Dixit, CJP 46, 2079 (1968).

PO (continued):

(19) Verma, CJP 48, 2391 (1970).
(20) Verma, Broida, CJP 48, 2991 (1970).
(20a) Verma, Dixit, Jois, Nagaraj, Singhal, CJP 49, 3180 (1971).
(21) Mulliken, Liu, JACS 93, 6738 (1971).
(22) Ackermann, Lefebvre-Brion, Roche, CJP 50, 692 (1972).
(23) Coquart, Couet, Guenebaut, Larzillière, Ngo, CJP 50, 1014 (1972).
(24) Guha, Jois, Verma, CJP 50, 1579 (1972).
(25) Coquart, Larzillière, Ngo, CJP 50, 2945 (1972).
(26) Coquart, Prudhomme, CR B 275, 383 (1972).
(27) Drowart, Myers, Szwarc, Vander Auwera-Mahieu, Uy, JCS FT II 68, 1749 (1972).
(28) Rai, Rai, Upadhya, JP B 5, 1038 (1972).
(29) Roche, Lefebvre-Brion, JCP 59, 1914 (1973).
(30) Tseng, Grein, JCP 59, 6563 (1973).
(31) Verma, Jois, CJP 51, 322 (1973).
(32) Prudhomme, Larzillière, Couet, CJP 51, 2464 (1973).
(33) Ngo, Da Paz, Coquart, Couet, CJP 52, 154 (1974).
(34) Coquart, Da Paz, Prudhomme, CJP 52, 177 (1974).
(35) Prudhomme, Coquart, CJP 52, 2150 (1974).
(36) Coquart, Da Paz, Prudhomme, CJP 53, 377 (1975).
(37) Verma, Singhal, CJP 53, 411 (1975).
(38) Zaidi, Verma, CJP 53, 420 (1975).
(39) Verma, Nagaraj, JMS 58, 301 (1975).
(40) Ghosh, Nagaraj, Verma, CJP 54, 695 (1976).
(41) Cornet, Dubois, Houbrechts, JP B 10, L415 (1977).

PO^+:
PO^-:
Po_2:
See p. 543.

System	ν_e	ω_e'		$\omega_e'x_e'$	ω_e''		$\omega_e''x_e''$	Remarks	Degrad.	ν_{00}		References
$^{141}Pr^{16}O$	$\mu = 14.3643666_1$	$D_0^0 = 7.7_4$ eV[a]			I.P. = 4.9_0 eV[b]							JUL 1977 A
The following classifications and analyses are by (8). Most of the bands are observed in emission and absorption.												
XXII									R	18961.2	H	(8)
XXI	18703.6	786.4	H	4.07	834.0	H	2.56	Different analysis by (1).	R	18679.5	H	(1)(2)* (8)
XX	17886.5	785.4	H	2.93	832.2	H	2.01		R	17863.0	H	(2)* (8)
XIX	17588.0	791.8	H	4.00	833.1	H	3.05		R	17567.1	H	(2)* (8)
XVIII									R	17345.8	H	(2)* (8)
XVII	16631.4	790.0	H	4.49	831.9	H	1.75		R	16609.6	H	(2)* (8)
XVI	15464.6	785.9	H	2.70	830.7	H	2.00		R	15442.0	H	(2)* (8)
XV	14461	[786.1]	H		[830.9]	H			R	14438.5	H	(2)* (8)
XIV	14436	[730.0]	H		835.8	H	2.4		R	14384.0	H	(2)* (8)
XIII	14112.8	789.6	H	2.76	835.0	H	2.25		R	14090.1	H	(8)
XII	14045	[754.0]	H		[830.8]	H			R	14006.1	H	(8)
XI	13678	[788.0]	H		[829.4]	H			R	13656.9	H	(2)* (8)
X	13079	[754.4]	H		835.8	H	2.35	Single P, Q, R branches; rot. analysis.[c]	R	13038.62	Z	(2)* (8)
IX[d]	12756.0	745.0	H	2.27	835.1	H	1.92		R	12710.8	H	(2)* (8)
VIII	11971	[733.5]	H		[831.0]	H			R	11922.0	H	(8)
VII[d]	11815.8	[741.03]	Z	2.16 H	[831.78]	Z	2.22 H	Single P, Q, R branches; rot. analysis.[c]	R	11770.33	Z	(2)*(7)* (8)
VI	11150	[754.0]	H		836.4	H	2.45		R	11109.3	H	(8)
V	11021	[742]	H		[830.7]	H			R	10976.2	H	(8)
IV	10482	[740.2]	H		[829.6]	H			R	10437.6	H	(8)
III	10240.2	764.5	H	2.60	835.8	H	3.25		R	10204.7	H	(8)
II	10048	[728.1]	H		[831.2]	H			R	9996.0	H	(8)
I	9281	[735.3]	H		[830.5]	H			R	9233.0	H	(8)
	From the IR spectrum of PrO in an Ar matrix: $\Delta G(\frac{1}{2}) = 817.1$ for $Pr^{16}O$, and 775.4 for $Pr^{18}O$.											(9)

PrO: [a]Thermochemical value (mass-spectrom.)(3)(4), recalculated (5)(6).

[b]Corrected electron impact appearance potential (10).

[c]Rotational constants obtained by (8):

System X: $B_0' = 0.3459$, $D_0' = 3.1 \times 10^{-7}$, $r_0' = 1.842$;
$B_0'' = 0.3620$, $D_0'' = 2.4 \times 10^{-7}$, $r_0'' = 1.801$.

System VII: $B_1' = 0.3418$, $D_1' = 2.4 \times 10^{-7}$;
$B_0' = 0.3414$, $D_0' = 1.1 \times 10^{-7}$, $r_0' = 1.854$;
$B_1'' = 0.3595$, $D_1'' = 2.6 \times 10^{-7}$;
$B_0'' = 0.3610$, $D_0'' = 2.6 \times 10^{-7}$, $r_0'' = 1.803$;

(B, D in cm^{-1}, r in Å). The upper state of system VII is perturbed. Different rotational constants and band origins have been obtained by (7).

[d](7) suggest that systems VII and IX form the two components of a doublet system.

(1) Watson, PR 53, 639 (1938).
(2) Gatterer, Junkes, Salpeter, Rosen, METOX (1957).
(3) Walsh, Dever, White, JPC 65, 1410 (1961).
(4) Ames, Walsh, White, JPC 71, 2707 (1967).
(5) Smoes, Coppens, Bergman, Drowart, TFS 65, 682 (1969).
(6) Bergman, Coppens, Drowart, Smoes, TFS 66, 800 (1970).
(7) Venkitachalam, Krishnamurty, Narasimham, PIAS A 76, 113 (1972).
(8) Shenyavskaya, Egorova, Lupanov, JMS 47, 355 (1973).
(9) Gabelnick, Reedy, Chasanov, JCP 60, 1167 (1974).
(10) Ackermann, Rauh, Thorn, JCP 65, 1027 (1976).

PO^+: [a]$D_0^0(PO) + I.P.(P) - I.P.(PO)$.

(1) Dressler, HPA 28, 563 (1955).

PO^-: [a]From $D_0^0(PO)$ and the electron affinities of PO and O.

[b]From the laser photoelectron spectrum of PO^- (2).

[c]Franck-Condon factor analysis of the PO^- photodetachment peaks (2).

(1) Boyd, Lipscomb, JCP 46, 910 (1967).
(2) Zittel, Lineberger, JCP 65, 1236 (1976).

Po_2: [a]Extrapolation of the lower state vibrational levels.

[b]$\omega_e y_e = - 0.0003226$.

(1) Charles, Timma, Hunt, Pish, JOSA 47, 291 (1957).

State	T_e	ω_e	$\omega_e x_e$	B_e	α_e	D_e ($10^{-7}cm^{-1}$)	r_e (Å)	Observed Transitions Design.	ν_{00}	References
$^{141}Pr^{(32)}S$		(μ = 26.0592154)		$D_0^0 = 5.2_1$ eV[a]						APR 1975
$^{31}P^{32}S$		μ = 15.7325008		$D_0^0 = 4.5_4$ eV[a]						JUL 1977
C $^2\Sigma$	34686.5	534.8 H^Q	3.31	0.2644_2[b]	0.00196	2.5	2.013_0	C→X, R	34263.5 34584.3 H^Q	(1)(2)(3)* (4)*
B $^2\Pi$	22987.7 22894.0	512.2 H	2.15					B→X, R	22553.7 22780.8 H	(2)(3)*
X $^2\Pi_r$	320.8[c] 0	739.1 H^Q	2.96	[0.2967_4] [0.2963_2]		[2.0]	[1.900_9]			
$^{31}P^{32}S^+$										JUL 1977
A ($^1\Sigma$)	40617.5	607.5 H	4.5					A→X, R	40498.7 H	(1)(2)
X $^1\Sigma^+$	0	844.6 H	3.3							
$^{31}P^{(80)}Se$		(μ = 22.3222031)		$D_0^0 = 3.7_3$ eV[a]						JUL 1977
$^{(194,195)}Pt_2$		(μ = 97.231234)								AUG 1976
A		218.4[a]	0.9					A←X,[a]	11248.7	(1)
X	0									
$^{(195)}Pt^{(11)}B$		(μ = 10.4208585)		$D_0^0 = 4.9_1$ eV[a]						MAY 1975
$^{195}Pt^{12}C$		μ = 11.30422952		D_0^0 = 6.28 eV[a]						MAY 1975
B $^1\Sigma^{(+)}$	(32779)	[843.8] H	bc	0.468[c]	0.006	7	1.78_5	B←X, R	32676.8 H	(3)* (5)(6)*
A $^1\Pi$	18627.01	818.74 Z	5.44	0.48023[d]	0.00411	6.7	1.7621_9	A←X, R	18510.67 Z	(1)(3)* (6)
A' $^1\Pi$	13262.8	[906.93] Z	5.6 H	0.50584[e]	0.00390	6.20[f]	1.7170_0	A'↔X, R	13196.14 Z	(2)(5)(6)
A" $^1\Sigma^{(+)}$	12697.16	943.40 Z	5.28	0.50957	0.00370_5	6.04[g]	1.7107_1	A"↔X, R	12643.19 Z	(5)(6)*
X $^1\Sigma^{(+)}$	0	1051.13 Z	4.86	0.53044	0.003273	5.46[h]	1.6767_2			

State	T_e	ω_e	$\omega_e x_e$	B_e	α_e	D_e (10^{-} cm^{-1})	r_e (Å)	Observed Transitions Design. ν_{00}	References
$^{31}P(^{130})Te$	(μ = 25.0104736)		$D_0^0 = 3.0_5$ eV[a]						JUL 1977

PrS: [a]Thermochemical value (1), recalculated (2).

(1) Cater, Holler, Fries, quoted in ref. (5) of PrO.
(2) See ref. (6) of PrO.

PS, PS^+:

[a]Thermochemical value (mass-spectrom.) (5).
[b]Spin doubling constant $\gamma = 0.015_2$.
[c]$A_0 = +321.9_3$, from the rotational analysis of the C→X 1-0 and 2-0 bands (4).

(1) Dressler, Miescher, PPS A 68, 542 (1955).
(2) Dressler, HPA 28, 563 (1955).
(3) Narasimham, Subramanian, JMS 29, 294 (1969).
(4) Narasimham, Balasubramanian, JMS 37, 371 (1971).
(5) Drowart, Myers, Szwarc, Vander Auwera-Mahieu, Uy, HTS 5, 482 (1973).

PSe: [a]Thermochemical value (mass-spectrom.)(1), based on $D_0^0(Se_2) = 3.2_9$ eV.

(1) See ref. (5) of PS.

Pt_2: [a]In Ar matrix at 12 K; not observed in the gas phase.

(1) Jansson, Scullman, JMS 61, 299 (1976).

PtB: [a]Thermochemical value (mass-spectrom.)(1).

(1) McIntyre, Vander Auwera-Mahieu, Drowart, TFS 64, 3006 (1968).

PtC: [a]Thermochemical value (mass-spectrom.) (4)(7).
[b]$\Delta G(3/2) \approx 808.9$.
[c]B $^1\Sigma$ is strongly perturbed.
[d]Λ-type doubling $\Delta\nu_{ef}(v=0) = +0.64 \times 10^{-4} J(J+1)$.
[e]Λ-type doubling $\Delta\nu_{ef}(v=0) = +2.18 \times 10^{-4} J(J+1)$.
[f]$H_0 = +7.0 \times 10^{-13}$.
[g]$\beta_e = +0.06 \times 10^{-7}$; $H_0 = +5.4 \times 10^{-13}$, $H_1 = +6.6 \times 10^{-13}$.
[h]$\beta_e = +0.03 \times 10^{-7}$; $H_0 \ldots H_2 (10^{-13} cm^{-1}) = +8.6, +9.9, +8.$

(1) Neuhaus, Scullman, Yttermo, ZN 20 a, 162 (1965).
(2) Appelblad, Barrow, Scullman, PPS 91, 260 (1967).
(3) Scullman, Yttermo, AF 33, 231 (1967).
(4) Vander Auwera-Mahieu, Drowart, CPL 1, 311 (1967).
(5) Scullman, Dissertation (Stockholm, 1971); see USIP Report 71-02.
(6) Appelblad, Nilsson, Scullman, PS 7, 65 (1973).
(7) Gingerich, CPL 23, 270 (1973).

PTe: [a]Thermochemical value (mass-spectrom.)(1), based on $D_0^0(Te_2) = 2.68$ eV.

(1) See ref. (5) of PS.

State	T_e	ω_e	$\omega_e x_e$	B_e	α_e	D_e ($10^{-4} cm^{-1}$)	r_e (Å)	Observed Transitions: Design.	Observed Transitions: ν_{00}	References
(195)Pt^1H		(μ = 1.00264229)		$D_0^0 \leq 3.44_0$ eV[a]						MAY 1975
IV $(^2\Sigma)$[b]		[2051] H		8.03	0.40	[4.4][c]	1.447	IV ← III, R	36504 H	(6)
III $(^2\Sigma)$[b]				[9.51]		[3.7]	[1.330]			
II $(^2\Sigma)$[b]				[5.22]		[4.1]	[1.79_5]	II ← I, R	30311 H	(6)
I $(^2\Sigma)$[b]				[7.13]		[2.8]	[1.536]			
B $(^2\Delta)_{5/2}$	(26962)	[1548.18][d] Z	(80)[e]	6.003[fg]	0.301	[3.15][h]	1.673_6	B ↔ X_1, R	26613.91[d] Z	(1)(3)* (4) (5)
B' $(^2\Phi)_{7/2}$	(24218)	[1428.7][d] Z	(74)	5.758	0.326[i]	[3.6][j]	1.708_8	B' → X_1, R	23806.48[d] Z	(5)*
A $(^2\Delta)_{5/2}$	(22311)	$1690._6$[d] Z	55.3[k]	[5.534][ℓ]	m	[3.55][n]	[1.743_0]	A ↔ X_1, R	21960.59[d] Z	(1)(2)(3)* (5)
A' $(^2\Delta)_{3/2}$	x_2+(19938)	[1500.08][d] Z	(58)	6.1103[o]	0.286_9	[4.71_7][p]	1.6587_9	A' ↔ X_2, R	19610.82[d] Z	(5)*
X_2 $(^2\Delta)_{3/2}$	x_2[q]	[2177.31][d] Z	(43)	7.2784[o]	0.2029	[2.83_4][r]	1.51987			
X_1 $(^2\Delta)_{5/2}$	0[s]	[2294.68][d] Z	(46)	7.1963	0.1996	[2.61_3][t]	1.52852			
(195)Pt^2H		(μ = 1.99350810)		$D_0^0 \leq 3.59$ eV[a]						MAY 1975
B $(^2\Delta)_{5/2}$	(26947)	$1211._8$[b] Z	$40._3$	3.039[cd]	0.111	0.77[e]	1.668_1	B ← X, R	26703.6[b] Z	(1)* (2)*
A $(^2\Delta)_{5/2}$	(22287)	$1198._5$[b] Z	$26._5$[f]	2.935[c]	0.109	0.76[g]	1.697_4	A ← X, R	22040.5[b] Z	(1)(2)
X $(^2\Delta)_{5/2}$	0	[1644.3][b] Z	(23)	3.640	0.071	0.66	1.524_2			

Pt^1H: [a]From the predissociation in v=0 of B $(^2\Delta)_{5/2}$, assuming that X_1 is the ground state.

[b]Preliminary data only.

[c]$D_1 = 4.9 \times 10^{-4}$.

[d]Band origins in the tables for Pt^1H and Pt^2H correspond to the energy of J'=0 relative to J"=0. Vibrational constants recalculated accordingly.

[e]From the corresponding value for Pt^2H.

Pt^1H (continued):

[f]Perturbations in both v=0 and v=1.

[g]Predissociation above v=0, J = 12.5, see (5).

[h]$D_1 = 3.94 \times 10^{-4}$.

[i]Perturbation in v=1 at J ≈ 6.5, see (6).

[j]$D_1 = 3.5 \times 10^{-4}$.

[k]$\omega_e y_e$ = - 3.88 (v=0, ..., 3).

[ℓ]Ω-type doubling; for details see (3)(5)(7).

Pt^1H (continued):

[m] B_1, B_2, B_3 = 5.244, 4.924, 4.517 (v=3 perturbed for $J \geq 10.5$).

[n] D_1, D_2, D_3 (10^{-4} cm^{-1}) = 3.68, 4.42, 8.8.

[o] Ω-type doubling; see (5).

[p] $D_1 = 6.08 \times 10^{-4}$.

[q] $x_2 \approx 1320$; see (5).

[r] $D_1 = 2.84 \times 10^{-4}$.

[s] Not certain that this is the ground state.

[t] $D_1 = 2.60_7 \times 10^{-4}$.

(1) Loginov, OS(Engl. Transl.) 16, 220 (1964).

(2) Neuhaus, Scullman, ZN 19 a, 659 (1964).

(3) Scullman, AF 28, 255 (1964).

(4) Loginov, OS(Engl. Transl.) 20, 88 (1966).

(5) Kaving, Scullman, CJP 49, 2264 (1971).

(6) Scullman, Dissertation (Stockholm, 1971); see USIP Report 71-02.

(7) Kaving, Scullman, PS 9, 33 (1974).

Pt^2H: [a] From the predissociation in v=2 of B $(^2\Delta)_{5/2}$, assuming that X is the ground state. From the value for Pt^1H: $D_0^0 \leq 3.48_2$ eV.

[b] See footnote [d] of Pt^1H.

[c] Perturbations.

[d] All lines of the 2-0 band are diffuse.

[e] $\beta_e \approx +0.10 \times 10^{-4}$.

[f] $\omega_e y_e = -1._5$ (v=0, ..., 3).

[g] $\beta_e \approx +0.06 \times 10^{-4}$.

(1) See ref. (6) of Pt^1H.

(2) See ref. (7) of Pt^1H.

State	T_e	ω_e	$\omega_e x_e$	B_e	α_e	D_e ($10^{-7}cm^{-1}$)	r_e (Å)	Observed Transitions Design.	ν_{00}	References
$^{195}Pt^{16}O$		$\mu = 14.7821844_2$		$D_0^0 = 3.8_2$ eV[a]						JAN 1976 A
D $^1\Sigma$	(24863)	[567.1] Z	b	[0.33671]	b	$[5.19]^c$	$[1.8403_5]$	D→X, R	24722.11 Z	(5)*
A $^1\Sigma$	16995.12	727.07 Z	5.42	0.35385	0.00291	$3.2_7{}^c$	1.7952_3	A→X, R	16932.99 Z	(1)(2)(4)*
(X) $^1\Sigma$	0	851.11 Z	4.98	0.38224	0.00283	$3.0_5{}^c$	1.7272_6			
$^{(195)}Pt^{(28)}Si$		(μ = 24.4661075)		$D_0^0 = 5.1_5$ eV[a]						MAY 1975
$^{(195)}Pt^{232}Th$		(μ = 105.946025)		D_0^0 = 5.7 eV[a]						MAY 1975
$^{(244)}Pu^{19}F$		(μ = 17.6263383)		$D_0^0 = 5.4_6$ eV[a]						MAY 1975
$^{(226)}Ra^{(35)}Cl$		(μ = 30.2836142)								JUL 1977
C $^2\Pi$	15386.5 14782.1	252.9 H 253.8 H	0.72 0.71					C→X, R	15384.8 14780.9 H	(1)
X $^2\Sigma$	0	256.2 H	0.71							
$^{(85)}Rb_2$		($\mu = 42.455899_5$)		D_0^0 = 0.49 eV[a]		$3.44 < I.P.(eV)^b \lesssim 3.95$				JUL 1977 A
		Unidentified features in the absorption spectrum of rubidium vapour at 37270 and 40590 cm^{-1}.								(9)
		Diffuse absorption bands corresponding to van der Waals molecules.								(4)
D	22777.5	40.42 H	0.0745^c					D^d← X, R	22769.1 H	(3)
C $^1\Pi_u$	20835.1	36.46 H	0.124	Predissociation[e]				C^d↔X,[f] R	20824.7 H	(3)
		"Quasicontinuous" emission 16400 - 18500 cm^{-1}.								(14)
B $^1\Pi_u$	14662.6	48.05 H	0.191					B↔X,[g] R	14657.9 H	(2)
A ($^1\Sigma_u^+$)		Unresolved band system 9200 - 12500 cm^{-1}.[h]						A↔X		(12)(13)(20)
X $^1\Sigma_g^+$	0	57.31 H	0.105					Mol. beam magn. reson.[i]		
$^{(85)}Rb_2{}^+$				$D_0^0 \gtrsim 0.72$ eV[b]						JUL 1977
X $^2\Sigma_g^+$	0						(3.94)[j]			

PtO: [a]Thermochemical value (mass-spectrom.)(3).

[b]Levels with v > 0 are perturbed.

[c]Also higher order constants.

(1) Feast, PPS A 63, 549 (1950).
(2) Raziunas, Macur, Katz, JCP 43, 1010 (1965).
(3) Norman, Staley, Bell, JPC 71, 3686 (1967); AdC No. 72, 101 (1968).
(4) Nilsson, Scullman, Mehendalé, JMS 35, 177 (1970).
(5) Scullman, Sassenberg, Nilsson, CJP 53, 1991 (1975).

PtSi: [a]Thermochemical value (mass-spectrom.)(1).

(1) Vander Auwera-Mahieu, Peeters, McIntyre, Drowart, TFS 66, 809 (1970).

PtTh: [a]Thermochemical value (mass-spectrom.)(1).

(1) Gingerich, CPL 23, 270 (1973).

PuF: [a]Thermochemical value (mass-spectrom.)(1).

(1) Kent, JACS 90, 5657 (1968).

RaCℓ: (1) Lagerqvist, AF 6, 141 (1953).

Rb_2, Rb_2^+:

[a]Spectroscopic value (3), extrapolation of vibrational levels in X, C, D.

[b]Associative photoionization of rubidium vapour by atomic line absorption (1)(7)(8).

[c]$\omega_e y_e = - 0.00144$.

[d]Lifetime measurements by (11) vary from 61 ns to 14 ns, the former attributed to the D state, the latter to C $^1\Pi_u$. The two states have also been observed in two-photon ionization of Rb_2 (18).

[e]The state responsible for the partial predissociation

Rb_2, Rb_2^+ (continued):

($\lesssim$ 25%) of C $^1\Pi_u$ correlates with the 5p $^2P_{3/2}$ state of Rb (14)(17); see also (16) whose observations of the diffuse and sharp series of Rb in two-photon ionization of rubidium vapour through intermediate continuum states of Rb_2 include only transitions from the $^2P_{3/2}$ component.

[f]Polarization studies of the fluorescence spectrum (17)(19) confirm its composition of P and R as well as Q lines, contrary to the conclusions of (14) that it consists of Q line progressions only.

[g]Magnetic rotation spectrum (2). The B→X (and A→X) emission observed by (14) is attributed to atomic recombination of Rb(5^2P) and Rb(5^2S), the former formed in the predissociation of Rb_2 C $^1\Pi_u$.

[h]Interference by the a $^3\Pi_u$ state may be responsible for irregularities in the spectrum at 9900 cm^{-1} (20).

[i]$g_J(^{85}Rb_2) = 0.00953\ \mu_N$ (6); $eqQ(^{85}Rb) = -1.10$ MHz (5).

[j]Rough estimate based on the analysis of charge exchange cross sections (10). Theoretical calculations predict $r_e = 4.45$ Å (15).

(1) Lawrence, Edlefsen, PR 34, 233 (1929).
(2) Kusch, PR 49, 218 (1936).
(3) Tsi-Ze, San-Tsiang, PR 52, 91 (1937).
(4) Tsi-Ze, Shang-Yi, JP(Paris) 9, 169 (1938).
(5) Logan, Cote, Kusch, PR 86, 280 (1952).
(6) Brooks, Anderson, Ramsey, PRL 10, 441 (1963); PR A 136, 62, (1964).
(7) Lee, Mahan, JCP 42, 2893 (1965).
(8) Hudson, JCP 43, 1790 (1965).
(9) Creek, Marr, JQSRT 8, 1431 (1968).
(10) Olson, PR 187, 153 (1969).
(11) Baumgartner, Demtröder, Stock, ZP 232, 462 (1970).

(continued p. 551)

State	T_e	ω_e	$\omega_e x_e$	B_e	α_e	D_e $(10^{-8}cm^{-1})$	r_e (Å)	Observed Transitions Design.	Observed Transitions ν_{00}	References
$^{(85)}Rb^{40}Ar$		(μ = 27.1735746) D_e^0 = (0.0054) eV[a]								JUL 1977
		Green emission "bands" near the Rb 6s-5s and 4d-5s forbidden transitions (8); similar features observed in absorption (2).								
B $^2\Sigma^+$ A $^2\Pi_{3/2}$ A $^2\Pi_{1/2}$		Potential curves for all four states have been constructed from studies of the far-wing emission spectra of the Rb resonance lines (4)(5). Near-wing intensities have been measured by (7). Only the A $^2\Pi$ curves have distinct potential wells with $D_e \approx 330$ cm^{-1} and $r_e \approx 3.35$ Å.								
X $^2\Sigma^+$	0			b			(5.2)[a]			
$^{85}Rb^{79}Br$		μ = 40.902717$_4$ D_0^0 = 3.9$_0$ eV[b] I.P. = 7.7$_5$ eV[b]								JUL 1977
		Absorption continua with maxima at 35700, 38800, 46700 cm^{-1}.[c]								(1)(2)(7)
A		Diffuse absorption (fluctuation) bands 26600 - 32100 cm^{-1}; the chemiluminescence spectrum consists of a long lower-state vibrational progression 19000 - 25000 cm^{-1}.						A[d] $\leftrightarrow$ X		(2)(10)
X $^1\Sigma^+$	0	169.46[e]	0.463[e]	0.04752798	0.00018596[f]	1.4959	2.94474$_4$	Microwave sp.[g]		(3)(6)(14)
$^{(85)}Rb^{(79)}Br^+$		D_0^0 = 0.3$_3$ eV[h]								JUL 1977
(D)	(116200) 114400 [i] 111800 [i] (109700)	Ionization from the metal 4p shell.[j]								
(C)	106700 [i] (104300)									
A ($\frac{1}{2}$)	3630 [i]	Ionization from the halogen 4p shell.								
X ($\frac{3}{2}$, $\frac{1}{2}$)	0 [i]									

Rb_2, Rb_2^+ (continued):

(12) Sorokin, Lankard, JCP 55, 3810 (1971).
(13) Kostin, Khodovoi, BASPS 37 (10), 69 (1973).
(14) Brom, Broida, JCP 61, 982 (1974).
(15) Bellomonte, Cavaliere, Ferrante, JCP 61, 3225 (1974).
(16) Collins, Curry, Johnson, Mirza, Chellehmalzadeh, Anderson, PR A 14, 1662 (1976).
(17) Feldman, Zare, CP 15, 415 (1976).
(18) Granneman, Klewer, Nygaard, Van der Wiel, JP B 9, 865, (1976).
(19) Tam, Happer, JCP 64, 4337 (1976).
(20) Drummond, Schlie, JCP 65, 2116 (1976).

RbAr: [a]See (1)(3).

[b]An average value $\bar{\gamma} = 3.5 \times 10^{-6} cm^{-1}$ for the spin-rotation interaction constant has been derived from relaxation measurements on optically polarized Rb atoms in Ar (6).

(1) Baylis, JCP 51, 2665 (1969).
(2) Besombes, Granier, Granier, OC 1, 161 (1969).
(3) Nikiforov, Shcherba, OS(Engl. Transl.) 32, 567 (1972).
(4) Drummond, Gallagher, JCP 60, 3426 (1974).
(5) Carrington, Gallagher, PR A 10, 1464 (1974).
(6) Bouchiat, Brossel, Mora, Pottier, JP(Paris) 36, 1075 (1975).
(7) Ottinger, Scheps, York, Gallagher, PR A 11, 1815 (1975).
(8) Tam, Moe, Park, Happer, PRL 35, 85 (1975)

RbBr, $RbBr^+$:

[a]Thermochem. value (4); 3.98 eV by flame photometry (5).

[b]Onset of the photoelectron spectrum (11), vertical transition at 8.17 eV.

RbBr, $RbBr^+$ (continued):

[c]Also observed in the electron energy loss spectrum (8), additional peaks at 7.4, 16.3, 19.8 eV. (7) give absorption cross sections from 30300 to 50000 cm^{-1}.

[d]There is no conclusive evidence yet for the identity of the upper states observed in the absorption and emission sp..

[e]Calculated from the rotational constants (6).

[f]$\gamma_e = + 2.1_4 \times 10^{-7}$ (14).

[g]$eqQ(^{85}Rb) = -47.2_0 + 0.28(v+\frac{1}{2})$ MHz
$eqQ(^{79}Br) = + 3.0_8 + 0.75(v+\frac{1}{2})$ MHz } (14);

$\mu_{e\ell} = 10.8_6$ D [molecular beam electric deflection (13)].

[h]From D_0^0(RbBr), I.P.(Rb), and I.P.(RbBr).

[i]From the maxima of the photoelectron peaks (11)(12); see also (9).

[j]The complexity of the metal 4p photoelectron spectrum is tentatively attributed to configuration interaction (12).

(1) Müller, AP(Leipzig) 82, 39 (1927).
(2) Barrow, Caunt, PRS A 219, 120 (1953).
(3) Honig, Mandel, Stitch, Townes, PR 96, 629 (1954).
(4) Brewer, Brackett, CRev 61, 425 (1961).
(5) Bulewicz, Phillips, Sugden, TFS 57, 921 (1961).
(6) Rusk, Gordy, PR 127, 817 (1962).
(7) Davidovits, Brodhead, JCP 46, 2968 (1967).
(8) Geiger, Pfeiffer, ZP 208, 105 (1968).
(9) Goodman, Allen, Cusachs, Schweitzer, JESRP 3, 289 (1974).
(10) Oldenborg, Gole, Zare, JCP 60, 4032 (1974).
(11) Potts, Williams, Price, PRS A 341, 147 (1974).
(12) Potts, Williams, JCS FT II 72, 1892 (1976).
(13) Story, Hebert, JCP 64, 855 (1976).
(14) Tiemann, Hölzer, Hoeft, ZN 32 a, 123 (1977).

State	T_e	ω_e	$\omega_e x_e$	B_e	α_e	D_e $(10^{-7}cm^{-1})$	r_e (Å)	Observed Transitions Design.	ν_{00}	References
$^{85}Rb^{35}Cl$		$\mu = 24.7685361$ $D_0^0 = 4.3_4$ eV[a] I.P. $= 8.2_6$ eV[b]								JUL 1977
		Continuous absorption with first maximum at ~ 40000 cm^{-1}.[c]								(1)(11)
A		Diffuse absorption bands (fluctuation b.) 30000 - 38300 cm^{-1}; long lower-state vibrational progression in chemiluminescence from 18000 to 26000 cm^{-1}.						$A^d \leftrightarrow X$,		(2)(4)(15)
X $^1\Sigma^+$	0	228[e]	(0.92)	0.0876404_1	0.0004536_5[f]	0.4947[g]	2.78673_6	Rot.-vibr. b.		(6)
								Rotation sp.[h]		(5)(9)
								Mol. beam rf electric[i]		(13)
								and magn. reson.[j]		(10)
$^{(85)}Rb^{(35)}Cl^+$		$D_0^0 = 0.2_6$ eV[k]								JUL 1977
(D)	108600 ℓ 106900 ℓ	Ionization from the metal 4p shell.								
(C)	(102000) ℓ 100300 ℓ									
X	0 ℓ	Ionization from the halogen 3p shell.								
$^{(85)}Rb^{133}Cs$		$(\mu = 51.810591_0)$								JUL 1977
		Evidence for an additional state in the 19000 - 22000 cm^{-1} region by two-photon ionization.								(3)
A	13747.2	38.46 H						$A \leftarrow X$, R	13741.7	(1)(2)
X $^1\Sigma^+$	0	49.41 H								
$^{85}Rb^{19}F$		$\mu = 15.5248345$ $D_0^0 = 5.0_0$ eV[a]								JUL 1977
		Continuous absorption with maximum at 46500 cm^{-1}.[b]								(1)(4)
A		Diffuse absorption bands 33500 - 42300 cm^{-1}.						$A \leftarrow X$		
X $^1\Sigma^+$	0	376[c]	(1.9)	0.2106640_1	0.0015227_9[d]	2.683_9[e]	2.270333	Rot.-vibr. b.		(13)
		SCF-LCAO-MO calculations (16).						Rotation sp.		(5)(8)
								Mol. beam rf electric[fg]		(2)(10)(11)
								and magn. reson.[hg]		(3)(9)

RbCℓ, RbCℓ^+:

[a]Thermochem. value (7); flame photom. value 4.40 eV (8).

[b]Onset of a broad unresolved photoelectron peak with maximum at 8.74 eV (14)(16).

[c]Additional peaks in the electron energy loss spectrum at 7.8, 15.5, 19.0, 23.0 eV (12). (11) give absorption cross sections from 34000 to 50000 cm^{-1}.

[d]See [d] of RbBr.

[e]From the IR spectrum (6). From the rotational constants (9) calculate $\omega_e = 233.34$, $\omega_e x_e = 0.856$.

[f]$\gamma_e = +7.0 \times 10^{-7}$.

[g]$\beta_e = +2._3 \times 10^{-11}$.

[h]$eqQ(^{85}Rb) = -52.675 + 0.38\,v$ MHz; $eqQ(^{35}C\ell) = +0.774 - 0.155\,v$ MHz; v = 0,1,2 (5); for constants of $^{87}Rb^{35}C\ell$ see (3)(5).

[i]$\mu_{e\ell}[D] = 10.483 + 0.054(v+\frac{1}{2})$, v = 0,1,2.

[j]$g_J = (-)0.018_3\ \mu_N$.

[k]From $D_0^0(RbC\ell)$ and the ionization potentials of Rb and RbCℓ.

[ℓ]From maxima of the photoelectron spectrum (16)(17); the halogen 3p spectrum is unresolved. See also [j] of $RbBr^+$.

(1) Müller, AP(Leipzig) 82, 39 (1927).
(2) Sommermeyer, ZP 56, 548 (1929).
(3) Bolef, Zeiger, PR 85, 799 (1952).
(4) Barrow, Caunt, PRS A 219, 120 (1953).
(5) Trischka, Braunstein, PR 96, 968 (1954).
(6) Rice, Klemperer, JCP 27, 573 (1957).
(7) Brewer, Brackett, CRev 61, 425 (1961).
(8) Bulewicz, Phillips, Sugden, TFS 57, 921 (1961).
(9) Clouser, Gordy, PR A 134, 863 (1964).
(10) Mehran, Brooks, Ramsey, PR 141, 93 (1966).

RbCℓ, RbCℓ^+ (continued):

(11) Davidovits, Brodhead, JCP 46, 2968 (1967).
(12) Geiger, Pfeiffer, ZP 208, 105 (1968).
(13) Hebert, Lovas, Melendres, Hollowell, Story, Street, JCP 48, 2824 (1968).
(14) Goodman, Allen, Cusachs, Schweitzer, JESRP 3, 289 (1974).
(15) Oldenborg, Gole, Zare, JCP 60, 4032 (1974).
(16) Potts, Williams, Price, PRS A 341, 147 (1974).
(17) Potts, Williams, JCS FT II 72, 1892 (1976).

RbCs:

(1) Loomis, Kusch, PR 46, 292 (1934).
(2) Kusch, PR 49, 218 (1936).
(3) Granneman, Klewer, Nygaard, Van der Wiel, JP B 9, 865 (1976).

RbF:

[a]Thermochemical value (6); flame photometry gives 5.2 eV (7).

[b]The electron energy loss spectrum has peaks at 4.9, 8.2, 9.4, 14.8, 18.4, 19.7, 21.0 eV (14).

[c]From the IR spectrum (13); good agreement with $\omega_e = 373.27$ and $\omega_e x_e = 1.80$ as calculated from the rot. constants (8).

[d]$\gamma_e = +3.30 \times 10^{-6}$. (5)(8) give constants for ^{87}RbF.

[e]$\beta_e = -0.003_7 \times 10^{-7}$ (8).

[f]$\mu_{e\ell}[D] = 8.5131 + 0.0665_0(v+\frac{1}{2}) + 0.0002_6(v+\frac{1}{2})^2$, v=0,1,2 (12)(15); $eqQ(^{85}Rb)[MHz] = -70.739 + 0.797_5(v+\frac{1}{2}) - 0.005_3(v+\frac{1}{2})^2$, v=0...4 (10)(11)(12).

[g](17) give constants for ^{87}RbF (eqQ, $\mu_{e\ell}$, g_J, etc.).

[h]$g_J = (-)0.044_1\ \mu_N$ (9); somewhat different values by (12) for v=0 (−0.05470) and v=1 (−0.05455).

(1) Caunt, Barrow, Nature 164, 753 (1949).
(2) Hughes, Grabner, PR 79, 314 (1950).

(continued p. 555)

State		T_e	ω_e	$\omega_e x_e$	B_e	α_e	D_e	r_e (Å)	Observed Transitions Design.	ν_{00}	References
$^{(85)}Rb^1H$			(μ = 0.99600357)								JUL 1977
A	$^1\Sigma^+$	$18219._8$	211.7 Z	-6.4_7[a]	1.112	-0.054[b]	0.000110[c]	3.901	A→X, R	$17862._3$ Z	(1)* (2)
X	$^1\Sigma^+$	0	936.9_4 Z	14.21[d]	3.020	0.072[e]	0.000123[f]	2.367			
$^{(85)}Rb^2H$			(μ = 1.96743480)								JUL 1977
A	$^1\Sigma^+$		$\Delta G(10\frac{1}{2})$ = 193.6 Z		{ 0.642 { 0.648	(v = 11) (v = 10)	0.000011 0.000026		A←X, R	ν(10-1) = 19084.9 Z	(2)
X	$^1\Sigma^+$	0			1.488	(v = 1)	0.00002_4	r_1 = 2.400			
$^{(85)}Rb^4He$			(μ = 3.8224206_1)		D_e^0 = (0.000115) eV[a]						JUL 1977
Green emission "bands" near the forbidden 6s-5s and 4d-5s transitions of Rb (4); similar features in absorption (2). Far-wing emission spectra of the Rb resonance lines (3), near-wing intensities (6). Predicted potential curves (1).											
X	$^2\Sigma^+$	0						(7.36)[a]			
$^{85}Rb^{127}I$			μ = 50.872801_7		D_0^0 = 3.3_0 eV[a]		I.P. = 7.1_2 eV[b]				JUL 1977
			Continuous absorption with maxima at 30700, 38100, 41800 cm^{-1}.[c]								(1)(2)(7)
A			Diffuse absorption (fluctuation) bands 23150 - 27550 cm^{-1}; long lower-state progression in chemiluminescence 17500 - 22500 cm^{-1}.						A[d]↔X		(2)(10)
X	$^1\Sigma^+$	0	138.51[e]	0.335[e]	0.03283293	0.00010946[f]	7.380[g] ($10^{-9}cm^{-1}$)	3.17687_9	Rotation sp.[h]		(3)(6)
$^{(85)}Rb^{127}I^+$					D_0^0 = 0.3_6 eV[i]						JUL 1977
(D)		119700[j] 118600[j]	Ionization from the metal 4p shell.								
(C)		113100[j] 111500[j]									
A ($\frac{1}{2}$)		7800[j]	Ionization from the halogen 5p shell.								
X ($\frac{3}{2}$,$\frac{1}{2}$)		0[j]									

RbF (continued):

(3) Bolef, Zeiger, PR 85, 799 (1952).
(4) Barrow, Caunt, PRS A 219, 120 (1953).
(5) Lew, Morris, Geiger, Eisinger, CJP 36, 171 (1958); Lew, CJP 42, 1004 (1964) (erratum).
(6) Brewer, Brackett, CRev 61, 425 (1961).
(7) Bulewicz, Phillips, Sugden, TFS 57, 921 (1961).
(8) Veazey, Gordy, PR A 138, 1303 (1965).
(9) Mehran, Brooks, Ramsey, PR 141, 93 (1966).
(10) Zorn, English, Dickinson, Stephenson, JCP 45, 3731 (1966).
(11) Bonczyk, Hughes, PR 161, 15 (1967).
(12) Gräff, Schönwasser, Tonutti, ZP 199, 157 (1967).
(13) Baikov, Vasilevskii, OS(Engl. Transl.) 22, 198 (1967).
(14) Geiger, Pfeiffer, ZP 208, 105 (1968).
(15) Hebert, Lovas, Melendres, Hollowell, Story, Street, JCP 48, 2824 (1968).
(16) Matcha, JCP 53, 4490 (1970).
(17) Heitbaum, Schönwasser, ZN 27 a, 92 (1972).

Rb^1H, Rb^2H:

[a] $\omega_e y_e = -0.25_3$, $\omega_e z_e = +0.0030$ $(v \leq 9)$.
[b] $-0.0054(v+\frac{1}{2})^2 + 0.00016_2(v+\frac{1}{2})^3$ $(v \leq 9)$.
[c] $\beta_e \approx -0.026 \times 10^{-4}$.
[d] $\omega_e y_e = +0.082$.
[e] $\gamma_e = +0.0003$.
[f] $\beta_e \approx -0.01_3 \times 10^{-4}$.

(1) Gaydon, Pearse, PRS A 173, 28 (1939).
(2) Bartky, JMS 21, 1 (1966).

RbHe: [a] Pseudopotential calculations of (1), corrected values quoted by (7). According to (7) the rotationless ground state potential supports one bound level, and a single quasibound level exists for N=1. The quasibound state has a natural lifetime of 0.6 ns [electron spin relaxation study of optically pumped Rb (5)].

RbHe (continued):

(1) Baylis, JCP 51, 2665 (1969).
(2) Besombes, Granier, Granier, OC 1, 161 (1969).
(3) Drummond, Gallagher, JCP 60, 3426 (1974).
(4) Tam, Moe, Park, Happer, PRL 35, 85 (1975).
(5) Franz, Volk, PRL 35, 1704 (1975).
(6) Ottinger, Scheps, York, Gallagher, PR A 11, 1815 (1975).
(7) Kiehl, PL A 56, 82 (1976).

RbI, RbI^+:

[a] Thermochemical value (4), 3.52 eV by flame photometry (5).
[b] Onset of the first photoelectron peak with maximum at 7.51 eV (11).
[c] Additional peaks in the electron energy loss spectrum at 6.4, 15.6, 19.2 eV (8). (7) give absorption cross sections from 26000 to 50000 cm^{-1}.
[d] See [d] of RbBr.
[e] Calculated from the rotational constants (6).
[f] $\gamma_e = +1.18 \times 10^{-7}$.
[g] $\beta_e = +0.005_3 \times 10^{-9}$.
[h] $eqQ(^{85}Rb) = -40.4_0$ MHz, $eqQ(^{127}I) = -59.8_9$ MHz, both for v=0 (14). $\mu_{el} = 11.4_8$ D [electric deflection method (13)].
[i] $D_0^0(RbI)$ + I.P.(Rb) - I.P.(RbI).
[j] From maxima of the photoelectron peaks (9)(11)(12). See also [j] of $RbBr^+$.

(1) Schmidt-Ott, ZP 69, 724 (1931).
(2) See ref. (4) of RbF.
(3) Honig, Mandel, Stitch, Townes, PR 96, 629 (1954).

(continued p. 557)

State	T_e	ω_e	$\omega_e x_e$	B_e	α_e	D_e $(10^{-}\ cm^{-1})$	r_e (Å)	Observed Transitions Design.	Observed Transitions ν_{00}	References
$^{(85)}Rb^{(84)}Kr$		$(\mu = 42.204344_4)$ $D_e^0 = 0.0091$ eV[a]								JUL 1977
		Theoretical potential energy curves correlated with Rb 5^2D, 7^2S, and 4^2F (7). Green emission "bands" near the Rb 6s-5s and 4d-5s forbidden transitions (11); similar features in absorption (3).								
B $^2\Sigma^+$		Potential energy curves for all four states have been constructed from studies of the far-wing emission spectra of the Rb resonance lines (8)(9); near-wing intensities measured by (10). Only the Π curves have deep potential wells with $D_e \approx 420$ cm^{-1} and $r_e \approx 3.5$ Å.								
A $^2\Pi_{3/2}$										
A $^2\Pi_{1/2}$										
X $^2\Sigma^+$	0			b			5.2_9[a]			
$^{(85)}Rb^{(20)}Ne$		$(\mu = 16.1823211)$ $D_e^0 = (0.00015)$ eV[a]								JUL 1977
		Green emission "bands" near the Rb 6s-5s and 4d-5s forbidden transitions (6); similar features in absorption (2).								
B $^2\Sigma^+$		Far-wing emission spectra of the Rb resonance lines studied by (3)(4), near-wing spectra by (5). The derived $^2\Pi$ potential curves are attractive with $D_e \approx 150$ cm^{-1} and $r_e \approx 3.4$ Å.								
A $^2\Pi_{3/2}$										
A $^2\Pi_{1/2}$										
X $^2\Sigma^+$	0						(7.8)[a]			
$^{(85)}Rb^{16}O$		$(\mu = 13.4595306_4)$								JUL 1977
A $^2\Pi_i$	(606)[a]	(389)[a]					(2.41)[a]			
X $^2\Sigma^+$	0	(433)[a]					(2.28)[a]			
$^{(85)}Rb^{(132)}Xe$		$(\mu = 51.657726_1)$ $D_e^0 = (0.0134)$ eV[a]								JUL 1977
		Green emission "bands" near the Rb 6s-5s and 4d-5s forbidden transitions (8); similar features in absorption (2).								
B $^2\Sigma^+$		Potential curves for all four states have been constructed from the study of the far-wing emission spectra of the Rb resonance lines (4)(6); near-wing intensities measured by (7). B $^2\Sigma$ is repulsive, A $^2\Pi$ attractive with $D_e \approx 650$ cm^{-1} and $r_e \approx 3.43$ Å. Teratomic recombination in A $^2\Pi_{1/2}$ (5)(9).								
A $^2\Pi_{3/2}$										
A $^2\Pi_{1/2}$										
X $^2\Sigma^+$	0			b			(4.97)[a]			

RbI, RbI^+ (continued):

(4) Brewer, Brackett, CRev 61, 425 (1961).
(5) Bulewicz, Phillips, Sugden, TFS 57, 921 (1961).
(6) Rusk, Gordy, PR 127, 817 (1962).
(7) Davidovits, Brodhead, JCP 46, 2968 (1967).
(8) Geiger, Pfeiffer, ZP 208, 105 (1968).
(9) Goodman, Allen, Cusachs, Schweitzer, JESRP 3, 289 (1974).
(10) Oldenborg, Gole, Zare, JCP 60, 4032 (1974).
(11) Potts, Williams, Price, PRS A 341, 147 (1974).
(12) Potts, Williams, JCS FT II 72, 1892 (1976).
(13) Story, Hebert, JCP 64, 855 (1976).
(14) Tiemann, Hoeft, ZN 31 a, 236 (1976).

RbKr: [a]From atomic scattering data (1), see also (2)(6).
[b]The existence of Rb-rare-gas van der Waals molecules has been shown in relaxation experiments on optically polarized Rb atoms; see e.g. (4)(5). Average values for the spin-rotation interaction have been derived from these observations: $\bar{\gamma} = 0.0000216$ for RbKr and $\bar{\gamma} \approx 0.00005_2$ for RbXe.

(1) Buck, Pauly, ZP 208, 390 (1968).
(2) Baylis, JCP 51, 2665 (1969).
(3) Besombes, Granier, Granier, OC 1, 161 (1969).
(4) Bouchiat, Brossel, Pottier, JCP 56, 3703 (1972).
(5) Bouchiat, Pottier, JP(Paris) 33, 213 (1972).
(6) Nikiforov, Shcherba, OS(Engl. Transl.) 32, 567 (1972).
(7) Pascale, Vandeplanque, JCP 60, 2278 (1974).
(8) Drummond, Gallagher, JCP 60, 3426 (1974).
(9) Carrington, Gallagher, PR A 10, 1464 (1974).
(10) Ottinger, Scheps, York, Gallagher, PR A 11, 1815 (1975).
(11) Tam, Moe, Park, Happer, PRL 35, 85 (1975).

RbNe: [a]Pseudopotential calculations (1).

(1) See ref. (2) of RbKr.
(2) See ref. (3) of RbKr.
(3) See ref. (8) of RbKr.
(4) See ref. (9) of RbKr.
(5) See ref. (10) of RbKr.
(6) See ref. (11) of RbKr.

RbO: [a]Ab initio calculations (2); experimental evidence for a $^2\Sigma$ ground state comes from the ESR spectrum of matrix isolated ^{87}RbO (1).

(1) Lindsay, Herschbach, Kwiram, JCP 60, 315 (1974).
(2) So, Richards, CPL 32, 227 (1975).

RbXe: [a]Pseudopotential calculations (1); see also (3)(4).
[b]See [b] of RbKr.

(1) See ref. (2) of RbKr.
(2) See ref. (3) of RbKr.
(3) See ref. (6) of RbKr.
(4) See ref. (8) of RbKr.
(5) Carrington, Gallagher, JCP 60, 3436 (1974).
(6) See ref. (9) of RbKr.
(7) See ref. (10) of RbKr.
(8) See ref. (11) of RbKr.
(9) Scheps, Gallagher, JCP 65, 859 (1976).

State	T_e	ω_e	$\omega_e x_e$	B_e	α_e	D_e $(10^{-6}cm^{-1})$	r_e (Å)	Observed Transitions Design.	Observed Transitions ν_{00}	References
$^{(187)}Re^{16}O$?		$(\mu = 14.7343266_2)$ Mostly R shaded band heads in the emission spectrum from 11500 to 17400 cm^{-1}. No analysis.								MAY 1975 (1)* (2)
$^{103}Rh_2$		$\mu = 51.452756_0$ $D_0^0 = 2.9_2$ eV[a]								MAY 1975
$^{103}Rh^{(11)}B$		$(\mu = 9.94531026_6)$ $D_0^0 = 4.8_9$ eV[a]								MAY 1975
$^{103}Rh^{12}C$		$\mu = 10.74679641$ $D_0^0 = 6.01$ eV[a]								MAY 1975 A
		Additional systems observed in matrix absorption.								(5)
D $^2\Sigma^{(-)}$	(21756)	(782) [b]		(0.482) [bc]			(1.80_4)			(2)
C $^2\Sigma^{(+)}$	21439.2	927.8 [d] Z	13.7_3	0.5510 [de]	0.0060	(1.0)	1.687_3	C ← X, R	21376.0 [d] Z	(1)(2)* (5)
B $^2\Sigma^{(+)}$				[0.5067] [fg]		(0.8)	$[1.759_5]$	B ← X, R	21361.0 [f] Z	(1)(2)(5)
A $^2\Pi_r$	10242.75 [h]	939.12 Z	5.48 [i]	0.57149 [j]	0.00428	0.832 [k]	1.655_4	A ↔ X, R	10187.24 [m] Z	(4)(5)
	9462.94	949.41 Z	5.357	0.57329 [j]	0.00426	0.826 [ℓ]			9412.60 [m] Z	
X $^2\Sigma^{(+)}$	0	1049.87 Z	4.937	0.6027 [n]	0.00396	0.78_3 [o]	1.613_3	ESR sp. [p]		
$^{103}Rh^{16}O$		$\mu = 13.8432212_7$ $D_0^0 = 4.2$ eV[a] Unclassified R shaded emission bands in the region 15150 - 17050 cm^{-1}.								MAY 1975 (2)
$^{103}Rh^{(28)}Si$		$(\mu = 21.9966876)$ $D_0^0 = 4.0_5$ eV[a]								MAY 1975
$^{103}Rh^{(48)}Ti$		$(\mu = 32.707955_2)$ $D_0^0 = 4.0_1$ eV[a]								MAY 1975

ReO: (1) Gatterer, Junkes, Salpeter, Rosen, METOX (1957).
(2) Raziunas, Macur, Katz, JCP 43, 1010 (1965).

Rh_2: [a]Thermochemical value (mass-spectrom.)(1)(2).

(1) Gingerich, Cocke, CC (1972), 536.
(2) Cocke, Gingerich, JCP 60, 1958 (1974).

RhB: [a]Thermochemical value (mass-spectrom.)(1).

(1) Vander Auwera-Mahieu, Peeters, McIntyre, Drowart, TFS 66, 809 (1970).

RhC: [a]Thermochemical value (mass-spectrom.)(3)(6).
[b]Vibrational numbering uncertain. All information based on perturbations in C $^2\Sigma^{(+)}$.
[c]Spin-splitting constant $\gamma \approx -1.6$.
[d]Strong perturbations produced by interaction with B $^2\Sigma^{(+)}$, and weaker perturbations produced by D $^2\Sigma^{(-)}$. The constants in the table are deperturbed values. The observed origin of the 0-0 band is at 21452.0 cm^{-1}.
[e]Spin-splitting constant $\gamma = -0.03$.
[f]Vibrational numbering uncertain. Strong interaction with v=0 of C $^2\Sigma^{(+)}$. The constants in the table are deperturbed values. The observed band origin is at 21285.0 cm^{-1}.
[g]Spin-splitting constant $\gamma = +1.00$.
[h]$A_v = +781.07 - 10.50(v+\frac{1}{2})$; also J-dependent terms.
[i]$\omega_e y_e = +0.021$.
[j]Λ-type doubling in $^2\Pi_{1/2}$:
$\Delta v_{fe} = (+)[0.0177 + 0.0016(v+\frac{1}{2})](J+\frac{1}{2})$.
[k]$\beta_e = +0.0032 \times 10^{-6}$.
[l]$\beta_e = -0.0019 \times 10^{-6}$.
[m]J'=0 relative to N"=0.

RhC (continued):
[n]Spin-splitting constant $\gamma = -0.065$.
[o]$\beta_e = +0.01_2 \times 10^{-6}$.
[p]In rare gas matrices at 4 K (5).

(1) Lagerqvist, Neuhaus, Scullman, ZN 20 a, 751 (1965).
(2) Lagerqvist, Scullman, AF 32, 479 (1966).
(3) Vander Auwera-Mahieu, Drowart, CPL 1, 311 (1967).
(4) Kaving, Scullman, JMS 32, 475 (1969).
(5) Brom, Graham, Weltner, JCP 57, 4116 (1972).
(6) Cocke, Gingerich, JCP 57, 3654 (1972); 60, 1958 (1974).

RhO: [a]Thermochemical value (mass-spectrom.)(1).

(1) Norman, Staley, Bell, JPC 68, 662 (1964); AdC No. 72, 101 (1968).
(2) See ref. (2) of ReO.

RhSi: [a]Thermochemical value (mass-spectrom.)(1).

(1) See ref. (1) of RhB.

RhTi: [a]Thermochemical value (mass-spectrom.)(1)(2).

(1)(2) See ref. (1)(2), resp., of Rh_2.

(Sub-) system	v_e	ω'_e ω''_e	$\omega'_e x'_e$ $\omega''_e x''_e$	B'_e B''_e	α'_e α''_e	D'_e D''_e (10^{-7}cm^{-1})	r'_e r''_e (Å)	$\Omega' \leftrightarrow \Omega''$	ν_{00}	References and Remarks
$^{(102)}$Ru$^{(11)}$B		(μ = 9.9358762$_1$)		D_0^0 = 4.6$_0$ eV[a]						MAY 1975
^{102}Ru^{12}C		μ = 10.73578136		D_0^0 = 6.6$_8$ eV[a]						MAY 1975
XI		Constants for v=1:		0.513$_3$ [0.587$_5$]		9.$_7$ [7.$_5$]	1.749 [1.635]	←	R 23802.00[b] Z	(5) P and R br.
X	23299	[743.25] Z 1038.8 Z	 4.6$_4$	0.5234 0.5878	0.0069 0.0036	[9.$_5$][c] 8.3	1.732$_1$ 1.634$_4$	4←4,[d]	R 23152.00 Z	(3)(5)* P and R br.
IX	22925	[775.25][e] Z [1018] H		[0.5273] [0.588$_5$]	0.0077[e]	[10] [7.$_5$]	[1.725$_6$] [1.633]	←	R 22803.25 Z	(3)(5) P and R br.
VIII		0-0 sequence only.		[0.5637] [0.5882]		[9.$_6$] [8.9]	[1.669$_0$] [1.633$_9$]	→	R 15344.82 Z	(4) P, Q, R br.
VII	13862.6$_4$	954.5$_6$ Z 1039.1$_4$ Z	5.39 4.7$_5$	0.5702 0.5882	0.0043 0.0040	9.0[f] 7.7	1.659$_5$ 1.633$_9$	4↔3,	R 13820.19 Z	(4)* (5) P, Q[g], R br.
VI	13353	[962] H [1043] H		[0.5701] [0.5887]		[10.$_2$] [9.$_6$]	[1.659$_6$] [1.633$_2$]	→	R 13312.69 Z	(4) P, Q, R br.
V	13328	[949] H [1032] H		[0.569$_7$] [0.5882]		[9.$_6$] [8.$_6$]	[1.660] [1.633$_9$]	→	R 13286.43 Z	(4) P, Q, R br.
IV	13138	960 H 1048 H	3 5.$_5$	[0.5698] [0.5879][h]		[8.$_4$] [8.1]	[1.660$_0$] [1.634$_3$]	2→1,	R 13094.87 Z	(4) P, Q, R br.
III	12913	[944] H [1020] H		[0.5691] [0.5882]		[9.$_4$] [9.$_0$]	[1.661$_1$] [1.633$_9$]	→	R 12875.23 Z	(4) P, Q, R br.
II		0-0 sequence only.		[0.5710] [0.5864]		[9.$_1$] [7.8]	[1.658$_3$] [1.636$_4$]	(2)→(3),	R 12658.26 Z	(4) P, Q, R br.
I	12664	(950) H (1030) H		[0.5653] [0.5870]		[7.$_5$] [6.$_6$]	[1.666$_6$] [1.635$_5$]	→	R 12624.28 Z	(4) P, Q, R br.

RuB: [a]Thermochemical value (mass-spectrom.)(1).

(1) See ref. (1) of RhB.

RuC: [a]Thermochemical value (mass-spectrom.)(1), revised (2); (6).

[b]Only 1-0 band analyzed.

[c]$D_1 = 8._0 \times 10^{-7}$.

[d]It is possible that the lower state of system X is identical with that of system VII. In this case, $\Omega' = \Omega'' = 3$.

[e]Assignment of 1-0 band not certain. Lower state constants for this band are $B''_0 = 0.5869$, $D''_0 = 7._0 \times 10^{-7}$.

[f]$\beta_e = -0.3 \times 10^{-7}$.

[g]Perturbations.

[h]The Ω-type doubling in the 0-0 band, $\Delta v \approx 0.00010_3 \times J(J+1)$, is believed to arise in the lower state.

RuC (continued):

(1) McIntyre, Vander Auwera-Mahieu, Drowart, TFS 64, 3006 (1968).
(2) See ref. (1) of RhB.
(3) Scullman, Dissertation (Stockholm, 1971); see USIP Report 71-02.
(4) Scullman, Thelin, PS 3, 19 (1971).
(5) Scullman, Thelin, PS 5, 201 (1972).
(6) Gingerich, CPL 25, 523 (1974).

State	T_e	ω_e	$\omega_e x_e$	B_e	α_e	D_e ($10^{-7}cm^{-1}$)	r_e (Å)	Observed Transitions: Design.	Observed Transitions: ν_{00}	References
$^{102}Ru^{16}O$		$\mu = 13.8249497_5$		$D_0^0 = 5.3$ eV[a]						MAY 1975
		Additional unclassified emission bands in the region 15500 - 17200 cm^{-1}.								(3)
B_3	$a_3 + 18121.4$	792.9 H	4.1	0.3818	0.0025	4	1.787_1	$B_3 \to A_3$,[b] R	18086.2 Z	(1)(3)*
B_2 ($\Omega = 3$)	$a_2 + 18101$	[783] H		[0.382][c]		[6][c]	[1.78_7]	$B_2 \to A_2$,[b] R	18065.1 Z	(1)(3)*
B_1				[(0.384)]			[(1.78)]	$B_1 \to A_1$,[b] R	18024 Z	(1)(3)
A_3	a_3[d]	863.5 H	4.6	0.4137	0.0028	3	1.716_8			
A_2 ($\Omega = 3$)	a_2[d]	[855] H		[0.4144][e]		[3.9]	[1.715_4]			
A_1	a_1[d]			[(0.414)]			[(1.72)]			
$^{(102)}Ru^{(28)}Si$		($\mu = 21.9505901$)		$D_0^0 = 4.0_8$ eV[a]						MAY 1975
$^{(102)}Ru^{232}Th$		($\mu = 70.807681_5$)		$D_0^0 = 6.1$ eV[a]						MAY 1975
$^{32}S_2$		$\mu = 15.9860364$		$D_0^0 = 4.3693$ eV[a]		I.P. = 9.36 eV[b]				JUL 1977
		Additional progressions and unassigned bands in the absorption spectrum 65700 - 71900 cm^{-1}.								(50)
F c	(66333) (66229)	[827][d] H						$F \leftarrow X$,	66384[d] H 66280[d] H	(36)(50)
E c	(65933) (65829)	[818][d] H						$E \leftarrow X$,	65980[d] H 65876[d] H	(36)(50)
j		Only v=0 observed.						$j \to (b)$,[e] V	56077.7 H	(25)*
i		[785.0] H						$i \to (b)$,[e] V	55099.3 H	(12)*
h		819.6 H	2.70					$h \to (b)$,[e] V	51461.4 H	(12)*
D $^3\Pi_{u,r}$	58978.7 58691.7 58518.3	793.8 H	4.00	[0.3073] [0.3066] [0.3059]		[1.85]	[1.854_6]	$D \leftrightarrow X$, V	59012.50 Z 58725.47 Z 58552.05 Z	(8)* (12) (31)
g $^1\Delta_u$	$x + 52187.7$	816.0 H	2.70	[0.3210]		(2.0)	[1.812_5]	$g \leftrightarrow a$,[f] V	52244.66 Z	(12)* (26)
$C'(^3\Sigma_u^-)$		v=0 only; system e − X of (12).						$C' \to X$, V	56621.6 H	(12)*

RuO: [a]Thermochemical value (mass-spectrom.)(2).

[b]R and P branches only.

[c]$B_1 = 0.381$, $D_1 = 9 \times 10^{-7}$. Perturbations.

[d]Relative position of these three states unknown.

[e]$B_1 = 0.413$.

(1) See ref. (2) of ReO.

(2) Norman, Staley, Bell, AdC No. 72, 101 (1968).

(3) Scullman, Thelin, JMS 56, 64 (1975).

RuSi: [a]Thermochemical value (mass-spectrom.)(1).

(1) See ref. (1) of RhB.

RuTh: [a]Thermochemical value (mass-spectrom.)(1).

(1) See ref. (6) of RuC.

S_2: [a]From the predissociation limit in B $^3\Sigma_u^-$ assuming dissociation at this limit into $^3P_2 + {}^3P_1$ (32). The value given here (35240.2 cm^{-1}) is relative to the lowest existing level in X $^3\Sigma_g^-(v=0)$, i.e. J=0, in accordance with the definition of D_0^0 but at variance with the value given by (32) which refers to a hypothetical level (N=0) of the F_2 component at 23.1 cm^{-1} above J=0. $D_0^0 = 4.38$ eV by photoionization mass-spectrometry (30), 4.4_1 eV from thermal measurements (27). See also (23).

S_2 (continued):

[b]By photoionization mass-spectrometry (28)(30).

[c]The two states E and F are believed to be members of two Rydberg series, one converging to X $^2\Pi_{1/2}$ of S_2^+ (C,E,...), the other to X $^2\Pi_{3/2}$ (F,...). The apparent doublet structure of the bands is tentatively attributed to (Ω_c,ω) coupling (36).

[d]Measurements of (50); assignments of higher members of the two progressions appear uncertain.

[e]The lower state(s) of the three systems could be either a or b; see (18)(26). Bands originating from the j level have double heads, all others single heads. No absorption corresponding to these transitions has been reported, although strong absorption from a $^1\Delta_g$ ($g \leftarrow a$, $f \leftarrow a$) has been seen in the flash photolysis of S_2Cl_2 (28a)(36) and COS (35). (12) use c and c' instead of h and i, respectively.

[f]Called $d \rightarrow x$ by (12). Observed in absorption in the flash photolysis of S_2Cl_2 (36).

S_2 (continued from p. 565):

[b']Spin splitting constants $\lambda_v = +11.82 + 0.05(v+\frac{1}{2}) + 0.0024 \times (v+\frac{1}{2})^2$, $\gamma_v = -0.00659 - 0.000126(v+\frac{1}{2})$ for $v \leq 27$, from (44) who give also data for $^{34}S_2$; see also (13)(18)(45). From the pure rotational Raman spectrum (46) obtain $B_0 = 0.29443$.

[c']$\gamma_e = -1.8_2 \times 10^{-6}$ (18).

[d']Raman spectra in solid matrices (37)(47) yield $\Delta G = 717$.

State	T_e	ω_e	$\omega_e x_e$	B_e	α_e	D_e $(10^{-7}cm^{-1})$	r_e (Å)	Observed Transitions: Design.	Observed Transitions: ν_{00}	References
$^{32}S_2$ (continued)										
C $^3\Sigma_u^-$	55581.7	829.1_5 Z	3.3_4	0.3219^{g}	0.0013_8^{h}	$[2.1_7]$	1.810_0	C↔X, V	55633.3^{i} Z	(3)(8)* (12)(29)*
f $^1\Delta_u$	x + 36875.45	438.32 Z	2.70^{j}	0.22704^{k}	0.00178	(2.43)	2.1551	f↔a,$^{\ell}$ R	36743.53 Z	(26)(35)
B' $^3\Pi_{g,i}$	-	-		$-^{n}$					-	
	z+(14504)	$[533.7]^{m}$ (Z)		$[0.244_1]^{n}$			[2.078]	B'→A, V	13451.9_5^{o} Z	(16)(24) (49)*
	z+(14295)	-		$[0.243_5]^{n}$					13320.6_4^{o} Z	
									-	
								B'→A', V	14144.3_7 Z	(10)(16) (24)(48)*
									14318.0_7 Z	
B $^3\Sigma_u^-$	31835	434.0^{p}	2.75^{p}	0.2239^{pqr}	0.0023^{p}	$[2.4]^{p}$	2.170	B^{s}↔X,^{t} R	31689^{u}	(6)(9)(18)
A $^3\Sigma_u^+$ 0_u^-	$z + 1078^{v}$	482.7_5 Z	2.5_8	0.230_1	0.0021		2.141			
1_u	$z + 1000.4_9$	482.1_5 Z	2.5_6	0.225_9^{w}	0.0014		2.161			
A' $^3\Delta_{u,i}$	-	-		-	-					
	z + 383	488.1_6 Z	2.5_1	0.228_5	0.001_4		2.148			
	z^{x}	488.2_5 Z	2.5_2	0.228_5	0.001_5					
b $^1\Sigma_g^+$	y	$(699.7)^{y}$	$(3.4)^{y}$							
a $^1\Delta_g$	x^{z}	702.35 Z	3.09	0.29262	0.00173	(2.01)	1.8983	Raman sp.$^{d'}$		(40)(46)
X $^3\Sigma_g^-$	$0^{a'}$	725.65 Z	2.844	$0.2954_7^{b'}$	$0.001570^{c'}$	$[1.9_0]$	1.8892	EPR and		(45)
								mol. beam rf sp.		(25a)

S_2 (continued):

^{g}Spin splitting constants $\lambda_0 = -11.6_1$, $\gamma_0 = +0.033$.

$^{h}\gamma_e = -0.00023$.

^{i}This number, given or implied by (18)(29), refers presumably to the F_2 levels in both upper and lower state.

$^{j}\omega_e y_e = -0.005$.

^{k}Breaking-off in emission above v'=10 (20)(22). In absorption (35) bands with v'=11 and 12 have been observed, the rotational lines being only very slightly broadened. Predissociation probably into $^3\Delta_u$ from $^3P + ^1D$.

$^{\ell}$First observed by (4)(15). Vibrational numbering established

S_2 (continued):

by isotope investigations (17)(21).

[m]Fragments of two V shaded emission bands at $v_0 = 13451.9$ and $13985.5\ cm^{-1}$ have been observed by (10) and assigned (18) to a $^1\Pi_g \rightarrow {}^1\Sigma_u^-$ transition later called e→c (26). The first band ($B' \approx 0.24_4$, $B'' \approx 0.22_9$, predissociated except for low J) is undoubtedly the 0-0 band of the $B'\ {}^3\Pi_{g1} \rightarrow A\ {}^3\Sigma_u^+(0_u^-)$ transition, the second presumably the corresponding 1-0 band since the $\Delta G(\frac{1}{2})$ value agrees fairly well with $\omega_e \approx 500$ as estimated from isotope shift studies (24). However, no emission from levels having $v' > 0$ was reported by other investigators.

[n]The last observed levels in emission are J'=33 and 15 in $^3\Pi_2$ and $^3\Pi_1$, respectively; higher levels, and presumably all levels of the unobserved $^3\Pi_0$ component, are predissociated.

[o]Origins of the $^3\Pi_{2g} \rightarrow {}^3\Sigma_u^+(1_u)$ and $^3\Pi_{1g} \rightarrow {}^3\Sigma_u^+(0_u^-)$ transitions.

[p]Vibrational constants from (6), rotational constants from (9). (18) give $B_0 = 0.2235$, $\alpha_e = 0.0018$ (i.e. $B_e = 0.2244$) without mentioning whether this is based on a revised analysis. This state is heavily perturbed by a $^3\Pi_u$ state (18); as a result none of the constants are very meaningful.

[q]Spin splitting constant $\lambda \approx -4.7$ for v=0,2,4 (v=6,7 are also inverted) but +9.5 for v=1,3,5 (43); see [p]. (18) give $\gamma = +0.05$ for v=0,1,4.

[r]Breaking-off in emission (at low pressure) above

$$\left.\begin{matrix} J=61 \\ J=58 \\ J=59 \end{matrix}\right\}\text{ in v=8, and }\left.\begin{matrix} J=37 \\ J=36 \\ J=35 \end{matrix}\right\}\text{ in v=9 }\left.\begin{matrix} (F_1\text{ component}) \\ (F_2\text{ component}) \\ (F_3\text{ component}) \end{matrix}\right\}(32).$$

These together with similar breaking-off points in $^{34}S_2$ and $^{32}S^{34}S$ yield a predissociation limit at $35636.3\ cm^{-1}$ above $X\ {}^3\Sigma_g^-(v=0,J=0)$ of $^{32}S_2$. Bands with $v' \geq 10$ are absent in emission [except at high pressure (1)(2)(11)] and broadened in absorption. Above v'=18 there is strongly increased diffuseness indicating a second predissociation (7). Pressure effects on the intensity distribution of the absorption bands (5)(14).

[s]Lifetime $\tau(v=0...3) = 17$ ns [phase shift method (33)];
$\tau(v=3,4) = 19._5$ ns [Hanle effect (41)].
The most recent measurements [single-photon time correlation (51)] give $\tau = 45.0$ ns.

[t]Secondary heads on the short-wavelength side of the bands are formed by the forbidden $^TR_{31}$ branches [(10), see also (39)]. Experimental Franck-Condon factors (v"=0...25) from resonance fluorescence series with v'=3,4 (42), see also (40). Theoretical Franck-Condon factors (14)(38) [see, however, [p]]. $^{32}S_2/^{34}S_2$ isotope shifts (34). Absorption in inert gas matrices at low temperature (19).

[u]The observed position of v'=0 relative to $X\ {}^3\Sigma_g^-(v''=0)$ is at $31659\ cm^{-1}$; strong vibrational perturbation.

[v]$\lambda = -39.0$ (and $\gamma = +0.008$) derived from the observed $F_1(N) - F_2(N)$ splittings (49).

[w]$B^+(F_2) - B^-(F_1) = +0.0021$.

[x]$z \approx 22000\ cm^{-1}$, very rough estimate based on the fact that the upper state of B'→A' is predissociated and, therefore, cannot lie below the dissociation limit $^3P + {}^3P$. A similar value is obtained by extrapolation of the vibrational levels in A' and X to their common limit $^3P + {}^3P$.

[y]Assuming that b is the lower state of the three singlet systems originating from h, i, j.

[z](35) estimate $x \approx 4700\ cm^{-1}$.

[a']Refers to the F_2 component.

(continued on p. 563, ref. on p. 567)

State	T_e	ω_e	$\omega_e x_e$	B_e	α_e	D_e (10^{-} cm^{-1})	r_e (Å)	Observed Transitions Design.	Observed Transitions ν_{00}	References
$^{32}S_2^+$		μ = 15.9858993		D_0^0 = 5.37 eV[a]						JUL 1977
Several additional unresolved photoelectron peaks with vertical I.P.'s of 15.58 ($^2\Pi_u$), 17.73, 18.10, 18.66, 23.33, 25.99 eV (2).										
B $^2\Sigma_g^-$	41820	540	All constants obtained by photoelectron spectroscopy (2); data of somewhat lower resolution are given by (1). Internuclear distances derived from Franck-Condon factor analyses of the photoelectron spectra (2).				(1.98_3)			
b $^4\Sigma_g^-$	30920	580					(1.93_6)			
A $^2\Pi_u$	22390	530					(2.04_2)			
a $^4\Pi_u$	17440	600					(2.05_8)			
X $^2\Pi_{g,r}$	470 0	790					(1.82_5)			
$^{32}S_2^-$		μ = 15.9861735		D_0^0 = 3.95_5 eV[a]		I.P. = 1.66_3 eV[b]				JUL 1977
A' ($^2\Pi_u$)	20220[c]	364.2_5[d]	2.00	Constants for S_2^- ions dissolved in KI crystals, from spectra studied at 2 K (5)(7); see also (8). Host crystals other than KI lead to somewhat different values.				A'← X,	20102	(7)
A ($^2\Pi_u$)	20143							A→X',	19452	(5)(8)
								A↔X,	20025	(2)(5)(7)(8)
X' ($^2\Pi_{g,i}$)	573[c]	600.8	3.01					Raman sp.		(3)(8)
X ($^2\Pi_{g,i}$)	0	600.8[e]	2.16					EPR sp.		(1)

S_2^+: [a]$D_0^0(S_2)$ + I.P.(S) - I.P.(S_2).

(1) Berkowitz, JCP 62, 4074 (1975).

(2) Dyke, Golob, Jonathan, Morris, JCS FT II 71, 1026 (1975).

S_2^-: [a]From $D_0^0(S_2)$ and the electron affinities of S_2 and S.

[b]From laser photodetachment experiments (6).

[c]The splitting is due to the crystal field, not spin-orbit coupling which in the $^2\Pi_g$ state amounts to approximately -420 cm^{-1} (1)(7).

[d]The Raman spectrum of ions pumped into this state by laser irradiation consists of a sharp line shifted by 362 cm^{-1} (4).

S_2^- (continued):

[e](3) predict a gas phase frequency of ~550 cm^{-1}.

(1) Vannotti, Morton, PR 161, 282 (1967).

(2) Rolfe, JCP 49, 4193 (1968).

(3) Holzer, Murphy, Bernstein, JMS 32, 13 (1969).

(4) Holzer, Racine, Cipriani, AdRS 1, 393 (1973).

(5) Ikezawa, Rolfe, JCP 58, 2024 (1973).

(6) Celotta, Bennett, Hall, JCP 60, 1740 (1974).

(7) Vella, Rolfe, JCP 61, 41 (1974).

(8) Sawicki, Fitchen, JCP 65, 4497 (1976).

S_2 (continued):

(1) Asundi, Nature 127, 93 (1931).
(2) Asundi, CS 3, 154 (1934).
(3) Wieland, Wehrli, Miescher, HPA 7, 843 (1934).
(4) Rosen, Désirant, BSRSL 4, 233 (1935).
(5) Kondratjew, Olsson, ZP 99, 671 (1936).
(6) Olsson, Thesis, Stockholm (1938).
(7) Herzberg, Mundie, JCP 8, 263 (1940).
(8) Maeder, HPA 21, 411 (1948).
(9) Ikenoue, SL 9, 79 (1960).
(10) Meakin, Barrow, CJP 40, 377 (1962).
(11) Sugden, Demerdache, Nature 195, 596 (1962).
(12) Tanaka, Ogawa, JCP 36, 726 (1962).
(13) Barrow, Ketteringham, CJP 41, 419 (1963).
(14) Herman, Felenbok, JQSRT 3, 247 (1963).
(15) Haranath, ZP 173, 428 (1963).
(16) Narasimham, CS 33, 261 (1964).
(17) Narasimham, Brody, PIAS A 59, 345 (1964).
(18) Barrow, du Parcq, in "Elemental Sulphur" (ed. Meyer), p. 251, New York Interscience (1965).
(19) Brewer, Brabson, Meyer, JCP 42, 1385 (1965).
(20) Asundi, JCP 43, S24 (1965).
(21) Narasimham, Bhagvat, PIAS A 61, 75 (1965).
(22) Narasimham, Gopal, CS 34, 454 (1965).
(23) Drowart, Goldfinger, QR 20, 545 (1966).
(24) Narasimham, Apparao, Nature 210, 1034 (1966).
(25) Lakshminarayana, Narasimham, CS 36, 533 (1967).
(25a) Channappa, Pendlebury, Smith, in "La Structure Hyperfine Magnétique des Atomes et des Molécules" (ed. Moser and Lefebvre), p. 73, éditions du CNRS (Paris, 1967).
(26) Barrow, du Parcq, JP B 1, 283 (1968).
(27) Budininkas, Edwards, Wahlbeck, JCP 48, 2859 (1968).
(28) Berkowitz, Lifshitz, JCP 48, 4346 (1968).
(28a) Donovan, Husain, Jackson, TFS 64, 1798 (1968).
(29) Barrow, du Parcq, Ricks, JP B 2, 413 (1969).
(30) Berkowitz, Chupka, JCP 50, 4245 (1969).
(31) Ricks, Barrow, JP B 2, 906 (1969).
(32) Ricks, Barrow, CJP 47, 2423 (1969).
(33) Smith, JQSRT 9, 1191 (1969).
(34) Chaudhry, Upadhya, Nair, IJPAP 8, 52 (1970).
(35) Carleer, Colin, JP B 3, 1715 (1970).
(36) Donovan, Husain, Stevenson, TFS 66, 1 (1970).
(37) Barletta, Claassen, McBeth, JCP 55, 5409 (1971).
(38) Smith, Liszt, JQSRT 11, 45 (1971).
(39) Tatum, Watson, CJP 49, 2693 (1971).
(40) Yee, Barrow, Rogstad, JCS FT II 68, 1808 (1972).
(41) Meyer, Crosley, JCP 59, 1933 (1973).
(42) Meyer, Crosley, JCP 59, 3153 (1973).
(43) Meyer, Crosley, CJP 51, 2119 (1973).
(44) Barrow, Yee, APH 35, 239 (1974).
(45) Wayne, Davies, Thrush, MP 28, 989 (1974).
(46) Freedman, Jones, Rogstad, JCS FT II 71, 286 (1975).
(47) Hopkins, Brown, JCP 62, 1598 (1975).
(48) Narasimham, Sethuraman, Apparao, JMS 59, 142 (1976).
(49) Narasimham, Apparao, Balasubramanian, JMS 59, 244 (1976).
(50) Mahajan, Lakshminarayana, Narasimham, IJPAP 14, 488 (1976).
(51) McGee, Weston, CPL 47, 352 (1977).

State	T_e	ω_e	$\omega_e x_e$	B_e	α_e	D_e $(10^{-9}cm^{-1})$	r_e (Å)	Observed Transitions Design.	Observed Transitions ν_{00}	References
$^{(121,123)}Sb_2$		$(\mu = 60.947907_4)$		$D_0^0 = 3.09$ eV[a]						JUL 1977
Fragments of other emission band systems 11900 - 13900 (V shaded), 23800 - 27800, 33300 - 34500 cm^{-1} (R shaded).										(6)
U	(70194)	$[272]^{b}$		Weak diffuse bands.				$U \leftarrow X$,	70195	(7)
M	(63258)	$[152]^{c}$ H		Three sharp bands.				$M \leftarrow X$,	63199 H	(7)
I	(59142)	$[210]^{d}$ H		Weak system.				$I \leftarrow X$,	59112 H	(7)
G	(53888)	$[185]^{e}$ H						$G \leftarrow X$,	R 53846 H	(7)* (10)(11)
E	(48645)	$[228]^{f}$ H		The bands appear diffuse.				$E \leftarrow X$,	48624 H	(7)* (10)
F	44780	226.0^{g} H	1.17					$F \leftarrow X$,	(R) 44758 H	(1)(10)
H	(44329)	$[479]^{h}$ H		Single progression, observed in $^{121}Sb_2$.				$H \leftarrow X$,	V $(44433)^{h}$ H	(11)
D	(32087)	$[209.6]^{i}$ H						$D \leftrightarrow X$,	R 32057 H	(2)(6)
B 0_u^+	19068.9	218.0_8 H	0.53_7	$B_2 = 0.044844^{j}$		$D_2 = 9.2$	$r_2 = 2.4835$	$B \leftrightarrow X$,	R 19043.0 H	(3)(4)(8)
A	14991	217.2 H	0.44					$A \leftrightarrow X$,	R 14965 H	(3)(4)
X $^1\Sigma_g^+$	0	269.9_8 H	0.58_8	$B_2 = 0.050447^{j}$		$D_2 = 9.5$	$r_2 = 2.3415$			
$^{121}Sb^{75}As$		$\mu = 46.257057_6$								JUL 1977
A $(^1\Pi)$	27366	[204.7] H						$A \rightarrow X$,	R 27297.1 H	(1)
X $^1\Sigma^+$	0	343.0 H	0.8							
$^{(121)}Sb^{209}Bi$		$(\mu = 76.592111_7)$								JUL 1977
A	40647^{a}	190.2 H	0.73					$A \leftarrow X$,	R 40632 H	(1)*
X $^1\Sigma^+$	0	220.0 H	0.50							
$^{(121)}Sb^{(79)}Br$		$(\mu = 47.750100_7)$								JUL 1977
		Two emission continua with maxima at 15400 and 19200 cm^{-1}.								(1)(4)
C $(^3\Pi)$		Strong diffuse absorption bands in the flash photolysis of $SbBr_3$, 41200 - 44200 cm^{-1}. $\Delta G' \approx 340$, $\Delta G'' \approx 258$; analysis seems doubtful.						$C \leftarrow X$,	V 44220 43269 42265	(2)

State	T_e	ω_e	$\omega_e x_e$	B_e	α_e	D_e (10^{-} cm^{-1})	r_e (Å)	Observed Transitions Design.	ν_{00}	References
$^{(121)}Sb^{(79)}Br$ (continued)										
B_2	(33740)	201.0 H	0.30					$B_2 \rightarrow X,^a$ R	33719 H	(1)(3)
B_1	(32372)	207.6 H	0.85					$B_1 \rightarrow X,^a$ R	32354 H	(1)(3)
A	(19749)	215.8 H	0.35					$A \rightarrow X,^a$ R	19736 H	(1)(4)
X_2 X_1 $(^3\Sigma^-)$ 1 0^+	0	242.7 H	0.5_6							

Sb_2: [a]Thermochemical value (mass-spectrom.)(5)(9).

[b]$\Delta G(3/2) = 254$.

[c]$\Delta G(3/2) = 149$.

[d]$\Delta G(3/2,5/2) = 216, 217$.

[e]$\Delta G(3/2,5/2) = 195, 205$ (10); (7) give also $\Delta G(7/2) = 196$. Bands of $^{121}Sb_2$ (11).

[f]$\Delta G(3/2...9/2) = 216, 217, 220, 246$ (7) [flash photolysis of SbH_3]; (10) [high temp. abs.] give no assignments.

[g](10) express doubts about the correctness of the analysis.

[h]$\Delta G(3/2...13/2) = 541, 528, 477, 517, 548, 472$. The constants assume v'=0,v"=2 for the first b. at 43897 cm^{-1}.

[i]Levels up to v'=17 have been observed; the ΔG values are irregular due to perturbations and vary between 208 and 228. (1) give different assignments and constants.

[j]Constants for $^{121,123}Sb_2$ from the rotational analysis of the B→X 2-2 band by (8) who give also constants for $^{121}Sb_2$ and $^{123}Sb_2$.

(1) Nakamura, Shidei, JJP 10, 11 (1935).
(2) Naudé, SAJS 32, 103 (1935).

Sb_2 (continued):
(3) Almy, JPC 41, 47 (1937).
(4) Almy, Schultz, PR 51, 62 (1937).
(5) De Maria, Drowart, Inghram, JCP 31, 1076 (1959).
(6) Mrozowski, Santaram, JOSA 57, 522 (1967).
(7) Donovan, Strachan, TFS 67, 3407 (1971).
(8) Sfeila, Perdigon, Martin, Fémelat, JMS 42, 239 (1972).
(9) Kordis, Gingerich, JCP 58, 5141 (1973).
(10) Topouzkhanian, Sibai, d'Incan, ZN 29 a, 436 (1974).
(11) Sibai, Topouzkhanian, d'Incan, CR B 280, 79 (1975).

SbAs: (1) Yee, Jones, CC (1969), 752.

SbBi: [a](1) give erroneously 40617.

(1) See ref. (1) of Sb_2.

SbBr: [a]Undecided whether the lower state is X_1 or X_2.

(1) Singh, Avasthi, IJPAP 1, 197 (1963).
(2) Danon, Chatalic, Deschamps, Pannetier, CR C 269, 1249 (1969).
(3) Avasthi, ZN 26 a, 250 (1971).
(4) Avasthi, Sharma, Sud, ZN 30 a, 695 (1975).

State	T_e	ω_e	$\omega_e x_e$	B_e	α_e	D_e ($10^{-7}cm^{-1}$)	r_e (Å)	Observed Transitions Design.	Observed Transitions ν_{00}	References
$^{(121)}Sb^{35}Cl$		(μ = 27.1238563)								JUL 1977
E	(47358)	[(446)] H						$E \leftarrow X$, V	47394 H	(4)
D	(45216)	[436] H						$D \leftarrow X$,[a] V	45247[b] H	(4)* (5)
C	(43069)	444 H	3					$C \leftarrow X$,[a] V	43103[b] H	(2)(4)* (5)
B	(41616)	448 H	4					$B \leftarrow X$,[a] V	41652 H	(5)
A_2	(25906)	$240._{9}$[c] H	0.8_{5}					$A_2 \rightarrow X$,[a] R	25839[c] H	(1)* (3)
A_1	(22178)	$237._{5}$[c] H	1.5					$A_1 \rightarrow X$,[a] R	22109[c] H	(1)* (3)
b 0^+	12148.7	382.1 H	0.6					$b \rightarrow X_2$,[d] V	12152.4 H^Q	(6)*
								$b \rightarrow X_1$,[e] V	12958.9 H^P	(6)*
X_2 ($^3\Sigma^-$) 1	$816._{3}$[f]	374.7 H	0.6							
X_1 0^+	0									
$^{121}Sb^{19}F$		μ = 16.4184644		D_0^0 = (4.4) eV						JUL 1977
C_3 1	44756.7	[696.2_{7}] Z	3.0 H	0.2983[a]	0.0029	(2.2)	1.855_{3}	$C_3 \rightarrow b$, V	31148.9 Z	(8)*
								$C_3 \rightarrow a$, V	37983.53 Z	(1a)(8)*
								$C_3 \leftrightarrow X_1$, V	44802.6 Z	(1a)(2)(8)*
C_2	44310.4	700.9 H	2.8_{0}					$C_2 \rightarrow X_2$, V	43558.2 H	(1a)(2)
A_3 1	28706.6	411.1 Z	1.67	0.2414	0.0017_{0}	(3.3)	2.062_{4}	$A_3 \rightarrow b$, R	14953.8[b] (Z)	(9)*
								$A_3 \rightarrow a$, R	21788.3[b] (Z)	(1)* (9)*
								$A_3 \rightarrow X_2$, R	27809.7[b] (Z)	(1a)(9)*
A_2 2	24788.4	420.3 Z	1.71	0.2411	0.0016_{5}	(3.2)	2.063_{6}	$A_2 \rightarrow a$, R	17874.6[c] (Z)	(9)*
								$A_2 \rightarrow X_2$, R	23896.1[c] (Z)	(1)* (9)*
A_1	22589.4	418.9 H	2.6_{4}					$A_1 \rightarrow X_1$, R	22493.7 H	(3)*
A 0^+	$21407._{5}$	416 H	2.1	[0.2385_{3}]		[3.2_{2}]	[2.074_{7}]	$A \rightarrow X_2$, R	20513.0[d] (Z)	(6)(9)*
								$A \rightarrow X_1$,[e] R	21311	(9)
b $^1\Sigma^+$	13651.1	615.5 H	2.8	0.2815	0.0025	(2.4)	1.909_{8}	$b \rightarrow X_2$,	12856.1_{5} Z	(7)*
								$b \rightarrow X_1$,	13653.6_{8} Z	(7)*

State	T_e	ω_e	$\omega_e x_e$	B_e	α_e	D_e (10^{-7}cm^{-1})	r_e (Å)	Observed Transitions Design.	ν_{00}	References
$^{121}Sb^{19}F$ (continued)										
a $^1\Delta$	6816.0	616.3 Z	2.7_0	0.2806	0.0018_7	(2.3)	1.912_9			
$X_2\,(^3\Sigma^-)$ 1	796.3	612.6 H	2.6_1	0.2805^f	0.0020	(2.4)	1.913_2			
X_1 0^+	0	[605.0] Z	2.6_2 H	0.2792	0.0020	(2.3)	1.917_7			

SbCℓ: [a]It is not decided whether the lower state is X_1 or X_2, both of the emission and of the absorption bands.

[b]Average values of constants given by different groups of authors.

[c]Constants of (3) who has observed a few more bands than (1) as well as Cℓ isotope shifts. The lower-state constants for these two systems agree poorly with those derived from the b→X bands (6).

[d]P and Q heads.

[e]Single sequence of P heads.

[f]From the P heads of the b→X transitions.

(1) Ferguson, Hudes, PR 57, 705 (1940).

(2) Basco, Yee, SpL 1, 19 (1968).

(3) Avasthi, SpL 3, 291 (1970).

(4) Danon, Chatalic, Pannetier, CR C 272, 1411 (1971).

(5) Briggs, Kemp, JCS FT II 68, 1083 (1972).

(6) de Bie - Prévot, Thèse (U. Libre de Bruxelles, 1974).

SbF: [a]Ω-type doubling $\Delta\nu_{ef}(v=0) = (+)8._9 \times 10^{-5} J(J+1)$. The rotational analyses by (4) and (5) are in disagreement with (8).

[b]Extrapolations from v'-0 bands having v' ≥ 2.

[c]Extrapolations from v'-0 ($A_2 \rightarrow X_2$) and 0-v" ($A_2 \rightarrow a$) bands having v' ≥ 1 and v" ≥ 2, respectively.

[d]Extrapolated from the 0-2 band, the only rotationally analysed band. The lower-state vibrational numbering of (6) has been increased by 1 (9).

[e]Band heads observed, but no details given (9).

[f]Ω-type doubling $\Delta\nu_{ef}(v=0) = +0.00031\,J(J+1)$ (9).

(1) Rochester, PR 51, 486 (1937).

(1a) Howell, Rochester, PPS 51, 329 (1939).

(2) Patel, Abraham, IJPAP 7, 641 (1969).

(3) Abraham, Patel, JP B 3, 882 (1970).

(4) Abraham, Patel, JP B 3, 1183 (1970).

(5) Abraham, Patel, JP B 4, 1398 (1971).

(6) Chakravorty, Abraham, Patel, JP B 6, 757 (1973).

(7) Wang, Jones, Prévot, Colin, JMS 49, 377 (1974).

(8) Prévot, Colin, Jones, JMS 56, 432 (1975).

(9) Vasudev, Jones, JMS 59, 442 (1976).

State	T_e	ω_e	$\omega_e x_e$	B_e	α_e	D_e (10^{-4} cm^{-1})	r_e (Å)	Observed Transitions: Design.	Observed Transitions: ν_{00}	References
$^{(121)}Sb^1H$		(μ = 0.99949368)								JUL 1977
$A_3(^3\Pi)$ 0^+	[30115.8]			[4.10]			[2.02_8]	$A_3 \leftarrow X_2$, R	29460.8 Z	
								$A_3 \leftarrow X_1$, R	30115.8 Z	
B 0^+	[29761.2]			[4.06]			[2.03_8]	$B \leftarrow X_2$, R	29106.2 Z	(1)(2)
								$B \leftarrow X_1$, R	29761.2 Z	
$A_2(^3\Pi)$ 1	[28541]							$A_2 \leftarrow X_2$, R	27886 H	
$A_1(^3\Pi)$ 2	[26901]	Very diffuse bands, rotational structure not resolved.						$A_1 \leftarrow X_2$, R	26246 H	
$X_2(^3\Sigma^-)$ 1	[654.97]									
X_1 0^+	[0]			[5.684][a]		[2.4]	[1.722_6]			
$^{(121)}Sb^2H$		(μ = 1.98109965)								JUL 1977
C 0^+	[30566.4]			[1.95]			[2.09]	$C \leftarrow X_2$, R	29906.4 Z	
$A_3(^3\Pi)$ 0^+	[30159.5]			[2.06]			[2.03]	$A_3 \leftarrow X_2$, R	29499.5 Z	
								$A_3 \leftarrow X_1$, R	30159.5 Z	
B 0^+	[29959.8]			[2.10]			[2.01]	$B \leftarrow X_2$, R	29299.8 Z	(1)
								$B \leftarrow X_1$, R	29959.8 Z	
$A_2(^3\Pi)$ 1	[28677]	Diffuse, partly resolved rotational structure.						$A_2 \leftarrow X_2$, R	28017.3 H	
$A_1(^3\Pi)$ 2	[27017.9]	Fragments only.		[2.53]		[10]	[1.83]	$A_1 \leftarrow X_2$, R	26357.9 Z	
$X_2(^3\Sigma^-)$ 1	[660.01]									
X_1 0^+	[0]			[2.8782][b]		[0.45]	[1.7194]			
$^{(121)}Sb^{127}I$		(μ = 61.915748_2)								JUL 1977
	(41909)	[264] H							41942 H	
C ($^3\Pi$)	(41387)	[247] H						$C \leftarrow X$, V	41411 H	(1)
	(40846)	[244] H							40869 H	
X	0	[198] H								

State	T_e	ω_e	$\omega_e x_e$	B_e	α_e	D_e $(10^{-7}cm^{-1})$	r_e (Å)	Observed Transitions Design.	ν_{00}	References
$^{(121)}Sb^{14}N$		(μ = 12.5495824$_2$)								JUL 1977 A
B ($^1\Pi$)	34465	830.7 H	6.0					B↔X, R	34409 H	(1)* (2)
X $^1\Sigma^+$	0	942.0 H	5.6							
$^{(121)}Sb^{16}O$		(μ = 14.1261081$_1$) $D_0^0 \leq$ 4.39 eV (see [c])								JUL 1977
E $^2\Sigma^{(+)}$	39785	849.$_4$[a] H^Q	2.9					E↔X, V	37531 39802 H^Q	(2)(5)
F ($^2\Delta$)	(38958)	[588.$_5$] H^Q						F↔X, R	36573.7 38844.$_5$ H^Q	(5)
G		Diffuse heads 35800 - 38950 cm^{-1}.						G↔X, (V)		(5)
D $^2\Pi_{\frac{1}{2}}$	34544	505.$_9$[b] H	3.0	B_1 = 0.2777[c]		D_1 = 4.0	r_1 = 2.073	D→X[d], R	34389.$_5$[e]	(3)*(5)*(7)*
C ($^2\Delta_r$)	30315 29747	568.$_1$ H 570.$_6$ H	3.2$_8$ 3.5$_2$	B_2 = 0.2991[f]		D_2 = 3.25	r_2 = 1.997	C→X, R	27919.6 29624.$_8$[g] H	(3)*(5)*(8)*
B $^2\Sigma$	26594	582.$_0$ H^Q	6.5					B→X, R	24204 26476 H^Q	(1)(3)* (4)*
A $^2\Pi_r$	20801 20668	569.$_0$ H	5.0					A→X, R	18405 20544 H	(1)
X $^2\Pi_r$	2272 0	816 H^Q	4.2	0.3580[fh]	0.0022	[2.7]	1.825$_8$			

Sb^1H, Sb^2H:

[a]Spin splitting constants λ = +333.39, γ = -0.44.

[b]Spin splitting constants λ = +333.01 (slight J dependence), γ = -0.19.

(1) Bollmark, Lindgren, CPL 1, 480 (1967); PS 10, 325 (1974).

(2) Basco, Yee, SpL 1, 13 (1968).

SbI: (1) Danon, Chatalic, Deschamps, Pannetier, CR C 269, 1249 (1969).

SbN: (1) Coy, Sponer, PR 58, 709 (1940).

(2) Ciach, Thistlethwaite, JCP 53, 3381 (1970).

SbO: [a]From the analysis by (5).

[b]Vibrational numbering of (5) increased by 1; see (6)(7).

[c]v=2 of D $^2\Pi_{\frac{1}{2}}$ predissociated at low J.

[d]X $^2\Pi_{\frac{1}{2}}$ component.

[e]Calculated from the origin of the 1-1 band.

[f]Rotational constants for ^{121}SbO.

(continued p. 575)

State	T_e	ω_e	$\omega_e x_e$	B_e	α_e	D_e ($10^{-7}cm^{-1}$)	r_e (Å)	Observed Transitions: Design.	Observed Transitions: ν_{00}	References
$^{121}Sb^{31}P$		μ = 24.6570049		D_0^0 = 3.68 eV[a]						JUL 1977
B $^1\Pi$	28136	[394.0][b] Z		0.127_7[b]	0.000_2	(3)	2.31_4	B→X, R	28083.69 Z	(1)* (3)*
X $^1\Sigma^+$	0	500.07 Z	1.632[c]	0.1406	0.0005	0.4	2.205			
$^{(121)}Sb^{32}S$		(μ = 25.2855155)								JUL 1977
		Additional unassigned emission bands (both V and R shaded) in the region 12000 – 28500 cm^{-1}.								(1)
G		(380) H	(4.4)					G→(X), R	$24626._0$ H $27026._6$ H	(1)
F		(296) H						F↔X, R	$27406._0$ H $28664._0$ H	(1)*
E		(442) H						E↔X, R	$26248._3$ H $26720._4$ H	(1)*
D		(389) H						D↔X, R	$24310._4$ H -	(1)*
C		(390) H						C→X, R	$23249._3$ H $25996._4$ H	(1)*
B		(397) H	(16.5)					B→(X), R	$22517._2$ H -	(1)*
A		(341) H						A↔X, R	$19472._6$ H $22278._9$ H	(1)*
X $^2\Pi_r$		(470) H (480) H	(1.6) (1.2)							
$^{(121)}Sb^{(80)}Se$		(μ = 48.113717_0)								JUL 1977
D	43756	365.7_4 H	0.76					D←X, V	43776 H	(1)
C	(41600)	Unassigned V shaded bands 40700 – 42500 cm^{-1}.						C←X, V		(1)
B	36041	418.0 H	0.48					B←X, V	36087 H	(1)
A	28965	221.8 H	1.0					A←X, R	28913 H	(1)
X	0	326.1 H	1.04							

State	T_e	ω_e	$\omega_e x_e$	B_e	α_e	D_e (10^{-} cm^{-1})	r_e (Å)	Observed Transitions Design.	ν_{00}	References
$^{(121)}Sb^{(130)}Te$		(μ = 62.621731$_9$)		D_0^0 = 2.8 eV[a]						JUL 1977
A	43553	314.5 H	0.48					A← X,	V 43568 H	(1)
X	0	284.4 H	0.20							

SbO (continued):

[g]Calculated from the origin of the 2-0 band (^{121}SbO).

[h]Λ-type doubling in $^2\Pi_{\frac{1}{2}}$: $\Delta\nu = 0.107(J+\frac{1}{2})$.

(1) Sen Gupta, IJP 13, 145 (1939).
(2) Sen Gupta, IJP 17, 216 (1943).
(3) Lakshman, ZP 158, 367 (1960).
(4) Lakshman, ZP 158, 386 (1960).
(5) Shimauchi, SL 9, 109 (1960).
(6) Rao, Rao, CS 37, 310 (1968).
(7) Rai, Upadhya, Rai, JP B 3, 1374 (1970).
(8) Rai, Rai, Rai, CJP 52, 592 (1974).

SbP: [a]Thermochemical value (mass-spectrom.) (2).

[b]$\Delta G(3/2) = 395.6$; strong perturbations in v=0 and 1 make the evaluation of equilibrium constants not very meaningful.

[c]$\omega_e y_e$ = +0.0025.

(1) Yee, Jones, Kopp, JMS 33, 119 (1970).
(2) Kordis, Gingerich, JPC 76, 2336 (1972); JCP 58, 5141 (1973).
(3) Jones, Flinn, Yee, JMS 52, 344 (1974).

SbS: (1) Shimauchi, Nishiyama, SL 17, 76 (1968).

SbSe, SbTe:

[a]Thermochemical value (mass-spectrom.) (2), recalculated with the revised value for the dissociation energy of Te_2.

(1) Sharma, PPS A 66, 1109 (1953).
(2) Porter, Spencer, JCP 32, 943 (1960).

State	T_e	ω_e	$\omega_e x_e$	B_e	α_e	D_e ($10^{-7} cm^{-1}$)	r_e (Å)	Observed Transitions Design.	Observed Transitions ν_{00}	References
$^{45}Sc_2$		μ = 22.477958$_7$		D_0^0 = 1.6$_5$ eV[a]						MAY 1975
$^{45}Sc^{35}Cl$		μ = 19.6692074		D_0^0 = (3.4) eV						MAY 1975
y	x + 27189.7	482.7 H^Q	2.5					y → x,[a] V	27202.0 H^Q	(2)
x	x	458.2 H^Q	2.8							
e ($^3\Pi$)	d + 22260.0	312.5 H^Q	0.55					e → d, V	22461.0 H^P 22361.5 H^P 22267.6[b] H^Q	(1)(2)
d ($^3\Sigma$)	d	297.3 H^Q	0.61							
c ($^3\Delta$)	a + 13113.8	355.9 H	2.18					c → a, R	13096.1 H 13092.4 H 13090.7 H	(1)(2)
b ($^3\Phi$)								b → a, R	12596.2 H 12567.6 H 12537.6 H	(1)(2)
a ($^3\Delta$)	a	398.3 H	1.36							
F ($^1\Pi$)	31249.9	364.7 H	1.0					F → X, R	31208.8 H	(1)(2)
E ($^1\Sigma^+$)	27033.3	472.1 H	1.32					E → X, R	27045.8 H	(1)(2)
D $^1\Pi$	21521.1	373.1 H^R	1.6	[0.1569][c]		[1.6]	[2.337]	D → X, R	21484.0 H^R	(1)(2)
B $^1\Pi$	17613.3	374.3 H^R	2.3	[0.1551][c]		[1.1]	[2.351]	B → X, R	17576.6 H^R	(1)(2)
A $^1\Sigma^+$	12431.2	373.9 H	0.9	[0.1574]		[1.1]	[2.333]	A → X, R	12394.7 H	(1)(2)
X $^1\Sigma^+$	0	447.4 H	1.8	0.1725	0.0010	[1.1]	2.229			

Sc_2: [a]Thermochemical value (mass-spectrom.)(1)(2).

(1) Verhaegen, Smoes, Drowart, JCP 40, 239 (1964).

(2) Verhaegen, PhD Thesis, University of Brussels (1965) [quoted by Drowart in "Phase Stability in Metals and Alloys"; Rudman, Stringer, Jaffee, Eds.; McGraw-Hill (1967)].

ScCℓ: [a]Unidentified system.

[b]P head at 22263.1 cm^{-1}.

[c]Small Λ-type doubling.

(1) Shenyavskaya, Mal'tsev, Gurvich, VMUK 22(4), 104 (1967)

(2) Shenyavskaya, Mal'tsev, Kataev, Gurvich, OS(Engl. Transl.) 26, 509 (1969).

State	T_e	ω_e	$\omega_e x_e$	B_e	α_e	D_e (10^{-} cm^{-1})	r_e (Å)	Observed Transitions Design.	ν_{00}	References
$^{45}Sc^{19}F$		μ = 13.3546988		$D_0^0 = 6.1_7$ eV[a]						MAY 1975
H $^1\Pi$				[0.3671][b]			[1.854_3]	H↔X, R	38806.1 Z	(5)(9)
		0-0 sequence of V shaded absorption bands. Complex structure.						V	35942.0 H[Q]	(5)(9)
G $^1\Pi$	35009	[570] H		[0.378][bc]			[1.82_7]	G↔X, R	34926.7[d] Z	(1)(2)(5)
$^3\Phi_4$				[0.3463]			[1.909_2]		27202.2 Z	
g $^3\Phi_3$				[0.3441]			[1.915_3]	g← a, R	27171.1 Z	(5)
$^3\Phi_2$				[0.3413]			[1.923_1]		27138.2 Z	
F $^1\Pi$	26891.5	[565.3] Z	(3.1)	0.3461[e]	0.0025		1.909_8	F←X, R	26809.6 Z	(2)(5)
		Complex group of R shaded absorption bands.						R	26300	(5)
		Strong 0-0 sequence of V shaded bands.[f]						V	21927 H	(5)(7)(9)
E $^1\Pi$	20383.5	622.1 Z	3.7	0.3630[bc]	0.00296		1.864_8	E↔X, R	20326.8 Z	(2)(4)(5)(7)
($^3\Pi_2$)				[0.3677][b]			[1.852_8]		18361.4 Z	
d ($^3\Pi_1$)								d← a, R	18336.0	(5)
C $^1\Sigma^+$	16164.7	589.6 Z	2.64	0.3473	0.0024		1.906_5	C↔X, R	16092.0 Z	(2)(4)(5)
$^3\Phi_4$	a_3 + 15356.9			0.3545	0.00310		1.887_0		15317.6 Z	
c $^3\Phi_3$ g	a_2 + 15316.8	570.4 Z	2.96	[0.3511]			[1.896_1]	c↔a, R	15277.5 Z	(3)(5)(7)
$^3\Phi_2$	a_1 + 15273.6			[0.3490]			[1.901_8]		15234.4 Z	
B $^1\Pi$ g	10735.4_9	586.25 Z	2.015[h]	0.3431_0[i]	0.0026_2[j]		1.9181	B↔X, R	10661.25 Z	(4)(5)(10)
$^3\Delta_3$	a_3 [k]	649.1_1 Z		0.3706	0.00258		1.845_6			
a $^3\Delta_2$ g	a_2	648.9_8 Z	3.03	0.3665	0.00254		1.855_9			
$^3\Delta_1$	a_1	648.9_1 Z		0.3623	0.00250		1.866_6			
X $^1\Sigma^+$	0 [ℓ]	735.6 Z	3.8	0.3950	0.00266		1.787_7			

ScF: [a]Thermochemical value (mass-spectrom.)(8).

[b]Perturbations.

[c]B values are for the f levels.

[d]From (2); 34920.7 in (5) appears to be a misprint.

[e]Λ-type doubling $\Delta v_{fe} = +\ 0.0021 \times J(J+1)$.

[f]The authors of (5) assume the bands to arise from a transition $^3\Delta \leftarrow a\ ^3\Delta$ with nearly equal coupling constants in both states. The upper state, then, has $\omega_e = 724.7$, $\omega_e x_e = 3.64$. The system was also observed in emission by (7) who interpreted it as $^3\Pi \rightarrow {}^3\Sigma$ with $\omega'_e = 591.0$, $\omega'_e x'_e = 2.73$, $\omega''_e = 514.1$, $\omega''_e x''_e = 1.98$. Neither (5) nor (7) give enough details.

[g]For a reassignment of molecular orbital configurations see (11).

[h]$\omega_e y_e = -\ 0.0633$.

[i]Λ-type doubling $\Delta v_{fe} \begin{Bmatrix} (v=0) = +\ 1.115 \\ (v=1) = +\ 0.714 \\ (v=2) = +\ 0.452 \\ (v=3) = +\ 0.315 \end{Bmatrix} \times 10^{-3} J(J+1) - \ldots$

[j]$\gamma_e = +\ 3.25 \times 10^{-5}$.

[k]Theoretical calculations by (6) put the $a\ ^3\Delta$ state at 461 cm^{-1} and the corresponding $^1\Delta$ state at 2612 cm^{-1} above $X\ ^1\Sigma^+$.

[ℓ]That $X\ ^1\Sigma^+$ is the ground state is strongly suggested by the fact that only $B \leftarrow X$, $C \leftarrow X$, and $E \leftarrow X$, but none of the triplet systems, were observed in matrix absorption at 4 K (4).

ScF (continued):

(1) Gurvich, Shenyavskaya, OS(Engl. Transl.) 14, 161 (1963).

(2) Barrow, Gissane, Le Bargy, Rose, Ross, PPS 83, 889 (1964).

(3) Barrow, Gissane, PPS 84, 615 (1964).

(4) McLeod, Weltner, JPC 70, 3293 (1966).

(5) Barrow, Bastin, Moore, Pott, Nature 215, 1072 (1967).

(6) Carlson, Moser, JCP 46, 35 (1967).

(7) See ref. (1) of ScCℓ.

(8) Zmbov, Margrave, JCP 47, 3122 (1967).

(9) Barrow, in DONNSPEC (1970).

(10) Barrow, Pedersen, JP B 4, L11 (1971).

(11) Scott, Richards, CPL 28, 101 (1974).

State	T_e	ω_e	$\omega_e x_e$	B_e	α_e	D_e $(10^{-7} cm^{-1})$	r_e (Å)	Observed Transitions Design.	Observed Transitions ν_{00}	References
$^{45}Sc^{1}H$		$\mu = 0.98572711$								OCT 1977
$^{45}Sc^{2}H$		$\mu = 1.92773633$								
		Weak emission bands in the green ($\sim$17900 cm^{-1}) and near infrared (11800 - 12300 cm^{-1});								(4)*
		absorption bands of complex structure in the region 17690 - 18350 cm^{-1}; no analysis.								(1)*
	Theoretical calculations (2)(3) predict a $^1\Sigma^+$ or $^3\Delta$ ground state, and a dissociation energy of $\sim$1.8 eV.									
$^{32}S^{(35)}Cl$		$(\mu = 16.7016919_7)$								JUL 1977
		A single progression of absorption bands starting at 22644 cm^{-1} with a spacing of 460 - 490 cm^{-1} and tentatively assigned to SCℓ has been observed in the flash photolysis of $S_2C\ell_2$ (1).								
$^{45}Sc^{16}O$		$\mu = 11.7974775_7$		$D_0^0 = 6.96$ eV[a]						JAN 1977 A
B $^2\Sigma^+$	20645.1	825.4_7 Z	4.2_1	[0.48308][bc]	0.0032	[6.74]	[1.7198_6]	B↔X,[d] R	20571.15 Z	(1)(2)(3)(4)* (5)* (6)* (7)* (8)(14)(15)*
A $^2\Pi_r$	16547.0[e]	876.0[f] H	5.0_0	[0.50277][g]	(0.0037)	[6.5_4]	[1.6858_5]	A↔X,[d] R	16498.13[h] Z	(1)(2)* (3) (4)* (5)* (6)* (7)* (8)(14) (16)(18)*
A' $^2\Delta_r$	$15135._9$ i $15029._8$	$845._9$ H	$4._9$					A'→X,[j] R	$15072._0$ $14965._9$ H	(21)
X $^2\Sigma^+$ k	0	[964.9_5] Z	4.2_0 H	[0.51343][ℓc]	0.0033	[5.85]	[1.6682_6]	ESR sp.[m]		
$^{45}Sc^{32}S$		$\mu = 18.6841468$		$D_0^0 = 4.9_2$ eV[a]						JAN 1977
B $^2\Sigma$	12497.6	488.2 H	2.0					B→X, R	12459.0 H	(4)
A $^2\Pi$		Emission bands in the region 11000 - 12600 cm^{-1}, v' numbering unknown. Progression of absorption bands in Ne matrix, 11100 - 12500 cm^{-1}, tentatively attributed to ScS (2).						A→X, R	(11150) (11040)	(4)
X $^2\Sigma$	0	565.2[b] H	1.8					ESR sp.[c]		

Sc^1H, Sc^2H:

(1) Smith, PRS A 332, 113 (1973).
(2) Scott, Richards, JP B 7, 1679 (1974).
(3) Kunz, Guse, Blint, JP B 8, L358 (1975).
(4) Bernard, Effantin, Bacis, CJP 55, 1654 (1977).

ScCl: (1) Donovan, Husain, Jackson, TFS 64, 1798 (1968).

ScO: [a]Thermochemical value (mass-spectrom.)(11)(12)(13).
[b]Spin-splitting constant γ_0 = (-)0.0670.
[c]Potential energy curves (20).
[d]Intensity distribution (19)(22).
[e]A_0 = +115.33; slight J dependence.
[f]Corrected for head-origin separations.
[g]Λ-type doubling in $^2\Pi_{1/2}$, $\Delta v_{fe}(v=0) = -0.0655(J+\frac{1}{2}) + \ldots$, and in $^2\Pi_{3/2}$ [see (18)].
[h]$J'=\frac{1}{2}$ (average of F_1 and $\{F_2\}$) relative to N"=0. The sub-band origins ($\{J'=0\}$ relative to N"=0) are at 16440.61 and 16554.83 cm^{-1}.
[i]T_e in good agreement with theoretical predictions (9)(21).
[j]Observed in the "single-collision" $Sc + O_2$ chemiluminescence spectrum.
[k]The $^2\Sigma$ character of the ground state was theoretically predicted by (9) and experimentally verified by (10)(14) [ESR spectrum and matrix absorption].
[ℓ]Large magnetic hyperfine structure, 4b = +0.254 (15)(18) [see also (10)(14)].
[m]In rare gas matrices (10)(14)(17).

(1) Johnson, Johnson, PRS A 133, 207 (1931).
(2) Meggers, Wheeler, JRNBS 6, 239 (1931).
(3) Jevons, "Band Spectra of Diatomic Molecules", Phys. Soc. London (1932).

ScO (continued):

(4) Piccardi, GCI 63, 127 (1933).
(5) Gatterer, RS 1, 153 (1942).
(6) Gatterer, Junkes, Salpeter, Rosen, METOX (1957).
(7) Åkerlind, AF 22, 41 (1962).
(8) Ortenberg, Glasko, Dimitriev, SAAJ 8, 258 (1964).
(9) Carlson, Ludeña, Moser, JCP 43, 2408 (1965).
(10) Kasai, Weltner, JCP 43, 2553 (1965).
(11) Ames, Walsh, White, JPC 71, 2707 (1967).
(12) Coppens, Smoes, Drowart, TFS 63, 2140 (1967).
(13) Drowart, Pattoret, Smoes, PBCS No. 8, 67 (1967).
(14) Weltner, McLeod, Kasai, JCP 46, 3172 (1967).
(15) Adams, Klemperer, Dunn, CJP 46, 2213 (1968).
(16) Athénour, Bacis, Féménias, Stringat, CR B 271, 567 (1970).
(17) McIntyre, Lin, Weltner, JCP 56, 5576 (1972).
(18) Stringat, Athénour, Féménias, CJP 50, 395 (1972).
(19) Féménias, Athénour, Stringat, CJP 52, 361 (1974); 53, 2353 (1975) (erratum).
(20) Narasimhamurthy, Murthy, Pramāṇa 2, 269 (1974).
(21) Chalek, Gole, JCP 65, 2845 (1976).
(22) Féménias, Goure, Stringat, CJP 55, 180 (1977).

ScS: [a]Thermochemical value (mass-spectrom.)(1); (3) give a slightly lower value.
[b]IR spectrum in Ar matrix at 4 K gives $\Delta G(\frac{1}{2}) = 554$ (2).
[c]In Ar matrix at 4 K (2).

(1) See ref. (12) of ScO.
(2) See ref. (17) of ScO.
(3) Tuenge, Laabs, Franzen, JCP 65, 2400 (1976).
(4) Stringat, Fenot, CJP 54, 2293 (1976).

State	T_e	ω_e	$\omega_e x_e$	B_e	α_e	D_e ($10^{-8}cm^{-1}$)	r_e (Å)	Observed Transitions Design.	ν_{00}	References
$^{45}Sc^{(80)}Se$		(μ = 28.771125_4)		$D_0^0 = 3.9_6$ eV[a]						MAY 1975
$^{45}Sc^{(130)}Te$		(μ = 33.398044_4)		$D_0^0 = 3.0_5$ eV[a]						MAY 1975
$^{80}Se_2$		$\mu = 39.958262_7$		D_0^0, see [a].		I.P. = 8.88 eV[b]				JUL 1977
F (1_u)	(55421)	[430.2] H						$(F \leftarrow X_2)$ V	54932.8 H	(13)
E 0_u^+	54752.5	403.9 H	1.3	0.0924	0.00033		2.13_7	$E \leftarrow X_2$,[c] V	54249	(13)(22)
								$E \leftarrow X_1$, V	54761.7 H	(13)
D (1_u)	(53075)	[426.2] H		[0.0965]			$[2.09_1]$	$D \leftarrow X_1$,[d] V	53096.1 H	(13)*
C_2 $^3\Sigma_u^-$ 0_u^+	(53324)	[414] H		Diffuse bands				$C_2 \leftarrow X_1$, V	53339 H	(13)
C_1 $^3\Sigma_u^-$ 1_u	53220.5	428.0 H	1.22	0.09664	0.00033_3		2.089_4	$C_1 \leftarrow X_2$, V	52730.9 Z	(13)*
B_2 $^3\Sigma_u^-$ 1_u	$26058._6$	246.42[e] Z	1.225	0.07086[fg]	0.00055_3	(2)	2.440_0	B_2[h] $\leftrightarrow X_2$,[i] R	$25478._2$[j]	(5)
								$B_2 \rightarrow X_1$, R	$25989._2$	(21)(22)
B_1 $^3\Sigma_u^-$ 0_u^+	25980.36	246.291[ek] Z	1.016[ℓ]	0.07048[m]	0.000345	(4)	2.446_6	$B_1 \rightarrow X_2$, R	25399.8	(21)(22)
								$B_1 \leftrightarrow X_1$,[i] R	25910.84 Z	(5)*
X_2 $^3\Sigma_g^-$ 1_g	510.0[n]	387.156[o] Z	0.9640[o]	0.09019[op]	0.000299[o]	(2)	2.162_8	t		
X_1 $^3\Sigma_g^-$ 0_g^+	0	385.303 Z	0.9636_3[q]	0.08992	0.000288[r]	2.4[s]	2.166_0			

ScSe: [a]Thermochemical value (mass-spectrom.)(1)(2).

(1) Bergman, Coppens, Drowart, Smoes, TFS 66, 800 (1970).
(2) Ni, Wahlbeck, HTS 4, 326 (1972).

ScTe: [a]Thermochemical value [(1), no details].

(1) See ref. (1) of ScSe.

Se_2: [a]From the predissociation in B_1 0_u^+ (see [m]) three possible spectroscopic values for the dissociation energy of $^{80}Se_2$, i.e. $D_0^0 = 3.410_5$, 3.163_8, 3.096_4 eV, can be derived depending on the assumed atomic states at the observed predissociation limits. (5) prefer $D_0^0 = 3.164$ eV on the basis of indirect spectroscopic arguments. However, both photoionization (10)(24) and thermochemical studies [mass-spec-

Se_2 (continued):

trometry (6)(9)(12), Knudsen-torsion effusion (8); see also (11)(17)] strongly favour the higher value D_0^0 = 3.411 eV.

[b]Photoionization mass-spectrometry (10). From the photoelectron spectrum (23) derive adiabatic and vertical ionization potentials of 8.70 and 8.89 eV, respectively.

[c]This transition is much weaker than $E \leftarrow X_1$.

[d]The assumption (13) that this transition is $C_1 \leftarrow X_1$ has been withdrawn (22) since it gives the wrong $X_1 - X_2$ splitting.

[e]Vibrational analysis confirmed by isotope investigations | (5).

[f]Average value, $B_e(F_3) - B_e(F_2) = +0.00038$.

[g]Rotational perturbations; a tentative analysis (18) of these and similar perturbations in the 0_u^+ component was based on an erroneous value of the $0_u^+ - 1_u$ splitting. Predissociation in v=5 at J=72(F_3) and 73(F_2); bands with v'=6 have not been seen.

[h]Lifetime τ(v=0,J=105) = 58 ns, from Hanle effect measurements (19)(20) combined with experimentally determined Landé g_J factors (20)(25).

[i]Various proposed other "systems" in the region 14500 - 18500 cm^{-1} (1)(4) have been shown (5) to belong to the main B−X system. (18) have extended the rotational analysis of the $^{78}Se_2$ bands to higher values of v".

[j]Extrapolated from bands having v" ≥ 8, using lower state constants of (14).

[k]The $B_1\ 0_u^+$ state has a substantial potential maximum arising from an avoided crossing with a repulsive 0_u^+ state. The interaction strongly affects vibrational levels above v=15; the theoretical discussion by (16) predicts irregular level shifts and widths above the crossing point.

[l]$\omega_e y_e = -0.00549$.

[m]Sharp predissociation limits occur for v=4(J=106), v=5(J=82), v=6(J=50), leading to a dissociation limit at 27508 cm^{-1} above $X_1\ 0_g^+$(v=0), and for v=13(J=96), v=14(J=78), v=15(J=50), leading to a dissociation limit at 29498 cm^{-1}. (5) attribute the former to $^3P_2 + {}^3P_1$, the latter to $^3P_1 + {}^3P_1$. There are many rotational perturbations and several accidental predissociations; see also [g].

[n]From (22) whose measurements of fluorescence series in the range 6 ≤ v" ≤ 12 lead to $\Delta v = +509.95 + 2.125_6(v+\frac{1}{2})$ for the separation of the rotationless 1_g and 0_g^+ substates.

[o]Constants apply to 8 ≤ v ≤ 29.

[p]Average of F_2 and F_3, $B_e(F_3) - B_e(F_2) = +0.00006$.

[q]$\omega_e y_e = -0.000818_4$.

[r]$\gamma_e = -6.1 \times 10^{-7}$.

[s]$\beta_e = -0.023 \times 10^{-8}$.

[t]For a theoretical calculation of the magnetic moment of Se_2, consistent with experimental results, see (20a).

(1) Rosen, Monfort, Physica 3, 257 (1936).
(2) Shin-Piaw, AP(Paris) (11) 10, 173 (1938).
(3) Migeotte, BSRSL 10 (12), 658 (1941).
(4) Leelavathi, Rao, IJP 29, 1 (1955).
(5) Barrow, Chandler, Meyer, PTRSL A 260, 395 (1966).
(6) Berkowitz, Chupka, JCP 45, 4289 (1966).
(7) Drowart, Goldfinger, QR 20, 545 (1966).
(8) Budininkas, Edwards, Wahlbeck, JCP 48, 2867 (1968).
(9) Colin, Drowart, TFS 64, 2611 (1968).
(10) Berkowitz, Chupka, JCP 50, 4245 (1969).
(11) Meschi, Searcy, JCP 51, 5134 (1969).
(12) Uy, Drowart, TFS 65, 3221 (1969).

(continued p. 585)

State	T_e	ω_e	$\omega_e x_e$	B_e	α_e	D_e (10^{-} cm^{-1})	r_e (Å)	Observed Transitions Design.	Observed Transitions ν_{00}	References
$^{(80)}Se_2^+$		(μ = 39.958125$_5$)		D_0^0 = 4.3$_8$ eV[a]						JUL 1977
D ($^2\Pi_u$)	41210	From the maxima of the observed photoelectron peaks (2); vibrational structure unresolved. Additional partly resolved peaks in the region 17.3 - 20.6 eV above X $^3\Sigma_g^-$ of Se_2 (2).								
C ($^2\Sigma_g^-$)	35650									
B ($^2\Pi_u$)	(31620)									
b ($^4\Sigma_g^-$)	27260									
A ($^2\Pi_u$)	19200									
a ($^4\Pi_u$)	14440									
X_2 $^2\Pi_{g\frac{3}{2}}$	1940									
X_1 $^2\Pi_{g\frac{1}{2}}$	0									
$^{80}Se_2^-$		μ = 39.958399$_8$								JUL 1977
A ($^2\Pi_u$)	(16173)	216.$_7$	0.61$_5$	Constants for Se_2^- ions dissolved in KI crystals, from fluorescence and absorption spectra studied at 2 K (4)(5).				A$\leftrightarrow$X,	16192[a] 16040	(2)(4)(5)
X ($^2\Pi_g$)	0	330.$_3$	0.86					Raman sp.[b]		(3)
								EPR sp.[b]		(1)
$^{(80)}Se^{(79)}Br$		(μ = 39.707146$_1$)								JUL 1977
B_2 ($^2\Pi$)	(x + 46158)	[390] H						$B_2 \leftarrow X_2$,[a] V	46195 H	(1)
B_1	47227.3	392.5 H	2.0					$B_1 \leftarrow X_1$,[a] V	47265 H	(1)
X_2 ($^2\Pi_i$)	x									
X_1	0	316.9 H	0.7							
$^{(80)}Se^{(35)}Cl$		(μ = 24.3250212)								JUL 1977
A ($^2\Pi_i$)		[595] H	Single progression consisting of seven bands[b]; vibrational numbering uncertain.					A$\leftarrow$X,[a] R	27116 H	(1)
X ($^2\Pi_i$)	0									

State	T_e	ω_e	$\omega_e x_e$	B_e	α_e	D_e (10^{-} cm^{-1})	r_e (Å)	Observed Transitions Design.	ν_{00}	References
$^{(80)}Se^{19}F$		(μ = 15.3494168)		$D_0^0 = (3.2_1)$ eV[a]						JUL 1977
X $^2\Pi_i$	0[b]	(757)[a]		[0.3624]			[1.7408]	ESR sp.[c]		(1)(3)

Se_2 (continued):

(13) Barrow, Burton, Callomon, TFS 66, 2685 (1970).
(14) Barrow, Beattie, Burton, Gilson, TFS 67, 583 (1971).
(15) Tatum, Watson, CJP 49, 2693 (1971).
(16) Atabek, Lefebvre, CPL 17, 167 (1972).
(17) Smoes, Mandy, Auwera-Mahieu, Drowart, BSCB 81, 45 (1972).
(18) Yee, Barrow, JCS FT II 68, 1181 (1972).
(19) Dalby, Vigué, Lehmann, CJP 53, 140 (1975).
(20) Gouedard, Lehmann, CR B 280, 471 (1975).
(20a) Büchler, Meschi, JCP 63, 3586 (1975).
(21) Gouedard, Lehmann, JP B 9, 2113 (1976).
(22) Greenwood, Barrow, JP B 9, 2123 (1976).
(23) Streets, Berkowitz, JESRP 9, 269 (1976).
(24) Radler, Berkowitz, JCP 66, 2176 (1977).
(25) Gouedard, Lehmann, JP(Paris) 38, L-85 (1977).

Se_2^+: [a]Value obtained by photoionization mass-spectrometry (1), consistent with the highest of three possible spectroscopic values for $D_0^0(Se_2)$ [see [a] of Se_2].

(1) See ref. (10) of Se_2.
(2) See ref. (23) of Se_2.

Se_2^-: [a]Splitting due to the crystal field, not spin-orbit coupling.
[b]In NaI and KI crystals.

(1) Vannotti, Morton, JCP 47, 4210 (1967).
(2) Rolfe, JCP 49, 4193 (1968).
(3) Holzer, Murphy, Bernstein, JMS 32, 13 (1969).
(4) Ikezawa, Rolfe, JCP 58, 2024 (1973).
(5) Vella, Rolfe, JCP 61, 41 (1974).

SeBr: [a]Observed in the flash photolysis of Se_2Br_2.

(1) Oldershaw, Robinson, TFS 67, 907 (1971).

SeCℓ: [a]Observed in the flash photolysis of $SeC\ell_2$. The assignment of this spectrum to SeCℓ is not certain; it is assumed that only one component of $^2\Pi - {}^2\Pi$ is observed because of large doublet splitting in the lower state.
[b]The first and last bands of the progression are diffuse.

(1) Ciach, Power, Thistlethwaite, CPL 9, 349 (1971).

SeF: [a]Theoretical estimate (4).
[b]A_0(observed) ≈ -560 (1), A_0(calculated) = -1790 (3).
[c]$\mu_{e\ell}$ = 1.52 D (2). Magnetic hfs parameter, rotational and third-order g factors (3).

(1) Carrington, Currie, Miller, Levy, JCP 50, 2726 (1969).
(2) Byfleet, Carrington, Russell, MP 20, 271 (1971).

SeF (cont'd):

(3) Brown, Byfleet, Howard, Russell, MP 23, 457 (1972).
(4) O'Hare, JCP 60, 4084 (1974).

State	T_e	ω_e	$\omega_e x_e$	B_e	α_e	D_e $(10^{-7}cm^{-1})$	r_e (Å)	Observed Transitions Design.	ν_{00}	References
$^{(80)}Se^{1}H$		(μ = 0.99527385)		D_0^0 = (3.2) eV[a]		I.P. = (9.8) eV[b]				JUL 1977
F				Weak diffuse band.				$F \leftarrow X_1$,	71190	(5)
E				Diffuse band.				$E \leftarrow X_1$,	69604	(5)
D				Strong diffuse band.				$D \leftarrow X_1$,	66814	(5)
C				Strong double-headed band.				$C \leftarrow X_1$, R	55797 H^Q	(5)
A $(^2\Sigma^+)$	(31500)	[1232] (H)	(172)[c]	Diffuse bands.				$A \leftarrow X_1$, R	31048 (H)	(2)
X_2 $^2\Pi_{1/2}$	(1815)[d]									
X_1 $^2\Pi_{3/2}$	0	(2400)[c]		[7.78][e]			$[1.47_5]$	ESR sp.[f]		(1)(3)(4)
$^{(80)}Se^{2}H$		(μ = 1.96458951)		D_0^0 = (3.2) eV[a]						JUL 1977
A $(^2\Sigma^+)$	(31490)	[959] (H)	(87)[c]	Diffuse bands.				$A \leftarrow X_1$, R	31178 (H)	(2)
X_2 $^2\Pi_{1/2}$	(1815)[d]									
X_1 $^2\Pi_{3/2}$	0	(1708)[c]		$[3.94_0]$[e]			$[1.47_6]$	ESR sp.[g]		(3)(4)
$^{(80)}Se^{1}H^-$						I.P. = 2.21 eV[d]				JUL 1977
$^{80}Se^{16}O$		μ = 13.3274823_8		$D_0^0 \leq 4.4_1$ eV[b]						OCT 1977
A large number of V shaded emission bands in the region 40000 - 55000 cm^{-1} has tentatively been assigned to five or six systems with upper state vibrational frequencies ranging from ~940 to ~1030 cm^{-1} and lower state constants similar to those of the ground and first excited states of SeO.										(2)* (8)*
C_3 (0)	35484	581 H	3.5					$C_3 \rightarrow X_2$, R	35156 H	
C_2 $(^3\Pi_i)$ (1)	35405	585 H	4.1					$C_3 \rightarrow X_1$, R	35313 H	(10)*
C_1 (2)	-	-	-					$C_2 \rightarrow X_2$, R	35075 H	
								$C_2 \rightarrow X_1$, R	35240 H	
B_2 $^3\Sigma^-$ {1	(34376)	[517.5] Z		[0.3417][cd]	0.040_7		1.869	$B_2 \rightarrow X_2$, R	34012.2 Z	
B_1 {0^+	34277.8	522.3 Z	$3._9$[a]	[0.3332][d]	0.029_2		1.907	$B_1 \rightarrow X_1$, R	34081.8 Z	(1)(3)
A'	(17364)[e]	$885._2$ H	5.8_5					$A' \rightarrow X$,[f] R	17349 H	(6)

State	T_e	ω_e	$\omega_e x_e$	B_e	α_e	D_e ($10^{-7}cm^{-1}$)	r_e (Å)	Observed Transitions: Design.	v_{00}	References
$^{80}Se^{16}O$ (continued)										
A_3 (2)	(16773)	[980] H						$A_3 \rightarrow X_2$, V	16640^h H	
$A_2(^3\Pi_r)$(1)	16458	996 H	7.0	$[(0.47_0)]^g$			[(1.64)]	$A_2 \rightarrow X_2$, V	16334^h H	(11)*
								$A_2 \rightarrow X_1$, V	16495^h H	
A_1 (0)	16144	994 H	6.5					$A_1 \rightarrow X_2$, V	16015^h H	
								$A_1 \rightarrow X_1$, V	16185^h H	
b $^1\Sigma^+$	9723.5	838.8_3 (Z)	5.11	[0.4560]	(0.0035)	[8]	$[1.665_5]$	$b \rightarrow X_2$, R	9519.1 Z	(7)* (9)(12)*
								$b \rightarrow X_1$, R	9685.4 Z	
a $^1\Delta$	$(5300)^i$			[0.461]			$[1.65_6]$	ESR sp.[j]		(4)
X_2 $^3\Sigma^-$ {1	165.9	915.43^k Z	4.52	0.4738^ℓ	0.00339	5	1.633_9	ESR sp.[m]		(4)
X_1 {0^+	0	914.69^k Z	4.52	0.4655	0.00323		1.648_4			

Se^1H, Se^2H, Se^1H^-:

[a]Extrapolation of the vibrational levels in A $^2\Sigma^+$ assuming dissociation of this state into $^2S + {}^1D$.

[b]From Rydberg assignments of (5); doubtful.

[c]From isotope relations between SeH and SeD.

[d]From the photodetachment spectra of SeH^- and SeD^- (6).

[e]Λ-doubling frequencies (3).

[f]$\mu_{e\ell}$ = 0.49 D (4). ^{77}Se hf coupling (3).

[g]$\mu_{e\ell}$ = 0.483 D (4). ^{77}Se hf coupling (3).

(1) Radford, JCP 40, 2732 (1964).
(2) Lindgren, JMS 28, 536 (1968).
(3) Carrington, Currie, Lucas, PRS A 315, 355 (1970).
(4) Byfleet, Carrington, Russell, MP 20, 271 (1971).
(5) Donovan, Little, Konstantatos, JCS FT II 68, 1812 (1972).
(6) Smyth, Brauman, JCP 56, 5993 (1972).

SeO: [a]Emission bands with v' > 2 have not been observed, probably owing to predissociation. The limit is at ~35600 cm^{-1} above X_1(v"=0).

[b]From the predissociation in B $^3\Sigma^-$; see [a].

[c]Average of F_2 and F_3, $B_0(F_3) - B_0(F_2)$ = +0.0052.

[d]Rotational perturbations (1).

[e](6) give $17338._5$ which does not agree with their v_{00} value. It is not clear whether X_1 or X_2 is the lower state of this system.

[f](6) considered the upper state to be b $^1\Sigma^+$, but v_{00} appears too high for this interpretation.

[g]From P,Q head separations.

[h]Extrapolated from bands having v"=3.

[i]Theoretical estimate (9).

[j]$\mu_{e\ell}(a^1\Delta)$ = 2.01 D, from Stark effect on the ESR sp. (5).

(continued p. 589)

State	T_e	ω_e	$\omega_e x_e$	B_e	α_e	D_e $(10^{-8}cm^{-1})$	r_e (Å)	Observed Transitions Design.	ν_{00}	References
$^{78}Se^{32}S$		μ = 22.6698689		D_0^0 = (3.7) eV[a]						JUL 1977
B_2 $^3\Sigma^-$ 1	(28330)	$[327.5_6]$	b	[0.1349]	c		[2.348]	$B_2 \leftarrow X_2$, R	28011	(2)
B_1 0^+	28248.2	[330.82][d] Z	(2.92)	0.1369[e]	0.00121		2.331	$B_1 \leftarrow X_1$, R	28138.46 Z	(2)
A 0^+	$27328._4$	332.1[df]	(2.66)	0.1186[ef]	0.00095		2.504	$A \leftarrow X_1$, R	27216.5 (Z)	(2)
X_2 $^3\Sigma^-$ 1	205[g]	556.26 Z	1.831	0.1812_1[h]	0.00089	7	2.025_7			
X_1 0^+	0	555.56[d] Z	1.848	0.1792_6	0.00076	6	2.036_7			
$^{80}Se^{32}S^-$		μ = 22.83624								JUL 1977
A ($^2\Pi$)				The constants refer to SeS^- ions dissolved in KI crystals; from fluorescence spectra studied at 2 K (4).				$A \rightarrow X$,	17768.1	(2)(4)
X ($^2\Pi$)	0	468.9[a]	1.66					Raman sp.[b]		(3)
								EPR sp.[b]		(1)
$^{32}S^{19}F$		μ = 11.9170627		$D_0^0 \lesssim$ 3.3 eV[a]						JUL 1977
A_2 $^2\Pi$ 1/2		483[b] H	2.6[b]					$A_2 \leftarrow X_2$, R	25205 H	(2)
A_1 3/2		488 H	3.1	(0.554)[c]	0.004		(1.59_8)	$A_2 \leftarrow X_1$, R	25606 H	
								$A_1 \leftarrow X_1$, R	24995 H	(2)*
X_2 $^2\Pi$ 1/2	(401)[d]									
X_1 3/2	0			$[0.552174_0]$			1.600574	Microwave sp.[e]		(4)
								ESR sp.		(1)(3)

SeO (continued):

[k]From $B \rightarrow X$ bands with $v'' \geq 5$; vibrational numbering confirmed by isotope studies.

[ℓ]Average of F_2 and F_3, $B_0(F_3) - B_0(F_2) = +0.0048$ (12).

[m]Hf coupling of ^{77}Se (4).

(1) Barrow, Deutsch, PPS 82, 548 (1963).
(2) Haranath, JMS 13, 168 (1964).
(3) Haranath, IJPAP 3, 75 (1965).
(4) Carrington, Currie, Levy, Miller, MP 17, 535 (1969).
(5) Byfleet, Carrington, Russell, MP 20, 271 (1971).
(6) Kushawaha, Pathak, SpL 5, 393 (1972).
(7) Azam, Reddy, CJP 51, 2166 (1973).
(8) Reddy, Azam, JMS 49, 461 (1974).
(9) Barrow, Lemanczyk, CJP 53, 553 (1975).
(10) Verma, Azam, Reddy, JMS 58, 367 (1975).
(11) Verma, Azam, Reddy, JMS 65, 289 (1977).
(12) Verma, Reddy, JMS 67, 360 (1977).

SeS: [a]Estimate based on a Birge-Sponer extrapolation of the ground state vibrational levels (2); see also (1).

[b]$\Delta G(3/2) = 328.9_5$; see [c].

[c]$B_1 = 0.130_5$, $B_2 = 0.123$. Extensive perturbations by states having $\Omega = 1$ or 2.

[d]Vibrational numbering confirmed by isotope studies.

[e]Strong mutual perturbations between B and A as well as other perturbations in B.

[f]Only v'=3 and 4 have been analyzed, $\Delta G(7/2) = 310.79$.

[g]Estimated from the magnitude of the Ω-type doubling; see [h].

[h]$B_e(F_3) - B_e(F_2) = +0.00063$.

SeS (continued):

(1) Drowart, Goldfinger, QR 20, 545 (1966).
(2) Ahmed, Barrow, JP B 7, 2256 (1974).

SeS^-: [a]A Raman frequency of 464 cm^{-1} in KI has been observed by (3).

[b]In KI and NaI crystals.

(1) Vannotti, Morton, JCP 47, 4210 (1967).
(2) Rolfe, JCP 49, 4193 (1968).
(3) Holzer, Murphy, Bernstein, JMS 32, 13 (1969).
(4) Ikezawa, Rolfe, JCP 58, 2024 (1973).

SF: [a]From the predissociation in the $A_1 \leftarrow X_1$ bands; see [c].

[b]From the $A_2 \leftarrow X_2$ progression; slightly different numbers are obtained from $A_2 \leftarrow X_1$. The assignments of both progressions are tentative.

[c]Only $B_3 = 0.540$ and $B_5 = 0.532$ have been measured. Individual rotational lines are diffuse for $v' \geq 3$. Bands with $v' \geq 7$ are very diffuse.

[d]Based on the assignments of progressions $A_2 \leftarrow X_2$ and $A_2 \leftarrow X_1$; from the ESR spectrum (1) derive $A_0 = -387 \pm 25$.

[e]$\mu_{e\ell}(v=0) = 0.79_4$ D (4); (3) obtain 0.87 D. Hfs parameter $a + \frac{1}{2}(b+c) = 428.6_0$ MHz (1)(4).

(1) Carrington, Currie, Miller, Levy, JCP 50, 2726 (1969).
(2) Di Lonardo, Trombetti, TFS 66, 2694 (1970).
(3) Byfleet, Carrington, Russell, MP 20, 271 (1971).
(4) Amano, Hirota, JMS 45, 417 (1973).

State		T_e	ω_e	$\omega_e x_e$	B_e	α_e	D_e $(10^{-4}cm^{-1})$	r_e (Å)	Observed Transitions Design.	Observed Transitions ν_{00}	References
$^{32}S^{1}H$			$\mu = 0.97702732$		$D_0^0 = 3.5_5$ eV[ab]		I.P. = 10.43 eV[cb]				JUL 1977
H	$^2\Delta$				[9.46]			$[1.35_1]$	H← X,	80847.5[d]	(7)
G	$^2\Delta$				[9.01]			$[1.38_4]$	G← X,	79343.3[d]	(7)
F	$^2\Delta$				[9.19]			$[1.37_0]$	F← X,	76707.9[d]	(7)
E	$^2\Sigma$				[9.076]			$[1.378_8]$	E← X,	71317.7[d]	(7)
D	$^2\Delta$				[9.215]		[6.1]	$[1.368_4]$	D← X,	71194.7_2[d]	(7)
C	$^2\Delta$		Weak diffuse band near 1561 Å; sharper in SD.						C← X,	(64060)	(7)
B	$^2\Sigma$	(59641)	[2557.03] Z	(56.8)[e]	8.785	0.259	[8.2]	1.401_4	B← X, R	59621.7_1 Z	(7)
A	$^2\Sigma^+$	(31038)	1979.8 Z	97.6_5	8.521[fg]	0.46_4[h]	6.3_6[i]	1.423_0	A↔ X,[j] R	30662.4_2 Z	(2)* (4)* (11)
X	$^2\Pi_i$	0[k]	$(2711._6)$[ℓ]	$(59._9)$[ℓ]	$[9.461_1]$[m]	(0.27_0)[ℓ]	$[4.8_0]$	1.340_9	Vibration sp.[n]		(12)
									Hf Λ-doubling sp.[op]		(18)(19)
									EPR sp.[q]		(5)(6)(13) (16)(17)
$^{32}S^{2}H$			$\mu = 1.89474169$		$D_0^0 = 3.6_0$ eV[ab]		I.P. = 10.43 eV[cb]				JUL 1977
H	$^2\Delta$								H← X,	80858[d]	(7)
G	$^2\Delta$				[4.96]			$[1.33_9]$	G← X,	79319.8[d]	(7)
F	$^2\Delta$				[4.74]			$[1.37_0]$	F← X,	76717.4[d]	(7)
E	$^2\Sigma$				[4.739]		$[1.3_5]$	$[1.370_2]$	E← X,	71327.7[d]	(7)*
D	$^2\Delta$				[4.745]		[0.70]	$[1.369_3]$	D← X,	71205.39[d]	(7)*
C	$^2\Delta$				[4.693]			$[1.376_9]$	C← X,	63872.4_0[d]	(7)
B	$^2\Sigma$	59581[r]	[1859.16] Z	(29.3)[e]	4.532	0.105	[1.35]	1.401_1	B← X, R	59566.39 Z	(7)*
A	$^2\Sigma^+$	31039	1417.0 Z	48.8_5	4.392[s]	0.172[t]	1.7_6[u]	1.423_3	A↔ X,[j] R	30769.4_9 Z	(2)(4)* (11)*
X	$^2\Pi_i$	0[v]	$[1885._5]$ Z	$(30._9)$[w]	$[4.900_3]$[m]	0.10_0	$[1.3_5]$	1.3406	Vibration sp.[n]		(12)
									Hf Λ-doubling sp.[x]		(19)
									EPR sp.		(5)

S^1H, S^2H:

[a]From an extrapolation of the vibrational levels in A $^2\Sigma^+$ to the limit $^1D + ^2S$ (4); consistent with the observed predissociation in A $^2\Sigma^+$. Photoionization mass-spectrometry of H_2S (9) gives $D_0^0(S^1H) = 3.67$ eV (recalc. using updated auxiliary data). See also (20).

[b]Both D_0^0 and I.P. refer to the lowest existing molecular level, in agreement with definitions but contrary to the values given by (4) and (7), respectively, which refer to the zero-point of the Hill-Van Vleck expression for the ground state.

[c]Extrapolation of a short Rydberg series (7).

[d]The v_{00} values refer to the zero-point of the Hill-Van Vleck expression for the lower state and are exclusive of the J independent terms $-B\Lambda^2$ in the upper states, contrary to definitions normally adopted in these tables. Only the X $^2\Pi_{3/2}$ subbands have been observed.

[e]From isotope shifts.

[f]Spin splitting constant $\gamma_0 = +0.313$.

[g]The rotational lines of bands having $v' \geq 1$ are increasingly diffuse.

[h]$\gamma_e = -0.02_2$.

[i]$\beta_e = +0.6_0 \times 10^{-4}$.

[j]Observed in absorption (flash photolysis of H_2S and D_2S) by (2)(4), in matrix absorption by (12), and in emission by (1)(11). Franck-Condon factors (3).

[k]$A_0 = -376.9_6$ (2). On the basis of certain higher order corrections (15) gives $A_0 = -378.5_3$ (and $B_0 = 9.465$, $D_0 = 4.7 \times 10^{-4}$).

[ℓ]From the constants for SD (11).

[m]Λ-doubling and hfs parameters (19).

[n]In argon matrices at 20.4 K.

[o]Molecular beam electric resonance study. The strongest Λ-doubling transitions in $^2\Pi_{3/2}$, J=3/2 occur at 111.4862 (F=1 − 1) and 111.5452 MHz (F=2 − 2), in J=5/2 at 442.4781 (F=2 − 2) and 442.6277 MHz (F=3 − 3); these observations supersede earlier predictions (6)(16)(17) from EPR measurements.

[p]$\mu_{e\ell}(v=0) = 0.7580$ D (18)(19). Stark effect in EPR spectrum (8)(13a) yields the less accurate value 0.62 D. Predicted dipole moment function (20). Theoretical charge distributions (10).

[q]For EPR sp. of ^{33}SH and ^{33}S hf interaction see (14).

[r]Large electronic isotope shift.

[s]Spin splitting constant $\gamma_0 = +0.163$. The lines of the 2-0 absorption band are diffuse.

[t]$\gamma_e = -0.005$.

[u]$\beta_e = +0.1_0 \times 10^{-4}$.

[v]$A_0 = -376.7_5$ (2). Taking into account higher order corrections (15) derives $A_0 = -378.32$ (and $B_0 = 4.899$, $D_0 = 1.3 \times 10^{-4}$).

[w]Estimates by (2) and (11).

[x]$\mu_{e\ell}(v=0) = 0.7571$ D (19).

(1) Leach, CR 230, 2181 (1950).
(2) Ramsay, JCP 20, 1920 (1952).
(3) Nicholls, Fraser, Jarmain, McEachran, ApJ 131, 399 (1960).
(4) Johns, Ramsay, CJP 39, 210 (1961).
(5) McDonald, JCP 39, 2587 (1963).
(6) Radford, Linzer, PRL 10, 443 (1963).
(7) Morrow, CJP 44, 2447 (1966).
(8) Carrington, Levy, Miller, JCP 47, 3801 (1967).
(9) Dibeler, Liston, JCP 49, 482 (1968).
(10) Cade, Bader, Henneker, Keaveny, JCP 50, 5313 (1969).
(11) Pathak, Palmer, JMS 32, 157 (1969).

(continued p. 593)

State		T_e	ω_e	$\omega_e x_e$	B_e	α_e	D_e ($10^{-5}cm^{-1}$)	r_e (Å)	Observed Transitions		References
									Design.	ν_{00}	
$^{32}S^1H^+$			$D_0^0 = 3.4_8$ eV[a]								JUL 1977
			Theoretical predictions of several excited states (3)(4).								
A	$^3\Pi_i$	b			$[7.474_7]$[c]		$[62._7]$	$[1.520]$[d]	$A^e \rightarrow X$, R	29911.7_1[f] Z	(1)
X	$^3\Sigma^-$	0			$[9.134_0]$[g]		$[48._9]$	$[1.3744]$			
$^{32}S^1H^-$			$D_0^0 = 3.7_9$ eV[a] I.P. = 2.31_9 eV[b]								JUL 1977
X	$^1\Sigma^+$		The analysis of the shape of the photodetachment cross section curve (2) leads to ground state constants which are indistinguishable from those of S^1H ($X^2\Pi$), confirming theoretical predictions by (1).								
$^{28}Si_2$			μ = 13.9884643		$D_0^0 = 3.21$ eV[a]						AUG 1977
P	$^3\Pi_{g,i}$	b	Only v=0 observed.		$[(0.224)]$			$[(2.32)]$	$P \leftarrow D$, R	53132.4 H 53173.5 H 53219.2 H	(4)*
O	$^3\Sigma_u^-$	53395.5_8	404.2 H	3.0	0.222_5[c]	0.003	$[0.05_0]$	2.327	$O \leftarrow X$, R	53341.94 Z	(4)*
N	$^3\Sigma_u^-$	46789.1_0	458.6 H	4.8	0.2193	0.0025	$[0.023]$	2.344	$N \leftarrow X$, R	46762.21 Z	(3)* (4)*
L	$^3\Pi_{g,i}$	x+28629[d]	[494] H		$[0.2370]$			$[2.255]$	$L \leftrightarrow D$,[e] R	28602.2[f]	(1)* (6)
K	$^3\Sigma_u^-$	30794.0	462.6 Z	5.95 H	0.2186	0.0031_6		2.348	$K \leftarrow X$,[g] R	30768.8[h]	(3)*
H	$^3\Sigma_u^-$	24429.1_5	275.30[i] Z	1.99	0.1712[ij]	0.0013_5	0.030	2.653	$H \leftrightarrow X$,[g] R	24311.3_2[i] Z	(1)* (3)* (6)
D	$^3\Pi_{u,i}$	x[k]	547.94 Z	2.43	0.2596	0.0015_5		2.155	$(D \leftarrow X)$[ℓ]	(34730)[ℓ]	(5)
X	$^3\Sigma_g^-$	0	510.98 Z	2.02	0.2390	0.0013_5	0.021	2.246			

S^1H, S^2H (continued):

(12) Acquista, Schoen, JCP 53, 1290 (1970).
(13) Uehara, Morino, JMS 36, 158 (1970).
(13a)Byfleet, Carrington, Russell, MP 20, 271 (1971).
(14) Miller, JCP 54, 1658 (1971).
(15) Veseth, JMS 38, 228 (1971).
(16) Brown, Thistlethwaite, MP 23, 635 (1972).
(17) Tanimoto, Uehara, MP 25, 1193 (1973).
(18) Meerts, Dymanus, ApJ 187, L45 (1974).
(19) Meerts, Dymanus, CJP 53, 2123 (1975).
(20) Meyer, Rosmus, JCP 63, 2356 (1975).

S^1H^+: [a] $D_0^0(S^1H)$ + I.P.(S) - I.P.(S^1H).
[b] A_0 = -216.5 (1); see also (2).
[c] Effective B value; for Λ-doubling constants see (1)(2).
[d] From the "true" $B_0 = 7.47_2$ (2).
[e] Lifetime $\tau(v=0)$ = 1090 ns (5).
[f] Effective value (1); the "true" origin is at 29911.28 cm^{-1} (2), the subband origins at 29675.55, 29912.81, 30141.71 cm^{-1} (1).
[g] Spin splitting constants $\lambda_0 = +5.71_0$, $\gamma_0 = -0.16_5$ cm^{-1}.

(1) Horani, Leach, Rostas, JMS 23, 115 (1967).
(2) Horani, Rostas, Lefebvre-Brion, CJP 45, 3319 (1967).
(3) Cade, CJP 46, 1989 (1968).
(4) Liu, Legentil, Verhaegen, in "Selected Topics in Molecular Physics" (ed. Clementi), p. 19, Chemie GmbH (1972).
(5) Brzozowski, Elander, Erman, Lyyra, PS 10, 241 (1974).

S^1H^-: [a] From $D_0^0(S^1H)$ and the electron affinities of S^1H and S.
[b] From the photodetachment cross section (2).

(1) Cade, JCP 47, 2390 (1967).
(2) Steiner, JCP 49, 5097 (1968).

Si_2: [a] From the observed predissociation in H (3) and thermochemical data [mass-spectrom.(2), recalc.(3a)].
[b] A = -28.2.
[c] Spin splitting constants λ_0 = -6.68 (slight J dependence), γ_0 = +0.030.
[d] A = $-22._6$ (from A" and the observed subband origins).
[e] The $^3\Pi_0$ subbands are essentially complete, but only fragments of the other subbands have been observed.
[f] Average of the $^3\Pi_0$ and $^3\Pi_2$ subband origins. (1) gives 28059.1 which refers to the 0-1 rather than 0-0 band.
[g] Also observed in rare gas matrices (3b)(5).
[h] Extrapolated from the origins of the 1-0 and 2-0 bands. The 0-0 band (v_H = 30771) is completely diffuse.
[i] Corrected vibrational numbering of (6).
[j] The rotational lines of absorption bands having v'=6 are diffuse, indicating predissociation above 25877 cm^{-1}. Higher levels have not been observed.
[k] A = -71.6 (from the effective B values).
[ℓ] Progression of absorption bands in argon matrix, 34700 - 36300 cm^{-1}; tentative interpretation (5).

(1) Douglas, CJP 33, 801 (1955).
(2) Drowart, De Maria, Inghram, JCP 29, 1015 (1958).
(3) Verma, Warsop, CJP 41, 152 (1963).
(3a)Verhaegen, Stafford, Drowart, JCP 40, 1622 (1964).
(3b)Weltner, McLeod, JCP 41, 235 (1964).
(4) Lagerqvist, Malmberg, PS 2, 45 (1970).
(5) Milligan, Jacox, JCP 52, 2594 (1970).
(6) Dubois, Leclercq, CJP 49, 3053 (1971).

$^{(28)}Si^{(79)}Br$

(μ = 20.6547281) SEP 1977

Fragments of two other emission systems (30100 - 31500, 34900 - 36900 cm^{-1}) tentatively assigned to SiBr or $SiBr^+$. (2)

Additional absorption bands at 45762, 46266, 46343, 46693, 47445 cm^{-1}. (9)

State	T_e	ω_e	$\omega_e x_e$	B_e	α_e	D_e ($10^{-7}cm^{-1}$)	r_e (Å)	Observed Transitions: Design.	Observed Transitions: ν_{00}	References
(F)		[505][a] H						(F)← X,	45017 H	(9)
E	(44560) 44521	552 H	1.5	Bands with v'=1 and 2 are diffuse.				E← X, V	(44201) 44585 H	(9)
D	44017	[565][b] H						D← X, V	44088 H	(9)
C ($^2\Pi$)	41060 41051	531[c] H	2.0					C← X, V	40690[c] 41104[c] H	(9)
B ($^2\Sigma$)	$33572._4$	$571._2$ H	2.4	de				B↔ X, V	$33223._1$ $33645._1$ H	(1)* (3)(5) (9)
B'($^2\Delta$)	23911 23889	395 H	4					B'→ X, R	23473 23874 H[Q]	(7)
A ($^2\Sigma$)	20937.6	$250._3$ H	0.5					A→ X, R	(20428) $20850._9$ H	(7)(8)
X $^2\Pi_r$	$423._1$[f] 0	$424._3$ H	1.5	e						

$^{(28)}Si^{79}Br^+$

SEP 1977

State	T_e	ω_e	$\omega_e x_e$	B_e	α_e	D_e	r_e	Design.	ν_{00}	References
A	29005.4	428.7 H	6.9					A→ X,[g] R	28950.5 H	(4)
X	0	535.8 H	1.6							

$^{28}Si^{12}C$

μ = 8.39792238 $D_0^0 = 4.6_4$ eV[a] AUG 1977 A

No spectra have yet been conclusively assigned to SiC. The following constants are from <u>ab initio</u> calculations (3):

State	T_e	ω_e	$\omega_e x_e$	B_e	α_e	D_e	r_e	Design.	ν_{00}	References
a $^1\Sigma^+$	(6628)	(1018)		(0.695)			(1.70)			
A $^3\Sigma^-$	(5597)	(606)		(0.556)			(1.90)			
X $^3\Pi_i$	0[b]	(983)		(0.606)			(1.82)[b]			

State	T_e	ω_e		$\omega_e x_e$	B_e	α_e	D_e ($10^{-7}cm^{-1}$)	r_e (Å)	Observed Transitions Design.		ν_{00}		References
$^{28}Si^{35}Cl$		μ = 15.5422824											SEP 1977
F									F← X,		45968 46179	H	(12)
E									E← X,		45169 45375	H	(12)
D	44943.$_9$	659.4	H	3.8					D↔X,	V	44798.$_3$ 45005.$_8$	H	(1)(12)
C $^2\Pi$	41177.$_2$ 41165.$_7$	674.2	H	2.20	[0.2888][a]	0.0009	2.4	1.936	C↔X,	V	41039.$_9$ 41235.$_0$	H	(1)* (10) (12)
B' $^2\Delta$	35631.0[b]	511.1	Z	5.6	0.2618$_7$	0.00243	[3.59][c]	2.035$_2$	B'↔X,	V_R	35411.2$_5$[d] 35617.8$_5$[d]	Z	(4)(6)* (9)

SiBr, $SiBr^+$:

[a]Only two bands; it is not certain that they belong to the same system.

[b]It is possible that the v'=1 progression is in fact a separate system D'← X.

[c]Vibrational numbering uncertain.

[d]Emission bands with v' > 2 have not been observed.

[e](6) report the following rotational constants for the B− X bands: $B_0' = 0.1771$, $B_0''(^2\Pi_{\frac{1}{2}}) = 0.1598$. See also (10) who give considerably different results.

[f]From (5); (3) prefer a doublet separation of 418.0 cm^{-1} leading to slightly different constants.

[g]Br isotope shifts clearly observed.

(1) Miescher, HPA 8, 587 (1935).

(2) Asundi, Karim, PIAS A 6, 281 (1937).

(3) Jevons, Bashford, PPS 49, 554 (1937).

(4) Kuznetsova, Kuzmenko, Kuzyakov, OS(Engl. Transl.) 24, 434 (1968).

SiBr, $SiBr^+$ (continued):

(5) Kuznetsova, Kuzmenko, Kuzyakov, VMUK 23(3), 30 (1968).

(6) Kuznetsova, Kuzyakov, VMUK 24(6), 103 (1969).

(7) Kuznetsova, Kuzyakov, ZPS 10(3), 413 (1969).

(8) Rao, Haranath, JP B 2, 1381 (1969).

(9) Oldershaw, Robinson, TFS 67, 1870 (1971).

(10) Mishra, Khanna, IJP 46, 1 (1972).

SiC: [a]Thermochemical value (mass-spectrom.)(1).

[b](2) assumes a $^1\Sigma$ ground state and estimates r_e = 1.65 Å from an extrapolation to SiC of the shortening of Si-O, Si-N single bonds in the corresponding diatomic molecules.

(1) Verhaegen, Stafford, Drowart, JCP 40, 1622 (1964).

(2) Lovas, ApJ 193, 265 (1974).

(3) Lutz, Ryan, ApJ 194, 753 (1974).

SiCℓ: [a]From (10); rotational analyses of a few additional sub-bands reported by (13).

(continued p. 597)

State	T_e	ω_e	$\omega_e x_e$	B_e	α_e	D_e ($10^{-7}cm^{-1}$)	r_e (Å)	Observed Transitions Design.	ν_{00}	References
$^{28}Si^{35}Cl$ (continued)										
B $^2\Sigma^+$	34108.6	706.6[e] H	3.9[e]	[0.2784][f]	0.0017[f]	1.8	1.971	B↔X,[g] V	33987.1 34193.6 Z	(1)* (2)(3) (4)(5)(12) (16)
A $^2\Sigma$	23113.9	[294.9$_5$] Z	0.73 H	0.1986	0.0007	[2.$_9$][h]	2.337	A→X, R	22788.0 22994.7 Z	(7)* (15)*
X $^2\Pi_r$	206.6[i] 0	535.60 Z	2.16$_8$[j]	0.2561[k]	0.0016	2.5[ℓ]	2.058			
$^{28}Si^{19}F$	μ = 11.3148108	D_0^0 = 5.5$_7$ eV[a]		I.P. = 7.2$_8$ eV[b]						SEP 1977
	Theoretical studies of low-lying valence states (22).									
I	[52834]							I→X, V	52325[c] 52489[c] H	(5)
H ($^2\Sigma^+$)	(52095)	[1022] H						H→X, V	52098[c] 52260[c] H	(5)
(G)	(51941)	[1008] H						(G→X) V	51938[c] 52098[c] H	(5)
F	[52195]							F→X, V	51685[c] 51851[c] H	(5)
E	[51650]							E→X, V	51143[c] 51302[c] H	(5)
	Unassigned V shaded bands in the region 47639 - 48325 cm^{-1}.									(4)
D $^2\Sigma^+$	47418.6	1003.2	5.64	0.625	0.005$_5$		1.54$_4$	D→X, V	47491.4[d]	(4)(5)*
D' $^2\Pi$	46612.5[e]	1032.9	5.28	0.6329[f]	0.0044		1.534$_3$	D'→B,	(12061.8)	(6)
								D'→A, V UV ε bands	23912.6	(1)(6)
								D'→X,	(46700.2)	(6)
C' $^2\Pi$	41964.7[g]	1031.8	4.45	0.6376[f]	0.0039		1.528$_6$	C'→A, V Green ε b.	19264.5	(1)(6)
								C'→X,	(42052.1)	(6)
C $^2\Delta$	39438.0[h]	[878.38] Z	5.8[i]	0.60338[j]	0.00539	[12.1][k]	1.5713$_7$	C→X,[ℓ] V γ bands	39454.1$_4$ Z	(2)(5)(11)*

State	T_e	ω_e	$\omega_e x_e$	B_e	α_e	D_e ($10^{-7}cm^{-1}$)	r_e (Å)	Observed Transitions: Design.	Observed Transitions: ν_{00}	References
$^{28}Si^{19}F$ (continued)										
B $^2\Sigma^+$	34561.5	1011.2_3 Z	4.825	0.62707^j	0.00462	$10._1$	1.5414_0	$B\rightarrow A$,[m] V $B\rightarrow X$,[n] V β bands	11850.8_4 Z 34638.5	(5)* (1)* (2)(12)
a $^4\Sigma^-$	29805.06	863.16 Z	5.370	$0.5786_2{}^o$	0.00502	10.5	1.6046	$a\rightarrow X$,[p] η bands	29807.89 Z	(7)* (20)*
A $^2\Sigma^+$	22858.3	718.5_8 Z	10.167^q	0.57839^{rj}	0.00941^s	$15._7$	1.6049_6	$A^t\leftrightarrow X$,[u] R α bands	22787.64 Z	(1)* (3)(5)
X $^2\Pi_r$	0^v	857.19 Z	4.735	$0.5812_1{}^{wj}$	0.00494	10.7	1.6011			

SiCℓ (continued):

[b]$A_0 = +3.772$ [or -2.729, see (6)].

[c]$D_1 = 3.90 \times 10^{-7}$.

[d]The origins refer to the zero-point of the Hill-Van Vleck expression for the upper state and to the hypothetical J"=0 levels in the lower state.

[e]Corrected using calculated head-origin separations (5).

[f]Average values of (3) and (16). See also (8)(11).

[g]Franck-Condon factors (14).

[h]$D_1 = 1._9 \times 10^{-7}$.

[i]$A_v = +207.21 - 0.32v$; small J dependent terms have been determined for v=5...10 (15).

[j]$\omega_e y_e = +0.0053$.

[k]Λ-type doubling $\Delta\nu(^2\Pi_{\frac{1}{2}}) = 0.005(J+\frac{1}{2})$ (15)(16).

[ℓ]$\beta_e = -0.09 \times 10^{-7}$.

(1) Jevons, PPS 48, 563 (1936).
(2) Barrow, Drummond, Walker, PPS A 67, 186 (1954).
(3) Ovcharenko, Tunitskii, Yakutin, OS(Engl. Transl.) 8, 393 (1960).
(4) Thrush, Nature 186, 1044 (1960).

SiCℓ (continued):

(5) Ovcharenko, Kuzyakov, OS(Engl. Transl.) 13, 362 (1962).
(6) Verma, CJP 42, 2345 (1964).
(7) Sanii, Verma, CJP 43, 960 (1965).
(8) Cordes, Gehrke, ZPC (Frankfurt am Main) 51, 281 (1966).
(9) Ovcharenko, Kuzyakov, Tatevskii, OS(Engl. Transl.) Suppl. 2, 6 (1966).
(10) Ovcharenko, Kuzyakov, OS(Engl. Transl.) 20, 14 (1966).
(11) Mishra, Khanna, CS 38, 361 (1969).
(12) Oldershaw, Robinson, JMS 38, 306 (1971).
(13) Pandey, Upadhya, Nair, IJPAP 9, 36 (1971).
(14) Singh, Dube, IJPAP 9, 164 (1971).
(15) Singhal, Verma, CJP 49, 407 (1971).
(16) Rai, Singh, Upadhya, Rai, JP B 7, 415 (1974).

SiF: [a]Thermochemical value (mass-spectrom.)(9).

[b]Extrapolation of a short Rydberg series B,D,H,... (5); the value of (5) has been slightly modified in accordance with the definition of the I.P.. (9) give an electron impact ap-

(continued p. 599)

State	T_e	ω_e	$\omega_e x_e$	B_e	α_e	D_e ($10^{-4} cm^{-1}$)	r_e (Å)	Observed Transitions: Design.	Observed Transitions: ν_{00}	References
$^{28}Si^{1}H$		μ = 0.97278226		$D_0^0 \leq 3.06_0$ eV[a]		I.P. $\leq$ 8.04 eV[b]				AUG 1977
E $^2\Sigma^+$	[53411.2]	Only v=0 observed		[7.528]		[3.92]	[1.517_2]	E← X, V	52399.19 Z	(10)*
D $^2\Delta$	[49522.1]	Only v=0 observed		[7.90][c]			[1.48_1]	D← X, V	48510.1 Z	(5)(10)*
B $^2\Sigma^+$	[31842.2]	Only v=0 observed		[6.62][de]			[1.61_8]	B← X, R	30830.2[d] Z	(5)* (13)
C $^2\Sigma^+$	[31832.4]	Only one level observed		[1.17][d]			[3.8_5]	C← X, R	30820.4[d] Z	(5)(13)
		"Slightly diffuse" weak absorption bands in the region 25600 - 26700 cm^{-1}.								(4)
A $^2\Delta$	24300.4[f]	1858.90[g] Z	99.17_5	7.4664[g]	0.3445[h]	[5.24][i]	1.5234_7	$A^j \leftrightarrow X$,[k] R	24193.04 Z	(1)* (2)(5)* (6)(10)
X $^2\Pi_r$	0[ℓ]	2041.80[g] Z	35.51	7.4996[gm]	0.2190[n]	[3.97]	1.5201_0	Extensive theoretical calculations[o]		

Si^1H, Si^2H:

[a]From the predissociation in B $^2\Sigma^+$ assuming dissociation into $^1D + ^2S$ at the predissociation limit (5). According to (2) extrapolation of the vibrational levels in A $^2\Delta$ gives very nearly the same limit.

[b]From $D_0^0(Si^1H)$, I.P.(Si), and $D_0^0(Si^1H^+)$ (12).

[c]Increasing diffuseness with increasing N on account of predissociation.

[d]Deperturbed constants of (13) whose T_0 values correspond to ν_{00} + 1079.5 [see (6)]. As in similar cases, ν_{00} refers to the zero-point of the Hill-Van Vleck expression for the ground state. Interaction parameter $H_{B,C} = 16.1\ cm^{-1}$. The v numbering of the C level is uncertain.

[e]Strongly predissociated above N=2.

[f]$A_0 = 3.58$, $A_1 = 3.11$, $A_2 = 2.59$ (6). Discussion of second order spin-orbit splittings (14).

[g]Recalculated (10) from data for v=0,1,2 (6).

[h]$\gamma_e = -0.0418_5$.

[i]$D_1 = 6.08 \times 10^{-4}$, $D_2 = 7.36 \times 10^{-4}$.

[j](11) gives a radiative lifetime of 0.7 μs for both SiH and SiD corresponding to $f_{00} = 0.0037$ (16); see also (17).

[k]Potential functions, Franck-Condon factors (15)(16). | (6).

[ℓ]$A_0 = +142.83$, $A_1 = +143.43$, $A_2 = +144.04$. Slight J dependence

[m]For Λ-doubling constants ($p_0 = 0.0819$, $q_0 = 0.00831$) see (6) (20); the extrapolated splitting of the v=0, J=½ level is $\Delta\nu_{fe} = +0.0978\ cm^{-1}$ (2932 ± 20 MHz) (20); ab initio calculations (19) predict 0.1057 cm^{-1} (3168 MHz).

[n]$\gamma_e = +0.0017$.

[o]Hartree-Fock wavefunctions and energies (8), charge distributions (9), spectroscopic constants (18).

(References on p. 601)

SiF (continued):

pearance potential of 7.5 eV; ab initio calculations predict 7.4 eV (19).

[c] Q_1 and P_2 heads. (The head designations in Table 4 of (5) are erroneous.)

[d] According to (4) the 0-0 Q_1 and P_2 heads are at 47569.2 and 47408.9 cm^{-1}, respectively.

[e] $A \approx 0$.

[f] (6) give Λ-doubling constants.

[g] $A = 16.5_4$.

[h] (11) assume a regular state with $A_0 = +2.46$ and $A_1 = +2.35$.

[i] From (5); Pekeris' relation gives 7.01 (11).

[j] RKR potential curves (10).

[k] $D_1 = 13.5 \times 10^{-7}$.

[l] Franck-Condon factors (14).

[m] Franck-Condon factors (18).

[n] Franck-Condon factors (13)(18); variation of transition moment with r (16).

[o] Spin splitting constants $\lambda_0 = +0.274$, $\gamma_0 = +0.00188$; (20) find that the $^4\Sigma$ levels can be fitted with one ρ-type doubling parameter γ rather than the two proposed by (8) to account for earlier results by (7).

[p] Two short 0-0 sequences of headless bands centred at 29728 and 29890 cm^{-1}. Franck-Condon factors (21).

[q] $\omega_e y_e = +0.157$.

[r] Spin splitting constant $\gamma_0 = -0.0062_5$.

[s] $\gamma_e = +0.00013$.

[t] Radiative lifetime $\tau(v=0) = 0.23\ \mu s$ (23).

[u] Observed in absorption in a shock tube experiment (15). Franck-Condon factors (18). Electronic transition moment (15), variation with r (17).

[v] $A_0 = +161.88$, $A_1 = +162.04$, $A_2 = +162.19$ (20), see also (11). Small J dependent terms have also been determined.

[w] Λ-doubling constants $p_0 = -0.00299$ [i.e. $\Delta v_{fe}(^2\Pi_{\frac{1}{2}}) \approx -0.0030(J+\frac{1}{2})$], $q_0 = -1.06 \times 10^{-5}$ (20).

(1) Johnson, Jenkins, PRS A 116, 327 (1927).
(2) Asundi, Samuel, PIAS A 3, 346 (1936).
(3) Eyster, PR 51, 1078 (1937).
(4) Dovell, Barrow, PPS A 64, 98 (1951).
(5) Johns, Barrow, PPS 71, 476 (1958).
(6) Barrow, Butler, Johns, Powell, PPS 73, 317 (1959).
(7) Verma, CJP 40, 586 (1962).
(8) Hougen, CJP 40, 598 (1962).
(9) Ehlert, Margrave, JCP 41, 1066 (1964).
(10) Singh, Rai, IJPAP 4, 102 (1966).
(11) Appelblad, Barrow, Verma, JP B 1, 274 (1968).
(12) Singh, Singh, CS 37, 8 (1968).
(13) Mohanty, Singh, IJPAP 7, 109 (1969).
(14) Singh, Maheshwari, IJPAP 7, 708 (1969).
(15) Kuzyakov, Ovcharenko, Kuz'menko, Kurdyumova, ZPS 12(3), 555 (1970).
(16) Kuz'menko, Kuzyakov, Smirnov, ZPS 13(4), 616 (1970).
(17) Kuz'menko, Smirnov, Kuzyakov, VMUK 25(3), 357 (1970).
(18) Wentink, Spindler, JQSRT 10, 609 (1970).
(19) O'Hare, Wahl, JCP 55, 666 (1971).
(20) Martin, Merer, CJP 51, 634 (1973).
(21) Singh, IJPAP 13, 204 (1975).
(22) Bialski, Grein, JMS 61, 321 (1976).
(23) Davis, Hadley, PR A 14, 1146 (1976).

State		T_e	ω_e	$\omega_e x_e$	B_e	α_e	D_e $(10^{-4}cm^{-1})$	r_e (Å)	Observed Transitions Design.	ν_{00}	References
$^{28}Si^2H$			μ = 1.87884152		$D_0^0 \leq 3.09_5$ eV[p]						AUG 1977
E	$^2\Sigma^+$	[53111.85]			[3.9161]		[1.028]	$[1.5136_5]$	E← X, V	52381.75 Z	(10)*
D	$^2\Delta$	[49255.6]			$[4.00_9]$		[1]	[1.496]	D← X, V	48525.5[q] Z	(5)*
C	$^2\Sigma^+$	[31728.2]			[1.09][r]			$[2.8_7]$	C← X, R	30998.1[r] Z	(13)
B	$^2\Sigma^+$	[31634.9]			[3.703][rs]		[3.4][r]	$[1.556_6]$	B← X, R	30904.8[r] Z	(5)* (13)
A	$^2\Delta$	24313.8[t]	1328.08[g] Z	48.11	3.8680[g]	0.1318[u]	[1.379][v]	1.5230_4	A↔ X, R	24235.66 Z	(5)* (6)
X	$^2\Pi_r$	0[w]	1469.32[g] Z	18.23	3.8840[gx]	0.0781[y]	[1.054]	1.5198_9			
$^{28}Si^1H^+$					D_0^0 = 3.17 eV[a]						AUG 1977
A	$^1\Pi$	$25846._1$	[390.17] Z	(72.0)[b]	4.9125[c]	0.7667	[19.92][d]	1.8782	A→ X,[eh] R	25025.20 Z	(1)[f]
X	$^1\Sigma^+$	0	2157.17 Z	34.24	7.6603	0.2096[g]	3.83[g]	1.5041			
$^{28}Si^1H^-$					$D_0^0 \leq 2.95_2$ eV[a]		I.P. = 1.277 eV[b]				AUG 1977
b	$^1\Sigma^+$	[8460]	(2100)[c]					(1.50)[c]			
a	$^1\Delta$	[4580]	(2100)[c]					(1.50)[c]			
X	$^3\Sigma^-$	[0]	(2175)[c]					(1.474)[c]			
$^{28}Si^{127}I$			μ = 22.9233294		D_0^0 = 3.0 eV[a]						SEP 1977
			Unassigned absorption bands in the region 41500 - 43600 cm^{-1}.								(4)
F									F← X_1, V	44995 H	(4)*
(E)			Diffuse bands, assignment uncertain.						(E← X)	(44104)	(4)*
D									D← X_1, V	42859 H	(4)*
C		42711	486 H	3.5	v=2 diffuse.				C← X_1, V	42772 H	(4)*
B	$^2\Sigma$	32380.3	471.7 H	0.9	All levels predissociated.[b]				B← X_1, V	32434.3 H	(1)(3)*
A	$^2\Sigma$	21204.9	208.6 H	1.66[c]	d				A← X_1, R	21127.2 H	(3)*
a	$(^4\Sigma^-_{\frac{1}{2}})$	x+20289.7	275.7 H	5.6	[(0.118)][e]			[(2.50)]	a↔ X_2,[f] R	20246.9 H	(2)* (3)*

Si^1H, Si^2H (continued):

[p]From the value for Si^1H, confirmed by the observed predissociation in B $^2\Sigma^+$.

[q]Recalculated from the data of (5).

[r]Deperturbed constants of (13) whose T_0 values correspond to v_{00} + 797.50 [see (6)]. Interaction parameter $H_{B,C}$ = 5.87 cm^{-1}. See [d].

[s]Increasing linewidth above N=8 indicating predissociation.

[t]$A_0 = 3.45$, $A_1 = 3.42$, $A_2 = 2.14$ (6); see also (14).

[u]$\gamma_e = -0.0101$.

[v]$D_1 = 1.524 \times 10^{-4}$, $D_2 = 1.708 \times 10^{-4}$.

[w]$A_0 = +142.73$, $A_1 = +143.10$, $A_2 = +143.76$; slight J dependence

[x](6) give Λ-doubling constants (p = 0.039, q = 0.0024). | (6).

[y]$\gamma_e = -0.0009$.

(1) Rochester, ZP 101, 769 (1936).
(2) Douglas, CJP 35, 71 (1957).
(3) Barrow, Deutsch, PCS (1960), p. 122.
(4) Thrush, Nature 186, 1044 (1960).
(5) Verma, CJP 43, 2136 (1965).
(6) Klynning, Lindgren, AF 33, 73 (1966).
(7) Cade, PPS 91, 842 (1967).
(8) Cade, Huo, JCP 47, 649 (1967).
(9) Cade, Bader, Henneker, Keaveny, JCP 50, 5313 (1969).
(10) Herzberg, Lagerqvist, McKenzie, CJP 47, 1889 (1969).
(11) Smith, JCP 51, 520 (1969).
(12) Douglas, Lutz, CJP 48, 247 (1970).
(13) Bollmark, Klynning, Pagès, PS 3, 219 (1971).
(14) Veseth, Physica 56, 286 (1971).
(15) Rao, Lakshman, Physica 56, 322 (1971).
(16) Smith, Liszt, JQSRT 11, 45 (1971).
(17) Grevesse, Sauval, JQSRT 11, 65 (1971).
(18) Meyer, Rosmus, JCP 63, 2356 (1975).

Si^1H, Si^2H (continued):

(19) Wilson, Richards, Nature 258, 133 (1975).
(20) Freedman, Irwin, AA 53, 447 (1976).

Si^1H^+: [a]From a short extrapolation of the vibrational levels in

[b]Estimated from Pekeris' relation (1). | A $^1\Pi$ (1).

[c]From average B_v values for the two Λ-doubling components; B(P,R) - B(Q) = +0.0062 and +0.0156 for v=0 and 1, resp..

[d]$D_1 = 17.89 \times 10^{-4}$. In both v=0 and 1, rotational levels having $J \geqq 9$ cannot be represented by short power series in J(J+1).

[e]Also observed in the solar spectrum (2). On the basis of this observation (3) obtain $f_{00} = 0.0005$; see, however, (5).

[f]Table IV as well as eqn.[3] and eqn.[5] of (1) contain several errors which were later corrected in an unpublished erratum.

[g]$\gamma_e = +0.0045_5$, $\beta_e = -0.05 \times 10^{-4}$.

[h]Potential functions, Franck-Condon factors (4)(5).

(1) Douglas, Lutz, CJP 48, 247 (1970).
(2) Grevesse, Sauval, AA 9, 232 (1970).
(3) Grevesse, Sauval, JQSRT 11, 65 (1971).
(4) Rao, Lakshman, Physica 56, 322 (1971).
(5) Liszt, Smith, JQSRT 12, 947 (1972).

Si^1H^-: [a]From $D_0^0(Si^1H)$ and the electron affinities of Si and Si^1H.

[b]From laser photoelectron spectroscopy (1). See also ref. (7) of Si^1H.

[c]Franck-Condon analysis of the photodetachment sp. (1).

(1) Kasdan, Herbst, Lineberger, JCP 62, 541 (1975).

SiI: [a]Extrapolation of the vibrational levels in A $^2\Sigma$ (3). An upper limit of 4.02 eV follows from the prediss. in B $^2\Sigma$.

(continued p. 603)

State	T_e	ω_e	$\omega_e x_e$	B_e	α_e	D_e ($10^{-6} cm^{-1}$)	r_e (Å)	Observed Transitions Design.	Observed Transitions ν_{00}	References
$^{28}Si^{127}I$ (continued)										
X_2 $^2\Pi_{3/2}$	x[g]	359.0 H	1.1							
X_1 $^2\Pi_{1/2}$	0	363.8 H	1.2_5	[(0.123)][h]			[(2.45)]			
$^{28}Si^{14}N$		$\mu = 9.3321339_9$								SEP 1977
		The emitter of the Woods band at 26017 cm^{-1} (3) has been shown (11) to be SiO^+ and not SiN as suggested by (9).								
L $^2\Pi_i$	a+(32661)[a]	[718][b] H		[0.549][b]			$[1.81_4]$[b]	L→A, R	32491[b] 32508[b] H	(14)
D $^2\Pi_i$	a+27865.6[c]	699.3_3 Z	3.48	0.5238[d]	0.0041	1.0	1.857_1	D→A, R	27693.8 Z	(1)(6)(13) (14)*
K $^2\Sigma$	$a+25718._2$	1142[b] Z	11.5	0.677_5[b]	0.005		1.633[b]	K→A, V	$25765._7$[b] Z	(14)*
B $^2\Sigma^+$	24299.2_1	1031.03 Z	16.85[e]	0.7238[fgh]	0.01048	1.5	1.579_8	B↔X,[i] R	24236.4_7 Z	(1)(2)* (4) (10)(14)
A $^2\Pi_i$	a[jk]	1044.4_1 Z	6.20[ℓ]	0.67516	0.00538	1.10	1.6357_0			
X $^2\Sigma^+$	0	1151.3_6 Z	6.47[m]	0.7311[noh]	0.00565	1.2	1.571_9			

SiN: [a]$A \approx -72$.

[b]Vibrational numbering unknown; the lowest observed level is arbitrarily assumed to have v=0.

[c]The observed spin-orbit coupling constants vary from $A_3 = -60.42$ to $A_7 = -63.49$ (14).

[d]Perturbations in v=3 by a Σ state (13)(14).

[e]$\omega_e y_e = +0.15$, $\omega_e z_e = -0.011$.

[f](11) give the spin splitting constant $\gamma_0 = +0.0020$ but (14) change this to -0.0034.

[g]Numerous perturbations; for a summary see (14).

[h]Potential curves (7), see also (14).

[i]The 0-0 sequence was wrongly attributed to SiO^+ by (8), see the correction by (9). Measured relative intensities, Franck-Condon factors, r dependence of the transition moment (5)(12).

[j](14) estimate $a \approx 8000$.

[k]The observed spin-orbit coupling constants $[A_1 \ldots A_5$ (14)] are approximately given by $A_v = -89.54 + 0.27(v+\frac{1}{2})$.

[ℓ]$\omega_e y_e = -0.011$.

[m]$\omega_e y_e = -0.007$.

SiN (continued):

[n](11) give the spin splitting constant $\gamma_0 = +0.0153$ [see also (9)(10)] but (14) change this to -0.0172.

[o]A small perturbation in v=8 probably arises from interaction with A $^2\Pi$ (14).

(1) Mulliken, PR 26, 319 (1925).
(2) Jenkins, de Laszlo, PRS A 122, 103 (1929).
(3) Woods, PR 63, 426 (1943).
(4) Thrush, Nature 186, 1044 (1960).
(5) Stevens, Ferguson, CJP 41, 240 (1963).
(6) Schofield, Broida, PP 4, 989 (1965).
(7) Singh, Rai, IJPAP 4, 102 (1966).
(8) Nagaraj, Verma, CJP 46, 1597 (1968).
(9) Dunn, Rao, Nagaraj, Verma, CJP 47, 2128 (1969).
(10) Dunn, Dunn, CJP 50, 860 (1972).
(11) Singh, Bredohl, Remy, Dubois, JP B 6, 2656 (1973).
(12) Gohel, Shah, IJPAP 13, 162 (1975).
(13) Linton, JMS 55, 108 (1975).
(14) Bredohl, Dubois, Houbrechts, Singh, CJP 54, 680 (1976).

SiI (continued):

[b]The bands become progressively more diffuse with increasing v' (3).

[c]$\omega_e y_e = +0.079$, $\omega_e z_e = -0.0055$.

[d]Only the 7-0 and 8-0 bands of $A \leftarrow X_1$ have been analyzed (3), $B' \approx 0.085$. Of the six expected branches only four have been observed; R_{11} and $^QP_{21}$ (i.e. R_{ee} and P_{ff}) lines are absent. Extensive perturbations by levels of a $^4\Sigma^-_{\frac{1}{2}}$ state; see [e].

[e]As a full rotational analysis of the emission bands [called $A' \rightarrow X$ by (3) and $A \rightarrow X$ by (2)] was not possible, the nature of the a state is not known. (3) suggests that this is the same $^4\Sigma^-_{\frac{1}{2}}$ state whose higher vibrational levels ($\Delta G \approx 176$, $B \approx 0.097$) have been identified in perturbations of A $^2\Sigma$(v=7,8).

[f]A much weaker system reported by (2) at 650 cm^{-1} to the red of the main system was not confirmed by (3). Transitions to or from X_1 $^2\Pi_{\frac{1}{2}}$ have not been observed.

[g](3) assumes $x \approx 700$ cm^{-1}. Tentative assignments of weak absorption bands would give $x = 649$ (3) or 757 cm^{-1} (1).

[h]Estimated by assuming the same percentage decrease in bond length from HSiI to SiI as from HSiCℓ to SiCℓ (3). The rotational analysis of the $A \leftarrow X_1$ 7-0 and 8-0 bands gives $B'' - |p''|/2 = 0.10987$ and $D'' = 2.1 \times 10^{-7}$.

(1) Oldershaw, Robinson, TFS 64, 2256 (1968).
(2) Lakshminarayana, Haranath, JP B 3, 576 (1970).
(3) Billingsley, JMS 43, 128 (1972).
(4) Oldershaw, Robinson, JMS 44, 602 (1972).

State	T_e	ω_e	$\omega_e x_e$	B_e	α_e	D_e $(10^{-6}cm^{-1})$	r_e (Å)	Observed Transitions		References
								Design.	ν_{00}	
$^{28}Si^{16}O$		$\mu = 10.1767076_1$ $D_0^0 = 8.26$ eV[a] I.P. $= 11.4_3$ eV[b]								AUG 1977 A
		Theoretical potential curves and spectroscopic constants for the ground and several valence-excited states (19a)(46).								
P $^1\Sigma^+$(5pσ)	82019	[1121.5] Z		[0.692]	c		$[1.54_7]$	P← X,[d] R	81960.3 Z	(24)*
O $^1\Pi$	[82208.1]			[0.556]	e		$[1.72_6]$	O← X, R	81588.8 Z	(24)
N $^1\Pi$	(81232)	[(1024)]		[0.635]	f		$[1.61_5]$	N← X, R	81124.8 Z	(24)
M $^1\Pi$	81203	[833]		[0.640]	g		$[1.60_9]$	M← X,[d] R	81000.5 Z	(24)
L $^1\Sigma^+$(5sσ)	80783	[1102.2] Z		[0.701]	h		$[1.53_7]$	L← X,[d] R	80715.3 Z	(24)*
K $^1\Sigma^+$	78369	[905.8] Z	i	[0.615]	i		$[1.64_1]$	K← X,[d] R	78202.6 Z	(24)
J $^1\Pi$ (4pπ)	76381	[1146]		[0.6983]	j		$[1.540_2]$	J← X,[d] R	76334.3 Z	(24)*
h $^3\Pi_i$	(70790)[k]	(431)[k]		$B_1 \approx 0.589$[k]			$r_1 \approx 1.67_7$			(33)
I $^1\Pi$	(70510)	[878.9] Z	ℓ	[0.614]	ℓ		$[1.64_3]$	I← X, R	70333.7 Z	(24)
H $^1\Sigma^+$(4pσ)	69727	$[1109.2_5]$ Z		0.7146	0.0094[m]		1.522_5	H← X,[d] R	69662.2_6 Z	(24)* (33)*
G $^1\Pi$	(69670)	$[862.8_8]$ Z	ℓ	[0.6292]	ℓ	[1.2]	$[1.622_6]$	G← X, R	69483.0_6 Z	(24)*
F $^1\Sigma^+$(4sσ)	68532.0	1120.0_0[n] Z	7.34_5[n]	[0.6938]	o		$[1.545_2]$	F← X,[d] R	68470.9_0 Z	(2)* (24)*
		Many unidentified emission bands in the region 21800 - 31500 cm^{-1}.								(47)
g $^3\Sigma^+$(4sσ)	[68091.4]			[0.71588]		[1.42]	$[1.5211_6]$	g → b, V	33567.9_5 Z 33639.0_1 Z 33711.5_9 Z	(35)*
f $^3\Pi_i$	59283.1 59260.8 59236.8	488.4[p]	3.4	0.586[p]	0.014_5	3	1.68_1	f → b, R	$25146._6$ H $25051._7$ H $24956._7$ H	(18)* (37) (47)*
c $^3\Sigma^{(+)}$	57551.3	949.10 H	17.30	0.6841[q]	0.0079	1.7	1.556_1	c → b, V_R	23498.0_5 Z 23569.1_1 Z 23641.6_9 Z	(11)* (35) (47)
E $^1\Sigma^+$	52860.9	675.52 Z	4.204	0.54727[r]	0.00555[s]	1.434[t]	1.7397_8	E[u] ↔ X,[v] R	52578.31 Z	(15)* (36) (47)*

SiO: [a]Average of several thermochemical determinations; see the summary in the Appendix of (30), also (5)(25).

[b]Average of the values obtained on the assumption that F,L,... and H,P,... are the first members of two Rydberg series converging to X $^2\Sigma^+$ of SiO^+ (24)(32). The electron impact appearance potential is 11.5_8 eV (8)(16).

[c]Only v=0 and 1 observed; $B_1 = 0.703$. Both levels are perturbed.

[d]Corresponding data for $Si^{18}O$ and confirmation of the vibrational numbering (32).

[e]Only one strongly perturbed level observed, vibrational numbering uncertain.

[f]v=1 diffuse. Vibrational numbering doubtful.

[g]Of the two observed levels only v=0 has been partially analyzed.

[h]v=0 strongly perturbed. Only fragments of the 1-0 band observed, $B_1 = 0.67$. The 2-0 band at 1207 Å (82850 cm^{-1}) has very broad lines.

[i]Higher ΔG and B_v values ($v \leq 5$) are irregular.

[j]v=0 strongly perturbed; v=1 observed but not analyzed.

[k]v=1 observed in a perturbation of H $^1\Sigma^+(v=1)$; $A_1 \approx -36$. Slightly different constants are given by (48a): T_e = 70826, $A_1 = -33.2$, $\omega_e = 414$, $B_1 = 0.5763$.

[ℓ]The G and I states interact strongly. Higher ΔG and B_v values are irregular. The vibrational assignments of (24) have been revised by (32) to account for isotope shifts; the vibrational numbering of the I state is still in doubt, see (48a). The linewidth in bands labelled G $-$ X (according to the new assignments of (32)] increases with v'. G(v=3) is perturbed by a $^3\Sigma^+$ state (probably v=4 of g $^3\Sigma^+$) having $B = 0.687$, $T_v = 71937$ cm^{-1}.

[m]v=1 perturbed by h $^3\Pi_i(v=1)$, see [k].

[n]These constants represent only v=0,1,2. There are vibrational perturbations for higher v which may also affect the lower levels.

[o]$B_1 = 0.6888$, $B_2 = 0.6785$.

[p]Approximate constants from a partial analysis of the 1-0, 3-0, 4-0 bands (18)(37); vibrational constants confirmed by (47). The v=0 level is perturbed (37); according to (48) the $^3\Pi_1$ component interacts strongly with A $^1\Pi(v=23)$. (48) give constants which differ considerably from those of (37).

[q]Spin splitting constants for v=1: $\lambda = 0.298$, $\gamma = -0.002$ (11).

[r]RKR potential curve (45); notice that the total energies of the potential curves calculated by (45) are unreliable since the T_e values used by them are, in fact, T_{00} values.

[s]γ_e = +0.000022.

[t]β_e = $+0.018 \times 10^{-6}$.

[u]Radiative lifetime $\tau(v=1...7)$ = 10.5 ns (23). These authors have calculated Franck-Condon factors and f values for bands of the v'-0 progression and have determined the r dependence of the transition moment.

[v]A new emission system reported by (22) has been shown (36) to be part of the $E \rightarrow X$ system with high v' and v"; see also (47).

(References on p. 607)

State	T_e	ω_e	$\omega_e x_e$	B_e	α_e	D_e $(10^{-6}cm^{-1})$	r_e (Å)	Observed Transitions: Design.	Observed Transitions: ν_{00}	References
$^{28}Si^{16}O$ (continued)										
A $^1\Pi$	42835.4	852.8 Z	6.43^a	0.6307^{bc}	0.00660	1.43	1.620_6	$A^d \leftrightarrow X,^e$ R	42640.71 Z	(1)* (9) (17)* (38)* (47)* (49)*
D $^1\Delta$	38823^f	730	3.9	$0.553_8{}^c$	0.0051_6		1.729			
C $^1\Sigma^-$	38624^f	740	4.27	$0.555_5{}^c$	(0.005_2)		1.727			
e $^3\Sigma^-$	38309^f	748	4.19	$0.556_3{}^c$	0.0051_5		1.726			
d $^3\Delta_r$	36487^{fg}	767	4.1	$0.563_1{}^c$	(0.005_2)		1.715			
b $^3\Pi_2$	34018.5			$[0.6892_0]^i$		$[2.2_2]$			R 33904.2_0 Z	
b $^3\Pi_1$	33947.4^h	1013.8 (Z)	7.5_7	$[0.6760_7]^{ic}$	0.0044_0	$[1.2_5]$	1.5624	b→X,	R 33833.1_4 Z	(27)(29)(39) (51)
b $^3\Pi_0$	33874.8			$[0.6638_6]^i$		[0.50]			R 33760.56 Z	
a $^3\Sigma^+$	(33630)	$(790)^j$	$(4.1)^j$	$(0.57)^{jc}$	$(0.0052)^j$		$(1.70)^j$	a→X,	R 33409 H	(29)(39)(51)
X $^1\Sigma^+$	0	1241.55_7 Z	5.966^k	0.7267512^c	0.0050377^{ℓ}	0.980	1.509739	Rot.-vibr. sp.[m]		(19)(26)
								Rotation sp.[n]		(7)(31)(34) (50)
								Mol. beam rf el. reson.[o]		(12)(28)

SiO: [a]$\omega_e y_e$ = +0.0238 (49).

[b]The A $^1\Pi$ state is extensively perturbed by levels of d $^3\Delta$, e $^3\Sigma^-$, C $^1\Sigma^-$, D $^1\Delta$; see (49) who give a very detailed analysis of these perturbations for $^{28}Si^{16}O$ and $^{28}Si^{18}O$. Selective enhancement of perturbed and corresponding "extra" lines in emission under fast pumping conditions (44).

[c]RKR potential energy curves for the eight lowest states (49); see also (4)(45) and [r] on p. 605 concerning the results of (45).

[d]Radiative lifetime $\tau = 9._6$ ns [phase-shift method (20)] corresponding to $f_{e\ell} = 0.13$; see also (10). Considerably smaller f values have been derived by (6)(13); for a summary of reported f values see (46) whose theoretical calculations predict τ = 32 ns.

[e]Franck-Condon factors (3)(21)(48a). Isotope shifts for $^{29}Si^{16}O$ and $^{30}Si^{16}O$ (42)(47).

[f]All constants for these states derived from perturbations in A $^1\Pi$ (1)(49).

[g]A = +8. There is a strong spin-orbit interaction between $^3\Delta_2$ and $^1\Delta_2$ which causes a large asymmetry in the spin splitting of d $^3\Delta$ (49).

[h]A_0 = +73.19, A_1 = +73.02 (11).

[i]Effective B_0 values (35); using instead sums of $\Delta_2F(J)$ for

SiO (continued):

the three components (11) obtain $B_0 = 0.6766_8$. Λ-type doubling in $^3\Pi_0$ and $^3\Pi_1$ (35).

[j]Predicted constants (49); see also (19a)(46). Only v'=0 has been observed in chemiluminescence. From partially resolved rotational structure (51) derive $B_0 \approx 0.59$.

[k]$\omega_e y_e = +0.0054557$; these are the constants of (36)(49) which are very similar to the old constants of (1) and the more recent ones of (26) (rotation-vibration sp.); see also (14)(17).

[l]$\gamma_e = +2.35 \times 10^{-6}$ (50).

[m]Observed in late-type stars (26)(43).

[n]Several microwave lines have been observed in interstellar space and some extended stellar atmospheres, see e.g. (40)(41). Maser action is prevalent.

[o]$\mu_{el}(v) = 3.0882 + 0.0197(v+\frac{1}{2})$ D for $v \leq 3$ (12); for g_J factors ($-0.1536\ \mu_N$ for v=0) and other magnetic properties see (28), also (31).

(1) Lagerqvist, Uhler, AF 6, 95 (1952).
(2) Barrow, Rowlinson, PRS A 224, 374 (1954).
(3) Nicholls, JRNBS A 66, 227 (1962).
(4) Nair, Singh, Rai, JCP 43, 3570 (1965).
(5) Coppens, Smoes, Drowart, TFS 63, 2140 (1967).
(6) Hooker, Main, JQSRT 8, 1527 (1968); Physica 41, 35 (1969).
(7) Törring, ZN 23 a, 777 (1968).
(8) Hildenbrand, Murad, JCP 51, 807 (1969).
(9) Singh, Upadhya, Nair, IJP 43, 665 (1969).
(10) Czernichowski, Zyrnicki, APP A 37, 865 (1970).
(11) Nagaraj, Verma, CJP 48, 1436 (1970).
(12) Raymonda, Muenter, Klemperer, JCP 52, 3458 (1970).
(13) Rusin, VMUK 25, 397, 526 (1970); 27, 196 (1972).
(14) Cornet, BCSARB (5) 57, 1069 (1971).
(15) Elander, Lagerqvist, PS 3, 267 (1971).
(16) Hildenbrand, IJMSIP 7, 255 (1971).
(17) Bosser, Lebreton, Marsigny, CR C 275, 531 (1972).
(18) Cornet, Dubois, CJP 50, 630 (1972).
(19) Hedelund, Lambert, ApL 11, 71 (1972).
(19a) Heil, Schaefer, JCP 56, 958 (1972).
(20) Smith, Liszt, JQSRT 12, 505 (1972).
(21) Liszt, Smith, JQSRT 12, 947 (1972).
(22) Bredohl, Cornet, Dubois, Remy, CJP 51, 2332 (1973).
(23) Elander, Smith, ApJ 184, 311 (1973).
(24) Lagerqvist, Renhorn, Elander, JMS 46, 285 (1973).
(25) Nagai, Niwa, Shinmei, Yokokawa, JCS FT I 69, 1628 (1973).
(26) Beer, Lambert, Sneden, PASP 86, 806 (1974).
(27) Bredohl, Cornet, Dubois, Remy, JP B 7, L66 (1974).
(28) Davis, Muenter, JCP 61, 2940 (1974).
(29) Hager, Wilson, Hadley, CPL 27, 439 (1974).
(30) Hildenbrand, Murad, JCP 61, 1232 (1974).
(31) Honerjäger, Tischer, ZN 29 a, 1695 (1974).
(32) Lagerqvist, Renhorn, JMS 49, 157 (1974).
(33) Lagerqvist, Renhorn, APH 35, 155 (1974).
(34) Lovas, Krupenie, JPCRD 3, 245 (1974).
(35) Singh, Bredohl, Remy, Dubois, CJP 52, 569 (1974).
(36) Barrow, Stone, JP B 8, L13 (1975).
(37) Bredohl, Cornet, Dubois, JP B 8, L16 (1975).
(38) Deutsch, Deutsch, Elander, Lagerqvist, PS 12, 248 (1975).
(39) Hager, Harris, Hadley, JCP 63, 2810 (1975).
(40) Kaifu, Buhl, Snyder, ApJ 195, 359 (1975).
(41) Buhl, Snyder, Lovas, Johnson, ApJ 201, L29 (1975).
(42) Podkorytova, OS(Engl. Transl.) 38, 637 (1975).
(43) Singh, AA 44, 411 (1975).

(continued p. 609)

State	T_e	ω_e	$\omega_e x_e$	B_e	α_e	D_e (10^{-7}cm^{-1})	r_e (Å)	Observed Transitions: Design.	ν_{00}	References
$^{28}Si^{16}O^+$				$D_0^0 = 4.9_8$ eV[a]						AUG 1977
B $^2\Sigma^+$				[0.7103][b]		[11.9]	[1.527_1]	B→X,[c]	R 26016.55 Z	(1)*(2)*(5)
X $^2\Sigma^+$	0			[0.7178][b]		[12.0]	[1.519_1]			
$^{28}Si^{31}P$		μ = 14.6995860		$D_0^0 = 3.7_3$ eV[a]						SEP 1977
$^{28}Si^{32}S$		μ = 14.9206889		$D_0^0 = 6.4_2$ eV[a]						AUG 1977 A
E $^1\Sigma^+$	41915.8	405.6 Z	1.60[b]	0.22137[c]	0.00139[d]		2.2591	E↔X,[e]	R 41744.0 Z	(2)* (6)(12)
I $^1\Sigma^-$	≤37462[f]	≥404.9	1.1_8	≥0.214_2	0.001_8		≤2.29_7			(5)
C $^1\Delta$	37269[f]	$439._9$	3.9_7	0.226_9	0.002_8		2.23_1			(5)
e $^3\Sigma^-$	≤35322[f]	≥$407._2$	1.7_7	≥0.223_0	0.002_9		≤2.25_1			(5)(3)
D $^1\Pi$	35026.8_6	513.1_2 Z	2.9_3	0.2664_7[gc]	0.0021_6	2.91[h]	2.059_1	D↔X,	R 34908.5_1 Z	(1)*(3)*(5)
a $^3\Pi_1$	[$30239._2$]	Only v'=0 observed		[0.28180]			[2.0023]	a→X,	R $29865._0$ Z	(14)
X $^1\Sigma^+$	0	749.64 Z	2.577[i]	0.30352788[c]	0.00147308[j]	2.01	1.929321[k]	Microwave sp.[ℓ]		(10)(15)
$^{28}Si^{(80)}Se$		(μ = 20.7224706)		$D_0^0 = 5.64$ eV[a]						AUG 1977
E	38505.9	308.8 H	1.95[b]					E←X,	R 38370.3 H	(2)
D $^1\Pi$	32450.3	399.8 H	1.93					D↔X,	R 32360.2 H	(1)* (2)
a ($^3\Pi_1$)	(25077)	[(412)] H						a→X,	R 24993.7_3 H	(5)
X $^1\Sigma^+$	0	580.0 H	1.78	0.1920117[c]	0.0007767[c]	0.842[c]	2.058324	Microwave sp.		(3)
$^{28}Si^{(130)}Te$		(μ = 23.0194111)		$D_0^0 = 4.64$ eV[a]						AUG 1977
E	33991	242 H	(3.63)[b]					E←X,	R 33871 H	(2)
D $^1\Pi$	28661.8	338.6 H	1.70					D↔X,	R 28590.4 H	(1)* (2)
X $^1\Sigma^+$	0	481.2 H	1.30							

SiO (continued):

(44) Bredohl, Remy, Cornet, JP B 9, 2307 (1976).

(45) Lakshman, Rao, Naidu, Pramāṇa 7, 369 (1976).

(46) Oddershede, Elander, JCP 65, 3495 (1976).

(47) Shanker, Linton, Verma, JMS 60, 197 (1976).

(48) Verma, Shanker, JMS 63, 553 (1976).

(48a) Renhorn, Dissertation (Stockholm, 1976).

(49) Field, Lagerqvist, Renhorn, PS 14, 298 (1976); JCP 66, 868 (1977).

(50) Manson, Clark, De Lucia, Gordy, PR A 15, 223 (1977).

(51) Linton, Capelle, JMS 66, 62 (1977).

SiO^+: [a]D_0^0(SiO) + I.P.(Si) - I.P.(SiO).

[b]Spin splitting constants $\gamma_0' = -0.0066$, $\gamma_0'' = +0.0028$.

[c]There has been some doubt whether this spectrum is due to SiO^+ or SiN (2)(3) but the recent work of (5) seems to settle this point in favour of SiO^+. Another extensive system of R shaded bands, also ascribed to B→X of SiO^+, has been observed (4) in discharges through $O_2 + SiC\ell_4$ and, with much higher intensity, from a hollow cathode containing a stoichiometric mixture of SiO_2 + Si: $\nu_{00} = 30500.1$,
$\omega_e' = 634.9_0$, $\omega_e'x_e' = 4.45$, $B_e' = 0.518_0$, $\alpha_e' = 0.0054$,
$\omega_e'' = 976.06$, $\omega_e''x_e'' = 5.57$, $B_e'' = 0.6103$, $\alpha_e'' = 0.0051$.
No doublet splitting was observed and, as pointed out by (5), the lower state B value does not fit with those of the Rydberg states of SiO.

(1) Woods, PR 63, 426 (1943).

(2) Nagaraj, Verma, CJP 46, 1597 (1968).

(3) Dunn, Rao, Nagaraj, Verma, CJP 47, 2128 (1969).

(4) Cornet, Dubois, Gerkens, Tripnaux, BSRSL 41, 183 (1972).

(5) Singh, Bredohl, Remy, Dubois, JP B 6, 2656 (1973).

SiP: [a]Thermochemical value (mass-spectrom.)(1).

(1) Smoes, Depière, Drowart, RIHTR 9, 171 (1972).

SiS: [a]From a short extrapolation of the vibrational levels of the E state (4) assuming dissociation into $^3P_2 + {}^3P_2$ (8).

[b]$\omega_e y_e = -0.028$.

[c]RKR potential curves (7)(13).

[d]$\gamma_e = -0.000018$.

[e]Bands in the near UV and visible region (3490 - 6170 Å) first described by (1) have been shown (12) to be part of the E→X system ($1 \leq v' \leq 17$, $20 \leq v'' \leq 51$). Franck-Condon factors (13).

[f]The constants for these states have been derived (5) from perturbations in D $^1\Pi$; the vibrational numbering is uncertain except for C $^1\Delta$.

[g]Numerous perturbations due to interactions with e $^3\Sigma^-$, C $^1\Delta$, and I $^1\Sigma^-$.

[h]$\beta_e = -0.03 \times 10^{-7}$.

[i]$\omega_e y_e = +0.00052$.

[j]$\gamma_e = -8.6 \times 10^{-8}$.

[k]From studies of four different isotopes the equilibrium internuclear distance at the minimum of the Born-Oppenheimer potential curve can be derived: $r_e^{(BO)} = 1.929264$ Å (10).

[ℓ]$\mu_{e\ell}(v=0) = +1.73$ D ($^+SiS^-$), see (7a)(9)(10); $g_J = -0.09097 - 0.00029_6(v+\frac{1}{2})$ (11); $Si^{33}S$ hyperfine structure (9); adiabatic and non-adiabatic corrections (10).

(References on p.611)

SiSe: }
SiTe: } See p.611 .

State	T_e	ω_e	$\omega_e x_e$	B_e	α_e	D_e (10^{-} cm^{-1})	r_e (Å)	Observed Transitions Design.	ν_{00}	References
$^{(152)}Sm^{(35)}Cl$	(μ = 28.4258078) $D_0^0 \geq 4.3_4$ eV[a] Bands in the region 12500 - 17200 cm^{-1} of the low-resolution Sm + $C\ell_2$ chemiluminescence spectrum have been attributed to SmCℓ.									SEP 1976 (1)
$^{(152)}Sm^{19}F$	(μ = 16.8866373) $D_0^0 = 5.4_6$ eV[a] Unresolved emission in the region 14000 - 33000, maximum at 24000 cm^{-1}.									SEP 1976 (2)(3)
$^{(152)}Sm^{16}O$	(μ = 14.4713000$_4$) D_0^0 = 5.90 eV[a] I.P. = 5.5$_5$ eV[b] Large number of unclassified, mostly R shaded emission bands from 13200 to 23300 cm^{-1}. Low-resolution chemiluminescence spectra have two broad maxima at 15600 and 21100 cm^{-1}.									SEP 1977 (1)(2)* (3) (6)(7)(9)

SmCℓ: [a]From the Sm + $C\ell_2$ chemiluminescence spectrum (1).

(1) Yokozeki, Menzinger, CP 14, 427 (1976).

SmF: [a]Thermochemical value (mass-spectrom.)(1); consistent with lower limits from the Sm + F_2 chemiluminescence spectrum (2)(3).

(1) Zmbov, Margrave, JINC 29, 59 (1967).
(2) Dickson, Zare, CP 7, 361 (1975).
(3) See ref. (1) of SmCℓ.

SmO: [a]Thermochemical value (mass-spectrom.)(10), compatible with (7) and superseding earlier results of (4)(5).
[b]Corrected electron impact appearance potential (8).

(1) Piccardi, AANL 21, 589 (1935); 25, 86 (1937).
(2) Gatterer, Junkes, Salpeter, Rosen, METOX (1957).
(3) Herrmann, Alkemade, "Chemical Analysis by Flame Photometry", 2nd rev. ed., Interscience (1963).
(4) Ames, Walsh, White, JPC 71, 2707 (1967).
(5) Smoes, Coppens, Bergman, Drowart, TFS 65, 682 (1969).
(6) Edelstein, Eckstrom, Perry, Benson, JCP 61, 4932 (1974).
(7) See ref. (2) of SmF.
(8) Ackermann, Rauh, Thorn, JCP 65, 1027 (1976).
(9) See ref. (1) of SmCℓ.
(10) Hildenbrand, CPL 48, 340 (1977).

SiS (continued):

(1) Barrow, Jevons, PRS A 169, 45 (1938).
(2) Vago, Barrow, PPS 58, 538 (1946).
(3) Lagerqvist, Nilheden, Barrow, PPS A 65, 419 (1952).
(4) Robinson, Barrow, PPS A 67, 95 (1954).
(5) Nilheden, AF 10, 19 (1955).
(6) Barrow, Deutsch, Lagerqvist, Westerlund, PPS 78, 1307 (1961).
(7) Nair, Singh, Rai, JCP 43, 3570 (1965).
(7a) Murty, Curl, JMS 30, 102 (1969).
(8) Barrow, DONNSPEC (1970), p. 323.
(9) Hoeft, Lovas, Tiemann, Törring, ZN 24 a, 1422 (1969); JCP 53, 2736 (1970).
(10) Tiemann, Renwanz, Hoeft, Törring, ZN 27 a, 1566 (1972).
(11) Honerjäger, Tischer, ZN 28 a, 1374 (1973).
(12) Bredohl, Cornet, Dubois, Wildéria, JP B 8, L259 (1975).
(13) Katti, Korwar, APH 39, 145 (1975).
(14) Bredohl, Cornet, Dubois, JP B 9, L207 (1976).
(15) Tiemann, JPCRD 5, 1147 (1976).

SiSe: [a]From an extrapolation of the vibrational levels of the E $^1\Sigma^+$ state assuming dissociation into $^3P_2 + {}^3P_2$ (2)(4).

[b]$\omega_e y_e = -0.032$.

[c]Rotational constants for $^{28}Si^{80}Se$; (3) gives constants for eight isotopic species.

(1) Barrow, PPS 51, 267 (1939).
(2) See ref. (2) of SiS.
(3) Hoeft, ZN 20 a, 1122 (1965).
(4) See ref. (8) of SiS.
(5) Lebreton, Bosser, Ferran, Marsigny, JP B 8, L141 (1975).

SiTe: [a]Average of two recent thermochemical determinations (3)(4), corrected for the new value of $D_0^0(Te_2)$; see also (5).

[b]$\omega_e y_e = (+0.13)$.

(1) See ref. (1) of SiSe.
(2) See ref. (2) of SiS.
(3) Exsteen, Drowart, Auwera-Mahieu, Callaerts, JPC 71, 4130 (1967).
(4) Brebrick, JCP 49, 2584 (1968).
(5) Barrow, DONNSPEC (1970), p. 323 and 367.

State	T_e	ω_e	$\omega_e x_e$	B_e	α_e	D_e $(10^{-}$ $cm^{-1})$	r_e (Å)	Observed Transitions Design.	ν_{00}	References
$^{(118,120)}Sn_2$		(μ = 59.446747$_4$) D_0^0 = 1.9$_9$ eV[a]								SEP 1977 A
$^{(120)}Sn^{(79)}Br$		(μ = 47.593081$_9$)								SEP 1977
		Unidentified absorption bands in the region 42300 - 45800 cm^{-1}.								(6)
E	(43570)	[320]						$E \leftarrow X_1$,	43607	(6)
D	(42742)	[269]						$D \leftarrow X_1$,	42753	(6)
C	[40869]							$C \leftarrow X_1$,	40746	(6)
B ($^2\Sigma$)	33062.$_3$	304.3 H	0.7$_1$	No emission from $v' \geq 1$.				$B \rightarrow X_2$, V $B \leftrightarrow X_1$,	(30628)[a] 33090.$_8$ H	(1)* (4)
A ($^2\Delta$)	(27063) 26695	[163.6] 169.1 H	6.8					$A \rightarrow X$, R	24557.9 26654.$_3$ H	(1)*
A' ($^2\Sigma$)	(18717)[b]	164.1 H	0.9					$A' \rightarrow X$[c], R	18675.$_8$ H	(2)* (3)
X_2 $^2\Pi_{3/2}$	(2463)	247.2[d] H	0.6$_3$[d]							
X_1 $^2\Pi_{1/2}$	0									
$^{120}Sn^{35}Cl$		μ = 27.0731195								SEP 1977
		Several weak diffuse absorption bands, 45400 - 47600 cm^{-1}.								(8)
		Strong absorption continua 29000 - 33000 and 40000 - 53000 cm^{-1} [see, however, (7)].								(3)
C	(43718) 43650	399.3[a] H	1.1					$C \leftarrow X$,[b] V	41384[a] 43674[a] H	(3)(8)
C'	41229	419.3[a] H	1.7					$C' \leftarrow X_1$,[b] V	41263[a] H	(8)
B $^2\Sigma^+$	33583.3	431.8 H^Q	1.25	0.1216	0.0006		2.263	$B \leftrightarrow X$, V	31265.4 33623.6 H^Q	(1)* (3)(4) (8)(9)(12) (14)
A c	28963 28692	303.3[a] 300.8[a] H^Q	3.7 4.0	c				$A \leftrightarrow X$, R	26580[a] 28666[a] H^Q	(1)* (2)(3) (4)
A' $^2\Sigma^+$	19418.4	232.3 H	0.71	[0.0908]			[2.619]	$A' \rightarrow X_1$, R	19359.1 H	(4)* (5)(6) (10)(12)
X_2 $^2\Pi_{3/2}$	2356.$_6$	354.4 H^Q	1.05	0.1122	0.0004		2.356			
X_1 $^2\Pi_{1/2}$	0	351.1 H^Q	1.06	0.1117[d]	0.0004		2.361			

Sn_2: [a]Thermochemical value (mass-spectrom.) (1)(2).

(1) Drowart, Honig, JPC 61, 980 (1957).
(2) Ackerman, Drowart, Stafford, Verhaegen, JCP 36, 1557 (1962).

SnBr: [a]0-0 head obscured by an atomic line; the v" numbering of this subsystem is uncertain.
[b]Assuming X_1 as the lower state of the $A' \rightarrow X$ bands; see [c].
[c]It is not clear whether the lower-state component is X_1 or X_2. (5) suggest X_1 and assign a number of weaker bands near 6300 Å to $A' \rightarrow X_2$.
[d]Average of the constants obtained by (1)(4).

(1) Jevons, Bashford, PPS 49, 554 (1937).
(2) Sarma, Venkateswarlu, JMS 17, 203 (1965).
(3) Naegeli, Palmer, JMS 21, 325 (1966).
(4) Oldershaw, Robinson, TFS 64, 616 (1968).
(5) Chatalic, Deschamps, Pannetier, CR C 269, 584 (1969).
(6) Oldershaw, Robinson, TFS 67, 2499 (1971).

SnCℓ: [a]Natural Sn isotopic mixture.
[b]According to (13) also observed in emission.
[c]Originally assumed to be $^2\Delta$, this state has recently been reinterpreted as $^4\Sigma^-$ ($\Omega = 1/2$ and $3/2$) on the basis of a tentative analysis of incompletely resolved rotational structure (11).
[d]Λ-type doubling $\Delta v_{fe} = -0.009(J+\frac{1}{2})$ (12).

(1) Jevons, PRS A 110, 365 (1926).
(2) Ferguson, PR 32, 607 (1928).
(3) Fowler, PR 62, 141 (1942).
(4) Sarma, Venkateswarlu, JMS 17, 252 (1965).
(5) See ref. (3) of SnBr.
(6) Pannetier, Deschamps, JCPPB 65, 1164 (1968).
(7) Hastie, Hauge, Margrave, JMS 29, 152 (1969).
(8) Oldershaw, Robinson, JMS 32, 469 (1969).
(9) Richter, ZPC (Frankfurt a. M.) 71, 303 (1970).
(10) Chatalic, Iacocca, Pannetier, JCPPB 69, 82 (1972).
(11) Chatalic, Iacocca, Pannetier, JCPPB 70, 481 (1973).
(12) Chatalic, Iacocca, Pannetier, JCPPB 70, 908 (1973).
(13) Katti, Korwar, PL A 48, 461 (1974); CS 43, 374 (1974).
(14) Katti, Korwar, IJPAP 13, 710 (1975).

State	T_e	ω_e	$\omega_e x_e$	B_e	α_e	D_e ($10^{-6} cm^{-1}$)	r_e (Å)	Observed Transitions Design.	ν_{00}	References
$^{118}Sn^{19}F$		μ = 16.3618865		$D_0^0 = 4.9_0$ eV[a]						SEP 1977
		Strong absorption continua with maxima at 41000 and 53000 cm^{-1}.								(1)
G $^2\Delta$	(46427) (46338)	$[609._9]$[b] H		$[0.2870]$[c]		$[0.22_5]$[c]	[1.895]	G↔X, V	44121.9 Z 46351.8[b] H	(1)(3)
F $(^2\Sigma)$	45500.5	688.2[d] H	4.6_5					F→A, V	25497.0	(3)
								F↔X, V	43233.1 45552.6[d] H	(1)(3)
E	42137.1 41856.1	677.0[d] H	3.0					E↔X, V	39864.5[d] H 41903.0[d] H	(1)(3)
D	(41341)	[622][d] H						D→X, V	39041.9 41361.4[d] H	(3)
C $^2\Delta$	40831 40760	[600][d] H	(5.4)	$[0.2856]$[c]		$[0.22_3]$[c]	[1.899]	C↔X, V	38524.7 Z 40772.8[d] H	(1)(3)(6)
B $^2\Sigma^+$	34109.0	677.6[d] H	2.74	[0.2896]		$[0.24_6]$	[1.886]	B→A, V	14100.9	(3)
								B↔X, V	31836.97 Z (34156.09)	(1)* (4)
B' $(^4\Sigma^-)$[e]								B'→X,	33039.8[d] H	(3)
A $^2\Sigma^+$	20136.9	[415.76] Z	2.20_6 H	0.2471[f]	0.0026	$[0.38_2]$[g]	2.042	A→X,[h] R	17736.08 Z 20055.58 Z	(2)* (4)(9)*
X $^2\Pi_{3/2}$	2316.9	[582.67] Z	2.82 H	0.2738	0.0014	$[0.22_5]$[i]	1.940			
X $^2\Pi_{1/2}$	0	$[577.6_4]$ Z	2.69 H	0.2727[jk]	0.0014	$[0.26_2]$[ℓ]	1.944			
$^{(120)}Sn^1H$		(μ = 0.99942466)		$D_0^0 \leq 2.73$ eV[a]						AUG 1977
A $^2\Delta$	b			[4.904][cd]		[433][e]	$[1.854_6]$	A↔X, R	23468.27 Z	(1)(4)*
a $^4\Sigma^-$				[5.3723][f]		[298][g]	[1.7719]	a↔X,[h] R	15439.27 Z	(2)(4)*
X $^2\Pi_r$	0[i]			[5.31488][j]		[207.5]	[1.78146]			
$^{(120)}Sn^2H$		(μ = 1.98082851)								AUG 1977
A $^2\Delta$	(23790)[k]	[(736)]		[2.5161][c]	(0.2)	[110.4][ℓ]	[1.8391]	A←X, R	23563.65 Z	(4)

State	T_e	ω_e	$\omega_e x_e$	B_e	α_e	D_e ($10^{-6}cm^{-1}$)	r_e (Å)	Observed Transitions		References
								Design.	ν_{00}	
$^{(120)}Sn^2H$ (continued)										
a $^4\Sigma^-$	(15580)	[992.5] Z		$[2.740]^m$	0.13	$[80]^n$	[1.762]	a← X,	R 15482.5 Z	(4)
X $^2\Pi_r$	0°	[1188.0] Z		$[2.6950]^p$	0.049_0	[53.4]	[1.7770]			

SnF: [a]Thermochemical value (mass-spectrom.)(7), recalculated with $D_0^0(CaF) = 5.4_8$ eV.

[b]From (3) (natural isotopic mixture); (1) give $\Delta G(\frac{1}{2})$ = 607 and 598 cm^{-1}.

[c]Only the $^2\Delta_{5/2}$ component has been analyzed (10).

[d]Natural isotopic mixture.

[e]Assignment by analogy with GeF (11).

[f]Spin doubling constant $\gamma = -0.0836$.

[g]$D_1 = 0.32_2 \times 10^{-6}$.

[h]Franck-Condon factors (8).

[i]$D_1 = 0.25_2 \times 10^{-6}$.

[j]Λ-type doubling $\Delta\nu_{fe} = -0.058_8(J+\frac{1}{2})$.

[k]RKR potential curve (5).

[ℓ]$D_1 = 0.33_6 \times 10^{-6}$.

(1) Jenkins, Rochester, PR 52, 1135 (1937).
(2) Yuasa, PPMSJ 21, 498 (1939).
(3) Barrow, Butler, Johns, Powell, PPS 73, 317 (1959).
(4) Barrow, Kopp, Merer, PPS 79, 749 (1962).
(5) Singh, Rai, IJPAP 4, 102 (1966).
(6) Uzikov, Kuzyakov, VMUK 23(5), 33 (1968).
(7) Zmbov, Hastie, Margrave, TFS 64, 861 (1968).
(8) Singh, Dube, IJPAP 9, 164 (1971).
(9) Rai, Singh, SpL 5, 155 (1972).
(10) Ram, Upadhya, Rai, JP B 6, L372 (1973).
(11) Merer, private communication.

Sn^1H, Sn^2H:

[a]From the predissociation in A $^2\Delta$ assuming dissociation into $^3P_2 + ^2S$.

[b]Spin coupling constants for v=0: A = 20.41, $A_J = 0.0625$, γ = 1.23 (4). For a more elaborate evaluation of these constants (including centrifugal distortion) see (8).

[c]Rotational constants of (4) based on a modified Hill-Van Vleck expression; see also (8). A case "c" treatment of the same data is given by (11).

[d]Broadening of absorption lines above N'≈14 (4), sharp cut-off in emission at N'=17 (1).

[e]$H_0 = -2.3 \times 10^{-7}$; see [c].

[f]Spin splitting constants for v=0: $\lambda = 45.78$, $\gamma_1 = 0.190$, γ_2 = 0.191. ΔB(difference in B values between the states $\Omega = 1/2$ and 3/2) = −0.0266. All constants derived by (4) according to the modified expressions [see (4)] for the energy levels of a $^4\Sigma$ state (3). For further refinements see (7)(9). The case "c" treatment of the same data by (11) gives $B_0 = 5.404$ and 5.339 for $\Omega = 1/2$ and 3/2, respectively. For a discussion of $^4\Sigma^-_{\frac{1}{2}}$ in terms of case "a'" coupling conditions see (6).

[g]$H_0 = -3.3 \times 10^{-8}$.

[h]Rotational intensity distribution (5)(6).

[i]$A_0 = +2178.88 + 0.01719\,J(J+1)$ (4); see, however, (8).

(continued p. 617)

State	T_e	ω_e	$\omega_e x_e$	B_e	α_e	D_e ($10^{-7}cm^{-1}$)	r_e (Å)	Observed Transitions Design.	ν_{00}	References
$^{(120)}Sn^{127}I$		(μ = 61.652004$_6$)								SEP 1977
		Fragments of several overlapping systems of absorption bands in the region 38800 - 46300 cm^{-1}.								(3)(4)
B ($^2\Sigma$)	32172.8	241.1 H	0.1$_7$					B← X, V	32193.9 H	(1)*
A ($^2\Sigma$)	17916.7	129.8 H	0.46					A→X, R	17880.8 H	(2)
X ($^2\Pi_{\frac{1}{2}}$)	0	199.0[a] H	0.5$_5$							
$^{120}Sn^{16}O$		μ = 14.1123342$_7$ D_0^0 = 5.49 eV[a]								SEP 1977
		Unidentified absorption bands, 64920 - 68360 cm^{-1}.								(5)
G	58806[b]	724 H	21					G← X, R	58752 H	(5)*
F	(57669)	[466] H						F← X, R	57492 H	(5)*
		Unidentified absorption bands, 53250 - 57670 cm^{-1}.								(5)
		Absorption continuum between 51630 and 51920 cm^{-1}.								(5)
E	36295	508.0 H	2.9[c]					E↔X, R	36138 H	(2)(4)*
D $^1\Pi$	29624	[573.6] Z	3.08[d] H	0.3145$_5$[ef]	0.0025	4.0	1.948$_7$	D↔X,[gh] R	29503.2 Z	(1)* (3)(6)*
		A large number of emission bands in the visible and near UV region has been observed by various methods (1)(10)(12). A satisfactory interpretation is still lacking.								
B (1)	(24890)	[560]		B_1 = 0.3010[i]			[1.992]	B← X, R	24760	(7)
A (0$^+$)	(24333)	[555]		[0.2964]			[2.008]	A← X, R	24200	(7)
X $^1\Sigma^+$	0	[814.6] Z	3.73 H	0.3557191[f]	0.002142$_9$[j]	(2.66)	1.832505	IR spectrum[k]		(14)
								Microwave sp.[ℓ]		(11)

SnI: [a]From the absorption spectrum (1). From the emission spectrum (2) obtain ω_e'' = 201.6, $\omega_e''x_e''$ = 0.53.

(1) Oldershaw, Robinson, TFS 64, 616 (1968).

SnI (continued):

(2) Murty, Haranath, Rao, IJP 45, 203 (1971).
(3) Iacocca, Chatalic, Pannetier, CR C 274, 1892 (1972).
(4) Oldershaw, Robinson, JMS 45, 489 (1973).

SnO: [a]Thermochemical value (8). Extrapolation of the E − X v"=0 progression gives a dissociation limit at 45770 cm^{-1} (5); if it corresponds to $^3P_1 + ^3P_1$ [see (15)] a dissociation energy of 5.45 eV would follow.

[b](5) give 58809.6 which does not fit with their observed (or calculated) v_{00}.

[c]$-5 \times 10^{-5}(v+\frac{1}{2})^5$ (valid for $v' \leq 17$).

[d]$\omega_e y_e = -0.135$.

[e]Small Λ-type doubling. Several perturbations, probably by a $^1\Pi$ (or $^1\Delta$) state (6).

[f]RKR potential functions (9).

[g]Franck-Condon factors, relative band intensities, variation of electronic transition moment with r (16).

[h]The observed ^{118}Sn and ^{116}Sn isotope shifts for the 0-1 band (6) seem to confirm the v' numbering. See, however, (12) who suggest that v' be increased by 1.

[i]Both v=1 and v=2 are extensively perturbed.

[j]$\gamma_e \approx -7 \times 10^{-7}$.

[k]In argon and nitrogen matrices.

[ℓ](11) gives rotational constants for seven isotopes. Dipole moment from Stark effect of rotation spectrum

SnO (continued):

$\mu_{e\ell}(v=0) = 4.3_2$ D (13). Zeeman effect (17), $g_J(v=0) = -0.1463$.

(1) Mahanti, ZP 68, 114 (1931).
(2) Loomis, Watson, PR 45, 805 (1934).
(3) Jevons, PPS 50, 910 (1938).
(4) Eisler, Barrow, PPS A 62, 740 (1949).
(5) Barrow, Rowlinson, PRS A 224, 374 (1954).
(6) Lagerqvist, Nilsson, Wigartz, AF 15, 521 (1959).
(7) Deutsch, Barrow, Nature 201, 815 (1964).
(8) Colin, Drowart, Verhaegen, TFS 61, 1364 (1965).
(9) Nair, Singh, Rai, JCP 43, 3570 (1965).
(10) Joshi, Yamdagni, IJP 41, 275 (1967).
(11) Törring, ZN 22 a, 1234 (1967).
(12) Smith, Meyer, JMS 27, 304 (1968).
(13) Hoeft, Lovas, Tiemann, Tischer, Törring, ZN 24 a, 1222 (1969).
(14) Ogden, Ricks, JCP 53, 896 (1970).
(15) Barrow, DONNSPEC (1970), p. 323.
(16) Dube, Rai, JP B 4, 579 (1971).
(17) Honerjäger, Tischer, ZN 28 a, 1372 (1973).

Sn^1H, Sn^2H (continued):

[j]Λ-type doubling in $^2\Pi_{\frac{1}{2}}(v=0)$, $\Delta v_{fe} = +1.003(J+\frac{1}{2}) - 1.76 \times 10^{-4}(J+\frac{1}{2})^3$ (4).

[k]Spin coupling constants for v=0: $A = 8.52$, $A_J = 0.0225$, $\gamma = 0.555$. The large difference from Sn^1H remains unexplained; see (10)(11).

[ℓ]$H_0 = -0.135 \times 10^{-7}$; see [c].

[m]Spin splitting constants for v=0: $\lambda = 45.0$, $\gamma_1 = 0.076$, $\gamma_2 = 0.0883$. ΔB(see [f]) = −0.012.

[n]$H_0 = -0.13 \times 10^{-8}$.

[o]$A_0 = 2177.05 + 0.0080\,J(J+1)$.

[p]$\Delta v_{fe}(^2\Pi_{\frac{1}{2}}, v=0) = +0.5076(J+\frac{1}{2}) - 0.000045(J+\frac{1}{2})^3$.

(1) Watson, Simon, PR 55, 358 (1939).
(2) Watson, Simon, PR 57, 708 (1940).
(3) Hougen, CJP 40, 598 (1962).
(4) Klynning, Lindgren, Åslund, AF 30, 141 (1965).
(5) Klynning, AF 31, 281 (1966).
(6) Kopp, Hougen, CJP 45, 2581 (1967).
(7) Kovács, Korwar, JP B 4, 759 (1971).
(8) Kovács, Vujisić, JP B 4, 1123 (1971).
(9) Kovács, Pacher, JP B 4, 1633 (1971).
(10) Veseth, Physica 56, 286 (1971).
(11) Veseth, JMS 48, 283 (1973).

State	T_e	ω_e		$\omega_e x_e$	B_e	α_e	D_e ($10^{-8}cm^{-1}$)	r_e (Å)	Observed Transitions		References
									Design.	ν_{00}	
$^{120}Sn^{32}S$		μ = 25.2414174			D_0^0 = 4.77 eV[a]						AUG 1977
		Continuous absorption from 56470 to 56887 cm^{-1}.									(2)
G[b]	[56171]	v=0 only.							G← X, R	55928 H	(2)
F[b]	(52257)	(408)	H	(4.5)					F← X, R	52217 H	(2)
E $^1\Sigma^+$	33037.0	294.25	H	1.15	B_{20} = 0.0889 B_{19} = 0.0915				E↔X, R	32941 H	(1)* (2)(3)
D $^1\Pi$	28336.6_0	331.35	Z	1.265	0.12023[cd]	0.00070		2.3569	D↔X, R	28258.67 Z	(1)* (3)
		The spectrum in the visible and near UV region is very complex, both in absorption and in emission (1)(3)(7); partial analysis by (3):									
C' 1	[23950.3]	Only one level[e] analyzed.			[0.1075]			[2.493]	C'↔X, R	23707.0 Z	(3)
B 1	(23589.8)	[366.60][f]	Z		[0.1214]	g		[2.345]	B↔X, R	23529.82 Z	(3)
A 0^+	[22915.97]	Only one level[h] analyzed.			[0.1184]			[2.375]	A↔X, R	22672.68 Z	(3)
a		Fluorescence bands in rare gas matrices, 18300 - 14300 cm^{-1}.							a→X,	(18300)	(8)
X $^1\Sigma^+$	0	487.26	Z	1.358	0.13686139[d]	0.00050563[i]	4.24	2.209026	IR spectrum[j]		(12)
									Microwave sp.[k]		(9)
$^{(120)}Sn^{(80)}Se$		(μ = 47.954301_7)			D_0^0 = 4.2_0 eV[a]						AUG 1977
F	(47850)	(290)							F← X, R	(47830) H	(3)*
E	30738.9	196.6	H	0.77[b]					E↔X, R	30671.6 H	(2)(3)*
D	27549.6	225.1	H	0.69					D↔X, R	27496.6 H	(1)(2)* (3)*
		A definitive analysis of the visible absorption bands (15800 - 23800 cm^{-1}) is still lacking. At least three or four systems seem to be present with upper state frequencies of ~226 and 218 cm^{-1}, but there is considerable disagreement concerning the assignments of the bands.									(1)(7)(8)
X $^1\Sigma^+$	0	331.2	H	0.736	0.0649977_7[c]	0.0001704_8[d]	1.1	2.325601	IR spectrum[e]		(9)
									Microwave sp.[f]		(5)

SnS: [a]Thermochemical value (4); agrees fairly well with values derived from the continuous absorption ($D_0^0 \leq$ 4.84 eV) and from a short extrapolation of the vibrational levels of the E state (D_0^0 = 4.81 eV) if dissociation into $^1D + ^1D$ and $^3P_1 + ^3P_1$, respectively, is assumed.

[b]p and q of (2).

[c]Perturbations.

[d]RKR potential functions (6).

[e]Called z by (3) and renamed C' by (5); v numbering uncertain. Low J lines ($J < 29$) are weak or absent causing the headless appearance of the bands.

[f]The two levels x and w of (3) are believed (5) to belong to the same electronic state; v numbering uncertain. Rotational perturbations in both observed levels.

[g]$B_1 = 0.1218$; see [f].

[h]Called y by (3), renamed A by (5); v numbering uncertain.

[i]$-1.8_7 \times 10^{-7}(v+\frac{1}{2})^2 - 4 \times 10^{-9}(v+\frac{1}{2})^3$.

[j]In argon matrix.

[k]$\mu_{e\ell} = 3.17 + 0.022(v+\frac{1}{2})$ D (9)(11); see also (10). ^{33}S hf structure (11).

(1) Rochester, PRS A 150, 668 (1935).
(2) Barrow, Drummond, Rowlinson, PPS A 66, 885 (1953).
(3) Douglas, Howe, Morton, JMS 7, 161 (1961).
(4) Colin, Drowart, JCP 37, 1120 (1962).
(5) Barrow, Fry, Le Bargy, PPS 81, 697 (1963).
(6) Nair, Singh, Rai, JCP 43, 3570 (1965).
(7) Yamdagni, Joshi, IJP 40, 495 (1966).
(8) Smith, Meyer, JMS 27, 304 (1968).
(9) Hoeft, Lovas, Tiemann, Tischer, Törring, ZN 24 a, 1222 (1969).

SnS (continued):

(10) Murty, Curl, JMS 30, 102 (1969).
(11) Hoeft, Lovas, Tiemann, Törring, JCP 53, 2736 (1970).
(12) Marino, Guerin, Nixon, JMS 51, 160 (1974).

SnSe: [a]From thermochemical data (mass-spectrom.)(4), recalculated with $D_0^0(Se_2) = 3.411$ eV (see [a] of Se_2). Assuming dissociation into $^3P_1 + ^3P_1$ at the dissociation limit of the E state [see [b] and (7)] gives $D_0^0 = 4.0_8$ eV.

[b]$\omega_e y_e = -0.0016$. Vibrational levels observed to v=29; a short extrapolation yields a dissociation limit at 35470 cm^{-1} while the banded absorption seems to go over into a continuum at 36570 cm^{-1}.

[c]Rotational constants for $^{120}Sn^{80}Se$; see (5) who gives similar data for 28 other isotopic molecules.

[d]$\gamma_e = -1._3 \times 10^{-7}$.

[e]In argon matrix.

[f]$\mu_{e\ell}(v=0) = 2.82$ D, from Stark effect of microwave sp. (6).

(1) Walker, Straley, Smith, PR 53, 140 (1938).
(2) Barrow, Vago, PPS 55, 326 (1943).
(3) Vago, Barrow, PPS 58, 707 (1946).
(4) Colin, Drowart, TFS 60, 673 (1964).
(5) Hoeft, ZN 21 a, 437 (1966).
(6) Hoeft, Lovas, Tiemann, Törring, ZN 24 a, 1843 (1969).
(7) Barrow, DONNSPEC (1970), p. 376/7.
(8) Yamdagni, JMS 33, 531 (1970).
(9) See ref. (12) of SnS.

State	T_e	ω_e	$\omega_e x_e$	B_e	α_e	D_e $(10^{-6}cm^{-1})$	r_e (Å)	Observed Transitions Design.	Observed Transitions ν_{00}	References
$^{(120)}Sn^{(130)}Te$		(μ = 62.351952$_6$)		D_0^0 = 3.69 eV[a]						AUG 1977
J	(47260)	(230) H						J← X,	R (47245) H	(4)(5)*
I	44033.5	229.7 H	1.25[b]					I← X,	R 44018.4 H	(4)(5)*
H	30818.3	201.0 H	0.6					H← X,	R 30789.0 H	(3)(4)
G	29071.8	200.8 H	0.3					G← X,	R 29042.5 H	(3)(4)
F	28545.9	98.0 H	1.0					F← X,	R 28465.0 H	(3)(4)
E	(27642.8)	(135.0) H	(2.5)					E↔ X,[c]	R (27580.0) H	(1)(3)(5)*
D	25444.3	179.1 H	0.40					D↔ X,	R 25404.1 H	(1)*(2)*(5)*
C	21418.6	218.1 H	0.98					C← X,	R 21397.8 H	(2)*
B	20394.9	230.3 H	1.53[d]					B← X,	R 20380.0 H	(2)*
A	16844.0	178.5 H	0.44					A← X,	R 16803.5 H	(2)*
X $^1\Sigma^+$	0	259.5 H	0.50	0.04247917[e]	0.00009543[f]	0.0055	2.522814	Microwave sp.[g]		(7)
$^{32}S^{16}O$		μ = 10.6613029$_2$		D_0^0 = 5.359 eV[a]		I.P. = 10.29 eV[b]				AUG 1977
		Potential energy diagram (17).								
		Fragment of another system near 70700 cm^{-1}.								(24)
E ($^3\Pi$)	(67884)	[1220]						E← X,	{68092, 67921, 67746}	(18)
D ($^3\Pi$)	(54340)	[1254]						D← X,	{54586, 54394, 54259}	(18)
C	(42200)	(170)	{State causing perturbations and predissociations in B $^3\Sigma^-$.				(2.2)			(17)
B $^3\Sigma^-$	41629	630.4[cd] H	4.79[c]	0.502$_0$[ef]	0.006$_2$	(1.28)	1.775	B[g]↔ X,[h]	R 41370	(1)(2)(2a) (17)*
A $^3\Pi$ 2	38622	412.7[i] Z	1.7	0.6164	0.0204[j]	(4.8)	1.601$_6$	A[n]↔ X,	R {38255[k]	(17)*
A $^3\Pi$ 1	38462	413.3[i] Z	1.6	0.6107[ℓ]	0.0194[m]	(4.0)	1.609$_1$		38095[k]	
A $^3\Pi$ 0	38306	415.2[i] Z	1.6	0.6067[o]	0.0194[m]	(3.7)	1.614$_4$		37940[k]}	

SnTe: [a]Thermochemical value (mass-spectrom.)(6), corrected for the new value of $D_0^0(Te_2)$.

[b]$\omega_e y_e$ = -0.003.

[c]According to (5) the assignments in this system are quite uncertain; (5) give $v_e \approx 28000$, $\omega_e' \approx 150$.

[d]$\omega_e y_e$ = -0.013.

[e]Rotational constants for $^{120}Sn^{130}Te$.

[f]$- 5._3 \times 10^{-8}(v+\frac{1}{2})^2 + 1._8 \times 10^{-9}(v+\frac{1}{2})^3$.

[g]Dipole moment of $^{120}Sn^{130}Te$: $\mu_{el}(v=0) = 2.19$ D, from Stark effect of microwave spectrum (8). Microwave spectra have been observed for 27 isotopic molecules (7).

(1) Barrow, PPS 52, 380 (1940).
(2) Barrow, Vago, PPS 56, 78 (1944).
(3) Sharma, PNASI A 14, 232 (1945).
(4) Sharma, Nature 157, 663 (1946).
(5) Vago, Barrow, PPS 58, 707 (1946).
(6) Colin, Drowart, TFS 60, 673 (1964).
(7) Hoeft, Tiemann, ZN 23 a, 1034 (1968).
(8) Hoeft, Lovas, Tiemann, Törring, ZN 24 a, 1843 (1969).

SO: [a]From the first predissociation limit in B $^3\Sigma^-$ (see [e]) assuming that it corresponds to dissociation into $^3P_2 + {}^3P_2$. A similar value ($D_0^0 = 5.36_4$ eV) follows from the assumption that the convergence limit of the B state (see [d]) corresponds to $S(^1D) + O(^3P_2)$. The value given here has been confirmed by thermochemical measurements (5); s. also (16).

[b]From the photoelectron spectrum (25)(29).

[c]The constants given represent the best approximation for $v \leq 6$ (2).

[d]Vibrational levels observed to v'=30, very close to the convergence limit $S(^1D) + O(^3P)$ at 52500 ± 100 cm^{-1} (2a) (17). A large drop in the value of ΔG occurs near v=16 owing to an avoided crossing with another $^3\Sigma^-$ state (3)

SO (continued):

(17). Lower vibrational levels appear to converge to a limit at ~59090 cm^{-1} corresponding to $S(^3P) + O(^1D)$.

[e]Strong rotational perturbations, particularly for v=1 and 2 (1)(4)(20) and v=7, 11 (17). Predissociation (breaking-off in emission) for v=0, 1, 2, 3 above N=66, 53, 39, 6, respectively (1), leading to a dissociation limit near 43224 cm^{-1}. A second predissociation (diffuseness in absorption) sets in above v=8 and reaches its maximum for v=14, 15, substantially above the corresponding dissociation limit $^3P + {}^3P$ (17).

[f]Spin splitting constants $\lambda_0 = 3.5$, $\gamma_0 = 0.010$ or -0.020 (6).

[g]Radiative lifetimes $\tau = 17._3$, $16._6$, $16._2$ ns for v=0, 1, 2, respectively (19).

[h]Franck-Condon factors (26)(30); measured relative intensities, variation of electronic transition moment with r (30).

[i]Constants derived from v=1...4; v=0 is perturbed. $\Delta G(\frac{1}{2})$ = 425, 416, 414 for $^3\Pi_0$, $^3\Pi_1$, $^3\Pi_2$, respectively. ΔG decreases rapidly above v=4.

[j]γ_e = +0.0010.

[k]Approximate origins for the deperturbed 0-0 band, by extrapolation from the unperturbed levels v=1...4; see [i].

[l]Λ-type doubling Δv_{fe} = +0.00031 J(J+1).

[m]γ_e = +0.0009.

[n]Radiative lifetime $\tau = 12._4$ ns corresponding to a band oscillator strength $f \approx 0.018$ near the Franck-Condon maximum (2-0 band); see (27).

[o]Λ-type doubling $\Delta v_{fe} \approx +1.2$ cm^{-1}, slightly dependent on J.

(continued p. 622)

State	T_e	ω_e	$\omega_e x_e$	B_e	α_e	D_e ($10^{-6}cm^{-1}$)	r_e (Å)	Observed Transitions: Design.	Observed Transitions: ν_{00}	References
$^{32}S^{16}O$ (continued)										
b $^1\Sigma^+$	10510.0	1068.6_6[p] Z	7.2_5	0.7026_2	0.0063_5	[1.20][q]	1.5001	b→X, R	10469.33 Z	(15)* (22)*
a $^1\Delta$	(6350)[r]			[0.7103383][s]		[1.168]	[1.491971]	Microwave sp.[t]		(21)(39)
								ESR sp.[u]		(8)(26a)(27a)
X $^3\Sigma^-$	0	1149.2_2[p] Z	5.6_3	0.7208171[v]	0.005736_7	[1.134]	1.481087	IR sp.[w]		(35)
								Microwave sp.[x]		(6)(7)(10) (31)(32)(39)
								ESR sp.		(9)(12)(34)
$^{32}S^{16}O^+$		D_0^0 = 5.43 eV[a]								AUG 1977
C ($^2\Pi$)	(77100)									
B ($^2\Sigma^-$)	49780	1000					(1.55)			
b $^4\Sigma^-$	37690	990					(1.53_5)			
A ($^2\Pi$)	(35600)									
a $^4\Pi$	(26170)	800					(1.64)			
X $^2\Pi_r$	340 0	1360					(1.42_4)			

All data obtained from the photoelectron spectrum of SO (1); bond lengths derived from Franck-Condon analyses of the observed spectra.

SO (continued):

[p]Vibrational constants of (22); the older values of (2) (15) are based on an extrapolation of the ground state vibrational levels (1) required by the revised v" numbering of (2). This change has been confirmed by isotope studies (14). IR fundamental in Ar matrix: $\Delta G(\frac{1}{2})$ = 1136.7 (35).

[q]$D_1 = 1.30 \times 10^{-6}$.

[r]From a comparison with O_2 and S_2 (15).

[s]"True" B_0 of (39); the effective value is 0.7103476.

[t]μ_{el} = 1.32 D [average of values obtained from Stark effect of ESR (11)(23) and microwave (21) spectra].

[u]The ^{33}S hf interaction has been studied by (25a).

[v]Spin splitting constants λ_0 = +5.2787981 cm^{-1}, λ_1 = +5.3105, γ_0 = -0.0056153 cm^{-1}, γ_1 = -0.00572 (39). For an improved representation of the rotational levels according to the intermediate case "c"- case "e" coupling model see (28)(33)(36); for an ab initio calculation of the spin-orbit part of λ (dominant contribution to the observed splitting) see (37). Rotational constants for $^{34}S^{16}O$, $^{32}S^{18}O$, $^{33}S^{16}O$ are

SO (continued):

given in (31)(32).

[w]In argon matrix.

[x]μ_{el} = 1.55 D from Stark effect of microwave sp. (6). The hf interaction due to ^{33}S has been studied by microwave (10) and ESR spectroscopy (13); see also (38).

(1) Martin, PR 41, 167 (1932).
(2) Norrish, Oldershaw, PRS A 249, 498 (1959).
(2a) McGrath, McGarvey, JCP 37, 1574 (1962); McGarvey, McGrath, PRS A 278, 490 (1964).
(3) Abadie, Herman, CR 257, 2820 (1963).
(4) Abadie, Herman, JQSRT 4, 195 (1964).
(5) Colin, Goldfinger. Jeunehomme, TFS 60, 306 (1964).
(6) Powell, Lide, JCP 41, 1413 (1964).
(7) Winnewisser, Sastry, Cook, Gordy, JCP 41, 1687 (1964).
(8) Carrington, Levy, Miller, PRS A 293, 108 (1966).
(9) Daniels, Dorain, JCP 45, 26 (1966).
(10) Amano, Hirota, Morino, JPSJ 22, 399 (1967).
(11) Carrington, Levy, Miller, JCP 47, 3801 (1967).
(12) Carrington, Levy, Miller, PRS A 298, 340 (1967).
(13) Carrington, Levy, Miller, MP 13, 401 (1967).
(14) Apparao, Narasimham, PIAS A 68, 173 (1968).
(15) Colin, CJP 46, 1539 (1968).
(16) Gaydon, DISSEN (1968).
(17) Colin, CJP 47, 979 (1969).
(18) Donovan, Husain, Jackson, TFS 65, 2930 (1969).
(19) Smith, JQSRT 9, 1191 (1969).
(20) Abadie, AP(Paris) 5, 227 (1970).
(21) Saito, JCP 53, 2544 (1970).
(22) Bouchoux, Marchand, Janin, SA A 27, 1909 (1971); Bouchoux, Marchand, SA A 28, 1771 (1972).
(23) Byfleet, Carrington, Russell, MP 20, 271 (1971).

SO (continued):

(24) Donovan, Little, SpL 4, 213 (1971).
(25) Jonathan, Smith, Ross, CPL 9, 217 (1971).
(25a) Miller, JCP 54, 1658 (1971).
(26) Smith, Liszt, JQSRT 11, 45 (1971).
(26a) Uehara, MP 21, 407 (1971).
(27) Smith, ApJ 176, 265 (1972).
(27a) Brown, Uehara, MP 24, 1169 (1972).
(28) Veseth, JP B 6, 1484 (1973).
(29) Dyke, Golob, Jonathan, Morris, Okuda, Smith, JCS FT II 70, 1818 (1974).
(30) Hébert, Hodder, JP B 7, 2244 (1974).
(31) Tiemann, JPCRD 3, 259 (1974).
(32) Tiemann, JMS 51, 316 (1974).
(33) Veseth, Lofthus, MP 27, 511 (1974).
(34) Davies, Wayne, Stone, MP 28, 1409 (1974).
(35) Hopkins, Brown, JCP 62, 2511 (1975).
(36) Veseth, MP 29, 321 (1975).
(37) Wayne, CPL 31, 97 (1975).
(38) Veseth, JMS 59, 51 (1976).
(39) Clark, De Lucia, JMS 60, 332 (1976).

SO^+: [a]D_0^0(SO) + I.P.(S) - I.P.(SO).

(1) See ref. (29) of SO.

State	T_e	ω_e	$\omega_e x_e$	B_e	α_e	D_e (10^{-} cm^{-1})	r_e (Å)	Observed Transitions Design.	Observed Transitions ν_{00}	References
$^{(88)}Sr^{79}Br$		(μ = 41.584947$_1$) D_0^0 = 3.4$_1$ eV[a]								SEP 1977
		V shaded emission bands in the region 17000 - 20000 cm^{-1} have been attributed to transitions from three additional states to A $^2\Pi$ and B $^2\Sigma$.								(5)
E $^2\Sigma^+$	32052.5	248.0 H	0.65					E→X, V	32068.2 H	(4)
D $^2\Sigma^+$	28958.2	247.8 H	0.55					D↔X, V	28973.8 H	(3)(11)
C $^2\Pi$	24665.8 24343.7	205.2 H	0.49					C[b]↔X, R	24660.2 H 24338.1 H	(1)(3)(10) (11)
B $^2\Sigma^+$	15352.0	222.0 H	0.55					B[b]↔X, V	15354.7 H	(2)(3)
A $^2\Pi$	15000.7 14699.4	222.1 H	0.53					A[b]↔X, V	15003.5 H 14702.2 H	(2)(3)
X $^2\Sigma^+$	0	216.5 H	0.51							
$^{(88)}Sr^{35}Cl$		(μ = 25.0170662) D_0^0 = 4.1$_6$ eV[a]								SEP 1977
H $^2\Sigma^+$	34256.7	364.6 H	1.08					H→B,	18562.7 H	(5)
								H↔X, V	34287.8 H	(4)(5)
G ($^2\Delta$)	34085.4 34059.6	356.7 H	1.0					G→A,	18996.4 H 19264.8 H	(5)
F ($^2\Pi$)	32974.8 32905.9	354.2 H	1.09					F→B,	17277.5 H 17208.4 H	(5)
								F→A,	17884.6 H 18109.9 H	(5)
E $^2\Sigma^+$	32201.8	346.3 H	1.10					E→A,	- 17396.1 H	(5)
								E↔X, V	32223.8 H	(4)(5)
D $^2\Sigma^+$	28822.9	344.8 H	1.04					D→B,	13123.7 H	(5)
								D→A,	13727.0 H 14021.8 H	(5)
								D↔X, V	28844.1 H	(4)(5)
C $^2\Pi$	25399.8 25244.6	283.4 H 282.1 H	0.92 0.89					C[b]↔X, R	25390.4 H 25234.5 H	(1)(2)(5)

State	T_e	ω_e	$\omega_e x_e$	B_e	α_e	D_e (10^{-} cm^{-1})	r_e (Å)	Observed Transitions Design.	ν_{00}	References
$^{(88)}Sr^{35}Cl$ (continued)										
B $^2\Sigma^+$	15719.5	306.4 H	0.98					$B^b \leftrightarrow X$, V	15721.5[c] H 15713.1[c] H	(1)(3)* (4) (7)
A $^2\Pi$	15112.6 14818.4	309.4 H	0.98					$A^b \leftrightarrow X$, V	15116.1 H 14821.9 H	(1)(2)(3)* (4)(5)(7)
X $^2\Sigma^+$	0	302.3 H	0.95							

SrBr: [a]Thermochemical value (mass-spectrom.) (12); in good agreement with the flame photometric value of (6)(7) but considerably below the lower limit to D_0^0 derived by (9) from the study of chemiluminescence spectra.

[b]The following radiative lifetimes have been measured by (8): A 34 ns, B 42 ns, C 29 ns.

(1) Olmsted, ZWP 7, 300 (1906).
(2) Hedfeld, ZP 68, 610 (1931).
(3) Harrington, Dissertation (U. of California, 1942).
(4) Reddy, Rao, IJPAP 4, 251 (1966).
(5) Reddy, Reddy, Ashrafunnisa, Rao, CS 40, 317 (1971).
(6) Gurvich, Ryabova, Khitrov, FSCS No. 8, 83 (1973).
(7) Khitrov, Ryabova, Gurvich, HT(USSR) 11, 1005 (1973).
(8) Dagdigian, Cruse, Zare, JCP 60, 2330 (1974).
(9) Menzinger, CJC 52, 1688 (1974).
(10) Joshi, Gopal, Pramāṇa 4, 276 (1975).
(11) Puri, Mohan, CS 44, 152 (1975).
(12) Hildenbrand, JCP 66, 3526 (1977).

SrCℓ: [a]Thermochemical value (mass-spectrom.)(6), in very good agreement with the most recent flame photometric results of (8)(10). (9) place a lower limit to the dissociation energy at 4.2_9 eV based on their study of the $Sr + C\ell_2$ chemiluminescent reaction.

[b](11) have measured the following radiative lifetimes: A 31 ns, B 39 ns, C 26 ns.

[c]Double heads on account of large spin doubling.

(1) Parker, PR 47, 349 (1935).
(2) More, Cornell, PR 53, 806 (1938).
(3) Gatterer, RS 1, 153 (1942).
(4) See ref. (3) of SrBr.
(5) Novikov, Gurvich, OS(Engl. Transl.) 19, 76 (1965).
(6) Hildenbrand, JCP 52, 5751 (1970).
(7) Singh, Nair, Upadhya, Rai, OPA 3, 76 (1970).
(8) Gurvich, Ryabova, Khitrov, Starovoitov, HT(USSR) 9, 261 (1971).
(9) Jonah, Zare, CPL 9, 65 (1971).
(10) See ref. (6) of SrBr.
(11) See ref. (8) of SrBr.

State	T_e	ω_e		$\omega_e x_e$	B_e	α_e	D_e $(10^{-5}cm^{-1})$	r_e (Å)	Observed Transitions Design.		ν_{00}		References
$^{(88)}Sr^{19}F$		(μ = 15.6221112)			D_0^0 = 5.58 eV[a]								OCT 1977
G ($^2\Pi$)	34758.9	573.9	H	1.28					G← X,	V	34795.4	H	(3)
F $^2\Sigma^+$	32823.7	598.5	H	3.42	[0.26966][b]	(0.00187)	[0.022_3]	[2.0004]	F⟷X,	V	32871.96	Z	(3)(8)(9)
E $^2\Pi$	31615 31529.1	- 564.4	H	- 3.20					E⟷X,	V	31646.5 31560.4	H H	(3)(9)
D $^2\Sigma^+$	28296.6	552.1	H	2.15					D⟷X,	V	28322.0	H	(3)(9)
C $^2\Pi$	27445 $27384._1$	- 450.5[c]	H	- 1.72[c]					C⟷X,	R	27419.3 27358.8	H H	(1)(3)(9)
B $^2\Sigma^+$	17267.42	495.8[d]	Z	2.3_4	0.249396[de]	0.001557	0.0252	2.08010	B⟷X,	R	17264.10	Z	(1)(2)* (3) (4)(13)
A $^2\Pi$	$15349._0$ $15068._3$	507.3[c] 507.9[c]	H H	2.18[c] 2.21[c]					A[f]⟷X,	V	15352.0 15071.6	H H	(1)(2)* (3) (9)
X $^2\Sigma^+$	0	502.4[dg]	Z	2.2_7	0.250533[dh]	0.001546	0.0249	2.07537	Rotation sp.[i]				(14)
									ESR sp.[j]				(11)
$^{(88)}Sr^{1}H$		(μ = 0.99640162)			$D_0^0 \leq$ 1.66 eV[a]								SEP 1977
		Fragments of several additional absorption systems above 30000 cm^{-1}; tentative rotational analyses.											(9)* (10) (16)* (17)*
F $^2\Sigma^+$	34096	[1337.0]	Z	(33)[b]	[4.0020][c]	0.09_2	[13.84]	[2.0561]	F← X,	V	34189.97	Z	(6)* (7)(9) (10)
C $^2\Sigma^+$	26230	1347	H	$23._5$	4.008[d]	0.132	14	2.055	C⟷X,	V	26298.7	Z	(2)(4)(5) (8)*
D $^2\Sigma^+$	20847.6	1014.1	Z	15.4	1.925[e]	0.024	3	2.965	D⟷X,[f]	R	20752.0	Z	(3)(5)
E $^2\Pi$					[3.869] [3.639][g]		[20] [10]	[2.091] [2.156]	E→X,	V	18960 18860	Z Z	(2)
B $^2\Sigma^+$	14340	[$1193._0$]	Z	(19)[b]	3.8788[h]	0.0930	17.3	2.0885	B→X,	V	14352.1	Z	(1)* (3)
A $^2\Pi_r$	i				[3.679] [3.668][j]		[11.3] [13.3]	[2.144] [2.148]	A→X,	V	13653 13360	Z Z	(1)*
X $^2\Sigma^+$	0	1206.2	Z	17.0	3.6751[k]	0.0814	[13.5][ℓ]	2.1456	ESR sp.[m]				(12)

SrF:

[a]Thermochemical value (mass-spectrom.)(10). Re-evaluation by (10) of the earlier data of (5)(6) gives $D_0^0 = 5.45$ eV; flame photometric value 5.7_2 eV (7).

[b]Spin doubling constant $|\gamma| = 0.043$ (8).

[c]Vibrational constants of (9).

[d]Constants for $^{88}Sr^{19}F$ (13)(14).

[e]Large spin doubling, $\gamma = -0.1353_4$ cm^{-1} (14).

[f]Radiative lifetime $\tau = 23$ ns (12).

[g]Similar constants are obtained from bandhead measurements in $D-X$, $E-X$, $F-X$: $\omega_e'' = 501.3$, $\omega_e''x_e'' = 2.2_5$ (averaged values).

[h]Spin-rotation interaction $\gamma = +0.00249$ (14).

[i]Microwave optical double resonance.

[j]In Ne and Ar matrices at 4 K.

(1) Johnson, PRS A 122, 161 (1929).
(2) Harvey, PRS A 133, 336 (1931).
(3) Fowler, PR 59, 645 (1941).
(4) Ahrens, PR 74, 74 (1948).
(5) Blue, Green, Ehlert, Margrave, Nature 199, 804 (1963).
(6) Ehlert, Blue, Green, Margrave, JCP 41, 2250 (1964).
(7) Ryabova, Gurvich, HT(USSR) 2, 749 (1964).
(8) Barrow, Beale, CC (1967), 606.
(9) Novikov, Gurvich, OS(Engl. Transl.) 22, 395 (1967).
(10) Hildenbrand, JCP 48, 3657 (1968).
(11) Knight, Easley, Weltner, Wilson, JCP 54, 322 (1971).
(12) Dagdigian, Cruse, Zare, JCP 60, 2330 (1974).
(13) Steimle, Domaille, Harris, JMS 68, 134 (1977).
(14) Domaille, Steimle, Harris, JMS 68, 146 (1977).

Sr^1H, Sr^2H:

[a]From the predissociation in C $^2\Sigma^+$ assuming dissociation into $^3P_0 + {}^2S$.

[b]From isotope relations.

[c]Spin doubling constant $\gamma_0 = +0.076$. All lines are diffuse owing to predissociation.

[d]Breaking-off in emission at low pressure above $N'=19$ and 4 in $v'=0$ and 1, respectively (8). Perturbations.

[e]Strong perturbations.

[f]Franck-Condon factors (11).

[g]Perturbations.

[h]Spin doubling constant $\gamma_0 = -3.81$ [for $N < 10$ (3)]. For a more elaborate evaluation of the spin splitting including higher-order correction terms see (14).

[i]$A_0 \approx +300$. See also [j].

[j]Very large Λ-type doubling, see (1)(15). The constants given here are effective values; see, however, (13) who has evaluated the true constants by taking into account several usually neglected corrections to the rotational energy formulae for a $^2\Pi$ state.

[k]From (3); spin doubling constant $\gamma_0 = +0.122$. (6) give $B_0 = 3.6336$, $D_0 = 13.39 \times 10^{-5}$.

[ℓ]$D_1 = 12.9 \times 10^{-5}$.

[m]In argon matrix at 4 K.

(continued p. 628)

State	T_e	ω_e	$\omega_e x_e$	B_e	α_e	D_e ($10^{-5}cm^{-1}$)	r_e (Å)	Observed Transitions Design.	ν_{00}	References
$^{(88)}Sr^2H$		(μ = 1.96898856)		$D_0^0 \leq$ 1.70 eV[a]						SEP 1977
		Fragments of additional absorption systems above 30000 cm^{-1}; tentative rotational analyses.								(9)(10)(16)* (17)*
F $^2\Sigma^+$	34097	[964.6] Z	(17)[b]	[2.0334][n]	0.032	[3.47]	[2.0519]	F← X, V	34164.87 Z	(6)(7)(9) (10)
C $^2\Sigma^+$	26226	[943] Z	(11.9)[o]	1.98[p]	0.03	[2.5]	2.08	C→X, V	26279 Z	(3)(8)*
B $^2\Sigma^+$	14335	[857] Z	(10)[b]	1.9426[q]	0.0349	[4.02][r]	2.0994	B→X, V	14343.9 Z	(3)
X $^2\Sigma^+$	0	[841] Z	(8.6)[o]	1.8609[s]	0.0292	[3.47][t]	2.1449			
$^{(88)}Sr^{127}I$		(μ = 51.932462$_5$)		$D_0^0 \geq$ 2.8$_2$ eV[a]						SEP 1977
F		210.4 H	0.40					F→A or B, V	19902.1 H 19599.3 H	(12)
D ($^2\Sigma$)	28944.0	200.4 H	0.50					D→X, V	28957.1 H	(8)* (10)
C ($^2\Pi$)	23223.4 22666.1	168.5 H 170.9 H	0.40 0.36					C[b]↔X, R	23220.6 H 22664.5 H	(1)(2)(5)*
B ($^2\Sigma$)[c]	14815.9	182.2 H	0.37					B[b]→X, V	14820.1 H	(9)*
A ($^2\Pi$)[c]	14748.8 14422.7	179.5 H 182.3 H	0.32 0.54					A[b]↔X, V	14751.7 H 14427.0 H	(1)(4)(6) (9)*
X $^2\Sigma^+$	0	174.1[d] H	0.35[d]							

Sr^1H, Sr^2H (continued):

[n]Spin doubling constant γ_0 = +0.035$_7$.
[o]From the value for Sr^1H.
[p]Breaking-off above N'=29 and 19 in v'=0 and 1, respectively (8). Strong perturbations.
[q]Spin doubling constant γ_0 = -2.01 [for N < 10 (3)]. See [h].
[r]D_1 = 3.81 x 10^{-5}.

Sr^1H, Sr^2H (continued):

[s]Spin doubling constant γ_0 = +0.061$_3$.
[t]D_1 = 3.64 x 10^{-5}.

(1) Watson, Fredrickson, PR 39, 765 (1932).
(2) Fredrickson, Hogan, Watson, PR 48, 602 (1935).
(3) Watson, Fredrickson, Hogan, PR 49, 150 (1936).
(4) Humphreys, Fredrickson, PR 50, 542 (1936).

Sr^1H, Sr^2H (continued):

(5) More, Cornell, PR 53, 806 (1938).
(6) Edvinsson, Kopp, Lindgren, Åslund, AF 25, 95 (1963).
(7) Khan, PPS 81, 1047 (1963).
(8) Khan, PPS 82, 564 (1963).
(9) Khan, PPS 89, 165 (1966).
(10) Khan, Butt, JP B 1, 745 (1968).
(11) Singh, Srivastava, JQSRT 8, 1443 (1968).
(12) Knight, Weltner, JCP 54, 3875 (1971).
(13) Veseth, JMS 38, 228 (1971).
(14) Veseth, MP 20, 1057 (1971).
(15) Veseth, MP 21, 287 (1971).
(16) Khan, Rafi, Hussainee, JP B 9, 1953 (1976).
(17) Khan, Rafi, Khan, Baig, JP B 9, 2313 (1976); 10, 111 (1977).

SrI: [a]From a reactive scattering study of the Sr + HI reaction by the crossed molecular beam technique (7). The ionic model calculations of (3) predict $D_0^0 \approx 2.92$ eV.
[b]Radiative lifetimes measured by (11): A 42 ns, B 46 ns,

SrI (continued):

C 36 ns. There seems to be some uncertainty concerning the correct identification of the B state emission.
[c]According to (9) the tentative assignments for these two states or substates (i.e. Σ and Π) should be reversed.
[d]Average values for the lower state constants of A...D → X.

(1) Walters, Barratt, PRS A 118, 120 (1928).
(2) Mesnage, AP(Paris) 12, 5 (1939).
(3) Krasnov, Karaseva, OS(Engl. Transl.) 19, 14 (1965).
(4) Shukla, IJPAP 8, 855 (1970).
(5) Reddy, Reddy, Rao, JP B 4, 574 (1971).
(6) Reddy, Reddy, Rao, Rao, CS 40, 186 (1971).
(7) Mims, Lin, Herm, JCP 57, 3099 (1972).
(8) Shah, Patel, Darji, JP B 6, 1344 (1973).
(9) Ashrafunnisa, Rao, Rao, JP B 6, 1503 (1973).
(10) Ashrafunnisa, Rao, Rao, JP B 6, 2653 (1973).
(11) Dagdigian, Cruse, Zare, JCP 60, 2330 (1974).
(12) Kamalasanan, IJPAP 13, 124 (1975).

SrS (continued from p. 631):

(1) Marquart, Berkowitz, JCP 39, 283 (1963).
(2) Colin, Goldfinger, Jeunehomme, TFS 60, 306 (1964).
(3) Cater, Johnson, JCP 47, 5353 (1967).
(4) Marcano, Barrow, TFS 66, 1917 (1970).

SrSe (Table on p. 630):

[a]Thermochemical value (mass-spectrom.)(1), corrected for new value of $D_0^0(Se_2)$.

(1) Berkowitz, Chupka, JCP 45, 4289 (1966).

State	T_e	ω_e	$\omega_e x_e$	B_e	α_e	D_e (10^{-7}cm^{-1})	r_e (Å)	Observed Transitions Design.	Observed Transitions ν_{00}	References
$^{88}Sr^{16}O$		$\mu = 13.5325861_5$		$D_0^0 = 4.8_8$ eV[a]						SEP 1977 A
C $^1\Sigma^+$	28632.7	480.2 Z	2.6	0.2742[b]	0.0021	3.5	2.131	C→X,[c]	R 28546.4$_3$ Z	(1)(9)*
B $^1\Pi$	24701.0	519.9$_1$ Z	3.2$_4$	0.2937	0.0015	5.$_0$	2.059	B→X,[c]	R 24634.4 Z	(1)(7)* (8)
		Unclassified bands in the regions 14700 - 15700 and 16600 - 16900 cm^{-1}. The emitter of these bands has long been in doubt (13)(32); in flames, and in arcs in water vapour, strong bands attributed to SrOH occur in the same regions (11)(14). The bands have recently been observed in dry Sr + N_2O + CO (or N_2^*) flames, both in emission and in absorption (33)(34)(35), and have tentatively been attributed to the transitions $^3\Delta \leftrightarrow a^3\Pi$ and $^1\Delta, ^1\Sigma^- \leftrightarrow A'^1\Pi$ of SrO.								(3)* (16)*
A $^1\Sigma^+$	10886.59	619.5$_8$ Z	0.8$_9$[d]	0.30471$_1$[e]	0.0011$_2$	3.2	2.021$_9$	A→X,[f]	R 10870.40 Z	(2)* (4)* (5) (12)* (16)* (21)
A' $^1\Pi$	9400	472.$_8$[g] H	2.06	0.256$_5$[h]	0.0017		2.20$_4$	A'→X,	R 9310 H	(29)(31)
a $^3\Pi_i$	9149[i]	463.5	1.61	0.2584	0.0020		2.196			
X $^1\Sigma^+$	0	653.4$_9$ Z	3.96	0.33798$_0$	0.00219$_4$	3.6	1.91983	IR spectrum[j]		(28)
								Microwave and rf sp.[kℓ]		(19)
$^{88}Sr^{32}S$		$\mu = 23.4449370$		$D_0^0 = 3.48$ eV[a]						SEP 1977
B $^1\Sigma^+$	39332.1	286.8$_0$ Z	0.84	0.10566[b]	0.00032$_0$	0.575	2.608$_7$	B←X,	R 39281.4$_2$ Z	(4)*
X $^1\Sigma^+$	0	388.38 Z	1.31	0.12072	0.00044$_0$	0.475	2.4405			
$^{(88)}Sr^{(80)}Se$		($\mu = 41.860458_9$)		$D_0^0 \approx 2.9$ eV[a]						SEP 1977

SrO: [a]From the appearance threshold of SrO in a crossed-beam study of the Sr + O_2 reaction (25); the interpretation of the chemiluminescence spectrum resulting from the reaction Sr + $C\ell O_2$ in a similar experiment (30) leads to $D_0^0 \geq 4.6_7$ eV. Good agreement with the flame-photometric results ($D_0^0 = 4.8_1$ eV) of (15) and (10), the latter recalculated for a $^1\Sigma$ state, and with the thermochemical value ($D_0^0 = 4.9_0$ eV) of (6). Flame-photometric values of (20)(26) are lower, even when recalculated with a $^1\Sigma$ ground state, and are close to the mass-spectrometric value of (18) ($D_0^0 = 4.4_3$ eV). Earlier references have been reviewed in these papers; see also (22).

SrO (continued):

[b]Several perturbations (9).

[c]RKR Franck-Condon factors (24).

[d]$\omega_e y_e = -0.05_4$. Bandheads in the region 8959 - 9166 cm^{-1}, originally (12) attributed to a new system of SrO, have been shown (21) to belong to the A→X 0-3 sequence.

[e]Numerous perturbations by levels of a $^3\Pi_i$ and A' $^1\Pi$ (5) (27).

[f]Franck-Condon factors (17).

[g]Vibrational numbering (29) confirmed by isotope studies (31).

[h]The rotational constants have been derived (29) with the help of information gained from the analysis of perturbations in A $^1\Sigma^+$ (5)(27).

[i]A = -70; all constants for this state derived by (27) from perturbations in A $^1\Sigma^+$ (5).

[j]In nitrogen matrices.

[k]By the molecular beam electric resonance method. $\mu_{e\ell}$ = 8.913 - 0.026(v+½) D (19).

[ℓ]Ab initio calculations of various ground state properties (23).

(1) Mahanti, PR 42, 609 (1932).
(2) Mahla, ZP 81, 625 (1933).
(3) Gatterer, RS 1, 153 (1942).
(4) Almkvist, Lagerqvist, AF 1, 477 (1949).
(5) Almkvist, Lagerqvist, AF 2, 233 (1950).
(6) Drummond, Barrow, TFS 47, 1275 (1951).
(7) Kovács, Budó, APH 1, 469 (1952); AP(Leipzig) 12, 17 (1953).
(8) Deézsi, Koczkás, Mátrai, APH 3, 95 (1953).
(9) Lagerqvist, Almkvist, AF 8, 481 (1954).
(10) Lagerqvist, Huldt, ZN 9 a, 991 (1954).
(11) Lagerqvist, Huldt, Naturw. 42, 365 (1955).

SrO (continued):

(12) Lagerqvist, Selin, AF 11, 323 (1956).
(13) Charton, Gaydon, PPS A 69, 520 (1956).
(14) Huldt, Lagerqvist, AF 11, 347 (1956).
(15) Veits, Gurvich, OS 1, 22 (1956); 2, 145 (1957); ZFK 31, 2306 (1957).
(16) Gatterer, Junkes, Salpeter, Rosen, METOX (1957).
(17) Nicholls, JRNBS A 66, 227 (1962).
(18) Drowart, Exsteen, Verhaegen, TFS 60, 1920 (1964).
(19) Kaufman, Wharton, Klemperer, JCP 43, 943 (1965).
(20) Kalff, Hollander, Alkemade, JCP 43, 2299 (1965).
(21) Brewer, Hauge, JMS 25, 330 (1968).
(22) Gaydon, DISSEN (1968), p. 242.
(23) Yoshimine, JPSJ 25, 1100 (1968).
(24) Liszt, Smith, JQSRT 11, 1043 (1971).
(25) Batalli-Cosmovici, Michel, CPL 16, 77 (1972).
(26) Kalff, Alkemade, JCP 59, 2572 (1973).
(27) Field, JCP 60, 2400 (1974).
(28) Ault, Andrews, JCP 62, 2320 (1975).
(29) Capelle, Broida, Field, JCP 62, 3131 (1975).
(30) Engelke, Sander, Zare, JCP 65, 1146 (1976).
(31) Hecht, JCP 65, 5026 (1976).
(32) Pearse, Gaydon, IDSPEC (1976), p. 339.
(33) Benard, Slafer, Hecht, JCP 66, 1012 (1977).
(34) Eckstrom, Barker, Hawley, Reilly, AO 16, 2102 (1977).
(35) Benard, Slafer, Love, Lee, AO 16, 2108 (1977).

SrS: [a]Thermochemical value (mass-spectrom.)(1)(2)(3).
[b]Extensive perturbations.

References on p. 629 .

SrSe: See p. 629 .

State	T_e	ω_e	$\omega_e x_e$	B_e	α_e	D_e (10^{-7}cm^{-1})	r_e (Å)	Observed Transitions Design.	ν_{00}	References
$^{181}Ta^{16}O$		$\mu = 14.6958722_5$		$D_0^0 = 8.2$ eV[a]						JUN 1975 A
		Weak emission bands at 35476 and 36379 cm^{-1}, no analysis.								(7)
V ($^2\Delta$) 5/2				[0.375][bc]		[3.3]	[1.74_9]	V→X_2, R	33280[b] (Z)	(7)
U ($^2\Delta$) 5/2				[0.3715][c]		[3.3]	[1.757_2]	U→X_2, R	33110 (Z)	(7)
T ($^2\Delta$) 5/2	35954	(891)		[0.37688][d]		[2.70]	[1.7446_1]	T→X_2, R	32380.7_7 Z	(7)
S ($^2\Delta$) 5/2	35864	(871)		[0.37536][d]		[2.79]	[1.7481_4]	S→X_2, R	32280.4_0 Z	(7)
R ($^2\Delta$) 3/2	32445	(885)		[0.38393][d]		[2.89][e]	[1.7285_2]	R→X_1, R	32373.60 Z	(7)
Q	(29306)	[(895)][f]						Q← X_1,	(29240)[f]	(6)
Q'($^2\Delta$) 5/2	27353.0	[896.1] Z	(4.07)	0.38183_4[g]	0.00219	[2.744]	1.73326	Q'→X_2, R	23785.2_0 Z	(7)
P ($^2\Delta$) 3/2	26736.19	902.68 Z	4.08	0.37750_0[h]	0.00181	[2.573]	1.74318	P↔X_1, R	26673.04 Z	(1)(2)* (4) (6)(7)
O	(26342)	(913)[f]	(4.5)					O← X_1,	(26284)[f]	(6)
O'($^2\Phi$) 7/2	26186	(899)		[0.381304]		[2.745]	[1.73446_2]	O'→X_2, R	22616.0_7 Z	(7)
N ($^2\Pi$) 3/2	25657	(900)		[0.37720_7]		[2.64_9]	[1.74386]	N→X_2, R	22087.7_0 Z	(7)
								N↔X_1, R	25593.1_3 Z	(6)(7)
M ($^2\Phi$) 5/2	24123.7	[890.31] Z	4.1 H	0.37706_4	0.00184	[2.635]	1.74419	M↔X_1, R	24058.42 Z	(2)* (4)(6) (7)
L ($^2\Pi$) 1/2	23408.3	[887.70] Z	4.1 H	0.37742_4[i]	0.00195	[2.70_6]	1.74335	L↔X_1, R	23341.74 Z	(2)* (4)(6) (7)
K'($^2\Phi$) 7/2	22981.5_8	903.06 Z	3.56	0.38081	0.00192	[2.756]	1.7355_9	K'→X_2, R	19413.32 Z	(7)
K	(22396)	($901._7$)[f]	(3.3_4)					K↔X_1,	(22333)[f]	(6)
J	(22196)	(892)[f]	(3)					J← X_1,	(22128)[f]	(6)
(I)	Ne matrix emission and absorption spectra suggest the existence of an additional state close to H.									(6)
H	(20868)	[(900)][j]						H↔X_1,	(20805)[j]	(6)
G								G← X_1,	(18007)[f]	(6)
F	(16770)	[(922)][f]						F← X_1,	(16718)[f]	(6)
		Incompletely analysed emission band.						R	16051 Z	(7)
E ($^2\Phi$) 5/2	15928	(935)[f]	(5)	[0.38618][k]		[3.2_6]	[1.7234_8]	E↔X_1 R	15880.6_2 Z	(6)(7)

State	T_e	ω_e	$\omega_e x_e$	B_e	α_e	D_e $(10^{-7}cm^{-1})$	r_e (Å)	Observed Transitions Design.	ν_{00}	References
$^{181}Ta^{16}O$ (continued)										
D	(14437)	$(943)^{\ell}$						$D\leftrightarrow X_1$,	$(14395)^{fm}$	(6)(7)
C $(^2\Delta)$ 3/2	13612	(942)		$[0.38754_7]$		$[2.62_4]$	[1.72044]	$C\leftrightarrow X_1$, R	13569.27 Z	(6)(7)
B $(^2\Phi)$ 5/2	12900	(931)		$[0.38685_1]$		$[2.67_4]$	[1.72198]	$B\leftrightarrow X_1$, R	12852.02 Z	(6)(7)
A	(12134)	$[(939)]^n$						$A\leftrightarrow X_1$,	$(12090)^o$	(6)(7)
A'$(^2\Pi)$ 1/2				[0.389]			$[1.71_7]$	$A'\rightarrow X_1$, R	11062 (Z)	(7)
A"$(^2\Delta)$ 3/2	10908	(933)		$[0.38729_1]$		$[2.66_8]$	[1.72100]	$A''\rightarrow X_1$, R	10860.95 Z	(7)
X_2 $(^2\Delta)$ 5/2	3504.39	1030.81 Z	3.59	0.40358_4	0.00187	[2.503]	1.68591			
X_1 3/2	0	1028.69 Z	3.51	0.40284_0	0.00182	[2.450]	1.68746	IR spectrum[p]		(6)
$^{181}Ta^{16}O^+$?										JUN 1975
A	17933	(952)		0.3906	0.0022	2.63	1.713_7	$A\rightarrow X$,[a] R	17880.41 Z	(1)
(X)	0	[1051.12] Z	(3.41)	0.4128_7	0.0018_1	2.52	1.6668			

TaO: [a]Average of two thermochemical values (mass-spectrom.) | (3)(5).

[b]This level could possibly be T(v=1).

[c]Extensive perturbations.

[d]Perturbations.

[e](7) give 2.289 which appears to be a misprint.

[f]From the Ne matrix absorption spectrum.

[g]Perturbations in v=0.

[h]v=1 perturbed by a state of higher B value.

[i]Λ-type doubling $\Delta\nu(v=0) = 0.0927(J+\frac{1}{2})$.

[j]From Ne matrix absorption and emission spectra.

[k]Perturbed by a state of larger B value.

[ℓ]Calculated from $Ta^{18}O$ frequency in neon.

[m]In the gas phase probably at 14362 cm^{-1}.

[n]From the Ar matrix absorption spectrum.

TaO (continued):

[o]From (7); the observed band is too weak for analysis.

[p]In rare gas matrices.

(1) Premaswarup, IJP 29, 109 (1955).

(2) Gatterer, Junkes, Salpeter, Rosen, METOX (1957).

(3) Inghram, Chupka, Berkowitz, JCP 27, 569 (1957).

(4) Premaswarup, Barrow, Nature 180, 602 (1957).

(5) Krikorian, Carpenter, JPC 69, 4399 (1965).

(6) Weltner, McLeod, JCP 42, 882 (1965).

(7) Cheetham, Barrow, TFS 63, 1835 (1967).

TaO^+: [a]P, Q, R branches, probably singlet system.

(1) See ref. (7) of TaO.

State	T_e	ω_e	$\omega_e x_e$	B_e	α_e	D_e ($10^{-8}cm^{-1}$)	r_e (Å)	Observed Transitions Design.	Observed Transitions ν_{00}	References
$^{159}Tb_2$		μ = 79.462693		$D_0^0 = 1.3_2$ eV[a]						JUN 1975
$^{159}Tb^{16}O$		μ = 14.5323214$_6$		$D_0^0 = 7.3_0$ eV[a]						JUN 1975 A
		Mostly R shaded emission bands in the region 15400 - 19000 cm^{-1}; no analysis.								(1)* (2)
X	0	[824.3][b]								(5)
$^{130}Te_2$		μ = 64.953116$_0$		$D_0^0 = 2.677$ eV[a]		I.P. = 8.29 eV[b]				NOV 1977
Large number of absorption bands, 40000 - 52000 cm^{-1} (7)(8); tentative assignments to several systems (8). There are rather frequent coincidences with bands attributed to TeS and TeSe, see (1) and (2) of TeS and TeSe, respectively.										
B_2 ($^3\Sigma_u^-$) 1_u				0.03016[c]				$B_2 \rightarrow X_2$, R	(21282.4)[c] Z	(22)*
B_1 ($^3\Sigma_u^-$) 0_u^+	22207.4	162.32[d] Z	0.453[e]	0.032535[f]	0.000125		2.8244	$B_1 \rightarrow X_2$, R	20188.6 Z	(19)(22)*
								$B_1 \leftrightarrow X_1$,[gh] R	22165.0 Z	(10)* (11) (15)* (17)
A 0_u^+	19450.8	143.588 Z	0.4543[i]	0.031238[j]	0.000130		2.8824	$A \leftrightarrow X_1$,[g] R	19399.1 Z	(17)(22)
X_2 ($^3\Sigma_g^-$) 1_g	1974.9$_7$	250.033 Z	0.5155	0.039820	0.0001002[k]		2.5530			
X_1 ($^3\Sigma_g^-$) 0_g^+	0	247.07 Z	0.5148[l]	0.039681[j]	0.000106	[0.44]	2.5574			
$^{(130)}Te_2^+$		(μ = 64.952978$_8$)		$D_0^0 = 3.40$ eV[m]						NOV 1977
D ($^2\Pi_u$)	33230	From the photoelectron spectrum (20)(23); the identifications of the peaks are tentative, no vibrational structure has been observed. There are additional peaks corresponding to higher ionization energies.								
C ($^2\Sigma_g^-$)	28790									
B ($^2\Pi_u$)	26450									
b ($^4\Sigma_g^-$)	21940									
A ($^2\Pi_u$)	14520									
a ($^4\Pi_u$)	9030									
X ($^2\Pi_{g,r}$)	3790 0									

Tb_2: [a]Thermochemical value (mass-spectrom.)(1).

(1) Kordis, Gingerich, Seyse, JCP 61, 5114 (1974).

TbO: [a]Thermochemical value (mass-spectrom.)(3), recalculated (4).

[b]From the IR sp. of $Tb^{16}O$ in argon at 15 K; for $Tb^{18}O$ at 781.7. Derived constants $\omega_e = 828._4$, $\omega_e x_e = 2.1$ (5).

(1) Gatterer, Junkes, Salpeter, Rosen, METOX (1957).
(2) Herrmann, Alkemade, "Chemical Analysis by Flame Photometry", 2nd rev. ed., Interscience (1963).
(3) Ames, Walsh, White, JPC 71, 2707 (1967).
(4) Smoes, Coppens, Bergman, Drowart, TFS 65, 682 (1969).
(5) Gabelnick, Reedy, Chasanov, JCP 60, 1167 (1974).

Te_2, Te_2^+:

[a]Weighted mean (17) of several values obtained by spectroscopic (14)(17) and thermochemical (12)(18) methods and from a photoionization threshold (13).

[b]By photoionization mass-spectrometry (13)(16). Similar values from the photoelectron spectrum (20)(23).

[c]Constants for the only observed vibrational level of this state, from laser-excited fluorescence spectra. The v numbering is unknown (probably $6 \lesssim v' \lesssim 10$).

[d]The vibrational numbering adopted here was first suggested by (4) and has recently been confirmed by isotope studies (14)(15)(17); v' has been increased by +3 units compared to the numbering used by (1)(9a)(10).

[e]$\omega_e y_e = -0.01109$ (for $v < 20$, see [f]).

[f]Numerous local perturbations. A much stronger homogeneous perturbation affects levels with $v \gtrsim 20$ (14)(15)(17). Predissociation above v=23 (2)(4), i.e. at ~ 25600 cm^{-1} above $X_1\ 0_g^+(v=0)$; induced predissociation below this limit (3)(4)(6); accidental prediss. of the vibrational type (9).

[g]Rot. analyses of bands of $^{128}Te_2$ (17), extended by (19).

Te_2, Te_2^+ (continued):

[h]Continuous emission with maximum at 19200 cm^{-1} has been interpreted as recombination continuum (5).

[i]$\omega_e y_e = -0.003892$.

[j]Potential curves (21).

[k]$\gamma_e = +2.2 \times 10^{-7}$.

[ℓ]$\omega_e y_e = -0.00055$ (17).

[m]From $D_0^0(Te_2)$ + I.P.(Te) - I.P.(Te_2).

(1) Rosen, ZP 43, 69 (1927).
(2) Hirschlaff, ZP 75, 315 (1932).
(3) Kondratjew, Lauris, ZP 92, 741 (1934).
(4) Olsson, ZP 95, 215 (1935); Dissertation (Stockholm, 1938).
(5) Rompe, ZP 101, 214 (1936); PZ 37, 807 (1936).
(6) Olsson, CR 204, 1182 (1937).
(7) Choong, AP(Paris) (11) 10, 173 (1938).
(8) Migeotte, MSRSL (4) 5, 3 (1942).
(9) Rosen, PR 68, 124 (1945).
(9a) Herzberg, MOLSPEC 1 (1950).
(10) Prasad, Rao, IJP 28, 549 (1954).
(11) du Parcq, Barrow, CC (1966), p. 270.
(12) Budininkas, Edwards, Wahlbeck, JCP 48, 2870 (1968); 51, 1686 (1969) (erratum).
(13) Berkowitz, Chupka, JCP 50, 4245 (1969).
(14) Jha, Rao, CPL 3, 175 (1969).
(15) Jha, Subbaram, Rao, JMS 32, 383 (1969).
(16) Guyon, Berkowitz, JCP 54, 1814 (1971).
(17) Barrow, du Parcq, PRS A 327, 279 (1972).
(18) Smoes, Mandy, Auwera-Mahieu, Drowart, BSCB 81, 45 (1972).
(19) Yee, Barrow, JCS FT II 68, 1397 (1972).
(20) Berkowitz, JCP 62, 4074 (1975).
(21) Rao, Rao, Rao, SpL 8, 745 (1975).
(22) Stone, Barrow, CJP 53, 1976 (1975).
(23) Streets, Berkowitz, JESRP 9, 269 (1976).

State	T_e	ω_e	$\omega_e x_e$	B_e	α_e	D_e (10^{-} cm^{-1})	r_e (Å)	Observed Transitions Design.	ν_{00}	References
$^{(130)}Te^{(79)}Br$		(μ = 49.093760$_6$)								NOV 1977
B	$(43125)^a$	314.2 H	0.5					B← X, V	41430 H 43148 H	(1)*
X ($^2\Pi_i$)	$(1718)^a$ 0	267.4 H	0.7							
$^{(130)}Te^{(35)}Cl$		(μ = 27.5522041)								NOV 1977
B	$(44262)^a$	458 H	1.$_5$					B← X, V	42624 H 44298 H	(1)*
X ($^2\Pi_i$)	$(1674)^a$ 0	[386] H								
$^{(130)}Te^1H$		(μ = 1.00006661)			I.P. = 9.0$_9$ eV[a]					NOV 1977
T								T← X_1,	69589	(2)*
S								S← X_1,	68729	(2)*
R								R← X_1,	67797	(2)*
Q								Q← X_1,	66711	(2)*
P								P← X_1,	62150	(2)*
O		Very strong.	Only single absorption bands have been observed without vibrational structure. According to (2) none of the bands shows any significant shift on deuteration.					O← X_1,	60753	(2)*
N								N← X_2, N← X_1,	56850 60680	(2)*
H								H← X_1,	58824	(2)*
D								D← X_1,	55006	(2)*
C		Very strong.						C← X_2, C← X_1,	45872 49702	(2)*
B								B← X_1,	46404	(2)*
X_2($^2\Pi_i$) X_1	$(3830)^b$ 0			[5.56]			[1.74$_1$]	EPR sp.		(1)

State	T_e	ω_e	$\omega_e x_e$	B_e	α_e	D_e (10^{-} cm^{-1})	r_e (Å)	Observed Transitions Design.	ν_{00}	References
$^{(130)}Te^{127}I$		(μ = 64.193905_3)								NOV 1977
C	43658	$251._8$ H	0.5					C← X, V	$43\bar{6}75$ H	(1)*
B	41057	$260._2$ H	1.4					B← X, V	$(39263)^a$ 41078 H	(1)*
X ($^2\Pi_i$)	$(1815)^a$ 0	$217._3$ H	0.6							

TeBr, TeCℓ:

[a]Assuming that there is no spin-orbit splitting in the B state.

(1) Oldershaw, Robinson, JMS 37, 314 (1971).

Te^1H: [a]From a tentative interpretation of several Rydberg series (e.g. C,P,Q,...) by (2).

[b]From the interpretation of the VUV absorption spectrum (2); from the paramagnetic resonance spectrum (1) obtains $A_0 = -2250$.

(1) Radford, JCP 40, 2732 (1964).
(2) Donovan, Little, Konstantatos, JCS FT II 68, 1812 (1972).

TeI: [a]Tentative, based on a single band head.

(1) Oldershaw, Robinson, TFS 67, 907 (1971).

TeSe (see p.639):

[a]Thermochemical value (mass-spectrom.)(1), corrected for new values of $D_0^0(Se_2)$ and $D_0^0(Te_2)$.

[b]The band structure is somewhat diffuse suggesting predissociation (2)(3).

[c]All constants for these states are derived from laser-excited fluorescence spectra. In carrying out the analysis it was necessary to assume the value of the rotational constant B_e for the ground state (3), see [f].

[d]From the $B_1 \rightarrow X_1$ and $B_1 \rightarrow X_2$ fluorescence series.

[e]Combined results from fluorescence and absorption spectra.

[f]Estimated value (3) assuming $r_e(TeSe) \approx \frac{1}{2}[r_e(Te_2)+r_e(Se_2)]$.

(1) Porter, Spencer, JCP 32, 943 (1960).
(2) Joshi, Sharma, PPS 90, 1159 (1967).
(3) Ahmed, Barrow, Yee, JP B 8, 649 (1975).

State		T_e	ω_e	$\omega_e x_e$	B_e	α_e	D_e ($10^{-7}cm^{-1}$)	r_e (Å)	Observed Transitions Design.	ν_{00}	References
$^{130}Te^{16}O$			$\mu = 14.2414174_5$		$D_0^0 \lesssim 3.90$ eV[a]		I.P. = 8.72 eV[b]				NOV 1977
			Fragment of an absorption system in the region 32700 - 36000 cm^{-1}; it is not certain that these bands are due to TeO.								(1)
A_2	1	[28719]	c		[0.2771][d]	(0.004)	[2]	[2.067]	$A_2 \rightarrow X_2$, R	27641.9_5 Z	(2)* (3)*
A_1	0^+	(28212)	[444.95] Z	e	0.2760[df]	0.0052	4	2.071	$A_1 \leftrightarrow X_1$, R	28037.0_4 Z	(2)* (3)*
			Unclassified emission bands 16000 - 20000 cm^{-1}.								(2)
X_2	1	679[g]	798.06 Z	4.00	0.3564[h]	0.00236	3	1.822_4			
X_1	0^+	0	797.11 Z	4.00	0.3554	0.00237	2.7	1.825_0			
$^{(130)}Te^{16}O^+$					$D_0^0 \lesssim 4.19$ eV[i]						NOV 1977
	($^2\Pi$)	(38470)	From the photoelectron spectrum (5). Identifications are tentative, no vibrational structure was observed. Numbers in parantheses refer to peaks overlapped by impurities.								
B	($^2\Sigma$)	26450									
b	($^4\Sigma$) }	(19760)									
A	($^2\Pi$) }										
a	($^4\Pi$)	16780									
X	($^2\Pi$) 3/2	(4840)									
	1/2	0									
$^{130}Te^{32}S$			$\mu = 25.6573697$		$D_0^0 = (3.5)$ eV[a]						NOV 1977
D_2		x + 42199	[526] H						$D_2 \leftarrow X_2$,[b] V	42226 H	(1)*
D_1		43283	524.2 H	0.8					$D_1 \leftarrow X_1$, V	43310 H	(1)*
C	0^+				[0.08115][c]		[3.5][c]	$[2.845_4]$[c]	$C \leftarrow X_1$, R	26886.96[c] Z	(3)
B	0^+	24530	250.3[d] Z	(3.37)	0.1027[d]	0.00120	(1.1)	2.529	$B \leftarrow X_1$, R	24419[d] Z	(3)
A	0^+	23549	204.2_3[d] Z	1.018	0.0976_2[d]	0.00075	(0.7)	2.594	$A \leftarrow X_1$, R	23415.4[d] Z	(3)
X_2	1	x[e]	(472) H								
X_1	0^+	0	471.18[f] Z	(1.57)	0.13216[g]	0.00050[g]	0.40	2.2297			

State	T_e	ω_e	$\omega_e x_e$	B_e	α_e	D_e ($10^{-7}cm^{-1}$)	r_e (Å)	Observed Transitions Design.	ν_{00}	References
$^{128}Te^{78}Se$		$\mu = 48.420396_1$		$D_0^0 = 3.0$ eV[a]						NOV 1977
		Unclassified absorption bands 43900 - 46500 cm^{-1}.								(2)
D_2	42232.3	353.3 H	1.0	Double-headed bands[b]				$D_2 \leftarrow X_2$, V	40703.0 H^Q	(2)* (3)
D_1	42168.4	355.6 H	1.0					$D_1 \leftarrow X_1$, V	42188.0 H^Q	(2)* (3)
C 1	[c]			$B_2 = 0.0477$			$r_2 = 2.70_2$	$C \rightarrow X_2$,	$\nu(2-0) = 22490$	(3)
B_2 1	23596 [c]	[191.5]		[0.048]			$[2.6_9]$	$B_2 \rightarrow X_2$,	21986	(3)
B_1 0^+	23393 [c]	[189]		0.0497			2.64_7	$B_1 \rightarrow X_2$,	21781	(3)
								$B_1 \rightarrow X_1$,	23329.1	(3)
A 0^+	[c]							$A \rightarrow X_1$,	$\nu(2-0) = 22052$	(3)
X_2 1	1547.2[d]	317.39[e]	0.7175	(0.0619)[f]			(2.37_2)			
X_1 0^+	0	316.24[e]	0.738	(0.0619)[f]	(0.00018)	(0.095)	(2.37_2)			

TeO, TeO^+:

[a]From the predissociation in A_1 0^+ (see [f]).

[b]From the photoelectron spectrum (5).

[c](3) give, without details, $\Delta G(\frac{1}{2}) \approx 458$ for ^{128}TeO. (2), by contrast, suggest $\omega_e = 408$, $\omega_e x_e = 4.0$. According to (3) this state is predissociated between v=2 and 3.

[d]Extensive perturbations.

[e]$\Delta G(3/2) = 409.9$, $\Delta G(5/2) = 393.4$, $\Delta G(7/2) = 364.53$. From v=3 to 10 the vibrational intervals are quite well represented by $\omega_e = 410.71$, $\omega_e x_e = 5.772$; large deviations occur below and above these limits. Vibrational levels observed to v=15. The v' numbering has been established from isotope shifts (3) and has been increased by +3 units compared to the numbering used by (1) and (2).

[f]Absorption lines become diffuse above J'≈42 and 14 in v'=9 and 10, respectively; the highest observed level in emission is v'=9. The predissociation limit (~31450 cm^{-1}) is

TeO, TeO^+ (continued):

assumed to correspond to normal atomic products.

[g]Estimated (4) from the magnitude of the Ω-type doubling in X_2 1 (see [h]).

[h]$B(1^+) - B(1^-) \approx +0.00066$ (4).

[i]$D_0^0(TeO) + I.P.(Te) - I.P.(TeO)$.

(1) Choong, AP(Paris) (11) 10, 173 (1938).
(2) Haranath, Rao, Sivaramamurty, ZP 155, 507 (1959).
(3) Chandler, Hurst, Barrow, PPS 86, 105 (1965).
(4) Barrow, Hitchings, JP B 5, L132 (1972).
(5) Potts, Williams, CPL 42, 550 (1976).

TeS: See p. 641 .
TeSe: See p. 637 .

State	T_e	ω_e	$\omega_e x_e$	B_e	α_e	D_e ($10^{-7}cm^{-1}$)	r_e (Å)	Observed Transitions Design.	ν_{00}	References
$^{232}Th^{(11)}B$		(μ = 10.5106174)		$D_0^0 = 3.0_3$ eV[a]						JUN 1975
$^{232}Th^{14}N$		μ = 13.2061097_7		$D_0^0 = 5.9_0$ eV[a]						JUN 1975
$^{232}Th^{16}O$		μ = 14.9634501_8		$D_0^0 = 9.00$ eV[a]		I.P. ≥ 6.0 eV[b]				JUN 1975
		A good reproduction of the arc spectrum from 11400 to 27800 cm^{-1} can be found in (1).								
		Sequence of R shaded bands; no analysis.							R 24291 H	(1)*
K $^1\Pi$	22683.4_8	[795.47] Z	(2.30)	0.318636[ce] 0.318642[de]	0.00124_4 0.00132_5	[2.007] [2.025]	1.88033	K→X,	R 22635.65 Z	(4)
M $^1\Pi$	21752.2	861.41 Z	5.27	0.325857[ce] 0.325754[de]	0.001390[f] 0.001398[g]	[2.059][f] [2.051][g]	1.8595_3	M→X,	R 21734.32 Z	(4)*
		Sequence of R shaded bands; no analysis.							R 21407 H	(1)*
I $^1\Pi$	19586.29	800.85 Z	1.47	0.330434[ch] 0.328921[dh]	0.00182_5[i] 0.00190_6[j]	2.360[i] 2.218[j]	1.848_6	I→X,	R 19539.06 Z	(3)* (6)
F $^1\Sigma^{(+)}$	(18406)	[757.36][k] Z		[0.321397]	k	[2.042][k]	[1.87224]	F→X,	R 18337.56 Z	(2)
G $(^1\Delta)^\ell$	18038[m]	816	2.4	0.318192[h]	0.00127_6	[1.936]	1.88165	G→H,	R 12693.35 Z	(2)* (3)* (6)
E $^1\Sigma^{(+)}$	16353.60	829.26 Z	2.30	0.323090	0.001303	[1.990]	1.86733	E→X,	R 16320.37 Z	(2)(6)
D $^1\Pi$	15974.5_3	[834.22] Z	(2.49)	0.325691[d] 0.321549[c]	0.001357 0.00129_8	[1.997] [1.850]	1.866	D→X,	R 15946.22 Z	(2)(6)
		Strong unidentified head at 15606 cm^{-1}.							R	(1)*
C $^1\Pi$	14520.3_5	[830.33] Z	(2.39)	0.322455[c] 0.321617[d]	0.001281 0.00128_7	[1.931] [1.873]	1.870_4	C→X,	R 14490.02 Z	(2)(6)
		Violet degraded bands in the region 13700 - 14100 cm^{-1}. No analysis.							V	(1)*
B $(^1\Pi)^\ell$	11155.57	842.80 Z	2.18	0.324973[c] 0.32364 [d]	0.001299 0.00129	[1.942] [1.882]	1.863_8	B→X,	R 11129.14 Z	(2)* (6)(7)
A $(^1\Sigma^+)^\ell$	10625.5_4	[841.48] Z	(2.44)	0.323044	0.001294	[1.866]	1.86746	A→X,	R 10600.82 Z	(2)* (6)
H $(^1\Phi)^\ell$	5321	(864)	(2.4)	0.326427	0.00125_8	[1.864]	1.85776			
X $^1\Sigma^{(+)}$	0	895.77 Z	2.39	0.332644	0.001302	[1.833]	1.84032	IR spectrum[n]		(9)

ThB: [a]Thermochemical value (mass-spectrom.)(1).

(1) Gingerich, HTS 1, 258 (1969).

ThN: [a]Thermochemical value (mass-spectrom.)(1).

(1) Gingerich, JCP 49, 19 (1968).

ThO: [a]Thermochemical value (mass-spectrom.)(8). (11) prefer 8.7_9 eV, the difference being largely due to the choice of free energy functions.

[b]By electron impact (10)(11).

[c]e levels.

[d]f levels.

[e]v=0 and 1 of K are perturbed by v=1 and 2 of M, respectively. The constants given are deperturbed values. An unidentified third state interacts strongly with M (4).

[f]$B_2 = 0.32174$; $D_1 = 2.178 \times 10^{-7}$, $D_2 = 3.02 \times 10^{-7}$.

[g]$B_2 = 0.32218$; $D_1 = 2.026 \times 10^{-7}$, $D_2 = 2.22 \times 10^{-7}$.

[h]Perturbations in v=0,1,2,... of I by v=2,3,4,... of G, respectively. All constants are deperturbed values.

[i]$\gamma_e = +0.000106$; $\beta_e = +0.055 \times 10^{-7}$.

[j]$\gamma_e = +0.000063$; $\beta_e = -0.015 \times 10^{-7}$.

ThO (continued):

[k]$F\,^1\Sigma$ is perturbed; $B_1 = 0.324261$, $D_1 = 3.20_5 \times 10^{-7}$.

[ℓ]According to (5) these states should be identified as H $^3\Delta_1$, A $^3\Pi_{0^+}$, B $^3\Pi_1$, G $^3\Phi_2$.

[m]From the analysis of perturbations; see [h].

[n]In Ar matrix at 15 K.

(1) Gatterer, Junkes, Salpeter, Rosen, METOX (1957).
(2) Edvinsson, Selin, Åslund, AF 30, 283 (1965).
(3) Edvinsson, von Bornstedt, Nylén, AF 38, 193 (1968).
(4) von Bornstedt, Edvinsson, PS 2, 205 (1970).
(5) Edvinsson, Thesis (Stockholm, 1971). USIP Report 71-09.
(6) Wentink, Spindler, JQSRT 12, 1569 (1972).
(7) Zare, Schmeltekopf, Harrop, Albritton, JMS 46, 37 (1973).
(8) Ackermann, Rauh, HTS 5, 463 (1973); JCP 60, 2266 (1974).
(9) Gabelnick, Reedy, Chasanov, JCP 60, 1167 (1974).
(10) Rauh, Ackermann, JCP 60, 1396 (1974).
(11) Hildenbrand, Murad, JCP 61, 1232 (1974).

TeS (Table p. 638):

[a]Estimated (2) by comparison with other molecules of this type.

[b]Interpretation by (4).

[c]Only one band observed; its isotope shift indicates a fairly high value of v'.

[d]Long extrapolation from high vibrational levels; only v=7,8 of B 0^+ $[\Delta G(7\frac{1}{2}) = 196.39]$ and v=11...15 of A 0^+ have been analyzed. The v numbering derives from isotope shifts of levels which are often perturbed.

[e]$x \gtrsim 1084$ according to the interpretation by (4) of the V shaded UV absorption bands.

TeS (continued):

[f]$\Delta G(\frac{1}{2}) = 468.04$.

[g](3) give $B_e = 0.13220$, $\alpha_e = 0.000545$ which do, however, not reproduce B_0 and B_1 as determined by them.

(1) Mohan, Majumdar, PPS 77, 147 (1961).
(2) Drowart, Goldfinger, QR 20, 545 (1966).
(3) Barrow, Dudley, Hitchings, Yee, JP B 5, L172 (1972).
(4) Barrow, Yee, APH 35, 239 (1974).

State	T_e	ω_e	$\omega_e x_e$	B_e	α_e	D_e (10^{-} cm^{-1})	r_e (Å)	Observed Transitions Design.	Observed Transitions ν_{00}	References
$^{232}Th^{31}P$		μ = 27.3261175 $D_0^0 = 3.8_6$ eV[a]								JUN 1975
$^{(48)}Ti_2$		(μ = 23.9739746) $D_0^0 = 1.3_0$ eV[a]								JUN 1975
$^{(48)}Ti^{(79)}Br$		(μ = 29.826460$_9$)								JUN 1975
		Complex spectrum consisting of a large number of bands in the regions 23000 - 24100, 24600 - 24900, 25700 - 26400 cm^{-1}; in emission and in absorption. Tentative vibrational analyses.								(1)(2)
$^{(48)}Ti^{(35)}Cl$		(μ = 20.2212912)								JUN 1975
		Fragments of additional systems in the region 24600 - 26400 cm^{-1}; in emission and absorption.								(2)(5)
		Complex spectrum of mostly V shaded bands, 23500 - 24700 cm^{-1}; in emission and absorption. Tentative vibrational analyses (2)(3) suggesting a $^4\Pi - ^4\Sigma$ transition are suspect in view of more recent work (7) which indicates that of several distinct band systems present in this region the most prominent one is $^2\Phi - ^2\Delta$.								(1)(2)(3)(4) (5)(6)
$^{(48)}Ti^{19}F$		(μ = 13.6069328) D_0^0 = (5.9) eV[a]								JUN 1975
		R shaded bands in the region 26450 - 27450 cm^{-1}; in emission and absorption.								(2)(3)
		Weak system of V shaded bands, 25600 - 25850 cm^{-1}; in emission and absorption.								(2)(3)
		V shaded bands in the region 24450 - 25400 cm^{-1}, tentatively attributed to a $^4\Pi - ^4\Sigma$ transition; both in emission and in absorption.								(2)(3)
$^{(48)}Ti^{1}H$		(μ = 0.98707768) D_0^0 = (1.6) eV[a]								JUN 1975
$^{(48)}Ti^{2}H$		(μ = 1.93290844)								
X ($^4\Phi$)[a]		Open but complex rotational structure in absorption in the regions 18200 - 19200[b], 19550 - 19900, 20800 - 21300[c] cm^{-1}.								(1)* (3)
$^{(48)}Ti^{127}I$		(μ = 34.799685$_2$)								JUN 1975
		Sequence of R shaded emission bands 24705 - 24748 cm^{-1}. V shaded em. b. 22200 - 22600 cm^{-1}.								(1)

ThP: [a]Thermochemical value (mass-spectrom.)(1).

(1) Gingerich, HTS 1, 258 (1969).

Ti_2: [a]Thermochemical value (mass-spectrom.)(1).

(1) Kant, Lin, JCP 51, 1644 (1969).

TiBr: (1) Chatalic, Deschamps, Pannetier, JCPPB 67, 316 (1970).

(2) Sivaji, Rao, JP B 3, 720 (1970).

TiCℓ: (1) More, Parker, PR 52, 1150 (1937).

(2) Rao, IJP 23, 535 (1949).

(3) Shenyavskaya, Kuzyakov, Tatevskii, OS(Engl. Transl.) 12, 197 (1962).

(4) Parkinson, Reeves, unpublished; quoted by Nicholls, Parkinson, Reeves, AO 2, 919 (1963).

(5) Chatalic, Deschamps, Pannetier, CR C 268, 1111 (1969).

(6) Diebner, Kay, JCP 51, 3547 (1969).

(7) Dunn, Lanini, work reported in progress in Newsletter No. 78 [ed. Phillips, Davis, U. of California (Berkeley, March 1972)].

TiF: [a]Estimated from thermochemical data (1).

(1) Zmbov, Margrave, JPC 71, 2893 (1967).

(2) See ref. (6) of TiCℓ.

(3) Chatalic, Deschamps, Pannetier, CR C 270, 146 (1970).

Ti^1H, Ti^2H:

[a]Theoretical predictions of (2).

[b]Partial rotational analyses (3) of two V shaded subbands of Ti^1H with P heads at 18576 and 18646 cm^{-1} give the lower state constants $B'' = 4.956$ and 5.22, respectively. The upper states are strongly perturbed. The bands have P, Q, and R branches.

[c]Diffuse.

(1) Smith, Gaydon, JP B 4, 797 (1971).

(2) Scott, Richards, JP B 7, 500 (1974).

(3) Gaydon, JP B 7, 2429 (1974).

TiI: (1) Sivaji, Rao, Rao, CS 39, 153 (1970).

TiO (continued from p. 647):

(54) Rauh, Ackermann, JCP 60, 1396 (1974).

(55) Collins, JP B 8, 304 (1975).

(56) Żyrnicki, JQSRT 15, 575 (1975).

(57) Liu, Wahlbeck, JCP 63, 1694 (1975).

(58) Brom, Broida, JCP 63, 3718 (1975).

(59) Feinberg, Bilal, Davis, Phillips, ApL 12, 147 (1976); Feinberg, Davis, JMS 65, 264 (1977).

(60) Hildenbrand, CPL 44, 281 (1976).

(61) Dubois, Gole, JCP 66, 779 (1977).

(62) Linton, Broida, JMS 64, 382, 389 (1977).

(63) Sheldon, Gilles, JCP 66, 3705 (1977); Wheatley, Sheldon, Gilles, JCP 66, 3712 (1977).

State	T_e	ω_e	$\omega_e x_e$	B_e	α_e	D_e ($10^{-7}cm^{-1}$)	r_e (Å)	Observed Transitions Design.	ν_{00}	References
$^{48}Ti^{14}N$		$\mu = 10.8378951_7$	$D_0^0 = 4.9$ eV[a]							MAY 1976
		Additional bands in the region 30400 - 33800 cm^{-1}; in absorption and in emission.								(1)
B $^2\Sigma$				$[0.574_5]$		[6.4]	$[1.645_4]$	B→X, R	23487.3 Z	(6)
A $^2\Pi_r$	b	c		[0.6103][d]		[13]	$[1.596_4]$	A→X, R	16197.25[e] Z	(4)*
X $^2\Sigma$	0	c		[0.6211]		[13]	$[1.582_5]$			
$^{(48)}Ti^{16}O$		($\mu = 11.9938851_9$)	$D_0^0 = 6.87$ eV[a]		I.P. = 6.4 eV[b]					NOV 1977 A
	Very weak progression ($\omega_e'' \approx 994$, $\omega_e''x_e'' \approx 5$) of R shaded bands in emission from flames.							R	(30422) H (30367) H	(34)(44)
D	(31920)	[(1040)]						D↔X,[c]	(31940)	(34)(36)(44)(61)
e $^1\Sigma^+$	a + 26598.1	[845.2] Z	4.2 H	0.4892[d]	0.0023	$[4._7]$[d]	1.695_0	e↔d, R	24297.5 Z	(37)* (42)* (52)
f $^1\Delta$	a+(19132)	(890)		[0.50221]		[6.4]	$[1.6729_2]$	f↔a,[e] R	19068.93 Z	(43)* (52)
c $^1\Phi$	a + 17890.2	[909.6] Z	4.1_9 H	0.5230[f]	0.0031_3	[3.9][f]	1.639_3	c[g]↔a,[eh] R β bands	17840.6 Z	(3)(6)* (14)* (29) (33)(41)(51)
C $^3\Delta_r$	19617.0 19525.5 19427.12	838.26 Z	4.76[i]	0.48989[jk]	0.00306[ℓ]	6.7	1.6938_3	C↔X,[mn] R α bands	19334.03 Z 19343.66 Z 19341.68 Z	(2)* (4)(9) (11)(14)* (20)(21)(31) (33)(45)(48) (49)
								C→a,[o]		(62)
B $^3\Pi_r$	$16331._3$ $16315._1$ $16293._5$	875 H	5	[0.50617][jp]		[6.86]	$[1.6663_6]$	B↔X,[m] R γ' bands	16066.7[q] Z 16151.6[q] Z 16226.4 Z	(5)(14)* (17)(18)(24) (31)* (38)* (48)(55)
b $^1\Pi$	a + 11322.0_3	[911.20] Z	(3.7_2)	0.51337[r]	0.0029_1	6.1	1.6546_4	b↔d,[e] R φ bands	9054.02 Z	(15)(19)(29) (30)(33)(49) (52)(58)
								b↔a,[e] R δ bands	11272.82 Z	(6)* (29)(30) (33)(52)(58)
								(b→X),[s]	(14710)	(58)

State	T_e	ω_e	$\omega_e x_e$	B_e	α_e	D_e ($10^{-7} cm^{-1}$)	r_e (Å)	Observed Transitions Design.	ν_{00}	References
$^{(48)}Ti^{16}O$ (continued)										
A $^3\Phi_r$	14431.0 14262.8 14089.91	867.78 Z	3.942	0.50739[j]	0.00315[t]	6.92[t]	1.6643_6	A↔X,[mn] R γ bands	14163.00 Z 14095.88 Z 14019.43 Z	(1)(3)(7)* (10)(11) (14)* (21) (31)(48)(49)
E $^3\Pi$	12025	$924._2$ H	$5._1$					E↔X,[u] ε bands	11871 H 11886 H 11899 H	(36)(62)
d $^1\Sigma^+$	a + 2215.6	[1014.6] Z	(4.6_4)	0.54922	0.00337	[6.0]	1.5997_2			
a $^1\Delta$	a[v]	[1009.3] H	3.9_3	0.53760	0.00298	5.9	1.6169_2			
X $^3\Delta_r$	197.5[w] 96.4[w] 0	1009.02 Z	4.498[x]	0.53541[j]	0.00301[y]	6.03[y]	1.6202_2	Rotation-vibr. sp.[z]		

TiN: [a]Thermochemical value (mass-spectrom.)(5). See also (2)(3).
[b]A_0 = +156.70; slight J dependence.
[c]0-0 sequence only; the R_2+Q_{21} and Q_1+R_{12} heads of the 0-0 band are at 16285.8 and 16125.8 cm^{-1}, the corresponding heads of the 1-1 band at 16193.2 and 16035.2.
[d]Λ-type doubling in $^2\Pi_{\frac{1}{2}}$, $\Delta v = 0.037(J+\frac{1}{2})$. Perturbations.
[e]Refers to the zero-point of the Hill-Van Vleck expression in the upper state. (4) give 16197.52, in poor agreement with our recalculated value 16197.21. We assume that the former is intended to be 16197.25.

(1) Parkinson, Reeves, CJP 41, 702 (1963).
(2) Carlson, Claydon, Moser, JCP 46, 4963 (1967).
(3) Gingerich, JCP 49, 19 (1968).
(4) Dunn, Hanson, Rubinson, CJP 48, 1657 (1970).
(5) Stearns, Kohl, HTS 2, 146 (1970).
(6) Bates, Ranieri, Dunn, CJP 54, 915 (1976).

TiO: [a]Thermochemical value (mass-spectrom.)(27)(35)(57)(60); different values have been proposed by (26)(28)(40). See also (12)(13)(47)(63). A lower bound of 6.9_3 eV for the dissociation energy has been deduced by (61) from the study of chemiluminescent spectra resulting from the reactions $Ti + O_2$ and $Ti + N_2O$.
[b]Electron impact appearance potential (54).
[c]Multiple heads in flames. Absorption in a neon matrix at 4 K. Analysis uncertain.
[d]Perturbations in v=0 and 1 by levels of smaller B values (52). $D_1 = 9._1 \times 10^{-7}$.
[e]Absorption in stellar atmospheres.
[f]The data suggest a slight perturbation of v=0 by an unidentified level of smaller B value and lower energy. $D_1 \ldots D_3 (10^{-7} cm^{-1})$ = 5.8, 6.9, 7.5; $H_0 = -2.3 \times 10^{-11}$.
[g]Radiative lifetime $\tau(v=0) = 17._5$ ns (59). See [h].
[h]The absolute transition probabilities of (56) are in gross
(continued p. 646)

State		T_e	ω_e	$\omega_e x_e$	B_e	α_e	D_e ($10^{-7}cm^{-1}$)	r_e (Å)	Observed Transitions Design.	ν_{00}	References
$^{(48)}Ti^{32}S$			(μ = 19.1816179)		$D_0^0 = 4.7_5$ eV[a]						JUN 1975
C $^3\Delta$	3	(11806)	484.12 Z	2.55	0.18905	0.00102[b]	[1.22]			11582.25 Z	
	2	(11716)	484.30 Z	2.51_5	0.18820[c]	0.00099[d]	[1.20]	2.1610	C↔X, R	11587.05 Z	(4)
	1	$11624._3$			[0.18684]	e	[1.06]			11585.31 Z	
(X) $^3\Delta$	3	(185)[f]	[558.17] Z		0.20344	0.00092	[1.12]				
	2	(90)[f]	[558.30] Z	(1.95)	0.20268	0.00092	$[1.09_5]$	2.0825			
	1	0	[558.37] Z		0.20180	0.00090	$[1.03_5]$				
$^{(48)}Ti^{(130)}Te$			($\mu = 35.021596_7$)		D_0^0 = 2.94 eV[a]						JUN 1975

TiS: [a]Thermochemical value (mass-spectrom.)(3); see also (1) [quoted in (2)].

[b]$\gamma_e = -0.00001_3$.

[c]Small perturbations in v=0.

[d]$\gamma_e = -0.00001_9$.

[e]Extensive perturbations.

[f]Estimated from the effective B values.

(1) Franzen, Thesis (U. of Kansas, 1962).

TiS (continued):

(2) Suzuki, Wahlbeck, JPC 70, 1914 (1966).

(3) Smoes, Coppens, Bergman, Drowart, TFS 65, 682 (1969).

(4) Clements, Barrow, TFS 65, 1163 (1969).

TiTe: [a]Thermochemical value (mass-spectrom.)(1); no details.

(1) Bergman, Coppens, Drowart, Smoes, TFS 66, 800 (1970).

TiO (continued):

disagreement with the lifetime measurements of (59).

[i]$\omega_e y_e = +0.047$.

[j]For spin coupling constants (spin-orbit, spin-spin, spin-rotation) see (48), also (23)(25)(32).

[k]Levels with $v \geq 4$ are perturbed (45)(48). The perturbing (singlet?) state has $B \approx 0.510$, $\omega \approx 900$.

[ℓ]$\gamma_e = -0.000030$.

[m]Absorption in stellar atmospheres. Also observed in absorption in rare gas matrices (24)(36).

TiO (continued):

[n]Electronic oscillator strengths for the α system (39)(50), for the γ bands (39).

[o]Only three lines [R(16), Q(17), P(18)] of the $^3\Delta_3 \rightarrow {}^1\Delta$ 2-0 band have been observed in laser-excited photoluminescence.

[p]The Λ-type doubling in $^3\Pi_0$, $\Delta v = 1.60$ cm^{-1}, is nearly constant up to J≈65, then diminishes and changes sign for J > 120.

[q]$\nu_{00}(^3\Pi_1 - {}^3\Delta_1) = 16248.0$, $\nu_{00}(^3\Pi_2 - {}^3\Delta_2) = 16167.8$.

[r]Λ-type doubling $\Delta\nu_{fe} = +0.00014\,J(J+1)$.

[s]In neon at 4 K.

[t] $\gamma_e = -0.000010$; $\beta_e = +0.02 \times 10^{-7}$.

[u] Absorption in a neon matrix at 4 K.

[v] $a = 3440 \pm 10$ cm^{-1} (62), based on the identification of three lines of the C→a 2-0 band (see [o]). A similar value ($a \approx 3500$) follows from the assignment (58) of the b→X intercombination transition in neon. Qualitative agreement with theoretical predictions (22). Earlier estimates (8)(18) are considerably lower.

[w] From the observation of two satellite bands of γ'(0-0) (38); see [q].

[x] $\omega_e y_e = -0.0107$.

[y] $\gamma_e = -0.000011$; $\beta_e = +0.03 \times 10^{-7}$.

[z] Not observed; for the predicted structure of the fundamental band see (53).

(1) Christy, ApJ 70, 1 (1929).
(2) Christy, PR 33, 701 (1929).
(3) Lowater, PPS 41, 557 (1929).
(4) Budó, ZP 98, 437 (1936).
(5) Coheur, BSRSL 12, 98 (1943).
(6) Phillips, ApJ 111, 314 (1950).
(7) Phillips, ApJ 114, 152 (1951).
(8) Phillips, ApJ 115, 567 (1952).
(9) Phillips, ApJ 119, 274 (1954).
(10) Fraser, Jarmain, Nicholls, ApJ 119, 286 (1954).
(11) Uhler, Dissertation (Stockholm, 1954).
(12) Groves, Hoch, Johnston, JPC 59, 127 (1955).
(13) Berkowitz, Chupka, Inghram, JPC 61, 1569 (1957).
(14) Gatterer, Junkes, Salpeter, Rosen, METOX (1957).
(15) Pettersson, AF 16, 185 (1959).
(16) Ortenberg, OS(Engl. Transl.) 9, 80 (1960).
(17) Pedoussaut, CR 252, 2819 (1961).
(18) Merrill, Deutsch, Keenan, ApJ 136, 21 (1962).
(19) Pettersson, Lindgren, AF 22, 491 (1962).
(20) Prasad, PPS 79, 1078 (1962); 82, 419 (1963).

(21) Ortenberg, Glasko, SAAJ 6, 714 (1963).
(22) Carlson, Nesbet, JCP 41, 1051 (1964); Carlson, Moser, JCP 46, 35 (1967).
(23) Kovács, JMS 18, 229 (1965).
(24) Weltner, McLeod, JPC 69, 3488 (1965).
(25) Törös, APH 20, 91 (1966).
(26) Wahlbeck, Gilles, JCP 46, 2465 (1967).
(27) Drowart, Coppens, Smoes, JCP 50, 1046 (1969).
(28) Gilles, Hampson, Wahlbeck, JCP 50, 1048 (1969).
(29) Linton, Nicholls, JP B 2, 490 (1969).
(30) Lockwood, ApJ 157, 275 (1969).
(31) Phillips, ApJ 157, 449 (1969).
(32) Kovács, Korwar, APH 29, 399 (1970).
(33) Linton, Nicholls, JQSRT 10, 311 (1970).
(34) Pathak, Palmer, JMS 33, 137 (1970).
(35) Hampson, Gilles, JCP 55, 3712 (1971).
(36) McIntyre, Thompson, Weltner, JPC 75, 3243 (1971).
(37) Phillips, Davis, ApJ 167, 209 (1971).
(38) Phillips, ApJ 169, 185 (1971).
(39) Price, Sulzmann, Penner, JQSRT 11, 427 (1971); 14, 1273 (1974).
(40) Balducci, De Maria, Guido, Piacente, JCP 56, 3422 (1972).
(41) Dube, IJPAP 10, 70 (1972).
(42) Lindgren, JMS 43, 474 (1972).
(43) Linton, CJP 50, 312 (1972).
(44) Palmer, Hsu, JMS 43, 320 (1972).
(45) Phillips, Davis, ApJ 175, 583 (1972).
(46) Wentink, Spindler, JQSRT 12, 1569 (1972).
(47) Wu, Wahlbeck, JCP 56, 4534 (1972).
(48) Phillips, ApJ(Suppl.) 26 (232), 313 (1973).
(49) Collins, Fay, JQSRT 14, 1259 (1974).
(50) Fairbairn, Wolnik, Berthel, ApJ 193, 273 (1974).
(51) Linton, JMS 50, 235 (1974).
(52) Linton, Singhal, JMS 51, 194 (1974).
(53) Phillips, ApJ(Suppl.) 27 (247), 319 (1974).

(continued p. 643)

State	T_e	ω_e	$\omega_e x_e$	B_e	α_e	D_e $(10^{-9}cm^{-1})$	r_e (Å)	Observed Transitions Design.	ν_{00}	References
$^{(205)}Tl_2$		$(\mu = 102.487219)$ $D_0^0 < 0.9$ eV[a]								NOV 1977
		Several diffuse emission bands and continua, 15000 - 36500 cm^{-1}.								(1)
		Strong R and V shaded absorption bands, 22900 - 23500 cm^{-1}.								(3)
		Weak R shaded bands in thermal emission, 15000 - 16000 cm^{-1}. $\omega' \approx 88$, $\omega'' \approx 136$ cm^{-1}.								(3)
$^{(205)}Tl^{40}Ar$		$(\mu = 33.4423676)$								NOV 1977
		The far-wing emission intensities of the Tℓ resonance lines (26478 and 18685 cm^{-1}) broadened by rare gases have been studied by (1); Franck-Condon analyses of the observed continua yield potential curves in the 3 - 4 Å region for the $X_1\ \frac{1}{2}$, $X_2\ \frac{3}{2}$, $A\ \frac{1}{2}$, $B\ ^2\Sigma$ states arising from the atomic limits $^2P_{\frac{1}{2}}$, $^2P_{\frac{3}{2}}$, $^2S_{\frac{1}{2}} + {}^1S$. Near-wing intensities have been investigated by (2) yielding additional information on the potentials at larger internuclear distances.								
$^{(205)}Tl^{75}As$		$(\mu = 54.866846_3)$ $D_0^0 = 2.0_2$ eV[a]								NOV 1977
$^{(205)}Tl^{209}Bi$		$(\mu = 103.479018)$ $D_0^0 = 1.2_1$ eV[a]								NOV 1977
$^{205}Tl^{81}Br$		$\mu = 58.014373$ $D_0^0 = 3.42$ eV[a] I.P. = 9.14 eV[b]								NOV 1977
D		Fragments of an emission system at ~25340 cm^{-1}.						$(D \rightarrow A)$, R		(2)* (4)*
		Continuous absorption for $\nu > 40000$ cm^{-1}.								(2)(11)
C $(^1\Pi)$		Continuous absorption with maximum at 37160 cm^{-1}.[d]						$C \leftarrow X$, [c]		(1)(2)(11)*
		Weak absorption continua at 31000 and 32040 cm^{-1}.						[c]		(11)*
		Absorption continua with maxima at 29540 and 30000 cm^{-1}.[d]						[c]		(1)* (2)(11)*
A $(^3\Pi_{0^+})$[e]	29191.5	108.32 H	5.15[f]	Bands having $v' \geq 3$ are diffuse.				$A \leftrightarrow X$, [c] R	29148.4 H	(1)* (2)* (3)(4)
X $^1\Sigma^+$	0	192.10 H	0.39	0.0423895[g]	0.0001275_5	(8.3)	2.61819_1	IR spectrum[h]		(13)
								Microwave sp.[i]		(5)(6)(16)
								Mol. beam el. reson.[j]		(12)

State	T_e	ω_e	$\omega_e x_e$	B_e	α_e	D_e (10^{-} cm^{-1})	r_e (Å)	Observed Transitions Design.	ν_{00}	References
$^{(205)}Tl^{(79)}Br^+$		$D_0^0 = 0.39$ eV[k]								NOV 1977
B $^2\Sigma$	(36500)	From the photoelectron spectrum (15), using vertical potentials for A and B and the adiabatic potential for X $^2\Pi$. The peak corresponding to the shallow $^2\Pi$ state is very broad and has a vertical potential which lies at ~0.35 eV above that of A $^2\Sigma$.								
A $^2\Sigma$	(2700)									
X $^2\Pi$	0									

$T\ell_2$: [a]Mass-spectrometric result of (2) who suggest the "preferred" value 0.6 eV.

(1) Hamada, PM 12, 50 (1931).
(2) Drowart, Honig, JPC 61, 980 (1957).
(3) Ginter, Ginter, Innes, JPC 69, 2480 (1965).

TℓAr:

(1) Cheron, Scheps, Gallagher, JCP 65, 326 (1976).
(2) Cheron, Scheps, Gallagher, PR A 15, 651 (1977).

TℓAs: [a]Thermochemical value (mass-spectrom.)(1).

(1) Piacente, Malaspina, JCP 56, 1780 (1972).

TℓBi: [a]Thermochemical value (mass-spectrom.)(1).

(1) De Maria, Malaspina, Piacente, JCP 56, 1978 (1972).

TℓBr, $T\ell Br^+$:

[a]Photoionization mass-spectrometry (10). In good agreement with spectroscopic and thermochemical results (7), flame photometry (8), and electron impact mass-spectrometry (9).

[b]Photoionization value (10). Notice that this is the adiabatic value corresponding to the second vertical potential (9.83 eV) observed in the photoelectron spectrum (15); first vertical I.P. at 9.48 eV.

[c]Absorption cross sections have been measured by (11).

TℓBr, $T\ell Br^+$ (continued):

[d](2) report emission continua with maxima at 30020 and 33770 cm^{-1}.

[e]Assigned by analogy with TℓCℓ; (2) assume $^3\Pi_1$.

[f]$\omega_e y_e = -0.22$.

[g](16) give rotational constants for four isotopes.

[h]In argon at 10 K.

[i]Dipole moment of $^{205}T\ell^{79}Br$: $\mu_{e\ell}(v=0) = 4.49$ D, from Stark effect of rotation sp. (14).

[j]Hfs constants for four isotopes in the first five vibrational states.

[k]$D_0^0(T\ell Br) + I.P.(T\ell) - I.P.(T\ell Br)$.

(1) Butkow, ZP 58, 232 (1929).
(2) Howell, Coulson, PPS 53, 706 (1941).
(3) Rao, IJP 23, 265 (1949).
(4) Rao, IJP 23, 425 (1949).
(5) Barrett, Mandel, PR 109, 1572 (1958).
(6) Fitzky, ZP 151, 351 (1958).
(7) Barrow, TFS 56, 952 (1960).
(8) Bulewicz, Phillips, Sugden, TFS 57, 921 (1961).
(9) Khvostenko, Sultanov, RJPC 39, 252 (1965).
(10) Berkowitz, Walter, JCP 49, 1184 (1968).
(11) Davidovits, Bellisio, JCP 50, 3560 (1969).

(continued p. 651)

State	T_e	ω_e	$\omega_e x_e$	B_e	α_e	D_e ($10^{-7}cm^{-1}$)	r_e (Å)	Observed Transitions Design.	ν_{00}	References
$^{205}Tl^{35}Cl$		$\mu = 29.8725631$ $D_0^0 = 3.82$ eV[a] I.P. = 9.70 eV[b]								NOV 1977
D		Complex system of emission bands; tentative vibrational analysis (3).						(D→A), R_V	(24628) (24040)	(2)* (3)*
		Continuous absorption for $\nu > 40000$ cm^{-1}.								(1)(16)
C ($^1\Pi$)		Continuous absorption with maximum at 39820 cm^{-1}, preceded by diffuse bands of longer wavelengths.						C← X, [c]		(1)* (16)*
		Very weak absorption continua with maxima at 32910 and 34830 cm^{-1}.						[c]		(1)* (16)*
		Absorption continua with maxima at 31690 and 32190 cm^{-1}.						[c]		(1)* (16)*
A $^3\Pi_{0^+}$	31049.$_4$	223.1[d] (Z)	11.4[e]	0.09227_0[f]	0.00131	0.7_5	2.4730	A↔X, [c] R	$31016._6$ (Z)	(1)* (2)* (6)
X $^1\Sigma^+$	0	283.7_5[d] (Z)	0.81_8	0.09139702_2	0.00039793_2[g]	0.375_3	2.484826	Vibration sp.[h]		(17)(21)
								Rotation sp.		(7)(8)(12) (20)
								Mol. beam electric and magnetic resonance[i]		(4)(15)(19) (5)
$^{(205)}Tl^{(35)}Cl^+$		$D_0^0 = 0.23$ eV[j]								NOV 1977
B $^2\Sigma$	(33800)	From the photoelectron spectrum (18), using vertical potentials for A and B and the adiabatic potential for X. The $^2\Pi$ peak is very broad and has a vertical potential which is 0.49 eV higher than that of A $^2\Sigma$.								
A $^2\Sigma$	(1600)									
X $^2\Pi$	0									
$^{205}Tl^{19}F$		$\mu = 17.3868727$ $D_0^0 = 4.57$ eV[a] I.P. = 10.5_2 eV[b]								NOV 1977
		(2) report emission bands in the region 28500 - 29500 cm^{-1}; uncertain.								
		Absorption continua at ~45400 and above 50000 cm^{-1}.								(1)*
C ($^1\Pi$)	(45546)	[346] H		Only v=0 and 1.[c]				C← X, R	45481 H	(1)*
		Absorption continuum at ~38400 cm^{-1}.								(1)*
B $^3\Pi_1$	36863.0_8	366.6_4	10.2_2[de]	0.2249_2	0.0030_7[f]	[3.9_2][g]	2.076_2	B↔X, R	36805.6_3 Z	(1)* (3)
A $^3\Pi_{0^+}$	35186.0_2	436.3	7.1[he]	0.2309_1	0.0027_4[i]	2.8_1[j]	2.049_1	A↔X, V_R	35164.3_1 Z	(1)* (3)
X $^1\Sigma^+$	0	477.3	2.3	0.223150163	0.001503850[k]	1.948	2.084438	Vibration sp.[ℓ]		(16)(21)
								Rotation sp.		(12)(15)(17) (19)(20)
								Mol. beam el. reson.[m]		(6)(7)(10) (11)

State	T_e	ω_e	$\omega_e x_e$	B_e	α_e	D_e (10^{-} cm^{-1})	r_e (Å)	Observed Transitions Design.	ν_{00}	References
(205)Tl19F+		$D_0^0 = 0.16$ eV[n]								NOV 1977
B $^2\Sigma$	(28500)	From the photoelectron spectrum (18) (adiabatic potentials). The peak corresponding to A is very broad.								
A $^2\Pi$	(5100)									
X $^2\Sigma$	0									

TℓBr, TℓBr+ (continued):

(12) Dickinson, Stephenson, Zorn, JCP 53, 1525 (1970).

(13) Brom, Franzen, JCP 54, 2874 (1971).

(14) Tiemann, ZN 26 a, 1809 (1971).

(15) Berkowitz, JCP 56, 2766 (1972).

(16) Lovas, Tiemann, JPCRD 3(3), 609 (1974).

TℓCℓ, TℓCℓ+:

[a]Photoionization mass-spectrometry (14); good agreement with thermochemical and spectroscopic values (9)(10), electron impact mass-spectrometry (13), and flame photometry (11).

[b]Photoionization value (14). As for TℓBr, this adiabatic potential corresponds to the second rather than first vertical potential observed by photoelectron spectroscopy (18).

[c]Absorption cross sections (16).

[d]Vibrational constants of (2) whose measurements at higher dispersion are more extensive than those of (1). Rotational analyses (6) give $\Delta G'(\frac{1}{2}) = 201.60$, $\Delta G''(\frac{1}{2}) = 281.88$.

[e]$\omega_e y_e = +1.33$, $\omega_e z_e = -0.115$.

[f]Predissociation for $v' \geq 4$; see (2).

[g]$\gamma_e = +4.17_0 \times 10^{-7}$; from (12) who gives rotational constants for all four TℓCℓ isotopes. See also (20).

TℓCℓ, TℓCℓ+ (continued):

[h]IR and Raman spectra in argon and krypton matrices.

[i]$\mu_{e\ell} = 4.515_3 + 0.0552_7(v+\frac{1}{2}) + 0.00007_8(v+\frac{1}{2})^2$ D; from (19) who give also hfs constants and magnetic parameters for $v = 0...3$.

[j]D_0^0(TℓCℓ) + I.P.(Tℓ) − I.P.(TℓCℓ).

(1) Howell, Coulson, PRS A 166, 238 (1938).

(2) Miescher, HPA 14, 148 (1941).

(3) Rao, IJP 23, 393 (1949).

(4) Carlson, Lee, Fabricand, PR 85, 784 (1952).

(5) Zeiger, Bolef, PR 85, 788 (1952).

(6) Barrow, PPS A 70, 622 (1957).

(7) Barrett, Mandel, PR 109, 1572 (1958).

(8) Fitzky, ZP 151, 351 (1958).

(9) Barrow, PPS 75, 933 (1960).

(10) Barrow, TFS 56, 952 (1960).

(11) Bulewicz, Phillips, Sugden, TFS 57, 921 (1961).

(12) De Wijn, Physica 31, 1193, 1557 (1965).

(13) Khovstenko, Sultanov, RJPC 39, 252 (1965).

(14) Berkowitz, Walter, JCP 49, 1184 (1968).

(15) Hammerle, Dickinson, VanAusdal, Stephenson, Zorn, JCP 50, 2086 (1969).

(continued p. 653)

TℓF, TℓF+: See p. 653 .

State	T_e	ω_e	$\omega_e x_e$	B_e	α_e	D_e ($10^{-4}cm^{-1}$)	r_e (Å)	Observed Transitions Design.	ν_{00}	References
$^{(205)}Tl^1H$		(μ = 1.00289416)		D_0^0 = 1.97 eV[a]						NOV 1977
		Absorption continua at 22720, 23590, 23920, 24550, 24930 cm^{-1}.								(3)
D $^3\Pi_2$ (2)	[24344.3]			[1.53][b]		[35]	[3.31]	C← X, R	23654.6 Z	(4)
C $^1\Pi$ (1)	(24181)	$[98.0_5]$ Z	c	[2.027][d]	e	[27][f]	[2.880]	D← X, R	23556.25 Z	(4)*
B $^3\Pi_1$ (1)		Unstable.								
A $^3\Pi_{0^+}(0^+)$	(17723)	[759.1] Z	gh	[4.617]	hij	$[5.6_8]$[k]	$[1.908_1]$	A↔X, R	17519.9 Z	(1)* (3)
X $^1\Sigma^+$	0	1390.7 Z	22.7	4.806[ℓ]	0.154[m]	[2.54]	1.870_2			
$^{(205)}Tl^2H$		(μ = 1.99450400)		D_0^0 = 2.00 eV[n]						NOV 1977
C $^1\Pi$ (1)		$\Delta G(3/2) = 57.56$		$B_2 = 0.820$ $B_1 = 0.973$		$D_2 = 16$ $D_1 = 11$	$r_1 = 2.94_7$	D← X,[o] R	v(1-0) = 23823.69 Z	(6)
A $^3\Pi_{0^+}(0^+)$	(17717)	[604.50] Z	ph	[2.380]	hqj	[1.14][k]	$[1.884_5]$	A← X R	17590.21 Z	(3)*
X $^1\Sigma^+$	0	987.7 Z	12.04[r]	2.419	0.057[s]	0.60	1.869_2			

$T\ell^1H$, $T\ell^2H$:

[a]From the highest observed level in C $^1\Pi$ (4); this state dissociates to $^2P_{3/2} + {}^2S$. Flame photometry (2) gives 1.9_1 eV.

[b]This level interacts with v=1 of C $^1\Pi$ (4).

[c]$\Delta G(3/2) = 56.2$.

[d]$B_0^+ - B_0^- = +0.006$.

[e]$B_1 = 1.84$, $B_2 = 0.7$.

[f]$D_1 = 60 \times 10^{-4}$.

[g]$\Delta G(3/2...11/2)$ = 474.9, 393.0, 402.2, 416.0, 426.1.

[h]Anomalous potential curve, see (5)(6).

[i]$B_1...B_6$ = 3.916, 3.201, 2.981, 2.856, 2.780, 2.692.

[j]Lines become increasingly diffuse with increasing rotation.

$T\ell^1H$, $T\ell^2H$ (continued):

[k]For higher D_v values see (1)(3).

[ℓ]RKR potential curve (5).

[m]$\gamma_e = +0.0044$.

[n]From the value for $T\ell^1H$.

[o]v'=0 not observed, possibly because of predissociation by the repulsive $^3\Pi_1$ state; see (6).

[p]$\Delta G(3/2...11/2)$ = 439.13, 314.61, 279.65, 281.23, 288.14.

[q]$B_1...B_6$ = 2.144, 1.842, 1.614, 1.515, 1.47, 1.43.

[r]$\omega_e y_e \approx +0.1$.

[s]$\gamma_e = +0.002$.

$T\ell^1H$, $T\ell^2H$ (continued):

(1) Grundström, Valberg, ZP 108, 326 (1938).
(2) Bulewicz, Sugden, TFS 54, 830 (1958).
(3) Neuhaus, Muld, ZP 153, 412 (1959).
(4) Larsson, Neuhaus, AF 23, 461 (1963).
(5) Ginter, Battino, JCP 42, 3222 (1965).
(6) Larsson, Neuhaus, AF 31, 299 (1966).

$T\ell C\ell$, $T\ell C\ell^+$ (continued):

(16) Davidovits, Bellisio, JCP 50, 3560 (1969).
(17) Brom, Franzen, JCP 54, 2874 (1971).
(18) Berkowitz, JCP 56, 2766 (1972).
(19) Ley, Schauer, ZN 27 a, 77 (1972).
(20) Lovas, Tiemann, JPCRD 3(3), 609 (1974).
(21) Lesiecki, Nibler, JCP 63, 3452 (1975).

$T\ell F$, $T\ell F^+$ (Table p. 650/1):

[a] Photoionization mass-spectrometry (14); good agreement with thermochemical results [(8) and ref. given in (14)]; similar results by equilibrium mass-spectrometry (13) and flame photometry (9).

[b] Adiabatic potential from the photoelectron spectrum (18).

[c] According to (8) [see Table 3, Note c of this reference] the vibrational analysis is uncertain and the 0-0 band may lie at 45010 cm^{-1}.

[d] $\omega_e y_e = -1.15_5$.

[e] Both A and B have small potential humps of ~ 0.18 eV.

[f] $\gamma_e = -0.00090_5$.

[g] $D_1 = 5.2_2 \times 10^{-7}$, $D_2 = 6.7_3 \times 10^{-7}$.

[h] $\omega_e y_e = -0.1$.

[i] $\gamma_e = -0.00013_5$.

[j] $\beta_e = +0.2_2 \times 10^{-7}$.

[k] $\gamma_e = +3.142 \times 10^{-6}$ (17).

[ℓ] IR and Raman spectra in argon and krypton matrices.

$T\ell F$, $T\ell F^+$ (continued):

[m] $\mu_{e\ell} = 4.193_6 + 0.069_1(v+\frac{1}{2})$ D; $g_J = -0.05356\ \mu_N$ (11).

[n] $D_0^0(T\ell F) + I.P.(T\ell) - I.P.(T\ell F)$.

(1) Howell, PRS A 160, 242 (1937).
(2) Rao, Rao, IJP 29, 20 (1955).
(3) Barrow, Cheall, Thomas, Zeeman, PPS 71, 128 (1958).
(4) Barrett, Mandel, PR 109, 1572 (1958).
(5) Fitzky, ZP 151, 351 (1958).
(6) Gräff, Paul, Schlier, ZP 153, 38 (1958).
(7) Gräff, ZP 155, 433 (1959).
(8) Barrow, TFS 56, 952 (1960).
(9) Bulewicz, Phillips, Sugden, TFS 57, 921 (1961).
(10) Drechsler, Gräff, ZP 163, 165 (1961).
(11) Boeckh, Gräff, Ley, ZP 179, 285 (1964).
(12) Ritchie, Lew, CJP 43, 1701 (1965).
(13) Murad, Hildenbrand, Main, JCP 45, 263 (1966).
(14) Berkowitz, Walter, JCP 49, 1184 (1968).
(15) Hoeft, Lovas, Tiemann, Törring, ZN 25 a, 1029 (1970).
(16) Brom, Franzen, JCP 54, 2874 (1971).
(17) Dijkerman, Flegel, Gräff, Mönter, ZN 27 a, 100 (1972).
(18) Dehmer, Berkowitz, Cusachs, JCP 58, 5681 (1973).
(19) Honerjäger, Tischer, ZN 28 a, 458 (1973).
(20) Lovas, Tiemann, JPCRD 3(3), 609 (1974).
(21) Lesiecki, Nibler, JCP 63, 3452 (1975).

State	T_e	ω_e	$\omega_e x_e$	B_e	α_e	D_e (10^{-9} cm^{-1})	r_e (Å)	Observed Transitions Design.	Observed Transitions ν_{00}	References
$^{(205)}Tl^{4}He$		(μ = 3.9259401$_4$) See TℓAr.								NOV 1977
$^{205}Tl^{127}I$		μ = 78.378506$_4$ D_0^0 = 2.76 eV[a] I.P. = 8.47 eV[b]								NOV 1977
		Continuous absorption at higher energies.						c		(12)*
C ($^1\Pi$)		Absorption continuum with maximum at 33050 cm^{-1}.						c		(2)(12)*
		Very weak continuous absorption with maxima at 28240, 28400 cm^{-1}.						c		(12)*
		Further emission bands have been reported by (3)(5) in the regions 15800 - 18000 and 27100 - 27900 cm^{-1}.								
A $^3\Pi_{0^+}$		(30)	Broad fluctuations in absorption.					A$\leftrightarrow$X,[c]	(26250)	(1)(2)(3)* (4)* (12)*
X $^1\Sigma^+$	0	(150)[d]		0.0271676$_1$	0.0000663$_5$	(3.6)	2.81367$_6$	IR spectrum[d]		(14)
								Microwave sp.[e]		(6)(7)(8)(17)
								Mol. beam rf el. reson.[f]		(13)
$^{(205)}Tl^{127}I^+$		D_0^0 = 0.40 eV[g]								NOV 1977
B	(37200) (34100)	From the photoelectron spectrum (16) (vertical potentials).								
A	(6500)									
X	0									
$^{(205)}Tl^{(84)}Kr$		(μ = 59.538077$_4$) See TℓAr.								
$^{(205)}Tl^{(20)}Ne$		(μ = 18.2157448) See TℓAr.								
$^{(205)}Tl^{(132)}Xe$		(μ = 80.257341$_5$) See TℓAr.								

$T\ell I$, $T\ell I^+$:

[a]Photoionization mass-spectrometry (11); similar values are obtained from thermochemical data (9), by flame photometry (10), and by photofragment spectroscopy (18).

[b]Photoionization mass-spectrometry (11) and photoelectron spectroscopy (16).

[c]Absorption cross sections (12).

[d]$\Delta G(\frac{1}{2}) = 143.0\ cm^{-1}$ in argon at 10 K (14).

[e]$\mu_{e\ell}(v=0) = 4.61$ D, from Stark effect of rotation sp.(15).

[f]Hfs constants.

[g]$D_0^0(T\ell I) + I.P.(T\ell) - I.P.(T\ell I)$.

(1) Butkow, Terenin, ZP 49, 865 (1928).
(2) Butkow, ZP 58, 232 (1929).
(3) Rao, Rao, IJP 23, 185 (1949).
(4) Rao, IJP 24, 434 (1950).
(5) Rao, Rao, IJP 29, 20 (1955).
(6) Happ, ZP 147, 567 (1957).
(7) Barrett, Mandel, PR 109, 1572 (1958).
(8) Fitzky, ZP 151, 351 (1958).
(9) Barrow, TFS 56, 952 (1960).
(10) Bulewicz, Phillips, Sugden, TFS 57, 921 (1961).
(11) Berkowitz, Chupka, JCP 45, 1287 (1966).
(12) Davidovits, Bellisio, JCP 50, 3560 (1969).
(13) Stephenson, Dickinson, Zorn, JCP 53, 1529 (1970).
(14) Brom, Franzen, JCP 54, 2874 (1971).
(15) Tiemann, ZN 26 a, 1809 (1971).
(16) Berkowitz, JCP 56, 2766 (1972).
(17) Lovas, Tiemann, JPCRD 3(3), 609 (1974).
(18) Kawasaki, Litvak, Bersohn, JCP 66, 1434 (1977).

State	T_e	ω_e	$\omega_e x_e$	B_e	α_e	D_e	r_e (Å)	Observed Transitions: Design.	Observed Transitions: ν_{00}	References
$^{169}Tm^{16}O$		$\mu = 14.6114809_2$ $D_0^0 = 5.7_6$ eV[a]								JUN 1975 A
		Unclassified, mostly R shaded emission bands in the regions 20000 - 21300 cm^{-1},								(1)(2)
		16700 - 19200 cm^{-1},								(1)* (2)
		11600 - 15600 cm^{-1}.								(3)
$^{238}U^{(11)}B$		($\mu = 10.5226566$) $D_0^0 = 3.3_0$ eV[a]								JUN 1975
$^{238}U^{12}C$		$\mu = 11.42411706$ $D_0^0 = 4.7_7$ eV[a]								JUN 1975
$^{238}U^{14}N$		$\mu = 13.2251213_5$ $D_0^0 = 5.5_1$ eV[a]								JUN 1975 A
$^{238}U^{16}O$		$\mu = 14.9878628_4$ $D_0^0 = 7.8_7$ eV[a] I.P. = 5.6_6 eV[b]								JUN 1975
		Very weak band structure in emission from an arc; no measurements.								(1)
X		[820.0][c]								
$^{238}U^{(32)}S$		($\mu = 28.1864180$) $D_0^0 = 5.3_8$ eV[a]								JUN 1975
$^{51}V_2$		$\mu = 25.4719822$ $D_0^0 = 2.4_8$ eV[a]								JUN 1975 A
$^{51}V^1H$		$\mu = 0.98827418$								OCT 1975
$^{51}V^2H$		$\mu = 1.93750185$								
		Absorption bands of complex structure in the region 20800 - 22700 cm^{-1} with strong features near 21230 and 21330 cm^{-1}. Theoretical calculations (2)(3) predict a $^5\Delta$ ground state.								(1)*

TmO: [a]Thermochemical value (mass-spectrom.)(4), recalc. (5).

(1) Gatterer, Junkes, Salpeter, Rosen, METOX (1957).
(2) Herrmann, Alkemade, "Chemical Analysis by Flame Photometry", Wiley-Interscience (1963).
(3) Mavrodineanu, Boiteux, "Flame Spectroscopy", Wiley (1965).
(4) Ames, Walsh, White, JPC 71, 2707 (1967).
(5) Smoes, Coppens, Bergman, Drowart, TFS 65, 682 (1969).

UB: [a]Thermochemical value (mass-spectrom.)(1).

(1) Gingerich, JCP 53, 746 (1970).

UC: [a]Thermochemical value (mass-spectrom.)(1).

(1) Gingerich, JCP 50, 2255 (1969).

UN: [a]Thermochemical value (mass-spectrom.)(1).

(1) Gingerich, JCP 47, 2192 (1967).

UO: [a]Thermochemical value (mass-spectrom.)(3)(4)(5).
[b]Electron impact appearance potential, average of two values (2)(8).
[c]IR spectrum of $U^{16}O$ in Ar matrix at 15 K (6)(7). For $U^{18}O$ at 776.3 cm^{-1}. Derived constants for $U^{16}O$ are $\omega_e = 825._0$, $\omega_e x_e = 2.5$.

(1) See ref. (1) of TmO.
(2) Mann, JCP 40, 1632 (1964).
(3) Drowart, Pattoret, Smoes, PBCS No. 8, 67 (1967).
(4) Coppens, Smoes, Drowart, TFS 64, 630 (1968).
(5) Pattoret, Drowart, Smoes, in "Thermodynamics of Nuclear Materials" (Proc. Symp. Thermodynamics, Vienna, 1967), p. 613. Publ. IAEA, Vienna (1968).
(6) Carstens, Gruen, Kozlowski, HTS 4, 436 (1972).
(7) Gabelnick, Reedy, Chasanov, CPL 19, 90 (1973); JCP 58, 4468 (1973); 60, 1167 (1974).
(8) Rauh, Ackermann, JCP 60, 1396 (1974).

US: [a]Thermochemical value (mass-spectrom.)(1).

(1) Cater, Rauh, Thorn, JCP 44, 3106 (1966); 48, 538 (1968)(Erratum).

V_2: [a]Thermochemical value (mass-spectrom.)(1).

(1) Kant, Lin, JCP 51, 1644 (1969).

V^1H, V^2H:

(1) Smith, PRS A 332, 113 (1973).
(2) Scott, Richards, JP B 7, L347 (1974).
(3) Henderson, Das, Wahl, JCP 63, 2805 (1975).

State	T_e	ω_e	$\omega_e x_e$	B_e	α_e	D_e ($10^{-7}cm^{-1}$)	r_e (Å)	Observed Transitions: Design.	Observed Transitions: ν_{00}	References
$^{51}V^{14}N$		$\mu = 10.9839053_0$		$D_0^0 = 4.9_1$ eV[a]						JUN 1975
$^{51}V^{16}O$		$\mu = 12.1729611_9$		$D_0^0 = 6.4_1$ eV[a]						JUN 1975
D	23980	[833]						D← X,[b]	23890	(18)
C $^4\Sigma^-$	17494.3	863.4[c] (Z)	5.35	0.4953[de]	0.0035	6	1.672_1	C[f]↔ X,[gh] R	17420.2 Z	(1)(3)* (6)* (11)(12)(15) (17)*
B $^4\Pi_{(r)}$	12760.8 12689 12625 12567	910.9	5.0	0.5246	0.004		1.625	B↔ X,[h] R	12710.6[i] 12638.6 H^Q 12574.7 12516.7	(2)(6)* (12) (15)(17)*
		[For a more rigorous treatment of the fine structure see (20).]								
		R shaded bands in the region 9470 - 9560 cm^{-1}.[j]								(4)
X $^4\Sigma^-$ [k]	0	1011.3[c] (Z)	4.86	0.54825[ℓ]	0.00352	6	1.5893_2	ESR sp.[m]		(10)
$^{51}V^{(32)}S$		($\mu = 19.6437769$)		$D_0^0 = 4.6_2$ eV[a]						JUN 1975
$^{51}V^{(80)}Se$		($\mu = 31.111488_5$)		$D_0^0 = 3.5_4$ eV[a]						JUN 1975
$^{(184)}W^{16}O$		($\mu = 14.7153823_1$)		$D_0^0 = 6.8$ eV[a]						JUL 1975
		R shaded bands in gas phase emission from 12900 to 24400 cm^{-1}.								(1)(2)(3)(5)
G		[933]		Observed in absorption in rare gas matrices (6). Frequencies are for solid neon, except for D→X which was also identified (1) in the gas phase spectrum.				G← X,	23794	(6)
F		[982][b]	c					F← X,	23366	(6)
E		[944][b]	c					E← X,	21509	(6)
D	$20834._2$	$993._0$ H	6.5					D↔X, R	20799.9 H	(1)(6)
C		[931]	c					C← X,	19190	(6)
B		[955][d]	c					B← X,	17283	(6)
A		[951][e]	c					A← X,	17132[e]	(6)
X	0	$1059._9$ H	3.1							

VN: [a]Thermochemical value (mass-spectrom.)(1).

(1) Farber, Srivastava, JCS FT I 69, 390 (1973).

VO: [a]Thermochemical value (mass-spectrom.)(5)(9)(14)(19).

[b]Low temp. matrix abs.; not observed in the gas phase.

[c]From (17). Slightly different constants in (15).

[d]Spin splitting constants (7)(15) for v=0: $\lambda = +0.53$, $\gamma_1 = -0.009$, $\gamma_2 = -0.007$.

[e]Perturbations.

[f]Radiative lifetime $\tau(v=0,1) = 0.41\ \mu s$ (16).

[g]Absolute transition moments, band oscillator strengths (17). This ref. contains a useful review and bibliography of earlier work.

[h]The existence of a new system between 13900 and 14500 cm^{-1}, tentatively identified by Grosjean and Rosen [see (6)], is doubtful since the 2-0 and 0-3 sequences of B−X and C−X, respectively, are expected to overlap in this region.

[i]Subband origin at 12706.8 cm^{-1}.

[j]It is not certain whether these bands form a new system or are part of B−X.

[k]The ESR spectrum of matrix isolated VO (10) is compatible only with a $^4\Sigma^-$ ground state, confirming theoretical predictions by (8).

[ℓ]Spin splitting constants (7)(15) for v=0: $\lambda = +1.371$, $\gamma_1 = +0.0112$, $\gamma_2 = +0.0111$. Large nuclear magnetic hfs, b = 0.080 cm^{-1} (10)(13). An unusual perturbation occurs in v=0,1 between F_2 and F_3 levels with $\Delta N = \Delta F = 0$ (13).

[m]In argon matrix.

(1) Mahanti, PPS 47, 433 (1935).
(2) Keenan, Schroeder, ApJ 115, 82 (1952).
(3) Lagerqvist, Selin, Naturw. 42, 65 (1955); AF 12, 553 (1957).
(4) Lagerqvist, Selin, AF 11, 429 (1956).
(5) Berkowitz, Chupka, Inghram, JCP 27, 87 (1957).
(6) Gatterer, Junkes, Salpeter, Rosen, METOX (1957).

VO (continued):

(7) Hougen, CJP 40, 598 (1962).
(8) Carlson, Moser, JCP 44, 3259 (1966).
(9) Coppens, Smoes, Drowart, TFS 63, 2140 (1967).
(10) Kasai, JCP 49, 4979 (1968).
(11) Laud, Kalsulkar, IJP 42, 61 (1968).
(12) Richards, Barrow, Nature 217, 842 (1968).
(13) Richards, Barrow, Nature 219, 1244 (1968).
(14) Frantseva, Semenov, HT(USSR) 7, 52 (1969).
(15) Richards, D. Phil. Thesis (Oxford, 1969); see DONNSPEC.
(16) Diebold, Wentink, unpubl.(1970); quoted in (17).
(17) Harrington, Seel, Hebert, Nicholls, IAMS 7 (1970).
(18) Weltner, unpubl.; quoted by Barrow in DONNSPEC.
(19) Farber, Uy, Srivastava, JCP 56, 5312 (1972).
(20) Veseth, PS 12, 125 (1975).

VS: [a]Thermochemical value (mass-spectrom.)(1). Slightly higher value in (2).

(1) Drowart, Pattoret, Smoes, PBCS No. 8, 67 (1967).
(2) Owzarski, Franzen, JCP 60, 1113 (1974).

VSe: [a]Thermochemical value (mass-spectrom.)(1); no details.

(1) Bergman, Coppens, Drowart, Smoes, TFS 66, 800 (1970).

WO: [a]Thermochemical value (mass-spectrom.)(4).

[b]Strong interaction between levels of E and F.

[c]Higher ΔG and deperturbed values may be found in (6).

[d]Strong interaction between levels of A and B.

[e]Lowest observed level (v=3?) and ΔG. Weak system; the intensities of the bands derive mostly from the strong A∼B interaction.

References on p. 661.

State	T_e	ω_e	$\omega_e x_e$	B_e	α_e	D_e (10^{-} cm^{-1})	r_e (Å)	Observed Transitions Design.	ν_{00}	References
$^{(129,132)}Xe_2$		(μ = 65.193611$_8$) D_0^0 = 0.0230 eV[a] I.P. = 11.12$_7$ eV[b]								NOV 1977
		Several diffuse emission bands and continua in the visible and near UV regions; interpretation and assignment to Xe_2 uncertain.								(3)(5)
K	85139.$_6$	26	1	Associated with $7s[1\frac{1}{2}]_2^o$ (85189.31 cm^{-1}) [c]				K← X,[d]	85141.9	(19)
		Unclassified abs. bands, associated with $5d[1\frac{1}{2}]_1^o$ (83890.47 cm^{-1}) [c]; $5d[2\frac{1}{2}]_3^o$ (82430.72 cm^{-1}) [c]						[e]		(19)(20)*
H	82001	[30.$_0$]		Associated with $5d[2\frac{1}{2}]_2^o$ (81926.04 cm^{-1}) [c]				H← X,[f]	82005.7	(19)
		Unclassified abs. bands, associated with $5d[1\frac{1}{2}]_2^o$ (80323.28 cm^{-1}) [c]; $6p[0\frac{1}{2}]_0$ (80119.47 cm^{-1}) [c]; $5d[0\frac{1}{2}]_1^o$ (79987.16 cm^{-1}) [c]; $6p[1\frac{1}{2}]_2$ (79212.97 cm^{-1}) [c]; $6p[1\frac{1}{2}]_1$ (78956.54 cm^{-1}) [c]; $6p[2\frac{1}{2}]_3$ (78403.56 cm^{-1}) [c]; $6p[2\frac{1}{2}]_2$ (78120.30 cm^{-1}) [c]						[e]; [g]		(9)(19) (20)*
		Unclassified abs. bands, associated with $6s'[0\frac{1}{2}]_1^o$ (77185.56 cm^{-1}) [c]						[h]		(14)
B ($^1\Sigma_u^+$) 0_u^+; A ($^3\Sigma_u^+$) 1_u		Unclassified absorption bands longward and shortward of the resonance line at 68045.66 cm^{-1}(1469.6 Å); pressure broadening of this line due to molecule formation. Continuous emission 68000 - 50000 cm^{-1}(1470 - 2000 Å) with maxima near 68030, 67070 cm^{-1}(1470, 1491 Å, "first continuum") and ~58800 cm^{-1}(1700 Å, "second cont.")						B^i↔ X,[j]; A^{ik}↔ X,[j]		(1)(2)(3)* (4)(6)(8) (13)(14)
X $^1\Sigma_g^+$	0	21.12 H	0.65[ℓ]				4.36$_1$[m]			

Xe_2: [a]Spectroscopic value, by extrapolation from the observed lowest ten vibrational levels of the ground state (20). The corresponding well depth is in very good agreement with D_e^0 = 0.0243 eV derived by (18) from bulk properties and differential scattering cross sections as well as spectroscopic data. Integral absorption measurements of the 1274.8 Å band (78444 cm^{-1}) as a function of temperature give D_0^0 = 0.029 eV (25).

Xe_2 (continued):

[b]Photoionization of Xe_2 (28). An earlier determination based on the formation of Xe_2^+ from Rydberg excited Xe (associative ionization) yielded 11.14 eV (7).

[c]Atomic energy levels relative to $5p^6\ ^1S$.

[d]Several v" progressions (presumably v'=0...4) near 1174 Å.

[e]Bands longward of 1192 and 1250 Å have been partially analyzed by (19)(20); the ground state vibrational levels of (20) are from this analysis.

[f]Two v" progressions shortward, and two broad bands longward, of 1221 Å.

[g](25) have determined the oscillator strength ($f = 0.03_9$) for a band at 1274.8 Å.

[h]Continuous absorption and ill defined vibrational structure longward of 1296 Å; small number of sharp bands at shorter wavelengths.

[i]Radiative lifetimes: A 100 ns [theoretical value of 23 ns predicted by (17)], B 6 ns (21)(26)(27). A and B are associated with the lowest excited states of Xe I, i.e. $6s[1\frac{1}{2}]_2^o$ (67068.05 cm^{-1}) and $6s[1\frac{1}{2}]_1^o$ (68045.66 cm^{-1}), respectively.

[j]The continuum is observed in discharges at high pressure as well as in fluorescence excited by the Xe I 1470 Å resonance line (12), by CO fourth positive bands (23), and by synchrotron radiation (24). There is good evidence that the "first continuum" is due to vibrationally unrelaxed molecules in the A and B states, while the "second continuum" corresponds to transitions from v'≈0 of these states to the steep repulsive part of the X $^1\Sigma_g^+$ potential curve (11)(22). A broad emission peak at 1725 Å, probably representing the same transition, has been observed by bombardment of Xe in a Ne matrix with α particles (16).

[k](22) estimates $r_e \approx 3.25$ Å.

[ℓ]$\omega_e y_e = +0.003$. Zero point energy 10.40 cm^{-1}; last stable vibrational level (extrapol.) v=25 (20).

[m]From bulk properties and differential scattering cross sections (18) [see (20), "note added in proof"]; see also (15). Electron diffraction (10) gives 4.41 Å.

(1) McLennan, Turnbull, PRS A 139, 683 (1933).
(2) Tanaka, Zelikoff, JOSA 44, 254 (1954).
(3) Wilkinson, Tanaka, JOSA 45, 344 (1955).
(4) Tanaka, JOSA 45, 710 (1955).
(5) Roth, Gloersen, JCP 29, 820 (1958).
(6) Huffman, Larrabee, Tanaka, AO 4, 1581 (1965).
(7) Samson, Cairns, JOSA 56, 1140 (1966).
(8) Wilkinson, JQSRT 6, 823 (1966).
(9) Herzberg, unpublished.
(10) Audit, JP(Paris) 30, 192 (1969).
(11) Mulliken, JCP 52, 5170 (1970).
(12) Freeman, McEwan, Claridge, Phillips, CPL 10, 530 (1971).
(13) Kosinskaya, Polozova, OS(Engl. Transl.) 30, 458 (1971).
(14) Castex, Damany, CPL 13, 158 (1972); 24, 437 (1974).
(15) Docken, Schafer, JMS 46, 454 (1973).

(continued p. 662)

WO (continued): (1) Gatterer, Krishnamurty, Nature 169, 543 (1952).
(2) Vittalachar, Krishnamurty, CS 23, 357 (1954).
(3) Gatterer, Junkes, Salpeter, Rosen, METOX (1957).
(4) DeMaria, Burns, Drowart, Inghram, JCP 32, 1373 (1960).
(5) Foster, Gaydon, quoted in IDSPEC (1963).
(6) Weltner, McLeod, JMS 17, 276 (1965).

State	T_e	ω_e	$\omega_e x_e$	B_e	α_e	D_e (10^{-} cm^{-1})	r_e (Å)	Observed Transitions Design.	ν_{00}	References
$^{(129,132)}Xe_2^+$		(μ = 65.193474$_6$) $D_0^0 = 1.02_6$ eV[a]								NOV 1977 A
		Diffuse emission bands 61690 - 63130 cm^{-1} in a condensed discharge through Xe, tentatively assigned to Xe_2^+.								(1)*
D ($^2\Sigma_g^+$) $\frac{1}{2}g$	(17500)	$D_0 \approx 0.17$ eV								
C ($^2\Pi_u$) $\frac{1}{2}u$	(16700)	$D_0 \approx 0.26$ eV								
$\frac{3}{2}u$	(8600)	Repulsive.								
B ($^2\Pi_g$) $\frac{1}{2}g$	(7000)	$D_0 \approx 0.16$ eV								
$\frac{3}{2}g$	(5300)	$D_0 \approx 0.37$ eV	Preliminary data from the photoelectron spectrum (6). For an evaluation of difference potentials from g-u oscillations in the differential elastic scattering cross section see (4)(4b). Absolute photodissociation cross sections 14400 - 17700 cm^{-1} (5).					B← X		(5)
X $^2\Sigma_u^+$	0						(3.25)[b]			
$^{(132)}Xe^{(79)}Br$		(μ = 49.376402$_4$)								NOV 1977
B ($^2\Sigma^+$)	36010[a]	120	(0.23)				(2.96)	B↔X,[b]		(1)(2)(6)
		Broad unclassified emission bands at 21500 and 28300 cm^{-1}.[c]								
X ($^2\Sigma^+$)	0	Repulsive state with van der Waals minimum of unknown depth.								(1)(2)
$^{(132)}Xe^{35}Cl$		(μ = 27.6410016) $D_0^0 = 0.030_0$ eV[a]								NOV 1977
D ($\frac{1}{2}$)								D↔X,[b]	42450	(8)
B ($^2\Sigma^+$)	32405.0	195.2 H	0.54				(2.94)[c]	B↔X,[d] V	32489.3 H	(2)(3)(4)(5)
X ($^2\Sigma^+$)	0	26.3 H	-0.28[e]				(3.18)[c]	ESR sp.[f]		(7)

Xe_2 (continued):

(16) Gedanken, Raz, Jortner, JCP 59, 1630 (1973).
(17) Weihofen, JCP 60, 445 (1974).
(18) Barker, Watts, Lee, Schafer, Lee, JCP 61, 3081 (1974).
(19) Castex, CP 5, 448 (1974).
(20) Freeman, Yoshino, Tanaka, JCP 61, 4880 (1974).
(21) Keto, Gleason, Walters, PRL 33, 1365 (1974).
(22) Mulliken, RR 59, 357 (1974).

Xe_2 (continued):

(23) Fink, Comes, CPL 30, 267 (1975).
(24) Brodmann, Zimmerer, Hahn, CPL 41, 160 (1976); Brodmann, Zimmerer, JP B 10, 3395 (1977).
(25) Chashchina, Shreider, ZPS 25, 163 (1976). | (1976).
(26) Keto, Gleason, Bonifield, Walters, Soley, CPL 42, 125
(27) Leichner, Palmer, Cook, Thieneman, PR A 13, 1787
(28) Ng, Trevor, Mahan, Lee, JCP 65, 4327 (1976) | (1976).

Xe_2^+: [a]$D_0^0(Xe_2)$ + I.P.(Xe) - I.P.(Xe_2). Values of 0.97 and 0.99 eV for D_e^0 have been derived by (3) and (4) from elastic scattering of Xe^+ by Xe using ground state r_e values which differ slightly from the one given in the Table.

[b]Estimated value (4a).

(1) Tanaka, JOSA 45, 710 (1955).
(2) Mulliken, JCP 52, 5170 (1970).
(3) Lorents, Olson, Conklin, CPL 20, 589 (1973).
(4) Mittmann, Weise, ZN 29 a, 400 (1974).
(4a) See ref. (22) of Xe_2.
(4b) Jones, Conklin, Lorents, Olson, PR A 10, 102 (1974).
(5) Miller, Ling, Saxon, Moseley, PR A 13, 2171 (1976).
(6) Dehmer, Dehmer, JCP 67, 1774 (1977).

XeBr: [a]Energy of the potential minimum relative to Xe(^1S) + Br($^2P_{3/2}$). All constants derived (6) by trial-and-error matching of observed and theoretically simulated spectra.

[b]Structured continuum with principal peak at 35480 cm^{-1}, observed in emission in the reaction of metastable ($^3P_2$) Xe atoms with Br_2, CH_2Br_2, and PBr_3 at ~1 torr (1)(3) or on electron bombardment of Ar-Xe-Br_2 or Xe-HBr mixtures at pressures of the order of 1 atm (2)(6). Under the latter conditions laser action has been found (4)(6). Observed in absorption in an Ar matrix at 34000 cm^{-1} (5).

[c](5) report emission peaks in argon matrices at 21200 and 26700 cm^{-1}.

(1) Velazco, Setser, JCP 62, 1990 (1975).
(2) Brau, Ewing, JCP 63, 4640 (1975).
(3) Golde, JMS 58, 261 (1975).
(4) Searles, Hart, APL 27, 243 (1975); 28, 602 (1976).

XeBr (continued):

(5) Ault, Andrews, JCP 65, 4192 (1976).
(6) Tellinghuisen, Hays, Hoffman, Tisone, JCP 65, 4473 (1976).

XeCℓ: [a]Extrapolation of the vibrational levels in X $^2\Sigma$ (5).

[b]Observed in a low-pressure discharge through Xe + $C\ell_2$ (8). Absorption bands at similar wavelengths are observed in inert gas matrices (6).

[c]Based on potential functions chosen for best representation of observed levels and intensities (5).

[d]Observed in emission in the reaction of metastable ($^3P_2$) Xe atoms with $C\ell_2$, NOCℓ, SOCℓ_2, CCℓ_4 (2)(3) or on electron bombardment of mixtures of Ar + Xe + $C\ell_2$ and Xe + HCℓ or $C\ell_2$ at high pressure (of the order of 1 atm) (4)(5). Under the latter conditions laser action has been detected (1)(5). Observed in absorption and emission in inert gas matrices (6).

[e]$\omega_e y_e = -0.06_7$.

[f]In argon at 4.2 K.

(1) Ewing, Brau, APL 27, 350 (1975).
(2) See ref. (3) of XeBr.
(3) See ref. (1) of XeBr.
(4) See ref. (2) of XeBr.
(5) Tellinghuisen, Hoffman, Tisone, Hays, JCP 64, 2484 (1976).
(6) See ref. (5) of XeBr.
(7) Adrian, Bowers, JCP 65, 4316 (1976).
(8) Shuker, APL 29, 785 (1976).

State	T_e	ω_e	$\omega_e x_e$	B_e	α_e	D_e (10^{-} cm^{-1})	r_e (Å)	Observed Transitions Design.	ν_{00}	References
$^{(132)}Xe^{19}F$		(μ = 16.6065342)		$D_0^0 = 0.13_3$ eV[a]		I.P. ≤ 10.23 eV[b]				NOV 1977 A
D ($^2\Pi_{1/2}$)	38057	351 H	$2._1$					D↔X,[c] V	38118 H	(2)(7)(13)(16)*
B ($^2\Pi_{1/2}$)	28826	308.6 (Z)	1.52					B[d]↔X,[e] V	28866	(2)(5)* (7)(9)(12)(13)
C ($^2\Pi_{3/2}$)	(25550)	(370)	Tentative interpretation (10) of observations in solid argon.					C→A,	(18600)	(10)
A ($^2\Pi_{3/2}$)		Unstable						C←X,	(25620)	
X ($^2\Sigma$)	0	234.2[f] (Z)	12.0					ESR sp.[g]		(1)(3)
$^{(132)}Xe^{19}F^+$				$D_0^0 \geq 2.03$ eV[a]						NOV 1977
$^{(132)}Xe^1H$		(μ = 1.00018323)		$D_e^0 = 0.0068_3$ eV[a]						NOV 1977
X $^2\Sigma$	0	Evidence for the existence of quasibound levels in the van der Waals well [orbiting resonances in the integral scattering cross section (3)].					3.9_4[a]			
$^{(132)}Xe^1H^+$				$D_e^0 = (4.32)$ eV[a]						NOV 1977
X $^1\Sigma$	0						(1.59)[a]			

XeF: [a]From (12) after correction for change in the v numbering of the ground state. Extensive ab initio calculations (4) suggested an essentially repulsive pot. for X $^2\Sigma$; much better agreement with the spectroscopic result is obtained (11) if, in addition, the contribution to binding from the dispersion interaction is taken into account.

[b]$D_0^0(XeF) + I.P.(Xe) - D_0^0(XeF^+)$.

[c]Observed in emission in the reaction of metastable rare gas atoms with fluorine containing molecules (13)(16) and in electron-beam excited mixtures of Ar + Xe + F_2 at high total pressure (7). Absorption in inert gas matrices at similar wavelengths (10). Laser-excited emission spectra in solid Ne and Ar (14); lifetimes of $12._5$ and $11._5$ ns, respectively, have been measured.

XeF (continued):

[d]Lifetime τ = 18.8 ns (15); (14) measure 6 ns in solid Ar.

[e]Observed in emission at low pressure in the reaction of metastable ($^3P_2$) Xe atoms with F_2, NOF, N_2F_4, CF_3OF [flowing afterglow (6)(9)] and at high pressure (of the order

XeF (continued):

of 1 atm) on electron bombardment of mixtures of Ar, Xe, and F_2 or NF_3 (5)(7)(12); two bands (0-3 and 1-4 according to the new v" numbering) at 3532 and 3511 Å show strong laser action (5)(8)(12). Also observed in fluorescence in the photolysis of XeF_2 (17). In emission and absorption in inert gas matrices (10)(14).

[f]The v" numbering in (12) has been changed in a "note added in proof". See also (16). Emission spectra in a neon matrix (14) give $\omega_e'' = 247$, $\omega_e''x_e'' = 10.2$.

[g]In γ-irradiated solid XeF_4 and XeF_2 (1)(3).

(1) Morton, Falconer, JCP 39, 427 (1963).
(2) Kuznetsova, Kuzyakov, Shpanskii, Khutoretskii, VMUK 19, 19 (1964).
(3) Eachus, Symons, JCS A (1971), p. 304.
(4) Liskow, Schaefer, Bagus, Liu, JACS 95, 4056 (1973).
(5) Ault, Bradford, Bhaumik, APL 27, 413 (1975).
(6) Velazco, Setser, JCP 62, 1990 (1975).
(7) Brau, Ewing, JCP 63, 4640 (1975).
(8) Brau, Ewing, APL 27, 435 (1975).
(9) Golde, JMS 58, 261 (1975).
(10) Ault, Andrews, JCP 64, 3075 (1976); 65, 4192 (1976).
(11) Krauss, Liu, CPL 44, 257 (1976).
(12) Tellinghuisen, Tisone, Hoffman, Hays, JCP 64, 4796 (1976).
(13) Velazco, Kolts, Setser, JCP 65, 3468 (1976).
(14) Goodman, Brus, JCP 65, 3808 (1976).
(15) Burnham, Harris, JCP 66, 2742 (1977).
(16) Velazco, Kolts, Setser, Coxon, CPL 46, 99 (1977).
(17) Brashears, Setser, Desmarteau, CPL 48, 84 (1977).

XeF^+: [a]From photoionization data (3). XeF^+ is formed in the photoionization of XeF_2 (1)(3) and in the reaction $Xe^+(^2P_{\frac{3}{2},\frac{1}{2}}) + F_2 = XeF^+ + F$ (2).

(1) Morrison, Nicholson, O'Donnell, JCP 49, 959 (1968).
(2) Berkowitz, Chupka, CPL 7, 447 (1970).
(3) Berkowitz, Chupka, Guyon, Holloway, Spohr, JPC 75, 1461 (1971).

Xe^1H: [a]Scattering of H atoms by Xe (2). The repulsive potential at higher energies has been determined by (1).

(1) Picot, Fink, JCP 56, 4241 (1972).
(2) Bickes, Lantzsch, Toennies, Walaschewski, FDCS No. 55, 167 (1973).
(3) Toennies, Welz, Wolf, JCP 61, 2461 (1974).

Xe^1H^+: [a]From the ab initio calculations of (2); D_e^0 refers to $Xe^+ + H$ and has been reduced by 1.47 eV from the value given in (2) [$Xe + H^+$]. See also (3). The proton scattering results of (1) are energy dependent, probably because of charge exchange in the curve crossing region.

(1) Weise, Mittmann, Ding, Henglein, ZN 26 a, 1122 (1971).
(2) Kubach, Sidis, JP B 6, L289 (1973).
(3) Gallup, Macek, JP B 10, 1601 (1977).

State	T_e	ω_e	$\omega_e x_e$	B_e	α_e	D_e $(10^{-}\ cm^{-1})$	r_e (Å)	Observed Transitions Design.	ν_{00}	References
$^{(132)}Xe^{127}I$		(μ = 64.678012$_1$)								NOV 1977
B ($^2\Sigma^+$)	40210[a]	112	(0.24)	Diffuse emission bands with principal peak at 39510 cm^{-1}.			(3.31)	B→X,[b]		(2)(3)(4)
		Broad diffuse bands at 27770 and 30760 cm^{-1}.						B→A,[b]		(2)(3)
A ($^2\Pi_{\frac{3}{2},\frac{1}{2}}$)		Unstable.								
X ($^2\Sigma^+$)	0	Repulsive state with van der Waals minimum of unknown depth.								
$^{(132)}Xe^{14}N$?		(μ = 12.6591650$_4$)								NOV 1977
		Emission system at 20300 cm^{-1}, observed in a d.c. discharge through Xe + N_2.								(1)
$^{(132)}Xe^{16}O$		(μ = 14.2651049$_3$)								NOV 1977
F		Broad emission continuum with maximum at 42200 cm^{-1}, in discharges through Xe + trace of O_2 and in chemiluminescence of Xe($^3P_2$) + N_2O.						F↔X,[a]		(1b)(4)(5)
		Narrow band group at 34300 cm^{-1} in discharges through Xe + O_2. Laser excitation produces a broad emission band at 34500 cm^{-1} in solid argon.						(F→A)		(2)(13)
d ($^1\Sigma^+$)	(33268)[b]	152 H	9.9	*Ab initio* potential functions (12). The molecular states arise from Xe(^1S) + O(^1S): d, Xe(^1S) + O(^1D): a,b,c, Xe(^1S) + O(^3P): X,A.				d→b,[c]	(17695)[d]	(8)(13)
								d[e]→a,[f] R	20097[g] H	(1)(1a)(1b) (3)* (13)
								(d→X)[h]	32460	(1b)(8)(13)
c ($^1\Delta$)		Repulsive state.								
b ($^1\Pi$)	(15600)[b]	97[d] H	8.7							
a ($^1\Sigma^+$)	(13068)[b]	372[g] H	12.0							
A ($^3\Sigma^-$)		Repulsive state.								
X ($^3\Pi$)		Repulsive state.								

XeI: [a]Energy of the potential minimum relative to $Xe(^1S) + I(^2P_{3/2})$. All constants obtained by trial-and-error matching of observed and calculated spectra (4).

[b]Observed in the reaction of metastable Xe $(^3P_2)$ atoms with I_2 and CF_3I (1) and when Xe + HI or Ar + Xe + I_2 mixtures at high pressure (~1 atm) are bombarded by electrons (2)(4).

(1) Velazco, Setser, JCP 62, 1990 (1975).
(2) Brau, Ewing, JCP 63, 4640 (1975).
(3) Ewing, Brau, PR A 12, 129 (1975).
(4) Tellinghuisen, Hays, Hoffman, Tisone, JCP 65, 4473 (1976).

XeN: (1) Herman, Herman, Nature 193, 156 (1962).

XeO: [a]Charge-transfer transition from a largely ionic upper state ($^+XeO^-$) to the largely atomic ground state; see (5). In argon matrix absorption at 44040 cm^{-1} (10).

[b]T_e relative to $Xe(^1S) + O(^3P_2)$.

[c]Observed in emission from electron-beam excited high-density Xe + O_2 mixtures (8) and as laser-excited fluorescence in rare gas matrices (13).

[d]The b state vibrational numbering (8) is uncertain; according to (13) it may have to be increased by +1.

[e]Lifetime $\tau(v=0,1) \approx 50$ μs (7); (9) suggest the much shorter value $\tau \approx 0.1$ μs.

[f]Observed in an ozonizer-type discharge through Xe (~1 atm) + traces of O_2 (3); partially resolved rotational structure. Laser-excited fluorescence in rare gas matrices (13). Laser oscillation observed using electron-beam excitation of high-pressure Xe containing small amounts of O_2 (6)(11).

XeO (continued):

[g]The a state vibrational numbering (3) may have to be raised by +3 units; an unambiguous numbering has not been established in spite of the availability of $Xe^{18}O$ data in solid argon (13).

[h]Narrow emission peak; in solid argon and neon at 31940 and 32670 cm^{-1}, respectively.

(1) Kenty, Aicher, Noel, Poritsky, Paolino, PR 69, 36 (1946).
(1a)Herman, CR 222, 492 (1946).
(1b)Herman, Herman, JPR 11, 69 (1950).
(2) Cuthbertson, Herman, CR 234, 1355 (1952).
(3) Cooper, Cobb, Tolnas, JMS 7, 223 (1961).
(4) Stedman, Setser, JCP 52, 3957 (1970).
(5) Golde, Thrush, CPL 29, 486 (1974).
(6) Powell, Murray, Rhodes, APL 25, 730 (1974).
(7) Tisone, JCP 60, 3716 (1974).
(8) Lorents, Huestis, in Lecture Notes in Physics, Vol. 43: Laser Spectroscopy, p. 100 (Springer, Berlin 1975).
(9) Aleksandrov, Vinogradov, Lugovskii, Podmoshenskii, OS (Engl. Transl.) 41, 224 (1976).
(10) Ault, Andrews, CPL 43, 350 (1976).
(11) Basov, Babeiko, Zuev, Mikheev, Orlov, Pogorelskii, Stavrovskii, Startsev, Yalovoi, SJQE 6, 505 (1976).
(12) Dunning, Hay, JCP 66, 3767 (1977).
(13) Goodman, Tully, Bondybey, Brus, JCP 66, 4802 (1977).

State	T_e	ω_e	$\omega_e x_e$	B_e	α_e	D_e ($10^{-7}cm^{-1}$)	r_e (Å)	Observed Transitions Design.	ν_{00}	References
$^{89}Y_2$		$\mu = 44.452933_4$		$D_0^0 = 1.6_2$ eV[a]						JUL 1975
$^{(172,174)}Yb_2$		($\mu = 86.465923$)		$D_0^0 = 0.1_7$ eV[a]						JUL 1975
$^{(174)}Yb^{(35)}Cl$		($\mu = 29.1154527$)								JUL 1975 A
B ($^2\Sigma$)[a]	19928.0	315.0 H	2.0	(fragment only)				B→X, V	19939.1 H	(1)
A_2 ($^2\Pi$)[a]	19369.3	314.6 H	1.38					A_2→X, V	19380.0 H	(1)*
A_1	17882.0[b]	314.4 H	1.12[c]					A_1→X, V	17892.1 H	(1)*
X ($^2\Sigma$)[a]	0	293.6[d] H	1.23[d]							
$^{(174)}Yb^{19}F$		($\mu = 17.1276445$)		$D_0^0 = 4.8$ eV[a]						SEP 1976 A
B ($^2\Sigma^+$) $\frac{1}{2}$	(21067)	[511.75] H	b	[0.2486][c]			[1.990]	B↔X, V	21074.25[d] H	(2)
A_2 ($^2\Pi$) $\frac{3}{2}$	(19460)	[540.35] H	b	[0.24863]		[3.16][e]	[1.9896]	A_2↔X, V	19470.99[f] Z	(2)*
A_1 $\frac{1}{2}$	(18090)	[473.72] H	b	[0.2470_4][g]		[2.1_4]	[1.996_0]	A_1↔X, V	18106.28[f] Z	(2)
X $^2\Sigma^+$	0	[501.91] Z	2.20_5 H	[0.24140]	0.0015	[2.18]	2.0161			

Y_2: [a]Thermochemical value (mass-spectrom.)(1).

(1) Verhaegen, Smoes, Drowart, JCP 40, 239 (1964).

Yb_2: [a]Thermochemical value (mass-spectrom.)(1).

(1) Guido, Balducci, JCP 57, 5611 (1972).

YbCℓ: [a]Assignments by analogy with YbF (2).
[b](1) give 17800.9 which does not fit their data.
[c]$\omega_e y_e = +0.0072$.
[d]Average of the constants from $A_1 - X$ and $A_2 - X$.

(1) Gatterer, Piccardi, Vincenzi, RS 1, 181 (1942).

(2) See ref. (2) of YbF.

YbF: [a]Corrected (1) linear Birge-Sponer extrapolation for the ground state (2). See, however, (3) who derive $D_0^0 \geq 5.3_6$ eV from the $Yb + F_2$ chemiluminescence spectrum.
[b]Higher ΔG values in (2); strong perturbations.
[c]Large perturbations.
[d]Head nearest to origin.
[e]$H_0 = +4._1 \times 10^{-12}$.
[f]$\{J'=0\}$ relative to $N''=0$.
[g]Ω-type doubling $\Delta\nu_{fe} = -0.40(J+\frac{1}{2})$.

(1) Hildenbrand, AdHTC 1, 193 (1967).

(2) Barrow, Chojnicki, JCS FT II 71, 728 (1975).

(3) Yokozeki, Menzinger, CP 14, 427 (1976).

State	T_e	ω_e	$\omega_e x_e$	B_e	α_e	D_e ($10^{-5}cm^{-1}$)	r_e (Å)	Observed Transitions Design.	Observed Transitions ν_{00}	References
$^{(174)}Yb^1H$		(μ = 1.00201939)		$D_0^0 \lessapprox$ (1.93) eV[a]						JAN 1976 A
F $\frac{1}{2}$[b]				[4.155][c]		[16.6]	[2.012]	F↔X, V	22002.39[d] Z	(2)(6)
E ($^2\Sigma^+$) $\frac{1}{2}$				[4.2497][e]		[19.7]	[1.9897]	E↔X, V	17822.56[d] Z	(2)* (3)(6)
D ($^2\Pi$) $\frac{3}{2}$	16780	[1319.7] Z	(19.4)	4.1609[f]	0.088_6	[14.68][g]	2.0108	D↔X, V	16834.72[d] Z	(1)(2)
D ($^2\Pi$) $\frac{1}{2}$	15325.6	[1307.5_4] Z	21.0_2[h]	4.1950[i]	0.089_8[h]	[15.33][j]	2.0026		15375.58[d] Z	(1)(2)* (3)(6)
C ($\Omega > \frac{1}{2}$)	≤ 15200	≥1500[k]	(30)	(4.4)[k]						(2)
B ($\frac{1}{2}$)	≲15400	≳1370[ℓ]		(3.2)[ℓ]						(2)(4)
a $^4\Sigma^+_{3/2}$, $^4\Sigma^+_{1/2}$		Not observed.		[4.672_3][m]		[$7._1$][n]	[1.8976]	a→X,	13183.92[d] Z	(5)(7)
X $^2\Sigma^+$	0	1249.54 Z	21.06	3.9930_5[o]	0.0956_5[p]	16.18[q]	2.0526			
$^{(174)}Yb^2H$		(μ = 1.99104715)								JAN 1976 A
F $\frac{1}{2}$[a]				[2.1289][b]		[4.24]	[1.9943]	F→X,[c] V	21986.20[d] Z	(2)* (3)(6)
E ($^2\Sigma^+$) $\frac{1}{2}$	17768.6	968.65 Z	24.41	[2.1425][e]	f	[-8][g]	[1.9879]	E→X, V	17806.16[d] Z	(2)(3)(6)
D ($^2\Pi$) $\frac{3}{2}$	16778	[945.50] Z	(15.6)	[2.09241][h]	0.0472	[3.962][i]	2.0003	D→X, V	16822.18[d] Z	(2)
D ($^2\Pi$) $\frac{1}{2}$	15326.4	957.9 Z	9.7	2.122[j]	0.039	[4.18][k]	1.997		15362.3[d] Z	(2)(7)
a $^4\Sigma^+_{3/2}$	13008.5	1084.4 Z	9.84	2.3516[ℓ]	0.0371	3.40[m]	1.8975	a→X, V	13107.60[d] Z	(5)(7)*
a $^4\Sigma^+_{1/2}$				[2.1931][n]		[4.8][o]	[1.9648]		12493.15[d] Z	(3)(4)* (7)
X $^2\Sigma^+$	0	886.6 Z	10.57	2.01162[p]	0.03425[q]	[4.160][r]	2.0516			

Yb^1H: [a]From the predissociation in D ½ (v=0). The fact that a $^4\Sigma_{1/2}$ was not observed for Yb^1H may be due to predissociation; in this case $D_0^0 \lessapprox 1.55$ eV.

[b]Veseth (6) suggests $\Omega = 3/2$.

[c]Ω-type doubling, $\Delta v_{fe} = - 0.23(J+\frac{1}{2}) + \ldots$ (2). Different constants in (6). Perturbations.

[d]$\{J'=0\}$ relative to N"=0.

[e]Large Ω-type doubling, $\Delta v_{fe} = + 13.60(J+\frac{1}{2}) - \ldots$. f levels perturbed by v=2 of D ½ (6).

[f]Perturbations. Ω-type doubling, $\Delta v_{fe}(v=0) = + 5.19 \times 10^{-4}(J-\frac{1}{2})(J+\frac{1}{2})(J+\frac{3}{2}) - \ldots$.

[g]$D_1 = 16.5 \times 10^{-5}$; $H_1 = + 2.8 \times 10^{-8}$.

[h]Data for v=2 (B = 3.958) from a perturbation in the f levels of E, v=0. See (6).

[i]Ω-type doubling, $\Delta v_{fe}(v=0) = - 4.89(J+\frac{1}{2}) + \ldots$. Perturbations. An accidental predissociation occurs in v=0 near the intersection with a level of B (½).

[j]$D_1 = 15.7 \times 10^{-5}$.

[k]From perturbations in the ½ and 3/2 components of D. Vibrational numbering unknown.

[ℓ]From perturbations in the e levels of D ½, v=0 and 1. Vibrational numbering unknown.

[m]Ω-type doubling, $\Delta v_{fe} = + 0.0285_8(J-\frac{1}{2})(J+\frac{1}{2})(J+\frac{3}{2}) - \ldots$.

[n]Also higher order constants.

[o]Spin splitting $\Delta v_{12}(v=0) = + 0.5688(N+\frac{1}{2}) - \ldots$.

[p]$\gamma_e = - 0.0010_8$.

[q]$\beta_e = + 0.18 \times 10^{-5}$; $H_0 = + 4.02 \times 10^{-9}$, $H_1 = + 4.39 \times 10^{-9}$, $H_2 = + 4.6 \times 10^{-9}$.

For ref. see Yb^2H.

Yb^2H: [a]See [b] of Yb^1H.

[b]Ω-type doubling, $\Delta v_{fe} = - 0.231(J+\frac{1}{2}) + \ldots$. Perturbations.

[c]One P, one R, and two Q branches.

[d]See [d] of Yb^1H.

[e]Ω-type doubling, $\Delta v_{fe}(v=0) = + 6.989(J+\frac{1}{2}) - \ldots$ (2). Slightly different constants in (6). Perturbations.

[f]$B_1 = 2.104$, $B_2 = 2.013$.

[g]$H_0 = - 34 \times 10^{-8}$. Additional D_v, H_v values in (2).

[h]Perturbations. Ω-type doubling, $\Delta v_{fe}(v=0) = + 0.682 \times 10^{-4}(J-\frac{1}{2})(J+\frac{1}{2})(J+\frac{3}{2}) - \ldots$; for v=1 see (2).

[i]$D_1 = 6.0 \times 10^{-5}$; $H_0 = + 5.0 \times 10^{-10}$, $H_1 = + 4.9 \times 10^{-8}$.

[j]Ω-type doubling, $\Delta v_{fe}(v=0) = - 2.590(J+\frac{1}{2}) + \ldots$. Perturbations by levels of $^4\Sigma^+_{3/2}$.

[k]D_1, $D_2(10^{-5}cm^{-1}) = 3.80$, 3.3_6; and higher order constants.

[ℓ]Ω-type doubling, $\Delta v_{fe}(v=0) = + 0.00363_4(J-\frac{1}{2})(J+\frac{1}{2})(J+\frac{3}{2}) - \ldots$.

[m]$\beta_e \approx - 0.09$; also higher order constants.

[n]Large Ω-type doubling, $\Delta v_{fe}(v=0) = - 9.051(J+\frac{1}{2}) + \ldots$

[o]Also higher order constants.

[p]Spin splitting $\Delta v_{12}(v=0) = + 0.2876(N+\frac{1}{2}) - \ldots$.

[q]$\gamma_e = - 0.000320$.

[r]$D_1 = 4.174 \times 10^{-5}$, $H_e \approx + 6.3 \times 10^{-10}$.

(1) Kopp, Naturw. 49, 202 (1962).
(2) Hagland, Kopp, Åslund, AF 32, 321 (1966).
(3) Kopp, Hougen, CJP 45, 2581 (1967).
(4) Hagland, Kopp, AF 39, 257 (1969).
(5) L. Hagland, Thesis (Stockholm, 1969), quoted in DONNSPEC (1970).
(6) Veseth, JP B 6, 1484 (1973).
(7) Kopp, Hagland, Rydh, CJP 53, 2242 (1975); USIP Report 75 - 15 (July 1975).

State	T_e	ω_e	$\omega_e x_e$	B_e	α_e	D_e ($10^{-7}cm^{-1}$)	r_e (Å)	Observed Transitions		References
								Design.	ν_{00}	
$^{(174)}Yb^{16}O$		(μ = 14.6479335$_3$) $D_0^0 \le 3.6_8$ eV[a] or ≥ 4.08 eV[b].								SEP 1976 A
		Unclassified, mostly R shaded emission bands in the region 17700 - 21500 cm^{-1}.								(1)* (2)*
$^{89}Y^{35}Cl$		μ = 25.0974229								JUL 1975
C $^1\Sigma$	14907.6	324.5 H	1.14	(0.1089)[a]	(0.0007)	(0.9)[a]	(2.48$_4$)	C→X, R	14879.5 H	(1)
X $^1\Sigma$ b	0	380.7 H	1.3	(0.1160)[a]	(0.0003)	(0.9)[a]	(2.40$_6$)			
$^{89}Y^{19}F$		μ = 15.6534084 D_0^0 = 6.2 eV[a]								DEC 1976 A
G $^1\Pi$	31253.6	[536.30] (Z)	2.1$_3$	0.27661	0.00233	[2.96]	1.9731	G↔X, R	31205.80 Z	(1)(2)(3)(6)
F $^1\Sigma^+$	28022.5	552.9 H	2.6$_9$	[0.27536]		[2.76]	[1.9776]	F↔X, R	27980.81 Z	(2)(4)(7)
E $^1\Pi$	b	[581.92] (Z)		[0.27090][c]	0.0024	[2.5]	[1.9930]	E↔X, (R)	25464.33 Z	(3)(4)(6)
D $^1\Pi$	b			[0.26805][d]		[3.3]	[2.0044]	D ←X, R	25324.90 Z	(3)(6)
C $^1\Sigma^+$	19242.4	[527.20] (Z)	2.45	0.26666	0.00177	[2.64]	2.0096	C↔X, R	19190.35 Z	(2)(3)(6)
b $^3\Phi_4$ e $^3\Phi_3$ $^3\Phi_2$	a_3 + 15051.7 a_2 + 14865.9 a_1 + 14658.1	536.1 H	2.41	[0.277]			[1.97$_2$]	b↔a, R	15028.0 H 14842.8 H 14635.5 H	(2)(3)
B $^1\Pi$	15934.2	[534.67] (Z)	2.35[f]	0.26709[g]	0.00156	[2.61]	2.0080	B↔X, R	15885.78 Z	(1)(2)(3)(4)(6)
a $^3\Delta_3$ e $^3\Delta_2$ $^3\Delta_1$	a_3 a_2 a_1	583.5 H 582.3 H 581.2 H	2.49 2.42 2.39	[0.285]			[1.94$_4$]			
X $^1\Sigma^+$	0	[631.29] (Z)	2.50	0.29042	0.00163	[2.37]	1.9257			
$^{89}Y^{19}F^+$										DEC 1976
$^2\Pi_{(r)}$	x_2 + 28062.4 x_1 + 27964.0	716.6 H [711.5$_3$] Z	2.38 2.41 H	[0.3065] [0.3055][h]	0.0015	[2.0] [2.2]	[1.876$_0$]	$^2\Pi \rightarrow {}^2\Delta$, V	28088.37 Z 27990.3$_4$ Z	(7)
$^2\Delta_{(r)}$	x_2 x_1	664.7 H 663.7 H	2.30 2.29	[0.2960] [0.2953]		[2.2] [2.2]	[1.908$_6$]			

YbO: [a]Thermochemical value (mass-spectrom.)(3), recalc. (4).
[b]From the $Yb + O_3$ chemiluminescence spectrum (5).

(1) See ref. (1) of YbCℓ.
(2) Gatterer, Junkes, Salpeter, Rosen, METOX (1957).
(3) Ames, Walsh, White, JPC 71, 2707 (1967).
(4) Smoes, Coppens, Bergman, Drowart, TFS 65, 682 (1969).
(5) See ref. (3) of YbF.

YCℓ: [a]Uncertain. The D values from the rotational analysis are approximately twice as large as the values calculated from $D = 4B^3/\omega^2$.
[b]Not certain that this is the ground state.

(1) Janney, JOSA 56, 1706 (1966).

YF, YF^+: [a]Thermochemical value (mass-spectrom.)(5).
[b]The absorption spectrum in this region contains several overlapping bands, and it is not certain how many states are involved (3)(6). Perturbations.
[c]Λ-type doubling (6), $\Delta\nu = 0.0007 \times J(J+1)$.
[d]Λ-type doubling (6), $\Delta\nu = 0.00049 \times J(J+1)$.
[e]Assignments by analogy with ScF.
[f]1.61 in (2)(4).
[g]Λ-type doubling (3)(6), $\Delta\nu_{ef} = (-)0.00013_2 \times J(J+1)$.
[h]Λ-type doubling, $\Delta\nu = 0.0079(J+\frac{1}{2})$.

(1) Barrow, Gissane, PPS 84, 615 (1964).
(2) Shenyavskaya, Mal'tsev, Gurvich, OS(Engl. Transl.) 21, 374 (1966).
(3) Barrow, Bastin, Moore, Pott, Nature 215, 1072 (1967).
(4) Shenyavskaya, Mal'tsev, Gurvich, VMUK 22(4), 104 (1967).
(5) Zmbov, Margrave, JCP 47, 3122 (1967).
(6) R. F. Barrow, in DONNSPEC (1970).
(7) Shenyavskaya, Ryabov, JMS 63, 23 (1976).

State	T_e	ω_e	$\omega_e x_e$	B_e	α_e	D_e ($10^{-7}cm^{-1}$)	r_e (Å)	Observed Transitions Design.	ν_{00}	References
$^{89}Y^{16}O$		$\mu = 13.5560646_9$		$D_0^0 = 7.2_9$ eV[a]						JUL 1975 A
B $^2\Sigma^+$	20791	765.5 H	$8._0$	[0.3722][b]		[3.8]	[1.827_9]	B↔X, R	20741.9_2 Z	(1)(2)(3)(4)* (5)* (6)* (7)* (9)(11)(16)(17)(18)(19)
A $^2\Pi_r$	16742.2 16315.0	822.7[c] 820.7[c]	3.9_7 3.5_0	[0.3857][d]	(0.0023)	[3.5]	[1.795_6]	A↔X, R	16722.7_5[e] Z 16294.7_2[e] Z	(1)(2)* (3)(4)* (5)* (6)* (7)* (9)(11)(17)(18)
X $^2\Sigma^+$	0	861.0[c]	2.9_3	[0.3881][f]	(0.0018)	[3.2]	[1.790_0]	ESR sp.[g]		
$^{89}Y^{(32)}S$		($\mu = 23.5154972$)		$D_0^0 = 5.4_5$ eV[a]						JUL 1975
A $^2\Pi$	13971 13462	[438] [447]						A← X,[b]	13949 13444	(2)
X $^2\Sigma$	0	[483]						b		
$^{89}Y^{(80)}Se$		($\mu = 42.085933_4$)		$D_0^0 = 4.4_9$ eV[a]						JUL 1975
$^{89}Y^{(130)}Te$		($\mu = 52.782392_8$)		$D_0^0 = 3.4_8$ eV[a]						JUL 1975

YO: [a]Thermochemical value (mass-spectrom.)(20), see also (8) (10)(12)(13)(14)(15).

[b]Spin-splitting constant γ_0 = (-)0.148.

[c]Redetermined, taking into account estimated head-origin separations.

[d]Λ-type doubling in $^2\Pi_{\frac{1}{2}}$, $\Delta v_{fe}(v=0) = -0.151(J+\frac{1}{2})$.

[e]{J'=0} relative to N"=0.

[f]Nuclear magnetic hyperfine structure; $b = -0.027$ cm^{-1} [from the ESR spectrum, (11)(18)].

[g]In rare gas matrices (11)(18).

(1) Johnson, Johnson, PRS A 133, 207 (1931).
(2) Meggers, Wheeler, JRNBS 6, 239 (1931).
(3) Jevons, "Band Spectra of Diatomic Molecules", Phys. Soc. London (1932).
(4) Piccardi, GCI 63, 127 (1933).
(5) Gatterer, RS 1, 153 (1942).
(6) See ref. (2) of YbO.
(7) Uhler, Åkerlind, AF 19, 1 (1961).
(8) White, Walsh, Ames, Goldstein, in "Thermodynamics of Nuclear Materials", p. 417. IAEA, Vienna (1962).
(9) Ortenberg, Glasko, SAAJ 6, 714 (1963); Ortenberg, Glasko, Dimitriev, SAAJ 8, 258 (1964).
(10) Ackermann, Rauh, Thorn, JCP 40, 883 (1964).
(11) Kasai, Weltner, JCP 43, 2553 (1965).
(12) Smoes, Drowart, Verhaegen, JCP 43, 732 (1965).
(13) See ref. (3) of YbO.
(14) Coppens, Smoes, Drowart, TFS 63, 2140 (1967).
(15) Drowart, Pattoret, Smoes, PBCS No. 8, 67 (1967).
(16) Murthy, Murthy, PPS 90, 881 (1967).
(17) Veits, Gurvich, DC 173, 377 (1967).
(18) Weltner, McLeod, Kasai, JCP 46, 3172 (1967).
(19) Dube, Rai, Singh, IJPAP 10, 87 (1972); Dube, IJPAP 10, 167 (1972).
(20) Ackermann, Rauh, JCP 60, 2266 (1974).

YS: [a]Thermochemical value (mass-spectrom.)(1).

[b]Visible, IR, and ESR spectra of YS in rare gas matrices at 4 K (2). Gas phase spectrum not observed.

(1) See ref. (14) of YO.
(2) McIntyre, Lin, Weltner, JCP 56, 5576 (1972).

YSe: [a]Thermochemical value (mass-spectrom.)(1).

(1) Ni, Wahlbeck, HTS 4, 326 (1972).

YTe: [a]Thermochemical value, no details (1).

(1) Uy, Drowart, HTS 2, 293 (1970).

State	T_e	ω_e	$\omega_e x_e$	B_e	α_e	D_e	r_e (Å)	Observed Transitions Design.	Observed Transitions ν_{00}	References
$^{(64,66)}Zn_2$		(μ = 32.456117$_9$) Large number of continua and diffuse bands in emission and absorption. See (1).								JUL 1975 A
$^{(64)}Zn^{(79)}Br$		(μ = 35.318658$_6$)								JUL 1975
C ($^2\Pi$)	32523 32125	358.0[a] H 350.0[a] H	2.00 2.00					C↔X, b	32543[a] H 32141[a] H	(1)(2)(3)(5) (7)(8)
B ($^2\Sigma$)		Unclassified bands from 11800 to 30000 cm^{-1}.[c]						B→X, R		(2)(4)(6)
X ($^2\Sigma$)	0	318.0[a] H	2.00							
$^{(64)}Zn^{(35)}Cl$		(μ = 22.6043892)		D_0^0 = 2.1 eV[a]						JUL 1975
E ($^2\Sigma$)	(48186.4)	(345.4)	(5.0)					E→X,[b] R	48163.0 (Z)	(4)
C ($^2\Pi$)	33977.9 33593.4	381.8 H 384.0 H	1.0 1.1					C↔X, R	33973.7 H 33590.2 H	(1)(4)
B ($^2\Sigma$)	(27316)	(185.$_0$) H	(0.5$_3$)					B→X, R	(27213) H	(2)(5)(6)
X ($^2\Sigma$)	0	390.5 H	1.5$_5$							
$^{(64)}Zn^{133}Cs$?		(μ = 43.165842$_1$) Diffuse V shaded absorption bands at 19363 and 19503 cm^{-1}.								JUL 1975 (1)
$^{(64)}Zn^{19}F$		(μ = 14.6459379)								JUL 1975
D ($^2\Sigma$)								D←X,[a] R	(38633) H	(1)
C ($^2\Pi$)	(37359) (36987)	([596.8]) H^Q ([601.5]) H^Q						C←X,[a] R	(37343.9) H^Q (36974.2) H^Q	(1)
X ($^2\Sigma$)	0	(628) H	(3.5)							

Zn_2: (1) W. Finkelnburg, "Kontinuierliche Spektren", Springer (Berlin, 1938).

ZnBr: [a]From the analysis (8) of the absorption spectrum. An earlier analysis (3) of absorption measurements by (1) gave $\omega' \approx 250$, $\omega'' \approx 220$.
[b]Shading uncertain.
[c]Suggested vibrational constants in (6).

(1) Walter, Barratt, PRS A 122,201 (1929).
(2) Wieland, HPA 2, 46, 77 (1929).
(3) Howell, PRS A 182, 95 (1943).
(4) Wieland, in "Contribution à l'Etude de la Structure Moléculaire", Vol. comm. Victor Henri, Ed. Desoer, Liège (1948); p. 229.
(5) Ramasastry, Sreeramamurty, PNISI 16, 305 (1950).
(6) Patel, Rajan, IJP 42, 125 (1968).
(7) Rajan, Shah, IJPAP 7, 61 (1969).
(8) Gosavi, Greig, Young, Strausz, JCP 54, 983 (1971).

$ZnC\ell$: [a]Na D line chemiluminescence in $Na/ZnC\ell_2$ flames [(3), revised].
[b]0-0 sequence only.

(1) See ref. (1) of ZnBr.
(2) See ref. (2) of ZnBr.
(3) Horn, Polanyi, Sattler, ZPC B 17, 220 (1932).
(4) Cornell, PR 54, 341 (1938).
(5) See ref. (4) of ZnBr.
(6) Patel, Rajan, IJPAP 5, 330 (1967).

ZnCs: (1) Barratt, TFS 25, 758 (1929).

ZnF: [a]Diffuse bands, predissociation.

(1) Rochester, Olsson, ZP 114, 495 (1939).

State	T_e	ω_e	$\omega_e x_e$	B_e	α_e	D_e $(10^{-4} cm^{-1})$	r_e (Å)	Observed Transitions Design.	ν_{00}	References
$^{64}Zn^1H$		μ = 0.99218372		$D_0^0 = 0.85_1$ eV[a]						JUL 1975 A
C $^2\Sigma^+$	41090	1824 H	48	$[7.23]$[b]		[4.7]	$[1.53_3]$	C←X, V	41200 H	(6)
B $^2\Sigma^+$	$27587._7$	$1020._7$ Z	$16._5$	[3.288][c]		[1.40]	[2.273]	B→X, R	27303.9 Z	(2)
A $^2\Pi_r$	23276.9[d]	1910.2 Z	40.8	7.433_2[e]	0.238_5	$[4.48_2]$[f]	1.5119	A→X, V	23431.8[g] Z	(1)(2)(3)(4)(5)(9)
X $^2\Sigma^+$	0	1607.6 Z	55.14[h]	6.6794[i]	0.2500[j]	[4.66][k]	1.5949_0	ESR sp.[ℓ]		
$^{(64)}Zn^2H$		(μ = 1.95258556)		$D_0^0 = 0.87_9$ eV[m]						JUL 1975 A
C $^2\Sigma^+$	(41110)	1313 H	24					C←X, V	41204 H	(6)
A $^2\Pi_r$	n			[3.736][o]		$[1.3_1]$[p]	$[1.520_2]$	A→X, V	23391.5[g] Z	(3)(9)
X $^2\Sigma^+$	0	[1072] H	(28)	$[3.349_7]$[q]		$[1.24_0]$	[1.6054]			
$^{(64)}Zn^1H^+$				D_0^0 = (2.5) eV						JUL 1975
A $^1\Sigma^+$	46700	1365 Z	15	5.767	0.105	4.0	1.716	A→X, R	46431 Z	(1)(2)
X $^1\Sigma^+$	0	1916 Z	39[a]	7.407	0.238	4.8	1.515			
$^{(64)}Zn^2H^+$										JUL 1975
A $^1\Sigma^+$	46693.9	974.4 Z	7.6	2.928	0.042	1.0	1.717	A→X, R	46501.7 Z	(3)
X $^1\Sigma^+$	0	1364.8 Z	19.8	3.766	0.107	1.0	1.514			

Zn^1H, Zn^2H:

[a]Short extrapolation for the ground state.

[b]All lines diffuse; predissociation.

[c]All observed vibrational levels of B $^2\Sigma$ show perturbations by A $^2\Pi$ levels with $v \geq 1$. The rotational constants in (2) are satisfactorily reproduced by $B_v = 3.304 - 0.033(v+\frac{1}{2}) + 0.00060(v+\frac{1}{2})^2 - 0.00024_5(v+\frac{1}{2})^3$, except for v=1, 2, and 3 all of which are close to the intersection of the two potential curves.

[d]$A_0 = +342.66$, $A_1 = +342.06$ [as recalculated in (9) from the data of (2)].

[e]Rotational constants for v=0 and 1 as recalculated in (9) from the data of (2). Λ-type doubling in $^2\Pi_{\frac{1}{2}}$, $\Delta v_{fe}(v=0) \approx +0.26_9(J+\frac{1}{2}) - \ldots$, and in $^2\Pi_{3/2}$ [see (3) and (2)].

[f]$D_1 = 4.46 \times 10^{-4}$; $H_0 = +1.3_4 \times 10^{-8}$, $H_1 = +0.9 \times 10^{-8}$.

[g]$J'=\frac{1}{2}$ (average of F_1 and $\{F_2\}$) relative to N"=0. A different definition was used in (2).

[h]$\omega_e y_e = +0.398$, $\omega_e z_e = -0.4339$. Very rapid convergence. The constants [from (2)] lead to a discrepancy of nearly 6 cm^{-1} for the highest observed level, v=5.

[i]Spin doubling, $\Delta v_{12}(v=0) = +0.254(N+\frac{1}{2}) - \ldots$ [(2), see also (7)], decreasing rapidly with increasing v.

[j]$- 0.03765(v+\frac{1}{2})^2 + 0.00897(v+\frac{1}{2})^3 - 0.001479(v+\frac{1}{2})^4$.

[k]$D_1, \ldots, D_5(10^{-4} cm^{-1}) = 5.00, 5.49, 6.58, 8.40, 10.5$. Also higher order terms, see (2).

[ℓ]In Ar matrix at 4 K (8).

[m]From the value for Zn^1H.

[n]$A_0 = +342.82$, as recalculated in (9) from the data of (3).

[o]Rotational constants recalculated in (9) from data in (3). Λ-type doubling in $^2\Pi_{\frac{1}{2}}$, $\Delta v_{fe} \approx +0.17_0(J+\frac{1}{2}) - \ldots$, see (3).

Zn^1H, Zn^2H (continued):

[p]$H_0 = +1._6 \times 10^{-8}$.

[q]Spin doubling $\Delta v_{12} = +0.131(N+\frac{1}{2})$, see (3).

(1) Watson, PR 36, 1134 (1930).

(2) G. Stenvinkel, Dissertation (Stockholm, 1936). Ref. to earlier work are reviewed in this paper.

(3) Fujioka, Tanaka, Sci. Pap. IPCR (Tokyo) 32, 143 (1937).

(4) Stenvinkel, Svensson, Olsson, AMAF 26, No. 10, 1 (1939).

(5) Mrozowski, PR 58, 597 (1940).

(6) Khan, PPS 80, 599 (1962).

(7) Veseth, JP B 3, 1677 (1970).

(8) Knight, Weltner, JCP 55, 2061 (1971).

(9) Veseth, JMS 38, 228 (1971).

Zn^1H^+, Zn^2H^+:

[a]$\omega_e y_e = -0.2$.

(1) Bengtsson, Grundström, ZP 57, 1 (1929).

(2) Bengtsson-Knave, NARSSU Ser. IV, 8, No.4 (1932).

(3) Gabel, Zumstein, PR 52, 726 (1937).

State	T_e	ω_e	$\omega_e x_e$	B_e	α_e	D_e	r_e (Å)	Observed Transitions Design.	Observed Transitions ν_{00}	References
$^{(64)}Zn^{127}I$		(μ = 42.512918$_7$)								JUL 1975
E	44115	142 H	3					E→X, R	44073 H	(5)
D	(39911)	(80)[a] H	(1.3)					D→X,	(39839)[a] H	(7)
C_2 ($^2\Pi$)	30125.8[b]	248.2 H	0.72					C_2↔X,[c] V	30138.2[b] H	(2)(8)*
C_1	29498.9	272.0 H	0.50					C_1←X, V	29523.2 H	(8)*
B ($^2\Sigma$)		Bands in the region 16000 - 28600 cm^{-1}; no analysis.						B→X, R		(1)(2)(3)(6)
X ($^2\Sigma$)	0	223.4 H	0.6$_3$[d]							
$^{(64)}Zn^{(115)}In$		(μ = 41.075781$_9$)								JUL 1975
		Two narrow emission continua near 22160 and 24370 cm^{-1}.								(1)
D ($^2\Sigma$)	c + 17732.0	107.0 H	1.0					D→C,[a] V	17757.2 H	(1)
C ($^2\Sigma$)	c	56.1 H	0.1							
B ($^2\Pi$)	a + 18831.2 18810.8	193.9[b] H 201.2 H	0.6 0.6					B→A, V	18854.8 H 18838.0 H	(1)*
A ($^2\Sigma$)	a	146.7 H	0.7							
$^{(64)}Zn^{(39)}K$?		(μ = 24.2088389)								JUL 1975
		V shaded diffuse absorption band at 24107 cm^{-1}.								(1)
$^{(64)}Zn^{16}O$		(μ = 12.7939099$_4$) $D_0^0 \leq 2.8_2$ eV[a]								JUL 1975
		Unclassified, mostly R shaded emission bands in the region 17300 - 20500 cm^{-1}.								(4)
$^{(64)}Zn^{(85)}Rb$?		(μ = 36.470733$_9$)								JUL 1975
		Unclassified absorption bands in the region 22500 - 24000 cm^{-1}.								(1)

ZnI: [a]Only bands with $v'' \geq 15$ and $v' \leq 2$; neither the analysis nor the assignment to ZnI appear certain.

[b]Wieland (2) gives T_e = 30117.6, ν_{00} = 30129.5.

[c]According to (8) most of the diffuse emission bands of system D of (4) in the region 30500 - 31300 cm^{-1} fit well into the $C_2 - X$ system.

[d]Average value from $C_1 - X$ and $C_2 - X$.

(1) Terenin, ZP 44, 713 (1927).
(2) See ref. (2) of ZnBr.
(3) Oeser, ZP 95, 699 (1935).
(4) Rao, Rao, IJP 20, 49 (1946).
(5) Ramasastry, IJP 22, 119 (1948)
(6) See ref. (4) of ZnBr.
(7) Ramasastry, IJP 23, 35 (1949).
(8) See ref. (8) of ZnBr.

ZnIn: [a]System of complex appearance; more recently (2) attributed to $ZnIn_2$.

[b]0-0 sequence only. It is assumed that $A(^2\Sigma)$ is the lower state.

(1) Santaram, Winans, PR A 136, 57 (1964).
(2) See ref. (1) of ZnTℓ.

ZnK, ZnRb:

(1) See ref. (1) of ZnCs.

ZnO: [a]Thermochemical value (mass-spectrom.)(1); (2)(3).

(1) Anthrop, Searcy, JPC 68, 2335 (1964).
(2) Hirschwald, Stolze, Stranski, ZPC (Frankfurt am Main) 42, 96 (1964).
(3) See ref. (4) of ZnS.
(4) Pešić, CCA 38, 313 (1966).

ZnRb: See ZnK.

State	T_e	ω_e	$\omega_e x_e$	B_e	α_e	D_e	r_e (Å)	Observed Transitions Design.	ν_{00}	References
$^{(64)}Zn^{(32)}S$	(μ = 21.3130477) $D_0^0 = 2.0_8$ eV[a]									JUL 1975
	The origin of two absorption continua with long wavelength limits at 35700 and 46500 cm^{-1}, formerly attributed to ZnS (1), appears doubtful; see (3)(4).									
$^{(64)}Zn^{(80)}Se$	(μ = 35.517196_3) $D_0^0 = 1.3_7$ eV[a]									JUL 1975
$^{(64)}Zn^{(130)}Te$	(μ = 42.844572_7) $D_0^0 = 0.9_5$ eV[a]									JUL 1975
	The origin of two absorption continua with long wavelength limits at 17830 and 31480 cm^{-1}, formerly attributed to ZnTe (1), appears doubtful; see (2).									
$^{(64)}Zn^{(205)}Tl$	(μ = 48.730625_4)									JUL 1975
	Emission continuum, 25400 - 26500 cm^{-1}.									(1)
	Broad, intense emission band at 21360 cm^{-1}, followed by weaker V shaded bands at 21440, 21550, 21660, 21770, 21880 cm^{-1}.									(1)
	Emission continuum, 17900 - 18700 cm^{-1}.									(1)
$^{(90)}Zr^{(79)}Br$	(μ = 42.027022_4)									JUL 1975
	Unclassified, mostly R shaded band heads in emission at 26600 - 26650 and 26910 - 26980 cm^{-1}.									(1)*
System B:	Four 0-0 sequences ($\omega'-\omega'' \approx -3.5$) of narrow R shaded heads; in em.. Possibly quartet system.								26481.0 H 26305.2 H 26133.9 H 25959.6 H	(1)*
System C:	Four groups of narrow R shaded heads, three of the groups having the appearance of long 0-0 sequences ($\omega'-\omega'' \approx +6$); in emission. $^4\Pi \rightarrow {}^4\Sigma$?								24396.0 H 24177.2 H 23936.3 H (23708)	(1)*

ZnS: [a]Thermochemical value (mass-spectrom.)(2)(3)(4).

(1) Sen Gupta, PRS A 143, 438 (1934).
(2) Colin, ICB 26, 1129 (1961).
(3) Marquart, Berkowitz, JCP 39, 283 (1963).
(4) De Maria, Goldfinger, Malaspina, Piacente, TFS 61, 2146 (1965).

ZnSe: [a]Thermochemical value (mass-spectrom.)(1).

(1) See ref. (4) of ZnS.

ZnTe: [a]Estimated thermochemical value (2).

(1) Mathur, IJP 11, 177 (1937).
(2) See ref. (4) of ZnS.

ZnTℓ: (1) Santaram, Vaidyan, Winans, JP B 4, 133 (1971).

ZrBr: (1) Sivaji, Rao, PRIA A 70, 1 (1970).

State	T_e	ω_e	$\omega_e x_e$	B_e	α_e	D_e	r_e (Å)	Observed Transitions Design.	ν_{00}	References
$^{(90)}Zr^{(35)}Cl$		(μ = 25.1763829)								JUL 1975
System A:		Three 0-0 sequences ($\omega'-\omega'' \approx +15$) of line-like heads. $\Omega'=\frac{1}{2},\frac{3}{2},\frac{5}{2} \rightarrow {}^4\Sigma$?						R	35231.7 H 34816.4 H 34354.6 H	(1)*
System B:		Very complex system of R and V shaded band heads in the region 26300 - 27600 cm^{-1}; in emission.							(26918)	(1)*
System C:		Four 0-0 sequences ($\omega'-\omega'' \approx +8$) of complex structure. $^4\Pi \rightarrow {}^4\Sigma$?						R	24704.6 H 24530.4 H 24343.8 H (24155)	(1)*
$^{(90)}Zr^{19}F$		(μ = 15.6840882)								JUL 1975
		The spectrum attributed to ZrF (1) was shown to be due to CuF (2).								
$^{(90)}Zr^{127}I$		(μ = 52.623739_5)								JUL 1975
		Groups of unclassified emission bands in the region 24800 - 26000 cm^{-1}.								(1)
System B:		Four 0-0 sequences ($\omega'-\omega'' \approx -2$) of line-like R shaded heads. Possibly quartet system.						R	25416.6 H 25276.6 H 25117.2 H 24958.6 H	(1)*
System C:		Four 0-0 sequences ($\omega'-\omega'' \approx +5$) of line-like band heads. $^4\Pi \rightarrow {}^4\Sigma$?						R	23564.8 H 23282.2 H 22995.4 H (22720)	(1)*
$^{90}Zr^{14}N$		μ = 12.1159579_3		D_0^0 = 5.8_1 eV^a						JUL 1975 A
A $^2\Pi_r$	b			[0.4798]			[1.702_9]	A→X,[c]	17701.8 H	(2)
X $^2\Sigma$	0			[0.4832]			[1.696_9]		17133.0 H	

ZrCℓ: (1) Carroll, Daly, PRIA A 61, 101 (1961).

ZrF: (1) Afaf, PPS A 63, 544, 1156 (1950).
(2) Carroll, Daly, PPS A 70, 549 (1957).

ZrI: (1) Sivaji, Rao, PRIA A 70, 7 (1970)

ZrN: [a]Thermochemical value (mass-spectrom.)(1).
[b]$A_0 \approx +567$.
[c]Perturbations.

(1) Gingerich, JCP 49, 14 (1968).
(2) J. K. Bates, T. M. Dunn (Dept. of Chemistry, U. of Michigan, Ann Arbor), abstract of paper Σ8 presented at the 27th Symposium on Molecular Structure and Spectra, Columbus, Ohio (1972).

State	T_e	ω_e	$\omega_e x_e$	B_e	α_e	D_e $(10^{-7}cm^{-1})$	r_e (Å)	Observed Transitions Design.	Observed Transitions ν_{00}	References
$^{90}Zr^{16}O$		$\mu = 13.5790678_9$ $D_0^0 = 7.85$ eV[a] I.P. = 6.1 eV[b]								AUG 1975 A
		Additional unidentified systems in gas phase emission:								
		a) Headless bands in the region 12600 - 13200 cm^{-1}; no analysis.								(6)(10)
		b) Isolated R shaded band (one R, one P branch) at 12203.1 cm^{-1}, accompanied by a weaker head at 12216.7 cm^{-1}.								(10)* (14) (16)*
		c) Mostly R shaded heads in the region 10500 - 11800 cm^{-1}; tentative analysis (10) gives $\omega_e' = 862.9$, $\omega_e'x_e' = 8.8$; $\omega_e'' = 945.4$, $\omega_e''x_e'' = 8.6$; uncertain.							10750.3 H 10731.4 H 10715.3 H 10700.1 H 10685.3 H	(2)(10)
		and in matrix absorption at 17025[c] and 19397[d] cm^{-1}.								(19)
f ($^3\Delta$)								f → a, R φ-system	33993.7 H 33888.4 H 33685.2 H	(7)(10)
e								e → (a), R δ-system	28780.3[e] H^Q 28620.1[e] H^Q 28501.7[e] H^Q	(7)(10)
E $^1\Sigma^+$	27212.4	843.2_7 Z	3.0_4	0.3951	0.0019	3.4	1.772_6	E ↔ X,[fg] R System A	27144.7_1 Z	(7)(10)(13)* (16)* (19) (21)(26)
D $^1\Delta$	y + 19321.5	[835.4] Z	2.5_6 H	0.3986	0.0021	3.8	1.764_8	D → A, R System B	19272.5_5 Z	(9)(14)* (16)* (21)
d $^3\Delta$	x + 22314.9 x + 21894.3 x + 21594.3	820.6 H	3.3_1	[0.3953] [0.3926] [0.3896]	(0.0021)	[2.4] [1.8] [3.5]	1.776_1	d ↔ a,[f] R α-system	21631.48 Z 21548.46 Z 21536.36 Z	(1)* (3)* (10)(11)* (16)* (17) (18)(20)(21) (32)
c $^3\Pi_r$	x + 18137.6 x + 18079.4 x + 18041.3	845.4 H^R	3.6_4	[0.4058][h] [0.4032] [0.3960]	(0.0023)	[3.9] [5.0] [2.4]	1.756	c ↔ a,[f] R β-system	17466.46[i] Z 17745.89[i] Z 17995.70 Z	(1)* (12)* (16)* (21) (29)(32)
b $^3\Phi_r$	x + 16700.5 x + 16070.4 x + 15468.0	853.9 H^R	3.1_4	[0.40438] [0.40368] [0.40307]	0.00191[j] 0.00210[j] 0.00198[j]	[3.64_2][k] [3.61_7][k] [3.56_2][k]	1.7514_3	b ↔ a,[f] R γ-system	16033.81[ℓ] Z 15741.31[ℓ] Z 15426.78[ℓ] Z	(1)* (4) (11)* (16)* (17)(18)(21) (22)(23)* (27)(32)

ZrO: [a]Thermochemical value (mass-spectrom.)(15)(24)(31)(34).

[b]By electron impact (30)(34).

[c]In a Ne matrix; $\Delta G'(\frac{1}{2}) = 872$. For tentative assignments of gas phase emission bands in the same region see (16).

[d]In a Ne matrix; $\Delta G'(\frac{1}{2}) = 836$.

[e]The stronger R heads are at 28512.0, 28630.4, and 28790.5 cm^{-1}.

[f]Absorption in stellar atmospheres (5)(8)(25).

ZrO (continued):

[g]Absorption in rare gas matrices (19).

[h]Λ-type doubling, see (29).

[i]$\nu_{00}(^3\Pi_1 - {}^3\Delta_1) = 18033.80$, $\nu_{00}(^3\Pi_2 - {}^3\Delta_2) = 17804.1$.

[j]$B_0 - B_1$; constants for v=2 in (27).

[k]D_v values for v=1, 2 in (27).

[ℓ]$\{J'=0\}$ relative to $\{J''=0\}$. A different definition was used in (27).

State	T_e	ω_e	$\omega_e x_e$	B_e	α_e	D_e ($10^{-7}cm^{-1}$)	r_e (Å)	Observed Transitions Design.	ν_{00}	References
$^{90}Zr^{16}O$ (continued)										
B $^1\Pi$	15443	859[m]	3	[0.40154][n]		[3.52]	[1.7583_2]	B ↔ X,[fg] R	15383.41° Z	(16)* (19) (28)*
A $^1\Delta$	y	938.1 H	1.8_0	[0.4167]	(0.0012)	[3.5]	[1.726_0]			
a $^3\Delta_r$	x + 625.5 x + 287.9_1[p] x	936.5 H^R	3.4_7	[0.41573] [0.41475] [0.41328]	0.00173[j] 0.00190[j] 0.00178[j]	[3.30_9][k] [3.26_9][k] [3.16_9][k]	1.7285			
X $^1\Sigma^+$	0	[969.7_6] Z	4.9_0	[0.42263]	0.0023	[3.19]	1.7116	IR fundamental[q]		

[m]From the matrix (Ne) absorption spectrum, in good agreement with ω = 858 from Kratzer's relation.

[n]Λ-type doubling, $\Delta\nu_{fe}$ = + 0.00024 x J(J+1).

[o]{J'=0} relative to J"=0. A different definition was used in (28).

[p]The energy of $a\ ^3\Delta$ above $X\ ^1\Sigma$ is estimated (33) as 1700 ± 250 cm^{-1} from the temperature dependence of the integral absorption coefficients for the 0-0 bands of E − X and d − a. The triplet splittings derive from the observation (29) of two satellite bands of β(0-0); see [i].

[q]In a Ne matrix at 4 K (19).

(1) Lowater, PPS 44, 51 (1932).
(2) Meggers, Kiess, JRNBS 9, 309 (1932).
(3) Lowater, PTRSL A 234, 355 (1935).
(4) Tanaka, Horie, PPMSJ 23, 464 (1941).
(5) Davis, ApJ 106, 28 (1947).
(6) Kiess, PASP 60, 252 (1948).
(7) Afaf, Nature 164, 752 (1949).
(8) Herbig, ApJ 109, 109 (1949).
(9) Afaf, PPS A 63, 674 (1950).
(10) Afaf, PPS A 63, 1156 (1950).
(11) Lagerqvist, Uhler, Barrow, AF 8, 281 (1954).
(12) Uhler, AF 8, 295 (1954).
(13) Uhler, Åkerlind, AF 10, 431 (1956).

ZrO (continued):

(14) Åkerlind, AF 11, 395 (1956).
(15) Chupka, Berkowitz, Inghram, JCP 26, 1207 (1957).
(16) Gatterer, Junkes, Salpeter, Rosen, METOX (1957).
(17) Ortenberg, SAAJ 5, 588 (1962).
(18) Ortenberg, Glasko, SAAJ 6, 714 (1963).
(19) Weltner, McLeod, Nature 206, 87 (1965); JPC 69, 3488 (1965).
(20) Singh, Pathak, PPS 90, 543 (1967).
(21) Nicholls, Tyte, PPS 91, 489 (1967).
(22) Singh, Pathak, PPS 91, 497 (1967).
(23) Schadee, Davis, ApJ 152, 169 (1968).
(24) Brewer, Rosenblatt, AdHTC 2, 1 (1969).
(25) Davis, Keenan, PASP 81, 230 (1969).
(26) Liszt, Smith, JQSRT 11, 1043 (1971).
(27) Tatum, Balfour, JMS 48, 292 (1973).
(28) Balfour, Tatum, JMS 48, 313 (1973).
(29) Lindgren, JMS 48, 322 (1973).
(30) Rauh, Ackermann, JCP 60, 1396 (1974).
(31) Ackermann, Rauh, JCP 60, 2266 (1974).
(32) Schoonveld, Sundaram, ApJ 192, 207 (1974).
(33) Veits, Gurvich, Kobylyanskii, Smirnov, Suslov, JQSRT 14, 221 (1974).
(34) Murad, Hildenbrand, JCP 63, 1133 (1975).

APPENDIX

Ag_2	(10) Brown, Ginter, JMS 69, 25 (1978).	Absorption spectrum.
AgBi	(4) Lochet, JP B 10, 277 (1977).	Laser excited fluorescence.
AgCd, AgHg	(1) Kasai, McLeod, JPC 79, 2324 (1975).	ESR sp. in rare gas matrices.
AgO	(5) Griffiths, Barrow, JP B 10, 925 (1977).	$A\,^2\Pi \rightarrow X\,^2\Pi$.
	(6) Griffiths, Barrow, JCS FT II 73, 943 (1977).	Electronic spectra in rare gas matrices.
AgZn	(1) Kasai, McLeod, JPC 79, 2324 (1975).	See AgCd.
AℓBr	(12) Ram, SpL 9, 435 (1976).	
	(13) Rosenwaks, JCP 65, 3668 (1976).	$a\,^3\Pi \rightarrow X\,^1\Sigma^+$ in chemiluminescence.
AℓCℓ	(13) Rosenwaks, JCP 65, 3668 (1976).	$a\,^3\Pi \rightarrow X\,^1\Sigma^+$, $b\,^3\Sigma^+ \rightarrow a\,^3\Pi$ in chemiluminescence.
AℓF	(30) Kopp, Lindgren, Malmberg, PS 14, 170 (1976).	$a\,^3\Pi \leftarrow X\,^1\Sigma^+$, rotational analysis.
	(31) Rosenwaks, JCP 65, 3668 (1976).	$a\,^3\Pi$, $A\,^1\Pi \rightarrow X\,^1\Sigma^+$; $b\,^3\Sigma^+$, $c\,^3\Sigma^+ \rightarrow a\,^3\Pi$ in chemiluminescence.
$A\ell^1H$, $A\ell^2H$	(29) Pelissier, Malrieu, JCP 67, 5963 (1977).	Theoretical calculation of the ground and valence excited states.
AℓI	(5) Rosenwaks, JCP 65, 3668 (1976).	See AℓBr.
AℓO	(53) Frank, Krauss, ZN 31 a, 1193 (1976).	$D_0^0 = 5.1_7$ eV.
	(54) Lindsay, Gole, JCP 66, 3886 (1977).	$B\,^2\Sigma^+ \rightarrow X\,^2\Sigma^+$ in chemiluminescence.
	(55) Sayers, Gole, JCP 67, 5442 (1977).	$A\,^2\Pi_i \rightarrow X\,^2\Sigma^+$ in chemiluminescence.
Ar_2	(25) Gillen, Saxon, Lorents, Ice, Olson, JCP 64, 1925 (1976).	$^3\Sigma_u^+$ well depth from scattering data, $D_e = 0.78$ eV, $r_e = 2.33$ Å.
	(26) Saxon, Liu, JCP 64, 3291 (1976).	_Ab initio_ calculations of $^3\Sigma_g^+$ and $^3\Sigma_u^+$ states.
	(27) Aziz, JCP 65, 490 (1976).	Ground state potential.
	(28) Ng, Trevor, Mahan, Lee, JCP 66, 446 (1977).	Photoionization of Ar_2, I.P. = 14.5_4 eV.
	(29) Aziz, Chen, JCP 67, 5719 (1977).	Ground state potential and molecular constants.
Ar_2^+	(6) Miller Ling, Saxon, Moseley, PR A 13, 2171 (1976).	Photodissociation cross sections, 5650 - 6950 Å.

Ar_2^+ (cont'd)	(7) Moseley, Saxon, Huber, Cosby, Abouaf, Tadjeddine, JCP 67, 1659 (1977).	Photofragment spectroscopy; $X\,^2\Sigma_u^+$, $^2\Sigma_g^+$, $^2\Pi_g$ potentials. $D_0^0 = 1.33$ eV.
	(8) Wadt, JCP 68, 402 (1978).	Calculated potential curves of low-lying states.
ArF, ArF^+	(3) Gardner, Karo, Wahl, JCP 65, 1222 (1976).	Hartree-Fock potential energy curves for $X\,^2\Sigma^+$ and $^2\Pi$ states of ArF.
Ar^1H, Ar^2H	(3) Bassi, Dondi, Tommasini, Torello, Valbusa, PR A 13, 584 (1976).	H - Ar potential from scattering data. $D_e^0 = 4.16$ meV, $r_e = 3.62$ Å.
ArKr	(4) Piper, Setser, Clyne, JCP 63, 5018 (1975).	Interaction of metastable Ar with Kr. Molecular emission at 7565 Å.
	(5) Bobetic, Barker, JCP 64, 2367 (1976).	Calculated ground state vibrational energy levels.
	(6) Nain, Aziz, MP 33, 303 (1977).	Ar - Kr intermolecular potential.
	(7) Buontempo, Cunsolo, Dore, Maselli, JCP 66, 1278 (1977).	Translational absorption spectrum.
	(8) Ng, Tiedemann, Mahan, Lee, JCP 66, 5737 (1977).	Photoionization of ArKr, I.P. = 13.42_5 eV.
ArO	(5) Julienne, Krauss, Stevens, CPL 38, 374 (1976).	Collision-induced oxygen $^1S_0 - {}^1D_2$ emission near 5577 Å in argon. Calculated bound - bound spectrum.
ArO^+	(1) Ding, Karlau, Weise, CPL 45, 92 (1977).	$^4\Sigma^-$ potential from O^+-on-Ar scattering data; $D_e = 0.68$ eV, $r_e = 2.02$ Å.
ArXe	(3) See ref. (5) of ArKr.	
	(4) Chashchina, Shreider, JAS 23, 914 (1975).	
	(5) Castex, JCP 66, 3854 (1977).	Absorption spectra 1150 - 1500 Å. Potential curves.
	(6) See ref. (8) of ArKr.	Photoionization of ArXe; I.P. = 11.98_5 eV.
AsF	(11) Vasudev, Jones, CJP 55, 337 (1977).	Rotational analysis of two overlapping systems $f\,0^+$, $e\,0^- \leftrightarrow X_2\,1$ (previously called $B \leftrightarrow X$).
AsN	(7) Krishnamurty, Thomas, IJPAP 14, 236 (1976).	Isotope shifts.
AsS	(6) Ashrafunnisa, Rao, Rao, IJP 49, 580 (1975).	$A_1' - X_2$ Franck-Condon factors, potential functions.
AuCa	(2) Coquant, Houdart, CR B 284, 171 (1977).	Rotational analysis of A - X, B - X; preliminary results.
AuLi	(2) Ihle, Langenscheidt, Zmbova, JCP 66, 5105 (1977).	IR sp. in rare gas matrices; extrapol. gas phase frequencies $\Delta G(\frac{1}{2}) = 746$ and 705 cm^{-1} for Au^6Li and Au^7Li, respectively.

AuO	(2) Griffiths, Barrow, JCS FT II 73, 943 (1977).	Absorption and emission spectra in rare gas matrices. Predicted gas phase constants $v_{00} \approx 25000$, $\omega' \approx 600$, $\omega'' \approx 675$.
AuSi	(6) Coquant, Houdart, CR B 284, 171 (1977).	Rotational analysis of $A-X_1$, preliminary constants.
BaBr	(9) Hildenbrand, JCP 66, 3526 (1977).	$D_0^0 = 3.71$ eV (mass-spectrometric value).
BaCℓ	(15) See ref. (9) of BaBr.	$D_0^0 = 4.48$ eV (mass-spectrometric value).
BaO	(41) Ackermann, Rauh, Thorn, JCP 65, 1027 (1976).	I.P. = 6.85 eV (electron impact).
	(42) Wormsbecher, Lane, Harris, JCP 66, 2745 (1977).	Microwave optical double resonance, Stark effect; dipole moment of A $^1\Sigma^+$, $\mu_{e\ell}(v=7) = 2.2_0$ D.
	(43) Hocking, Pearson, Creswell, Winnewisser, JCP 68, 1128 (1978).	Millimeter wave spectrum.
BaS	(6) Tiemann, Ryzlewicz, Törring, ZN 31 a, 128 (1976).	Rotation sp. of $^{138}Ba^{32}S$; $B_e = 0.10331430$, $\alpha_e = 0.00031509_5$, $\gamma_e = -4.47 \times 10^{-7}$, $D_e = 3.09_5 \times 10^{-8}$.
$BeAr^+$	(3) Subbaram, Coxon, Jones, CJP 54, 1535 (1976).	$A\ ^2\Pi_r \rightarrow X\ ^2\Sigma^+$, high resolution analysis; molecular constants.
BeF	(17) Gohel, Shah, IJP 48, 932 (1974).	A − X RKR Franck-Condon factors.
$BeKr^+$	(2) Coxon, Jones, Subbaram, CJP 55, 254 (1977).	$A\ ^2\Pi_r \rightarrow X\ ^2\Sigma^+$, high resolution analysis.
B^1H	(24) Stern, Kaldor, JCP 64, 2002 (1976).	Many-body perturbation theory applied to eight states.
	(25) Pyper, Gerratt, PRS A 355, 407 (1977).	Spin-coupled theory of molecular wavefunctions [$X\ ^1\Sigma^+$].
Bi_2	(12) Süzer, Lee, Shirley, JCP 65, 412 (1976).	Photoelectron sp., I.P. = 7.53, 8.94 eV ($X\ ^2\Pi_{ui}$) and 9.30 eV ($A\ ^2\Sigma_g^+$).
	(13) Teichman, Nixon, JCP 67, 2470 (1977).	New electronic states of matrix-isolated Bi_2.
BiCℓ	(12) Kuijpers, Törring, Dymanus, CP 18, 401 (1976).	Microwave sp.; constants for $Bi^{35}C\ell$ $B_e = 0.09212553$, $\alpha_e = 0.00040203$, ..., also hfs constants.
BiF	(16) Kuijpers, Dymanus, CP 24, 97 (1977).	Microwave sp.; $B_e = 0.22998897$, $\alpha_e = 0.00150262$, ...; also hfs constants.
BO, BO^+, BO^-	(23) Griffing, Simons, JCP 64, 3610 (1976).	Theoretical study of BO^-, I.P. = 2.79 eV.
Br_2	(44) Le Roy, Macdonald, Burns, JCP 65, 1485 (1976).	B and C state repulsive potential curves from a fit of calculated continuum absorption coefficients to experimental data; B − X and C − X transition moment functions.

Br_2 (cont'd)	(45) Zaraga, Nogar, Moore, JMS 63, 564 (1976).	B – X transition moment from absolute line absorption measurements. Estimated B state radiative lifetime 20 μs; predissociation quantum yields near unity.
	(46) Chang, Hwang, JMS 65, 430 (1977).	Resonance fluorescence with argon-ion laser excitation.
	(47) Chang, Hwang, JCP 67, 3624 (1977).	Resonance Raman study; fundamental and eight overtones investigated with high resolution.
	(48) Tellinghuisen, CPL 49, 485 (1977).	Reassignment of the "E→B" system to the transition $D'(^3\Pi_g)2_g \rightarrow A'(^3\Pi_u)2_u$. Vibrational constants.
	(49) Lindeman, Wiesenfeld, CPL 50, 364 (1977).	Laser induced photodissociation; B← X and C← X relative absorptivities.
	(50) de Vlieger, Eisendrath, JP B 10, L463 (1977).	Hanle effect studies of several rotational levels of $B(^3\Pi_u)0_u^+$, v'=40,45.
BrCℓ	(15) Wright, Spates, Davis, JCP 66, 1566 (1977).	Laser-excited fluorescence from B $^3\Pi_{0^+}$; lifetime $18._5$ μs.
BrF, BrF^+	(12) Clyne, Curran, Coxon, JMS 63, 43 (1976).	Laser excitation spectrum of BrF $B\,^3\Pi_{0^+} \leftarrow X\,^1\Sigma^+$. Rotational analysis. Predissociation.
	(13) Clyne, McDermid, JCS FT II 73, 1094 (1977).	Laser-excited fluorescence from B $^3\Pi_{0^+}$; lifetime $24._7$ μs.
BS	(9) Bell, McLean, JMS 63, 521 (1976).	Two new emission systems, $E\,^2\Sigma^+ \rightarrow X\,^2\Sigma^+$ (v_{00} = 47933.6) and $E\,^2\Sigma^+ \rightarrow A\,^2\Pi_i$. Rotational and vibrational constants.
C_2	(60) Tatarczyk, Fink, Becker, CPL 40, 126 (1976).	Lifetimes of single vibrational levels of $d\,^3\Pi_g$; τ = 120 ns corresponding to $f_{00} \approx 0.025$ for the Swan system.
	(61) Cooper, Nicholls, SpL 9, 139 (1976).	Transition probability data for seven band systems.
	(62) Curtis, Engman, Erman, PS 13, 270 (1976).	High resolution lifetime studies of $d\,^3\Pi_g$, $C\,^1\Pi_g$, $D\,^1\Sigma_u^+$, $\tau(v'=0) = 123, 31._1, 18._1$ ns, respectively. Corresponding oscillator strength for the Swan bands $f_{00} = 0.026_1$.
	(63) Leach, Velghe, JQSRT 16, 861 (1976).	Laser pyrolysis; rad. lifetime of $d\,^3\Pi_g(v=0) \approx 240$ ns.
	(64) Bell, Branch, ApJ 212, 591 (1977).	Phillips system; rot. dependence of Franck-Condon factors.
	(65) Swamy, O'Dell, ApJ 216, 158 (1977).	Franck-Condon factors for the Fox-Herzberg and Ballik-Ramsay bands.
	(66) Kini, Savadatti, JP B 10, 1139 (1977).	Swan bands; relative intensities in discharge through CO.
	(67) Chauville, Maillard, Mantz, JMS 68, 399 (1977).	Line positions and molecular constants for $A\,^1\Pi_u - X\,^1\Sigma_g^+$. Rotational perturbations in $A\,^1\Pi_u(v=1)$ by a previously unob-

C_2 (cont'd)		served level of c $^3\Sigma_u^+$ leading to revised constants for the c state ($T_e = 9227.4$, etc.).
	(68) Langhoff, Sink, Pritchard, Kern, Strickler, Boyd, JCP 67, 1051 (1977).	*Ab initio* study of perturbations between X $^1\Sigma_g^+$ and b $^3\Sigma_g^-$.
C_2^-	(13) Cederbaum, Domcke, Niessen, JP B 10, 2963 (1977).	Theoretical calculation of I.P. = 3.60 eV.
Ca_2	(5) Sakurai, Broida, JCP 65, 1138 (1976).	Laser photoluminescence.
	(6) Miller, Andrews, CPL 50, 315 (1977); Miller, Ault, Andrews, JCP 67, 2478 (1977).	Absorption and emission spectra in solid argon and krypton. Resolved vibrational structure 14000 - 16000 cm^{-1}, $\omega' \approx$ 111(Ar), 117(Kr); $\omega'' \approx 78$(Kr).
CaBr	(11) Puri, Mohan, IJPAP 14, 512 (1976).	Extension of the C→X and D→X systems.
	(12) Hildenbrand, JCP 66, 3526 (1977).	$D_0^0 = 3.18$ eV (mass-spectrometric value).
CaCℓ	(15) Domaille, Steimle, Wong, Harris, JMS 65, 354 (1977).	B $^2\Sigma^+$ – X $^2\Sigma^+$, rotational analysis. See (16).
	(16) Domaille, Steimle, Harris, JMS 66, 503 (1977).	X $^2\Sigma^+$ rotation spectrum using laser microwave optical double resonance. Rotational constants for X: $B_e'' = 0.152233$, $\alpha_e'' = 0.000800$, and B: $B_e' = 0.154700$, $\alpha_e' = 0.000889$. Spin-splitting constants $\gamma_0'' = +0.00136$, $\gamma_0' = -0.0652$.
	(17) Brinkmann, Telle, JP B 10, 133 (1977).	Ca* + HCℓ(or $C\ell_2$) chemiluminescence, A→X and B→X. $D_0^0 \geq$ 4.28 eV.
Ca^1H	(25) Berg, Klynning, Martin, OC 17, 320 (1976).	Laser excitation spectroscopy of the B $^2\Sigma$ – X $^2\Sigma$ transition.
Ca^2H	(9) Kaving, Lindgren, PS 13, 39 (1976).	Absorption spectrum 2800 - 3200 Å. Rotational analyses of the 5p complex ($K^2\Sigma$, $L^2\Pi$) and of the 4d complex ($G^2\Sigma$, $J^2\Pi$ only).
CaMg	(1) Miller, Ault, Andrews, JCP 67, 2478 (1977).	Absorption in solid argon and krypton, vibrational structure 17300 - 19000 cm^{-1}.
CaO	(22) Andrews, Ault, JMS 68, 114 (1977).	IR spectra of matrix-isolated CaO; four isotopes.
	(23) Benard, Slafer, Love, Lee, AO 16, 2108 (1977).	Modulated transmission spectroscopy of chemi-excited CaO. Absorption from a $^3\Pi$ and A' $^1\Pi$.
	(24) Creswell, Hocking, Pearson, CPL 48, 369 (1977).	Pure rotational spectrum, $B_e = 0.4444527_4$, $\alpha_e = 0.0033126_6$ for $^{40}Ca^{16}O$. See (25).
	(25) Hocking, Pearson, Creswell, Winnewisser, JCP 68, 1128 (1978).	Millimeter wave spectrum.

CaSr	(1) Miller, Ault, Andrews, JCP 67, 2478 (1977).	Absorption in solid argon, vibr. structure 14400 - 15200 cm^{-1}.
Cd_2	(6) Ault, Andrews, JMS 65, 102 (1977).	UV absorption in solid argon and krypton, vibr. structure 35000 - 36700 cm^{-1}.
Cd^1H, Cd^2H	(11) Jourdan, Negre, Dufayard, Nedelec, JP(Paris) 37, L-29 (1976).	Lifetime of A $^2\Pi$, $\tau \approx 70$ ns.
	(12) Dufayard, Nedelec, JP(Paris) 38, 449 (1977).	Lifetimes, Λ-doubling, hfs of A $^2\Pi$.
CeI	(1) Chantry, JCP 65, 4412 (1976).	Negative ion formation in CeI_3. D(Ce-I) $\geq 3.7_7$ eV.
	(2) Chantry, JCP 65, 4421 (1976).	Positive ions by electron impact on CeI_3. I.P.(CeI) = 5.9 eV.
CeO	(9) Ackermann, Rauh, Thorn, JCP 65, 1027 (1976).	I.P. = 4.90 eV (electron impact).
C^1H	(50) Evenson, Radford, Moran, APL 18, 426 (1971).	Pure rotational transition by IR laser magnetic resonance.
	(51) Carozza, Anderson, JOSA 67, 118 (1977).	Radiative lifetime of A $^2\Delta$, $\tau(v=0,1) = 511$ ns.
	(52) Levy, Hinze, ApJ 211, 980 (1977).	Predicted Λ-doubling spectrum of ^{13}CH.
	(53) Hammersley, Richards, ApJ 214, 951 (1977).	Ab initio calculation of the Λ-doubling in excited rotational levels of CH and CD.
C^2H	(9) See ref. (53) of C^1H.	
C^1H^+	(23) Erman, ApJ 213, L89 (1977).	$f_{00} = 0.0072$ for A − X, from improved lifetime measurements.
	(24) Elander, Oddershede, Beebe, ApJ 216, 165 (1977).	Theoretical spectroscopic constants and lifetimes for A − X.
C^2H^+	(3) See ref. (24) of C^1H^+.	
$C\ell_2$	(44) Wells, Zipf, JCP 66, 5828 (1977).	Translational spectroscopy, dissociative excitation of $C\ell_2$.
	(45) Chang, Hwang, JCP 67, 4777 (1977).	Resonance Raman spectrum of gaseous chlorine, $\Delta v \leq 8$.
	(46) Hwang, Chang, JMS 69, 11 (1978).	Resonance fluorescence of gaseous chlorine.
	(47) Coxon, Shanker, JMS 69, 109 (1978).	Rotational analysis of $B\,^3\Pi_{0^+} \rightarrow X\,^1\Sigma^+$ of $^{35}C\ell_2$ and $^{35,37}C\ell_2$ in the chlorine afterglow emission spectrum. Constants for $^{35}C\ell_2$: $T_e = 17817.67$, $\omega_e' = 255.38$, $\omega_e'x_e' = 4.59$, $\omega_e'y_e' = -0.038$, $B_e' = 0.16313$, $\alpha_e' = 0.00242$, $\gamma_e' = -0.000057$. Reinterpretation of the $0_g^+ \leftarrow 0_u^+$ system observed by (13).
$C\ell_2^-$	(8) Martinez de Pinillos, Weltner, JCP 65, 4256 (1976).	ESR sp. of $C\ell_2^-$ in various ion pairs at 4 K.
	(9) Sullivan, Freiser, Beauchamp, CPL 48, 294 (1977).	Gas phase photodissociation of $C\ell_2^-$; broad peak at 3500 Å attributed to $^2\Sigma_g^+ \leftarrow X\,^2\Sigma_u^+$.

$C\ell F$, $C\ell F^+$	(19) Fabricant, Muenter, JCP 66, 5274 (1977).	Molecular beam electric resonance; hf, Stark, Zeeman properties. $\mu_{e\ell}(^+C\ell F^-) = +0.8_5$ D.
$C\ell O$, $C\ell O^-$	(18) Bulgin, Dyke, Jonathan, Morris, MP 32, 1487 (1976).	Photoelectron sp., I.P. = 10.87 eV, $\omega(C\ell O^+) \approx 1040$ in $X\,^3\Sigma^-$.
	(19) Amano, Hirota, JMS 66, 185 (1977).	Microwave sp. in $X\,^2\Pi(v=1)$.
	(20) Arnold, Whiting, Langhoff, JCP 66, 4459 (1977).	MCSCF+IC wavefunctions, properties of $X\,^2\Pi$, $A\,^2\Pi$.
	(21) Langhoff, Dix, Arnold, Nicholls, Danylewych, JCP 67, 4306 (1977).	Theoretical intensity parameters for vibration-rotation bands.
	(22) Mandelman, Nicholls, JQSRT 17, 483 (1977).	$A-X$ band oscillator strengths, continuum absorption cross sections, electronic transition moment.
	(23) Cooper, JQSRT 17, 543 (1977).	$A-X$ electronic transition moment.
	(24) Menzies, Margolis, Hinkley, Toth, AO 16, 523 (1977).	Fundamental vibration-rotation spectrum.
	(25) Rigaud, Leroy, Le Bras, Poulet, Jourdain, Combourieu, CPL 46, 161 (1977).	$A-X$ absorption cross sections.
	(26) Wine, Ravishankara, Philen, Davis, Watson, CPL 50, 101 (1977).	$A-X$ high resolution absorption cross sections.
CN	(74) Miller, Freund, Field, JCP 65, 3790 (1976).	Identification and characterization of $a\,^4\Sigma^+$ by anticrossing spectroscopy.
	(75) Bailey, ApJ 211, 596 (1977).	Abs. coefficients for the fundamental and first overtone vibration-rotation bands.
	(76) Bacis, Cerny, d'Incan, Guelachvili, Roux, ApJ 214, 946 (1977).	$A\,^2\Pi - X\,^2\Sigma$; absolute wavenumber measurements by Fourier spectrometry, analysis of the 0-0 band. Rotational constants.
	(77) Dixon, Woods, JCP 67, 3956 (1977).	Laboratory microwave spectrum. Rotational and hfs constants for $X\,^2\Sigma^+(v=0,1)$.
CN^-	(3) Pacansky, Liu, JCP 66, 4818 (1977).	Hartree-Fock electron affinity of CN 3.29 eV.
CO	(163) Rao, in "Spectroscopy", International Review of Science, Physical Chemistry (2) 3 (ed. Ramsay), Butterworths, London (1976), p. 317.	Wavenumbers for the 1-0 and 2-0 vibr.-rot. bands of $^{12}C^{16}O$, CO laser lines. Improved Dunham coefficients.
	(164) Fleming, JQSRT 16, 63 (1976).	Line strength and halfwidth measurements from far-infrared absorption spectra.

CO (cont'd)	(165) Bouanich, JQSRT 16, 1119 (1976).	X $^1\Sigma$ dipole moment function and transition moments.
	(166) Ogilvie, Koo, JMS 61, 332 (1976).	Dunham potential energy coefficients.
	(167) Todd, Clayton, Telfair, McCubbin, Pliva, JMS 62, 201 (1976).	Infrared emission; precise line positions for the 1-0, 2-1, 2-0 transitions of $^{12}C^{16}O$. $\Delta v=1$ sequences measured up to $v'=20,11,4$ for $^{12}C^{16}O$, $^{13}C^{16}O$, $^{12}C^{18}O$, respectively. Dunham coefficients for all three molecules.
	(168) Chackerian, Valero, JMS 62, 338 (1976).	Absolute intensities for the 4-0 vibr.-rot. band.
	(169) Banna, Shirley, JESRP 8, 255 (1976).	Photoelectron spectrum at 132.3 eV.
	(170) Hamnett, Stoll, Brion, JESRP 8, 367 (1976).	PE branching ratios, partial ionization cross sections.
	(171) Phillips, Lee, Judge, JCP 65, 3118 (1976).	d $^3\Delta_i \rightarrow$ X $^1\Sigma$ fluorescence, $\tau(v'=3) = 4.7$ μs. Electronic transition moment variation, (d − X)/(d − a) branching ratios.
	(172) Chackerian, JCP 65, 4228 (1976).	X $^1\Sigma^+$ dipole moment function.
	(173) Corvilain-Berger, Verhaegen, CPL 50, 468 (1977).	Theoretical $1\sigma(1s_O)$ and $2\sigma(1s_C)$ ionization potentials.
	(174) Domin, Domin, Rytel, APP A 51, 783 (1977).	Analysis of the 4th pos. system of $^{12}C^{16}O$ in the near UV.
	(175) Huffaker, JMS 65, 1 (1977).	X $^1\Sigma^+$ RKR potential.
	(176) Kirschner, Le Roy, Ogilvie, Tipping, JMS 65, 306 (1977).	Ground state dipole moment function.
	(177) Kay, Van der Leeuw, Van der Wiel, JP B 10, 2513 (1977).	Energy loss spectra near the carbon K edge.
	(178) Marcoux, Piper, Setser, JCP 66, 351 (1977).	IR radiative decay constants for the vibrational levels of a $^3\Pi$.
	(179) Ermler, Mulliken, Wahl, JCP 66, 3031 (1977).	Molecular orbital correlation diagrams.
	(180) Phillips, Lee, Judge, JCP 66, 3688 (1977).	A $\rightarrow$ X fluorescence from photodissociation of CO_2; transition moment variation.
	(181) Kirby-Docken, Liu, JCP 66, 4309 (1977).	Ground state dipole moment function (theor.).
	(182) Hemminger, Cavanagh, Lisy, Klemperer, JCP 67, 4952 (1977).	Stark effect in d $^3\Delta_i(v=4) \leftarrow$ a $^3\Pi_r(v=0)$, $\mu_{e\ell}(d^3\Delta, v=4) =$ 0.42 D ($^-CO^+$).
	(183) Provorov, Stoicheff, Wallace, JCP 67, 5393 (1977).	Laser excited fluorescence from A $^1\Pi$; lifetimes of unperturbed rotational levels ~10.9 ns, longer for perturbed levels.
	(184) Möhlmann, de Heer, CP 21, 119 (1977).	Dependence on r of the electronic transition moment for the 3rd pos. gr. (b $\rightarrow$ a).

CO (cont'd)	(185) Meerts, de Leeuw, Dymanus, CP 22, 319 (1977).	Molecular beam electric resonance. Stark-Zeeman spectra in v=0,J=1 of $^{12}C^{16}O$, $^{13}C^{16}O$, $^{12}C^{18}O$.
	(186) van Sprang, Möhlmann, de Heer, CP 24, 429 (1977).	Radiative lifetimes of $a'\,^3\Sigma^+(v=4...9)$ 10.24...6.82 μs, $b\,^3\Sigma^+(v=0)$ 56 ns, $C\,^3\Pi(v=0)$ 16 ns, $B\,^1\Sigma^+(v=0)$ 34 ns, $d\,^3\Delta(v=1...16)$ 7.30 ... 2.94 μs
	(187) Norbeck, Merkel, Certain, MP 34, 589 (1977).	Theor. dipole moment functions for a $^3\Pi$ and A $^1\Pi$.
	(188) Plummer, Gustafsson, Gudat, Eastman, PR A 15, 2339 (1977).	Partial photoionization cross sections 18 - 50 eV.
CO^+, CO^{++}	(41) Janjic, Pesic, ApJ 209, 642 (1976).	Rotational analysis of the 1st neg. system of $^{12}C^{18}O^+$.
	(42) Gagnaire, Goure, CJP 54, 2111 (1976).	Rot. anal. of the CO^+ A→X 2-0 band; Λ doubling constants.
	(43) Möhlmann, de Heer, CPL 43, 170 (1976).	Radiative lifetimes for A $^2\Pi_i(v=0...9)$ $3.8_2 ... 2.1_1$ μs.
	(44) Carrington, Sarre, MP 33, 1495 (1977).	$A\,^2\Pi \leftarrow X\,^2\Sigma$ absorption in an ion beam.
	(45) See ref. (179) of CO.	
	(46) Kay, Van der Leeuw, Van der Wiel, JP B 10, 2521 (1977).	Auger decay of carbon-K ionized CO.
	(47) Lee, JP B 10, 3033 (1977).	Cross sections for the production of B→X fluorescence by photoionization.
	(48) Möhlmann, de Heer, CP 21, 119 (1977).	Electronic transition moment function for the 1st negative system (B−X).
CO^-	(9) Zubek, Szmytkowski, JP B 10, L27 (1977).	Comparison of experimental and calculated cross sections for resonant vibrational excitation of CO by electron scattering; derived ground state constants $\omega = 1895$ cm^{-1}, $r_0 = 1.223$ Å, I.P. = -1.52 eV, Γ = 0.80 eV.
	(10) King, McConkey, Read, JP B 10, L541 (1977).	Negative-ion resonances associated with K-shell-excited states of CO.
CP	(8) Murthy, Gowda, Narasimhamurthy, Pramana 6, 25 (1976).	$B\,^2\Sigma - A\,^2\Pi$ RKR Franck-Condon factors.
CrMo	(1) Efremov, Samoilova, Gurvich, CPL 44, 108 (1976).	Absorption spectrum near 4800 Å.
CS, CS^+	(30) Coxon, Marcoux, Setser, CP 17, 403 (1976).	$A\,^1\Pi \rightarrow X\,^1\Sigma$, $A\,^2\Pi \rightarrow X\,^2\Sigma$; intensity and wavelength measurements.
	(31) Domcke, Cederbaum, von Niessen, Kraemer, CPL 43, 258 (1976).	Calculated He I photoelectron spectrum of CS.

CS, CS^+ (cont'd)		
	(32) Robbe, Schamps, JCP 65, 5420 (1976).	*Ab initio* calculation of CS valence states and perturbation parameters.
	(33) Black, Sharpless, Slanger, JCP 66, 2113 (1977).	$CS(a\,^3\Pi)$ produced in photodissociation of CS_2. Radiative lifetime $\tau(a\,^3\Pi) \approx 16$ ms.
	(34) Chong, Takahata, JESRP 10, 137 (1977).	Theoretical calculations of photoelectron spectra.
	(35) Todd, JMS 66, 162 (1977).	High resolution measurement of the 2-0 vibr.-rot. band of $^{12}C^{32}S$ and $^{12}C^{34}S$; improved vibr. and rot. constants.
	(36) Cossart, Horani, Rostas, JMS 67, 283 (1977).	$a\,^3\Pi_r \rightarrow X\,^1\Sigma^+$; rotational analysis, perturbations.
	(37) Gauyacq, Horani, CJP, in press (1978).	$A\,^2\Pi_i \rightarrow X\,^2\Sigma^+$ of CS^+; rotational analysis, perturbations.
Cs_2, Cs_2^+	(23) Benedict, Drummond, Schlie, JCP 66, 4600 (1977).	Absorption spectra 8000 - 13000 Å, interpreted as $A\,^1\Sigma_u^+ \leftarrow X\,^1\Sigma_g^+$ and $b\,^3\Pi_u \leftarrow X\,^1\Sigma_g^+$.
	(24) Klewer, Beerlage, Los, Van der Wiel, JP B 10, 2809 (1977).	Ion-pair formation in two-photon absorption of molecular cesium (5630 - 6350 Å).
	(25) Toader, Collins, Johnson, Mirza, PR A 16, 1490 (1977).	Molecular satellites of the $6^2S - 7^2P$ doublet of cesium.
	(26) Gupta, Happer, Wagner, Wennmyr, JCP 68, 799 (1978).	Absorption studies in the visible and near visible region.
CsAr	(5) Sayer, Ferray, Lozingot, Berlande, JP B 9, L293 (1976).	Emission spectra of discharges in Cs-rare gas mixtures.
$CsBr, CsBr^+$	(22) Potts, Williams, JCS FT II 72, 1892 (1976).	He II photoelectron spectrum.
CsCℓ	(23) See ref. (22) of $CsBr, CsBr^+$.	
	(24) Cederberg, JCP 66, 5247 (1977).	Molecular beam electric resonance; hfs constants.
CsHe	(2) See ref. (5) of CsAr.	
CsI, CsI^+	(20) See ref. (22) of $CsBr, CsBr^+$.	
	(21) Grossman, Hurst, Payne, Allman, CPL 50, 70 (1977).	Saturated photodissociation of CsI.
CsKr, CsNe	(3) See ref. (5) of CsAr.	
CsXe	(6) Marek, Niemax, PL A 57, 414 (1976).	Emission bands 3500 - 6400 Å.
	(7) Marek, JP B 10, L325 (1977).	Time-resolved fluorescence.

Cu_2	(14) Steele, JMS 61, 477 (1976).	Photoluminescence, lifetimes, and discharge excitation.
CuI	(13) Nair, Tiemann, Hoeft, ZN 32 a, 1053 (1977).	On the hfs in the rotational spectrum.
CuO	(14) Appelblad, Lagerqvist, PS 13, 275 (1976).	Rotational analyses of $E\,^2\Delta \rightarrow X\,^2\Pi$, $H\,^2\Pi \rightarrow X\,^2\Pi$, $I\,^2\Pi \rightarrow X\,^2\Pi$, $P\,^2\Pi \rightarrow X\,^2\Pi$. Molecular constants, perturbations.
	(15) Griffiths, Barrow, JCS FT II 73, 943 (1977).	Electronic spectra in rare gas matrices.
	(16) Lefebvre, Pinchemel, Schamps, JMS 68, 81 (1977).	Rotational analysis of a new system $A'\,^2\Sigma^+ \rightarrow X\,^2\Pi_i$. Upper state constants $T_0 = 15531.90$, $\omega_e = 614.0$, $B_0 = 0.4382_8$; large spin doubling.
	(17) Pinchemel, Lefebvre, Schamps, JP B 10, 3215 (1977).	New $^2\Delta_{5/2} \rightarrow X\,^2\Pi_{3/2}$ transition. Upper state constants $T_0 = 15317.24$, $B_0 = 0.4253_6$.
CuS	(6) Biron, CR B 283, 209 (1976).	Rotational analysis of $A \rightarrow X$ 0-0 band.
DyO	(6) See ref. (9) of CeO.	I.P. = 6.08 eV (electron impact).
ErO	(5) See ref. (9) of CeO.	I.P. = 6.30 eV (electron impact).
EuO	(8) Dirscherl, Michel, CPL 43, 547 (1976).	Dependence of reaction cross sections of $Eu + O_2$ on collisional energy; $D_0^0 = 4.97$ eV.
	(9) See ref. (9) of CeO.	I.P. = 6.48 eV (electron impact).
	(10) Murad, Hildenbrand, JCP 65, 3250 (1976).	$D_0^0 = 4.8_5$ eV (equilibrium mass-spectrometry).
	(11) Balducci, Gigli, Guido, JCP 67, 147 (1977).	$D_0^0 = 4.84$ eV (equilibrium mass-spectrometry).
F_2	(21) Hay, Cartwright, CPL 41, 80 (1976).	Rydberg, ionic, and valence interactions in the excited states of F_2.
F_2^+	(8) Banyard, Ellis, Tait, Dixon, JP B 7, 1411 (1974).	Ab initio studies of potential curves for low-lying states of F_2^+ and F_2^{++}.
	(9) Guyon, Spohr, Chupka, Berkowitz, JCP 65, 1650 (1976).	Threshold photoelectron spectrum of F_2. Constants for $X\,^2\Pi_g$, $A = 346$, $\omega_e = 1113$, $\omega_e x_e = 9.75$.
Fe^1H, Fe^2H	(6) McCormack, O'Connor, AA(Suppl.) 26, 373 (1976).	Wavelengths and line intensities for the 4920, 5320, and 8690 Å bands.
	(7) Wing, Cohen, Brault, ApJ 216, 659 (1977).	Confirmation of the presence of FeH in sunspots and cool stars.
GdO	(9) See ref. (9) of CeO.	I.P. = 5.75 eV (electron impact).
	(10) Yadav, Rai, Rai, CJP 54, 2429 (1976).	Emission spectrum 19400 - 22500 cm^{-1}.

GeO	(27) Mummigatti, Jyoti, APH 42, 99 (1977).	A – X RKRV Franck-Condon factors.
$^1H_2$	(170) Dehmer, Chupka, JCP 65, 2243 (1976).	High resolution measurement of relative photoabsorption and photoionization cross sections at 78 K, 715 - 805 Å.
	(171) Jette, JCP 65, 4325 (1976); 67, 2934 (1977) (erratum).	Ab initio calculation of spin-rotation coupling in c $^3\Pi_u$.
	(172) Lee, Carlson, Judge, JQSRT 16, 873 (1976).	Absolute absorption cross sections of H_2 and D_2, 180 - 780 Å.
	(173) Miller, Freund, JMS 63, 193 (1976).	Singlet-triplet anticrossings between B' $^1\Sigma_u^+(3p\sigma)$ and f $^3\Sigma_u^+(4p\sigma)$.
	(174) Wolniewicz, JMS 63, 537 (1976).	Non-adiabatic corrections to the rotational energies of X $^1\Sigma_g^+(v=0...3)$.
	(175) Anderson, Watson, Sharpton, JOSA 67, 1641 (1977).	Electron impact excitation of G $^1\Sigma_g^+$; lifetime ~30 ns.
	(176) Bunker, McLarnon, Moss, MP 33, 425 (1977).	Application of the effective vibration-rotation hamiltonian to H_2 and D_2; calculated vibration-rotation energies for H_2 ($v \leq 14$, $J \leq 5$).
	(177) Glover, Weinhold, JCP 66, 303 (1977).	Calculated potential curve for the G,K $^1\Sigma_g^+$ double minimum state.
	(178) Borrell, Guyon, Glass-Maujean, JCP 66, 818 (1977).	Predissociation, 680 - 860 Å.
	(179) Jungen, Atabek, JCP 66, 5584 (1977).	Rovibronic interactions in the absorption spectrum; an application of multichannel quantum defect methods.
	(180) Ford, Greenawalt, Browne, JCP 67, 983 (1977).	Computation of nuclear motion and mass polarization adiabatic energy corrections for X $^1\Sigma_g^+$, B $^1\Sigma_u^+$, C $^1\Pi_u$, D $^1\Pi_u$.
	(181) Prochaska, Andrews, JCP 67, 1139 (1977).	Vibrational-rotational and pure rotational laser Raman spectrum in solid inert gas matrices at 12 K.
	(182) Bhattacharyya, Chiu, JCP 67, 5727 (1977).	Theoretical radiative lifetime of c $^3\Pi_u(v=0)$ = 1.76 ms.
	(183) Kolos, Rychlewski, JMS 66, 428 (1977).	Ab initio potential energy curves for c $^3\Pi_u$, I $^1\Pi_g$, i $^3\Pi_g$.
	(184) Wolniewicz, Dressler, JMS 67, 416 (1977).	The E,F and G,K $^1\Sigma_g^+$ states of hydrogen. Adiabatic calculation of vibronic states.
	(185) Reddy, Varghese, Prasad, PR A 15, 975 (1977).	Collision-induced fundamental IR absorption band.
	(186) Rumble, Sims, Purdy, JP B 10, 2553 (1977).	Calculated oscillator strengths for the $C^1\Pi_u - X^1\Sigma_g^+$ Werner b..
	(187) Turner, Kirby-Docken, Dalgarno, ApJ(Suppl.) 35(3), 281 (1977).	Calculated electric quadrupole vibration-rotation transition probabilities for the ground state ($v \leq 14$, $J \leq 20$).
	(188) Vogler, Meierjohann, PRL 38, 57 (1977).	Predissociation of c $^3\Pi_u$.

$^1H_2$ (cont'd)	(189) Glass-Maujean, Breton, Guyon, PRL 40, 181 (1978).	Accidental predissociation of $D'\,^1\Pi_u^+(4p\pi)$.
	(190) Kligler, Rhodes, PRL 40, 309 (1978).	Two-photon excitation of E,F $^1\Sigma_g^+$.
	(191) Wicke, JCP 68, 337 (1978).	G,K double-minimum state.
$^1H^2H$	(52) Smirnov, Mazurenko, OS(Engl. Transl.) 41, 566 (1976).	Calculation of the permanent dipole moment.
	(53) See ref. (174) of $^1H_2$.	
	(54) Prasad, Reddy, JCP 66, 707 (1977).	IR absorption; collision-induced fundamental band in HD + Kr and HD + Xe mixtures at room temperature.
	(55) Reddy, Prasad, JCP 66, 5259 (1977).	IR absorption; collision-induced fundamental band of the pure gas.
	(56) Alemar-Rivera, Ford, JMS 67, 336 (1977).	Non-adiabatic effects in the B, C, and E,F states of HD.
	(57) See ref. (184) of $^1H_2$.	
	(58) See ref. (181) of $^1H_2$.	
$^2H_2$	(61) See ref. (174) of $^1H_2$.	
	(62) See ref. (172) of $^1H_2$.	
	(63) See ref. (178) of $^1H_2$.	
	(64) See ref. (179) of $^1H_2$.	
	(65) See ref. (180) of $^1H_2$.	
	(66) See ref. (181) of $^1H_2$.	
	(67) See ref. (184) of $^1H_2$.	
$^1H_2^+$	(28) van Asselt, Maas, Los, CP 5, 429 (1974).	Laser-induced photodissociation of H_2^+ and D_2^+ in ion beams.
	(29) Colbourn, CPL 44, 374 (1976).	The use of the electron reduced mass in the electronic Schrödinger equation for H_2^+ and its isotopes.
	(30) Ozenne, Durup, Odom, Pernot, Tabché-Fouhaillé, Tadjeddine, CP 16, 75 (1976).	Laser photodissociation of H_2^+, D_2^+, HD^+ in ion beams. Comparison between experimental and *ab initio* computed fragment kinetic energy spectra.
	(31) Bishop, JCP 66, 3842 (1977).	Relativistic corrections for H_2^+ and its isotopes.
	(32) Bishop, Cheung, PR A 16, 640 (1977).	Calculated IR transition frequencies for H_2^+, D_2^+, HD^+, HT^+, DT^+.
	(33) Cohen, McEachran, Schlifer, CPL 49, 374 (1977).	Hyperfine structure of H_2^+.
$^1H^2H^+$, $^1H^3H^+$	(7) Colbourn, Bunker, JMS 63, 155 (1976).	Accurate theoretical vibration-rotation energies and transition moments for HD^+, HT^+, DT^+.
	(8) See ref. (30) of $^1H_2^+$.	

$^1H^2H^+$, $^1H^3H^+$ (cont'd)

	(9) See ref. (31) of $^1H_2^+$.	
	(10) Thomas, Dale, Paulson, JCP 67, 793 (1977).	Photodissociation spectra of HD^+ and D_2^+.
	(11) See ref. (32) of $^1H_2^+$.	
	(12) Ray, Certain, PRL 38, 824 (1977).	Calculated hyperfine structure in the IR spectrum of HD^+.
$^2H_2^+$, $^2H^3H^+$, $^3H_2^+$	(13) See ref. (28) of $^1H_2^+$.	
	(14) See ref. (7) of $^1H^2H^+$, $^1H^3H^+$.	
	(15) See ref. (30) of $^1H_2^+$.	
	(16) See ref. (32) of $^1H_2^+$.	
	(17) See ref. (10) of $^1H^2H^+$, $^1H^3H^+$.	
$^1H_2^-$	(7) Buckley, Bottcher, JP B 10, L635 (1977).	Feshbach projection-operator calculation of the resonant states of H_2^-.
^1HBr	(48) Johnson, Ramsey, JCP 67, 941 (1977).	Molecular beam electric resonance, Stark hf structure.
	(49) Niay, Bernage, Coquant, Fayt, CJP 55, 1829 (1977).	Dunham potential coefficients.
^2HBr	(14) Bernage, Niay, JMS 63, 317 (1976).	High resolution measurements of the 5-0 IR band; Dunham coefficients.
	(15) Fayt, Van Lerberghe, Guelachvili, Amiot, Bernage, Niay, MP 32, 955 (1976).	1-0, 2-0, 3-0, 4-0 IR absorption bands of $D^{79}Br$ and $D^{81}Br$. Equilibrium constants; precise calculated wavenumbers for observed chemical laser lines.
	(16) See ref. (48) of ^1HBr.	
^3HBr	(4) Bernage, Niay, CJP 55, 1016 (1977).	Calculated equilibrium constants from the constants for HBr and DBr.
$^1HBr^+$, $^2HBr^+$	(8) Möhlmann, de Heer, CP 17, 147 (1976).	Radiative lifetimes of A $^2\Sigma^+$, $\tau(v=0,1,2) = 4.5$, 4.0, and 3.6 (DBr^+) μs.
$^1HC\ell$	(65) Guelachvili, OC 19, 150 (1976).	Absolute wavenumber measurements of the 2-0 IR absorption band of $H^{35}C\ell$ and $H^{37}C\ell$ by Fourier spectrometry. Band origins and rotational constants.
$^2HC\ell$	(28) Niay, Coquant, Bernage, Bocquet, JMS 65, 388 (1977).	IR 3-0 absorption band, high resolution measurements. Dunham coefficients. Extrapolated constants for $^3HC\ell$.

$^1HC\ell^+$, $^2HC\ell^+$	(6) Brown, Watson, JMS 65, 65 (1977).	Spin-orbit and spin-rotation coupling in doublet states; application to $HC\ell^+$.
	(7) Möhlmann, Bhutani, de Heer, CP 21, 127 (1977).	Radiative lifetimes of $HC\ell^+$ $A\,^2\Sigma^+$, $\tau(v=0...6) = 2.58 ... 1.85\,\mu s$. Dependence on r of the electronic transition moment.
$^1HC\ell^-$	(3) Gianturco, Thompson, JP B 10, L21 (1977).	Theoretical evidence for short-lived resonances in electron scattering.
	(4) Taylor, Goldstein, Segal, JP B 10, 2253 (1977).	Computation of resonant states of $HC\ell^-$.
	(5) Goldstein, Segal, Wetmore, JCP 68, 271 (1978).	CI calculation on the resonant states of $HC\ell^-$.
He_2	(68) Brutschy, Haberland, PRL 38, 686 (1977).	Crossed-atomic-beam study of the long-range parts of $A\,^1\Sigma_u^+$ and $C\,^1\Sigma_g^+$. The A state has a potential hump of 0.047 eV at 3.14 Å.
	(69) Dacre, CPL 50, 147 (1977).	Calculation of the He-He well depth (0.000908 eV at 3.03 Å).
	(70) Ermler, Mulliken, Wahl, JCP 66, 3031 (1977).	Molecular orbital correlation diagrams.
	(71) Orth, Ginter, JMS 64, 223 (1977).	$i\,^3\Pi_g \rightarrow a\,^3\Sigma_u^+$, $\ell\,^3\Pi_g \rightarrow a\,^3\Sigma_u^+$; constants for i, ℓ. Perturbations.
	(72) Saxon, Gillen, Liu, PR A 15, 543 (1977).	*Ab initio* potential curves for $^1\Sigma_g^+$ states.
	(73) Orth, Brown, Ginter, JMS 69, 53 (1978).	Characterization of the $^1\Pi$ states associated with the UAO's 4p ...9p . Wavenumber tables for transitions to $A\,^1\Sigma_u^+$ and molecular constants.
He_2^+, He_2^{++}	(15) Blint, PR A 14, 2055 (1976).	*Ab initio* CI calculation of $^2\Sigma_g^+$, $^2\Sigma_u^+$, $^2\Pi_g$, $^2\Pi_u$ potential curves for He_2^+.
	(16) Maas, Van Asselt, Nowak, Los, Peyerimhoff, Buenker, CP 17, 217 (1976).	*Ab initio* calculation of $X\,^2\Sigma_u^+$ and adjustment based on measurements of the rotational predissociation of $^4He_2^+$, $^3He_2^+$, $^3He^4He^+$.
	(17) See ref. (70) of He_2.	
	(18) Yagisawa, Sato, Watanabe, PR A 16, 1352 (1977).	Accurate adiabatic potential curves for the ground state and eleven excited states of He_2^{++}.
HeAr	(7) Smith, Rulis, Scoles, Aziz, Nain, JCP 67, 152 (1977).	Crossed-beam differential scattering cross sections. Interatomic potential, $D_e = 0.0026_0$ eV, $r_e = 3.46$ Å.
He^1H	(6) Toennies, Welz, Wolf, CPL 44, 5 (1976).	He-H potential from low-energy elastic scattering data; $D_e = 0.00046$ eV, $r_e = 3.72$ Å.

He^1H^+	(13) Stewart, Watson, Dalgarno, JCP 65, 2104 (1976).	Hartree-Fock calculation of excitation energies, oscillator strengths, etc..
	(14) Price, MP 33, 559 (1977).	Calculation of eigen-energies and widths of the quasi-bound states (orbiting resonances) of HeH^+.
He^1H^{++}	(3) Winter, Duncan, Lane, JP B 10, 285 (1977).	Exact eigenvalues, electronic wavefunctions and their derivatives with respect to r for the lowest twenty states.
HeKr	(3) See ref. (7) of HeAr.	$D_e = 0.0026_0$ eV, $r_e = 3.67$ Å.
HeNe	(5) Grace, Johnson, Skofronick, JCP 67, 2443 (1977).	Observation of orbiting resonances in He + Ne scattering.
HeXe	(4) See ref. (7) of HeAr.	$D_e = 0.0024_1$ eV, $r_e = 3.95$ Å.
^1HF	(46) Guelachvili, OC 19, 150 (1976).	Absolute wavenumber measurements of the 1-0 and 2-0 IR absorption bands by Fourier spectrometry.
	(47) Chan, Rao, ZN 32 a, 897 (1977).	Dipole moment function.
	(48) Huffaker, JMS 65, 1 (1977).	$X\ ^1\Sigma^+$ RKR potential.
	(49) Pyper, Gerratt, PRS A 355, 407 (1977).	Spin-coupled theory of molecular wave functions; application to HF $(X\ ^1\Sigma^+)$.
HfO	(10) Ackermann, Rauh, Thorn, JCP 65, 1027 (1976).	First ionization potentials of the lanthanide monoxides.
Hg_2	(14) Siara, Krause, PR A 11, 1810 (1975).	Formation and decay of Hg_2 excimers in Hg-Ar mixtures.
	(15) Drullinger, Hessel, Smith, JCP 66, 5656 (1977).	Experimental studies of mercury molecules.
	(16) Smith, Drullinger, Hessel, Cooper, JCP 66, 5667 (1977).	A theoretical analysis of mercury molecules.
	(17) Komine, Byer, JCP 67, 2536 (1977).	Optically pumped Hg_2 studies.
HgAr	(11) Woodworth, JCP 66, 754 (1977).	Electron-beam excited fluorescence from mixtures of Hg with high-pressure Xe, Kr, Ar.
HgBr	(14) Wilcomb, Bernstein, JMS 62, 442 (1976).	$D_0^0 = 0.74_4$ eV, from the Le Roy-Bernstein modified Birge-Sponer extrapolation for $X\ ^2\Sigma^+$.
	(15) Djeu, Mazza, CPL 46, 172 (1977).	$B \rightarrow X$ in laser-excited fluorescence; $\tau = 23$ ns.
HgCℓ	(14) See ref. (14) of HgBr.	$D_0^0 = 1.06_6$ eV.
HgCs	(6) Düren, JP B 10, 3467 (1977).	Interpretation of experimental scattering cross sections with pseudopotential calculations.

Hg^1H	(16) Hehenberger, Laskowski, Brändas, JCP 65, 4559, 4571 (1976); Hehenberger, JCP 67, 1710 (1977).	Weyl's theory applied to predissociation by rotation.
HgI	(17) See ref. (14) of HgBr.	D_0^0 = 0.387 eV.
HgK	(6) See ref. (6) of HgCs.	
HgKr	(1) See ref. (11) of HgAr.	
HgNa	(5) See ref. (6) of HgCs.	
HgRb	(3) See ref. (6) of HgCs.	
HgTℓ	(3) Drummond, Schlie, JCP 65, 3454 (1976).	Fluorescence spectra of bands in the blue and red regions. Potential of the Tℓ-Hg system as a laser.
HgXe	(1) Nikiforov, Plimak, Predtechenski, Shcherba, OS(Engl. Transl.) 41, 195 (1976).	Formation of excited HgXe molecules.
	(2) See ref. (11) of HgAr.	
^1HI	(33) Eland, Berkowitz, JCP 67, 5034 (1977).	Photoionization mass-spectrometry. I.P.($^2\Pi_{\frac{3}{2}}$) = 10.386 eV, I.P.($^2\Pi_{\frac{1}{2}}$) = 11.049$_5$ eV.
	(34) Ungemach, Schaefer, Liu, JMS 66, 99 (1977).	X $^1\Sigma^+$, theoretical dipole moment function.
^2HI	(10) See ref. (33) of ^1HI.	I.P.($^2\Pi_{\frac{3}{2}}$) = 10.387 eV, I.P.($^2\Pi_{\frac{1}{2}}$) = 11.0505 eV.
HoO	(6) See ref. (9) of CeO.	I.P. = 6.17 eV (electron impact).
I_2	(93) Glozman, Zakharenko, Melnikov, Tkachenko, Fofanov, OS(Engl. Transl.) 41, 519 (1976).	Saturated absorption in I_2.
	(94) Koo, Newton, Smith, Andrews, PL A 58, 449 (1976).	Lifetimes of R(127) hf components in the B – X 11-5 band (~380 ns).
	(95) Landsberg, CPL 43, 102 (1976).	Nuclear hf splittings in the B – X system; no evidence for magnetic octupole coupling.
	(96) Callear, Erman, Kurepa, CPL 44, 599 (1976).	Lifetime of D $^1\Sigma_u^+$, τ = 15.5 ns.
	(97) Lehmann, SJQE 6, 442 (1976).	Prediss. in B $^3\Pi_{0^+u}$; dependence on rotation and vibration.
	(98) Yee, JCS FT II 72, 2113 (1976).	B $^3\Pi_{0^+u} \leftrightarrow$ X $^1\Sigma_g^+$ of $^{127}I_2$ and $^{129}I_2$. Rotational distortion constants. Identification and assignment of laser-excited molecular transitions.
	(99) Dalby, Petty-Sil, Pryce, Tai, CJP 55, 1033 (1977).	Nonlinear resonant photoionization. New Rydberg 1_g state with T_e = 53562.75, ω_e = 241.41, B_e = (0.04029).

I_2 (cont'd)	(100) Danyluk, King, CP 22, 59 (1977).	Two-photon sequential absorption spectroscopy. Constants for five excited electronic states (T_e = 40389...42600 cm^{-1}).
	(101) Vigué, Broyer, Lehmann, JP B 10, L379 (1977).	Fluorescence yield of individual hf components in the B state. A test of hyperfine predissociation.
	(102) Williams, Fernández, Rousseau, CPL 47, 150 (1977).	$B''^1\Pi_{1u}$ repulsive potential function from resonance Raman scattering data.
	(103) Tellinghuisen, CPL 49, 485 (1977).	Reassignment of the "H→B" system to the transition G→A'. Vibrational constants.
IBr	(25) Child, MP 32, 1495 (1976).	Predissociation and photodissociation of IBr. Potential curves for B and B'; new constants for B', T_e = 17053, ω_e = 123, $\omega_e x_e$ = 4.4, B_e = 0.0323.
	(26) Clyne, McDermid, JCS FT II 72, 2242 (1976).	B–X, spectroscopic constants, RKR potentials, Franck-Condon factors.
	(27) Tiemann, Dreyer, CP 23, 231 (1977).	Ground state dipole moment $\mu_{e\ell}(v=0) = 0.73_7$ D.
	(28) Wright, Havey, JCP 68, 864 (1978).	B state lifetime $\tau(v=2,3,4)$ = 0.54 μs.
ICℓ	(26) Clyne, McDermid, JCS FT II 72, 2242, 2252 (1976).	See ref. (26) of IBr. B–X laser excitation spectrum, first observation of v'=0. New improved constants for $B\ ^3\Pi_{0^+}$.
	(27) Clyne, McDermid, JCS FT II 73, 1094 (1977).	$B\ ^3\Pi_{0^+}$ fluorescence decay lifetimes.
IF	(9) See ref. (26) of IBr and ICℓ.	B–X excitation spectra; onset of predissociation in v'=10 at J'=12, leading to $D_0^0 \leq 2.814_5$ eV.
	(10) See ref. (27) of ICℓ.	
InBr	(7) Tiemann, Köhler, Hoeft, ZN 32 a, 6 (1977).	Hfs of J=2←1 transition; new determination of the quadrupole coupling constants.
K_2, K_2^+	(29) Herrmann, Leutwyler, Schumacher, Wöste, CPL 52, 418 (1977).	Two-photon ionization in a molecular beam.
	(30) Lemont, Giniger, Flynn, JCP 66, 4509 (1977).	$B\ ^1\Pi_u$ radiative lifetime 12.2 ns.
KAr, KAr^+	(11) Budenholzer, Gislason, Jorgensen, JCP 66, 4832 (1977).	Potassium ion - rare gas potentials from total cross-section measurements.
KHe^+, KKr^+, KNe^+	(1) See ref. (11) of KAr, KAr^+.	

Kr_2, Kr_2^+	(15) Miller, Ling, Saxon, Moseley, PR A 13, 2171 (1976).	Photodissociation cross sections of Kr_2^+, 5650 - 6950 Å.
	(16) Ng, Trevor, Mahan, Lee, JCP 66, 446 (1977).	Photoionization in a molecular beam, I.P. = 12.87 eV.
	(17) Wadt, JCP 68, 402 (1978).	Calculated potential curves for low-lying states of Kr_2^+.
KrF, KrF^+	(13) Hay, Dunning, JCP 66, 1306 (1977).	*Ab initio* CI calculation of states dissociating to Kr + F, $Kr^+ + F^-$, $Kr(^3P) + F$.
	(14) Krauss, JCP 67, 1712 (1977).	Theory, electronic structure.
	(15) Burnham, Searles, JCP 67, 5967 (1977).	KrF B state radiative lifetime 9.0 ns.
KrH^+	(4) Benoit, Kubach, Sidis, Pommier, Barat, JP B 10, 1661 (1977).	Charge-exchange processes in collisions between H^+ and Kr.
KrO	(3) Dunning, Hay, JCP 66, 3767 (1977).	*Ab initio* CI calculation of low-lying electronic states.
KrXe, $KrXe^+$	(7) Castex, JCP 66, 3854 (1977).	Absorption spectra in the far UV region.
	(8) Ng, Tiedemann, Mahan, Lee, JCP 66, 5737 (1977).	Photoionization study; I.P. = 11.757 eV.
	(9) Morlais, Rupin, Robin, CR B 284, 385 (1977).	
KXe^+	(1) See ref. (11) of KAr, KAr^+.	
LaO	(33) Ackermann, Rauh, Thorn, JCP 65, 1027 (1976).	First ionization potentials of the lanthanide monoxides.
	(34) Gole, Chalek, JCP 65, 4384 (1976).	Chemiluminescent spectra. $D_0^0 \geq 8.19$ eV.
	(35) Behere, Sardesai, Pramāṇa 8, 108 (1977).	Potential energy curves.
Li_2, Li_2^+, Li_2^-		
	(36) Lam, Gallagher, Hessel, JCP 66, 3550 (1977).	Intensity distribution in the A - X system.
	(37) Kusch, Hessel, JCP 67, 586 (1977).	Analysis of $A\,^1\Sigma_u^+ - X\,^1\Sigma_g^+$. New constants.
	(38) Konowalow, Olson, JCP 67, 590 (1977).	Theoretical vibrational energy levels for $A\,^1\Sigma_u^+$.
	(39) Olson, Konowalow, CP 21, 393; 22, 29 (1977).	Calculated potential energy curves.
	(40) Pyper, Gerratt, PRS A 355, 407 (1977).	Spin-coupled theory of molecular wavefunctions, application to Li_2 ($X\,^1\Sigma_g^+$).
	(41) Wine, Melton, CPL 45, 509 (1977).	$A\,^1\Sigma_u^+$ radiative lifetime $18._0$ ns.
	(42) Watson, Cerjan, Guberman, Dalgarno, CPL 50, 181 (1977).	Calculated potential energy curves.
Li^1H	(23) Pyper, Gerratt, PRS A 355, 407 (1977).	Spin-coupled theory of molecular wavefunctions, application to LiH ($X\,^1\Sigma^+$).

LiHe, $LiHe^+$	(10) Gatland, Morrison, Ellis, Thackston, McDaniel, Alexander, Viehland, Mason, JCP 66, 5121 (1977).	Li^+ - He interaction potential.
LiK	(5) Zmbov, Wu, Ihle, JCP 67, 4603 (1977).	Mass-spectrometric study, $D_0^0 = 0.81$ eV, I.P. = 4.69 eV.
LiNa, $LiNa^+$	(10) See ref. (5) of LiK.	$D_0^0 = 0.90$ eV, I.P. = 4.94 eV (constants for LiNa).
LuF	(4) Athénour, Féménias, Effantin, JP B 9, 2893 (1976).	Constants for X,A,B,E,F.
	(5) Effantin, Wannous, d'Incan, CJP 55, 64 (1977).	Classification of bands belonging to eight systems in the region 3000 - 6500 Å.
	(6) Effantin, Bacis, d'Incan, PR A 15, 1053 (1977).	$B\ ^1\Pi \rightarrow X\ ^1\Sigma$; hyperfine structure.
LuO	(11) See ref. (9) of CeO.	I.P. = 6.79 eV (electron impact).
Mg_2	(12) Scheingraber, Vidal, JCP 66, 3694 (1977).	A - X discrete and continuous Franck-Condon factors.
	(13) Stevens, Krauss, JCP 67, 1977 (1977).	Electronic structure of the ground and excited states of Mg_2 and Mg_2^+.
	(14) Miller, Ault, Andrews, JCP 67, 2478 (1977).	Absorption in solid argon and krypton, resolved vibrational structure 22300 - 23000 and 25000 - 28000 cm^{-1}.
Mg_2^+	(1) See ref. (13) of Mg_2.	
MgO	(28) Evans, Mackie, JMS 65, 169 (1977).	Triplet-triplet transitions and intensities.
	(29) Ikeda, Wong, Harris, Field, JMS 68, 452 (1977).	Laser-induced photoluminescence spectra; $B\rightarrow X$, $B\rightarrow A$. Analysis of $a\ ^3\Pi_i \sim X\ ^1\Sigma^+$ perturbations.
MgSr	(1) Miller, Ault, Andrews, JCP 67, 2478 (1977).	Absorption in solid argon and krypton; vibrational structure 16000 - 17100 cm^{-1}.
MnO	(14) Pinchemel, Schamps, CP 18, 481 (1976).	Electronic structure.
	(15) Ferrante, Wilkerson, Graham, Weltner, JCP 67, 5904 (1977).	ESR sp. at 4 K; $^6\Sigma^+$ ground state.
N_2	(200) Becker, Engles, Tatarczyk, ZN 31 a, 673 (1976).	$C\ ^3\Pi_u \leftrightarrow B\ ^3\Pi_g (v'' \leq 13)$
	(201) Dehmer, Dill, JCP 65, 5327 (1976).	K-shell spectrum of N_2.
	(202) Hillier, Kendrick, JESRP 8, 239 (1976).	CI calc. of satellite peaks in core and valence PE spectra.
	(203) Banna, Shirley, JESRP 8, 255 (1976).	Photoelectron spectrum at 132.3 eV.
	(204) Kotlikov, OS(Engl.Transl.) 41, 434 (1976).	Hanle effect on a rotational level of $B\ ^3\Pi_g$. | 50 eV.
	(204a) Hamnett, Stoll, Brion, JESRP 8, 367 (1976).	PE branching ratios, partial ionization cross sections 18 -

N_2 (cont'd)	(205) Wells, Borst, Zipf, PR A 14, 695 (1976).	Translational spectroscopy of metastable fragments produced by dissociative excitation.
	(206) Ermler, Mulliken, Wahl, JCP 66, 3031 (1977).	Molecular orbital correlation diagrams.
	(207) Brennen, Shuman, JCP 66, 4248 (1977).	Spectrum of the Lewis-Rayleigh nitrogen afterglow.
	(208) Chutjian, Ajello, JCP 66, 4544 (1977).	Threshold photoelectron spectrum, 15.5 - 19.0 eV (800 - 650 Å).
	(209) Hummer, Burns, JCP 67, 4062 (1977).	Lifetimes of $x\ ^1\Sigma_g^-$ (23.1 ns) and $y\ ^1\Pi_g$ (19.9 ns).
	(210) Gürtler, Saile, Koch, CPL 48, 245 (1977).	High-resolution absorption spectrum, 990 - 440 Å. Rydberg-series leading to X,A,B,C of N_2^+. Abs. cross sections.
	(211) Rescigno, Langhoff, CPL 51, 65 (1977).	K-shell photoionization.
	(212) Becker. Engels, Tatarczyk, CPL 51, 111 (1977).	$C\ ^3\Pi_u$ radiative lifetimes, $\tau(v=0) = 36.6$ ns, $\tau(v=4) = 36.5$ ns.
	(213) Plummer, Gustafsson, Gudat, Eastman, PR A 15, 2339 (1977).	Partial photoionization cross sections, 18 - 40 eV.
	(214) Rastogi, Lowndes, JP B 10, 495 (1977).	Collision-induced translational-rotational far-infrared absorption; quadrupole moment of N_2, $\|Q\| = 1.21 \times 10^{-26}$ esu cm^2.
	(215) Samson, Haddad, Gardner, JP B 10, 1749 (1977).	Total and partial photoionization cross sections from threshold to 100 Å.
	(216) Kay, Van der Leeuw, Van der Wiel, JP B 10, 2513 (1977).	Energy loss spectra and absolute oscillator strengths for shape resonances near the nitrogen K edge.
	(217) Woodruff, Marr, PRS A 358, 87 (1977).	Photoelectron spectrum; partial cross sections as a function of energy from 16 to 40 eV.
	(218) Yoshino, Tanaka, JMS 66, 219 (1977).	High-resolution VUV absorption spectrum; homogeneous perturbation $c_4'\ ^1\Sigma_u^+(v=0) \sim b'\ ^1\Sigma_u^+(v=1)$.
N_2^+	(75) Erman, PS 14, 51 (1976).	Predissociation of $C\ ^2\Sigma_u^+$; lifetimes $\tau(v=2) = 78._9$ ns, $\tau(v \geqq 3) \approx 4$ to 5 ns.
	(76) Bouchoux, Goure, CJP 55, 1492 (1977).	$B\ ^2\Sigma_u^+ \rightarrow X\ ^2\Sigma_g^+$; perturbation $B(v=1) \sim A(v=11)$.
	(77) See ref. (206) of N_2.	
	(78) Lee, JP B 10, 3033 (1977).	Cross sections for the production of $B \rightarrow X$ fluorescence by photoionization.
	(79) Dick, Benesch, Crosswhite, Tilford, Gottscho, Field, JMS 69, 95 (1978).	High-resolution study of $B \rightarrow X$ bands.
N_2^{++}	(7) Stockdale, JCP 66, 1792 (1977).	High kinetic energy N^+ ions from decay of doubly ionized N_2.

N_2^-	(11) King, McConkey, Read, JP B 10, L541 (1977).	Negative-ion resonances associated with K-shell-excited states of N_2.
	(12) Veillette, Marchand, CJP 55, 2134 (1977).	Negative ion resonances by electron impact in the 10.5-15.5 eV region.
Na_2	(41) Lam, Gallagher, Hessel, JCP 66, 3550 (1977).	Intensity distribution in the A − X system.
	(42) Woerdman, CPL 50, 41 (1977).	Two-photon absorption.
	(43) See ref. (29) of K_2, K_2^+.	
	(44) Woerdman, CPL 53, 219 (1978).	Experimental Franck-Condon factors for the A − X system from a comparative study of the saturation behaviour of Na_2 and Na transitions.
NaAr	(6) Smalley, Auerbach, Fitch, Levy, Wharton, JCP 66, 3778 (1977).	Analysis of the A − X fluorescence excitation spectrum; vibrational and rotational constants. $D_e^0 = 0.0048_2$ eV, $r_e(X^2\Sigma^+) = 4.991$ Å.
	(7) Saxon, Olson, Liu, JCP 67, 2692 (1977).	Ab initio calculations for X $^2\Sigma$, A $^2\Pi$, B $^2\Sigma$.
NaI, NaI^+	(22) Anderson, Wilson, Ross, CPL 48, 284 (1977).	Photodissociation in a molecular beam using linearly polarized light.
NaK	(9) Zmbov, Wu, Ihle, JCP 67, 4603 (1977).	Mass-spectrometric study, $D_0^0 = 0.63$ eV, I.P. = 4.57 eV.
NaNe	(4) Ahmad-Bitar, Lapatovich, Pritchard, Renhorn, PRL 39, 1657 (1977).	A − X laser excitation spectrum; spectroscopic constants. $D_e^0 = 0.0010_0$ eV, $r_e(X^2\Sigma^+) = 5.2_9$ Å.
NdO	(6) See ref. (9) of CeO.	I.P. = 4.97 eV (electron impact).
NeF	(2) Winter, Bender, Rescigno, JCP 67, 3122 (1977).	Potential energy curves and predicted fluorescence.
NeKr, $NeKr^+$ NeXe, $NeXe^+$	(5) Castex, JCP 66, 3854 (1977).	NeXe, absorption spectra in the far UV region.
N^1H	(56) Banerjee, Grein, JCP 66, 1054, 2589 (1977).	Calculated potential energy curves and spectroscopic constants for b $^1\Sigma^+$, d $^1\Sigma^+$.
	(57) Hsu, Smith, JCP 66, 1835 (1977).	Lifetime of d $^1\Sigma^+(v=0)$ 46 ns (NH) and 62 ns (ND).
N^2H	(17) See ref. (57) of N^1H.	
Ni_2	(3) Moskovits, Hulse, JCP 66, 3988 (1977).	Matrix spectra in the visible and UV regions.
	(4) Anderson, JCP 66, 5108 (1977).	Theory of UV spectra of Ni_2.

NO	(210) Brzozowski, Erman, Lyyra, PS 14, 290 (1976).	Lifetimes for A,B,B',C,D,F. Predissociation rates and perturbations.
	(211) Amiot, JP B 10, L317 (1977).	Spectral coincidences between CO laser lines and absorption lines of NO.
	(212) Sukumar, Bottcher, JP B 10, L335 (1977).	Interpretation of experimental results on preionization and predissociation via high-lying molecular states.
	(213) Freedman, CJP 55, 1387 (1977).	C $^2\Pi(v=0) \leftarrow X\ ^2\Pi(v=0)$ two-photon absorption spectrum.
	(214) Goodman, Brus, JCP 67, 933 (1977).	The A $^2\Sigma^+(3s\sigma)$ Rydberg state in solid rare gases.
	(215) Lee, JCP 67, 3998 (1977).	Energy loss spectrum.
	(216) Möhlmann, de Heer, CPL 49, 588 (1977).	Radiative lifetimes (6.4_3 and 5.7_7 μs for v=1 and 2, resp.) and excitation energy (6.4 eV for v'=2) of b $^4\Sigma^-$.
	(217) Patel, Kerl, Burkhardt, PRL 38, 1204 (1977).	Opto-acoustic high-resolution spectroscopy of the $v=2\leftarrow 1$ vibration-rotation transition.
	(218) Takezawa, JMS 66, 121 (1977).	Rydberg series converging to b $^3\Pi$, A $^1\Pi$, w $^3\Delta$ of NO^+.
	(219) Kristiansen, JMS 66, 177 (1977).	On the determination of molecular parameters for NO (X $^2\Pi$). Term values for v=0.
	(220) Dale, Johns, McKellar, Riggin, JMS 67, 440 (1977).	High-resolution laser magnetic resonance and infrared-radiofrequency double resonance spectroscopy of NO and its isotopes near 5.4 μm.
	(221) Amiot, Bacis, Guelachvili, CJP 56, 251 (1978).	IR study of the X $^2\Pi$ v=0,1,2 levels of $^{14}N^{16}O$. Preliminary results for v=0,1 of $^{14}N^{17}O$, $^{14}N^{18}O$, $^{15}N^{16}O$.
	(222) Frueholz, Rianda, Kuppermann, JCP 68, 775 (1978).	$a^4\Pi \leftarrow X^2\Pi$ and $b^4\Sigma^- \leftarrow X^2\Pi$ by electron impact spectroscopy. Excitation energies of $a^4\Pi(v=4)$ 5.22 eV and of $b^4\Sigma^-(v=0)$ 5.70 eV.
NO^+, NO^{++}	(45) Shimauchi, SL 25, 1 (1976).	NO^+ A – X Franck-Condon factors.
	(46) Natalis, Delwiche, Collin, Caprace, Praet, CPL 49, 177 (1977).	Ne I photoelectron sp. of NO; NO^+ ground state levels observed to v=32.
O_2	(199) Galkin, OS(Engl. Transl.) 42, 486 (1977).	The b← X 0-1 band in the absorption spectrum of the earth's atmosphere.
	(200) Guberman, JCP 67, 1125 (1977).	Accurate ab initio potential curve for X $^3\Sigma_g^-$.
	(201) Goodman, Brus, JCP 67, 1482 (1977).	Electronic spectroscopy and dynamics of the A, A', and c states in van der Waals solids.

O_2 (cont'd)	(202) Katayama, Ogawa, Ogawa, Tanaka, JCP 67, 2132 (1977).	VUV absorption spectra from b $^1\Sigma_g^+$ and a $^1\Delta_g$.
	(203) Lee, JCP 67, 3998 (1977).	Energy loss spectrum.
	(204) Saxon, Liu, JCP 67, 5432 (1977).	Ab initio CI study of the valence states.
	(205) Lee, Slanger, Black, Sharpless, JCP 67, 5602 (1977).	Quantum yields for the production of O(^{1}D) by photodissociation of O_2 at 1160 - 1770 Å.
O_2^+	(53) Albritton, Schmeltekopf, Harrop, Zare, Czarny, JMS 67, 157 (1977).	Analysis of the b $^4\Sigma_g^- \rightarrow$ a $^4\Pi_u$ 1st negative band system.
	(54) Carrington, Roberts, Sarre, MP 34, 291 (1977).	Laser-induced predissociation spectrum of O_2^+, b$\leftarrow$a. The b state predissociates via b' $^4\Pi_g$.
	(55) Lambert, Goure, Albritton, CJP 55, 1842 (1977).	Comparison of experimental and theoretical line intensities in the b$\rightarrow$a 0-0 band.
	(56) Samson, Gardner, JCP 67, 755 (1977).	Vibrational energy levels and dissociation energy of X $^2\Pi_g$.
	(57) Banna, Shirley, JESRP 8, 255 (1976).	Photoelectron spectroscopy at 132.3 eV.
O^1H	(127) Destombes, Marliere, Rohart, JMS 67, 93 (1977).	Molecular constants for X $^2\Pi$ and A $^2\Sigma$ from the analysis of the Λ-doubling hfs, the pure rotational sp., and the A$-$X 0-0 band.
	(128) German, JCP 67, 5411 (1977).	Predissociation and tunneling in the A $^2\Sigma^+$ state.
	(129) Palmer, JCP 67, 5413 (1977).	Predissociation and tunneling in the A $^2\Sigma^+$ state.
P_2	(28) Chong, Takahata, JESRP 10, 137 (1977).	Theoretical calculations of photoelectron spectra.
	(29) Malicet, Brion, Daumont, CR C 284, 175 (1977).	b $^3\Pi_g \rightarrow$ a $^3\Sigma_u^+$; perturbations in b(v=6,7) are attributed to a new $^5\Sigma_g^+$ state.
	(30) Brion, Malicet, Daumont, CR C 284, 647 (1977).	A $^1\Pi_g \rightarrow$ X $^1\Sigma_g^+$, new data.
	(31) Effantin, Bacis, Amiot, Verges, JMS 69, 79 (1978).	High-resolution Fourier spectrometry of P_2 infrared emission. Analysis of a new system A $^1\Pi_g \rightarrow$ A' $^1\Sigma_u^-$. Constants for A': $T_e = 28521.79$, $\omega_e = 608.095$, $\omega_e x_e = 2.357$, $\omega_e y_e = 0.0016$, $B_e = 0.258731$, $\alpha_e = 0.001396$.
P_2^-	(1) Cederbaum, Domcke, Niessen, JP B 10, 2963 (1977).	Theoretical calculation of I.P. = 0.30 eV.
Pb_2	(8) Bondybey, English, JCP 67, 3405 (1977).	Emission and laser excitation spectra in rare gas solids. Reassignment of gas phase data.

$PC\ell$	(3) Coxon, Wickramaaratchi, JMS 68, 372 (1977).	Visible and near infrared emission excited in the reaction $Ar(^3P_{2,0}) + PC\ell_3$. Vibrational analyses of A $^3\Pi \rightarrow$ X $^3\Sigma^-$ and b $^1\Sigma^+ \rightarrow$ X $^3\Sigma^-$. Vibrational constants for all three states ($P^{35}C\ell$ and $P^{37}C\ell$).
P^1H, P^2H	(16) Di Stefano, Lenzi, Margani, Xuan, JCP 68, 959 (1978).	b$\rightarrow$X observed in the VUV photolysis of PH_3; $T_0 = 14340$ cm^{-1}.
PN, PN^+	(13) Atkins, Timms, SA A 33, 853 (1977).	Matrix IR spectrum of PN.
	(14) Bulgin, Dyke, Morris, JCS FT II 73, 983 (1977).	He I photoelectron spectrum; X $^2\Sigma^+$, A $^2\Pi$, B $^2\Sigma^+$ of PN^+.
	(15) Chong, Takahata, JESRP 10, 137 (1977).	Theoretical calculations of photoelectron spectra.
$PtA\ell$?	(1) Scullman, Cederbalk, JP B 10, 3659 (1977).	Unclassified emission bands 16700 - 18300 cm^{-1}.
PtO	(6) Jansson, Scullman, JMS 61, 299 (1976).	Optical absorption spectra in rare gas matrices.
Rb_2, $Rb_2{}^+$	(21) Gupta, Happer, Wagner, Wennmyr, JCP 68, 799 (1978).	Absorption studies in the visible and near visible regions.
RhC	(7) Shadrin, Zhirnov, OS(Engl. Transl.) 38, 367 (1975).	C - X Franck-Condon factors.
S_2	(52) Leone, Kosnik, APL 30, 346 (1977).	Laser action on the B - X transition.
SbN	(3) Jenouvrier, Daumont, Pascat, CJP 56, 30 (1978).	Electronic spectrum 3300 - 4000 Å. Analysis of the A $^1\Pi \rightarrow$ X $^1\Sigma^+$ system of four isotopes. Upper state strongly perturbed. Constants for X $^1\Sigma^+$ of $^{121}Sb^{14}N$: $\omega_e = 864.80$, $\omega_e x_e = 4.75$, $B_e = 0.3988$, $\alpha_e = 0.0026$, $r_e = 1.835_3$. The spectrum previously attributed to SbN is due to Sb_2.
ScO	(23) Rao, Rao, Rao, Physica C 81, 392 (1976).	A - X and B - X Franck-Condon factors.
SiC	(4) Graeffe, Juslén, Karras, JP B 10, 3219 (1977).	Si K_α X-ray emission spectrum from SiC.
SiO	(52) Chong, Takahata, JESRP 10, 137 (1977).	Theoretical calculations of photoelectron spectra.
SiS	(16) Atkins, Timms, SA A 33, 853 (1977).	Matrix IR spectrum.
Sn_2	(3) Teichman, Epting, Nixon, JCP 68, 336 (1978).	Raman sp. in solid argon; $\omega_e \approx 188$, $\omega_e x_e \approx 0.53$.
Sr_2	(1) Miller, Ault, Andrews, JCP 67, 2478 (1977).	Absorption in solid argon and krypton, vibrational structure from 13400 to 14600 cm^{-1}.

SrO	(36) Hocking, Pearson, Creswell, Winnewisser, JCP 68, 1128 (1978).	Millimeter wave spectrum.
TaO	(8) See ref. (9) of CeO.	I.P. = 7.92 eV (electron impact).
TbO	(6) See ref. (9) of CeO.	I.P. = 5.62 eV (electron impact).
TiO	(64) Kovács, El Agrab, APH 42, 67 (1977).	On the anomalous triplet splitting of C $^3\Delta_r$.
	(65) Steele, Linton, JMS 69, 66 (1978).	Flame spectroscopy; C – X radiative lifetimes and oscillator strengths, τ = 37, 29, 28 ns for v'=0, 1, 2, respectively.
TmO	(6) See ref. (9) of CeO.	I.P. = 6.44 eV (electron impact).
U_2	(1) Gorokhov, Emel'yanov, Khodeev, HT(USSR) 12, 1156 (1974).	Mass-spectroscopic investigation; $D_0^0 = 2.2_5$ eV.
UN	(2) Green, Reedy, JCP 65, 2921 (1976).	Identification in argon matrices.
V_2	(2) Ford, Huber, Klotzbücher, Kündig, Moskovits, Ozin, JCP 66, 524 (1977).	Matrix study.
Xe_2^+	(7) Wadt, JCP 68, 402 (1978).	Calculated potential curves for low-lying states.
XeF	(18) Smith, Kobrinsky, JMS 69, 1 (1978).	B← X; $T_e = 28813.9$, $\omega_e' = 308.2$, $\omega_e'x_e' = 1.43$, $\omega_e'' = 225._1$, $\omega_e''x_e'' = 10.6$.
YbCℓ	(3) Lee, Zare, JMS 64, 233 (1977).	Chemiluminescent spectrum, $A_1, A_2, B \rightarrow X$. Vibrational const..
YbF	(4) See ref. (3) of YbCℓ.	Vibr. constants for $A_2 \rightarrow X$; A_1 and B strongly perturbed.
Yb^1H, Yb^2H	(8) Van Zee, Seely, Weltner, JCP 67, 861 (1977).	ESR and optical spectroscopy in argon matrices at 4 K.
YbO	(6) See ref. (9) of CeO.	I.P. = 6.55 eV (electron impact).
YF, YF^+	(8) Shenyavskaya, Gurvich, JMS 68, 41 (1977).	Rotational analysis of b $^3\Phi \rightarrow$ a $^3\Delta$ of YF. New rotational constants.
Y^1H, Y^2H	(1) Bernard, Bacis, CJP 55, 1322 (1977).	Analysis of singlet and triplet systems of YH and YD.
YO	(21) Chalek, Gole, JCP 65, 2845 (1976).	Chemiluminescence spectrum. Analysis of A' $^2\Delta \rightarrow$ X $^2\Sigma^+$; $T_e = \begin{cases}14870.4\\14531.2\end{cases}$ $\omega_e' = \begin{cases}794.9\\794.0\end{cases}$, $\omega_e'' = 862.0$.
Zn_2	(2) Hay, Dunning, Raffenetti, JCP 65, 2679 (1976).	Ab initio calculation of electronic states dissociating to $^1S + ^1S, ^3P, ^1P$.

Zn_2 (cont'd)	(3) Ault, Andrews, JMS 65, 102 (1977).	UV absorption in solid argon and krypton; vibrational structure 38300 - 40000 cm^{-1}.
Zn^1H, Zn^2H	(10) Dufayard, Nedelec, JP(Paris) 38, 449 (1977).	Lifetimes ($\approx$ 75 ns), Λ-doubling, hfs in A $^2\Pi$.
ZrN	(3) Kabankova, Moskvitina, Kuzyakov, VMUK 16(5), 620 (1975). For engl. transl. see MUCB 30(5), 80 (1975).	Electronic absorption spectrum.
	(4) Bates, Dunn, CJP 54, 1216 (1976).	The yellow and violet emission systems.
ZrO	(35) Phillips, Davis, ApJ 206, 632 (1976).	A new $^1\Sigma \rightarrow {}^1\Sigma$ system of ZrO; v_{00} = 17050.47, B_0' = 0.40479 [observed in matrix absorption (19) at 17025 cm^{-1}, see c on p. 687].
	(36) Phillips, Davis, ApJ(Suppl.) 32(3), 537 (1976).	B$\rightarrow$X system; 18 additional bands, improved upper and lower state constants.
	(37) Lauchlan, Brom, Broida, JCP 65, 2672 (1976).	Laser photoluminescence in neon at 4 K. Tentative identification of two new infrared transitions, B$\rightarrow$A and b' $^3\Pi \rightarrow$ a $^3\Delta$.